# 1992
NATIONAL CONSTRUCTION ESTIMATOR

40th Edition

Includes 1992 Estimate Writer Disk

Edited by

Martin D. Kiley

and

William M. Moselle

Craftsman Book Company

6058 Corte del Cedro, P.O. Box 6500, Carlsbad, CA 92018-6500

Contents

A complete index begins on page 569

How to Use This Book	3-4
Craft Codes and Crews	5-7
Residential Division Labor Cost	8
Adjusting Labor Costs	9
Area Modification Factors	10-11
Credits and Acknowledgements	12
Residential Division	**13-218**
Adhesives	13
Architectural illustrations	15
Bathroom accessories	18
Building paper	21
Cabinets	22
Carpentry	24
Ceiling materials	39
Closet door systems	42
Concrete	45
Demolition	54
Doors	59
Electrical work	70
Elevators	73
Excavation	75
Fans	78
Fences	79
Fireplace components	83
Flooring	85
Garage doors	92
Glass, glazing	92
Gutters and downspouts	94
Gypsum drywall	95
Hardboard	96
Hardware	98
Heating and cooling	100
Insulation	104
Insurance	106
Landscaping	109
Lath	112
Lighting fixtures	114
Lumber	117
Markup	130
Masonry	131
Medicine cabinets	141
Mouldings	144
Nails	148
Paint	151
Paneling	157
Paving	159
Plastering	160
Plumbing	160
Plywood	168
Range hoods	170
Roofing	171
Sheet metal	181
Siding	189
Skylights	196
Stairs	200
Taxes	203
Tile	204
Wallboard	210
Windows	212

Construction Economics Division	219
Abbreviations and Symbols	220
Industrial & Commercial Division Labor Cost	221
Industrial & Commercial Division	**222-552**
General Requirements 1	222-240
Markup & overhead	222
Equipment rental rates	232
Site Work 2	240-297
Demolition	241
Excavation and grading	251
Pile foundations	259
Utilities	264
Paving and walkways	272
Concrete 3	297-324
Formwork	298
Concrete placing	306
Tilt-up	316
Masonry 4	324-332
Clay brick	324
Concrete block	326
Metals 5	333-340
Structural steel	336
Wood and Plastic 6	340-347
Rough carpentry	340
Panelized roof	343
Thermal & Moisture Protection 7	348-360
Insulation	348
Roofing	351
Doors and Windows 8	360-369
Doors	360
Storefront	365
Glazing	368
Finishes 9	370-383
Gypsum wallboard	372
Flooring	376
Painting	379
Specialties 10	383-395
Equipment 11	395-405
Furnishings 12	405-408
Special Construction 13	408-413
Conveying Systems 14	413-416
Mechanical 15	416-500
Steel pipe	416
Cast iron soil pipe	428
Copper tube	436
Hangers and supports	445
Valves	447
Plumbing rough-in and fixtures	461
Heating, ventilating, air conditioning	466
Duct work	494
Electrical 16	500-552
Electrical duct and conduit	500
Wire	519
Outlets & junction boxes	524
Switches	526
Motor connections	528
Lighting fixtures	539
Security and alarm systems	548
How to Use Estimate Writer	**553-568**
Index	569

Copyright 1991 Craftsman Book Company ISBN 0-934041-68-7 1st printing November 1991 for the year 1992

This Book Is An Encyclopedia Of 1992 Building Costs

We'll Answer Your Questions about any part of this book and explain how to apply these costs. Free telephone assistance is available from 9 a.m. until noon and from 1 p.m. until 5 p.m. Pacific time Monday thru Friday except holidays. Phone 619-438-7816. We don't accept collect calls and won't estimate the job for you. But if you need clarification on something in this manual, we can help.

The 1992 National Construction Estimator lists estimated construction costs to general contractors as of mid-1992.

This Manual Has Two Parts; the Residential Construction Division begins on page 13. Use the figures in this division when estimating the cost of homes and apartments with a wood or masonry frame. The Industrial and Commercial Division begins on page 222 and can be used to estimate costs for nearly all construction not covered by the Residential Division.

The Residential Construction Division is arranged in alphabetical order by construction trade and type of material. The Industrial and Commercial Division follows the 16 section Construction Specification Format. A complete index begins on page 569.

 **Estimate Writer** (inside the back cover) is an electronic version of this book. Instructions for using Estimate Writer begin on page 553.

Material Costs for each item are listed in the column headed "Material." These are neither retail nor wholesale prices. They are estimates of what most contractors who buy in moderate volume will pay as of mid-1992.

Add Delivery Expense to the material cost for other than local delivery of reasonably large quantities. Cost of delivery varies with the distance from source of supply, method of transportation, and quantity to be delivered. But most material dealers absorb the delivery cost on local delivery (5 to 15 miles) of larger quantities to good customers. Add the expense of job site delivery when it is a significant part of the material cost.

Add Sales Tax when sales tax will be charged to the contractor buying the materials.

Equipment Cost is included with the material cost in this manual unless specifically excluded or listed separately.

Mobilization Cost on site is included in this book. But the cost of moving major pieces of equipment to the site or transporting work crews to remote locations must be added.

Waste and Coverage loss is included in the installed material cost. The cost of many materials per unit after installation is greater than the purchase price for the same unit because of waste, shrinkage or coverage loss during installation. For example, about 120 square feet of nominal 1" x 4" square edge boards will be needed to cover 100 square feet of floor or wall. There is no coverage loss with plywood sheathing, but waste due to cutting and fitting will average about 6%.

Costs in the "Material" column of this book assume normal waste and coverage loss. Small and irregular jobs may require a greater waste allowance. Materials priced without installation (with no labor cost) do not include an allowance for waste and coverage except as noted.

Labor Costs for installing the material or doing the work described are listed in the column headed "Labor." The labor cost per unit is the labor cost per hour multiplied by the time needed to complete the task. Labor cost includes the basic wage, the employer's contribution to welfare, pension, vacation and apprentice funds and all tax and insurance charges based on wages. Hourly labor costs for the various Crafts are listed on page 8 (for the Residential Division) and page 221 (for the Industrial and Commercial Division).

Hourly labor costs used in the Industrial and Commercial Division are higher than those used in the Residential Division, reflecting the fact that craftsmen on industrial and commercial jobs are often paid more than craftsmen on residential jobs.

Supervision Expense to the general contractor is not included in the labor cost. The cost of supervision and non-productive labor varies widely from job to job. Calculate the cost of supervision and non-productive labor and add this to the estimate.

Taxes and Insurance included in the labor cost are itemized in the sections beginning on pages 106 and 203.

Manhours per Unit and the Craft performing the work are listed in the "Craft@Hrs" column. Divide the manhours per unit into 8 to find the units of work done per man in an 8-hour day. Multiply by the number of crew members to find the units per crew in an 8-hour day.

Adjust the Labor Cost to the job you are figuring when your actual hourly labor cost is known or can be estimated. The labor costs listed on pages 8 and 221 will apply within a few percent on many jobs. But labor costs may be much higher or much lower on the job you are estimating.

If the hourly wage rates listed on page 8 or page 221 are not accurate, divide your known or estimated cost per hour by the listed cost per hour. The result is your adjustment for any figure in the "Labor" column for that craft. See page 9 for more information on adjusting labor costs.

Adjust for Unusual Labor Productivity. Costs in the labor column are for normal conditions: experienced Craftsmen working on reasonably well planned and managed new construction with fair to good productivity. Labor estimates assume that materials are standard grade, appropriate tools are on hand, work done by other Crafts is adequate, layout and installation are relatively uncomplicated, and working conditions don't slow progress.

Labor costs will be higher where these conditions don't apply. Only experience and careful analysis can help you predict the cost of unusual labor productivity.

Use an Area Modification Factor from page 10 or 11 if your hourly labor cost is unknown and can't be estimated.

Here's how: Use the labor and material costs in this manual without modification. Then multiply the factor from page 10 or 11 times the estimate total to find your local estimated cost.

The "Total" Column sums figures in the "Material" and "Labor" (and "Equipment" when shown) columns.

The "Unit" Column identifies the unit of measure for each material. Abbreviations are defined on page 220.

Subcontracted Work includes the subcontractor's overhead and profit. Many sections in the Residential Division and all of the Industrial and Commercial Division give "subcontract" costs—what a general contractor would expect to pay a subcontractor-specialist for this work. Subcontract costs include the subcontractor's supervision expense, contingency allowance, overhead and profit.

Costs in the Residential Division that aren't identified as "subcontract" assume the general contractor does the work with his own crews. Work done by the general contractor's Craftsmen may cost more or less than the same work done by a subcontractor. Specialists may do work at a lower cost, even after the subcontractor's overhead and profit have been added.

A Subcontractor's Markup has been added to both material and labor costs in the Industrial and Commercial Division. A markup of 30% has been added to the labor costs. Most material costs include a 15% markup, but lower markups have been used on major items of equipment which involve relatively little labor cost.

The General Contractor's Markup is not included in any costs in this book. On page 129 we suggest a 20% markup on the contract price for general contractors handling residential construction. Apply this markup or some figure you select to all costs, including both subcontract items and work done by your own crews. To realize a markup of 20% on the contract price, you'll have to add 25% to your costs. See page 130 for an example of how markup is calculated. Markup is the most difficult item to estimate.

Labor and Material Costs Change. These costs were compiled in the fall of 1991 and projected to mid-1992 by adding 2 to 4%. This estimate will be accurate for some materials but inaccurate for others. No one can predict material price changes accurately.

How Accurate Are These Figures? As accurate as possible considering that the estimators who wrote this book don't know your subcontractors or material suppliers, haven't seen the plans or specifications, don't know what building code applies or where the job is, had to project material costs at least 6 months into the future, and had no record of how much work the crew that will be assigned to the job can handle.

You wouldn't bid a job under those conditions. And we don't claim that all construction is done at these prices.

Estimating Is an Art, not a science. On many jobs the range between high and low bid will be 20% or more. There's room for legitimate disagreement on what the correct costs are, even when complete plans and specifications are available, the date and site are established, and labor and material costs are identical for all bidders.

No cost fits all jobs. Good estimates are custom made for a particular project and a single contractor through judgment, analysis and experience.

This book is not a substitute for judgment, analysis and sound estimating practice. It's an aid in developing an informed opinion of cost. If you're using this book as your sole cost authority for contract bids, you're reading more into these pages than the editors intend.

Use These Figures to compile preliminary estimates, to check your costs and subcontract bids and when no actual costs are available. This book will reduce the chance of error or omission on bid estimates, speed "ballpark" estimates, and be a good guide when there's no time to get a quote.

Where Do We Get These Figures? From the same sources all professional estimators use: contractors and subcontractors, architectural and engineering firms, material suppliers, material price services, analysis of plans, specifications, estimates and completed project costs, and both published and unpublished cost studies. In addition, we conduct nationwide mail and phone surveys and have the use of several major national estimating data bases.

Craft Codes, Hourly Costs and Crew Compositions

Both the Residential Division and Commercial and Industrial Division of this book include a column titled *Craft@Hrs*. Letters and numbers in this column show our estimates of:

- Who will do the work (the craft code)
- An @ symbol which means *at*
- How long the work will take (manhours).

For example, on page 34 you'll find estimates for installing CD plywood wall sheathing by the square foot. The *Craft@Hrs* column opposite 1/2" plywood wall sheathing shows:

B1@.016

That means we estimate the installation rate for crew B1 at .016 manhours per square foot. That's the same as 16 manhours per 1,000 square feet.

The table that follows defines each of the craft codes used in this book. Notice that crew B1 is composed of two craftsmen: one laborer and one carpenter. To install 1,000 square feet of 1/2" CD wall sheathing at .016 manhours per square foot, that crew would need 16 manhours (one 8-hour day for a crew of two).

Notice also in the table below that the cost per manhour for crew B1 is listed as $21.19. That's the average for a building laborer (listed at $18.80 per hour on page 8) and a building carpenter (listed at $23.59 per hour on page 8): $18.80 plus $23.59 is $42.39. Divide by 2 to get $21.19, the average cost per manhour for crew B1.

In the table below, the manhour average cost is the sum of hourly costs of all crew members divided by the number of crew members.

Costs in the *Labor* column in this book are the product of the installation time (in manhours) multiplied by the cost per manhour. For example, on page 34 the labor cost listed for 1/2" CD wall sheathing is $.34 per square foot. That's the installation time (.016 manhours per square foot) multiplied by $21.19, the average cost per manhour for crew B1.

Craft Code	Manhour Avg. Cost	Crew Composition
A1	$40.70	1 asbestos worker, 1 laborer
AT	34.92	1 air tool operator
AW	47.29	1 asbestos worker
B1	21.19	1 laborer, 1 carpenter
B2	21.99	1 laborer, 2 carpenters
B3	20.41	2 laborers, 1 carpenter
B4	23.19	1 laborer, 1 operating engineer, 1 reinforcing ironworker
B5	23.07	1 laborer, 1 carpenter, 1 cement mason, 1 operating engineer, 1 reinforcing ironworker
B6	20.49	1 laborer, 1 cement mason
B7	19.74	1 laborer, 1 truck driver
B8	22.36	1 laborer, 1 operating engineer
B9	20.55	1 bricklayer, 1 bricklayer's tender
BB	$23.38	1 bricklayer
BC	23.59	1 carpenter
BE	25.28	1 electrician
BF	22.45	1 floor layer
BG	21.97	1 glazier
BH	17.71	1 bricklayer's tender (Residential)
BL	18.80	1 laborer
BM	45.89	1 boilermaker
BR	22.92	1 lather
BS	20.90	1 marble setter
BT	32.47	1 bricklayer's tender (Commercial & Industrial)
C1	34.80	4 laborers, 1 truck driver
C2	39.05	1 laborer, 2 truck drivers, 2 tractor operators
C3	38.22	1 laborer, 1 truck driver, 1 tractor operator
C4	35.26	2 laborers, 1 truck driver

Craft Code	Manhour Avg. Cost	Crew Composition
C5	$37.19	2 laborers, 1 tractor operator, 1 truck driver
C6	36.57	6 laborers, 2 truck drivers, 2 tractor operators
C7	38.25	2 laborers, 3 truck drivers, 1 crane operator, 1 tractor operator
C8	38.47	1 laborer, 1 carpenter
C9	38.96	1 laborer, 1 crane operator
CB	42.44	1 bricklayer
CC	42.84	1 carpenter
CD	42.54	1 drywall installer
CE	46.33	1 electrician
CF	22.17	1 concrete mason
CG	43.73	1 glazier
CL	34.10	1 laborer
CM	40.14	1 cement mason
CO	43.82	1 crane operator
CT	21.79	1 mosaic & terrazzo worker
D1	23.47	1 drywall installer, 1 drywall taper
D2	38.32	1 drywall installer, 1 laborer
D3	42.83	1 laborer, 1 ironworker (structural), 1 millwright
D4	39.04	1 millwright, 1 laborer
D5	39.99	1 boilermaker, 1 laborer
D6	43.24	2 millwrights, 1 tractor operator
D7	37.35	1 painter, 1 laborer
D8	42.05	2 elevator constructors, 1 laborer
D9	40.28	2 millwrights, 1 laborer
DI	24.03	1 drywall installer
DT	22.91	1 drywall taper
E1	40.77	2 electricians, 2 laborers, 1 tractor operator
E2	38.18	1 electrician, 2 laborers
E3	40.88	2 electricians, 2 carpenters, 2 laborers
E4	40.22	1 electrician, 1 laborer
EC	46.02	1 elevator constructor
F5	39.34	3 carpenters, 2 laborers
F6	$39.30	2 carpenters, 2 laborers, 1 tractor operator
F7	40.69	2 carpenters, 1 laborer, 1 tractor operator
F8	39.05	2 plasterers, 1 plasterer's helper
F9	38.09	1 floor layer, 1 laborer
FL	42.08	1 floor layer
G1	38.91	1 glazier, 1 laborer
H1	42.74	1 carpenter, 1 laborer, 1 ironworker (structural), 1 tractor operator
H2	40.70	1 crane operator, 1 truck driver
H3	37.76	1 carpenter, 3 laborers, 1 crane operator, 1 truck driver
H4	48.44	1 crane operator, 6 ironworkers (structural), 1 truck driver
H5	42.81	1 crane operator, 2 laborers, 2 ironworkers (structural)
H6	42.56	1 ironworker (structural), 1 laborer
H7	48.62	1 crane operator, 2 ironworkers (structural)
H8	47.58	1 crane operator, 4 ironworkers (structural), 1 truck driver
H9	46.15	1 electrician, 1 sheet metal worker
HC	17.71	1 plasterer's helper (Residential)
IW	51.02	1 ironworker (structural)
LA	40.99	1 lather
M1	37.45	1 bricklayer, 1 bricklayer's tender
M2	37.01	1 carpenter, 2 laborers
M3	37.65	1 plasterer, 1 plasterer's helper
M4	38.44	1 laborer, 1 marble setter
M5	40.06	1 pipefitter, 1 laborer
M6	42.47	1 asbestos worker, 1 laborer, 1 pipefitter
M7	42.05	1 laborer, 2 pipefitters

Craft Code	Manhour Avg. Cost	Crew Composition
M8	$43.05	1 laborer, 3 pipefitters
M9	46.18	1 electrician, 1 pipefitter
MR	23.74	1 millwright (Residential)
MS	42.78	1 marble setter
MT	39.74	1 mosaic & terrazzo worker
MW	43.37	1 millwright (Commercial & Industrial)
OE	25.92	1 operating engineer
P1	21.80	1 plumber, 1 laborer (Residential)
P2	20.73	1 painter, 1 laborer
P5	36.57	3 laborers, 1 tractor operator, 1 truck driver
P6	40.87	1 plumber, 1 laborer (Commercial & Industrial)
P7	38.54	1 carpenter, 3 laborers, 1 painter, 2 tractor operators, 2 truck drivers
P8	37.12	1 laborer, 1 cement mason
P9	39.03	1 carpenter, 1 laborer, 1 cement mason
PA	40.60	1 painter (Commercial & Industrial)
PD	46.12	1 pile driver
PF	46.03	1 pipefitter
PH	33.47	1 plasterer's helper (Commercial & Industrial)
PL	47.64	1 plumber (Commercial & Industrial)
PM	24.79	1 plumber (Residential)
PR	23.38	1 plasterer (Residential)
PS	41.84	1 plasterer (Commercial & Industrial)
PT	22.66	1 painter (Residential)
R1	22.29	1 roofer, 1 laborer
R3	42.16	2 roofers, 1 laborer
RB	46.73	1 reinforcing ironworker (Commercial & Industrial)
RF	46.20	1 roofer (Commercial & Industrial)
RI	24.85	1 reinforcing ironworker (Residential)
RR	25.78	1 roofer (Residential)
S1	38.54	1 laborer, 1 tractor operator
S2	$40.02	1 laborer, 2 tractor operators
S3	40.28	1 tractor operator, 1 truck driver
S4	34.10	3 laborers
S5	35.99	5 laborers, 1 crane operator, 1 truck driver
S6	37.06	2 laborers, 1 tractor operator
S7	38.54	3 laborers, 3 tractor operators
S8	40.69	2 pile drivers, 2 laborers, 1 crane operator, 1 tractor operator, 1 truck driver
S9	38.98	1 pile driver, 2 laborers, 1 tractor operator, 1 truck driver
SM	45.97	1 sheet metal worker (Commercial & Industrial)
SP	48.05	1 sprinkler fitter
SW	25.18	1 sheet metal worker (Residential)
T1	20.47	1 tile layer, 1 laborer
T2	42.30	3 laborers, 3 carpenters, 3 ironworkers (structural), 1 crane operator, 1 truck driver
T3	40.41	1 laborer, 1 reinforcing ironworker
T4	36.92	1 laborer, 1 mosaic & terrazzo worker
T5	40.04	1 sheet metal worker, 1 laborer
T6	42.01	2 sheet metal workers, 1 laborer
TD	37.59	1 truck driver/teamster/oiler (Commercial & Industrial)
TL	22.14	1 tile layer
TO	42.98	1 tractor operator
TR	20.69	1 truck driver/teamster/oiler (Residential)
U1	39.71	1 plumber, 2 laborers, 1 tractor operator
U2	38.61	1 plumber, 2 laborers

Residential Division

Hourly Labor Cost

Craft	1 Base wage per hour	2 Taxable fringe benefits (@5.15% of base wage)	3 Insurance and employer taxes(%)	4 Insurance and employer taxes($)	5 Non-taxable fringe benefits (@4.55% of base wage)	6 Total hourly cost used in this book
Bricklayer	$16.50	$0.85	30.4%	$5.28	$0.75	$23.38
Bricklayer Helper	12.50	0.64	30.4	4.00	0.57	17.71
Building Laborer	13.00	0.67	33.2	4.54	0.59	18.80
Building Carpenter	16.25	0.84	33.7	5.76	0.74	23.59
Cement Mason	16.00	0.82	27.5	4.62	0.73	22.17
Drywall Installer	16.25	0.84	36.3	6.20	0.74	24.03
Drywall Taper	16.25	0.84	29.7	5.08	0.74	22.91
Electrician	18.50	0.95	25.6	4.98	0.84	25.28
Floor layer	16.25	0.84	27.1	4.62	0.74	22.45
Glazier	15.50	0.80	30.4	4.96	0.71	21.97
Lather	17.00	0.88	23.9	4.27	0.77	22.92
Marble Setter	14.75	0.76	30.4	4.72	0.67	20.90
Millwright	17.00	0.88	28.5	5.09	0.77	23.74
Mosaic & Terrazzo Worker	15.75	0.81	27.3	4.52	0.72	21.79
Operating Engineer	18.40	0.95	29.6	5.73	0.84	25.92
Painter	16.00	0.82	30.3	5.10	0.73	22.66
Plasterer	16.50	0.85	30.4	5.28	0.75	23.38
Plasterer Helper	12.50	0.64	30.4	4.00	0.57	17.71
Plumber	18.00	0.93	26.7	5.04	0.82	24.79
Reinforcing Ironworker	16.25	0.84	41.1	7.02	0.74	24.85
Roofer	16.50	0.85	44.3	7.68	0.75	25.78
Sheet Metal Worker	18.00	0.93	28.7	5.43	0.82	25.18
Tile Layer	16.00	0.82	27.3	4.59	0.73	22.14
Truck Driver	14.50	0.75	31.4	4.78	0.66	20.69

The labor costs shown in Column 6 were used to compute the manhour costs for crews on pages 5 to 7 and the figures in the "Labor" column of the Residential Division of this manual. Figures in the "Labor" column of the Industrial and Commercial Division of this book were computed using the hourly costs shown on page 221. All labor costs are in U.S. dollars per manhour.

It's important that you understand what's included in the figures in each of the six columns above. Here's an explanation:

Column 1, the base wage per hour, is the craftsman's hourly wage. These figures are representative of what many contractors will be paying craftsmen working on residential construction in 1992.

Column 2, taxable fringe benefits, includes vacation pay, sick leave and other taxable benefits. These fringe benefits average 5.15% of the base wage for many construction contractors. This benefit is in addition to the base wage.

Column 3, insurance and employer taxes in percent, shows the insurance and tax rate for construction trades. The cost of insurance in this column includes workers' compensation and contractor's casualty and liability coverage. Insurance rates vary widely from state to state and depend on a contractor's loss experience. Typical rates are shown in the Insurance section of this manual beginning on page 106. Taxes are itemized in the section on page 203. Note that taxes and insurance increase the hourly labor cost by 25 to 30% for most trades. There is no legal way to avoid these costs.

Column 4, insurance and employer taxes in dollars, shows the hourly cost of taxes and insurance for each construction trade. Insurance and taxes are paid on the costs in both columns 1 and 2.

Column 5, non-taxable fringe benefits, includes employer paid non-taxable benefits such as medical coverage and tax-deferred pension and profit sharing plans. These fringe benefits average 4.55% of the base wage for many construction contractors. The employer pays no taxes or insurance on these benefits.

Column 6, the total hourly cost in dollars, is the sum of columns 1, 2, 4, and 5.

These hourly labor costs will apply within a few percent on many jobs. But wage rates may be much higher or lower in some areas. If the hourly costs shown in column 6 are not accurate for your work, develop modification factors that you can apply to the labor costs in this book. The following paragraphs explain the procedure.

Adjusting Labor Costs

Here's how to customize the labor costs in this book if your wage rates are different from the wage rates shown on page 8 or 221.

Start with the taxable benefits you offer. Assume craftsmen on your payroll get one week of vacation each year and one week of sick leave each year. Convert these benefits into hours. Your computation might look like this:

```
   40 vacation hours
+  40 sick leave hours
   80 taxable leave hours
```

Then add the regular work hours for the year:

```
   2,000 regular hours
+     80 taxable benefit hours
   2,080 total hours
```

Multiply these hours by the base wage per hour. If you pay carpenters $8.00 per hour, the calculation would be:

```
   2,080 hours
x  $8.00 per hour
   $16,640 per year
```

Next determine the tax and insurance rate for each trade. If you know the rates that apply to your jobs, use those rates. If not, use the rates in column 3 on page 8. Continuing with our example, we'll use 33.7%, the rate for carpenters in column 3 on page 8. To increase the annual taxable wage by 33.7%, we'll multiply by 1.337:

```
   $16,640 per year
x    1.337 tax & insurance rate
   $22,248 annual cost
```

Then add the cost of non-taxable benefits. Suppose your company has no pension or profit sharing plan but does provide medical insurance for employees. Assume that the cost for your carpenter is $113 per month or $1,356 per year.

```
   $ 1,356 medical plan
+   22,248 annual cost
   $23,604 total annual cost
```

Divide this total annual cost by the actual hours worked in a year. This gives the contractor's total hourly labor cost including all benefits, taxes and insurance. Assume your carpenter will work 2,000 hours a year:

$$\frac{\$23,604}{2,000} = \$11.80 \text{ per hour}$$

Finally, find your modification factor for the labor costs in this book. Divide your total hourly labor cost by the total hourly cost shown on page 8. For the carpenter in our example, the figure in column 8 is $23.59:

$$\frac{\$11.80}{\$23.59} = .50$$

Your modification factor is 50%. Multiply any building carpenter (Craft Code BC) labor costs in the Residential Division of this book by .50 to find your estimated cost. For example, notice on page 19 that the labor cost for installing an 18" long towel bar is $6.60 per each bar. If installed by your carpenter working at $8.00 per hour, your estimated cost would be one-half of $6.60 or $3.30. The manhours would remain the same @ .280.

If the Labor Rate Is Unknown

On some estimates you may not know what labor rates will apply. In that case, use both labor and material figures in this book without making any adjustment. When all labor, equipment and material costs have been compiled, multiply the totals by the appropriate factor in the area modification table on page 10 or 11.

Adjusting the labor costs in this book will make your estimates much more accurate.

Area Modification Factors

Construction costs are higher in some cities than in other cities. Use the factors on this page or page 11 to adapt the costs listed in this book to your job site. Multiply your estimated total project cost by the factor listed for the appropriate city in this table to find your estimated building cost.

These factors were compiled by comparing the actual construction cost of residential, institutional and commercial buildings in 402 communities throughout the United States. Because these factors are based on completed project costs, they consider all construction cost variables, including labor, equipment and material cost, labor productivity, climate, job conditions and markup.

Use the factor for the nearest or most comparable city. If the city you need is not listed in the table, use the factor for the appropriate state.

Note that these location factors are composites of many costs and will not necessarily be accurate when estimating the cost of any particular part of a building. But when used to modify all estimated costs on a job, they should improve the accuracy of your estimates.

Location	Factor
Alabama	**0.80**
Birmingham	0.82
Dothan	0.76
Florence	0.78
Huntsville	0.80
Mobile	0.83
Montgomery	0.80
Tuscaloosa	0.79
Alaska	**2.21**
Anchorage	1.47
Fairbanks	1.57
Juneau	1.51
Kenai	1.35
Ketchikan	1.51
Sitka	1.53
Arizona	**0.97**
Casa Grande	1.01
Flagstaff	0.94
Kingman	0.94
Phoenix	1.03
Sierra Vista	0.92
Tucson	1.00
Yuma	0.94
Arkansas	**0.82**
Fayetteville	0.81
Fort Smith	0.81
Jonesboro	0.80
Little Rock	0.84
Texarkana	0.80
California	**1.20**
Arrowhead	1.24
Bakersfield	1.18
Barstow	1.22
Big Bear	1.25
Desert Center	1.31
El Cajon	1.20
Eureka	1.17
Fresno	1.09
Inyokern	1.20
Lancaster	1.28
Los Angeles	1.32
Modesto	1.08
Mojave	1.23
Needles	1.23
Oakland	1.34
Ojai	1.22
Oxnard	1.25
Paso Robles	1.22
Piru	1.21
Placerville	1.09
Redding	1.05
Ridgecrest	1.17
Sacramento	1.08
San Bernardino	1.23
San Diego	1.19
San Francisco	1.47
San Jose	1.32
Santa Ana	1.31
Santa Barbara	1.26
Santa Cruz	1.32
Santa Maria	1.24
Santa Rosa	1.19
South Lake Tahoe	1.10
Tehachapi	1.22
Ventura	1.25
Victorville	1.23
Yreka	1.05
Colorado	**1.02**
Aspen-Vail	1.14
Denver	0.99
Grand Junction	1.02
Connecticut	**1.21**
Bridgeport	1.24
Hartford	1.22
New Haven	1.20
New London	1.18
New Milford	1.21
Norwich	1.17
Ridgefield	1.25
Windam	1.17
Delaware	**1.07**
Dover	1.06
Wilmington	1.08
District of Columbia	**1.22**
Washington D.C.	1.22
Florida	**0.90**
Jacksonville	0.85
Key West	0.94
Miami	0.93
Orlando	0.99
Pensacola	0.83
Tampa	0.91
Georgia	**0.82**
Albany	0.81
Atlanta	0.84
Augusta	0.81
Brunswick	0.80
Columbus	0.80
Macon	0.80
Rome	0.79
Savannah	0.80
Valdosta	0.80
Guam	**1.39**
Agana	1.39
Hawaii	**1.49**
Hilo	1.60
Honolulu	1.52
Kauai	1.63
Kona	1.62
Maui	1.59
Idaho	**0.99**
Boise	0.98
Coeur D'Alene	1.04
Idaho Falls	0.98
Pocatello	1.02
Illinois	**1.05**
Bellville	1.04
Chicago	1.13
East St. Louis	1.06
Moline	1.06
Springfield	1.03
Indiana	**0.96**
Bloomington	0.94
Evansville	0.92
Fort Wayne	0.92
Gary	1.02
Hammond	1.00
Indianapolis	0.93
Lafayette	0.94
South Bend	0.95
Terre Haute	0.96
Iowa	**0.92**
Bettendorf	0.97
Cedar Rapids	0.93
Council Bluffs	0.89
Davenport	0.95
Des Moines	0.92
Dubuque	0.92
Mason City	0.91
Sioux City	0.90
Waterloo	0.92
Kansas	**0.86**
Garden City	0.81
Kansas City	0.96
Pittsburg	0.80
Salina	0.82
Topeka	0.88
Wichita	0.82
Kentucky	**0.92**
Ashland	0.96
Covington	0.98
Louisville	0.89
Middlesboro	0.93
Owensboro	0.90
Paducah	0.89
Louisiana	**0.85**
Alexandria	0.80
Baton Rouge	0.89
Houma	0.87
Lafayette	0.85
Lake Charles	0.88
Marshall	0.88
Monroe	0.81
New Orleans	0.88
Shreveport	0.83
Maine	**1.06**
Augusta	1.10
Bangor	1.04
Brunswick	1.13
Lewiston	1.11
Portland	1.11
Waterville	1.06
Maryland	**1.00**
Baltimore	1.02
Baltimore City	1.05
Hagerstown	0.95
Salisbury	0.98
Waldorf	1.00
Massachusetts	**1.23**
Boston	1.36
Fall River	1.25
Worchester	1.22
Michigan	**1.00**
Ann Arbor	1.04
Battle Creek	0.96
Benton Harbor	0.99
Detroit	1.06
Flint	1.03
Grand Rapids	0.95
Jackson	1.01
Lansing	1.04
Marquette	1.00
Mt. Pleasant	0.99
Muskegon	0.94
Saginaw	0.98
Ypsilanti	1.07
Minnesota	**1.02**
Duluth	1.04
Mankato	1.00
Minneapolis	1.07
Rochester	1.02
St. Cloud	1.01
Worthington	0.95
Mississippi	**0.80**
Corinth	0.78
Greenville	0.79
Greenwood	0.78

Area Modification Factors

Gulfport	0.80
Hattisburg	0.77
Jackson	0.78
Southaven	0.78
Missouri	**0.95**
Cape Girardeau	0.95
Columbia	0.94
Joplin	0.89
Kansas City	1.00
Kirksville	0.95
Rolla	0.89
St. Joseph	0.95
St. Louis	1.04
Sedalia	0.93
Springfield	0.92
Montana	**0.96**
Billings	0.96
Great Falls	0.97
Helena	0.96
Missoula	0.96
Nebraska	**0.89**
Grand Island	0.91
Lincoln	0.88
Omaha	0.91
Norfolk	0.91
North Platte	0.87
Scottsbluff	0.87
Nevada	**1.13**
Las Vegas	1.16
Reno	1.15
New Hampshire	**1.07**
Concord	1.07
Dover	1.07
Keene	1.05
Manchester	1.07
Nashua	1.09
Portsmouth	1.09
New Jersey	**1.21**
Ashbury Park	1.23
Atlantic City	1.25
Burlington	1.22
Camden	1.23
Freehold	1.23
Gloucester	1.22
Newark	1.30
North Bergen	1.30
Trenton	1.23
Vineland	1.21
New Mexico	**0.89**
Albuquerque	0.84
Clovis	0.83
Santa Fe	0.87
Silver City	0.85
Taos	0.93
New York	**1.21**
Albany	1.07
Binghamton	1.05
Buffalo	1.06
Elmira	1.02
Jamestown	0.98
Nassau County	1.41
N.Y. City (Metro)	1.46
N.Y. City (Inner)	1.52
Orange County	1.35
Plattsburgh	1.01
Poughkeepsie	1.21
Rochester	1.04
Rockland County	1.36
Suffolk County	1.27
Syracuse	1.06
Westchester County	1.42
North Carolina	**0.81**
Asheville	0.81
Charlotte	0.81
Durham	0.82
Elizabeth City	0.82
Fayetteville	0.78
Greensboro	0.79
Greenville	0.78
Raleigh	0.83
Wilmington	0.79
Winston-Salem	0.79
North Dakota	**0.93**
Bismarck	0.96
Dickinson	0.95
Fargo	0.92
Ohio	**1.07**
Akron	1.12
Cincinnati	1.01
Cleveland	1.15
Columbus	1.00
Dayton	1.00
Findlay	1.03
Lorain	1.11
Mansfield	1.03
Toledo	1.12
Youngstown	1.06
Oklahoma	**0.84**
Ada	0.85
Ardmore	0.84
Bartlesville	0.84
Enid	0.83
Guymon	0.85
Lawton	0.83
McAlester	0.85
Muskogee	0.86
Oklahoma City	0.84
Shawnee	0.85
Stillwater	0.85
Tulsa	0.84
Woodward	0.84
Oregon	**1.02**
Bend	1.00
Coos Bay	1.03
Eugene	0.99
Portland	1.02
Pennsylvania	**1.06**
Allentown	1.17
Altoona	1.06
Bellefonte	1.08
Erie	1.09
Harrisburg	1.04
Johnstown	1.06
Lancaster	1.08
Philadelphia	1.21
Pittsburgh	1.10
Reading	1.06
Scranton	1.08
Wellsboro	1.07
York	1.03
Puerto Rico	**0.88**
Arecibo	0.88
Mayaguez	0.88
Old San Juan	0.89
Ponce	0.88
San Juan	0.87
Rhode Island	**1.16**
Providence	1.18
South Carolina	**0.81**
Aiken	0.80
Anderson	0.80
Beaufort	0.82
Charleston	0.84
Columbia	0.80
Florence	0.79
Greenville	0.81
Greenwood	0.80
Myrtle Beach	0.82
North Augusta	0.82
Orangeburg	0.80
Rockhill	0.80
Spartansburg	0.81
South Dakota	**0.95**
Pierre	0.98
Rapid City	0.96
Sioux Falls	0.98
Tennessee	**0.81**
Chattanooga	0.83
Clarksville	0.79
Columbia	0.81
Jackson	0.82
Johnson City	0.79
Kingsport	0.80
Knoxville	0.82
Memphis	0.82
Nashville	0.80
Oak Ridge	0.81
Texas	**0.83**
Abilene	0.85
Amarillo	0.87
Austin	0.81
Beaumont	0.85
Bryan	0.85
Corpus Christi	0.77
Dallas	0.77
Eagle Pass	0.78
El Campo	0.85
El Paso	0.82
Fort Worth	0.85
Harlingen	0.77
Houston	0.84
Junction	0.79
Laredo	0.77
Lubbock	0.88
Lufkin	0.84
Midland	0.90
Odessa	0.90
San Angelo	0.84
San Antonio	0.79
Sherman	0.84
Texas City	0.85
Tyler	0.82
Waco	0.81
Wichita Falls	0.85
Victoria	0.79
Utah	**0.94**
Cedar City	0.95
Salt Lake City	0.98
Vernal	0.97
Vermont	**1.07**
Bennington	1.09
Brattleboro	1.09
Burlington	1.06
Montpelier	1.06
Rutland	1.06
Virginia	**0.92**
Charlottesville	0.92
Harrisonburg	0.88
Newport News	0.91
Norfolk	0.93
Norton	0.92
Richmond	0.94
Virgin Islands	**1.16**
St. Thomas	1.17
St. Croix	1.15
Washington	**1.06**
Aberdeen	1.06
Bellingham	1.06
Cheney	1.05
Kennewick	1.07
Longview	1.05
Olympia	1.07
Port Angeles	1.07
Pullman	1.06
Seattle	1.07
Spokane	1.04
Yakima	1.06
West Virginia	**1.00**
Bluefield	0.96
Charleston	1.02
Fairmont	1.03
Huntington	0.99
Martinsburg	0.96
Parkersburg	1.01
Point Pleasant	1.03
Wheeling	1.08
Wisconsin	**1.00**
Eau Claire	1.01
Green Bay	0.97
Madison	1.01
Milwaukee	1.05
Reedsville	0.99
Superior	1.03
Wausau	0.98
Wyoming	**0.94**
Casper	0.95
Cheyenne	0.93
Cody	0.96

Credits and Acknowledgments

This book has over 30,000 cost estimates for 1992. To develop these estimates, the editors relied on information supplied by hundreds of construction cost authorities. We offer our sincere thanks to the contractors, engineers, design professionals, construction estimators, material suppliers, and manufacturers who, in the spirit of cooperation, have assisted in the preparation of this Fortieth Edition of the *National Construction Estimator*. Many of the cost authorities who supplied information for this volume are listed below.

Gina Adams, CSI, R.C.P. Brick Co.; **Mike Allen**, Bayside Specialties; **Don Altevers**, S.D.G. & E.; **Neal Ames**, Castec, Inc.; **Jeff Andrews**, Delta Pacific Builders; ; **Richard V. Aronson**, Norwesco Inc.; **David D. Backer**, David Backer Ltd.; **Bill Baker**, Stagecraft Industries; **Doug Barkee**, Weyerhaeuser Co.; **Mel Baugh**, Hafer Steel Co.; **Mel Beard**, U. S. Tile Co.; **Alec W. Bechtold**, Ventarama Skylights; **Greg Becker**, Cement Cutting, Inc.; **Doug Bench**, J.H. Baxter Co.; **Emil Bonfini**, Home Federal Insurance; **Bill Boren**, Arizona Rebar; **Robert Branlett**, Potlatch Corp.; **Brent Brewer**, Cabrillo Crane & Rigging Co.; **Tom Briggs**, ICBO; **Ed Brock**, Stucco Stone Products; **Mike Bunke**, Lasco Panels; **Bill Burkhart**, Peerless Products, Inc.; **Everett Castillo**, Otis Elevator Co.; **Claude Coker**, Fortifiber Corp.; **Gerry Colburn**, Rheem Manufacturing; **Mark Costello**, Superior Fireplace Co.; **John L. Cousins**, Flexi-Wall Systems; **Bob Crede**, Inclinator Co.; **Ben Crocker**, Iron-A-Way Co.; **Dick Degreffinreid**, Soil Stabilization Mixes; **Mike Depin**, Tate Access Floors; **Doug Didion**, Ultraflo Corp.; **Bob DiNello**, Specialty Wood Co.; **Jack Eagen**, Moore & Taber Consultants; **Tom Fitts**, System Parking, Inc.; **Tom Forsyth**, Florentine Co.; **James H. Francis**, Kirsch Co.; **Anthony Galeota**, Area Home Inspection; **Jack Garroute**, Rebar Engrg. Inc.; **Kathy Gayhart**, Stark Ceramics; **Alex Gere**, Stone Tech Co.; **Craig Gilbert**, Armor Safe Co.; **Charlie Gindele**, Unique Building Products; **Bill Godfrey**, American Boiler Works; **Steve Gold**, Gold Electric; **Estelle Gore**, Wasco Products; **Ed Gregg**, Roppe Rubber Corp.; **Linda Grundman**, El Cajon Electric; **G. O. Guericke**, Georgia Pacific Co.; **Ed Gunderson**, Plastmo Raingutters; **John Gunderson**, Nasscor; **T. Gunusky**, Perma Grain Prod.; **Andrew Gutt**, A&A Drafting Service; **Robert Hammond**, Cad-Cor Corp.; **Mark Harbold**, Vintage Lumber & Const.; **James Hay**, Henry Co.; **Larry Hoagland**, Professional Photographic Srvs.; **Michael Hoffman**, Wilkinson Co.; **Mary Holman**, Koffler Sales Corp.; **Jeff Horn**, Aged Woods; **Morris Hudspeth**, Bessler Stairways; **Bob Johnson**, Allied Barricade; **Ned Jones**, Pioneer Manufacturing; **E. C. Kamps**, San Diego Acoustics; **G. Kellett**, Superior Stakes; **Jim Kneaper**, West Coast Lumber Inspection Bureau; **John Kosar**, Construction Specialties; **John Kratz**, Multi Water Systems; **Jerry Leavitt**, M. H. Loe Co.; **Jim LeBeau**, LeBeau Building Specialties; **Peter Malek**, A-1 Security Systems; **Kelly McBride**, Sealant Specialties; **Diane McCann**, San Diego Aerial Equipment Co.; **Lewis Mitchell**, Shakespeare Co.; **Warren Monsees**, Putnam Rolling Ladder Co.; **Donna Morgan**, Classic Ceilings; **Reid Neslage**, H&H Specialties; **Doug Niel**, Builder's Carpet Mart; **Mel Nortley**, Mel-Nor; **Walter O'Brien**; **Al Panton**, Kyocera, Solar Division; **Ted H. Pope**, Mirafi Inc.; **David Pyman**, Bilco Co.; **Marty Ratner**, Nuclear Associates; **Thomas D. Reider**, Jewett Refrigerator Co.; **L. A. Richards**, Giles and Kendall, Inc.; **Ryan Roulette**, Roulette Photography; **Mike Ryan**, Sedgwick Lifts Inc.; **Toby Sanders**, Sanders Hydroseeding; **James Sauder**, Alumax Magnolia Div.; **Steve Scholl**, National Schedule Masters; **Michael Sedano**, C. R. Laurence Co., Inc.; **Edward F. Sedlak**, Abitibi Price; **Bill Sharp**, Western Exterminators; **Susan Smallwood**, Sau-Sea Swimming Pool Products; **Ray Smith**, Koppers Co.; **Pat Stebbins**, Stebbins Co.; **John Stern**, Kentucky Wood Floors; **Jim Stevens**, Atascadero Glass; **Don Strawbridge**, Vital Services Co.; **Reino J. Tarkianen**, Finlandia Sauna; **Clayton Taylor**, California Pneumatic Systems; **John Tefler**, Asphalt Service Co.; **Roland Temple**, Velux-America; **Fling Traylor**, El-Do Tile; **T. Vanderhoof**, Ted's Quality Roofing; **Bob Warren**, Creative Ceilings; **L. D. Warren**, Tarkett Corp.; **Brice Westphall**, Hydrodynamics Assoc.; **Bruce Wickern**, Architectural Illustration; **D. Widders**, Lane Stanton Vance Lmbr; **Eleanor Wiese**, Centennial Flag Pole Co.; **Charles Williamson**, Sico, Inc.; **J. Zanger**, Walker & Zanger; **Joe Zavada**, Pittsburgh Corning Co.

A special thanks to the following people. Production and layout: **Ted Fukumoto, Karen Sheffield, Shanna Walborn**. Editorial and proofreading: **Laurence Jacobs, Alice Nolan**. Research: **John Rowen, John Svalina**. Indexing: **Rozanne Bryson, Charlene Jenia, Genie Runyon, Claudia Stebelski**. Cover photography: **Ed Kessler**

Adhesives

Adhesives. See also Carpet, Caulking, Flooring, Glazing, Gypsum, and Tile in the Residential Division of this book.

Panel adhesives Better quality, gun applied in continuous bead to wood or metal framing or furring members, material only, add labor below. Per 100 SF of wall, floor, or ceiling including 6% waste.

		Bead diameter			
		1/8"	1/4"	3/8"	1/2"
Subfloor adhesive (at $.13 per oz), on floors					
12" OC members	CSF	5.40	12.20	13.75	21.90
16" OC members	CSF	4.20	9.90	10.90	17.45
20" OC members	CSF	3.80	8.45	9.75	14.95
Wall sheathing or shear panel adhesive (at $.14 per oz), on walls					
12" OC members	CSF	3.12	12.60	28.85	51.50
16" OC members	CSF	2.90	11.60	26.30	46.75
20" OC members	CSF	2.80	11.20	25.65	44.60
24" OC members	CSF	2.60	11.10	24.10	42.65
Polystyrene or polyurethane foam panel adhesive (at $.23 per oz), on walls					
12" OC members	CSF	2.96	12.50	28.10	50.20
16" OC members	CSF	2.74	11.50	25.60	45.70
20" OC members	CSF	2.63	10.90	24.60	43.70
24" OC members	CSF	2.63	10.50	23.50	41.60
Gypsum drywall adhesive (at $.19 per oz), on ceilings					
12" OC members	CSF	1.78	7.50	16.60	29.40
16" OC members	CSF	1.34	5.60	12.70	22.40
20" OC members	CSF	1.22	4.70	10.40	18.30
24" OC members	CSF	1.12	4.02	9.00	16.00
Gypsum drywall adhesive (at $.19 per oz), on walls					
12" OC members	CSF	2.53	10.00	22.50	40.00
16" OC members	CSF	2.22	9.08	20.50	36.30
20" OC members	CSF	2.11	8.65	20.60	34.60
24" OC members	CSF	2.11	8.34	18.70	33.25
Hardboard or plastic panel adhesive (at $.19 per oz), on walls					
12" OC members	CSF	2.43	10.35	23.35	41.50
16" OC members	CSF	2.32	9.40	21.20	37.80
20" OC members	CSF	2.22	9.00	20.30	35.90
24" OC members	CSF	2.01	8.65	19.40	34.60

Labor to apply adhesive to framing members, 1/8" to 1/2" bead diameter, no material included

	Craft@Hrs	Unit	Material	Labor	Total
Floor or ceiling joists					
12" OC members	BC@.075	CSF	--	1.77	--
16" OC members	BC@.056	CSF	--	1.32	--
20" OC members	BC@.052	CSF	--	1.23	--
24" OC members	BC@.042	CSF	--	.99	--
Interior and exterior wall members					
12" OC members	BC@.100	CSF	--	2.36	--
16" OC members	BC@.090	CSF	--	2.12	--
20" OC members	BC@.084	CSF	--	1.98	--
24" OC members	BC@.084	CSF	--	1.98	--

	29 oz cartridge	1 pint can	1 quart can	1 gallon can
General purpose adhesives				
Acoustic tile adhesive, solvent base, waterproof, sound deadening type	--	--	5.60	16.70
Aliphatic resin woodworking glue	--	5.80	8.50	23.60
Carpet cement, latex, multi-purpose	--	--	6.30	18.20

Adhesives

	29 oz cartridge	1 pint can	1 quart can	1 gallon can
Contact cement, latex base, water cleanup, waterproof, bonds veneers to plywood, particleboard, wallboard	--	7.70	11.70	35.60
Contact cement, solvent base, solvent cleanup waterproof, bonds metal, linoleum, rubber, veneers, plastic laminates, leather, plywood	--	6.10	9.70	26.50
Contact cement solvent	--	8.90	12.90	--
Foam panel insulation adhesive, waterproof, for polyurethane & polystyrene panels, bonds to wood, metal, masonry & concrete	8.80	--	--	27.90
Gypsum drywall adhesive, waterproof, bonds to wood, metal, masonry and concrete	6.15	--	--	23.00
Hardboard, fiberboard, and plastic panel adhesive, waterproof, bonds to wood, plaster, concrete, masonry and gypsum	--	--	8.00	23.80
Parquet tile adhesive, solvent base	--	--	--	30.00
Plastic resin glue, marine grade	--	5.90	--	--
Tile cement, solvent base, for installation of ceramic, quarry, slate and marble tiles	--	--	7.90	24.30
Resilient flooring adhesive, latex base, adheres to concrete, plywood, felt, sheet flooring	--	--	7.00	21.70
Resorcinol glue, waterproof, 2 part	--	21.50	34.50	--
Subfloor and general purpose construction adhesive, waterproof, specification grade, bonds to wood, metal, concrete and masonry	7.40	--	5.60	15.00

	Craft@Hrs	Unit	Material	Labor	Total
Aggregate. Typical prices, 5 mile haul, 24 ton minimum.					
Crushed stone (1.4 tons per CY)					
3/4" (Number 3)	--	Ton	12.30	--	--
1-1/2" (Number 2)	--	Ton	12.30	--	--
Crushed slag, typical prices where available					
3/4"	--	Ton	8.00	--	--
1-1/2"	--	Ton	8.10	--	--
Gravel (1.4 tons per CY)					
3/4"	--	Ton	12.50	--	--
1-1/2"	--	Ton	12.75	--	--
Sand (1.35 tons per CY)	--	Ton	9.95	--	--
Add per ton less than 24 tons	--	Ton	4.65	--	--
Add for delivery over 5 miles, one way	--	Mile	2.60	--	--

Appraisal Fees. Costs for determining the value of existing buildings, land, and equipment. Actual fee charged is based on the level of difficulty and the time spent on appraisal plus travel to location and cost of support services, if any. Costs include research and report by a professional appraiser. Client may request an appraisal on a "fee not to exceed" basis. Fees shown are averages and are not quoted as a percentage of value. Costs shown are typical fees.

Residences, single family dwellings (not including contents)					
To $75,000 valuation	--	LS	--	--	300.00
Over $75,000 to $300,000 valuation	--	LS	--	--	350.00
Over $300,000 to $500,000 valuation	--	LS	--	--	450.00
Over $500,000 to $1,000,000 valuation	--	LS	--	--	650.00
Over $1,000,000 to $3,000,000 valuation	--	LS	--	--	2,500.00

Appraisal Fees

	Craft@Hrs	Unit	Material	Labor	Total
Apartment houses, commercial and industrial buildings					
To $300,000 valuation	--	LS	--	--	3,000.00
Over $300,000 to $1,000,000	--	LS	--	--	4,000.00
Over $1,000,000 to $3,000,000	--	LS	--	--	6,000.00
Over $3,000,000 to $5,000,000	--	LS	--	--	7,500.00
Machinery (Fee is based on value of each piece of equipment appraised.) Charges listed are per $1,000 of valuation, $500.00 minimum fee					
To $30,000 valuation	--	Ea	--	--	25.00
Over $ 30,000 to $100,000	--	Ea	--	--	40.00
Over $100,000 to $400,000	--	Ea	--	--	45.00
Over $400,000 to $500,000	--	Ea	--	--	50.00
Over $500,000 to $1,000,000	--	Ea	--	--	60.00
Appraisals per 1/2 day (typical)	--	Rate	--	--	350.00
Appraisals per day (typical)	--	Rate	--	--	650.00
Court testimony (excluding preparation)	--	Day	--	--	1,000.00

Arbitration. These are administrative fees paid to the American Arbitration Association (AAA). Includes appointment of arbitrator(s) and use of hearing rooms. Arbitrators are chosen from the National Panel of Construction Arbitrators and are paid a fee negotiated by the parties. The fee is waived for the first day of service. Legal representation, if desired (although not necessary), is at the expense of each party. A refund schedule is applicable to settled or withdrawn cases.

Administrative AAA filing fee to initiate disclosed case is $300.00
Balance of the fee is based on the amount of each claim and counterclaim
  $10,000 to $25,000, 3%
  $25,000 to $50,000, $750 + 2% of amount over $25,000
  $50,000 to $100,000, $1,250 + 1% of amount over $50,000
  $100,000 to $500,000, $1,750 + .5% of amount over $100,000
  $500,000 to $5,000,000, $3,750 + .25% of amount over $500,000
  $5,000,000 to $50,000,000, $15,000 + .10% of amount over $50,000,000
The fee is increased 10% for each additional party represented in excess of two.
Fee charged to party causing a postponement of any scheduled arbitration hearing

	Craft@Hrs	Unit	Material	Labor	Total
Sole-arbitrator cases					
First occurrence	--	LS	--	--	100.00
Second occurrence	--	LS	--	--	200.00
Three-arbitrator cases					
First occurrence	--	LS	--	--	150.00
Second occurrence	--	LS	--	--	300.00

Architectural Illustrations. Full color painting on watercolor board with matted frame with title and credit on matte. Glass and elaborate framing are extra. Costs for pen and ink illustrations with color mylar overlay are similar to cost for watercolor illustrations. Typical fee

	Craft@Hrs	Unit	Material	Labor	Total
Custom home, eye level view					
Simple rendering	--	LS	--	--	850.00
Complex rendering	--	LS	--	--	1,250.00
Custom home, bird's eye view					
Simple rendering	--	LS	--	--	1,000.00
Complex rendering	--	LS	--	--	1,500.00
Tract homes in groups of five or more (single floor plans with multiple elevations), eye level view					
Simple rendering	--	LS	--	--	500.00
Complex rendering	--	LS	--	--	700.00

Architectural Illustrations

	Craft@Hrs	Unit	Material	Labor	Total
Tract homes in groups of five or more (single floor plans with multiple elevations), bird's eye view					
Simple rendering	--	LS	--	--	650.00
Complex rendering	--	LS	--	--	1,000.00
Tract homes or condominium project, overall bird's eye view,					
10-25 homes or living units	--	LS	--	--	3,250.00
Typical commercial structure					
Eye level view	--	LS	--	--	1,400.00
Bird's eye view	--	LS	--	--	1,600.00
Complex commercial structure					
Eye level view	--	LS	--	--	2,100.00
Bird's eye view	--	LS	--	--	2,750.00
Deduct for pen and ink drawings (no color)	--	%	--	--	-33.0
Computer generated perspective drawings using CAD system for design studies					
Custom home	--	LS	--	--	550.00
Large condo or apartment projects	--	LS	--	--	1,400.00
Tract homes		LS	--	--	350.00
Commercial structure, line drawing with hidden lines removed	--	LS	--	--	1,000.00

Awnings and Canopies for Doors and Windows. Costs for awnings include all hardware. All have adjustable support arms to control angle height and preferred amount of window coverage. For larger size aluminum awnings, price two awnings and add for splice kit below.

	Craft@Hrs	Unit	Material	Labor	Total
Natural aluminum ribbed awning with clear, weather-resistant finish and 26" arms					
36" wide x 30" long	SW@1.45	Ea	47.00	36.50	83.50
48" wide x 30" long	SW@1.86	Ea	62.00	46.80	108.80
60" wide x 30" long	SW@2.07	Ea	75.00	52.10	127.10
72" wide x 30" long	SW@2.27	Ea	85.00	57.20	142.20
Add for door canopy with 17" drop sides	--	%	58.0	--	--
Custom colored window awnings in stripes or solids, with baked enamel finish and ventilation panels					
30" wide x 24" high	SW@1.45	Ea	77.00	36.50	113.50
36" wide x 36" high	SW@1.45	Ea	116.00	36.50	152.50
48" wide x 48" high	SW@1.87	Ea	200.00	47.10	247.10
48" wide x 60" high	SW@1.87	Ea	285.00	47.10	332.10
60" wide x 72" high	SW@2.07	Ea	315.00	52.10	367.10
72" wide x 72" high	SW@2.26	Ea	420.00	56.90	476.90
Add for splice kit with overlap slats	SW@.218	Ea	15.00	5.49	20.49
Security roll-up awning with pull cord assembly and folding arms, clear weather-resistant finish. Awning rolls down to cover whole window. 48" long, 24" arms					
36" wide	SW@1.52	Ea	145.00	38.30	183.30
48" wide	SW@1.94	Ea	170.00	48.80	218.80
Plastic awning with baked-on acrylic finish, ventilated side panels, and reinforced metal frame, hardware included. 24" drop, 24" projection					
36" wide	BC@1.58	Ea	80.00	37.30	117.30
42" wide	BC@1.75	Ea	92.00	41.30	133.30
48" wide	BC@2.02	Ea	100.00	47.70	147.70
60" wide	BC@2.26	Ea	120.00	53.30	173.30
72" wide	BC@2.47	Ea	126.00	58.30	184.30
96" wide	BC@2.79	Ea	162.00	65.80	227.80
Plastic door canopy with 36" projection					
42" wide	BC@1.80	Ea	195.00	42.50	237.50

Awnings and Canopies

	Craft@Hrs	Unit	Material	Labor	Total
Traditional fabric awning, with waterproof, acrylic duck, colorfast fabric, double stitched seams, and tubular metal framing and pull cord assembly. 24" drop, 24" projection					
30" wide	BC@1.35	Ea	37.00	31.80	68.80
36" wide	BC@1.58	Ea	42.00	37.30	79.30
42" wide	BC@1.80	Ea	50.00	42.50	92.50
48" wide	BC@2.02	Ea	55.00	47.70	102.70
Add for 30" drop, 30" projection	--	%	10.0	20.0	--
Cloth canopy patio cover, with front bar and tension support rafters, 9" valance and 8' projection					
8' x 10'	BC@2.03	Ea	220.00	47.90	267.90
8' x 15'	BC@2.03	Ea	320.00	47.90	367.90

Barricade, Construction Safety. Purchase prices except as noted.
Reflectorized plywood barricade, 8" to 12" wide rail, 4" to 6" wide stripes, 45" 14 gauge steel legs, no light

	Craft@Hrs	Unit	Material	Labor	Total
Type I, 2' wide, 3' high, 1 reflectorized rail each side					
To 100 units	--	Ea	17.50	--	--
Over 100 units	--	Ea	16.00	--	--
Add for 12 gauge steel legs	--	%	15.0	--	--
Type II, 2' wide, 3' high, 2 reflectorized rails each side					
To 100 units	--	Ea	26.80	--	--
Over 100 units	--	Ea	24.70	--	--
Type III, 4' wide, 5' high, 3 reflectorized rails each side, wood & steel legs					
To 100 units	--	Ea	61.80	--	--
Over 100 units	--	Ea	58.70	--	--
Add for all wood legs	--	%	45.0	--	--
Utility barricade, 2" x 2" legs, 1" x 6" face, no light					
To 100 units	--	Ea	12.40	--	--
Over 100 units	--	Ea	10.80	--	--
Add for lighted units without batteries (batteries last 2 months)					
To 100 units	--	Ea	14.90	--	--
Over 100 units	--	Ea	13.90	--	--
Batteries, 6 volt (2 needed)	--	Ea	1.65	--	--
Lighted units, rental, per day	--	Ea	.45	--	--
Unlighted units, rental per day	--	Ea	.35	--	--
Add for pickup and delivery, per trip	--	Ea	--	--	15.00
"Road Closed", reflectorized, 30" x 48"	--	Ea	53.60	--	--
"Construction Zone", 4' x 4'	--	Ea	89.60	--	--
High-rise tripod with 3 orange flags	--	Ea	82.40	--	--
High-rise sign holder, wind resistant	--	Ea	134.00	--	--
Traffic cones, PVC					
18" high	--	Ea	3.80	--	--
28" high	--	Ea	8.24	--	--
Add for better quality rubber cones	--	%	20.0	--	--
Lane delineator, 42" orange plastic cylinder with 2 reflectors on a 12 pound rubber base	--	Ea	12.40	--	--
Mesh signs, orange, 48" x 48", includes brace and clamp	--	Ea	56.70	--	--
Hand-held traffic paddles, "Stop" and "Slow"	--	Ea	9.79	--	--
Typical labor cost, place and remove, any of above per use	BL@.160	Ea	--	2.88	--

Basement Doors

	Craft@Hrs	Unit	Material	Labor	Total

Basement Doors. Good quality 12 gauge primed steel, center opening basement doors. Costs include assembly and installation hardware. No concrete, masonry, anchor placement or finish painting included. Bilco Company

Doors (overall dimensions)

	Craft@Hrs	Unit	Material	Labor	Total
30" H, 47" W, 58" L	BC@3.41	Ea	245.00	80.40	325.40
22" H, 51" W, 64" L	BC@3.41	Ea	250.00	80.40	330.40
19-1/2" H, 55" W, 72" L	BC@3.41	Ea	271.00	80.40	351.40
52" H, 51" W, 43-1/4" L	BC@3.41	Ea	290.00	80.40	370.40

Door extensions (available for 19-1/2" H, 55" W, 72" L door only)

	Craft@Hrs	Unit	Material	Labor	Total
6" deep	BC@1.71	Ea	65.00	40.30	105.30
12" deep	BC@1.71	Ea	81.00	40.30	121.30
18" deep	BC@1.71	Ea	100.00	40.30	140.30
24" deep	BC@1.71	Ea	118.00	40.30	158.30

Stair stringers, steel, pre-cut for 2" x 10" wood treads (without treads)

	Craft@Hrs	Unit	Material	Labor	Total
32" to 39" stair height	BC@1.71	Ea	50.00	40.30	90.30
48" to 55" stair height	BC@1.71	Ea	66.00	40.30	106.30
56" to 64" stair height	BC@1.71	Ea	76.00	40.30	116.30
65" to 72" stair height	BC@1.71	Ea	84.00	40.30	124.30
73" to 78" stair height	BC@1.71	Ea	115.00	40.30	155.30
81" to 88" stair height	BC@1.71	Ea	125.00	40.30	165.30
89" to 97" stair height	BC@1.71	Ea	133.00	40.30	173.30

Bathroom Accessories. Average quality. Better quality brass accessories cost 75% to 100% more. See also Medicine Cabinets and Vanities.

	Craft@Hrs	Unit	Material	Labor	Total
Cup and toothbrush holder, chrome	BC@.258	Ea	7.50	6.09	13.59
Cup holder, chrome, surface mounted	BC@.258	Ea	9.15	6.09	15.24
Cup, toothbrush & soap holder, recessed	BC@.258	Ea	19.00	6.09	25.09
Toothbrush holder, chrome, surface mount	BC@.258	Ea	8.00	6.09	14.09
Robe hook, double, chrome	BC@.258	Ea	7.25	6.09	13.34
Soap holder, chrome, with drain holes, surface mounted	BC@.258	Ea	8.00	6.09	14.09
Toilet roll holder, chrome, recessed	BC@.258	Ea	22.60	6.09	28.69
Recessed tissue holder, stainless steel	BC@.258	Ea	32.30	6.09	38.39

Electrical plates, chrome plated

	Craft@Hrs	Unit	Material	Labor	Total
Switch plate, single	BE@.154	Ea	6.00	3.89	9.89
Switch plate, double	BE@.154	Ea	7.60	3.89	11.49
Duplex receptacle plate	BE@.154	Ea	5.90	3.89	9.79
Duplex receptacle and switch	BE@.154	Ea	5.80	3.89	9.69

Grab bars, tubular stainless steel, with anchor plates

	Craft@Hrs	Unit	Material	Labor	Total
Straight bar, 16"	BC@.414	Ea	23.70	9.77	33.47
Straight bar, 24"	BC@.414	Ea	26.00	9.77	35.77
Straight bar, 32"	BC@.414	Ea	29.00	9.77	38.77
"L" shaped bar, 16" x 32"	BC@.620	Ea	55.00	14.60	69.60

Mirrors, stainless steel framed, surface mount, no light or cabinet

	Craft@Hrs	Unit	Material	Labor	Total
16" high x 20" wide	BG@.420	Ea	53.00	9.23	62.23
18" high x 24" wide	BG@.420	Ea	60.00	9.23	69.23
18" high x 36" wide	BG@.420	Ea	87.00	9.23	96.23
24" high x 36" wide	BG@.420	Ea	111.00	9.23	120.23
48" high x 24" wide	BG@.420	Ea	136.00	9.23	145.23

Mirrors, wood framed, surface mount, better quality

	Craft@Hrs	Unit	Material	Labor	Total
18" x 29" rectangular	BG@.420	Ea	70.00	9.23	79.23
20" x 27" oval, oak	BG@.420	Ea	99.00	9.23	108.23

Bathroom Accessories

	Craft@Hrs	Unit	Material	Labor	Total
Shower curtain rods, chrome plated					
60", recessed	BC@.730	Ea	26.90	17.20	44.10
66", recessed	BC@.730	Ea	29.60	17.20	46.80
Towel bars, polished chrome, 3/4" round bar					
18" long	BC@.280	Ea	14.50	6.60	21.10
24" long	BC@.280	Ea	15.60	6.60	22.20
30" long	BC@.280	Ea	17.20	6.60	23.80
36" long	BC@.280	Ea	18.40	6.60	25.00
Deduct for square bar units	--	%	-25.0	--	--
Swing-arm towel rack, chrome, 3 bars, 12" L	BC@.280	Ea	18.40	6.60	25.00
Towel shelf, chrome, 24" L with bar below	BC@.280	Ea	40.30	6.60	46.90
Towel rings, chrome	BC@.343	Ea	13.00	8.09	21.09
Towel rings, chrome base, clear plastic ring	BC@.343	Ea	12.40	8.09	20.49

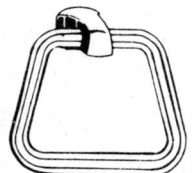

Beds, Folding. Concealed-in-wall type. Steel framed, folding wall bed system. Bed requires 18-5/8" or 22" deep recess. Includes frame, lift mechanism, all hardware. Installed in framed opening. Padded vinyl headboard. Bed face panel accepts paint, wallpaper, vinyl or laminate up to 1/4" thick. Add for box spring and mattress. Sico Incorporated

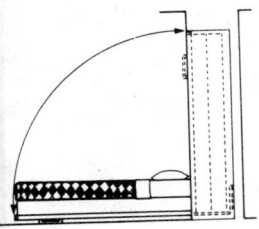

	Craft@Hrs	Unit	Material	Labor	Total
Twin, 41" W x 83-1/2" H x 19" D	B1@5.16	Ea	835.00	109.00	944.00
Box spring and mattress	--	Ea	275.00	--	--
Twin-long, 41" W x 89-1/2" H x 19" D	B1@5.41	Ea	860.00	115.00	975.00
Box spring and mattress	--	Ea	295.00	--	--
Double, 56" W x 83-1/2" H x 19" D	B1@5.41	Ea	835.00	115.00	950.00
Box spring and mattress	--	Ea	350.00	--	--
Double-long, 56" W x 89-1/2" H x 19" D	B1@5.41	Ea	860.00	115.00	975.00
Box spring and mattress	--	Ea	365.00	--	--
Queen, 62-1/2" W x 89-1/2" H x 19" D	B1@5.41	Ea	910.00	115.00	1,025.00
Box spring and mattress	--	Ea	410.00	--	--
King, 79" W x 89-1/2" H x 19" D	B1@5.41	Ea	1,750.00	115.00	1,865.00
Box spring and mattress	--	Ea	575.00	--	--
Add for chrome frame, any bed above	--	LS	125.00	--	--
Add for brass frame, any bed above	--	LS	135.00	--	--

Blueprinting. (Reproduction only) Assumes original is on semi-transparent drafting paper or film. See also Architectural Illustration and Drafting. Small quantities. Cost per sheet (ISO standard sizes).

		Unit			Total
Erasable vellum, made from blueprint	--	SF	--	--	1.50
Blue or black line prints on white paper					
46-3/4" x 33-1/8"	--	Ea	--	--	1.62
39-3/8" x 27-7/8"	--	Ea	--	--	1.17
33-1/8" x 23-3/8"	--	Ea	--	--	.83
23-3/8" x 16-1/2"	--	Ea	--	--	.42
16-1/2" x 11-3/4"	--	Ea	--	--	.21
8-1/4" x 11-3/4"	--	Ea	--	--	.15
Brown line opaque prints on white paper					
46-3/4" x 33-1/8"	--	Ea	--	--	6.00
39-3/8" x 27-7/8"	--	Ea	--	--	4.00
33-1/8" x 23-3/8"	--	Ea	--	--	3.00
23-3/8" x 16-1/2"	--	Ea	--	--	1.50
16-1/2" x 11-3/4"	--	Ea	--	--	.75
8-1/4" x 11-3/4"	--	Ea	--	--	.37

Blueprinting

	Craft@Hrs	Unit	Material	Labor	Total
Erasable sepia on white paper					
46-3/4" x 33-1/8"	--	Ea	--	--	9.00
39-3/8" x 27-7/8"	--	Ea	--	--	6.00
33-1/8" x 23-3/8"	--	Ea	--	--	4.50
23-3/8" x 16-1/2"	--	Ea	--	--	2.30
16-1/2" x 11-3/4"	--	Ea	--	--	1.15
8-1/4" x 11-3/4"	--	Ea	--	--	.60
Diazo mylar on white paper					
46-3/4" x 33-1/8"	--	Ea	--	--	29.10
39-3/8" x 27-7/8"	--	Ea	--	--	20.60
33-1/8" x 23-3/8"	--	Ea	--	--	14.50
23-3/8" x 16-1/2"	--	Ea	--	--	7.25
16-1/2" x 11-3/4"	--	Ea	--	--	3.65
8-1/4" x 11-3/4"	--	Ea	--	--	1.82
Add for local pickup and delivery, round trip	--	LS	--	--	5.00
Add for stapled edge sets, per set	--	Ea	--	--	.20
Add for edge bound sets, per set	--	Ea	--	--	.30
Deduct for jobs over 500 SF of print	--	%	--	--	-10.0
Deduct for jobs over 1,000 SF of print	--	%	--	--	-20.0

Building Inspection Service. (Home inspection service) Inspection of all parts of building by qualified engineer or certified building inspection technician. Includes written report covering all doors and windows, electrical system, foundation, heating and cooling system, insulation, interior and exterior surface conditions, landscaping, plumbing system, roofing, and structural integrity.

	Craft@Hrs	Unit	Material	Labor	Total
Single-family residence					
Base fee (up to 2,500 SF)	--	LS	--	--	200.00
Add for additional 1,000 SF or fraction	--	LS	--	--	60.00
Add for out buildings (each)	--	LS	--	--	25.00
Add for houses over 50 years old	--	LS	--	--	50.00
Add for houses with over 10 rooms	--	LS	--	--	50.00
Add for houses with over 15 rooms	--	LS	--	--	65.00
Add for swimming pool, spa or sauna	--	LS	--	--	25.00
Add for soil testing	--	LS	--	--	150.00
Add for water testing (coliform only)	--	LS	--	--	65.00
Add for warranty protection					
Houses to 10 rooms & 50 years old	--	LS	--	--	250.00
Houses over 50 years old	--	LS	--	--	270.00
Houses over 10 rooms	--	LS	--	--	270.00
Multi-family structures					
Two family residence base fee	--	LS	--	--	220.00
Apartment or condominium base fee	--	LS	--	--	160.00
Warranty protection (base cost)	--	LS	--	--	250.00
Add for each additional unit	--	LS	--	--	20.00
Add for each family living unit					
Standard inspection	--	LS	--	--	50.00
Detailed inspection	--	LS	--	--	60.00
Add for swimming pool, spa, sauna	--	LS	--	--	50.00
Add for water quality testing	--	LS	--	--	225.00
Add for water quantity test, per well	--	LS	--	--	150.00
Add for soil testing	--	LS	--	--	1,500.00

Building Inspection Service

	Craft@Hrs	Unit	Material	Labor	Total
Hazards testing for single and multi-family dwellings					
Lead paint testing	--	LS	--	--	350.00
Urea-formaldehyde insulation testing	--	LS	--	--	75.00
Asbestos testing	--	LS	--	--	110.00
Radon gas testing	--	LS	--	--	125.00

Building Paper. See also Roofing for roof applications and Polyethylene Film. Costs include 7% coverage allowance for 2" lap and 5% waste allowance. See installation costs at the end of this section.

	Craft@Hrs	Unit	Material	Labor	Total
Fortifiber products					
Asphalt felt					
15 lb, (324 SF roll at $11.20), 36" x 108'	--	SF	.04	--	--
30 lb, (216 SF roll at $14.40), 36" x 72'	--	SF	.08	--	--
Asphalt shake felt					
30 lb, (180 SF roll at $14.80), 18" x 120'	--	SF	.09	--	--
Double kraft (Aquabar)					
Class A, 36" wide, 30-50-30, (500 SF roll at $15.95)	--	SF	.04	--	--
Class B, 36" wide, 30-30-30, (500 SF roll at $14.25)	--	SF	.03	--	--
"Jumbo Tex" gun grade sheathing paper, 40" wide, (324 SF roll at $6.50)	--	SF	.02	--	--
Rattan black building paper, 36", 40" wide, (500 SF roll at $10.15)	--	SF	.02	--	--
Red rosin sized sheathing (duplex sheathing) 36" wide, (500 SF roll at $9.75)	--	SF	.02	--	--
Flashing paper					
Kraft paper, 6" wide x 150' long (at $5.35 per 75 SF roll)	--	SF	.08	--	--
Copper armored Sisalkraft, 10" wide x 120' long					
1 oz per SF (at $80.00 per 120 LF roll)	--	LF	.75	--	--
2 oz per SF (at $95.00 per 120 LF roll)	--	LF	.89	--	--
3 oz per SF (at $155.00 per 120 LF roll)	--	LF	1.45	--	--
Fortifiber products					
Above grade vapor barriers, no waste or coverage allowance included.					
Vaporstop 298, (fiberglass reinforcing and asphaltic adhesive between 2 layers of kraft), 32" x 405' roll (1,080 SF roll at $52.50)	--	SF	.05	--	--
Pyro-Kure 600, (2 layers of heavy kraft with fire retardant adhesive, edge reinforced with fiberglass strands) 32" x 405' roll (1,080 SF roll at $100.00)	--	SF	.09	--	--
Foil Barrier 718 (fiberglass reinforced aluminum foil and non-asphaltic adhesive between 2 layers of kraft) 52" x 231' roll (1,000 SF roll at $144.00)	--	SF	.14	--	--
Below grade vapor barriers					
Moistop (fiberglass reinforced kraft between 2 layers of polyethylene) 8' x 250' roll (2,000 SF roll at $142.00)	--	SF	.07	--	--
Concrete curing papers					
Orange Label Sisalkraft (fiberglass reinforcing and asphaltic adhesive between 2 layers of kraft), 4.8 lbs per CSF, 48" x 125' roll, (500 SF roll at $35.00)	--	SF	.07	--	--
Sisalkraft SK-10, economy curing papers (fiberglass reinforcing and asphaltic adhesive between 2 layers of kraft) 4.2 lbs per CSF, 48" x 300' roll (1,200 SF roll at $41.00)	--	SF	.03	--	--

Building Paper

	Craft@Hrs	Unit	Material	Labor	Total
Protective papers					
Seekure (fiberglass reinforcing strands and nonstaining adhesive between 2 layers of kraft) 48" x 300' roll, (1,200 SF roll at $62.00)	--	SF	.05	--	--
Tyvek™ housewrap by DuPont					
Air infiltration barrier (high-density polyethylene fibers in sheet form) 3' x 195' rolls (585 SF @ $50.00) or 9' x 195' rolls (1755 SF @ $145.00)	--	SF	.09	--	--
Labor to install building papers					
Felts, vapor barriers, infiltration barriers, building papers on walls, ceilings, roofs					
Heavy stapled, typical	BC@.003	SF	--	.08	--
Tack stapled, typical	BC@.002	SF	--	.05	--
Curing papers, protective papers and vapor barriers, minimal fasteners	BC@.001	SF	--	.03	--
Flashing papers, 6" to 8" wide	BC@.010	LF	--	.24	--

Building Permit Fees. These fees are based on recommendations in the Uniform Building Code and will apply in many areas. Building Departments publish fee schedules. The permit fee will be doubled when work is started without getting a permit. When the valuation of the proposed construction exceeds $1,000, plans are often required. Estimate the plan check fee at 65% of the permit fee. The fee for reinspection is usually $30. Inspections outside normal business hours are $45 per hour with a two hour minimum.

The minimum fee for construction values to $500 is $15.00
Over $500 to $2,000 the fee is $15.00 for the first $500
  plus $2.00 for each additional $100 or fraction thereof, up to $2,000.
Over $2,000 to $25,000 the fee is $45.00 for the first $2,000
  plus $9.00 for each additional $1,000 or fraction thereof to $25,000.
Over $25,000 to $50,000 the fee is $250.00 for the first $25,000
  plus $6.50 for each additional $1,000 or fraction thereof to $50,000.
Over $50,000 to $100,000 the fee is $425.00 for the first $50,000
  plus $4.50 for each additional $1,000 or fraction thereof to $100,000.
Over $100,000 to $500,000 the fee is $650.00 for the first $100,000
  plus $3.50 for each additional $1,000 or fraction thereof to $500,000.
Over $500,000 to $1,000,000 the fee is $2,050.00 for the first $500,000
  plus $3.00 for each additional $1,000 or fraction thereof to $1,000,000.
Over $1,000,000 the fee is $3,550.00 for the first $1,000,000
  plus $2.00 for each additional $1,000 or fraction thereof.

Cabinets, Kitchen. See also Vanities. Good quality mill-made modular units with solid hardwood face frames, hardwood door frames and drawer fronts, hardwood veneer on raised door panels (front and back), glued mortise, dowel, and dado joint construction, full backs (1/8" vinyl laminated plywood), vinyl laminated cabinet interiors, vinyl laminated composition drawer bodies with nylon and metal guides. Includes self-closing hinges, door and drawer pulls, mounting hardware and adjustable shelves. See illustrations for unit types. See the price adjustments at the end of this section for pricing of other units. No tops included. See Countertops.

> Kitchen cabinet costs vary widely. The prices listed in this section are for standard grade cabinets. Add 65% to material costs for premium grade cabinets with solid hardwood fronts and frames, mitered corners and solid wood drawer bodies with steel guides and ball bearings. Deduct 45% from material costs for economy grade cabinets, melamine laminated to particleboard, in textured colors or woodgrain print finish.

Cabinets, Kitchen

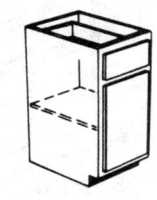
One door base

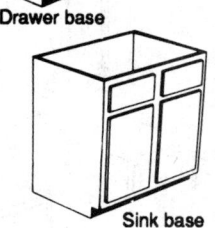
Drawer base

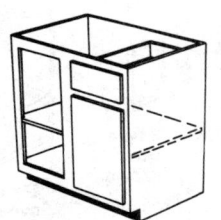
Sink base

	Craft@Hrs	Unit	Material	Labor	Total
Cabinets, Kitchen. (See the note on the preceding page concerning cabinet costs.)					
One door base cabinets, 34-1/2" high, 24" deep					
9" wide, tray divider	BC@.461	Ea	96.00	10.90	106.90
12" wide, 1 door, 1 drawer	BC@.461	Ea	132.00	10.90	142.90
15" wide, 1 door, 1 drawer	BC@.638	Ea	142.00	15.10	157.10
18" wide, 1 door, 1 drawer	BC@.766	Ea	152.00	18.10	170.10
21" wide, 1 door, 1 drawer	BC@.766	Ea	163.00	18.10	181.10
24" wide, 1 door, 1 drawer	BC@.911	Ea	170.00	21.50	191.50
Drawer base cabinets, 34-1/2" high, 24" deep					
15" wide, 4 drawers	BC@.638	Ea	150.00	15.10	165.10
18" wide, 4 drawers	BC@.766	Ea	160.00	18.10	178.10
24" wide, 4 drawers	BC@.911	Ea	181.00	21.50	202.50
Sink base cabinets, 34-1/2" high, 24" deep					
24" wide, 1 door, 1 drawer front	BC@.740	Ea	142.00	17.50	159.50
30" wide, 2 doors, 2 drawer fronts	BC@.766	Ea	175.00	18.10	193.10
33" wide, 2 doors, 2 drawer fronts	BC@.766	Ea	183.00	18.10	201.10
36" wide, 2 doors, 2 drawer fronts	BC@.766	Ea	190.00	18.10	208.10
42" wide, 2 doors, 2 drawer fronts	BC@.911	Ea	206.00	21.50	227.50
48" wide, 2 doors, 2 drawer fronts	BC@.911	Ea	225.00	21.50	246.50
Two door base cabinets, 34-1/2" high, 24" deep					
27" wide, 2 doors, 2 drawers	BC@1.25	Ea	227.00	29.50	256.50
30" wide, 2 doors, 2 drawers	BC@1.25	Ea	235.00	29.50	264.50
33" wide, 2 doors, 2 drawers	BC@1.25	Ea	245.00	29.50	274.50
36" wide, 2 doors, 2 drawers	BC@1.35	Ea	256.00	31.80	287.80
42" wide, 2 doors, 2 drawers	BC@1.50	Ea	278.00	35.40	313.40
48" wide, 2 doors, 2 drawers	BC@1.71	Ea	300.00	40.30	340.30
Blind corner base cabinets, 34-1/2" high					
Minimum 36", maximum 39" at wall	BC@1.39	Ea	163.00	32.80	195.80
Minimum 39", maximum 42" at wall	BC@1.50	Ea	180.00	35.40	215.40
45-degree corner base, revolving, 34-1/2" high,					
36" wide at each wall	BC@2.12	Ea	254.00	50.00	304.00
Corner sink front, 34-1/2" high,					
40" wide at walls	BC@2.63	Ea	144.00	62.00	206.00
Wall cabinets, adjustable shelves, 30" high, 12" deep					
9" wide, 1 door	BC@.461	Ea	89.00	10.90	99.90
12" wide, 1 door	BC@.461	Ea	94.00	10.90	104.90
15" wide, 1 door	BC@.461	Ea	102.00	10.90	112.90
18" wide, 1 door	BC@.638	Ea	111.00	15.10	126.10
21" wide, 1 door	BC@.638	Ea	119.00	15.10	134.10
24" wide, 1 door	BC@.766	Ea	128.00	18.10	146.10
27" wide, 2 doors	BC@.766	Ea	151.00	18.10	169.10
30" wide, 2 doors	BC@.911	Ea	159.00	21.50	180.50
33" wide, 2 doors	BC@.911	Ea	166.00	21.50	187.50
36" wide, 2 doors	BC@1.03	Ea	174.00	24.30	198.30
42" wide, 2 doors	BC@1.03	Ea	191.00	24.30	215.30
48" wide, 2 doors	BC@1.16	Ea	209.00	27.40	236.40
Above-appliance wall cabinets, 12" deep					
12" high, 30" wide, 2 doors	BC@.461	Ea	99.00	10.90	109.90
15" high, 30" wide, 2 doors	BC@.461	Ea	111.00	10.90	121.90
15" high, 33" wide, 2 doors	BC@.537	Ea	116.00	12.70	128.70
15" high, 36" wide, 2 doors	BC@.638	Ea	122.00	15.10	137.10
18" high, 18" wide, 2 doors	BC@.537	Ea	91.00	12.70	103.70
18" high, 30" wide, 2 doors	BC@.766	Ea	119.00	18.10	137.10
18" high, 36" wide, 2 doors	BC@.911	Ea	128.00	21.50	149.50

45 degree corner base cabinet

1 door wall cabinet

Above appliance wall cabinet

Cabinets, Kitchen

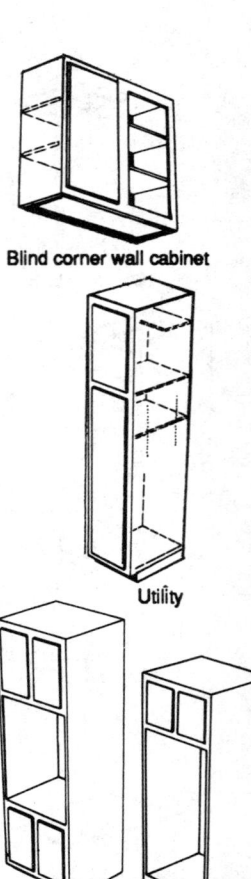

Blind corner wall cabinet

Utility

Single oven cabinet

Double oven cabinet

	Craft@Hrs	Unit	Material	Labor	Total
Corner wall cabinet, 30" high, 12" deep					
24" at each wall, fixed shelves	BC@1.03	Ea	163.00	24.30	187.30
24" at each wall, revolving shelves	BC@1.03	Ea	217.00	24.30	241.30
Blind corner wall cabinet, 30" high					
24" minimum, 1 door	BC@1.03	Ea	119.00	24.30	143.30
36" minimum, 1 door	BC@1.32	Ea	154.00	31.10	185.10
42" minimum, 2 doors	BC@1.20	Ea	198.00	28.30	226.30
Utility cabinet, 66" high, 12" deep, no shelves					
18" wide	BC@1.32	Ea	198.00	31.10	229.10
24" wide	BC@1.71	Ea	215.00	40.30	255.30
Utility cabinet, 66" high, 24" deep, add shelf cost below					
18" wide	BC@1.24	Ea	209.00	29.30	238.30
24" wide	BC@1.71	Ea	244.00	40.30	284.30
Add for utility cabinet revolving shelves, includes mounting hardware					
18" wide x 24" deep	BC@.360	Ea	177.00	8.49	185.49
24" wide x 24" deep	BC@.360	Ea	204.00	8.49	212.49
Add for utility cabinet plain shelves					
18" wide x 24" deep	BC@.541	Ea	57.00	12.80	69.80
24" wide x 24" deep	BC@.541	Ea	60.00	12.80	72.80
Oven cabinets, 66" high, 24" deep					
27" wide, single oven	BC@2.19	Ea	239.00	51.70	290.70
27" wide, double oven	BC@2.19	Ea	183.00	51.70	234.70
Microwave cabinet, with trim,					
21" high, 20" deep, 30" wide	BC@.986	Ea	144.00	23.30	167.30

Carpentry. See finish carpentry items in separate sections: Cabinets, Ceilings, Closet Door Systems, Countertops, Cupolas, Doors, Entrances, Flooring, Hardboard, Hardware, Lumber, Moulding, Paneling, Shutters, Siding, Skylights, Soffits, Stairs, Thresholds, and Weatherstripping.

Carpentry rule of thumb Rough carpentry (framing) cost per square foot of living area. These figures will apply on many types of residential jobs. The six lines of cost data in the paragraph that follows these estimates show the cost breakdown for framing a medium cost house. Labor on tract homes may run 50% less and complex framing jobs will cost more.

Low cost house	B1@.188	SF	2.82	3.98	6.80
Medium cost house	B1@.224	SF	3.37	4.74	8.11
Better quality house	B1@.305	SF	4.59	6.46	11.05

Carpentry cost breakdown Costs per square foot of living area for single-story residences based on lumber at $450 per MBF. These figures were used to compile the framing cost for a medium cost home in the paragraph above.

Sills, pier blocks, floor beams					
(145 BF per 1,000 SF)	B1@.017	SF	.07	.36	.43
Floor joists, blocking, bridging					
(1,480 BF per 1,000 SF)	B1@.028	SF	.67	.59	1.26
Subfloor, 5/8" CD plywood (at $460 per MSF)	B1@.011	SF	.53	.23	.76
(1,150 SF per 1,000 SF)					
Studs, sole plates, top plates, header and end joists, blocking, bracing and framing for openings					
(2,250 BF per 1,000 SF)	B1@.093	SF	1.01	1.97	2.98
Ceiling joists, header and end joists, blocking and bracing					
(1,060 BF per 1,000 SF)	B1@.045	SF	.48	.95	1.43
Rafters, collar beams, ridge boards, 2" x 8" rafters					
16" OC, (1,340 BF per 1000 SF)	B1@.030	SF	.61	.64	1.25

Carpentry

	Craft@Hrs	Unit	Material	Labor	Total

Carpentry cost detail Material costs shown here can be adjusted to reflect your actual lumber cost: divide your actual lumber cost (per MBF) by the cost listed in parentheses (per MBF). Then multiply the cost in the material column by this adjustment factor. No waste included.

Backing, for appliances and fixtures. See also Nailers in this section.

	Craft@Hrs	Unit	Material	Labor	Total
1" x 4", Std & Btr (at $405 per MBF)	B1@.012	LF	.15	.25	.40
1" x 4", Std & Btr (at $405 per MBF)	B1@.035	BF	.41	.74	1.15
1" x 6", Std & Btr (at $505 per MBF)	B1@.018	LF	.26	.38	.64
1" x 6", Std & Btr (at $505 per MBF)	B1@.035	BF	.51	.74	1.25
2" x 4", Std & Btr (at $430 per MBF)	B1@.024	LF	.30	.51	.81
2" x 4", Std & Btr (at $430 per MBF)	B1@.035	BF	.43	.72	1.15
2" x 6", #2 grade (at $420 per MBF)	B1@.035	LF	.42	.72	1.14
2" x 6", #2 grade (at $420 per MBF)	B1@.035	BF	.42	.74	1.16

Bracing. See also sheathing for plywood bracing and shear panels.

Let-in wall bracing, using standard and better lumber

	Craft@Hrs	Unit	Material	Labor	Total
1" x 4" at $405 per MBF	B1@.021	LF	.15	.45	.60
1" x 4" at $405 per MBF	B1@.054	BF	.41	1.14	1.55
1" x 6" at $505 per MBF	B1@.027	LF	.26	.57	.83
1" x 6" at $505 per MBF	B1@.053	BF	.51	1.12	1.63
2" x 4" at $430 per MBF	B1@.035	LF	.30	.74	1.04
2" x 4" at $430 per MBF	B1@.053	BF	.43	1.12	1.55

Steel strap bracing, 1-1/4" wide,

	Craft@Hrs	Unit	Material	Labor	Total
9'6" or 11'6" lengths	B1@.010	LF	.40	.21	.61

Steel "V" bracing, 3/4" x 3/4",

	Craft@Hrs	Unit	Material	Labor	Total
9'6" or 11'6" lengths	B1@.010	LF	.55	.21	.76

Temporary wood frame wall bracing, assumes salvage at 50% and 3 uses

	Craft@Hrs	Unit	Material	Labor	Total
1" x 4" Std & Btr (at $405 per MBF)	B1@.006	LF	.04	.13	.17
1" x 4" Std & Btr (at $405 per MBF)	B1@.018	BF	.10	.38	.48
1" x 6" Std & Btr (at $505 per MBF)	B1@.010	LF	.07	.21	.28
1" x 6" Std & Btr (at $505 per MBF)	B1@.018	BF	.13	.38	.51
2" x 4" utility (at $370 per MBF)	B1@.012	LF	.06	.25	.31
2" x 4" utility (at $370 per MBF)	B1@.018	BF	.10	.38	.48
2" x 6" utility (at $340 per MBF)	B1@.018	LF	.09	.38	.47
2" x 6" utility (at $340 per MBF)	B1@.018	BF	.09	.38	.47

Bridging, material only, including 5% waste, based on 10' bays. Add the labor costs that follow.

		Center to center spacing			
	Unit	12"	16"	20"	24"
2" x 6" members					
1" x 3" cross, Std & Btr (at $500 per MBF)	Bay	.26	.34	.42	.52
1" x 4" cross, Std & Btr (at $405 per MBF)	Bay	.29	.37	.45	.55
2" x 2" cross, Std & Btr (at $450 per MBF)	Bay	.32	.43	.54	.64
2" x 6" solid, Std & Btr (at $450 per MBF)	Bay	.46	.64	.82	1.02
2" x 8" members					
1" x 3" cross, Std & Btr (at $500 per MBF)	Bay	.28	.35	.44	.52
1" x 4" cross, Std & Btr (at $405 per MBF)	Bay	.31	.40	.48	.56
2" x 2" cross, Std & Btr (at $450 per MBF)	Bay	.35	.46	.56	.65

Carpentry

	Unit	\-\-Center to center spacing\-\-			
		12"	16"	20"	24"
Bridging, material only, including 5% waste, based on 10' bays. Add labor costs that follow.					
2" x 8" solid, Std & Btr					
(at $450 per MBF)	Bay	.66	.90	1.14	1.39
Steel, no nail type, cross	Bay	.98	1.07	--	1.36
2" x 10" members					
1" x 3" cross, Std & Btr					
(at $500 per MBF)	Bay	.31	.29	.45	.55
1" x 4" cross, Std & Btr					
(at $405 per MBF)	Bay	.35	.43	.52	.61
2" x 2" cross, Std & Btr					
(at $450 per MBF)	Bay	.39	.48	.58	.68
2" x 10" solid, Std & Btr					
(at $610 per MBF)	Bay	.94	1.31	1.70	2.05
Steel, no nail type, cross	Bay	.98	1.07	--	1.36
2" x 12" members					
1" x 3" cross, Std & Btr					
(at $500 per MBF)	Bay	.34	.41	.49	.56
1" x 4" cross, Std & Btr					
(at $405 per MBF)	Bay	.37	.44	.52	.60
2" x 2" cross, Std & Btr					
(at $450 per MBF)	Bay	.43	.51	.61	.70
2" x 12" solid, Std & Btr					
(at $595 per MBF)	Bay	1.14	1.58	2.03	2.45
Steel, no nail type	Bay	.98	1.07	--	1.36

	Craft@Hrs	Unit	Material	Labor	Total
Labor to install bridging, per bay					
1" x 3", 1" x 4", 2" x 2" cross, pre-cut	B1@.034	Ea	--	.72	--
2" x 6", 2" x 8", solid, job cut	B1@.042	Ea	--	.89	--
2" x 10", 2" x 12", solid, job cut	B1@.057	Ea	--	1.21	--
Steel, no nail type, cross	B1@.020	Ea	--	.42	--
Cant strips					
2" x 4" beveled	B1@.020	LF	.35	.42	.77
For short or cut-up runs	B1@.030	LF	.40	.64	1.04
Catwalks (walkways), 1" x 6" solid planking					
(at $550 per MBF, 1 BF per SF)	B1@.017	SF	.55	.36	.91
Ceiling beams, to 4" x 12",					
(at $900 per MBF, 4 BF per LF)	B1@.053	LF	3.60	1.12	4.72

Ceiling Joists. Per SF of area covered. Figures in parentheses indicate board feet per square foot of ceiling including end joists, header joists, and 5% waste. No beams, bridging, or blocking included. Deduct for openings over 25 SF. Costs shown are based on a job with 1,000 SF of area covered. For scheduling purposes, estimate that a two-man crew can complete 650 SF of area per 8-hour day for 12" center to center framing; 800 SF for 16" OC; 950 SF for 20" OC; or 1,100 SF for 24" OC.

	Craft@Hrs	Unit	Material	Labor	Total
2" x 4" (at $430 per MBF), Std & Btr grade					
12" centers (.78 BF per SF)	B1@.023	SF	.35	.49	.84
16" centers (.59 BF per SF)	B1@.020	SF	.27	.42	.69
20" centers (.48 BF per SF)	B1@.016	SF	.22	.34	.56
24" centers (.42 BF per SF)	B1@.014	SF	.18	.30	.48
2" x 6" (at $450 per MBF), Std & Btr grade					
12" centers (1.15 BF per SF)	B1@.026	SF	.55	.55	1.10
16" centers (.88 BF per SF)	B1@.020	SF	.42	.42	.84

Carpentry

	Craft@Hrs	Unit	Material	Labor	Total
2" x 6" (at $450 per MBF), Std & Btr grade					
20" centers (.72 BF per SF)	B1@.017	SF	.34	.36	.70
24" centers (.63 BF per SF)	B1@.015	SF	.31	.32	.63
2" x 8" (at $450 per MBF), Std & Btr grade					
12" centers (1.53 BF per SF)	B1@.028	SF	.73	.59	1.32
16" centers (1.17 BF per SF)	B1@.022	SF	.55	.47	1.02
20" centers (.96 BF per SF)	B1@.018	SF	.45	.38	.83
24" centers (.84 BF per SF)	B1@.016	SF	.40	.34	.74
2" x 10" (at $590 per MBF), Std & Btr grade					
12" centers (1.94 BF per SF)	B1@.030	SF	1.21	.64	1.85
16" centers (1.47 BF per SF)	B1@.023	SF	.91	.49	1.40
20" centers (1.21 BF per SF)	B1@.020	SF	.75	.42	1.17
24" centers (1.04 BF per SF)	B1@.017	SF	.64	.36	1.00
2" x 12" (at $600 per MBF), Std & Btr grade					
12" centers (2.30 BF per SF)	B1@.031	SF	1.46	.66	2.12
16" centers (1.76 BF per SF)	B1@.025	SF	1.11	.53	1.64
20" centers (1.44 BF per SF)	B1@.020	SF	.92	.42	1.34
24" centers (1.26 BF per SF)	B1@.018	SF	.80	.38	1.18

Ceiling Assemblies. Costs for wood framed ceiling joists with ceiling finish and fiberglass insulation, based on performing the work at the construction site. These costs include the ceiling joists, ceiling finish as described, blocking, nails and 3-1/2" thick R-11 fiberglass insulation batts between the ceiling joists. Figures in parentheses indicate board feet per square foot of ceiling framing including end joists and typical header joists. No beams included. Ceiling joists and blocking are based on Std & Btr grade lumber. Costs shown are per square foot of area covered and include normal waste. Deduct for openings over 25 SF. Ceiling joists with regular gypsum drywall taped and sanded smooth finish, ready for paint.

	Craft@Hrs	Unit	Material	Labor	Total
2" x 4" ceiling joists at 16" on center (.59 BF per SF at $430 per MBF),					
with insulation and 1/2" gypsum drywall	B1@.053	SF	.72	1.12	1.84
2" x 6" ceiling joists at 16" on center (.88 BF per SF at $450 per MBF),					
with insulation and 1/2" gypsum drywall	B1@.055	SF	.89	1.17	2.06
2" x 8" ceiling joists at 16" on center (1.17 BF per SF at $450 per MBF),					
with insulation and 1/2" gypsum drywall	B1@.057	SF	1.02	1.21	2.23
For spray applied plaster finish (cottage cheese)					
Textured, add	DT@.002	SF	.05	.05	.10
Acoustic, add	DT@.003	SF	.06	.07	.13
For different type gypsum drywall with taped and sanded smooth finish, ready for paint					
1/2" fire rated type X board, add	--	SF	.03	--	--
1/2" moisture resistant greenboard, add	--	SF	.09	--	--
5/8" thick plain board, add, add	--	SF	.02	--	--
5/8" fire rated type X board, add	--	SF	.04	--	--
5/8" moisture resistant greenboard, add	--	SF	.11	--	--
For different ceiling joist center to center dimensions					
2" x 4" ceiling joists					
12" on center, add	B1@.004	SF	.08	.08	.16
20" on center, deduct	--	SF	-.05	-.06	-.11
24" on center, deduct	B1@.005	SF	-.09	-.11	-.20
2" x 6" ceiling joists					
12" on center, add	B1@.006	SF	.13	.13	.26
20" on center, deduct	--	SF	-.09	-.06	-.15
24" on center, deduct	--	SF	-.11	-.11	-.22

Carpentry

	Craft@Hrs	Unit	Material	Labor	Total
Ceiling Assemblies, continued					
2" x 8" ceiling joists					
12" on center, add	B1@.006	SF	.18	.13	.31
20" on center, deduct	--	SF	-.10	-.08	-.18
24" on center, deduct	--	SF	-.15	-.13	-.28
For different type insulation					
Fiberglass batts					
6-1/4" thick R-19, add	--	SF	.11	--	--
10" thick R-30, add	--	SF	.32	--	--
12" thick R-38, add	--	SF	.53	--	--
Blown-in fiberglass					
8" thick R-19, add	--	SF	.04	--	--
Collar beams, 2" x 6" (at $450 per MBF),					
Std & Btr grade, includes 10% waste	B1@.013	LF	.50	.28	.78
Collar ties, 1" x 6" (at $500 per MBF),					
Std & Btr grade, includes 10% waste	B1@.006	LF	.28	.13	.41

Columns, S4S, green. Material costs include 10% waste. For scheduling purposes, estimate that a two-man crew can complete 100 to 125 LF per 8-hour day.

	Craft@Hrs	Unit	Material	Labor	Total
4" x 8" (at $630 per MBF)	B1@.143	LF	1.86	3.03	4.89
4" x 10" (at $630 per MBF)	B1@.140	LF	2.31	2.97	5.28
4" x 12" (at $640 per MBF)	B1@.145	LF	2.81	3.07	5.88
6" x 10" (at $800 per MBF)	B1@.147	LF	4.40	3.12	7.52
6" x 12" (at $800 per MBF)	B1@.150	LF	5.28	3.18	8.46
8" x 10" (at $800 per MBF)	B1@.155	LF	5.86	3.28	9.14
8" x 12" (at $800 per MBF)	B1@.166	LF	7.05	3.52	10.57

Curbs (see Trimmers below)

Door opening framing in wall studs. Figures in parentheses indicate typical board feet per SF of opening area, based on walls 8' in height. Costs shown are per SF of door opening and include header of appropriate size, double vertical studs each side of the opening, double top plates, cripples, blocking, nails and normal waste.

2" x 4" wall studs (at $450 per MBF), Std & Btr, opening size as shown

	Craft@Hrs	Unit	Material	Labor	Total
up to 18 SF (2 BF per SF)	B1@.100	SF	.99	2.12	3.11
over 18 SF to 24 SF (1.5 BF per SF)	B1@.067	SF	.74	1.42	2.16
over 24 SF to 36 SF (1.3 BF per SF)	B1@.050	SF	.64	1.06	1.70
over 36 SF to 54 SF (1.2 BF per SF)	B1@.040	SF	.59	.85	1.44
over 54 SF to 67 SF (1.1 BF per SF)	B1@.035	SF	.54	.74	1.28
over 67 SF (1.0 BF per SF)	B1@.030	SF	.50	.64	1.14
Add per foot of height, walls over 8' in height	--	LF	1.20	--	--
Deduct for 2" x 3" wall studs	--	%	-15.0	--	--
Add for 2" x 6" wall studs	--	%	30.0	--	--

Dormer studs (2" x 4"), Std & Btr

	Craft@Hrs	Unit	Material	Labor	Total
(at $430 per MBF), including 10% waste	B1@.033	BF	.47	.70	1.17

Fascia material, material column includes 10% waste

Engelmann spruce, S4S, dry, # 2 and Btr

	Craft@Hrs	Unit	Material	Labor	Total
1" x 4" (at $555 per MBF)	B1@.044	LF	.20	.93	1.13
1" x 6" (at $625 per MBF)	B1@.044	LF	.34	.93	1.27
1" x 8" (at $640 per MBF)	B1@.051	LF	.47	1.08	1.55

Hem-fir, S4S, dry, Std and Btr

	Craft@Hrs	Unit	Material	Labor	Total
2" x 6" (at $440 per MBF)	B1@.044	LF	.49	.93	1.42
2" x 8" (at $455 per MBF)	B1@.051	LF	.67	1.08	1.75
2" x 10" (at $550 per MBF)	B1@.051	LF	1.01	1.08	2.09

Carpentry

	Craft@Hrs	Unit	Material	Labor	Total
Redwood, S4S kiln dried, select all heart					
1" x 4" (at $2,300 per MBF)	B1@.044	LF	.84	.93	1.77
1" x 6" (at $2,280 per MBF)	B1@.044	LF	1.25	.93	2.18
1" x 8" (at $2,400 per MBF)	B1@.051	LF	1.76	1.08	2.84
2" x 4" (at $2,680 per MBF)	B1@.044	LF	1.97	.93	2.90
2" x 6" (at $2,570 per MBF)	B1@.044	LF	2.83	.93	3.76
2" x 8" (at $2,690 per MBF)	B1@.051	LF	3.94	1.08	5.02

Fireblocks, installed in wood frame walls, per LF of wall to be blocked. Figures in parentheses indicate board feet of fire blocking per linear foot of wall including 10% waste. See also Backing, Bridging, and Nailers in this section.

	Craft@Hrs	Unit	Material	Labor	Total
2" x 3" blocking (at $450 per MBF), Std & Btr					
12" OC members (.48 BF per LF)	B1@.020	LF	.22	.42	.64
16" OC members (.50 BF per LF)	B1@.015	LF	.23	.32	.55
20" OC members (.51 BF per LF)	B1@.012	LF	.23	.25	.48
24" OC members (.52 BF per LF)	B1@.010	LF	.23	.21	.44
2" x 4" blocking (at $430 per MBF), Std & Btr					
12" OC members (.64 BF per LF)	B1@.020	LF	.28	.42	.70
16" OC members (.67 BF per LF)	B1@.015	LF	.29	.32	.61
20" OC members (.68 BF per LF)	B1@.012	LF	.29	.25	.54
24" OC members (.69 BF per LF)	B1@.010	LF	.30	.21	.51
2" x 6" blocking (at $450 per MBF), Std & Btr					
12" OC members (.96 BF per LF)	B1@.021	LF	.43	.45	.88
16" OC members (1.00 BF per LF)	B1@.016	LF	.45	.34	.79
20" OC members (1.02 BF per LF)	B1@.012	LF	.46	.25	.71
24" OC members (1.03 BF per LF)	B1@.010	LF	.46	.21	.67

Floor joists. Per SF of area covered. Figures in parentheses indicate board feet per square foot of floor including box or band joist, typical double joists, and 6% waste. No beams, blocking or bridging included. Deduct for openings over 25 SF. Costs shown are based on a job with 1,000 SF of area covered. For scheduling purposes, estimate that a two-man crew can complete 750 SF of area per 8-hour day for 12" center to center framing; 925 SF for 16" OC; 1,100 SF for 20" OC; or 1,250 SF for 24" OC.

	Craft@Hrs	Unit	Material	Labor	Total
2" x 6" (at $450 per MBF), Std & Btr					
12" centers (1.28 BF per SF)	B1@.021	SF	.58	.44	1.02
16" centers (1.02 BF per SF)	B1@.017	SF	.46	.36	.82
20" centers (.88 BF per SF)	B1@.014	SF	.40	.30	.70
24" centers (.73 BF per SF)	B1@.013	SF	.33	.28	.61
2" x 8" (at $450 per MBF), Std & Btr					
12" centers (1.71 BF per SF)	B1@.023	SF	.77	.49	1.26
16" centers (1.36 BF per SF)	B1@.018	SF	.61	.38	.99
20" centers (1.17 BF per SF)	B1@.015	SF	.53	.32	.85
24" centers (1.03 BF per SF)	B1@.014	SF	.46	.30	.76
2" x 10" (at $590 per MBF), Std & Btr					
12" centers (2.14 BF per SF)	B1@.025	SF	1.26	.53	1.79
16" centers (1.71 BF per SF)	B1@.020	SF	1.01	.42	1.43
20" centers (1.48 BF per SF)	B1@.016	SF	.87	.34	1.21
24" centers (1.30 BF per SF)	B1@.016	SF	.77	.34	1.11
2" x 12" (at $600 per MBF), Std & Btr					
12" centers (2.56 BF per SF)	B1@.026	SF	1.54	.55	2.09
16" centers (2.05 BF per SF)	B1@.021	SF	1.23	.44	1.67
20" centers (1.77 BF per SF)	B1@.018	SF	1.06	.38	1.44
24" centers (1.56 BF per SF)	B1@.017	SF	.94	.36	1.30

Carpentry

	Craft@Hrs	Unit	Material	Labor	Total

Floor Assemblies. Costs for wood framed floor joists with subflooring and R-19 insulation, based on performing the work at the construction site. These costs include the floor joists, subflooring as described, blocking, nails and 6-1/4" thick R-19 fiberglass insulation between the floor joists. Figures in parentheses indicate board feet per square foot of floor framing including box or band joists and typical double joists. No beams included. Floor joists and blocking are based on Std & Btr grade lumber at $450 per MBF. Plywood is based on CD standard interior sheathing, plugged and touch sanded with 1/2" thick at $11.00 per 4' x 8' sheet and 3/4" thick at $17.50 per 4' x 8' sheet. 1" x 6" planked subflooring is based on Std & Btr grade lumber at $500 per MBF with 1.24 BF per square foot of floor. Costs shown are per square foot of area covered and include normal waste. Deduct for openings over 25 SF.

	Craft@Hrs	Unit	Material	Labor	Total
Floor joists 16" O.C., R-19 insulation and plywood subflooring					
2" x 6" (1.02 BF per SF) and 1/2" plywood	B1@.040	SF	1.23	.85	2.08
2" x 8" (1.36 BF per SF) and 1/2" plywood	B1@.041	SF	1.39	.87	2.26
2" x 6" (1.02 BF per SF) and 3/4" plywood	B1@.042	SF	1.45	.89	2.34
2" x 8" (1.36 BF per SF) and 3/4" plywood	B1@.042	SF	1.61	.89	2.50
Floor joists 16" O.C., R-19 insulation and planked subflooring					
2" x 6" (1.02 BF per SF) and 1" x 6" planking	B1@.047	SF	1.49	1.00	2.49
2" x 8" (1.36 BF per SF) and 1" x 6" planking	B1@.048	SF	1.65	1.02	2.67
For different floor joist center to center dimensions					
2" x 6" floor joists					
12" on center, add	B1@.004	SF	.12	.09	.21
20" on center, deduct	--	SF	-.06	-.06	-.12
24" on center, deduct	--	SF	-.13	-.19	-.32
2" x 8" floor joists					
12" on center, add	B1@.005	SF	.16	.11	.27
20" on center, deduct	--	SF	-.08	-.06	-.14
24" on center, deduct	--	SF	-.15	-.09	-.24
For different type insulation					
Fiberglass batts					
10" thick R-30, add	--	SF	.32	--	--
12" thick R-38, add	--	SF	.53	--	--

Furring. Per SF of surface area to be covered. Figures in parentheses show coverage including 7% waste, typical job

	Craft@Hrs	Unit	Material	Labor	Total
Over masonry, lumber at $315 per MBF, Utility grade					
12" OC, 1" x 2" (.24 BF per SF)	B1@.025	SF	.08	.53	.61
16" OC, 1" x 2" (.20 BF per SF)	B1@.020	SF	.06	.42	.48
20" OC, 1" x 2" (.17 BF per SF)	B1@.018	SF	.05	.38	.43
24" OC, 1" x 2" (.15 BF per SF)	B1@.016	SF	.05	.34	.39
12" OC, 1" x 3" (.36 BF per SF)	B1@.025	SF	.11	.53	.64
16" OC, 1" x 3" (.29 BF per SF)	B1@.020	SF	.09	.42	.51
20" OC, 1" x 3" (.25 BF per SF)	B1@.018	SF	.08	.38	.46
24" OC, 1" x 3" (.22 BF per SF)	B1@.016	SF	.07	.34	.41
Over wood frame, lumber at $315 per MBF, Utility grade					
12" OC, 1" x 2" (.24 BF per SF)	B1@.016	SF	.08	.34	.42
16" OC, 1" x 2" (.20 BF per SF)	B1@.013	SF	.06	.28	.34
20" OC, 1" x 2" (.17 BF per SF)	B1@.011	SF	.05	.23	.28
24" OC, 1" x 2" (.15 BF per SF)	B1@.010	SF	.05	.21	.26
12" OC, 1" x 3" (.36 BF per SF)	B1@.016	SF	.11	.34	.45
16" OC, 1" x 3" (.29 BF per SF)	B1@.013	SF	.09	.28	.37
20" OC, 1" x 3" (.25 BF per SF)	B1@.011	SF	.08	.23	.31
24" OC, 1" x 3" (.22 BF per SF)	B1@.010	SF	.07	.21	.28

Carpentry

	Craft@Hrs	Unit	Material	Labor	Total

Over wood subfloor, 1" lumber at $315 per MBF, 2" lumber at $450 per MBF

12" OC, 1" x 2" utility (.24 BF per SF)	B1@.007	SF	.08	.15	.23
16" OC, 1" x 2" utility (.20 BF per SF)	B1@.028	SF	.06	.59	.65
20" OC, 1" x 2" utility (.17 BF per SF)	B1@.024	SF	.05	.51	.56
24" OC, 1" x 2" utility (.15 SF per SF)	B1@.021	SF	.05	.44	.49
12" OC, 2" x 2" Std & Btr (.48 BF per SF)	B1@.033	SF	.22	.70	.92
16" OC, 2" x 2" Std & Btr (.39 BF per SF)	B1@.028	SF	.18	.59	.77
20" OC, 2" x 2" Std & Btr (.34 BF per SF)	B1@.024	SF	.15	.51	.66
24" OC, 2" x 2" Std & Btr (.30 BF per SF)	B1@.021	SF	.14	.44	.58

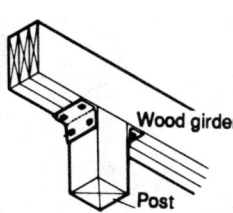

Girders, built-up from 2" lumber, Std & Btr, first floor work. Figures in parentheses show board feet per linear foot of girder, including 7% waste

4" x 6" (2.15 BF per LF, $450 per MBF)	B1@.034	LF	.97	.72	1.69
4" x 8" (2.85 BF per LF, $450 per MBF)	B1@.045	LF	1.28	.95	2.23
4" x 10" (3.58 BF per LF, $590 per MBF)	B1@.057	LF	2.11	1.20	3.31
4" x 12" (4.28 BF per LF, $600 per MBF)	B1@.067	LF	2.57	1.42	3.99
6" x 6" (3.21 BF per LF, $450 per MBF)	B1@.051	LF	1.44	1.08	2.52
6" x 8" (4.28 BF per LF, $450 per MBF)	B1@.067	LF	1.93	1.42	3.35
6" x 10" (5.35 BF per LF, $590 per MBF)	B1@.083	LF	3.16	1.76	4.92
6" x 12" (6.42 BF per LF, $600 per MBF)	B1@.098	LF	3.85	2.08	5.93
8" x 8" (5.71 BF per LF, $450 per MBF)	B1@.088	LF	2.57	1.86	4.43
Add for work above first floor	--	%	--	60.0	--

Grounds, Std & Btr, 1" x 2",
(at $400 per MBF, .18 BF per LF) including 10% waste

Over masonry	B1@.047	LF	.07	1.00	1.07
Over wood	B1@.038	LF	.07	.81	.88

Headers, #2 & Btr. Installed over wall openings and around floor, ceiling and roof openings, including 10% waste

2" x 6" (at $450 per MBF, 1.10 BF per LF)	B1@.028	LF	.50	.59	1.09
2" x 8" (at $450 per MBF, 1.47 BF per LF)	B1@.037	LF	.66	.78	1.44
2" x 10" (at $590 per MBF, 1.83 BF per LF)	B1@.046	LF	1.08	.97	2.05
2" x 12" (at $600 per MBF, 2.20 BF per LF)	B1@.057	LF	1.32	1.20	2.52
4" x 6" (at $625 per MBF, 2.20 BF per LF)	B1@.057	LF	1.38	1.20	2.58
4" x 8" (at $630 per MBF, 2.93 BF per LF)	B1@.073	LF	1.85	1.55	3.40
4" x 10" (at $630 per MBF, 3.67 BF per LF)	B1@.094	LF	2.31	1.99	4.30
4" x 12" (at $640 per MBF, 4.40 BF per LF)	B1@.112	LF	2.82	2.37	5.19

Lally columns (residential basement column) 3-1/2" diameter, concrete filled steel tube, material only

	--	LF	6.00	--	--
Add for base or cap plate	--	Ea	2.25	--	--
Add for installation, including cut per installed column, to 12' high	B1@.458	Ea	--	9.70	--

Ledger strips, Std & Btr nailed to faces of beams, girders, joists, etc. See also Ribbons in this section for let-in type. Figures in parentheses indicate board feet per LF including 10% waste

1" x 2" (at $440 per MBF .18 BF per LF)	B1@.010	LF	.08	.21	.29
1" x 3" (at $500 per MBF .28 BF per LF)	B1@.010	LF	.14	.21	.35
1" x 4" (at $440 per MBF .37 BF per LF)	B1@.010	LF	.16	.21	.37
2" x 2" (at $450 per MBF .37 BF per LF)	B1@.010	LF	.17	.21	.38
2" x 3" (at $450 per MBF .55 BF per LF)	B1@.010	LF	.25	.21	.46
2" x 4" (at $430 per MBF .73 BF per LF)	B1@.010	LF	.31	.21	.52

Add for one 1/2" or 3/4" diameter bolt, drilled and tightened

1-1/2" to 3" bolt with hex nut & washer	B1@.067	Ea	.55	1.42	1.97
3-1/4" to 5" bolt with hex nut & washer	B1@.073	Ea	.65	1.55	2.20
5-1/4" to 7" bolt with hex nut & washer	B1@.073	Ea	.80	1.55	2.35

Carpentry

	Craft@Hrs	Unit	Material	Labor	Total

Nailers, #2 & Btr, for wall finishes, trim, etc. See also Backing, Bridging and Fireblocking in this section. Figures in parentheses show board feet per LF including 10% waste

	Craft@Hrs	Unit	Material	Labor	Total
1" x 4" (at $555 per MBF, .37 BF per LF)	B1@.011	LF	.21	.23	.44
1" x 6" (at $625 per MBF, .55 BF per LF)	B1@.017	LF	.34	.36	.70
1" x 8" (at $640 per MBF, .73 BF per LF)	B1@.022	LF	.47	.47	.94
2" x 4" (at $430 per MBF, .73 BF per LF)	B1@.023	LF	.31	.49	.80
2" x 6" (at $450 per MBF, 1.10 BF per LF)	B1@.034	LF	.50	.72	1.22
2" x 8" (at $450 per MBF, 1.47 BF per LF)	B1@.045	LF	.66	.95	1.61
2" x 10" (at $590 per MBF, 1.83 BF per LF)	B1@.057	LF	1.08	1.21	2.29

Pier pads, 2" x 6", treated, #2 & Btr (lumber at $450 per MBF plus treatment

	Craft@Hrs	Unit	Material	Labor	Total
at $150 per MBF, 1.10 BF per LF)	B1@.034	LF	.66	.72	1.38

Plates, (wall plates), #2 & Btr, untreated. See also Sill Plates in this section. Figures in parentheses indicate board feet per LF including 10% waste

	Craft@Hrs	Unit	Material	Labor	Total
2" x 3" (at $450 per MBF, .55 BF per LF)	B1@.010	LF	.25	.21	.46
2" x 4" (at $430 per MBF, .73 BF per LF)	B1@.012	LF	.31	.25	.56
2" x 6" (at $450 per MBF, 1.10 BF per LF)	B1@.018	LF	.50	.38	.88

Posts, 4" x 4" material costs including 10% waste (1.47 BF per LF). See also Lally columns in this section and Posts in Lumber section.

	Craft@Hrs	Unit	Material	Labor	Total
Fir, S4S, green, select structural ($850 MBF)	--	LF	1.25	--	--
Fir, S4S, green, #1 structural ($715 MBF)	--	LF	1.05	--	--
Fir, rough, green, construction ($755 MBF)	--	LF	1.11	--	--
Red cedar, rough, green, construction ($635 MBF)	--	LF	.95	--	--
Redwood, S4S, green, construction ($850 MBF)	--	LF	1.25	--	--
Redwood, green, construction heart ($925 MBF)	--	LF	1.36	--	--
Southern yellow pine, treated ($680 MBF)	--	LF	1.00	--	--
Labor to install post, 4" x 4"					
In wall framing	B1@.064	LF	--	1.36	--
In deck railings	B1@.084	LF	--	1.78	--

Posts set on precast concrete pier block, including pier block with anchor (at $4.50) placed on existing grade, temporary 1" x 6" bracing (8 LF at $.25 = $2.00) and stakes (2 at $.35 =$.70). Cost is for each post set. Add for excavation if required.

	Craft@Hrs	Unit	Material	Labor	Total
Heights to 8', cost of post not included	BL@.166	Ea	7.20	3.13	10.33

Posts set in a hole with concrete, including temporary 1" x 6" bracing (8 LF at $.25 = $2.00) and stakes (2 at $.35 = $.70). Heights to 8'. Cost of post not included. Excavation and concrete assume posthole is 1' x 1' x 3' deep. Costs shown are for each post set.

	Craft@Hrs	Unit	Material	Labor	Total
Excavate posthole, by hand (normal soil)	BL@.250	Ea	--	4.70	--
Set post	BL@.125	Ea	2.70	2.35	5.05
Concrete for post, pre-mixed posthole mix	BL@.125	Ea	2.30	2.35	4.65

Purlins (perling), #2 & Btr, installed below roof rafters. Figures in parentheses indicate board feet per LF including 5% waste

	Craft@Hrs	Unit	Material	Labor	Total
2" x 4" (at $430 per MBF, .70 BF per LF)	B1@.012	LF	.30	.25	.55
2" x 6" (at $450 per MBF, 1.05 BF per LF	B1@.017	LF	.47	.36	.83
2" x 8" (at $450 per MBF, 1.40 BF per LF)	B1@.023	LF	.63	.49	1.12
4" x 6" (at $625 per MBF, 2.10 BF per LF)	B1@.034	LF	1.31	.72	2.03
4" x 8" (at $630 per MBF, 2.80 BF per LF)	B1@.045	LF	1.76	.95	2.71

Rafters, flat, shed, or gable roofs, up to 5 in 12 slope (5/24 pitch), maximum 25' span. Figures in parentheses indicate board feet per SF of actual roof surface area (not roof plan area), including rafters, ridge boards, collar beams and normal waste, but no blocking, bracing, purlins, curbs, or gable walls

Carpentry

OBTAINING ROOF AREA FROM PLAN AREA

Rise	Factor	Rise	Factor
3"	1.031	8"	1.202
3½"	1.042	8½"	1.225
4"	1.054	9"	1.250
4½"	1.068	9½"	1.275
5"	1.083	10"	1.302
5½"	1.100	10½"	1.329
6"	1.118	11"	1.357
6½"	1.137	11½"	1.385
7"	1.158	12"	1.414
7½"	1.179		

When a roof has to be figured from a plan only, and the roof pitch is known, the roof area may be fairly accurately computed from the table above. The horizontal or plan area (including overhangs) should be multiplied by the factor shown in the table opposite the rise, which is given in inches per horizontal foot. The result will be the roof area.

	Craft@Hrs	Unit	Material	Labor	Total
Rafters, continued					
2" x 4" (at $450 per MBF), Std & Btr					
12" center (.89 BF per SF)	B1@.021	SF	.40	.45	.85
16" center (.71 BF per SF)	B1@.017	SF	.32	.36	.68
24" center (.53 BF per SF)	B1@.013	SF	.24	.28	.52
2" x 6" (at $450 per MBF), Std & Btr					
12" center (1.29 BF per SF)	B1@.028	SF	.58	.59	1.17
16" center (1.02 BF per SF)	B1@.023	SF	.46	.49	.95
24" center (.75 BF per SF)	B1@.017	SF	.34	.36	.70
2" x 8" (at $450 per MBF), Std & Btr					
12" center (1.71 BF per SF)	B1@.036	SF	.77	.76	1.53
16" center (1.34 BF per SF)	B1@.030	SF	.61	.64	1.25
24" center (1.12 BF per SF)	B1@.024	SF	.51	.51	1.02
2" x 10" (at $590 per MBF), Std & Btr					
12" center (2.12 BF per SF)	B1@.045	SF	1.25	.95	2.20
16" center (1.97 BF per SF)	B1@.041	SF	1.16	.87	2.03
24" center (1.21 BF per SF)	B1@.026	SF	.71	.55	1.26
2" x 12" (at $600 per MBF), Std & Btr					
12" center (2.52 BF per SF)	B1@.050	SF	1.51	1.06	2.57
16" center (1.97 BF per SF)	B1@.040	SF	1.18	.85	2.03
24" center (1.43 BF per SF)	B1@.030	SF	.86	.64	1.50
Add for hip roof	--	%	--	15.0	--
Add for cut-up roof	--	%	5.0	35.0	--
Add for slope over 5 in 12	--	%	--	25.0	--
Deduct for small or no overhang	--	%	--	-10.0	--
Ribbons (ribbands), let-in to wall framing. See also Ledgers in this section. Figures in parentheses indicate board feet per LF including 10% waste					
1" x 3", Std & Btr (.28 BF per LF)					
(at $500 per MBF)	B1@.020	LF	.14	.42	.56
1" x 4", Std & Btr (.37 BF per LF)					
(at $440 per MBF)	B1@.020	LF	.16	.42	.58
1" x 6", Std & Btr (.55 BF per LF)					
(at $500 per MBF)	B1@.030	LF	.28	.64	.92
2" x 3", Std & Btr (.55 BF per LF)					
(at $450 per MBF)	B1@.041	LF	.25	.87	1.12
2" x 4", Std & Btr (.73 BF per LF)					
(at $430 per MBF)	B1@.041	LF	.31	.87	1.18
2" x 6", #2 & Btr (1.10 BF per LF)					
(at $450 per MBF)	B1@.045	LF	.50	.95	1.45
2" x 8", #2 & Btr (1.47 BF per LF)					
(at $450 per MBF)	B1@.045	LF	.66	.95	1.61
Roof decking. See also Sheathing in this section. Flat, shed, or gable roofs to 5 in 12 slope (5/24 pitch). Figures in parentheses indicate board feet per SF of actual roof area (not roof plan area), including 5% waste. These material costs are based on T&G fir, commercial grade.					
2" x 6" (2.28 BF per SF, $525 per MBF)	B1@.043	SF	1.48	.91	2.39
2" x 8" (2.25 BF per SF, $525 per MBF)	B1@.043	SF	1.47	.91	2.38
3" x 6" (3.43 BF per SF, $725 per MBF)	B1@.047	SF	3.09	1.00	4.09
Add for steep pitch roof	--	%	--	40.0	--
Roof trusses, 2" x 4" top and bottom chord					
Fink truss "W"					
24' span, 3 in 12 slope	B1@.811	Ea	40.80	17.20	58.00
24' span, 4 in 12 slope	B1@.771	Ea	42.70	16.30	59.00

Carpentry

	Craft@Hrs	Unit	Material	Labor	Total
28' span, 5 in 12 slope	B1@.958	Ea	48.60	20.30	68.90
32' span, 5 in 12 slope	B1@1.26	Ea	63.20	26.70	89.90
40' span, 5 in 12 slope	B1@1.70	Ea	74.00	36.00	110.00
Gable truss					
28' span, 5 in 12 slope	B1@.958	Ea	60.70	20.30	81.00
32' span, 5 in 12 slope	B1@1.26	Ea	81.90	26.70	108.60
40' span, 5 in 12 slope	B1@1.73	Ea	119.00	36.70	155.70

Sheathing, roof, per SF of roof surface including 10% waste
CD plywood sheathing, rough, interior grade

	Craft@Hrs	Unit	Material	Labor	Total
1/2" (at $ 8.80 per 4' x 8' sheet)	B1@.013	SF	.30	.28	.58
5/8" (at $14.70 per 4' x 8' sheet)	B1@.013	SF	.50	.28	.78
Add for hip roof	--	%	5.0	35.0	--
Add for steep pitch or cut-up roof	--	%	--	80.0	--

Board sheathing, 1" x 6" or 1" x 8" utility T&G laid diagonal (at $340 per MBF)

	Craft@Hrs	Unit	Material	Labor	Total
(1.13 BF per SF)	B1@.026	SF	.73	.55	1.28
Add for hip roof	--	%	5.0	35.0	--
Add for cut-up steep roofs	--	%	--	80.0	--

Sheathing, wall, per SF of wall surface, including 5% waste
CD plywood sheathing, plugged and touch sanded, interior grade

	Craft@Hrs	Unit	Material	Labor	Total
5/16" (at $ 8.80 per 4' x 8' sheet)	B1@.013	SF	.29	.28	.57
3/8" (at $ 9.60 per 4' x 8' sheet)	B1@.015	SF	.32	.32	.64
1/2" (at $11.00 per 4' x 8' sheet)	B1@.016	SF	.36	.34	.70
5/8" (at $14.70 per 4' x 8' sheet)	B1@.018	SF	.48	.38	.86
3/4" (at $17.50 per 4' x 8' sheet)	B1@.020	SF	.57	.42	.99

Board sheathing 1" x 6" or 1" x 8" utility T&G (at $340 per MBF)

	Craft@Hrs	Unit	Material	Labor	Total
(1.13 BF per SF)	B1@.020	SF	.73	.42	1.15
Add for diagonal patterns	--	%	--	10.0	--

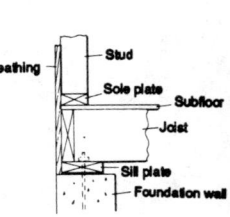

Sill plates (at foundation), pressure treated lumber, drilled and installed with foundation bolts at 48" OC, no bolts, nuts or washers included. Figures in parentheses indicate board feet per LF of foundation, including 5% waste and treatment at $150 per MBF, using

	Craft@Hrs	Unit	Material	Labor	Total
2" x 3", Std & Btr (at $600 per MBF)					
(.53 BF per LF)	B1@.020	LF	.32	.42	.74
2" x 4", Std & Btr (at $580 per MBF)					
(.70 BF per LF)	B1@.023	LF	.49	.49	.98
2" x 6", #2 & Btr (at $600 per MBF)					
(1.05 BF per LF)	B1@.024	LF	.64	.51	1.15
2" x 8", #2 & Btr (at $600 per MBF)					
(1.40 BF per LF)	B1@.031	LF	.84	.66	1.50

Sleepers, #2 & Btr pressure treated lumber at $450 per MBF and treatment at $150 per MBF, including 5% waste, using

	Craft@Hrs	Unit	Material	Labor	Total
2" x 6" ($600 per MBF)	B1@.017	BF	.63	.36	.99
Add for taper cuts on sleepers	--	%	--	100.0	--

Studding, per board foot of studs and plates. As determined by detailed material take-off. No waste and no backing, bracing, fireblocking, headers, nailers or foundation plates included. 12", 16", 20", or 24" spacing, including normal waste

	Craft@Hrs	Unit	Material	Labor	Total
2" x 3" members, Std & Btr (at $450 per MBF)	B1@.027	BF	.45	.57	1.02
2" x 4" members, Std & Btr (at $430 per MBF)	B1@.025	BF	.43	.53	.96
2" x 6" members, Std & Btr (at $450 per MBF)	B1@.022	BF	.45	.47	.92

Studding, per square foot of wall area. Figures in parentheses indicate typical board feet per SF of wall area measured on one side. Costs include studding, single bottom plate, double top plate, fireblocking, nails and normal waste. Also see door and window opening framing, backing, let-in bracing and sheathing for shear walls.

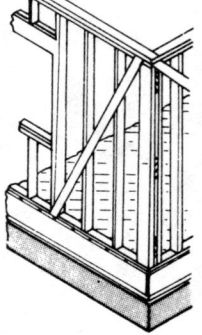

Carpentry

	Craft@Hrs	Unit	Material	Labor	Total

Studding, per square foot of wall area, continued
Wall stud framing
  2" x 3" (at $450 per MBF), Std & Btr

	Craft@Hrs	Unit	Material	Labor	Total
12" center (.73 BF per SF)	B1@.020	SF	.33	.42	.75
16" center (.68 BF per SF)	B1@.018	SF	.31	.38	.69
20" center (.65 BF per SF)	B1@.018	SF	.29	.38	.67
24" center (.62 BF per SF)	B1@.017	SF	.28	.36	.64

  2" x 4" (at $430 per MBF), Std & Btr

	Craft@Hrs	Unit	Material	Labor	Total
12" center (.96 BF per SF)	B1@.024	SF	.41	.51	.92
16" center (.92 BF per SF)	B1@.023	SF	.40	.49	.89
20" center (.88 BF per SF)	B1@.022	SF	.38	.47	.85
24" center (.82 BF per SF)	B1@.021	SF	.35	.45	.80

  2" x 6" (at $450 per MBF), Std & Btr

	Craft@Hrs	Unit	Material	Labor	Total
12" center (1.45 BF per SF)	B1@.032	SF	.65	.68	1.33
16" center (1.32 BF per SF)	B1@.029	SF	.59	.62	1.21
20" center (1.26 BF per SF)	B1@.028	SF	.57	.59	1.16
24" center (1.21 BF per SF)	B1@.027	SF	.54	.57	1.11

Subflooring
Board sheathing, 1" x 6" Std & Btr, (at $500 per MBF, 1.24 BF per SF)

	Craft@Hrs	Unit	Material	Labor	Total
includes 12% shrinkage and 5% waste & nails	B1@.020	SF	.62	.42	1.04
Add for diagonal patterns	--	%	5.0	10.0	--

Plywood sheathing, CD standard interior grade, plugged and touch sanded.
Includes 5% waste & fasteners

	Craft@Hrs	Unit	Material	Labor	Total
5/16" (at $ 8.80 per 4' x 8' sheet)	B1@.011	SF	.29	.23	.52
3/8" (at $ 9.60 per 4' x 8' sheet)	B1@.011	SF	.32	.23	.55
1/2" (at $11.00 per 4' x 8' sheet)	B1@.012	SF	.36	.25	.61
5/8" (at $14.70 per 4' x 8' sheet)	B1@.012	SF	.48	.25	.73
3/4" (at $17.50 per 4' x 8' sheet)	B1@.013	SF	.57	.28	.85
1-1/8" (2-4-1) underlayment, T&G, (at $27.50 per 4' x 8' sheet)	B1@.020	SF	.90	.42	1.32

Trimmers, at stairwells, skylights, dormers, etc. Figures in parentheses show board feet per LF including 10% waste

	Craft@Hrs	Unit	Material	Labor	Total
2" x 4", Std & Btr (at $430 per MBF) (.73 BF per LF)	B1@.018	LF	.31	.38	.69
2" x 6", #2 & Btr (at $450 per MBF) (1.10 BF per LF)	B1@.028	LF	.50	.59	1.09
2" x 8", #2 & Btr (at $450 per MBF) (1.47 BF per LF)	B1@.038	LF	.66	.81	1.47
2" x 10", #2 & Btr (at $590 per MBF) (1.83 BF per LF)	B1@.047	LF	1.08	1.00	2.08
2" x 12", #2 & Btr (at $600 per MBF) (2.20 BF per LF)	B1@.057	LF	1.32	1.21	2.53

Window openings, framing in wall studs, based on walls 8' in height. Figures in parentheses indicate typical board feet per opening. Costs shown are per window opening and include header of appropriate size, double vertical studs each side of the opening, double top plates, single bottom plate, cripples, blocking, nails and normal waste.
  2" x 4" wall studs (at $430 per MBF), Std & Btr, opening size shown is width/height.

	Craft@Hrs	Unit	Material	Labor	Total
2'0"/2'0" or 2'0"/3'0" (45 BF)	B1@1.75	Ea	21.30	37.10	58.40
3'0"/2'0", 3'0"/3'0" or 3'0"/4'0" (47 BF)	B1@2.00	Ea	22.20	42.40	64.60
4'0"/2'0", 4'0"/3'0" or 4'0"/4'0" (50 BF)	B1@2.25	Ea	23.70	47.70	71.40
4'0"/5'0" or 5'0"/5'0" (55 BF)	B1@2.50	Ea	26.00	53.00	79.00
6'0"/3'0", 6'0"/4'0" or 6'0"/6'0" (70 BF)	B1@2.75	Ea	33.10	58.30	91.40
7'0"/2'0", 7'0"/3'0" or 7'0"/4'0" (85 BF)	B1@3.00	Ea	40.20	63.60	103.80

Carpentry

	Craft@Hrs	Unit	Material	Labor	Total
Window openings, framing, continued					
8'0"/3'0", 8'0"/4'0" or 8'0"/6'0" (90 BF)	B1@3.25	Ea	42.60	68.90	111.50
10'0"/3'0", 10'0"/4'0" or 10'0"/6'0" (115 BF)	B1@3.50	Ea	54.40	74.20	128.60
Add per foot of height, walls over 8' high,	--	LF	1.15	--	--
Deduct for 2" x 3" wall studs	--	%	-15.0	--	--
Add for 2" x 6" wall studs	--	%	30.0	--	--

Wall Assemblies. Costs for 2" x 4" wood framed stud walls with wall finish treatment on both sides, based on performing the work at the construction site. These costs include 2" x 4" wall studs at 16" center to center, double top plates, single bottom plates, fire blocking, nails and wall finish treatment as described. Costs shown are per SF or LF of wall measured on one face and include normal waste. Lumber is Std & Btr at $430 per MBF with 1.12 BF per SF of wall measured on one side.

Interior wall assemblies

	Craft@Hrs	Unit	Material	Labor	Total
Stud walls with 1/2" gypsum drywall both sides, ready for painting					
Cost per square foot	B1@.059	SF	.74	1.25	1.99
Cost per running foot, for 8' high walls	B1@.472	LF	5.92	10.00	15.92
Stud walls with 5/8" gypsum drywall both sides, ready for painting					
Cost per square foot	B1@.062	SF	.78	1.31	2.09
Cost per running foot, for 8' high walls	B1@.496	LF	6.24	10.51	16.75

Exterior wall assemblies

Drywall interior, wood siding exterior
Stud walls with 1/2" gypsum drywall on inside face ready for painting, over 3-1/2" R-11 insulation with 5/8" thick rough sawn T-1-11, exterior grade plywood siding (4 ply, 4' x 8' panels at $700 per MSF) on the outside face.

	Craft@Hrs	Unit	Material	Labor	Total
Cost per square foot	B1@.081	SF	1.76	1.72	3.48
Cost per running foot, for 8' high walls	B1@.648	LF	14.08	13.76	27.84

Drywall interior, 1" x 6" drop siding exterior
Stud walls with 1/2" gypsum drywall on inside face ready for painting, over 3-1/2" R-11 insulation with 1" x 6" southern yellow pine drop siding, D grade, (1.19 BF per SF with 5-1/4" exposure, at $760 per MBF) on the outside face.

	Craft@Hrs	Unit	Material	Labor	Total
Cost per square foot	B1@.086	SF	1.89	1.82	3.71
Cost per running foot, for 8' high wall	B1@.688	LF	15.12	14.56	29.68

Drywall interior, stucco exterior
Stud walls with 1/2" gypsum drywall on inside face ready for painting, over 3-1/2" R-11 insulation and a three-coat exterior plaster (stucco) finish with integral color on the outside face.

	Craft@Hrs	Unit	Material	Labor	Total
Cost per square foot	B1@.063	SF	3.01	1.33	4.34
Cost per running foot, for 8' high wall	B1@.504	SF	24.08	10.64	34.72

Add for different type drywall gypsum board

	Craft@Hrs	Unit	Material	Labor	Total
1/2" or 5/8" moisture resistant greenboard					
Cost per SF, greenboard per side, add	--	SF	.09	--	--
Cost per running foot, for 8' high walls, per side, add	--	LF	.72	--	--
5/8" thick gypsum regular drywall					
Cost per square foot, per side, add	--	SF	.02	--	--
Cost per running foot, for 8' high walls, per side, add	--	LF	.16	--	--

Carpeting, Subcontract. Costs listed are average prices for complete residential jobs and include consultation, measurement, pad, carpet, and professional installation of pad and carpet using tack strips and hot melt tape on seams. Prices can be expected to vary, up or down, by 50% depending on quantity and quality of actual materials installed.

Carpeting

	Craft@Hrs	Unit	Material	Labor	Total
Minimum quality, 25 to 35 oz face weight nylon carpet installed over a 1/2" (4 lb density) rebond pad	--	SY	--	--	18.50
Medium quality, 35 to 50 oz face weight nylon carpet installed over a 1/2" (6 lb density) rebond pad	--	SY	--	--	28.00
Better quality, 50 (plus) oz face weight nylon carpet installed over a 1/2" (6 lb density) rebond pad	--	SY	--	--	36.00
Wool Berber carpet (large loop) installed	--	SY	--	--	52.50
Add for waterfall (box steps) stairways	--	Riser	--	--	3.00
Add for wrapped steps (open riser), sewn	--	Step	--	--	7.00
Add for sewn edge treatment on one side	--	Riser	--	--	7.00
Add for circular stair steps, depending on frame	--	Step	--	--	14.00

Carports. Material costs include all hardware. Labor is based on bolting posts to existing concrete slab.

Single carport, 6 posts, 8' high
  Natural aluminum finish, posts attach to galvanized steel beams

	Craft@Hrs	Unit	Material	Labor	Total
8' wide x 16' long, 40 PSF	B1@7.96	Ea	535.00	169.00	704.00
10' wide x 20' long, 20 PSF	B1@7.96	Ea	685.00	169.00	854.00

White enamel finish, baked-on enamel roof panels, gutter-mounted posts, 10' wide x 20' long

	Craft@Hrs	Unit	Material	Labor	Total
20 PSF	B1@7.96	Ea	795.00	169.00	964.00
40 PSF	B1@7.96	Ea	840.00	169.00	1,009.00
60 PSF	B1@7.96	Ea	940.00	169.00	1,109.00

Double carport, 10' high, with steel posts and aluminum gutters, roof pans and supports

	Craft@Hrs	Unit	Material	Labor	Total
20' x 20', 4 posts, 40 PSF	B1@8.80	Ea	1,840.00	186.00	2,026.00
20' x 20', 6 posts, 60 PSF	B1@9.34	Ea	2,400.00	198.00	2,598.00
24' x 24', 4 posts, 40 PSF	B1@8.80	Ea	2,900.00	186.00	3,086.00
24' x 24', 6 posts, 60 PSF	B1@9.34	Ea	3,200.00	198.00	3,398.00
Deduct for steel beams	--	%	-25.0	--	--

Caulking. Material costs are typical costs for bead diameter listed. Figures in parentheses indicate approximate coverage including 5% waste. Labor costs are for good quality application on smooth to slightly irregular surfaces. Per LF bead length.

Acoustical caulk, flexible, sound deadening (@ $.26 per fluid oz)

	Craft@Hrs	Unit	Material	Labor	Total
1/8" (11.6 LF per fluid oz)	BC@.018	LF	.02	.43	.45
1/4" (2.91 LF per fluid oz)	BC@.025	LF	.09	.59	.68
3/8" (1.29 LF per fluid oz)	BC@.030	LF	.20	.71	.91
1/2" (.728 LF per fluid oz)	BC@.033	LF	.36	.78	1.14

Butyl caulk, premium quality, 20 year life expectancy (@ $.33 per fluid oz)

	Craft@Hrs	Unit	Material	Labor	Total
1/8" (11.6 LF per fluid oz)	BC@.018	LF	.03	.43	.46
1/4" (2.91 LF per fluid oz)	BC@.025	LF	.12	.59	.71
3/8" (1.29 LF per fluid oz)	BC@.030	LF	.26	.71	.97
1/2" (.728 LF per fluid oz)	BC@.033	LF	.46	.78	1.24

Butyl caulk, good quality, 10 year life expectancy (@ $.32 per fluid oz)

	Craft@Hrs	Unit	Material	Labor	Total
1/8" (11.6 LF per fluid oz)	BC@.018	LF	.03	.43	.46
1/4" (2.91 LF per fluid oz)	BC@.025	LF	.11	.59	.70
3/8" (1.29 LF per fluid oz)	BC@.030	LF	.25	.71	.96
1/2" (.728 LF per fluid oz)	BC@.033	LF	.44	.78	1.22

Latex, vinyl acrylic caulk, good quality, 10 year life (@ $.35 per fluid oz)

	Craft@Hrs	Unit	Material	Labor	Total
1/8" (11.6 LF per fluid oz)	BC@.018	LF	.03	.43	.46
1/4" (2.91 LF per fluid oz)	BC@.025	LF	.12	.59	.71
3/8" (1.29 LF per fluid oz)	BC@.030	LF	.27	.71	.98
1/2" (.728 LF per fluid oz)	BC@.033	LF	.48	.78	1.26

Caulking

	Craft@Hrs	Unit	Material	Labor	Total
Latex caulk, economy grade (@$.31 per fluid oz)					
1/8" (11.6 LF per fluid oz)	BC@.018	LF	.03	.43	.46
1/4" (2.91 LF per fluid oz)	BC@.025	LF	.11	.59	.70
3/8" (1.29 LF per fluid oz)	BC@.030	LF	.24	.71	.95
1/2" (.728 LF per fluid oz)	BC@.033	LF	.43	.78	1.21
Oil base caulk, good quality (@$.27 per fluid oz)					
1/8" (11.6 LF per fluid oz)	BC@.018	LF	.02	.43	.45
1/4" (2.91 LF per fluid oz)	BC@.025	LF	.09	.59	.68
3/8" (1.29 LF per fluid oz)	BC@.030	LF	.21	.71	.92
1/2" (.728 LF per fluid oz)	BC@.033	LF	.37	.78	1.15
Oil base caulk, economy grade (@$.25 per fluid oz)					
1/8" (11.6 LF per fluid oz)	BC@.018	LF	.02	.43	.45
1/4" (2.91 LF per fluid oz)	BC@.025	LF	.09	.59	.68
3/8" (1.29 LF per fluid oz)	BC@.030	LF	.19	.71	.90
1/2" (.728 LF per fluid oz)	BC@.033	LF	.34	.78	1.12
Peel-off caulking for temporary use (@$.23 per fluid oz)					
1/8" (11.6 LF per fluid oz)	BC@.018	LF	.02	.43	.45
1/4" (2.91 LF per fluid oz)	BC@.022	LF	.08	.52	.60
3/8" (1.29 LF per fluid oz)	BC@.027	LF	.18	.64	.82
1/2" (.728 LF per fluid oz)	BC@.030	LF	.32	.71	1.03
Siliconized acrylic caulk, premium quality, 20 year life (@$.25 per fluid oz)					
1/8" (11.6 LF per fluid oz)	BC@.018	LF	.03	.43	.46
1/4" (2.91 LF per fluid oz)	BC@.025	LF	.09	.59	.68
3/8" (1.29 LF per fluid oz)	BC@.030	LF	.19	.71	.90
1/2" (.728 LF per fluid oz)	BC@.033	LF	.34	.78	1.12
Silicon rubber sealant, premium quality, 20 year life (@$.53 per fluid oz)					
1/8" (11.6 LF per fluid oz)	BC@.043	LF	.05	1.01	1.06
1/4" (2.91 LF per fluid oz)	BC@.048	LF	.18	1.13	1.31
3/8" (1.29 LF per fluid oz)	BC@.056	LF	.41	1.32	1.73
1/2" (.728 LF per fluid oz)	BC@.060	LF	.73	1.42	2.15
Tub caulk, white siliconized (@$.38 per fluid oz)					
1/8" (11.6 LF per fluid oz)	BC@.018	LF	.03	.43	.46
1/4" (2.91 LF per fluid oz)	BC@.022	LF	.13	.52	.65
3/8" (1.29 LF per fluid oz)	BC@.027	LF	.29	.64	.93
1/2" (.728 LF per fluid oz)	BC@.030	LF	.52	.71	1.23
Anti-algae and mildew-resistant tub caulk, premium quality white or clear silicone (@$.42 per fluid oz)					
1/8" (11.6 LF per fluid oz)	BC@.018	LF	.04	.43	.47
1/4" (2.91 LF per fluid oz)	BC@.022	LF	.14	.52	.66
3/8" (1.29 LF per fluid oz)	BC@.027	LF	.33	.64	.97
1/2" (.728 LF per fluid oz)	BC@.030	LF	.58	.71	1.29
Urethane one-part caulk, good quality, 10 year life (@$.44 per fluid oz)					
1/8" (11.6 LF per fluid oz)	BC@.043	LF	.04	1.01	1.05
1/4" (2.91 LF per fluid oz)	BC@.048	LF	.15	1.13	1.28
3/8" (1.29 LF per fluid oz)	BC@.056	LF	.34	1.32	1.66
1/2" (.728 LF per fluid oz)	BC@.060	LF	.60	1.42	2.02
Urethane two-part caulk, top quality, 20 year life (@$.55 per fluid oz)					
1/8" (11.6 LF per fluid oz)	BC@.043	LF	.05	1.01	1.06
1/4" (2.91 LF per fluid oz)	BC@.052	LF	.19	1.23	1.42
3/8" (1.29 LF per fluid oz)	BC@.056	LF	.43	1.32	1.75
1/2" (.728 LF per fluid oz)	BC@.066	LF	.76	1.56	2.32
Add for irregular surfaces such as vertical masonry or lap siding	--	%	5.0	25.0	--

Caulking

	Craft@Hrs	Unit	Material	Labor	Total
Caulking gun, professional type, heavy duty					
Bulk caulking gun	--	Ea	30.00	--	--
Caulking gun, economy grade					
11-oz cartridge	--	Ea	5.00	--	--
29-oz cartridge	--	Ea	10.00	--	--

Cedar Closet Lining. Material costs are per SF of floor, wall, ceiling or door of closet including 7% waste. Labor is for good quality installation in typical residential closet over studs or gypsum wallboard.

	Craft@Hrs	Unit	Material	Labor	Total
Compressed cedar chip panels, class C flamespread rating, 4' x 8' panels					
3/16" thick ($21.00 per 4' x 8' panel)	BC@.016	SF	.71	.38	1.09
1/4" thick ($23.50 per 4' x 8' panel)	BC@.016	SF	.79	.38	1.17
Cedar boards, 3/8" x 3 1/2", random lengths, T&G, end matched					
($25.00 per 20 SF bundle)	BC@.050	SF	1.34	1.18	2.52

Ceiling Domes. Enveldomes, complete ceiling lighted dome kits, includes grid, hardware, panels and crating, from flat to 30 degree pitch in 5 degree increments, Envel Design Corp.

	Craft@Hrs	Unit	Material	Labor	Total
10' diameter octagon	B1@10.7	Ea	3,250.00	227.00	3,477.00
16' diameter octagon	B1@17.2	Ea	6,050.00	364.00	6,414.00
24' diameter, round	B1@24.7	Ea	12,550.00	523.00	13,073.00

Ceiling Panel Suspension Grids. Typical costs, plain white. Add panel costs below.

	Craft@Hrs	Unit	Material	Labor	Total
Main runner, 12' long (@$3.75 ea)	--	LF	.31	--	--
Cross tee, 48" long (@$1.00 ea)	--	LF	.25	--	--
Cross tee, 24" long (@$.56 ea)	--	LF	.29	--	--
Wall mould, 12' long (@$2.30 ea)	--	LF	.19	--	--
Add for walnut colored	--	%	10.0	--	--
Add for fire-rated white	--	%	5.0	--	--
Add for chrome or gold color grid	--	%	40.0	--	--
Total grid system including runners, tees, wall mould, hooks and hanging wire (but no tile or lighting fixtures)					
2' x 2' grid	BC@.014	SF	.38	.33	.71
2' x 4' grid	BC@.012	SF	.30	.28	.58
Add for under 400 SF job	BC@.007	SF	--	.17	--
Lighting fixtures for 2' x 4' grid, includes installation, connection, lens and tubes					
2 tube fixtures	BE@1.00	Ea	45.60	25.30	70.90
4 tube fixtures	BE@1.00	Ea	68.90	25.30	94.20
Light-diffusing panels for installation in 2' x 4' suspension system under fluorescent lighting					
Acrylic, prismatic clear					
.095" thick	BC@.021	Ea	5.30	.50	5.80
.110" thick	BC@.021	Ea	6.36	.50	6.86
.125" thick	BC@.021	Ea	7.95	.50	8.45
Acrylic, dual white					
.080" thick	BC@.021	Ea	13.80	.50	14.30
.125" thick	BC@.021	Ea	20.70	.50	21.20
Acrylic frost overlay, clear or white					
.040" thick	BC@.021	Ea	7.42	.50	7.92
Eggcrate louvers, 1/2" x 1/2" x 1/2" cell size					
White styrene	BC@.188	Ea	9.01	4.44	13.45
White acrylic	BC@.188	Ea	13.80	4.44	18.24
Mill aluminum	BC@.188	Ea	25.40	4.44	29.84
White aluminum	BC@.188	Ea	29.70	4.44	34.14
Anodized aluminum	BC@.188	Ea	33.90	4.44	38.34

Ceiling Grids

	Craft@Hrs	Unit	Material	Labor	Total
Eggcrate louvers, 1/2" x 1/2" x 3/8" cell size					
White styrene	BC@.188	Ea	6.36	4.44	10.80
White acrylic	BC@.188	Ea	12.20	4.44	16.64
Black styrene	BC@.188	Ea	9.01	4.44	13.45
Parabolic louvers, 5/8" x 5/8" x 7/16" cell size					
Silver styrene	BC@.188	Ea	22.30	4.44	26.74
Silver acrylic	BC@.188	Ea	37.10	4.44	41.54
Gold styrene	BC@.188	Ea	23.30	4.44	27.74
Gold acrylic	BC@.188	Ea	38.20	4.44	42.64
Polystyrene, clear or white					
.095", .070" thick	BC@.056	Ea	4.24	1.32	5.56
.040" thick, smooth white	BC@.056	Ea	4.77	1.32	6.09
Stained glass look, one dimension, 2' x 2'	BC@.033	Ea	27.60	.78	28.38
Stained glass look, one dimension, antiqued colors					
2' x 2'	BC@.035	Ea	37.10	.83	37.93

Ceiling Panels, Lay-in Type. Costs per SF of ceiling area covered. No waste included. Add suspension grid cost above.

	Craft@Hrs	Unit	Material	Labor	Total
24" x 24" non-acoustic					
Smooth, plain white	BC@.007	SF	1.02	.17	1.19
Smooth, marbleized or simulated wood	BC@.007	SF	1.75	.17	1.92
Textured	BC@.005	SF	1.36	.12	1.48
24" x 24" acoustic,					
Deep textured, 3/4" reveal edge	BC@.008	SF	1.20	.19	1.39
Random pinhole, 5/8" square edge	BC@.004	SF	.36	.09	.45
24" x 48" acoustic					
Deep textured, 3/4" reveal edge	BC@.008	SF	1.27	.19	1.46
Fissured, 5/8" square edge	BC@.004	SF	.36	.09	.45
Random pinhole, 5/8" square edge	BC@.004	SF	.36	.09	.45
24" x 48" fire retardant, acoustic,					
Deep textured, 3/4" reveal edge	BC@.010	SF	1.56	.24	1.80
Fissured, 5/8" square edge	BC@.004	SF	.45	.09	.54
Embossed parquet, pattern, grid hiding	BC@.004	SF	1.07	.09	1.16
Insulating (R-12), acoustic, textured	BC@.004	SF	2.05	.09	2.14
Fiberglass reinforced 24" x 24" gypsum panels for T-bar systems. Weight is 3 to 4 pounds each, depending on pattern					
White painted finish	BC@.015	SF	6.89	.35	7.24
Wood grain finish	BC@.020	SF	7.95	.47	8.42
Multi-dimensional faceted and beveled ceiling panels,					
2' x 2' (Envel Design)	BC@.488	Ea	75.90	11.50	87.40

Ceiling, Metal Pans. Flat, 2' x 2' pans.

	Craft@Hrs	Unit	Material	Labor	Total
Aluminum with concealed grid					
2' x 2' painted	BC@.070	SF	3.71	1.65	5.36
2' x 2' polished	BC@.070	SF	7.95	1.65	9.60
Stainless steel with concealed grid					
2' x 2', polished clear mirror	BC@.082	SF	11.70	1.93	13.63
Stainless steel with exposed grid					
2' x 2', polished clear mirror	BC@.060	SF	6.89	1.42	8.31

Ceiling Tile. 12" x 12" tile nailed, glued or stapled to ceilings. Costs per SF of ceiling area covered including fasteners and trim. No waste included.

	Craft@Hrs	Unit	Material	Labor	Total
Non-acoustic					
Economy, embossed texture pattern	BC@.022	SF	.56	.52	1.08

Ceiling Tile

	Craft@Hrs	Unit	Material	Labor	Total
Smooth	BC@.022	SF	.66	.52	1.18
Textured	BC@.023	SF	.75	.54	1.29
Acoustic					
Random pinhole	BC@.030	SF	.91	.71	1.62
Heavily textured	BC@.030	SF	1.45	.71	2.16
Fire retardant					
Non-acoustic simulated cork	BC@.034	SF	1.90	.80	2.70
Acoustic, textured	BC@.030	SF	1.48	.71	2.19

Ceilings, Tin. Various plain and intricate patterns, unfinished.

	Craft@Hrs	Unit	Material	Labor	Total
Embossed panels					
2' x 2' square panels, cost each	B1@.147	Ea	12.00	3.12	15.12
3' x 3' square panels, cost each	B1@.314	Ea	64.00	6.65	70.65
Embossed plates					
12" x 12" plates, cost per SF	B1@.042	SF	5.00	.89	5.89
12" x 24" plates, cost per SF	B1@.040	SF	5.70	.85	6.55
24" x 24" plates, cost per SF	B1@.037	SF	3.05	.78	3.83
24" x 48" plates, cost per SF	B1@.035	SF	3.05	.74	3.79
Embossed moulding, 4' lengths					
3" and 4" widths, cost per linear foot	B1@.055	LF	1.57	1.17	2.74
6" widths, cost per linear foot	B1@.067	LF	2.32	1.42	3.74
12" widths cost per linear foot	B1@.088	LF	3.27	1.87	5.14
Embossed moulding crosses, tees or ells					
For 3", 4" and 6" wide moulding, cost each	B1@.181	Ea	4.18	3.84	8.02
For 12" wide moulding, cost each	B1@.228	Ea	6.75	4.83	11.58
Perimeter and beam cornice moulding, 4' lengths					
2" projection x 2" deep, per LF	B1@.180	LF	1.97	3.81	5.78
3" projection x 4" deep, per LF	B1@.197	LF	2.08	4.17	6.25
4" projection x 6" deep, per LF	B1@.197	LF	2.22	4.17	6.39
6" projection x 8" deep, per LF	B1@.218	LF	2.63	4.62	7.25
6" projection x 10" deep, per LF	B1@.218	LF	2.63	4.62	7.25
10" projection x 12" deep, per LF	B1@.238	LF	3.54	5.04	8.58
12" projection x 18" deep, per LF	B1@.238	LF	4.40	5.04	9.44
Mitered corners for cornice moulding					
2" x 2"	B1@.155	Ea	7.77	3.28	11.05
3" x 4"	B1@.184	Ea	7.77	3.90	11.67
4" x 6"	B1@.184	Ea	8.30	3.90	12.20
6" x 8"	B1@.230	Ea	8.80	4.87	13.67
6" x 10"	B1@.230	Ea	9.50	4.87	14.37
10" x 12"	B1@.280	Ea	10.00	5.93	15.93
12" x 18"	B1@.280	Ea	11.35	5.93	17.28
Frieze moulding, 4' lengths					
6" width, per LF	B1@.067	LF	2.35	1.42	3.77
12" and 14" widths per LF	B1@.088	LF	3.25	1.87	5.12
18" width per LF	B1@.110	LF	4.85	2.33	7.18
Foot moulding, 4' lengths					
4" width, per LF	B1@.055	LF	1.82	1.17	2.99
6" width, per LF	B1@.067	LF	2.35	1.42	3.77
Nosing, 4' lengths					
1-1/4" width, per LF	B1@.047	LF	1.65	1.00	2.65
4-1/2" width, per LF	B1@.055	LF	1.65	1.17	2.82

Cement

	Craft@Hrs	Unit	Material	Labor	Total
Cement. See also Adhesives, Aggregates, and Concrete.					
Portland cement, regular 94 lb sack	--	Sa	6.50	--	--
Plastic cement, 94 lb sack	--	Sa	7.00	--	--
High early strength cement, 94 lb sack	--	Sa	7.50	--	--
Concrete mix, 90 lb sack	--	Sa	3.10	--	--
Topping mix, 60 lb sack	--	Sa	2.85	--	--
Mortar mix, 70 lb sack	--	Sa	4.00	--	--
Exterior stucco mix, 60 lb sack	--	Sa	4.00	--	--
Plaster patch, 25 lb sack	--	Sa	7.50	--	--
Asphalt mix, 60 lb sack	--	Sa	4.00	--	--
White cement, 94 lb sack	--	Sa	12.75	--	--
Deduct for pallet quantities	--	%	-13.0	--	--

Closet Door Systems

Louver-over-panel wood doors. Solid pine, unfinished, for 80" to 81" heights, hardware included

	Craft@Hrs	Unit	Material	Labor	Total
Two-section, folding door or two door set					
24" wide	BC@1.20	Ea	45.00	28.30	73.30
30" wide	BC@1.20	Ea	58.00	28.30	86.30
32" wide	BC@1.50	Ea	64.00	35.40	99.40
36" wide	BC@1.50	Ea	69.00	35.40	104.40
Four-section (two 2-section doors)					
24" wide	BC@1.80	Ea	64.00	42.50	106.50
30" wide	BC@1.80	Ea	72.00	42.50	114.50
32" wide	BC@1.80	Ea	78.00	42.50	120.50
36" wide	BC@2.09	Ea	84.00	49.30	133.30
Add for colonial (raised panel) styles	--	LS	15.00	--	--

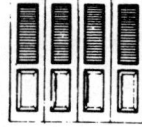

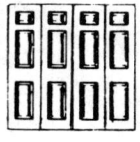

Folding hollow core, wood doors, 1-3/8" thick, with hardware. Add labor below.

	Hardboard	Lauan	Birch	Ash
6'8" high (2 door units)				
2'0" wide	41.00	44.00	50.00	51.00
2'6" wide	44.00	48.00	56.00	57.00
3'0" wide	48.00	53.00	62.00	63.00
6'8" high (4 door units)				
4'0" wide	71.00	78.00	92.00	94.00
5'0" wide	76.00	88.00	104.00	106.00
6'0" wide	82.00	97.00	115.00	118.00
7'0" wide	113.00	123.00	145.00	155.00
8'0" wide	118.00	130.00	154.00	164.00
8'0" high (4 door units)				
4'0" wide	101.00	116.00	137.00	138.00
5'0" wide	109.00	130.00	155.00	155.00
6'0" wide	118.00	144.00	172.00	173.00
7'0" wide	166.00	185.00	220.00	221.00
8'0" wide	166.00	185.00	220.00	221.00

	Craft@Hrs	Unit	Material	Labor	Total
Labor to install folding hollow core closet doors					
2'0", 2'6" or 3'0" wide x 6'8" high	BC@1.30	Ea	--	30.70	--
2'0", 2'6" or 3'0" wide x 8'0" high	BC@1.50	Ea	--	35.40	--
4'0" or 5'0" wide x 6'8" high	BC@1.60	Ea	--	37.70	--
4'0" or 5'0" wide x 8'0" high	BC@1.80	Ea	--	42.50	--
6'0", 7'0" or 8'0" wide x 6'8" high	BC@1.90	Ea	--	44.80	--
6'0", 7'0" or 8'0" wide x 8'0" high	BC@2.10	Ea	--	49.50	--

Closet Doors

Folding doors, pine panel and louver, 1-3/8" thick, 6'8" high, with hardware. Add labor below.

	Full louver	Louver panel	2 panel design	3 panel design
2'0" (2 door unit)	87.00	101.00	113.00	127.00
2'6" (2 door unit)	93.00	109.00	128.00	141.00
2'8" (2 door unit)	96.00	114.00	133.00	146.00
3'0" (2 door unit)	100.00	120.00	142.00	157.00
4'0" (4 door unit)	159.00	185.00	224.00	246.00
5'0" (4 door unit)	171.00	200.00	250.00	275.00
6'0" (4 door unit)	183.00	216.00	276.00	304.00
7'0" (4 door unit)	208.00	247.00	306.00	337.00
8'0" (4 door unit)	218.00	263.00	330.00	362.00

Folding doors, pine panel and louver, 1-1/8" thickness, with hardware. Add labor below

	Full louver	Louver panel	2 panel design	3 panel design
6'8" high (2 door units)				
2'0" wide	57.00	66.00	113.00	127.00
2'6" wide	62.00	73.00	128.00	141.00
2'8" wide	67.00	77.00	133.00	146.00
3'0" wide	69.00	82.00	142.00	157.00
6'8" high (4 door units)				
4'0" wide	159.00	185.00	224.00	246.00
5'0" wide	171.00	200.00	249.00	275.00
6'0" wide	183.00	216.00	276.00	304.00
7'0" wide	208.00	247.00	306.00	337.00
8'0" wide	218.00	263.00	330.00	362.00
8'0" high (2 door units)				
2'0" wide	86.00	93.00	--	--
2'6" wide	94.00	105.00	--	--
3'0" wide	101.00	114.00	--	--
8'0" high (4 door units)				
4'0" wide	143.00	154.00	--	--
5'0" wide	157.00	157.00	--	--
6'0" wide	168.00	168.00	--	--
7'0" wide	188.00	188.00	--	--
8'0" wide	209.00	209.00	--	--

	Craft@Hrs	Unit	Material	Labor	Total
Labor to install folding panel and louver closet doors					
2'0", 2'6" or 3'0" wide x 6'8" high	BC@1.30	Ea	--	30.70	--
2'0", 2'6" or 3'0" wide x 8'0" high	BC@1.50	Ea	--	35.40	--
4'0" or 5'0" wide x 6'8" high	BC@1.60	Ea	--	37.70	--
4'0" or 5'0" wide x 8'0" high	BC@1.80	Ea	--	42.50	--
6'0", 7'0" or 8'0" wide x 6'8" high	BC@1.90	Ea	--	44.80	--
6'0", 7'0" or 8'0" wide x 8'0" high	BC@2.10	Ea	--	49.50	--

Mirror panels for closet doors

Mirror panels for folding closet doors, polished edge

	Craft@Hrs	Unit	Material	Labor	Total
12" x 6'8", for 24" bi-fold	BG@.406	Pr	85.00	8.92	93.92
15" x 6'8", for 30" bi-fold	BG@.406	Pr	100.00	8.92	108.92
18" x 6'8", for 36" bi-fold	BG@.406	Pr	115.00	8.92	123.92

Mirror panels for hinged doors, 3/16" float plate glass with 1/2" bevel on all edges. Includes mounting clips. Price per each

	Craft@Hrs	Unit	Material	Labor	Total
14" x 50"	BG@.208	Ea	14.00	4.57	18.57
16" x 56"	BG@.208	Ea	20.00	4.57	24.57
16" x 60"	BG@.208	Ea	24.00	4.57	28.57
18" x 60"	BG@.208	Ea	27.00	4.57	31.57

Closet Doors

	Craft@Hrs	Unit	Material	Labor	Total
18" x 68"	BG@.208	Ea	28.00	4.57	32.57
20" x 68"	BG@.208	Ea	33.00	4.57	37.57
22" x 68"	BG@.208	Ea	36.00	4.57	40.57
24" x 68"	BG@.208	Ea	45.00	4.57	49.57

By-passing, hollow core wood closet doors, 1-3/8" thick. Includes hardware
Add labor below

	Hardboard	Lauan	Birch	Ash	Red oak
6'8" high					
4'0" (2 door unit)	62.00	71.00	86.00	86.00	103.00
5'0" (2 door unit)	67.00	82.00	100.00	101.00	120.00
6'0" (2 door unit)	76.00	94.00	115.00	117.00	139.00
7'0" (2 door unit)	94.00	131.00	160.00	161.00	217.00
8'0" (2 door unit)	102.00	144.00	177.00	178.00	224.00
9'0" (3 door unit)	117.00	144.00	177.00	179.00	255.00
10'0" (3 door unit)	143.00	197.00	240.00	243.00	326.00
12'0" (3 door unit)	159.00	222.00	271.00	274.00	341.00
8'0" high					
4'0" (2 door unit)	91.00	120.00	130.00	134.00	--
5'0" (2 door unit)	96.00	128.00	146.00	147.00	--
6'0" (2 door unit)	104.00	143.00	167.00	168.00	--
7'0" (2 door unit)	124.00	181.00	215.00	220.00	--
8'0" (2 door unit)	132.00	192.00	232.00	235.00	--
9'0" (3 door unit)	159.00	217.00	254.00	354.00	--
10'0" (3 door unit)	187.00	272.00	324.00	324.00	--
12'0" (3 door unit)	208.00	298.00	358.00	358.00	--

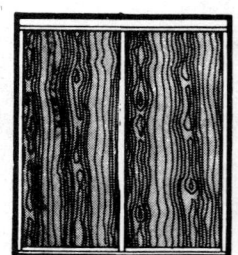

	Craft@Hrs	Unit	Material	Labor	Total
Labor to install by-passing closet doors					
4'0" or 5'0" wide x 6'8" high	BC@1.95	Set	--	46.00	--
4'0" or 5'0" wide x 8'0" high	BC@2.25	Set	--	53.10	--
6'0", 7'0", or 8'0" wide x 6'8" high	BC@2.42	Set	--	57.10	--
6'0", 7'0", or 8'0" wide x 8'0" high	BC@2.82	Set	--	66.50	--
9'0", 10'0", or 12'0" wide x 6'8" high	BC@2.69	Set	--	63.50	--
9'0", 10'0", or 12'0" wide x 8'0" high	BC@3.10	Set	--	73.10	--

Sliding by-passing closet door units, steel frame, includes hardware and track, hollow core, opening sizes, 2 doors per set

	Craft@Hrs	Unit	Material	Labor	Total
Vinyl over hardboard doors, 6'8" high					
4'0"	BC@2.07	Set	85.00	48.80	133.80
6'0"	BC@2.58	Set	100.00	60.90	160.90
8'0"	BC@2.58	Set	120.00	60.90	180.90
10'0"	BC@3.01	Set	160.00	71.00	231.00
Mirror doors, 3/16" sheet glass over hardboard, 6'8" high					
4'0"	BC@2.08	Set	155.00	49.10	204.10
5'0"	BC@2.08	Set	185.00	49.10	234.10
6'0"	BC@2.58	Set	200.00	60.90	260.90
7'0"	BC@2.58	Set	255.00	60.90	315.90
8'0"	BC@2.58	Set	285.00	60.90	345.90
9'0"	BC@3.02	Set	345.00	71.20	416.20
10'0"	BC@3.02	Set	380.00	71.20	451.20
Mirror doors, 3/16" sheet glass over hardboard, 8'0" high					
5'0"	BC@2.08	Set	190.00	49.10	239.10
6'0"	BC@2.58	Set	225.00	60.90	285.90
7'0"	BC@2.58	Set	247.00	60.90	307.90
8'0"	BC@2.58	Set	295.00	60.90	355.90

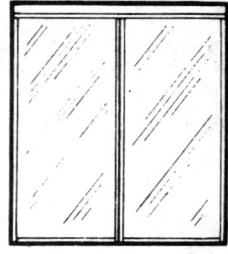

Closet Doors

	Craft@Hrs	Unit	Material	Labor	Total
9'0"	BC@3.02	Set	350.00	71.20	421.20
10'0"	BC@3.02	Set	380.00	71.20	451.20
Add for safety glass and better quality hardware	--	%	50.0	--	--

Clothesline Units
Retractable type, indoor-outdoor, one end is wall-mounted, 5 lines, 34' long each, lines retract into metal case 34" L, 6-1/4" W, 5" H, mounting hardware included,

	Craft@Hrs	Unit	Material	Labor	Total
Cost per unit as described	B1@.860	Ea	37.00	18.20	55.20
Add for galvanized steel post, if required	B1@1.11	Ea	16.00	23.50	39.50

T-shaped galvanized steel pole type, including line,
7 hook, 6' x 2" post, 36" cross arm width,

	Craft@Hrs	Unit	Material	Labor	Total
complete with two posts, set in concrete	BL@1.91	Ea	43.00	35.90	78.90

5 hook, 6' x 1-1/2" post, 30" cross arm width,

	Craft@Hrs	Unit	Material	Labor	Total
(set of two posts)	BL@1.91	Ea	35.00	35.90	70.90

Umbrella type, 8' (6' above ground) galvanized steel center post, requires 10' x 10' area, folds for storage,

	Craft@Hrs	Unit	Material	Labor	Total
includes line, and pole set in concrete	BL@.885	Ea	32.00	16.60	48.60

Columns and Porch Posts. Colonial style.
Round wood, standard cap and base

	Craft@Hrs	Unit	Material	Labor	Total
8" x 8'	B1@5.15	Ea	102.00	109.00	211.00
10" x 10'	B1@5.15	Ea	153.00	109.00	262.00
Add for fluted wood, 8" diameter	--	Ea	16.00	--	--
Add for fluted wood, 10" diameter	--	Ea	19.00	--	--

Round aluminum, standard cap and base

	Craft@Hrs	Unit	Material	Labor	Total
8" x 9'	B1@5.15	Ea	123.00	109.00	232.00
10" x 10'	B1@5.15	Ea	173.00	109.00	282.00
12" x 12'	B1@5.15	Ea	300.00	109.00	409.00
Add per foot over 12' high	B1@.363	Ea	--	7.69	--
Add per inch over 12" diameter	B1@.186	Ea	--	3.94	--
Add for ornamental wood cap, typical price	--	Ea	355.00	--	--
Add for ornamental aluminum cap, typical	--	Ea	305.00	--	--

Porch posts, clear laminated west coast hemlock, solid turned

	Craft@Hrs	Unit	Material	Labor	Total
3-1/4" x 8'	B1@1.04	Ea	36.00	22.00	58.00
4-1/4" x 8'	B1@1.18	Ea	55.00	25.00	80.00
5-1/4" x 8'	B1@1.28	Ea	80.00	27.10	107.10

Concrete. Ready-mix delivered by truck. Typical prices for most cities. Includes delivery up to 20 miles for 10 CY or more, 3" to 4" slump. Material cost only, no placing or pumping included. See forming and finishing costs on the following pages.
Footing and foundation concrete, 1-1/2" aggregate

	Craft@Hrs	Unit	Material	Labor	Total
2,000 PSI, 4.8 sack mix	--	CY	46.20	--	--
2,500 PSI, 5.2 sack mix	--	CY	47.90	--	--
3,000 PSI, 5.7 sack mix	--	CY	50.40	--	--
3,500 PSI, 6.3 sack mix	--	CY	53.30	--	--
4,000 PSI, 6.9 sack mix	--	CY	56.20	--	--

Slab, sidewalk, and driveway concrete, 1" aggregate

	Craft@Hrs	Unit	Material	Labor	Total
2,000 PSI, 5.0 sack mix	--	CY	46.90	--	--
2,500 PSI, 5.5 sack mix	--	CY	49.30	--	--
3,000 PSI, 6.0 sack mix	--	CY	50.60	--	--
3,500 PSI, 6.6 sack mix	--	CY	54.70	--	--
4,000 PSI, 7.1 sack mix	--	CY	56.70	--	--

Concrete

	Craft@Hrs	Unit	Material	Labor	Total
Pea-gravel pump mix / grout mix, 3/8" aggregate					
2,000 PSI, 6.0 sack mix	--	CY	54.80	--	--
2,500 PSI, 6.5 sack mix	--	CY	57.30	--	--
3,000 PSI, 7.2 sack mix	--	CY	60.90	--	--
3,500 PSI, 7.9 sack mix	--	CY	64.40	--	--
5,000 PSI, 8.5 sack mix	--	CY	67.50	--	--
Extra costs for ready-mix concrete					
Add for delivery over 20 miles	--	Mile	.60	--	--
Add for standby charge in excess of 5 minutes per CY delivered, per minute of extra time	--	Ea	1.00	--	--
Add for super plasticized mix, 7"-8" slump	--	%	8.0	--	--
Add for high early strength concrete					
5 sack mix	--	CY	7.00	--	--
6 sack mix	--	CY	8.80	--	--
Add for lightweight aggregate, typical	--	CY	29.00	--	--
Add for lightweight aggregate, pump mix	--	CY	29.00	--	--
Add for granite aggregate, typical	--	CY	2.90	--	--
Add for white cement (architectural)	--	CY	35.00	--	--
Add for 1% calcium chloride	--	CY	1.10	--	--
Add for chemical compensated shrinkage	--	CY	11.50	--	--
Add for less than 9 CY per load	--	CY	10.00	--	--
Add for colored concrete, in 25 pound sacks					
Blended red	--	Lb	1.70	--	--
Adobe	--	Lb	1.65	--	--
Yellow	--	Lb	1.60	--	--
Black	--	Lb	1.65	--	--
Brown	--	Lb	1.70	--	--
Green	--	Lb	4.00	--	--

Concrete Expansion Joints. Fiber or Manville asphalt felt sided. Per linear foot.

Width (thickness)	1/4"	3/8"	1/2"
3"	.18	.18	.20
3-1/2"	.19	.20	.22
4"	.21	.22	.24
6"	.29	.31	.34

	Craft@Hrs	Unit	Material	Labor	Total
Labor to install concrete expansion joints, 3" to 6" width, 1/4" to 1/2" thick	B1@.030	LF	--	.64	--

Concrete Footing Form Ties. Cartons of 200.

	Craft@Hrs	Unit	Material	Labor	Total
6" thru 12"	--	Ea	.31	--	--
14"	--	Ea	.33	--	--
16"	--	Ea	.36	--	--
18"	--	Ea	.38	--	--
20"	--	Ea	.40	--	--
22"	--	Ea	.43	--	--
24"	--	Ea	.45	--	--

Concrete Form Excavation for wall footings, grade beams and column footings. Loosening and one throw only, no disposal or backfilling included. Per CY

	Craft@Hrs	Unit	Material	Labor	Total
Light soil	BL@1.10	CY	--	20.70	--
Average soil	BL@1.65	CY	--	31.00	--
Heavy soil or loose rock	BL@2.14	CY	--	40.20	--
Add for wet soil	--	%	--	75.0	--

Concrete

	13/16" x 13/16"	1/2" x 1"
Concrete Form Stakes		
12" long	1.60	1.60
18" long	1.70	1.70
24" long	1.85	2.10
30" long	2.40	2.55
36" long	2.70	2.70

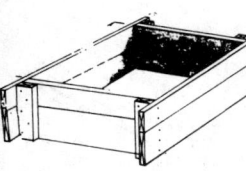

Concrete Formwork. Multiple use of forms assumes that forms can be removed, cleaned and reused without being completely disassembled. Material costs in this section can be adjusted to reflect your actual cost of lumber. Here's how: Divide your actual lumber cost per MBF by the assumed cost (listed in parentheses). Then multiply the cost in the material column by this adjustment factor.

	Craft@Hrs	Unit	Material	Labor	Total

Board forming for wall footings, grade beams, column footings, site curbs and steps. Includes 5% waste. No stripping included. Per SF of contact area.
1" thick forms and bracing, using 1.64 BF of Std & Btr lumber per SF (@ $515 per MBF). Includes nails, ties, and form oil (@ $.20 per SF)

	Craft@Hrs	Unit	Material	Labor	Total
1 use	B2@.100	SF	1.05	2.20	3.25
3 use	B2@.100	SF	.63	2.20	2.83
5 use	B2@.100	SF	.53	2.20	2.73

2" thick forms, 2.85 BF of forming lumber (@$430 per MBF) per SF. Includes nails, ties, and form oil (@ $.20 per SF)

	Craft@Hrs	Unit	Material	Labor	Total
1 use	B2@.115	SF	1.43	2.53	3.96
3 use	B2@.115	SF	.77	2.53	3.30
5 use	B2@.115	SF	.65	2.53	3.18

Add for keyway beveled on two edges, 1 use. No stripping included

	Craft@Hrs	Unit	Material	Labor	Total
2" x 4", Std & Btr (@$430 per MBF)	B2@.027	LF	.27	.59	.86
2" x 6", Std & Btr (@$450 per MBF)	B2@.027	LF	.41	.59	1.00

Driveway and walkway edge forms Material costs include stakes, nails, and form oil (@$.16 per LF) and 5% waste. No stripping included. Per LF of edge form.
2" x 4" form, Std & Btr (@$430 per MBF, .7 BF per LF)

	Craft@Hrs	Unit	Material	Labor	Total
1 use	B2@.050	LF	.46	1.10	1.56
3 use	B2@.050	LF	.25	1.10	1.35
5 use	B2@.050	LF	.19	1.10	1.29

2" x 6" form, Std & Btr (@$450 per MBF, 1.05 BF per LF)

	Craft@Hrs	Unit	Material	Labor	Total
1 use	B2@.050	LF	.63	1.10	1.73
3 use	B2@.050	LF	.35	1.10	1.45
5 use	B2@.050	LF	.27	1.10	1.37

2" x 8" form, Std & Btr (@$450 per MBF) (1.4 BF per LF)

	Craft@Hrs	Unit	Material	Labor	Total
1 use	B2@.055	LF	.79	1.21	2.00
3 use	B2@.055	LF	.44	1.21	1.65
5 use	B2@.055	LF	.33	1.21	1.54

2" x 10" form, Std & Btr (@$590 per MBF) (1.75 BF per LF)

	Craft@Hrs	Unit	Material	Labor	Total
1 use	B2@.055	LF	1.19	1.21	2.40
3 use	B2@.055	LF	.66	1.21	1.87
5 use	B2@.055	LF	.50	1.21	1.71

2" x 12" form, Std & Btr (@$600 per MBF) (2.1 BF per LF)

	Craft@Hrs	Unit	Material	Labor	Total
1 use	B2@.055	LF	1.42	1.21	2.63
3 use	B2@.055	LF	.78	1.21	1.99
5 use	B2@.055	LF	.60	1.21	1.81

Concrete

	Craft@Hrs	Unit	Material	Labor	Total

Plywood forming for foundation walls, building walls and retaining walls, using 3/4" plyform (@$710 per MSF) with 10% waste, and 2" bracing (@$345 per MBF) with 20% waste. All material costs include nails, ties, clamps and form oil (@$.20 per SF). No stripping included. Per SF of contact area.

Walls up to 4' high (1.10 SF of plywood and .42 BF of bracing per SF of form area)

	Craft@Hrs	Unit	Material	Labor	Total
1 use	B2@.051	SF	1.21	1.12	2.33
3 use	B2@.051	SF	.67	1.12	1.79
5 use	B2@.051	SF	.55	1.12	1.67

Walls 4' to 8' high (1.10 SF of plywood and .60 BF of bracing per SF of form area)

1 use	B2@.060	SF	1.28	1.32	2.60
3 use	B2@.060	SF	.70	1.32	2.02
5 use	B2@.060	SF	.58	1.32	1.90

Walls 8' to 12' high (1.10 SF of plywood and .90 BF of bracing per SF of form area)

1 use	B2@.095	SF	1.38	2.09	3.47
3 use	B2@.095	SF	.76	2.09	2.85
5 use	B2@.095	SF	.62	2.09	2.71

Walls 12' to 16' high (1.10 SF of plywood and 1.05 BF of bracing per SF of form area)

1 use	B2@.128	SF	1.43	2.82	4.25
3 use	B2@.128	SF	.79	2.82	3.61
5 use	B2@.128	SF	.64	2.82	3.46

Walls over 16' high (1.10 SF of plywood and 1.20 BF of bracing per SF of form area)

1 use	B2@.153	SF	1.48	3.36	4.84
3 use	B2@.153	SF	.82	3.36	4.18
5 use	B2@.153	SF	.67	3.36	4.03

Form stripping Labor to remove forms and bracing, clean, and stack on job site.
Board forms at wall footings, grade beams and column footings. Per SF of contact area

	Craft@Hrs	Unit	Material	Labor	Total
1" thick lumber	BL@.010	SF	--	.19	--
2" thick lumber	BL@.012	SF	--	.23	--
Keyways, 2" x 4" and 2" x 6"	BL@.006	LF	--	.11	--

Slab edge forms. Per LF of edge form

2" x 4" to 2" x 6"	BL@.012	LF	--	.23	--
2" x 8" to 2" x 12"	BL@.013	LF	--	.24	--

Walls, plywood forms. Per SF of contact area

To 4' high	BL@.017	SF	--	.32	--
Over 4' to 8' high	BL@.018	SF	--	.34	--
Over 8' to 12' high	BL@.020	SF	--	.38	--
Over 12' to 16' high	BL@.027	SF	--	.51	--
Over 16' high	BL@.040	SF	--	.75	--

Steel reinforcing bars Based on a minimum of 5,000 lb job. Material costs are for deformed steel reinforcing bars, including 10% lap allowance, cutting and heavy bending. Labor costs are based on installation in walls, footings, and grade beams. Costs per lb

	Craft@Hrs	Unit	Material	Labor	Total
1/4" diameter, #2 bar (.17 lb per LF)	RI@.015	Lb	.53	.37	.90
3/8" diameter, #3 bar (.38 lb per LF)	RI@.011	Lb	.29	.27	.56
1/2" diameter, #4 bar (.67 lb per LF)	RI@.010	Lb	.25	.25	.50
5/8" diameter, #5 bar (1.04 lb per LF)	RI@.009	Lb	.25	.22	.47
3/4" diameter: #6 bar (1.50 lb per LF)	RI@.008	Lb	.25	.20	.45
7/8" diameter #7 bar (2.04 lb per LF)	RI@.008	Lb	.25	.20	.45
1" diameter, #8 bar (2.67 lb per LF)	RI@.008	Lb	.25	.20	.45
1-1/8" diameter, #9 bar (3.40 lb per LF)	RI@.008	Lb	.25	.20	.45
1-1/4" diameter, #10 bar (4.30 lb per LF)	RI@.007	Lb	.25	.17	.42
1-3/8" diameter, #11 bar (5.31 lb per LF)	RI@.007	Lb	.25	.17	.42
Add for less than 5,000 lb job	--	Lb	.05	.10	.15
Add for light bending	--	%	--	10.0	--
Deduct for structural slabs	--	%	--	-10.0	--

Concrete

	Craft@Hrs	Unit	Material	Labor	Total

Steel reinforcing mesh
Material costs are for welded wire mesh including 15% waste and overlap. Labor costs are for installation in slabs

	Craft@Hrs	Unit	Material	Labor	Total
6" x 6", W1.4 x W1.4 (#10 x #10) (@$90 per MSF)	RI@.003	SF	.10	.08	.18
6" x 6", W2.0 x W2.0 (#8 x #8) (@$130 per MSF)	RI@.003	SF	.15	.08	.23
6" x 6", W2.9 x W2.9 (#6 x #6) (@$185 per MSF)	RI@.003	SF	.21	.08	.29

Footings and grade beams
Concrete placing only. Material costs are for 3,000 PSI, 5.7 sack mix, with 1-1/2" aggregate (@$50.40 per CY), no waste included. Labor costs are for placing concrete only using earth forming. Add the cost of excavation, wood forming (if required), and steel reinforcing.

	Craft@Hrs	Unit	Material	Labor	Total
6" D x 12" W (1.85 CY per CLF)	BL@.024	LF	.93	.45	1.38
8" D x 12" W (2.47 CY per CLF)	BL@.032	LF	1.25	.60	1.85
8" D x 16" W (3.29 CY per CLF)	BL@.043	LF	1.66	.81	2.47
8" D x 18" W (3.70 CY per CLF)	BL@.048	LF	1.87	.90	2.77
10" D x 12" W (3.09 CY per CLF)	BL@.036	LF	1.56	.68	2.24
10" D x 16" W (4.12 CY per CLF)	BL@.050	LF	2.08	.94	3.02
10" D x 18" W (4.63 CY per CLF)	BL@.055	LF	2.33	1.03	3.36
12" D x 12" W (3.70 CY per CLF)	BL@.044	LF	1.87	.83	2.70
12" D x 16" W (4.94 CY per CLF)	BL@.063	LF	2.49	1.18	3.67
12" D x 20" W (6.17 CY per CLF)	BL@.066	LF	3.11	1.24	4.35
12" D x 24" W (7.41 CY per CLF)	BL@.081	LF	3.74	1.52	5.26
Typical cost per CY	BL@1.26	CY	50.40	23.70	74.10

Column Footings
Concrete placing only. Material costs are for 3,000 PSI, 5.7 sack mix, with 1-1/2" aggregate (@$50.40 per CY), no waste included. Labor costs are for placing concrete only in an existing form. Add for excavation, board forming and steel reinforcing.

	Craft@Hrs	Unit	Material	Labor	Total
Typical cost per CY	B1@.875	CY	50.40	18.50	68.90

Concrete Foundations, for building walls or retaining walls
Material costs are for 2,500 PSI, 6.5 sack mix, with 3/8" aggregate (@$57.30 per CY), no waste included. Pumping (@$9.00 per CY based on a 25 CY job) included in material cost where noted. Labor costs are for placing only. Add the cost of excavation, formwork, steel reinforcing, finishes and curing. Square foot costs are based on SF of wall measured on one face only.

	Craft@Hrs	Unit	Material	Labor	Total
4" thick walls (1.23 CY per CSF)					
To 4' high, direct from chute	B1@.013	SF	.71	.28	.99
4' to 8' high, pumped	B3@.015	SF	.82	.31	1.13
8' to 12' high, pumped	B3@.017	SF	.82	.35	1.17
12' to 16' high, pumped	B3@.018	SF	.82	.37	1.19
16' high, pumped	B3@.020	SF	.82	.41	1.23
6" thick walls (1.85 CY per CSF)					
To 4' high, direct from chute	B1@.020	SF	1.06	.42	1.48
4' to 8' high, pumped	B3@.022	SF	1.23	.45	1.68
8' to 12' high, pumped	B3@.025	SF	1.23	.51	1.74
12' to 16' high, pumped	B3@.027	SF	1.23	.55	1.78
16' high, pumped	B3@.030	SF	1.23	.61	1.84

Concrete

	Craft@Hrs	Unit	Material	Labor	Total
Concrete Foundations, building walls or retaining walls, continued					
8" thick walls (2.47 CY per CSF)					
To 4' high, direct from chute	B1@.026	SF	1.42	.55	1.97
4' to 8' high, pumped	B3@.030	SF	1.64	.61	2.25
8' to 12' high, pumped	B3@.033	SF	1.64	.67	2.31
12' to 16' high, pumped	B3@.036	SF	1.64	.74	2.38
16' high, pumped	B3@.040	SF	1.64	.82	2.46
10" thick walls (3.09 CY per CSF)					
To 4' high, direct from chute	B1@.032	SF	1.77	.68	2.45
4' to 8' high, pumped	B3@.037	SF	2.05	.76	2.81
8' to 12' high, pumped	B3@.041	SF	2.05	.84	2.89
12' to 16' high, pumped	B3@.046	SF	2.05	.94	2.99
16' high, pumped	B3@.050	SF	2.05	1.02	3.07
12" thick walls (3.70 CY per CSF)					
To 4' high, placed direct from chute	B1@.040	SF	2.12	.85	2.97
4' to 8' high, pumped	B3@.045	SF	2.45	.92	3.37
8' to 12' high, pumped	B3@.050	SF	2.45	1.02	3.47
12' to 16' high, pumped	B3@.055	SF	2.45	1.12	3.57
16' high, pumped	B3@.060	SF	2.45	1.23	3.68

Concrete Foundations, footings, grade beams. Use the figures below for preliminary estimates. Concrete costs are based on 2,000 PSI, 4.8 sack mix with 1-1/2" aggregate placed directly from the chute of a ready-mix truck at $46.20 per CY. Figures in parentheses show the cubic yards of concrete per linear foot of foundation (including 5% waste). Costs shown include concrete, 60 pounds of reinforcing per CY of concrete, and typical excavation with a 3/4 CY backhoe spreading excess backfill on site.

Footings and grade beams, cast directly against the earth, no forming or finishing required. For scheduling purposes estimate that a crew of 3 can lay out, excavate, place and tie the reinforcing steel and place 13 CY of concrete in an 8-hour day. Use $250.00 as a minimum cost for this type work.

	Craft@Hrs	Unit	Material	Labor	Equipment	Total
Typical cost per CY	B4@1.80	CY	67.70	41.70	2.85	112.25
12" W x 18" D (0.06 CY per LF)	B4@.108	LF	4.06	2.51	.18	6.75
18" W x 24" D (0.12 CY per LF)	B4@.216	LF	8.13	5.01	.34	13.48
24" W x 24" D (0.16 CY per LF)	B4@.288	LF	10.90	6.68	.46	18.04
24" W x 30" D (0.19 CY per LF)	B4@.342	LF	12.90	7.93	.54	21.37
24" W x 36" D (0.23 CY per LF)	B4@.414	LF	15.60	9.60	.65	25.85
30" W x 36" D (0.29 CY per LF)	B4@.521	LF	19.70	12.10	.83	32.63
30" W x 42" D (0.34 CY per LF)	B4@.611	LF	23.00	14.20	.97	38.17
36" W x 48" D (0.47 CY per LF)	B4@.845	LF	31.85	19.60	1.34	52.79

Tricks of the trade: To quickly estimate the cost of sizes not shown, multiply the width in inches by the depth in inches and divide the result by 3700. This is the CY of concrete per LF of footing including 5% waste. Multiply this quantity by the "Typical cost per CY" shown above to find the cost per LF.

Tip: When using the Estimate Writer computerized estimating program, calculate the quantity (CY) of concrete required as described and select the "Typical cost per CY" shown above.

Continuous concrete footing with foundation stem wall. These figures assume the foundation stem wall projects 24" above the finished grade and extends into the soil 18" to the top of the footing. Costs shown include forming both sides of the foundation wall and the footing, based on three uses of the forms. Use $1,250.00 as a minimum cost for this type work.

	Craft@Hrs	Unit	Material	Labor	Equipment	Total
Typical cost per CY	B5@7.16	CY	134.00	165.00	4.55	303.55

Concrete

	Craft@Hrs	Unit	Material	Labor	Equipment	Total

Continuous concrete footing with foundation stem wall, continued
Typical single story structure, footing 12" W x 8" D,
wall 6" T x 42" D (.10 CY per LF)   B5@.716   LF   13.40   16.50   .46   30.36
Typical two story structure, footing 18" W x 10" D,
wall 8" T x 42" D (.14 CY per LF)   B5@1.00   LF   18.75   23.10   .64   42.49
Typical three story structure, footing 24" W x 12" D,
wall 10" T x 42" D (.19 CY per LF)   B5@1.36   LF   25.45   31.40   .87   57.72

	Craft@Hrs	Unit	Material	Labor	Total

Slabs, Walks and Driveways
Concrete placing only. Material costs are for 3,000 PSI, 6 sack mix, with 1" aggregate (@$50.60 per CY), no waste included. Labor costs are for placing and screeding only. Add for excavation, fine grading, edge forming, sand and gravel base, vapor barrier, thickened edges, steel reinforcing and finishing.
    2" thick (.617 CY per CSF)   BL@.011   SF   .31   .21   .52
    2-1/2" thick (.772 CY per CSF)   BL@.012   SF   .39   .23   .62
    3" thick (.926 CY per CSF)   BL@.013   SF   .47   .24   .71
    3-1/2" thick (1.08 CY per CSF)   BL@.014   SF   .54   .26   .80
    4" thick (1.23 CY per CSF)   BL@.015   SF   .62   .28   .90
    4-1/2" thick (1.39 CY per CSF)   BL@.015   SF   .70   .28   .98
    5" thick (1.54 CY per CSF)   BL@.017   SF   .78   .32   1.10
    6" thick (1.85 CY per CSF)   BL@.018   SF   .93   .34   1.27
    8" thick (2.47 CY per CSF)   BL@.020   SF   1.25   .38   1.63
    10" thick (3.09 CY per CSF)   BL@.024   SF   1.56   .45   2.01
    12" thick (3.70 CY per CSF)   BL@.027   SF   1.87   .51   2.38

Slab Base
Aggregate base for slabs. No waste included. Labor costs are for spreading aggregate from piles only. Add for fine grading, using hand tools.
Crushed stone base at $17.20 per cubic yard. 1.4 tons equal one cubic yard
    1" base (.309 CY per CSF)   BL@.001   SF   .05   .02   .07
    2" base (.617 CY per CSF)   BL@.003   SF   .11   .06   .17
    3" base (.926 CY per CSF)   BL@.004   SF   .16   .08   .24
    4" base (1.23 CY per CSF)   BL@.006   SF   .21   .11   .32
    5" base (1.54 CY per CSF)   BL@.007   SF   .26   .13   .39
    6" base (1.85 CY per CSF)   BL@.008   SF   .32   .15   .47
Sand fill base at $13.40 per cubic yard. 1.35 tons equal one cubic yard
    1" fill (.309 CY per CSF)   BL@.001   SF   .04   .02   .06
    2" fill (.617 CY per CSF)   BL@.002   SF   .08   .03   .11
    3" fill (.926 CY per CSF)   BL@.003   SF   .12   .06   .18
    4" fill (1.23 CY per CSF)   BL@.004   SF   .16   .08   .24
    5" fill (1.54 CY per CSF)   BL@.006   SF   .21   .11   .32
    6" fill (1.85 CY per CSF)   BL@.007   SF   .25   .13   .38
Add for fine grading for slab on grade   BL@.008   SF   --   .15   --

Slab membrane Material costs are for 10' wide membrane with 6" laps. Labor costs are to lay membrane over a level slab base.
    4 mil polyethylene (@$2.60 per CSF)   BL@.002   SF   .03   .04   .07
    6 mil polyethylene (@$4.20 per CSF)   BL@.002   SF   .04   .04   .08
    Reinforced kraft and polyethylene laminate
    (@$8.30 per CSF)   BL@.002   SF   .08   .04   .12
    Add for adhesive cemented seams   BL@.001   SF   .01   .02   .03

Concrete

	Craft@Hrs	Unit	Material	Labor	Total

Curbs and gutters Small quantities. Material costs are for 2,000 PSI, 5.0 sack mix with 1" aggregate (@$46.90 per CY). No waste included. Labor cost is for placing and finishing concrete only. Add for excavation, wood forming, steel reinforcing.

Per cubic yard of concrete	B6@1.61	CY	46.90	33.00	79.90

Steps on grade Material costs are for 2,000 PSI, 5.0 sack mix, with 1" aggregate (@$46.90 per CY), no waste included. Labor cost is for placing concrete only. Add for excavation, wood forming, steel reinforcing, and finishing. See also Concrete in the Industrial Division.

Per cubic yard of concrete	BL@1.07	CY	46.90	20.10	67.00

Concrete Slab Finishes

	Craft@Hrs	Unit	Material	Labor	Total
Float finish	B6@.008	SF	.06	.16	.22
Trowel finishing					
Steel, machine work	B6@.011	SF	.07	.23	.30
Steel, hand work	B6@.014	SF	--	.29	--
Broom finish	B6@.010	SF	.05	.21	.26
Scoring concrete surface, hand work	B6@.005	LF	--	.10	--
Exposed aggregate (washed, including finishing), no disposal of slurry	B6@.017	SF	.09	.35	.44
Non-metallic color and hardener, troweled on, 2 applications, red, gray or black					
60 pounds per 100 SF	B6@.021	SF	.25	.43	.68
100 pounds per 100 SF	B6@.025	SF	.39	.51	.90
Add for other standard colors	--	%	50.0	--	--
Liquid curing and sealing compound, spray-on, 400 SF and $20 per gallon,					
Mastercure	B6@.335	CSF	5.00	6.86	11.86
Sweep, scrub and wash down	B6@.552	CSF	1.30	11.30	12.60
Finish treads and risers					
No abrasives, no plastering, per LF of tread	B6@.042	LF	--	.86	--
With abrasives, plastered, per LF of tread	B6@.063	LF	.49	1.29	1.78

Concrete Wall Finishes

	Craft@Hrs	Unit	Material	Labor	Total
Cut back ties and patch	B6@.011	SF	.09	.23	.32
Remove fins	B6@.008	LF	.03	.16	.19
Grind smooth	B6@.021	SF	.06	.43	.49
Sack, burlap grout rub	B6@.013	SF	.03	.27	.30
Wire brush, green	B6@.015	SF	.03	.31	.34
Wash with acid and rinse	B6@.004	SF	.20	.08	.28
Break fins, patch voids, Carborundum rub	B6@.035	SF	.04	.72	.76
Break fins, patch voids, burlap grout rub	B6@.026	SF	.05	.53	.58

Specialty Concrete Finishes

	Craft@Hrs	Unit	Material	Labor	Total
Monolithic natural aggregate topping					
1/16"	B6@.006	SF	.05	.12	.17
3/16"	B6@.020	SF	.16	.41	.57
1/2"	B6@.022	SF	.12	.45	.57
Integral colors, typical, figure 8 pounds per sack of cement	--	Lb	--	--	1.60
Stamped finish (embossed concrete)					
Diamond, square, octagonal patterns	B6@.047	SF	--	.96	--
Spanish paver pattern	B6@.053	SF	--	1.09	--
Add for grouting of joints, if required	B6@.023	SF	.17	.47	.64

Countertops

	Craft@Hrs	Unit	Material	Labor	Total

Countertops. One-piece custom tops, 25"-36" deep, with 4" backsplash, shop assembled. Laminated plastic on composition base. Costs are per LF of back edge and for solid colors. Tops surfaced with Formica, Textolite, or Wilsonart, $35 minimum price per LF.

	Craft@Hrs	Unit	Material	Labor	Total
Bar tops					
Square edge	B1@.181	LF	28.90	3.84	32.74
Bevel edge	B1@.181	LF	40.60	3.84	44.44
Rolled drip edge (post formed)	B1@.181	LF	19.20	3.84	23.04
Sink tops					
Square edge	B1@.181	LF	28.80	3.84	32.64
Square edge with woodgrain at edge	B1@.181	LF	23.50	3.84	27.34
Bevel edge with woodgrain at edge	B1@.181	LF	58.80	3.84	62.64
Rolled drip edge (post formed)	B1@.181	LF	19.20	3.84	23.04
Vanity tops					
Square edge	B1@.181	LF	28.90	3.84	32.74
Square edge with woodgrain at edge	B1@.181	LF	24.60	3.84	28.44
Bevel edge with woodgrain at edge	B1@.181	LF	59.90	3.84	63.74
Rolled drip edge (post formed)	B1@.181	LF	19.20	3.84	23.04
Cultured marble vanity tops and integral sink, 22" deep					
25" long	B1@.654	Ea	73.00	13.90	86.90
31" long	B1@.742	Ea	82.00	15.70	97.70
37" long	B1@.830	Ea	93.00	17.60	110.60
43" long	B1@.924	Ea	108.00	19.60	127.60
49" long	B1@1.02	Ea	116.00	21.60	137.60
Corian vanity tops and integral sink, 3/4" thick, 22" deep					
25" long	B1@.654	Ea	286.00	13.90	299.90
31" long	B1@.742	Ea	335.00	15.70	350.70
37" long	B1@.830	Ea	385.00	17.60	402.60
43" long	B1@.924	Ea	440.00	19.60	459.60
49" long	B1@1.02	Ea	497.00	21.60	518.60
Add for square endsplash, attached	--	Ea	--	--	13.50
Add for square endsplash, loose	--	Ea	--	--	12.00
Add for contour endsplash, loose	--	Ea	--	--	15.80
Add for mitered corners	--	Ea	--	--	17.90
Add for seamless tops	--	Ea	--	--	26.00
Deduct for omitting end cap	--	Ea	--	--	-6.00
Add for sink, range or vanity cutout	--	Ea	--	--	6.70
Add for drilling 3 plumbing fixture holes	--	LS	--	--	9.00
Add for quarter round corner	--	Ea	--	--	18.20
Add for half round corner	--	Ea	--	--	28.00
Add for textured surfaces	--	%	14.0	--	--
Deduct for patterns and wood grains	--	%	-7.0	--	--

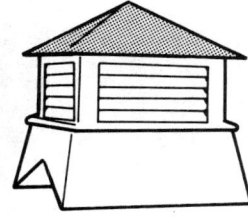

Cupolas. Finished redwood. Add the cost of weather vanes below.

	Craft@Hrs	Unit	Material	Labor	Total
Aluminum roof covering					
22" x 22" x 9" base, 25" high	B1@.948	Ea	148.00	20.10	168.10
29" x 29" x 11-1/4" base, 30" high	B1@.948	Ea	210.00	20.10	230.10
35" x 35" x 11-1/4" base, 33" high	B1@1.43	Ea	295.00	30.30	325.30
47" x 47" x 11-1/4" base, 39" high	B1@2.02	Ea	465.00	42.80	507.80
Copper roof covering					
22" x 22" x 9" base, 25" high	B1@.948	Ea	170.00	20.10	190.10
25" x 25" x 9" base, 26" high	B1@.948	Ea	190.00	20.10	210.10
29" x 29" x 11-1/4" base, 30" high	B1@.948	Ea	245.00	20.10	265.10
35" x 35" x 11-1/4" base, 33" high	B1@1.43	Ea	350.00	30.30	380.30
41" x 41" x 11-1/4" base, 35" high	B1@1.59	Ea	420.00	33.70	453.70

Cupolas

	Craft@Hrs	Unit	Material	Labor	Total
47" x 47" x 11-1/4" base, 39" high	B1@2.02	Ea	540.00	42.80	582.80
Weather vanes for cupolas					
Standard aluminum, baked black finish					
26" height	B1@.275	Ea	38.00	5.83	43.83
Deluxe model, gold-bronze aluminum, black finish					
36" height, bronze ornament	B1@.368	Ea	125.00	7.80	132.80

Decking, Lumber Patio and Yard Deck. These costs assume 2" x 4" redwood deck supported by doubled 2" x 6" redwood beams 24" on center. Beams are supported by 4" x 4" posts set 5' on center in concrete. The deck requires 2.6 board feet of 2" x 4" per square foot of surface, 2 board feet of 2" x 6" per square foot of surface, and .64 board feet of 4" x 4" post per square foot of deck. Post length is assumed to be 6 feet. Fasteners are 16d galvanized nails, and galvanized steel joist hangers and post anchors. Costs assume that decking is predrilled for nailing and that 2" x 4" decking is spaced at 1/8" to permit drainage. Costs are based on clear select redwood 2" x 4" at $1,200 per MBF ($.80 per LF), clear select redwood 2" x 6" at $1,350 per MBF ($1.35 per LF), construction heart redwood 4" x 4" at $1,600 per MBF ($2.15 per LF) and anchors, hangers and nails averaging $.65 per SF.

Deck, beams, posts and hardware	B1@.036	SF	7.50	.76	8.26
Add for premixed concrete at each post (.5 CF)					
based on posts at 10' OC both ways	B1@.010	SF	.04	.21	.25
Add for diagonal decking	B1@.007	SF	1.50	.15	1.65
Add for stain and sealer finish	B1@.006	SF	.07	.13	.20
Add for bleached redwood finish	B1@.005	SF	.05	.11	.16

Note: For decks made with other materials, see the Lumber section for pricing information and then use the board foot per square foot equation to calculate the total square foot cost.

Demolition. Itemized costs for demolition of building components when building is being remodeled, repaired or rehabilitated, not completely demolished. Costs are to break out the items listed and pile debris on site only. No hauling, dump fees or salvage value included. Figures in parentheses give the approximate "loose" volume of the materials (volume after being demolished).

	Craft@Hrs	Unit	Material	Labor	Total
Brick walls, cost per in-place SF, using pneumatic breaker					
4" walls (80 SF per CY)	BL@.061	SF	--	1.15	--
8" walls (54 SF per CY)	BL@.110	SF	--	2.07	--
12" walls (20 SF per CY)	BL@.133	SF	--	2.50	--
Brick sidewalks, cost per in-place SF					
By hand (120 SF per CY)	BL@.003	SF	--	.06	--
Ceiling tile, cost per in-place SF					
Stapled or glued 12" x 12" (150 SF per CY)	BL@.015	SF	--	.28	--
Suspended panels 2' x 4' (175 SF per CY)	BL@.015	SF	--	.28	--
Concrete masonry walls, cost per in-place SF, using pneumatic breaker					
4" (60 SF per CY)	BL@.066	SF	--	1.24	--
6" (45 SF per CY)	BL@.075	SF	--	1.41	--
8" (30 SF per CY)	BL@.098	SF	--	1.84	--
12" (20 SF per CY)	BL@.140	SF	--	2.63	--
Add for reinforced and grouted block walls	--	%	--	50.0	--
Concrete foundations, steel reinforced, cost per in-place CY,					
Using pneumatic breaker (.75 CY per CY)	BL@3.96	CY	--	74.40	--
Concrete footings, steel reinforced using pneumatic breaker, cost per in-place LF					
6" W x 12" D (35 LF per CY)	BL@.075	LF	--	1.41	--
8" W x 12" D (30 LF per CY)	BL@.098	LF	--	1.84	--
8" W x 16" D (20 LF per CY)	BL@.133	LF	--	2.50	--
8" W x 18" D (18 LF per CY)	BL@.147	LF	--	2.76	--

Demolition

	Craft@Hrs	Unit	Material	Labor	Total
10" W x 12" D (21 LF per CY)	BL@.121	LF	--	2.28	--
10" W x 16" D (16 LF per CY)	BL@.165	LF	--	3.10	--
10" W x 18" D (14 LF per CY)	BL@.185	LF	--	3.48	--
12" W x 12" D (20 LF per CY)	BL@.147	LF	--	2.76	--
12" W x 16" D (13 LF per CY)	BL@.196	LF	--	3.69	--
12" W x 20" D (10 LF per CY)	BL@.245	LF	--	4.61	--
12" W x 24" D (9 LF per CY)	BL@.294	LF	--	5.53	--
Concrete sidewalks, to 4" thick, cost per in-place SF					
Non-reinforced, by hand (60 SF per CY)	BL@.050	SF	--	.94	--
Reinforced, using pneumatic breaker (55 SF per CY)	BL@.033	SF	--	.62	--
Concrete slabs, non-reinforced, per in-place CY					
Using pneumatic breaker, (.80 CY per CY)	BL@3.17	CY	--	59.60	--
Concrete slabs, non-reinforced, cost per in-place SF of slab area					
Using pneumatic breaker					
3" slab thickness (90 SF per CY)	BL@.030	SF	--	.56	--
4" slab thickness (60 SF per CY)	BL@.040	SF	--	.75	--
5" slab thickness (50 SF per CY)	BL@.050	SF	--	.94	--
6" slab thickness (45 SF per CY)	BL@.056	SF	--	1.05	--
8" slab thickness (30 SF per CY)	BL@.092	SF	--	1.73	--
Concrete slabs, steel reinforced, cost per in-place CY,					
Using pneumatic breaker (.75 CY per CY)	BL@3.96	CY	--	74.40	--
Concrete slabs, steel reinforced, cost per in-place SF of slab area					
Using pneumatic breaker					
3" slab thickness (80 SF per CY)	BL@.036	SF	--	.68	--
4" slab thickness (55 SF per CY)	BL@.050	SF	--	.94	--
5" slab thickness (45 SF per CY)	BL@.061	SF	--	1.15	--
6" slab thickness (40 SF per CY)	BL@.075	SF	--	1.41	--
8" slab thickness (25 SF per CY)	BL@.098	SF	--	1.84	--
Concrete walls, steel reinforced, cost per in-place CY					
Using pneumatic breaker (.75 CY per CY)	BL@4.76	CY	--	89.50	--
Concrete walls, steel reinforced, cost per in-place SF of wall area					
Using pneumatic breaker (deduct openings over 25 SF)					
3" wall thickness (80 SF per CY)	BL@.045	SF	--	.85	--
4" wall thickness (55 SF per CY)	BL@.058	SF	--	1.09	--
5" wall thickness (45 SF per CY)	BL@.075	SF	--	1.41	--
6" wall thickness (40 SF per CY)	BL@.090	SF	--	1.69	--
8" wall thickness (25 SF per CY)	BL@.120	SF	--	2.26	--
10" wall thickness (20 SF per CY)	BL@.147	SF	--	2.76	--
12" wall thickness (15 SF per CY)	BL@.176	SF	--	3.31	--
Curbs and 18" gutter, concrete, cost per in-place LF					
Using pneumatic breaker (8 LF per CY)	BL@.422	LF	--	7.93	--
Doors, typical cost to remove door, frame and hardware					
3' x 7' metal (2 doors per CY)	BL@1.00	Ea	--	18.80	--
3' x 7' wood (2 doors per CY)	BL@.497	Ea	--	9.34	--
Flooring, cost per in-place SY of floor area					
Ceramic or quarry tile (25 SY per CY)	BL@.263	SY	--	4.94	--
Hardwood, nailed (25 SY per CY)	BL@.290	SY	--	5.45	--
Hardwood strip, glued (25 SY per CY)	BL@.503	SY	--	9.46	--
Linoleum and sheet flooring (30 SY per CY)	BL@.056	SY	--	1.05	--
Carpet (40 SY per CY)	BL@.028	SY	--	.53	--
Add for carpet pad (35 SY per CY)	BL@.014	SY	--	.26	--
Resilient tile (30 SY per CY)	BL@.070	SY	--	1.32	--
Terrazzo (25 SY per CY)	BL@.286	SY	--	5.38	--

Demolition

	Craft@Hrs	Unit	Material	Labor	Total

Pavement, asphaltic concrete (bituminous) to 6" thick, cost per in-place SY, no disposal included (see also Concrete Slabs this section)
   Using pneumatic pick (4 SY per CY)   BL@.250   SY   --   4.70   --

Plaster walls or ceilings, cost per in-place SF
   By hand (90 SF per CY)   BL@.015   SF   --   .28   --

Roofing, cost per in-place Sq, removed using hand tools
   Built-up, 5 ply   (1.25 Sq per CY)   BL@2.77   Sq   --   52.10   --
   Shingles, asphalt (2.50 Sq per CY)   BL@1.33   Sq   --   25.00   --
   Shingles, slate   (1.00 Sq per CY)   BL@1.79   Sq   --   33.70   --
   Shingles, wood   (1.66 Sq per CY)   BL@2.02   Sq   --   38.00   --

Sheathing, up to 1" thick, cost per in-place SF, removed using hand tools
   Gypsum (250 SF per CY)   BL@.017   SF   --   .32   --
   Plywood (200 SF per CY)   BL@.017   SF   --   .32   --

Siding, up to 1" thick, cost per in-place SF, removed using hand tools
   Metal (200 SF per CY)   BL@.027   SF   --   .51   --
   Plywood (200 SF per CY)   BL@.017   SF   --   .32   --
   Wood, boards (250 SF per CY)   BL@.030   SF   --   .56   --

Trees. See Excavation

Wallboard up to 1" thick, cost per in-place SF, removed using hand tools
   Gypsum, walls or ceilings (500 SF per CY)   BL@.018   SF   --   .34   --
   Plywood or insulation board (500 SF per CY)   BL@.018   SF   --   .34   --

Windows, typical cost to remove window, frame and hardware
   3' x 4' metal window (2 per CY)   BL@.696   Ea   --   13.10   --
   3' x 4' wood window (3 per CY)   BL@.780   Ea   --   14.70   --

Windows, typical cost to remove window, frame and hardware, cost per in-place SF of window
   Metal (24 SF per CY)   BL@.058   SF   --   1.09   --
   Wood (36 SF per CY)   BL@.063   SF   --   1.18   --

Wood framing demolition. Typical labor costs for demolition of wood frame structural components. Labor (Craft) is based on using Building Laborers with a cost of $18.80 per manhour. The manhours can be determined by dividing the cost per in-place SF by the Craft rate of $18.80. (Add for demolition of flooring, roofing, sheathing, etc.)

	Center to center spacing			
Ceiling joists, cost per in-place SF	12"	16"	20"	24"
2" x 4" (100 SF per CY)	.23	.17	.14	.13
2" x 6" (70 SF per CY)	.33	.25	.20	.18
2" x 8" (50 SF per CY)	.44	.34	.27	.24
2" x 10" (40 SF per CY)	.55	.42	.35	.31
2" x 12" (35 SF per CY)	.66	.51	.42	.36
Floor joists, cost per in-place SF of floor area				
2" x 6" (70 SF per CY)	.37	.30	.25	.21
2" x 8" (50 SF per CY)	.50	.40	.34	.30
2" x 10" (40 SF per CY)	.59	.50	.43	.37
2" x 12" (35 SF per CY)	.74	.58	.48	.46
Rafters, cost per in-place SF of actual roof area				
2" x 4" (100 SF per CY)	.26	.20	.18	.16
2" x 6" (70 SF per CY)	.37	.30	.26	.21
2" x 8" (50 SF per CY)	.50	.38	.35	.33
2" x 10" (40 SF per CY)	.60	.49	.43	.35
2" x 12" (35 SF per CY)	.73	.56	.49	.42
Stud walls, interior or exterior, cost per in-place SF of wall area				
2" x 3" (100 SF per CY)	.24	.23	.21	.20
2" x 4" (70 SF per CY)	.32	.31	.29	.28
2" x 6" (50 SF per CY)	.47	.44	.42	.41

Demolition

	Craft@Hrs	Unit	Material	Labor	Total
Wood stairs, cost per in-place SF or LF					
Risers (25 SF per CY)	BL@.116	SF	--	2.18	--
Landings (50 SF per CY)	BL@.021	SF	--	.40	--
Handrails (100 LF per CY)	BL@.044	LF	--	.83	--
Posts (200 LF per CY)	BL@.075	LF	--	1.41	--

Demolition, Subcontract. Costs for demolishing an entire building. Includes loading and hauling up to 5 miles, but no dump fees. Costs are for above ground structure only, per in-place SF of total floor area. Up to 8' high ceilings. No salvage value assumed. Add foundation demolition. Figures in parentheses give the approximate "loose" volume of the materials (volume after being demolished).

	Craft@Hrs	Unit	Material	Labor	Total
Light wood frame structures, cost per in-place SF floor area					
One story (75 SF per CY)	--	SF	--	--	1.95
Two story (80 SF per CY)	--	SF	--	--	2.60
Three story (80 SF per CY)	--	SF	--	--	3.45
Masonry building, cost per in-place SF floor area					
based on 5,000 SF job (50 SF per CY)	--	SF	--	--	4.10
Concrete building, cost per in-place SF floor area					
based on 5,000 SF job (30 SF per CY)	--	SF	--	--	4.95
Steel reinforced concrete building, cost per in-place SF floor area					
based on 5,000 SF job (20 SF per CY)	--	SF	--	--	5.60

Door Chimes. Electric, typical installation, add transformer and wire below.

	Craft@Hrs	Unit	Material	Labor	Total
Surface mounted, 2 notes					
Simple white plastic chime, 4" x 7"	BE@3.50	Ea	17.00	88.50	105.50
Modern oak grained finish, plastic, 8" x 5"	BE@3.50	Ea	32.00	88.50	120.50
Walnut finish, plastic, cloth grille, 8" x 5"	BE@3.50	Ea	32.00	88.50	120.50
Designer, wood-look plastic strips across a cloth grille, 9" x 7"	BE@3.50	Ea	45.00	88.50	133.50
Modern real oak wood, with brass plated chimes, 9" x 7"	BE@3.50	Ea	55.00	88.50	143.50
Surface mounted, 4 to 8 notes, solid wood door chime					
Three 12" long tubes	BE@3.64	Ea	85.00	92.00	177.00
Four 55" long tubes	BE@3.64	Ea	85.00	92.00	177.00
Add for chime transformer	--	Ea	--	--	7.00
Add for bell wire, 50' per coil	--	Ea	--	--	4.50
Add for lighted push buttons	--	Ea	--	--	5.00
Non-electric, push button type, door mounted					
Modern design, 2-1/2" x 2-1/2"	BE@.497	Ea	15.00	12.60	27.60
With peep sight, 7" x 3-1/2"	BE@.497	Ea	17.00	12.60	29.60

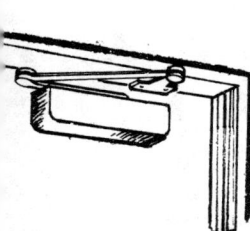

Door Closers

	Craft@Hrs	Unit	Material	Labor	Total
Standard duty door closers, hydraulic, bronze or aluminum finish					
To 4'0" door interior, or 3'0" door exterior	BC@1.04	Ea	68.00	24.50	92.50
Add for adapter plate	--	Ea	9.60	--	--
Light duty model, pneumatic for interior doors to 85 lb	BC@1.04	Ea	36.00	24.50	60.50
Light duty model, pneumatic, hinge mounted	BC@.390	Ea	10.20	9.20	19.40
Screen door closer, brass finish					
Hydraulic closer, medium duty	BC@.515	Ea	8.20	12.10	20.30
Pneumatic closer, heavy duty	BC@.546	Ea	12.00	12.90	24.90
Sliding door closer, pneumatic, aluminum finish					
Screen	BC@.250	Ea	12.00	5.90	17.90
Glass	BC@.250	Ea	24.00	5.90	29.90

Door Frames

	Craft@Hrs	Unit	Material	Labor	Total

Door Frames
Exterior frames with 1-3/4" x 4", 4-3/8" or 4-3/4" head and side jambs, and brick or stucco mould casing. For doors up to 3'0" x 6'8"

	Craft@Hrs	Unit	Material	Labor	Total
Select finger-jointed pine					
No sill	BC@1.08	Ea	41.20	25.50	66.70
Oak sill	BC@1.65	Ea	61.00	38.90	99.90
Aluminum sill	BC@1.46	Ea	63.80	34.40	98.20
Solid stock fir					
No sill	BC@1.08	Ea	53.30	25.50	78.80
Oak sill	BC@1.65	Ea	76.70	38.90	115.60
Aluminum sill	BC@1.46	Ea	79.40	34.40	113.80
Add for widths over 3'0"					
No sill	--	LF	3.10	--	--
Oak sill	--	LF	7.80	--	--
Aluminum sill	--	LF	6.65	--	--
Add for 7'0" high frame	--	Ea	6.55	--	--
Add for 8'0" high frame	--	Ea	13.30	--	--
Add for multiple frames					
For double frames, add the cost of two frames, plus	--	LS	9.00	--	--
For triple frames, add the cost of three frames, plus	--	LS	18.00	--	--
Add for solid mullions					
3" x 6", per mullion	--	Ea	52.00	--	--
4" x 6", per mullion	--	Ea	75.00	--	--

Interior sliding frames, pocket type, door and casing not included, 6'8" high, complete frame and hardware, 4-5/8" or 4-3/8" jamb, 2" x 4" studs

	Craft@Hrs	Unit	Material	Labor	Total
2'0" wide	BC@.981	Ea	45.00	23.10	68.10
2'4"	BC@.981	Ea	45.00	23.10	68.10
2'6"	BC@.981	Ea	45.00	23.10	68.10
2'8"	BC@.981	Ea	45.00	23.10	68.10
3'0"	BC@.981	Ea	45.00	23.10	68.10
3'6"	BC@.981	Ea	56.00	23.10	79.10
4'0"	BC@.981	Ea	56.00	23.10	79.10
Add for 7'0" height	--	Ea	9.80	--	--
Add for 8'0" height	--	Ea	19.40	--	--
Add for 2" x 6" stud wall	--	Ea	16.30	--	--
Add for solid jambs	--	Ea	17.40	--	--
Sliding door hardware set only	--	Ea	7.50	--	--

Door Jambs, Exterior
Select finger-joint, kiln dried pine, 6'9-1/2" high

	Craft@Hrs	Unit	Material	Labor	Total
4-1/8" set	BC@.800	Ea	22.50	18.90	41.40
4-1/2" set	BC@.800	Ea	23.70	18.90	42.60
4-3/4" set	BC@.800	Ea	24.60	18.90	43.50
5-1/4" set	BC@.850	Ea	24.60	20.10	44.70

Solid stock kiln dried fir, 6'9-1/2" high

	Craft@Hrs	Unit	Material	Labor	Total
4-1/8" set	BC@.800	Ea	31.00	18.90	49.90
4-1/2" set	BC@.800	Ea	34.40	18.90	53.30
4-3/4" set	BC@.800	Ea	36.00	18.90	54.90
5-1/4" set	BC@.850	Ea	39.30	20.10	59.40

Door Jambs

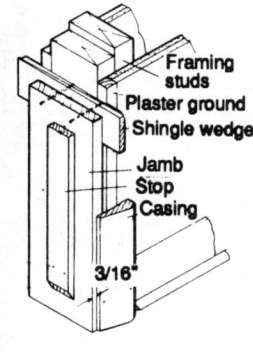

	Craft@Hrs	Unit	Material	Labor	Total

Door Jambs, Interior. Flat jamb with square cut heads and rabbeted sides, pine, 6'9-1/2" high.

	Craft@Hrs	Unit	Material	Labor	Total
3-9/16" set	BC@.800	Ea	9.70	18.90	28.60
4-9/16" set	BC@.800	Ea	11.90	18.90	30.80
4-11/16" set	BC@.800	Ea	12.90	18.90	31.80
5-1/4" set	BC@.850	Ea	14.80	20.10	34.90
Add for solid jambs	--	Ea	17.50	--	--
Add for special width jambs (to next larger width)	--	Ea	9.80	--	--
Add for 7' high	--	Ea	9.80	--	--
Add for 8' high	--	Ea	19.50	--	--
Add for round jamb without stop	--	Ea	8.80	--	--
Deduct for unfinished lauan	--	Ea	-1.30	--	--
Add for prefinished lauan	--	Ea	2.50	--	--
Add for fir or hemlock jamb sets	--	%	40.0	--	--

Door Trim Sets. Costs listed are per set (one head casing and two side casings), 6'8" height, ponderosa pine. One side of door opening only.

Casing size 11/16" x 2-1/4"

	Craft@Hrs	Unit	Material	Labor	Total
3'0" W, ranch style, unfinished	BC@.510	Ea	9.00	12.00	21.00
3'0" W, ranch style, primed	BC@.510	Ea	6.20	12.00	18.20
6'0" W, ranch style, unfinished	BC@.612	Ea	13.00	14.40	27.40
6'0" W, ranch style, primed	BC@.612	Ea	7.50	14.40	21.90
3'0" W, colonial style, unfinished	BC@.510	Ea	10.00	12.00	22.00
3'0" W, colonial style, primed	BC@.510	Ea	5.50	12.00	17.50
6'0" W, colonial style, unfinished	BC@.612	Ea	12.00	14.40	26.40
6'0" W, colonial style, primed	BC@.612	Ea	7.50	14.40	21.90

Casing size 1/2" x 2", embossed honey oak color

	Craft@Hrs	Unit	Material	Labor	Total
3'0" W, ranch style, prefinished light	BC@.510	Ea	7.50	12.00	19.50
6'0" W, ranch style, prefinished light	BC@.616	Ea	8.60	14.50	23.10
3'0" W, colonial style, prefinished light	BC@.510	Ea	7.75	12.00	19.75
6'0" W, colonial style, prefinished light	BC@.612	Ea	9.40	14.40	23.80

Casing size 3/4" x 3-1/2", ponderosa pine, with corner and base blocks

	Craft@Hrs	Unit	Material	Labor	Total
6'0" W, butterfly, prefinished dark	BC@.510	Ea	45.00	12.00	57.00
6'0" W, fluted, prefinished dark	BC@.612	Ea	50.00	14.40	64.40

Casing size 3/4" x 5-1/4", ponderosa pine, with corner and base blocks

	Craft@Hrs	Unit	Material	Labor	Total
6'0" W, butterfly, prefinished dark	BC@.510	Ea	66.00	12.00	78.00
6'0" W, fluted, prefinished dark	BC@.612	Ea	66.00	14.40	80.40

Door Index

Item	Page
Bar doors	59-60
Combination doors	60
Dutch doors	60
Fire doors	60
Flush doors	61-63
Folding doors	64
French doors	64
Louver doors	64-65
Panel doors	65
Pet doors	65
Prehung doors	66-67
Screen doors	67

Doors. See also Basement Doors, Closet Doors, Door Frames, Door Jambs, Door Closers, Door Trim, Entrances, Garage Doors, Hardware (door), Mouldings (astragal, sill, threshold), Shower and Tub Doors, Thresholds, and Weatherstripping listed alphabetically in the Residential division of this book.

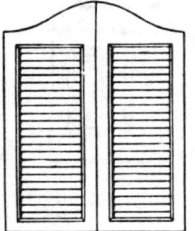

Bar Doors (cafe doors)
Louvered 1-1/8", 4' high, ponderosa pine, opening sizes, hardware included

	Craft@Hrs	Unit	Material	Labor	Total
2'6" unfinished	BC@.686	Pr	45.00	16.20	61.20
2'8" unfinished	BC@.686	Pr	48.00	16.20	64.20
3'0" unfinished	BC@.686	Pr	52.00	16.20	68.20
2'6" prefinished	BC@.686	Pr	60.00	16.20	76.20
2'8" prefinished	BC@.686	Pr	63.00	16.20	79.20
3'0" prefinished	BC@.686	Pr	67.00	16.20	83.20

Doors

	Craft@Hrs	Unit	Material	Labor	Total
Spindle over raised panel design, 1-1/8", 4' high, hemlock, pair opening sizes					
2'6" unfinished	BC@.686	Pr	71.00	16.20	87.20
2'8" unfinished	BC@.686	Pr	75.00	16.20	91.20
3'0" unfinished	BC@.686	Pr	79.00	16.20	95.20
Add for hinges					
Gravity pivot hinges	--	Pr	3.50	--	--
Spring hinges, polished brass	--	Pr	35.00	--	--

Combination Doors (storm and screen doors)
Wood construction, western hemlock, full view type, with aluminum screen cloth, tempered glass storm insert, 4-1/2" top rail and stiles, 9-1/4" bottom rail. Unfinished. No frame or hardware included

	Craft@Hrs	Unit	Material	Labor	Total
2'8" x 6'8"	BC@1.25	Ea	114.00	29.50	143.50
3'0" x 6'8"	BC@1.25	Ea	118.00	29.50	147.50

Wood construction, western hemlock, colonial style with half height crossbuck, aluminum screen cloth, tempered glass storm insert, 4-1/4" top rail and stiles, 9-1/4" bottom rail. Unfinished, no frame or hardware included

	Craft@Hrs	Unit	Material	Labor	Total
2'8" x 6'8"	BC@1.25	Ea	140.00	29.50	169.50
3'0" x 6'8"	BC@1.25	Ea	145.00	29.50	174.50

Aluminum frame, fiberglass screen cloth, latch, keyed lock, pneumatic closer, weatherstripping, and glass insert panel. Frame included. Available for openings 6'8" to 7'0" high, 2'8" to 3'0" wide

	Craft@Hrs	Unit	Material	Labor	Total
Better quality, half-height crossbuck with upper window or full length window styles, baked enamel finish and full weatherstripping	BC@1.01	Ea	243.00	23.80	266.80
Better quality with 3/4 length jalousie storm insert and foam-filled double wall kick plate	BC@1.05	Ea	322.00	24.80	346.80
Good quality with stationary upper glass panel, large aluminum kick plate, horizontal pushbar, and baked enamel finish	BC@1.05	Ea	147.00	24.80	171.80

Dutch Doors. Fir, 1-3/4" thick stiles and rails, single pane tempered glazing. No frame, jambs, moulding, or hardware included. Add one top section and one bottom section to find the door cost. See the labor cost to install two-section dutch doors below. 2'6", 2'8" or 3'0" sizes

Bottom sections	Craft@Hrs	Unit	Material	Labor	Total
Two vertical panel style	--	Ea	122.00	--	--
Cross buck panel style	--	Ea	123.00	--	--
Top sections					
Full lite style	--	Ea	213.00	--	--
Four lite style	--	Ea	248.00	--	--
Six lite style	--	Ea	244.00	--	--
Nine lite style	--	Ea	220.00	--	--
Seven lite diagonal style	--	Ea	339.00	--	--
Twelve lite diagonal style	--	Ea	248.00	--	--
Add for dutch door shelf at lock rail	--	Ea	35.40	--	--
Labor to install two-section dutch door	BC@2.95	Ea	--	69.60	--

Fire Doors. Class B label, noncombustible mineral core, unfinished, labor hanging door only, 1-3/4" thick, natural birch
1 hour rating, 6'8" high

	Craft@Hrs	Unit	Material	Labor	Total
2'0" wide	BC@1.36	Ea	116.00	32.10	148.10
2'4" wide	BC@1.41	Ea	127.00	33.30	160.30
2'6" wide	BC@1.41	Ea	133.00	33.30	166.30

Doors

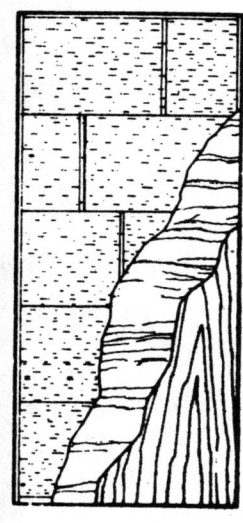

	Craft@Hrs	Unit	Material	Labor	Total
2'8" wide	BC@1.41	Ea	139.00	33.30	172.30
3'0" wide	BC@1.46	Ea	152.00	34.40	186.40
3'6" wide	BC@1.46	Ea	168.00	34.40	202.40
4'0" wide	BC@1.46	Ea	182.00	34.40	216.40
1 hour rating, 7'0" high					
2'6" wide	BC@1.46	Ea	139.00	34.40	173.40
2'8" wide	BC@1.46	Ea	146.00	34.40	180.40
3'0" wide	BC@1.57	Ea	159.00	37.00	196.00
3'6" wide	BC@1.57	Ea	176.00	37.00	213.00
4'0" wide	BC@1.57	Ea	190.00	37.00	227.00
1-1/2 hour rating, 6'8" high					
2'6" wide	BC@1.36	Ea	188.00	32.10	220.10
2'8" wide	BC@1.46	Ea	196.00	34.40	230.40
3'0" wide	BC@1.46	Ea	213.00	34.40	247.40
3'6" wide	BC@1.46	Ea	237.00	34.40	271.40
4'0" wide	BC@1.46	Ea	257.00	34.40	291.40
1-1/2 hour rating, 7'0" high					
2'6" wide	BC@1.46	Ea	196.00	34.40	230.40
2'8" wide	BC@1.57	Ea	206.00	37.00	243.00
3'0" wide	BC@1.57	Ea	224.00	37.00	261.00
3'6" wide	BC@1.57	Ea	248.00	37.00	285.00
4'0" wide	BC@1.57	Ea	368.00	37.00	405.00
Add for 10" x 10" lite, wired glass	--	Ea	--	--	45.00

Flush Doors, Hollow Core

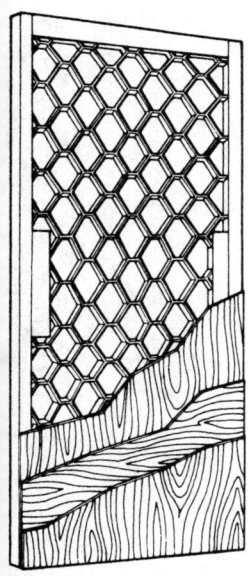

6'8" high x 1-3/8" thick	Hardboard	Lauan	Birch	Ash	Red oak
1'3" wide	23.00	28.00	35.00	36.00	44.00
1'4" wide	23.00	28.00	35.00	36.00	44.00
1'6" wide	23.00	28.00	35.00	36.00	44.00
1'8" wide	24.00	28.00	35.00	36.00	44.00
1'10" wide	24.00	28.00	36.00	36.00	44.00
2'0" wide	24.00	28.00	36.00	36.00	44.00
2'2" wide	25.00	31.00	40.00	39.00	49.00
2'4" wide	25.00	31.00	40.00	39.00	49.00
2'6" wide	25.00	33.00	42.00	42.00	52.00
2'8" wide	27.00	35.00	44.00	45.00	54.00
2'10" wide	28.00	35.00	48.00	49.00	60.00
3'0" wide	28.00	37.00	48.00	49.00	60.00
3'6" wide	36.00	54.00	69.00	70.00	97.00
4'0" wide	39.00	60.00	76.00	77.00	100.00
6'8" high x 1-3/4" thick	Hardboard	Lauan	Birch	Ash	
2'6" wide	31.00	38.00	47.00	48.00	
2'8" wide	32.00	40.00	49.00	50.00	
3'0" wide	34.00	43.00	53.00	54.00	
3'6" wide	42.00	--	74.00	--	
7'0" high x 1-3/8" thick		Hardboard	Lauan	Birch	
2'6" wide		31.00	38.00	47.00	
2'8" wide		32.00	40.00	49.00	
3'0" wide		34.00	43.00	53.00	
7'0" high x 1-3/4" thick	Hardboard	Lauan	Birch	Ash	
2'0" wide	40.00	--	52.00	--	
2'4" wide	41.00	--	56.00	--	
2'6" wide	36.00	44.00	53.00	--	
2'8" wide	37.00	46.00	55.00	--	

Doors

	Hardboard	Lauan	Birch	Ash
3'0" wide	39.00	48.00	59.00	60.00
3'6" wide	54.00	--	87.00	88.00
4'0" wide	57.00	--	94.00	--

8'0" high x 1-3/8" thick

	Hardboard	Birch
1'6" wide	36.00	57.00
2'0" wide	38.00	58.00
2'2" wide	39.00	62.00
2'4" wide	39.00	62.00
2'6" wide	40.00	65.00
2'8" wide	40.00	68.00
3'0" wide	42.00	74.00
3'6" wide	51.00	97.00
4'0" wide	53.00	104.00

8'0" high x 1-3/4" thick

	Hardboard	Birch
2'6" wide	46.00	71.00
2'8" wide	47.00	74.00
3'0" wide	48.00	80.00
3'6" wide	57.00	103.00
4'0" wide	60.00	110.00

Flush Doors, Solid Core

6'8" high x 1-3/8" thick

	Hardboard	Lauan	Birch	Ash	Red oak
2'0" wide	43.00	48.00	55.00	56.00	64.00
2'4" wide	44.00	51.00	59.00	60.00	69.00
2'6" wide	45.00	53.00	61.00	62.00	71.00
2'8" wide	46.00	54.00	64.00	64.00	74.00
3'0" wide	48.00	57.00	68.00	68.00	80.00

6'8" high x 1-3/4" thick

	Hardboard	Lauan	Birch	Ash	Red oak
2'6" wide	50.00	58.00	67.00	67.00	77.00
2'8" wide	52.00	60.00	69.00	70.00	79.00
3'0" wide	53.00	63.00	73.00	74.00	85.00
3'6" wide	68.00	--	100.00	--	--

7'0" high x 1-3/8" thick

	Hardboard	Lauan	Birch	Ash	Red oak
2'6" wide	50.00	58.00	67.00	--	--
2'8" wide	52.00	60.00	69.00	--	--
3'0" wide	53.00	63.00	73.00	--	--

7'0" high x 1-3/4" thick

	Hardboard	Lauan	Birch	Ash	Red oak
2'0" wide	60.00	65.00	72.00	--	--
2'4" wide	61.00	67.00	76.00	--	--
2'6" wide	56.00	64.00	72.00	--	82.00
2'8" wide	57.00	65.00	75.00	--	85.00
3'0" wide	59.00	68.00	79.00	79.00	90.00
3'6" wide	81.00	99.00	113.00	114.00	--
4'0" wide	83.00	104.00	120.00	--	--

8'0" high x 1-3/4" thick

	Hardboard	Birch	Red oak
2'6" wide	71.00	97.00	116.00
2'8" wide	72.00	100.00	117.00
3'0" wide	74.00	106.00	118.00
3'6" wide	85.00	130.00	--
4'0" wide	88.00	138.00	--

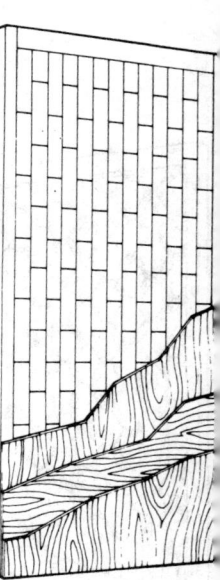

Doors

	Craft@Hrs	Unit	Material	Labor	Total
Flush Doors, Additional Costs, hollow core and solid core					
Non-stock and odd sizes, add to next larger size, hollow or solid core	--	Ea	28.00	--	--
Exterior glue, Type I waterproof					
Hardboard faces	--	Ea	5.00	--	--
Hardwood faces	--	Ea	11.00	--	--
Cut-outs, no stops included					
Hollow core, edge blocked	--	Ea	19.00	--	--
Solid core	--	Ea	15.00	--	--
Cut-out stops, two sides					
Softwood	--	Ea	12.00	--	--
Birch or mahogany	--	Ea	15.00	--	--
Ash, oak or walnut	--	Ea	28.00	--	--
Glazed openings, per SF or fraction of SF. No cut-outs or stops included					
Tempered clear, 3/16" thick	--	SF	11.00	--	--
Tempered bronze, 3/16" thick	--	SF	13.00	--	--
Tempered grey, 3/16" thick	--	SF	13.00	--	--
Tempered obscure, 7/32" thick	--	SF	13.00	--	--
Tempered Flemish amber	--	SF	17.00	--	--
Flemish acrylics, 3/16" thick	--	SF	9.00	--	--
Flemish acrylics, 1/8" thick	--	SF	6.50	--	--
Clear safety laminate 1/4" thick	--	SF	14.00	--	--
Polished wire plate 1/4" thick	--	SF	16.00	--	--
Vents, no cut-outs or stops included					
Fiber cane, 12" x 8"	--	Ea	10.00	--	--
Metal cane, 10" x 10"	--	Ea	29.00	--	--
Prefitting flush doors. Includes beveled lock stile, lockset bore, and hinge mortising. No hardware included					
6'8" to 7'0" high x 1-3/8" or 1-3/4" thick					
To 3'0" wide, hollow core	--	Ea	2.20	--	--
Over 3'0" to 4'0", hollow core	--	Ea	4.30	--	--
To 3'0" wide, solid core	--	Ea	4.30	--	--
Over 3'0" to 4'0" wide, solid core	--	Ea	7.10	--	--
8'0" high x 1-3/8" or 1-3/4" thick					
To 3'0" wide, hollow core	--	Ea	7.10	--	--
Over 3'0" to 4'0", hollow core	--	Ea	7.90	--	--
To 3'0" wide, solid core	--	Ea	7.90	--	--
Over 3'0" to 4'0" wide, solid core	--	Ea	11.50	--	--
Stave lumber cores for solid core doors					
To 3'0" x 7'0" sizes	--	Ea	43.00	--	--
Over 3'0" x 7'0 to 4'0" x 8'0" sizes	--	Ea	54.00	--	--
Special construction, 1-3/8" or 1-3/4" thick doors					
5" top and bottom rails	--	Ea	5.50	--	--
6" lock blocks	--	Ea	5.50	--	--
2-1/2" wide stiles	--	Ea	19.00	--	--
2-1/2" wide center stile	--	Ea	11.00	--	--
Labor to Hang Flush Doors. Typical cost per door. No frame, trim or finish. Add for hardware as required.					
Hollow core, to 3'0" x 7'0"	BC@.941	Ea	--	22.20	--
Hollow core, over 3'0" x 7'0" to 4'0" x 8'0"	BC@1.08	Ea	--	25.50	--
Solid core, to 3'0" x 7'0"	BC@1.15	Ea	--	27.10	--
Solid core, over 3'0" x 7'0" to 4'0" x 8'0"	BC@1.32	Ea	--	31.10	--

Doors, Folding

Folding Doors. 4" wide hardwood panels with embossed, prefinished wood grain print, 3/8" thick, includes track and hardware.

	Walnut or oak 6'8"	Red oak 6'8"	Pine 8'0"	Birch 8'0"
2'8" wide	133.00	211.00	159.00	258.00
3'0" wide	144.00	228.00	172.00	279.00
3'6" wide	155.00	243.00	185.00	297.00
4'0" wide	174.00	275.00	208.00	330.00
4'6" wide	243.00	340.00	257.00	415.00
5'0" wide	225.00	357.00	270.00	436.00
5'6" wide	236.00	374.00	283.00	456.00
6'0" wide	275.00	437.00	330.00	534.00
7'0" wide	315.00	500.00	377.00	611.00
8'0" wide	356.00	564.00	426.00	689.00
9'0" wide	400.00	631.00	477.00	771.00
10'0" wide	437.00	639.00	524.00	850.00
11'0" wide	472.00	947.00	565.00	915.00
12'0" wide	549.00	974.00	657.00	1,067.00
13'0" wide	592.00	983.00	708.00	1,140.00
14'0" wide	630.00	1,000.00	754.00	1,222.00
15'0" wide	673.00	1,065.00	806.00	1,300.00

	Craft@Hrs	Unit	Material	Labor	Total
Labor to install wood panel folding doors, per LF of width	BC@.647	LF	--	15.30	--

French Doors (sash doors). Douglas fir or hemlock, with tempered glass set in putty, 6'8" high.

	Craft@Hrs	Unit	Material	Labor	Total
1 lite, 1-3/8" thick sash					
2'0" to 2'8" wide	BC@1.45	Ea	137.00	34.20	171.20
3'0" wide	BC@1.45	Ea	142.00	34.20	176.20
5 lites, 5 high, 1-3/8" thick sash					
2'0" to 2'8" wide	BC@1.45	Ea	120.00	34.20	154.20
3'0" wide	BC@1.45	Ea	124.00	34.20	158.20
10 lites, 5 high, 1-3/8" thick sash					
2'0" to 2'8" wide	BC@1.45	Ea	133.00	34.20	167.20
3'0" wide	BC@1.45	Ea	137.00	34.20	171.20
1 lite, 1-3/4" thick sash					
2'0" to 2'8" wide	BC@1.45	Ea	138.00	34.20	172.20
3'0" wide	BC@1.45	Ea	141.00	34.20	175.20
5 lites, 5 high, 1-3/4" thick sash					
2'0" to 2'8" wide	BC@1.45	Ea	130.00	34.20	164.20
3'0" wide	BC@1.45	Ea	134.00	34.20	168.20
10 lites, 5 high, 1-3/4" thick sash					
2'0" to 2'8" wide	BC@1.45	Ea	133.00	34.20	167.20
3'0" wide	BC@1.45	Ea	137.00	34.20	171.20

Louver Doors, with full length slats, pine, 1-3/8", unfinished, 6'8" height

	Craft@Hrs	Unit	Material	Labor	Total
1'0"	BC@.750	Ea	38.10	17.70	55.80
1'3"	BC@.750	Ea	40.80	17.70	58.50
1'4"	BC@.750	Ea	41.90	17.70	59.60
1'6"	BC@.750	Ea	45.00	17.70	62.70
1'8"	BC@.750	Ea	55.30	17.70	73.00

Doors

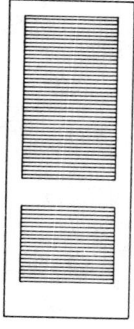

	Craft@Hrs	Unit	Material	Labor	Total
2'0"	BC@.750	Ea	60.00	17.70	77.70
2'4"	BC@.750	Ea	66.70	17.70	84.40
2'6"	BC@.750	Ea	67.90	17.70	85.60
2'8"	BC@.750	Ea	71.10	17.70	88.80
3'0"	BC@.750	Ea	76.10	17.70	93.80
Add for prefinished doors	--	%	25.0	--	--
Add for half louver with raised lower panel	--	%	15.0	--	--
Add for false louvers	--	%	200.0	--	--
Add for 8'0" height, any design	--	%	145.0	--	--

Panel (Colonial) Doors, 1-3/8" thick with raised panels 7/16" thick, 6'8" high, primed, labor hanging door only. No frame, trim, finish or hardware. Add for same as required.

Primed hardboard, hollow core

	Craft@Hrs	Unit	Material	Labor	Total
1'2"	BC@.827	Ea	29.90	19.50	49.40
1'4"	BC@.827	Ea	29.90	19.50	49.40
1'6"	BC@.827	Ea	29.90	19.50	49.40
2'0"	BC@.827	Ea	30.80	19.50	50.30
2'4"	BC@.827	Ea	33.30	19.50	52.80
2'6"	BC@.827	Ea	33.30	19.50	52.80
2'8"	BC@.827	Ea	34.50	19.50	54.00
3'0"	BC@.827	Ea	37.10	19.50	56.60

Ponderosa pine, stile and rail, 6 panel

	Craft@Hrs	Unit	Material	Labor	Total
1'2"	BC@.827	Ea	52.90	19.50	72.40
1'3"	BC@.827	Ea	54.70	19.50	74.20
1'4"	BC@.827	Ea	57.90	19.50	77.40
1'6"	BC@.827	Ea	72.40	19.50	91.90
1'8"	BC@.827	Ea	75.90	19.50	95.40
1'10"	BC@.827	Ea	76.80	19.50	96.30
2'0"	BC@.827	Ea	77.70	19.50	97.20
2'2"	BC@.827	Ea	83.90	19.50	103.40
2'4"	BC@.827	Ea	83.90	19.50	103.40
2'6"	BC@.827	Ea	83.90	19.50	103.40
2'8"	BC@.827	Ea	95.50	19.50	115.00
3'0"	BC@.827	Ea	101.00	19.50	120.50
Add for prefinished panel doors	--	%	20.0	--	--

Pet Doors

Entrance doors, lockable, swinging, aluminum frame with security panel

	Craft@Hrs	Unit	Material	Labor	Total
For cats and miniature dogs, 5" x 7-1/2"	BC@.800	Ea	37.00	18.90	55.90
For small dogs, 8-1/2" x 12-1/2"	BC@.800	Ea	58.00	18.90	76.90
For medium dogs, 11-1/2" x 16-1/2"	BC@.800	Ea	78.00	18.90	96.90
For large dogs, 15" x 20"	BC@.850	Ea	103.00	20.10	123.10

Sliding screen or patio doors, adjustable full length 1/2" panel, 80" high, "Lexan" plastic above lockable, swinging PVC door, aluminum frame

	Craft@Hrs	Unit	Material	Labor	Total
For cats and miniature dogs, panel is 11-1/2" wide with 5" x 7-1/2" door	BC@.500	Ea	160.00	11.80	171.80
For small dogs, panel is 15" wide with 8-1/2" x 12-1/2" door	BC@.500	Ea	185.00	11.80	196.80
For medium dogs, panel is 18" wide with 11-1/2" x 16-1/2" door	BC@.500	Ea	210.00	11.80	221.80

Doors

	Prefinished embossed oak tone, flush	Natural birch, flush	Unfinished panel pine

Prehung Hollow Core Interior Doors. Primed split pine jamb, matching stop and casing, hinges, door sizes, 6'8" high, 1-3/8" thick, add for lockset.

Size	Prefinished embossed oak tone, flush	Natural birch, flush	Unfinished panel pine
1'6"	64.10	76.70	126.00
2'0"	65.10	78.80	131.00
2'4"	68.30	80.90	139.00
2'6"	68.30	80.90	139.00
2'8"	69.30	83.00	151.00
3'0"	70.40	84.00	156.00

Labor to install prehung hollow core doors	Craft@Hrs	Unit	Material	Labor	Total
To 2'0" wide	BC@.581	Ea	--	13.70	--
Over 2'0" wide	BC@.827	Ea	--	19.50	--

Prehung Solid Core Exterior Wood Doors. 1-3/4" thick, bored for lockset with finger-jointed pine frame, exterior casing, aluminum sill and 4-1/2" jamb. Assembled, primed and weatherstripped, 6'8" high. 5'4" and 6'0" widths are double doors. Add for interior trim and lockset.

	Single door unit		Double door unit	
	2'8"	3'0"	5'4"	6'0"
Flush birch	203.00	207.00	435.00	444.00
6 panel, no glass, fir	305.00	313.00	639.00	655.00
French, 12 SF insulating glass	336.00	--	701.00	--
4 panel, 4 lite fan-shaped glass upper panels	--	334.00	--	718.00
Solid wood lower panel, 9 lite uninsulated glass upper panels	--	450.00	--	929.00
Mediterranean detail, no glass	--	400.00	--	830.00

Labor to install prehung solid core doors	Craft@Hrs	Unit	Material	Labor	Total
2'8" or 3'0" wide single door units	BC@.750	Ea	--	17.70	--
5'4" or 6'0" wide double door units	BC@1.00	Ea	--	23.60	--

Sidelights for packaged prehung doors, 1-3/4" thick, with insulated glass, assembled.

	Craft@Hrs	Unit	Material	Labor	Total
6'8" x 1'2", fir or hemlock	BC@1.36	Ea	219.00	32.10	251.10

Prehung Steel-clad Exterior Doors. 24 gauge steel cladding, 1-3/4" door bored for lockset, with wood frame, exterior casing only, polystyrene core, 1-1/2 hour (Class B) fire rating, weatherstripped, with hinges, assembled, primed, with aluminum sill, threshold. 6'8" high. Add for interior trim and lockset.

	Single doors			Double doors	
	2'8"	3'0"	3'6"	5'0"	5'4"	6'0"
Flush doors	143.00	146.00	194.00	359.00	361.00	366.00
Flush with up to 3 SF glass lite per door	182.00	185.00	195.00	447.00	447.00	453.00
Flush with 12 SF insulating glass per door	249.00	251.00	295.00	--	550.00	556.00
Doors with decorative insulating glass and moulding face	311.00	314.00	314.00	673.00	674.00	680.00

Labor to install prehung steel-clad doors	Craft@Hrs	Unit	Material	Labor	Total
2'8" to 3'6" wide single door units	BC@.750	Ea	--	17.70	--
5'0" to 6'0" wide double door units	BC@1.00	Ea	--	23.60	--

Doors

	Craft@Hrs	Unit	Material	Labor	Total
Prehung Steel-clad Exterior Doors, continued. Add for:					
Single door steel frame	--	Ea	22.90	--	--
Double door steel frame	--	Ea	25.70	--	--
Sidelights, typical, 1'2" wide	BC@1.22	Ea	143.00	28.80	171.80
Interior trim sets					
2'6" to 2'8"	BC@.471	Ea	8.46	11.10	19.56
3'0" to 3'6"	BC@.471	Ea	9.31	11.10	20.41
5'0" to 6'0"	BC@.471	Ea	9.50	11.10	20.60
Mail slot (cut out only)	--	Ea	10.70	--	--
Peephole (hole only)	--	Ea	3.04	--	--
Extension jamb, 4-9/16" to 5-1/4"					
Single door	--	Ea	11.40	--	--
Double door	--	Ea	19.00	--	--

Screen Doors

Wood screen doors, western hemlock, 1-3/8" thick, aluminum screen cloth, 3-1/8" stiles, 3-1/8" top and center rails, 5" bottom rail. No frame or hardware included.

	Craft@Hrs	Unit	Material	Labor	Total
2'8" x 6'9"	BC@1.25	Ea	57.00	29.50	86.50
3'0" x 6'9"	BC@1.25	Ea	59.00	29.50	88.50

Aluminum screen doors, fiberglass screen cloth, latch, lock, pneumatic closer, and hinges. Available for openings 6'8" to 6'9" high, 2'5" to 3'1" wide

	Craft@Hrs	Unit	Material	Labor	Total
Better quality full view with ornamental grill, tempered glass, kick plate, and deadbolt lock	BC@1.00	Ea	205.00	23.60	228.60
Better quality, full length side lite with two decorative acrylic side panels, tempered glass, deadbolt lock	BC@1.00	Ea	212.00	23.60	235.60
Good quality, half height grill, with horizontal pushbar, kick plate, and baked enamel finish	BC@1.00	Ea	150.00	23.60	173.60
Minimum quality, center pushbar, satin aluminum finish	BC@1.00	Ea	78.00	23.60	101.60

Sliding 6'8" Aluminum Patio Doors. 3/16" single pane tempered safety glass, aluminum frame with baked-on enamel, weatherstripped, includes adjustable locking device, screen included.

	Craft@Hrs	Unit	Material	Labor	Total
1 sliding door, 1 stationary lite					
5'0" wide	BC@2.08	Ea	246.00	49.10	295.10
6'0" wide	BC@2.58	Ea	268.00	60.90	328.90
7'0" wide	BC@2.58	Ea	313.00	60.90	373.90
8'0" wide	BC@3.16	Ea	313.00	74.50	387.50
10'0" wide	BC@3.16	Ea	415.00	74.50	489.50
1 sliding door, 2 stationary lites					
9'0" wide	BC@3.16	Ea	360.00	74.50	434.50
12'0" wide	BC@3.65	Ea	425.00	86.10	511.10
15'0" wide	BC@4.14	Ea	578.00	97.70	675.70
2 sliding doors, 2 stationary lites					
12'0" wide	BC@3.65	Ea	482.00	86.10	568.10
16'0" wide	BC@4.14	Ea	575.00	97.70	672.70

Sliding 8'0" Aluminum Patio Doors. 3/16" single pane tempered safety glass, 5-1/2" anodized frame, weatherstripped, screen included.

	Craft@Hrs	Unit	Material	Labor	Total
1 sliding door, 1 stationary lite					
5'0" wide	BC@2.25	Ea	280.00	53.10	333.10
6'0" wide	BC@2.75	Ea	307.00	64.90	371.90
8'0" wide	BC@3.50	Ea	355.00	82.60	437.60
10'0" wide	BC@3.50	Ea	461.00	82.60	543.60

Doors

	Craft@Hrs	Unit	Material	Labor	Total
1 sliding door, 2 stationary lites					
9'0" wide	BC@3.50	Ea	409.00	82.60	491.60
12'0" wide	BC@4.00	Ea	481.00	94.40	575.40
15'0" wide	BC@4.50	Ea	641.00	106.00	747.00
2 sliding doors, 2 stationary lites					
12'0" wide	BC@4.00	Ea	546.00	94.40	640.40
16'0" wide	BC@4.50	Ea	645.00	106.00	751.00
Add for tinted grey glass, average	--	%	50.0	--	--
Add for bronze finish frame	--	%	10.0	--	--
Add for installation in masonry or concrete	--	%	--	10.0	--
Add for key lock	--	Ea	16.20	--	--

Sliding Aluminum-clad Wood Patio Doors. Aluminum clad on exterior surfaces, 7/8" insulating glass, ball bearing rollers, weatherstripped, with deadbolt security lock, screen included.

	Craft@Hrs	Unit	Material	Labor	Total
1 sliding door, 1 stationary lite					
5'0" wide	BC@2.00	Ea	664.00	47.20	711.20
6'0" wide	BC@2.75	Ea	694.00	64.90	758.90
8'0" wide	BC@3.25	Ea	810.00	76.70	886.70
1 sliding door, 2 stationary lites					
9'0" wide	BC@3.25	Ea	960.00	76.70	1,036.70
12'0" wide	BC@3.75	Ea	1,390.00	88.50	1,478.50
2 sliding doors, 2 stationary lites					
12'0" wide	BC@3.75	Ea	1,180.00	88.50	1,268.50
16'0" wide	BC@4.25	Ea	1,440.00	100.00	1,540.00
Add for installation in masonry or concrete	--	%	--	10.0	--
Add for key lock	--	Ea	20.00	--	--

Sliding Wood Patio Doors. 5/8" tempered insulating glass, treated pine frame, double weatherstripped, with screen, key lock, and deadbolt, 6'8" high.

	Craft@Hrs	Unit	Material	Labor	Total
5' wide, 1 sliding door, 1 stationary lite	BC@2.00	Ea	936.00	47.20	983.20
6' wide, 1 sliding door, 1 stationary lite	BC@2.75	Ea	1,010.00	64.90	1,074.90
8' wide, 1 sliding door, 1 stationary lite	BC@2.75	Ea	1,160.00	64.90	1,224.90
9' wide, 1 sliding door, 2 stationary lites	BC@3.50	Ea	1,240.00	82.60	1,322.60
12' wide, 1 sliding door, 2 stationary lites	BC@3.75	Ea	1,800.00	88.50	1,888.50
Add for colonial grille per door section	BC@.500	LS	103.00	11.80	114.80

Swinging Steel Patio Doors. 1" tempered insulating glass, foam core, single lite, magnetic weatherstripping, with sliding screen, lock and deadbolt, 6'8" high.

	Craft@Hrs	Unit	Material	Labor	Total
5'4" wide, 1 swinging door, 1 fixed lite	BC@2.00	Ea	608.00	47.20	655.20
6'0" wide, 1 swinging door, 1 fixed lite	BC@2.75	Ea	608.00	64.90	672.90
7'0" wide, 2 swinging doors	BC@2.75	Ea	631.00	64.90	695.90
9'0" wide, 1 swinging door, 2 fixed lites	BC@3.25	Ea	940.00	76.70	1,016.70
Add for 15 lite door	--	Ea	37.70	--	--

Drafting. Per SF of floor area, architectural only, not including engineering fees. See also Architectural Illustrations and Blueprinting. Typical prices.

	Craft@Hrs	Unit	Material	Labor	Total
Apartments	--	SF	--	--	1.10
Warehouses and storage buildings	--	SF	--	--	.75
Office buildings	--	SF	--	--	.85

Drafting

	Craft@Hrs	Unit	Material	Labor	Total
Residences					
Minimum quality tract work	--	SF	--	--	1.50
Typical work	--	SF	--	--	1.75
Better work (exposed woods, detail work, hillsides)	--	SF	--	--	2.50

Drainage Piping. Perforated or solid corrugated plastic drainage pipe installed in trenches or at foundation footings. No excavation, gravel or backfill included. Available in 10' to 250' lengths. Snap fittings. Per linear foot.

	Craft@Hrs	Unit	Material	Labor	Total
3" diameter pipe	BL@.010	LF	.40	.19	.59
4" diameter pipe	BL@.010	LF	.50	.19	.69
5" diameter pipe	BL@.010	LF	1.05	.19	1.24
6" diameter pipe	BL@.012	LF	1.10	.23	1.33
8" diameter pipe	BL@.012	LF	1.90	.23	2.13
Reducer fittings					
4" x 3"	BL@.050	Ea	2.40	.94	3.34
5" x 4"	BL@.050	Ea	2.60	.94	3.54
6" x 4"	BL@.050	Ea	2.70	.94	3.64
8" x 6"	BL@.060	Ea	4.30	1.13	5.43
Adapters, end caps, split couplers					
3"	BL@.050	Ea	1.20	.94	2.14
4"	BL@.050	Ea	1.60	.94	2.54
5"	BL@.050	Ea	2.10	.94	3.04
6"	BL@.060	Ea	2.55	1.13	3.68
Ells, wyes, tees					
3"	BL@.050	Ea	3.70	.94	4.64
4"	BL@.050	Ea	6.10	.94	7.04
5"	BL@.050	Ea	6.90	.94	7.84
6"	BL@.060	Ea	7.95	1.13	9.08

Rock or sand fill for drainage systems. Labor cost is for spreading base and covering pipe. 3 mile haul dumped on site, and placed by wheelbarrow

	Craft@Hrs	Unit	Material	Labor	Total
Crushed stone (1.4 tons per CY)					
3/4" (Number 3)	BL@.700	CY	12.30	13.20	25.50
1-1/2" (Number 2)	BL@.700	CY	12.30	13.20	25.50
Crushed slag (1.86 tons per CY)					
3/4"	BL@.700	CY	8.00	13.20	21.20
1-1/2"	BL@.700	CY	8.10	13.20	21.30
Sand (1.35 tons per CY)	BL@.360	CY	9.95	6.77	16.72

Draperies, Custom, Subcontract. Custom draperies include 4" double hems and headers, 1-1/2" side hems, and weighted corners. Finished sizes allow 4" above, 4" below (except for sliding glass openings), and 5" on each side of wall openings. Prices are figured at one width, 48" of fabric, pleated to 16". Costs listed include fabric, full liner, manufacturing, all hardware, and installation. Add 50% per LF of width for jobs outside working range of metropolitan centers.

Minimum quality, 200% fullness, average is 6 panels of fabric, limited choice of fabric styles, colors, and textures

	Craft@Hrs	Unit	Material	Labor	Total
To 95" high	--	LF	--	--	11.90
To 84" high	--	LF	--	--	11.10
To 68" high	--	LF	--	--	10.20
To 54" high	--	LF	--	--	9.60
To 44" high	--	LF	--	--	8.65

Draperies

	Craft@Hrs	Unit	Material	Labor	Total

Good quality, fully lined, 250% fullness, average is 5 panels of fabric, better selection of fabric styles, colors, and textures, weighted seams

To 95" high	--	LF	--	--	44.70
To 84" high	--	LF	--	--	41.20
To 68" high	--	LF	--	--	37.30
To 54" high	--	LF	--	--	34.40
To 44" high	--	LF	--	--	32.40
Deduct for tract or multi-unit jobs	--	%	--	--	-25.0

Add for insulated fabrics or liners

Pleated shade liner	--	SY	--	--	7.85
Thermal liner	--	SY	--	--	8.65

Better quality, fully lined, choice of finest fabric styles, colors, and textures. (Does not include special treatments such as elaborate swags, or custom fabrics)

To 95" high	--	LF	--	--	64.00
To 84" high	--	LF	--	--	58.20
To 68" high	--	LF	--	--	50.90
To 54" high	--	LF	--	--	46.50
To 44" high	--	LF	--	--	42.20
Add for each visit to job for measuring	--	LS	--	--	27.00
Add per job for delivery	--	LS	--	--	27.00

Electrical Work, Subcontract
Costs listed below are for wiring new residential and light commercial buildings with Romex cable and assume circuit lengths averaging 40 feet. If flex cable is required, add $10 for 15 amp circuits and $25 for 20 amp circuits. No fixtures or appliances included except as noted. Work on second and higher floors may cost 25% more. Work performed by a qualified subcontractor.

Rule of thumb: Total cost for electrical work, performed by a qualified subcontractor, per SF of floor area. Includes service entrance, outlets, switches, basic lighting fixtures and connecting appliances only

All wiring and fixtures	--	SF	--	--	3.00
Lighting fixtures only (no wiring)	--	SF	--	--	1.00

Air conditioners (with thermostat)

Central, 2 ton (220 volt)	--	LS	--	--	300.00
First floor room (115 volt)	--	LS	--	--	125.00
Second floor room (115 volt)	--	LS	--	--	200.00
Add for thermostat on second floor	--	LS	--	--	50.00

Alarms. See also Security Alarms

Fire or smoke detector, wiring and outlet box only	--	LS	--	--	50.00
Add for detector unit	--	LS	--	--	24.00

Bathroom fixtures, wiring only, no fixtures or equipment included

Mirror lighting, (valance or side lighted mirrors)	--	LS	--	--	51.00
Sauna heater (40 amp branch circuit)	--	LS	--	--	175.00
Steam bath generator	--	LS	--	--	175.00
Sunlamps	--	LS	--	--	100.00
Whirlpool bath system/wall switch	--	LS	--	--	155.00
Clock outlets (recessed)	--	LS	--	--	40.00

Closet lighting

Including ceramic "pull chain" fixture	--	LS	--	--	45.00
Including ceramic switch-operated fixture, add for wall switch below	--	LS	--	--	45.00

Electrical

	Craft@Hrs	Unit	Material	Labor	Total

Electrical Work, Subcontract, continued

Clothes dryers
  Motor-driven gas dryer, receptacle only -- LS -- -- 60.00
  Direct connection for up to 5,760
    watt (30 amp) electric dryer, 220 volt -- LS -- -- 130.00
  Direct connection for over 5,760
    watt (40 amp) electric dryer, 220 volt -- LS -- -- 155.00
Clothes washers (115 volt) -- LS -- -- 79.00
Dishwashers -- LS -- -- 79.00
Door bells (rough wiring, front and rear with transformer,
no chime or bell switch included) -- LS -- -- 77.00
Fans (exhaust)
  Attic (wiring only) -- LS -- -- 87.00
  Bathroom (includes 70 CFM fan) -- LS -- -- 102.00
  Garage (wiring only) -- LS -- -- 70.00
  Kitchen (includes 225 CFM fan) -- LS -- -- 160.00
Freezers, branch circuit wiring only
  Upright (115 volt) -- LS -- -- 75.00
  Floor model -- LS -- -- 56.00
Furnace, wiring and blower hookup only -- LS -- -- 85.00
Garage door opener, wiring and hookup
only -- LS -- -- 87.00
Garbage disposers (wiring, switch and connection
only, no equipment included) -- LS -- -- 116.00
Grounding devices (see also Lightning Protection Systems below)
  Entire electrical system -- LS -- -- 56.00
  Single appliance -- LS -- -- 56.00
Ground fault circuit interrupter, rough and finish wiring,
including outlet box, and GFCI -- LS -- -- 69.00
Heaters
  Baseboard (115 volt) per branch circuit -- LS -- -- 81.00
  Bathroom (ceiling type) wiring, switch
    connection (with GFCI) only -- LS -- -- 112.00
  Ceiling heat system (radiant-resistance type) 1,000 watt, 120 volt,
    per branch circuit, including thermostat -- LS -- -- 214.00
  Space (flush in-wall type) up to 2,000
    watt, 220 volt -- LS -- -- 198.00
Humidifiers, central -- LS -- -- 75.00
Lamp posts, wiring and hookup only, to 25' with
buried wire and conduit -- LS -- -- 173.00
Lighting fixture outlets (rough wiring and box only)
  Ceiling -- Ea -- -- 40.00
  Floor -- Ea -- -- 58.00
  Set in concrete -- Ea -- -- 75.00
  Set in masonry -- Ea -- -- 75.00
  Underground -- Ea -- -- 103.00
  Valance -- Ea -- -- 41.00
  Wall -- Ea -- -- 36.00
Lightning protection systems (static electric grounding system)
  Residential, ridge protection including one chimney and connected garage
    (houses with cut-up roofs will cost more) -- LS -- -- 1,500.00
  Barns and light commercial buildings -- LS -- -- 3,500.00

Electrical

	Craft@Hrs	Unit	Material	Labor	Total
Electrical Work, Subcontract, continued					
Ovens, wall type (wiring and hookup only), 220 volt					
To 4,800 watts (20 amp)	--	LS	--	--	117.00
4,800 to 7,200 watts (30 amp)	--	LS	--	--	153.00
7,200 to 9,600 watts (40 amp)	--	LS	--	--	179.00
Ranges, (wiring and hookup only), 220 volt					
Countertop type					
To 4,800 watts (20 amp)	--	LS	--	--	112.00
4,800 to 7,200 watts (30 amp)	--	LS	--	--	143.00
7,200 to 9,600 watts (40 amp)	--	LS	--	--	173.00
Freestanding type (50 amp)	--	LS	--	--	214.00
Receptacle outlets (rough wiring, box, receptacle, and finish wiring with plate)					
Ceiling	--	Ea	--	--	40.00
Counter top wall	--	Ea	--	--	40.00
Floor	--	Ea	--	--	71.00
Split-wired	--	Ea	--	--	77.00
Standard indoor, duplex wall outlet	--	Ea	--	--	40.00
Waterproof, with ground fault circuit	--	Ea	--	--	68.00
Refrigerator or freezer wall outlet	--	Ea	--	--	41.00
Service entrance connections, complete (panel box hookup but no wiring)					
100 amp service including meter, main switch, GFCI, and 5 single pole breakers in 20 breaker space exterior panel box	--	LS	--	--	750.00
200 amp service including meter, main switch, two GFCI and 15 single pole breakers in 40 breaker space exterior panel box	--	LS	--	--	1,250.00
Sub panel connections (panel box hookup but no wiring)					
40 amp circuit panel including 12 breaker indoor panel box and circuit breakers	--	LS	--	--	200.00
50 amp circuit panel including 16 breaker indoor panel box and circuit breakers	--	LS	--	--	250.00
Sump pump connection, with 15' of underground cable	--	Ea	--	--	153.00
Switches (includes rough and finish wiring)					
Dimmer	--	Ea	--	--	50.00
Lighted	--	Ea	--	--	37.00
Quiet	--	Ea	--	--	37.00
Mercury	--	Ea	--	--	37.00
Standard indoor wall switch	--	Ea	--	--	37.00
Stair-wired (3 way)	--	Ea	--	--	90.00
Waterproof	--	Ea	--	--	50.00
Television outlet wiring (300 ohm wire)	--	Ea	--	--	36.00
Vacuuming system, central					
Central unit hookup	--	LS	--	--	94.00
Remote vacuum outlets (receptacle wiring only)	--	Ea	--	--	37.00
Water heaters (connection only)	--	LS	--	--	150.00
Water pumps (domestic potable water, connection only)	--	LS	--	--	94.00

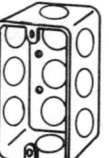

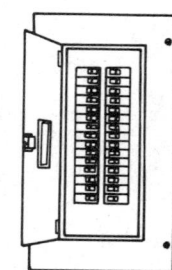

Elevators

	Craft@Hrs	Unit	Material	Labor	Total

Elevators and Lifts, Subcontract
Elevators for apartments, hydraulic vertical cab, meets code requirements for public buildings, 2,000 pound capacity, to 13 passengers, 6' x 5' platform. Includes side opening sliding door, illuminated controls, emergency light, alarm, code-approved fire service operation, motor control unit, wiring in hoistway, cab and door design options. Basic 2 stop prices

	Craft@Hrs	Unit	Material	Labor	Total
100 feet per minute	--	LS	--	--	36,000.00
125 feet per minute	--	LS	--	--	38,500.00
150 feet per minute	--	LS	--	--	41,500.00
Add per stop to 5 stops	--	LS	--	--	3,900.00
Add for photoelectric safety door opener	--	Ea	--	--	1,270.00
Add for hall position indicator	--	Ea	--	--	360.00
Add for car position indicator	--	Ea	--	--	360.00
Add for car direction light & tone	--	Ea	--	--	255.00
Add for hall lantern with audible tone	--	Ea	--	--	360.00

Elevators for private residences, electric cable, motor driven, 500 pound capacity, includes complete installation of elevator, controls and safety equipment. Meets code requirements for residences

2 stops (up to 10' rise)	--	LS	--	--	15,700.00
3 stops (up to 20' rise)	--	LS	--	--	18,500.00
4 stops (up to 30' rise)	--	LS	--	--	21,000.00
5 stops (up to 40' rise)	--	LS	--	--	23,300.00
Add for 700 pound capacity	--	LS	--	--	4,150.00
Add for additional gate	--	Ea	--	--	630.00

Dumbwaiters, electric, 2 stop

100 pound, 24" x 24" x 36", tray type	--	LS	--	--	8,400.00
Add for each extra stop	--	Ea	--	--	1,670.00
300 pound, 30" x 30" x 36", tray type	--	LS	--	--	12,600.00
Add for each extra stop	--	Ea	--	--	2,100.00
500 pound, 23" x 56" x 48", cart type	--	LS	--	--	14,200.00
Add for each extra stop	--	Ea	--	--	2,100.00

Stairlifts (incline lifts), single passenger, indoor. Lift to 18' measured diagonally. 300 pound capacity. Includes top and bottom call button, safety chair, track, power unit, wiring in hoistway and installation by a licensed elevator contractor. Single family residential use

Straight stairs	--	LS	--	--	4,000.00
Curved or turning stairs, costs vary widely	--	LS	--	--	14,000.00

Hydraulic elevators, 450 pound capacity. Includes controls and safety equipment, with standard unfinished birch plywood cab and one gate.

2 stops (up to 15' rise)	--	LS	--	--	21,000.00
3 stops (up to 22' rise)	--	LS	--	--	23,000.00
4 stops (up to 28' rise)	--	LS	--	--	25,000.00
5 stops (up to 34' rise)	--	LS	--	--	26,800.00
Add for adjacent opening	--	Ea	--	--	750.00
Add for additional gate	--	Ea	--	--	1,200.00
Add for plastic laminate cab	--	Ea	--	--	475.00
Add for unfinished oak picture frame cab	--	Ea	--	--	1,500.00
Add for unfinished oak raised panel cab	--	Ea	--	--	2,500.00
Add for finished oak cabs	--	%	--	--	20.0

Wheelchair lift (porch lift), screw driven, fully enclosed sides. Typical electrical connections and control included. Up to 5' rise

Residential use model	--	LS	--	--	10,000.00
Public use (code approved)	--	LS	--	--	11,000.00

Engineering

	Craft@Hrs	Unit	Material	Labor	Total

Engineering Fees. Typical cost for consulting work.
Acoustical engineering (Based on an hourly wage of $65.00 to $125.00)

Environmental noise survey of a lot; includes technician with measuring equipment, data reduction and analysis, and written report with recommendations	--	LS	--	--	1,200.00
Exterior to interior noise analysis	--	LS	--	--	400.00
Measure "Impact Insulation Class" and "Sound Transmission Class" in existing buildings, (walls, floors, and ceilings), minimum cost	--	LS	--	--	800.00
Priced by wall, floor, or ceiling sections, approx. 100 SF	--	Ea	--	--	400.00
Analyze office, conference room, or examining room to assure acoustical privacy	--	LS	--	--	350.00
Prepare acoustical design for church, auditorium, or lecture hall, (fees vary greatly with complexity of acoustics desired)	--	LS	--	--	1,500.00
Evaluate noise problems from proposed building on surrounding environment	--	LS	--	--	800.00
Evaluate noise problems from surrounding environment on proposed building	--	LS	--	--	650.00

Front end scheduling. Coordinate architectural, engineering and other consultant services. Plan check times and other approval times for residential and commercial

projects	--	LS	--	--	4,000.00

Project scheduling using CPM (Critical Path Method). Includes consultation, review of construction documents, development of construction logic, and graphic schedule. Comprehensive schedules to meet government or owner specifications will cost more.

Wood frame buildings, 1 or 2 stories	--	LS	--	--	2,000.00
Structural engineering plan check, typical 2-story, 8 to 10 unit apartment building	--	LS	--	--	775.00
Add for underground parking	--	LS	--	--	670.00

Entrances, Colonial
Single door entrance, pine, primed (no doors, frames, or sills included)

Without pediment, 3'0" x 6'8"	BC@1.19	Ea	129.00	28.10	157.10
With pediment, 3'0" x 6'8"	BC@1.19	Ea	150.00	28.10	178.10
With arch pediment, 3'0" x 6'8"	BC@1.19	Ea	258.00	28.10	286.10
Add for exterior door frames, 3'0" x 6'8" x 1-3/4"					
4-9/16" jamb width	BC@1.25	Ea	39.50	29.50	69.00

Double door entrance, pine, primed (no doors, frames, or sills included)
Without pediment, two door (individual width)

2'6" x 6'8"	BC@1.45	Ea	152.00	34.20	186.20
2'8" x 6'8"	BC@1.45	Ea	164.00	34.20	198.20
3'0" x 6'8"	BC@1.45	Ea	173.00	34.20	207.20

With pediment, two doors (individual width)

2'6" x 6'8"	BC@1.45	Ea	169.00	34.20	203.20
2'8" x 6'8"	BC@1.45	Ea	205.00	34.20	239.20
3'0" x 6'8"	BC@1.45	Ea	268.00	34.20	302.20

With arched pediment, 7'6-1/2" x 1'9" (above door). Opening width of 5'7-1/4" will also fit 3'0" door with two 14" sidelights

3'0" x 6'8" (two doors)	BC@1.45	Ea	383.00	34.20	417.20

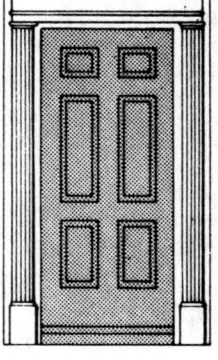

Entrances

	Craft@Hrs	Unit	Material	Labor	Total
Add for exterior door frames, 1-3/4" thick, 5-1/4" jamb					
2'8" x 6'8"	BC@1.25	Ea	53.40	29.50	82.90
3'0" x 6'8"	BC@1.25	Ea	53.50	29.50	83.00
6'0" x 6'8"	BC@2.75	Ea	64.30	64.90	129.20
Decorative split columns (with cap and base) alone, no pediment					
8" x 8' to 12'	BC@.800	Ea	24.50	18.90	43.40
10" x 8' to 12'	BC@.800	Ea	27.80	18.90	46.70
12" x 8' to 12'	BC@.800	Ea	31.00	18.90	49.90
10" x 14' to 16'	BC@1.00	Ea	49.00	23.60	72.60

Excavation and Backfill by Hand

	Craft@Hrs	Unit	Material	Labor	Total
General excavation (loosening and one throw)					
Light soil	BL@1.10	CY	--	20.70	--
Average soil	BL@1.70	CY	--	32.00	--
Heavy soil or loose rock	BL@2.25	CY	--	42.30	--
Add for wet soil	--	%	--	75.0	--
Backfilling (one throw from stockpile)					
Sand	BL@.367	CY	--	6.90	--
Average soil	BL@.467	CY	--	8.78	--
Rock or clay	BL@.625	CY	--	11.80	--
Add for compaction	--	%	--	100.0	--
Fine grading	BL@.008	SF	--	.15	--
Footings, average soil					
6" deep x 12" wide (1.85 CY per CLF)	BL@.034	LF	--	.64	--
8" deep x 12" wide (2.47 CY per CLF)	BL@.050	LF	--	.94	--
8" deep x 16" wide (3.29 CY per CLF)	BL@.055	LF	--	1.03	--
8" deep x 18" wide (3.70 CY per CLF)	BL@.060	LF	--	1.13	--
10" deep x 12" wide (3.09 CY per CLF)	BL@.050	LF	--	.94	--
10" deep x 16" wide (4.12 CY per CLF)	BL@.067	LF	--	1.26	--
10" deep x 18" wide (4.63 CY per CLF)	BL@.075	LF	--	1.41	--
12" deep x 12" wide (3.70 CY per CLF)	BL@.060	LF	--	1.13	--
12" deep x 16" wide (4.94 CY per CLF)	BL@.081	LF	--	1.52	--
12" deep x 20" wide (6.17 CY per CLF)	BL@.100	LF	--	1.88	--
16" deep x 16" wide (6.59 CY per CLF)	BL@.110	LF	--	2.07	--
12" deep x 24" wide (7.41 CY per CLF)	BL@.125	LF	--	2.35	--
Loading trucks (shoveling, one throw)					
Average soil	BL@1.45	CY	--	27.30	--
Rock or clay	BL@2.68	CY	--	50.40	--
Pits to 6'					
Light soil	BL@1.34	CY	--	25.20	--
Average soil	BL@2.00	CY	--	37.60	--
Heavy soil	BL@2.75	CY	--	51.70	--
For each additional lift add	--	%	--	100.0	--
Shaping trench bottom for pipe					
To 10" pipe	BL@.024	LF	--	.45	--
12" to 20" pipe	BL@.076	LF	--	1.43	--
Shaping embankment slopes					
Up to 1 in 4 slope	BL@.060	SY	--	1.13	--
Over 1 in 4 slope	BL@.075	SY	--	1.41	--
Add for top crown or toe	--	%	--	50.0	--
Add for swales	--	%	--	90.0	--
Spreading material piled on site					
Average soil	BL@.367	CY	--	6.90	--
Stone or clay	BL@.468	CY	--	8.80	--

Excavation Factors

Depth	CY per SF
2"	.00617
4"	.01235
6"	.01852
8"	.02469
10"	.03086
1' 0"	.03704
1' 6"	.05555
2' 0"	.07407
2' 6"	.09259
3' 0"	.11111
3' 6"	.13333
4' 0"	.14815
4' 6"	.16667
5' 0"	.18519
5' 6"	.20370
6' 0"	.22222
6' 6"	.24074
7' 0"	.25926
7' 6"	.27777
8' 0"	.29630
8' 6"	.31481
9' 0"	.33333
9' 6"	.35185
10' 0"	.37037

Multiply the excavation factor by the area in square feet to find the cubic yards of soil to be excavated.
Example: Assume an excavation 24 ft. x 30 ft. and 6 ft. deep. 24 x 30 = 720. In the table the 6 ft. depth has a factor of .22222 (the number of cu. yd. in an excavation 1 ft. square and 6 ft. deep).
720 x .22222 = 160 Cu.Yds.

Excavation

	Craft@Hrs	Unit	Material	Labor	Total
Strip and pile top soil	BL@.024	SF	--	.45	--
Tamping, hand tamp only	BL@.612	CY	--	11.50	--
Trenches to 5', soil piled only					
Light soil	BL@1.13	CY	--	21.20	--
Average soil	BL@1.84	CY	--	34.60	--
Heavy soil or loose rock	BL@2.86	CY	--	53.80	--
Add for depth over 5' to 9'	BL@.250	CY	--	4.70	--

Excavation with Heavy Equipment. These figures assume a crew of one operator unless noted otherwise. Only labor costs are included here. See equipment rental costs at the end of this section. Use the productivity rates listed here to determine the number of hours or days that the equipment will be needed. For larger jobs and commercial work, see excavation costs under Site Work in the Commercial and Industrial division of this book.

	Craft@Hrs	Unit	Material	Labor	Total
Excavation rule of thumb for small jobs	--	CY	--	--	2.85
Backhoe, operator and one laborer, 3/4 CY bucket					
Light soil (13.2 CY per hour)	B8@.152	CY	--	3.40	--
Average soil (12.5 CY per hour)	B8@.160	CY	--	3.58	--
Heavy soil (10.3 CY per hour)	B8@.194	CY	--	4.34	--
Rock (16 CY per hour)	B8@.125	CY	--	2.80	--
Add when using 1/2 CY bucket	--	%	--	25.0	--
Bulldozer, 50 HP unit					
Backfill (36 CY per hour)	OE@.027	CY	--	.70	--
Clearing brush (900 SF per hour)	OE@.001	SF	--	.26	--
Add for thick brush	--	%	--	300.0	--
Spread dumped soil (42 CY per hour)	OE@.024	CY	--	.62	--
Strip top soil (17 CY per hour)	OE@.059	CY	--	1.53	--
Bulldozer, 66 HP unit					
Backfill (40 CY per hour)	OE@.025	CY	--	.65	--
Clearing brush (1,000 SF per hour)	OE@.001	SF	--	.26	--
Add for thick brush	--	%	--	300.0	--
Spread dumped soil (45 CY per hour)	OE@.023	CY	--	.60	--
Strip top soil (20 CY per hour)	OE@.050	CY	--	1.30	--
Bulldozer, 120 HP unit					
Backfill (53 CY per hour)	OE@.019	CY	--	.49	--
Clearing brush (1,250 SF per hour)	OE@.001	SF	--	.02	--
Add for thick brush	--	%	--	200.0	--
Spread dumped soil (63 CY per hour)	OE@.016	CY	--	.42	--
Strip top soil (25 CY per hour)	OE@.040	CY	--	1.04	--
Dump truck. Spot, load, unload, travel					
3 CY truck					
Short haul (9 CY per hour)	B7@.222	CY	--	4.38	--
2-3 mile haul (6 CY per hour)	B7@.333	CY	--	6.57	--
4 mile haul (4 CY per hour)	B7@.500	CY	--	9.87	--
5 mile haul (3 CY per hour)	B7@.666	CY	--	13.10	--
4 CY truck					
Short haul (11 CY per hour)	B7@.182	CY	--	3.59	--
2-3 mile haul (7 CY per hour)	B7@.286	CY	--	5.65	--
4 mile haul (5.5 CY per hour)	B7@.364	CY	--	7.19	--
5 mile haul (4.5 CY per hour)	B7@.444	CY	--	8.77	--
5 CY truck					
Short haul (14 CY per hour)	B7@.143	CY	--	2.82	--
2-3 mile haul (9.5 CY per hour)	B7@.211	CY	--	4.17	--
4 mile haul (6.5 CY per hour)	B7@.308	CY	--	6.08	--
5 mile haul (5.5 CY per hour)	B7@.364	CY	--	7.19	--

Excavation

	Craft@Hrs	Unit	Material	Labor	Total
Jackhammer, one laborer (per CY of unloosened soil)					
Average soil (2-1/2 CY per hour)	BL@.400	CY	--	7.52	--
Heavy soil (2 CY per hour)	BL@.500	CY	--	9.40	--
Igneous or dense rock (.7 CY per hour)	BL@1.43	CY	--	26.90	--
Most weathered rock (1.2 CY per hour)	BL@.833	CY	--	15.70	--
Soft sedimentary rock (2 CY per hour)	BL@.500	CY	--	9.40	--
Air tamp (3 CY per hour)	BL@.333	CY	--	6.26	--
Loader (tractor shovel)					
Piling earth on premises, 1 CY bucket					
Light soil (43 CY per hour)	OE@.023	CY	--	.60	--
Heavy soil (35 CY per hour)	OE@.029	CY	--	.75	--
Loading trucks, 1 CY bucket					
Light soil (53 CY per hour)	OE@.019	CY	--	.49	--
Average soil (43 CY per hour)	OE@.023	CY	--	.60	--
Heavy soil (40 CY per hour)	OE@.025	CY	--	.65	--
Deduct for 2-1/4 CY bucket	--	%	--	-43.0	--
Sheepsfoot roller (18.8 CSF per hour)	OE@.001	SF	--	.01	--
Sprinkling with truck, (62.3 CSF per hour)	TR@.001	SF	--	.01	--
Tree and brush removal, labor only (clear and grub, one operator and one laborer)					
Light brush	B8@10.0	Acre	--	224.00	--
Heavy brush	B8@12.0	Acre	--	268.00	--
Wooded	B8@64.0	Acre	--	1,430.00	--
Tree removal, cutting trees, removing branches, cutting into short lengths with chain saws and axes, by tree diameter, labor only					
8" to 12" (2.5 manhours per tree)	BL@2.50	Ea	--	47.00	--
13" to 18" (3.5 manhours per tree)	BL@3.50	Ea	--	65.80	--
19" to 24" (5.5 manhours per tree)	BL@5.50	Ea	--	103.00	--
25" to 36" (7.0 manhours per tree)	BL@7.00	Ea	--	132.00	--
Tree stump removal, using a small dozer, by tree diameter, operator only					
6" to 10" (1.6 manhours per stump)	OE@1.60	Ea	--	41.50	--
11" to 14" (2.1 manhours per stump)	OE@2.10	Ea	--	54.40	--
15" to 18" (2.6 manhours per stump)	OE@2.60	Ea	--	67.40	--
19" to 24" (3.1 manhours per stump)	OE@3.10	Ea	--	80.40	--
25" to 30" (3.3 manhours per stump)	OE@3.30	Ea	--	85.50	--
Trenching machine, crawler-mounted					
Light soil (27 CY per hour)	OE@.037	CY	--	.96	--
Heavy soil (21 CY per hour)	OE@.048	CY	--	1.24	--
Add for pneumatic-tired machine	--	%	--	15.0	--

Excavation Equipment Costs. Typical cost not including fuel, delivery or pickup charges. Add labor costs from the previous section. Half day minimums.

	1/2 day	Day	Week
Backhoe, wheeled type			
1/2 CY bucket	133.50	222.00	731.00
3/4 CY bucket	127.50	212.00	701.00
Add for each delivery or pickup	85.00	85.00	85.00
Bulldozer, crawler tractor			
60 HP unit	148.00	246.00	807.00
75 HP unit	172.00	286.00	931.00
135 HP unit	354.00	590.00	1,818.00
Add for each delivery or pickup	85.00	85.00	85.00
Dump truck, on-highway type			
3 CY	85.20	131.00	378.00
5 CY	90.50	138.00	554.00
10 CY	125.00	192.00	735.00

Excavation

	1/2 day	Day	Week
Dump truck, on-highway type, continued			
Add for mileage charge, per mile driven	.35	.38	.40
Loader tractor, wheeled type			
1 CY bucket	132.00	202.00	703.00
2-1/2 CY bucket	291.00	447.00	1,490.00
Add for each delivery or pickup	100.00	100.00	100.00
Sheepsfoot roller, towed type, 40" diameter, 48" wide, double drum	48.00	80.00	243.00
Trenching machine, 4' depth, 12" width capacity, crawler-mounted type	74.50	124.00	446.00
Add for each delivery or pickup	100.00	100.00	100.00
Compactor, 32" plate width, manually guided, gasoline engine, vibratory plate type	56.00	93.00	302.00
Chain saw, 18" bar, gasoline engine	24.30	40.50	131.00
Breakers, pavement, medium duty, with blade and 50' hose	25.00	41.00	113.00
Add for trailer-mounted air compressor, gas powered, silenced			
100 CFM	36.00	60.00	195.00
150 CFM	40.00	66.00	212.00
175 CFM	46.00	76.00	250.00

Fans

Attic fans, indoor type (see also Roof-mounted fans below), belt drive, overall dimensions, labor installing and wiring.

	Craft@Hrs	Unit	Material	Labor	Total
Two speed, horizontal, 1/3 HP					
30" x 30" (5,100/3,400 CFM)	BE@3.80	Ea	495.00	96.10	591.10
36" x 36' (7,400/4,900 CFM)	BE@3.80	Ea	525.00	96.10	621.10
42" x 42" (9,750/6,500 CFM)	BE@3.80	Ea	595.00	96.10	691.10
Variable speed, vertical or horizontal, 1/4 HP, with ceiling shutters					
24" x 24" (to 5,400 CFM)	BE@3.80	Ea	330.00	96.10	426.10
30" x 30" (to 8,150 CFM)	BE@3.80	Ea	355.00	96.10	451.10
Add for time switch	BE@.950	Ea	30.00	24.00	54.00
Automatic ceiling shutters for exhaust fans, white aluminum,					
24"	BE@.667	Ea	70.00	16.90	86.90
30"	BE@.667	Ea	80.00	16.90	96.90
36"	BE@.750	Ea	95.00	19.00	114.00
Window fans, 3 speed, white/gray finish					
12", 25" to 38" panel (2,000 CFM)	BE@1.50	Ea	55.00	37.90	92.90
16", 22-1/2" to 33" panel (6,000 CFM)	BE@1.50	Ea	70.00	37.90	107.90
20", 26-1/2" to 36" panel (6,500 CFM)	BE@1.50	Ea	75.00	37.90	112.90

Bathroom exhaust fans, ceiling-mounted, labor wiring (with one wall switch), installing and connecting. See also Heaters

	Craft@Hrs	Unit	Material	Labor	Total
10" x 10" grille, 3" diameter duct (60 CFM)	BE@2.00	Ea	30.00	50.60	80.60
10" x 10" grille, with light, 4" diameter duct (50 CFM)	BE@2.00	Ea	45.00	50.60	95.60
Add for duct to 8', wall cap, and wire	BE@.417	LS	25.00	10.50	35.50

Bathroom wall-mounted direct exhaust fan. Labor wiring, installing and connecting, includes wall vent, switch, and wire

	Craft@Hrs	Unit	Material	Labor	Total
8" x 8" (110 CFM)	BE@2.00	Ea	35.00	50.60	85.60

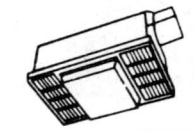

Kitchen exhaust fans, circular, aluminum, labor wiring, installing and connecting

	Craft@Hrs	Unit	Material	Labor	Total
200 CFM ceiling fan (with switch, wire, duct at $18.90 and wall cap at $11.60)	BE@2.00	Ea	121.00	50.60	171.60
300 CFM direct vent wall fan (with switch)	BE@2.00	Ea	110.00	50.60	160.60

Fans

	Craft@Hrs	Unit	Material	Labor	Total

Roof-mounted power ventilators, direct connected, with thermostat and exhaust hood. Labor installing, wiring and connecting

	Craft@Hrs	Unit	Material	Labor	Total
26" diameter roof vent flange, 1,200 CFM, 1/10 HP (for 4,800 to 7,200 CF attics)	BE@3.45	Ea	84.00	87.20	171.20
28" diameter vent, 1,600 CFM, 1/5 HP (for 7,200 to 9,600 CF attics)	BE@3.45	Ea	120.00	87.20	207.20
Add for automatic humidity sensor	--	Ea	20.00	--	--

Ceiling fans, decorative, 3-speed reversible motors, reversible wood blades, on/off pull chain, with white enamel, polished or antique brass motor housing, lighting kits optional (see below). Labor includes hanging and connecting only

	Craft@Hrs	Unit	Material	Labor	Total
42" four-blade plain fan with heavy duty motor housing	BE@1.50	Ea	150.00	37.90	187.90
42" five solid plywood blades, modern styles, tarnish, scratch resistant finish, 5 year motor warranty	BE@1.50	Ea	100.00	37.90	137.90
52" four plywood blades with decorator cane insert	BE@1.50	Ea	106.00	37.90	143.90
52" walnut finish blades, with lighted stained glass housing (takes four bulbs)	BE@1.50	Ea	120.00	37.90	157.90
52" four cabinet-grade hardwood blades with decorator cane insert, wobble-free motor housing, scratch resistant finish, lifetime motor warranty	BE@1.50	Ea	140.00	37.90	177.90
Optional lighting fixtures for ceiling fans					
Standard globe dome light, 6" H, 10" Dia	BE@.283	Ea	20.00	7.15	27.15
Old fashioned four-light 9-1/4" H, 20" W	BE@.283	Ea	65.00	7.15	72.15
Beveled smoke plate glass lamp, 10-3/4" H, 10-1/2" W	BE@.283	Ea	70.00	7.15	77.15
Add for tassel pull chain	BE@.017	Ea	1.50	.43	1.93

Fencing, Chain Link

Fence, 11-1/2 gauge steel, 2" x 2" galvanized fabric and framework, including top rail, ties, 1-5/8" OD line post at 10', 1-5/8" top rail and sleeves. Material includes 2/3 CF sack of concrete per post.

	Craft@Hrs	Unit	Material	Labor	Total
36" high	BL@.125	LF	2.00	2.35	4.35
42" high	BL@.125	LF	2.30	2.35	4.65
48" high	BL@.125	LF	2.45	2.35	4.80
60" high	BL@.125	LF	2.70	2.35	5.05
72" high	BL@.125	LF	3.00	2.35	5.35
Add for 9 gauge galvanized fabric	--	%	25.0	--	--
Add for 9 gauge vinyl coated fabric	--	%	20.0	--	--
Add for redwood filler strips	BL@.024	LF	2.50	.45	2.95
Add for aluminum filler strips	BL@.018	LF	2.70	.34	3.04
Add for plastic tubing filler strips	BL@.018	LF	2.80	.34	3.14

Gates, driveway or walkway. 11-1/2 gauge steel, 2" x 2" galvanized chain link with frame and tension bars. Costs shown are per square foot of frame area. Add for hardware and gate posts from below.

	Craft@Hrs	Unit	Material	Labor	Total
With 1-3/8" diameter pipe frame	BL@.025	SF	2.15	.47	2.62
With 1-5/8" diameter pipe frame	BL@.025	SF	2.50	.47	2.97
Add for hardware, per gate					
Walkway gate	BL@.250	Ea	10.00	2.40	12.40
Driveway gate	BL@.250	Ea	25.00	2.40	27.40

Fences

	Craft@Hrs	Unit	Material	Labor	Total

Fencing, Chain Link, continued
Corner, end or gate posts, complete, heavyweight, 2-1/2" outside diameter, with fittings and brace. Costs include 2/3 CF sack of concrete per post

	Craft@Hrs	Unit	Material	Labor	Total
36" high fence	BL@.612	Ea	12.00	11.50	23.50
42" high fence	BL@.612	Ea	14.00	11.50	25.50
48" high fence	BL@.612	Ea	15.00	11.50	26.50
60" high fence	BL@.668	Ea	17.00	12.60	29.60
72" high fence	BL@.735	Ea	19.00	13.80	32.80
Deduct for 2" outside diameter	--	%	-15.0	--	--

Chain link fence pet enclosures, galvanized steel, 11-1/2 gauge, 2" mesh, 1-3/8" tubular steel frame, 6' high, lockable gate

	Craft@Hrs	Unit	Material	Labor	Total
4' x 10'	BL@4.00	Ea	320.00	75.20	395.20
4' x 15'	BL@5.00	Ea	398.00	94.00	492.00
8' x 10'	BL@4.75	Ea	398.00	89.30	487.30
4' x 10', with roof	BL@5.00	Ea	342.00	94.00	436.00

Fencing, Chain Link, Subcontract. Work performed by a fence contractor. Galvanized steel fabric and framework, posts set in concrete, 10' maximum line post spacing. Based on minimum hand labor, non-rocky soil and post locations accessible to a trailered mechanical post hole digger. These costs include the subcontractor's overhead and profit.
11-1/2 gauge fence, 2" x 2" galvanized fabric and frame with 1-5/8" OD top rail, 1-5/8" OD line posts, and 2-1/2" OD terminal posts. Add for other posts as shown

	Craft@Hrs	Unit	Material	Labor	Total
48" high fence	--	LF	--	--	6.60
72" high fence	--	LF	--	--	7.20
Add for single strand barbed wire at top	--	LF	--	--	.25
Add for corner, end or gate posts					
48" high fence	--	Ea	--	--	35.70
72" high fence	--	Ea	--	--	43.90
Add for gates, walkway or driveway	--	SF	--	--	3.62
Add for walkway gate hardware, per gate	--	Ea	--	--	15.00
Add for driveway gate hardware, per gate	--	Ea	--	--	32.60
Add for vinyl coated fence and posts	--	%	--	--	10.0
Add for 9 gauge chain link fabric	--	%	--	--	12.0

Fence, Galvanized Wire Mesh with 1-3/8" diameter steel posts, set without concrete 10' OC
12-1/2 gauge steel, 12" vertical wire spacing

	Craft@Hrs	Unit	Material	Labor	Total
36" high, 6 horizontal wires	BL@.127	LF	.55	2.39	2.94
48" high, 8 horizontal wires	BL@.127	LF	.60	2.39	2.99
48" high, 8 horizontal wires (11 gauge)	BL@.127	LF	.72	2.39	3.11

11 gauge fence fabric, 9" vertical wire spacing

	Craft@Hrs	Unit	Material	Labor	Total
42" high, 8 horizontal wires	BL@.127	LF	.73	2.39	3.12
48" high, 9 horizontal wires	BL@.127	LF	.77	2.39	3.16

14 gauge fence fabric, 6" vertical wire spacing

	Craft@Hrs	Unit	Material	Labor	Total
36" high, 8 horizontal wires	BL@.127	LF	.56	2.39	2.95
48" high, 9 horizontal wires	BL@.127	LF	.74	2.39	3.13
60" high, 10 horizontal wires	BL@.127	LF	.94	2.39	3.33

Components for electrified galvanized wire mesh fence

	Craft@Hrs	Unit	Material	Labor	Total
15 gauge aluminum wire, single strand	BL@.050	LF	.03	.94	.97

Charging units for electric fence, solar powered, 25 mile range, with built-in

	Craft@Hrs	Unit	Material	Labor	Total
6-volt battery for night use	--	Ea	210.00	--	--

Battery powered charging unit, indoor or outdoor mounting, no battery included

	Craft@Hrs	Unit	Material	Labor	Total
25 mile range, add 12-volt battery	--	Ea	80.00	--	--
15 mile range, add 6-volt battery	--	Ea	55.00	--	--

Fences

	Craft@Hrs	Unit	Material	Labor	Total
12-1/2 gauge barbed wire, 4 barbs per LF	BL@.050	LF	.04	.94	.98
Fence stays for barbed wire, 48" high	--	Ea	.35	--	--
Add for electric fence warning sign, yellow with black lettering	--	Ea	.32	--	--
Add for polyethylene insulators	--	Ea	.06	--	--
Add for non-conductive gate fastener	--	Ea	1.58	--	--
Add for lightning arrestor kit	--	Ea	4.08	--	--

Fence, Wood. Costs shown include posts set in concrete. Gates have 2" x 4" frame and are 3' wide. Add for concrete footings, gate hardware or painting as required from costs shown at the end of this section.

Basketweave fence, redwood, "B" grade, 1" x 6" boards, 2" x 4" stringers or spreaders, 4" x 4" posts

	Craft@Hrs	Unit	Material	Labor	Total
Tight weave, 4' high	B1@.090	LF	11.40	1.90	13.30
Tight weave, 6' high	B1@.095	LF	17.90	2.01	19.91
Wide span, 4' high	B1@.090	LF	15.80	1.90	17.70
Wide span, 8' high	B1@.095	LF	17.00	2.01	19.01

Board fence, 1" x 6" boards, 2" x 4" rails, 4" x 4" posts

	Craft@Hrs	Unit	Material	Labor	Total
Fir or larch					
4' high, 3 rail	B1@.110	LF	3.70	2.33	6.03
6' high, 3 rail	B1@.119	LF	4.90	2.52	7.42
4' high gate	B1@.846	Ea	24.50	17.90	42.40
6' high gate	B1@1.05	Ea	32.50	22.20	54.70
Red cedar, #2 & better					
4' high, 3 rail	B1@.112	LF	4.80	2.37	7.17
6' high, 3 rail	B1@.119	LF	6.40	2.52	8.92
4' high gate	B1@.846	Ea	31.50	17.90	49.40
6' high gate	B1@1.05	Ea	42.20	22.20	64.40
Redwood, B grade					
4' high, 3 rail	B1@.110	LF	14.50	2.33	16.83
6' high, 3 rail	B1@.119	LF	19.00	2.52	21.52
4' high gate	B1@.846	Ea	52.50	17.90	70.40
6' high gate	B1@1.05	Ea	75.00	22.20	97.20

Board and batten fence, 1" x 6" boards, 1" x 2" battens, 2" x 4" rails, 4" x 4" posts

	Craft@Hrs	Unit	Material	Labor	Total
Fir or larch					
4' high, 3 rail	B1@.124	LF	4.25	2.62	6.87
6' high, 3 rail	B1@.130	LF	5.75	2.75	8.50
4' high gate	B1@.846	Ea	28.20	17.90	46.10
6' high gate	B1@1.05	Ea	38.00	22.20	60.20
Red cedar, #2 grade					
4' high, 3 rail	B1@.124	LF	5.60	2.62	8.22
6' high, 3 rail	B1@.130	LF	8.95	2.75	11.70
4' high gate	B1@.846	Ea	37.10	17.90	55.00
6' high gate	B1@1.05	Ea	59.00	22.20	81.20
Redwood, B grade					
4' high, 3 rail	B1@.124	LF	16.30	2.62	18.92
6' high, 3 rail	B1@.130	LF	21.90	2.75	24.65
4' high gate	B1@.846	Ea	107.00	17.90	124.90
6' high gate	B1@1.05	Ea	145.00	22.20	167.20

Split rail fence, red cedar

	Craft@Hrs	Unit	Material	Labor	Total
Split 2" x 4" rails, 3' high, 2 rail	B1@.095	LF	2.05	2.01	4.06
Split 2" x 4" rails, 4' high, 3 rail	B1@.102	LF	2.35	2.16	4.51
Rustic 4" round, 3' high, 2 rail	B1@.095	LF	2.50	2.01	4.51
Rustic 4" round, 4' high, 3 rail	B1@.102	LF	2.80	2.16	4.96

Fences

	Craft@Hrs	Unit	Material	Labor	Total
Picket fence, 1" x 2" pickets, 2" x 4" rails, 4" x 4" posts					
Fir or larch					
3' high, 2 rail	B1@.105	LF	3.00	2.22	5.22
5' high, 3 rail	B1@.113	LF	4.70	2.39	7.09
3' high gate	B1@.770	Ea	20.50	16.30	36.80
5' high gate	B1@.887	Ea	31.10	18.80	49.90
Red cedar, #2 and better					
3' high, 2 rail	B1@.105	LF	3.75	2.22	5.97
5' high, 3 rail	B1@.113	LF	5.85	2.39	8.24
3' high gate	B1@.770	Ea	24.80	16.30	41.10
5' high gate	B1@.887	Ea	38.70	18.80	57.50
Redwood, B grade					
3' high, 2 rail	B1@.105	LF	8.60	2.22	10.82
5' high, 3 rail	B1@.105	LF	13.40	2.22	15.62
3' high gate	B1@.770	Ea	30.30	16.30	46.60
5' high gate	B1@.887	Ea	45.30	18.80	64.10
Concrete fence post footing,					
8" diameter, 2' deep, typical	B1@.136	Ea	3.00	2.88	5.88
Gate hardware					
Latch, standard duty	B1@.281	Ea	2.75	5.95	8.70
Latch, heavy duty	B1@.281	Ea	6.40	5.95	12.35
Hinges, standard, per pair	B1@.374	Pr	5.10	7.93	13.03
Hinges, heavy duty, per pair	B1@.374	Pr	12.30	7.93	20.23
No-sag cable kit	B1@.281	Ea	7.25	5.95	13.20
Self closing spring	B1@.281	Ea	6.40	5.95	12.35
Paint or stain fence or gate with a roller, 2 coats, 2 sides					
3' high fence	PT@.093	LF	.55	2.11	2.66
4' high fence	PT@.115	LF	.70	2.61	3.31
5' high fence	PT@.128	LF	.85	2.90	3.75
6' high fence	PT@.171	LF	1.00	3.87	4.87
3' or 4' high gate (includes brushwork)	PT@.043	LF	1.45	.97	2.42
5' or 6' high gate (includes brushwork)	PT@.147	LF	2.40	3.33	5.73

Fiberglass Panels. Corrugated, 8', 10', 12' panels, 2-1/2" corrugations, nailed on wood frame, colors

	Craft@Hrs	Unit	Material	Labor	Total
5 oz, 26" or 27-1/2" wide	BC@.012	SF	1.10	.28	1.38
6 oz, 26" or 51" wide	BC@.012	SF	1.30	.28	1.58
8 oz, 26" or 27" wide	BC@.012	SF	1.60	.28	1.88
Flat panels, clear, green and white					
.06" flat sheets, 4' x 8', 10', 12'	BC@.012	SF	1.60	.28	1.88
.03" rolls, 24", 36", 48" x 50'	BC@.012	SF	1.35	.28	1.63
.037" rolls, 24", 36", 48" x 50'	BC@.012	SF	1.50	.28	1.78
Accessories					
Nails, ring shank with rubber washer, 1-3/4",					
100 per box (130 SF)	--	Box	3.00	--	--
Self-tapping screws, 1-1/4",					
100 per box (130 SF)	--	Box	21.00	--	--
Horizontal closure strips, corrugated					
Redwood, 2-1/2" x 1-1/2"	--	LF	.40	--	--
Polyfoam, 1" x 1"	--	LF	.30	--	--
Rubber, 1" x 1"	--	LF	.40	--	--
Vertical crown moulding					
Redwood, 1-1/2" x 1" or Polyfoam, 1" x 1"	--	LF	.40	--	--
Rubber, 1" x 1"	--	LF	.75	--	--

Fire Sprinkler Systems

	Craft@Hrs	Unit	Material	Labor	Total

Fire Sprinkler Systems. The following gives typical costs for systems installed on a subcontract basis in a single family residence not over two stories high in accord with NFPA Section 13D. Cost shown includes connection to domestic water line inside garage (1" pipe size minimum), exposed copper riser with shut-off and drain valve, CPVC concealed distribution piping and residential type sprinkler heads below finished ceilings. Crew is based on 1 sprinkler pipe installer and 1 helper with an average rate of $25.00 per manhour. These costs include hydraulic design calculations and inspection of completed system for compliance with design requirements. Costs shown are per SF of protected area. When estimating the square footage of protected area include non-inhabited areas such as bathrooms, closets and attached garages. For scheduling purposes, estimate that a crew of 2 men can install the rough-in piping for 1500 to 1600 SF of protected area in an 8-hour day and about the same amount of finish work in another 8-hour day.

	Craft@Hrs	Unit	Material	Labor	Total
Fabricate, install, and test system	--	SF	--	--	1.50
Deduct for tract work	--	%	--	--	-15.0
Add for condominiums or apartments	--	%	--	--	40.0

Fireplace Components. See also Masonry. Superior Fireplace Products

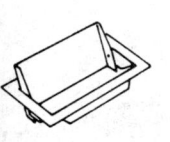

Angle iron (lintel support), lengths 25" to 96"

	Craft@Hrs	Unit	Material	Labor	Total
3" x 3" x 3/16"	B9@.167	LF	3.40	3.43	6.83

Ash drops, 16-3/4" x 12-1/2" cast iron top, galvanized container

	Craft@Hrs	Unit	Material	Labor	Total
8" deep	B9@.750	Ea	49.00	15.40	64.40
14" deep	B9@.750	Ea	59.00	15.40	74.40

Ash dumps, 4-1/2" x 8-1/4", 12 gauge

	Craft@Hrs	Unit	Material	Labor	Total
stamped steel	B9@.500	Ea	5.70	10.30	16.00

Chimney anchors, per pair

	Craft@Hrs	Unit	Material	Labor	Total
48" x 1-1/2" x 3/16"	--	Pr	7.50	--	--
72" x 1-1/2" x 3/16"	--	Pr	10.00	--	--

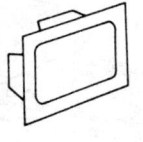

Cleanout doors, 14 gauge stamped steel

	Craft@Hrs	Unit	Material	Labor	Total
8" x 8"	B9@.500	Ea	9.90	10.30	20.20
8" x 10"	B9@.500	Ea	10.50	10.30	20.80
Combustion air inlet kit for masonry fireplaces	--	Ea	85.00	--	--

Dampers for single opening firebox, heavy steel with rockwool insulation

	Craft@Hrs	Unit	Material	Labor	Total
14" high, 25" to 30" wide, 10" deep, 25 lbs	B9@1.33	Ea	33.00	27.30	60.30
14" high, 36" wide, 10" deep, 31 lbs	B9@1.33	Ea	35.50	27.30	62.80
14" high, 42" wide, 10" deep, 38 lbs	B9@1.33	Ea	43.00	27.30	70.30
14" high, 48" wide, 10" deep, 43 lbs	B9@1.75	Ea	56.00	36.00	92.00
16" high, 54" wide, 13" deep, 66 lbs	B9@2.50	Ea	111.00	51.40	162.40
16" high, 60" wide, 13" deep, 74 lbs	B9@2.50	Ea	115.00	51.40	166.40
16" high, 72" wide, 13" deep, 87 lbs	B9@2.50	Ea	138.00	51.40	189.40

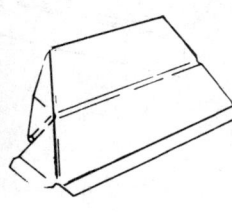

Dampers for multiple opening or corner opening firebox, high capacity, with steel downdraft shelf support

	Craft@Hrs	Unit	Material	Labor	Total
35" wide	B9@1.33	Ea	134.00	27.30	161.30
41" wide	B9@1.33	Ea	153.00	27.30	180.30
47" wide	B9@1.75	Ea	196.00	36.00	232.00

Fuel grates

	Craft@Hrs	Unit	Material	Labor	Total
21" x 16"	--	Ea	32.00	--	--
29" x 16"	--	Ea	41.00	--	--
40" x 30"	--	Ea	62.00	--	--
Gas valve, log lighter and key	--	LS	49.00	--	--

Fireplace Forms. Heavy gauge steel heat circulating forms for masonry fireplaces. Finished opening dimensions. Superior Heatform

Single front opening

	Craft@Hrs	Unit	Material	Labor	Total
31" x 25", 4,300 CF capacity	B9@3.50	Ea	355.00	71.90	426.90

Fireplaces

	Craft@Hrs	Unit	Material	Labor	Total
Fireplace Forms, continued					
34" x 25", 5,000 CF capacity	B9@3.50	Ea	380.00	71.90	451.90
37" x 29", 5,750 CF capacity	B9@3.50	Ea	430.00	71.90	501.90
42" x 29", 6,500 CF capacity	B9@3.50	Ea	565.00	71.90	636.90
48" x 31", 7,500 CF capacity	B9@4.00	Ea	715.00	82.20	797.20
60" x 31", 8,500 CF capacity	B9@4.00	Ea	1,090.00	82.20	1,172.20
72" x 31", 9,500 CF capacity	B9@5.50	Ea	1,480.00	113.00	1,593.00
Front and end opening (either right or left)					
38" wide, 27" high	B9@3.50	Ea	815.00	71.90	886.90
Open on both faces (see-through), including grate, 23" high					
33" wide	B9@3.50	Ea	810.00	71.90	881.90
43" wide	B9@4.00	Ea	930.00	82.20	1,012.20
Add for 2 air inlet grilles, 8" x 11"	--	LS	37.00	--	--
Add for 1 air outlet grille, 6" x 32"	--	LS	33.00	--	--
Add for 2 fan-forced air inlets with grilles and motor	--	LS	93.00	--	--

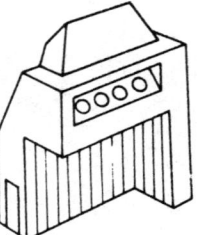

Fireplaces, Prefabricated. Zero clearance factory built fireplaces, including metal fireplace body, refractory interior and fuel grate. UL listed. Add for flues, doors and blower below. Listed by screen size.

	Craft@Hrs	Unit	Material	Labor	Total
Radiant heating, open front only, with fuel grate					
33" wide	B9@5.00	Ea	310.00	103.00	413.00
38" wide	B9@5.00	Ea	390.00	103.00	493.00
43" wide	B9@5.00	Ea	500.00	103.00	603.00
Radiant heating, open front and one end,					
38" wide	B9@5.00	Ea	630.00	103.00	733.00
Radiant heating, three sides open, with brass and glass doors and refractory floor					
36" wide (long side)	B9@6.00	Ea	1,345.00	123.00	1,468.00
Radiant heating traditional masonry look					
45" wide	B9@5.00	Ea	930.00	103.00	1,033.00
Convection-type heat circulating, open front only					
33" screen size	B9@5.50	Ea	355.00	113.00	468.00
38" screen size	B9@5.50	Ea	445.00	113.00	558.00
43" screen size	B9@5.50	Ea	560.00	113.00	673.00
Forced air heat circulating, open front only (add for blower below)					
33" screen size	B9@5.50	Ea	365.00	113.00	478.00
39" screen size	B9@5.50	Ea	465.00	113.00	578.00
43" screen size	B9@5.50	Ea	570.00	113.00	683.00
Forced air deluxe heat circulating, open front only (add for two blowers below)					
38" screen size	B9@5.50	Ea	610.00	113.00	723.00
43" screen size	B9@5.50	Ea	730.00	113.00	843.00
Heat circulating, see-through with brass and glass doors, designer model					
38" screen size	B9@6.50	Ea	1,430.00	134.00	1,564.00
Header plate, connects venting system to gas fireplace	B9@.167	Ea	43.00	3.43	46.43
Fireplace doors, complete set including frame and hardware, antique brass					
33", for radiant heat units	B9@1.25	Ea	124.00	25.70	149.70
38", for radiant heat units	B9@1.25	Ea	155.00	25.70	180.70
43", for radiant heat units	B9@1.25	Ea	180.00	25.70	205.70
33", for convection units	B9@1.25	Ea	124.00	25.70	149.70
38", for convection units	B9@1.25	Ea	155.00	25.70	180.70
43", for convection units	B9@1.25	Ea	180.00	25.70	205.70
33", for forced air units	B9@1.25	Ea	124.00	25.70	149.70
38", for forced air units	B9@1.25	Ea	155.00	25.70	180.70

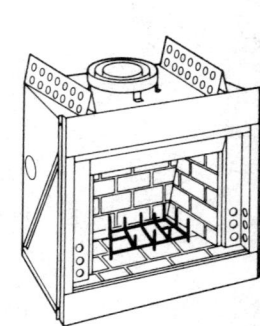
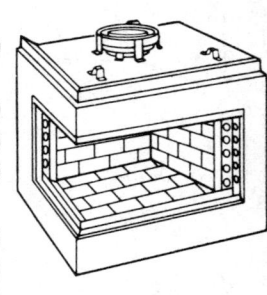
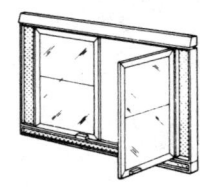

Fireplaces

	Craft@Hrs	Unit	Material	Labor	Total
43", for forced air units	B9@1.25	Ea	180.00	25.70	205.70
For corner open model	B9@1.25	Ea	205.00	25.70	230.70
Blower for forced air fireplaces, requires electrical receptacle outlet (no electrical wiring included)	BE@.500	Ea	77.00	12.60	89.60
Fireplace flues (chimney), 8" inside diameters, double wall, typical cost for straight vertical installation	B9@.167	LF	19.50	3.43	22.93
Add for 10" inside diameter	--	%	33.0	--	--
Add for stabilizer, one required per chimney	B9@.250	Ea	26.00	5.14	31.14
Add for flue spacers, for passing through ceiling, 1" or 2" clearance	B9@.167	Ea	9.00	3.43	12.43
Add for typical flue offset	B9@.250	Ea	41.30	5.14	46.44
Add for flashing and storm collar, one required per roof penetration:					
Flat to 6/12 pitch roofs	B9@3.25	Ea	36.00	66.80	102.80
6/12 to 12/12 pitch roofs	B9@3.25	Ea	57.00	66.80	123.80
Add for spark arrestor top	B9@.500	Ea	55.00	10.30	65.30
Wood burning stove/fireplace, forced air type, with cast iron doors	B1@6.00	Ea	1,550.00	127.00	1,677.00
Natural gas fireplace, 16,000 Btu (non-wood burning), 36" wide, no gas piping included	B1@6.00	Ea	465.00	127.00	592.00
Trim and hood assembly, for use with gas burning fireplaces	B1@.500	Ea	65.00	10.60	75.60

Flooring, Resilient. Includes typical waste and adhesive (1 gallon at $16 covers 250 SF or 27.7 SY).

	Craft@Hrs	Unit	Material	Labor	Total
Asphalt tile, 9" x 9" x 1/8" thick, marbleized					
B grade colors, dark	BF@.021	SF	1.10	.47	1.57
C grade colors, medium	BF@.021	SF	1.25	.47	1.72
D grade colors, light	BF@.021	SF	1.25	.47	1.72
Sheet vinyl, no wax, Armstrong					
Solarian Supreme (.095" gauge)	BF@.286	SY	38.00	6.42	44.42
Designer Solarian II (.090" gauge)	BF@.286	SY	36.00	6.42	42.42
Designer Solarian (.070" gauge)	BF@.286	SY	28.00	6.42	34.42
Imperial Accotone (.065" gauge)	BF@.243	SY	9.00	5.46	14.46
Sundial Solarian (.077" gauge)	BF@.243	SY	14.00	5.46	19.46
Sheet vinyl, no wax, Mannington					
Granite Run	BF@.247	SY	35.00	5.55	40.55
Country Quilt	BF@.247	SY	33.00	5.55	38.55
Sonesta	BF@.247	SY	28.00	5.55	33.55
St. Moritz	BF@.247	SY	24.00	5.55	29.55
Sheet vinyl, Corlon, Armstrong					
Heatherdale (.065" gauge)	BF@.247	SY	25.00	5.55	30.55
Sheet vinyl, no wax, Armstrong					
Timespan (.080" gauge)	BF@.247	SY	12.00	5.55	17.55
Cambray (.065" gauge)	BF@.238	SY	8.00	5.34	13.34
Royelle (.050" gauge)	BF@.238	SY	6.50	5.34	11.84
Sheet vinyl					
Vega (.088" gauge), Mannington	BF@.247	SY	8.00	5.55	13.55
Vinyl Ease (.073" gauge), Mannington	BF@.200	SY	7.00	4.49	11.49
Preference (.065" gauge), Tarkett	BF@.238	SY	6.50	5.34	11.84
Softred (.062" gauge), Tarkett	BF@.238	SY	8.00	5.34	13.34
Vinyl tile, self-adhesive, Armstrong Solarian no wax					
Century Solarian (3/32")	BF@.023	SF	2.00	.52	2.52

Flooring

	Craft@Hrs	Unit	Material	Labor	Total
Flooring, Resilient, continued					
Premier Solarian (.080")	BF@.023	SF	1.15	.52	1.67
Excelon (1/8")	BF@.023	SF	.80	.52	1.32
Excelon (3/32")	BF@.023	SF	.70	.52	1.22
Stylistik (.050")	BF@.018	SF	.95	.40	1.35
Vernay (.045")	BF@.015	SF	.50	.34	.84
Vinyl tile, including adhesive, National Floor Products					
Vega II, 1/16" x 12" x 12"	BF@.018	SF	1.00	.40	1.40
Designer slate, .080" x 12" x 12"	BF@.018	SF	2.15	.40	2.55
Majestic slate, 1/8" x 12" x 12"	BF@.018	SF	2.15	.40	2.55
Terrazzo, 1/8" x 12" x 12"	BF@.018	SF	2.05	.40	2.45
Mediterranean marble, 1/8" x 12" x 12"	BF@.018	SF	2.35	.40	2.75
Embassy oak, 1/8" x 12" x 12"	BF@.018	SF	2.65	.40	3.05
Pecan, 1/8" x 12" x 12"	BF@.018	SF	2.75	.40	3.15
Plymouth plank, 1/8" x 12" x 12"	BF@.018	SF	3.00	.40	3.40
Teak, 1/8" x 12" x 12"	BF@.018	SF	2.55	.40	2.95
Plaza brick, 1/8" x 9" x 9"	BF@.018	SF	4.15	.40	4.55
Vinyl tile, self-stick, Tarkett					
Elite, .080" x 12" x 12"	BF@.023	SF	1.60	.52	2.12
Proclaim, 1/16" x 12" x 12"	BF@.023	SF	1.05	.52	1.57
Stylglo, .080" x 12" x 12"	BF@.023	SF	1.20	.52	1.72
Decorator 1/16" x 12" x 12"	BF@.023	SF	.80	.52	1.32
Composition vinyl tile, 12" x 12", Armstrong					
3/32" standard colors (marbleized)	BF@.023	SF	1.00	.52	1.52
3/32" designer colors	BF@.023	SF	.90	.52	1.42
1/8" standard colors (marbleized)	BF@.023	SF	1.15	.52	1.67
1/8" designer colors, pebbled	BF@.023	SF	1.70	.52	2.22
1/8" solid black or white	BF@.015	SF	1.60	.34	1.94
1/8" solid colors	BF@.015	SF	1.75	.34	2.09
Accessories					
Adhesive (250 SF per gallon)	--	Gal	16.00	--	--
Latex floor adhesive (150 SF per gallon)	--	Gal	16.00	--	--
Vinyl seam sealer (seals 50 LF)	--	Pint	8.00	--	--
Cove base, colors, Flexco Rubber					
2-1/2" high	BF@.025	LF	.50	.56	1.06
4"	BF@.025	LF	.60	.56	1.16
6"	BF@.025	LF	.84	.56	1.40
Cove base cement					
(200 LF of 4" base per gallon)	--	Gal	16.00	--	--
Felt underlay	BF@.006	SF	.25	.14	.39
Accent strip, 1/8" x 24" long					
1/2" wide	BF@.025	LF	.35	.56	.91
1"	BF@.025	LF	.44	.56	1.00
2"	BF@.025	LF	.82	.56	1.38
Vinyl edging strip, 1" wide					
1/16" thick	--	LF	.45	--	--
1/8" thick	--	LF	.58	--	--

TILE WASTE ALLOWANCES

1 to 50 sq. ft.	14%
50 to 100 sq. ft.	10%
100 to 200 sq. ft.	8%
200 to 300 sq. ft.	7%
300 to 1000 sq. ft.	5%
Over 1000 sq. ft.	3%

Flooring, Wood

	3/4" x 2-1/4" 138.3 BF/ 100 SF	3/4" x 1-1/2" 155 BF/ 100 SF	1/2" x 2" 130 BF/ 100 SF	1/2 x 1-1/2 138 BF/ 100 SF

Flooring, Wood Plank and Strip. Includes typical waste and coverage loss when installed over a prepared subfloor but no sanding or finishing.

Unfinished oak strip flooring, tongue and groove edge. These are material costs per square foot for a 1,000 SF job. (See labor costs below.)

	3/4" x 2-1/4"	3/4" x 1-1/2"	1/2" x 2"	1/2 x 1-1/2
Clear plain white or red	3.10	3.10	3.00	3.00
Select plain white or red	2.00	1.90	1.90	1.90
Number 1 common white	1.75	1.75	1.75	1.75
Number 1 common red	1.75	1.75	1.75	1.75
Number 2 common red & white	1.50	1.50	1.50	1.50
Number 1 common & better, short	1.50	1.50	1.50	1.50

Unfinished antique tongue and groove plank flooring, 25/32" thick, random widths and lengths. Vintage Lumber

	Craft@Hrs	Unit	Material	Labor	Total
Chestnut, American, 3" to 7"	BF@.090	SF	6.39	2.02	8.41
Chestnut, distressed, 3" to 7"	BF@.090	SF	7.28	2.02	9.30
Chestnut, distressed, 6" and up	BF@.090	SF	10.11	2.02	12.13
Hemlock, 3" to 7"	BF@.090	SF	4.48	2.02	6.50
Hemlock, 6" and up	BF@.090	SF	7.28	2.02	9.30
Oak, 3" to 7"	BF@.090	SF	8.00	2.02	10.02
Oak, 6" and up	BF@.090	SF	10.11	2.02	12.13
Oak, distressed, 3" to 7"	BF@.090	SF	6.23	2.02	8.25
Oak, distressed, 6" and up	BF@.090	SF	8.94	2.02	10.96
Poplar, 3" to 8"	BF@.090	SF	3.68	2.02	5.70
Poplar, 8" and up	BF@.090	SF	4.92	2.02	6.94
White pine, 3" to 7"	BF@.090	SF	4.90	2.02	6.92
White pine, distressed, 3" to 7"	BF@.090	SF	5.60	2.02	7.62
White pine, distressed, 6" and up	BF@.090	SF	8.26	2.02	10.28
Yellow pine, 3" to 5"	BF@.090	SF	4.80	2.02	6.82
Yellow pine, 6" to 10"	BF@.090	SF	6.40	2.02	8.42

Unfinished tongue and groove plank flooring, 25/32" thick, random widths and random 3' to 16' lengths. Vintage Lumber

	Craft@Hrs	Unit	Material	Labor	Total
Ash, select, 3" to 7"	BF@.090	SF	3.44	2.02	5.46
Cherry, #1, 3" to 7"	BF@.090	SF	3.75	2.02	5.77
Cherry, #1, 6" and up	BF@.090	SF	5.00	2.02	7.02
Hickory, #1, 3" to 7"	BF@.090	SF	3.00	2.02	5.02
Oak, #1, red or white, 3" to 7"	BF@.090	SF	3.44	2.02	5.46
Oak, #1, red or white, 6" and up	BF@.090	SF	5.00	2.02	7.02
Maple, FAS, 3" to 7"	BF@.090	SF	3.13	2.02	5.15
Walnut, #3, 3" to 7"	BF@.090	SF	3.44	2.02	5.46
White pine, knotty grade, 12" and up	BF@.090	SF	2.69	2.02	4.71

Prefinished oak strip flooring, tongue and groove, 25/32" thick

	Craft@Hrs	Unit	Material	Labor	Total
2-1/4" wide, standard & better, red and white oak	BF@.036	SF	6.00	.81	6.81
Random widths 3", 5" and 7", walnut plugs, clear and select grade	BF@.042	SF	7.20	.94	8.14
Random plank 2-1/4" to 3-1/4" wide, dark number 1 to C and better	BF@.032	SF	6.10	.72	6.82
Adhesive for oak plank	--	SF	.75	--	--

Unfinished northern hard maple strip flooring 25/32" x 2-1/4" (138.3 BF per 100 SF)

	Craft@Hrs	Unit	Material	Labor	Total
1st grade	BF@.036	SF	3.90	.81	4.71
2nd grade	BF@.036	SF	3.55	.81	4.36
3rd grade	BF@.036	SF	3.00	.81	3.81

Flooring, Wood

	Craft@Hrs	Unit	Material	Labor	Total

Unfinished northern hard maple strip flooring, continued
 25/32" x 1-1/2" (155 BF per 100 SF)

	Craft@Hrs	Unit	Material	Labor	Total
1st grade	BF@.040	SF	4.50	.90	5.40
2nd grade	BF@.040	SF	4.20	.90	5.10
3rd grade	BF@.040	SF	3.60	.90	4.50

33/32" x 1-1/2" (170 BF per 100 SF)

	Craft@Hrs	Unit	Material	Labor	Total
1st grade	BF@.036	SF	5.00	.81	5.81
2nd grade	BF@.036	SF	4.80	.81	5.61
3rd grade	BF@.036	SF	4.10	.81	4.91

Edge-wired "Worthwood" strip block flooring by Oregon Lumber Company. End grain unfinished kiln-dried fir, hemlock or alder in flexible strips. Costs include minimal cutting and fitting, installation in mastic but no subfloor preparation
 4-5/8" wide strips 12" to 33" long

	Craft@Hrs	Unit	Material	Labor	Total
1" thick	BF@.055	SF	3.00	1.24	4.24
1-1/2" thick	BF@.055	SF	3.60	1.24	4.84
2" thick	BF@.055	SF	3.85	1.24	5.09
2-1/2" thick	BF@.055	SF	4.30	1.24	5.54
Add for extensive cutting and fitting	--	%	--	100.0	--

Unfinished long plank flooring, random 3" to 8" widths, 4' to 16' lengths, 3/4" thick, including 8% waste, Specialty Wood Floors

	Craft@Hrs	Unit	Material	Labor	Total
Select grade, white oak	BF@.062	SF	5.00	1.39	6.39
Select grade, red oak	BF@.062	SF	4.80	1.39	6.19
Select grade, rock maple	BF@.062	SF	4.05	1.39	5.44
Select grade, figured maple	BF@.062	SF	5.00	1.39	6.39
Select grade, ash	BF@.062	SF	5.00	1.39	6.39
Select grade, cherry	BF@.062	SF	5.40	1.39	6.79
Common grade, cherry	BF@.062	SF	4.35	1.39	5.74

Unfinished wide plank flooring, random 7" to 14" widths, 4' to 16' lengths, 3/4" thick, including 8% waste, Specialty Wood Floors

	Craft@Hrs	Unit	Material	Labor	Total
Select grade, red oak	BF@.074	SF	5.10	1.66	6.76
Select grade, white oak	BF@.074	SF	5.60	1.66	7.26
Select grade, cherry	BF@.074	SF	5.90	1.66	7.56
Select grade, maple	BF@.074	SF	4.60	1.66	6.26
Knotty grade, pine	BF@.074	SF	3.00	1.66	4.66
Adhesive for 3/4" plank	--	SF	.45	--	--

Labor installing hardwood strip flooring, nailed, including felt

	Craft@Hrs	Unit	Material	Labor	Total
1-1/2" width	BF@.066	SF	--	1.48	--
2"	BF@.062	SF	--	1.39	--
2-1/4"	BF@.062	SF	--	1.39	--
Sanding, 3 passes (60/80/100 grit)	BF@.023	SF	.05	.52	.57
2 coats of stain/sealer	BF@.010	SF	.09	.23	.32
2 coats of urethane	BF@.012	SF	.11	.27	.38
Sand, fill and 2 coats of lacquer	BF@.047	SF	.20	1.06	1.26

Flooring, Wood Parquet Blocks. Includes typical waste and coverage loss when installed over a prepared subfloor but no sanding or finishing.
Unfinished parquet block flooring, 3/4" thick, 10" x 10" to 39" x 39" blocks, includes 5% waste. Cost varies by pattern with larger blocks costing more. These are material costs only. Add labor cost below. Kentucky Wood Floors

	Craft@Hrs	Unit	Material	Labor	Total
Ash or quartered oak, select	--	SF	14.60	--	--
Ash or quartered oak, natural	--	SF	10.60	--	--
Cherry or walnut select grade	--	SF	15.80	--	--
Cherry or walnut natural grade	--	SF	12.60	--	--
Oak, select plain	--	SF	11.40	--	--

Flooring, Wood

	Craft@Hrs	Unit	Material	Labor	Total
Oak, natural plain	--	SF	10.10	--	--

Prefinished parquet block flooring, 5/16" thick, select grade, specified sizes, includes 5% waste. Cost varies by pattern and size, with larger blocks costing more. These are material costs only. Add labor cost below. Kentucky Wood Floors.

	Craft@Hrs	Unit	Material	Labor	Total
Cherry or walnut					
9" x 9"	--	SF	4.20	--	--
13" x 13"	--	SF	8.00	--	--
Red oak					
9" x 9"	--	SF	3.00	--	--
13" x 13"	--	SF	6.60	--	--
Teak, 12" x 12"	--	SF	3.20	--	--

Unfinished parquet block flooring, select grade, 2-1/4" x 6-3/4" to 19" x 19" blocks, 5/16" to 3/4" thick, includes 5% waste. Cost varies by pattern and size, with larger blocks costing more. These are material costs only. Add labor cost below. Kentucky Wood Floors

	Craft@Hrs	Unit	Material	Labor	Total
Ash, walnut, or cherry	--	SF	10.30	--	--
Plain oak	--	SF	5.30	--	--
Quartered oak	--	SF	6.10	--	--
Red oak	--	SF	2.80	--	--
Add for beveled edges	--	SF	.60	--	--

Prefinished crossband 3-ply birch parquet flooring. Costs include cutting, fitting and 5% waste, but no subfloor preparation. 6" to 12" wide by 12" to 48" blocks, Oregon Lumber, "Saima", 9/16" thick

	Craft@Hrs	Unit	Material	Labor	Total
	--	SF	3.45	--	--
Bonded blocks, 12" x 12", 1/8" thick, laid in mastic					
Vinyl bonded cork	BF@.050	SF	3.60	1.12	4.72
Vinyl bonded domestic woods	BF@.062	SF	7.10	1.39	8.49
Vinyl bonded exotic woods.	BF@.062	SF	9.75	1.39	11.14
Acrylic impregnated wood, includes adhesive, 5/16" thick,					
12" x 12" block laid in mastic					
Oak	BF@.057	SF	6.09	1.28	7.37
Ash	BF@.057	SF	6.87	1.28	8.15
Cherry	BF@.057	SF	7.30	1.28	8.58
2-3/4" x RL plank laid in mastic					
Oak	BF@.057	SF	6.51	1.28	7.79
Maple	BF@.057	SF	7.12	1.28	8.40
Tupelo	BF@.057	SF	6.64	1.28	7.92
Lindenwood	BF@.057	SF	7.29	1.28	8.57
Labor and material installing parquet block flooring					
Most parquet laid in mastic	BF@.055	SF	.26	1.24	1.50
Edge-glued over high density foam	BF@.052	SF	.43	1.17	1.60
Labor to install custom and irregular-shaped					
tiles, including adhesive	BF@.074	SF	.45	1.66	2.11
Add for installations under 1,000 SF	--	%	--	40.0	--

Flooring, Wood, Softwood. Including typical waste and coverage loss when laid at right angles but no sanding or finishing.

Douglas fir, "C" and better, dry, vertical grain, including 5% waste

	Craft@Hrs	Unit	Material	Labor	Total
1" x 3"	BF@.037	SF	2.70	.83	3.53
1" x 4"	BF@.037	SF	2.40	.83	3.23
1-1/4" x 4"	BF@.037	SF	2.85	.83	3.68
Deduct for "D" grade	--	%	-35.0	--	--
Add for "B" grade and better	--	%	25.0	--	--
Add for specific lengths	--	%	3.0	--	--
Deduct for flat grain	--	%	-20.0	--	--
Add for diagonal patterns	--	%	--	10.0	--

Flooring, Wood

	Craft@Hrs	Unit	Material	Labor	Total

Flooring, Wood, Softwood, continued
Select kiln dried heart pine, 97% dense heartwood, random lengths, including 5% waste, 1" to 1-1/4" thick

	Craft@Hrs	Unit	Material	Labor	Total
4" wide	BF@.042	SF	5.25	.94	6.19
5" wide	BF@.042	SF	5.40	.94	6.34
6" wide	BF@.042	SF	5.60	.94	6.54
8" wide	BF@.046	SF	6.30	1.03	7.33
10" wide	BF@.046	SF	6.95	1.03	7.98

Unfinished plank flooring, 3/4" thick, 3" to 8" widths, 2' to 12' lengths, Aged Woods

	Craft@Hrs	Unit	Material	Labor	Total
Antique white pine	BF@.100	SF	6.70	2.25	8.95
Antique yellow pine	BF@.100	SF	5.70	2.25	7.95
Milled heart pine	BF@.100	SF	6.80	2.25	9.05
Distressed oak	BF@.100	SF	9.00	2.25	11.25
Milled oak	BF@.100	SF	7.70	2.25	9.95
Distressed American chestnut	BF@.100	SF	10.50	2.25	12.75
Milled American chestnut	BF@.100	SF	9.00	2.25	11.25
Aged hemlock/fir	BF@.100	SF	5.70	2.25	7.95
Antique poplar	BF@.100	SF	5.80	2.25	8.05
Aged cypress	BF@.100	SF	3.50	2.25	5.75
Add for 9" to 15" widths	--	%	15.0	--	--

Flooring, Subcontract. These costs include the subcontractor's overhead and profit.
Hardwood flooring
   Strip flooring, T&G, unfinished, 2-1/4" wide, 25/32" thick oak

	Craft@Hrs	Unit	Material	Labor	Total
Clear quartered	--	SF	--	--	6.90
Select quartered	--	SF	--	--	7.00
No. 1 common	--	SF	--	--	6.60
No. 2 common	--	SF	--	--	6.10
Deduct for 1/2" thick	--	%	--	--	-20.0
Deduct for tract work	--	%	--	--	-10.0

   Parquet blocks, set in mastic, prefinished

	Craft@Hrs	Unit	Material	Labor	Total
Oak, 3/16", select	--	SF	--	--	7.70
Oak, 5/16", select	--	SF	--	--	8.65
Oak, 3/8", select	--	SF	--	--	9.00
Oak, 25/32", select	--	SF	--	--	9.75
Maple, 5/16", select	--	SF	--	--	8.55
Walnut, 5/16"	--	SF	--	--	10.95
Teak, 5/16", typical	--	SF	--	--	16.30

   Ranch plank flooring, red oak, 3", 5," 7" widths,
   walnut pegs, finished — SF — — 10.50

   Additions
      For sanding and finishing add

	Craft@Hrs	Unit	Material	Labor	Total
Light finishes	--	SF	--	--	1.90
Dark finishes	--	SF	--	--	2.45
Custom fancy, bleached	--	SF	--	--	3.25

Resilient flooring
   Asphalt tile, 1/8", laid with underfelt, standard or wood designs

	Craft@Hrs	Unit	Material	Labor	Total
B grade (dark colors)	--	SF	--	--	1.85
C grade (medium colors)	--	SF	--	--	2.05
D grade (light colors)	--	SF	--	--	2.05
Cork tile, standard, 3/16"	--	SF	--	--	2.90
Cork tile, 5/16"	--	SF	--	--	3.95
Rubber tile, 1/8"	--	SF	--	--	3.45

Flooring, Resilient

	Craft@Hrs	Unit	Material	Labor	Total
Vinyl composition tile					
1/16" standard	--	SF	--	--	1.37
3/32" standard	--	SF	--	--	1.49
1/8" standard	--	SF	--	--	1.68
1/16" metallic accent	--	SF	--	--	1.32
Vinyl tile, solid vinyl, Azrock					
1/16" standard	--	SF	--	--	3.47
1/8" standard	--	SF	--	--	4.52
Vinyl sheet flooring, including rubber cove base					
.065", Corlon or equal	--	SF	--	--	2.65
.065", metallic accent	--	SF	--	--	2.85
.040", standard	--	SF	--	--	2.35
.090", top quality custom	--	SF	--	--	5.40
Base for resilient flooring					
Rubber, 6"	--	LF	--	--	1.72
Rubber, 4"	--	LF	--	--	1.48
Rubber, 2-1/2"	--	LF	--	--	1.37
Stair treads, 12" width, Flexco					
Rubber, molded, 3/4"	--	LF	--	--	10.55
Rubber, molded, 1/8"	--	LF	--	--	8.45
Rubber, grit safety tread, 1/4"	--	LF	--	--	12.35
Rubber, grit safety tread, 1/8"	--	LF	--	--	10.10
Vinyl, molded, 1/8"	--	LF	--	--	5.55
Vinyl, molded, 1/4"	--	LF	--	--	6.75
Stair risers, 7" high, 1/8" thick					
Rubber	--	LF	--	--	3.50
Vinyl	--	LF	--	--	2.75
Slate, random rectangles, typical	--	SF	--	--	18.50
Particleboard underlayment, 3/8"	--	SF	--	--	.68

Foundation Bolts. With nuts attached, full bag quantities.

	Craft@Hrs	Unit	Material	Labor	Total
1/2" x 6", .34 lb, bag of 300	B1@.107	Ea	.20	2.27	2.47
1/2" x 8", .44 lb, bag of 250	B1@.107	Ea	.23	2.27	2.50
1/2" x 10", .51 lb, bag of 200	B1@.107	Ea	.27	2.27	2.54
5/8" x 6", .62 lb, bag of 50	B1@.107	Ea	.56	2.27	2.83
5/8" x 8", .78 lb, bag of 50	B1@.107	Ea	.66	2.27	2.93
5/8" x 10", .88 lb, bag of 50	B1@.107	Ea	.75	2.27	3.02
5/8" x 12", 1.1 lb, bag of 50	B1@.121	Ea	.80	2.56	3.36
3/4" x 8", 1.2 lb, bag of 50	B1@.107	Ea	.85	2.27	3.12
3/4" x 10", 1.4 lb, bag of 50	B1@.107	Ea	1.00	2.27	3.27
3/4" x 12", 1.7 lb, bag of 50	B1@.121	Ea	1.12	2.56	3.68
Add for wrought, cut washers, plain					
1/2"	--	Ea	.09	--	--
5/8"	--	Ea	.14	--	--
3/4"	--	Ea	.20	--	--

Garage Door Openers. Radio controlled, electric, single or double door, includes typical electric hookup to adjacent 110 volt outlet.

	Craft@Hrs	Unit	Material	Labor	Total
Screw worm drive, 1/3 HP opener, safety stop device, receiver and one transmitter, multiple frequency, lighted	BC@3.90	LS	125.00	92.00	217.00
Deluxe model, includes 1/2 HP opener, receiver, transmitter, automatic light, built-in time delay, safety stop device					
Chain drive, not for vault-type garages	BC@4.04	Ea	180.00	95.30	275.30
Add for additional transmitter or key switch	--	Ea	31.50	--	--

Garage Doors

Garage Doors. Unfinished, with hardware, see labor costs below.

Type		Single 8' x 7'	9' x 7'	Double 16' x 7'
Aluminum framed fiberglass, sectional, white, with key lock	Ea	360.00	380.00	605.00
Steel garage doors				
One piece, primed	Ea	155.00	165.00	--
Sectional, continuous hinges	Ea	280.00	295.00	510.00
Sectional, with glass panels	Ea	305.00	310.00	530.00
Wood garage doors, sectional, without insulation				
Sectional door	Ea	240.00	260.00	445.00
Door with two small lites	Ea	260.00	300.00	520.00
Door with plain panels	Ea	275.00	290.00	575.00
Door with raised redwood panels	Ea	590.00	650.00	1,210.00
Door with detailed face	Ea	370.00	400.00	770.00
Honeycomb core wood door	Ea	270.00	295.00	--
Styrofoam core hardboard door	Ea	295.00	305.00	510.00
Styrofoam core textured wood door	Ea	295.00	310.00	525.00
Wood doors with from 4 to 12 better-quality glass panel inserts				
Cathedral style	Ea	295.00	310.00	650.00
Sunburst style	Ea	325.00	325.00	695.00
Add for low headroom hardware	Ea	27.50	27.50	35.50

Labor installing garage doors	Craft@Hrs	Unit	Material	Labor	Total
8' or 9' wide doors	B1@4.42	Ea	--	93.70	--
16' wide doors	B1@5.90	Ea	--	125.00	--

Garage Doors, Subcontract. Complete installed costs including hardware.

		Unit	Material	Labor	Total
Aluminum (.019 gauge) on wood frame, jamb hinge					
8' x 7'		--	Ea	--	--	360.00
16' x 7'		--	Ea	--	--	500.00
Plywood (3/8") on wood frame, flush face, jamb hinge					
8' x 7'		--	Ea	--	--	260.00
16' x 7'		--	Ea	--	--	360.00
Add for moulding on face		--	LF	--	--	1.40
Board on wood frame, jamb hinge					
Plain face, 8' x 7'		--	Ea	--	--	405.00
Plain face, 16' x 7'		--	Ea	--	--	725.00
Pattern face, 8' x 7'		--	Ea	--	--	415.00
Pattern face, 16' x 7'		--	Ea	--	--	745.00
Sectional upward acting garage doors					
Aluminum, 8' x 7'		--	Ea	--	--	505.00
Aluminum, 16' x 7'		--	Ea	--	--	980.00
Steel, 8' x 7'		--	Ea	--	--	460.00
Steel, 16' x 7'		--	Ea	--	--	750.00
Wood, 8' x 7'		--	Ea	--	--	415.00
Wood, 16' x 7'		--	Ea	--	--	1,020.00
Add for garage door operators, radio controlled					
Single door, 1 transmitter		--	LS	--	--	315.00
Double door, 2 transmitters		--	LS	--	--	395.00

Glass. Material only. See labor costs below

SSB glass (single strength, "B" grade)		Unit	Material	Labor	Total
60" width plus length	--	SF	1.85	--	--
70" width plus length	--	SF	1.90	--	--

Glass

	Craft@Hrs	Unit	Material	Labor	Total
80" width plus length	--	SF	2.10	--	--
100" width plus length	--	SF	2.15	--	--
DSB glass (double strength, "B" grade)					
60" width plus length	--	SF	2.45	--	--
70" width plus length	--	SF	2.50	--	--
80" width plus length	--	SF	2.60	--	--
100" width plus length	--	SF	2.70	--	--
Non-glare glass	--	SF	4.10	--	--
Float glass, 1/4" thick					
To 25 SF	--	SF	3.00	--	--
25 to 35 SF	--	SF	3.25	--	--
35 to 50 SF	--	SF	3.50	--	--
Insulating glass, clear					
1/2" thick, 1/4" airspace, up to 25 SF	--	SF	6.00	--	--
5/8" thick, 1/4" airspace, up to 35 SF	--	SF	7.10	--	--
1" thick, 1/2" airspace, up to 50 SF	--	SF	7.80	--	--
Tempered glass, clear, 1/4" thick					
To 25 SF	--	SF	6.15	--	--
25 to 50 SF	--	SF	7.40	--	--
50 to 75 SF	--	SF	8.00	--	--
Tempered glass, clear, 3/16" thick					
To 25 SF	--	SF	5.35	--	--
25 to 50 SF	--	SF	6.20	--	--
50 to 75 SF	--	SF	6.90	--	--
Laminated safety glass, 1/4" thick, ASI safety rating					
Up to 25 SF	--	SF	6.25	--	--
25 to 35 SF	--	SF	6.70	--	--
Labor setting glass in wood or metal sash with vinyl bed					
To 15 SF, 1 hour charge	BG@1.77	LS	--	38.90	--
15 to 25 SF, 2 hour charge	BG@3.53	LS	--	77.60	--
25 to 35 SF, 3 hour charge	BG@5.30	LS	--	116.00	--
Beveled edge mirrors, unframed	--	SF	3.20	--	--
Labor to bevel edges, per lineal inch					
1/2" beveled edge	BG@.013	Inch	--	.29	--
1" beveled edge	BG@.063	Inch	--	1.38	--

Glazing, Subcontract

Hack out broken lite, wood sash (minimum cost is $35 to hack out and reglaze)

	Craft@Hrs	Unit	Material	Labor	Total
To 40" width plus length	--	Ea	--	--	12.50
41" to 50" width plus length	--	Ea	--	--	13.50
51" to 60" width plus length	--	Ea	--	--	15.50
61" to 70" width plus length	--	Ea	--	--	17.30
71" to 80" width plus length	--	Ea	--	--	23.90
81" to 90" width plus length	--	Ea	--	--	37.70
Add for metal sash with putty or wraparound rubber bed	--	%	--	--	12.0
Reglaze at job site, wood sash, no glass included					
To 40" width plus length	--	Ea	--	--	12.50
41" to 50"	--	Ea	--	--	13.50
51" to 60"	--	Ea	--	--	15.50
61" to 70"	--	Ea	--	--	17.30
71" to 80"	--	Ea	--	--	23.90
81" to 90"	--	Ea	--	--	37.70

Glazing

	Craft@Hrs	Unit	Material	Labor	Total
Reglaze at job site, wood sash, no glass included, continued					
Add for metal sash with putty or rubber wraparound bed,	--	%	--	--	20.0
Add for metal sash, angles and putty	--	%	--	--	20.0
Add for monitoring or work above first floor	--	%	--	--	50.0
Additional cost items					
Edge polishing	--	Inch	--	--	.25
Miter storefront corner	--	LF	--	--	11.70
Showcase cementing	--	LF	--	--	8.00
Gutters and Downspouts					
Galvanized steel, "K" style box (10' lengths)					
Gutter, 4", 28 gauge	SW@.065	LF	.87	1.64	2.51
Gutter, 4", 26 gauge	SW@.065	LF	1.48	1.64	3.12
Gutter, 5", 28 gauge	SW@.070	LF	1.59	1.76	3.35
Gutter, 5", 26 gauge	SW@.070	LF	2.00	1.76	3.76
Add for fittings (connectors, hangers, etc)	--	LF	1.03	--	--
Add for inside or outside corner, 4"	SW@.143	Ea	5.40	3.60	9.00
Add for inside or outside corner, 5"	SW@.143	Ea	12.30	3.60	15.90
Downspouts, 2" x 3", 28 gauge, square, corrugated	SW@.054	LF	1.13	1.36	2.49
Add for drop outlet, 3 ells and strap, per downspout	--	Ea	15.00	--	--
Gutter, 4", and average number of downspouts, subcontract, per LF of gutter and downspout	--	LF	--	--	5.60
Add for galvanized steel gutter with white enamel finish	--	%	50.0	--	--
Galvanized steel, half round gutter (10' lengths), single bead					
28 gauge, 4"	SW@.065	LF	1.90	1.64	3.54
28 gauge, 5"	SW@.070	LF	2.05	1.76	3.81
28 gauge, 6"	SW@.078	LF	2.66	1.96	4.62
26 gauge, 4"	SW@.065	LF	2.25	1.64	3.89
26 gauge, 5"	SW@.070	LF	2.30	1.76	4.06
26 gauge, 6"	SW@.078	LF	3.28	1.96	5.24
Add for double bead gutter	--	%	25.0	--	--
Add for fittings (connectors, hangers, etc)	--	LF	1.03	--	--
Add for inside or outside corner	SW@.143	Ea	9.70	3.60	13.30
Downspouts, square, plain					
2" x 3", 28 gauge	SW@.054	LF	1.75	1.36	3.11
2" x 3", 26 gauge	SW@.054	LF	1.95	1.36	3.31
2" x 4", 28 gauge	SW@.054	LF	2.15	1.36	3.51
2" x 4", 26 gauge	SW@.054	LF	1.95	1.36	3.31
3" x 4", 28 gauge	SW@.054	LF	2.36	1.36	3.72
3" x 4", 26 gauge	SW@.054	LF	2.51	1.36	3.87
Downspouts, round					
2", 28 gauge	SW@.054	LF	2.20	1.36	3.56
2", 26 gauge	SW@.054	LF	2.77	1.36	4.13
3", 28 gauge	SW@.054	LF	1.38	1.36	2.74
3", 26 gauge	SW@.054	LF	1.79	1.36	3.15
4", 28 gauge	SW@.054	LF	1.69	1.36	3.05
4", 26 gauge	SW@.054	LF	2.21	1.36	3.57
Add for downspout fittings, per downspout	--	Ea	16.40	--	--

Gutters

	Craft@Hrs	Unit	Material	Labor	Total
Vinyl, white PVC, (Plastmo Vinyl), half round gutter, (10' lengths)					
Gutter, 4"	SW@.054	LF	.52	1.36	1.88
Add for fittings (connectors, hangers, etc)	--	LF	.53	--	--
Add for inside or outside corner	SW@.091	Ea	4.30	2.29	6.59
Add for bonding kit (covers 150 LF)	--	Ea	3.38	--	--
Downspouts, 3" round	SW@.022	LF	.89	.55	1.44
Add for 3 ells and 2 clamps, per downspout	--	Ea	19.10	--	--

Gypsum Drywall and Accessories

Material cost for each 1,000 square feet of board includes 5 gallons of premixed joint compound (at $14 per 5 gallons), 380 linear feet of 2" perforated joint tape (at $5 per 500' roll) and 4-1/2 pounds of drywall screws (at $2 per pound). Labor costs are for good quality hanging and taping with smooth sanded joints but no texturing. Material costs include 6% waste. Costs per square foot of wall or ceiling covered.

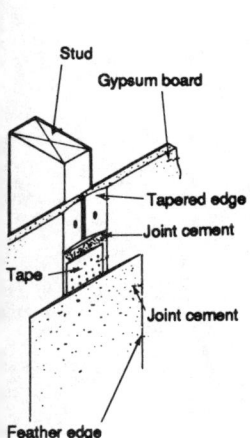

	Craft@Hrs	Unit	Material	Labor	Total
1/4" plain board at $141 per MSF					
Walls	D1@.016	SF	.17	.38	.55
Ceilings	D1@.021	SF	.17	.49	.66
3/8" plain board at $141 per MSF					
Walls	D1@.017	SF	.17	.40	.57
Ceilings	D1@.023	SF	.17	.54	.71
1/2" plain board at $141 per MSF					
Walls	D1@.018	SF	.17	.42	.59
Ceilings	D1@.024	SF	.17	.56	.73
5/8" plain board at $158 per MSF					
Walls	D1@.020	SF	.19	.47	.66
Ceilings	D1@.025	SF	.19	.59	.78
3/8" foil backed board at $187 per MSF					
Walls	D1@.017	SF	.22	.40	.62
Ceilings	D1@.023	SF	.22	.54	.76
1/2" foil backed board at $192 per MSF					
Walls	D1@.018	SF	.23	.42	.65
Ceilings	D1@.024	SF	.23	.56	.79
5/8" foil backed board at $227 per MSF					
Walls	D1@.020	SF	.26	.47	.73
Ceilings	D1@.025	SF	.26	.59	.85
1/2" moisture resistant greenboard at $227 per MSF					
Walls	D1@.018	SF	.26	.42	.68
Ceilings	D1@.024	SF	.26	.56	.82
1/2" fire rated type X board at $167 per MSF					
Walls	D1@.018	SF	.20	.42	.62
Ceilings	D1@.024	SF	.20	.56	.76
5/8" fire rated type X board at $172 per MSF					
Walls	D1@.020	SF	.21	.47	.68
Ceilings	D1@.025	SF	.21	.59	.80
1/2" fire rated type X foil backed insulating board at $231 per MSF					
Walls	D1@.018	SF	.27	.42	.69
Ceilings	D1@.024	SF	.27	.56	.83
5/8" fire rated type X foil backed insulating board at $236 per MSF					
Walls	D1@.020	SF	.28	.47	.75
Ceilings	D1@.025	SF	.28	.59	.87
3/8" plain backerboard at $197 per MSF					
Walls	D1@.017	SF	.23	.40	.63
Ceilings	D1@.023	SF	.23	.54	.77

Gypsum

	Craft@Hrs	Unit	Material	Labor	Total
1/2" plain backerboard at $202 per MSF					
Walls	D1@.018	SF	.24	.42	.66
Ceilings	D1@.024	SF	.24	.56	.80
5/8" plain backerboard at $236 per MSF					
Walls	D1@.020	SF	.27	.47	.74
Ceilings	D1@.025	SF	.27	.59	.86
Sheathing board, 1/2" thickness	D1@.010	SF	.39	.24	.63
"V" joint sheathing, 5/8" thickness	D1@.010	SF	.38	.24	.62
Drywall corner trim	D1@.030	LF	.19	.70	.89
Beadex outside corner	D1@.030	LF	.19	.70	.89
Beadex inside corner	D1@.030	LF	.19	.70	.89
Corner guard	D1@.030	LF	.19	.70	.89
Drywall edge trim					
"L" casing	D1@.016	LF	.20	.38	.58
Clip on casing	D1@.016	LF	.25	.38	.63
Beadex casing B-8	D1@.016	LF	.28	.38	.66
Beadex casing B-9	D1@.016	LF	.23	.38	.61
Perforated paper tape (figure 380 LF of tape per 1,000 SF of board)					
500 LF roll	--	Ea	5.00	--	--
250 LF roll	--	Ea	3.50	--	--
75 LF roll	--	Ea	2.50	--	--
Perfatape joint system (figure 50 lbs of finishing compound per 1,000 SF of board)					
60' tape, 4-1/2 lbs compound, knife	--	Ea	5.00	--	--
250' tape, 18 lbs compound, knife	--	Ea	14.00	--	--
5 lb bag taping compound	--	Ea	4.00	--	--
25 lb bag all-purpose compound	--	Ea	7.50	--	--
5 gal pail ready-mix cement	--	Ea	14.00	--	--
Triple "T" joint compound, 25 lb bag	--	Ea	11.00	--	--
Topping compound, 50 lb carton	--	Ea	9.00	--	--
Deduct if taping and finishing are not required					
Walls	--	SF	--	-.21	--
Ceilings	--	SF	--	-.23	--
Add for spray applied plaster finish (cottage cheese)					
Textured	DT@.226	CSF	5.00	5.18	10.18
Acoustic	DT@.272	CSF	6.00	6.23	12.23

Hardboard. See also Siding for hardboard siding.

	Craft@Hrs	Unit	Material	Labor	Total
Plastic coated decorative tileboard, Bestile					
Solid colors, 1/8" x 4' x 8' sheets	--	SF	.80	--	--
Patterns, 1/8" x 4' x 8' sheets	--	SF	.80	--	--
Adhesive, gallon covers 80 SF at $20.00 per gallon	--	SF	.25	--	--
Sealer for edges, 11 oz cartridge at $6.00 covers 130 LF	--	LF	.05	--	--
Labor applying panels with adhesive in bathrooms	B1@.043	SF	--	.91	--
Mouldings for bathroom tileboard panels, cost per LF					
Aluminum	B1@.034	LF	1.00	.72	1.72
PVC	B1@.034	LF	.75	.72	1.47
Tub kits, various colors, Bestile (includes two 5' x 5' panels, adhesive, caulking, and moulding)	--	Ea	100.00	--	--
Standard hardboard, untempered, surfaced one side					
1/8" x 4' x 8'	B1@.034	SF	.21	.72	.93
1/4" x 4' x 8'	B1@.034	SF	.32	.72	1.04

Hardboard

	Craft@Hrs	Unit	Material	Labor	Total
Tempered hardboard					
1/8" x 4' x 8'	B1@.034	SF	.28	.72	1.00
1/8" x 4' x 10'	B1@.034	SF	.38	.72	1.10
1/4" x 4' x 8'	B1@.034	SF	.41	.72	1.13
1/4" x 4' x 10'	B1@.034	SF	.62	.72	1.34
Surfaced two sides					
1/8" x 4' x 8'	B1@.034	SF	.41	.72	1.13
1/8" x 4' x 10'	B1@.034	SF	.51	.72	1.23
1/4" x 4' x 8'	B1@.034	SF	.62	.72	1.34
1/4" x 4' x 10'	B1@.034	SF	.74	.72	1.46
Pegboard, holes 1" on centers					
1/8" standard, 3/16" holes, 4' x 8'	B1@.034	SF	.42	.72	1.14
1/4" standard, 9/32" holes, 4' x 8'	B1@.034	SF	.45	.72	1.17
1/8" tempered	B1@.034	SF	.46	.72	1.18
1/4" tempered	B1@.034	SF	.58	.72	1.30
Underlayment, 4' x 8' particleboard					
3/8" thickness	B1@.011	SF	.24	.23	.47
1/2" thickness	B1@.011	SF	.25	.23	.48
5/8" thickness	B1@.011	SF	.32	.23	.55
3/4" thickness	B1@.011	SF	.37	.23	.60
Marlite plastic coated panels, 1/8" x 4' x 8'					
High gloss	B1@.043	SF	2.00	.91	2.91
Satin gloss	B1@.043	SF	2.00	.91	2.91

Hardboard, Masonite

Decorative hardboard panels for bathroom applications, adhesive applied. Add for adhesive below.

	Craft@Hrs	Unit	Material	Labor	Total
Royalclad, solid colors, 1/4" x 4' x 8' sheets	B1@.043	SF	.41	.91	1.32
Royalclad, patterns, 1/4" x 4' x 8' sheets	B1@.043	SF	.41	.91	1.32
Add for panel adhesive (@$16.80 per gal)					
(80 SF per gal)	--	SF	.21	--	--
Add for edge sealer (@$7.00 per 11 oz cartridge)					
(130 LF per cartridge)	--	LF	.05	--	--
Interior hardboard paneling, nail applied. Add for adhesive below					
Standard Presdwood, untempered, surfaced one side					
1/8" x 4' x 8' sheets	B1@.034	SF	.28	.72	1.00
1/4" x 4' x 8' sheets	B1@.034	SF	.40	.72	1.12
Royalcote panels, smooth and grooved, 1/4" x 4' x 8'					
Standard woodgrain	B1@.034	SF	.42	.72	1.14
Somerset	B1@.034	SF	.43	.72	1.15
Hanover	B1@.034	SF	.47	.72	1.19
Royalcote panels, textured, 1/4" x 4' x 8'					
Woodfield	B1@.034	SF	.49	.72	1.21
Naturel	B1@.034	SF	.49	.72	1.21
Gallon of adhesive at $17 covers 80 SF	B1@.004	SF	.20	.09	.29
Pegboard, standard					
1/8" x 4' x 8' sheets	B1@.034	SF	.32	.72	1.04
1/4" x 4' x 8' sheets	B1@.034	SF	.37	.72	1.09
Pegboard, tempered					
1/8" thick	B1@.034	SF	.35	.72	1.07
1/4" thick	B1@.034	SF	.49	.72	1.21

Hardware

	Craft@Hrs	Unit	Material	Labor	Total

Hardware. See also Bath Accessories, Door Chimes, Door Closers, Foundation Bolts, Garage Door Openers, Joist Bridging, Joist Hangers, Mailboxes, Nails, Screen Wire, Sheet Metal, Timber Connectors, and Weatherstripping.

Rule of thumb: Cost of finish hardware (including door hardware, catches, stops, racks, brackets and pulls) per house

Low to medium cost house (1600 SF)	BC@16.0	LS	325.00	377.00	702.00
Better quality house (2000 SF)	BC@20.0	LS	825.00	472.00	1,297.00

Door exterior locksets

Cabinet locks (commercial quality), keyed, 6 pin, including strike, 1/2" or 1" throw — BC@.572 Ea 48.50 13.50 62.00

Deadbolt locks, keyed cylinder outside, thumb turn inside, including strike, strike reinforcer, and strike box

Standard duty, 5 pin, 1" throw	BC@1.04	Ea	39.00	24.50	63.50
Heavy duty (commercial quality), 5 pin, 1" throw, steel reinforced	BC@1.04	Ea	52.00	24.50	76.50
Extra heavy duty, 6 pin, 1" throw	BC@1.04	Ea	86.50	24.50	111.00

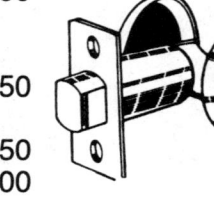

Entry locks, bored type, including trim, strike, and strike box

Heavy duty, 6 pin tumbler, 1/2" throw, knob and escutcheon trim and strike reinforcer	BC@1.04	Ea	68.00	24.50	92.50
Heavy duty, 6 pin tumbler, 1/2" throw, lever and escutcheon trim and strike reinforcer	BC@1.04	Ea	74.50	24.50	99.00
Standard, 5 pin tumbler, 1/2" throw, knob and escutcheon trim and strike reinforcer	BC@1.04	Ea	32.50	24.50	57.00
Standard, 5 pin tumbler, 1/2" throw, lever and escutcheon trim and strike reinforcer	BC@1.10	Ea	38.00	25.90	63.90
Minimum quality, 5 pin tumbler, 1/2" throw, knob and escutcheon trim	BC@1.00	Ea	21.00	23.60	44.60
Add for construction keying system (minimum order 20 keyed entry locks), per lock	--	Ea	2.85	--	--
Add for electrically operated latch release, no switches or wiring included	--	Ea	103.00	--	--

Grip handle entry lock, deluxe trim, 5 pin cylinder, includes strike, strike reinforcer, and strike box — BC@1.04 Ea 135.00 24.50 159.50

Mortise type locks, heavy duty, 6 pin tumbler, with integral deadbolt, strike, strike reinforcer, strike box and escutcheon trim

Residential, knob type	BC@1.30	Ea	145.00	30.70	175.70
Residential, lever type	BC@1.30	Ea	146.00	30.70	176.70
Apartment, lever type	BC@1.30	Ea	173.00	30.70	203.70
Retail store, lever type	BC@1.30	Ea	173.00	30.70	203.70
Add for construction keying system (minimum order of 20 keyed entry locks), per lock	--	Ea	2.71	--	--
Add for armored fronts	--	Ea	65.00	--	--
Add for electrically operated locks, no switches or wiring included	--	Ea	316.00	--	--
Add for monitor switch for electrically operated locks	--	Ea	54.00	--	--

Security entrance (cylinder/deadbolt combination), 5 pin cylinder and 1" throw, 5 pin deadbolt — BC@1.30 Ea 80.00 30.70 110.70

Hardware

	Craft@Hrs	Unit	Material	Labor	Total
Door passage locksets, keyless					
Bored type, including knob trim, non-locking					
Standard residential latchset	BC@.572	Ea	14.00	13.50	27.50
Apartment/commercial latchset	BC@.572	Ea	30.50	13.50	44.00
Heavy duty commercial latchset	BC@.572	Ea	86.50	13.50	100.00
Standard residential button locking	BC@.572	Ea	16.50	13.50	30.00
Apartment/commercial button locking	BC@.572	Ea	37.00	13.50	50.50
Heavy duty commercial button locking	BC@.572	Ea	103.00	13.50	116.50
Dummy knobs, one side only					
Residential dummy knob	BC@.428	Ea	5.30	10.10	15.40
Apartment/commercial dummy knob	BC@.428	Ea	12.50	10.10	22.60
Heavy duty commercial dummy knob	BC@.428	Ea	33.50	10.10	43.60
Add for lever trim	--	%	45.0	--	--
Mortise type latchsets					
Passage (non-locking) knob trim	BC@1.04	Ea	159.00	24.50	183.50
Passage (non-locking) lever trim	BC@1.04	Ea	204.00	24.50	228.50
Privacy (thumb turn lock), knob trim	BC@1.04	Ea	187.00	24.50	211.50
Privacy (thumb turn lock), lever trim	BC@1.04	Ea	210.00	24.50	234.50
Office privacy lock, keyed, with lever and escutcheon trim	BC@1.04	Ea	210.00	24.50	234.50
Add for lead X-ray shielding	--	Ea	82.00	--	--
Add for electricity operated locks, no switches or wiring included	--	Ea	319.00	--	--
Miscellaneous door hardware					
Hasps, safety type, heavy duty zinc plated					
2-1/2"	--	Ea	2.21	--	--
4-1/2"	--	Ea	4.67	--	--
6"	--	Ea	7.14	--	--
7-1/2"	--	Ea	13.20	--	--
Hinges					
Butt, heavy steel, full mortise					
2" x 2", loose pin	--	Ea	3.89	--	--
3" x 3", loose pin	--	Ea	4.67	--	--
4" x 4", loose pin	--	Ea	7.14	--	--
4" x 4", ball bearing	--	Ea	24.20	--	--
Butt, light steel, brass finish, full mortise					
3" x 3", loose pin	--	Ea	1.37	--	--
4" x 4", loose pin	--	Ea	2.10	--	--
Screen door					
Brass finish	--	Ea	5.20	--	--
Prime pointed finish	--	Ea	5.78	--	--
Strap, heavy zinc-plated steel					
4" each side	--	Ea	3.89	--	--
6" each side	--	Ea	5.51	--	--
8" each side	--	Ea	8.82	--	--
10" each side	--	Ea	16.00	--	--
12" each side	--	Ea	33.60	--	--
Tee, heavy zinc-plated steel					
4"	--	Ea	4.94	--	--
6"	--	Ea	6.62	--	--
8"	--	Ea	11.00	--	--
10"	--	Ea	17.60	--	--
12"	--	Ea	41.00	--	--

Hardware

	Craft@Hrs	Unit	Material	Labor	Total
Door knockers, good quality polished brass					
3-1/2" x 7", traditional style	BC@.258	Ea	34.70	6.09	40.79
4" x 8-1/2", ornate style	BC@.258	Ea	46.20	6.09	52.29
Door pull, brass finish, 4" x 16"	BC@.258	Ea	36.80	6.09	42.89
Door push plates					
Brass finish, 3" x 12"	BC@.258	Ea	6.20	6.09	12.29
Brass finish, 4" x 16"	BC@.258	Ea	9.35	6.09	15.44
Door kick plate					
Aluminum finish					
6" x 30", to fit 2'6" and 2'8" doors	BC@.258	Ea	12.10	6.09	18.19
6" x 34", to fit 3'0" and 3'6" doors	BC@.258	Ea	13.20	6.09	19.29
Polished brass, good quality					
6" x 30", to fit 2'6" and 2'8" doors	BC@.258	Ea	38.90	6.09	44.99
6" x 34", to fit 3'0" and 3'6" doors	BC@.258	Ea	43.10	6.09	49.19
Door stops					
Rigid baseboard type, medium quality	BC@.084	Ea	.74	1.98	2.72
Rigid polished brass, good quality	BC@.084	Ea	3.41	1.98	5.39
Flexible spring baseboard type	BC@.084	Ea	1.10	1.98	3.08
Angle floor type, polished brass	BC@.084	Ea	.89	1.98	2.87
Butt hinge type	BC@.084	Ea	.78	1.98	2.76
Door peep sight, 5/8" diameter	BC@.258	Ea	3.26	6.09	9.35
House numbers, good quality polished brass					
3" high	BC@.103	Ea	9.65	2.43	12.08
5" high	BC@.103	Ea	12.50	2.43	14.93
Garage door hardware, jamb type, with springs and slide bolt (@$5.00)					
8' to 10' wide door	BC@4.36	Ea	113.00	103.00	216.00
10' to 16' wide door, with truss	BC@5.50	Ea	175.00	130.00	305.00
Truss assembly for 16' door	BC@.493	Ea	29.10	11.60	40.70
Garage door hardware, track type, with springs, 7' height, with lock					
7' to 9' wide door	BC@4.36	Ea	169.00	103.00	272.00
10' to 16' double track	BC@5.50	Ea	296.00	130.00	426.00

Heating and Cooling Equipment. No wiring included except as noted. See the Electrical Work section. Labor includes installation and connection only.

Baseboard electric heaters
Baseboard, convection type, surface mounted, 20 gauge steel, labor includes connecting and installation, 7" high, 3-1/4" deep, 3.41 Btu/hr/watt, 240 volt. Low or medium density

	Craft@Hrs	Unit	Material	Labor	Total
2'6" long, 500 watt (1,700 Btu)	BE@1.70	Ea	28.00	43.00	71.00
3'0" long, 750 watt (2,225 Btu)	BE@1.81	Ea	30.00	45.80	75.80
4'0" long, 1,000 watt (3,400 Btu)	BE@1.81	Ea	38.00	45.80	83.80
5'0" long, 1,250 watt (4,250 Btu)	BE@1.81	Ea	43.00	45.80	88.80
6'0" long, 1,500 watt (5,100 Btu)	BE@1.89	Ea	48.50	47.80	96.30
8'0" long, 2,000 watt (6,800 Btu)	BE@2.04	Ea	59.00	51.60	110.60
10'0" long, 2,500 watt (10,200 Btu)	BE@2.32	Ea	70.00	58.70	128.70
Add for line voltage, wall mount thermostat, good quality	BE@1.62	Ea	30.00	41.00	71.00
Add for integral thermostat	BE@.094	Ea	22.00	2.38	24.38

Hot water circulating baseboard heaters, 240 volts, 9-1/2" high, 3-1/4" deep. Labor includes connecting and installation.

39" long, 500 watts	BE@1.62	Ea	120.00	41.00	161.00
47" long, 750 watts	BE@1.70	Ea	138.00	43.00	181.00
59" long, 1,000 watts	BE@1.77	Ea	159.00	44.70	203.70
71" long, 1,380 watts	BE@1.79	Ea	182.00	45.30	227.30

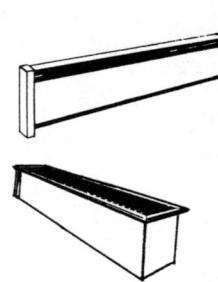

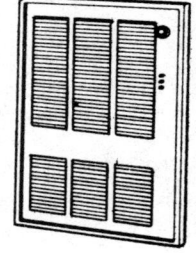

Heating

	Craft@Hrs	Unit	Material	Labor	Total
Baseboard electric heaters					
83" long, 1,500 watts	BE@1.88	Ea	198.00	47.50	245.50
107" long, 2,000 watts	BE@2.03	Ea	234.00	51.30	285.30
Add for line voltage wall thermostat	BE@1.62	Ea	26.00	41.00	67.00

Recessed water circulating baseboard heaters, 240 volt floor mounted, 10" high, 6" wide. Labor includes connecting and installation.

	Craft@Hrs	Unit	Material	Labor	Total
3'0" long, 400 watts	BE@2.27	Ea	179.00	57.40	236.40
4'0" long, 600 watts	BE@2.27	Ea	188.00	57.40	245.40
5'0" long, 800 watts	BE@2.36	Ea	198.00	59.70	257.70
6'0" long, 1,000 watts	BE@2.59	Ea	240.00	65.50	305.50
8'0" long, 1,400 watts	BE@2.82	Ea	265.00	71.30	336.30
Add for wall thermostat	BE@1.62	Ea	26.00	41.00	67.00

Bathroom electric heaters

Ceiling blower type, snap-on grille, prewired rocker switch and plate, with light and air circulator, 14" long, 8" wide, 7" high, 1,320 watts, 120 volts, (with vent

	Craft@Hrs	Unit	Material	Labor	Total
kit at $30.00)	BE@2.00	Ea	174.00	50.60	224.60

Wall mount forced-air chrome wall unit, 11" x 8" x 4" rough-in, includes rough-in box

	Craft@Hrs	Unit	Material	Labor	Total
Automatic, recessed, 1,000 watts	BE@2.00	Ea	80.00	50.60	130.60
Manual, surface mount, 750 watts	BE@2.00	Ea	72.50	50.60	123.10

Circular ceiling unit, fan-forced, 8-1/2" diameter, 3-1/2" deep, chrome, surface mount, 1,500

	Craft@Hrs	Unit	Material	Labor	Total
watts, with switch	BE@2.00	Ea	100.00	50.60	150.60

Radiant ceiling unit, 1,500 watts, 120 volts, with bath exhaust fan

	Craft@Hrs	Unit	Material	Labor	Total
and light	BE@2.25	Ea	175.00	56.90	231.90
Add for switch (4 position)	--	Ea	30.00	--	--

Radiant wall unit, steel grille, built-in switch, 14" wide, 25" high, 4" deep, 120 volts

	Craft@Hrs	Unit	Material	Labor	Total
1,000 watt, 3,400 Btu, with blower	BE@2.75	Ea	66.00	69.50	135.50
1,200 watt, 4,095 Btu	BE@1.95	Ea	55.00	49.30	104.30
4,000/2,500 watt, with blower	BE@2.23	Ea	133.00	56.40	189.40

Wall and ceiling electric heaters

Wall heater, forced-air type, with built-in thermostat, and automatic safety shut-off switch, 11" high, 8" wide, 4" deep, recessed

	Craft@Hrs	Unit	Material	Labor	Total
2,000 watt, 240 volt	BE@1.78	Ea	108.00	45.00	153.00
1,500 watt, 120 volt	BE@1.78	Ea	107.00	45.00	152.00

Floor units, drop-in type, 14" long x 7-1/4" wide, includes fan motor assembly, housing and grille

	Craft@Hrs	Unit	Material	Labor	Total
120 volts, 375 or 750 watts	BE@1.76	Ea	181.00	44.50	225.50
277 volts, 750 watts	BE@1.76	Ea	221.00	44.50	265.50
Add for wall thermostat kit	BE@1.62	Ea	44.00	41.00	85.00
Add for concrete accessory kit, housing cover and 2 stakes	--	LS	17.00	--	--

Infrared quartz tube heaters, indoor or outdoor, chromed guard, polished aluminum reflector. Includes modulating control

	Craft@Hrs	Unit	Material	Labor	Total
1,500 watts, 33" long, 120 or 240 volts	BE@4.14	Ea	224.00	105.00	329.00
2,000 watts, 33" long, 240 volts	BE@4.14	Ea	224.00	105.00	329.00
3,000 watts, 58-1/2" long, 240 volts	BE@4.14	Ea	295.00	105.00	400.00
4,000 watts, 58-1/2" long, 240 volts	BE@4.14	Ea	295.00	105.00	400.00
Add for wall or ceiling brackets (pair)	--	LS	13.00	--	--
Deduct for indoor use only	--	%	-30.0	--	--

Suspension blower heaters, 208 or 240 volts, single phase, propeller type

	Craft@Hrs	Unit	Material	Labor	Total
3,000 watts	BE@4.80	Ea	195.00	121.00	316.00
5,000 watts	BE@4.80	Ea	200.00	121.00	321.00
7,500 watts	BE@4.80	Ea	325.00	121.00	446.00

Heating

	Craft@Hrs	Unit	Material	Labor	Total
Wall and ceiling electric heaters					
Suspension blower heaters, 208 or 240 volts, single phase, propeller type					
10,000 watts	BE@4.80	Ea	350.00	121.00	471.00
Add for wall mounting bracket	--	LS	9.20	--	--
Add for line voltage wall thermostat					
Single pole	--	LS	24.50	--	--
Double pole	--	LS	36.00	--	--
Wall mount fan-forced downflow insert heaters, heavy duty, built-in thermostat, 240 volts, 14" wide, 20" high, 4" deep rough-in, 3.41 Btu/hr/watt					
1,500 watts	BE@3.95	Ea	165.00	99.90	264.90
2,000 watts	BE@3.95	Ea	170.00	99.90	269.90
3,000 watts	BE@3.95	Ea	185.00	99.90	284.90
4,000 watts	BE@3.95	Ea	200.00	99.90	299.90
Add for surface mounting kit	--	LS	19.00	--	--
Gas and oil furnaces Complete costs including installation, connection and vent.					
Floor gas furnaces, with vent at $82, valves and wall thermostat. Pilot ignition					
35 MBtu input, 24 MBtu output	PM@7.13	Ea	637.00	177.00	814.00
50 MBtu input, 35 MBtu output	PM@8.01	Ea	698.00	199.00	897.00
65 MBtu input, 45 MBtu output	PM@8.92	Ea	753.00	221.00	974.00
Add for dual room floor furnaces	--	%	30.0	--	--
Add for spark ignition furnaces	--	LS	47.30	--	--
Wall gas furnaces, 65" high, 14" wide, includes typical piping and valve at $39 and built-in remote dial thermostat, pilot ignition, blower, and vent at $79					
25 MBtu input, 17.5 MBtu output	PM@4.67	Ea	491.00	116.00	607.00
35 MBtu input, 24.5 MBtu output	PM@4.67	Ea	512.00	116.00	628.00
50 MBtu input, 40 MBtu output	PM@4.67	Ea	615.00	116.00	731.00
65 MBtu input, 48 MBtu output	PM@4.67	Ea	648.00	116.00	764.00
Add for wall thermostat	PM@.972	Ea	--	24.10	--
Add for spark ignition	--	LS	37.10	--	--
Deduct for no circulating fan	--	Ea	-77.00	--	--
Add for free standing vent trim kit	PM@.254	LS	40.00	6.30	46.30
Add for 2 wall unit (rear registers)	PM@.455	LS	40.00	11.30	51.30
Direct vent through-the-wall gas furnaces, including valve at $36, pilot ignition, wall-mounted thermostat, wall vent cap and guard					
15 MBtu output	PM@2.98	Ea	233.00	73.90	306.90
35 MBtu output	PM@2.98	Ea	399.00	73.90	472.90
Add for vent kit	PM@.455	Ea	29.00	11.30	40.30
Deduct for built-in thermostat	--	Ea	-22.50	-24.80	-47.30
Deduct for no blower	--	Ea	-100.00	--	--
Forced air, upflow gas furnaces with electronic ignition. Includes vent at $89, typical piping at $39, thermostat at $32.50 and distribution plenum at $98. No ducting included					
50 MBtu input, 48 MBtu output	PM@7.13	Ea	1,820.00	177.00	1,997.00
75 MBtu input, 70 MBtu output	PM@7.13	Ea	1,940.00	177.00	2,117.00
100 MBtu input, 92 MBtu output	PM@7.80	Ea	2,060.00	193.00	2,253.00
125 MBtu input, 113 MBtu output	PM@7.80	Ea	2,180.00	193.00	2,373.00
Add for liquid propane models	--	LS	--	--	212.00
Add for time control thermostat	--	Ea	81.80	--	--
Forced air oil furnaces, with warm air distribution plenum. Use costs for forced air gas furnace above, plus	--	%	10.0	--	--
Add for slab-mounted 275 gallon tank and piping	PM@7.80	LS	400.00	193.00	593.00

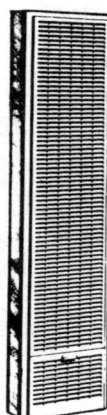

Heating

	Craft@Hrs	Unit	Material	Labor	Total
Add for 10' Class "A" 8" triple wall vent, flue top, fittings and supports	PM@4.92	LS	600.00	122.00	722.00

Ductwork for forced air furnaces

	Craft@Hrs	Unit	Material	Labor	Total
Sheet metal, 6 duct and 2 air return system elbows, attaching boots, 4" x 12" registers, and 2" R-5 insulation	SW@22.1	LS	455.00	556.00	1,011.00
Add for each additional duct run	SW@4.20	Ea	92.20	106.00	198.20
Fiberglass flex duct, insulated, 8" dia.	SW@.012	LF	2.34	.30	2.64

Heating and cooling combinations

Evaporative coolers, no wiring or duct work, roof installation, with switch, electrical and water connection at $89, recirculation pump, and mounting hardware at $65.

	Craft@Hrs	Unit	Material	Labor	Total
4,000 CFM, 1/3 HP	PM@11.1	Ea	620.00	275.00	895.00
4,500 CFM, 1/2 HP	PM@11.1	Ea	675.00	275.00	950.00
5,200 CFM, 1/2 HP	PM@11.1	Ea	790.00	275.00	1,065.00
6,600 CFM, 3/4 HP	PM@11.1	Ea	905.00	275.00	1,180.00

Window type air conditioning/cooling units, high-efficiency units, adjustable thermostat, labor is for installation in an existing opening

	Craft@Hrs	Unit	Material	Labor	Total
5,600 Btu/hr	PM@.751	Ea	415.00	18.60	433.60
7,000 Btu/hr	PM@.751	Ea	475.00	18.60	493.60
8,000 Btu/hr	PM@.751	Ea	495.00	18.60	513.60
14,000 Btu/hr	PM@1.00	Ea	700.00	24.80	724.80
18,000 Btu/hr	PM@1.00	Ea	755.00	24.80	779.80
21,000 Btu/hr	PM@1.50	Ea	845.00	37.20	882.20

Split system heat pump, one outdoor unit, with indoor evaporator coil, direct drive multi-speed blower, filter, and fiberglass cabinet insulation. No electrical or concrete work included

	Craft@Hrs	Unit	Material	Labor	Total
18 MBtu cooling, 17.6 MBtu heating	PM@9.59	Ea	1,180.00	238.00	1,418.00
25 MBtu cooling, 23.8 MBtu heating	PM@11.3	Ea	1,310.00	280.00	1,590.00
30 MBtu cooling, 29.6 MBtu heating	PM@12.3	Ea	1,440.00	305.00	1,745.00
34 MBtu cooling, 34 MBtu heating	PM@13.1	Ea	1,680.00	325.00	2,005.00
42 MBtu cooling, 40 MBtu heating	PM@14.3	Ea	1,920.00	354.00	2,274.00
Add for electrical connection	PM@7.34	LS	56.50	182.00	238.50
Add for drain pan	--	LS	45.00	--	--

Packaged split system electric heat pumps, indoor and outdoor units including compressor, coils, blower section with electric insert heater (kw as shown), 35' of tubing, two thermostats and control package. Installed on concrete pad. No electrical, concrete, or duct work included

Cooling	Heating	Craft@Hrs	Unit	Material	Labor	Total
20,300 Btu	18,300 Btu, 10 kw	PM@10.9	Ea	2,650.00	270.00	2,920.00
24,400 Btu	24,600 Btu, 15 kw	PM@12.0	Ea	3,010.00	297.00	3,307.00
30,000 Btu	32,000 Btu, 15 kw	PM@12.9	Ea	3,130.00	320.00	3,450.00
34,600 Btu	37,600 Btu, 30 kw	PM@14.0	Ea	3,300.00	347.00	3,647.00
40,500 Btu	42,000 Btu, 10 kw	PM@14.9	Ea	3,480.00	369.00	3,849.00
46,500 Btu	48,000 Btu, 30 kw	PM@16.3	Ea	3,940.00	404.00	4,344.00
Add for electrical connections		PM@14.0	LS	113.00	347.00	460.00

Single unit heat pumps, including heating and cooling coil, compressor, two thermostats and a 14 kw electric insert heater. No concrete, electrical or duct work included

Cooling	Heating	Craft@Hrs	Unit	Material	Labor	Total
24.8 MBtu	25.2 MBtu	PM@6.46	Ea	2,610.00	160.00	2,770.00
29.6 MBtu	29.0 MBtu	PM@7.59	Ea	2,930.00	188.00	3,118.00
35.2 MBtu	32.0 MBtu	PM@8.47	Ea	3,050.00	210.00	3,260.00
39.4 MBtu	38.0 MBtu	PM@9.59	Ea	3,210.00	238.00	3,448.00

Insulation

	Craft@Hrs	Unit	Material	Labor	Total

Insulation. See also Building Paper and Polyethylene Film. Coverage allows for studs and joists.

Aluminum foil insulation, wall and ceiling applications, including 10% waste and overlap, 36" width on 40 lb kraft paper, 250 SF per roll
1 side foil ($11.00 per roll)	BC@.006	SF	.05	.14	.19
2 sides foil ($17.50 per roll)	BC@.006	SF	.07	.14	.21

Fiberglass insulation, wall and ceiling application, no waste included
  Kraft paper-faced, roll and batt insulation
    16" OC framing members
3-1/2" (R-11)	BC@.007	SF	.25	.17	.42
6-1/4" (R-19)	BC@.007	SF	.37	.17	.54
10" (R-30)	BC@.008	SF	.59	.19	.78
12" (R-38)	BC@.008	SF	.81	.19	1.00

    24" OC framing members
3-1/2" (R-11)	BC@.004	SF	.25	.09	.34
6-1/4" (R-19)	BC@.005	SF	.37	.12	.49
10" (R-30)	BC@.005	SF	.59	.12	.71
12" (R-38)	BC@.005	SF	.81	.12	.93

  Foil faced, roll and batt insulation
    16" OC framing members
3-1/2" (R-11)	BC@.007	SF	.28	.17	.45
6-1/2" (R-19)	BC@.007	SF	.42	.17	.59

    24" OC framing members
3-1/2" (R-11)	BC@.004	SF	.28	.09	.37
6-1/4" (R-19)	BC@.005	SF	.42	.12	.54

  Unfaced, roll and batt insulation
    16" OC framing members
3-1/2" (R-11)	BC@.004	SF	.26	.09	.35
6-1/4" (R-19)	BC@.005	SF	.38	.12	.50
10" (R-30)	BC@.005	SF	.59	.12	.71
12" (R-38)	BC@.006	SF	.77	.14	.91

    24" OC framing members
3-1/2" (R-11)	BC@.003	SF	.26	.07	.33
6-1/4" (R-19)	BC@.003	SF	.38	.07	.45
10" (R-30)	BC@.003	SF	.59	.07	.66
12" (R-38)	BC@.004	SF	.77	.09	.86

Polystyrene insulation board, No. 100 regular, white, applied over walls with adhesive, 2' x 8' panels, including 5% waste
1" thick (R-5)	BC@.010	SF	.39	.24	.63
2" thick (R-8)	BC@.011	SF	.74	.26	1.00
Add for adhesive	--	SF	.07	--	--

Foil faced, two sides, rigid foam (isocyanurate) insulation board, 4' x 8', including 5% waste
1/2" (R-4)	BC@.010	SF	.31	.24	.55
3/4" (R-4)	BC@.010	SF	.38	.24	.62
1-1/2" (R-8)	BC@.010	SF	.59	.24	.83

Foil faced rigid urethane board, no paper backing, 4' x 8', including 5% waste
1-1/2" (R-10)	BC@.011	SF	.69	.26	.95
2" (R-14)	BC@.011	SF	.84	.26	1.10

Add for nails, 50 lb cartons (large square heads)
2-1/2", for 1-1/2" boards	--	LS	85.00	--	--
3", for 2" boards	--	LS	85.00	--	--

Insulation Recommendations
Walls, using 2" x 6" framing
R-19: R-19 (6") fiberglass batts
Walls, using 2" x 4" framing
R-19: R-11 (3½") fiberglass batts and 1" polystyrene
R-13: R-13 (3⅝") fiberglass batts
R-11: R-11 (3½") fiberglass batts
Floors
R-22: R-22 (6½") mineral fiber
R-19: R-19 (6") mineral fiber
R-13: R-13 (3⅝") mineral fiber
R-11: R-11 (3½") mineral fiber
Ceilings, double layers of batts
R-38: Two layers of R-19 (6") mineral fiber
R-33: One layer of R-22 (6½") and one layer of R-11 (3½") mineral fiber
R-30: One layer of R-19 (6") and one layer of R-11 (3½") mineral fiber
R-26: Two layers of R-13 (3⅝") mineral fiber
Ceilings, loose fill mineral wool and batts
R-38: R-19 (6") mineral fiber and 20 bags of wool per 1,000 S.F. (8¾")
R-33: R-22 (6½") mineral fiber and 11 bags of wool per 1,000 S.F. (5")
R-30: R-19 (6") mineral fiber and 11 bags of wool per 1,000 S.F. (5")
R-26: R-19 (6") mineral fiber and 8 bags of wool per 1,000 S.F. (3¼")

Insulation

	Craft@Hrs	Unit	Material	Labor	Total
Vermiculite insulation, poured over ceilings					
Vermiculite, 25 lb sack (3 CF)	--	Ea	8.00	--	--
At 3" depth (60 sacks per 1,000 SF)	BL@.007	SF	.48	.13	.61
At 4" depth (72 sacks per 1,000 SF)	BL@.007	SF	.58	.13	.71
Roof insulation, 24" x 48", perlite board					
3/4" thick (R-2.08)	RR@.007	SF	.31	.18	.49
1" thick (R-2.78)	RR@.007	SF	.38	.18	.56
1-1/2" thick (R-4.17)	RR@.007	SF	.55	.18	.73
2" thick (R-5.26)	RR@.007	SF	.67	.18	.85
Cant strips for insulated roof board, one piece wood, tapered					
2"	RR@.018	LF	.19	.46	.65
3"	RR@.020	LF	.39	.52	.91
4"	RR@.018	LF	.68	.46	1.14
Sill sealer, 1" x 50', 6" wide	RR@.003	LF	.29	.08	.37
Sound control batts, unfaced, 96" long					
4" x 16", 24"	BC@.007	SF	.45	.17	.62
2-3/4" x 16", 24"	BC@.007	SF	.38	.17	.55
Masonry batts, 1-1/2" x 15" x 48"	B9@.008	SF	.21	.16	.37
Masonry fill, 3 CF bag @ $5.50, perlite poured in concrete block cores					
4" wall, 8.1 SF per CF	B9@.006	SF	.25	.12	.37
6" wall, 5.4 SF per CF	B9@.006	SF	.36	.12	.48
8" wall, 3.6 SF per CF	B9@.006	SF	.53	.12	.65
Asphalt sheathing board on walls, nailed, 4' x 8'					
1/2", standard	BC@.013	SF	.27	.31	.58
1/2", intermediate	BC@.013	SF	.31	.31	.62
1/2", nail base	BC@.013	SF	.37	.31	.68
25/32", standard	BC@.014	SF	.36	.33	.69
Sound board, on wall, 1/2"	BC@.013	SF	.27	.31	.58
Add for ceiling applications	BC@.004	SF	--	.09	--

Insulation, Subcontract. New construction, coverage includes framing.

	Craft@Hrs	Unit	Material	Labor	Total
Batts, walls and ceilings to 10' high, fiberglass, kraft faced					
3-1/2" (R-11) in walls	--	MSF	--	--	315.00
3-1/2" (R-11) in ceilings	--	MSF	--	--	345.00
6-1/4" (R-19)	--	MSF	--	--	465.00
10" (R-30)	--	MSF	--	--	715.00
Blown rockwool					
3-1/2" (R-11)	--	MSF	--	--	440.00
6" (R-19)	--	MSF	--	--	595.00
Blown cellulose					
3-1/2" (R-12)	--	MSF	--	--	425.00
6" (R-21)	--	MSF	--	--	605.00
Blown fiberglass, ceilings					
5" (R-11)	--	MSF	--	--	370.00
8" (R-19)	--	MSF	--	--	490.00
Sprayed-on urethane foam (rigid)					
1" thick	--	MSF	--	--	1,050.00
Add for acrylic waterproofing	--	MSF	--	--	765.00
Add for urethane rubber waterproofing	--	MSF	--	--	1,210.00
Add for cleaning rock off existing roof	--	MSF	--	--	75.00

Insulation

	Craft@Hrs	Unit	Material	Labor	Total

Insulation, Subcontract, continued
Fiberglass blown into walls in an existing structure, includes coring hole in exterior or interior of wall, filling cavity, sealing hole and paint to match existing surface.
R-12 to R-14 rating depending on cavity thickness

	Craft@Hrs	Unit	Material	Labor	Total
Stucco (exterior)	--	MSF	--	--	1,240.00
Wallboard (interior)	--	MSF	--	--	1,240.00
Plywood or wood siding	--	MSF	--	--	1,640.00

Thermal analysis (infrared thermography). Includes infrared video inspection of entire building, written report by qualified technician or engineer, and thermograms (infrared photographs) of all problem areas of building. Inspection is done using high-quality thermal sensitive instrumentation. Costs listed are for exterior facade shots, or interior (room by room) shots. All travel expenses for inspector are extra

Residence, typically 4 hours required	--	Hr	--	--	180.00
Commercial structure ($800.00 minimum)	--	Hr	--	--	105.00

Add per day lost due to weather, lack of preparation by building occupants, etc., when inspector and equipment are at building site prepared to commence
| inspection | -- | LS | -- | -- | 300.00 |

Add per night for accommodations when overnight stay is required. This is common when inspection must be performed at night to get accurate results.
| Typical cost, per night | -- | LS | -- | -- | 90.00 |

Insurance and Bonding. Typical rates. Costs vary by state and class of construction.
Rule of thumb: Employer's cost for payroll taxes, and insurance.
| Per $100 (C$) of payroll | -- | C$ | -- | -- | 30.00 |

Rule of thumb: Complete insurance program (comprehensive general liability policy, truck, auto, and equipment floaters and fidelity bond), contractor's typical cost
| Per $100 of payroll | -- | C$ | -- | -- | 6.75 |

Liability insurance Comprehensive contractor's liability insurance, including operations, completed operations, bodily injury and property damage, protective and contractual coverages, $1,000,000 policy limit. Minimum annual premium will be between $5,000 and $8,000. Rates vary by state and with the contractor's loss experience. Typical costs per $100 (C$) of payroll for each trade employed. (Calculate and add for each category separately.)

General contractors	--	C$	--	--	1.58
Carpentry	--	C$	--	--	3.74
Concrete, formed or flat	--	C$	--	--	4.94
Drywall hanging and finishing	--	C$	--	--	3.37
Electrical wiring	--	C$	--	--	3.37
Floor covering installation	--	C$	--	--	4.47
Glaziers	--	C$	--	--	3.37
Heating, ventilating, air conditioning	--	C$	--	--	4.63
Insulation	--	C$	--	--	4.16
Masonry, tile	--	C$	--	--	3.58
Painting	--	C$	--	--	4.89
Plastering and stucco	--	C$	--	--	3.58
Plumbing	--	C$	--	--	5.79

Workers' compensation coverage Rates vary from state to state and from year to year. Coverage cost per $100 (C$) of base payroll excluding fringe benefits.

Bricklayer	--	C$	--	--	14.05
Carpenter					
1 and 2 family dwellings	--	C$	--	--	14.32
Multiple units and commercial	--	C$	--	--	21.96

Insurance

	Craft@Hrs	Unit	Material	Labor	Total
Workers' compensation coverage, continued					
Coverage cost per $100 (C$) of base payroll excluding fringe benefits.					
Clerical (office worker)	--	C$	--	--	.90
Concrete					
1 and 2 family dwellings	--	C$	--	--	9.33
Other concrete	--	C$	--	--	17.78
Construction laborer	--	C$	--	--	14.32
Drywall taper	--	C$	--	--	12.03
Electrical wiring	--	C$	--	--	7.82
Elevator erectors	--	C$	--	--	5.92
Executive supervisors	--	C$	--	--	4.98
Excavation, grading	--	C$	--	--	7.55
Excavation, rock (no tunneling)	--	C$	--	--	11.90
Glazing	--	C$	--	--	12.94
Insulation work	--	C$	--	--	22.76
Iron or steel erection work	--	C$	--	--	24.76
Lathing	--	C$	--	--	8.13
Operating engineers	--	C$	--	--	9.60
Painting and paperhanging	--	C$	--	--	14.90
Pile driving	--	C$	--	--	19.03
Plastering and stucco	--	C$	--	--	15.92
Plumbing	--	C$	--	--	10.36
Reinforcing steel installation (concrete)	--	C$	--	--	11.19
Roofing	--	C$	--	--	36.32
Sewer construction	--	C$	--	--	13.58
Sheet metal work (on site)	--	C$	--	--	13.11
Steam and boiler work	--	C$	--	--	8.73
Tile, stone, and terrazzo work	--	C$	--	--	7.68
Truck driver	--	C$	--	--	16.80
Tunneling	--	C$	--	--	15.68
Payment and performance bonds Cost per $1,000 (M$) of final contract price.					
Rates depend on the experience, credit, and net worth of the applicant.					
Preferred Rates, contract coverage for:					
First $100,000	--	M$	--	--	14.40
Next $2,400,000	--	M$	--	--	8.70
Next $2,500,000	--	M$	--	--	6.90
Standard Rates, contract coverage for:					
First $100,000	--	M$	--	--	30.00
Next $400,000	--	M$	--	--	18.00
Next $2,000,000	--	M$	--	--	12.00
Next $2,500,000	--	M$	--	--	9.00
Next $2,500,000	--	M$	--	--	8.40
Over $7,500,000	--	M$	--	--	7.80
Substandard Rates, contract coverage for:					
First $100,000	--	M$	--	--	36.00
Balance of contract	--	M$	--	--	24.00

Intercom System, Subcontract. Costs include all labor and equipment but assume that wiring is done before interior wall and ceiling finish is supplied.
Central system with AM/FM music, hands-free talk back, privacy feature and room monitoring. Includes 1 master station, 5 selective call remote stations and 2 door stations

	Craft@Hrs	Unit	Material	Labor	Total
Basic system as described above	--	LS	--	--	790.00

Intercom Systems

	Craft@Hrs	Unit	Material	Labor	Total
Intercom System, Subcontract, continued					
Add cassette tape player	--	LS	--	--	160.00
Add door and gate release	--	Ea	--	--	150.00
Add telephone answering feature, per station	--	Ea	--	--	59.00
Add telephone handset stations, per station	--	Ea	--	--	35.00

Ironing Centers, Built-In
Wood frame with hinged door, can be recessed or surface mounted. Includes one 110 volt control panel where shown, receptacle outlet, timed shutoff switch, spotlight, storage shelf, and steel ironing board. No electrical work included. Iron-A-Way Products

	Craft@Hrs	Unit	Material	Labor	Total
Unit type 1, no electric control panel, non-adjustable 42" board, 48" high, 15" wide, 6" deep	BC@.994	Ea	125.00	23.40	148.40
Unit type 2, with electric control panel, non-adjustable 42" board, 48" high, 15" wide, 6" deep	BC@.994	Ea	175.00	23.40	198.40
Unit type 3, with electric control panel, adjustable 46" board, 61" high, 15" wide, 6" deep	BC@.994	Ea	245.00	23.40	268.40
Added costs, any of above units					
Add for sleeveboard with holder	--	Ea	16.00	--	--
Add for door mirror	--	Ea	28.00	--	--
Add for steam/dry iron, G.E.	--	Ea	30.00	--	--
Add for oak paneled door front	--	Ea	35.00	--	--

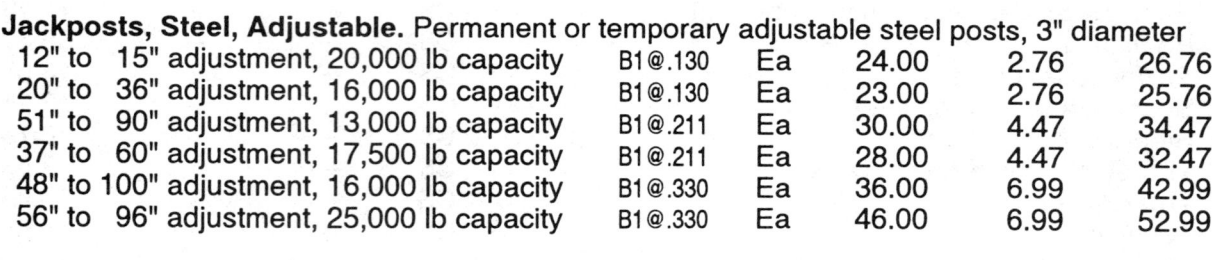

Jackposts, Steel, Adjustable. Permanent or temporary adjustable steel posts, 3" diameter

	Craft@Hrs	Unit	Material	Labor	Total
12" to 15" adjustment, 20,000 lb capacity	B1@.130	Ea	24.00	2.76	26.76
20" to 36" adjustment, 16,000 lb capacity	B1@.130	Ea	23.00	2.76	25.76
51" to 90" adjustment, 13,000 lb capacity	B1@.211	Ea	30.00	4.47	34.47
37" to 60" adjustment, 17,500 lb capacity	B1@.211	Ea	28.00	4.47	32.47
48" to 100" adjustment, 16,000 lb capacity	B1@.330	Ea	36.00	6.99	42.99
56" to 96" adjustment, 25,000 lb capacity	B1@.330	Ea	46.00	6.99	52.99

Joist Bridging. See also Carpentry, Joist Hangers, and Timber Connectors.
Cross bridging, steel, per pair, no nails required.

	Craft@Hrs	Unit	Material	Labor	Total
12" joist spacing on center					
2" x 8" to 2" x 16" joists	BC@.022	Pr	1.42	.52	1.94
16" joist spacing on center					
2" x 8" to 2" x 10" joists	BC@.022	Pr	1.44	.52	1.96
2" x 12" to 2" x 16" joists	BC@.022	Pr	1.52	.52	2.04
24" joist spacing on center					
2" x 8" to 2" x 10" joists	BC@.022	Pr	1.90	.52	2.42
2" x 12" to 2" x 14" joists	BC@.022	Pr	1.92	.52	2.44
2" x 16" joists	BC@.022	Pr	1.94	.52	2.46

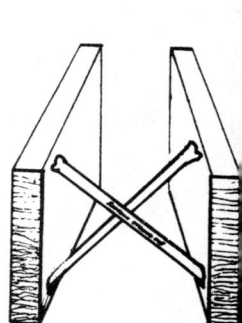

Joist Hangers. See also Timber Connectors and Plywood Framing Clips. Costs shown are based on carton lots. "U" type or formed seat type, standard duty, 16 gauge, galvanized

Nominal size	Craft@Hrs	Unit	Material	Labor	Total
2" x 4"	BC@.040	Ea	.72	.94	1.66
2" x 6" or 2" x 8"	BC@.046	Ea	.82	1.09	1.91
2" x 10" or 2" x 12"	BC@.046	Ea	1.20	1.09	2.29
2" x 14"	BC@.052	Ea	1.66	1.23	2.89
2" x 16"	BC@.052	Ea	1.66	1.23	2.89
3" x 4"	BC@.046	Ea	1.58	1.09	2.67
3" x 6" or 3" x 8"	BC@.046	Ea	1.85	1.09	2.94
3" x 10" or 3" x 12"	BC@.052	Ea	2.31	1.23	3.54

Joist Hangers

	Craft@Hrs	Unit	Material	Labor	Total
Joist Hangers					
3" x 14" or 3" x 16"	BC@.070	Ea	2.56	1.65	4.21
4" x 4"	BC@.046	Ea	1.33	1.09	2.42
4" x 6" or 4" x 8"	BC@.052	Ea	1.42	1.23	2.65
4" x 10" or 4" x 12"	BC@.070	Ea	2.12	1.65	3.77
4" x 14" or 4" x 16"	BC@.087	Ea	3.08	2.05	5.13
6" x 6" or 6" x 8"	BC@.052	Ea	4.60	1.23	5.83
6" x 10" or 6" x 12"	BC@.087	Ea	6.14	2.05	8.19
Landscaping					
Ground cover. Costs include plants and hand labor to plant					
English Ivy, 2-3/4" pots	BL@.098	Ea	.37	1.84	2.21
Liriope Muscari, 1 quart	BL@.133	Ea	2.10	2.50	4.60
Periwinkle, 2-3/4" pots	BL@.095	Ea	.39	1.79	2.18
Purple-leaved Wintercreeper, 1 quart	BL@.128	Ea	1.24	2.41	3.65
Plants and shrubs. Costs include plants, planting by hand, fertilization					
Abelia, gaucheri-pink, 5 gal.	BL@.835	Ea	8.91	15.70	24.61
Acacia, Fern leaf, 5 gal.	BL@.868	Ea	10.30	16.30	26.60
Andorra Juniper, 18" - 24"	BL@.868	Ea	7.98	16.30	24.28
Andorra Juniper, 24" - 30"	BL@1.19	Ea	9.17	22.40	31.57
Australian Brush Cherry, 5 gal.	BL@.868	Ea	10.30	16.30	26.60
Azalea, Indica or Kaepferi, 18" - 24"	BL@.868	Ea	13.40	16.30	29.70
Azalea, Kurume types, 18" - 24"	BL@.868	Ea	10.90	16.30	27.20
Begonia, Angel Wing	BL@.835	Ea	8.91	15.70	24.61
Blue Hibiscus, 5 gal.	BL@.835	Ea	8.91	15.70	24.61
Bougainvillea, all varieties, 5 gal.	BL@.868	Ea	10.30	16.30	26.60
Bridal Wreath, 30" - 36"	BL@1.44	Ea	8.24	27.10	35.34
Burford Holly, 30" - 36"	BL@1.44	Ea	16.00	27.10	43.10
Camellia, Sasanqua, 5 gal.	BL@.835	Ea	11.90	15.70	27.60
Cherry Laurel, 36" - 48"	BL@1.53	Ea	29.70	28.80	58.50
Cleyera Japonica, 24" - 30"	BL@1.14	Ea	10.80	21.40	32.20
Dwarf Japanese Holly, 15" - 18"	BL@.696	Ea	9.22	13.10	22.32
Dwarf Lily of the Nile, 5 gal.	BL@.835	Ea	8.91	15.70	24.61
Dwarf Yaupon Holly, 15" - 18"	BL@.668	Ea	9.22	12.60	21.82
Gardenia, all varieties, 5 gal.	BL@.835	Ea	8.91	15.70	24.61
Geranium, all varieties, 5 gal.	BL@.835	Ea	8.91	15.70	24.61
Gold Dust Aucuba, 24" - 30"	BL@1.19	Ea	10.04	22.40	32.44
Glossy Abelia, 24" - 36"	BL@1.31	Ea	9.00	24.60	33.60
Grey Lavender Cotton, 12" - 15"	BL@.601	Ea	2.83	11.30	14.13
Lily of the Nile, 5 gal.	BL@.835	Ea	8.91	15.70	24.61
Night Blooming Jessamine, 5 gal.	BL@.835	Ea	8.91	15.70	24.61
Pampas Grass, 2 gal.	BL@.417	Ea	5.97	7.84	13.81
Pink Escallonia, 5 gal.	BL@.835	Ea	8.91	15.70	24.61
Poinciana Paradise Shrub, 5 gal.	BL@.868	Ea	10.30	16.30	26.60
Prickly Pear Cactus, 2 gal.	BL@.417	Ea	8.03	7.84	15.87
Red Leaved Japanese Barberry, 18" - 24"	BL@.868	Ea	7.83	16.30	24.13
Rhododendron, 24" - 30"	BL@.835	Ea	27.30	15.70	43.00
Rubber Plant, 5 gal.	BL@.835	Ea	13.50	15.70	29.20
Sprengers Ash, 5 gal.	BL@.907	Ea	10.50	17.10	27.60
Shore Juniper, 18" - 24"	BL@.868	Ea	8.75	16.30	25.05
Tamarix Juniper, 18" - 24"	BL@.835	Ea	8.96	15.70	24.66
Thunbergi Spiraea, 24" - 30"	BL@1.14	Ea	7.83	21.40	29.23
Unedo- Strawberry Tree, 5 gal.	BL@.835	Ea	8.91	15.70	24.61
Vanhoutte Spiraea, 24" - 36"	BL@1.31	Ea	8.86	24.60	33.46

Landscaping

	Craft@Hrs	Unit	Material	Labor	Total

Landscaping, continued

Plants and shrubs. Costs include plants, planting by hand, fertilization

	Craft@Hrs	Unit	Material	Labor	Total
Veronica, all varieties, 5 gal.	BL@.835	Ea	8.91	15.70	24.61
Viburnum, all varieties, 5 gal.	BL@.835	Ea	8.91	15.70	24.61
Willow Leaf Japanese Holly, 24" - 30"	BL@.442	Ea	11.00	8.31	19.31
Wintergreen Barberry, 18" - 24"	BL@.868	Ea	10.09	16.30	26.39
Yew, Hicks, 24" - 30"	BL@1.31	Ea	16.50	24.60	41.10

Trees. Costs include trees, planting, fertilization, backfill, and support as required. Add machine excavation costs that follow

	Craft@Hrs	Unit	Material	Labor	Total
American Holly, 5' - 6'	BL@1.75	Ea	85.00	32.90	117.90
American Planetree Sycamore, 14' - 16'	BL@3.66	Ea	172.00	68.80	240.80
American Elm, 12' - 14'	BL@3.66	Ea	100.90	68.80	169.70
Armstrong Juniperus, 5 gal.	BL@.835	Ea	8.91	15.70	24.61
Australian Tree Fern, 5 gal.	BL@.835	Ea	10.50	15.70	26.20
Austrian Pine, 8' - 10'	BL@1.99	Ea	140.00	37.40	177.40
Bamboo multiplex, Golden Goddess, 5 gal.	BL@2.08	Ea	38.10	39.10	77.20
Bradford Pear, 8' - 10',	BL@2.25	Ea	124.00	42.30	166.30
Camellia, common type, 4' - 5'	BL@1.10	Ea	26.10	20.70	46.80
Colorado Blue Spruce, 4' - 5'	BL@1.99	Ea	48.00	37.40	85.40
Cape Myrtle, 8' - 10'	BL@2.25	Ea	77.30	42.30	119.60
Deodara Cedar, 6' - 8'	BL@3.66	Ea	77.30	68.80	146.10
Eastern Redbud, 6' - 8'	BL@2.00	Ea	36.00	37.60	73.60
Foster's American Holly, 5' - 6'	BL@1.75	Ea	55.00	32.90	87.90
Flowering Crabapple, 8' - 10'	BL@2.25	Ea	67.60	42.30	109.90
Fuchsia, all varieties, 5 gal.	BL@.835	Ea	8.91	15.70	24.61
Goldenrain Tree, 8' - 10'	BL@2.25	Ea	75.20	42.30	117.50
Green Ash, 11' - 13'	BL@2.00	Ea	63.90	37.60	101.50
Greenspire Linden, 10' - 12'	BL@3.05	Ea	148.00	57.30	205.30
Hibiscus, all varieties, 5 gal.	BL@.835	Ea	9.48	15.70	25.18
Japanese Privet, 4' - 5'	BL@1.46	Ea	19.30	27.40	46.70
Kwanzan Cherry, 6' - 8'	BL@3.05	Ea	48.60	57.30	105.90
Prince of Wales Juniperus, 5 gal	BL@.835	Ea	8.91	15.70	24.61
Psidium (guava), 5 gal.	BL@.835	Ea	8.91	15.70	24.61
Pyracantha Firethorn, 3' - 4'	BL@1.10	Ea	14.20	20.70	34.90
Live Oak, 10' - 12'	BL@2.25	Ea	172.00	42.30	214.30
Maidenhair Tree, 10' - 12'	BL@3.66	Ea	124.00	68.80	192.80
Northern Red Oak, 11' - 13'	BL@2.00	Ea	107.00	37.60	144.60
Pin Oak, 10' - 12'	BL@3.48	Ea	100.90	65.40	166.30
Purpleleaf Flowering Plum, 6' - 8'	BL@2.00	Ea	48.00	37.60	85.60
Red Maple, 11' - 13'	BL@3.05	Ea	100.60	57.30	157.90
Southern Magnolia, 6' - 8'	BL@2.00	Ea	82.90	37.60	120.50
Star Magnolia, 4' - 5'	BL@1.46	Ea	82.00	27.40	109.40
Sugar Maple, 11' - 13'	BL@2.00	Ea	124.00	37.60	161.60
Thornless Honeylocust, 10' - 12'	BL@2.25	Ea	100.70	42.30	143.00
Washington Hawthorne, 8' - 10'	BL@2.25	Ea	123.00	42.30	165.30
Weeping Fig, 5 gal.	BL@.835	Ea	13.50	15.70	29.20
Weeping Forsythia, 3' - 4'	BL@1.10	Ea	8.55	20.70	29.25
White Flowering Dogwood, 5' - 6'	BL@1.75	Ea	40.00	32.90	72.90
White Pine, 6' - 8'	BL@2.00	Ea	48.00	37.60	85.60
Willow Oak, 10' - 12'	BL@2.74	Ea	107.00	51.50	158.50
Wisconsin Weeping Willow, 8' - 10'	BL@2.00	Ea	59.20	37.60	96.80

Landscaping

	Craft@Hrs	Unit	Material	Labor	Total
Palm trees, based on West Coast prices					
Mexican Fan Palm, 5 gal.	BL@.835	Ea	8.91	15.70	24.61
Mediterranean Fan Palm, 5 gal.	BL@.835	Ea	10.80	15.70	26.50
Pigmy Date Palm, 5 gal.	BL@.835	Ea	19.00	15.70	34.70
Common King Palm, 15 gal.	BL@2.25	Ea	43.30	42.30	85.60

Machine excavation for planting trees. Typical excavation only, no backfill or planting included. Cost by tree height not including root ball. Assumes 8 to 10 trees on the job site

	Craft@Hrs	Unit	Material	Labor	Total
3' to 4' tree	--	Ea	--	--	11.50
4' to 5' tree	--	Ea	--	--	15.10
5' to 6' tree	--	Ea	--	--	18.30
6' to 8' tree	--	Ea	--	--	20.90
8' to 10' tree	--	Ea	--	--	23.40
10' to 12' tree	--	Ea	--	--	31.40
11' to 13' tree	--	Ea	--	--	34.20
12' to 14' tree	--	Ea	--	--	36.40
Relocate shrub up to 4' high, includes hand and machine excavation, lifting, 100' haul, and replanting	--	Ea	--	--	62.50
Relocate tree, 2" - 5" caliper, includes hand and machine excavation, lifting, 100' haul, and replanting	--	Ea	--	--	160.00
Seeding, level area (general mixture at $3.20 per lb)					
Seeding preparation, grade, rake, clean	BL@.003	SY	.10	.06	.16
Fine grade, seed (with grass), and lime	BL@.018	SY	.70	.34	1.04
Mechanical seeding (1 lb per 22 SY)	BL@.005	SY	.15	.09	.24
Hand seeding (1 lb per 10 SY)	BL@.006	SY	.32	.11	.43
Fertilizer, pre-planting,					
12-nitrogen, 4-phosphorus, 6-potassium (1 lb per 12 SY)	BL@.002	SY	.31	.04	.35
Liming (1 lb per 1.6 SY)	BL@.002	SY	.12	.04	.16
Placing topsoil, delivered to site, 1 CY covers 81 SF at 4" depth					
With equipment, level site	BL@.092	CY	16.50	1.73	18.23
With equipment, sloped site	BL@.107	CY	17.00	2.01	19.01
By hand, level site	BL@.735	CY	16.50	13.80	30.30
By hand, sloped site	BL@.946	CY	16.50	17.80	34.30
Placing, by hand spreading					
Wood chip mulch, pine bark	BL@1.27	CY	30.00	23.90	53.90
Bale of straw, spread, 90 lb bale	BL@.200	Ea	5.00	3.76	8.76
Landscape stepping stones, concrete					
12" round	BL@.133	Ea	.90	2.50	3.40
14" round	BL@.133	Ea	2.60	2.50	5.10
18" round	BL@.153	Ea	4.60	2.88	7.48
24" round	BL@.153	Ea	8.50	2.88	11.38
Add for square shapes	--	%	30.0	--	--
18" or 24" diameter natural tree ring	BL@.133	Ea	1.75	2.50	4.25
Redwood benderboard, staked redwood					
5/16" x 4"	BL@.011	LF	.35	.21	.56
1" x 6"	BL@.016	LF	1.20	.30	1.50
1" x 8"	BL@.016	LF	1.50	.30	1.80
2" x 8" rough	BL@.020	LF	5.10	.38	5.48
Rototilling light soil					
To 4" depth	BL@.584	CSY	6.50	11.00	17.50
Sodding, no soil preparation, nursery sod, Bermuda or Blue Rye Mix					
Per SF (500 SF minimum charge)	BL@.013	SF	.38	.24	.55

Lath

	Craft@Hrs	Unit	Material	Labor	Total
Lath					
Gypsum lath, perforated or plain, nailed to walls and ceilings					
3/8" x 16" x 48"	BR@.076	SY	2.10	1.74	3.84
1/2" x 16" x 48"	BR@.083	SY	2.07	1.90	3.97
Add for foil back insulating lath	--	SY	.50	--	--
Steel lath					
Diamond lath (junior mesh), 27" x 96" sheets, nailed to walls					
2.5 lb black painted	BR@.076	SY	1.50	1.74	3.24
2.5 lb galvanized	BR@.076	SY	2.00	1.74	3.74
1.75 lb black painted	BR@.076	SY	1.50	1.74	3.24
3.4 lb black painted	BR@.076	SY	2.00	1.74	3.74
3.4 lb galvanized	BR@.076	SY	2.25	1.74	3.99
Z-Riblath, 1/8" to wood frame (flat rib)					
2.75 lb painted, 24", 27" x 96" sheets	BR@.076	SY	2.70	1.74	4.44
3.4 lb painted, 24", 27" x 96" sheets	BR@.088	SY	3.30	2.02	5.32
Riblath, 3/8" to wood frame (high rib)					
3.4 lb painted, 24", 27" x 96" sheets	BR@.088	SY	2.25	2.02	4.27
3.4 lb galvanized, 24", 27" x 96" sheets	BR@.088	SY	2.60	2.02	4.62
3.4 lb painted paperback, 27" x 96" sheets	BR@.088	SY	2.50	2.02	4.52
Add for ceiling applications	--	%	--	25.0	--
Corner beads, 26 gauge, 8' to 12' lengths					
Flexible all-purpose bead	BR@2.74	CLF	40.00	62.80	102.80
Expansion, 2-1/2" flange	BR@2.74	CLF	42.00	62.80	104.80
Cornerite or cornalath					
2" x 2" x 4", steel-painted copper alloy	BR@2.74	CLF	12.50	62.80	75.30
Corneraid (Stockton), 2-1/2" smooth wire	--	CLF	23.00	--	--
Casing beads					
Square nose, short flange, 1/2" or 3/4"	BR@2.54	CLF	30.00	58.20	88.20
Quarter round, short flange, 1/2" or 3/4"	BR@2.54	CLF	29.00	58.20	87.20
Square nose, expansion, 1/2" x 3/4"	BR@2.54	CLF	36.00	58.20	94.20
Quarter round, expansion, 1/2" x 3/4"	BR@2.54	CLF	36.00	58.20	94.20
Archbead (flexible plastic nose) for curves	BR@2.54	CLF	50.00	58.20	108.20
Base screed,					
flush or curved point	BR@1.82	CLF	35.00	41.70	76.70
Expansion joint, 26 gauge, 1/2" ground	BR@2.54	CLF	90.00	58.20	148.20
Archaid (for curves and arches), 8' length	BR@.028	LF	.37	.64	1.01
Galvanized stucco netting					
1" x 20 gauge x 48", 450 SF/roll	--	Roll	90.00	--	--
1" x 18 gauge x 48", 450 SF/roll	--	Roll	140.00	--	--
1" x 18 gauge x 48", 600 SF/roll	--	Roll	175.00	--	--
1-1/2" x 17 gauge x 36", 150 SF/roll	--	Roll	50.00	--	--
Self-furring	--	Roll	50.00	--	--
Self-furring, paperback, 100 SF/roll	--	Roll	55.00	--	--
Wood lath. See Lumber, Redwood					
Lathing, Subcontract					
Metal lath, nailed on wood studs					
2.5 lb diamond mesh, copper alloy	--	SY	--	--	4.20
2.5 lb galvanized	--	SY	--	--	4.70
3.4 lb painted copper alloy	--	SY	--	--	5.05
Add for ceiling applications	--	%	--	--	15.0

Lathing

	Craft@Hrs	Unit	Material	Labor	Total
Lathing, Subcontract					
Gypsum lath, perforated or plain, nailed on wood studs					
3/8"	--	SY	--	--	4.85
1/2"	--	SY	--	--	5.20
Add for ceiling applications	--	%	--	--	15.0
Add for foil facing	--	SY	--	--	.80
15 lb felt and stucco netting	--	SY	--	--	2.75
Lawn Sprinkler System					
PVC pipe, Schedule 40, including typical hand trenching, plastic fittings (but no valves or sprinkler heads), installed					
1/2" pipe at $.15/LF, with 1 fitting each					
10 LF at $.35	BL@.040	LF	.19	.75	.94
3/4" pipe at $.18/LF, with 1 fitting each					
10 LF at $.45	BL@.040	LF	.23	.75	.98
1" pipe at $.27/LF, with 1 fitting each					
10 LF at $.55	BL@.040	LF	.33	.75	1.08
Deduct for Class 315 pipe	--	%	-40.0	--	--
Lawn Sprinklers, individual components					
Trenching for pipe installation, by hand					
Light soil, 8" wide					
12" deep	BL@.027	LF	--	.51	--
18" deep	BL@.040	LF	--	.75	--
24" deep	BL@.053	LF	--	1.00	--
Average soil, 8" wide					
12" deep	BL@.043	LF	--	.81	--
18" deep	BL@.065	LF	--	1.22	--
24" deep	BL@.087	LF	--	1.64	--
Heavy soil or loose rock, 8" wide					
12" deep	BL@.090	LF	--	1.69	--
18" deep	BL@.100	LF	--	1.88	--
24" deep	BL@.134	LF	--	2.52	--
Impact sprinkler heads, riser mounted, with nozzle, full circle					
Brass, with diffuser	BL@.422	Ea	12.00	7.93	19.93
Plastic, with diffuser	BL@.422	Ea	6.00	7.93	13.93
Impact sprinkler, pop-up type					
Brass and stainless steel	BL@.422	Ea	34.00	7.93	41.93
Plastic	BL@.422	Ea	17.00	7.93	24.93
Rotor pop-up,					
Full circle	BL@.422	Ea	80.00	7.93	87.93
Rotor pop-up, with electric valve in head,					
Full circle	BL@2.52	Ea	110.00	47.40	157.40
Spray head pop-up type, plastic, with nozzle					
2" height	BL@.422	Ea	2.50	7.93	10.43
4" height	BL@.422	Ea	4.50	7.93	12.43
6" height	BL@.422	Ea	8.00	7.93	15.93
Spray head, brass, for riser mounting					
Flat	BL@.191	Ea	2.50	3.59	6.09
Standard	BL@.191	Ea	2.80	3.59	6.39
Stream bubbler, all patterns	BL@.191	Ea	4.00	3.59	7.59

Lawn Sprinklers

	Craft@Hrs	Unit	Material	Labor	Total
Lawn Sprinklers, individual components, continued					
Sprinkler head risers, 1/2" or 3/4", installed to main or branch supply					
6" riser with three fittings	BL@.056	Ea	7.00	1.05	8.05
8" riser with three fittings	BL@.056	Ea	7.00	1.05	8.05
12" riser with three fittings	BL@.056	Ea	8.00	1.05	9.05
18" riser with three fittings	BL@.056	Ea	12.00	1.05	13.05
24" riser with three fittings	BL@.056	Ea	14.00	1.05	15.05
Valves, non-siphon, manual, brass					
3/4"	BL@1.03	Ea	20.00	19.40	39.40
1"	BL@1.38	Ea	22.00	25.90	47.90
Valves, electric solenoid, brass, with siphon breaker, hookup but no wire included					
3/4"	BL@2.52	Ea	42.00	47.40	89.40
1"	BL@3.53	Ea	46.00	66.40	112.40

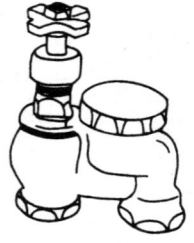

Valve control wire, 16 gauge low voltage direct burial cable, laid with pipe. No trenching or end connections included

	Craft@Hrs	Unit	Material	Labor	Total
2 wire	BL@.003	LF	.11	.06	.17
3 wire	BL@.003	LF	.12	.06	.18
4 wire	BL@.003	LF	.14	.06	.20
6 wire	BL@.003	LF	.20	.06	.26
8 wire	BL@.003	LF	.26	.06	.32

Automatic irrigation control stations, add for electrical service connection below, no valve, or wiring included

	Craft@Hrs	Unit	Material	Labor	Total
Electro-mechanical type, controls up to					
6 separate electric valves	BL@1.83	LS	115.00	34.40	149.40
Programmable computerized type, controls up to					
8 separate electric valves	BL@1.83	LS	170.00	34.40	204.40
Add for 120 volt electrical connection	--	LS	--	--	85.00

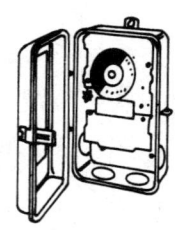

Typical subcontract price, residential system for 1,700 SF area using PVC Schedule 40 pipe, 340 LF of 3/4" pipe, 35 LF of 1" pipe, four 1" valves, 30 brass, full circle heads and 6" risers on each head, based on heads spaced at 10'.

	Craft@Hrs	Unit	Material	Labor	Total
Large, regular shaped areas	--	SF	--	--	.46
Narrow and irregular areas	--	SF	--	--	.51
Add for freezing zones	--	SF	--	--	.06

Layout. Foundation layout, medium to large size residence, 6 to 8 outside corners, includes shooting elevation from nearby reference, stakes and batterboards as required. No surveying or clearing included

	Craft@Hrs	Unit	Material	Labor	Total
Per residence	B1@7.86	LS	35.00	167.00	202.00

Lighting Fixtures. See also Electrical Work. Costs are to hang and connect fixtures only. No wiring included.

Surface mounted fluorescent fixtures Including ballasts but no lamps.
High level output, wrap around acrylic diffuser, black metal pan

	Craft@Hrs	Unit	Material	Labor	Total
14-1/8" wide x 26-5/8" long x 4" deep (requires four 20 watt tubes)	BE@1.60	Ea	127.00	40.40	167.40
13-1/2" wide x 50" long x 4" deep (requires two 40 watt tubes)	BE@1.60	Ea	90.50	40.40	130.90
26" wide x 26" long x 4" deep (requires two 40 watt U/6 tubes)	BE@1.60	Ea	105.00	40.40	145.40
25" wide x 50" long x 4" deep (requires four 40 watt tubes)	BE@1.60	Ea	155.00	40.40	195.40

Lighting

	Craft@Hrs	Unit	Material	Labor	Total

Surface mounted fluorescent fixtures
Decorative model with vinyl trim on metal frame and acrylic diffuser.
Available in oak, walnut or butcher block woodgrained vinyl

	Craft@Hrs	Unit	Material	Labor	Total
9-1/4" wide x 24" long x 4" deep (requires two 20 watt tubes)	BE@1.60	Ea	41.50	40.40	81.90
9-1/4" wide x 48" long x 4" deep (requires two 40 watt tubes)	BE@1.60	Ea	52.50	40.40	92.90
15" wide x 15" long x 4" deep (one 22/32 watt circline tube)	BE@1.60	Ea	53.00	40.40	93.40
16" wide x 25" long x 4" deep (requires four 20 watt tubes)	BE@1.60	Ea	64.50	40.40	104.90
14" wide x 48" long x 4" deep (requires four 40 watt tubes)	BE@1.60	Ea	87.00	40.40	127.40
19" wide x 19" long x 4" deep (one 32/40 watt circline tube)	BE@1.60	Ea	56.00	40.40	96.40
24" wide x 24" long x 4" deep (requires two 40 watt U/6 tubes)	BE@1.60	Ea	68.00	40.40	108.40

Decorative, wrap around acrylic diffusers, solid oak wood ends

	Craft@Hrs	Unit	Material	Labor	Total
8-3/4" wide x 24" long x 2-1/2" deep (requires two 20 watt tubes)	BE@1.60	Ea	36.00	40.40	76.40
13-3/8" wide x 48" long x 2-1/2" deep (requires two 40 watt tubes)	BE@1.60	Ea	69.50	40.40	109.90

Economy, wrap around acrylic diffuser, basic metal ends

	Craft@Hrs	Unit	Material	Labor	Total
8-3/4" wide x 24" long x 2-1/2" deep (requires two 20 watt tubes)	BE@1.60	Ea	38.00	40.40	78.40
8-3/4" wide x 48" long x 2-1/2" deep (requires two 40 watt tubes)	BE@1.60	Ea	48.50	40.40	88.90

Decorative circular fixtures for circline tubes
Ornate simulated glass crystal with antiqued metal band,

	Craft@Hrs	Unit	Material	Labor	Total
13" diameter x 7-1/2" high (requires one 34 watt tube)	BE@.696	Ea	76.40	17.60	94.00
Ornate metal shade with wood finial, 17" diameter x 8-1/2" high (requires one 22 watt and one 32 watt tube)	BE@.696	Ea	78.80	17.60	96.40
Natural oak finish with glass diffusers, 14-1/2" diameter x 5-1/4" high (requires one 22 watt and one 32 watt tube)	BE@.696	Ea	66.80	17.60	84.40
Basic fixture with frosted glass diffuser, 15-3/4" diameter x 4-1/2" high (requires one 32 watt and one 40 watt tube)	BE@.696	Ea	53.80	17.60	71.40

Utility type circular fixtures without diffusers

	Craft@Hrs	Unit	Material	Labor	Total
Two tube fixture, 16" diameter x 3-1/2" high (requires one 32 and one 40 watt circline tube)	BE@.696	Ea	36.70	17.60	54.30
Two tube fixture, 12-1/4" diameter x 3-1/2" high (requires one 22 and one 32 watt circle tube)	BE@.696	Ea	23.10	17.60	40.70
One tube fixture, 12-1/2" diameter x 3-1/2" high (requires one 22 watt circline tube)	BE@.696	Ea	20.40	17.60	38.00

Low profile rectangular wall or ceiling mount fixture, metal housing, white acrylic diffuser, 15-1/2" x 5",

	Craft@Hrs	Unit	Material	Labor	Total
(requires two 9 watt energy saver bulbs)	BE@.696	Ea	48.00	17.60	65.60

Rectangular indoor/outdoor model with discoloration resistant/impact resistant acrylic diffuser and cold weather ballast

	Craft@Hrs	Unit	Material	Labor	Total
9-7/8" wide x 9-7/8" long x 4-3/4" deep (requires one 22 watt tube)	BE@.696	Ea	52.50	17.60	70.10
12" wide x 12" long x 5" deep (requires one 34 watt circline tube)	BE@.696	Ea	68.50	17.60	86.10

Lighting

	Craft@Hrs	Unit	Material	Labor	Total

Surface mounted fluorescent fixtures, continued

Hall lantern type, including tubes

| 4-1/4" wide x 12-1/8" long x 4" deep | BE@.696 | Ea | 33.00 | 17.60 | 50.60 |

Decorative valance type, with solid oak frame and acrylic diffuser

11-11/16" wide x 25" long x 4" deep (requires two 20 watt tubes)	BE@2.06	Ea	90.30	52.10	142.40
25" wide x 25" long x 4" deep (requires two 40 watt U/6 tubes)	BE@1.59	Ea	112.00	40.20	152.20
11-11/16" wide x 49-1/8" long x 4" deep (requires two 40 watt tubes)	BE@1.59	Ea	102.00	40.20	142.20

Economy valance type, with metal ends. Round or square models available. All models are 2-5/8" wide x 4-1/4" deep

24-3/4" L (requires one 15 watt tube)	BE@1.59	Ea	23.50	40.20	63.70
24-3/4" L (requires one 20 watt tube)	BE@1.59	Ea	26.50	40.20	66.70
36-3/4" L (requires one 30 watt tube)	BE@1.59	Ea	34.50	40.20	74.70
36-3/4" L (requires two 30 watt tubes)	BE@1.59	Ea	42.00	40.20	82.20
48-3/4" L (requires one 40 watt tube)	BE@1.59	Ea	47.50	40.20	87.70
48-3/4" L (requires two 40 watt tubes)	BE@1.59	Ea	53.50	40.20	93.70

Under shelf/cabinet type for wiring to wall switch

18" long x 5" wide x 1-1/2" deep (requires one 15 watt tube)	BE@.696	Ea	12.50	17.60	30.10
24" long x 5-1/4" wide x 1-3/4" deep (requires one 20 watt tube)	BE@.696	Ea	17.00	17.60	34.60
36" long x 5-1/4" wide x 1-3/4" deep (requires one 30 watt tube)	BE@.696	Ea	27.00	17.60	44.60
48" long x 5-1/4" wide x 1-3/4" deep (requires one 40 watt tube)	BE@.696	Ea	30.00	17.60	47.60

	Cool white	Deluxe cool white	Energy saving cool white	Warm white
Fluorescent tubes, add labor to install below				
8 watt (12" long)	5.15	8.70	--	7.55
13 watt (21" long)	6.40	--	--	--
15 watt (18" long)	4.70	7.40	--	6.00
20 watt (24" long)	4.95	8.00	7.55	5.30
30 watt (36" long)	6.55	9.80	5.20	7.90
40 watt (48" long)	3.50	--	--	--
Slimline straight tubes				
20 watt (24" long)	12.40	--	--	--
30 watt (36" long)	12.40	--	9.95	14.40
40 watt (48" long)	8.30	11.30	17.00	10.20
Circular (circline) tubes				
20 watt (6-1/2" diameter)	8.10	--	--	8.70
22 watt (8-1/4" diameter)	7.10	9.10	--	8.60
32 watt (12" diameter)	7.50	11.00	--	9.70
40 watt (16" diameter)	10.60	--	--	13.00
"U" shaped tubes (3" spacing x 12" long)				
U-3 40 watt	11.20	15.60	--	11.20
U-6 40 watt	11.20	15.60	--	11.20
Labor to install tubes	1.16	1.16	1.16	1.16

Lighting

	Craft@Hrs	Unit	Material	Labor	Total
Incandescent lighting fixtures					
Outdoor fixtures, bulbs not included					
Post lantern type fixture, cast aluminum, with aluminum 8' post, no digging included	BE@2.00	Ea	83.00	50.60	133.60
Modern 7-1/2" x 7-1/2" porch ceiling fixture, 3" drop, single light	BE@2.15	Ea	19.50	54.40	73.90
Flood light fixture, zinc mounting					
1 lite (no lamps)	BE@2.15	Ea	8.61	54.40	63.01
2 lite (no lamps)	BE@2.15	Ea	11.00	54.40	65.40
Path lighting, low voltage, light kits including transformer, timer, 100' of wire, 12 volt lights, hardware, and typical installation					
8 flood lights, 18-1/4" high x 6" wide	BE@3.18	Ea	82.80	80.40	163.20
6 flood lights, 8-1/2" high x 5-3/4" wide	BE@2.98	Ea	62.10	75.30	137.40
6 path lights, 11" high x 66" diameter	BE@2.63	Ea	62.10	66.50	128.60
Recessed fixtures, including housing, incandescent					
Wired ceiling fixtures, chrome frame, pre-wired housing					
5" shower light, with opal lens, 60 watt	BE@1.72	Ea	30.50	43.50	74.00
5" mini-light for restricted spaces, 40 watt	BE@1.72	Ea	30.50	43.50	74.00
7" frame, drop lens or stepped baffle	BE@1.72	Ea	30.50	43.50	74.00
7" frame, flush lens, 75 watt	BE@1.72	Ea	28.50	43.50	72.00
7" clear aluminum reflector, lights large area	BE@1.72	Ea	38.00	43.50	81.50
8-1/8" white enamel finish eyeball spotlight, 100 watt	BE@1.72	Ea	34.00	43.50	77.50
12" long x 10" wide square fixture, with diffusing glass lens	BE@1.72	Ea	40.00	43.50	83.50
Incandescent ceiling fixtures, using 60 watt standard bulbs, pre-wired housing					
15-1/2" x 7", acrylic cut crystal-look fixture (requires 3 bulbs)	BE@1.59	Ea	24.50	40.20	64.70
11-1/4" x 6-1/2", with frosted glass globe and brass plated band (requires 2 bulbs)	BE@1.59	Ea	36.50	40.20	76.70
12-1/2" brass base, designer fixture with 3 enameled metal swivel heads (requires 3 bulbs)	BE@1.59	Ea	48.50	40.20	88.70
14" x 6", frosted glass globe with floral design, antique brass plated base (requires 3 bulbs)	BE@1.59	Ea	61.00	40.20	101.20

Lumber. Prices for lumber vary with changes in supply and demand and increase with increases in distance from the mill or source of supply. Lumber costs in this section are based on an average of West Coast and Midwest prices as of summer 1991 and have been projected to mid 1992 by adding approximately 5%. Carload lots will cost 15% to 20% less and retail purchases will cost from 20% to 25% more. Generally, lumber prices will be 10% to 20% lower in the Southern U.S. and slightly higher in the Northeast.

		Construction Green	Std. & Btr K.D.	Utility K.D.
Douglas Fir S4S, 8' to 20' lengths				
2" x 4"	MBF	490.00	445.00	370.00
2" x 6"	MBF	495.00	450.00	340.00
2" x 8"	MBF	490.00	465.00	325.00
2" x 10"	MBF	635.00	610.00	320.00
2" x 12"	MBF	615.00	605.00	330.00
Fir & Larch S4S, 8' to 20' lengths				
2" x 4"	MBF	435.00	430.00	--
2" x 6"	MBF	440.00	430.00	--
2" x 8"	MBF	440.00	430.00	--

Lumber

		Construction Green	Std. & Btr K.D.	Utility K.D.
Fir & Larch S4S, 8' to 20' lengths				
2" x 10"	MBF	600.00	610.00	--
2" x 12"	MBF	600.00	595.00	--
Hemlock - Fir S4S, 8' to 20' lengths				
2" x 4"	MBF	--	440.00	--
2" x 6"	MBF	--	430.00	--
2" x 8"	MBF	--	455.00	--
2" x 10"	MBF	--	550.00	--
2" x 12"	MBF	--	545.00	--
Ponderosa Pine S4S, 8' to 20' lengths				
2" x 4"	MBF	--	430.00	--
2" x 6"	MBF	--	430.00	--
2" x 8"	MBF	--	430.00	--
2" x 10"	MBF	--	445.00	--
2" x 12"	MBF	--	465.00	--
Red Cedar S4S, 8' to 20' lengths				#2 & Btr
2" x 4"	MBF	--	--	570.00
2" x 6"	MBF	--	--	620.00
2" x 8"	MBF	--	--	610.00
2" x 10"	MBF	--	--	670.00
2" x 12"	MBF	--	--	705.00
Southern Yellow Pine S4S, 8' to 20' lengths		#1	#2	
2" x 4"	MBF	420.00	385.00	--
2" x 6"	MBF	420.00	395.00	--
2" x 8"	MBF	420.00	395.00	--
2" x 10"	MBF	490.00	480.00	--
2" x 12"	MBF	585.00	570.00	--
Spruce & Pine S4S, 8' to 20' lengths			Std. & Btr K.D.	
2" x 4"	MBF	--	370.00	--
2" x 6"	MBF	--	380.00	--
2" x 8"	MBF	--	370.00	--
2" x 10"	MBF	--	395.00	--
2" x 12"	MBF	--	445.00	--
Studs, random lengths, S4S, 2" x 4" or 2" x 6"		Construction Green	Std. & Btr K.D.	
Hemlock - Fir	MBF	--	430.00	--
Douglas Fir	MBF	475.00	455.00	--
Fir & Larch	MBF	455.00	420.00	--
White Woods	MBF	--	430.00	--
Spruce or Pine	MBF	--	370.00	--
Southern Yellow Pine	MBF	--	400.00	--
Economy grade, non-code	MBF	155.00	--	--
Studs, precut lengths, S4S, 2" x 4" or 2" x 6"				
Hemlock - Fir	MBF	--	425.00	--
Douglas Fir	MBF	465.00	450.00	--
Fir & Larch	MBF	450.00	415.00	--
White Woods	MBF	--	420.00	--
Spruce or Pine	MBF	--	365.00	--
Southern Yellow Pine	MBF	--	390.00	--

Lumber Index

Item	Page
Douglas fir	117-118
Dowels	125
Englemann spruce	119-120
Fir and larch	118, 120
Flooring	121
Hardwood lumber	123
Hemlock	118
Idaho white pine	119
Particleboard	121
Poles and posts	123
Ponderosa pine	118-119
Plywood	121-122
Red cedar	118, 119
Redwood	120
Redwood, common	119
Redwood, clear	120
Sheathing	122
Siding, board	125-127
Southern pine	120
Southern yellow pine	121
Specialty lumber	121
Spruce and pine	118
Spruce Lodgepole	119
beams and timbers	119
Stakes	127
Strandboard	122
Studs	118
Veneers	124-125

Lumber

		Construction Green	Std. & Btr K.D.	
White Woods - Spruce Lodgepole S4S, 8' to 20' lengths				
2" x 4"	MBF	--	420.00	
2" x 6"	MBF	--	410.00	

Beams and Timbers

Douglas Fir, rough, 8' to 20' lengths
4" x 4"	MBF	755.00	--
4" x 6"	MBF	755.00	--
4" x 8"	MBF	755.00	--
4" x 10"	MBF	755.00	--
4" x 12"	MBF	765.00	--
6" x 6"	MBF	1,000.00	--

Red Cedar, rough, 8' to 20' lengths
4" x 4"	MBF	635.00	--
4" x 6"	MBF	660.00	--
4" x 8"	MBF	660.00	--
4" x 10"	MBF	705.00	--
4" x 12"	MBF	725.00	--
6" x 6"	MBF	710.00	--

Ponderosa Pine Boards, 8' to 20' lengths		C & Btr	D & Btr
1" x 4"	MBF	1,590.00	1,010.00
1" x 6"	MBF	2,040.00	1,330.00
1" x 8"	MBF	1,815.00	1,330.00
1" x 10"	MBF	1,665.00	1,590.00
1" x 12"	MBF	2,340.00	1,960.00

Ponderosa Pine Boards, 8' to 20' lengths		#2 & Btr	#3 & Btr	#4 & Btr
1" x 4"	MBF	580.00	410.00	295.00
1" x 6"	MBF	590.00	425.00	310.00
1" x 8"	MBF	635.00	410.00	310.00
1" x 10"	MBF	640.00	455.00	325.00
1" x 12"	MBF	885.00	580.00	330.00

Idaho White Pine Boards, 8' to 20' lengths		Choice & Btr	Quality	Sterling
1" x 4"	MBF	1,735.00	1,025.00	875.00
1" x 6"	MBF	2,260.00	1,390.00	875.00
1" x 8"	MBF	2,035.00	1,285.00	875.00
1" x 10"	MBF	1,850.00	1,700.00	875.00
1" x 12"	MBF	2,410.00	2,075.00	890.00

Idaho White Pine Boards, 8' to 20' lengths		Standard	Utility	
1" x 4"	MBF	405.00	285.00	--
1" x 6"	MBF	505.00	295.00	--
1" x 8"	MBF	460.00	295.00	--
1" x 10"	MBF	460.00	300.00	--
1" x 12"	MBF	575.00	340.00	--

Engelmann Spruce Boards, 8' to 20' lengths		D & Btr	#2 & Btr	
1" x 4"	MBF	815.00	555.00	--
1" x 6"	MBF	1,310.00	625.00	--
1" x 8"	MBF	1,310.00	640.00	--

Lumber

		D & Btr	#2 & Btr	
Engelmann Spruce Boards, 8' to 20' lengths				
1" x 10"	MBF	1,345.00	640.00	
1" x 12"	MBF	1,870.00	940.00	

		#3 & Btr	#4 & Btr	
Engelmann Spruce Boards, 8' to 20' lengths				
1" x 4"	MBF	395.00	290.00	
1" x 6"	MBF	415.00	300.00	
1" x 8"	MBF	405.00	300.00	
1" x 10"	MBF	440.00	315.00	
1" x 12"	MBF	575.00	340.00	

		D & Btr	#3 & Btr	#4 & Btr
Fir & Larch Boards, 8' to 20' lengths				
1" x 4"	MBF	875.00	385.00	325.00
1" x 6"	MBF	895.00	415.00	305.00
1" x 8"	MBF	910.00	350.00	300.00
1" x 10"	MBF	915.00	380.00	305.00
1" x 12"	MBF	1,050.00	435.00	325.00

		#2 & Btr	#3 & Btr	
Red Cedar Boards, S1S2E, 8' to 20' lengths				
1" x 4"	MBF	570.00	660.00	--
1" x 6"	MBF	620.00	905.00	--
1" x 8"	MBF	610.00	930.00	--
1" x 10"	MBF	670.00	955.00	--
1" x 12"	MBF	705.00	1,015.00	--

Redwood, random lengths 6' to 20', S4S kiln dried. West Coast prices, transportation costs will add from 10% to 30% to these costs in other areas

Size		Select All Heart	Clear Select	B Grade
1" x 2"	MBF	2,180.00	1,750.00	1,350.00
1" x 3"	MBF	2,250.00	1,550.00	1,950.00
1" x 4"	MBF	2,250.00	1,380.00	1,350.00
1" x 6"	MBF	2,280.00	1,480.00	1,650.00
1" x 8"	MBF	2,350.00	1,850.00	2,250.00
1" x 10"	MBF	2,500.00	1,800.00	2,400.00
1" x 12"	MBF	2,750.00	1,650.00	2,450.00
2" x 4"	MBF	2,650.00	1,150.00	1,900.00
2" x 6"	MBF	2,500.00	1,300.00	1,700.00
2" x 8"	MBF	2,650.00	1,500.00	2,000.00
2" x 10"	MBF	3,100.00	1,800.00	2,750.00
2" x 12"	MBF	3,700.00	1,600.00	3,000.00
4" x 4" to 6"	MBF	3,700.00	--	1,500.00
Add for specified lengths	%	20.0	20.0	20.0

		C & Btr	D	
Southern Pine Boards, 8' to 20' lengths				
1" x 4"	MBF	785.00	630.00	--
1" x 6"	MBF	935.00	780.00	--
1" x 8"	MBF	980.00	815.00	--
1" x 10"	MBF	955.00	825.00	--
1" x 12"	MBF	1,290.00	970.00	--

		#2 & Btr	#3	
Southern Pine Boards, 8' to 20' lengths				
1" x 4"	MBF	390.00	310.00	--
1" x 6"	MBF	490.00	320.00	--

Lumber

		#2 & Btr	#3
Southern Pine Boards, 8' to 20' lengths			
1" x 8"	MBF	475.00	325.00
1" x 10"	MBF	470.00	340.00
1" x 12"	MBF	515.00	355.00
Southern Yellow Pine Decking, 8' to 16' lengths			
5/4" x 6"	MBF	--	845.00

		C & Btr	D Grade
Fir Flooring, 4' to 20' random lengths, T&G			
1" x 3" flooring, vertical grain	MBF	2,500.00	1,800.00
1" x 4" flooring, vertical grain	MBF	2,500.00	1,800.00
1" x 3" flooring, flat grain	MBF	1,450.00	1,150.00
1" x 4" flooring, flat grain	MBF	1,450.00	1,150.00

			D Grade
Southern Yellow Pine Flooring, random lengths			
1" x 4"	MBF	--	630.00
1" x 6"	MBF	--	780.00
1" x 8"	MBF	--	815.00
1" x 10"	MBF	--	825.00
1" x 12"	MBF	--	970.00

Specialty Lumber

Douglas fir, finish and clear, S4S, dry, 8' to 20' specified lengths, flat or vertical grain. West Coast prices, transportation costs will add from 10% to 30% to these costs in other areas.

Size		C & Btr (V/G)	C & Btr (F/G)
1" x 2"	MBF	1,850.00	1,600.00
1" x 3"	MBF	1,850.00	1,600.00
1" x 4"	MBF	1,850.00	1,600.00
1" x 6"	MBF	2,250.00	2,000.00
1" x 8"	MBF	2,750.00	2,500.00
1" x 10"	MBF	2,750.00	2,500.00
1" x 12"	MBF	4,150.00	3,900.00
2" x 2"	MBF	2,050.00	1,750.00
2" x 3"	MBF	2,350.00	2,100.00
2" x 4"	MBF	1,800.00	1,575.00
2" x 6"	MBF	2,250.00	2,000.00
2" x 8"	MBF	2,400.00	2,175.00
2" x 10"	MBF	2,900.00	2,650.00
2" x 12"	MBF	4,250.00	4,000.00
Add for each 2" width over 12"	MBF	50.00	50.00
Add for 22' and 24' lengths	MBF	20.00	20.00

Particleboard, underlayment		Standard	Industrial
3/8"	MSF	200.00	220.00
1/2"	MSF	220.00	265.00
5/8"	MSF	260.00	290.00
3/4"	MSF	340.00	305.00

Plywood
Sanded, exterior grade

		AA	AB	AC	BC
1/4"	MSF	620.00	595.00	435.00	360.00
3/8"	MSF	735.00	715.00	550.00	420.00
1/2"	MSF	815.00	795.00	635.00	525.00
5/8"	MSF	930.00	910.00	745.00	640.00
3/4"	MSF	1,080.00	1,060.00	895.00	785.00

Lumber

Plywood
Sanded, interior grade

		AA	AB	AD	BD
1/4"	MSF	605.00	580.00	420.00	345.00
3/8"	MSF	720.00	700.00	535.00	400.00
1/2"	MSF	805.00	785.00	620.00	510.00
5/8"	MSF	915.00	895.00	730.00	620.00
3/4"	MSF	1,065.00	1,040.00	880.00	765.00

Sheathing
Exterior grade

		CD	CC	Mill Grade
5/16"	MSF	235.00	270.00	215.00
3/8"	MSF	260.00	310.00	225.00
1/2"	MSF	325.00	--	255.00
1/2" 4 ply	MSF	345.00	420.00	220.00
1/2" 5 ply	MSF	395.00	--	--
5/8"	MSF	375.00	--	--
5/8" 4 ply	MSF	450.00	510.00	330.00
5/8" 5 ply	MSF	480.00	--	--
3/4"	MSF	500.00	590.00	415.00

Interior grade

		CD	CC	Mill Grade
5/16"	MSF	315.00	--	--
3/8"	MSF	345.00	--	--
1/2"	MSF	395.00	--	--
5/8"	MSF	530.00	--	--
3/4"	MSF	625.00	--	--

Strand board, oriented grain

		Sq. Edged	T & G	
1/4"	MSF	185.00	--	--
3/8"	MSF	240.00	--	--
7/16"	MSF	290.00	--	--
1/2"	MSF	310.00	--	--
5/8"	MSF	--	405.00	--
3/4"	MSF	--	465.00	--

Underlayment, X-Band, C Grade

5/8" 4 ply	MSF	--	465.00	--
3/4" 4 ply	MSF	--	555.00	--

Concrete form (Plyform)

5/8" 4 ply	MSF	620.00	--	--
3/4" 4 ply	MSF	710.00	--	--

Sidings, rough sawn, 4 ply

		T-1-11	RBB	
3/8" thick, 4' x 8' panels	MSF	525.00	--	--
5/8" thick, 4' x 8' panels	MSF	705.00	760.00	--
3/8" thick, 4' x 9' panels	MSF	565.00	--	--
5/8" thick, 4' x 9' panels	MSF	740.00	800.00	--

	Craft@Hrs	Unit	Material	Labor	Total
Lath, Redwood "A" & better, 50 per bundle, nailed to walls for architectural treatment					
3/8" x 1-1/2" x 4'	BC@.500	Bdle	15.00	11.80	26.80
3/8" x 1-1/2" x 6'	BC@.625	Bdle	24.00	14.70	38.70
3/8" x 1-1/2" x 8'	BC@.750	Bdle	27.50	17.70	45.20

Lumber

Poles, select appearance grade building poles. Poles have an average taper of 1" in diameter per 10' of length. Ponderosa pine treated with Penta-Dow or ACZA (.60 lbs/CF), meets AWPA-C 23-77. Costs based on quantity of 8 to 10 poles. Add for delivery below. Add for labor following Posts. Costs are per linear foot.

		8" butt size	10" butt size	12" butt size
To 15' long pole	LF	5.75	7.75	10.90
To 20' long pole	LF	5.25	7.15	10.20
To 25' long pole	LF	4.75	6.75	9.80
Add for Douglas fir	LF	1.75	1.75	1.75
Add for delivery, typical	LF	2.55	2.55	2.55

Posts

		6' high	7' high	8' high	10' high
Sawn posts, merchantable, redwood, 4" x 4"	Ea	9.50	10.50	12.00	14.00
Yellow pine penta treated post					
3" x 3"	Ea	3.15	4.35	5.00	6.00
3-1/2" x 3-1/2"	Ea	4.50	5.00	5.75	7.00
4" x 4"	Ea	5.50	6.50	7.25	9.00
4-1/2" x 4-1/2"	Ea	5.50	7.00	8.25	9.50
5" x 5"	Ea	7.50	8.50	9.50	12.00
6" x 6"	Ea	9.00	10.00	11.00	13.00
7" x 7"	Ea	11.00	13.50	14.80	17.00

Labor to Install Poles and Posts
Costs shown are per post and include measuring, cutting one end, drilling two holes at the top and bottom and installation. Add for hardware and bolts as needed.

	Craft@Hrs	Unit	Material	Labor	Total
3" x 3", 4" x 4" or 5" x 5" posts	B1@.333	Ea	--	7.06	--
6" x 6" or 7" x 7" posts	B1@.500	Ea	--	10.60	--
8" or 10" butt diameter poles	B1@.666	Ea	--	14.10	--
12" butt diameter poles	B1@.750	Ea	--	15.90	--

Hardwood Lumber. See also Paneling and Flooring.
Kiln dried dimension hardwood lumber, rough, random widths and lengths, first and second grade. Costs per board foot.

	Thickness				
	1/2"	3/4"	1"	1-1/4"	1-3/4"
Alder	--	1.73	2.10	2.15	2.34
Ash, Eastern domestic	--	2.62	3.35	3.55	3.16
Beech, Japanese	--	--	3.00	--	3.50
Birch, Eastern domestic	--	2.01	2.50	2.55	2.87
Cherry, American Eastern	--	3.24	4.05	--	3.94
Elm, American	--	--	3.75	--	4.70
Koa	--	8.10	5.75	--	9.00
Mahogany					
Philippine, Light Red	1.77	2.04	2.75	2.75	2.77
Genuine Honduras	--	3.25	4.40	4.65	4.65
Maple, Eastern hard	--	2.08	2.50	2.65	2.80
Oak					
Appalachian white	--	2.11	2.95	3.15	4.42
Domestic red	--	2.50	3.65	2.98	3.85
Poplar	1.73	1.73	1.90	2.05	2.10
Shina (basswood), Japanese, select	--	--	2.65	--	2.90
Sitka spruce, vertical grain, clear	--	--	2.65	--	3.10
Teak, not kiln dried	--	7.20	7.25	7.80	8.10
Walnut, South American	--	--	7.10	--	9.10

Lumber

	Thickness			
	1"	1-1/4"	1-1/2"	2"

Hardwood Lumber, continued. See also Paneling and Flooring.
Kiln dried dimension hardwood lumber, rough, random widths and lengths, first and second grade. Costs per board foot.

	1"	1-1/4"	1-1/2"	2"
Walnut, Eastern Black	3.26	4.10	4.25	4.80
Willow	--	4.00	--	4.35

Veneers, architectural grade, 1/42" thick, 8' lengths, random widths, 4" to 12". Costs are based on greater than 20 SF and less than 100 SF quantities. Material costs include 5% for waste but no substrate under the veneer. Labor shown includes layout fit-up, trim, glue and clamp, remove clamps and trim edges with a router. For scheduling purposes, estimate that a two-man crew will install 65 SF of veneer per 8-hour day.

	Craft@Hrs	Unit	Material	Labor	Total
Acacia	B1@.250	SF	.95	5.30	6.25
Afromosia, Africa	B1@.250	SF	.65	5.30	5.95
Ash, brown	B1@.250	SF	.60	5.30	5.90
Ash, white	B1@.250	SF	.70	5.30	6.00
Avodire	B1@.250	SF	1.35	5.30	6.65
Beech	B1@.250	SF	.55	5.30	5.85
Benin (tigerwood), stripe, Africa	B1@.250	SF	.60	5.30	5.90
Birch, natural	B1@.250	SF	.55	5.30	5.85
Birch, white	B1@.250	SF	.80	5.30	6.10
Bubinga, Africa, stripe	B1@.250	SF	.75	5.30	6.05
Butternut	B1@.250	SF	1.30	5.30	6.60
Cedar	B1@.250	SF	.50	5.30	5.80
Cherry	B1@.250	SF	.60	5.30	5.90
Chestnut, wormy	B1@.250	SF	1.00	5.30	6.30
East Indian laurel	B1@.250	SF	1.40	5.30	6.70
Elm	B1@.250	SF	.45	5.30	5.75
Ebony, Maccassur (India)	B1@.250	SF	3.70	5.30	9.00
English brown oak	B1@.250	SF	2.40	5.30	7.70
European white oak	B1@.250	SF	.90	5.30	6.20
French olive ash (benge)	B1@.250	SF	1.65	5.30	6.95
Goncalo alves (Brazil)	B1@.250	SF	1.00	5.30	6.30
Harewood (English)	B1@.250	SF	1.70	5.30	7.00
Kevazingo fig (Africa)	B1@.250	SF	.90	5.30	6.20
Lacewood-silky oak (Australia)	B1@.250	SF	.90	5.30	6.20
Limba (Korina) Africa	B1@.250	SF	.65	5.30	5.95
Mahogany, stripe	B1@.250	SF	.60	5.30	5.90
Mahogany, plain (Africa)	B1@.250	SF	.70	5.30	6.00
Makore, African cherry	B1@.250	SF	1.30	5.30	6.60
Maple, white	B1@.250	SF	.50	5.30	5.80
Maple, fiddleback	B1@.250	SF	1.30	5.30	6.60
Oak, white plain	B1@.250	SF	.60	5.30	5.90
Oak, white rift	B1@.250	SF	.70	5.30	6.00
Oak, red	B1@.250	SF	.55	5.30	5.85
Oriental wood, Australia	B1@.250	SF	.65	5.30	5.95
Padouk, (Vermillion), India	B1@.250	SF	.75	5.30	6.05
Paldao, Australia	B1@.250	SF	1.30	5.30	6.60
Pearwood, Swiss	B1@.250	SF	2.65	5.30	7.95
Pine, southern yellow	B1@.250	SF	.55	5.30	5.85
Pecan	B1@.250	SF	.45	5.30	5.75
Poplar	B1@.250	SF	.50	5.30	5.80
Prima vera, Central American	B1@.250	SF	1.35	5.30	6.65

Lumber

	Craft@Hrs	Unit	Material	Labor	Total
Veneers, continued					
Purpleheart, Dutch Guiana	B1@.250	SF	.90	5.30	6.20
Rosewood, Brazilian	B1@.250	SF	8.10	5.30	13.40
Rosewood, East Indian	B1@.250	SF	3.50	5.30	8.80
Sapele, stripe, Africa	B1@.250	SF	.80	5.30	6.10
Satinwood, stripe	B1@.250	SF	2.70	5.30	8.00
Satinwood, bee wing	B1@.250	SF	2.70	5.30	8.00
Sen, Japanese	B1@.250	SF	.85	5.30	6.15
Sycamore	B1@.250	SF	2.65	5.30	7.95
Tamo, peanut fig, Japanese	B1@.250	SF	3.80	5.30	9.10
Teak (Indian) plain	B1@.250	SF	1.65	5.30	6.95
Teak (Indian) stripe	B1@.250	SF	1.20	5.30	6.50
Walnut, plain, USA	B1@.250	SF	.75	5.30	6.05
Walnut, stripe, USA	B1@.250	SF	.85	5.30	6.15
Wenge, plain	B1@.250	SF	.90	5.30	6.20
Zebra stripe, Africa	B1@.250	SF	.65	5.30	5.95
Deduct for over 100 SF of same type veneer	--	%	-7.0	--	--

		Unit	Oak	Birch
Dowels, birch and oak.				
1/8"		Ea	.42	.17
3/16"		Ea	.54	.24
1/4"		Ea	.65	.31
5/16"		Ea	.76	.42
3/8"		Ea	.87	.54
7/16"		Ea	1.00	.73
1/2"		Ea	1.43	.91
5/8"		Ea	1.80	1.24
3/4"		Ea	3.00	2.18
7/8"		Ea	3.33	2.95
1"		Ea	3.83	3.69

	Craft@Hrs	Unit	Material	Labor	Total
Dowel pins, birch, fluted, per bag of 125					
3/8" x 1-1/2"	--	Bag	2.50	--	--
3/8" x 2"	--	Bag	3.00	--	--
3/8" x 2-1/2"	--	Bag	3.50	--	--
7/16" x 2"	--	Bag	4.00	--	--
1/2" x 2"	--	Bag	4.50	--	--

Board siding See also Siding and Hardboard.

Douglas fir C and better siding, flat grain, coverage and 5% waste included, drop, rustic or shiplap patterns

	Craft@Hrs	Unit	Material	Labor	Total
1" x 6" ($2.25/BF, 1.14 BF per SF)	BC@3.16	Sq	255.00	74.50	329.50
1" x 8" ($2.50/BF, 1.12 BF per SF)	BC@3.16	Sq	280.00	74.50	354.50

Douglas fir D grade siding, flat grain, coverage and 5% waste included, drop, rustic or shiplap patterns

1" x 6" ($2.00/BF, 1.14 BF per SF)	BC@3.16	Sq	230.00	74.50	304.50
1" x 8" ($2.10/BF, 1.12 BF per SF)	BC@3.16	Sq	235.00	74.50	309.50

Redwood siding, clear all heart, kiln dried, random lengths, vertical grain, includes normal waste and coverage

Resawn redwood bevel siding

1/2" x 4" ($2,250/MBF, 1.51 BF per SF, 2-3/4" exposure)	BC@2.77	Sq	340.00	65.30	405.30
5/8" x 6" ($2,000/MBF, 1.31 BF per SF, 4-3/4" exposure)	BC@2.68	Sq	260.00	63.20	323.20

Lumber, Sidings

	Craft@Hrs	Unit	Material	Labor	Total
Redwood siding, clear all heart, kiln dried, random lengths, vertical grain, includes normal waste and coverage					
Resawn redwood bevel siding,					
5/8" x 8" ($2,350/MBF, 1.23 BF per SF, 6-3/4" exposure)	BC@2.68	Sq	290.00	63.20	353.20
3/4" x 6" ($2,390/MBF, 1.31 BF per SF, 4-3/4" exposure)	BC@2.68	Sq	315.00	63.20	378.20
3/4" x 8" ($2,750/MBF, 1.23 BF per SF, 6-3/4" exposure)	BC@2.68	Sq	340.00	63.20	403.20
3/4" x 10" ($2,750/MBF, 1.19 BF per SF, 8-3/4" exposure)	BC@2.58	Sq	330.00	60.90	390.90
Deduct for clear all heart, flat grain	--	%	-15.0	--	--
Deduct for clear flat grain	--	%	-20.0	--	--
Deduct for green	--	Sq	-20.00	--	--
Deduct for clear vertical grain	--	%	-15.0	--	--
Rustic, anzac or drop redwood siding, clear, all-heart, vertical grain, kiln dried					
1" x 4" ($2,660/MBF, 1.25 BF per SF)	BC@2.48	Sq	335.00	58.50	403.50
1" x 6" ($2,920/MBF, 1.17 BF per SF)	BC@2.30	Sq	340.00	54.30	394.30
1" x 8" ($2,960/MBF, 1.12 BF per SF)	BC@2.30	Sq	330.00	54.30	384.30
1" x 10" ($2,675/MBF, 1.11 BF per SF)	BC@2.30	Sq	295.00	54.30	349.30
1" x 12" ($3,080/MBF, 1.09 BF per SF)	BC@2.30	Sq	335.00	54.30	389.30
Deduct for clear all-heart flat grain	--	%	-10.0	--	--
Deduct for clear vertical grain	--	%	-15.0	--	--
Add for resawn face	--	Sq	10.00	--	--
Add for specified lengths	--	Sq	10.00	--	--
Western red cedar face, bevel siding, dry, includes waste and coverage allowance					
Clear grade red cedar siding					
1/2" x 4" (1.51 BF per SF at 2-3/4" exposure, $1,900/MBF)	BC@2.77	Sq	285.00	65.30	350.30
1/2" x 6" (1.31 BF per SF at 4-3/4" exposure, $1,970/MBF)	BC@2.68	Sq	260.00	63.20	323.20
1/2" x 8" (1.23 BF per SF at 6-3/4" exposure, $2,000/MBF)	BC@2.68	Sq	245.00	63.20	308.20
3/4" x 8" (1.23 BF per SF at 6-3/4" exposure, $2,065/MBF)	BC@2.68	Sq	255.00	63.20	318.20
3/4" x 10" (1.19 BF per SF at 8-3/4" exposure, $2,135/MBF)	BC@2.68	Sq	255.00	63.20	318.20
"A" grade red cedar siding					
1/2" x 4" (1.51 BF per SF at 2-3/4" exposure, $2,180/MBF)	BC@2.77	Sq	330.00	65.30	395.30
1/2" x 6" (1.31 BF per SF at 4-3/4" exposure, $2,015/MBF)	BC@2.68	Sq	265.00	63.20	328.20
1/2" x 8" (1.23 BF per SF at 6-3/4" exposure, $2,025/MBF)	BC@2.68	Sq	250.00	63.20	313.20
3/4" x 8" (1.23 BF per SF at 6-3/4" exposure, $2,190/MBF)	BC@2.68	Sq	270.00	63.20	333.20
3/4" x 10" (1.19 BF per SF at 8-3/4 "exposure, $2,130/MBF)	BC@2.58	Sq	255.00	60.90	315.90
"B" grade red cedar siding					
1/2" x 6" (1.31 BF per SF at 4-3/4" exposure, $1,675/MBF)	BC@2.68	Sq	220.00	63.20	283.20
1/2" x 8" (1.23 BF per SF at 6-3/4" exposure, $1,735/MBF)	BC@2.68	Sq	215.00	63.20	278.20

Lumber, Sidings

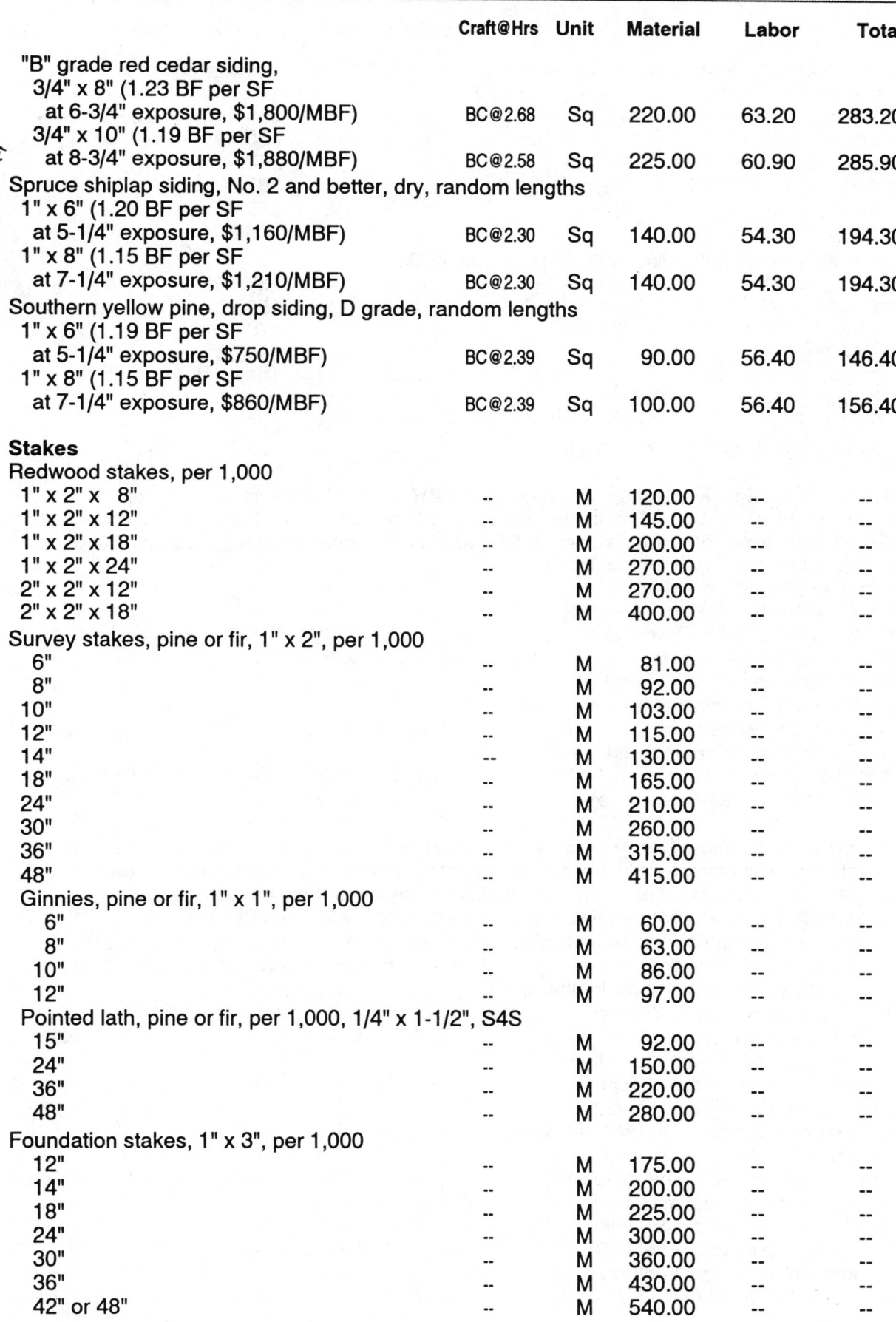

	Craft@Hrs	Unit	Material	Labor	Total
"B" grade red cedar siding,					
3/4" x 8" (1.23 BF per SF					
at 6-3/4" exposure, $1,800/MBF)	BC@2.68	Sq	220.00	63.20	283.20
3/4" x 10" (1.19 BF per SF					
at 8-3/4" exposure, $1,880/MBF)	BC@2.58	Sq	225.00	60.90	285.90
Spruce shiplap siding, No. 2 and better, dry, random lengths					
1" x 6" (1.20 BF per SF					
at 5-1/4" exposure, $1,160/MBF)	BC@2.30	Sq	140.00	54.30	194.30
1" x 8" (1.15 BF per SF					
at 7-1/4" exposure, $1,210/MBF)	BC@2.30	Sq	140.00	54.30	194.30
Southern yellow pine, drop siding, D grade, random lengths					
1" x 6" (1.19 BF per SF					
at 5-1/4" exposure, $750/MBF)	BC@2.39	Sq	90.00	56.40	146.40
1" x 8" (1.15 BF per SF					
at 7-1/4" exposure, $860/MBF)	BC@2.39	Sq	100.00	56.40	156.40
Stakes					
Redwood stakes, per 1,000					
1" x 2" x 8"	--	M	120.00	--	--
1" x 2" x 12"	--	M	145.00	--	--
1" x 2" x 18"	--	M	200.00	--	--
1" x 2" x 24"	--	M	270.00	--	--
2" x 2" x 12"	--	M	270.00	--	--
2" x 2" x 18"	--	M	400.00	--	--
Survey stakes, pine or fir, 1" x 2", per 1,000					
6"	--	M	81.00	--	--
8"	--	M	92.00	--	--
10"	--	M	103.00	--	--
12"	--	M	115.00	--	--
14"	--	M	130.00	--	--
18"	--	M	165.00	--	--
24"	--	M	210.00	--	--
30"	--	M	260.00	--	--
36"	--	M	315.00	--	--
48"	--	M	415.00	--	--
Ginnies, pine or fir, 1" x 1", per 1,000					
6"	--	M	60.00	--	--
8"	--	M	63.00	--	--
10"	--	M	86.00	--	--
12"	--	M	97.00	--	--
Pointed lath, pine or fir, per 1,000, 1/4" x 1-1/2", S4S					
15"	--	M	92.00	--	--
24"	--	M	150.00	--	--
36"	--	M	220.00	--	--
48"	--	M	280.00	--	--
Foundation stakes, 1" x 3", per 1,000					
12"	--	M	175.00	--	--
14"	--	M	200.00	--	--
18"	--	M	225.00	--	--
24"	--	M	300.00	--	--
30"	--	M	360.00	--	--
36"	--	M	430.00	--	--
42" or 48"	--	M	540.00	--	--

Lumber, Stakes

	Craft@Hrs	Unit	Material	Labor	Total
Hubs, pine or fir, 2" x 2", per 1,000					
6"	--	M	195.00	--	--
8"	--	M	215.00	--	--
10"	--	M	225.00	--	--
12"	--	M	240.00	--	--
14"	--	M	285.00	--	--
18"	--	M	325.00	--	--
24"	--	M	445.00	--	--
Form stakes (wedges), pine or fir, 5" point, per 1,000					
2" x 2" x 6"	--	M	135.00	--	--
2" x 2" x 8"	--	M	150.00	--	--
2" x 2" x 10"	--	M	170.00	--	--
2" x 4" x 6"	--	M	165.00	--	--
2" x 4" x 8"	--	M	185.00	--	--
2" x 4" x 10"	--	M	200.00	--	--
2" x 4" x 12"	--	M	225.00	--	--
Feather wedges	--	M	92.00	--	--

Lumber Grading. Quality inspector daily rate: $210.00. Half day rate: $115.00 plus 30 cents per mile from office, $50.00 minimum. For lumber in remote areas, the daily rate applied is $280.00 plus travel plus subsistence of $70.00 per day. Rates assume adequate lifting equipment and a helper are available.

	Craft@Hrs	Unit	Material	Labor	Total
1" lumber (20 MBF per day)	--	MBF	--	--	12.50
2" x 4", 6" lumber (23 MBF per day)	--	MBF	--	--	11.50
2" x 8", 10", 12" (33 MBF per day)	--	MBF	--	--	10.00
4" x 4", 6" (33 MBF per day)	--	MBF	--	--	10.00
4" x 8" and over (28 MBF per day)	--	MBF	--	--	11.50
Lumber inventory verification, not including travel and expenses	--	Day	--	--	210.00
Court appearance, not including preparation, travel or expenses	--	Day	--	--	400.00
Lumber grading training (on site)	--	Day	--	--	250.00

Lumber Preservation Treatments, Subcontract. See also Soil Treatments. Costs assume treatment at a treatment plant, not on the construction site. Treatments listed below are designed to protect wood and wood products from decay and destruction caused by moisture (fungi), wood boring insects and organisms. All retention rates listed (in parentheses) are in pounds per cubic foot and are the minimum recommended by the Society of American Wood Preservers. Costs are for treatment only and assume treating of 10,000 board feet or more. No lumber or transportation included.

	Craft@Hrs	Unit	Material	Labor	Total
Lumber submerged in or frequently exposed to salt water					
ACZA treatment (2.5)	--	MBF	--	--	425.00
ACZA treatment for structural lumber					
Piles and building poles (2.5)	--	CF	--	--	5.50
Board lumber and timbers (2.5)	--	MBF	--	--	425.00
Creosote treatment for structural lumber					
Piles and building poles (20.0)	--	CF	--	--	4.00
Board lumber and timbers (25.0)	--	MBF	--	--	350.00
Lumber in contact with fresh water or soil, ACZA treatment					
Building poles, structural (.60)	--	CF	--	--	3.80
Boards, timbers, structural (.60)	--	MBF	--	--	165.00
Boards, timbers, nonstructural (.40)	--	MBF	--	--	135.00
Plywood, structural (.40), 1/2"	--	MSF	--	--	179.00

Lumber Preservation Treatments

	Craft@Hrs	Unit	Material	Labor	Total
Lumber Preservation Treatments, Subcontract					
Posts, guardrail, signs (.60)	--	MBF	--	--	240.00
All weather foundation KD lumber (.60)	--	MBF	--	--	245.00
All weather foundation, 1/2" plywood (.60)	--	MSF	--	--	194.00
Lumber in contact with fresh water or soil, creosote treatment					
Piles, building poles, structural (12.0)	--	CF	--	--	4.20
Boards, timbers, structural (12.0)	--	MBF	--	--	220.00
Boards, timbers, nonstructural (10.0)	--	MBF	--	--	200.00
Plywood, structural, 1/2" (10.0)	--	MSF	--	--	204.00
Posts (fence, guardrail, sign) (12.0)	--	MBF	--	--	220.00
Lumber used above ground, ACZA treatment					
Boards, timbers, structural (.25)	--	MBF	--	--	120.00
Boards, timbers, nonstructural (.25)	--	MBF	--	--	120.00
Plywood, structural 1/2" (.25)	--	MSF	--	--	165.00
Lumber used above ground, creosote treatment					
Boards, timbers, structural (8.0)	--	MBF	--	--	230.00
Boards, timbers, nonstructural (8.0)	--	MBF	--	--	230.00
Plywood, structural (8.0), 1/2"	--	MSF	--	--	240.00
Lumber in contact with man, animals and plants, ACZA treatment					
Boards, structural (.25)	--	MBF	--	--	120.00
Boards, timbers, nonstructural (.25)	--	MSF	--	--	120.00
Plywood and other 1/2" sheets (.25)	--	MSF	--	--	175.00
Millwork (.25)	--	MBF	--	--	122.00
Fire retardant treatments					
Structural use (interior only), "D-Blaze"					
Surfaced lumber	--	MBF	--	--	220.00
Rough lumber	--	MBF	--	--	220.00
Architectural use (interior and exterior), "NCX"					
Surfaced lumber	--	MBF	--	--	510.00
Rough lumber	--	MBF	--	--	510.00

Mailboxes

Apartment type, meets current postal regulations, tumbler locks, aluminum or gold finish, price per unit, recessed

	Craft@Hrs	Unit	Material	Labor	Total
3 box unit, 5-1/2" wide x 16" high	B1@1.88	Ea	78.00	39.80	117.80
4 box unit, 5-1/2" wide x 16" high	B1@1.88	Ea	103.00	39.80	142.80
5 box unit, 5-1/2" wide x 16" high	B1@1.88	Ea	128.00	39.80	167.80
6 box unit, 5-1/2" wide x 16" high	B1@1.88	Ea	154.00	39.80	193.80
7 box unit, 5-1/2" wide x 16" high	B1@1.88	Ea	180.00	39.80	219.80
Add for surface mounted	--	%	20.0	--	--

Office type, horizontal, meets postal regulations, tumbler locks, aluminum or brass finish

	Craft@Hrs	Unit	Material	Labor	Total
For front load, per door					
6" wide x 5" high x 15" deep	B1@1.02	Ea	33.00	21.60	54.60
Rear load, per door					
6" wide x 5" high x 15" deep	B1@1.02	Ea	30.00	21.60	51.60
Rural, #1 size, aluminum finish, 19" long x 6-1/2" wide x 9" high, with flag,					
post mount	B1@.776	Ea	10.50	16.40	26.90
Rural, #1-1/2 size, aluminum finish, 19" long x 6-1/2" wide x 11" high, with flag,					
post mount	B1@.776	Ea	24.00	16.40	40.40
Suburban style, finished, 16" long x 3" wide x 6" deep,					
with flag, post mount	B1@.776	Ea	11.50	16.40	27.90
Add for redwood or wrought iron post	--	Ea	11.00	--	--

Mailboxes

	Craft@Hrs	Unit	Material	Labor	Total
Thru-the-wall type, 13" long x 3" high,					
gold or aluminum finish	B1@1.00	Ea	26.00	21.19	47.19
Wall-mounting security type, lockable, 15" wide x 14" high x 8" deep,					
black plastic	B1@1.25	Ea	67.00	26.49	93.49
Mantels, Ponderosa Pine, Unfinished					
Modern design, 50" x 37" opening,					
paint grade, 11" x 77" shelf	B1@3.81	Ea	375.00	80.70	455.70
Colonial design, 50" x 39" opening,					
stain grade, 7" x 72" shelf	B1@3.81	Ea	575.00	80.70	655.70
Ornate design, 51-1/2" x 39" opening,					
with detailing and scroll work, stain grade,					
11" x 68" shelf	B1@3.81	Ea	1,050.00	80.70	1,130.70
Mantel moulding (simple trim for masonry fireplace openings),					
15 LF, for opening 6'0" x 3'6" or less	B1@1.07	Set	120.00	22.70	142.70
Mantel shelf only, with iron or wood support brackets, prefinished					
4" x 8" x 8"	B1@.541	Ea	32.00	11.50	43.50
4" x 8" x 6"	B1@.541	Ea	25.00	11.50	36.50
4" x 10" x 10"	B1@.541	Ea	60.00	11.50	71.50

Marble Setting, Subcontract. See also Tile section. Marble prices vary with quality, quantity, and availability of materials. Check prices with local suppliers. These prices are based on 200 SF quantities.

Marble tile flooring, tiles or slabs
 3/8" x 12" x 12", thin set over concrete slab, based on $7.00 to $10.00 per SF material cost
 and $10.50 to $12.50 per SF labor cost

Total cost per SF of area covered	--	SF	--	--	20.00

3/4" slabs set in mortar bed, based on $40.00 to $50.00 per SF material cost
 and $15.00 per SF labor cost

Total cost per SF of area covered	--	SF	--	--	60.00

Marble bathroom vanity 6' x 2' wide, with 1 sink cutout and 4" splash on 3 sides
 One per home, based on $600.00 to $900.00 per unit material cost
 and $180.00 per unit labor cost

Total cost per unit	--	Ea	--	--	900.00

Four per home, based on $600.00 to $900.00 per unit material cost
 and $80.00 per unit labor cost

Total cost per unit	--	Ea	--	--	800.00

Fireplace hearth and facing, one per home, based on
 $415.00 to $740.00 per unit material cost and $340.00 to $545.00 per unit labor cost

Total cost per unit	--	Ea	--	--	1,100.00

Markup

Typical markup for light construction, percentage of gross contract price. No insurance or taxes included. See also Insurance and Taxes sections. Markup is the amount added to an estimate after all job costs have been accounted for: labor and materials, job fees, permits, bonds and insurance, job site supervision and similar job-related expenses. Profit is the return on money invested in the construction business.

Contingency					
(varies widely from job to job)	--	%	--	--	2.0
Overhead (office, administrative and					
general business expense)	--	%	--	--	10.0
Profit	--	%	--	--	8.0

Markup

	Craft@Hrs	Unit	Material	Labor	Total

Markup, continued
Total markup (add 25% to the total job cost to
attain 20% of gross contract price) -- % -- -- 20.0

Note that costs listed in this book do not include the general contractor's markup. However, sections identified as "subcontract" include the subcontractor's markup.

For example, note how markup is calculated on a project with these costs:

Project Costs and Markup

Material	$50,000
Labor (with taxes & insurance)	40,000
Subcontract	10,000
Total job costs	100,000
Markup (25% of $100,000)	**25,000**
Contract price	$125,000

Markup is $25,000 which is 25% of the total job costs ($100,000) but only 20% of the contract price ($25,000 divided by $125,000 is .20).

For general contractors handling larger commercial and industrial projects, a markup of 23.8% on the total job cost is suggested. See the paragraph General Contractor Markup at the beginning of the section General Requirements in the Industrial and Commercial Division of this manual. For example, note how markup is calculated on a project with these costs:

Material	$4,000,000
Labor (with taxes and insurance)	5,000,000
Equipment	1,000,000
Total job cost	$10,000,000
Indirect overhead (8%)	**800,000**
Direct overhead (5.8%)	**580,000**
Contingency (2%)	**200,000**
Profit (8%)	**800,000**
Contract price	$12,380,000

The total job cost shown above is the cost that a general contractor would expect to pay subcontractor-specialists for this work performed on a subcontract basis including the subcontractors' supervision expense, contingency allowance, overhead and profit.

Note that markup for the general contractor (overhead, contingency and profit) is $2,380,000 or 23.8% of the total job costs but only 19.22% of the contract price (2,380,000 divided by 12,380,000 is .1922).

Masonry, Face Brick. Foundations, running bond, 3/8" concave joints, waste, and wall ties not included (add mortar below). Includes delivery up to 30 miles with a 5 ton minimum.
Coverage (units per SF) includes 3/8" mortar joints but the SF cost does not include mortar cost. Add for mortar below.

	Craft@Hrs	Unit	Material	Labor	Total
Standard brick, 2-1/4" high x 3-3/4" deep x 8" long (6.45 units per SF)					
Brown (at $480 per M)	B9@.211	SF	3.10	4.34	7.44
Burnt oak (at $480 per M)	B9@.211	SF	3.10	4.34	7.44
Flashed (at $440 per M)	B9@.211	SF	2.84	4.34	7.18
Old English (at $450 per M)	B9@.211	SF	2.90	4.34	7.24
Smooth Red (at $360 per M)	B9@.211	SF	2.32	4.34	6.66
Norman brick, 2-1/2" high x 3-1/2" deep x 11-1/2" long (4.22 units per SF)					
Brown (at $680 per M)	B9@.211	SF	2.87	4.34	7.21
Burnt Oak (at $680 per M)	B9@.211	SF	2.87	4.34	7.21
Flashed (at $660 per M)	B9@.211	SF	2.79	4.34	7.13
Red (at $630 per M)	B9@.211	SF	2.66	4.34	7.00

Masonry

	Craft@Hrs	Unit	Material	Labor	Total
Modular, 3-1/2" high x 3" deep x 11-1/2" long (3.13 units per SF)					
Brown (at $830 per M)	B9@.156	SF	2.60	3.21	5.81
Burnt oak (at $830 per M)	B9@.156	SF	2.60	3.21	5.81
Flashed (at $710 per M)	B9@.156	SF	2.22	3.21	5.43
Red (at $660 per M)	B9@.156	SF	2.07	3.21	5.28
Jumbo (filler brick) 3-1/2" high x 3" deep x 11-1/2" long (3.13 units per SF)					
Fireplace, cored (at $310 per M)	B9@.156	SF	.97	3.21	4.18
Fireplace, solid (at $330 per M)	B9@.156	SF	1.04	3.21	4.25
Mission, cored (at $345 per M)	B9@.156	SF	1.08	3.21	4.29
Mission, solid (at $365 per M)	B9@.156	SF	1.14	3.21	4.35
Padre brick, 4" high x 3" deep					
7-1/2" long cored, 4.2 per SF, $.70 each	B9@.156	SF	2.94	3.21	6.15
7-1/2" long solid, 4.2 per SF, $.87 each	B9@.156	SF	3.65	3.21	6.86
11-1/2" long cored, 2.8 per SF, $.88 each	B9@.177	SF	2.46	3.64	6.10
11-1/2" long solid, 2.8 per SF, $1.09 each	B9@.177	SF	3.05	3.64	6.69
15-1/2" long cored, 2.1 per SF, $1.16 each	B9@.211	SF	2.44	4.34	6.78
15-1/2" long solid, 2.1 per SF, $1.45 each	B9@.211	SF	3.05	4.34	7.39
Commercial, 3-1/2" high x 3" deep					
Smooth, 8" long, 4.4 per SF, $.37 each	B9@.156	SF	1.63	3.21	4.84
Colonial, 10" long, 3.6 per SF, $.45 each	B9@.177	SF	1.62	3.64	5.26

Mortar for brick Based on an all-purpose mortar mix costing $5.70 per cubic foot (60 lb. bag @ $2.85) used in single wythe walls laid with running bond. Includes normal waste. Costs are per SF of wall face using the brick sizes (width x height x length) listed.

	Craft@Hrs	Unit	Material	Labor	Total
4" x 2-1/2" to 2-5/8" x 8" bricks					
3/8" joints (5.5 CF per CSF)	--	SF	.31	--	--
1/2" joints (7.0 CF per CSF)	--	SF	.40	--	--
4" x 3-1/2" to 3-5/8" x 8" bricks					
3/8" joints (4.8 CF per CSF)	--	SF	.27	--	--
1/2" joints (6.1 CF per CSF)	--	SF	.35	--	--
4" x 4" x 8" bricks					
3/8" joints (4.2 CF per CSF)	--	SF	.24	--	--
1/2" joints (5.3 CF per CSF)	--	SF	.30	--	--
4" x 5-1/2" to 5-5/8" x 8" bricks					
3/8" joints (3.5 CF per CSF)	--	SF	.20	--	--
1/2" joints (4.4 CF per CSF)	--	SF	.25	--	--
4" x 2" x 12" bricks					
3/8" joints (6.5 CF per CSF)	--	SF	.37	--	--
1/2" joints (8.2 CF per CSF)	--	SF	.47	--	--
4" x 2-1/2" to 2-5/8" x 12" bricks					
3/8" joints (5.1 CF per CSF)	--	SF	.29	--	--
1/2" joints (6.5 CF per CSF)	--	SF	.37	--	--
4" x 3-1/2" to 3-5/8" x 12" brick					
3/8" joints (4.4 CF per CSF)	--	SF	.25	--	--
1/2" joints (5.6 CF per CSF)	--	SF	.32	--	--
4" x 4" x 12" bricks					
3/8" joints (3.7 CF per CSF)	--	SF	.21	--	--
1/2" joints (4.8 CF per CSF)	--	SF	.27	--	--
4" x 5-1/2" to 5-5/8" x 12" brick					
3/8" joints (3.0 CF per CSF)	--	SF	.17	--	--
1/2" joints (3.9 CF per CSF)	--	SF	.22	--	--
6" x 2-1/2" to 2-5/8" x 12" bricks					

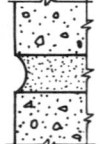
Concave joint

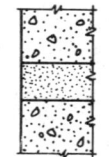
Flush joint

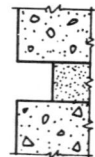
Raked joint

Masonry

	Craft@Hrs	Unit	Material	Labor	Total
3/8" joints (7.9 CF per CSF)	--	SF	.45	--	--
1/2" joints (10.2 CF per CSF)	--	SF	.58	--	--
6" x 3-1/2" to 3-5/8" x 12" bricks					
3/8" joints (6.8 CF per CSF)	--	SF	.39	--	--
1/2" joints (8.8 CF per CSF)	--	SF	.50	--	--
6" x 4" x 12" bricks					
3/8" joints (5.6 CF per CSF)	--	SF	.32	--	--
1/2" joints (7.4 CF per CSF)	--	SF	.42	--	--
Add for 8" 22 gauge galvanized wall ties (50 per 100 SF)	B9@.011	SF	.04	.23	.27
Add for raked joint	--	%	--	15.0	--
Add for flush joint	--	%	--	40.0	--
Add for header course every 6th course	--	%	20.0	5.0	--
Add for Flemish bond with header course every 6th course	--	%	12.0	20.0	--

Flemish

Mini-brick veneer strips Figures in parentheses indicate coverage using 1/2" grouted joints. No waste included.

	Craft@Hrs	Unit	Material	Labor	Total
2-3/16" high x 7/16" thick x 7-1/2" long (6.75 units per SF)					
Unglazed, standard colors, $.21 each	B9@.300	SF	1.42	6.17	7.59
Unglazed, flashed, $.39 each	B9@.300	SF	2.63	6.17	8.80
Glazed, all colors, $.57 each	B9@.300	SF	3.85	6.17	10.02
2-3/16" high x 7/16" thick x 11-1/2" long (4.50 units per SF)					
Unglazed, standard colors, $.47 each	B9@.200	SF	2.12	4.11	6.23
Unglazed, flashed, $.77 each	B9@.200	SF	3.47	4.11	7.58
Glazed, all colors, $.95 each	B9@.200	SF	4.28	4.11	8.39
Mini-brick veneer strip corners					
Unglazed, standard colors	B9@.077	Ea	.81	1.58	2.39
Unglazed, flashed	B9@.077	Ea	1.18	1.58	2.76
Glazed, all colors	B9@.077	Ea	1.44	1.58	3.02

Firebrick Used in chimneys and boilers, delivered to 30 miles, typical costs.

	Craft@Hrs	Unit	Material	Labor	Total
Standard backs, 9" x 4-1/2" x 2-1/2"	B9@30.9	M	725.00	635.00	1,360.00
Split backs, 9" x 4-1/2" x 1-1/2"	B9@23.3	M	725.00	479.00	1,204.00
Fireclay, Dosch clay					
10 lb sack	--	Ea	1.20	--	--
50 lb sack	--	Ea	4.40	--	--

Brick pavers Costs assume simple pattern with 1/2" mortar joints. Delivered to 30 miles, 10 ton minimum. Costs in material column include 8% for waste and breakage.

	Craft@Hrs	Unit	Material	Labor	Total
Adobe paver (Fresno)					
6" x 12" (2 units per SF, $.43 each)	B9@.193	SF	.93	3.97	4.90
12" x 12" (1 units per SF, $.81 each)	B9@.135	SF	.87	2.77	3.64
California paver, 1-1/4" thick, brown					
3-5/8" x 7-5/8" (4.5 units per SF, $.46 each)	B9@.340	SF	2.24	6.99	9.23
3-5/8" x 11-5/8" (3 units per SF, $.70 each)	B9@.222	SF	2.27	4.56	6.83
Giant mission paver, 2-3/8" thick x 5-5/8" x 11-5/8"					
Red (2 units per SF, $.66 each)	B9@.193	SF	1.43	3.97	5.40
Redskin (2 units per SF, $.77 each)	B9@.193	SF	1.66	3.97	5.63
Mini-brick paver, 7/16" thick x 3-5/8" x 7-5/8"					
Unglazed, standard colors (4.5 units per SF, $.37 each)	B9@.340	SF	1.80	6.99	8.79

Masonry

	Craft@Hrs	Unit	Material	Labor	Total
Unglazed, flashed (4.5 units per SF $.66 each)	B9@.340	SF	3.21	6.99	10.20
Glazed, all colors (4.5 units per SF $.65 each)	B9@.340	SF	3.16	6.99	10.15
Norman paver, 2-1/2" thick x 3-1/2" x 11-1/2" (3 units per SF, $.59 each)	B9@.222	SF	1.91	4.56	6.47
Padre paver, 1-1/4" thick x 7-1/2" x 7-1/2" (2.5 units per SF, $1.07 each)	B9@.211	SF	2.89	4.34	7.23
Mortar joints for pavers, 1/2" typical	B9@.010	SF	.43	.21	.64

Masonry, Concrete Block. Foundations not included. Includes local delivery and 8% waste but no cutting. Natural gray, medium weight, no reinforcing included.

	Craft@Hrs	Unit	Material	Labor	Total
4" wide units					
8" x 16" ($.69 ea), 1.125 per SF	B9@.085	SF	.84	1.75	2.59
8" x 12", corner	B9@.054	Ea	.69	1.11	1.80
8" x 8" ($.68 ea), 2.25 per SF	B9@.093	SF	1.65	1.91	3.56
4" x 16" ($.52 ea), 2.25 per SF	B9@.093	SF	1.26	1.91	3.17
4" x 12", corner	B9@.052	Ea	.52	1.07	1.59
4" x 8" ($.51 ea), 4.5 per SF (1/2 block)	B9@.106	SF	2.48	2.18	4.66
6" wide units					
8" x 16" ($.79 ea), 1.125 per SF	B9@.095	SF	.96	1.95	2.91
8" x 14" corner	B9@.058	Ea	.79	1.19	1.98
8" x 8" ($.78 ea), 2.25 per SF, 1/2 block	B9@.103	SF	1.90	2.12	4.02
4" x 16" ($.70 ea), 2.25 per SF	B9@.103	SF	1.70	2.12	3.82
4" x 14" corner	B9@.070	Ea	.71	1.44	2.15
4" x 8" ($.69 ea), 4.5 per SF, 1/2 block	B9@.114	SF	3.35	2.34	5.69
4" x 12" (3/4 block)	B9@.052	Ea	.69	1.07	1.76
8" wide units					
4" x 16" ($.88 ea), 2.25 per SF	B9@.114	SF	2.14	2.34	4.48
8" x 16" ($.98 ea), 1.125 per SF	B9@.111	SF	1.19	2.28	3.47
6" x 8" ($.86 ea), 3 per SF	B9@.137	SF	2.79	2.82	5.61
8" x 12" (3/4 block)	B9@.074	Ea	.98	1.52	2.50
8" x 8" (half)	B9@.052	Ea	1.06	1.07	2.13
8" x 8" (lintel)	B9@.052	Ea	1.12	1.07	2.19
6" x 16" ($.93 ea), 1.5 per SF	B9@.117	SF	1.51	2.40	3.91
4" x 8" (half block)	B9@.053	Ea	.94	1.09	2.03
12" wide units					
8" x 16" ($1.63 ea), 1.125 per SF	B9@.148	SF	1.98	3.04	5.02
8" x 8" (half) ($1.62 ea)	B9@.082	Ea	1.75	1.69	3.44
8" x 8" (lintel) ($1.68 ea)	B9@.082	Ea	1.81	1.69	3.50
Add for high strength units	--	%	27.0	--	--
Add for lightweight block, typical	--	Ea	.18	--	--
Deduct for lightweight block, labor	--	%	--	-7.0	--
Add for 1/2" rebars 24" OC	B9@.004	SF	.11	.08	.19
Add for other than running bond	--	%	--	20.0	--
Add for 2" caps, natural color					
4" x 16" ($.67 ea)	B9@.027	LF	.51	.56	1.07
6" x 16" ($.80 ea)	B9@.038	LF	.59	.78	1.37
8" x 16" ($.93 ea)	B9@.046	LF	.69	.95	1.64
Add for color block					
Light colors	--	%	12.0	--	--

Masonry

	Craft@Hrs	Unit	Material	Labor	Total
Medium colors	--	%	18.0	--	--
Dark colors	--	%	25.0	--	--

Add for mortar (at $5.70 per CF). Cost in material column includes normal waste

	Craft@Hrs	Unit	Material	Labor	Total
8" x 8" x 16" block (5.23 CF per CSF)	--	SF	.30	--	--
4" x 8" x 16" block (3.23 CF per CSF)	--	SF	.19	--	--
6" x 8" x 16" block (4.3 CF per CSF)	--	SF	.25	--	--

Add for grout cores at 36" interval (at $2.00 per CF), poured by hand, no waste included

	Craft@Hrs	Unit	Material	Labor	Total
4" wide wall (16.1 SF/CF grout)	B9@.007	SF	.13	.14	.27
6" wide wall (10.7 SF/CF grout)	B9@.007	SF	.19	.14	.33
8" wide wall (8.0 SF/CF grout)	B9@.007	SF	.25	.14	.39
12" wide wall (5.4 SF/CF grout)	B9@.010	SF	.37	.21	.58
Add for cores grouted at 24" interval	--	%	25.0	25.0	--
Deduct for cores grouted at 48" interval	--	%	-18.0	-18.0	--

Detailed concrete block Foundations not included. Lightweight, 3/8" score, including local delivery and 8% waste.

8" x 8" x 16" (1.125 units per SF)

	Craft@Hrs	Unit	Material	Labor	Total
Single score, one side					
Natural ($1.45 ea)	B9@.111	SF	1.76	2.28	4.04
Tan ($1.63 ea)	B9@.111	SF	1.98	2.28	4.26
Single score, two sides					
Natural ($1.56 ea)	B9@.111	SF	1.90	2.28	4.18
Tan ($1.74 ea)	B9@.111	SF	2.12	2.28	4.40
Three or five score, one side					
Natural ($1.64 ea)	B9@.111	SF	2.00	2.28	4.28
Tan ($1.83 ea)	B9@.111	SF	2.23	2.28	4.51
Three or five score, 2 side					
Natural ($1.95 ea)	B9@.111	SF	2.37	2.28	4.65
Tan ($2.13 ea)	B9@.111	SF	2.59	2.28	4.87

12" x 8" x 16" (1.125 units per SF)

	Craft@Hrs	Unit	Material	Labor	Total
Single score one side					
Natural ($2.13 ea)	B9@.151	SF	2.59	3.10	5.69
Tan ($2.40 ea)	B9@.151	SF	2.92	3.10	6.02
Single score 2 side					
Natural (at $2.19 ea)	B9@.151	SF	2.66	3.10	5.76
Tan ($2.48 ea)	B9@.151	SF	3.01	3.10	6.11
Three or five score, one side					
Natural ($2.31 ea)	B9@.151	SF	2.81	3.10	5.91
Tan ($2.58 ea)	B9@.151	SF	3.14	3.10	6.24
Three or five score, two sides					
Natural ($2.62 ea)	B9@.151	SF	3.18	3.10	6.28
Tan ($2.88 ea)	B9@.151	SF	3.50	3.10	6.60
Add for reinforcement	B9@.004	SF	.11	.08	.19
Deduct for medium weight block	--	%	-15.0	--	--
Add for medium weight block	--	%	--	6.0	--

Add for mortar at $5.70/CF

	Craft@Hrs	Unit	Material	Labor	Total
(7 CF per 100 SF)	--	SF	.40	--	--

Add for grouting of cores ($2.00/CF), poured by hand, no waste included

	Craft@Hrs	Unit	Material	Labor	Total
Solid grout (3.2 SF per CF grout)	B9@.015	SF	.63	.31	.94
24" interval (6.5 SF per CF grout)	B9@.008	SF	.31	.16	.47
36" interval (8.0 SF per CF grout)	B9@.006	SF	.25	.12	.37
48" interval (9.9 SF per CF grout)	B9@.004	SF	.21	.08	.29

Masonry

	Craft@Hrs	Unit	Material	Labor	Total

Split face concrete block Textured block, medium weight, split one side, structural. Includes 8% waste and local delivery.

	Craft@Hrs	Unit	Material	Labor	Total
8" wide, 4" x 16", ($1.25 ea)	B9@.111	SF	3.04	2.28	5.32
8" wide, 8" x 16", ($1.40 ea)	B9@.111	SF	1.70	2.28	3.98
12" wide, 8" x 8", ($1.75 ea)	B9@.117	SF	3.94	2.40	6.34
12" wide, 8" x 16", ($1.85 ea)	B9@.117	SF	2.25	2.40	4.65
Add for lightweight block	--	%	12.0	--	--
Add for color	--	SF	.30	--	--
Add for mortar (at $5.70 per cubic foot and 7 cubic feet per 100 square feet of wall)	--	SF	.40	--	--

Concrete pavers Natural concrete, including local delivery, 3/4" thick.

	Craft@Hrs	Unit	Material	Labor	Total
6" x 6"	--	Ea	2.80	--	--
6" x 12"	--	Ea	2.85	--	--
12" x 12"	--	Ea	2.65	--	--
12" x 18"	--	Ea	2.80	--	--
18" x 18"	--	Ea	2.85	--	--
Add for light colors	--	%	15.0	--	--
Add for beach pebble surface	--	SF	1.20	--	--
Add for dark colors	--	%	30.0	--	--
Labor installing pavers in concrete	B9@.150	SF	--	3.08	--

Concrete screen block Foundations and supports not included. Including local delivery and 8% waste, natural color.
Textured screen (no corebar defacements)

	Craft@Hrs	Unit	Material	Labor	Total
4" x 12" x 12", $1.55 ea, 1 per SF	B9@.211	SF	1.67	4.34	6.01
4" x 6" x 6", $.51 ea, 3.68 per SF	B9@.407	SF	2.03	8.36	10.39
4" x 8" x 8", $1.23 ea, 2.15 per SF	B9@.285	SF	2.85	5.86	8.71

Four sided sculpture screens

	Craft@Hrs	Unit	Material	Labor	Total
4" x 12" x 12", $1.55 ea, 1 per SF	B9@.312	SF	1.67	6.41	8.08

Sculpture screens (patterns face, rug texture reverse)

	Craft@Hrs	Unit	Material	Labor	Total
4" x 12" x 12", $2.05 ea, 1 per SF	B9@.211	SF	2.21	4.34	6.55
4" x 16" x 16", $3.61 ea, .57 per SF	B9@.285	SF	2.22	5.86	8.08
4" x 16" x 12", $2.88 ea, .75 per SF	B9@.285	SF	2.33	5.86	8.19

Add for channel steel support (high rise)

	Craft@Hrs	Unit	Material	Labor	Total
2" x 2" x 3/16"	--	LF	1.70	--	--
3" x 3" x 3/16"	--	LF	2.05	--	--
4" x 4" x 1/4"	--	LF	4.05	--	--
6" x 4" x 3/8"	--	LF	7.90	--	--

Concrete slump block Natural concrete color, 4" widths are veneer, no foundation or reinforcing included. Cost in material column includes 8% for breakage, waste, and local delivery.

	Craft@Hrs	Unit	Material	Labor	Total
4" x 4" x 8", $.51 ea, 4.5 per SF	B9@.267	SF	2.39	5.49	7.88
4" x 4" x 12", $.52 ea, 3 per SF	B9@.267	SF	1.66	5.49	7.15
4" x 4" x 16", $.53 ea, 2.25 per SF	B9@.267	SF	1.24	5.49	6.73
6" x 4" x 16", $.69 ea, 2.25 per SF	B9@.243	SF	1.63	4.99	6.62
6" x 6" x 16", $.78 ea, 1.5 per SF	B9@.217	SF	1.22	4.46	5.68
8" x 4" x 16", $.90 ea, 2.25 per SF	B9@.164	SF	2.11	3.37	5.48
8" x 6" x 16", $1.01 ea, 1.5 per SF	B9@.243	SF	1.58	4.99	6.57
12" x 4" x 16", $1.50 ea, 2.25 per SF	B9@.267	SF	3.52	5.49	9.01
12" x 6" x 16", $1.67 ea, 1.5 per SF	B9@.260	SF	2.61	5.34	7.95

Add for caps or slabs, 2" thick, 16" long, .75 units per LF. No waste included

Masonry

	Craft@Hrs	Unit	Material	Labor	Total
4" width ($.65 ea)	B9@.160	LF	.49	3.29	3.78
6" width ($.82 ea)	B9@.160	LF	.62	3.29	3.91
8" width ($.94 ea)	B9@.160	LF	.71	3.29	4.00
12" width ($1.83 ea)	B9@.160	LF	1.37	3.29	4.66
Add for curved block					
4" width, 2' or 4' diameter circle	--	%	125.0	25.0	--
8" width, 2' or 4' diameter circle	--	%	150.0	20.0	--
Add for light colors	--	%	13.0	--	--
Add for darker colors	--	%	25.0	--	--
Add for mortar ($5.70 per CF). Cost includes normal waste					
4" x 4" x 8" (4 CF per CSF)	--	SF	.23	--	--
6" x 6" x 16" (4 CF per CSF)	--	SF	.23	--	--
8" x 6" x 16" (5 CF per CSF)	--	SF	.29	--	--
12" x 6" x 16" (7.5 CF per CSF)	--	SF	.43	--	--
Add for slump one side units	--	%	15.0	--	--

Adobe block Walls to 8', not including reinforcement or foundations, plus delivery, 4" high, 16" long, 2.25 blocks per SF, including 10% waste.

	Craft@Hrs	Unit	Material	Labor	Total
4" wide, ($.57 ea)	B9@.206	SF	1.41	4.23	5.64
6" wide, ($.69 ea)	B9@.233	SF	1.71	4.79	6.50
8" wide, ($.93 ea)	B9@.238	SF	2.30	4.89	7.19
12" wide, ($1.50 ea)	B9@.267	SF	3.71	5.49	9.20
Add for reinforcement (24" OC both ways)	--	SF	.23	--	--
Pavers, 12" x 12" x 2-1/2"	--	SF	.92	--	--

Flagstone Including concrete bed.

	Craft@Hrs	Unit	Material	Labor	Total
Veneer on walls, 3" thick	B9@.480	SF	3.72	9.86	13.58
Walks and porches, 2" thick	B9@.248	SF	1.86	5.10	6.96
Coping, 4" x 12"	B9@.272	SF	5.90	5.59	11.49
Steps					
Risers	B9@.434	LF	2.10	8.92	11.02
Treads	B9@.455	LF	4.21	9.35	13.56

Flue lining Pumice, 1' lengths, delivered to 30 miles, including 8% waste.

	Craft@Hrs	Unit	Material	Labor	Total
8-1/2" x 17"	B9@.148	Ea	4.65	3.04	7.69
10" x 14"	B9@.206	Ea	5.05	4.23	9.28
10" x 17-1/2"	B9@.228	Ea	5.50	4.69	10.19
10" x 21"	B9@.243	Ea	6.50	4.99	11.49
13" x 13"	B9@.193	Ea	5.80	3.97	9.77
13" x 17"	B9@.193	Ea	5.80	3.97	9.77
13" x 20"	B9@.228	Ea	8.20	4.69	12.89
17" x 17"	B9@.228	Ea	10.20	4.69	14.89
17" x 20"	B9@.254	Ea	12.20	5.22	17.42
21" x 20"	B9@.265	Ea	16.20	5.45	21.65
Deduct for 2' lengths	--	%	-5.0	-20.0	--

Glass block Costs include 6% for waste and local delivery. Add mortar and reinforcing costs below. Labor cost includes caulking, cleaning with sponge and fiber brush, and continuous tuckpointing. Based on Pittsburgh Corning Glass, costs are for a 100 SF job.

3-7/8" thick block, smooth or irregular faces (Argus, Decora, Essex AA or Vue)

	Craft@Hrs	Unit	Material	Labor	Total
4" x 8", 4.5 per SF, $3.00 each	B9@.392	SF	14.30	82.06	96.36
6" x 6", 4 per SF, $3.00 each	B9@.376	SF	12.70	7.73	20.43
8" x 8", 2.25 per SF, $4.00 each	B9@.296	SF	9.55	6.08	15.63
12" x 12", 1 per SF, $10.50 each	B9@.211	SF	11.70	4.34	16.04

Masonry, Glass Block

	Craft@Hrs	Unit	Material	Labor	Total
Add for heat and glare reducing insert	--	%	75.0	--	--
3-7/8" thick block, solar reflective bronze or grey tint, one side, smooth finish					
8" x 8", 2.25 per SF, $9.75 each	B9@.296	SF	23.30	6.08	29.38
Add for solar reflective two sides, grey	--	%	7.0	--	--
3-7/8" thick corner units	B9@.125	Ea	14.25	2.70	16.95
3-7/8" thick end units	B9@.125	Ea	14.25	2.70	16.95
3" thick block, solid glass, coated or uncoated, smooth faces (Vistabrick)					
8" x 8", 2.25 per SF, $19.50 each	B9@.376	SF	46.50	7.73	54.23
3-1/8" thick block, smooth, irregular, or diamond relief faces (Vue, Decora, or Delphi thinline)					
6" x 6", 4 per SF, $2.95 each	B9@.376	SF	12.50	7.73	20.23
8" x 8", 2.25 per SF, $3.05 each	B9@.296	SF	7.25	6.08	13.33
4" x 8", 4.5 per SF, $2.95 each	B9@.391	SF	14.10	8.04	22.14
6" x 8", 3 per SF, $2.95 each	B9@.344	SF	9.40	7.07	16.47
Add for heat and glare reducing insert	--	%	75.0	--	--
Deduct for over 100 SF to 500 SF job	--	%	-5.0	--	--
Deduct for over 500 SF job to 1,000 SF job	--	%	-10.0	--	--
Deduct for over 1,000 SF job	--	%	-12.0	--	--
Add for wall ties	--	Ea	1.90	--	--
Add for expansion strips, fiberglass	--	Ea	1.10	--	--
Add for panel anchors, 20 gauge perforated steel, 1-3/4" x 24"	--	LF	1.42	--	--
Add for panel reinforcing, 10' lengths, 3" wide	--	LF	.25	--	--
Mortar for glass block					
White cement mix (assumes 1/2 part hydrated lime, 1 part white cement, 2-1/4 to 3 parts #20 or #30 silica sand)	--	CF	7.15	--	--
Portland cement mix (same as above except portland cement replaces white cement, use when adding colors)	--	CF	4.91	--	--
1/4" mortar joints, 3-7/8" thick block, white cement mix					
6" x 6" block (4.3 CF/CSF)	--	SF	.31	--	--
8" x 8" block (3.2 CF/CSF)	--	SF	.23	--	--
12" x 12" block (2.2 CF/CSF)	--	SF	.16	--	--
1/4" mortar joints, 3" and 3-1/8" thick block, white cement mix					
6" x 6" block (3.5 CF/CSF)	--	SF	.25	--	--
8" x 8" block (2.6 CF/CSF)	--	SF	.19	--	--
Deduct for portland cement mix	--	%	-30.0	--	--

Masonry accessories

	Craft@Hrs	Unit	Material	Labor	Total
Asphalt emulsion (50 SF per gallon) for below grade waterproofing					
5 gallon can	--	Ea	18.50	--	--
1 gallon can	--	Ea	5.00	--	--
Glass reinforcing mesh	--	SF	.11	--	--
Lime					
Hydrated (builders)					
50 lb sack	--	Sack	6.45	--	--
25 lb sack	--	Sack	3.35	--	--
10 lb sack	--	Sack	1.66	--	--
Pebbled lime, 100 lb sack	--	Sack	11.00	--	--
Processed or ground lime, 60 lb sack	--	Sack	7.50	--	--
Silica sand, #30, 100 lb sack	--	Sack	4.00	--	--
White cement, 94 lb sack	--	Sack	12.75	--	--
Portland cement, 94 lb sack	--	Sack	6.50	--	--

Masonry, Accessories

	Craft@Hrs	Unit	Material	Labor	Total
Mortar colors, 8 pounds of color per sack of cement or lime.					
Red (light or dark)	--	Lb	1.70	--	--
Yellow	--	Lb	1.60	--	--
Brown	--	Lb	1.70	--	--
Black	--	Lb	1.65	--	--
Green	--	Lb	4.00	--	--
Trowel-ready (factory prepared, ready to use) mortar, 30 hour life, deposit required on container, local delivery included.					
Type "S" natural mortar, 1/3 CY	--	Ea	43.00	--	--
Type "S" white mortar, 1/3 CY	--	Ea	56.00	--	--
Type "M" natural mortar, 1/3 CY	--	Ea	50.00	--	--
Add for colored mortar	--	%	18.0	--	--
Trowel-ready mortar, cost per square foot of wall, 3/8" joint					
4" x 4" x 8" brick wall	--	SF	.29	--	--
8" x 8" x 16" block wall	--	SF	.26	--	--
Hi-rise channel reinforcing steel					
2" x 2" x 3/16"	--	LF	1.70	--	--
3" x 3" x 3/16"	--	LF	2.05	--	--
4" x 4" x 1/4"	--	LF	4.05	--	--
6" x 4" x 3/8"	--	LF	7.90	--	--
Welding	--	Ea	--	--	4.80
Cutting	--	Ea	--	--	2.40
Flanges, drill one hole, weld	--	LS	--	--	7.40
Additional holes	--	Ea	--	--	.75
Expansion joints					
1/4" x 2"	--	LF	.10	--	--
1/4" x 3-1/2"	--	LF	.17	--	--
1/2" x 3-1/2"	--	LF	.19	--	--
Wall anchors, 1-3/4" x 24"	--	Ea	1.98	--	--
Dur-O-Wal reinforcing, Ladur or standard					
3" wide	--	LF	.12	--	--
4" wide	--	LF	.12	--	--
6" wide	--	LF	.13	--	--
8" wide	--	LF	.14	--	--
10" wide	--	LF	.15	--	--
12" wide	--	LF	.15	--	--
Wired adobe blocks					
2" x 2"	--	Ea	.11	--	--
3" x 3"	--	Ea	.23	--	--
Tie wire, 16 gauge					
Roll, 400'	--	Ea	2.20	--	--
Box, 20 rolls	--	Ea	52.50	--	--
Wall ties, 22 gauge	--	M	35.00	--	--
Wall ties, 28 gauge	--	M	23.00	--	--
Angle iron					
3" x 3" x 3/16"	--	LF	2.20	--	--
4" x 4" x 1/4"	--	LF	3.80	--	--
6" x 4" x 3/8"	--	LF	8.00	--	--
Post anchors (straps)					
4" x 4" x 10"	--	Ea	2.00	--	--
4" x 4" x 16"	--	Ea	2.75	--	--

Masonry, Accessories

	Craft@Hrs	Unit	Material	Labor	Total

Masonry, Subcontract
Ashlar veneer, typical prices

4" random	--	SF	--	--	14.10
4" limestone	--	SF	--	--	17.80

Face brick

Common	--	SF	--	--	7.05
Roman	--	SF	--	--	8.50
Norman	--	SF	--	--	7.10
Jumbo	--	SF	--	--	5.22

Jumbo, double

6" x 4" x 12"	--	SF	--	--	10.10
8" x 4" x 12"	--	SF	--	--	12.80
Select modular, 3-3/4" x 2-1/4" x 7-5/8"	--	SF	--	--	9.50
Glazed brick veneer, standard	--	SF	--	--	11.60
Add for reinforcing, typical	--	SF	--	--	.27
Light sandblast	--	SF	--	--	.27

Brick walls, common brick, reinforced with 1/2" bars 24" OC

10" thick wall	--	SF	--	--	15.30
13" thick wall	--	SF	--	--	17.50
16" thick wall	--	SF	--	--	20.30
20" thick wall	--	SF	--	--	23.00

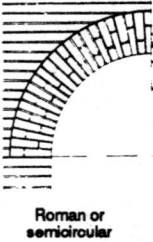

Roman or semicircular

Add for:

1/3 running bond	--	%	--	--	8.0
Weave bond	--	%	--	--	20.0
Flemish bond	--	%	--	--	25.0
Soldier or herringbone bond	--	%	--	--	20.0
Stacked bond	--	%	--	--	20.0
Short runs or cut up work	--	%	--	--	15.0
Wall over one story, per story	--	%	--	--	10.0
Modular brick	--	%	--	--	15.0
Arches, typical	--	LF	--	--	19.60
Caps and coping	--	LF	--	--	7.40
Cuts	--	LF	--	--	5.60
Jambs	--	LF	--	--	6.00
Sills, cut	--	LF	--	--	6.70

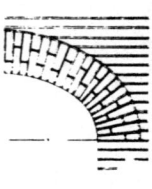

Elliptical

Masonry fill insulation for cavity walls, 4" thick (80 sacks of 4 CF each covers 1,000 SF)

	--	SF	--	--	.76
Deduct for tract work	--	%	--	--	-15.0

Concrete block, standard natural units, reinforced each 24"

4" wide block	--	SF	--	--	4.85
6" wide block	--	SF	--	--	5.50

8" wide block

8" x 16"	--	SF	--	--	5.80
4" x 16" all cells filled	--	SF	--	--	8.80
4" x 16" split face	--	SF	--	--	9.80
Add for colors	--	SF	--	--	.16
Add for waterproof, one side	--	SF	--	--	.33
Light sandblast	--	SF	--	--	.37
Heavy architectural sandblast	--	SF	--	--	.84

Fireplaces and chimneys, masonry only. See also Fireplace Materials

5' chimney (3' opening)	--	Ea	--	--	1,690.00
13' chimney (3' opening)	--	Ea	--	--	2,000.00

Masonry, Subcontract

	Craft@Hrs	Unit	Material	Labor	Total
Floors					
Flagstone	--	SF	--	--	8.10
Patio tile	--	SF	--	--	6.10
Pavers	--	SF	--	--	6.59
Quarry tile, unglazed	--	SF	--	--	9.15

Mats and Treads. Koffler Products

	Craft@Hrs	Unit	Material	Labor	Total
Entrance mats					
One piece, corrugated rubber mats, black					
3/8" thick, stock sizes up to 3' x 6'	B1@.125	SF	12.65	2.65	15.30
Link mats, open link, vinyl, 1/2" thick, including nosing					
Solid black	B1@.167	SF	12.40	3.54	15.94
Solid colors	B1@.167	SF	14.30	3.54	17.84
Any two colors	B1@.167	SF	14.30	3.54	17.84
For 9" to 12" letters add	--	Ea	9.30	--	--
For closed link add	--	%	25.0	--	--
Steel link mats, 3/8" deep, 1" mesh					
16" x 24"	B1@.501	Ea	16.00	10.60	26.60
22" x 36"	B1@.501	Ea	30.00	10.60	40.60
30" x 48"	B1@.501	Ea	52.00	10.60	62.60
36" x 54"	B1@.501	Ea	62.00	10.60	72.60
36" x 72"	B1@.501	Ea	84.00	10.60	94.60
40" x 84"	B1@.501	Ea	104.00	10.60	114.60
Non-slip strips, press-on abrasive, black, 1/16" thick					
3/4" x 24", box of 50 at $30.00	B1@.080	Ea	.64	1.70	2.34
6" x 24" cleats, box of 50 at $180.00	B1@.080	Ea	3.90	1.70	5.60
Corrugated rubber runner, cut lengths, 2', 3', 4' widths					
1/8" thick, black	B1@.060	SF	1.83	1.27	3.10
3/16" thick, black	B1@.060	SF	4.05	1.27	5.32
1/4" thick, black	B1@.060	SF	4.70	1.27	5.97
Deduct for full rolls (75 or 150 LF)	--	%	-10.0	--	--
Corrugated or round ribbed vinyl runner, 1/8" thick, 2', 3', or 4' wide, cut lengths, flame resistant					
Black or brown	B1@.060	SF	1.95	1.27	3.22
Terra cotta, green or grey	B1@.060	SF	2.15	1.27	3.42
White, 3' widths only	B1@.060	SF	2.38	1.27	3.65
Deduct for full rolls (100 LF)	--	%	-10.0	--	--
Linked hardwood strip mats, 3/4" thick					
Widths to 42"	B1@.125	SF	10.90	2.65	13.55
Over 42" wide to 60"	B1@.125	SF	11.90	2.65	14.55
Stair treads, homogeneous rubber, 1/4" thick, 1-1/2" nosing, 12" deep					
Diamond design, marbleized	B1@.080	LF	9.70	1.70	11.40
Diamond design, black	B1@.080	LF	9.70	1.70	11.40
Plain, with abrasive strips	B1@.080	LF	8.80	1.70	10.50

Medicine Cabinets. No electrical work included.

	Craft@Hrs	Unit	Material	Labor	Total
Recessed cabinets, hinged door, no lighting					
Minimum quality, 14" W x 20" H x 3" D, two fixed shelves, plastic body, mirror door, stainless steel frame	BC@.850	Ea	30.50	20.10	50.60
Minimum quality, 16" W x 22" H x 3" D, two fixed shelves, polystyrene body, two mirror doors with stainless steel frames	BC@.850	Ea	31.00	20.10	51.10

Medicine Cabinets

	Craft@Hrs	Unit	Material	Labor	Total
Average quality, 17" W x 25" H x 3" D, adjustable shelves, polystyrene body with vinyl shutter door	BC@.850	Ea	39.00	20.10	59.10
Average quality, 16" W x 24" H x 3" D, adjustable shelves, polystyrene body with unfinished wood shutter door	BC@.850	Ea	44.50	20.10	64.60
Good quality, 21" W x 31" H x 3" D, adjustable shelves, polystyrene body, oval mirror with chrome frame, decorator model	BC@.941	Ea	73.00	22.20	95.20
Good quality, 14" W x 24" H x 3" D, adjustable shelves, steel body, mirror door with chrome frame	BC@.850	Ea	52.50	20.10	72.60
Good quality, 18" W x 26" H x 3" D, oak frame, adjustable shelves, mirror door with magnetic catch	BC@.850	Ea	73.00	20.10	93.10
Better quality, 21" W x 31" H x 3" D, adjustable shelves, polystyrene body, oak-framed oval mirror	BC@.941	Ea	235.00	22.20	257.20

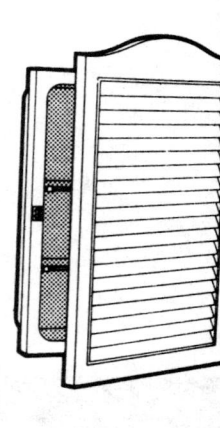

Recessed cabinet, hinged door, with lighting

	Craft@Hrs	Unit	Material	Labor	Total
Better quality, 20" W x 33" H x 3" D, adjustable shelves, polystyrene body, oak framed, brass and stained glass inserts, light strip above for 5 light bulbs	BC@1.18	Ea	310.00	27.80	337.80

Surface mounted cabinets with hinged door, no lighting

	Craft@Hrs	Unit	Material	Labor	Total
Good quality, corner models, sizes vary up to 16" W x 36" H x 3" D, adjustable shelves, fabricated for corner installation, mirror door with chrome frame, 14" W x 22" H x 3" D	BC@.850	Ea	73.00	20.10	93.10
Good quality, 18" W x 36" H x 5" D, adjustable shelves, polystyrene body, vinyl shutter	BC@.941	Ea	95.00	22.20	117.20
Good quality, 20" W x 30" H x 5" D, adjustable shelves, steel body, two mirror doors with steel frames	BC@.941	Ea	110.00	22.20	132.20
Better quality, 30" W x 30" H x 5" D, adjustable shelves, steel body, oak-framed mirrors, 1 fixed center door, 2 hinged side doors	BC@1.18	Ea	250.00	27.80	277.80

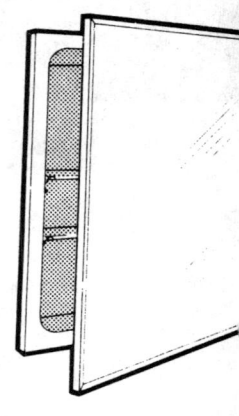

Surface mounted cabinets with sliding, by-passing mirror doors, no lighting

	Craft@Hrs	Unit	Material	Labor	Total
Minimum quality, 24" W x 16" H x 5" D, fixed shelves, polystyrene body with stainless steel frame	BC@.850	Ea	51.50	20.10	71.60
Good quality, 38" W x 31" H x 6" D, adjustable shelves, steel body with chrome frame	BC@.941	Ea	250.00	22.20	272.20

Surface mounted lighted cabinets with by-passing mirrored doors, four 60 watt bulbs, with plastic light diffusers, no electric work,

	Craft@Hrs	Unit	Material	Labor	Total
18" W x 27" H x 5" D, fixed shelves, steel body with oak-framed door	BC@1.18	Ea	52.50	27.80	80.30
16" W x 22" H x 5" D, fixed shelves, steel body with steel framed door	BC@1.18	Ea	23.00	27.80	50.80
26" W x 24" H x 6" D, two shelves, steel body, oak framed, light switch, grounded outlet	BC@1.42	Ea	112.00	33.50	145.50
24' W x 20" H x 9" D, three shelves, steel body, steel framed, light switch, grounded outlet	BC@1.42	Ea	65.00	33.50	98.50
Add for hardwood and glass valance lighting units, decorator style, no electrical work	BC@.377	Ea	88.00	8.89	96.89
Add for wiring and connecting lighted units (including switch)	BE@1.00	Ea	20.42	25.28	45.70

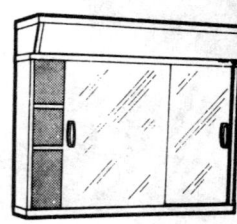

Meter Boxes

Precast concrete, FOB manufacturer (Brooks), inside dimension, approximately 100 to 190 lbs each, 11" deep

Meter Boxes

	Craft@Hrs	Unit	Material	Labor	Total

Meter Boxes
13-1/4" x 19-1/8"
With 1 piece concrete cover	BL@.334	Ea	18.00	6.28	24.28
With metal cover (cast iron)	BL@.334	Ea	31.00	6.28	37.28

14-5/8" x 19-3/4"
With 1 piece concrete cover	BL@.334	Ea	20.50	6.28	26.78
With metal cover (cast iron)	BL@.334	Ea	39.00	6.28	45.28

15-3/4" x 22-3/4"
With 1 piece concrete cover	BL@.334	Ea	25.50	6.28	31.78
With metal cover (cast iron)	BL@.334	Ea	44.50	6.28	50.78
Add for hinged or reading lid in 1 piece concrete cover	--	Ea	8.00	--	--
Deduct for non-bolting covers	--	%	-33.0	--	--

Plastic meter and valve boxes (Brooks), bolting plastic covers
Standard 16" x 10-3/4" x 12" deep	BL@.167	Ea	16.00	3.14	19.14
Jumbo 20-1/4" x 14-1/2" x 12" deep	BL@.167	Ea	28.50	3.14	31.64
Round pit box, 6" diameter, 9-1/16"	BL@.167	Ea	6.00	3.14	9.14

Mirrors. See also Bathroom Accessories and Glass.
Wall mirrors, distortion-free float glass, 1/8" x 3/16" thick
Cut and polished edges	--	SF	3.00	--	--
Beveled edges	--	SF	5.15	--	--

Decorative mirrors
Mirror strips, 8" x 72" (4 SF)
Clear	BG@.568	Ea	15.00	12.50	27.50
Clear with beveled edge	BG@.568	Ea	22.00	12.50	34.50
Bronze with beveled edge	BG@.568	Ea	24.00	12.50	36.50

Mirror arches (radius, top edge), 68" high
16" wide, clear (8.25 SF)	BG@1.05	Ea	21.00	23.10	44.10
20" wide, clear (10.9 SF)	BG@1.38	Ea	26.00	30.30	56.30
24" wide, clear (10.9 SF)	BG@1.38	Ea	70.00	30.30	100.30

Labor installing mirrors
To 5 SF	BG@.142	SF	--	3.12	--
5 to 15 SF	BG@.127	SF	--	2.79	--
15 to 25 SF	BG@.113	SF	--	2.48	--
25 to 50 SF	BG@.093	SF	--	2.04	--

Framed mirrors
Traditional rectangle with urethane corner and side ornamentation
31" x 42"	BG@.331	Ea	68.50	7.27	75.77
35" x 49"	BG@.331	Ea	89.50	7.27	96.77

Traditional hexagon mirror with urethane frame
32" x 42"	BG@.331	Ea	93.00	7.27	100.27
38" x 48"	BG@.331	Ea	100.00	7.27	107.27

Door mirrors
14" x 50"	BG@.251	Ea	26.50	5.51	32.01
16" x 56"	BG@.251	Ea	31.50	5.51	37.01
18" x 60"	BG@.301	Ea	37.00	6.61	43.61
18" x 68"	BG@.331	Ea	42.00	7.27	49.27
24" x 68"	BG@.331	Ea	47.50	7.27	54.77

Solid oak frame, plate glass
18" x 29"	BG@.331	Ea	52.50	7.27	59.77
24" x 24"	BG@.331	Ea	62.00	7.27	69.27

Mirrors

	Craft@Hrs	Unit	Material	Labor	Total
30" x 24"	BG@.331	Ea	72.50	7.27	79.77
36" x 26"	BG@.450	Ea	84.00	9.89	93.89
48" x 26"	BG@.450	Ea	105.00	9.89	114.89

Mouldings. Unfinished. Material costs include 5% waste. Labor costs are for typical installation only. For installation costs of Hardwood Moulding, see Softwood Moulding section.

Hardwood moulding

	Unit	Oak	Paint Grade Poplar	Birch
Base				
1/2" x 1-1/2"	LF	.80	.61	.69
1/2" x 2-1/2"	LF	.96	.72	.80
Casing				
1/2" x 1-1/2"	LF	.80	.61	.72
5/8" x 1-5/8"	LF	.87	.66	.75
Cap moulding, 1/2" x 1-1/2"	LF	.75	.64	.68
Crown, 1/2" x 2-1/4"	LF	.91	.69	.81
Cove				
1/2" x 1/2"	LF	.41	.35	.39
3/4" x 3/4	LF	.47	.38	.42
Corner mould				
3/4" x 3/4"	LF	.47	.35	.40
1" x 1"	LF	.62	.48	.54
Round edge stop, 3/8" x 1-1/4"	LF	.54	.36	.46
Chair rail, 1/2" x 2"	LF	1.06	1.46	1.17
Quarter round				
1/4" x 1/4"	LF	.52	.54	.47
1/2" x 1/2"	LF	.51	.48	.50
3/4" x 3/4"	LF	.57	.48	.50
Base shoe, 3/8" x 3/4"	LF	.46	.41	.44
Battens, 1/4" x 3/4"	LF	.32	.24	.26
Door trim sets, 16'8" total length				
1/2" x 1-5/8" casing	Ea	18.00	16.00	17.00
5/8" x 1-5/8" casing	Ea	20.00	18.00	19.00
3/8" x 1-1/4" round edge stop	Ea	12.00	10.00	11.00

Oak threshold, LF of threshold	Craft@Hrs	Unit	Material	Labor	Total
5/8" x 3-1/2"	BC@.174	LF	3.20	4.11	7.31
3/4" x 3-1/2"	BC@.174	LF	3.20	4.11	7.31
3/4" x 5-1/2"	BC@.174	LF	3.90	4.11	8.01
7/8" x 3-1/2"	BC@.174	LF	3.20	4.11	7.31

Redwood moulding

	Craft@Hrs	Unit	Material	Labor	Total
Band moulding, 3/4" x 5/16"	BC@.030	LF	.60	.71	1.31
Battens and lattice					
5/16" x 1-1/4"	BC@.026	LF	.30	.61	.91
5/16" x 1-5/8"	BC@.026	LF	.31	.61	.92
5/16" x 2-1/2"	BC@.026	LF	.46	.61	1.07
5/16" x 3-1/2"	BC@.026	LF	.49	.61	1.10
3/8" x 2-1/2"	BC@.026	LF	.47	.61	1.08
Brick mould, 1-1/2" x 1-1/2"	BC@.026	LF	1.17	.61	1.78
Drip cap-water table					
1-5/8" x 2-1/2" (no lip)	BC@.060	LF	1.46	1.42	2.88
1-5/8" x 2-1/2"	BC@.060	LF	1.46	1.42	2.88

Mouldings, Redwood

	Craft@Hrs	Unit	Material	Labor	Total
Quarter round, 11/16" x 11/16"	BC@.026	LF	.39	.61	1.00
Rabbeted siding mould, 1-1/8" x 1-5/8"	BC@.026	LF	.87	.61	1.48
S4S (rectangular)					
11/16" x 11/16"	BC@.026	LF	.40	.61	1.01
1" x 2"	BC@.026	LF	.59	.61	1.20
1" x 3"	BC@.026	LF	.96	.61	1.57
1" x 4"	BC@.026	LF	1.20	.61	1.81
Stucco mould					
13/16" x 1-5/16"	BC@.026	LF	.66	.61	1.27
13/16" x 1/2"	BC@.026	LF	1.00	.61	1.61
Window sill, 2" x 8"	BC@.043	LF	4.18	1.01	5.19
Softwood moulding, pine					
Astragal moulding					
For 1-3/8" x 7' doors	BC@.032	LF	2.56	.76	3.32
For 1-3/4" x 7' doors	BC@.032	LF	2.71	.76	3.47
Base (all patterns)					
7/16" x 7/16"	BC@.016	LF	.40	.38	.78
3/8" x 2-1/4"	BC@.016	LF	.60	.38	.98
7/16" x 1-5/8"	BC@.016	LF	.62	.38	1.00
1/2" x 2-1/4"	BC@.016	LF	.90	.38	1.28
1/2" x 2-1/2"	BC@.016	LF	.93	.38	1.31
1/2" x 3-1/2"	BC@.016	LF	1.20	.38	1.58
Base (combination or cove)					
7/16" x 1-5/8"	BC@.016	LF	.57	.38	.95
1/2" x 2-1/2"	BC@.016	LF	.98	.38	1.36
1/2" x 3-1/2"	BC@.016	LF	1.26	.38	1.64
Base shoe					
3/8" x 11/16"	BC@.016	LF	.26	.38	.64
7/16" x 3/4"	BC@.016	LF	.34	.38	.72
Blind stop					
3/8" x 1/2"	BC@.015	LF	.20	.35	.55
3/8" x 3/4"	BC@.015	LF	.21	.35	.56
3/8" x 1-1/4"	BC@.015	LF	.40	.35	.75
1/2" x 3/4"	BC@.015	LF	.41	.35	.76
1/2" x 1-1/4"	BC@.015	LF	.46	.35	.81
1/2" x 1-1/2"	BC@.015	LF	.56	.35	.91
Casing (all patterns)					
7/16" x 1-1/2"	BC@.023	LF	.37	.54	.91
9/16" x 1-5/8"	BC@.023	LF	.61	.54	1.15
1/2" x 1-5/8"	BC@.023	LF	.69	.54	1.23
5/8" x 2-1/2"	BC@.023	LF	.82	.54	1.36
5/8" x 3-1/2"	BC@.023	LF	1.52	.54	2.06
Chair rail					
1/2" x 1-5/8"	BC@.021	LF	.60	.50	1.10
5/8" x 2-1/2"	BC@.021	LF	.82	.50	1.32
Chamfer strip, 3/4" x 3/4"	BC@.016	LF	.42	.38	.80
Corner bead (outside corner moulding)					
3/4" x 3/4"	BC@.016	LF	.42	.38	.80
15/16" x 15/16"	BC@.016	LF	.82	.38	1.20
1-5/16" x 1-5/16"	BC@.016	LF	1.00	.38	1.38
1" x 2"	BC@.016	LF	.96	.38	1.34

Mouldings, Softwood

	Craft@Hrs	Unit	Material	Labor	Total
Cove moulding, solid					
3/8" x 3/8"	BC@.020	LF	.17	.47	.64
1/2" x 1/2"	BC@.020	LF	.26	.47	.73
5/8" x 5/8"	BC@.020	LF	.40	.47	.87
3/4" x 3/4"	BC@.020	LF	.41	.47	.88
15/16" x 15/16"	BC@.020	LF	.71	.47	1.18
Cove moulding, sprung					
3/4" x 1-5/8"	BC@.020	LF	.60	.47	1.07
3/4" x 2-1/4"	BC@.020	LF	.80	.47	1.27
3/4" x 3-1/2"	BC@.020	LF	1.18	.47	1.65
Crown or bed moulding					
3/4" x 3/4"	BC@.044	LF	.45	1.04	1.49
9/16" x 1-5/8"	BC@.044	LF	.62	1.04	1.66
9/16" x 2-1/4"	BC@.044	LF	.85	1.04	1.89
3/4" x 3-1/2"	BC@.044	LF	1.10	1.04	2.14
5/8" x 4-1/4"	BC@.044	LF	1.50	1.04	2.54
Drip cap (water table moulding or bar nosing)					
1-1/2" x 2-5/16" with lip	BC@.060	LF	1.53	1.42	2.95
1-1/2" x 2-5/16" no lip	BC@.060	LF	1.53	1.42	2.95
Drip moulding, 3/4" x 1-1/4"	BC@.030	LF	.47	.71	1.18
Glass bead, 5/8" x 3/4", 3/8" x 5/8"	BC@.016	LF	.31	.38	.69
Handrail, fir					
2" x 2" (oval)	BC@.060	LF	2.00	1.42	3.42
1" x 2-1/2" (detailed)	BC@.060	LF	1.90	1.42	3.32
Lattice					
1/4" x 1-1/4"	BC@.016	LF	.20	.38	.58
1/4" x 1-5/8"	BC@.016	LF	.25	.38	.63
1/4" x 2-1/2"	BC@.016	LF	.42	.38	.80
1/4" x 3-1/2"	BC@.016	LF	.53	.38	.91
Mullion casing, 1/4" x 3-1/2"	BC@.016	LF	.72	.38	1.10
Panel moulding					
3/8" x 5/8"	BC@.030	LF	.16	.71	.87
5/8" x 3/4"	BC@.030	LF	.28	.71	.99
3/4" x 1"	BC@.030	LF	.42	.71	1.13
3/4" x 1-5/8"	BC@.030	LF	.46	.71	1.17
1/2" x 7/8"	BC@.030	LF	.35	.71	1.06
Parting bead, 3/8" x 3/4"	BC@.016	LF	.19	.38	.57
Picture frame mould					
11/16" x 13/16"	BC@.040	LF	.58	.94	1.52
3/4" x 1-5/8"	BC@.040	LF	.86	.94	1.80
7/8" x 2-1/2"	BC@.040	LF	1.70	.94	2.64
Plaster ground					
3/4" x 7/8"	BC@.010	LF	.33	.24	.57
Quarter round					
1/4" x 1/4"	BC@.016	LF	.17	.38	.55
3/8" x 3/8"	BC@.016	LF	.18	.38	.56
1/2" x 1/2"	BC@.016	LF	.20	.38	.58
5/8" x 5/8"	BC@.016	LF	.25	.38	.63
3/4" x 3/4"	BC@.016	LF	.29	.38	.67
1" x 1"	BC@.016	LF	.62	.38	1.00
Half round					
1/4" x 1/2"	BC@.016	LF	.16	.38	.54
5/16" x 5/8"	BC@.016	LF	.17	.38	.55
3/8" x 3/4"	BC@.016	LF	.22	.38	.60

Mouldings, Softwood

	Craft@Hrs	Unit	Material	Labor	Total
1/2" x 1"	BC@.016	LF	.36	.38	.74
3/4" x 1-1/2"	BC@.016	LF	.51	.38	.89
Full round					
1/2"	BC@.016	LF	.42	.38	.80
3/4"	BC@.016	LF	.43	.38	.81
1"	BC@.016	LF	.68	.38	1.06
1-1/4"	BC@.016	LF	.85	.38	1.23
1-3/8"	BC@.016	LF	.47	.38	.85
1-5/8"	BC@.016	LF	.80	.38	1.18
2"	BC@.016	LF	1.60	.38	1.98
S4S (rectangular)					
1/2" x 2-1/2"	BC@.016	LF	.62	.38	1.00
1/2" x 3-1/2"	BC@.016	LF	.99	.38	1.37
1/2" x 5-1/2"	BC@.016	LF	1.50	.38	1.88
1/2" x 7-1/4"	BC@.016	LF	2.20	.38	2.58
1/2" x 3/4"	BC@.016	LF	.20	.38	.58
3/4" x 3/4"	BC@.016	LF	.31	.38	.69
3/4" x 1-1/4"	BC@.016	LF	.45	.38	.83
3/4" x 1-5/8"	BC@.016	LF	.51	.38	.89
3/4" x 2-1/2"	BC@.016	LF	.87	.38	1.25
3/4" x 3-1/2"	BC@.016	LF	1.26	.38	1.64
1" x 1"	BC@.016	LF	.65	.38	1.03
1-1/2" x 1-1/2"	BC@.016	LF	1.10	.38	1.48
Sash bar, 7/8" x 1-3/8"	BC@.016	LF	.75	.38	1.13
Screen moulding					
1/4" x 3/4" beaded	BC@.016	LF	.24	.38	.62
1/4" x 3/4" flat or insert	BC@.016	LF	.23	.38	.61
3/8" x 3/4" clover leaf	BC@.016	LF	.26	.38	.64
Square moulding					
1/2" x 1/2"	BC@.016	LF	.25	.38	.63
3/4" x 3/4"	BC@.016	LF	.35	.38	.73
1" x 1"	BC@.016	LF	.58	.38	.96
2" x 2"	BC@.016	LF	1.20	.38	1.58
Stops, round edge					
3/8" x 1/2"	BC@.025	LF	.19	.59	.78
3/8" x 3/4"	BC@.025	LF	.19	.59	.78
3/8" x 1"	BC@.025	LF	.27	.59	.86
3/8" x 1-1/4"	BC@.025	LF	.35	.59	.94
3/8" x 1-5/8"	BC@.025	LF	.34	.59	.93
1/2" x 3/4"	BC@.025	LF	.28	.59	.87
1/2" x 1-1/4"	BC@.025	LF	.41	.59	1.00
1/2" x 1-5/8"	BC@.025	LF	.50	.59	1.09
Wainscot cap					
5/8" x 1-1/4"	BC@.016	LF	.57	.38	.95
1/2" x 1-1/2"	BC@.016	LF	.48	.38	.86
Window stool, beveled					
7/8" x 2-1/4"	BC@.037	LF	.90	.87	1.77
15/16" x 2-1/2"	BC@.037	LF	1.37	.87	2.24
1" x 3-1/2"	BC@.037	LF	2.20	.87	3.07
1" x 5-1/2"	BC@.037	LF	3.49	.87	4.36
Window stool, flat					
11/16" x 4-5/8"	BC@.037	LF	1.40	.87	2.27
1" x 5-1/2"	BC@.037	LF	3.20	.87	4.07

Mouldings, Polystyrene

	Craft@Hrs	Unit	Material	Labor	Total
Polystyrene moulding Primed, ready for painting, Abitibi-Price					
Base, 9/16" x 3-5/16"	BC@.012	LF	1.04	.28	1.32
Casing, fluted, 5/8" x 3-1/4"	BC@.016	LF	1.06	.37	1.43
Cove, 13/16" x 3-3/4"	BC@.025	LF	1.04	.56	1.60
Chair rail, 5/8" x 3-1/8"	BC@.015	LF	.94	.34	1.28
Crown, 3/4" x 3-13/16"	BC@.025	LF	1.04	.56	1.60
Casing corner block, 3/4" x 3-1/2" x 3-1/2"	BC@.025	Ea	3.35	.56	3.91
Plinth block, 3/4" x 3-1/2" x 5-3/4"	BC@.025	Ea	3.71	.56	4.27
Base corner block					
Inside, 1-1/16" x 1-1/16" x 5-3/4"	BC@.025	Ea	2.26	.56	2.82
Outside, 1-5/16" x 1-5/16" x 5-3/4"	BC@.025	Ea	2.78	.56	3.34
Ceiling corner block					
Inside, 2-7/8" x 2-7/8" x 3-3/8"	BC@.025	Ea	3.67	.56	4.23
Outside, 3-1/2" x 3-1/2" x 3-3/8"	BC@.025	Ea	4.45	.56	5.01

Nails. Cost per pound in 50 lb cartons. Nails per pound is in parentheses.
Add 33% for 5 pound packages
Add 50% for 1 pound packages

	Sinkers cement coated	Common bright	Common galvanized
2 penny	.65 (1500)	--	--
3 penny	.62 (923)	.66 (568)	.84 (511)
4 penny	.61 (527)	.62 (294)	.81 (294)
5 penny	.51 (387)	.55 (271)	.75 (271)
6 penny	.46 (293)	.56 (167)	.71 (167)
7 penny	.46 (223)	.50 (161)	.71 (144)
8 penny	.44 (153)	.46 (101)	.68 (101)
10 penny	.44 (111)	.49 (66)	.67 (66)
12 penny	.44 (81)	.53 (61)	.67 (61)
16 penny	.43 (64)	.38 (47)	.65 (47)
20 penny	.44 (40)	.49 (29)	.66 (29)
30 penny	.46 (30)	.51 (22)	.71 (22)
40 penny	.48 (23)	.51 (17)	.71 (17)
50 penny	--	.50 (13)	.66 (13)
60 penny	--	.50 (10)	.66 (10)

Size & Kind of Material / Size of Nail / Quantity Required

Size & Kind of Material	Size of Nail	Quantity Required
1 x 2 furring on brick walls	20d common cut nails	5¼ lbs. / 100 lin. ft.
Base nailed 16"ctrs.	8d finish	1 lb. / 100 lin. ft.
Battens	4d finish	½ lb. / 100 lin. ft.
Sides of trim	4d, 6d, 8d finish	½ lb. / side of trim
48" wood lath	3d fine	7 lb. / 100 lath
Metal lath	7/16" hd. 1½" Roof Nail	17½ lb. / 100 sq. yds.
3/8" rock lath	Plaster board nail	9 lb. / 100 Sq. Yd.
¼" plaster board	4d cement coated	4½ lbs. / 1000 sq. ft.
3/8" plaster board	4d cement coated	4½ lbs. / 1000 sq. ft.
½" plaster board	6d cement coated	6 lbs. / 1000 sq. ft.
3/16" wallboard	3d cement coated	5 lbs. / 1000 sq. ft.
¼" wallboard	4d cement coated	9 lbs. / 1000 sq. ft.
3/8" wallboard	4d cement coated	9 lbs. / 1000 sq. ft.

NAILS FOR PARTICLE BOARD UNDERLAYMENT
3/8", ½" and 5/8" 6d cement coated box nails-10½ lbs. per 1000 square feet.
¾" 8d cement coated box nails - 17 lbs. per 1000 square feet.
Space nails at 6" centers around edges and 10: centers in interior area along line of floor joists. Edge nailing to be at a minimum of ½" from edge of panel.
(Approx. 100 nails per 4 x 8' panel.)

CEMENT COATED NAILS FOR PLYWOOD

Thickness		Lbs. / 1000 S. F.
¼"	4d	9
5/16"	6d	11
3/8"	6d	11
½"	7d	12
5/8"	8d	17
¾"	10d	21

Note: ¼" plywood will require about 5 lbs. per 1000 sq. ft. of 4d finish nails or 9 lbs. of 4d flat head nails when battens are not used figure 4½ lbs. of 4d flat head nails together with 3 lbs. of 4d finish nails when battens are used to cover joints spaced 4' 0" apart.

Box nails	Bright	Galvanized	Electro galv.	Cement coated
3 penny	.71 (588)	.95 (588)	.70 (511)	.66 (950)
4 penny	.69 (453)	.90 (453)	.66 (294)	.63 (710)
5 penny	.72 (389)	.91 (389)	.64 (271)	.56 (536)
6 penny	.62 (225)	.80 (225)	.61 (167)	.54 (306)
7 penny	.63 (210)	.87 (200)	.62 (144)	.55 (268)
8 penny	.50 (140)	.78 (136)	.58 (101)	.51 (186)
10 penny	.60 (94)	.81 (90)	.60 (66)	.53 (118)
12 penny	.60 (80)	.81 (76)	.60 (61)	.53 (105)
16 penny	.48 (71)	.76 (67)	.56 (47)	.50 (90)
20 penny	.59 (50)	.80 (50)	.59 (29)	.53 (65)

	Craft@Hrs	Unit	Material	Labor	Total
Box, nails, blued					
2 penny (890)	--	Lb	.61	--	--
3 penny (614)	--	Lb	.62	--	--
4 penny (473)	--	Lb	.64	--	--
Casing nails, bright					
3 penny (635)	--	Lb	.66	--	--

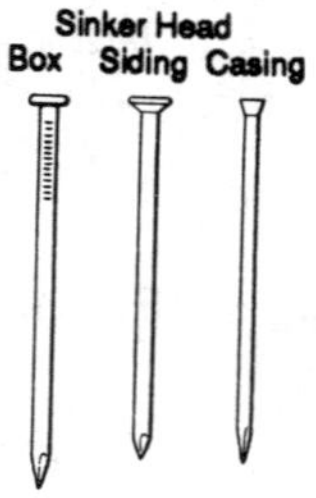

Nails

	Craft@Hrs	Unit	Material	Labor	Total
4 penny (473)	--	Lb	.59	--	--
5 penny (406)	--	Lb	.56	--	--
6 penny (236)	--	Lb	.56	--	--
8 penny (145)	--	Lb	.55	--	--
10 penny (94)	--	Lb	.55	--	--
16 penny (71)	--	Lb	.55	--	--
Concrete nails, (priced per pound in 25 lb carton)					
1/2" x 9 gauge (375)	--	Lb	.89	--	--
3/4" x 9 gauge (238)	--	Lb	.84	--	--
1" x 9 gauge (185)	--	Lb	.77	--	--
1-1/2" x 9 gauge (128)	--	Lb	.77	--	--
2" x 9 gauge (91)	--	Lb	.83	--	--
2-1/2" x 9 gauge (67)	--	Lb	.83	--	--
3" x 9 gauge (44)	--	Lb	.83	--	--
Duplex head nails (scaffold and form nails)					
6 penny (150)	--	Lb	.56	--	--
8 penny (88)	--	Lb	.55	--	--
10 penny (62)	--	Lb	.54	--	--
16 penny (44)	--	Lb	.52	--	--
20 penny (29)	--	Lb	.53	--	--
Finishing nails, bright					
2 penny (1351)	--	Lb	.82	--	--
3 penny (807)	--	Lb	.71	--	--
4 penny (584)	--	Lb	.57	--	--
5 penny (500)	--	Lb	.56	--	--
6 penny (309)	--	Lb	.53	--	--
8 penny (189)	--	Lb	.53	--	--
10 penny (121)	--	Lb	.55	--	--
12 penny (113)	--	Lb	.55	--	--
16 penny (90)	--	Lb	.52	--	--
20 penny (62)	--	Lb	.58	--	--
Finishing nails, galvanized					
3 penny (775)	--	Lb	.82	--	--
4 penny (560)	--	Lb	.71	--	--
6 penny (295)	--	Lb	.73	--	--
8 penny (180)	--	Lb	.64	--	--
16 penny (86)	--	Lb	.66	--	--
20 penny (60)	--	Lb	.69	--	--
Furring nails, 3/8" pads, 5,000 nails per case					
1-1/4"	--	Case	40.40	--	--
1-1/2"	--	Case	44.10	--	--
1-3/4"	--	Case	50.60	--	--
Lath nails, blued					
2 penny (1351)	--	Lb	.86	--	--
3 penny (778)	--	Lb	.86	--	--
3 penny (light) (1015)	--	Lb	.86	--	--
Plasterboard nails, 13 gauge, blued					
3/8" head, 1-1/8" (449)	--	Lb	.80	--	--
3/8" head, 1-1/4" (405)	--	Lb	.79	--	--
3/8" head, 1-1/2" (342)	--	Lb	.78	--	--
19/64" head, 1-1/8" (473)	--	Lb	.71	--	--
19/64" head, 1-1/4" (425)	--	Lb	.66	--	--
Ring shank nails, bright, 19/64" head, 12-1/2 gauge					

Nails

	Craft@Hrs	Unit	Material
1-1/4"	--	Lb	.71
1-3/8"	--	Lb	.71
1-1/2"	--	Lb	.71
1-5/8"	--	Lb	.71
Dritite nails, bright, 1/4" head			
1-3/8", 14 gauge	--	Lb	.66
1-5/8", 13-1/2 gauge	--	Lb	.66
1-7/8", 13 gauge	--	Lb	.66
2-3/8", 11-1/2 gauge	--	Lb	.66
Roofing nails, steel, large flat head, barbed, 7/16" head, 11 gauge			
Galvanized			
1/2" (355)	--	Lb	.74
3/4" (315)	--	Lb	.68
7/8" (280)	--	Lb	.66
1" (255)	--	Lb	.65
1-1/4" (210)	--	Lb	.65
1-1/2" (180)	--	Lb	.64
1-3/4" (150)	--	Lb	.63
2" (138)	--	Lb	.64
Electro galvanized			
1/2" (380)	--	Lb	.89
3/4" (315)	--	Lb	.76
7/8" (280)	--	Lb	.74
1" (255)	--	Lb	.66
1-1/4" (210)	--	Lb	.66
1-1/2" (180)	--	Lb	.64
1-3/4" (150)	--	Lb	.63
2" (138)	--	Lb	.60
Aluminum roofing nails, 7/16" head with 3/8" neoprene washer attached			
Screw thread, 10 gauge			
1-3/4", 525 per can, 1.87 lbs	--	Can	10.90
2", 525 per can, 2.60 lbs	--	Can	11.70
2-1/2", 315 per can, 3.40 lbs	--	Can	8.98
Screw shank, twin thread, 10 gauge			
1-3/4", 525 per can, 2.00 lbs	--	Can	10.90
2", 525 per can, 2.25 lbs	--	Can	11.70
2-1/2", 315 per can, 1.75 lbs	--	Can	9.77
Siding nails, board siding, plain shank			
6 penny (283), 2"	--	Lb	1.31
7 penny (248), 2-1/4"	--	Lb	1.29
8 penny (189), 2-1/2"	--	Lb	1.24
9 penny (171), 2-3/4"	--	Lb	1.28
10 penny (153), 3"	--	Lb	1.26
12 penny (113), 3-1/4"	--	Lb	1.22
16 penny (104), 3-1/2"	--	Lb	1.21
Add for anchor shank	--	%	25.0
Siding nails, hardboard siding, plain shank			
6 penny (181), 2"	--	Lb	1.18
8 penny (146), 2-1/2"	--	Lb	1.16
9 penny (136), 2-3/4"	--	Lb	1.20
10 penny (98), 3"	--	Lb	1.16
12 penny (90), 3-1/4"	--	Lb	1.20
16 penny (82), 3-1/2"	--	Lb	1.19

Nails For Asphalt Shingles

Type of shingle	Shingles / Sq. In.	Nails / Sq.
290 lb. 3-tab strip	80	400
255 lb. 3-tab strip	80	400
Sq. butts or thick butts	80	320
Dutch lap or angle lap	112	224
3-tab hexagon	86	344
Hip and ridge	30 strips, 120 shingles / bdl.	240 nails / bdl. 50 ft.

New work: Approx. 2 lbs. per square 7/8" barbed roofing nails.
Old work: Approx. 3 lbs. per square 1-3/4" barbed roofing nails.

Nails For Asbestos Roofing Shingles

Extra heavy American Colonial and American Colonial.
New work requires 1-3/4 lbs. 1-1/4" galvanized roofing nails per square.
Re-roofing requires 2-1/2 lbs. 2" galvanized roofing nails per square.

Ranch design.
New work requires 1 lb. 1-1/4" galvanized nails and 80 storm anchors per square.
Re-roofing requires 1-1/2 lbs. 2" galvanized nails and 80 storm anchors per square.

Lbs. of Nails Required per 1000 Ft. of Board Measure

Size & Kind of material	Size of nail	12"Ctrs.	16" Ctrs.	24" Ctrs.	# of nails to ea. bearing
Boards & Shiplap siding	**Common**				
1 x 4	8d	60	48	30	2
1 x 6	8d	40	32	20	2
1 x 8	8d	31	26	16	2
1 x 10	8d	25	20	13	2
1 x 12	8d	31	24	16	3
2 x 4	20d		82		2
2 x 6	20d		55		2
2 x 8	20d		41		2
2 x 10	20d		50		3
2 x 12	20d		41		3
Soft wood flooring	**Casing**				
1 x 3	8d	42	32	21	1
1 x 4	8d	30	24	16	1
	Common				
1 x 6	8d	22	16	11	1
Hard wood flooring	**Casing**				
3/8 x 1-1/2	4d	13	10		1
3/8 x 2	4d	11	8		1
25/32 x 2-1/4	8d	20	14		1
25/32 x 1-1/2	8d	27	20		1
Framing built-up girders	**Common**				
	20d	20	15	10	
2"	16d	10	8	6	
Blocking	10d	8	6	4	
Bridging	**Common**				4 per piece
	8d	1.5	1.2	.9	
Bevel siding	**Cem. coated**				
1/2 x 4	6d	23	18		1
1/2 x 6	6d	15	15		1
1/2 x 8	6d	12	10		1
3/4 x 10	7d	45	35		2
3/4 x 12	7d	60	50		3
Drop siding	**Cem. Coated**				
1 x 4	7d	45	35	22	2
1 x 6	7d	30	23	23	2
Ceiling mtrl.	**Finish**				
1/2, 5/8 x 4	6d	11	8	6	1
3/4 x 4	8d	18	14	9	1
Finish lumber	**Finish**				
7/8"	8d	25	12	13	2
1-1/4"	10d	12	10	6	2

Shingle Nails For Wood Roof Shingles
Approximate Number Required, and Weight of Rust - Resistant or Zinc Coated Nails Per Square.

	No. 1 Grade		No. 2 Grade		No. 3 Grade	
16 inch	#	WT. lbs	#	WT. lbs	#	WT. lbs
3-1/2" Exp.	1471	3-1/8	1872	4	2206	4-3/4
4" Exp.	1287	2-3/4	1637	3-1/2	1931	4-1/8
4-1/2" Exp.	1144	2-1/2	1454	3-1/8	1715	3-3/4
5" Exp.	1030	2-1/4	1310	2-3/4	1545	3-3/8
18 Inch						
4" Exp.	1287	2-3/4	1637	3-1/2	1931	4-1/8
4-1/2" Exp.	1144	2-1/2	1454	3-1/8	1715	3-3/4
5" Exp.	1030	2-1/4	1310	2-3/4	1545	3-3/8
5-1/2" Exp.	933	2	1190	2-5/8	1348	2-7/8
24 Inch						
6" Exp.	852	1-7/8	1060	2-3/8	1245	2-3/4
6-1/2" Exp.	784	1-3/4	974	2-1/8	1052	2-3/8
7" Exp.	760	1-3/4	945	2-1/8	1020	2-1/4
7-1/2" Exp.	716	1-5/8	885	2	955	2-1/8

Figures are for new roofs, on slat or solid decks.

For Over-Roofing, as larger nails are used, increase above weight two-thirds for 16" and 18" shingles and three-fourths for 24" shingles.

The above table allows a reasonable wastage of nails and fewer nails may be needed on some jobs.

For New Roofing: 16" and 18" shingles use 3d x 14 gauge shingle nails, approximate 466 per pound.

24" shingles use 4d galvanized box nails, approximate 453 per pound.

For Over Roofing: 16" and 18" shingles use 5d galvanized box nails, approximate 389 per pound.

24" shingles use 6d galvanized box nails, approximate 225 per pound.

Nails

	Craft@Hrs	Unit	Material	Labor	Total
20 penny (73), 4"	--	Lb	1.10	--	--
Add for anchor shank	--	%	17.0	--	--
Siding nails, for aluminum, steel and vinyl siding, plain shank					
1-1/4" (270)	--	Lb	1.19	--	--
1-1/2" (228)	--	Lb	1.16	--	--
2" (178)	--	Lb	1.11	--	--
2-1/2" (123)	--	Lb	1.13	--	--
3" (103)	--	Lb	1.21	--	--
Add for screw shank	--	%	30.0	--	--
Sinkers, green vinyl coated					
8 penny (90)	--	Lb	.35	--	--
16 penny (42)	--	Lb	.32	--	--
Spikes					
7" x 5/16" (6)	--	Lb	.19	--	--
8" x 5/16" (4)	--	Lb	.18	--	--
8" x 3/8" (4)	--	Lb	.18	--	--
10" x 3/8" (3)	--	Lb	.20	--	--
12" x 3/8" (2-1/2)	--	Lb	.31	--	--
Underlayment nails, ring grip,					
1-1/4" x 13 gauge (535)	--	Lb	.98	--	--
Shingle nails, 3d, electro galvanized	--	Lb	.61	--	--

NAILS FOR METAL ROOFING
Corrugated and 5-V crimp roofing - approx. 105 nails per square. Galvanized - approx. 1 lb. of spring head or approx. 1½ lbs. of lead head galvanized nails. Aluminum-approx. 1/3 lb. of aluminum roofing nails with neoprene washers.
Kaiser aluminum diamond-rib and twin-rib roofing. Approx. 90 nails per square.

Ornamental Iron, Columns and Railings Subcontract. Typical prices.
Handrail, 1-1/2" x 1" rectangular top rail tubing 1" square bottom rail tubing, pickets, 1/2" square with 5-1/2" space between.

	Craft@Hrs	Unit	Material	Labor	Total
42" high apartment style railing	--	LF	--	--	15.00
36" high railing for single family dwelling	--	LF	--	--	18.50
Add for bias	--	LF	--	--	3.50
Pool fence, minimum 5' high, 4-1/2" centers	--	LF	--	--	18.50
Add for gate, with self-closer and latch	--	LS	--	--	85.00

Paints, Coatings, and Supplies

Costs listed are for good to better quality coatings purchased in 1 to 4 gallon retail quantities. General contractor prices are generally discounted 10% to 15% on residential job quantities. Painting contractors receive discounts of 20% to 25% on steady volume accounts. Figures in parentheses are typical coverages per coat. Labor costs are listed at the end of this section. Add 30% for premium quality paints, deduct 30% for minimum quality paints.

	Craft@Hrs	Unit	Material	Labor	Total
Oil base primer					
Alkyd primer/sealer, interior (500 SF/Gal @ $15.75 per Gal)	--	SF	.03	--	--
Alkyd wood primer, exterior (400 SF/Gal @ $15.75 per Gal)	--	SF	.04	--	--
Prime coat oil wood primer (310 SF/Gal @ $21.00 per Gal)	--	SF	.07	--	--
Epoxy primer					
Catalyzed epoxy primer, two component type (425 SF/Gal @ $34.60 per Gal)	--	SF	.08	--	--
Cement base primer, per 25 lb sack, dry mix	--	Sack	15.75	--	--
Alkali resistant concrete/masonry primer/conditioner interior/exterior walls (225 SF/Gal @ $27.30 per Gal)	--	SF	.12	--	--
Latex primers					
Acrylic flat exterior, fast drying wood undercoater/ backprimer					

Paint

	Craft@Hrs	Unit	Material	Labor	Total
(400 SF/Gal @ $15.75 per Gal)	--	SF	.04	--	--
Acrylic (vinyl acrylic) block filler					
(50 SF/Gal @ $13.65 per Gal)	--	SF	.27	--	--
Acrylic (vinyl acrylic) rust inhibiting metal primer					
(500 SF/Gal @ $24.00 per Gal)	--	SF	.05	--	--
Acrylic heavy duty wall primer/sealer, low odor type, alkali-resistant					
(800 SF/Gal @ $22.40 per Gal)	--	SF	.03	--	--
Acrylic (vinyl acrylic) undercoater, all purpose, interior/exterior					
(500 SF/Gal @ $22.00 per Gal)	--	SF	.04	--	--
Latex wood primer, exterior (300 SF/Gal					
@ $22.60 per Gal)	--	SF	.08	--	--
Linseed primers					
Linseed (oil base) exterior primer					
(300 SF/Gal @ $14.70 per Gal)	--	SF	.05	--	--
Linseed (oil base) interior primer					
(300 SF/Gal @ $14.70 per Gal)	--	SF	.05	--	--
Red alkyd primer					
(500 SF/Gal @ $26.25 per Gal)	--	SF	.05	--	--
Wood sealer/primer, interior, clear					
(600 SF/Gal @ $15.75 per Gal)	--	SF	.03	--	--
Zinc chromate primer					
(300 SF/Gal @ $12.60 per Gal)	--	SF	.04	--	--
Zinc dust/zinc oxide primer					
(300 SF/Gal @ $29.40 per Gal)	--	SF	.10	--	--
Aluminum fiber roof coat (50 SF/Gal on					
composition roof @ $10.50 per Gal)	--	SF	.21	--	--
Anti-rust enamel, 500 SF/Qt @ $8.40 per Qt	--	SF	.02	--	--
Asphalt-fiber roof and foundation coating @ $5.25 per Gal					
75 SF per Gal on composition roof	--	SF	.07	--	--
100 SF per Gal on metal or concrete	--	SF	.05	--	--
Concrete enamel epoxy (350 SF per Gal					
@ $31.50 per Gal)	--	SF	.09	--	--
Concrete acrylic sealer (400 SF per Gal @ $17.90					
per Gal), rolled on concrete slab	--	SF	.05	--	--
Driveway coating					
Tar emulsion (120 SF per Gal @ $5.25/ Gal)	--	SF	.05	--	--
Acrylic latex (250 SF per Gal @ $8.40/ Gal)	--	SF	.04	--	--
Gutter paint (100 LF per Qt at $6.30 per Qt)	--	LF	.06	--	--
House paint, exterior					
Oil base, gloss enamel					
(400 SF per Gal @ $23.10 per Gal)	--	SF	.06	--	--
One coat, gloss oil base					
(400 SF per Gal @ $26.30 per Gal)	--	SF	.07	--	--
Latex, semi-gloss,					
(400 SF per Gal @ $15.75 per Gal)	--	SF	.04	--	--
Trim paint, flat latex,					
(500 SF per Gal @ $20.00 per Gal)	--	SF	.04	--	--
Latex flat paint,					
(300 SF per Gal @ $17.90 per Gal)	--	SF	.06	--	--
House paint, interior					
Latex flat paint					
(300 SF per Gal @ $17.90 per Gal)	--	SF	.06	--	--

Paint

	Craft@Hrs	Unit	Material	Labor	Total
Latex semi-gloss enamel, (300 SF per Gal @ $20.00 per Gal)	--	SF	.07	--	--
Latex fire-retardant flat paint (200 SF per Gal @ $36.80 per Gal)	--	SF	.18	--	--
Ceiling white latex flat, (350 SF per Gal @ $12.60 per Gal)	--	SF	.04	--	--
Linseed oil, boiled or raw @ $14.70 per Gal					
New exterior wood (400 SF per Gal)	--	SF	.04	--	--
Exterior brick, common, (200 SF per Gal)	--	SF	.07	--	--
Exterior concrete, stucco, brick, and shingle paint (300 SF per Gal @ $21.00 per Gal)	--	SF	.07	--	--
Masonry paint, two component type @ $18.90 per Gal					
Concrete floors or walls (200 SF per Gal)	--	SF	.09	--	--
Masonry block walls (120 SF per Gal)	--	SF	.16	--	--
Masonry waterproofing (100 SF per Gal @ $17.90 per Gal)	--	SF	.18	--	--
Paint remover, methylene chloride, (200 SF per Gal @ $19.00 per Gal)	--	SF	.10	--	--
Patio and floor paint					
latex (300 SF per Gal @ $17.90 per Gal)	--	SF	.06	--	--
acrylic (400 SF per Gal @ $23.10 per Gal)	--	SF	.06	--	--
Sealer					
Thompson's Water Seal (300 SF per Gal @ $21.00 for 2 Gallon can)	--	SF	.04	--	--
Spackling compounds (vinyl-paste), DAP					
Exterior	--	Qt	6.30	--	--
Exterior	--	Gal	15.75	--	--
Interior	--	Qt	5.80	--	--
Stain					
Latex redwood stain (water base) (300 SF per Gal @ $18.90 per Gal)	--	SF	.06	--	--
Redwood oil stain (300 SF per Gal @ $17.90 per Gal)	--	SF	.06	--	--
Creosote shingle stain, colors (200 SF per Gal @ $18.90 per Gal)	--	SF	.10	--	--
Swimming pool enamel					
Sau-Sea Type R (rubber base) (250 SF per Gal @ $40.00 per Gal)	--	SF	.16	--	--
Sau-Sea Plaster Pool Coating (250 SF per Gal @ $40.00 per Gal)	--	SF	.16	--	--
Thinner (Sau-Sea #3050-R)	--	Gal	14.00	--	--
Patching Plastic (150 LF per Gal @ $42.00 per Gal)	--	LF	.28	--	--
Texture paint, sand texture latex, armor coat (100 SF per Gal @ $6.80 per Gal)	--	SF	.07	--	--
Thinners					
Paint thinner	--	Gal	3.20	--	--
Shellac or lacquer thinner	--	Gal	9.50	--	--
Traffic paint, alkyd latex					
Black	--	Gal	18.90	--	--
White	--	Gal	20.00	--	--

Paint

	Craft@Hrs	Unit	Material	Labor	Total
Colors	--	Gal	27.30	--	--
Turpentine	--	Gal	12.60	--	--
Varnish, marine, clear	--	Gal	28.40	--	--
Wood filler paste	--	Gal	18.90	--	--
Wood preservative					
Pentachlorophenol, general purpose, 40%					
(150 SF per Gal @ $13.70 Gal)	--	SF	.09	--	--
Creosote (500 SF per Gal @ $7.35 per Gal)	--	SF	.02	--	--

Stains and wood preservatives	Quart	1 Gal.	5 Gal.
Original P.A.R., penetrating water repellent, redwood hue or clear	6.30	20.00	79.00
Woodlife wood preservative, clear	6.30	17.90	68.00
Patiolife redwood stain	7.40	21.00	84.00
Cuprolignum green wood preservative	8.40	18.90	76.50
Cuprinol stains, semi-transparent and solid colors	9.50	21.00	88.00
Wrought iron paint, black	23.00	--	--

Painting, Labor. Single coat applications except as noted. Not including equipment rental costs. These figures are based on hand work (roller, and brush where required) on residential jobs. Where spray equipment can be used to good advantage, reduce these costs 30% to 40%. Spray painting will increase paint requirements by 30% to 60%. Protection of adjacent materials is included in these costs but little or no surface preparation is assumed. Per SF of surface area to be painted.

	Craft@Hrs	Unit	Material	Labor	Total
Exterior surfaces, single coats					
Walls, siding	PT@.006	SF	--	.14	--
Doors and frames	PT@.007	SF	--	.16	--
Stucco	PT@.007	SF	--	.16	--
Trim, posts, and rails	PT@.012	LF	--	.27	--
Exterior surfaces, three coats					
Siding	PT@.015	SF	--	.34	--
Shingles	PT@.016	SF	--	.36	--
Doors and frames	PT@.020	SF	--	.45	--
Windows, SF of opening	PT@.038	SF	--	.86	--
Add for light sanding	PT@.011	SF	--	.25	--
Interior surfaces, single coat					
Flat painting	PT@.006	SF	--	.14	--
Enameling	PT@.007	SF	--	.16	--
Trim, filling and puttying	PT@.012	LF	--	.27	--
Staining, general	PT@.006	SF	--	.14	--
Shellacking, general	PT@.007	SF	--	.16	--
Varnishing, general	PT@.007	SF	--	.16	--
Interior surfaces, three coats					
Wallboard walls	PT@.015	SF	--	.34	--
Plaster walls	PT@.016	SF	--	.36	--
Wood walls	PT@.017	SF	--	.39	--
Wallboard ceilings	PT@.017	SF	--	.39	--
Plaster ceilings	PT@.018	SF	--	.41	--
Wood ceilings	PT@.021	SF	--	.48	--
Cabinet and case exteriors	PT@.025	SF	--	.57	--
Floors, wood					
Filling and puttying	PT@.007	SF	--	.16	--
Staining	PT@.006	SF	--	.14	--

Painting

	Craft@Hrs	Unit	Material	Labor	Total
Shellacking	PT@.006	SF	--	.14	--
Varnishing	PT@.006	SF	--	.14	--
Waxing, machine	PT@.006	SF	--	.14	--
Polishing, machine	PT@.008	SF	--	.18	--
Sanding, hand	PT@.012	SF	--	.27	--
Sanding, machine	PT@.007	SF	--	.16	--
Brick and concrete wall					
Oiling and sizing	PT@.007	SF	--	.16	--
Sealer coat	PT@.007	SF	--	.16	--
Masonry, finish coat	PT@.006	SF	--	.14	--
Concrete, finish coat	PT@.010	SF	--	.23	--
Concrete walls, 3 coats	PT@.017	SF	--	.39	--
Concrete steps, 3 coats	PT@.026	SF	--	.59	--

Paint Spraying Equipment Rental
Complete spraying outfit including paint cup, hose, spray gun, and conventional pump.

	Day	Week	Month
8 CFM electric compressor	37.00	100.00	290.00
17 CFM gas-driven compressor	47.00	155.00	410.00
Professional airless, complete outfit	55.00	200.00	600.00

Painting, Subcontract -- Rule of Thumb

	Craft@Hrs	Unit	Material	Labor	Total

Typical cost per square foot of floor area for painting the interior and exterior of residential buildings, including only minimum surface preparation

	Craft@Hrs	Unit	Material	Labor	Total
Economy, 1 or 2 coats, little brushwork	--	SF	--	--	1.40
Good quality, 2 or 3 coats	--	SF	--	--	3.21

Painting, Subcontract -- Itemized Costs
The painting costs that follow are based on the surface area painted.

Surface preparation
Light cleaning to remove surface dust and stains

	Craft@Hrs	Unit	Material	Labor	Total
Concrete or masonry surfaces	--	SF	--	--	.19
Gypsum or plaster surfaces	--	SF	--	--	.19
Wood surfaces	--	SF	--	--	.19

Normal painting preparation including scraping, patching and puttying

	Craft@Hrs	Unit	Material	Labor	Total
Concrete or masonry surface					
Painted	--	SF	--	--	.19
Unpainted	--	SF	--	--	.15
Gypsum or plaster surfaces, painted	--	SF	--	--	.19
Gypsum or plaster surfaces, unpainted	--	SF	--	--	.08
Metal surfaces, light sanding	--	SF	--	--	.08
Wood surfaces					
Painted	--	SF	--	--	.19
Unpainted	--	SF	--	--	.08
Puttying or reglazing windows	--	LF	--	--	1.04

Exterior painting Hand work, costs will be 15% to 30% lower when spray equipment can be used to good advantage.
Brick or concrete, including acrylic latex masonry primer and latex paint

	Craft@Hrs	Unit	Material	Labor	Total
1 coat latex	--	SF	--	--	.44
2 coats latex	--	SF	--	--	.62

Painting, Exterior

	Craft@Hrs	Unit	Material	Labor	Total
Concrete floors, etch and epoxy enamel	--	SF	--	--	.29
Columns and pilasters per coat	--	SF	--	--	.36
Cornices, per coat	--	SF	--	--	.91
Doors, including trim, exterior only					
3 coats	--	Ea	--	--	28.60
2 coats	--	Ea	--	--	20.60
Downspouts and gutters per coat	--	LF	--	--	.72
Eaves					
No rafters, per coat	--	SF	--	--	.32
With rafters, per coat, brush	--	SF	--	--	.62
Fences (gross area)					
Plain, 2 coats (per side)	--	SF	--	--	.42
Plain, 1 coat (per side)	--	SF	--	--	.22
Picket, 2 coats (per side)	--	SF	--	--	.38
Picket, 1 coat (per side)	--	SF	--	--	.20
Lattice work, 1 coat 1 side, gross area	--	SF	--	--	.38
Metal, typical					
1 coat	--	SF	--	--	.29
2 coats	--	SF	--	--	.54
Porch rail and balusters, per coat					
Gross area	--	SF	--	--	.78
Handrail only	--	LF	--	--	.53
Roofs, shingle, 1 coat stain plus 1 coat sealer					
Flat	--	SF	--	--	.36
4 in 12 pitch	--	SF	--	--	.46
8 in 12 pitch	--	SF	--	--	.49
12 in 12 pitch	--	SF	--	--	.60
Roofs, shingle, 1 coat stain plus 2 coats sealer					
Flat roof	--	SF	--	--	.60
4 in 12 pitch	--	SF	--	--	.69
8 in 12 pitch	--	SF	--	--	.76
12 in 12 pitch	--	SF	--	--	.82
Siding, plain					
Sanding and puttying, typical	--	SF	--	--	.21
Siding and trim, 3 coats	--	SF	--	--	.74
Siding and trim, 2 coats	--	SF	--	--	.55
Trim only, 2 coats	--	SF	--	--	.59
Trim only, 3 coats	--	SF	--	--	.79
Siding, shingles, including trim					
2 coats oil paint	--	SF	--	--	.56
1 coat oil paint	--	SF	--	--	.33
2 coats stain	--	SF	--	--	.53
1 coat stain	--	SF	--	--	.28
Spray painting, most surfaces, 2 coats	--	SF	--	--	.27
Steel sash, 3 coats (per side)	--	SF	--	--	.70
Stucco					
2 coats, rough surface	--	SF	--	--	.65
1 coat, rough surface	--	SF	--	--	.39
2 coats, smooth surface	--	SF	--	--	.54
1 coat, smooth surface	--	SF	--	--	.34
Window frames and sash, 3 coats					
One lite	--	Ea side	--	--	22.40
Add for each additional lite	--	Ea side	--	--	1.20

Painting, Interior

	Craft@Hrs	Unit	Material	Labor	Total

Interior painting Hand work (roller with some brush cut-in), costs will be 15% to 30% lower when spray equipment can be used to good advantage. These costs include only minimum of surface preparation.

Cabinets, bookcases, cupboards, per SF of face of cabinets

	Craft@Hrs	Unit	Material	Labor	Total
No doors, 3 coats	--	SF	--	--	1.73
Stain, shellac, varnish or plastic	--	SF	--	--	1.79
With doors, 3 coats	--	SF	--	--	2.85
Stain, shellac, varnish or plastic	--	SF	--	--	2.97
Concrete block or brick, primer and 1 coat latex	--	SF	--	--	.41
Doors, including trim					
3 coats (per side)	--	Ea	--	--	23.60
2 coats (per side)	--	Ea	--	--	16.20
Plaster or drywall walls, latex. Add for ceilings below					
Prime coat or sealer	--	SF	--	--	.20
1 coat smooth surface	--	SF	--	--	.24
1 coat rough surface	--	SF	--	--	.27
1 coat sealer, 1 coat flat					
Smooth surface	--	SF	--	--	.35
Rough surface	--	SF	--	--	.47
1 coat sealer, 2 coats flat					
Smooth surface	--	SF	--	--	.48
Rough surface	--	SF	--	--	.65
1 coat sealer, 2 coats gloss or semi-gloss					
Smooth surface	--	SF	--	--	.46
Rough surface	--	SF	--	--	.64
Add for ceilings	--	%	--	--	45.0
Stairs, including risers					
3 coats paint	--	SF	--	--	1.21
Stain, shellac, varnish	--	SF	--	--	1.24
Woodwork, painting					
Priming	--	SF	--	--	.26
2 coats	--	SF	--	--	.46
2 coats, top workmanship	--	SF	--	--	.59
3 coats	--	SF	--	--	.75
3 coats, top workmanship	--	SF	--	--	.98
Sanding and puttying	--	SF	--	--	.27
Woodwork	--	SF	--	--	.21
Staining					
Filling and paste wood filler	--	SF	--	--	.51
Woodwork, apply wax and polish	--	SF	--	--	.27

Paneling. See also Hardwood under Lumber and Hardboard.

Solid plank paneling These costs assume 5% cutting waste.

Prefinished, V joint or T&G, random lengths

	Craft@Hrs	Unit	Material	Labor	Total
Tennessee clear cedar, 3/8" x 4" or 6" wide	BC@.050	SF	1.35	1.18	2.53
Select tight knot cedar, 3/4" x 4" or 6" wide	BC@.050	SF	1.40	1.18	2.58
Knotty pine, 5/16" x 4" or 6" wide	BC@.050	SF	1.20	1.18	2.38
Redwood, clear, 5/16" x 4" wide	BC@.050	SF	2.60	1.18	3.78
Townsend hardwood by Potlatch, 1/2" x 3-1/4", 4" or 5-1/4" wide					
Burley pecan	BC@.063	SF	4.20	1.49	5.69
Colonial or Colonial ash	BC@.063	SF	2.90	1.49	4.39

Paneling

	Craft@Hrs	Unit	Material	Labor	Total
Colonial cherry	BC@.063	SF	3.62	1.49	5.11
Select red oak	BC@.063	SF	3.00	1.49	4.49
Wattled walnut	BC@.063	SF	4.30	1.49	5.79

Unfinished antique tongue and groove plank paneling, 25/32" thick, various widths and lengths, Vintage Lumber.

	Craft@Hrs	Unit	Material	Labor	Total
Chestnut, American, 3" to 7"	BC@.090	SF	6.35	2.12	8.47
Chestnut, distressed, 3" to 7"	BC@.090	SF	7.28	2.12	9.40
Chestnut, distressed, 6" and up	BC@.090	SF	10.11	2.12	12.23
Hemlock, 3" to 7"	BC@.090	SF	4.48	2.12	6.60
Hemlock, 6" and up	BC@.090	SF	7.28	2.12	9.40
Oak, 3" to 7"	BC@.090	SF	5.38	2.12	7.50
Oak, 6" and up	BC@.090	SF	8.00	2.12	10.12
Oak, distressed, 3" to 7"	BC@.090	SF	6.23	2.12	8.35
Oak, distressed, 6" and up	BC@.090	SF	8.94	2.12	11.06
Poplar, 3" to 8"	BC@.090	SF	3.68	2.12	5.80
Poplar, 8" and up	BC@.090	SF	4.92	2.12	7.04
White pine, 3" to 7"	BC@.090	SF	4.90	2.12	7.02
White pine, distressed, 3" to 7"	BC@.090	SF	5.60	2.12	7.72
White pine, distressed, 6" and up	BC@.090	SF	8.26	2.12	10.38
Yellow pine, 3" to 5"	BC@.090	SF	4.80	2.12	6.92
Yellow pine, 6" to 10"	BC@.090	SF	6.40	2.12	8.52

Unfinished tongue and groove planks, 25/32" thick, various widths and lengths, Aged Woods.

	Craft@Hrs	Unit	Material	Labor	Total
Cherry, #1, 3" to 7"	BC@.090	SF	3.80	2.12	5.92
Cherry, #1, 8" and up	BC@.090	SF	5.10	2.12	7.22
Maple, FAS, 3" to 7"	BC@.090	SF	3.25	2.12	5.37
Walnut, #3, 3" to 7"	BC@.090	SF	3.55	2.12	5.67
White pine, knotty grade, 12" and up	BC@.090	SF	2.85	2.12	4.97

Plywood and hardboard paneling Prefinished, 4' x 8' x 1/4" panels, including nails and adhesive at $.02 per SF and 5% waste, Georgia-Pacific.

	Craft@Hrs	Unit	Material	Labor	Total
Front Street (paper overlay)	BC@.030	SF	.50	.71	1.21
Concepts (direct print)	BC@.030	SF	.50	.71	1.21
Hillside birch, Lionite, Styleboard, Woodgrains, Terrace, or McKenzie birch	BC@.030	SF	.57	.71	1.28
Valley Forge, Barnplank, Millplank, Renaissance, Firelight, or Woodglen	BC@.030	SF	.85	.71	1.56
Timber Ridge	BC@.030	SF	1.00	.71	1.71
Bridgeport	BC@.030	SF	1.05	.71	1.76
Estate Oak	BC@.030	SF	1.20	.71	1.91
Firelight walnut	BC@.030	SF	1.35	.71	2.06
Estate cherry	BC@.030	SF	1.50	.71	2.21

Hardwood plywood paneling Unfinished, 4' x 8', including 5% waste, Georgia Pacific

Rotary cut ash, imported, flush face

	Craft@Hrs	Unit	Material	Labor	Total
1/4"	BC@.032	SF	.59	.76	1.35
1/2"	BC@.040	SF	1.38	.94	2.32
3/4", lumber core	BC@.042	SF	1.56	.99	2.55

Rotary cut birch, imported

	Craft@Hrs	Unit	Material	Labor	Total
1/4"	BC@.032	SF	.59	.76	1.35
1/2"	BC@.040	SF	1.03	.94	1.97
3/4"	BC@.042	SF	1.31	.99	2.30

Paneling

	Craft@Hrs	Unit	Material	Labor	Total
Plain slice white oak, imported, natural, flush face					
1/4"	BC@.032	SF	.72	.76	1.48
1/2"	BC@.040	SF	1.63	.94	2.57
3/4", lumber core	BC@.042	SF	1.75	.99	2.74
Rotary cut maple, natural					
1/4"	BC@.032	SF	.54	.76	1.30
1/2"	BC@.040	SF	.94	.94	1.88
3/4"	BC@.042	SF	1.19	.99	2.18
Plain slice walnut, flush face					
1/4"	BC@.032	SF	1.00	.76	1.76
1/2"	BC@.040	SF	2.21	.94	3.15
3/4", lumber core	BC@.042	SF	2.72	.99	3.71
Rotary cut lauan, natural oak, imported					
1/4"	BC@.032	SF	.38	.76	1.14
1/2"	BC@.040	SF	.69	.94	1.63
3/4"	BC@.042	SF	1.00	.99	1.99
Plain slice Honduras mahogany					
1/4"	BC@.032	SF	1.21	.76	1.97
1/2"	BC@.040	SF	1.54	.94	2.48
3/4", veneer core	BC@.042	SF	2.09	.99	3.08
Vinyl-clad hardboard paneling 1/4" vinyl-clad film over hardboard, Abitibi Price					
White or Aegean Gold	BC@.032	SF	.34	.76	1.10
Blue marble	BC@.032	SF	.38	.76	1.14
Harvest pattern	BC@.032	SF	.75	.76	1.51
Grooved paneling 5/32", grooves 1/2" OC "Tambour", Walker and Zanger					
Oak, teak, ash, unfinished veneer on plywood	BC@.032	SF	3.15	.76	3.91
Oak, teak, ash, prefinished veneer on plywood	BC@.032	SF	3.70	.76	4.46
Walnut, unfinished veneer on plywood	BC@.032	SF	4.00	.76	4.76
Walnut, prefinished veneer on plywood	BC@.032	SF	4.20	.76	4.96
Bright or brushed aluminum on hardboard	BC@.032	SF	5.25	.76	6.01
Bright or brushed brass on hardboard	BC@.032	SF	5.25	.76	6.01
White or almond enameled aluminum on hardboard	BC@.032	SF	5.25	.76	6.01
Plain unfinished hardboard	BC@.032	SF	2.10	.76	2.86
Z-brick paneling Non-ceramic mineral brick-like veneer. Interior or exterior. 2-3/8" x 8-1/8" x 5/16". Including adhesive/mortar and sealer	BC@.035	SF	3.18	.83	4.01
Paving, Subcontract. Small areas such as walks and driveways around residential buildings.					
Asphalt, including oil seal coat					
2"	--	SF	--	--	.83
3"	--	SF	--	--	1.05
4"	--	SF	--	--	1.28
4" base and fine grading	--	SF	--	--	.51
6" base and fine grading	--	SF	--	--	.57
Asphalt seal coats, applied with broom or squeegee					
Fog seal, thin layer of diluted SS1H oil	--	SF	--	--	.05

Paving

	Craft@Hrs	Unit	Material	Labor	Total
Seal and sand, thin layer of oil with sand laid on top	--	SF	--	--	.08
Simco "walk-top" or other quick dry creamy oil	--	SF	--	--	.12
Brick					
On concrete, grouted, laid flat, including concrete base	--	SF	--	--	7.90
On concrete, 8" x 2-1/4" exposure, no grouting, including concrete base	--	SF	--	--	9.10
On sand bed, flat	--	SF	--	--	6.35
On sand bed, 8" x 2-1/4" exposure	--	SF	--	--	7.60
Concrete (including fine grading)					
3" concrete, unreinforced	--	SF	--	--	1.78
4" concrete, unreinforced	--	SF	--	--	2.10
4" concrete, mesh reinforcing	--	SF	--	--	2.20
4" concrete, seeded aggregate finished, typical	--	SF	--	--	4.10
Add for colored concrete, most colors	--	SF	--	--	.27

Plastering, Subcontract. See also Lathing.

Stucco, not including lath, exterior walls, 3 coats, 7/8"

	Craft@Hrs	Unit	Material	Labor	Total
Textured finish	--	SY	--	--	10.00
Float finish	--	SY	--	--	11.90
Dash finish	--	SY	--	--	13.15
Add for soffits	--	%	--	--	15.0

Stucco, including lath, exterior, on wood frame, 7/8", portland cement plaster on metal lath, nailed, 3 coats

	Craft@Hrs	Unit	Material	Labor	Total
Walls	--	SY	--	--	18.15
Soffits	--	SY	--	--	19.10
Deduct for paper back lath	--	SY	--	--	-.16

Plaster on concrete or masonry, with bonding agent

	Craft@Hrs	Unit	Material	Labor	Total
1/2" sanded gypsum plaster	--	SY	--	--	11.90

Interior, on wood frame walls, no lath included

	Craft@Hrs	Unit	Material	Labor	Total
Gypsum plaster, 7/8", 3 coats	--	SY	--	--	11.75
Keenes cement finish, 3 coats	--	SY	--	--	14.95
Cement plaster finish, 3 coats	--	SY	--	--	14.20
Acoustical plaster, 1/4" including brown coat, 1-1/4" total	--	SY	--	--	17.20
Scratch coat, for tile	--	SY	--	--	6.75
Brown coat and scratch coat, for tile	--	SY	--	--	10.55
Gypsum plaster, 2 coats, tract work	--	SY	--	--	9.50
Add for ceiling work	--	%	--	--	15.0

Plumbing Fixtures and Equipment. Costs are for good to better quality fixtures and trim installed with the accessories listed. All vitreous china, enameled steel cast iron, and fiberglass fixtures are white. Colored acrylic fiberglass fixtures will cost about 6% more. Colored enameled steel, cast iron, or vitreous fixtures will cost about 25% more. No rough plumbing included. Rough plumbing is the drain, waste and vent (DWV) piping and water supply piping that isn't exposed when construction is complete. See the costs for rough plumbing at the end of this section. Minimum quality fixtures and trim will cost 20% to 50% less than the prices listed here. Deluxe fixtures will cost much more.

Plumbing

	Craft@Hrs	Unit	Material	Labor	Total

Bathtubs

Acrylic fiberglass bathtub, 5' L, 30" W, 16" H, (at $146) Includes polished chrome pop-up drain/overflow (at $84). Mixing valve, tub filler, and 2.5 GPM shower head (at $115) — P1@5.62, Ea, 345.00, 123.00, 468.00

Acrylic fiberglass drop-in whirlpool bath, slip resistant surface, 5' L, 42" W, 18-1/2" H (at $1,690), includes chrome pop-up drain/overflow (at $84), mixing valve and tub filler (at $118) — P1@5.62, Ea, 1,892.00, 123.00, 2,015.00

Enameled cast iron bathtub, 5' L, 32" W, 15" H (at $540), includes polished chrome pop-up drain/overflow (at $84), mixing valve, tub filler and 2.5 GPM shower head (at $115) — P1@7.63, Ea, 739.00, 166.00, 905.00

Enameled cast iron bathtub with chrome grab rails, 5' L, 32" W, 16" H, (at $730), includes polished chrome pop-up drain/overflow (at $84), mixing valve, tub filler, and 2.5 GPM shower head at $115 — P1@7.63, Ea, 929.00, 166.00, 1,095.00

Enameled cast iron bathtub, 5' L, 36" W, 20" H, built-in whirlpool with integral drain (at $2,820), includes mixing valve and tub filler (at $118) — P1@7.63, Ea, 2,938.00, 166.00, 3,104.00

Add for whirlpool electric connection — --, Ea, --, --, 170.00

Enameled steel bathtub, 5' L, 30" W, 15" H (at $125) includes polished chrome pop-up drain/overflow (at $84), mixing valve, tub filler, and 2.5 GPM shower head (at $115) — P1@6.10, Ea, 324.00, 133.00, 457.00

Add for shower curtain rod, installed — P1@.572, Ea, 21.50, 12.50, 34.00

Fiberglass whirlpool bath, 72" L x 42" W x 20" H (at $1,770), chrome drain kit (at $105) — P1@10.2, Ea, 1,875.00, 222.00, 2,097.00

Add for electrical connection — --, Ea, --, --, 165.00

Showers

Acrylic fiberglass shower stall, 32" W, 32" D, 72" H (at $220), includes mixing valve and 2.5 GPM shower head (at $84) — P1@5.15, Ea, 304.00, 112.00, 416.00

Acrylic fiberglass shower stall with integral soap dishes, grab bar, and drain, 48" W, 35" D, 72" H (at $260), includes mixing valve and 2.5 GPM shower head (at $84) — P1@5.62, Ea, 344.00, 123.00, 467.00

Acrylic fiberglass shower stall with integral soap dishes, grab bar, drain, and seat, 60" W, 35" D, 72" H (at $280), includes mixing valve and 2.5 GPM shower head (at $84) — P1@6.10, Ea, 364.00, 133.00, 497.00

Add for shower curtain rod, installed — P1@.572, Ea, 21.50, 12.50, 34.00

Tub and shower combinations

Acrylic fiberglass tub and shower with integral soap dishes and high and low grab bars, 5' W, 32" D, 72" H (at $310), includes polished chrome pop-up drain/overflow (at $32) mixing valve, tub filler, and 2.5 GPM shower head (at $93) — P1@7.87, Ea, 435.00, 172.00, 607.00

Handicapped/elderly access fiberglass tub and shower with integral soap dishes, seat and four grab bars, 64" L, 66" D, 82" H, with top (at $3,135), body washer shower (at $63), polished chrome pop-up drain/overflow (at $32), mixing valve and tub filler (at $70) — P1@9.72, Ea, 3,300.00, 212.00, 3,512.00

Add for shower curtain rod, installed — P1@.572, Ea, 21.50, 12.50, 34.00

Water closets

Floor-mounted, tank type vitreous china closet with elongated bowl (at $167) valved closet supply (at $22), and toilet seat with cover (at $43) — P1@2.65, Ea, 232.00, 57.80, 289.80

Plumbing

	Craft@Hrs	Unit	Material	Labor	Total
Wall-hung, tank type vitreous china closet with elongated bowl (at $425), valved closet supply (at $22), and toilet seat with cover (at $43)	P1@3.52	Ea	490.00	76.70	566.70
One piece, floor-mounted vitreous china closet with elongated bowl and toilet seat (at $627), valved closet supply (at $22)	P1@3.32	Ea	649.00	72.40	721.40
Deduct for round bowl	--	%	-25.0	--	--

Bidets

	Craft@Hrs	Unit	Material	Labor	Total
Standard vitreous china bidet with floor-mounted hardware (at $260), hot and cold valved supplies (at $27) and polished chrome fittings (at $260)	P1@3.67	Ea	547.00	80.00	627.00
Deluxe, designer type, vitreous china with floor mounting hardware (at $420), hot and cold valved supplies (at $27) and polished gold fittings (at $1,150)	P1@3.67	Ea	1,597.00	80.00	1,677.00

Lavatories

	Craft@Hrs	Unit	Material	Labor	Total
Oval, self-rimming vitreous china lavatory, 19" L, 17" D, with overflow (at $125), hot and cold valved supplies (at $35), faucets and pop-up drain fitting (at $110), and trap assembly (at $37)	P1@2.65	Ea	307.00	57.80	364.80
Rectangular, self-rimming vitreous china lavatory, 22" L, 19" D, with overflow, (at $156), hot and cold valved supplies (at $35), faucets and pop-up drain fitting (at $110) and trap assembly (at $37)	P1@2.65	Ea	338.00	57.80	395.80
Rectangular, wall-hung vitreous china lavatory, 20" L, 18" D, with overflow and wall mounting bracket (at $105), hot and cold valved supplies (at $35), faucet and pop-up drain fitting (at $110) and trap assembly (at $37)	P1@2.65	Ea	287.00	57.80	344.80
Oval, pedestal-mounted vitreous china lavatory with overflow, 24" L, 19" D (at $355), hot and cold valved supplies (at $35), faucets and pop-up drain fitting (at $110) and trap assembly (at $37)	P1@3.58	Ea	537.00	78.00	615.00
Rectangular, pedestal-mount vitreous china lavatory with overflow, 28" L, 22" D, (at $625), hot and cold valved supplies (at $35), faucets and pop-up drain fitting (at $110) and trap assembly (at $37)	P1@3.58	Ea	807.00	78.00	885.00
Wheelchair access lavatory, wall-hung vitreous china, 20" L, 27" D (at $260) insulated hot and cold valved supplies (at $45), chrome gooseneck faucet with wrist handles, pop-up drain (at $210) and insulated trap (at $46)	P1@4.29	Ea	561.00	93.50	654.50

Bar sinks

	Craft@Hrs	Unit	Material	Labor	Total
Acrylic bar sink, 15" L, 15" D, self-rimming (at $105), with chrome faucet (at $83), and strainer (at $16)	P1@2.56	Ea	204.00	55.80	259.80
Enameled cast iron bar sink, 16" L, 19" D, with strainer type drain fitting and chrome faucet (at $260), hot and cold water valved supplies (at $21) and trap assembly (at $32)	P1@2.56	Ea	313.00	55.80	368.80
Stainless steel bar sink with strainer type drain fitting, 15" L, 15" D (at $68), chrome bar faucet (at $83) hot and cold valved supplies (at $21), and trap assembly (at $16)	P1@2.56	Ea	188.00	55.80	243.80

Kitchen sinks

	Craft@Hrs	Unit	Material	Labor	Total
Single bowl, enameled cast iron, self-rimming, 24" L, 21" D (at $157), chrome faucet with sprayer (at $105), strainer drain fitting (at $16), hot and cold valved supplies (at $21) and trap (at $30)	P1@2.65	Ea	329.00	57.80	386.80

Plumbing

	Craft@Hrs	Unit	Material	Labor	Total

Single bowl, stainless steel, self-rimming sink, 24" L, 21" D (at $73), polished chrome faucet with sprayer (at $105), strainer type drain fitting (at $16), hot and cold valved supplies (at $21) and trap assembly (at $30) P1@2.65 Ea 245.00 57.80 302.80

Double bowl, enameled cast iron, self-rimming sink, 33" L, 22" D, (at $250), polished chrome faucet, with sprayer (at $105), two strainer type drain fittings (at $32), hot and cold valved supplies (at $21), and trap assembly (at $30) P1@3.06 Ea 438.00 66.70 504.70

Double bowl, stainless steel, self-rimming sink, 32" L, 22" D, (at $126), polished chrome faucet with sprayer (at $105), two strainer type drain fittings (at $32), hot and cold valved supplies (at $21) and trap assembly (at $30) P1@3.06 Ea 314.00 66.70 380.70

Triple bowl, enameled cast iron, self-rimming sink, 43" L, 22" D (at $315), polished chrome faucet with sprayer (at $105), three strainer type drain fittings (at $48), hot and cold valved supplies (at $21) and trap assembly (at $30) P1@4.03 Ea 519.00 87.90 606.90

Laundry sinks
Enameled cast iron laundry sink, 24" L, 20" D (at $210), service sink faucet (at $68), valved supplies (at $21), strainer drain fittings (at $16) and trap (at $30) P1@2.76 Ea 345.00 60.20 405.20

Service sinks
Enameled cast iron service sink, wall hung, 22" L, 18" D, with vinyl rim guard (at $260), service sink faucet (at $53), hot and cold valved supplies (at $21), strainer type drain fitting (at $16) and brass trap assembly (at $30) P1@3.67 Ea 380.00 80.00 460.00

Drinking fountains and electric water coolers
Wall-hung vitreous china fountain, 14" W, 13" H, complete with fittings and 1-1/4" trap P1@5.86 Ea 315.00 128.00 443.00

Semi-recessed vitreous china, 15" W, 27" H, with fittings, wall cleanout, and 1-1/2" trap P1@5.62 Ea 420.00 123.00 543.00

Wall-hung electric water cooler, 13 gallon capacity, complete with trap and fittings (requires electrical outlet for plug-in connection, not included) P1@5.86 Ea 445.00 128.00 573.00

Wall-hung wheelchair access electric water cooler, 7.5 gallon capacity, complete with trap and fittings (requires electrical outlet for plug-in connection, not included) P1@5.86 Ea 630.00 128.00 758.00

Fully recessed electric water cooler, 11.5 gallon capacity, complete with trap and fittings (requires electrical outlet for plug-in connection, not included) P1@5.62 Ea 775.00 123.00 898.00

Garbage disposers Add for electrical connection below.
Small, 1/3 HP kitchen disposer, (at $43), including fittings (at $30), no waste pipe included P1@2.39 Ea 73.00 52.10 125.10

Standard, 1/2 HP kitchen disposer (at $58), including fittings (at $30), no waste pipe included P1@2.39 Ea 88.00 52.10 140.10

Standard, 3/4 HP kitchen disposer (at $157), including fittings ($30), no waste pipe included P1@2.39 Ea 187.00 52.10 239.10

Plumbing

	Craft@Hrs	Unit	Material	Labor	Total
Large capacity, 1 HP kitchen disposer, stainless steel, sound-deadening insulation (at $200), including fittings (at $30), no waste pipe included	P1@2.45	Ea	230.00	53.40	283.40
Add for electrical connection and switch	BE@1.00	Ea	20.00	25.28	45.28

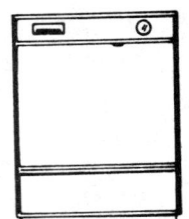

Dishwashers Under counter installation. Add for electrical connection below.

	Craft@Hrs	Unit	Material	Labor	Total
Deluxe	P1@4.00	Ea	600.00	87.20	687.20
Standard	P1@4.00	Ea	470.00	87.20	557.20
Economy	P1@4.00	Ea	260.00	87.20	347.20
Add for electrical connection	BE@1.00	Ea	25.00	25.28	50.28

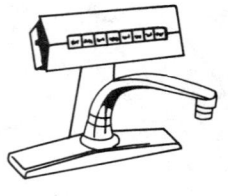

Ultraflo water systems. Single line, push button, centralized water distribution system, includes centralized mixing and metering valves, control wire, spouts, and push button fixture controls

	Craft@Hrs	Unit	Material	Labor	Total
Kitchen and one full bath system	P1@7.63	Ea	585.00	166.00	751.00
Kitchen and two full bath system	P1@11.5	Ea	935.00	251.00	1,186.00

Electric water heaters Includes labor and material to connect to existing water supply piping. 3,800 watts, 10 year guarantee, with temperature and pressure relief valve (at $10). All meet ASHRAE 9A-1980 efficiency standards.

	Craft@Hrs	Unit	Material	Labor	Total
30 gallon (at $200)	P1@5.20	LS	210.00	114.00	324.00
40 gallon (at $240)	P1@5.20	LS	250.00	114.00	364.00
52 gallon (at $250)	P1@5.20	LS	260.00	114.00	374.00
82 gallon (at $320)	P1@5.20	LS	330.00	114.00	444.00
Add for 5,500 watt heaters	--	LS	11.00	--	--
Add for foam insulation jacket	P1@.192	LS	16.00	4.19	20.19

Gas water heaters Includes labor and material to connect to existing water supply piping and gas line. 8 year guarantee, with temperature and pressure relief valve (at $10). Labor column includes cost of installing heater and making the gas connection and installing the flue (at $90). No gas piping included.

	Craft@Hrs	Unit	Material	Labor	Total
30 gallon (at $250)	P1@5.95	LS	350.00	130.00	480.00
40 gallon (at $260)	P1@5.95	LS	360.00	130.00	490.00
50 gallon (at $310)	P1@5.95	LS	410.00	130.00	540.00
65 gallon (at $350)	P1@5.95	LS	450.00	130.00	580.00
Add for foam insulation jacket	P1@.192	Ea	16.00	4.19	20.19

Solar water storage tanks Compatible with most closed-loop solar water heating systems, heat transfer by insulated all-copper heat exchanger, double wall construction and back-up electric heating element. Rheem

	Craft@Hrs	Unit	Material	Labor	Total
80 gallon, 59-1/2" high x 24" deep	P1@8.20	Ea	580.00	179.00	759.00
100 gallon, 53" high x 28" deep	P1@8.44	Ea	660.00	184.00	844.00
120 gallon, 62" high x 28" deep	P1@8.68	Ea	740.00	189.00	929.00

Solar water heaters

	Craft@Hrs	Unit	Material	Labor	Total
Passive thermosiphon domestic hot water system with integral 80 gallon water tank, retrofitted to an existing single story structure, with 5 year limited warranty	--	Ea	--	--	3,080.00
Active solar hot water system for 80 gallon water supply tank, fitted with automatic drain down, retrofitted to existing single story, single family unit 5 year warranty	--	Ea	--	--	3,400.00
Add for strengthening existing roof structure under solar collectors, per SF of supporting roof	--	SF	--	--	1.00

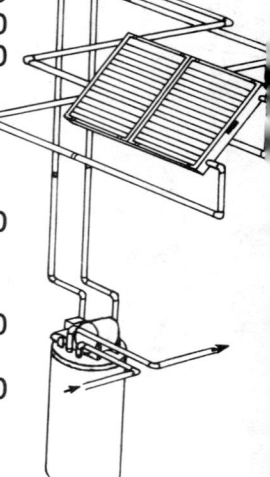

Plumbing

	Craft@Hrs	Unit	Material	Labor	Total
Active solar hot water system for new one-story home, including design and engineering	--	Ea	--	--	2,960.00
Add for custom-designed systems in homes with over 3,000 SF	--	%	--	--	20.0
Custom wood holding structure for solar collectors, mounted on concrete piers, 450 SF area	--	SF	--	--	3.35
Add for structure design	--	Ea	--	--	184.00
For designs requiring special cosmetic treatment such as concealment of pipe, wire or collectors add to total direct cost of system	--	%	--	--	15.0
Solar pool heater, active heater for pools with freeze tolerant rubber collector, roof mounted, SF of collector area	--	SF	--	--	4.75

Water softeners

9 gallons per minute, automatic	--	Ea	--	--	920.00
12 gallons per minute, automatic	--	Ea	--	--	980.00

Plumbing System Components. See following pages for detail cost breakdown.

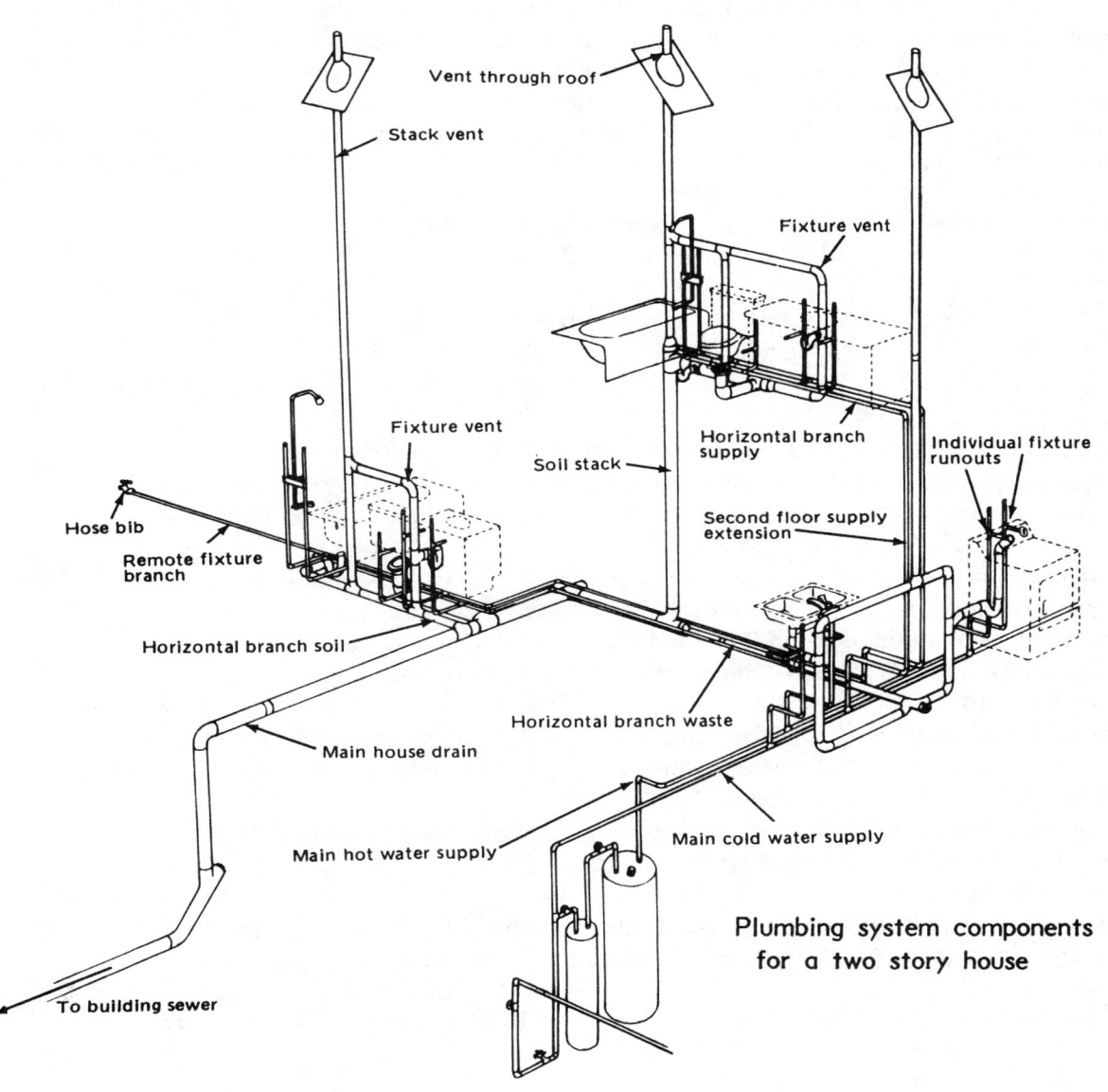

Plumbing system components for a two story house

Plumbing

	Craft@Hrs	Unit	Material	Labor	Total

Plumbing Rough-In Example. Rough plumbing is the drain, waste and vent (DWV) piping and water supply piping that isn't visible when construction is complete. Rough plumbing costs vary with the number of fixtures and length of the pipe runs. The figures in this section show typical rough plumbing costs for a small home as shown in the illustration on the preceding page. No plumbing fixtures or equipment are included in these costs.

Single story buildings
Supply plumbing	P1@41.4	LS	270.00	903.00	1,173.00
Drain, waste, vent plumbing	P1@25.2	LS	360.00	549.00	909.00
Single story, total rough plumbing costs	P1@66.6	LS	630.00	1,452.00	2,082.00

Two story buildings
Supply plumbing	P1@62.9	LS	385.00	1,370.00	1,755.00
Drain, waste, vent plumbing	P1@44.8	LS	603.00	977.00	1,580.00
Two story, total rough plumbing costs	P1@108.	LS	988.00	2,347.00	3,335.00

Plumbing Rough-In. Labor and material costs for installation of typical supply, waste, and vent piping in new construction only. Use these costs to estimate rough plumbing costs in residential buildings up to 4 units. Add supply, waste and vent piping to find the cost of the entire rough plumbing system. Figures in parentheses are material costs.

First floor main supply pipe Piping from the most remote downstream gate valve in a utility area. These figures assume that the main water supply line entrance is at a perimeter utility area of the building. All labor and material needed to install copper supply pipe with soldered brass fittings and pipe straps are included.

Hot supply, 3/4" type L copper tubing to centralized point in building, includes 22' of tubing (at $23.00), 4 typical joint fittings (at $3.10) and solder, flux, and pipe straps (at $1.60)	P1@6.01	LS	27.70	131.00	158.70
Cold supply, 3/4" type L copper tubing to centralized point in building, includes 22' of tubing (at $23.00), 4 typical joint fittings (at $3.10) and solder, flux, and pipe straps (at $1.60)	P1@6.01	LS	27.70	131.00	158.70
Add for 3/4" system drain gate valve	P1@.345	Ea	11.80	7.52	19.32

Second floor supply main supply (Vertical piping between floors)

Hot supply extension, 3/4" type L copper tubing from main supply to a centralized point on the second floor, includes an additional 12' run of tubing (at $12.55), 3 joint fittings (at $2.30), solder, flux and pipe straps (at $1.25)	P1@3.29	LS	16.10	71.70	87.80
Cold supply extension, 3/4" type L copper tubing from main supply to a centralized point on the second floor, includes an additional 12' run of tubing (at $12.55), 3 typical joint fittings (at $2.30), solder, flux and pipe straps (at $1.25)	P1@3.29	LS	16.10	71.70	87.80

Horizontal branch supply (First and second floor) Piping from the central upstream termination of the main supply pipe. Includes labor and material to install copper supply pipe with soldered brass fittings and pipe straps.

Hot supply Includes 3/4" type L copper tubing to bathroom, kitchen, or utility area, 15' of tubing (at $15.70), 4 typical fittings (at $3.10), solder, flux and pipe straps (at $1.60)	P1@4.11	LS	20.40	89.60	110.00
Cold supply. Includes 3/4" type L copper tubing to bathroom, kitchen, or utility area, 15' of tubing (at $15.70), 4 typical fittings (at $3.10), solder, flux and pipe straps (at $1.60)	P1@4.11	LS	20.40	89.60	110.00
Add for in-line shut-off	P1@.345	Ea	8.00	7.52	15.52

Plumbing

	Craft@Hrs	Unit	Material	Labor	Total

Fixture supply runouts (First and second floor) From the nearest point on the main or branch supply pipe to a wall or floor line at a fixture location. Includes labor and material to install copper supply pipe with soldered brass fittings and pipe straps. One needed for each fixture connection.

Hot supply, 1/2" type L copper tubing to each fixture, includes 5' of tubing (at $3.50), adapter (at $.85), 2 typical fittings (at $1.60), air chamber (at $4.00), solder, flux and pipe straps (at $1.25)	P1@1.36	LS	11.20	29.60	40.80
Cold supply, 1/2" type L copper tubing to individual fixture, includes 5' of tubing (at $3.50), adapter (at $.85), 2 typical fittings (at $1.60) air chamber (at $4.00) solder, flux and pipe straps (at $1.25)	P1@1.36	LS	11.20	29.60	40.80
Add for 1/2" in-line shutoff valve for dishwashers, clothes washers, ice makers, etc.	P1@.345	Ea	8.00	7.52	15.52

Remote fixture branch supply For hose bibs, etc. on building exterior.

Piping from the nearest point on the main or branch supply pipe to the fixture location. Includes 3/4" type L copper tubing (at $1.05/LF), soldered fittings every 10' (at $.11/LF), solder, flux, and pipe straps (at $.06/LF)	P1@.276	LF	1.22	6.02	7.24
Add for 3/4" frost-free hose bib with threaded outlet (at $12.00) and vacuum breaker/backflow preventer (at $13.00)	P1@.915	Ea	25.00	19.90	44.90

Soil piping Includes labor and materials to install solvent welded plastic soil piping including typical fittings (wyes, bends, couplings, and cleanouts).

Main house drain, 4" schedule 40 PVC DWV piping, from 5' outside the building line to a central location in the building, includes 30' of pipe (at $70) and 7 typical fittings (at $60)	P1@10.1	LS	130.00	220.00	350.00
Soil stack, 4" schedule 40 PVC DWV piping, from nearest point on main house drain to fixture location on second floor, includes 10' of pipe (at $23.40) and 2 fittings (at $20). Generally allow one soil stack per fixture area.	P1@3.37	LS	43.40	73.50	116.90
Horizontal branch soil pipe, 4" schedule 40 PVC DWV piping from main soil to remote fixture areas, includes 15' pipe (at $35) and 4 typical fittings (at $40), per branch	P1@4.34	LS	75.00	94.60	169.60

Individual fixture soil runouts, schedule 40 PVC DWV piping from the main or branch soil pipe to a fixture waste connection at the floor or wall line.

1-1/2" pipe connection for all fixtures except water closets. Includes 4" pipe (at $2.40) and 3 fittings (at $5.50)	P1@.825	LS	7.90	18.00	25.90
4" pipe connection to water closets, includes 6' of pipe (at $14) 2 typical fittings (at $20) and closet flange (at $7.00)	P1@.772	LS	41.00	16.80	57.80

Vent piping Includes labor and materials to install solvent-welded plastic vent piping with typical fittings (wyes, bends, couplings, and cleanouts).

Stack vents through roof, PVC DWV piping (generally allow one stack vent per fixture area, such as kitchen, bath, and utility rooms)

Single level vent, 2", includes 16' of pipe (at $12.55), 2 fittings (at $5.00) and galvanized steel roof penetration flashing (at $4.00), per vent stack	P1@1.70	LS	21.55	37.10	58.65

Plumbing

	Craft@Hrs	Unit	Material	Labor	Total
Vent piping, continued					
Two level vent, 2", includes 25' of pipe (at $19.60), 3 fittings (at $7.50) and galvanized steel roof penetration flashing (at $4.00), per vent stack	P1@4.25	LS	31.10	92.70	123.80
Individual fixture connections to stack vent, PVC DWV piping					
1-1/2" pipe connection for all fixtures except water closets, includes 4' of pipe (at $2.35) and 2 fittings (at $4.00), per fixture connection	P1@.548	LS	6.35	11.90	18.25
2" pipe connection for water closets, includes 6' of pipe (at $4.70), and 3 fittings (at $7.50)	P1@1.16	LS	12.20	25.30	37.50

Plywood. See also Lumber. Note: Softwood plywood prices change rapidly. The following prices are representative for fir plywood as of fall, 1990. Regional prices will vary by 5% to 30% or more. When lumber is purchased in carload lots, deduct 15% from these prices. Retail prices may be 15% higher than the prices listed here.

	Unit	AA	AB	AC	BC
Exterior grade, sanded					
1/4"	MSF	620.00	595.00	435.00	360.00
3/8"	MSF	735.00	715.00	550.00	420.00
1/2"	MSF	815.00	795.00	635.00	525.00
5/8"	MSF	930.00	910.00	745.00	640.00
3/4"	MSF	1,065.00	1,040.00	895.00	785.00

	Unit	AA	AB	AD	BD
Interior grade, sanded					
1/4"	MSF	605.00	580.00	420.00	345.00
3/8"	MSF	720.00	700.00	535.00	400.00
1/2"	MSF	805.00	785.00	620.00	510.00
5/8"	MSF	915.00	895.00	730.00	620.00
3/4"	MSF	1,065.00	1,040.00	880.00	765.00

	Unit	CD	CC	Mill Grade
Exterior grade, plain				
5/16"	MSF	235.00	270.00	215.00
3/8"	MSF	260.00	310.00	225.00
1/2"	MSF	325.00	--	255.00
1/2" 4 ply	MSF	345.00	420.00	220.00
1/2" 5 ply	MSF	395.00	--	--
5/8"	MSF	375.00	--	--
5/8" 4 ply	MSF	450.00	510.00	330.00
5/8" 5 ply	MSF	480.00	--	--
3/4"	MSF	500.00	590.00	415.00
Interior grade, plain				
5/16"	MSF	275.00	--	--
3/8"	MSF	300.00	--	--
1/2"	MSF	345.00	--	--
5/8"	MSF	460.00	--	--
3/4"	MSF	545.00	--	--

Identification Index: A set of two numbers separated by a slash that appears in the grade trademarks on standard sheathing, Structural I, Structural II and C-C grades. Number on left indicates recommended maximum spacing in inches for supports when panel is used for roof decking. Number on right shows maximum recommended spacing in inches for supports when the panel is used for subflooring.

A-A EXT-DFPA
Designed for use in exposed applications where the appearance of both sides is important. Fences, wind screens, exterior cabinets and built-ins, signs, boats, commercial refrigerators, shipping containers, tanks, ducts.

A-A · G-2 · EXT-DFPA · PS 1-66
Typical edge-mark

A-B EXT-DFPA
For uses similar to A-A EXT but where the appearance of one side is a little less important than the face.

A-B · G-3 · EXT-DFPA · PS 1-66
Typical edge-mark

	Unit	3/8"	1/2"	5/8"	3/4"
B-B plyform, Class 1, sanded, oiled and edge sealed	MSF	--	--	620.00	710.00
Add for 6', 7', 9' or 10' length	MSF	--	--	60.00	70.00
Add for T&G long sides	MSF	--	--	20.00	20.00
Particleboard, 40 lb interior underlayment	MSF	220.00	265.00	275.00	305.00

	Craft@Hrs	Unit	Material	Labor	Total
2-4-1 subfloor, 1-1/8", T&G 4 edges	--	MSF	946.00	--	--

Plywood

A-D INT-DFPA
For interior uses where the appearance of only one side is important. For built-ins, paneling, shelving, partitions, products made for sale, flow racks and other industrial uses.

Typical grade-trademark

A-C EXT-DFPA
Use in exposed applications where the appearance of only one side is important. Siding, soffits, fences, windscreens, structural uses, boxcar and truck lining, containers, tanks, agricultural equipment, commercial refrigerators, controlled atmosphere installations.

Typical grade-trademark

2·4·1 INT-DFPA
Combination subfloor-underlayment panel for use with supports on 4' centers. Good base for direct application of resilient flooring. Available square-edged or tongue and grooved 2 or 4 sides as specified. Use 2·4·1 with exterior glue in areas subject to excessive moisture. Unsanded or touch sanded as specified.

Typical grade-trademark

TEXTURE 1-11
Exterior unsanded type panel. Parallel grooves 1/4" deep, 3/8" wide 2" or 4" o.c. Long edges shiplapped to provide for continuous visual pattern. For siding, accent paneling, fences, parts and display cabinets. Available in 8' and 10' lengths and sanded or with MD Overlay.

T 1-11 · G-4 · EXT-DFPA · PS 1-66
Typical edge-mark

Yellow pine plywood sheathing (CD)	Unit	3/8"	1/2"	5/8"	3/4"
	MSF	260.00	325.00	375.00	500.00

Plywood Framing Clips, extruded aluminum, purchased in cartons of 500, material only

	Craft@Hrs	Unit	Material	Labor	Total
3/8"	--	Ea	.06	--	--
1/2"	--	Ea	.06	--	--
5/8"	--	Ea	.07	--	--
3/4"	--	Ea	.07	--	--

Cost of plywood clips per 100 SF of plywood surface area, including 5% waste

Bay spacing	Clips per bay	Clips per 100 SF	Unit	3/8"	1/2"	5/8"	3/4"
12" OC	1	26	CSF	1.64	--	--	--
16" OC	1	20	CSF	1.26	1.26	1.47	--
16" OC	2	40	CSF	2.52	--	--	--
20" OC	1	16	CSF	1.01	1.01	1.18	1.18
20" OC	2	32	CSF	2.02	2.02	2.35	--
24" OC	1	13	CSF	--	--	.96	.96
24" OC	2	26	CSF	1.64	1.64	1.91	1.91
32" OC	1	10	CSF	--	.63	.74	.74
32' OC	2	20	CSF	1.26	1.26	1.47	1.47
48" OC	1	7	CSF	--	--	.51	.51
48" OC	2	14	CSF	--	.88	1.03	1.03

Polyethylene Film, Clear or Black. Material costs include 5% for waste and 10% for laps. Labor shown is for installation on grade.

	Craft@Hrs	Unit	Material	Labor	Total
4 mil (.004" thick)					
50 LF rolls, 3', 4', 6', 8', 10', 12', 16' or 20' wide	BL@2.23	MSF	26.00	41.90	67.90
100 LF rolls, 3', 4', or 6' wide	BL@2.23	MSF	21.50	41.90	63.40
100 LF rolls, 8', 10', 12', 14', 16', or 20' wide	BL@2.23	MSF	22.00	41.90	63.90
6 mil (.006" thick)					
50 LF rolls, 6', 10', 12', 14', 16', 20', 24', 28', 32', or 40' wide	BL@2.23	MSF	43.00	41.90	84.90
100 LF rolls, 6' wide	BL@2.23	MSF	60.00	41.90	101.90
100 LF rolls, 10', 12', 14', 16', 20', 24', 28' or 32' wide	BL@2.23	MSF	33.00	41.90	74.90
100 LF rolls, 40' wide	BL@2.23	MSF	42.00	41.90	83.90
Add for installation on walls, ceilings and roofs, tack stapled, typical	--	%	--	60.0	--

Railings, Steel, Prefabricated. See also Ornamental Iron. Porch or step rail, site fitted, 2'10" high, 6" OC 1/2" solid twisted pickets, 4' or 6' sections

	Craft@Hrs	Unit	Material	Labor	Total
1" x 1/2" rails	RI@.171	LF	2.40	4.25	6.65
1" x 1" rails	RI@.180	LF	3.25	4.47	7.72
Add for each newel post 36" high with mounting hardware	--	Ea	9.00	--	--
Lamb's tongue for end post	--	Ea	3.50	--	--

Range Hoods

	Craft@Hrs	Unit	Material	Labor	Total

Range Hoods. Material costs are for ducted or ductless range hood only, no electrical wiring included. Add for ducting or charcoal filters below. Labor costs are for carpentry installation of range hood only. All models are steel construction with baked enamel finish in various colors.

Minimum quality, 160 CFM, 2 speed fan, 75 watt worklight, 9 RHP factor, .64 SF grease filter, 5.5 sone, 110 V, UL listed, front-mounted controls, top or back ducting

Size	Craft@Hrs	Unit	Material	Labor	Total
24" x 17-1/2" x 6"	SW@1.42	Ea	36.00	35.80	71.80
30" x 17-1/2" x 6"	SW@1.42	Ea	40.00	35.80	75.80
36" x 17-1/2" x 6"	SW@1.42	Ea	42.00	35.80	77.80
42" x 17-1/2" x 6"	SW@1.42	Ea	48.00	35.80	83.80

Good quality, 200 CFM, variable speed fan, 75 watt worklight, 23 RHP factor 1.04 SF grease filter, 5.5 sone, 110 V, UL listed, front-mounted controls

Size	Craft@Hrs	Unit	Material	Labor	Total
24" x 17-1/2" x 6"	SW@1.42	Ea	136.00	35.80	171.80
30" x 17-1/2" x 6"	SW@1.42	Ea	142.00	35.80	177.80
36" x 17-1/2" x 6"	SW@1.42	Ea	192.00	35.80	227.80
42" x 17-1/2" x 6"	SW@1.42	Ea	204.00	35.80	239.80
30" x 17-1/2" x 6", stainless steel	SW@1.42	Ea	163.00	35.80	198.80
36" x 17-1/2" x 6", stainless steel	SW@1.42	Ea	189.00	35.80	224.80

Better quality, 360 CFM, variable speed fan, two 75 watt worklights, 59 RHP factor, nightlight, 1.56 SF grease filter, 5.5 sone, 110 V, UL listed, front-mounted controls

Size	Craft@Hrs	Unit	Material	Labor	Total
30" x 20" x 9"	SW@1.42	Ea	210.00	35.80	245.80
36" x 20" x 9"	SW@1.42	Ea	236.00	35.80	271.80
30" x 20" x 9", stainless steel	SW@1.42	Ea	231.00	35.80	266.80
36" x 20" x 9", stainless steel	SW@1.42	Ea	257.00	35.80	292.80
42" x 20" x 9", stainless steel	SW@1.42	Ea	278.00	35.80	313.80

Deluxe quality, 460 CFM, variable speed fan, auto heat sensor control, high heat alarm, two 75 watt worklights, nightlight, 1.56 SF grease filter, 6.0 sone, 110 V, UL listed, front-mounted controls, stainless steel finish

Size	Craft@Hrs	Unit	Material	Labor	Total
30" x 20" x 9"	SW@1.50	Ea	252.00	37.80	289.80
36" x 20" x 9", stainless steel	SW@1.50	Ea	282.00	37.80	319.80

Add for ducting, including cap with back draft preventer at exterior wall or roof

Description	Craft@Hrs	Unit	Material	Labor	Total
Back ducted straight through wall, complete	SW@.526	Ea	14.50	13.20	27.70
Top or back ducted through wall with single bend, complete, to 4'	SW@.785	Ea	48.00	19.80	67.80
Top ducted straight through roof, to 4'	SW@.657	Ea	40.00	16.50	56.50
Top or back ducted through roof with single bend, to 4'	SW@.801	Ea	53.00	20.20	73.20
Add for each additional LF of straight duct over 4'	SW@.136	Ea	6.00	3.42	9.42
Add for each additional bend	SW@.136	Ea	14.50	3.42	17.92
Add for charcoal filters for ductless installation	--	Ea	9.50	--	--

Ranges, Built-In. Material costs are for good to better quality built-in appliances. Labor costs are for carpentry installation only. Add for gas and electric runs below. Note that all cooktops with grills require exhaust venting. See Range Hoods above.

Counter-mounted cooktops
 Gas, pilot free ignition, chrome plated, 3" high

Model	Craft@Hrs	Unit	Material	Labor	Total
Standard model, 30" x 21"	P1@1.92	Ea	262.00	41.90	303.90
Better unit, 36" x 20", griddle	P1@1.92	Ea	336.00	41.90	377.90

Ranges

	Craft@Hrs	Unit	Material	Labor	Total
Electric					
Smooth ceramic cooktop with one 9" element, one 8" element and two 6" elements					
30" x 21" x 3-3/8"	BE@.903	Ea	346.00	22.80	368.80
Induction type with smooth ceramic cooktop, three 1,300 watt elements and one 1,700 watt element,					
34-5/8" x 21-1/4" x 5-1/2"	BE@.903	Ea	810.00	22.80	832.80
Conventional coil type cooktop with two 8" and two 6" elements, 30-1/4" x 21-1/4" x 3"					
Baked enamel finish, various colors	BE@.903	Ea	200.00	22.80	222.80
Chrome plated finish	BE@.903	Ea	230.00	22.80	252.80
Conventional coil type cooktop, two 8" and two 6" elements, with built-in griddle, 35-1/2" x 20-1/2" x 3"					
Baked enamel finish, various colors	BE@.903	Ea	290.00	22.80	312.80
Chrome plated finish	BE@.903	Ea	315.00	22.80	337.80
Deluxe quality kitchen countertop range/grill combination, includes one 8" and one 6" element, grill module, chrome finish, built-in exhaust system below and 2 speed fan					
30" x 23" x 6"	BE@1.44	Ea	505.00	36.40	541.40
Wall ovens, with light and clock					
Gas, electronic ignition, continuous cleaning, with broiler below, 39-5/16" high, 23-3/4" wide, 26-3/8" deep, installation includes electric run (at $35.00)					
Porcelain enamel black glass door	P1@1.92	Ea	505.00	41.90	546.90
Electric, single oven 29" high, 27" deep, black glass door					
24" wide, continuous cleaning	BE@1.67	Ea	520.00	42.20	562.20
24" wide, self-cleaning	BE@1.67	Ea	650.00	42.20	692.20
24" wide, porcelain enameled	BE@1.67	Ea	460.00	42.20	502.20
Electric, double oven, 50" high, 27" deep, black glass doors					
24" wide, continuous cleaning	BE@1.67	Ea	700.00	42.20	742.20
24" wide, self-cleaning	BE@1.67	Ea	945.00	42.20	987.20
27" wide, self-cleaning	BE@1.67	Ea	965.00	42.20	1,007.20
27" wide, microwave upper, self-cleaning oven lower	BE@1.67	Ea	1,250.00	42.20	1,292.20
Range and oven combinations, automatic controls, with light and clock					
Drop-in models, 4 unit burners, black glass doors					
Gas, pilot-free ignition, continuous cleaning, timer					
30" wide x 27" deep x 32" high	P1@1.92	Ea	725.00	41.90	766.90
Electric, self-cleaning, chrome-plated top					
30" wide x 27" deep x 31" high	BE@2.00	Ea	715.00	50.60	765.60
Add for electric circuit for electric ovens or ranges, typical cost	BE@2.00	LS	90.00	50.60	140.60
Add for electrical circuit for clocks on gas ovens or oven & range combinations, typical cost	BE@.500	LS	18.00	12.64	30.64
Add for gas piping run for gas ovens or ranges, typical cost	P1@2.00	LS	31.40	43.60	75.00

Roof Coatings and Adhesives

Roofing asphalt, 100 lb carton (160 SF per carton)					
145 to 165 degree (low melt)	--	Ea	13.00	--	--
185 degree, certified (high melt)	--	Ea	13.00	--	--
Asphalt emulsion, Henry #107, (35 SF per gallon)					
1 gallon	--	Ea	4.60	--	--
5 gallons	--	Ea	17.00	--	--
55 gallons	--	Ea	150.00	--	--

Roof Coatings

	Craft@Hrs	Unit	Material	Labor	Total
Asphalt emulsion, Henry #106, for use with polyester (35 SF per gallon)					
5 gallons	--	Ea	18.00	--	--
55 gallons	--	Ea	140.00	--	--
Asphalt primer, Henry #105, (150 to 300 SF per gallon)					
1 gallon	--	Ea	7.50	--	--
5 gallons	--	Ea	31.00	--	--
55 gallons	--	Ea	300.00	--	--
Asphalt roof coating, Henry #101, plain, economy, (75 SF per gallon)					
1 gallon	--	Ea	5.00	--	--
5 gallons	--	Ea	21.00	--	--
55 gallons	--	Ea	190.00	--	--
Coal tar wet patch, Western Colloid #101, (12.5 SF per gallon at 1/8" thick)					
3 gallons	--	Ea	21.00	--	--
5 gallons	--	Ea	30.00	--	--
Cold application cement, Henry #203, (67 SF per gallon)					
1 gallon	--	Ea	6.50	--	--
5 gallons	--	Ea	27.00	--	--
55 gallons	--	Ea	240.00	--	--
Fiber roof coating, Henry #201, inert mineral fiber, (50 SF per gallon)					
1 gallon	--	Ea	5.00	--	--
5 gallons	--	Ea	19.00	--	--
55 gallons	--	Ea	200.00	--	--
Flashing compound, Western Colloid #106, (12-5 SF per gallon at 1/8" thick)					
3 gallons	--	Ea	15.00	--	--
5 gallons	--	Ea	22.00	--	--
Lap cement, Henry #108, (280 LF at 2" lap per gallon)					
1 gallon	--	Ea	5.00	--	--
5 gallons	--	Ea	23.00	--	--
Plastic roof cement, Henry #204, (12-1/2 SF @1/8" thick per gallon)					
1 gallon	--	Ea	5.50	--	--
5 gallons	--	Ea	18.00	--	--
55 gallons	--	Ea	190.00	--	--
Reflective non-fibered aluminum coating, Henry #120, (200 to 400 SF per gallon)					
1 gallon	--	Ea	13.00	--	--
5 gallons	--	Ea	56.00	--	--
55 gallons	--	Ea	580.00	--	--
Reflective fibered aluminum coating, Henry #220, (60 to 85 SF per gallon)					
1 gallon	--	Ea	12.50	--	--
5 gallons	--	Ea	56.00	--	--
55 gallons	--	Ea	560.00	--	--
Reflective white acrylic roof coating, Henry #280, (65 to 100 SF per gallon)					
1 gallon	--	Ea	16.50	--	--
5 gallons	--	Ea	82.00	--	--
55 gallons	--	Ea	850.00	--	--
Wet patch, Henry #208, (12-1/2 SF @1/8" thick per gallon)					
11 oz cartridge for caulking gun	--	Ea	1.40	--	--
1 gallon	--	Ea	6.50	--	--
5 gallons	--	Ea	26.00	--	--
Polyester fabric					
3 oz per SY, 36" x 375'	--	Sq	10.50	--	--

Roofing

	Craft@Hrs	Unit	Material	Labor	Total

OBTAINING ROOF AREA FROM PLAN AREA

Rise	Factor	Rise	Factor
3"	1.031	8"	1.202
3½"	1.042	8½"	1.225
4"	1.054	9"	1.250
4½"	1.068	9½"	1.275
5"	1.083	10"	1.302
5½"	1.100	10½"	1.329
6"	1.118	11"	1.357
6½"	1.137	11½"	1.385
7"	1.158	12"	1.414
7½"	1.179		

When a roof has to be figured from a plan only, and the roof pitch is known, the roof area may be fairly accurately computed from the table above. The horizontal or plan area (including overhangs) should be multiplied by the factor shown in the table opposite the rise, which is given in inches per horizontal foot. The result will be the roof area.

Roofing rule of thumb. Cost per square to install a roof on a non-tract house, not over two stories in height with 1,500 to 2,000 SF of floor area. Costs shown are typical subcontract prices. See detailed labor and material costs and built-up roofing costs in the following sections. Note: Many communities ban or sharply restrict using wood roofing products.

	Craft@Hrs	Unit	Material	Labor	Total
Asphalt shingles (depends on grade)	--	Sq	--	--	120.00
Cedar shingles, #3	--	Sq	--	--	170.00
Cedar shingles, #2	--	Sq	--	--	190.00
Cedar shingles, #1	--	Sq	--	--	225.00
Clay tiles	--	Sq	--	--	230.00
Concrete tiles	--	Sq	--	--	250.00
Metal tiles	--	Sq	--	--	230.00
Perlite shakes	--	Sq	--	--	240.00

Roofing, Built-Up, Subcontract

	Craft@Hrs	Unit	Material	Labor	Total
3 ply asphalt and gravel over wood roof deck (1 ply 30 lb felt, 2 plies 15 lb felt, 3 mop coats of asphalt, 400 lbs gravel)	--	Sq	--	--	120.00
3 ply asphalt and gravel over concrete or gypsum roof deck (3 plies 15 lb felt, 3 mop coats of asphalt and 400 lbs roofing gravel)	--	Sq	--	--	130.00
3 ply asphalt (2 plies 15 lb felt with 90 lb cap sheet, 2 mop coats)	--	Sq	--	--	115.00
4 ply asphalt and gravel over wood roof deck (4 plies 15 lb felt, 4 mop coats of asphalt, 400 lbs roofing gravel)	--	Sq	--	--	135.00
4 ply asphalt and gravel over concrete or gypsum roof deck (4 plies 15 lb felt, 5 mop coats of asphalt, 400 lbs roof gravel)	--	Sq	--	--	150.00
Cold process built-up roofing					
3 plies 25 lb fiberglass base sheet, 2 coats #203 cold application cement, one coat #107 asphalt emulsion, and #120 aluminum coating	--	Sq	--	--	120.00
1 ply 25 lb fiberglass base sheet, 2 plies #184 Rufon polyester fabric embedded in and top coated with #106 asphalt emulsion, and #120 aluminum coating, Henry	--	Sq	--	--	130.00

Roofing, Composition Shingle. Material costs include 10% waste. Labor costs include roof loading and typical cutting and fitting for roofs of average complexity.

	Craft@Hrs	Unit	Material	Labor	Total
Asphalt shingles					
Celotex Dimensional 4 (355 lb, 25 year)	R1@2.05	Sq	85.00	45.70	130.70
Certainteed Hallmark (340 lb, 30 year)	R1@2.05	Sq	75.00	45.70	120.70
Flintkote Estate (300 lb, 25 year)	R1@2.05	Sq	61.00	45.70	106.70
Fiberglass roofing shingles. Add for hip and ridge below.					
Bird Architect 80 (25 year class A)	R1@1.71	Sq	47.20	38.10	85.30
Celotex Big D (225 lb, 20 year)	R1@1.59	Sq	25.00	35.40	60.40
Celotex Big D 25 (300 lb, 25 year)	R1@1.83	Sq	42.00	40.80	82.80
Celotex Presidential (35 year)	R1@1.83	Sq	65.00	40.80	105.80
GAF Sentinel (225 lb, 20 year class A)	R1@1.59	Sq	24.00	35.40	59.40
GAF Royal Sovereign (25 year)	R1@1.59	Sq	32.00	35.40	67.40
GAF Woodline (30 year)	R1@1.83	Sq	37.00	40.80	77.80
GS Firescreen/Brigade (20 year)	R1@1.59	Sq	25.00	35.40	60.40
GS Firescreen/Brigade (25 year)	R1@1.59	Sq	31.00	35.40	66.40
GS Firehalt (25 year)	R1@1.59	Sq	37.00	35.40	72.40

Roofing

	Craft@Hrs	Unit	Material	Labor	Total
GS High Sierra (40 year)	R1@1.59	Sq	59.00	35.40	94.40
Pabco HM 30 (30 year)	R1@1.83	Sq	53.00	40.80	93.80
Pabco Horizon (25 year glass)	R1@1.59	Sq	40.00	35.40	75.40
Elk Prestique I (320 lb, 30 year class A)	R1@1.83	Sq	50.00	40.80	90.80
Elk Prestique II (240 lb, 25 year class A)	R1@1.59	Sq	40.00	35.40	75.40
Elk Prestique Plus (40 year, class A)	R1@1.83	Sq	55.00	40.80	95.80
Masonite Woodruf					
Rustic	R1@1.83	Sq	83.60	40.80	124.40
Traditional	R1@1.83	Sq	83.60	40.80	124.40
Rustic, flame test class C	R1@1.83	Sq	127.00	40.80	167.80
Rustic, flame test class B	R1@1.83	Sq	158.00	40.80	198.80
Add for hip and ridge units	R1@.020	LF	.56	.45	1.01

Roofing, Crushed Rock. Standard crush, 1/8" to 7/16", 10' lift and spread in 60 lb per 100 SF roofing asphalt at $11.00 per 100 lbs, typical costs based on rock at $85 per ton for 2 ton quantities.

	Craft@Hrs	Unit	Material	Labor	Total
Arctic white crushed tile, 220 lbs/sq	R1@.340	Sq	18.00	7.58	25.58
Canyon red crushed tile, 240 lbs/sq	R1@.340	Sq	19.00	7.58	26.58
Desert bronze, natural rock, 260 lbs/sq	R1@.340	Sq	20.00	7.58	27.58
Desert green, igneous rock, 250 lbs/sq	R1@.340	Sq	19.50	7.58	27.08
Glacier white, crushed porcelain, 200 lbs/sq	R1@.340	Sq	17.30	7.58	24.88
Laguna surf, desert rock, 260 lbs/sq	R1@.340	Sq	20.00	7.58	27.58
Calico, crushed firebrick, 400 lbs/sq	R1@.340	Sq	17.30	7.58	24.88
Deduct for full truckload quantities	--	%	-10.0	--	--

Roofing Papers. See also Building Paper. Labor laying paper on 3 in 12 pitch roof.

Asphalt felt (108 SF covers 100 SF), labor to roll and mop

	Craft@Hrs	Unit	Material	Labor	Total
15 lb (432 SF roll at $12.00)	R1@.360	Sq	2.80	8.02	10.82
30 lb (216 SF roll at $12.00)	R1@.495	Sq	5.60	11.00	16.60
Roofing asphalt, 25 lbs/sq	--	Sq	2.75	--	--

Mineral surfaced with fixtures (at $12.00/sq)

	Craft@Hrs	Unit	Material	Labor	Total
90 lb, 36" x 36' ($14.00 per roll)	R1@1.03	Sq	26.00	23.00	49.00
90 lb, 36" x 32' ($13.00 per roll)	R1@1.03	Sq	25.00	23.00	48.00

Smooth (roll covers 100 SF), with fixtures (at $11.00)

	Craft@Hrs	Unit	Material	Labor	Total
50 lb medium ($15.50 per roll)	R1@.738	Sq	26.50	16.50	43.00
65 lb extra heavy ($17.50 per roll)	R1@.738	Sq	28.50	16.50	45.00

Roofing Shakes and Shingles. Red cedar. No pressure treating.

Shakes

	Craft@Hrs	Unit	Material	Labor	Total
Shake hip and ridge units, 20 per bundle, covers 16.7 LF at 10" exposure and 20 LF at 12" exposure, per bundle	R1@.500	Ea	32.00	11.10	43.10
Sawn shakes, sawn 1 side, class "C" fire retardant					
1/2" to 3/4" x 24" (4 bdle/sq at 10" exp.)	R1@2.77	Sq	112.00	61.70	173.70
3/4" to 5/4" x 24" (5 bdle/sq at 10" exp.)	R1@3.28	Sq	155.00	73.10	228.10
Taper split shakes, 1/2" x 24"					
(4 bdle/sq at 10" exp.)	R1@2.77	Sq	94.00	61.70	155.70
Shake felt, 30 lb (100 SF roll at $15.00)	R1@.160	Sq	16.00	3.57	19.57
Add for pressure treated shakes	--	%	60.0	--	--

Shingles (red cedar)

Perfects 5X, green, 16", 5" exposure, 5 shingles are 2" thick (5/2)

	Craft@Hrs	Unit	Material	Labor	Total
#1 (for houses)	R1@2.77	Sq	110.00	61.70	171.70
#2, red label (houses or garages)	R1@2.77	Sq	63.00	61.70	124.70
#3 (for garages)	R1@2.77	Sq	53.00	61.70	114.70

Roofing

	Craft@Hrs	Unit	Material	Labor	Total
#4 (undercourse grade)	R1@1.09	Sq	17.50	24.30	41.80

Perfections, dry, 18", 5-1/2" exposure, 5 shingles are 2-1/4" thick, (5/2-1/4), fire treated

	Craft@Hrs	Unit	Material	Labor	Total
#1 (houses)	R1@2.77	Sq	125.00	61.70	186.70
#2, red label (houses or garages)	R1@2.77	Sq	115.00	61.70	176.70
#3 (garages)	R1@2.77	Sq	96.00	61.70	157.70

Hip and ridge units, 40 pieces per bundle (20 LF at 6" exp., 16-2/3 LF at 5" exposure, 15 LF at 4-1/2" exposure) per bundle

	Craft@Hrs	Unit	Material	Labor	Total
#1	R1@.500	Ea	27.00	11.10	38.10
#2	R1@.500	Ea	24.00	11.10	35.10
Add for pitch over 6" in 12"	--	%	--	40.0	--

Roofing Sheets. See also Fiberglass Panels.

Aluminum roofing and siding
  4-V corrugated, 2-1/2" corrugations, 5-V crimp, plain or embossed finish, on wood frame, includes 15% loss for coverage

	Craft@Hrs	Unit	Material	Labor	Total
.017" x 26" x 6' to 24'	R1@.022	SF	1.47	.49	1.96
.019" x 26" x 6' to 24'	R1@.022	SF	1.54	.49	2.03
Add for painted finish	--	%	15.0	--	--
Add for installation on metal frame	--	%	--	50.0	--

Ridge cap

	Craft@Hrs	Unit	Material	Labor	Total
12" x 10', plain	R1@.024	LF	1.40	.54	1.94
10" x 28' formed	R1@.024	LF	1.33	.54	1.87

Flashing, corrugated, embossed

	Craft@Hrs	Unit	Material	Labor	Total
End wall, 10" x 52"	R1@.047	Ea	2.60	1.05	3.65
Plain side wall, 7-1/2" x 10'	R1@.014	LF	1.16	.31	1.47
Rubber filler strip, 3/4" x 7/8" x 6'	--	Ea	1.34	--	--

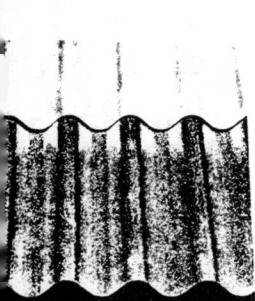

Galvanized corrugated steel sheets, on wood frame
  28 gauge x 27-1/2" wide (includes 20% coverage loss)

	Craft@Hrs	Unit	Material	Labor	Total
6' to 12' standard lengths	R1@.027	SF	.72	.60	1.32

26 gauge x 26" wide (include 15% coverage loss)

	Craft@Hrs	Unit	Material	Labor	Total
6' to 12' standard lengths	R1@.027	SF	.76	.60	1.36
Ridge roll, plain, 28 gauge, 10" wide	R1@.030	LF	.77	.67	1.44
Ridge cap, formed, plain, 28 gauge, 2-1/2"	R1@.061	LF	1.20	1.36	2.56

Sidewall flashing, 28 gauge, plain

	Craft@Hrs	Unit	Material	Labor	Total
3" x 4" x 10'	R1@.035	LF	.66	.78	1.44

Endwall flashing, 28 gauge

	Craft@Hrs	Unit	Material	Labor	Total
2-1/2" corrugated, 10" x 28"	R1@.035	LF	1.25	.78	2.03
Twin rib, 10" x 52"	R1@.035	LF	1.49	.78	2.27

Wood filler strip, corrugated

	Craft@Hrs	Unit	Material	Labor	Total
7/8" x 7/8" x 6'	--	Ea	2.50	--	--
Galvanized flat sheets, 26 gauge	--	SF	.90	--	--

Roofing Slate. Local delivery included. Costs will be higher where slate is not mined. Add freight cost at 800 to 1,000 pounds per 100 SF. Includes 20" long random width slate, 3/16" thick, 7-1/2" exposure. Meets Standard SS-S-451.

	Craft@Hrs	Unit	Material	Labor	Total
Semi-weathering green and gray	R1@11.3	Sq	300.00	252.00	552.00
Vermont black and gray black	R1@11.3	Sq	350.00	252.00	602.00
China black or gray	R1@11.3	Sq	300.00	252.00	552.00
Unfading and variegated purple	R1@11.3	Sq	400.00	252.00	652.00
Unfading mottled green and purple	R1@11.3	Sq	330.00	252.00	582.00
Unfading green	R1@11.3	Sq	330.00	252.00	582.00
Red slate	R1@13.6	Sq	1,500.00	303.00	1,803.00
Add for other specified widths and lengths	--	%	20.0	--	--

Roofing

	Craft@Hrs	Unit	Material	Labor	Total
Supra-Slate (Supradur) manufactured slate, 220 pieces per square, 500 lbs per square at 7" exposure, various colors, 16" x 9-1/2", 1/4" thick	R1@10.5	Sq	205.00	234.00	439.00

Roofing Tile. Clay roof tile. Material costs include felts and flashing. No freight or waste included. Costs will be higher where clay tile is not manufactured locally. U.S. Tile Co.

Spanish tile, "S" shaped, 88 pieces per square, (900 lbs per square at 11" centers and 15" exposure)

	Craft@Hrs	Unit	Material	Labor	Total
Red clay tile ($57 per square)	R1@3.46	Sq	74.00	77.10	151.10
Add for coloring	--	Sq	10.00	--	--
Add for hip/ridge units					
Red	R1@.047	LF	.65	1.05	1.70
Colors	R1@.047	LF	.88	1.05	1.93
Add for rake units					
Red	R1@.047	LF	1.17	1.05	2.22
Colors	R1@.047	LF	1.28	1.05	2.33

Red clay mission tile, 2-piece, 83 pans and 83 tops per square, 7-1/2" x 18" x 8-1/2" tiles at 11" centers and 15" exposure

	Craft@Hrs	Unit	Material	Labor	Total
Red clay tile ($107 per square)	R1@5.84	Sq	138.00	130.00	268.00
Add for coloring (costs vary widely)	--	Sq	36.00	--	--
Add for hip/ridge					
Red	R1@.047	LF	.65	1.05	1.70
Colors	R1@.047	LF	.88	1.05	1.93
Add for rake units					
Red	R1@.047	LF	1.17	1.05	2.22
Colors	R1@.047	LF	1.28	1.05	2.33

Flat shingles, 107 pieces per square (1,030 lbs/sq), at 9" centers and 15" exposure. Price depends on color

	Craft@Hrs	Unit	Material	Labor	Total
18" x 9" tiles (costs vary widely)	--	Sq	--	--	125.00
Add for ridge and rake units	R1@.051	LF	2.50	1.14	3.64
Labor to install flat shingles	R1@6.31	Sq	--	141.00	--

Concrete roof tile, material costs include felt, nails, flashing and 10% waste. No delivery or roof loading included.

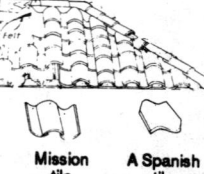

Mission tile   A Spanish tile

Ro-Tile, 150 pieces per square

	Craft@Hrs	Unit	Material	Labor	Total
Riviera French (750 lbs per square at $69.00/Sq)	R1@3.53	Sq	100.00	78.70	178.70
Tudor, Neo-Classic, Rustic Shake (900 lbs per square at $69.00/Sq)	R1@3.59	Sq	100.00	80.00	180.00
Concrete tile	R1@6.40	Sq	85.00	143.00	228.00
Hip starter pieces	--	Ea	4.40	--	--
Trim units	--	LF	1.80	--	--
Shed Ridge	--	Ea	6.60	--	--

Concrete roof tile, material includes felt, nails, and flashing, approximately 90 pieces per square, Monier Tile Co.

	Craft@Hrs	Unit	Material	Labor	Total
Homestead, slurry coat (at $56.00/Sq)	R1@3.69	Sq	92.00	82.30	174.30
Homestead, thru color (at $62.00/Sq)	R1@3.69	Sq	98.00	82.30	180.30
Mission "S", slurry coated (at $53.00/Sq)	R1@3.59	Sq	89.00	80.00	169.00
Mission "S", thru color (at $61.00/Sq)	R1@3.59	Sq	94.00	80.00	174.00
Slate, slurry coated (at $54.00/Sq)	R1@3.69	Sq	89.00	82.30	171.30
Shake, thru color (at $56.00/Sq)	R1@3.69	Sq	94.00	82.30	176.30
Villa, Roma, Classic "100", slurry coat (at $50.00/Sq)	R1@3.69	Sq	85.00	82.30	167.30
Villa, Roma, thru color (at $58.00/Sq)	R1@3.69	Sq	92.00	82.30	174.30

Roofing

	Craft@Hrs	Unit	Material	Labor	Total
Add for custom coloring	--	Sq	5.00	--	--
Trim tile					
Mansard "V" ridge or rake, slurry coated	--	Ea	1.10	--	--
Mansard, ridge or rake, thru color	--	Ea	1.20	--	--
Hipstarters, slurry coated	--	Ea	5.20	--	--
Hipstarters, thru color	--	Ea	6.20	--	--
Accessories					
Eave closure or birdstop	R1@.026	LF	.50	.58	1.08
Hurricane or wind clips	R1@.026	Ea	.50	.58	1.08
Underlayment or felt	R1@.051	Sq	18.00	1.14	19.14
Metal flashing and nails	R1@.306	Sq	11.00	6.82	17.82
Add for roof loading	R1@.822	Sq	--	18.30	--

Roofing, Tin Shingles. Material costs shown include 5% for waste and laps and an allowance of 2% for fasteners. The 5" x 7" require 411 per Sq and the 8" x 12" require 150 per Sq

	Craft@Hrs	Unit	Material	Labor	Total
Painted					
5" x 7"	R1@3.40	Sq	59.00	75.80	134.80
8" x 12"	R1@1.55	Sq	72.00	34.60	106.60
Galvanized					
5" x 7"	R1@3.40	Sq	119.00	75.80	194.80
8" x 12"	R1@1.55	Sq	130.00	34.60	164.60

Safes, Residential
"Safe-N-Sekure" wall storage safe, 14-1/2' wide, 9" high, 4" deep, with heavy gauge steel body, and dead bolt combination lock, self-trimming

	Craft@Hrs	Unit	Material	Labor	Total
lock, self-trimming	BC@.788	Ea	73.00	18.60	91.60

Sandblasting and Waterblasting, Subcontract. The minimum charge for on-site (truck-mounted) sandblasting and waterblasting will usually be about $250.00. Add the cost of extra liability insurance and physical barriers to protect adjacent swimming pools, residences, process plants, etc. No scaffolding included. Based on sandblasting with 450 CFM, 125 PSI compressor. Portable sandblasting equipment will increase costs about 35%.

	Craft@Hrs	Unit	Material	Labor	Total
Sandblasting most surfaces					
Water soluble paints	--	SF	--	--	.60
Oil paints	--	SF	--	--	.68
Heavy mastic	--	SF	--	--	.90
Sandblasting masonry					
Brick	--	SF	--	--	.60
Block walls, most work	--	SF	--	--	.65
Remove surface grime	--	SF	--	--	.60
Heavy, exposing aggregate	--	SF	--	--	.96
Sandblasting concrete tilt-up panels					
Light blast	--	SF	--	--	.84
Medium, exposing aggregate	--	SF	--	--	.90
Heavy, exposing aggregate	--	SF	--	--	.96
Sandblasting steel					
New, uncoated (commercial grade)	--	SF	--	--	.85
New, uncoated (near white grade)	--	SF	--	--	1.01
Epoxy coated (near white grade)	--	SF	--	--	1.58
Sandblasting wood					
Medium blast, clean and texture	--	SF	--	--	.96

Sandblasting

	Craft@Hrs	Unit	Material	Labor	Total
Waterblast (hydroblast) with mild detergent					
To 5,000 PSI blast (4 hour minimum)	--	Hr	--	--	105.00
5,000 to 10,000 PSI blast (8 hour min.)	--	Hr	--	--	125.00
Over 10,000 PSI blast (8 hour minimum)	--	Hr	--	--	140.00
Wet sandblasting (4 hour minimum)	--	Hr	--	--	115.00

Sauna Rooms. Costs listed are for labor and materials to install sauna rooms as described. Units are shipped as a disassembled kit complete with kiln dried softwood paneling boards, sauna heater, temperature and humidity controls and monitors, pre-assembled benches, prehung door, water bucket and dipper, and interior light. No electrical work included. Add freight cost from factory.

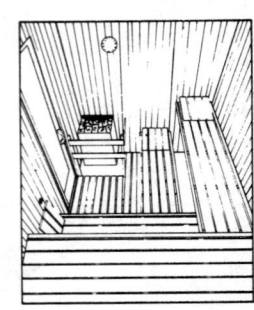

	Craft@Hrs	Unit	Material	Labor	Total
Personal sauna, 48" x 42" x 74"	B1@4.00	Ea	2,200.00	84.80	2,284.80

Pre-cut western red cedar sauna room package with heater, controls, duckboard floor and accessories, ready for nailing to framed and insulated walls with exterior finish, Amerec

	Craft@Hrs	Unit	Material	Labor	Total
4' x 4' room	B1@8.01	Ea	1,950.00	170.00	2,120.00
4' x 6' room	B1@8.01	Ea	2,200.00	170.00	2,370.00
6' x 6' room	B1@10.0	Ea	2,500.00	212.00	2,712.00
6' x 8' room	B1@12.0	Ea	3,100.00	254.00	3,354.00
8' x 8' room	B1@14.0	Ea	3,400.00	297.00	3,697.00
8' x 10' room	B1@15.0	Ea	3,800.00	318.00	4,118.00
8' x 12' room	B1@16.0	Ea	4,300.00	339.00	4,639.00
Add for electrical service connection and wiring of sauna	--	LS	--	--	293.00

Modular sauna room assembly. Includes the wired and insulated panels to construct a free-standing sauna on a waterproof floor. With heater, rocks, door, benches and accessories, Amerec.

	Craft@Hrs	Unit	Material	Labor	Total
4' x 4' room	B1@4.00	Ea	2,600.00	84.80	2,684.80
4' x 6' room	B1@4.00	Ea	3,100.00	84.80	3,184.80
6' x 6' room	B1@6.00	Ea	3,600.00	127.00	3,727.00
6' x 8' room	B1@8.01	Ea	4,300.00	170.00	4,470.00
6' x 10' room	B1@8.01	Ea	4,900.00	170.00	5,070.00
8' x 8' room	B1@10.0	Ea	4,600.00	212.00	4,812.00
8' x 10' room	B1@11.0	Ea	5,100.00	233.00	5,333.00
8' x 12' room	B1@12.0	Ea	6,000.00	254.00	6,254.00
Add for glass panel, 22-1/2" or 24"	--	Ea	70.00	--	--
Add for electrical service connection	--	LS	--	--	225.00

Pre-cut 100% vertical grain redwood sauna room package with heater, controls, duckboard floor and accessories (including non-skid removable rubberized flooring). Ready for nailing to preframed and insulated walls with exterior finish, Finlandia

	Craft@Hrs	Unit	Material	Labor	Total
3' x 5' or 4' x 4' room	B1@6.68	Ea	1,575.00	142.00	1,717.00
4' x 6' room	B1@6.68	Ea	1,790.00	142.00	1,932.00
5' x 6' room	B1@6.68	Ea	1,980.00	142.00	2,122.00
6' x 6' room	B1@8.01	Ea	2,155.00	170.00	2,325.00
6' x 8' room	B1@10.7	Ea	2,610.00	227.00	2,837.00
8' x 8' room	B1@10.7	Ea	3,055.00	227.00	3,282.00
8' x 10' room	B1@10.7	Ea	3,540.00	227.00	3,767.00
8' x 12' room	B1@10.7	Ea	4,230.00	227.00	4,457.00
10' x 10' room	B1@10.7	Ea	4,080.00	227.00	4,307.00
10' x 12' room	B1@10.7	Ea	4,595.00	227.00	4,822.00
Add for electrical service connection and wiring of sauna	--	LS	--	--	293.00

Saunas

	Craft@Hrs	Unit	Material	Labor	Total

Modular sauna room assembly. Completely pre-wired and pre-insulated panels to construct a free-standing sauna on waterproof floor. Includes heater, rocks, door, benches and accessories, Finlandia

	Craft@Hrs	Unit	Material	Labor	Total
3' x 4' room	B1@1.34	Ea	2,075.00	28.40	2,103.40
4' x 6' room	B1@1.34	Ea	2,635.00	28.40	2,663.40
5' x 6' room	B1@2.68	Ea	2,745.00	56.80	2,801.80
6' x 6' room	B1@2.68	Ea	3,110.00	56.80	3,166.80
6' x 8' room	B1@2.68	Ea	3,540.00	56.80	3,596.80
8' x 8' room	B1@4.02	Ea	4,375.00	85.20	4,460.20
8' x 10' room	B1@4.02	Ea	5,005.00	85.20	5,090.20
8' x 12' room	B1@4.02	Ea	5,615.00	85.20	5,700.20
10' x 10' room	B1@4.02	Ea	5,490.00	85.20	5,575.20
10' x 12' room	B1@4.02	Ea	6,155.00	85.20	6,240.20
Add for electrical service connection	--	LS	--	--	230.00
Add for thermometer	--	Ea	17.10	--	--
Add for sand timer	--	Ea	26.00	--	--
Add for delivery (typical)	--	%	--	--	5.0

Screen Wire
Cost per 100 LF roll.

	Alum. mesh	Galv. mesh	Bronze mesh	Fiber mesh
18" wide roll	40.80	45.10	--	--
20" wide roll	45.10	50.20	--	--
22" wide roll	49.30	55.30	--	--
24" wide roll	50.20	60.40	174.00	31.00
26" wide roll	59.50	64.60	189.00	40.00
28" wide roll	61.20	69.70	204.00	40.00
30" wide roll	65.50	74.80	218.00	42.00
32" wide roll	70.60	79.90	233.00	44.00
34" wide roll	74.80	81.60	248.00	48.00
36" wide roll	79.10	84.20	262.00	48.00
38" wide roll	87.00	89.30	--	49.00
40" wide roll	90.10	100.00	--	54.00
42" wide roll	91.80	105.00	306.00	59.00
44" wide roll	96.00	110.00	--	--
46" wide roll	116.00	116.00	--	--
48" wide roll	119.00	120.00	349.00	64.00
54" wide roll	122.00	--	--	73.00
60" wide roll	137.00	--	--	80.00
72" wide roll	163.00	--	--	97.00
84" wide roll	--	--	--	112.00

Security Alarms, Subcontract

	Craft@Hrs	Unit	Material	Labor	Total

Costs are for securing building perimeter and all accessible openings with electronic alarm system, including all labor and materials. Costs assume that no interior wall, floor, or ceiling finishes are in place. All equipment is good to better quality.

	Craft@Hrs	Unit	Material	Labor	Total
Alarm control panel	--	Ea	--	--	420.00
Wiring, detectors per opening, with switch	--	Ea	-	--	65.00
Monthly monitoring charge	--	Mo	--	--	25.00
Alarm options (installed)					
Audio detectors	--	Ea	--	--	150.00
Communicator (central station service)	--	Ea	--	--	205.00
Digital touch pad control	--	Ea	--	--	140.00
Entry/exit delay	--	Ea	--	--	45.00
Fire/smoke detectors	--	Ea	--	--	170.00

Security Alarms

	Craft@Hrs	Unit	Material	Labor	Total
Interior/exterior sirens	--	Ea	--	--	75.00
Panic button	--	Ea	--	--	48.00
Passive infrared detector	--	Ea	--	--	265.00
Pressure mat	--	LF	--	--	75.00
Security light control	--	Ea	--	--	455.00

Add for wiring detectors and controls in existing buildings where wall and ceiling finishes are already in place

	--	%	--	--	25.0

Security Guards, Subcontract

Construction site security guards, unarmed, in uniform with two-way radio, backup patrol car, bond, and liability insurance. Per manhour.

Short term (1 night to 1 week)	--	Hr	--	--	17.50
Medium duration (1 week to 1 month)	--	Hr	--	--	14.50
Long term (1 to 6 months)	--	Hr	--	--	13.00
Add for licensed armed guard	--	Hr	--	--	1.75
Add for holidays	--	%	--	--	50.0

Construction site guard dog service for sites with totally enclosed perimeter, 24 hour per day, 7 day per week service. Local business codes may restrict or prohibit unattended guard dogs. Includes dog handling training session, liability insurance, on-site kennel. Cost per month

Contractor feeding and tending dog	--	Mo	--	--	500.00
Daily delivery of guard dog at quitting time and pickup at starting time by guard dog service (may require advance notice for special working hours)	--	Mo	--	--	525.00

Construction site man and dog guard team, includes two-way radio communication with guard service office, backup patrol car, and liability insurance. Guard service may require that a telephone and guard shack be provided by the contractor. Cost for man and dog per hour

Short term (1 week to 1 month)	--	Hr	--	--	16.50
Long term (1 to 6 months)	--	Hr	--	--	14.50

Septic Sewer Systems, Subcontract

Soils testing (percolation test) by qualified engineer (required by some communities when applying for septic system permit). Does not include application fee.

Minimum cost	--	Ea	--	--	500.00

Residential septic sewer tanks (costs include excavation for tank with good site conditions, placing of tank, inlet and outlet fittings, and backfill after hookup)

Steel reinforced concrete tanks

1,000 gallons (3 bedroom house)	--	Ea	--	--	1,550.00
1,250 gallons (4 bedroom house)	--	Ea	--	--	1,600.00
1,500 gallons (5 or 6 bedroom house)	--	Ea	--	--	1,700.00

Fiberglass tanks

1,000 gallons (3 bedroom house)	--	Ea	--	--	1,600.00
1,250 gallons (4 bedroom house)	--	Ea	--	--	1,650.00
1,500 gallons (5 or 6 bedroom house)	--	Ea	--	--	1,750.00

Polyethylene tanks, 4" lines, 20" manhole diameter

500 gallons, 42" H x 51" W x 100" L	--	Ea	--	--	755.00
750 gallons, 57" H x 52" W x 95" L	--	Ea	--	--	840.00
1,000 gallons, 66" H x 52" W x 100" L	--	Ea	--	--	1,025.00
1,250 gallons, 68" H x 56" W x 112" L	--	Ea	--	--	1,115.00
1,500 gallons, 65" H x 55" W x 143" L	--	Ea	--	--	1,385.00

Septic Sewer Systems

	Craft@Hrs	Unit	Material	Labor	Total

Residential septic sewer drain fields (leach lines). Costs include labor, unsaturated paper, piping, gravel, excavation with good site conditions, backfill, and disposal of excess soil. 4" PVC pipe (ASTM D2729) laid in 3' deep by 1' wide trench

With 12" gravel base	--	LF	--	--	6.00
With 24" gravel base	--	LF	--	--	7.00
With 36" gravel base	--	LF	--	--	8.00
Add for pipe laid 6' deep	--	LF	--	--	2.00

Add for piping from house to septic tank, and from septic tank to remote drain fields (4" PVC Schedule 40) -- LF -- -- 4.00

Low cost pumping systems for residential wastes, including fiberglass basin, pump, installation from septic tank or sewer line, 40' pipe run, and automatic float switch (no electric work or pipe included)

To 15' head	--	LS	--	--	1,950.00
To 25' head	--	LS	--	--	2,300.00
To 30' head	--	LS	--	--	3,000.00
Add for high water/pump failure alarm	--	LS	--	--	425.00

Better quality pump system with two alternating pumps, 700 to 800 gallon concrete or fiberglass basin, automatic float switch, indoor control panel, high water/pump failure alarm, explosion proof electrical system but no electric work

or pipe -- LS -- -- 6,000.00

Pipe locating service to detect sewage and water leaks, breaks, and stoppages in pipes with diameter of 2" or more. Equipment can't be used where pipe has 90 degree turns. Maximum depth is 18' for cast iron and 30' for vitreous clay

Residential 1/2 day rate	--	LS	--	--	275.00
Commercial 1/2 day rate	--	LS	--	--	375.00
Add for each additional pipe entered	--	LS	--	--	35.00
Plus per LF of pipe length	--	LF	--	--	1.50
Add for travel over 15 miles, per mile	--	Ea	--	--	.25

Sewer Connections, Subcontract. Including typical excavation and backfill.

4" vitrified clay pipeline, house to property line

Long runs	--	LF	--	--	15.00
Short runs	--	LF	--	--	16.00

6" vitrified clay pipeline

Long runs	--	LF	--	--	17.00
Short runs	--	LF	--	--	20.00
Street work	--	LF	--	--	42.00

4" PVC sewer pipe, house to property line

Long runs	--	LF	--	--	13.00
Short runs	--	LF	--	--	15.30

6" PVC sewer pipe, property line to main line

Long runs	--	LF	--	--	16.00
Short runs	--	LF	--	--	19.50

Sheet Metal Access Doors

Attic access doors

22" x 22"	SW@.363	Ea	43.00	9.14	52.14
22" x 30"	SW@.363	Ea	56.00	9.14	65.14
30" x 30"	SW@.363	Ea	72.00	9.14	81.14

Wall access doors, nail-on type, galvanized or painted

24" x 18"	SW@.363	Ea	40.00	9.14	49.14
24" x 24"	SW@.363	Ea	47.00	9.14	56.14

Sheet Metal

	Craft@Hrs	Unit	Material	Labor	Total
Wall access doors, screen only					
24" x 18"	--	Ea	12.00	--	--
24" x 24"	--	Ea	13.50	--	--
Tub access doors					
14" x 14", plaster type	SW@.363	Ea	27.00	9.14	36.14
14" x 14", drywall type	SW@.363	Ea	32.00	9.14	41.14

Sheet Metal Area Walls. Provides light well for basement windows, galvanized, corrugated.

	Craft@Hrs	Unit	Material	Labor	Total
12" deep, 6" projection, 37" wide	SW@.410	Ea	14.00	10.30	24.30
18" deep, 6" projection, 37" wide	SW@.410	Ea	18.70	10.30	29.00
24" deep, 6" projection, 37" wide	SW@.410	Ea	24.10	10.30	34.40
30" deep, 6" projection, 37" wide	SW@.410	Ea	29.10	10.30	39.40
Add for steel grille cover	--	Ea	32.40	--	--

Sheet Metal Flashing

Counter flashing, in 200' pack

	Craft@Hrs	Unit	Material	Labor	Total
1/2" x 2"	SW@.020	LF	.32	.50	.82
1/2" x 3"	SW@.020	LF	.37	.50	.87
1/2" x 4"	SW@.020	LF	.47	.50	.97
1/2" x 5"	SW@.020	LF	.53	.50	1.03
1/2" x 6"	SW@.020	LF	.62	.50	1.12

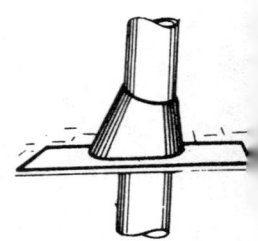

Flashing, all pitch, for plumbing vent pipes

	Unit	Galvanized	FHA painted
1/2", 3/4", 1	Ea	4.60	6.50
1-1/4", 1-1/2", 2"	Ea	3.70	5.65
3" conductor	Ea	5.80	8.60
3" amerivent, transite, soil	Ea	6.30	8.70
4" conductor	Ea	6.85	9.70
4" transite	Ea	8.30	11.20
4" amerivent, soil	Ea	6.85	9.00
5" conductor	Ea	10.30	13.80
5" transite	Ea	13.40	14.50
5" amerivent	Ea	11.20	14.10
6" conductor	Ea	14.20	17.10
6" transite	Ea	15.70	18.90
6" amerivent	Ea	15.70	18.90
7" conductor	Ea	17.70	18.90
7" transite	Ea	19.30	22.10
8" conductor	Ea	22.20	23.50
8" transite	Ea	24.10	23.50
9" conductor	Ea	24.10	23.50
10" conductor or transite	Ea	31.80	32.50
12" conductor or transite	Ea	35.50	36.50
14" conductor or transite	Ea	63.50	64.60

Flashing, all pitch, oval, galvanized

	Unit	Material	Total
4" opening	Ea	9.10	12.40
5" opening	Ea	13.30	18.60

Flashing, all pitch, tee top

	Unit	Material	Total
3" opening	Ea	7.55	9.10
4" opening	Ea	8.00	9.80
5" opening	Ea	13.00	15.90
6" opening	Ea	15.40	18.90

Sheet Metal Flashing

	Craft@Hrs	Unit	Material	Labor	Total
Flashing labor					
To 3" diameter	SW@.136	Ea	--	3.42	--
4" to 8" diameter	SW@.270	Ea	--	6.80	--
Gravel guard					
4-1/2"	SW@.030	LF	.36	.76	1.12
6"	SW@.030	LF	.48	.76	1.24
7-1/4"	SW@.030	LF	.53	.76	1.29
Nosing, roof edging, at 90 degree or 105 degree angle, galvanized, 10' lengths					
3/4" x 3/4"	SW@.040	LF	.20	1.01	1.21
1" x 1"	SW@.040	LF	.20	1.01	1.21
1" x 2"	SW@.040	LF	.24	1.01	1.25
1-1/2" x 1-1/2"	SW@.040	LF	.24	1.01	1.25
2" x 2"	SW@.040	LF	.30	1.01	1.31
2" x 3"	SW@.040	LF	.40	1.01	1.41
2" x 4"	SW@.040	LF	.48	1.01	1.49
3" x 3"	SW@.040	LF	.55	1.01	1.56
3" x 4"	SW@.040	LF	.66	1.01	1.67
3" x 5"	SW@.040	LF	.78	1.01	1.79
4" x 4"	SW@.043	LF	.78	1.08	1.86
4" x 5"	SW@.043	LF	.88	1.08	1.96
4" x 6"	SW@.043	LF	.98	1.08	2.06
5" x 5"	SW@.043	LF	.98	1.08	2.06
6" x 6"	SW@.043	LF	1.16	1.08	2.24
Add for 90 degree kickout nosing	--	%	5.0	--	--
Rain diverter, 1" x 3", galvanized, unpainted					
4' length	SW@.136	Ea	2.85	3.42	6.27
5' length	SW@.136	Ea	3.15	3.42	6.57
10' length	SW@.265	Ea	5.30	6.67	11.97
Roof edging (eave drip, with 1/4" kickout, galvanized)					
1" x 2"	SW@.040	LF	.31	1.01	1.32
1-1/2" x 1-1/2"	SW@.040	LF	.37	1.01	1.38
2" x 2"	SW@.040	LF	.37	1.01	1.38
Sill pans, galvanized, 5", 5-1/2" or 6", per pair	--	Pr	.68	--	--
Valley flashing, roll type, 50' lengths					
Tin galvanized, 28 gauge, galvanized					
8" wide	SW@.020	LF	.69	.50	1.19
14" wide	SW@.020	LF	1.14	.50	1.64
20" wide	SW@.020	LF	1.60	.50	2.10
Tin seamless, painted					
8" wide	SW@.020	LF	.58	.50	1.08
14" wide	SW@.020	LF	.89	.50	1.39
20" wide	SW@.020	LF	1.26	.50	1.76
Aluminum seamless, .016 gauge					
4" wide	SW@.020	LF	.42	.50	.92
6" wide	SW@.020	LF	.58	.50	1.08
8" wide	SW@.020	LF	.68	.50	1.18
10" wide	SW@.020	LF	.80	.50	1.30
12" wide	SW@.020	LF	.99	.50	1.49
14" wide	SW@.020	LF	1.16	.50	1.66
20" wide	SW@.020	LF	1.63	.50	2.13
Valley flashing, "W" type					
18", 26 gauge	SW@.030	LF	1.48	.76	2.24
24", 26 gauge	SW@.030	LF	1.93	.76	2.69

Sheet Metal Flashing

	Craft@Hrs	Unit	Material	Labor	Total
Vertical chimney flashing, 2" x 3", galvanized	SW@.050	LF	.50	1.26	1.76
Window flashing, painted tin					
3/4" x 1" x 28" long	SW@.020	LF	.21	.50	.71
1" x 1" x 28" long	SW@.020	LF	.23	.50	.73
"Z" bar flashing, galvanized					
Standard	SW@.015	LF	.47	.38	.85
Old style	SW@.015	LF	.47	.38	.85
For plywood siding	SW@.015	LF	.33	.38	.71
Sheet Metal Vents					
Attic and gable vents, opening sizes, louvers					
8" x 12"	SW@.448	Ea	9.90	11.30	21.20
14" x 12"	SW@.448	Ea	10.40	11.30	21.70
14" x 18"	SW@.448	Ea	15.50	11.30	26.80
14" x 24"	SW@.448	Ea	19.50	11.30	30.80
Chimney caps					
3" diameter	SW@.324	Ea	5.05	8.16	13.21
4" diameter	SW@.324	Ea	5.20	8.16	13.36
5" diameter	SW@.324	Ea	6.65	8.16	14.81
6" diameter	SW@.324	Ea	8.00	8.16	16.16
7" diameter	SW@.324	Ea	10.45	8.16	18.61
8" diameter	SW@.324	Ea	17.05	8.16	25.21
Clothes dryer vent set, aluminum, with hood and inside plate					
3" or 4" diameter	SW@.440	Ea	5.40	11.10	16.50
Flex hose, 4" diameter	--	LF	.96	--	--
Flex hose clamps, 4" diameter	--	Ea	1.28	--	--
Crawl hole vents, screen, painted	SW@.448	Ea	20.50	11.30	31.80
Cowl caps					
3" diameter	SW@.324	Ea	8.80	8.16	16.96
4" diameter	SW@.324	Ea	9.10	8.16	17.26
5" diameter	SW@.324	Ea	10.70	8.16	18.86
6" diameter	SW@.324	Ea	11.90	8.16	20.06
7" diameter	SW@.324	Ea	14.70	8.16	22.86
8" diameter	SW@.493	Ea	21.50	12.40	33.90
9" diameter	SW@.493	Ea	33.60	12.40	46.00
10" diameter	SW@.493	Ea	38.60	12.40	51.00
12" diameter	SW@.493	Ea	43.60	12.40	56.00
Dormer louvers, half round, galvanized, 1/4" mesh screen					
18" x 9", 3/12 pitch	SW@.871	Ea	37.80	21.90	59.70
24" x 12", 3/12 pitch	SW@.871	Ea	45.70	21.90	67.60
18" x 9", 5/12 pitch	SW@.871	Ea	31.70	21.90	53.60
24" x 12", 5/12 pitch	SW@.871	Ea	39.30	21.90	61.20
Add 1/8" mesh	--	%	3.0	--	--
Foundation vents, galvanized, with screen, no louvers					
6" x 14" stucco	SW@.255	Ea	2.20	6.42	8.62
8" x 14" stucco	SW@.255	Ea	2.70	6.42	9.12
6" x 14" 2-way	SW@.255	Ea	2.60	6.42	9.02
6" x 14" for siding (flat type)	SW@.255	Ea	2.20	6.42	8.62
6" x 14" louver type	SW@.255	Ea	2.50	6.42	8.92
6" x 14" louver type with mesh	SW@.255	Ea	3.20	6.42	9.62
6" x 14" foundation insert	SW@.255	Ea	2.20	6.42	8.62
6" x 14" vent with ears	SW@.255	Ea	2.20	6.42	8.62
6" x 14" ornamental grill & louvers	SW@.255	Ea	5.90	6.42	12.32

Sheet Metal Vents

	Craft@Hrs	Unit	Material	Labor	Total
Heater closet door vents, louver					
70 sq inches, louvers only	SW@.196	Ea	3.40	4.94	8.34
70 sq inches, louvers and screen	SW@.196	Ea	6.30	4.94	11.24
38 sq inches, louvers only	SW@.203	Ea	3.30	5.11	8.41
38 sq inches, louvers and screen	SW@.196	Ea	5.00	4.94	9.94
Midget louvers, aluminum, pack of 50					
1"	SW@.255	Ea	1.50	6.42	7.92
2"	SW@.255	Ea	2.20	6.42	8.62
3"	SW@.255	Ea	2.90	6.42	9.32
Rafter vents, 1/8" mesh, galvanized, louver					
Hand-formed rafter vents					
3" x 14"	SW@.220	Ea	1.85	5.54	7.39
3" x 22"	SW@.220	Ea	2.95	5.54	8.49
Plaster ground rafter vent four sides					
3" x 14"	SW@.220	Ea	1.90	5.54	7.44
3" x 22"	SW@.220	Ea	2.95	5.54	8.49
Flat type for siding					
3" x 14"	SW@.220	Ea	1.85	5.54	7.39
3" x 22"	SW@.220	Ea	2.95	5.54	8.49
Rafter insert					
3" x 14"	SW@.220	Ea	1.85	5.54	7.39
3" x 22"	SW@.220	Ea	2.95	5.54	8.49
Rafter vent with ears					
3" x 14"	SW@.220	Ea	2.55	5.54	8.09
3" x 22"	SW@.220	Ea	2.90	5.54	8.44
6" x 22"	SW@.220	Ea	3.55	5.54	9.09
For FHA painted, add	--	%	5.0	--	--
For galvanized 1/4" mesh, deduct	--	%	-5.0	--	--
Ridge ventilators, 1/8" louver opening, aluminum					
Mill finish	SW@.086	LF	2.95	2.17	5.12
Black or brown	SW@.086	LF	3.40	2.17	5.57
Add for end fitting	--	Ea	3.15	--	--
Roof jacks, round					
7" diameter	SW@.186	Ea	8.80	4.68	13.48
8" diameter	SW@.186	Ea	18.90	4.68	23.58
9" diameter	SW@.186	Ea	22.90	4.68	27.58
Roof vents, with mesh screen					
Square cap type, 26" L x 23" D x 5" H	SW@.448	Ea	38.30	11.30	49.60
Dormer type, 20" L x 10" D x 6" H	SW@.448	Ea	24.70	11.30	36.00
Dormer type, louvers, 19" L x 27" D x 18" H	SW@.448	Ea	59.00	11.30	70.30
Rotary roof ventilators, with base, by turbine diameter					
6" diameter	SW@.694	Ea	140.00	17.50	157.50
8" diameter	SW@.694	Ea	149.00	17.50	166.50
10" diameter	SW@.694	Ea	164.00	17.50	181.50
12" diameter	SW@.694	Ea	194.00	17.50	211.50
14" diameter	SW@.694	Ea	210.00	17.50	227.50
16" diameter	SW@.930	Ea	260.00	23.40	283.40
18" diameter	SW@1.21	Ea	300.00	30.50	330.50
20" diameter	SW@1.21	Ea	350.00	30.50	380.50
Round louver vents, 1/8" mesh, galvanized					
12" diameter	SW@.440	Ea	29.00	11.10	40.10
14" diameter	SW@.440	Ea	30.20	11.10	41.30
16" diameter	SW@.440	Ea	32.70	11.10	43.80

Sheet Metal Vents

	Craft@Hrs	Unit	Material	Labor	Total
18" diameter	SW@.440	Ea	37.60	11.10	48.70
24" diameter	SW@.661	Ea	69.00	16.60	85.60
Deduct for 1/4" mesh	--	%	-3.0	--	--

Sheet Metal Hoods
Entrance hoods, galvanized

	Craft@Hrs	Unit	Material	Labor	Total
48" x 22" x 12", double mold	SW@1.00	Ea	77.00	25.20	102.20
60" x 24" x 12", double mold	SW@1.00	Ea	73.00	25.20	98.20
48" x 22" x 12", scalloped edge	SW@1.00	Ea	94.00	25.20	119.20
60" x 24" x 12", scalloped edge	SW@1.00	Ea	100.00	25.20	125.20

Side hoods, galvanized, square

	Craft@Hrs	Unit	Material	Labor	Total
4" x 4"	SW@.190	Ea	7.20	4.78	11.98
6" x 6"	SW@.216	Ea	12.20	5.44	17.64
7" x 7"	SW@.272	Ea	13.80	6.85	20.65

Shelving
Wood shelving, based on using 1" thick #3 & Btr paint grade board lumber. Prices shown in parentheses are before allowance for accessories and waste. Costs shown include an allowance for ledger boards, nails and normal waste. Painting not included.

Closet shelves, with 1" diameter clothes pole with brackets 3' OC

	Craft@Hrs	Unit	Material	Labor	Total
10" wide shelf ($450 per MBF = $.37 per LF)	BC@.080	LF	.50	1.90	2.40
12" wide shelf ($580 per MBF = $.58 per LF)	BC@.080	LF	.75	1.90	2.65
18" wide shelf ($495 per MBF = $.74 per LF)	BC@.100	LF	.90	2.35	3.25

Linen cabinet shelves, cost of linen closet not included

	Craft@Hrs	Unit	Material	Labor	Total
18" wide shelf ($495 per MBF = $.74 per LF)	BC@.125	LF	1.00	2.95	3.95
24" wide shelf ($580 per MBF = $1.16 per LF)	BC@.125	LF	1.45	2.95	4.40

Utility shelves, laundry room walls, garage walls, etc.

	Craft@Hrs	Unit	Material	Labor	Total
18" wide shelf ($495 per MBF = $.74 per LF)	BC@.100	LF	.85	2.35	3.20
24" wide shelf ($580 per MBF = $1.16 per LF)	BC@.112	LF	1.35	2.65	4.00

Shower and Tub Doors
Swinging shower doors, with hardware, anodized aluminum frame, tempered safety glass

Door 64" high, good quality, silver or gold tone

	Craft@Hrs	Unit	Material	Labor	Total
24" to 26" wide	BG@1.57	Ea	130.00	34.50	164.50
26" to 28" wide	BG@1.57	Ea	140.00	34.50	174.50

Door 66" high, better quality, silver or brass frame

	Craft@Hrs	Unit	Material	Labor	Total
27" to 31" wide	BG@1.57	Ea	150.00	34.50	184.50
31" to 36" wide	BG@1.57	Ea	165.00	34.50	199.50

Door 73" high, better quality, silver or gold frame

	Craft@Hrs	Unit	Material	Labor	Total
32" to 42" wide	BG@1.57	Ea	205.00	34.50	239.50
42" to 48" wide	BG@1.57	Ea	230.00	34.50	264.50

Bi-folding doors, 70" high, 46" to 48"

	Craft@Hrs	Unit	Material	Labor	Total
total width	BG@2.41	Ea	185.00	52.90	237.90

Sliding shower doors, 2 panels, tempered glass, anodized aluminum frame, 46" total width, 70" high, outside towel bar

	Craft@Hrs	Unit	Material	Labor	Total
Both panels opaque	BG@1.57	Ea	175.00	34.50	209.50
1 panel double mirrored, 1 bronze tinted	BG@1.57	Ea	280.00	34.50	314.50
Snap-on plastic trim kit	BG@.151	Ea	22.00	3.32	25.32

Tub doors, sliding, tempered safety glass, anodized aluminum frame

Three panels, 56" H x 60" W opening

	Craft@Hrs	Unit	Material	Labor	Total
Mirrored center panel	BG@2.02	Ea	300.00	44.40	344.40
Reflective glass panels	BG@2.02	Ea	330.00	44.40	374.40

Shower Doors

	Craft@Hrs	Unit	Material	Labor	Total
Two panels, 56" H x 60" W opening					
Shatter-resistant plastic panels	BG@1.77	Ea	70.00	38.90	108.90
Semi-clear, tempered pebble pattern	BG@1.77	Ea	105.00	38.90	143.90
Semi-clear, with design	BG@1.77	Ea	175.00	38.90	213.90
One semi-clear, one mirrored	BG@1.77	Ea	145.00	38.90	183.90
Reflective glass panels	BG@1.77	Ea	195.00	38.90	233.90
Accordion folding tub and shower doors, vinyl, chrome frame					
60" x 58" high, tub enclosure	BG@1.57	Ea	70.00	34.50	104.50
30" to 32", 69" high, shower enclosure	BG@1.57	Ea	65.00	34.50	99.50
33" to 36", 69" high, shower enclosure	BG@1.57	Ea	70.00	34.50	104.50
46" to 48", 69" high, shower enclosure	BG@1.57	Ea	75.00	34.50	109.50

Shower and Tub Enclosures, Subcontract. No plumbing included. See also Plumbing Fixtures.

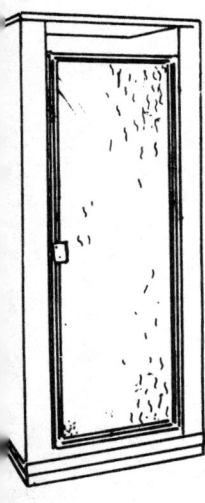

Shower enclosures, 48" W x 34" D x 66" H, tempered glass, installed, with composition receptor, door and towel bar

	Craft@Hrs	Unit	Material	Labor	Total
Aluminum frame, opaque glass	--	Ea	--	--	425.00
Gold frame, clear glass	--	Ea	--	--	445.00
Bronze frame, bronze tint	--	Ea	--	--	480.00
Corner shower enclosures, aluminum frame, plastic panels, installed, with composition receptor, door and towel bar					
30" to 34" W, 72" H	--	Ea	--	--	485.00
35" to 45" W, 72" H	--	Ea	--	--	410.00
45" to 60" W, 72" H	--	Ea	--	--	445.00
Add for tempered glass panels	--	Ea	--	--	100.00
"Neo-Angle" shower cabinets, tempered glass					
3' W x 3' D x 72" H, hammered glass	--	Ea	--	--	460.00
3'4" W x 3'4" D x 72" H, bronze frame	--	Ea	--	--	605.00
Shower doors, tempered opaque glass, installed					
Aluminum frame, 27" W x 66" H	--	Ea	--	--	210.00
Add for extra towel bars, installed	--	Ea	--	--	14.20
Add for clear or tinted glass	--	Ea	--	--	35.20
Tub enclosures, fiberglass, installed, 70" H, 60" W Plexiglass sliding door					
Semi-clear	--	Ea	--	--	210.00
Colors, tints, designs	--	Ea	--	--	240.00
Tub enclosures, tempered reflective glass, installed, 70" H, two panel, aluminum frame	--	Ea	--	--	345.00

Shower Receptors. Moulded fiberglass, corner style

	Craft@Hrs	Unit	Material	Labor	Total
33" x 34"	PM@.480	Ea	105.00	11.90	116.90
36" x 34"	PM@.480	Ea	110.00	11.90	121.90
42" x 34"	PM@.480	Ea	135.00	11.90	146.90
48" x 34"	PM@.480	Ea	140.00	11.90	151.90
54" x 34"	PM@.480	Ea	160.00	11.90	171.90
60" x 34"	PM@.480	Ea	170.00	11.90	181.90
37" x 37" neo angle	PM@.480	Ea	125.00	11.90	136.90
Drain, trap, cover and fittings	PM@1.00	LS	30.00	24.80	54.80

Shutters

Shutters, Exterior. Louvered, price per pair. Add hinges and installation below.

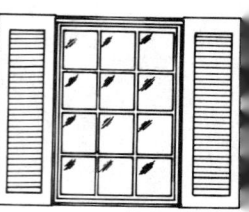

Height	Unfinished 7/8" pond. pine 15" wide	Painted 1-1/8" polypropylene 16" wide	Painted 1-1/8" polypropylene 15" wide	Painted 3/4" polystyrene 15" wide
23" to 25"	27.00	24.00	18.00	11.00
35"	28.00	31.00	24.00	15.00
39"	29.00	33.00	26.00	17.00
43"	31.00	35.00	29.00	19.00
47"	34.00	37.00	30.00	20.00
51"	37.00	39.00	32.00	22.00
55"	40.00	42.00	33.00	23.00
59"	43.00	46.00	35.00	25.00
63"	47.00	48.00	37.00	26.00
67"	50.00	51.00	38.00	--
71"	53.00	53.00	44.00	31.00

The shutters above may be installed as fixed units or as hinged units.
Installing operable shutters takes longer and requires hinges.

Labor installing shutters on hinges. Figures in the material column show the cost of anchors and hinges.

	Craft@Hrs	Unit	Material	Labor	Total
Wood frame construction, 1-3/8" throw					
25" to 51" high	BC@.400	Pr	16.50	9.44	25.94
52" to 80" high	BC@.634	Pr	23.70	15.00	38.70
Masonry construction, 4-1/4" throw					
25" to 51" high	BC@.598	Pr	24.75	14.10	38.85
52" to 80" high	BC@.827	Pr	35.00	19.50	54.50
Installation of fixed exterior shutters, wood frame or masonry, all sizes	BC@.390	Pr	2.50	9.20	11.70

Shutters, Interior

Movable 1-1/4" louver, 3/4" thick, kiln dried pine, with knobs and hooks, panels attached with hanging strips hinged to panels, finished, priced per set of either two or four panels

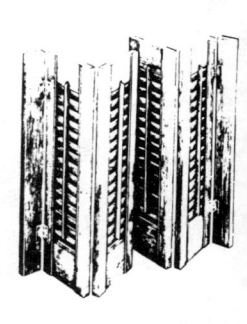

Width and Height	Craft@Hrs	Unit	Material	Labor	Total
23" x 20", 2 panels	BC@.250	Ea	44.00	5.90	49.90
23" x 24", 2 panels	BC@.300	Ea	49.00	7.08	56.08
23" x 28", 2 panels	BC@.300	Ea	56.00	7.08	63.08
23" x 32", 2 panels	BC@.333	Ea	63.00	7.86	70.86
23" x 36", 2 panels	BC@.367	Ea	67.00	8.66	75.66
27" x 20", 4 panels	BC@.250	Ea	61.00	5.90	66.90
27" x 24", 4 panels	BC@.300	Ea	68.00	7.08	75.08
27" x 28", 4 panels	BC@.300	Ea	76.00	7.08	83.08
27" x 32", 4 panels	BC@.333	Ea	82.00	7.86	89.86
27" x 36", 4 panels	BC@.367	Ea	92.00	8.66	100.66
31" x 20", 4 panels	BC@.250	Ea	65.00	5.90	70.90
31" x 24", 4 panels	BC@.300	Ea	71.00	7.08	78.08
31" x 28", 4 panels	BC@.300	Ea	81.00	7.08	88.08
31" x 32", 4 panels	BC@.367	Ea	90.00	8.66	98.66
31" x 36", 4 panels	BC@.367	Ea	100.00	8.66	108.66
33" x 20", 4 panels	BC@.250	Ea	70.00	5.90	75.90
33" x 24", 4 panels	BC@.300	Ea	77.00	7.08	84.08
33" x 28", 4 panels	BC@.300	Ea	86.00	7.08	93.08
33" x 32", 4 panels	BC@.333	Ea	94.00	7.86	101.86
33" x 36", 4 panels	BC@.367	Ea	106.00	8.66	114.66
35" x 20", 4 panels	BC@.250	Ea	81.00	5.90	86.90
35" x 24", 4 panels	BC@.300	Ea	89.00	7.08	96.08

Shutters

	Craft@Hrs	Unit	Material	Labor	Total
35" x 28", 4 panels	BC@.300	Ea	95.00	7.08	102.08
35" x 32", 4 panels	BC@.367	Ea	103.00	8.66	111.66
35" x 36", 4 panels	BC@.367	Ea	110.00	8.66	118.66
39" x 20", 4 panels	BC@.250	Ea	90.00	5.90	95.90
39" x 24", 4 panels	BC@.300	Ea	95.00	7.08	102.08
39" x 28", 4 panels	BC@.300	Ea	103.00	7.08	110.08
39" x 32", 4 panels	BC@.315	Ea	108.00	7.43	115.43
39" x 36", 4 panels	BC@.315	Ea	115.00	7.43	122.43
47" x 24", 4 panels	BC@.300	Ea	101.00	7.08	108.08
47" x 28", 4 panels	BC@.300	Ea	109.00	7.08	116.08
47" x 32", 4 panels	BC@.367	Ea	119.00	8.66	127.66
47" x 36", 4 panels	BC@.367	Ea	128.00	8.66	136.66

Siding. See also Siding in the Lumber Section, Plywood and Hardboard.

Mineral fiber siding 3/16" thick

	Craft@Hrs	Unit	Material	Labor	Total
Straight edge shingles, 12" x 24", 4.3 bundles per square	B1@2.85	Sq	90.00	60.40	150.40
Wavy butt shingles, 12" x 24", 4.3 bundles per square	B1@2.85	Sq	90.00	60.40	150.40
Siding corners, 12" long	--	Ea	1.00	--	--
Starter track, window molding, 10'	--	Ea	2.50	--	--

Composition siding 1/8" to 1/4" thick, prefinished. Includes 5% waste, Supradur products

	Craft@Hrs	Unit	Material	Labor	Total
12" x 24", 11" exposure, 3 bundles per square, various patterns	B1@1.88	Sq	55.00	39.80	94.80
14-1/2" x 25", 13" exposure, 4 bundles per square, thatched edge					
"Cedatex"	B1@1.88	Sq	60.00	39.80	99.80
"Pocono"	B1@1.88	Sq	75.00	39.80	114.80
14-1/2" x 32 x 13-1/2" exposure, 3 bundles per square, "Magna-Tones"	B1@1.88	Sq	55.00	39.80	94.80
9" x 32", 8" exposure, 3 bundles per square, "Stride" and "Narro-Shake"	B1@1.88	Sq	60.00	39.80	99.80

Hardboard siding, Abitibi-Price 7/16" thick. Includes 10% waste.

	Craft@Hrs	Unit	Material	Labor	Total
Lap siding, 8" or 12" x 16', primed	B1@.050	SF	.45	1.06	1.51
Plain panel, 4' x 8' or 9' primed	B1@.025	SF	.45	.53	.98
Vertical grooved, 8" OC, 4' x 8' or 9' primed	B1@.025	SF	.50	.53	1.03
Smooth beaded lap, 8" or 9" x 16' primed	B1@.050	SF	.52	1.06	1.58
Textured beaded lap, 8" or 9" x 16' primed	B1@.050	SF	.54	1.06	1.60
Cedar lap, 8" or 12" x 16' primed	B1@.050	SF	.53	1.06	1.59
Cedar lap, 8" x 16' prefinished	B1@.066	SF	.75	1.40	2.15
Cedar panels					
4' x 8' primed	B1@.025	SF	.53	.53	1.06
4' x 9' primed	B1@.025	SF	.54	.53	1.07
4' x 8' prefinished	B1@.025	SF	.75	.53	1.28
4' x 9' prefinished	B1@.025	SF	.77	.53	1.30
Cross sawn fir, reverse board and batten, 12" OC					
4' x 8' unprimed	B1@.025	SF	.51	.53	1.04
4' x 9' unprimed	B1@.025	SF	.52	.53	1.05
4' x 8' primed	B1@.025	SF	.53	.53	1.06
4' x 9' primed	B1@.025	SF	.54	.53	1.07

Siding

	Craft@Hrs	Unit	Material	Labor	Total
Cross sawn fir, vertical groove, 4" OC					
4' x 8' primed	B1@.050	SF	.57	1.06	1.63
4' x 9' primed	B1@.050	SF	.58	1.06	1.64
4' x 8' prefinished	B1@.050	SF	.79	1.06	1.85
Redwood panels					
4' x 8' primed	B1@.050	SF	.57	1.06	1.63
4' x 9' primed	B1@.050	SF	.58	1.06	1.64
4' x 8' prefinished	B1@.025	SF	.79	.53	1.32
4' x 9' prefinished	B1@.025	SF	.83	.53	1.36
Stucco panel					
4' x 8' primed	B1@.025	SF	.71	.53	1.24
4' x 9' primed	B1@.025	SF	.72	.53	1.25
4' x 8' prefinished	B1@.025	SF	.73	.53	1.26
4' x 9' prefinished	B1@.025	SF	.74	.53	1.27
"Great Random Shakes"					
1' x 4' unprimed	B1@.075	SF	.60	1.59	2.19
1' x 4' prefinished	B1@.075	SF	.82	1.59	2.41

Hardboard siding, Georgia-Pacific 7/16" thick, primed. Includes 10% waste.

	Craft@Hrs	Unit	Material	Labor	Total
Provincetown beaded lap, 8" x 16'	B1@.050	SF	.60	1.06	1.66
Sundance lap, 8" x 16'	B1@.050	SF	.59	1.06	1.65
Sundance lap, 12" x 16'	B1@.050	SF	.57	1.06	1.63
Yorktown textured beaded lap, 8" x 16'	B1@.050	SF	.64	1.06	1.70
V-8 Cadence, smooth, 4' x 8'	B1@.025	SF	.60	.53	1.13
Sundance (grooved) 4' x 8', or 9'	B1@.025	SF	.62	.53	1.15
1/2" Fairfax, 4" or 6" multilap x 16'	B1@.050	SF	.81	1.06	1.87
1/2" Sturbridge, 4" or 6" multilap x 16'	B1@.050	SF	.80	1.06	1.86
Windridge lap, 8" x 16', deep texture	B1@.050	SF	.65	1.06	1.71
Windridge, 4' x 8', deep texture	B1@.025	SF	.68	.53	1.21
Summerwood lap, 8" x 16', beaded or sq. edge	B1@.050	SF	.68	1.06	1.74
Summerwood, 4' x 8'	B1@.025	SF	.80	.53	1.33
Cadence lap, 8" or 12" x 16', smooth	B1@.050	SF	.55	1.06	1.61

Hardboard siding, Masonite 7/16" thick. Includes 10% waste.

	Craft@Hrs	Unit	Material	Labor	Total
"Cedar" lap, 8" x 16' primecoated	B1@.050	SF	.62	1.06	1.68
"Colorlok" lap					
8" x 16'	B1@.050	SF	1.05	1.06	2.11
12" x 16'	B1@.050	SF	1.05	1.06	2.11
"Ruf-X" panel, 4' x 8' or 9'					
Plain panel, primecoated	B1@.025	SF	.59	.53	1.12
Panelgroove and reverse batten,					
4" OC, primecoated	B1@.025	SF	.58	.53	1.11
"Ruf-X" lap, 12" x 16' primecoated	B1@.050	SF	.58	1.06	1.64
Panelgroove and reverse batten,					
4" OC, primecoated	B1@.025	SF	.58	.53	1.11
"Ruf-X" lap, 12" x 16' primecoated	B1@.050	SF	.58	1.06	1.64
"Ruf-X" beaded, 8" x 16' primecoated	B1@.050	SF	.59	1.06	1.65
"Stuccato" panel, 4' x 8' or 9'					
Prefinished	B1@.025	SF	.84	.53	1.37
Primecoated	B1@.025	SF	.59	.53	1.12
"Woodsman" lap, 8" x 16'					
Prestained	B1@.050	SF	.87	1.06	1.93
Primecoated	B1@.050	SF	.61	1.06	1.67

Siding

	Craft@Hrs	Unit	Material	Labor	Total
"Woodsman" lap, 12" x 16'					
Prestained	B1@.050	SF	.87	1.06	1.93
Primecoated	B1@.050	SF	.61	1.06	1.67
"Woodsman" panel, grooved 8" OC, 4' x 8' or 9'					
Prestained	B1@.025	SF	.87	.53	1.40
Primecoated	B1@.025	SF	.61	.53	1.14

Hardboard siding, Masonite Superside 1/2" thick. Includes 10% waste.

	Craft@Hrs	Unit	Material	Labor	Total
"Cedarside," 12" x 16', scaled only	B1@.050	SF	1.15	1.06	2.21
"Dropside" lap, 12" x 16'					
Smooth, primed	B1@.050	SF	.90	1.06	1.96
Textured, primed	B1@.050	SF	.90	1.06	1.96
Lap siding					
8" x 16' primed	B1@.050	SF	.75	1.06	1.81
12" x 16' primed	B1@.050	SF	.75	1.06	1.81
"Pineridge," 12" x 16' primed	B1@.050	SF	.90	1.06	1.96
Splined lap siding					
8" x 16' primed	B1@.050	SF	.95	1.06	2.01
12" x 16' primed	B1@.050	SF	.92	1.06	1.98
"V-Side," 8" x 16' unprimed	B1@.050	SF	.99	1.06	2.05
"Woodridge," 12" x 16' primed	B1@.050	SF	.90	1.06	1.96
"Woodsman" grooved, 4" OC					
4' x 8' primed	B1@.025	SF	.90	.53	1.43
4' x 9' primed	B1@.025	SF	.90	.53	1.43
"Woodsman" lap					
8" x 16' primed	B1@.050	SF	.96	1.06	2.02
12" x 16' primed	B1@.050	SF	.96	1.06	2.02
"Woodsman" plain panel					
4' x 8' primed	B1@.025	SF	.96	.53	1.49
4' x 9' primed	B1@.025	SF	.96	.53	1.49
"Woodsman" splined lap					
8" x 16' primed	B1@.050	SF	.98	1.06	2.04
12" x 16' primed	B1@.050	SF	.96	1.06	2.02

Hardboard siding, Weyerhaeuser Includes 10% waste.

	Craft@Hrs	Unit	Material	Labor	Total
Old Mill texture lap, 7/16" thick					
6" x 16' primed	B1@.050	SF	.53	1.06	1.59
8" x 16' primed	B1@.050	SF	.50	1.06	1.56
8" x 16' prefinished	B1@.050	SF	.65	1.06	1.71
9-1/2" x 16' primed	B1@.050	SF	.51	1.06	1.57
9-1/2" x 16' prefinished	B1@.050	SF	.66	1.06	1.72
12" x 16' primed	B1@.050	SF	.50	1.06	1.56
12" x 16' prefinished	B1@.050	SF	.65	1.06	1.71
Old Mill texture panel, 8" OC, 3/4" groove, 7/16" thick					
4' x 7' primed	B1@.025	SF	.35	.53	.88
4' x 7' prefinished	B1@.025	SF	.50	.53	1.03
4' x 8' primed	B1@.025	SF	.46	.53	.99
4' x 8' prefinished	B1@.025	SF	.61	.53	1.14
4' x 9' primed	B1@.025	SF	.51	.53	1.04
4' x 9' prefinished	B1@.025	SF	.66	.53	1.19
Adobe panels, shiplap edge, 7/16" thick					
4' x 8' primed	B1@.025	SF	.58	.53	1.11
4' x 8' prefinished	B1@.025	SF	.73	.53	1.26

Siding

	Craft@Hrs	Unit	Material	Labor	Total
Adobe panels, shiplap edge, 7/16" thick					
4' x 9' primed	B1@.025	SF	.68	.53	1.21
4' x 9' prefinished	B1@.025	SF	.83	.53	1.36
Smooth lap, 7/16" thick					
8" x 16' primed	B1@.050	SF	.50	1.06	1.56
9-1/2" x 16' primed	B1@.050	SF	.51	1.06	1.57
12" x 16' primed	B1@.050	SF	.50	1.06	1.56
Smooth panel, square edged, 7/16" thick					
4' x 7' primed	B1@.025	SF	.35	.53	.88
4' x 8' primed	B1@.025	SF	.46	.53	.99
4' x 9' primed	B1@.025	SF	.51	.53	1.04
Adobe panels, shiplap edge, 1/2" thick					
4' x 8' primed	B1@.025	SF	.66	.53	1.19
4' x 8' prefinished	B1@.025	SF	.82	.53	1.35
4' x 9' primed	B1@.025	SF	.79	.53	1.32
4' x 9' prefinished	B1@.025	SF	.93	.53	1.46
Cedar Shake lap, 1/2" thick					
10-1/2" x 16' primed	B1@.025	SF	.93	.53	1.46
Designer Shake lap, 1/2" thick					
9-1/2" x 4' primed	B1@.050	SF	1.60	1.06	2.66
Old Mill texture panel, 8" OC, 3/4" groove; 404 or 808, 1/2" thick					
4' x 7' primed	B1@.025	SF	.41	.53	.94
4' x 7' prefinished	B1@.025	SF	.56	.53	1.09
4' x 8' primed	B1@.025	SF	.55	.53	1.08
4' x 8' prefinished	B1@.025	SF	.70	.53	1.23
4' x 9' primed	B1@.025	SF	.61	.53	1.14
4' x 9' prefinished	B1@.025	SF	.76	.53	1.29
Old Mill and Smooth texture lap products, 1/2" thick, various patterns and groove sizes					
16" x 16' primed	B1@.050	SF	.70	1.06	1.76
16" x 16' prefinished	B1@.050	SF	.86	1.06	1.92
24" x 16' primed	B1@.050	SF	.73	1.06	1.79
Old Mill texture lap, square edge or self-aligning, 1/2" thick					
6" x 16" primed	B1@.025	SF	.63	.53	1.16
8" x 16' Sure-lock, primed	B1@.025	SF	.67	.53	1.20
8" x 16' Sure-lock, prefinished	B1@.025	SF	.82	.53	1.35
8" x 16' primed	B1@.025	SF	.59	.53	1.12
8" x 16' prefinished	B1@.025	SF	.74	.53	1.27
9-1/2" x 16' primed	B1@.025	SF	.61	.53	1.14
9-1/2" x 16' prefinished	B1@.025	SF	.76	.53	1.29
12" x 16' primed	B1@.025	SF	.59	.53	1.12
12" x 16' prefinished	B1@.025	SF	.74	.53	1.27
Smooth lap, square edge or self-aligning					
8" x 16' Sure-lock, primed	B1@.025	SF	.67	.53	1.20
8" x 16' primed	B1@.025	SF	.59	.53	1.12
9-1/2" x 16' primed	B1@.025	SF	.61	.53	1.14
12" x 16' primed	B1@.025	SF	.59	.53	1.12

Plywood siding, Douglas fir Plain or patterns, rough sawn. Includes 6% waste.

	Craft@Hrs	Unit	Material	Labor	Total
3/8" premium grade					
4' x 8'	B1@2.18	Sq	50.00	46.20	96.20
4' x 9'	B1@2.18	Sq	62.00	46.20	108.20
4' x 10'	B1@2.18	Sq	80.00	46.20	126.20

Siding

	Craft@Hrs	Unit	Material	Labor	Total

Plywood siding, Douglas fir Plain or patterns, rough sawn. Includes 6% waste.

	Craft@Hrs	Unit	Material	Labor	Total
3/8" shop grade					
4' x 8'	B1@2.18	Sq	30.00	46.20	76.20
4' x 9'	B1@2.18	Sq	32.00	46.20	78.20
4' x 10'	B1@2.18	Sq	35.00	46.20	81.20
5/8" premium grade					
4' x 8'	B1@2.45	Sq	56.00	51.90	107.90
4' x 9'	B1@2.45	Sq	63.00	51.90	114.90
4' x 10'	B1@2.45	Sq	70.00	51.90	121.90
5/8" shop grade					
4' x 8'	B1@2.45	Sq	36.00	51.90	87.90
4' x 9'	B1@2.45	Sq	39.00	51.90	90.90
4' x 10'	B1@2.45	Sq	42.00	51.90	93.90
Add for inverted batten patterns	--	Sq	5.00	--	--
Add for water repellent treated	--	Sq	6.00	--	--

Plywood siding, fiber bonded to exterior grade plywood. Includes 6% waste and coverage.

	Craft@Hrs	Unit	Material	Labor	Total
3/8" thick, 4' x 8' rough sawn texture	B1@2.18	Sq	60.00	46.20	106.20
3/8" thick, 4' x 9' rough sawn texture	B1@2.18	Sq	68.00	46.20	114.20
19/32" thick, patterns, 4", 6", 8" OC					
4' x 8'	B1@2.45	Sq	69.00	51.90	120.90
4' x 9'	B1@2.45	Sq	80.00	51.90	131.90
19/32" thick, inverted batten					
4' x 8'	B1@2.45	Sq	72.00	51.90	123.90
4' x 9'	B1@2.45	Sq	81.00	51.90	132.90

Plywood siding, redwood Water repellent treated. Includes 6% waste.

	Craft@Hrs	Unit	Material	Labor	Total
Premium grade, plain, rough sawn					
3/8" thick					
4' x 8'	B1@2.18	Sq	77.00	46.20	123.20
4' x 9', 10'	B1@2.18	Sq	88.00	46.20	134.20
5/8" thick					
4' x 8'	B1@2.45	Sq	81.00	51.90	132.90
4' x 9', 10'	B1@2.45	Sq	90.00	51.90	141.90
Premium grade, patterns, 4", 6", 8" OC, 5/8" thick					
4' x 8'	B1@2.45	Sq	90.00	51.90	141.90
4' x 9', 10'	B1@2.45	Sq	102.00	51.90	153.90
Premium grade, inverted batten, 8" and 12" OC, 5/8" thick					
4' x 8'	B1@2.45	Sq	96.00	51.90	147.90
4' x 9', 10'	B1@2.45	Sq	109.00	51.90	160.90
Select grade, plain, rough sawn					
3/8" thick					
4' x 8'	B1@2.18	Sq	50.00	46.20	96.20
4' x 9', 10'	B1@2.18	Sq	56.00	46.20	102.20
5/8" thick					
4' x 8'	B1@2.45	Sq	56.00	51.90	107.90
4' x 9', 10'	B1@2.45	Sq	64.00	51.90	115.90
Select grade, patterns 4", 6" 8" OC, rough sawn, 5/8" thick					
4' x 8'	B1@2.45	Sq	50.00	51.90	101.90
4' x 9', 10'	B1@2.45	Sq	59.00	51.90	110.90
Select grade, inverted batten, rough sawn, 5/8" thick					
4' x 8'	B1@2.45	Sq	72.00	51.90	123.90
4' x 9', 10'	B1@2.45	Sq	84.00	51.90	135.90

Siding

	Craft@Hrs	Unit	Material	Labor	Total
Plywood siding, western red cedar Includes 6% waste.					
3/8" thick, select grade, plain or patterns					
4' x 8'	B1@2.18	Sq	50.00	46.20	96.20
4' x 9', 10'	B1@2.18	Sq	55.00	46.20	101.20
1/2" thick, select grade, plain or patterns					
4' x 8'	B1@2.30	Sq	62.00	48.70	110.70
4' x 9', 10'	B1@2.30	Sq	70.00	48.70	118.70
5/8" thick, select grade, plain or patterns					
4' x 8'	B1@2.45	Sq	64.00	51.90	115.90
4' x 9', 10'	B1@2.45	Sq	72.00	51.90	123.90
3/8" thick, cabin grade (repaired and tight knots)					
4' x 8'	B1@2.18	Sq	36.00	46.20	82.20
4' x 9', 10'	B1@2.18	Sq	41.00	46.20	87.20
1/2" thick, cabin grade					
4' x 8'	B1@2.30	Sq	45.00	48.70	93.70
4' x 9', 10'	B1@2.30	Sq	50.00	48.70	98.70
5/8" thick, cabin grade					
4' x 8'	B1@2.45	Sq	56.00	51.90	107.90
4' x 9', 10'	B1@2.45	Sq	61.00	51.90	112.90
Plywood siding, stone covered Stonecast natural stone chips adhered to exterior grade plywood. United Panel.					
3/8" plywood substrate, regular aggregate					
47-3/4" x 8'	B1@.047	SF	2.40	1.00	3.40
47-3/4" x 9' or 10'	B1@.047	SF	2.65	1.00	3.65
3/8" plywood substrate, large aggregate					
47-3/4" x 8'	B1@.047	SF	3.20	1.00	4.20
47-3/4" x 9' or 10'	B1@.047	SF	3.35	1.00	4.35
1/2" plywood shiplap substrate, regular aggregate					
47-3/4" x 8'	B1@.047	SF	2.70	1.00	3.70
47-3/4" x 9' or 10'	B1@.047	SF	2.95	1.00	3.95
1/2" plywood shiplap substrate, large aggregate					
47-3/4" x 8'	B1@.047	SF	3.40	1.00	4.40
47-3/4" x 9'	B1@.047	SF	3.85	1.00	4.85
Cement board siding Stonecast fire-rated class 1 siding. 4' x 8', 10' panels. Includes 6% waste. United Panel.					
8mm (5/16"), regular texture	B1@.075	SF	3.45	1.59	5.04
8mm (5/16"), large texture	B1@.075	SF	4.00	1.59	5.59
10mm (3/8"), regular texture	B1@.083	SF	4.00	1.76	5.76
10mm (3/8"), large texture	B1@.083	SF	4.25	1.76	6.01
Fiberglass composite siding 4' x 8', 10', 12' panels. Stone chips adhered to integrated layers of fiberglass, includes 6% waste. United Panel.					
5/16", regular aggregate	B1@.075	SF	3.60	1.59	5.19
3/8", large aggregate	B1@.075	SF	4.00	1.59	5.59
3/16" fire-rated smooth panels	B1@.075	SF	3.80	1.59	5.39
Sheet metal siding, aluminum					
4-V corrugated, 2-1/2" corrugations, 5-V crimp, plain or embossed finish on wood frame. Includes 15% loss for coverage.					
17 gauge, 26" x 6' to 24'	B1@.034	SF	1.48	.72	2.20
19 gauge, 26" x 6' to 24'	B1@.034	SF	1.55	.72	2.27
Add for installation on metal frame	--	%	--	50.0	--

Siding

	Craft@Hrs	Unit	Material	Labor	Total
Flashing for corrugated aluminum siding, embossed					
End wall, 10" x 52"	B1@.048	Ea	2.58	1.02	3.60
Side wall, 7-1/2" x 10'	B1@.015	LF	1.17	.32	1.49
Rubber filler strip, 1" x 24"	B1@.132	Ea	.77	2.80	3.57
Aluminum panel siding, 24 gauge, smooth, horizontal patterns, non-insulated					
8" or double 4" widths, acrylic finish	B1@2.77	Sq	100.10	58.70	158.80
12" widths, bonded vinyl finish	B1@2.77	Sq	102.20	58.70	160.90
Add for foam backing	--	Sq	20.40	--	--
Starter strip	B1@.030	LF	.24	.64	.88
Inside and outside corners	B1@.033	LF	1.09	.70	1.79
Casing and trim	B1@.033	LF	.25	.70	.95
Drip cap	B1@.044	LF	.24	.93	1.17

Sheet metal siding, galvanized steel On wood frame.

	Craft@Hrs	Unit	Material	Labor	Total
28 gauge, 27-1/2" wide (includes 20% coverage loss)					
6' to 12' standard lengths	B1@.034	SF	.83	.72	1.55
For specific lengths	--	LF	1.73	--	--
26 gauge, 26" wide (includes 15% coverage loss)					
6' to 12' standard lengths	B1@.034	SF	.90	.72	1.62
For specific lengths	--	LF	1.67	--	--
Add for installation on metal frame	--	%	--	50.0	--

Shingle siding

Red cedar perfects 5X green, 16" long, 7-1/2" exposure, 5 shingles are 2" thick (5/2). 3 bundles cover 1 Sq including waste. Cost shown includes fasteners.

	Craft@Hrs	Unit	Material	Labor	Total
#1 ($30.00 per bundle)	B1@3.86	Sq	90.00	81.80	171.80
#2 red label ($23.00 per bundle)	B1@3.86	Sq	69.00	81.80	150.80
#3 ($14.00 per bundle)	B1@3.86	Sq	42.00	81.80	123.80

Red cedar perfections, dry, 18" long, 8" exposure, 5 shingles are 2-1/4" thick (5/2-1/4). 3 bundles cover 1 Sq including waste. Cost shown includes fasteners.

	Craft@Hrs	Unit	Material	Labor	Total
#1 ($30.00 per bundle)	B1@3.75	Sq	90.00	79.50	169.50
#2 red label ($24.00 per bundle)	B1@3.75	Sq	72.00	79.50	151.50
#3 ($20.00 per bundle)	B1@3.75	Sq	60.00	79.50	139.50

Panelized shingles, kiln dried, regraded shingle on plywood backing, no waste included. Shakertown

	Craft@Hrs	Unit	Material	Labor	Total
8' long x 7" wide single course					
Colonial 1	B1@.020	SF	1.50	.42	1.92
Barn shake	B1@.020	SF	2.00	.42	2.42
Cascade	B1@.020	SF	1.30	.42	1.72
8' long x 14" wide, single course					
Colonial 1	B1@.010	SF	1.50	.21	1.71
Barn shake	B1@.010	SF	2.00	.21	2.21
8' long x 14" wide, double course					
Colonial 2	B1@.010	SF	1.85	.21	2.06
Cascade Classic	B1@.010	SF	1.45	.21	1.66
Shingle grooved	B1@.010	SF	2.00	.21	2.21
Barn shake	B1@.010	SF	2.00	.21	2.21

Decorative shingles, 5" wide x 18" long, Shakertown, fancy cuts, 9 hand-shaped patterns,

	Craft@Hrs	Unit	Material	Labor	Total
7-1/2" exposure	B1@.027	SF	2.12	.57	2.69
Deduct for 10" exposure	--	SF	-.32	--	--

Siding

	Craft@Hrs	Unit	Material	Labor	Total
Vinyl siding Solid vinyl .024 gauge, woodgrained or colors.					
10" or double 5" panels	B1@3.34	Sq	64.00	70.80	134.80
Starter strip	B1@.030	LF	.22	.64	.86
J channel for corners	B1@.033	LF	.23	.70	.93
Outside corner	B1@.033	LF	.82	.70	1.52
Casing and trim	B1@.033	LF	.23	.70	.93
Deduct for white color	--	%	-5.0	--	--
Add for Amicor insulation (R-3)	B1@.350	Sq	17.00	7.42	24.42

Skylights

Polycarbonate dome, clear transparent or tinted, roof opening sizes, price each, Crestline Skylights, flush mount, self-flashing

	22" x 22"	30" x 30"	46" x 46"	22" x 46"	30" x 46"
Single dome	34.00	61.00	181.00	73.00	107.00
Double dome	48.00	94.00	212.00	104.00	137.00
Triple dome	68.00	120.00	248.00	146.00	183.00
Curb mount					
Single dome	34.00	61.00	181.00	73.00	107.00
Double dome	48.00	94.00	212.00	104.00	137.00
Triple dome	68.00	120.00	248.00	146.00	183.00
Fiberglass reinforced plastic curb mount, no skylight included					
Double dome	91.00	103.00	143.00	109.00	122.00
Triple dome	96.00	149.00	240.00	166.00	208.00
Ventilating model, fiberglass reinforced plastic curb mount, self-flashing					
Double dome	170.00	227.00	352.00	236.00	270.00
Triple dome	182.00	239.00	370.00	244.00	284.00

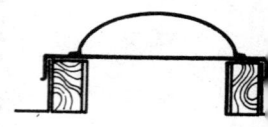

Labor installing self-flashing skylights, installation in new construction. No carpentry or roofing included.

	Craft@Hrs	Unit	Material	Labor	Total
Less than 4 SF	BG@1.56	Ea	--	34.30	--
4 SF to 10 SF	BG@2.58	Ea	--	56.70	--
Over 11 SF	BG@2.84	Ea	--	62.40	--

Tempered glass, double insulated, including prefabricated flashing, and roller blind. For installation on roofs with 10 to 85 degree slope. Labor costs are for installation of skylight, flashing and roller shade. No carpentry or roofing included.

	Craft@Hrs	Unit	Material	Labor	Total
Operable skylight, Velux model VSHH					
21-1/2" x 38-1/2"	BG@3.42	Ea	314.00	75.10	389.10
21-1/2" x 46-1/2"	BG@3.42	Ea	340.00	75.10	415.10
21-1/2" x 55"	BG@3.42	Ea	353.00	75.10	428.10
30-5/8" x 38-1/2"	BG@3.42	Ea	350.00	75.10	425.10
30-5/8" x 55"	BG@3.52	Ea	402.00	77.30	479.30
44-3/4" x 27-1/2"	BG@3.52	Ea	380.00	77.30	457.30
44-3/4" x 46-1/2"	BG@3.52	Ea	453.00	77.30	530.30
Add for prefabricated flashing for roofs with 10 to 18-1/2 degree slopes	--	%	45.0	--	--
Fixed ventilation, with removable filter on ventilation flap, Velux model FSFHH					
21-1/2" x 27-1/2"	BG@3.14	Ea	190.00	69.00	259.00
21-1/2" x 38-1/2"	BG@3.14	Ea	214.00	69.00	283.00
21-1/2" x 46-1/2"	BG@3.14	Ea	233.00	69.00	302.00
21-1/2" x 55"	BG@3.39	Ea	245.00	74.50	319.50
30-5/8" x 38-1/2	BG@3.14	Ea	252.00	69.00	321.00
30-5/8" x 55"	BG@3.39	Ea	289.00	74.50	363.50
44-3/4" x 27-1/2"	BG@3.36	Ea	289.00	73.80	362.80
44-3/4" x 46-1/2"	BG@3.39	Ea	335.00	74.50	409.50

Skylights

	Craft@Hrs	Unit	Material	Labor	Total
Add for prefabricated flashing for roofs with 10 to 18-1/2 degree slopes	--	%	45.0	--	--

Low profile, acrylic double domes with molded edge (low E insulated glass also available). Preassembled units include plywood curb-liner, 16 oz. copper flashing for pitched or flat roof, screen, and all hardware. Ventarama

	Craft@Hrs	Unit	Material	Labor	Total
Ventilating type					
30" x 72"	BG@2.58	Ea	505.00	56.70	561.70
45-1/2" x 45-1/2"	BG@2.58	Ea	455.00	56.70	511.70
45-1/2" x 30"	BG@2.58	Ea	380.00	56.70	436.70
30" x 45-1/2"	BG@2.58	Ea	380.00	56.70	436.70
30" x 30"	BG@2.07	Ea	310.00	45.50	355.50
30" x 22"	BG@2.07	Ea	280.00	45.50	325.50
22" x 30"	BG@2.07	Ea	280.00	45.50	325.50
22" x 45-1/2"	BG@2.07	Ea	340.00	45.50	385.50
Add for insulated glass	--	%	30.0	--	--
Add for bronze-tinted outer dome	--	Ea	33.00	--	--
Add for white inner dome	--	Ea	18.00	--	--
Add for roll shade					
22" width	--	Ea	105.00	--	--
30" width	--	Ea	115.00	--	--
45-1/2" width	--	Ea	125.00	--	--
Add for storm panel with weatherstripping					
30" x 30"	--	Ea	53.00	--	--
45-1/2" x 45-1/2"	--	Ea	72.00	--	--
Add for adjustable pole	--	Ea	37.00	--	--
Add for motorization	BE@3.23	Ea	225.00	81.70	306.70
Fixed type					
30" x 72"	BG@2.58	Ea	400.00	56.70	456.70
45-1/2" x 45-1/2"	BG@2.58	Ea	345.00	56.70	401.70
45-1/2" x 30"	BG@2.58	Ea	305.00	56.70	361.70
30" x 45-1/2"	BG@2.58	Ea	305.00	56.70	361.70
30" x 30"	BG@2.07	Ea	245.00	45.50	290.50
30" x 22"	BG@2.07	Ea	220.00	45.50	265.50
22" x 30"	BG@2.07	Ea	220.00	45.50	265.50
22" x 45-1/2"	BG@2.07	Ea	275.00	45.50	320.50
Add for insulated laminated glass	--	%	30.0	--	--

Skywindow, E-Class. Ultraseal Self-Flashing Units. Insulated glass, flexible flange, no mastic or step flashing required. Integral skyshade optional. Wasco

	Craft@Hrs	Unit	Material	Labor	Total
22-1/2" x 22-1/2", fixed	BG@0.75	Ea	180.00	16.50	196.50
22-1/2" x 30-1/2", fixed	BG@0.75	Ea	205.00	16.50	221.50
22-1/2" x 46-1/2", fixed	BG@0.75	Ea	240.00	16.50	256.50
22-1/2" x 70-1/2", fixed	BG@0.75	Ea	335.00	16.50	351.50
30-1/2" x 30-1/2", fixed	BG@0.75	Ea	245.00	16.50	261.50
30-1/2" x 46-1/2", fixed	BG@0.75	Ea	295.00	16.50	311.50
46-1/2" x 46-1/2", fixed	BG@0.75	Ea	360.00	16.50	376.50
22-1/2" x 22-1/2", venting	BG@0.75	Ea	340.00	16.50	356.50
22-1/2" x 30-1/2", venting	BG@0.75	Ea	365.00	16.50	381.50
22-1/2" x 46-1/2", venting	BG@0.75	Ea	415.00	16.50	431.50
30-1/2" x 30-1/2", venting	BG@0.75	Ea	400.00	16.50	416.50
30-1/2" x 46-1/2", venting	BG@0.75	Ea	450.00	16.50	466.50

Skywindow, Excel-10. Step Flash/Pan Flashed Units. Insulated clear tempered glass. Deck mounted step flash unit. Does not include Heat Mirror or laminated glass. Wasco

	Craft@Hrs	Unit	Material	Labor	Total
22-1/2" x 22-1/2", fixed	BG@2.35	Ea	165.00	51.60	216.60
22-1/2" x 30-1/2", fixed	BG@2.35	Ea	170.00	51.60	221.60

Skywindows

	Craft@Hrs	Unit	Material	Labor	Total
Skywindow, Excel-10					
22-1/2" x 46-1/2", fixed	BG@2.35	Ea	180.00	51.60	231.60
22-1/2" x 70-1/2", fixed	BG@2.35	Ea	285.00	51.60	336.60
30-1/2" x 30-1/2", fixed	BG@2.35	Ea	195.00	51.60	246.60
30-1/2" x 46-1/2", fixed	BG@2.35	Ea	240.00	51.60	291.60
22-1/2" x 22-1/2", vented	BG@2.35	Ea	275.00	51.60	326.60
22-1/2" x 30-1/2", vented	BG@2.35	Ea	300.00	51.60	351.60
22-1/2" x 46-1/2", vented	BG@2.35	Ea	350.00	51.60	401.60
30-1/2" x 30-1/2", vented	BG@2.35	Ea	345.00	51.60	396.60
30-1/2" x 46-1/2", vented	BG@2.35	Ea	380.00	51.60	431.60
Add for flashing kit	--	Ea	60.00	--	--

Skywindow, Genrae-1. Low profile insulated glass skywindow. Deck mount, tempered glass, self-flashing models. Does not include Heat Mirror or laminated glass. Wasco

	Craft@Hrs	Unit	Material	Labor	Total
22-1/2" x 22-1/2", fixed	BG@2.35	Ea	150.00	51.60	201.60
22-1/2" x 30-1/2", fixed	BG@2.35	Ea	170.00	51.60	221.60
22-1/2" x 46-1/2", fixed	BG@2.35	Ea	210.00	51.60	261.60
22-1/2" x 70-1/2", fixed	BG@2.35	Ea	285.00	51.60	336.60
30-1/2" x 30-1/2", fixed	BG@2.35	Ea	200.00	51.60	251.60
30-1/2" x 46-1/2", fixed	BG@2.35	Ea	245.00	51.60	296.60
46-1/2" x 46-1/2", fixed	BG@2.35	Ea	305.00	51.60	356.60
22-1/2" x 22-1/2", vented	BG@2.35	Ea	300.00	51.60	351.60
22-1/2" x 30-1/2", vented	BG@2.35	Ea	325.00	51.60	376.60
22-1/2" x 46-1/2", vented	BG@2.35	Ea	380.00	51.60	431.60
30-1/2" x 30-1/2", vented	BG@2.35	Ea	355.00	51.60	406.60
30-1/2" x 46-1/2", vented	BG@2.35	Ea	415.00	51.60	466.60
46-1/2" x 46-1/2", vented	BG@2.35	Ea	515.00	51.60	566.60

Roof windows, aluminum-clad wood frame, double insulated tempered glass, including prefabricated flashing, exterior awning, interior roller blind and insect screen. Sash rotates 180 degrees. For installation on roofs with 10 to 85 degree slope. Labor costs are for installation of roof window, flashing and sun screening accessories. No carpentry or roofing work other than curb included. Add for interior trim. Listed by actual unit dimensions, top hung, Velux model TPS

	Craft@Hrs	Unit	Material	Labor	Total
21-1/2" wide x 38-1/2" high	BG@3.42	Ea	442.00	75.10	517.10
27-1/2" wide x 46-1/2" high	BG@3.52	Ea	505.00	77.30	582.30
30-5/8" wide x 38-1/2" high	BG@3.42	Ea	496.00	75.10	571.10
30-5/8" wide x 55" high	BG@3.52	Ea	552.00	77.30	629.30
44-3/4" wide x 46-1/2" high	BG@3.52	Ea	616.00	77.30	693.30
Add for prefabricated flashing for roofs with 10 to 18-1/2 degree slope	--	%	45.0	--	--

Soffit Systems

Baked enamel finish, 6" fascia, J-channel, 8" perforated or solid soffit, .019 gauge aluminum

	Craft@Hrs	Unit	Material	Labor	Total
12" soffit	B1@.047	LF	2.75	1.00	3.75
18" soffit	B1@.055	LF	3.45	1.17	4.62
24" soffit	B1@.060	LF	4.00	1.27	5.27
Aluminum fascia alone					
4"	B1@.030	LF	.65	.64	1.29
6"	B1@.030	LF	.80	.64	1.44
8"	B1@.030	LF	.95	.64	1.59

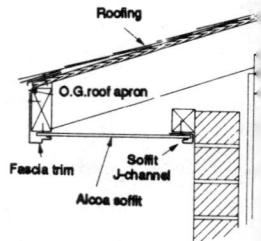

Soil Testing

	Craft@Hrs	Unit	Material	Labor	Total
Soil Testing					
Field observation and testing					
Soil technician (average)	--	Hr	--	--	50.00
Maximum density/optimum moisture test					
4" or 6" mold	--	Ea	--	--	120.00
Expansion index test	--	Ea	--	--	100.00
Report on compacted fill (costs vary widely)	--	Ea	--	--	1,500.00

Foundation investigation (bearing capacity, lateral loads, piles, seismic, stability, settlement for design purposes) and preliminary soils investigation (depth of fill, classification and profile of soil, expansion, shrinkage, grading, drainage recommendations). Drilling and sampling costs

	Craft@Hrs	Unit	Material	Labor	Total
Drill rig (driller and helper)	--	Hr	--	--	100.00
Staff engineer	--	Hr	--	--	60.00
Project engineer, RCE	--	Hr	--	--	70.00
Supervising engineer	--	Hr	--	--	100.00
Transportation	--	Mile	--	--	.50
Transportation	--	Hr	--	--	5.75
Laboratory testing (general)	--	Hr	--	--	50.00
Laboratory testing (individual basis)	--	Ea	--	--	100.00
Report preparation (conclusions, analysis, recommendations)	--	Ea	--	--	2,000.00
Foundation investigation, preliminary, soils	--	Ea	--	--	1,500.00
Depositions and court testimony	--	Hr	--	--	200.00
Add for hazardous studies	--	%	--	20.0	--

Soil Treatments, Subcontract. Costs based on total square footage inside of, and including, foundation. Includes inside and outside perimeter of foundation, interior foundation walls, and underneath all slabs. Three-plus year re-application guarantee against existence of termites. Aldrin, Chlordane, Dieldren, Heptachlor, or Heptrechlor/Chlordane.

	Craft@Hrs	Unit	Material	Labor	Total
Mix applied as per manufacturers' and state and federal specifications, complete application	--	SF	--	--	.20

Solar Photovoltaic Electrical Systems, Subcontract. Costs are for solar electric generating systems for use where commercial power is not available. These estimates assume 8 clear sunlight hours per day. Costs will be higher where less sun time is available. Kyocera Solar.

	Craft@Hrs	Unit	Material	Labor	Total
Small remote vacation home system (12 volt DC). Includes two 51 watt solar panels, two storage batteries (200 amp hours), voltage regulator, mounting rack and wiring. System yields 400 watt-hours per day of 12 volt DC power, enough to run 4 high-efficiency fluorescent lights and a DC stereo	--	LS	--	--	1,090.00
Larger remote primary home system (120 volt AC). Includes ten 51 watt solar panels, 10 storage batteries (1,000 amp hours), voltage regulator, mounting rack, wiring and a 2,000 watt DC to AC inverter. System produces 2,000 watt-hours per day of 120 volt AC power, enough to operate 6 incandescent fixtures, 8 fluorescent lights, color TV, stereo, vacuum cleaner, blender and ceiling fans. Refrigerator-freezer not included	--	LS	--	--	6,620.00
Deduct for no DC to AC converter	--	LS	--	--	-2,000.00
Add for portable backup generator when supplementary power is required	--	Ea	--	--	1,000.00

Solar Electric Systems

	Craft@Hrs	Unit	Material	Labor	Total

Basic home solar power package Simpler Solar Systems. Meets most strict building codes. Includes 2 batteries, 6 solar panels, adjustable aluminum mounting frame, DC load center, fuse box, wire harness, inverter, and voltage regulator.
Expandable. -- LS -- -- 3,100.00

Solar power inverter system Simpler Solar Systems. Includes 2000 watt inverter with built-in digital volt meter, frequency counter, in-line voltage meter, and optional 86 amp battery charger, five 60 watt panels, 2 steel case batteries, voltage regulator, digital computer volt meter and hardware -- LS -- -- 7,200.00

Solar powered lighting systems Simpler Solar Systems. Includes three 60 watt panels, voltage regulator, two 240 A/H batteries, two 15 watt fluorescent lights, 1 porch light with photocell, 4 walkway lights, two 40 watt fluorescent lights,
and hardware. -- LS -- -- 3,600.00

Solar lighting and electrical package Simpler Solar Systems. Includes fifteen 40 watt solar panels with mounting frame or 12 solar panels on a pole mounted tracker, 1200 amp hour battery, AC & DC circuit breaker box, 25 light fixtures utilizing 5" recessed cans and under the counter cove lights, 2.3 kw inverter, 600 watt computer inverter, 60 amp voltage regulator with monitor. Based on 2500 SF home. Handles all lighting, ceiling fans, computer, TV, stereo system, intercom and alarm system, garage door opener, and other necessary loads.
Expandable. -- LS -- -- 13,000.00

Solar refrigeration system Simpler Solar Systems. Includes six 60 watt PV panels, three 240 amp hour batteries, 30 amp voltage regulator, alarm, digital computer volt meter, 12 CF refrigerator and hardware. -- LS -- -- 7,400.00

Solar hot water system Simpler Solar Systems. Includes 80 gallon storage tank, 4' x 10' collector panel, drain down valve with DC brushless circulating pump,
and hardware. -- LS -- -- 2,600.00
  Deduct for free flow passive system -- LS -- -- -600.00

Solar water pumping system Simpler Solar Systems. Including two 60 watt panels, 30 amp voltage regulator, DC pump, 80 gallon holding tank, submersible pump, 2 batteries and hardware. 600 gallons per day. -- LS -- -- 3,100.00

Stairs. See also Basement Doors.
Factory cut and assembled open rise basement
stairway, 3' wide, 8' rise | B1@2.14 | LS | 165.00 | 45.30 | 210.30
  Add for handrail | B1@.161 | LF | 7.50 | 3.41 | 10.91
Factory cut and assembled straight closed box stairs, unfinished, 36" width
  Oak treads with 7-1/2" risers, price per riser
    3'0" wide | B1@.324 | Ea | 77.00 | 6.87 | 83.87
    3'6" wide | B1@.324 | Ea | 95.00 | 6.87 | 101.87
    4'0" wide | B1@.357 | Ea | 105.00 | 7.57 | 112.57
  Deduct for pine treads and plywood risers | -- | % | -22.0 | -- | --
  Add for prefinished assembled stair rail with
  balusters and newel, per riser | B1@.257 | Ea | 40.00 | 5.45 | 45.45
  Add for prefinished handrail, with
  brackets and balusters | B1@.135 | LF | 11.50 | 2.86 | 14.36
Basement stairs, open riser, delivered to job site assembled, 12 treads, 2'11-3/4" wide.
  13 risers at 7-13/16" riser height | B1@4.00 | Ea | 175.00 | 84.80 | 259.80
  13 risers at 8" riser height | B1@4.00 | Ea | 180.00 | 84.80 | 264.80
  13 risers at 8-3/16" riser height | B1@4.00 | Ea | 185.00 | 84.80 | 269.80

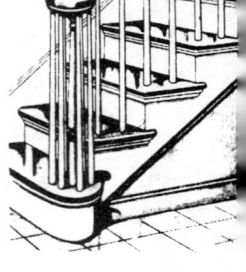

Stairs

	Craft@Hrs	Unit	Material	Labor	Total
Curved stair, clear oak, self-supporting, 8'6" radii, unfinished, unassembled (FOB factory prices)					
Open one side (1 side against wall)					
8'9" to 9'4" rise	B1@24.6	Ea	4,660.00	521.00	5,181.00
9'5" to 10'0" rise	B1@24.6	Ea	5,000.00	521.00	5,521.00
10'1" to 10'8" rise	B1@26.5	Ea	5,320.00	562.00	5,882.00
10'9" to 11'4" rise	B1@26.5	Ea	5,650.00	562.00	6,212.00
Open two sides					
8'9" to 9'4" rise	B1@30.0	Ea	8,350.00	636.00	8,986.00
9'5" to 10'0" rise	B1@30.0	Ea	8,920.00	636.00	9,556.00
10'1" to 10'8" rise	B1@31.9	Ea	9,510.00	676.00	10,186.00
10'9" to 11'4" rise	B1@31.9	Ea	10,100.00	676.00	10,776.00
Add for newel posts	B1@.334	Ea	54.00	7.08	61.08
Spiral stairs, aluminum, delivered to job site unassembled, 7-3/4" or 8-3/4" riser heights, 5' diameter					
85-1/4" to 96-1/4" with 10 treads	B1@9.98	Ea	1,640.00	211.00	1,851.00
93" to 105" with 11 treads	B1@10.7	Ea	1,790.00	227.00	2,017.00
100-3/4" to 113-3/4" with 12 treads	B1@11.3	Ea	1,940.00	239.00	2,179.00
108-1/2" to 122-1/2" with 13 treads	B1@12.0	Ea	2,090.00	254.00	2,344.00
116-1/4" to 131-1/4" with 14 risers	B1@12.6	Ea	2,230.00	267.00	2,497.00
124" to 140" with 15 risers	B1@12.8	Ea	2,390.00	271.00	2,661.00
131-3/4" to 148-3/4" with 16 risers	B1@13.5	Ea	2,530.00	286.00	2,816.00
138-1/2" to 157-1/2" with 17 risers	B1@14.2	Ea	2,680.00	301.00	2,981.00
147-1/4" to 166-1/4" with 18 risers	B1@14.8	Ea	2,830.00	314.00	3,144.00
Deduct for 4' diameter	--	%	-20.0	--	--
Deduct for 4'6" diameter	--	%	-10.0	--	--
Add for additional aluminum treads	--	Ea	149.00	--	--
Add for oak tread inserts	--	Ea	45.00	--	--
Spiral stairs, red oak, delivered to job site assembled, double handrails, 5' diameter					
74" to 82" with 9 risers	B1@16.0	Ea	2,600.00	339.00	2,939.00
82" to 90" with 10 risers	B1@16.0	Ea	2,800.00	339.00	3,139.00
90" to 98" with 11 risers	B1@16.0	Ea	2,850.00	339.00	3,189.00
98" to 106" with 12 risers	B1@16.0	Ea	2,900.00	339.00	3,239.00
106" to 114" with 13 risers	B1@20.0	Ea	2,950.00	424.00	3,374.00
114" to 122" with 14 risers	B1@20.0	Ea	3,100.00	424.00	3,524.00
122" to 130" with 15 risers	B1@20.0	Ea	3,200.00	424.00	3,624.00
Add for 6' diameter	--	%	20.0	--	--
Add for 8'6" diameter	--	%	30.0	--	--
Add for mahogany, any of above	--	Ea	150.00	--	--
Add for straight rail, unbored	B1@.060	LF	10.00	1.27	11.27
Add for straight nosing, unbored	B1@.060	LF	5.50	1.27	6.77
Add for balusters, 7/8" x 42"	B1@.334	Ea	3.25	7.08	10.33
Add for newel posts 1-7/8" x 44"	B1@.334	Ea	22.00	7.08	29.08
Spiral stairs, steel, tubular steel handrail, composition board treads, delivered unassembled, 5' diameter					
86" to 95" with 10 risers	B1@9.98	Ea	914.00	211.00	1,125.00
95" to 105" with 11 risers	B1@10.7	Ea	995.00	227.00	1,222.00
105" to 114" with 12 risers	B1@11.3	Ea	1,080.00	239.00	1,319.00
114" to 124" with 13 risers	B1@12.0	Ea	1,150.00	254.00	1,404.00
Deduct for 4' diameter	--	%	-12.0	--	--
Add for 6' diameter	--	%	24.0	--	--
Add for oak treads	--	%	25.0	--	--
Add for oak handrail	--	%	40.0	--	--

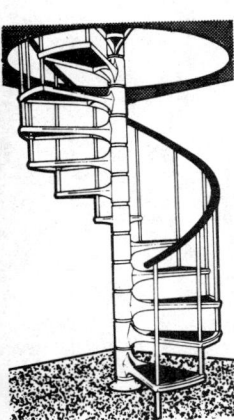

Stairs

	Craft@Hrs	Unit	Material	Labor	Total
Straight stairs, red oak, delivered to job site assembled, 36" tread, includes two 1-1/2" x 3-1/2" handrails, 7/8" round balusters					
70" to 77" with 10 risers	B1@12.0	Ea	1,300.00	254.00	1,554.00
77" to 85" with 11 risers	B1@12.0	Ea	1,400.00	254.00	1,654.00
85" to 93" with 12 risers	B1@12.0	Ea	1,500.00	254.00	1,754.00
93" to 100" with 13 risers	B1@14.0	Ea	1,600.00	297.00	1,897.00
100" to 108" with 14 risers	B1@14.0	Ea	1,800.00	297.00	2,097.00
108" to 116" with 15 risers	B1@14.0	Ea	1,900.00	297.00	2,197.00
116" to 124" with 16 risers	B1@16.0	Ea	2,100.00	339.00	2,439.00
124" to 131" with 17 risers	B1@16.0	Ea	2,200.00	339.00	2,539.00
131" to 139" with 18 risers	B1@16.0	Ea	2,400.00	339.00	2,739.00
Add for 44" tread	--	%	4.0	--	--
Add for straight rail, unbored	B1@.060	LF	10.00	1.27	11.27
Add for straight nosing	B1@.060	LF	5.50	1.27	6.77
Add for round balusters, 7/8" x 42"	B1@.334	Ea	3.25	7.08	10.33
Add for square balusters, 1-1/8" x 1-1/8" x 42"	B1@.334	Ea	4.25	7.08	11.33
Add for newel posts, 3" x 3" x 44"	B1@.334	Ea	26.00	7.08	33.08

Job-built stairways, from precut tread and riser material, closed stringers, cost per 7-1/2" rise, 36" wide, oak treads, pine risers

Straight, one flight	B1@1.99	Ea	81.50	42.20	123.70
"L" or "U" shape	B1@2.36	Ea	91.50	50.00	141.50
Deduct for pine or plywood treads	--	%	-22.0	--	--
Add for redwood lumber	--	%	20.0	--	--
Semi-circular, repetitive tract work	B1@2.52	Ea	86.00	53.40	139.40
Landings, framed and hardwood surfaced, per SF of landing surface	B1@.270	SF	13.00	5.72	18.72

Stairs, Disappearing. Bessler Stairways
Good quality folding stairway, full width main hinge, fir door panel, molded handrail, preassembled, 1" x 6" treads, 1" x 5" stringer and frame of select yellow pine, "space saver" for narrow openings, 22", 25-1/2", or 30" wide by 54" long,

8'9" or 10' height	B1@2.20	Ea	200.00	46.60	246.60

Better quality solid one-piece stringer sliding stairway, stringer and treads of clear yellow pine, unassembled, full-width piano hinge at head of stairs, with springs and cables.

Compact unit, for a 2' x 5' opening, 7'7" to 9'3" height	B1@3.33	Ea	340.00	70.60	410.60
Standard unit, for openings from 2'0" x 5'6" to 2'6" x 6'0", 7'7" to 10'10" heights	B1@3.86	Ea	369.00	81.80	450.80
Heavy-duty unit, for openings from 2'6" x 5'10" to 2'6" x 8'0", 7'7" to 12'10" heights	B1@4.56	Ea	773.00	96.60	869.60

Steam Generators
Material costs are for "Amerec" steam bath equipment including steam generator, timer switch and chrome-plated steam outlet. Labor costs include installation of steam unit and connections to enclosure, no rough plumbing or electrical included. Installed in existing, adequately sealed, tub enclosure. Maximum enclosure size.

60 cubic feet (4,000 watt generator)	P1@10.4	Ea	670.00	227.00	897.00
80 cubic feet (5,000 watt generator)	P1@10.4	Ea	690.00	227.00	917.00
135 cubic feet (6,000 watt generator)	P1@10.4	Ea	740.00	227.00	967.00
190 cubic feet (7,000 watt generator)	P1@10.4	Ea	830.00	227.00	1,057.00
270 cubic feet (8,500 watt generator)	P1@10.4	Ea	870.00	227.00	1,097.00
355 cubic feet (10,000 watt generator)	P1@10.4	Ea	940.00	227.00	1,167.00

Steam Generators

	Craft@Hrs	Unit	Material	Labor	Total
Add for individual valved hot water supply run out from hot water supply main to steam unit	--	LS	--	--	50.00
Add for thermostat	--	Ea	86.00	--	--
Add for chrome air switch control	--	Ea	120.00	--	--
Add for brass air switch control	--	Ea	143.00	--	--
Add for gold air switch control	--	Ea	145.00	--	--
Add for brass or chrome steam head	--	Ea	27.00	--	--
Add for gold steam head	--	Ea	64.60	--	--

Tarpaulins
Water resistant, with grommets, cut sizes

	Unit	Vinyl-coated nylon	Cotton duck	Canvas	Poly-ethylene
6'6" x 8'6"	Ea	36.00	--	--	--
8' x 10'	Ea	58.00	--	--	--
8'6" x 12'	Ea	82.00	--	--	--
10' x 12'	Ea	93.00	--	--	--
11'6" x 18'	Ea	158.00	--	--	--
15' x 20'	Ea	215.00	--	--	--
4'8" x 6'8"	Ea	--	--	19.00	8.00
5'8" x 7'8"	Ea	--	32.00	21.00	9.00
7'6" x 9'6"	Ea	--	47.00	35.00	13.00
9'6" x 11'6"	Ea	--	63.00	53.00	17.00
9'6" x 15'6"	Ea	--	--	68.00	24.50
15'6" x 19'6"	Ea	--	--	--	46.00

Taxes

	Craft@Hrs	Unit	Material	Labor	Total
Rule of thumb: Employer's cost for payroll taxes and insurance, expressed as a percent of total payroll. For each $1 paid in wages, the employer must pay an additional $.30 in taxes and insurance. See Insurance for specific costs on insurance coverage	--	%	--	--	30.0
Taxes, percent of total payroll					
State unemployment insurance (typical)	--	%	--	--	3.7
FICA (Social Security 6.20% and Medicare 1.45%)	--	%	--	--	7.65
Federal Unemployment Insurance (FUTA)	--	%	--	--	.80

Thresholds. See the Mouldings Section for oak threshold.
Aluminum with vinyl insert
 1-1/8" high, low rug, 3-1/4" wide

	Craft@Hrs	Unit	Material	Labor	Total
36" long	BC@.515	Ea	13.70	12.10	25.80
48" long	BC@.515	Ea	18.70	12.10	30.80
60" long	BC@.730	Ea	23.80	17.20	41.00
72" long	BC@.730	Ea	27.10	17.20	44.30

 1-1/8" high, 1-3/4" wide

	Craft@Hrs	Unit	Material	Labor	Total
30" long	BC@.515	Ea	15.50	12.10	27.60
48" long	BC@.515	Ea	22.00	12.10	34.10
60" long	BC@.730	Ea	27.30	17.20	44.50
72" long	BC@.730	Ea	31.00	17.20	48.20

Combination threshold, vinyl door shoe and drip cap
 Low rug

	Craft@Hrs	Unit	Material	Labor	Total
32" long	BC@.730	Ea	13.70	17.20	30.90
36" long	BC@.730	Ea	14.80	17.20	32.00
42" long	BC@.730	Ea	18.00	17.20	35.20
48" long	BC@.730	Ea	20.60	17.20	37.80
60" long	BC@.941	Ea	25.80	22.20	48.00
72" long	BC@.941	Ea	31.00	22.20	53.20

Thresholds

	Craft@Hrs	Unit	Material	Labor	Total
Combination threshold, vinyl door shoe and drip cap					
High rug					
32" long	BC@.730	Ea	17.30	17.20	34.50
36" long	BC@.730	Ea	19.00	17.20	36.20
42" long	BC@.730	Ea	23.20	17.20	40.40
48" long	BC@.730	Ea	26.60	17.20	43.80
60" long	BC@.941	Ea	33.10	22.20	55.30
72" long	BC@.941	Ea	39.70	22.20	61.90
Aluminum sill and interlocking threshold, low rug					
32"	BC@1.05	Ea	12.40	24.80	37.20
36"	BC@1.05	Ea	13.70	24.80	38.50
42"	BC@1.27	Ea	16.10	30.00	46.10
72"	BC@1.27	Ea	27.70	30.00	57.70
Door bottom caps, surface type, aluminum and vinyl					
32"	BC@.515	Ea	5.90	12.10	18.00
36"	BC@.515	Ea	6.00	12.10	18.10
42"	BC@.515	Ea	7.70	12.10	19.80
48"	BC@.515	Ea	8.80	12.10	20.90
Rain drip					
32"	BC@.230	Ea	3.80	5.43	9.23
36"	BC@.230	Ea	3.90	5.43	9.33
42"	BC@.230	Ea	5.30	5.43	10.73
48"	BC@.230	Ea	6.00	5.43	11.43

Tile. Figures in the material column are typical costs including 10% allowance for breakage and cutting waste. See labor costs below.

	Craft@Hrs	Unit	Material	Labor	Total
Glazed ceramic wall tile (5/16" thickness)					
2" x 2"					
Minimum quality	--	SF	2.04	--	--
Good quality	--	SF	4.50	--	--
Better quality	--	SF	7.91	--	--
3" x 3"					
Minimum quality	--	SF	1.78	--	--
Good quality	--	SF	3.93	--	--
Better quality	--	SF	7.05	--	--
4-1/2" x 4-1/2"					
Minimum quality	--	SF	2.40	--	--
Good quality	--	SF	3.93	--	--
Better quality	--	SF	5.76	--	--
6" x 6"					
Minimum quality	--	SF	1.68	--	--
Good quality	--	SF	3.82	--	--
Better quality	--	SF	6.43	--	--
8" x 8"					
Minimum quality	--	SF	2.30	--	--
Good quality	--	SF	4.30	--	--
Better quality	--	SF	7.00	--	--
Ceramic mosaic tile (1/4" thickness)					
1" hexagonal					
Minimum quality	--	SF	1.53	--	--
Good quality	--	SF	2.95	--	--
Better quality	--	SF	5.05	--	--

Tile

	Craft@Hrs	Unit	Material	Labor	Total
2" hexagonal					
Minimum quality	--	SF	1.68	--	--
Good quality	--	SF	3.78	--	--
Better quality	--	SF	6.58	--	--
3/4" penny round					
Minimum quality	--	SF	1.23	--	--
Good quality	--	SF	2.81	--	--
Better quality	--	SF	4.40	--	--
7/8" x 7/8" square					
Minimum quality	--	SF	1.95	--	--
Good quality	--	SF	4.30	--	--
Better quality	--	SF	7.50	--	--
1-1/2" x 1-1/2" square					
Minimum quality	--	SF	1.75	--	--
Good quality	--	SF	3.65	--	--
Better quality	--	SF	6.38	--	--
2" x 2" square					
Minimum quality	--	SF	2.15	--	--
Good quality	--	SF	4.60	--	--
Better quality	--	SF	7.86	--	--
Quarry tile (1/2" or 3/4" thickness)					
4" x 4", 6" x 6" or 8" x 8"					
Minimum quality	--	SF	2.05	--	--
Good quality	--	SF	4.00	--	--
Better quality	--	SF	8.40	--	--
Add for abrasive grain surfaces	--	SF	.65	--	--
Paver tile (1" thickness)					
4" x 4" or 6" x 6"					
Minimum quality	--	SF	2.60	--	--
Good quality	--	SF	6.02	--	--
Better quality	--	SF	10.50	--	--
12" x 12"					
Minimum quality	--	SF	2.50	--	--
Good quality	--	SF	5.11	--	--
Better quality	--	SF	8.85	--	--
Mexican red pavers, 12" x 12"	--	SF	1.18	--	--
Saltillo paver, 12" x 12"	--	SF	1.80	--	--

Marble and granite tile, "Martile". Note that prices for marble and granite tile will fluctuate greatly during the year.

	Craft@Hrs	Unit	Material	Labor	Total
Marble, 3/8" x 12" x 12", polished on one face	--	SF	5.95	--	--
Granite, natural, 3/8" x 12" x 12", polished on one face	--	SF	9.55	--	--
Marble, 3/4", cut to size, imported	--	SF	20.00	--	--
Granite, 3/4", cut to size, imported	--	SF	27.50	--	--
Travertine beige marble tile, 3/8" x 12" x 12", imported	--	SF	4.30	--	--
Carrera white marble tile, 3/8" x 12" x 12", imported	--	SF	4.95	--	--

Installation of tile, with organic adhesive and grout. No tile or surface preparation included. Good quality workmanship.

	Craft@Hrs	Unit	Material	Labor	Total
Countertops					
Ceramic mosaic	TL@.203	SF	.60	4.49	5.09
4-1/4" x 4-1/4" to 6" x 6" glazed	TL@.180	SF	.50	3.99	4.49

Tile

	Craft@Hrs	Unit	Material	Labor	Total
Floors					
Ceramic mosaic	TL@.121	SF	.60	2.68	3.28
4-1/4" x 4-1/4" to 6" x 6" glazed	TL@.110	SF	.50	2.44	2.94
Walls					
Ceramic mosaic	TL@.143	SF	.60	3.17	3.77
4-1/4" x 4-1/4" to 6" x 6" glazed	TL@.131	SF	.50	2.90	3.40

Installation of tile in a conventional mortar bed, with grout. No tile or surface preparation included. Good quality workmanship.

	Craft@Hrs	Unit	Material	Labor	Total
Countertops					
Ceramic mosaic	TL@.407	SF	1.60	9.01	10.61
4-1/4" x 4-1/4" to 6" x 6" glazed	TL@.352	SF	1.70	7.79	9.49
Floors					
Ceramic mosaic	TL@.238	SF	1.80	5.27	7.07
4-1/4" x 4-1/4" to 6" x 6" glazed	TL@.210	SF	1.50	4.65	6.15
Quarry or paver	TL@.167	SF	1.50	3.70	5.20
Marble or granite, 3/8" thick	TL@.354	SF	2.15	7.84	9.99
Marble or granite, 3/4" thick	TL@.591	SF	2.15	13.10	15.25
Walls					
Ceramic mosaic	TL@.315	SF	1.45	6.97	8.42
4-1/4" x 4-1/4" to 6" x 6" glazed	TL@.270	SF	1.55	5.98	7.53
Add for hand-made tile	--	%	--	15.0	--
Quarter round trim	TL@.020	LF	--	.44	--

Tile Backer Board (Durock™ or Wonderboard™). Water-resistant underlayment for ceramic tile on floors, walls, countertops and other interior wet areas. Material cost for 100 square feet (CSF) of board includes the backer board (at $1.00 per SF), 50 pounds of job mixed latex mortar for the joints and surface skim coat (at $20 per 50 pound sack), 75 lineal feet of joint tape (at $3.00 per 75' roll), 1/2 pound of 1-1/2" galvanized roofing nails (at $0.65 per pound) and 10% for waste. For scheduling purposes, estimate that a crew of 2 can install, tape and apply the skim coat on the following quantity of backer board in an 8-hour day: countertops 180 SF, floors 525 SF and walls 350 SF. Use $100.00 as a minimum cost for this type work.

1/2" thick backer board at $171.00 per CSF

	Craft@Hrs	Unit	Material	Labor	Total
Floors	T1@.030	SF	1.71	.61	2.32
Walls	T1@.045	SF	1.71	.92	2.63
Countertops	T1@.090	SF	1.71	1.84	3.55

Timber Connectors. See also Joist Hangers and Nails.

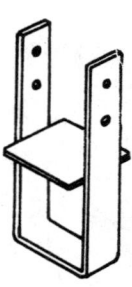

Rule of thumb: Cost of timber connectors (anchors, braces, caps, clips, hangers, and straps), based on total board footage of rough carpentry lumber in the structure

	Craft@Hrs	Unit	Material	Labor	Total
	--	MBF	21.00	--	--
Angle clips, 16 gauge, galvanized, no nails included					
3" (6 nails)	BC@.027	Ea	.40	.64	1.04
5" (6 nails)	BC@.027	Ea	.52	.64	1.16
7" (8 nails)	BC@.027	Ea	1.25	.64	1.89
9" (8 nails)	BC@.027	Ea	1.45	.64	2.09
Brick wall ties, 28 gauge, per 1,000					
3/4" x 6", corrugated	--	M	35.00	--	--
Column bases, 12 gauge, galvanized					
4" x 4"	--	Ea	4.15	--	--
4" x 6"	--	Ea	5.45	--	--
6" x 6"	--	Ea	6.65	--	--
Add for rough lumber sizes	--	%	15.0	--	--

Timber Connectors

	Craft@Hrs	Unit	Material	Labor	Total
Column bases, heavy duty					
4" x 4", 4 lbs	--	Ea	7.60	--	--
4" x 6", 5 lbs	--	Ea	8.10	--	--
4" x 8", 6 lbs	--	Ea	8.60	--	--
6" x 6", 7 lbs	--	Ea	10.90	--	--
6" x 8", 8 lbs	--	Ea	11.70	--	--
6" x 10", 9 lbs	--	Ea	12.90	--	--
6" x 12", 10 lbs	--	Ea	15.90	--	--
8" x 8", 13 lbs	--	Ea	15.90	--	--
8" x 10", 14 lbs	--	Ea	16.90	--	--
8" x 12", 15 lbs	--	Ea	18.15	--	--
10" x 10", 16 lbs	--	Ea	27.00	--	--
10" x 12", 17 lbs	--	Ea	29.50	--	--
12" x 12", 20 lbs	--	Ea	34.70	--	--
Concrete angles					
11" x 3-5/8", 2" wide	--	Ea	1.62	--	--
Concrete form ties, 18 gauge, per box of 250					
6"	--	Ea	.09	--	--
8"	--	Ea	.10	--	--
10"	--	Ea	.12	--	--
12"	--	Ea	.13	--	--
14"	--	Ea	.31	--	--
16"	--	Ea	.35	--	--
Wedges, 3-3/4", 14 gauge	--	Ea	.06	--	--
Corner braces, 12 gauge					
6" x 6", 1-1/2" wide	--	Ea	1.40	--	--
8" x 8", 2" wide	--	Ea	1.65	--	--
Delta R hangers					
2" x 6"	BC@.046	Ea	1.65	1.09	2.74
2" x 8"	BC@.046	Ea	2.05	1.09	3.14
2" x 10"	BC@.046	Ea	2.20	1.09	3.29
2" x 12"	BC@.046	Ea	2.35	1.09	3.44
2" x 14"	BC@.052	Ea	2.70	1.23	3.93
Framing anchors, 18 gauge, galvanized					
3" x 5"	BC@.046	Ea	.34	1.09	1.43
3" x 4"	BC@.046	Ea	.31	1.09	1.40
Framing hangers, box of 300					
2" x 4"	--	Ea	.40	--	--
Header hangers, 16 gauge, galvanized					
4"	BC@.046	Ea	3.90	1.09	4.99
6"	BC@.046	Ea	6.40	1.09	7.49
L straps, 12 gauge					
4" x 4", 1" wide	BC@.132	Ea	.79	3.11	3.90
6" x 6", 1-1/2" wide	BC@.132	Ea	1.40	3.11	4.51
8" x 8", 2" wide	BC@.132	Ea	1.75	3.11	4.86
12" x 12", 2" wide	BC@.165	Ea	3.35	3.89	7.24
Add for heavy duty straps, 3/16" thick	--	%	100.0	--	--
Add for heavy duty straps, 1/4" thick	--	%	150.0	--	--
Long straps, 12 gauge					
27"	BC@.165	Ea	3.50	3.89	7.39
36"	BC@.165	Ea	3.95	3.89	7.84
48"	BC@.265	Ea	4.85	6.25	11.10
60"	BC@.265	Ea	6.00	6.25	12.25

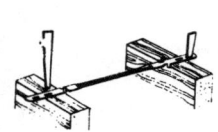

Timber Connectors

	Craft@Hrs	Unit	Material	Labor	Total
Mud sill anchors					
3-5/8"	BC@.046	Ea	.90	1.09	1.99
5-5/8"	BC@.046	Ea	1.15	1.09	2.24
Plate and stud straps, galvanized, 1-1/2" wide, 12 gauge					
6"	BC@.100	Ea	.57	2.36	2.93
9"	BC@.132	Ea	.66	3.11	3.77
12"	BC@.132	Ea	1.00	3.11	4.11
18"	BC@.132	Ea	1.21	3.11	4.32
24"	BC@.165	Ea	1.50	3.89	5.39
Plywood clips, aluminum					
3/8", 7/16", 1/2"	BC@.563	C	7.70	13.30	21.00
5/8"	BC@.563	C	8.75	13.30	22.05
Post anchors for decks, 12 gauge, galvanized					
4" x 4"	BC@.046	Ea	4.10	1.09	5.19
4" x 6"	BC@.046	Ea	5.45	1.09	6.54
6" x 4	BC@.070	Ea	6.65	1.65	8.30
6" x 6"	BC@.070	Ea	6.65	1.65	8.30
Add for rough lumber sizes	--	%	35.0	--	--
Post anchors, adjustable, 16 gauge, galvanized					
4" x 4"	BC@.046	Ea	3.85	1.09	4.94
4" x 6"	BC@.046	Ea	7.25	1.09	8.34
6" x 6"	BC@.070	Ea	11.00	1.65	12.65
8" x 8"	BC@.070	Ea	15.00	1.65	16.65
Add for rough lumber sizes	--	%	15.0	--	--
Post top tie or anchor, 18 gauge, galvanized					
4" x 4"	BC@.047	Ea	2.95	1.11	4.06
4" x 6"	BC@.047	Ea	3.50	1.11	4.61
6" x 6"	BC@.073	Ea	5.60	1.72	7.32
Post top tie plate, 16 gauge, galvanized					
4" x 4"	BC@.046	Ea	1.50	1.09	2.59
6" x 6"	BC@.046	Ea	2.05	1.09	3.14
Post cap and base combination, 18 gauge, galvanized					
4" x 4"	BC@.037	Ea	2.95	.87	3.82
4" x 6"	BC@.037	Ea	4.75	.87	5.62
6" x 6"	BC@.054	Ea	5.60	1.27	6.87
8" x 8"	BC@.054	Ea	14.85	1.27	16.12
Post caps, "P" style, 12 gauge, galvanized					
4" x 4"	BC@.052	Ea	11.15	1.23	12.38
4" x 6"	BC@.052	Ea	14.85	1.23	16.08
4" x 8"	BC@.052	Ea	20.75	1.23	21.98
6" x 4"	BC@.052	Ea	14.85	1.23	16.08
6" x 6"	BC@.077	Ea	20.25	1.82	22.07
6" x 8"	BC@.077	Ea	20.75	1.82	22.57
8" x 8"	BC@.077	Ea	25.00	1.82	26.82
Purlin hangers, without nails, heavy					
4" x 6"	BC@.052	Ea	6.90	1.23	8.13
4" x 8"	BC@.052	Ea	7.15	1.23	8.38
4" x 10"	BC@.070	Ea	7.45	1.65	9.10
4" x 12"	BC@.070	Ea	7.75	1.65	9.40
4" x 14"	BC@.087	Ea	7.75	2.05	9.80
4" x 16"	BC@.087	Ea	8.30	2.05	10.35
6" x 6"	BC@.052	Ea	10.50	1.23	11.73
6" x 8"	BC@.070	Ea	11.35	1.65	13.00

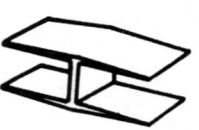

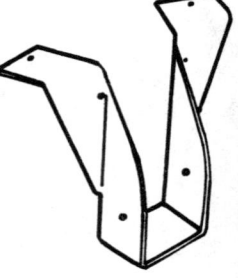

Timber Connectors

	Craft@Hrs	Unit	Material	Labor	Total
6" x 10"	BC@.087	Ea	11.90	2.05	13.95
6" x 12"	BC@.087	Ea	12.75	2.05	14.80
6" x 14"	BC@.117	Ea	13.55	2.76	16.31
6" x 16"	BC@.117	Ea	14.65	2.76	17.41
8" x 8"	BC@.070	Ea	14.00	1.65	15.65
8" x 10"	BC@.087	Ea	14.85	2.05	16.90
8" x 12"	BC@.117	Ea	15.65	2.76	18.41
8" x 14"	BC@.117	Ea	16.50	2.76	19.26
8" x 16"	BC@.117	Ea	17.30	2.76	20.06
Shear plates					
2-5/8"	BC@.068	Ea	2.20	1.60	3.80
4"	BC@.087	Ea	4.50	2.05	6.55
Split ring connectors					
2-5/8"	BC@.047	Ea	1.35	1.11	2.46
4"	BC@.054	Ea	2.10	1.27	3.37
T straps, 1/8" thick, 12 gauge					
6" x 6", 1-1/2" wide	BC@.132	Ea	1.40	3.11	4.51
8" x 8", 2" wide	BC@.132	Ea	2.05	3.11	5.16
12" x 8", 2" wide	BC@.132	Ea	2.30	3.11	5.41
12" x 12", 2" wide	BC@.132	Ea	3.35	3.11	6.46
Add for heavy-duty straps, 3/16" thick	--	%	100.0	--	--
Add for heavy-duty straps, 1/4" thick	--	%	150.0	--	--
Tie straps, no nails					
2-5/16" x 9-1/2"	BC@.046	Ea	.82	1.09	1.91
2-5/16" x 13"	BC@.046	Ea	.92	1.09	2.01
3/4" x 16-1/4"	BC@.046	Ea	.60	1.09	1.69
2-5/16" x 16-1/4"	BC@.070	Ea	1.20	1.65	2.85
2-5/16" x 23-1/4"	BC@.070	Ea	1.70	1.65	3.35
2-5/16" x 34"	BC@.087	Ea	2.55	2.05	4.60
Timber clips, 18 gauge					
2-3/4" x 1-3/8"	--	Ea	.20	--	--
5" x 1-1/4" x 1-1/4"	--	Ea	.29	--	--
5" x 2" x 1-1/4"	--	Ea	.32	--	--
7" x 2" x 2"	--	Ea	.56	--	--
9" x 2" x 2"	--	Ea	.65	--	--
11" x 2" x 2"	--	Ea	.88	--	--
13" x 2" x 2"	--	Ea	1.00	--	--
Toothed ring connectors					
2-5/8"	BC@.044	Ea	1.25	1.04	2.29
4"	BC@.047	Ea	1.40	1.11	2.51
Twist straps, 16 gauge, galvanized					
9-5/8"	--	Ea	.43	--	--
13-1/8"	--	Ea	.58	--	--
16-5/8"	--	Ea	.84	--	--
23-5/8"	--	Ea	1.12	--	--
Wall braces, diagonal, 12 gauge, 1-1/2" wide					
10' long, corrugated or flat	BC@.101	Ea	7.00	2.38	9.38
12' long, flat	BC@.101	Ea	8.50	2.38	10.88
14' long, flat	BC@.101	Ea	10.30	2.38	12.68
16' long, flat	BC@.101	Ea	12.50	2.38	14.88

Vacuum Cleaning Systems

	Craft@Hrs	Unit	Material	Labor	Total

Vacuum Cleaning Systems, Central, Subcontract
Costs listed are for better quality central vacuum power unit, piping to 4 remote outlets, hose and attachments (adequate for typical 2,000 SF home). Assumes that all vacuum tubes are installed before any interior finishes are applied.

	Craft@Hrs	Unit	Material	Labor	Total
Basic package with 23' hose	--	LS	--	--	1,000.00
Additional remote outlets	--	Ea	--	--	85.00
Add for electric or turbo power head	--	Ea	--	--	120.00

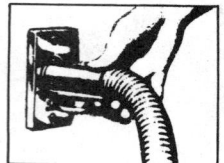

Vanities, Good Quality. No countertops, lavatory bowls or plumbing included. See also Countertops and Marble Setting.

Lavatory sink bases
  21" deep x 31-1/2" high

	Craft@Hrs	Unit	Material	Labor	Total
24" wide, 2 doors	B1@1.00	Ea	128.00	21.20	149.20
30" wide, 2 doors	B1@1.37	Ea	146.00	29.00	175.00
36" wide, 2 doors	B1@1.49	Ea	165.00	31.60	196.60
42" wide, 3 drawers over 3 doors	B1@1.65	Ea	233.00	35.00	268.00
48" wide, 3 drawers over 3 doors	B1@1.65	Ea	245.00	35.00	280.00

  18" deep x 31-1/2" high

	Craft@Hrs	Unit	Material	Labor	Total
24" wide, 2 doors	B1@1.00	Ea	128.00	21.20	149.20
30" wide, 2 doors	B1@1.37	Ea	146.00	29.00	175.00
16" deep x 31-1/2" high x 18" wide, 1 door	B1@1.00	Ea	113.00	21.20	134.20

Drawer bases
  21" deep x 31-1/2" high

	Craft@Hrs	Unit	Material	Labor	Total
12" wide, 3 drawers	B1@.703	Ea	116.00	14.90	130.90
15" wide, 3 drawers	B1@.703	Ea	122.00	14.90	136.90
18" wide, 3 drawers	B1@.845	Ea	130.00	17.90	147.90

Wallboard Partitions, Subcontract. Costs for interior partitions in one and two story buildings. Two 1/4" gypsum wallboard on 2 sides with "V-grooved" edges and plastic molding at corners fastened to 25 gauge metal framing at 24" OC, including framing, per LF of wall

	Craft@Hrs	Unit	Material	Labor	Total
Vinyl covered 8' high	--	LF	--	--	31.40
Vinyl covered 10' high	--	LF	--	--	42.10
Paintable paper-covered 8' high	--	LF	--	--	22.90
Paintable paper-covered 10' high	--	LF	--	--	33.60
Add for glued vinyl base on two sides	--	LF	--	--	1.90

Wallcoverings. Costs listed are for installation on a clean, smooth surface, and include adhesive, edge trimming, pattern matching, and normal amount of cutting and fitting and waste. Surface preparation such as patching and priming are extra. Typical roll is 8 yards long and 18" wide (approximately 36 square feet). Good to better material styles and quality.

Blank stock (underliner)

	Craft@Hrs	Unit	Material	Labor	Total
Bath, kitchen, and laundry room	P2@.668	Roll	8.25	13.80	22.05
Other rooms	P2@.558	Roll	8.25	11.60	19.85

Grasscloth or strings, vertical pattern

	Craft@Hrs	Unit	Material	Labor	Total
Bath, kitchen, and laundry rooms	P2@.995	Roll	43.30	20.60	63.90
Other rooms	P2@.830	Roll	43.30	17.20	60.50

Papers, vinyl-coated papers, trimmed, and pre-pasted

	Craft@Hrs	Unit	Material	Labor	Total
Bath, kitchen, and laundry rooms	P2@.884	Roll	18.50	18.30	36.80
Other rooms	P2@.780	Roll	18.50	16.20	34.70

Foil with or without pattern

	Craft@Hrs	Unit	Material	Labor	Total
Bath, kitchen, and laundry rooms	P2@1.12	Roll	26.00	23.20	49.20
Other rooms	P2@.844	Roll	26.00	17.50	43.50

Wallcoverings

	Craft@Hrs	Unit	Material	Labor	Total
Designer paper and hand-painted prints					
Bath, kitchen, and laundry rooms	P2@1.68	Roll	62.00	34.80	96.80
Other rooms	P2@1.41	Roll	62.00	29.20	91.20
Vinyls, no pattern					
Bath, kitchen, and laundry rooms	P2@.730	Roll	26.00	15.10	41.10
Other rooms	P2@.618	Roll	26.00	12.80	38.80
Vinyls, patterned					
Bath, kitchen, and laundry rooms	P2@.844	Roll	26.00	17.50	43.50
Other rooms	P2@.730	Roll	26.00	15.10	41.10
Vinyls, textured, no pattern, double cut seams, cloth backing	P2@.844	Roll	31.00	17.50	48.50
Coordinating borders, vinyl coated (roll is 4" to 6" x 12' to 15')	P2@.017	LF	.57	.35	.92
Add for imported papers	--	%	10.0	--	--
Add for sizing (preparatory coating, gallon covers approximately 800 SF)	--	Gal	13.40	--	--
Wallpaper adhesive (gallon covers approximately 300 SF)	--	Gal	11.30	--	--
Add for jobs $100 or less	--	%	--	--	10.0
Add heavy cut and trim jobs	--	%	--	--	10.0
Deduct for jobs over 50 rolls (2,000 SF)	--	%	--	--	-5.0
Deduct for jobs over 100 rolls (4,000 SF)	--	%	--	--	-10.0
Deduct for jobs over 200 rolls (8,000 SF)	--	%	--	--	-15.0

Waterproofing

Below grade waterproofing, applied to concrete or masonry walls, no surface preparation, excavation, or backfill included, average job

	Craft@Hrs	Unit	Material	Labor	Total
Bentonite waterproofing, one layer, nailed in place	BL@.081	SF	.93	1.52	2.45
Cementitious waterproofing with protection board, spray or brush applied, 2 coats	BL@.072	SF	.67	1.35	2.02
Crystalline waterproofing, spray or brush applied, 3 coats	BL@.110	SF	.39	2.07	2.46
Elastomeric waterproofing, sprayed, troweled, or rolled, 2 coats, 30 mm per coat	BL@.063	SF	.64	1.18	1.82

Above grade waterproofing, applied to smooth concrete or plywood deck, no surface preparation included

	Craft@Hrs	Unit	Material	Labor	Total
Between slabs membrane, sprayed, troweled or rolled, 2 coats	PT@.038	SF	.64	.86	1.50
Pedestrian walking surface, elastomeric membrane applied 4 coats thick with aggregate	PT@.065	SF	1.10	1.47	2.57
Chloroprene/Neoprene latex deck surfacing	PT@.144	SF	1.63	3.26	4.89

Weatherstripping Materials

Door bottom seals

	Craft@Hrs	Unit	Material	Labor	Total
Spring brass in pinch grip carrier (Kel-eez)					
32" x 1-3/4"	BC@.250	Ea	7.00	5.90	12.90
36" x 1-3/4"	BC@.250	Ea	7.90	5.90	13.80
Polyflex PVC, self adhesive, 8' lengths	BC@.250	Ea	3.80	5.90	9.70

Weatherstripping Materials

	Craft@Hrs	Unit	Material	Labor	Total
Door frame seal sets					
Spring bronze, nailed every 1 inch					
36" x 6'8" door	BC@.500	Ea	12.20	11.80	24.00
36" x 7'0" door	BC@.500	Ea	12.70	11.80	24.50
Flexible PVC (Draftshield)	BC@.084	LF	.51	1.98	2.49
Jamb side stripping, metal nailed every 1 inch, with rubber bead					
36" x 6'8" door, aluminum stripping	BC@.500	Ea	12.50	11.80	24.30
36" x 6'8" door, bronze stripping	BC@.500	Ea	17.00	11.80	28.80
Window seal sets					
Spring bronze, nailed every 1/2 or 1 inch					
Double hung window, 2 sides, head and sill, 18 LF, including removal of window sash	BC@.968	Ea	15.00	22.80	37.80
Casement window, 2 sides, head and sill, coil strip, 18 LF	BC@.500	Ea	9.00	11.80	20.80
Flexible PVC (Draftshield)	BC@.084	LF	.51	1.98	2.49
Garage door top and side seal					
Nylon gasket in aluminum frame (Kel-eez)					
9' x 7' door	BC@1.00	Ea	26.00	23.60	49.60
16' x 7' door	BC@1.25	Ea	31.00	29.50	60.50
Flexible PVC (Draftshield)	BC@.050	LF	1.06	1.18	2.24
Garage door bottom seal					
Soft rubber, set in adhesive					
9'	BC@.383	Ea	16.00	9.04	25.04
16'	BC@.500	Ea	27.00	11.80	38.80
Flexible PVC (Draftshield)	BC@.050	LF	1.06	1.18	2.24

Well Drilling, Subcontract. Costs are based on well depth and include welded steel casing (or continuous PVC casing). The amount of casing needed depends on soil conditions. Some types of rock require casings at the well surface only. Your local health department can probably provide information on the average well depth and subsurface conditions in the community. Note that unforeseen subsurface conditions can increase drilling costs substantially. Most subcontractors will estimate costs on an hourly basis.

	Craft@Hrs	Unit	Material	Labor	Total
Well hole with 4" ID PVC casing	--	LF	--	--	21.50
Well hole with 6" ID steel casing	--	LF	--	--	31.00
Well hole with 8" ID steel casing	--	LF	--	--	42.00
6" well hole in sturdy rock (no casing)	--	LF	--	--	12.50
Add per well for 6" diameter bottom filter screen where needed due to subsurface conditions					
Stainless steel	--	LF	--	--	95.00
Low carbon steel	--	LF	--	--	60.00
Add per well for drive shoe	--	LS	--	--	80.00
Add per well for surface seal (20' to 50' depth) where needed or required by law or site conditions	--	LS	--	--	400.00
Add per well for well drilling permit (fee will vary, typical fee in most municipalities.)	--	LS	--	--	200.00
Add for automatic electric pumping system (domestic use), including 87 gallon pressure tank, installed (no electrical hookup included), typical cost	--	LS	--	--	3,000.00
Add for electrical service drop pole	--	LS	--	--	750.00

Window Sills. Cultured marble, 1/2" radius front edge

	Craft@Hrs	Unit	Material	Labor	Total
3-1/2" sill depth, 2" x 4" stud walls	BS@.118	LF	3.00	2.47	5.47
4-7/8" sill depth, 2" x 6" stud walls	BS@.131	LF	3.30	2.74	6.04
5-1/2" sill depth	BS@.147	LF	4.00	3.07	7.07

Window

	Craft@Hrs	Unit	Material	Labor	Total

Window Treatments. Blinds and shades, stock sizes except as noted. See also Draperies.

Mini-blinds

	Craft@Hrs	Unit	Material	Labor	Total
Aluminum, 1"	BC@.050	SF	3.33	1.18	4.51
Aluminum, 1", tempered slats, custom sizes and colors	BC@.050	SF	4.70	1.18	5.88
Fabric, vertical, 3-1/2" wide, fire retardant, custom sizes and colors	BC@.083	SF	4.15	1.96	6.11
PVC, vertical, 3-1/2" wide, fire retardant, custom sizes and colors	BC@.075	SF	3.02	1.77	4.80

Polyester fabric blinds

Continuous pleated	BC@.066	SF	3.20	1.56	4.76
Wood blinds, 1" prefinished	BC@.075	SF	6.20	1.77	7.97
Vertical blinds, 3-1/2" vanes	BC@.083	SF	5.00	1.96	6.96

Steel blinds

1" enamel finish	BC@.050	SF	3.60	1.18	4.78
2" white enamel only	BC@.045	SF	2.55	1.06	3.61

Traverse rods for draperies

Steel	BC@.075	LF	4.42	1.77	6.19
Extruded aluminum, hand operated	BC@.083	LF	6.65	1.96	8.61
Extruded aluminum, cord operated	BC@.083	LF	7.75	1.96	9.71

Windows

Awning windows, aluminum Glazed, with hardware, no trim or surround needed, weatherstripped, natural finish, DSB glass with fiberglass screens, overall dimensions.

	Width				Mull bars
	19"	26"	37"	53"	
26" high, 2 lites high	43.00	52.50	60.00	72.00	5.10
38-1/2" high, 3 lites high	51.50	61.50	68.50	83.00	6.50
50-1/2" high, 4 lites high	61.50	78.50	82.00	105.00	9.10
63" high, 5 lites high	71.00	95.00	109.00	125.00	8.50

	Craft@Hrs	Unit	Material	Labor	Total
Add for storm sash, per SF of glass	--	SF	6.00	--	--
Add for enamel finish	--	%	10.0	--	--

Labor setting aluminum awning windows, no carpentry included

To 8 SF opening size	BG@1.00	Ea	--	22.00	--
Over 8 SF to 16 SF opening size	BG@1.50	Ea	--	33.00	--
Over 16 SF opening size	BG@2.00	Ea	--	43.90	--

Awning windows, pine Treated and primed, includes weatherstripping and exterior trim, glazed with 3/4" insulating glass, no drip cap, 1-3/4" sash, frame for 4-9/16" wall, each unit has 1 ventilating lite for each lite of width, overall dimensions

	One 32" lite wide	Two 32" lites wide	Three 32" lites wide
20" high, 1 lite high	171.00	305.00	440.00
40" high, 2 lites high	240.00	444.00	663.00
60" high, 3 lites high	353.00	631.00	1,530.00
24" high, 1 lite high	162.00	342.00	513.00
48" high, 2 lites high	296.00	572.00	861.00
72" high, 3 lites high	417.00	820.00	1,230.00
30" high, 1 lite high	177.00	337.00	492.00
60" high, 2 lites high	268.00	500.00	738.00
90" high, 3 lites high	375.00	705.00	1,050.00

Windows

	One 40" lite wide	Two 40" lites wide	Three 40" lites wide
20" high, 1 lite high	180.00	355.00	525.00
40" high, 2 lites high	300.00	575.00	860.00
60" high, 3 lites high	415.00	815.00	1,200.00
24" high, 1 lite high	195.00	375.00	560.00
48" high, 2 lites high	325.00	635.00	955.00
72" high, 3 lites high	450.00	910.00	1,350.00
30" high, 1 lite high	205.00	405.00	600.00
60" high, 2 lites high	355.00	685.00	1,050.00
90" high, 3 lites high	500.00	995.00	1,450.00

	One 48" lite wide	Two 48" lites wide	Three 48" lites wide
20" high, 1 lite high	155.00	275.00	435.00
40" high, 2 lites high	230.00	440.00	650.00
60" high, 3 lites high	315.00	600.00	910.00
24" high, 1 lite high	155.00	300.00	450.00
48" high, 2 lites high	265.00	570.00	770.00
72" high, 3 lites high	415.00	730.00	1,100.00
30" high, 1 lite high	175.00	315.00	495.00
60" high, 2 lites high	295.00	575.00	860.00
90" high, 3 lites high	415.00	825.00	1,230.00

Window Index

Item	Page
Aluminum	217-21
Awning windows	213-12
Bay windows	214-21
Double hung	217-21
Louvered	21
Picture windows	21
Window treatments	21

	Craft@Hrs	Unit	Material	Labor	Total
Deduct for awning windows with single sheet					
DSB glass	--	%	-25.0	--	--
Add for screen, per lite					
32" wide lites	--	Ea	5.05	--	--
40" wide lites	--	Ea	6.25	--	--
48" wide lites	--	Ea	6.65	--	--
Add for storm panels					
32" wide lites	--	Ea	21.50	--	--
40" wide lites	--	Ea	23.00	--	--
48" wide lites	--	Ea	24.50	--	--
Add for operating sash	--	Ea	34.00	--	--
Labor installing wood awning windows, no carpentry included					
To 8 SF opening size	BG@1.00	Ea	--	22.00	--
Over 8 SF to 16 SF opening size	BG@1.50	Ea	--	33.00	--
Over 16 SF to 32 SF opening size	BG@2.00	Ea	--	43.90	--
Over 32 SF to 48 SF opening size	BG@3.00	Ea	--	65.90	--
Over 48 SF opening size	BG@5.00	Ea	--	110.00	--

Bay windows Labor is for installing the window, no carpentry included.
Angle bay double hung units, 30 degree angle, pine, assembled, aluminum cladding, hardware, 2 ventilating sash, weatherstripped, double hung sides with picture center sash, insulated glass, white or bronze finish, 4-9/16" wall thickness, overall unit dimensions, 4'8" high

	Craft@Hrs	Unit	Material	Labor	Total
6'3"	BG@4.00	Ea	700.00	87.90	787.90
6'7"	BG@4.00	Ea	710.00	87.90	797.90
7'6"	BG@4.50	Ea	775.00	98.90	873.90
8'4"	BG@4.50	Ea	765.00	98.90	863.90
8'11"	BG@4.50	Ea	780.00	98.90	878.90
9'6"	BG@5.00	Ea	800.00	110.00	910.00
10'3"	BG@5.00	Ea	880.00	110.00	990.00

Windows

	Craft@Hrs	Unit	Material	Labor	Total
Add for 45 degree angle windows	--	Ea	10.00	--	--
Add for screens or grilles	--	Ea	12.50	--	--

Casement bow window unit, glazed with 1/2" insulated glass, assembled, screens, pine, fixed sash, 22" x 56" lites, weatherstripped, unfinished, 16" projection, 4-7/16" jambs, listed by actual unit dimensions

	Craft@Hrs	Unit	Material	Labor	Total
7'8" wide x 4'10" (4 lites, none venting)	BG@4.75	Ea	475.00	104.00	579.00
7'8" wide x 4'10" (4 lites, 2 venting)	BG@4.75	Ea	740.00	104.00	844.00
9'5" wide x 4'10" (5 lites, none venting)	BG@4.75	Ea	615.00	104.00	719.00
9'5" wide x 4'10" (5 lites, 2 venting)	BG@4.75	Ea	885.00	104.00	989.00
8'4" wide x 5'0" (4 lites, 2 venting)	BG@4.75	Ea	760.00	104.00	864.00
10'3" wide x 5'0" (5 lites, 2 venting)	BG@4.75	Ea	990.00	104.00	1,094.00
Add for interior and exterior trim					
3 or 4 lites	BG@2.00	Set	21.00	43.90	64.90
5 lites	BG@2.00	Set	25.00	43.90	68.90
Add for roof framing kit, 5 lites	--	Ea	84.00	--	--

Divided lite bow units, pine, insulated glass, 22" x 18", assembled, with hardware and screen, opening sizes, fixed sash, overall unit dimensions

	Craft@Hrs	Unit	Material	Labor	Total
7'8" x 4'9" (4 lites long x 3 lites high)	BG@4.75	Ea	530.00	104.00	634.00
Add for interior trim	BG@2.00	Set	21.00	43.90	64.90
Add for each ventilating sash	--	Ea	74.00	--	--
Add for roof framing kit	--	Ea	63.50	--	--

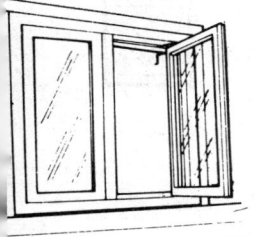

Casement windows, wood 3/4" insulated glass, with hardware, primed wood exterior frames and sash, weatherstripped, with drip cap, overall unit dimensions

15" wide glass		3'4"	4'0"	5'0"	6'0"
20" 1 ventilating lite		138.00	160.00	182.00	210.00
40" 2 ventilating lites		265.00	310.00	354.00	409.00
60" 2 ventilating, 1 fixed		380.00	450.00	512.00	595.00
80" 2 ventilating, 2 fixed		490.00	600.00	680.00	790.00
100" 3 ventilating, 2 fixed		615.00	750.00	850.00	990.00
19" wide glass	2'8"	3'4"	4'0"	5'0"	6'0"
24" 1 ventilating lite	140.00	151.00	165.00	187.00	219.00
48" 2 ventilating lites	272.00	292.00	320.00	365.00	429.00
72" 2 ventilating, 1 fixed	383.00	415.00	455.00	519.00	615.00
96" 2 ventilating, 2 fixed	500.00	545.00	597.00	680.00	807.00
120" 3 ventilating, 2 fixed	640.00	687.00	753.00	858.00	1,020.00
25" wide glass		3'4"	4'0"	5'0"	6'0"
30" 1 ventilating lite		164.00	180.00	215.00	230.00
60" 2 ventilating lites		316.00	355.00	419.00	450.00
90" 2 ventilating, 1 fixed		450.00	509.00	600.00	645.00
120" 2 ventilating, 2 fixed		592.00	670.00	788.00	850.00
150" 3 ventilating, 2 fixed		745.00	843.00	991.00	1,070.00

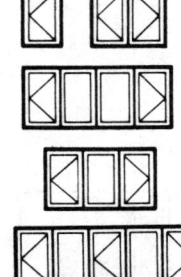

	Craft@Hrs	Unit	Material	Labor	Total
Add for screens					
32" height	--	Ea	6.10	--	--
40" height	--	Ea	6.50	--	--
48" height	--	Ea	7.55	--	--
60" height	--	Ea	9.00	--	--
72" height	--	Ea	9.35	--	--
Add for storm panels					
32" height	--	Ea	19.55	--	--
40" height	--	Ea	21.65	--	--

Windows

	Craft@Hrs	Unit	Material	Labor	Total
48" height	--	Ea	24.70	--	--
60" height	--	Ea	30.90	--	--
72" height	--	Ea	38.10	--	--
Deduct for single sheet glazing, SF of glass	--	SF	-2.85	--	--
Labor setting casement windows, no carpentry included.					
To 8 SF	BG@1.00	Ea	--	22.00	--
Over 8 SF to 16 SF	BG@1.50	Ea	--	33.00	--
Over 16 SF	BG@2.00	Ea	--	43.90	--

Double hung wood windows With primed exterior aluminum cladding, no screens, white or bronze finish, 1/2" insulating glass, weatherstripped, two lites per sash, listed by overall size (width x length). Labor is for installing the window, no carpentry included.

	Craft@Hrs	Unit	Material	Labor	Total
1'10" x 4'8" (16" x 24" lite)	BG@1.00	Set	200.00	22.00	222.00
2'2" x 3'4" (20" x 16" lite)	BG@1.00	Set	175.00	22.00	197.00
2'2" x 4'4" (20" x 24" lite)	BG@1.00	Set	210.00	22.00	232.00
2'2" x 5'4" (20" x 28" lite)	BG@1.50	Set	230.00	33.00	263.00
2'2" x 6'4" (20" x 34" lite)	BG@1.50	Set	265.00	33.00	298.00
2'6" x 3'4" (24" x 16" lite)	BG@1.00	Set	190.00	22.00	212.00
2'6" x 4'0" (24" x 20" lite)	BG@1.50	Set	210.00	33.00	243.00
2'6" x 4'8" (24" x 24" lite)	BG@1.50	Set	230.00	33.00	263.00
2'10" x 3'4" (28" x 16" lite)	BG@1.00	Set	200.00	22.00	222.00
2'10" x 4'0" (28" x 20" lite)	BG@1.50	Set	225.00	33.00	258.00
2'10" x 4'8" (28" x 24" lite)	BG@1.50	Set	240.00	33.00	273.00
2'10" x 5'4" (28" x 28" lite)	BG@1.50	Set	265.00	33.00	298.00
3'2" x 3'4" (32" x 16" lite)	BG@1.50	Set	215.00	33.00	248.00
3'2" x 4'0" (32" x 20" lite)	BG@1.50	Set	240.00	33.00	273.00
3'2" x 4'8" (32" x 24" lite)	BG@1.50	Set	265.00	33.00	298.00
3'2" x 5'4" (32" x 28" lite)	BG@1.50	Set	280.00	33.00	313.00
3'6" x 4'0" (36" x 20" lite)	BG@1.50	Set	255.00	33.00	288.00
3'6" x 4'8" (36" x 24" lite)	BG@2.00	Set	280.00	43.90	323.90
3'6" x 5'4" (36" x 28" lite)	BG@2.25	Set	295.00	49.40	344.40
Add for interior trim set, per SF of opening	--	SF	1.20	--	--
Labor to install interior trim set, any size opening listed above	BG@1.00	Set	--	22.00	--
Add for removable grille (Colonial), per SF of opening	--	SF	2.80	--	--
Deduct for single sheet glazing, SF of glass	--	SF	-3.00	--	--
Add for better quality aluminum or vinyl clad double hung wood windows, 7/16" insulating glass	--	%	40.0	--	--
Add for white aluminum screen with fiberglass screen cloth, per SF of glass	--	SF	1.75	--	--

Picture windows Listed by overall rough opening sizes. Labor is for installing the window, no carpentry included.

Picture windows with two double hung flankers, includes hardware and screens, primed, weatherstripped. Glazed with 1/2" insulated glass

	Craft@Hrs	Unit	Material	Labor	Total
8'3" x 4'8" (48" x 49" center sash)	BG@2.75	Ea	525.00	60.40	585.40
8'11" x 4'8" (48" x 46" center sash)	BG@2.75	Ea	545.00	60.40	605.40
9'4" x 4'8" (48" x 57" center sash)	BG@2.75	Ea	565.00	60.40	625.40
Add for 5'4" window height	--	%	12.0	15.0	--

Picture windows with two casement flankers, weatherstripped, with hardware, casing, and moulding, unit dimension sizes. Glazed with 3/4" insulated center sash and 1/2" insulated casements

Windows

		Width			
Height	7'4"	8'0"	8'4"	9'0"	10'0"
3'4"	505.00	530.00	525.00	550.00	575.00
4'0"	590.00	545.00	570.00	610.00	650.00
5'0"	625.00	660.00	680.00	745.00	820.00
6'0"	--	--	--	885.00	--

	Craft@Hrs	Unit	Material	Labor	Total
Add for screen, per flanker	--	Ea	7.80	--	--
Labor to hang and fit picture casement windows, no carpentry included					
To 8 SF opening size	BG@1.00	Ea	--	22.00	--
Over 8 SF to 16 SF opening size	BG@1.50	Ea	--	33.00	--
Over 16 SF to 32 SF opening size	BG@2.00	Ea	--	43.90	--
Over 32 SF opening size	BG@2.50	Ea	--	54.90	--

Sliding aluminum windows including bronze or white finish and hardware. One horizontal rolling sash and one stationary sash, double glazed.

			Width			
Height	2'0"	3'0"	3'4"	4'0"	5'0"	6'0"
1'0"	--	36.00	--	42.00	--	--
1'6"	--	41.00	--	47.00	--	--
2'0"	39.00	46.00	49.00	53.00	60.00	67.00
2'6"	--	52.00	54.00	53.00	66.00	74.00
3'0"	49.00	57.00	60.00	65.00	73.00	81.00
3'6"	--	62.00	66.00	71.00	80.00	89.00
4'0"	--	68.00	71.00	77.00	87.00	95.00
5'0"	--	86.00	--	77.00	113.00	125.00

Two horizontal rolling sashes and one stationary center sash, double glazed

			Width		
Height	6'0"	7'0"	8'0"	9'0"	10'0"
2'0"	82.00	89.00	95.00	--	157.00
2'6"	91.00	--	106.00	--	--
3'0"	99.00	108.00	117.00	131.00	139.00
3'6"	--	124.00	134.00	149.00	159.00
4'0"	125.00	134.00	144.00	161.00	167.00
5'0"	157.00	168.00	179.00	--	225.00

Single and double hung aluminum windows, bronze or white finish, vertical slide with screen, double glazed

			Width			
Height	1'6"	2'0"	2'6"	3'0"	3'6"	4'0"
Single hung, two lites						
2'0"	--	44.00	--	54.00	--	--
2'6"	--	48.00	--	--	--	--
3'0"	45.00	51.00	57.00	63.00	--	75.00
3'6"	--	55.00	61.00	67.00	73.00	81.00
4'0"	--	59.00	65.00	71.00	78.00	84.00
4'6"	--	62.00	69.00	75.00	82.00	90.00
5'0"	--	66.00	73.00	80.00	87.00	95.00
6'0"	--	74.00	81.00	88.00	--	--
Double hung, two lites						
2'0"	--	76.00	--	85.00	--	--
2'6"	--	82.00	--	--	--	--
3'0"	79.00	89.00	98.00	108.00	--	128.00
3'6"	--	94.00	105.00	115.00	125.00	136.00
4'0"	--	101.00	112.00	123.00	134.00	145.00

Windows

Height	1'6"	2'0"	2'6"	Width 3'0"	3'6"	4'0"
4'6"	--	108.00	119.00	130.00	142.00	155.00
5'0"	--	114.00	126.00	137.00	150.00	163.00
6'0"	--	126.00	139.00	152.00	--	--

Louvered aluminum windows With 4" glass louvers, including hardware

Height	1'6"	2'0"	2'6"	Width 3'0"	3'6"
1'0"	16.00	18.00	20.00	22.00	25.00
1'4"	21.00	23.00	27.00	29.00	32.00
1'6", 8"	23.00	27.00	31.00	34.00	38.00
1'10"	28.00	31.00	37.00	40.00	46.00
2'0"	31.00	35.00	41.00	45.00	52.00
2'4", 6"	34.00	39.00	46.00	51.00	58.00
2'8", 10"	41.00	46.00	55.00	60.00	68.00
3'0"	44.00	50.00	59.00	65.00	74.00
3'4", 6"	48.00	55.00	65.00	71.00	81.00
4'0"	55.00	63.00	75.00	82.00	94.00
4'6"	63.00	72.00	85.00	94.00	107.00
5'0"	71.00	81.00	96.00	106.00	121.00
5'6"	76.00	87.00	103.00	113.00	129.00
6'0"	84.00	96.00	113.00	125.00	143.00

	Craft@Hrs	Unit	Material	Labor	Total
Add for bronze finish on aluminum window frames	--	%	15.0	--	--
Deduct for single glazing on aluminum windows, per SF of glass	--	SF	-3.55	--	--
Labor setting aluminum windows, no carpentry included					
To 8 SF opening size	BG@1.00	Ea	--	22.00	--
Over 8 SF to 16 SF opening size	BG@1.50	Ea	--	33.00	--
Over 16 SF to 32 SF opening size	BG@2.00	Ea	--	43.90	--
Over 32 SF opening size	BG@3.00	Ea	--	65.90	--

Construction Economics Division
Construction Cost Index for New Single Family Homes

Period	% of 1967 Cost	$ Per SF of Floor	Period	% of 1967 Cost	$ Per SF of Floor	Period	% of 1967 Cost	$ Per SF of Floor
1974			**1980**			**1986**		
1st quarter	152.0	21.28	1st quarter	297.6	41.65	1st quarter	376.3	52.34
2nd quarter	157.3	22.02	2nd quarter	300.2	42.01	2nd quarter	383.5	53.67
3rd quarter	161.4	22.60	3rd quarter	302.6	42.34	3rd quarter	385.5	53.95
4th quarter	163.9	22.50	4th quarter	312.6	43.74	4th quarter	386.1	54.03
1975			**1981**			**1987**		
1st quarter	170.8	23.91	1st quarter	318.1	44.51	1st quarter	387.1	54.17
2nd quarter	172.5	24.15	2nd quarter	325.7	45.58	2nd quarter	388.2	54.32
3rd quarter	174.6	24.44	3rd quarter	331.5	46.30	3rd quarter	392.3	54.89
4th quarter	179.4	25.12	4th quarter	338.8	47.41	4th quarter	398.6	55.78
1976			**1982**			**1988**		
1st quarter	184.2	25.79	1st quarter	336.9	47.14	1st quarter	398.6	55.78
2nd quarter	189.5	26.53	2nd quarter	339.3	47.48	2nd quarter	399.6	55.92
3rd quarter	194.1	27.14	3rd quarter	342.3	47.90	3rd quarter	402.8	56.37
4th quarter	228.6	31.99	4th quarter	344.3	48.18	4th quarter	407.1	56.97
1977			**1983**			**1989**		
1st quarter	205.1	28.70	1st quarter	350.8	49.09	1st quarter	407.5	57.02
2nd quarter	212.9	29.79	2nd quarter	354.3	49.57	2nd quarter	411.9	57.64
3rd quarter	216.1	30.24	3rd quarter	363.9	50.92	3rd quarter	422.6	59.13
4th quarter	228.6	31.99	4th quarter	370.1	51.79	4th quarter	429.0	60.02
1978			**1984**			**1990**		
1st quarter	230.7	32.28	1st quarter	372.0	52.06	1st quarter	432.4	60.49
2nd quarter	241.9	33.85	2nd quarter	373.0	52.20	2nd quarter	434.2	60.74
3rd quarter	248.6	34.79	3rd quarter	375.1	52.49	3rd quarter	435.3	60.89
4th quarter	259.8	36.36	4th quarter	372.6	52.14	4th quarter	435.2	60.74
1979			**1985**			**1991**		
1st quarter	262.0	36.66	1st quarter	367.8	51.47	1st quarter	433.0	60.62
2nd quarter	278.5	38.97	2nd quarter	374.1	52.35	2nd quarter	432.6	60.56
3rd quarter	286.3	40.06	3rd quarter	376.2	52.64	3rd quarter	433.2	60.65
4th quarter	289.7	40.54	4th quarter	374.0	52.34	4th quarter	--	--

The figures under the column "$ Per SF of Floor" show construction costs for building a good quality home in a suburban area under competitive conditions in each calendar quarter since 1974. This home is described below under the section "Residential Rule of Thumb." These costs include the builder's overhead and profit and a 450 square foot garage but no basement. The cost of a finished basement per square foot will be approximately 40% of the square foot cost of living area. If the garage area is more than 450 square feet, use 50% of the living area cost to adjust for the larger or smaller garage. To find the total construction cost of the home, multiply the living area (excluding the garage) by the cost in the column "$ per SF of Floor."

Deduct for rural areas	5.0%
Add for 1,800 SF house (better quality)	4.0
Add for 2,000 SF house (better quality)	3.0
Deduct for over 2,400 SF house	3.0
Add for split level house	3.0
Add for 3-story house	10.0
Add for masonry construction	9.0

Construction costs are higher in some cities and lower in others. Square foot costs listed in the table above are national averages. To modify these costs to your job site, multiply by the appropriate area modification factor from either page 10 or 11 of this manual. But note that area modifications on pages 10 and 11 are based on recent construction and may not apply to work that was completed many years ago.

Residential Rule of Thumb

Construction Costs

The following figures are percentages of total construction cost for a good quality single family residence: 1,600 square foot single story non-tract 3-bedroom, 1¾ bath home plus attached two car (450 SF) garage. Costs assume a conventional floor, stucco exterior, wallboard interior, shake roof, single fireplace, forced air heat, copper water supply, ABS drain lines, range with double oven, disposer, dishwasher and 900 SF of concrete flatwork.

Item	Percent	Item	Percent	Item	Percent	Item	Percent	Item	Percent
Excavation	1.2	Finish lumber	.5	Painting	3.6	Doors	1.9	Wiring (Romex)	3.2
Flatwork (drive & walk)	2.4	Rough carpentry labor	8.5	Shower & tub enclosure	.5	Garage door	.4	Lighting fixtures	1.2
Foundation, slab, piers	3.7	Finish carpentry labor	1.7	Prefabricated fireplace	.9	Alum. windows, door	1.2	Insurance, payroll tax	2.8
Brick hearth & veneer	.7	Countertops	1.5	Bath accessories	.7	Exterior stucco	6.4	Plans and specs	.4
Rough hardware	.3	Cabinets	3.7	Built-in appliances	1.6	Gypsum wallboard	4.4	Permits & utilities	1.7
Finish hardware	.2	Insulation (R19 ceiling)	2.3	Heating and plumbing	4.9	Resilient flooring	2.0	Final cleanup	.4
Rough lumber	7.4	Roofing	5.5	Plumb. & sewer conn.	7.3	Carpeting	2.4	Overhead & profit	12.5

Selling Price

Construction costs are usually about 50% of the sale price of a house and lot. The following percentages of selling price are typical for single or two family residences: material 34%, on site labor 16%, land and site improvements 25%, sales and marketing 5%, financing 8%, overhead 7%, profit 5%.

Abbreviations

AASHO	American Assn. of State Highway Officials	FAA	Federal Aviation Administration	OC	spacing from center to center
ABS	acrylonitrile butadiene styrene	FICA	Federal Insurance Contributions Act (Social Security, Medicare tax)	OD	outside diameter
AC	alternating current			OS & Y	outside screw & yoke
AISC	American Institute of Steel Construction Inc.			oz	ounce
APP	attactic polypropylene	FOB	freight on board	perf	perforated
ASHRAE	American Society of Heating, Refrigerating and Air Conditioning Engineers	FPM	feet per minute	Pr	pair
		FRP	fiberglass reinforced plastic	PSF	pounds per square foot
		FS	Federal Specification	PSI	pounds per square inch
		FUTA	Federal Unemployment Compensation Act Tax	PV	photovoltaic
ASME	American Society of Mechanical Engineers			PVC	polyvinyl chloride
		Gal	gallon	Qt	quart
ASTM	American Society for Testing Materials	GFCI	ground fault circuit interruptor	R	thermal resistance
				R/L	random length(s)
AWPA	American Wood Products Association	GPH	gallon(s) per hour	R/W/L	random widths and lengths
		GPM	gallon(s) per minute	RPM	revolutions per minute
AWWA	American Water Works Association	H	height	RSC	rigid steel conduit
		HP	horsepower	S1S2E	surfaced 1 side, 2 edges
Ba	bay	Hr(s)	hour(s)	S2S	surfaced 2 sides
Bdle	bundle	IMC	intermediate metal conduit	S4S	surfaced 4 sides
BF	board foot	ID	Inside diameter	Sa	sack
BHP	boiler horsepower	KD	kiln dried or knocked down	SBS	styrene butyl styrene
Btr	better	KSI	kips per square inch	SDR	size to diameter ratio
Btu	British thermal unit	KV	kilovolt(s)	SF	square foot
B & W	black & white	KVA	1,000 volt amps	SFCA	square feet of form in contact with concrete
C	thermal conductance	kw	kilowatt(s)		
C	one hundred	kwh	kilowatt hour	Sq	100 square feet
CF	cubic foot	L	length	SSB	single strength B quality glass
CFM	cubic feet per minute	Lb(s)	pound(s)	STC	sound transmission class
CLF	100 linear feet	LF	linear foot	Std	standard
cm	centimeter	LP	liquified propane	SY	square yard
CPE	chlorinated polyethylene	LS	lump sum	T	thick
CPM	cycles per minute	M	one thousand	T&G	tongue & groove edge
CPVC	chlorinated polyvinyl chloride	Mb	million bytes (characters)	TV	television
		MBF	1,000 board feet	UBC	Uniform Building Code
CSPE	chloro sulphinated polyethylene	MBtu	1,000 British thermal units	UL	Underwriter's Laboratory
		MCM	1,000 circular mils	USDA	United States Dept. of Agriculture
CSF	100 square feet	MDO	medium density overlaid		
CY	cubic yard	MH	manhour	VLF	vertical linear foot
d	penny	Mi	mile	W	width
D	depth	MLF	1,000 linear feet	Wk	week
DC	direct current	MPH	miles per hour	W/	with
dia	diameter	mm	millimeter(s)	x	by or times
DSB	double strength B quality glass	Mo	month		
		MSF	1,000 square feet		**Symbols**
DWV	drain, waste, vent piping	NEMA	National Electrical Manufacturer's Association	/	per
Ea	each			—	through or to
EMT	electric metallic tube			@	at
EPDM	ethylene propylene diene monomer	NFPA	National Fire Protection Association	%	per 100 or percent
				$	U.S. dollars
equip.	equipment	No.	number	'	feet
exp.	exposure	NRC	noise reduction coefficient	"	inches
F	Fahrenheit			#	pound or number

Industrial and Commercial Division
Hourly Labor Costs

The hourly labor costs shown in the column headed "Hourly Cost Including Subcontractor's 30% Markup" have been used to compute the manhour costs for the crews on pages 5 to 7 and the costs in the "Labor" column on pages 222 to 552 of this book. All figures are in U.S. dollars per hour.

"Hourly Wage and Benefits" includes the wage, welfare, pension, vacation, apprentice and other mandatory contributions. The "Employer's Burden" is the cost to the contractor for Unemployment Insurance (FUTA), Social Security and Medicare (FICA), state unemployment insurance, workers' compensation insurance and liability insurance. Tax and insurance expense included in these labor-hour costs are itemized in the sections beginning on pages 106 and 203.

These hourly labor costs will apply within a few percent on many jobs. But wages may be much higher or lower on the job you are estimating. If the hourly cost on this page is not accurate, use the labor cost adjustment procedure on page 9.

If your hourly labor cost is not known and can't be estimated, use both labor and material figures in this book without adjustment. When all material and labor costs have been compiled, multiply the total by the appropriate figure in the area modification table on page 10 or 11.

Craft	Hourly Wage and Benefits ($)	Typical Employer Burden (%)	Employer's Burden Per Hour ($)	Hourly Cost ($)	Hourly Cost ($) Including Subcontractor 30% Markup
Air Tool Operator	$21.00	27.9%	$5.86	$26.86	$34.92
Asbestos Worker	28.30	28.5	8.07	36.38	47.29
Boilermaker	28.10	25.6	7.20	35.30	45.89
Bricklayer	25.30	29.1	7.35	32.65	42.44
Bricklayer Tender	19.35	29.1	5.62	24.98	32.47
Building Laborer	19.80	32.5	6.43	26.23	34.10
Carpenter	24.85	32.6	8.10	32.95	42.84
Cement Mason	24.40	26.5	6.48	30.88	40.14
Crane Operator	26.35	27.9	7.35	33.71	43.82
Drywall Installer	25.40	28.8	7.32	32.72	42.54
Electrician	29.00	22.9	6.64	35.64	46.33
Elevator Constructor	28.30	25.1	7.10	35.40	46.02
Floor Layer	25.90	25.0	6.47	32.37	42.08
Glazier	26.00	29.4	7.63	33.63	43.73
Iron Worker (Structural)	26.85	46.2	12.40	39.25	51.02
Lather	25.00	26.1	6.53	31.53	40.99
Marble Setter	25.50	29.1	7.41	32.91	42.78
Millwright	25.85	29.1	7.51	33.36	43.37
Mosaic & Terrazzo Worker	24.50	24.8	6.07	30.57	39.74
Painter	24.30	28.5	6.93	31.23	40.60
Pile Driver	24.80	43.1	10.68	35.48	46.12
Pipefitter	28.50	24.2	6.91	35.41	46.03
Plasterer	25.00	28.7	7.18	32.18	41.84
Plasterer Helper	20.00	28.7	5.75	25.75	33.47
Plumber	29.50	24.2	7.15	36.65	47.64
Reinforcing Ironworker	27.30	31.7	8.64	35.94	46.73
Roofer	24.50	45.0	11.03	35.54	46.20
Sheet Metal Worker	27.95	26.5	7.42	35.36	45.97
Sprinkler Fitter	29.75	24.2	7.21	36.96	48.05
Tractor Operator	25.85	27.9	7.21	33.06	42.98
Truck Driver	22.35	29.4	6.56	28.91	37.59

General Requirements 1

	Craft@Hrs	Unit	Material	Labor	Total

General Contractor's Markup. Costs listed in the Commercial and Industrial Division of this book include the subcontractor's markup (estimated at 15% of the material cost and 30% of the labor cost) but not the General Contractor's markup. The figures below show a typical markup for General Contractors handling commercial and industrial projects. The two sections that follow (*Indirect Overhead* and *Direct Overhead*) give a more detailed breakdown. For estimating purposes, include subcontractors' work as a "material" cost.

	Craft@Hrs	Unit	Material	Labor	Total
Indirect overhead (home office overhead)	--	%	2.0	6.0	8.0
Direct overhead (job site overhead)	--	%	0.9	4.9	5.8
Contingency	--	%	--	--	2.0
Profit	--	%	--	--	8.0
Total General Contractor's markup	--	%	--	--	**23.8**

Indirect Overhead (home office overhead). The figures below show typical indirect overhead costs per $1,000 of total contract price for a general contractor handling $5,000,000 to $10,000,000 commercial and industrial projects. Indirect overhead costs vary widely but will be about 8% of gross for most profitable firms. Total indirect overhead cost ($20.00 per $1,000 for materials and $60.00 per $1,000 for labor) appears as 2.0% and 6.0% in the section *General Contractor's Markup* at the top of this page.

	Craft@Hrs	Unit	Material	Labor	Total
Rent, supplies, utilities, equipment, etc.	--	Per M$	20.00	--	--
Office salaries and professional fees	--	Per M$	--	60.00	--
Total General Contractor's indirect overhead	--	Per M$	**20.00**	**60.00**	**80.00**

Direct Overhead (job site overhead). The figures below show typical direct overhead costs per $1,000 of total contract price for a general contractor handling a $5,000,000 to $10,000,000 project that requires 12 to 18 months for completion. Use these figures for preliminary estimates and to check final bids. Add the cost of mobilization, watchmen, fencing, hoisting, permits, bonds, scaffolding and testing. Total direct overhead cost ($8.72 per $1,000 for materials and $48.80 per $1,000 for labor) appears as .9% and 4.9% in the section *General Contractor's Markup* at the top of this page.

	Craft@Hrs	Unit	Material	Labor	Total
Project manager ($5,000 per month)	--	Per M$	--	10.00	--
Job superintendent ($4,500 per month)	--	Per M$	--	9.00	--
Assistant superintendent ($3,500 per month)	--	Per M$	--	7.00	--
Job site engineer ($3,000 per month)	--	Per M$	--	6.00	--
Office manager ($1,800 per month)	--	Per M$	--	3.60	--
Clerical help ($1,300 per month)	--	Per M$	--	2.60	--
Temporary power ($500 per month)	--	Per M$	1.00	--	--
Temporary water ($150 per month)	--	Per M$	.30	--	--
Temporary phone ($200 per month)	--	Per M$	.40	--	--
Temporary light and heat ($250 per month)	--	Per M$	.50	--	--
Job site toilet ($320 per month)	--	Per M$	.64	--	--
Job office ($300 per month)	--	Per M$	.60	--	--
Storage bin and tool shed ($250 per month)	--	Per M$	.50	--	--
Pickup for superintendent ($400 per month)	--	Per M$	.80	--	--
Progressive cleanup ($700 per month)	--	Per M$	--	1.40	--
Final cleanup ($6,000)	--	Per M$	--	.80	--
Debris removal ($500 per month)	--	Per M$	1.00	--	--
Glass cleaning ($2,000)	--	Per M$	--	.27	--
Job signs ($750)	--	Per M$	.10	--	--
Safety & first aid equipment ($90 per month)	--	Per M$	.18	--	--
Small tools, supplies ($1,000 per month)	--	Per M$	2.00	--	--
CPM scheduling ($350 per month)	--	Per M$	.70	--	--
Taxes and insurance on wages (at 20%)	--	Per M$	--	8.13	--
Total General Contractor's direct overhead	--	Per M$	**8.72**	**48.80**	--

General Requirements 1

	Craft@Hrs	Unit	Material	Labor	Total

Project Financing. Construction loans are usually made for a term of up to 18 months. The maximum loan will be based on the value of the land and building when completed. Typically this maximum is 75% for commercial, industrial and apartment buildings and 77.5% for tract housing. The initial loan disbursement will be about 50% of the land cost. Periodic disbursements are based on the percentage of completion as verified by voucher or inspection. The last 20% of loan proceeds is disbursed when an occupancy permit is issued. Fund control fees cover the cost of monitoring disbursements. Typical loan fees, loan rates and fund control fees are listed below.

	Craft@Hrs	Unit	Material	Labor	Total
Loan origination fee, based on amount of loan					
Tracts	--	%	--	--	2.5
Apartments	--	%	--	--	2.0
Commercial, industrial buildings	--	%	--	--	3.0
Loan interest rate, prime rate plus					
Tracts, commercial, industrial	--	%	--	--	2.0
Apartments	--	%	--	--	1.5
Fund control fees (costs per $1,000 disbursed)					
To $200,000	--	LS	--	--	950.00
Over $200,00 to $500,000, $950 plus	--	Per M$	--	--	1.50
Over $500,000 to $1,000,000, $1,400 plus	--	Per M$	--	--	1.00
Over $1,000,000 to $3,000,000, $1,200 plus	--	Per M$	--	--	.85
Over $3,000,000 to $5,000,000, $1,100 plus	--	Per M$	--	--	.75
Over $5,000,000.00, $1,000 plus	--	Per M$	--	--	.55

Project Scheduling by CPM (Critical Path Method). Includes consultation, review of construction documents, development of construction logic, and graphic schedule.

	Craft@Hrs	Unit	Material	Labor	Total
Wood frame buildings, one or two stories					
Simple schedule	--	LS	--	--	1,500.00
Complex schedule	--	LS	--	--	2,000.00
Tilt-up concrete buildings, 10,000 to 50,000 SF					
Simple schedule	--	LS	--	--	1,200.00
Complex schedule	--	LS	--	--	1,700.00
Two to five story buildings (mid rise)					
Simple schedule	--	LS	--	--	2,000.00
Complex schedule	--	LS	--	--	3,000.00
Five to ten story buildings (high rise)					
Simple schedule	--	LS	--	--	3,500.00
Complex schedule	--	LS	--	--	7,250.00
Manufacturing plants and specialized use low rise buildings					
Simple schedule	--	LS	--	--	1,550.00
Complex schedule	--	LS	--	--	2,000.00

Comprehensive schedules to meet government or owner specifications may cost 20% to 40% more. Daily schedule updates can increase costs by 40% to 50%.

Sewer Connection Fees. Check with the local sanitation district for actual charges. These costs are typical for work done by city or county crews and include excavation, recompaction and repairs to the street. Note that these costs will vary widely depending on local government policy.

	Craft@Hrs	Unit	Material	Labor	Total
Typical charges based on main up to 8' deep and lateral up to 5' deep					
4" connection and 40' run to main in street	--	LS	--	--	1,650.00
Add for additional run	--	LF	--	--	37.00
6" connection and 40' run to main in street	--	LS	--	--	2,500.00
Add for additional run	--	LF	--	--	63.00
4" connection and 15' run to main in alley	--	LS	--	--	1,150.00
Add for additional run	--	LF	--	--	37.00

General Requirements 1

	Craft@Hrs	Unit	Material	Labor	Total
6" connection and 15' run to main in alley	--	LS	--	--	2,000.00
Add for additional run	--	LF	--	--	63.00
Manhole cut-in	--	Ea	--	--	350.00
Add for main depths over 5'					
Over 5' to 8'	--	%	--	--	30.0
Over 8' to 11'	--	%	--	--	60.0
Over 11'	--	%	--	--	100.0

These costs do not include capacity fees or other charges which are often levied on connection to new construction. Cost per living unit may be $1,000 or more. Note that these costs will vary widely depending on local government policy. For commercial buildings, figure 20 fixture units equal one living unit. A 6" sewer connection will be required for buildings that include more than 216 fixture units. (Bathtubs, showers and sinks are 2 fixture units, water closets are 4, clothes washers are 3 and lavatories are 1.) The costs shown above assume that work will be done by sanitation district crews. Similar work done by a private contractor under a public improvement permit may cost 50% less.

Water Meters. Check with the local water district for actual charges. These costs are typical for work done by city or county crews and include excavation and pipe to 50' from the main, meter, vault, recompaction and repairs to the street. These costs do not include capacity fees or other charges which are often levied on connection to new construction. The capacity fee per living unit may be $500 or more. For commercial buildings, figure 20 fixture units equal one living unit. (Bathtubs, showers and sinks are 2 fixture units, water closets are 4, clothes washers are 3 and lavatories are 1.) The cost for discontinuing service will usually be about the same as the cost for starting new service, but no capacity fee will be charged. Add the backflow device and capacity fee, if required.

	Craft@Hrs	Unit	Material	Labor	Total
1" service, 3/4" or 1" meter	--	LS	--	--	1,350.00
Add per LF over 50'	--	LF	--	--	31.50
2" service, 1-1/2" or 2" meter	--	LS	--	--	2,250.00
Add per LF over 50'	--	LF	--	--	42.00
Two 2" service lines and meter manifold	--	LS	--	--	5,200.00
Add per LF over 50'	--	LF	--	--	70.00

Backflow preventer with single check valve. Includes pressure vacuum breaker (PVB), two ball valves, one air inlet and one check valve

5/8" or 3/4" pipe	--	LS	--	--	195.00
1"	--	LS	--	--	205.00
1-1/2"	--	LS	--	--	285.00
2"	--	LS	--	--	400.00

Backflow preventer with double check valves. Includes pressure vacuum breaker (PVB), two ball or gate valves, one air inlet and two check valves

3/4"	--	LS	--	--	250.00
1"	--	LS	--	--	260.00
1-1/2"	--	LS	--	--	300.00
2"	--	LS	--	--	425.00

Reduced pressure backflow preventer. Includes pressure reducing vacuum breaker (RPVB), two ball or gate valves, one air inlet and two check valves and a relief valve

3/4"	--	LS	--	--	500.00
1"	--	LS	--	--	560.00
1-1/2"	--	LS	--	--	600.00
2"	--	LS	--	--	800.00

Surveying
Surveying party, 2 or 3 technicians

Typical cost	--	Day	--	--	900.00
Higher cost	--	Day	--	--	1,150.00
Data reduction and drafting	--	Hr	--	--	47.50

General Requirements 1

	Craft@Hrs	Unit	Material	Labor	Total

Surveys. Including wood hubs or pipe markers as needed. These figures assume that recorded monuments are available adjacent to the site.

Residential lot, tract work, 4 corners	--	LS	--	--	545.00
Residential lot, individual, 4 corners	--	LS	--	--	975.00
Commercial lot, based on 30,000 SF lot	--	LS	--	--	1,000.00
Over 30,000 SF, add per acre	--	Acre	--	--	150.00

Lots without recorded markers cost more, depending on distance to the nearest recorded monument.

Add up to	--	%	--	--	100.0

Some states require that a corner record be prepared and filed with the county surveyor's office when new monuments are set. The cost of preparing and filing a corner record will be about $275.00 to $300.00.

Aerial Mapping. Typical costs based on a scale of 1" to 40'.

Back up ground survey including flagging model	--	Ea	--	--	1,500.00

(Typical "model" is approximately 750' by 1,260' or 25 acres)

Aerial photo flight, plane and crew, per local flight. One flight can cover several adjacent models	--	Ea	--	--	275.00
Vertical photos (2 required per model)	--	Ea	--	--	26.50
Data reduction and drafting, per hour	--	Hr	--	--	39.50
Complete survey with finished drawing, per model, including both field work and flying, typical cost	--	Ea	--	--	2,450.00
Deduct if property boundaries are marked on the ground by others	--	LS	--	--	-900.00
Typical mapping cost per acre (including aerial control, flight and compilation, based on 25 acre coverage)	--	Acre	--	--	100.00

Aerial photos, oblique (minimum 4 per job), one 8" x 10" print

Black and white	--	Ea	--	--	34.00
Color	--	Ea	--	--	70.00

Quantitative studies for earthmoving operations to evaluate soil quantities moved

Ground control and one flight	--	Ea	--	--	1,800.00
Each additional flight	--	Ea	--	--	500.00

Engineering Fees. Typical billing rates.
Registered engineers (calculations, design and drafting)

Assistant engineer	--	Hr	--	--	50.00
Senior engineer	--	Hr	--	--	74.00

Principal engineers (cost control, project management, client contact)

Assistant principal engineer	--	Hr	--	--	81.50
Senior principal engineer	--	Hr	--	--	100.00
Job site engineer (specification compliance)	--	Hr	--	--	47.50

Estimating Cost. Compiling detailed labor and material cost estimate for commercial and heavy construction, in percent of job cost. Actual fees are based on time spent plus cost of support services.
Most types of buildings, typical fees

Under $100,000 (range of .75 to 1.5%)	--	%	--	--	1.1
Over $100,000 to $500,000 (range of .5 to 1.25%)	--	%	--	--	0.9
Over $500,000 to $1,000,000 (range of .5 to 1.0%)	--	%	--	--	0.8
Over $1,000,000 (range of .25 to .75%)	--	%	--	--	0.5

Complex, high-tech, research or manufacturing facilities or other buildings with many unique design features, typical fees

Under $100,000 (range of 2.3 to 3%)	--	%	--	--	2.7
Over 100,000 to $500,000 (range of 1.5 to 2.5%)	--	%	--	--	2.0

General Requirements 1

	Craft@Hrs	Unit	Material	Labor	Total
Estimating Cost, continued					
Over $500,000 to $1,000,000 (range of 1.0 to 2.0%)	--	%	--	--	1.5
Over $1,000,000 (range of .5 to 1.0%)	--	%	--	--	0.8
Typical estimator's fee per hour	--	Hr	--	--	40.00

Specifications Fee. Preparing detailed construction specifications to meet design requirements for commercial and industrial buildings, in percent of job cost. Actual fees are based on time spent plus cost of support services, if any.

	Craft@Hrs	Unit	Material	Labor	Total
Typical fees for most jobs	--	%	--	--	0.5
Small or complex jobs					
Typical cost	--	%	--	--	1.0
High cost	--	%	--	--	1.5
Large jobs or repetitive work					
Typical cost	--	%	--	--	0.2
High cost	--	%	--	--	0.5

Job Layout. Setting layout stakes for the foundation from monuments identified on the plans. Per square foot of finished building floor area. Typical costs. Actual costs will be based on time spent.

	Craft@Hrs	Unit	Material	Labor	Total
Smaller or more complex jobs	--	SF	--	--	.45
Large job, easy work	--	SF	--	--	.15

General Non-Distributable Supervision and Expense

	Craft@Hrs	Unit	Material	Labor	Total
Project manager, typical	--	Mo	--	--	5,000.00
Job superintendent	--	Mo	--	--	4,500.00
Factory representative	--	Day	--	--	500.00
Timekeeper	--	Mo	--	--	1,610.00
Job site engineer for layout	--	Mo	--	--	3,000.00
Office manager	--	Mo	--	--	1,800.00
Office clerk	--	Mo	--	--	1,300.00

Mobilization. Typical costs for mobilization and demobilization within a 50 mile radius as a percent of total contract price. Includes cost of moving general contractor-owned equipment and job office to the site, setting up a fenced material yard, and removing equipment, office and fencing at job completion.

	Craft@Hrs	Unit	Material	Labor	Total
	--	%	--	--	0.5

Aggregate Testing. Add $34.50 per hour for sample preparation.

	Craft@Hrs	Unit	Material	Labor	Total
Sieve analysis, processed (each size), ASTM C 136	--	Ea	--	--	60.00
Sieve analysis, pit run aggregate	--	Ea	--	--	65.00
Specific gravity, coarse, ASTM C 127	--	Ea	--	--	55.00
Specific gravity, fine, ASTM C 127	--	Ea	--	--	75.00
Absorption, ASTM C 127, 128	--	Ea	--	--	44.50
Unit weight per cubic foot, ASTM C 29	--	Ea	--	--	22.00
Organic impurities, ASTM C 40	--	Ea	--	--	27.50
Clay lumps and friable particles, ASTM C 142	--	Ea	--	--	66.50
Material finer than #200 sieve, ASTM C 117	--	Ea	--	--	32.50
Coal and lignite, ASTM C 123	--	Ea	--	--	100.00
Soft particles, ASTM C 235	--	Ea	--	--	66.50
Acid solubility	--	Ea	--	--	33.50
Soundness, sodium or magnesium, 5 cycle, per series, ASTM C 88	--	Ea	--	--	120.00
Abrasion, L.A. Rattler, 100 and 500 cycles, ASTM C 131	--	Ea	--	--	100.00

General Requirements 1

	Craft@Hrs	Unit	Material	Labor	Total
Potential reactivity, chemical method, 3 determinations per series, ASTM C 289	--	Ea	--	--	375.00
Potential reactivity, mortar bar method, ASTM C 227	--	Ea	--	--	440.00
Percent flat or elongated particles, CRD C 119	--	Ea	--	--	100.00
Percent crushed particles, CAL 205	--	Ea	--	--	66.50
Aggregate test for mix design, including sieve analysis, specific gravity, number 200 wash, organic impurities and weight per cubic foot	--	Ea	--	--	75.00

Asphaltic Concrete Testing

	Craft@Hrs	Unit	Material	Labor	Total
Stability, premixed, Marshall, ASTM D 1559	--	Ea	--	--	150.00
Stability, lab mixed, Marshall, ASTM D 1559	--	Ea	--	--	200.00
Maximum density, premixed, Marshall, 1559	--	Ea	--	--	100.00
Maximum density, lab mixed, Marshall, 1559	--	Ea	--	--	120.00
Gradation on extracted sample (including wash), ASTM C 136	--	Ea	--	--	75.00
Maximum theoretical unit weight (Rice Gravity), ASTM 2041	--	Ea	--	--	75.00
Extraction, percent asphalt, ASTM D 2172 (Method B) excluding ash correction	--	Ea	--	--	60.00
Penetration, ASTM D 5	--	Ea	--	--	58.50
Asphaltic core density	--	Ea	--	--	22.50
Asphaltic concrete core drilling, field or laboratory, per linear inch	--	Ea	--	--	3.50
Add for equipment and technician for core drilling	--	Hr	--	--	66.50

Concrete Testing. Minimum fee is usually $25.00

	Craft@Hrs	Unit	Material	Labor	Total
Cylinder pickup within 40 miles of laboratory, cost per cylinder (minimum of 3)	--	Ea	--	--	6.50
Compression test, 6" x 12" cylinders, including mold ASTM C 39	--	Ea	--	--	11.00
Weight per cubic foot determination of lightweight concrete cylinders	--	Ea	--	--	3.50
Compression test, 2", 4", or 6" cores (excludes sample preparation), ASTM C 42	--	Ea	--	--	33.50
Flexure test, 6" x 6" beams, ASTM C 78	--	Ea	--	--	43.50
Splitting tensile, 6" x 12" cylinder, ASTM C 496	--	Ea	--	--	28.00
Modulus of elasticity test, static, ASTM C 469	--	Ea	--	--	71.50
Core cutting in laboratory	--	Ea	--	--	38.50
Laboratory trial, concrete batch, ASTM C 192	--	Ea	--	--	325.00
Length change (3 bars, 4 readings, up to 90 days), ASTM C 157 modified	--	LS	--	--	245.00
Additional readings, per set of 3 bars	--	LS	--	--	33.50
Storage over 90 days, per set of 3 bars, per month	--	Mo	--	--	33.50
Pickup or delivery of shrinkage molds, per trip	--	Ea	--	--	28.00
Aggregate tests for concrete mix designs only, including sieve analysis, specific gravity, number 200 wash, organic impurities, weight per cubic foot, per aggregate size	--	Ea	--	--	75.00
Mix design, determination of proportions	--	Ea	--	--	71.50
Amend or re-type existing mix designs not involving calculations	--	Ea	--	--	40.00
Review mix design prepared by others	--	Ea	--	--	57.50
Prepare special strength documentation for mix design	--	Ea	--	--	82.50

General Requirements 1

	Craft@Hrs	Unit	Material	Labor	Total
Sampling aggregate for mix design or laboratory batch, within 40 mile radius of laboratory, per trip	--	Ea	--	--	69.50
Proportional analysis, cement factor and percent of aggregate	--	Ea	--	--	250.00

Masonry and Tile Testing
Brick, ASTM C 67

	Craft@Hrs	Unit	Material	Labor	Total
Modulus of rupture (flexure)	--	Ea	--	--	21.50
Compressive strength of brick	--	Ea	--	--	21.50
Absorption, 5 hour or 24 hour	--	Ea	--	--	16.50
Boil, 1, 2, or 5 hours	--	Ea	--	--	16.50
Initial rate of absorption	--	Ea	--	--	16.50
Efflorescence	--	Ea	--	--	21.50
Dimensions, overall, coring, shell and web	--	Ea	--	--	16.50
Compression of core	--	Ea	--	--	33.50
Cores, shear, 6" and 8" diameter, 2 faces	--	Ea	--	--	46.50

Concrete block, ASTM C 140

	Craft@Hrs	Unit	Material	Labor	Total
Moisture content as received	--	Ea	--	--	16.50
Absorption	--	Ea	--	--	24.50
Compression	--	Ea	--	--	27.50
Tension	--	Ea	--	--	66.00
Shrinkage, modified British, ASTM C 426	--	Ea	--	--	66.00
Compression, 4", 6", 8" cores	--	Ea	--	--	33.50

Masonry prisms, ASTM E 447

	Craft@Hrs	Unit	Material	Labor	Total
Compression test, grouted prisms	--	Ea	--	--	120.00
Pickup prisms, within 40 miles of lab	--	Ea	--	--	38.50

Mortar and grout, UBC Standard 24-22

	Craft@Hrs	Unit	Material	Labor	Total
Compression test, 2" x 4" mortar cylinder	--	Ea	--	--	11.00
Compression test, 3" x 6" grout prisms	--	Ea	--	--	16.50
Compression test, 2" cubes, ASTM C 109	--	Ea	--	--	38.50

Gypsum roof fill, ASTM C 495

	Craft@Hrs	Unit	Material	Labor	Total
Compression test	--	Ea	--	--	16.50
Density	--	Ea	--	--	11.00

Gunite

	Craft@Hrs	Unit	Material	Labor	Total
Compression, 2", 4", 6" cores, ASTM C 42	--	Ea	--	--	33.50
Pickup gunite field sample (40 mile radius)	--	Ea	--	--	27.50

Roofing tile

	Craft@Hrs	Unit	Material	Labor	Total
Roofing tile breaking strength, UBC	--	Ea	--	--	18.00
Roofing tile absorption	--	Ea	--	--	18.00
Fireproofing, oven dry density	--	Ea	--	--	27.50

Slate

	Craft@Hrs	Unit	Material	Labor	Total
Modulus of rupture	--	Ea	--	--	33.50
Modulus of elasticity	--	Ea	--	--	55.00
Water absorption	--	Ea	--	--	27.50

Reinforcement Testing

	Craft@Hrs	Unit	Material	Labor	Total
Tensile test, number 11 bar or smaller, ASTM 615	--	Ea	--	--	22.00
Bend test, number 11 bar or smaller, ASTM 615	--	Ea	--	--	12.50
Tensile test, number 14 bar or smaller, ASTM 615	--	Ea	--	--	54.50
Tensile test, number 18 bar, ASTM 615	--	Ea	--	--	66.50

General Requirements 1

	Craft@Hrs	Unit	Material	Labor	Total
Sampling at fabricator's plant (within 40 miles of lab), $34.50 minimum					
During business hours, per sample	--	Ea	--	--	12.50
Other than business hours, per sample	--	Ea	--	--	21.50
Receive and distribute mill certificates	--	Ea	--	--	7.50
Tensile test, welded number 11 bar or smaller	--	Ea	--	--	31.50
Tensile test, welded number 14 bar	--	Ea	--	--	66.50
Tensile test, welded number 18 bar	--	Ea	--	--	82.50
Tensile test, mechanically spliced bar	--	Ea	--	--	94.50
Nick-break test, welded bar	--	Ea	--	--	32.50
Tensile and elongation test for prestress strands					
24" test, ASTM A 416	--	Ea	--	--	82.50
10" test, ASTM A 421	--	Ea	--	--	44.50
Modulus of elasticity for prestressing wire	--	Ea	--	--	94.50

Soils and Aggregate Base Testing

	Craft@Hrs	Unit	Material	Labor	Total
Sieve analysis, coarse, ASTM C 136	--	Ea	--	--	60.00
Sieve analysis, fine, ASTM C 136, C 177	--	Ea	--	--	77.50
Hydrometer analysis, ASTM D 422	--	Ea	--	--	71.50
Specific gravity					
Bulk, SSD, coarse, ASTM C 127, C 177	--	Ea	--	--	44.50
Fine, ASTM C 128	--	Ea	--	--	77.50
Atterberg limits, ASTM D 4318	--	Ea	--	--	94.50
Moisture-density relation, ASTM D 698, D 1557	--	Ea	--	--	120.00
Consolidation test, ASTM D 2435	--	Ea	--	--	165.00
Direct shear test, ASTM D 3080					
Unconsolidated, undrained, 1 point	--	Ea	--	--	55.00
Consolidated, undrained, 1 point	--	Ea	--	--	71.50
Consolidated, drained, 1 point	--	Ea	--	--	110.00
Unconfined compression test, ASTM D 2166	--	Ea	--	--	55.50
Expansion index, UBC 29-2	--	Ea	--	--	94.50
Permeability, constant or falling head, ASTM 2434	--	Ea	--	--	165.00
In place density, nuclear ASTM D2922-71					
COE0223-2000	--	Ea	--	--	45.00
Sand cone, ASTM D1556-064 D2922-71					
COE0223-2000	--	Ea	--	--	45.00

Steel Testing. $100.00 minimum for machine time.

	Craft@Hrs	Unit	Material	Labor	Total
Photo micrographs					
20x to 2000x	--	Ea	--	--	20.00
With negative	--	Ea	--	--	24.50
Tensile test, under 100,000 lbs	--	Ea	--	--	29.00
Electric extensometer	--	Ea	--	--	50.00
Hardness tests, Brinell or Rockwell	--	Ea	--	--	15.50
Charpy impact test, room temperature	--	Ea	--	--	10.50
Charpy impact test, reduced temperature	--	Ea	--	--	22.50
Welder Qualification Testing. Add $35 minimum per report					
Tensile test	--	Ea	--	--	22.50
Bend test	--	Ea	--	--	20.00
Macro etch	--	Ea	--	--	44.50
Fracture test	--	Ea	--	--	24.50
Machining for weld tests, 1/2" and less	--	Ea	--	--	24.50
Machining for weld tests, over 1/2"	--	Ea	--	--	39.50

General Requirements 1

	Craft@Hrs	Unit	Material	Labor	Total

Construction Photography. Professional quality work, 2-1/4" x 2-1/4" negatives, hand printed, labeled, numbered and dated.
Job progress photos, includes shots of all building corners, two 8" x 10" black and white prints of each shot, per visit

	Craft@Hrs	Unit	Material	Labor	Total
Typical cost	--	LS	--	--	300.00
High cost	--	LS	--	--	400.00

Periodic survey visits to job site to figure angles (for shots of all building corners, minimum two visits), per visit

	--	Ea	--	--	185.00

Exterior completion photographs, two 8" x 10" prints using 4" x 5" negatives

Black and white	--	Ea	--	--	200.00
Color	--	Ea	--	--	300.00

Interior completion photographs, with lighting setup, two 8" x 10" prints of each shot (minimum three shots)

Black and white	--	Ea	--	--	200.00
Color	--	Ea	--	--	300.00

Reprints or enlargements from existing negative (machine prints)

8" x 10", black and white	--	Ea	--	--	20.00
11" x 14", black and white	--	Ea	--	--	30.00
8" x 10", color	--	Ea	--	--	30.00
11" x 14", color	--	Ea	--	--	40.00

Signs and Markers
Pipe markers, snap-on, one-piece, acrylic plastic, by overall outside diameter of pipe or covering, meets OSHA and ANSI requirements. Costs are based on standard wording and colors and include the subcontractor's overhead and profit.

	Craft@Hrs	Unit	Material	Labor	Total
1" diameter	PF@.077	Ea	3.45	3.54	6.99
2"	PF@.077	Ea	4.50	3.54	8.04
3" to 5"	PF@.077	Ea	8.90	3.54	12.44
6"	PF@.095	Ea	9.50	4.37	13.87
8"	PF@.095	Ea	12.50	4.37	16.87
10"	PF@.125	Ea	15.30	5.75	21.05

"Caution," "Danger" and other stock signs, rectangular, 30 gauge, baked enamel

	Craft@Hrs	Unit	Material	Labor	Total
7" x 10"	PA@.301	Ea	16.60	12.20	28.80
10" x 14"	PA@.301	Ea	20.50	12.20	32.70
14" x 20"	PA@.455	Ea	29.50	18.50	48.00
20" x 28"	PA@.455	Ea	53.60	18.50	72.10
Add for double face signs	--	%	50.0	--	--
Add for 20 gauge, baked enamel	--	%	25.0	--	--
Deduct for semi-rigid butyrate or pressure sensitive signs	--	%	-25.0	--	--
Add for custom wording on signs	--	%	50.0	--	--
Add for pole-mounted signs	CL@1.00	Ea	12.00	34.10	46.10

Temporary Utilities. No permit fees included.
Drop pole for electric service
  Power pole installation, wiring, 6 months rent and pickup

100 amp meter pole	--	Ea	--	--	375.00
200 amp meter pole	--	Ea	--	--	490.00

Electric drop line and run to contractor's panel (assuming proper phase and voltage and service are available within 100' of site)
  Overhead run to contractor's "T" pole (including removal)

Single phase, 100 amps	--	LS	--	--	155.00
Single phase, over 100 to 200 amps	--	LS	--	--	235.00
Three phase, to 200 amps	--	LS	--	--	280.00

General Requirements 1

	Craft@Hrs	Unit	Material	Labor	Total

Temporary Utilities, continued
  Underground run in contractor's trench and conduit (including removal)

	Craft@Hrs	Unit	Material	Labor	Total
Single phase, 100 amps, "USA" wire	--	LS	--	--	245.00
Single phase, 101-200 amps, "USA" wire	--	LS	--	--	300.00
Three phase, to 200 amps, "USA" wire	--	LS	--	--	365.00
Typical power cost, per kilowatt hour ($5.00 minimum)	--	Ea	--	--	.09
Typical power cost, per 1000 SF of building per month of construction	--	MSF	--	--	1.75

Temporary water service, meter installed on fire hydrant. Add the cost of distribution hose or piping.

	Craft@Hrs	Unit	Material	Labor	Total
Meter fee, 2" meter (excluding $360 deposit)	--	LS	--	--	55.00
Water cost, per 1,000 gallons	--	M	--	--	1.50
Job phone, installation and 3 month service, 1 line, excluding message unit charges	--	LS	--	--	400.00

Portable job site toilets, including delivery, weekly service, and pickup, per 28 day month

	Craft@Hrs	Unit	Material	Labor	Total
Good quality (polyethylene)	--	Mo	--	--	87.50
Deluxe quality with sink, soap dispenser, towel dispenser and mirror	--	Mo	--	--	110.00

Trailer-mounted male and female compartment double, 2 stalls in each, sink, soap dispenser, mirror and towel dispenser in each compartment.

	Craft@Hrs	Unit	Material	Labor	Total
Rental per week	--	Wk	--	--	275.00

Construction Elevators and Hoists
Hoisting towers to 200', monthly rent including operator

	Craft@Hrs	Unit	Material	Labor	Total
Material hoist	CO@176.	Mo	4,600.00	7,710.00	12,310.00
Personnel hoist	CO@176.	Mo	2,000.00	7,710.00	9,710.00

Tower cranes, monthly rent including operator

	Craft@Hrs	Unit	Material	Labor	Total
4 ton lift at 130' radius, one operator	CO@176.	Mo	14,000.00	7,710.00	21,710.00
With two operators	CO@352.	Mo	14,000.00	15,400.00	29,400.00
6 ton lift at 130' radius, one operator	CO@176.	Mo	16,000.00	7,710.00	23,710.00
With two operators	CO@352.	Mo	16,000.00	15,400.00	31,400.00

Mobilize, erect and dismantle tower cranes

	Craft@Hrs	Unit	Material	Labor	Total
Good access	--	LS	--	--	35,000.00
Average access	--	LS	--	--	40,000.00
Poor access	--	LS	--	--	45,000.00

Hydraulic Truck Cranes. Includes equipment rental and operator. Four hour minimum. Time charge begins at the rental yard. By rated crane capacity.

	Craft@Hrs	Unit	Material	Labor	Total
5 ton	--	Hr	--	--	85.00
10 ton	--	Hr	--	--	106.00
14-1/2 ton	--	Hr	--	--	115.00
15 ton, rough terrain	--	Hr	--	--	115.00
30 ton	--	Hr	--	--	160.00
50 ton	--	Hr	--	--	190.00
65 ton	--	Hr	--	--	200.00
75 ton	--	Hr	--	--	212.00
Add for clamshell or breaker ball work	--	%	--	--	10.0

Conventional Cable Cranes. Including equipment rental and operator. Four hour minimum. Time charge begins at the rental yard. By rated crane capacity with 80' boom.

	Craft@Hrs	Unit	Material	Labor	Total
50 tons	--	Hr	--	--	185.00
70 tons	--	Hr	--	--	210.00
90 tons	--	Hr	--	--	220.00
115 tons	--	Hr	--	--	230.00
140 tons	--	Hr	--	--	250.00

General Requirements 1

	Craft@Hrs	Unit	Material	Labor	Total
Add for additional boom length (including jib) over 80'					
80 to 100'	--	Hr	--	--	3.00
Over 100 to 120'	--	Hr	--	--	4.00
Over 120 to 160'	--	Hr	--	--	5.00
Over 160 to 220'	--	Hr	--	--	8.00
Over 220 to 260'	--	Hr	--	--	10.00
Over 260 to 300'	--	Hr	--	--	14.00
Add for clamshell or breaker ball work	--	%	--	--	10.0

Equipment Rental Rates. Costs are for equipment in good condition but exclude accessories. Add the costs of damage waiver (about 10% of the rental cost) insurance, an equipment operator, pickup, return to yard, fuel, oil, and repairs, if needed. Delivery and pickup within 30 miles of the rental yard are usually offered at an additional cost of one day's rent. A rental "day" is assumed to be one 8-hour shift and begins when the equipment leaves the rental yard. A "week" is 40 hours in five consecutive days. A "month" is 176 hours in 30 consecutive days. Costs shown **do not** include subcontractor's overhead and profit.

Air equipment rental	Unit	Day	Week	Month
Air compressors, wheel-mounted				
15 CFM, shop type, electric	Ea	28.00	105.00	278.00
50 CFM, shop type, electric	Ea	55.00	166.00	499.00
90 CFM, gasoline unit	Ea	45.00	160.00	426.00
150 CFM, gasoline unit	Ea	60.00	192.00	560.00
175 CFM, gasoline unit	Ea	69.00	225.00	645.00
190 CFM, gasoline unit	Ea	78.00	240.00	700.00
Air compressors, wheel-mounted diesel units				
100 CFM	Ea	58.00	180.00	518.00
150 CFM	Ea	62.50	196.00	570.00
175 CFM	Ea	78.50	242.00	692.00
250 CFM	Ea	114.50	360.00	1,032.00
300 CFM	Ea	134.00	428.00	1,232.00
425 CFM	Ea	167.00	515.00	1,484.00
600 CFM	Ea	211.00	660.00	1,885.00
800 CFM	Ea	260.00	812.00	2,318.00
Paving breakers (no bits included)				
30 lb, light	Ea	20.00	62.00	173.00
70 lb, medium	Ea	22.50	69.00	190.00
90 lb, heavy	Ea	24.00	74.00	203.00
Paving breaker and jackhammer bits				
Moil points, 15" to 18"	Ea	5.10	12.50	29.00
Chisels, 3"	Ea	5.50	13.50	30.00
Clay spades, 5-1/2"	Ea	8.00	20.50	51.00
Asphalt cutters, 5"	Ea	8.00	19.00	46.00
Tamping pads	Ea	20.50	43.00	81.25
Rock drills, jackhammers				
30 lb, light	Ea	29.00	91.00	259.00
50 lb, medium	Ea	27.50	92.00	251.00
65 lb, heavy	Ea	26.00	85.00	244.00
Rock drill bits, hexagonal				
7/8", 2' long	Ea	6.50	15.00	36.00
1", 6' long	Ea	8.00	21.00	49.00
Pneumatic chippers, medium weight, 10 lb	Ea	24.00	74.50	195.50
Tampers, medium weight, 30 lb	Ea	24.00	75.00	208.00
Pneumatic grinders, hand held, 7 lb	Ea	20.00	71.50	205.00

General Requirements 1

	Unit	Day	Week	Month
Air hose, 50 LF section				
5/8"	Ea	4.00	10.50	26.00
3/4"	Ea	6.00	16.00	40.50
1"	Ea	7.00	20.50	57.00
1-1/2"	Ea	14.50	44.00	115.50
Compaction equipment rental				
Vibro plate, 300 lb, 24" plate width, gas	Ea	69.00	206.00	570.00
Vibro plate, 600 lb, 32" plate width, gas	Ea	93.50	302.00	847.00
Rammer, 60 CPM, 200 lb, gas powered	Ea	53.00	179.00	536.00
Rammer, 500 CPM, 150 lb, gas powered	Ea	60.00	196.00	544.00
Rollers, two axle steel drum, road type				
1 ton, gasoline	Ea	88.50	252.00	727.00
4 ton, gasoline	Ea	182.00	560.00	1,530.00
8 ton, diesel	Ea	303.50	961.00	2,827.00
Rollers, rubber tired, self propelled				
12 ton, 9 wheel, gasoline	Ea	225.00	668.00	2,010.00
15 ton, 9 wheel, diesel	Ea	236.00	800.00	2,400.00
30 ton, 11 wheel, diesel	Ea	460.00	1,240.00	4,250.00
Rollers, towed, sheepsfoot type				
40" diameter, double drum, 4' wide	Ea	80.00	243.00	725.00
60" diameter, double drum, 5' wide	Ea	100.00	289.00	1,032.00
Roller, self-propelled vibratory dual sheepsfoot drum, gas				
8 HP, 26" wide roll	Ea	105.00	310.00	910.00
12 HP, 32" wide roll	Ea	120.00	370.00	1,080.00
Rollers, vibrating steel drum, gas powered				
7 HP, 1,000 lb, walk behind, one drum	Ea	92.00	288.00	792.00
10 HP, 2,000 lb, walk behind, two drum	Ea	105.00	345.00	942.00
25 HP, 4,000 lb, towed, single drum	Ea	158.00	464.00	1,320.00
Rollers, self-propelled riding type, vibrating steel drum				
12 HP, 4,000 lb, gasoline, two drum	Ea	120.00	376.00	1,128.00
35 HP, 10,000 lb, diesel, two drum	Ea	233.00	681.00	1,994.00
100 HP, 20,000 lb, diesel, two drum	Ea	448.00	1,398.00	4,128.00
Concrete equipment rental	**Unit**	**Day**	**Week**	**Month**
Buggies, push type, 7 CF	Ea	18.00	44.00	103.00
Buggies, walking type, 12 CF	Ea	77.00	236.00	647.00
Buggies, riding type, 14 CF	Ea	72.00	218.00	649.00
Buggies, riding type, 18 CF	Ea	103.00	298.00	876.00
Vibrator, electric, 3 HP, flexible shaft	Ea	35.00	109.00	294.00
Vibrator, gasoline, 6 HP, flexible shaft	Ea	45.00	135.00	391.00
Troweling machine, 36", 4 paddle	Ea	45.50	143.00	392.00
Troweling machine, 48", riding type	Ea	145.00	440.00	1,390.00
Float finisher, 24" diameter	Ea	68.00	195.00	570.00
Concrete saws, gas powered, excluding blade cost				
10 HP, push type	Ea	54.00	175.00	488.00
15 HP, self propelled	Ea	64.50	216.50	602.00
30 HP, self propelled	Ea	98.50	316.00	906.00
Concrete bucket, bottom dump 1 CY	Ea	45.00	130.00	342.00
Concrete mixer, gas, trailer mount, 6 CF	Ea	47.00	152.00	420.00
Plaster mixer, gas, portable, 5 CF	Ea	44.00	144.00	402.00

General Requirements 1

	Unit	Day	Week	Month
Concrete conveyor, belt type, portable, gas powered				
16" wide, 32' long	Ea	206.00	523.00	1,194.00
18" wide, 56' long	Ea	305.00	878.00	2,428.00
Vibrating screed, 16', single beam	Ea	33.00	99.00	306.00
Column clamps, to 48", per set	Ea	--	--	3.25
Crane and pile driver rental	Unit	Day	Week	Month
Pile drivers, hammers				
MKT diesel, model DA 35	Ea	--	1,725.00	4,069.00
MKT steam, double acting, model 7	Ea	--	620.50	1,520.00
MKT vibratory, model V-20	Ea	--	3,652.00	9,791.00
MKT extractor, model E4	Ea	--	709.00	1,942.50
Cable controlled crawler-mounted lifting cranes, diesel				
35 ton	Ea	--	2,100.00	6,200.00
60 ton	Ea	781.00	2,120.00	6,110.00
100 ton	Ea	--	2,835.00	8,350.00
150 ton	Ea	--	3,485.00	8,765.00
Cable controlled, truck-mounted lifting cranes, diesel				
60 ton	Ea	--	2,400.00	7,700.00
75 ton	Ea	--	2,500.00	7,800.00
90 ton	Ea	742.00	2,560.00	7,785.00
Hydraulic controlled truck-mounted lifting cranes				
10 ton, gas	Ea	375.00	1,205.00	3,500.00
15 ton, diesel	Ea	440.00	1,500.00	4,350.00
35 ton, diesel	Ea	--	1,690.00	6,270.00
50 ton, diesel	Ea	650.00	1,910.00	7,150.00
Excavation equipment rental	Unit	Day	Week	Month
Clamshell excavators, diesel, crawler mounted				
25 ton	Ea	--	1,920.00	6,000.00
35 ton	Ea	--	2,700.00	8,200.00
Add for 1 CY clamshell bucket, general purpose	Ea	71.00	226.00	631.00
Dragline excavators, diesel, crawler mounted				
25 ton	Ea	--	1,900.00	6,000.00
45 ton	Ea	--	2,600.00	8,000.00
Add for 1 CY dragline bucket	Ea	49.00	159.00	452.00
Backhoes, crawler mounted, diesel				
3/8 CY	Ea	226.00	878.00	2,422.00
1/2 CY	Ea	374.00	1,166.00	3,322.00
3/4 CY	Ea	420.00	1,360.00	3,970.00
1 CY	Ea	485.00	1,546.00	4,574.00
1-1/2 CY	Ea	612.00	2,000.00	5,450.00
2 CY	Ea	884.00	3,000.00	8,634.00
Crawler tractors, with dozer blade				
55 to 65 HP, D-3	Ea	246.00	810.00	2,400.00
75 HP, D-4	Ea	286.00	930.00	2,775.00
105 HP, D-5	Ea	468.50	1,450.00	4,365.00
130 to 140 HP, D-6	Ea	590.00	1,818.00	5,527.00
190 to 200 HP, D-7	Ea	957.00	2,950.00	8,928.00
335 HP, D-8	Ea	1,100.00	3,770.00	11,820.00
460 HP, D-9	Ea	1,688.00	6,100.00	19,000.00

General Requirements 1

	Unit	Day	Week	Month
Crawler loaders, diesel				
3/4 CY	Ea	190.00	650.00	2,018.00
1-1/2 CY	Ea	366.00	1,260.00	3,768.00
2 CY	Ea	540.00	1,750.00	5,100.00
Gradall, truck mounted, diesel				
3/4 CY bucket	Ea	900.00	2,710.00	7,650.00
1 CY bucket	Ea	--	2,900.00	7,900.00
Wheel loaders, with integral backhoe				
1/2 CY bucket capacity, 55 HP	Ea	220.00	730.00	2,100.00
1 CY bucket capacity, 65 HP	Ea	210.00	700.00	2,100.00
1-1/4 CY bucket capacity, 75 HP	Ea	250.00	820.00	2,400.00
1-1/2 CY bucket capacity, 100 HP	Ea	300.00	1,050.00	3,000.00
Wheel tractor loaders, diesel				
3/4 CY bucket, 2 wheel drive	Ea	168.00	600.00	1,800.00
1 CY bucket, 2 wheel drive	Ea	200.00	670.00	2,000.00
2-1/2 CY bucket, 4 wheel drive, articulated	Ea	460.00	1,470.00	4,300.00
3-1/4 CY bucket, 4 wheel drive	Ea	--	2,400.00	6,825.00
5 CY bucket, articulated	Ea	1,000.00	3,100.00	8,800.00
Motor scraper-hauler, single engine drive, 2 wheel tractor				
200 HP, 12 CY capacity	Ea	970.00	3,150.00	9,250.00
300 HP, 24 CY capacity	Ea	1,350.00	4,200.00	12,000.00
450 HP, 24 CY capacity	Ea	1,850.00	6,150.00	16,500.00
200 HP, 15 CY, self loading	Ea	690.00	2,350.00	6,900.00
Motor graders, diesel				
10,000 lb	Ea	280.00	880.00	2,700.00
26,000 lb	Ea	520.00	1,600.00	4,700.00
30,000 lb	Ea	605.00	1,830.00	5,200.00
Trenchers, chain boom type, pneumatic wheels				
15 HP, 12" wide, 48" max. depth, walking	Ea	124.00	446.00	1,218.00
20 HP, 12" wide, 66" max. depth, riding	Ea	162.00	515.00	1,388.00
45 HP, 18" wide, 72" max. depth, riding	Ea	267.00	932.00	2,560.00
100 HP, 72" x 96" deep, crawler mount	Ea	490.00	1,560.00	4,850.00
Steel shores, to 12' width, adjustable	Ea	--	--	5.25
Hydraulic trench braces, 7' long, to 42" wide	Ea	16.50	48.00	145.00
Track-mounted wagon drills, self propelled				
4" drifter, straight boom	Ea	260.00	810.00	2,950.00
4-1/2" drifter, hydraulic swing boom	Ea	420.00	1,100.00	3,290.00
Dump trucks, rental rate plus mileage				
3 CY ($.38 mileage charge)	Ea	135.00	385.00	1,160.00
5 CY ($.42 mileage charge)	Ea	145.00	560.00	1,740.00
10 CY ($.46 mileage charge)	Ea	190.00	745.00	2,160.00
25 CY off highway	Ea	--	2,680.00	7,930.00

Forklifts and fork loaders	Unit	Day	Week	Month
Rough terrain, towable, gas powered				
4,000 lb capacity	Ea	158.00	535.50	1,572.00
8,000 lb capacity	Ea	250.00	828.50	2,500.00
Extension boom, 2 wheel steering				
4,000 lb, 30' lift	Ea	200.00	644.00	1,900.00
6,000 lb, 35' lift	Ea	175.00	620.00	2,250.00
8,000 lb, 40' lift, 4 wheel steering	Ea	318.00	1,030.00	2,871.00

General Requirements 1

	Unit	Day	Week	Month
Lifts, manual				
Rolling scissor lifts, manual				
15' high, 2' x 3' platform, 500 lbs	Ea	55.00	140.00	400.00
25' high, 2' x 3' platform, 400 lbs	Ea	100.00	232.00	638.00
Platforms, aerial, hydraulic	Unit	Day	Week	Month
Rolling scissor lifts, 2 wheel drive, electric				
16' high, 30" x 76" platform, 750 lbs	Ea	88.00	265.00	735.00
20' high, 29" x 92" platform, 1,000 lbs	Ea	88.00	265.00	770.00
25' high, 57" x 118" platform, 1,000 lbs	Ea	89.00	270.00	775.00
25', 68" x 161", 2,000 lbs, rough terrain	Ea	130.00	390.00	1,100.00
35', 82" x 168", 2,000 lbs, rough terrain	Ea	170.00	530.00	1,550.00
40', 72" x 144", 2,000 lbs, rough terrain	Ea	210.00	620.00	1,800.00
Boomlifts, telescoping and articulating booms, self propelled, gas or diesel				
34' high, 36" x 60" platform, 600 lbs	Ea	190.00	580.00	1,710.00
40' high, 43" x 66" platform, 500 lbs	Ea	223.00	695.00	2,050.00
60' high, 30" x 60" platform, 500 lbs	Ea	375.00	1,150.00	3,360.00
80' high, 36" x 96" platform, 500 lbs	Ea	757.00	2,410.00	7,000.00
Pump rental, hoses not included	Unit	Day	Week	Month
Pneumatic sump pumps, heavy duty				
Diaphragm	Ea	58.00	150.00	425.00
Impeller type	Ea	45.00	120.00	356.00
Centrifugal pumps				
1/2 HP electric, submersible, 80 GPM, 1-1/2"	Ea	25.00	83.50	216.00
2 HP electric, submersible, 120 GPM, 2"	Ea	38.00	120.00	335.00
Trash pumps, 3", gas	Ea	45.00	140.00	384.00
Diaphragm pumps				
2", gas, single action	Ea	34.00	108.00	305.00
4", gas, double action	Ea	87.00	250.00	725.00
Hose, coupled, 25 LF sections				
2" suction line	Ea	8.40	29.00	82.50
1-1/2 discharge line	Ea	6.50	18.50	45.50
3" discharge line	Ea	9.50	26.00	62.00
Sandblasting equipment rental	Unit	Day	Week	Month
Compressor and hopper				
To 250 PSI	Ea	42.00	130.00	318.00
Over 250 to 300 PSI	Ea	63.00	196.50	534.00
Over 600 PSI	Ea	57.50	175.50	486.00
Accessories				
50' of 1" air hose, coupled	Ea	14.00	32.50	76.50
Nozzle with 50' of 1" sandblast hose and 10' of 3/4" whip hose	Ea	29.50	66.00	165.00
Valve, remote control	Ea	29.00	50.00	145.00
Air-fed hood	Ea	24.00	65.00	166.00
Street and highway equipment rental	Unit	Day	Week	Month
Drophammer breaker, gas powered, 1,500 lb	Ea	500.00	1,435.00	4,300.00
Paving breaker, backhoe mount, pneumatic, 1,000 lb	Ea	235.00	735.00	1,640.00
Paving machine, diesel, 10' width, self propelled	Ea	786.50	3,140.00	9,350.00
Distribution truck for asphalt prime coat, 3,000 gallon	Ea	746.00	2,250.00	6,640.00

General Requirements 1

	Unit	Day	Week	Month
Pavement striper, 1 line, walk behind	Ea	21.00	63.00	--
Water blast truck for 1,000 PSI pavement cleaning	Ea	1,125.00	3,270.00	10,100.00
Router-groover for joint or crack sealing in pavement	Ea	168.00	572.00	1,720.00
Sealant pot for joint or crack sealing	Ea	198.00	620.00	1,850.00
Water truck, 6,000 gallon	Ea	328.00	1,000.00	3,000.00
Self-propelled street sweeper, vacuum (3 CY max.)	Ea	--	2,230.00	6,700.00

	Unit	Day	Week	Month
Miscellaneous construction equipment rental				
Brush chipper, trailer mount	Ea	230.00	780.00	2,340.00
Chain saws, 18", gasoline	Ea	40.00	130.00	350.00
Circular hand saws, 8", gas powered	Ea	41.00	135.50	355.00
Chippers, electric, 60 lb	Ea	47.00	158.00	446.00
Generators, electric, portable				
2.5 kw, gasoline	Ea	40.00	127.00	362.00
5 kw, gasoline	Ea	60.50	189.00	525.00
15 kw, diesel	Ea	100.00	300.00	840.00
60 kw, diesel	Ea	170.00	490.00	1,370.00
Grinders, hand held, electric	Ea	16.50	53.00	150.00
Heaters, portable, oil or gasoline, fan forced				
To 200 MBtu	Ea	31.00	102.50	280.00
Over 200 to 300 MBtu	Ea	40.50	150.00	390.00
Over 500 to 1,000 MBtu	Ea	70.00	228.00	600.00
Salamanders, LP gas	Ea	26.00	82.00	200.00
Laser pipe levels	Ea	83.00	240.00	700.00
Light towers, trailer mounted, gas powered				
1,000 watt	Ea	100.00	320.00	890.00
4,000 watt	Ea	150.00	390.00	1,175.00
Masonry saws, excluding blade cost				
Wall saw	Ea	40.00	132.00	380.00
Table saw, 2 HP	Ea	40.00	128.00	350.00
Post hole digger, gas powered, 12" auger	Ea	55.00	190.00	520.00
Power washer (water blaster), gas powered	Ea	54.00	168.00	430.50
Sander, belt, 3" width	Ea	15.50	58.00	152.00
Sander, disc, 9" diameter	Ea	16.50	47.00	120.00
Welding machine, with helmet				
300 amp, gas powered	Ea	60.00	220.00	550.00
300 amp, electric powered	Ea	38.50	122.00	327.00
Wheelbarrows, contractor style	Ea	9.00	27.00	65.50

Concrete Pumping. Small jobs, including truck rent and operator. Based on 6 sack concrete mix designed for pumping. No concrete included.

	Craft@Hrs	Unit	Material	Labor	Total
3/8" aggregate mix (pea gravel), using hose to 200'					
First 7 CY (7 CY minimum)	--	CY	--	--	14.50
Additional cubic yards over 7	--	CY	--	--	6.00
Add for hose over 200', per LF	--	LF	--	--	1.00
3/4" aggregate mix, using hose to 200'					
First 10 CY (10 CY minimum)	--	CY	--	--	24.00
Additional cubic yards over 10	--	CY	--	--	6.00
Add for hose over 200', per LF	--	LF	--	--	1.00

General Requirements 1

	Craft@Hrs	Unit	Material	Labor	Total

Concrete Pumping with a Boom Truck. Includes truck rent, operator, local travel but no concrete. Add costs equal to 1 hour for equipment setup and 1 hour for cleanup. Use 4 hours as the minimum cost for 28 and 32 meter boom trucks and 5 hours as the minimum cost for 36 thru 52 meter boom trucks. Estimate the actual pour rate at 70% of the pump rated capacity on thicker slabs and 50% of the truck rated capacity on most other work. Boom lengths over 36 meters include both an operator (at $50 per hour) and an oiler (at $46 per hour). Costs shown do not include subcontractor's markup.

	Craft@Hrs	Unit	Material	Labor	Total
28 meter boom (92'), 70 CY per hour rating	--	Hr	--	--	55.00
Add for each CY pumped	--	CY	--	--	1.25
32 meter boom (105'), 90 CY per hour rating	--	Hr	--	--	65.00
Add for each CY pumped	--	CY	--	--	1.50
36 meter boom (120'), 90 CY per hour rating	--	Hr	--	--	80.00
Add for each CY pumped	--	CY	--	--	1.75
42 meter boom (138'), 100 CY per hour rating	--	Hr	--	--	115.00
Add for each CY pumped	--	CY	--	--	2.00
52 meter boom (170'), 100 CY per hour rating	--	Hr	--	--	150.00
Add for each CY pumped	--	CY	--	--	2.50

Temporary Enclosures

Rented chain link fence and accessories. Costs are a one-time charge for up to six months usage on a rental basis. Costs include installation and one trip for removal and assume level site with truck access along pre-marked fence line. Add for gates and barbed wire as shown. Minimum charge is $250.

	Craft@Hrs	Unit	Material	Labor	Total
Chain link fence, 6' high					
Less than 200 feet	--	LF	--	--	.95
201 to 400 feet	--	LF	--	--	.90
401 to 600 feet	--	LF	--	--	.80
601 to 800 feet	--	LF	--	--	.75
801 to 1,000 feet	--	LF	--	--	.70
Over 1,000 feet	--	LF	--	--	.65
Add for gates					
6' x 10' single	--	Ea	--	--	75.00
6' x 12' single	--	Ea	--	--	95.00
6' x 15' single	--	Ea	--	--	125.00
6' x 20' double	--	Ea	--	--	145.00
6' x 24' double	--	Ea	--	--	195.00
6' x 30' double	--	Ea	--	--	235.00
Add for barbed wire					
per strand, per lineal foot	--	LF	--	--	.20
Contractor furnished items, installed and removed, based on single use and no salvage value					
Railing on stairway, 2" x 4"	CL@.121	LF	.90	4.13	5.03
Guardrail at second and higher floors (toe rail, mid rail and top rail)	CL@.124	LF	1.75	4.23	5.98
Plywood barricade fence bolted to pavement	CL@.430	LF	6.30	14.70	21.00
Plywood fence, post in ground, 8' high	CL@.202	LF	7.10	6.89	13.99
Plywood fence with sidewalk cover, 8' high	CL@.643	LF	17.00	21.90	38.90
For two uses, any above, deduct, each use	--	%	-30.0	-20.0	--

Scaffolding Rental

Exterior scaffolding, tubular steel, 60" wide, with 6'4" open end frames set at 7', with cross braces, base plates, mud sills, adjustable legs, post-mounted guardrail, climbing ladders and landings, brackets, clamps and building ties. Including two 2" x 10" scaffold planks on all side brackets and six 2" x 10" scaffold planks on scaffold where indicated. Including local delivery and pickup.

General Requirements 1

	Craft@Hrs	Unit	Material	Labor	Total

Costs shown are rental costs per square foot of building wall covered, per month. Minimum rental is one month.

	Craft@Hrs	Unit	Material	Labor	Total
With plank on all side brackets only	--	SF	.40	--	--
With plank on side brackets and scaffold	--	SF	.75	--	--
Deduct for rental over 2 months	--	%	-20.0	--	--

Caster-mounted scaffolding, 30" wide by 7' or 10' long, monthly rental. Minimum charge is one month.

4' to 6' high	--	Mo	73.50	--	--
7' to 11' high	--	Mo	94.50	--	--
12' to 16' high	--	Mo	100.00	--	--
17' to 21' high	--	Mo	150.00	--	--
22' to 26' high	--	Mo	165.00	--	--
27' to 30' high	--	Mo	180.00	--	--
Hook-end scaffold plank, rental, each	--	Mo	8.00	--	--

Plain-end micro-lam scaffold plank, rental each

9' length	--	Mo	3.50	--	--
12' length	--	Mo	5.00	--	--

Add for erection and dismantling, level ground, truck accessible

With plank on all side brackets only	CL@.023	SF	--	.78	--
With planks on side brackets and scaffold	CL@.036	SF	--	1.23	--
Safety nets nylon, 4" mesh, purchase	CL@.004	SF	1.50	.14	1.64
Safety nets nylon, 1/2" mesh, purchase	CL@.004	SF	1.70	.14	1.84
Swinging stage, 10', complete, motor operated, purchase	CL@4.00	Ea	1,150.00	136.00	1,286.00

Heavy Duty Shoring. Adjustable vertical tower type shoring. Rated capacity 5.5 tons per leg. Base frames are 4' wide by 6' high or 2' wide by 5' high. Extension frames are 4' wide by 5'4" high or 2' wide by 4'4" high. Frames are erected in pairs using cross-bracing to form a tower. Combinations of base frames and extension frames are used to reach the height required. Screw jacks with base plates are installed in the bottom of each leg. Similar screw jacks with "U" bracket heads are installed in the top of each leg. Material costs shown are rental rates based on a 1-month minimum. Add the cost of delivery and pickup. For scheduling purposes, estimate that a crew of 2 can unload, handle and erect 8 to 10 frames (including typical jacks and heads) per hour. Dismantling, handling and moving or loading will require nearly the same time.

Frames (any size), including braces and assembly hardware, rental per frame per month	--	Mo	7.75	--	
Screw jacks (four required per frame), with base plate or "U" head, rental per jack per month	--	Mo	3.00	--	
Add for erecting each frame	CL@.245	Ea	--	8.36	
Add for dismantling each frame	CL@.197	Ea	--	6.72	

Temporary Structures. Portable job site office trailers with electric air conditioners and baseboard heat, built-in desks, plan tables, insulated walls, tiled floors, paneled walls, locking doors and windows with screens, closet, fluorescent lights, and electrical receptacles. Add $30 per month for units with toilet and lavatory. Monthly rental based on 6 month minimum. Dimensions show length, width and height overall and floor area (excluding hitch area).

16' x 8' x 7' high (108 SF)	--	Mo	--	--	120.00
20' x 8' x 7' high (160 SF)	--	Mo	--	--	125.00
24' x 8' x 7' high (192 SF)	--	Mo	--	--	140.00
30' x 8' x 7' high (240 SF)	--	Mo	--	--	175.00
36' x 10' x 10' high (360 SF)	--	Mo	--	--	260.00
44' x 10' x 10' high (440 SF)	--	Mo	--	--	290.00
50' x 10' x 10' high (500 SF)	--	Mo	--	--	310.00
50' x 12' x 10' high (552 SF)	--	Mo	--	--	360.00

General Requirements 1

	Craft@Hrs	Unit	Material	Labor	Total
60' x 12' x 10' high (720 SF)	--	Mo	--	--	400.00
60' x 14' x 10' high (840 SF)	--	Mo	--	--	500.00
Add for skirting, per LF of perimeter	--	LF	--	--	8.50
Deduct for unit without heating & cooling	--	Mo	--	--	-21.00
Add for delivery and setup within 15 miles					
Typical cost	--	LS	--	--	150.00
High cost	--	LS	--	--	317.00
Add for typical pickup within 15 miles	--	LS	--	--	148.00
Add for delivery or pickup over 15 miles, per mile					
8' wide units	--	Mile	--	--	1.45
10' wide units	--	Mile	--	--	2.00
12' wide units	--	Mile	--	--	2.30
14' wide units	--	Mile	--	--	2.85

Portable steel storage containers (lockable), suitable for job site storage of tools and materials, based on 6 month rental

	Craft@Hrs	Unit	Material	Labor	Total
50' x 12' x 8'	--	Mo	--	--	227.00
26' x 8' x 8'	--	Mo	--	--	97.00
22' x 8' x 7'	--	Mo	--	--	92.00
16' x 8' x 7'	--	Mo	--	--	86.50
Add for typical pickup or delivery	--	LS	--	--	79.00

Portable job site shacks with lights, power receptacles, locking door and window

	Craft@Hrs	Unit	Material	Labor	Total
12' x 8' x 8'	--	Mo	--	--	65.50
8' x 4' x 8'	--	Mo	--	--	59.00
Add for typical pickup or delivery	--	LS	--	--	79.00

Cleanup

Progressive "broom clean" cleanup, as necessary, per 1,000 SF of floor per month

	Craft@Hrs	Unit	Material	Labor	Total
Typical cost	--	MSF	--	--	5.00
High cost	--	MSF	--	--	15.00
Final, total floor area (no glass cleaning)	--	MSF	--	--	45.00
Dumpster, 3 CY trash bin, emptied weekly	--	Mo	--	--	43.50

Glass cleaning, per 1,000 square feet of glass cleaned on one side. (Double these figures when both sides are cleaned.) Add the cost of staging or scaffolding when needed

	Craft@Hrs	Unit	Material	Labor	Total
Cleaning glass with sponge and squeegee	CL@1.34	MSF	9.00	45.70	54.70
Cleaning glass and window trim with solvent, towel, sponge and squeegee	CL@5.55	MSF	12.50	189.00	201.50
Removing paint and concrete splatter from windows and trim by scraping and solvent	CL@15.9	MSF	125.00	542.00	667.00
Mop resilient floor by hand	CL@1.20	MSF	1.10	40.90	42.00

Site Work 2

Structure Moving. Up to 5 mile haul, not including new foundation costs, utility hookup or finishing, per SF of ground floor. These figures assume no obstruction from trees or utility lines and that adequate right of way is available. Fees and charges imposed by government are not included. These costs include the subcontractor's overhead and profit.

	Craft@Hrs	Unit	Material	Labor	Equipment	Total
Concrete structures						
1,000 to 2,000 SF	C1@.168	SF	--	5.85	3.85	9.70
2,000 to 4,000 SF	C1@.149	SF	--	5.19	3.60	8.79

Site Work 2

	Craft@Hrs	Unit	Material	Labor	Equipment	Total
Wood frame						
One story, 1,000 to 2,000 SF	C1@.142	SF	--	4.94	3.40	8.34
One story, 2,000 to 3,000 SF	C1@.131	SF	--	4.56	3.45	8.01
Two story, 1,000 to 4,000 SF	C1@.086	SF	--	2.99	3.40	6.39
Masonry, 1,000 to 2,000 SF	C1@.168	SF	--	5.85	3.85	9.70
Steel frame, 2,000 to 4,000 SF	C1@.199	SF	--	6.93	5.15	12.08

Site Clearing. Using a dozer as described below, a 3/4 CY tractor loader and two 12 CY dump trucks. Clearing costs will be higher on smaller jobs and where the terrain limits production. These costs include hauling 6 miles to a legal dump and the subcontractor's overhead and profit but no dump fees. Add the cost of mobilization and demobilization. Quantity shown in parentheses gives the approximate "loose" volume of the materials removed (volume being hauled to the dump).

	Craft@Hrs	Unit	Material	Labor	Equipment	Total
Clear light brush and grub roots,						
Using a 105 HP D-5 dozer (350 CY)	C2@34.8	Acre	--	1,360.00	780.00	2,140.00
Clear medium brush and small trees, grub roots,						
Using a 130 HP D-6 dozer (420 CY)	C2@41.8	Acre	--	1,630.00	1,010.00	2,640.00
Clear brush and trees to 6" trunk diameter,						
Using a 335 HP D-8 dozer (455 CY)	C2@45.2	Acre	--	1,770.00	1,540.00	3,310.00
Clear wooded area, pull stumps,						
Using a 460 HP D-9 dozer (490 CY)	C2@48.7	Acre	--	1,900.00	2,230.00	4,130.00
Protect existing mature trees	CL@2.06	Ea	--	70.20	33.50	103.70
Large orchard						
With tap root trees to 15' high	C3@1.07	Ea	--	40.90	19.50	60.40
Add for stripping and removing palm fronds for palms over 30'						
high, per palm tree	C3@2.00	Ea	--	76.40	29.90	106.30

For manual production or hand clearing, allow 2 times any of the labor costs above.

Paving and Curb Demolition. No salvage of materials. These costs include the subcontractor's overhead and profit and the cost of loading and hauling to a legal dump within 6 miles but no dump fees. Equipment cost per hour is $43.00 and includes one air compressor, one paving breaker and jackhammer bits, one 55 HP wheel loader with integral backhoe and one 5 CY dump truck.

	Craft@Hrs	Unit	Material	Labor	Equipment	Total
Bituminous paving, depths to 3"						
Large area, with a wheel loader	C3@.028	SY	--	1.07	.40	1.47
Strips 24" wide for utility lines	C3@.029	SF	--	1.11	.42	1.53
Add for jobs under 500 SF	--	%	--	--	--	35.0
Bituminous curbs, to 12" width	C3@.028	LF	--	1.07	.40	1.47
Concrete paving and slabs on grade						
4" concrete without rebars	C3@.325	CY	--	12.40	4.65	17.05
4" concrete without rebars	C3@.037	SY	--	1.41	.53	1.94
6" concrete without rebars	C3@.421	CY	--	16.10	6.03	22.13
6" concrete without rebars	C3@.071	SY	--	2.71	1.02	3.73
7" to 9" concrete without rebars	C3@.809	CY	--	30.90	11.60	42.50
7" to 9" concrete without rebars	C3@.180	SY	--	6.88	2.58	9.46
12" to 18" concrete without rebars	C3@1.37	CY	--	52.40	19.64	72.04
12" to 18" concrete without rebars	C3@.574	SY	--	21.90	8.23	30.13
Add for bar reinforced paving	--	%	--	--	--	15.0
Add for jobs under 500 SF	--	%	--	--	--	35.0
Concrete curbs						
Curb and 24" monolithic gutter	C3@.100	LF	--	3.82	1.43	5.25
Planter and batter type curbs, 6" width	C3@.043	LF	--	1.64	.62	2.26

Site Work 2

	Craft@Hrs	Unit	Material	Labor	Equipment	Total
Surface milling of pavement, per inch of cut depth						
Asphalt pavement	C3@.014	SY	--	.54	.20	.74
Concrete	C3@.027	SY	--	1.03	.39	1.42
Removal of pavement markings by water blasting. Equipment cost is $20.00 per hour.						
4" wide strips	CL@.032	LF	--	1.09	.64	1.73
Per square foot	CL@.098	SF	--	3.34	1.96	5.30

Fence and Guardrail Demolition. No salvage except as noted. These costs include the subcontractor's overhead and profit and the cost loading and hauling to a legal dump within 6 miles but no dump fees.

Fencing demolition. Equipment cost per hour is $15.00 and includes one 5 CY dump truck.

	Craft@Hrs	Unit	Material	Labor	Equipment	Total
Remove and dispose chain link, 6' high	C4@.015	LF	--	.53	.08	.61
Remove and salvage chain link, 6' high	C4@.037	LF	--	1.31	.19	1.50
Remove and dispose board fence, 6' high	C4@.016	LF	--	.56	.08	.64
Highway type guardrail demolition, small quantities						
Remove and dispose	C4@.043	LF	--	1.52	.22	1.74
Remove in salvage condition	C4@.074	LF	--	2.61	.37	2.98

Remove and dispose guard posts. Equipment cost per hour is $43.00 and includes one air compressor, one paving breaker and jackhammer bits, one 55 HP wheel loader with integral backhoe and one 5 CY dump truck.

	Craft@Hrs	Unit	Material	Labor	Equipment	Total
Cost to remove and dispose of guard posts	C3@.245	Ea	--	9.36	3.51	12.87

Manhole and Piping Demolition. No salvage of materials except as noted. These costs include the subcontractor's overhead and profit and the cost of loading and hauling to a legal dump within 6 miles but no dump fees. Equipment cost per hour is $43.00 and includes one air compressor, one paving breaker and jackhammer bits, one 55 HP wheel loader with integral backhoe and one 5 CY dump truck.

Manhole and catch basin demolition

	Craft@Hrs	Unit	Material	Labor	Equipment	Total
Precast, break below collar and plug	C3@6.64	Ea	--	254.00	95.20	349.20
Brick, break below collar and plug	C3@5.16	Ea	--	197.00	74.00	271.00
Add for sand fill, either of above	C3@.125	CY	11.00	4.78	1.79	17.57
Remove and dispose masonry manhole or catch basin to 10' deep	C3@5.33	Ea	--	204.00	76.40	280.40
Remove and dispose reinforced concrete manhole or catch basin to 10' deep	C3@6.78	Ea	--	259.00	97.20	356.20
Remove frame and cover from manhole or catch basin and salvage	C3@1.45	Ea	--	55.40	20.80	76.20
Remove and reset frame and cover from manhole or catch basin	C3@3.99	Ea	--	152.00	57.20	209.20
Fire hydrant demolition						
Remove and dispose	C3@5.57	Ea	--	213.00	79.80	292.80
Break out storm or sewer pipe, including disposal but no excavation						
Up to 12" diameter	C3@.114	LF	--	4.36	1.63	5.99
15" to 18"	C3@.134	LF	--	5.12	1.92	7.04
21" to 24"	C3@.154	LF	--	5.89	2.20	8.09
27" to 36"	C3@.202	LF	--	7.72	2.90	10.62
Remove welded steel pipe for salvage, no excavation included						
4" diameter or smaller	C3@.111	LF	--	4.24	1.59	5.83
6" to 10" diameter	C3@.202	LF	--	7.72	2.90	10.62

Remove and dispose of an empty underground liquid storage tank and backfill with sand. (Draining and disposing of hazardous liquid in a tank may require special waste handling equipment. Cost of draining not included.)

	Craft@Hrs	Unit	Material	Labor	Equipment	Total
50 to 250 gallon tank	C3@4.24	Ea	--	162.00	60.80	222.80
Over 250 to 600 gallon tank	C3@11.7	Ea	--	447.00	168.00	615.00

Site Work 2

	Craft@Hrs	Unit	Material	Labor	Equipment	Total
Over 600 to 1,000 gallon tank	C3@22.4	Ea	--	856.00	321.00	1,177.00
Add for sand fill, any of above	C3@.125	CY	11.00	4.78	1.80	17.58

Miscellaneous Sitework Demolition. No salvage of materials except as noted. These costs include the subcontractor's overhead and profit and the cost of loading and hauling to a legal dump within 6 miles but no dump fees.

Railroad demolition, siding quantities. Equipment cost per hour is $43.00 and includes one air compressor, one pneumatic breaker and jackhammer bits, one 55 HP wheel loader with integral backhoe and one 5 CY dump truck.

	Craft@Hrs	Unit	Material	Labor	Equipment	Total
Remove track and ties for scrap	C5@.508	LF	--	18.90	5.46	24.36
Remove and dispose of ballast stone	C5@.111	CY	--	4.13	1.19	5.32
Remove wood ties alone	C5@.143	Ea	--	5.32	1.54	6.86

Remove light standards, flagpoles, playground poles, up to 30' high, including foundations. Equipment cost per hour is $23.00 and includes one air compressor, one pneumatic breaker and jackhammer bits and one 5 CY dump truck.

	Craft@Hrs	Unit	Material	Labor	Equipment	Total
Remove item, in salvage condition	C4@6.24	Ea	--	220.00	47.80	267.80
Remove item, no salvage	C4@.746	Ea	--	26.30	5.72	32.02

Torch cutting steel plate. Equipment cost per hour is $2.50 for an oxygen-acetylene manual welding and cutting torch with gases, regulator and goggles.

	Craft@Hrs	Unit	Material	Labor	Equipment	Total
To 3/8" thick	CL@.073	LF	--	2.49	1.46	3.95

Building Demolition. Costs for demolishing an entire building. Includes loading and hauling up to 6 miles but no dump fees. By square foot of floor area based on 8' ceiling height. No salvage value assumed. These costs include the subcontractor's overhead and profit.

Light wood-frame structures. Based on 2,500 SF job. No basements included. Equipment cost per hour is $35.00 and includes one 55 HP wheel loader with integral backhoe and one 5 CY dump truck.

	Craft@Hrs	Unit	Material	Labor	Equipment	Total
One story	C5@.038	SF	--	1.41	.33	1.74
Two story	C5@.053	SF	--	1.97	.46	2.43
Three story	C5@.070	SF	--	2.60	.61	3.21

Concrete and masonry structures

Demolition with pneumatic tools. Based on 5,000 SF job. Equipment cost per hour is $86.00 and includes one air compressor, three paving breaker and jackhammer bits, two 55 HP wheel loaders with integral backhoes and two 5 CY dump trucks.

	Craft@Hrs	Unit	Material	Labor	Equipment	Total
Concrete building	C6@.089	SF	--	3.26	.77	4.03
Reinforced concrete building	C6@.101	SF	--	3.69	.87	4.56
Masonry building	C6@.074	SF	--	2.71	.64	3.35

Demolition with crane and headache ball. Based on a 10,000 SF job. Equipment cost per hour is $113.00 and includes one air compressor, one pneumatic paving breaker and jackhammer bits, one 55 HP wheel loader with integral backhoe, one 15-ton hydraulic crane with headache ball and three 5 CY dump trucks.

	Craft@Hrs	Unit	Material	Labor	Equipment	Total
Concrete building	C7@.031	SF	--	1.19	.50	1.69
Reinforced concrete building	C7@.037	SF	--	1.42	.60	2.02
Masonry building	C7@.026	SF	--	1.00	.42	1.42

Concrete foundation and footing demolition. Equipment cost per hour is $73.00 and includes one air compressor, one pneumatic paving breaker and jackhammer bits, one D-6 crawler tractor with attachments and one 5 CY dump truck.

	Craft@Hrs	Unit	Material	Labor	Equipment	Total
Non-reinforced concrete	C3@1.10	CY	--	42.00	26.80	68.80
Up to double curtain reinforcing	C3@1.58	CY	--	60.40	38.40	98.80
Chip out concrete (no dozer used)	CL@.203	CF	--	6.92	4.67	11.59

Site Work 2

	Craft@Hrs	Unit	Material	Labor	Equipment	Total

Gutting a building Interior finishes stripped back to the structural walls. Building structure to remain. No allowance for salvage value. These costs include loading and hauling up to 6 miles but no dump fees. Costs shown are per square foot of floor area based on 8' ceiling height. Costs will be about 50% less if a small tractor can be used and up to 25% higher if debris must be carried to ground level in an elevator. These costs include the subcontractor's overhead and profit. Equipment cost per hour is $27.00 and includes one air compressor, two pneumatic breakers with jackhammer bits and one 5 CY dump truck.

	Craft@Hrs	Unit	Material	Labor	Equipment	Total
Residential buildings	C4@.265	SF	--	9.34	2.39	11.73
Commercial buildings	C4@.243	SF	--	8.57	2.19	10.76

Partition wall demolition Building structure to remain. No allowance for salvage value. No disposal or dump fees included except as noted. Costs shown are per square foot of wall removed (as measured on one side). These costs include the subcontractor's overhead and profit.
Knock down with pneumatic jackhammers and pile adjacent to site but no removal. Equipment cost per hour is $12.00 and includes one air compressor and two pneumatic breakers with jackhammer bits.

Brick or block partition demolition

	Craft@Hrs	Unit	Material	Labor	Equipment	Total
4" thick partition	CL@.040	SF	--	1.36	.24	1.60
8" thick partition	CL@.057	SF	--	1.94	.34	2.28
12" thick partition	CL@.073	SF	--	2.49	.44	2.93

Concrete partition demolition

	Craft@Hrs	Unit	Material	Labor	Equipment	Total
Non-reinforced	CL@.273	CF	--	9.31	1.64	10.95
Reinforced	CL@.363	CF	--	12.40	2.18	14.58

Knock down with hand tools and pile adjacent to site but no removal

Stud partition demolition

	Craft@Hrs	Unit	Material	Labor	Equipment	Total
Gypsum or terra cotta on metal lath	CL@.105	SF	--	3.58	--	--
Drywall on metal or wood studs	CL@.040	SF	--	1.36	--	--
Plaster on metal studs	CL@.075	SF	--	2.56	--	--

Ceiling demolition Building structure to remain. No allowance for salvage value. No disposal or dump fees included except as noted. Costs shown are per square foot of ceiling (as measured on one side). These costs include the subcontractor's overhead and profit.
Knocked down with hand tools to remove ceiling finish on framing at heights to 9'.

Plaster ceiling

	Craft@Hrs	Unit	Material	Labor	Equipment	Total
Including lath and furring	CL@.025	SF	--	.85	--	--
Including suspended grid	CL@.020	SF	--	.68	--	--

Acoustic tile ceiling

	Craft@Hrs	Unit	Material	Labor	Equipment	Total
Including suspended grid	CL@.010	SF	--	.34	--	--
Including grid in salvage condition	CL@.019	SF	--	.65	--	--
Including strip furring	CL@.014	SF	--	.48	--	--
Tile glued to ceiling	CL@.013	SF	--	.44	--	--

Drywall ceiling

	Craft@Hrs	Unit	Material	Labor	Equipment	Total
Nailed or attached with screws to joists	CL@.020	SF	--	.68	--	--
Including strip furring	CL@.025	SF	--	.85	--	--

Roof and floor demolition Building structure to remain. No allowance for salvage value. No disposal or dump fees included except as noted. Costs shown are per square foot of area removed (as measured on one side). These costs include the subcontractor's overhead and profit.

Roof surface removal, using hand tools

	Craft@Hrs	Unit	Material	Labor	Equipment	Total
Asphalt shingles	CL@.707	Sq	--	24.10	--	--
Built-up, including sheathing and gravel	CL@2.91	Sq	--	99.20	--	--
Clay or concrete tile	CL@1.03	Sq	--	35.10	--	--
Concrete plank without covering	CL@1.10	Sq	--	37.50	--	--
Gypsum plank without covering	CL@.820	Sq	--	28.00	--	--

Site Work 2

	Craft@Hrs	Unit	Material	Labor	Equipment	Total
Metal deck without covering	CL@2.27	Sq	--	77.40	--	--
Wood shingles	CL@.756	Sq	--	25.80	--	--
Add for removing gravel stop or flashing	CL@.070	LF	--	2.39	--	--

Floor slab demolition, using pneumatic jackhammers and piled adjacent to site but no removal. Equipment cost per hour is $12.00 and includes one air compressor and two pneumatic breakers with jackhammer bits.

	Craft@Hrs	Unit	Material	Labor	Equipment	Total
Slab on grade, 4" to 6" thick,						
reinforced with wire mesh	CL@.057	SF	--	1.94	.34	2.28
reinforced with number 4 bars	CL@.063	SF	--	2.15	.38	2.53
Slab, suspended, 6" to 8" thick, free fall	CL@.092	SF	--	3.14	.55	3.69
Slab fill, lightweight concrete fill						
on metal deck	CL@.029	SF	--	.99	.17	1.16
Topping, insulating, no sandblasting	CL@.023	SF	--	.78	.14	.92

Floor covering demolition, using pneumatic jackhammers and piled adjacent to site but no removal. Equipment cost per hour is $12.00 and includes one air compressor and two pneumatic breakers with jackhammer bits.

	Craft@Hrs	Unit	Material	Labor	Equipment	Total
Ceramic tile, quarry tile, brick	CL@.027	SF	--	.92	.16	1.08
Resilient materials only	CL@.006	SF	--	.21	.04	.25
Terrazzo	CL@.043	SF	--	1.47	.26	1.73
Wood blocks	CL@.015	SF	--	.51	.09	.60
Wood, residential strip floor	CL@.009	SF	--	.31	.05	.36

Debris removal These costs include the subcontractor's overhead and profit. Break demolition debris into manageable size with hand tools, load into wheelbarrow, move to chute and dump. Per cubic foot of material dumped

	Craft@Hrs	Unit	Material	Labor	Equipment	Total
Haul 50' to trash chute and dump	CL@.018	CF	--	.61	--	--
Haul 100' to trash chute and dump	CL@.022	CF	--	.75	--	--
Haul 50' to elevator, descend 10						
floors to chute and dump	CL@.024	CF	--	.82	--	--

Load demolition debris on truck and haul 6 miles to a dump site. Includes truck cost but no dump fees. Equipment cost per hour is based on $15.00 per hour for a 5 CY dump truck and $20.00 per hour for a 55 HP wheel loader with integral backhoe where shown.

	Craft@Hrs	Unit	Material	Labor	Equipment	Total
Per cubic yard of debris						
Loaded by hand	C4@.828	CY	--	29.20	4.14	33.34
Loaded using a wheel loader	C5@.367	CY	--	13.60	3.21	16.81

Cutting openings in frame walls, using hand tools. Cost per square foot of wall measured on one face.

	Craft@Hrs	Unit	Material	Labor	Equipment	Total
Metal stud wall with stucco or plaster	CL@.094	SF	--	3.21	--	3.21
Wood stud wall with drywall	CL@.051	SF	--	1.74	--	1.74
Dust control partitions, 6 mil plastic and frame	C8@.011	SF	.25	.42	--	.67

Trash chutes. Prefabricated steel chute installed and removed in a multi-story building. One-time charge for up to six-months usage.

	Craft@Hrs	Unit	Material	Labor	Equipment	Total
18" diameter	CL@.479	LF	--	16.30	16.50	32.80
36" diameter	CL@.634	LF	--	21.60	25.50	47.10

Removal of interior items. Based on removal using hand tools. Items removed in salvageable condition do not include an allowance for salvage value. These costs include the subcontractor's overhead and profit.

Removal, in non-salvageable condition
  Hollow metal door and frame in a masonry wall

	Craft@Hrs	Unit	Material	Labor	Equipment	Total
Single door to 4' x 7'	CL@1.56	Ea	--	53.20	--	--
Two doors, per opening to 8' x 7'	CL@1.99	Ea	--	67.90	--	--

Site Work 2

	Craft@Hrs	Unit	Material	Labor	Equipment	Total
Wood door and frame in a wood-frame wall						
Single door to 4' x 7'	CL@.832	Ea	--	28.40	--	--
Two doors, per opening to 8' x 7'	CL@1.56	Ea	--	53.20	--	--
Removal, in salvageable condition						
Door and frame						
Hollow metal door to 4' x 7' in masonry	C8@3.55	Ea	--	137.00	--	--
Wood door to 4' x 7' in frame wall	C8@2.21	Ea	--	85.00	--	--
Frame for resilient mat, metal						
per SF of mat area	C8@.054	SF	--	2.08	--	--
Lockers, metal 12" W, 60" H, 15" D	C8@1.13	Ea	--	43.50	--	--
Sink and soap dispenser, wall-hung	C8@.410	Ea	--	15.80	--	--
Toilet partitions, wood or metal	C8@1.52	Ea	--	58.50	--	--
Urinal screens, wood or metal	C8@.564	Ea	--	21.70	--	--
Window and frame, wood or metal	C8@.076	SF	--	2.92	--	--
Remove and reset airlock doors	C8@5.56	Ea	--	214.00	--	--

Asbestos Removal and Disposal. Typical costs including site preparation, monitoring, equipment, removal and disposal of waste. See detailed costs below. These costs include the subcontractor's overhead and profit.

Ceiling insulation in containment structure						
500 to 5,000 SF job	--	SF	--	--	--	30.00
5,000 to 20,000 SF job	--	SF	--	--	--	20.00
Pipe insulation in containment structure						
100 to 1,000 LF of 6" pipe	--	LF	--	--	--	55.00
1,000 to 3,000 LF of 6" pipe	--	LF	--	--	--	35.00
Pipe insulation using glove bags						
100 to 1,000 LF of 6" pipe	--	LF	--	--	--	45.00
1,000 to 3,000 LF of 6" pipe	--	LF	--	--	--	35.00

Asbestos hazard surveys Building inspection, hazard identification, sampling and ranking of asbestos risk. These costs include the subcontractor's overhead and profit.

Asbestos hazard survey and sample collection, (10,000 SF per hour), four hour minimum	--	Hour	--	--	--	52.50
Sample analysis (usually one sample per 1,000 SF), ten sample minimum	--	Ea	--	--	--	36.50
Report writing (per 1,000 SF of floor), $100 minimum	--	MSF	--	--	--	2.00

Asbestos removal site preparation Preparing an unoccupied building for friable asbestos abatement, including erection and dismantling of enclosures and barriers. These figures do not include the cost of removing movable furnishings and equipment from the containment area, modifying electrical and mechanical equipment in the containment area, or restoring surfaces after asbestos is removed. Costs will be higher when work is done in an occupied building and in localities with special site preparation or monitoring requirements. These figures include the cost of small tools and consumable supplies needed for abatement and the subcontractor's overhead and profit.

Videotape the work area (5,000 SF of floor area per hour), two hour minimum	--	Hour	--	--	--	52.50
Plastic cover for surfaces in containment area, includes pre-cleaning	CL@5.00	CSF	10.80	171.00	--	181.80
Enclosure walls on 2" x 4" framing, per 100 SF of enclosure wall						
Plywood	CC@3.25	CSF	46.50	139.00	--	185.50
Vinyl-coated gypboard	CC@7.99	CSF	82.50	342.00	--	424.50

Site Work 2

	Craft@Hrs	Unit	Material	Labor	Equipment	Total
Public access tunnel enclosures, vinyl-coated gypboard on 2" x 4" frame, plywood roof, 8' high, 2'6" wide, with flush-mount lights, per LF of tunnel, based on three uses of materials	CC@1.04	LF	19.00	44.60	--	63.60
Containment perimeter barrier, plastic attached to the building or enclosure walls, floor and ceiling						
1 layer (walls, ceiling)	CC@2.00	CSF	6.20	85.70	--	91.90
2 layers (floor)	CC@4.98	CSF	12.50	213.00	--	225.50
Tunnels	CC@.450	LF	1.40	19.30	--	20.70
Scaffolding, tubular steel, rental cost including erection and dismantling, per SF of wall area covered. Cost per SF per month used.						
In an open area	CL@.035	SF	--	1.19	2.15	3.34
Congested area	CL@.070	SF	--	2.39	2.15	4.54
Rolling tower, 4' x 6' to 15' high	CL@2.00	Ea	--	68.20	120.00	188.20
Decontamination unit (equipment, shower & clean rooms), plastic on wood frame, 5-man capacity, includes shower, heater, centrifugal pump, cartridge filter and piping.						
Prefabricated unit, delivered ready-to-use	--	Ea	1,500.00	--	--	--

Asbestos removal monitoring These costs include the subcontractor's overhead and profit.

	Craft@Hrs	Unit	Material	Labor	Equipment	Total
Daily air sampling (8 samples per shift) using phase contrast microscopy (PCM) at $20 per sample, per shift	--	Ea	--	--	--	160.00
Final air test (one sample per 5,000 SF), 3 sample minimum, PCM method, per sample	--	Ea	--	--	--	20.00
Final air test by transmission electron microscopy (TEM), if required, typically 5 to 13 samples per project, per sample	--	Ea	--	--	--	200.00
Industrial hygiene technician on site for 100% of project duration	--	Hour	--	--	--	47.50
Senior industrial hygienist on site, usually 50% of project duration	--	Hour	--	--	--	67.50
Industrial hygiene project director on site, usually 10% of project duration	--	Hour	--	--	--	95.00

Asbestos removal equipment These costs include the subcontractor's overhead and profit.

	Craft@Hrs	Unit	Material	Labor	Equipment	Total
Negative air vent system for a containment structure. Allow one fan per each 30,000 CF of enclosure with a two fan minimum, four air changes per hour						
Rental per day	--	Ea	--	--	30.00	--
Filters for vent fans, cost per day	--	Ea	31.50	--	--	--
High efficiency particulate air filter (HEPA) vacuum cleaner						
Rental per day	--	Ea	--	--	14.50	--
Vacuum cleaner maintenance, cost per day	--	Ea	--	--	19.00	--
Tyvek disposable protective suits, allow 4 suits per worker per shift, cost per suit	--	Ea	4.50	--	--	--
Respirator cartridges, per set of two, 4 sets per worker per shift, cost per set	--	Ea	3.75	--	--	--

Asbestos removal Costs are based on work done in a containment structure with standing headroom. See the section that follows for work done in crawl spaces. The crew is three laborers for each 1,000 SF of containment floor area. Work includes cutting, scraping and brushing away moistened friable asbestos with hand tools and bagging waste materials for disposal. Work is usually done on two 12-hour shifts per day. Add the cost of site preparation, monitoring, equipment, waste disposal, and restoring surfaces after asbestos is removed. These costs include the subcontractor's overhead and profit.

	Craft@Hrs	Unit	Material	Labor	Equipment	Total
Pipe insulation						
1/2" diameter pipe (72 LF per hour)	CL@.014	LF	--	.48	--	--
3/4" diameter pipe (59 LF per hour)	CL@.017	LF	--	.58	--	--

Site Work 2

	Craft@Hrs	Unit	Material	Labor	Equipment	Total
Asbestos removal, Pipe insulation, continued						
1" diameter pipe (46 LF per hour)	CL@.022	LF	--	.75	--	--
2" diameter pipe (26 LF per hour)	CL@.038	LF	--	1.30	--	--
3" diameter pipe (18 LF per hour)	CL@.055	LF	--	1.88	--	--
4" diameter pipe (14 LF per hour)	CL@.072	LF	--	2.46	--	--
6" diameter pipe (9.1 LF per hour)	CL@.110	LF	--	3.75	--	--
8" diameter pipe (7.1 LF per hour)	CL@.141	LF	--	4.81	--	--
10" diameter pipe (5.6 LF per hour)	CL@.179	LF	--	6.10	--	--
12" diameter pipe (4.8 LF per hour)	CL@.209	LF	--	7.13	--	--
Pipe fittings (for each fitting use the cost for 3 LF of pipe of the same diameter)						
Steel beams (costs for steel columns will be 20% less)						
W6x12 (5.0 LF per hour)	CL@.200	LF	--	6.82	--	--
W8x17 (3.8 LF per hour)	CL@.266	LF	--	9.07	--	--
W8x31 (3.3 LF per hour)	CL@.303	LF	--	10.30	--	--
W10x19 (3.7 LF per hour)	CL@.270	LF	--	9.21	--	--
W10x49 (2.4 LF per hour)	CL@.415	LF	--	14.20	--	--
W12x16.5 (3.4 LF per hour)	CL@.294	LF	--	10.00	--	--
W12x40 (2.6 LF per hour)	CL@.384	LF	--	13.10	--	--
W12x58 (2.3 LF per hour)	CL@.436	LF	--	14.90	--	--
W14x26 (2.9 LF per hour)	CL@.345	LF	--	11.80	--	--
W16x31 (2.6 LF per hour)	CL@.384	LF	--	13.10	--	--
W18x45 (2.1 LF per hour)	CL@.476	LF	--	16.20	--	--
W24x55 (1.8 LF per hour)	CL@.555	LF	--	18.90	--	--
W24x130 (1.4 LF per hour)	CL@.713	LF	--	24.30	--	--
W30x108 (1.4 LF per hour)	CL@.713	LF	--	24.30	--	--
W36x150 (1.1 LF per hour)	CL@.713	LF	--	24.30	--	--
Deduct for steel columns	--	%	--	-20.0	--	--
Asbestos overspray (2.7 SF per hour)	CL@.372	SF	--	12.70	--	--
Boiler surfaces, breaching (10 SF per hour)	CL@.100	SF	--	3.41	--	--
Concrete, smooth (7.7 SF per hour)	CL@.130	SF	--	4.43	--	--
Corrugated surfaces (9 SF per hour)	CL@.111	SF	--	3.79	--	--
Duct insulation (9 SF per hour)	CL@.109	SF	--	3.72	--	--
Plaster, smooth (6.7 SF per hour)	CL@.149	SF	--	5.08	--	--
Steel, smooth (10 SF per hour)	CL@.100	SF	--	3.41	--	--
Tanks (10 SF per hour)	CL@.100	SF	--	3.41	--	--

Asbestos removal in crawl spaces Costs assume work is done in a containment structure with less than 3' of headroom. Crew size is three laborers for each 1,000 SF of containment floor area. Work includes cutting, scraping and brushing away moistened friable asbestos with hand tools and bagging waste materials for disposal. Add the cost of site preparation, monitoring, equipment, waste disposal, and restoring surfaces after asbestos is removed. These costs include the subcontractor's overhead and profit.

	Craft@Hrs	Unit	Material	Labor	Equipment	Total
Job preparation and cleanup, per SF of containment area						
Pre-clean floors and walls (6.3 SF per hour)	CL@.159	SF	--	5.42	--	--
Cover adjacent surfaces with plastic (5.9 SF per hour)	CL@.170	SF	--	5.80	--	--
Remove and bag plastic cover (50 SF per hour)	CL@.020	SF	--	.68	--	--
Pipe insulation						
1" diameter pipe (6.3 LF per hour)	CL@.159	LF	--	5.42	--	--
1-1/2" diameter pipe (2.1 LF per hour)	CL@.476	LF	--	16.20	--	--
2" diameter pipe (2.0 LF per hour)	CL@.500	LF	--	17.10	--	--
3" diameter pipe (1.8 LF per hour)	CL@.555	LF	--	18.90	--	--

Site Work 2

	Craft@Hrs	Unit	Material	Labor	Equipment	Total
4" diameter pipe (1.7 LF per hour)	CL@.588	LF	--	20.10	--	--
6" diameter pipe (1.3 LF per hour)	CL@.768	LF	--	26.20	--	--
8" diameter pipe (1.2 LF per hour)	CL@.835	LF	--	28.50	--	--
Pipe fittings (for each fitting use the cost for 3 LF of pipe of the same diameter)						
Ductwork (1.0 SF of duct per hour)	CL@1.00	SF	--	34.10	--	--

Asbestos removal with glove bags Costs are based on work done in a containment structure. A negative pressure air system and a decontamination unit are required when glove bags are used. The crew is three laborers (wearing disposable protective suits and respirators) for each 1,000 SF of containment floor area. Work includes cutting, scraping and brushing friable asbestos with hand tools into glove bags. Bags cost about $5 each and come with the scraping knife already in the bag. Add the cost of site preparation, monitoring, waste disposal, and restoring surfaces after asbestos is removed. These costs include the subcontractor's overhead and profit.

Pipe insulation

	Craft@Hrs	Unit	Material	Labor	Equipment	Total
1/2" diameter pipe (3.3 LF per hour)	CL@.303	LF	--	10.30	--	--
3/4" diameter pipe (2.9 LF per hour)	CL@.345	LF	--	11.80	--	--
1" diameter pipe (2.5 LF per hour)	CL@.399	LF	--	13.60	--	--
1-1/2" diameter pipe (2.2 LF per hour)	CL@.454	LF	--	15.50	--	--
2" diameter pipe (2.0 LF per hour)	CL@.500	LF	--	17.10	--	--
2-1/2" diameter pipe (1.9 LF per hour)	CL@.527	LF	--	18.00	--	--
3" diameter pipe (1.8 LF per hour)	CL@.555	LF	--	18.90	--	--
4" to 6" diameter pipe (1.4 LF per hour)	CL@.713	LF	--	24.30	--	--
8" to 12" diameter pipe (1.1 LF per hour)	CL@.909	LF	--	31.00	--	--

Pipe fittings (for each fitting use the cost for 3 LF of pipe of the same diameter)

Asbestos waste disposal Typical costs for removing bags from the containment structure, loading into fiber drums, hauling to an approved dump site, and dump fees. Costs are per cubic yard of asbestos waste removed and disposed. Local regulations may or may not require that disposal bags be buried in fiber drums used for transportation. These costs include the subcontractor's overhead and profit.

	Craft@Hrs	Unit	Material	Labor	Equipment	Total
Disposal bags (1.5 CF, $.70 each)	--	CY	12.60	--	--	--
Remove bags from containment structure (.05 manhours per bag, 18 bags per CY)	CL@.900	CY	--	30.70	--	--
Fiber drums, $12.50 each, holds 3 disposal bags, six bags per CY	--	CY	25.00	--	--	--
Hauling waste to a legal dump ($5 per drum)	--	CY	--	--	--	10.00
Dump fees ($1 per drum)	--	CY	--	--	--	2.00

Asbestos encapsulation Spraying sealant on asbestos-based acoustical plaster or asbestos residue after removal. The crew is one painter working with an airless spray rig (at $25 per day). No containment structure or decontamination unit are needed, but add the cost of videotaping, protecting adjacent areas, monitoring and cleaning overspray. These costs include the subcontractor's overhead and profit.

	Craft@Hrs	Unit	Material	Labor	Equipment	Total
Penetrating sealant (at $7.50 per gallon)						
On walls and ceilings (25 SF per gallon)	PA@.040	SF	.30	1.62	.15	2.07
On columns and beams (25 SF per gallon)	PA@.070	SF	.30	2.84	.25	3.39
On pipe insulation, to 12" diameter (20 LF per gallon)	PA@.170	LF	.38	6.90	.55	7.83
Bridging sealant (at $20.00 per gallon)						
On walls and ceilings (31 SF per gallon)	PA@.040	SF	.65	1.62	.15	2.42
On columns and beams (31 SF per gallon)	PA@.070	SF	.65	2.84	.25	3.74
On pipe insulation, to 12" diameter (16 LF per gallon)	PA@.170	LF	1.25	6.90	.55	8.70
Protective covering for adjacent surfaces	PA@.050	SF	.05	2.03	--	2.08

Site Work 2

Wall Sawing, Subcontract. Using an electric saw. Per LF of cut including overcuts at each corner equal to the depth of the cut. Costs shown assume electric power is available. Minimum cost will be $250.

	Unit	Brick or block	Concrete w/#4 bar	Concrete w/#7 bar
To 4" depth	LF	7.15	8.30	9.35
To 5" depth	LF	8.30	9.35	11.00
To 6" depth	LF	9.35	11.00	13.20
To 7" depth	LF	11.00	13.20	15.40
To 8" depth	LF	13.20	15.40	17.60
To 10" depth	LF	15.40	17.60	22.10
To 12" depth	LF	17.60	22.10	24.20
To 14" depth	LF	22.10	24.20	37.50
To 16" depth	LF	24.20	37.50	46.30
To 18" depth	LF	28.70	46.30	50.50

Concrete Core Drilling, Subcontract. Using an electric drill. Costs shown are per LF of depth for diameter shown. Minimum cost will be $170.

	Craft@Hrs	Unit	Material	Labor	Total
1" diameter	--	LF	--	--	30.00
1-1/2"	--	LF	--	--	33.00
2"	--	LF	--	--	35.00
2-1/2"	--	LF	--	--	37.00
3"	--	LF	--	--	39.00
3-1/2"	--	LF	--	--	44.00
4"	--	LF	--	--	46.00
5"	--	LF	--	--	55.00
6"	--	LF	--	--	66.00
8"	--	LF	--	--	80.00
10"	--	LF	--	--	99.00
12"	--	LF	--	--	130.00
14"	--	LF	--	--	130.00
16"	--	LF	--	--	210.00
18"	--	LF	--	--	230.00
20"	--	LF	--	--	275.00
24"	--	LF	--	--	330.00
Hourly rate, 2 hour minimum	--	Hr	--	--	85.00

Prices are based on one mat reinforcing with 3/8" to 5/8" bars and include cleanup. Difficult drill setups are higher. Prices assume availability of 110 volt electricity. Figure travel time at $85 per hour.

Concrete Slab Sawing, Subcontract. Using a gasoline powered saw. Costs per linear foot for cured concrete assuming a level surface with good access and saw cut lines laid out and pre-marked by others. Costs include local travel time. Minimum cost will be $130.

Depth	Under 200'	200'-400'	400'-600'	600'-1000'	Over 1000'
1"	.65	.55	.50	.40	.35
1-1/2"	.70	.65	.60	.50	.45
2"	.80	.80	.75	.65	.60
2-1/2"	.90	.85	.80	.70	.65
3"	.95	.90	.85	.75	.70
3-1/2"	1.10	1.05	1.00	.90	.85
4"	1.25	1.15	1.10	1.00	.90
5"	1.55	1.45	1.40	1.30	1.25
6"	1.85	1.70	1.65	1.55	1.50

Site Work 2

Green (5 days old) concrete sawing will usually cost 25 to 50% less. Work done on an hourly basis will cost $88 per hour for slabs up to 4" thick and $100 per hour for 5" or 6" thick slabs. A two hour minimum charge will apply on work done on an hourly basis.

Asphalt Sawing, Subcontract. Using a gasoline powered saw. Minimum cost will be $130. Cost per linear foot of green or cured asphalt, assuming level surface with good access and saw cut lines laid out and pre-marked by others. Costs include local travel time.

Depth	Under 450'	450'-600'	600'-1000'	Over 1000'
1"	.22	.20	.16	.14
1-1/2"	.26	.24	.20	.18
2"	.32	.30	.26	.24
2-1/2"	.34	.32	.28	.26
3"	.36	.35	.30	.28
3-1/2"	.42	.40	.36	.34
4"	.46	.44	.41	.36
5"	.58	.56	.52	.50
6"	.68	.66	.62	.60

Work done on an hourly basis will cost $88 per hour for asphalt up to 4" thick and $100 per hour for 5" or 6" thick asphalt. A two hour minimum charge will apply on work done on an hourly basis.

Rock Excavation. These costs include the subcontractor's overhead and profit.

Rock drilling costs. Based on using a pneumatic track-mounted wagon drill. Equipment cost per hour is $62.50 and includes one compressor and one 4" pneumatic wagon drill with drill bits. Costs shown are per linear foot of hole drilled. Add blasting costs below.

LF per hour shown is for a 2-man crew	Craft@Hrs	Unit	Material	Labor	Equipment	Total
Easy work, 38 LF per hour	S1@.052	LF	--	2.00	1.64	3.64
Moderate work, 30 LF per hour	S1@.067	LF	--	2.58	2.11	4.69
Most work, 25 LF per hour	S1@.080	LF	--	3.08	2.52	5.60
Hard work, 20 LF per hour	S1@.100	LF	--	3.85	3.13	6.98
Dense rock, 15 LF per hour	S1@.132	LF	--	5.09	4.16	9.25
Drilling 2-1/2" hole for rock bolts, 24 LF/hour	S1@.086	LF	--	3.31	2.70	6.01

Blasting costs. Based on two cycles per hour and 20 linear foot lifts. These costs assume 75% fill of holes with explosives costing $2.00 per pound. Equipment cost per hour is $12.50 and includes one flatbed truck. Costs shown are per cubic yard of area blasted. Loading or hauling of blasted material not included

Load explosives in holes and detonate, by pattern spacing. CY per hour shown is for a 5-man crew						
6' x 6' pattern (26 CY per hour)	C1@.192	CY	6.10	6.68	.48	13.26
7' x 7' pattern (35 CY per hour)	C1@.144	CY	4.42	5.01	.36	9.79
7' x 8' pattern (47 CY per hour)	C1@.106	CY	3.41	3.69	.27	7.37
9' x 9' pattern (60 CY per hour)	C1@.084	CY	3.04	2.92	.21	6.17
10' x 10' pattern (72 CY per hour)	C1@.069	CY	2.42	2.40	.17	4.99
12' x 12' pattern (108 CY per hour)	C1@.046	CY	1.67	1.60	.12	3.39
14' x 14' pattern (147 CY per hour)	C1@.034	CY	1.20	1.18	.09	2.47

Ripping rock. Based on using a tractor-mounted ripper. Equipment cost is $110 per hour for a D-8 tractor and $175 per hour for a D-9 tractor. Costs shown are per CY of area ripped. Loading or hauling ripped material not included.

Clay or glacial tills, D-9 cat tractor with 2-shank ripper (500 CY per hour)	T0@.002	CY	--	.09	.35	.44
Clay or glacial tills, D-8 cat tractor with 1-shank ripper (470 CY per hour)	T0@.002	CY	--	.09	.23	.32
Shale, sandstone, or limestone, D-9 cat tractor with 2-shank ripper (125 CY per hour)	T0@.008	CY	--	.34	1.40	1.74

Site Work 2

	Craft@Hrs	Unit	Material	Labor	Equipment	Total

Ripping rock, continued

Shale, sandstone, or limestone, D-8 cat tractor with

	Craft@Hrs	Unit	Material	Labor	Equipment	Total
1-shank ripper (95 CY per hour)	TO@.011	CY	--	.47	1.16	1.63

Slate, metamorphic rock, D-9 cat tractor with

	Craft@Hrs	Unit	Material	Labor	Equipment	Total
2-shank ripper (95 CY per hour)	TO@.011	CY	--	.47	1.84	2.31

Slate, metamorphic rock, D-8 cat tractor with

	Craft@Hrs	Unit	Material	Labor	Equipment	Total
1-shank ripper (63 CY per hour)	TO@.016	CY	--	.69	1.75	2.44

Granite or basalt, D-9 cat tractor with 2-shank

	Craft@Hrs	Unit	Material	Labor	Equipment	Total
ripper (78 CY per hour)	TO@.013	CY	--	.56	2.24	2.80

Granite or basalt, D-8 cat tractor with 1-shank

	Craft@Hrs	Unit	Material	Labor	Equipment	Total
ripper (38 CY per hour)	TO@.027	CY	--	1.16	2.89	4.05

Rock loosening. Based on using pneumatic jackhammers. Equipment cost per hour is $12.00 and includes one compressor and two jackhammers with drill points. Costs shown are per CY of area loosened. Loading and hauling not included.

	Craft@Hrs	Unit	Material	Labor	Equipment	Total
Igneous or dense rock (.7 CY per hour)	CL@1.43	CY	--	48.80	8.58	57.38
Most weathered rock (1.2 CY per hour)	CL@.835	CY	--	28.50	5.01	33.51
Soft sedimentary rock (2 CY per hour)	CL@.500	CY	--	17.10	3.00	20.10

Site Grading. These costs include the subcontractor's overhead and profit. Layout, staking, flagmen, lights, watering, ripping, rock breaking, loading or hauling not included. Based on a 10 acre site. (One acre is 43,560 SF or 4,840 SY.)

Site grading based on using one 125 HP (10,000 lb) motor grader. Equipment cost is $28 per hour. Acres per hour shown are based on a 2-man crew.

	Craft@Hrs	Unit	Material	Labor	Equipment	Total
General area rough grading (.7 acres per hour)	S1@2.86	Acre	--	109.00	40.00	149.00
Fine grading subgrade to 1/10' (.45 acres per hour)	S1@4.44	Acre	--	172.00	62.20	234.20
Cut slope or shape embankment to 2' high (.25 acres per hour)	S1@8.00	Acre	--	308.00	112.00	420.00
Rough grade sub-base course on roadway (.2 acres per hour)	S1@10.0	Acre	--	385.00	140.00	525.00
Finish grade base or leveling course on roadway (.17 acres per hour)	S1@11.8	Acre	--	455.00	165.00	620.00
Finish grading for building slab, windrow excess (1,100 SF per hour)	S1@1.82	MSF	--	70.10	25.50	95.60

Site grading based on using one crawler tractor. Equipment cost per hour is as shown.

General area rough grading with 335 HP D-8 tractor at $110 per hour

	Craft@Hrs	Unit	Material	Labor	Equipment	Total
(.50 acres per hour)	TO@2.00	Acre	--	86.00	220.00	306.00

General area rough grading with 75 HP D-4 tractor at $28 per hour

	Craft@Hrs	Unit	Material	Labor	Equipment	Total
(.25 acres per hour)	TO@4.00	Acre	--	172.00	117.00	289.00

Embankment Grading. Earth embankment spreading, shaping and compacting. Based on a project with 2,000 CY or more of earth embankment work. These costs include the subcontractor's overhead and profit. Loading or hauling of earth fill not included.

Spreading and shaping

Spread and shape earth from loose piles, based on using a D-8 tractor at $110 per hour

	Craft@Hrs	Unit	Material	Labor	Equipment	Total
6" to 10" lifts (164 CY per hour)	TO@.006	CY	--	.26	.67	.93

Compacting and watering (add for cost of water below). CY per hour shown is for a 3-man crew Based on using a self-propelled 100 HP vibratory roller and a 6,000 gallon truck, equipment cost is $71.50 per hour.

	Craft@Hrs	Unit	Material	Labor	Equipment	Total
8" lifts (572 CY per hour)	C3@.005	CY	--	.19	.12	.31

Site Work 2

	Craft@Hrs	Unit	Material	Labor	Equipment	Total

Based on using a sheepsfoot roller towed behind a D-7 tractor and a 6,000 gallon truck, equipment cost is $120 per hour. Productivity assumes 3 passes at 5' wide

	Craft@Hrs	Unit	Material	Labor	Equipment	Total
6" lifts, (183 CY per hour)	C3@.016	CY	--	.61	.66	1.27
8" lifts (244 CY per hour)	C3@.012	CY	--	.46	.49	.95

Based on using a self-propelled 170 HP sheepsfoot roller and a 6,000 gallon truck, equipment cost is $87 per hour. Productivity assumes 3 passes at 7' wide

	Craft@Hrs	Unit	Material	Labor	Equipment	Total
6" lifts (513 CY per hour)	C3@.006	CY	--	.23	.17	.40
8" lifts (687 CY per hour)	C3@.004	CY	--	.15	.13	.28
10" lifts (817 CY per hour)	C3@.004	CY	--	.15	.11	.26

Cost of water for compacted earth embankments
Based on water at $1.50 per 1,000 gallons and 66 gallons per cubic yard of compacted material. Assumes optimum moisture at 10%, natural moisture of 2% and evaporation of 2%. Placed in conjunction with compaction shown above

	Craft@Hrs	Unit	Material	Labor	Equipment	Total
Cost per CY of compacted embankment	--	CY	--	--	--	.10

Finish shaping Earth embankment slopes and swales up to 1 in 4 incline.
Based on using a 125 HP grader and a 15 ton self-propelled rubber tired roller, equipment cost is $52 per hour

	Craft@Hrs	Unit	Material	Labor	Equipment	Total
(200 SY per hour based on a 3-man crew)	S2@.016	SY	--	.64	.28	.92

Based on using a D-8 tractor, equipment cost is $110 per hour

	Craft@Hrs	Unit	Material	Labor	Equipment	Total
(150 SY per hour)	TO@.007	SY	--	.30	.73	1.03

Finish shaping of embankment slopes and swales by hand.

	Craft@Hrs	Unit	Material	Labor	Equipment	Total
Slopes up to 1 in 4 (16 SY per hour)	CL@.063	SY	--	2.15	--	--
Slopes over 1 in 4 (12.5 SY per hour)	CL@.080	SY	--	2.73	--	--

Trench Excavation and Backfill. These costs and productivity are based on utility line trenches and continuous footings where the spoil is piled adjacent to the trench. Cubic yards (CY) or square feet (SF) per hour shown are based on a 2-man crew. Reduce productivity by 10% to 25% when spoil is loaded in trucks. Shoring, dewatering or unusual conditions are not included. These costs include the subcontractor's overhead and profit.

Excavation, using equipment shown
Truck-mounted Gradall with 1 CY bucket at $86 per hour

	Craft@Hrs	Unit	Material	Labor	Equipment	Total
Light soil (40 CY per hour)	S3@.050	CY	--	2.05	2.15	4.20
Medium soil (34 CY per hour)	S3@.059	CY	--	2.34	2.53	4.87
Heavy or wet soil (27 CY per hour)	S3@.074	CY	--	2.98	3.18	6.16
Loose rock (23 CY per hour)	S3@.087	CY	--	3.50	3.74	7.24

Crawler-mounted hydraulic backhoe with 3/4 CY bucket at $43 per hour

	Craft@Hrs	Unit	Material	Labor	Equipment	Total
Light soil (33 CY per hour)	S1@.061	CY	--	2.35	1.30	3.65
Medium soil (27 CY per hour)	S1@.074	CY	--	2.85	1.59	4.44
Heavy or wet soil (22 CY per hour)	S1@.091	CY	--	3.50	1.95	5.45
Loose rock (18 CY per hour)	S1@.112	CY	--	4.32	2.39	6.71

Crawler-mounted hydraulic backhoe with 1 CY bucket at $48 per hour

	Craft@Hrs	Unit	Material	Labor	Equipment	Total
Light soil (65 CY per hour)	S1@.031	CY	--	1.19	.74	1.93
Medium soil (53 CY per hour)	S1@.038	CY	--	1.46	.91	2.37
Heavy or wet soil (43 CY per hour)	S1@.047	CY	--	1.81	1.12	2.93
Loose rock (37 CY per hour)	S1@.055	CY	--	2.12	1.30	3.42
Blasted rock (34 CY per hour)	S1@.058	CY	--	2.24	1.42	3.66

Crawler-mounted hydraulic backhoe with 1-1/2 CY bucket at $62 per hour

	Craft@Hrs	Unit	Material	Labor	Equipment	Total
Light soil (83 CY per hour)	S1@.024	CY	--	.92	.74	1.66
Medium soil (70 CY per hour)	S1@.029	CY	--	1.12	.90	2.02
Heavy or wet soil (57 CY per hour)	S1@.035	CY	--	1.35	1.09	2.44
Loose rock (48 CY per hour)	S1@.042	CY	--	1.62	1.29	2.91
Blasted rock (43 CY per hour)	S1@.047	CY	--	1.81	1.44	3.25

Site Work 2

	Craft@Hrs	Unit	Material	Labor	Equipment	Total

Trench excavation, using equipment shown, continued
  Crawler-mounted hydraulic backhoe with 2 CY bucket at $89 per hour

	Craft@Hrs	Unit	Material	Labor	Equipment	Total
Light soil (97 CY per hour)	S1@.021	CY	--	.81	.92	1.73
Medium soil (80 CY per hour)	S1@.025	CY	--	.96	1.11	2.07
Heavy or wet soil (65 CY per hour)	S1@.031	CY	--	1.19	1.37	2.56
Loose rock (55 CY per hour)	S1@.036	CY	--	1.39	1.62	3.01
Blasted rock (50 CY per hour)	S1@.040	CY	--	1.54	1.78	3.32

Crawler-mounted hydraulic backhoe with 2-1/2 CY bucket at $105 per hour

	Craft@Hrs	Unit	Material	Labor	Equipment	Total
Light soil (122 CY per hour)	S1@.016	CY	--	.62	.86	1.48
Medium soil (100 CY per hour)	S1@.020	CY	--	.77	1.05	1.82
Heavy or wet soil (82 CY per hour)	S1@.024	CY	--	.93	1.28	2.21
Loose rock (68 C Y per hour)	S1@.029	CY	--	1.12	1.54	2.66
Blasted rock (62 CY per hour)	S1@.032	CY	--	1.23	1.69	2.92

Chain-boom Ditch Witch digging trench to 12" wide and 5'6" deep at $14 per hour

	Craft@Hrs	Unit	Material	Labor	Equipment	Total
Light soil (10 CY per hour)	S1@.200	CY	--	7.71	1.40	9.11
Most soils (8.5 CY per hour)	S1@.235	CY	--	9.06	1.65	10.71
Heavy soil (7 CY per hour)	S1@.286	CY	--	11.00	2.00	13.00

Crawler-mounted wheel type 80 HP trencher digging trench to 18" wide and 6' deep at $53 per hour

	Craft@Hrs	Unit	Material	Labor	Equipment	Total
Light soil (62 CY per hour)	S1@.032	CY	--	1.23	.86	2.09
Most soils (49 CY per hour)	S1@.041	CY	--	1.58	1.08	2.66
Heavy soil (40 CY per hour)	S1@.050	CY	--	1.93	1.33	3.26

Crawler-mounted wheel type 150 HP trencher digging trench to 24" wide and 9' deep at $63 per hour

	Craft@Hrs	Unit	Material	Labor	Equipment	Total
Light soil (155 CY per hour)	S1@.013	CY	--	.50	.41	.91
Most soils (125 CY per hour)	S1@.016	CY	--	.62	.50	1.12
Heavy soil (100 CY per hour)	S1@.020	CY	--	.77	.63	1.40

Trim trench bottom to 1/10'

	Craft@Hrs	Unit	Material	Labor	Equipment	Total
By hand (200 SF per hour)	CL@.010	SF	--	.34	--	--

Backfill trenches from loose material piled adjacent to trench. No compaction included.
  Soil, previously excavated

	Craft@Hrs	Unit	Material	Labor	Equipment	Total
By hand (8 CY per hour)	CL@.250	CY	--	8.53	--	--
D-3 crawler dozer ($24 & 25 CY per hour)	S1@.080	CY	--	3.08	.96	4.04
3/4 CY crawler loader ($21 & 33 CY per hour)	S1@.061	CY	--	2.35	.64	2.99
D-7 crawler dozer ($87 & 130 CY per hour)	S1@.015	CY	--	.58	.65	1.23

Sand or gravel bedding

	Craft@Hrs	Unit	Material	Labor	Equipment	Total
3/4 CY wheel loader ($23 & 80 CY per hour)	S1@.025	CY	15.00	.96	.29	16.25
Fine grade bedding by hand (22 SY per hr)	CL@.090	SY	--	3.07	--	--

Compaction of soil in trenches in 8" layers.

	Craft@Hrs	Unit	Material	Labor	Equipment	Total
Pneumatic tampers ($16 & 40 CY per hour)	CL@.050	CY	--	1.71	.40	2.11
Vibrating rammers ($10 & 20 CY per hour)	CL@.100	CY	--	3.41	.50	3.91

Dragline Excavation, for basements, footings or foundations. Costs shown include casting the excavated soil adjacent to the excavation or loading it into trucks. Hauling costs are not included. Cubic yards (CY) per hour are as measured in an undisturbed condition (bank measure) and are based on a 2-man crew. These costs include the subcontractor's overhead and profit.

  1-1/2 CY dragline with a 45 ton crawler crane at $84 per hour

	Craft@Hrs	Unit	Material	Labor	Equipment	Total
Loam or light clay (73 CY per hour)	H2@.027	CY	--	1.10	1.16	2.26
Sand or gravel (67 CY per hour)	H2@.030	CY	--	1.22	1.25	2.47
Heavy clay (37 CY per hour)	H2@.054	CY	--	2.20	2.27	4.47
Unclassified soil (28 CY per hour)	H2@.071	CY	--	2.89	2.99	5.88

2 CY dragline with a 60 ton crawler crane at $105 per hour

	Craft@Hrs	Unit	Material	Labor	Equipment	Total
Loam or light clay (88 CY per hour)	H2@.023	CY	--	.94	1.20	2.14
Sand or gravel (85 CY per hour)	H2@.024	CY	--	.98	1.24	2.22
Heavy clay (48 CY per hour)	H2@.042	CY	--	1.71	2.18	3.89

Site Work 2

	Craft@Hrs	Unit	Material	Labor	Equipment	Total
Unclassified soil (36 CY per hour)	H2@.056	CY	--	2.28	2.92	5.20
2-1/2 CY dragline with a 100 ton crawler crane at $147 per hour						
Loam or light clay (102 CY per hour)	H2@.019	CY	--	.77	1.44	2.21
Sand or gravel (98 CY per hour)	H2@.021	CY	--	.85	1.50	2.35
Heavy clay (58 CY per hour)	H2@.035	CY	--	1.42	2.53	3.95
Unclassified soil (44 CY per hour)	H2@.045	CY	--	1.83	3.34	5.17
3 CY dragline with a 150 ton crawler crane at $210 per hour						
Loam or light clay (116 CY per hour)	H2@.017	CY	--	.69	1.81	2.50
Sand or gravel (113 CY per hour)	H2@.018	CY	--	.73	1.86	2.59
Heavy clay (70 CY per hour)	H2@.029	CY	--	1.18	3.00	4.18
Unclassified soil (53 CY per hour)	H2@.038	CY	--	1.55	3.96	5.51

Backhoe Excavation, for basements, footings or foundations. Costs shown include casting the excavated soil adjacent to the excavation or loading it into trucks. Hauling costs are not included. Cubic yards (CY) per hour are as measured in an undisturbed condition (bank measure) and are based on a 2-man crew. Equipment costs are based on using a crawler-mounted hydraulic backhoe. These costs include the subcontractor's overhead and profit.

	Craft@Hrs	Unit	Material	Labor	Equipment	Total
1 CY backhoe at $48 per hour						
Light soil (69 CY per hour)	S1@.029	CY	--	1.12	.70	1.82
Most soils (57 CY per hour)	S1@.035	CY	--	1.35	.84	2.19
Wet soil, loose rock (46 CY per hour)	S1@.043	CY	--	1.66	1.03	2.69
1-1/2 CY backhoe at $62 per hour						
Light soil (90 CY per hour)	S1@.022	CY	--	.85	.68	1.53
Most soils (77 CY per hour)	S1@.026	CY	--	1.00	.81	1.81
Wet soil, loose rock (60 CY per hour)	S1@.033	CY	--	1.27	1.02	2.29
Blasted rock (54 CY per hour)	S1@.044	CY	--	1.70	1.36	3.06
2 CY backhoe at $89 per hour						
Light soil (100 CY per hour)	S1@.020	CY	--	.77	.89	1.66
Most soils (85 CY per hour)	S1@.023	CY	--	.89	1.02	1.91
Wet soil, loose rock (70 CY per hour)	S1@.029	CY	--	1.12	1.29	2.41
Blasted rock (54 CY per hour)	S1@.037	CY	--	1.43	1.65	3.08

Moving and Loading Excavated Materials. Costs shown include moving the material 50' and dumping it into piles or loading it into trucks. Add 25% for each 50' of travel beyond the first 50'. Hauling costs are not included. Cubic yards (CY) per hour are for material in a loose condition (previously excavated) and are based on a 2-man crew performing the work. Equipment costs are based on using a wheel-mounted front end loader.
These costs include the subcontractor's overhead and profit.

	Craft@Hrs	Unit	Material	Labor	Equipment	Total
3/4 CY loader ($23 and 32 CY per hour)	S1@.062	CY	--	2.39	.71	3.10
1-1/2 CY loader ($28 and 55 CY per hour)	S1@.036	CY	--	1.39	.50	1.89
2-1/2 CY loader ($49 and 90 CY per hour)	S1@.022	CY	--	.85	.54	1.39
3-1/2 CY loader ($65 and 125 CY per hour)	S1@.016	CY	--	.62	.52	1.14
6 CY loader ($100 and 218 CY per hour)	S1@.009	CY	--	.35	.45	.80

Hauling excavated material, using trucks. Costs shown include 4 minutes for loading, 3 minutes for dumping and travel time based on the one-way distance, speed and cycles per hour as noted. Costs for equipment to excavate and load the material are not included. Truck capacity shown is based on loose cubic yards of material. Allow the following percentage amounts when estimating quantities based on undisturbed (bank measure) materials for swell when exporting from the excavation site or shrinkage when importing from borrow site: clay (33%), common earth (25%), granite (65%), mud (21%), sand or gravel (12%). These costs include the subcontractor's overhead and profit.

	Craft@Hrs	Unit	Material	Labor	Equipment	Total
8 CY dump truck at $26 per hour						
1 mile haul at 20 MPH						
(4.2 cycles and 34 CY per hour)	TD@.029	CY	--	1.09	.76	1.85
3 mile haul at 30 MPH						
(2.9 cycles and 23 CY per hour)	TD@.044	CY	--	1.65	1.13	2.78

Site Work 2

	Craft@Hrs	Unit	Material	Labor	Equipment	Total
Hauling excavated material, using trucks, continued						
6 mile haul at 40 MPH						
(2.1 cycles and 17 CY per hour)	TD@.058	CY	--	2.18	1.53	3.71
12 CY dump truck at $30 per hour						
1 mile haul at 20 MPH						
(4.2 cycles and 50 CY per hour)	TD@.020	CY	--	.75	.60	1.35
3 mile haul at 30 MPH						
(2.9 cycles and 35 CY per hour)	TD@.028	CY	--	1.05	.86	1.91
6 mile haul at 40 MPH						
(2.1 cycles and 25 CY per hour)	TD@.041	CY	--	1.54	1.20	2.74
16 CY off-highway dump truck at $48 per hour						
1 mile haul at 20 MPH						
(4.2 cycles and 67 CY per hour)	TD@.015	CY	--	.56	.72	1.28
3 mile haul at 30 MPH						
(2.9 cycles and 46 CY per hour)	TD@.022	CY	--	.83	1.04	1.87
6 mile haul at 40 MPH						
(2.1 cycles and 34 CY per hour)	TD@.029	CY	--	1.09	1.41	2.50
25 CY off-highway dump truck at $65 per hour						
1 mile haul at 20 MPH						
(4.2 cycles and 105 CY per hour)	TD@.009	CY	--	.34	.61	.95
3 mile haul at 30 MPH						
(2.9 cycles and 73 CY per hour)	TD@.014	CY	--	.53	.88	1.41
6 mile haul at 40 MPH						
(2.1 cycles and 53 CY per hour)	TD@.019	CY	--	.71	1.21	1.92

Dozer Excavation. Mass excavation using a crawler tractor with a dozing blade attached. Costs shown include excavation and pushing soil 150' to stockpile based on good access and a maximum of 5% grade. Increase or reduce costs 20% for each 50' of push more or less than 150'. Cubic yards (CY) per hour shown are undisturbed bank measure and are based on a 1-man crew. These costs include the subcontractor's overhead and profit.

	Craft@Hrs	Unit	Material	Labor	Equipment	Total
Gravel or loose sand						
75 HP D-4 dozer with "S" blade at $28 per hour						
(29 CY per hour)	TO@.034	CY	--	1.46	.95	2.41
140 HP D-6 dozer with "S" blade at $57 per hour						
(56 CY per hour)	TO@.018	CY	--	.77	1.03	1.80
200 HP D-7 dozer with "S" blade at $87 per hour						
(78 CY per hour)	TO@.013	CY	--	.56	1.13	1.69
300 HP D-8 dozer with "U" blade at $110 per hour						
(140 CY per hour)	TO@.007	CY	--	.30	.77	1.07
460 HP D-9 dozer with "U" blade at $180 per hour						
(205 CY per hour)	TO@.005	CY	--	.22	.90	1.12
Loam or soft clay						
75 HP D-4 dozer with "S" blade at $28 per hour						
(26 CY per hour)	TO@.038	CY	--	1.63	1.06	2.69
140 HP D-6 dozer with "S" blade at $57 per hour						
(50 CY per hour)	TO@.020	CY	--	.86	1.14	2.00
200 HP D-7 dozer with "S" blade at $87 per hour						
(70 CY per hour)	TO@.014	CY	--	.60	1.22	1.82
300 HP D-8 dozer with "U" blade at $110 per hour						
(125 CY per hour)	TO@.008	CY	--	.34	.88	1.22
460 HP D-9 dozer with "U" blade at $180 per hour						
(185 CY per hour)	TO@.005	CY	--	.22	.90	1.12

Site Work 2

	Craft@Hrs	Unit	Material	Labor	Equipment	Total
Shale, sandstone or blasted rock						
75 HP D-4 dozer with "S" blade at $28 per hour						
(20 CY per hour)	TO@.051	CY	--	2.19	1.43	3.62
140 HP D-6 dozer with "S" blade at $57 per hour						
(38 CY per hour)	TO@.026	CY	--	1.12	1.48	2.60
200 HP D-7 dozer with "S" blade at $87 per hour						
(53 CY per hour)	TO@.019	CY	--	.82	1.65	2.47
300 HP D-8 dozer with "U" blade at $110 per hour						
(95 CY per hour)	TO@.011	CY	--	.47	1.21	1.68
460 HP D-9 dozer with "U" blade at $180 per hour						
(140 CY per hour)	TO@.007	CY	--	.30	1.26	1.56

Scraper-hauler Excavation. Mass excavation using a self-propelled scraper-hauler. Equipment costs include the scraper-hauler and a crawler tractor with a dozer blade attached pushing it 10 minutes each hour. Cubic yards (CY) per hour shown are undisturbed bank measure. These costs include the subcontractor's overhead and profit. Work done in clay, shale or soft rock will cost 10% to 25% more.

	Craft@Hrs	Unit	Material	Labor	Equipment	Total
12 CY elevating 200 HP scraper-hauler & D-8 300 HP tractor pushing ($108 per hour)						
1,000' haul (9 cycles and 99 CY per hr)	S2@.022	CY	--	.88	.79	1.67
2,500' haul (6 cycles and 66 CY per hr)	S2@.033	CY	--	1.32	1.19	2.51
4,000' haul (4.5 cycles and 50 CY per hr)	S2@.043	CY	--	1.72	1.55	3.27
15 CY self-propelled 330 HP scraper-hauler & D-8 300 HP tractor pushing ($143 per hour)						
1,000' haul (9 cycles and 135 CY per hr)	S2@.016	CY	--	.64	.76	1.40
2,500' haul (6 cycles and 90 CY per hr)	S2@.024	CY	--	.96	1.14	2.10
4,000' haul (4.5 cycles and 68 CY per hr)	S2@.032	CY	--	1.28	1.53	2.81
24 CY self-propelled 450 HP scraper-hauler & D-9 460 HP tractor pushing ($210 per hour)						
1,000' haul (9 cycles and 225 CY per hr)	S2@.010	CY	--	.40	.70	1.10
2,500' haul (6 cycles and 150 CY per hr)	S2@.014	CY	--	.56	.98	1.54
4,000' haul (4.5 cycles and 113 CY/hr)	S2@.019	CY	--	.76	1.33	2.09
35 CY self-propelled scraper-hauler with D-9 460 HP tractor pushing ($330 per hour)						
1,000' haul (9 cycles and 315 CY per hr)	S2@.007	CY	--	.28	.77	1.05
2,500' haul (6 cycles and 219 CY per hr)	S2@.010	CY	--	.40	1.10	1.50
4,000' haul (4.5 cycles and 158 CY/hr)	S2@.014	CY	--	.56	1.54	2.40
43 CY self-propelled scraper-hauler with D-9 460 HP tractor pushing ($380 per hour)						
1,000' haul (9 cycles and 378 CY per hr)	S2@.006	CY	--	.24	.76	1.00
2,500' haul (6 cycles and 258 CY per hr)	S2@.008	CY	--	.32	1.01	1.33
4,000' haul (4.5 cycles and 194 CY/hr)	S2@.011	CY	--	.44	1.39	1.83

Roadway and Embankment Earthwork Cut & Fill. Mass excavating, hauling, placing and compacting (cut & fill) using a 200 HP, 12 CY capacity self-propelled scraper-hauler with attachments. Equipment cost is $95.00 per hour. Productivity is based on a 1,500' haul with 6" lifts compacted to 95% per AASHO requirements. Cubic yards (CY) or square yards (SY) per hour shown are bank measurement. These costs include the subcontractor's overhead and profit and are typical for 50,000 CY jobs.

	Craft@Hrs	Unit	Material	Labor	Equipment	Total
Cut, fill and compact, clearing or finishing not included						
Most soil types (36 CY per hour)	TO@.028	CY	--	1.20	2.66	3.86
Rock and earth mixed (33 CY per hour)	TO@.030	CY	--	1.29	2.85	4.14
Rippable rock (19 CY per hour)	TO@.053	CY	--	2.28	5.04	7.32
Earth banks and levees, clearing or finishing not included						
Cut, fill and compact (67 CY per hour)	TO@.015	CY	--	.64	1.43	2.07
Channelizing compacted fill (44 CY per hour)	TO@.023	CY	--	.99	2.19	3.18
Roadway subgrade preparation						
Scarify & compact (91 CY per hour)	TO@.011	CY	--	.47	1.05	1.52
Roll & compact base course (83 CY per hour)	TO@.012	CY	--	.52	1.14	1.66
Excavation, open ditches (38 CY per hour)	TO@.026	CY	--	1.12	2.47	3.59

Site Work 2

	Craft@Hrs	Unit	Material	Labor	Equipment	Total

Roadway and Embankment Earthwork, continued
Trimming and finishing

| banks, swales and ditches (111 CY per hour) | TO@.009 | CY | -- | .39 | .86 | 1.25 |

Scarify existing asphalt pavement

| (38 SY per hour) | TO@.026 | SY | -- | 1.12 | 2.47 | 3.59 |

Slope Protection. These costs include the subcontractor's overhead and profit. See also, Soil Stabilization.

Riprap, dumped from trucks and placed using a 15 ton hydraulic crane. Riprap stone prices are FOB quarry and will vary widely. Typical riprap cost is $15.00 per cubic yard ($10.00 per ton) including subcontractor's markup. Add $.70 per CY per mile for trucking to the site. Cubic (CY) yards per hour shown are based on a 7-man crew. Equipment cost is $45.00 per hour.

| 5 to 7 CF pieces (9.4 CY per hour) | S5@.745 | CY | 15.00 | 26.80 | 4.78 | 46.58 |

5 to 7 CF pieces, sacked and placed,

| excluding bag cost (3 CY per hour) | S5@2.33 | CY | 15.00 | 83.90 | 15.00 | 113.90 |

PVC coated nylon riprap bags, quantities of 50 to 100

5' x 7', 1.50 CY capacity	--	Ea	45.00	--	--	--
5' x 10', 2.25 CY capacity	--	Ea	60.00	--	--	--
5' x 13', 3.00 CY capacity	--	Ea	73.00	--	--	--
Deduct for over 100 bags	--	%	-25.0	--	--	--

Loose riprap stone, under 90 lbs each,

| hand placed (2 CY per hour) | CL@.500 | CY | 15.00 | 17.10 | -- | 32.10 |

Erosion and drainage control armored woven fabric, (Mirafi #700X)

| for use under riprap, hand placed | CL@.011 | SY | 1.50 | .38 | -- | 1.88 |

Ornamental large rock. Rock prices include local delivery of 10 CY minimum. Cubic (CY) yards per hour shown are based on a 7-man crew. Equipment cost is $45.00 per hour based on using a 15 ton hydraulic crane.

Volcanic cinder, 1,200 lbs per CY

| (8 CY per hour) | S5@.875 | CY | 200.00 | 31.50 | 5.62 | 237.12 |

Featherock, 1,000 lbs per CY

| (12 CY per hour) | S5@.583 | CY | 150.00 | 21.00 | 3.74 | 174.74 |

Rock fill, dumped from trucks and placed with a 75 HP D-4 tractor. Rock prices include local delivery of 10 CY minimum. Cubic (CY) yards per hour shown are based on a 3-man crew. Equipment cost is $28 per hour.

Drain rock, 3/4" to 1-1/2" (12 CY per hour)	S6@.250	CY	17.00	9.27	2.33	28.60
Bank run gravel (12 CY per hour)	S6@.250	CY	15.00	9.27	2.33	26.60
Pea gravel (12 CY per hour)	S6@.250	CY	16.50	9.27	2.33	28.10

Straw bales secured to ground with reinforcing bars, bale is 3'6" long, 2' wide, 1'6" high. Straw at $100

| per ton (25 bales) and rebars at $1.00 | CL@.166 | Ea | 5.00 | 5.66 | -- | 10.66 |

Sedimentation control fence, installed vertically at bottom of slope,

| 36" high woven fabric (Mirafi 100X) | CL@.003 | LF | .20 | .10 | -- | .30 |

Add to fence above for #6 rebar stakes 6'0" long at 10'0" OC hand-driven 36" into the ground,

| per LF of fence | CL@.050 | LF | .21 | 1.71 | -- | 1.92 |

Manual Excavation. Based on one throw (except where noted) of shoveled earth from trench or pit. Shoring or dewatering not included. These costs include the subcontractor's overhead and profit.

Trench or pit in light soil (silty sand or loess, etc.)

Up to 4' deep (.82 CY per hour)	CL@1.22	CY	--	41.60	--	--
Over 4' to 6' deep (.68 CY per hour)	CL@1.46	CY	--	49.80	--	--
Over 6' deep, two throws (.32 CY per hour)	CL@3.16	CY	--	108.00	--	--

Trench or pit in medium soil (sandy clay, clayey sand, etc.)

Up to 4' deep (.68 CY per hour)	CL@1.46	CY	--	49.80	--	--
Over 4' to 6' deep (.54 CY per hour)	CL@1.86	CY	--	63.40	--	--
Over 6' deep, two throws (.27 CY per hour)	CL@3.66	CY	--	125.00	--	--

Site Work 2

	Craft@Hrs	Unit	Material	Labor	Equipment	Total
Trench or pit in heavy soil (clayey materials, shales, caliche, etc.)						
Up to 4' deep (.54 CY per hour)	CL@1.86	CY	--	63.40	--	--
Over 4' to 6' deep (.46 CY per hour)	CL@2.16	CY	--	73.70	--	--
Over 6' deep, 2 throws (.25 CY per hour)	CL@3.90	CY	--	133.00	--	--
Shovel from bank, wheel 300 feet in a 5 CF wheelbarrow on firm ground and dump						
Light soil (.74 CY per hour)	CL@1.35	CY	--	46.00	--	--
Medium soil (.65 CY per hour)	CL@1.54	CY	--	52.50	--	--
Heavy soil (.56 CY per hour)	CL@1.79	CY	--	61.00	--	--
Loose rock (.37 CY per hour)	CL@2.70	CY	--	92.10	--	--
Hand trim and shape						
Around utility lines (15 CF hour)	CL@.065	CF	--	2.22	--	--
For slab on grade (63 SY per hour)	CL@.016	SY	--	.55	--	--
Trench bottom (100 SF per hour)	CL@.010	SF	--	.34	--	--

Soil Stabilization. These costs include the subcontractor's overhead and profit. See also, Slope Protection.

Lime slurry injection treatment. Equipment cost is $84.50 per hour based on using a 125 HP grader, a 2 ton self-propelled steel roller and a 10 ton 100 HP vibratory steel roller.

	Craft@Hrs	Unit	Material	Labor	Equipment	Total
Cost per CY of soil treated (25 CY per hour)	S7@.240	CY	9.00	9.25	3.38	21.63

Vibroflotation treatment. Equipment cost is $130.00 per hour based on using a 50 ton hydraulic crane, a 1 CY wheel loader for 15 minutes of each hour, and a vibroflotation machine with pumps.

	Craft@Hrs	Unit	Material	Labor	Equipment	Total
Typical cost per CY of soil treated						
Low cost (92 CY per hour)	S7@.065	CY	--	2.51	1.41	3.92
High cost (46 CY per hour)	S7@.130	CY	--	5.01	2.82	7.83

Soil cement treatment. Includes materials and placement for 7% cement mix. Equipment cost is $100.00 per hour based on using a cross shaft mixer, a 2 ton 3 wheel steel roller and a vibratory roller. Cement priced at $100 per ton. Cost per CY of soil treated

	Craft@Hrs	Unit	Material	Labor	Equipment	Total
On level ground (5 CY per hour)	S7@1.20	CY	7.00	46.20	20.00	73.20
On slopes under 3 in 1 (4.3 CY per hour)	S7@1.40	CY	7.00	54.00	23.30	84.30
On slopes over 3 in 1 (4 CY per hour)	S7@1.50	CY	7.00	57.80	25.00	89.80

Construction fabrics (geotextiles). Mirafi products. Costs shown include 10% for lapover.

	Craft@Hrs	Unit	Material	Labor	Equipment	Total
Drainage fabric (Mirafi 140N), used to line trenches, allows water passage but prevents soil migration into drains	CL@.016	SY	.55	.55	--	1.10
Stabilization fabric (Mirafi 500X), used to prevent mixing of unstable soils with road base aggregate	CL@.008	SY	.58	.27	--	.85
Prefabricated drainage fabric (Miradrain), used to drain subsurface water from structural walls	CL@.006	SF	.80	.21	--	1.01
Erosion control mat (Miramat #1800), used to reduce surface soil erosion on slopes or in ditches while promoting seed growth	CL@.020	SY	6.10	.68	--	6.78

Pile Foundations. These costs assume solid ground for equipment support and a minimum of 4,000 LF of piling. Add the cost of mobilization and testing on all jobs. Costs shown per linear foot (LF) are per vertical foot of pile depth. Standby or idle time, access roads, rig mats or special engineering required by unusual conditions are not included. These costs include the subcontractor's overhead and profit.

Mobilization and demobilization. Typical costs

	Craft@Hrs	Unit	Material	Labor	Equipment	Total
Truck crane pile driver	--	LS	--	--	--	6,730.00
35 ton crawler crane pile driver	--	LS	--	--	--	7,800.00
65 ton crawler crane pile driver	--	LS	--	--	--	8,930.00
100 ton crawler crane driver	--	LS	--	--	--	14,300.00

Site Work 2

	Craft@Hrs	Unit	Material	Labor	Equipment	Total

Pile testing only. Testing costs will vary with the soil type, method of testing, and seismic conditions. Typical costs

	Craft@Hrs	Unit	Material	Labor	Equipment	Total
50 to 100 ton range	S8@204.	Ea	--	8,300.00	5,200.00	13,500.00
Over 100 ton to 200 ton range	S8@240.	Ea	--	9,770.00	6,220.00	15,990.00
Over 200 ton to 300 ton range	S8@272.	Ea	--	11,100.00	6,730.00	17,830.00

Prestressed concrete piles, square, to 60' long, predrilled for first 10', truck delivery

	Craft@Hrs	Unit	Material	Labor	Equipment	Total
10" (94 LF per hour)	S8@.064	LF	8.03	2.60	1.05	11.68
12" (66 LF per hour)	S8@.091	LF	11.20	3.70	1.45	16.35
14" (92 LF per hour)	S8@.096	LF	15.00	3.91	1.50	20.41
16" (80 LF per hour)	S8@.104	LF	18.70	4.23	1.60	24.53
18" (69 LF per hour)	S8@.112	LF	22.10	4.56	1.75	28.41
20" (57 LF per hour)	S8@.120	LF	25.50	4.88	1.90	32.28
24" (53 LF per hour)	S8@.132	LF	32.50	5.37	2.00	39.87
Add for predrilling in excess of first 10 LF	S8@.177	LF	--	7.20--	2.80	10.00
Add for piles over 60' long	--	%	10.0	--	--	--
Deduct for octagonal prestressed concrete piles	--	%	-5.0	-6.0	-6.0	--
Deduct for straight circular prestressed concrete piles	--	%	-10.0	-12.0	-12.0	--

Steel "H" section piles

	Craft@Hrs	Unit	Material	Labor	Equipment	Total
8" x 8", 36 lbs per LF	S8@.115	LF	11.20	4.68	2.15	18.03
10" x 10", 57 lbs per LF	S8@.120	LF	17.70	4.88	2.30	24.88
12" x 12", 74 lbs per LF	S8@.133	LF	22.90	5.41	2.60	30.91
14" x 14", 89 lbs per LF	S8@.147	LF	27.60	5.98	3.20	36.78

Steel pipe piles, concrete filled, including all materials

	Craft@Hrs	Unit	Material	Labor	Equipment	Total
8"	S8@.100	LF	11.00	4.07	2.40	17.47
10"	S8@.101	LF	13.70	4.11	2.50	20.31
12"	S8@.112	LF	18.10	4.56	2.70	25.36
14"	S8@.131	LF	19.70	5.33	3.10	28.13
16"	S8@.139	LF	23.20	5.66	3.30	32.16
18"	S8@.149	LF	26.80	6.06	3.60	36.46

Steel pipe piles, non-filled

	Craft@Hrs	Unit	Material	Labor	Equipment	Total
8"	S8@.085	Ea	9.59	3.46	2.10	15.15
10"	S8@.088	Ea	12.00	3.58	2.20	17.78
12"	S8@.096	Ea	15.60	3.91	2.40	21.91
14"	S8@.112	Ea	16.30	4.56	2.70	23.56
16"	S8@.120	Ea	18.50	4.88	2.90	26.28
18"	S8@.128	Ea	20.80	5.21	3.15	29.16

Splices for steel "H" section and pipe piles

	Craft@Hrs	Unit	Material	Labor	Equipment	Total
8"	S8@1.12	Ea	29.80	45.60	2.50	77.90
10"	S8@1.55	Ea	38.10	63.10	3.30	104.50
12"	S8@2.13	Ea	56.70	86.70	4.60	148.00
14"	S8@2.25	Ea	64.10	91.60	4.95	160.65
16"	S8@2.39	Ea	70.00	97.20	5.25	172.45
18"	S8@2.52	Ea	80.90	103.00	5.70	189.60

Standard points for steel "H" section and pipe piles

	Craft@Hrs	Unit	Material	Labor	Equipment	Total
8"	S8@1.49	Ea	42.20	60.60	3.40	106.20
10"	S8@1.68	Ea	49.80	68.40	3.85	122.05
12"	S8@1.84	Ea	58.80	74.90	4.15	137.85
14"	S8@2.03	Ea	63.50	82.60	4.65	150.75
16"	S8@2.23	Ea	67.50	90.70	5.05	163.25
18"	S8@2.52	Ea	75.80	103.00	5.60	184.40
Add for heavy duty points for steel pipe piles	--	%	30.0	20.0	--	--

Site Work 2

	Craft@Hrs	Unit	Material	Labor	Equipment	Total
Step tapered steel piles, concrete filled, by tip size						
8"	S8@.091	LF	9.10	3.70	2.90	15.70
10"	S8@.097	LF	10.10	3.95	3.10	17.15
12"	S8@.107	LF	12.10	4.35	3.35	19.80
14"	S8@.113	LF	14.10	4.60	3.60	22.30
Wood piles, treated with creosote						
To 30' long	S8@.077	LF	6.40	3.15	1.45	11.00
Over 30' to 40'	S8@.069	LF	5.40	2.80	1.35	9.55
Over 40' to 50'	S8@.067	LF	5.40	2.75	1.29	9.44
Over 50' to 60'	S8@.061	LF	6.40	2.50	1.19	10.09
Over 60' to 80'	S8@.059	LF	7.60	2.40	1.14	11.14
Add for drive shoe, per pile	--	Ea	27.80	--	--	--
Add for load test	--	Ea	--	--	--	8,600.00
Deduct for untreated piles	--	%	-33.0	--	--	--

Caissons. Not including waste disposal. Equipment is a flatbed truck with a power take-off, an auger and an A-frame hoist at $125.00 per hour. These costs include the subcontractor's overhead and profit.

	Craft@Hrs	Unit	Material	Labor	Equipment	Total
Auger holes, drilling only, in stable ground						
16" (57 LF per hour)	S1@.035	LF	--	1.35	2.30	3.65
24" (47 LF per hour)	S1@.042	LF	--	1.62	2.70	4.32
36" (36 LF per hour)	S1@.055	LF	--	2.12	3.60	5.72
Rock drilling, accessible, no hard rock, large quantity						
24" diameter (6 LF per hour)	S1@.348	LF	--	13.40	22.60	36.00
48" diameter (3 LF per hour)	S1@.687	LF	--	26.50	44.70	71.20
Caissons, accessible, drilling only						
Drilled, 16", no casing (55 LF per hour)	S1@.036	LF	--	1.39	2.35	3.74
Drilled, 24", no casing (45 LF per hour)	S1@.044	LF	--	1.70	2.90	4.60
Drilled, 16", casing removed (8 LF per hour)	S1@.252	LF	--	9.71	16.40	26.11
Drilled, 24", casing removed (7 LF per hour)	S1@.297	LF	--	11.40	19.30	30.70
Drilled and left cased, 16" (7 LF per hour)	S1@.258	LF	13.50	9.94	16.80	40.24
Drilled and left cased, 24" (6 LF per hour)	S1@.310	LF	40.50	11.90	20.20	72.60
Open caissons to 50', including shoring	S1@4.89	CY	--	188.00	318.00	506.00
Open caissons below 50', including shoring	S1@5.78	CY	--	223.00	376.00	599.00
Add for 16" concrete filling	S1@.075	LF	3.45	2.89	4.90	11.24
Add for 24" concrete filling	S1@.080	LF	8.90	3.08	5.20	17.18
Bell footings, accessible, typical costs, adjust for unusual hazards						
24" consisting of						
5' shaft plus 5' diameter bell	S1@1.86	Ea	--	71.70	121.00	192.70
Add for concrete filled footing	S1@1.54	Ea	216.00	59.40	100.00	375.40
24" consisting of						
6' shaft plus 6' diameter bell	S1@1.72	Ea	--	66.30	112.00	178.30
Add for concrete filled footing	S1@1.78	Ea	227.00	68.60	116.00	411.60
36" consisting of						
10' shaft plus 8' diameter bell	S1@3.10	Ea	--	119.00	202.00	321.00
Add for concrete filled footing	S1@3.98	Ea	260.00	153.00	259.00	672.00
Slurry trench in wet ground, based on 30' trench. Costs include reinforcing for hydrostatic forces and 3,000 PSI concrete, but no excavation, mats or bridging						
Cost per SF of trench, typical prices	--	SF	--	--	--	36.90
Pressure injected foundations (pressure grouting), per vertical LF, typical prices						
Uncased, 20' x 18", 75 tons	--	LF	--	--	--	39.00
Uncased, 20' x 24", 150 tons	--	LF	--	--	--	49.00
Cased, 25' x 14", 75 tons	--	LF	--	--	--	40.00

Site Work 2

	Craft@Hrs	Unit	Material	Labor	Equipment	Total

Shoring, Bulkheads and Underpinning. These costs include the subcontractor's overhead and profit. Steel sheet piling, driven and pulled, based on good driving conditions and using sheets 30' to 50' in length with 32 lbs of 38.5 KSI steel per SF. Material costs are based on steel at $790 per ton. Equipment cost for driving and pulling averages $210 per ton. Material cost assumes the piling is salvaged for $590 per ton

	Craft@Hrs	Unit	Material	Labor	Equipment	Total
Up to 20' deep	S8@.096	SF	3.20	3.91	3.50	10.61
Over 20' to 35' deep	S8@.088	SF	3.20	3.58	3.20	9.98
Over 35' to 50' deep	S8@.087	SF	3.20	3.54	3.15	9.89
Over 50' deep	S8@.085	SF	3.20	3.46	3.10	9.76

Additive and deductive costs for sheet steel piling
Add for:

	Craft@Hrs	Unit	Material	Labor	Equipment	Total
50' to 65' lengths	--	%	2.0	--	--	--
65' to 100' lengths	--	%	5.0	--	--	--
Epoxy coal tar 16 mil factory finish	--	SF	1.60	--	--	--
32 lb piling material left in place	--	SF	9.00	--	--	--
38 lbs per SF (MZ 38) steel	--	%	20.0	--	--	--
50 KSI high-strength steel	--	%	11.0	--	--	--
Marine grade 50 KSI high strength steel	--	%	23.0	--	--	--

Deduct for:

	Craft@Hrs	Unit	Material	Labor	Equipment	Total
28 lbs per SF (PS 28) steel	--	%	-12.0	--	--	--
23 lbs per SF (PS 23) steel	--	%	-25.0	--	--	--
Deduct labor & equipment cost when piling is left in place	--	%	--	-35.0	-15.0	--

Timber trench sheeting and bracing per square foot of trench wall measured on one side of trench. Costs include pulling and salvage, depths to 12'. Equipment cost is $42.50 per hour for a 10 ton hydraulic crane and a 1 ton flat-bed truck.

	Craft@Hrs	Unit	Material	Labor	Equipment	Total
Less than 6' wide, open bracing	S5@.053	SF	.95	1.91	.32	3.18
6' to 10' wide, open bracing	S5@.064	SF	1.00	2.30	.39	3.69
11' to 16' wide, open bracing	S5@.079	SF	1.20	2.84	.48	4.52
Over 16' wide, open bracing	S5@.091	SF	1.30	3.28	.55	5.13
Less than 6' wide, closed sheeting	S5@.114	SF	1.10	4.10	.69	5.89
6' to 10' wide, closed sheeting	S5@.143	SF	1.20	5.15	.87	7.22
11' to 16' wide, closed sheeting	S5@.163	SF	1.40	5.87	.99	8.26
Over 16' wide, closed sheeting	S5@.199	SF	1.60	7.16	1.21	9.97

Additional costs for bracing on trenches over 12' deep, add to costs above

	Craft@Hrs	Unit	Material	Labor	Equipment	Total
Normal bracing, to 15' deep	S5@.050	SF	.20	1.80	.30	2.30
One line of bracing, over 15' to 22' deep	S5@.053	SF	.25	1.91	.32	2.48
Two lines of bracing, over 22' to 35' deep	S5@.061	SF	.40	2.20	.37	2.97
Three lines of bracing, over 35' to 45' deep	S5@.064	SF	.45	2.30	.39	3.14

Bulkheads and tieback walls, (soldier beams), not including hydrostatic heads, 10" "H" soldier piles at 8' spacing, 3" lagging solid, 2 rows, raker bracing, SF of wall area, typical prices

	Craft@Hrs	Unit	Material	Labor	Equipment	Total
10' to 15' excavation	--	SF	--	--	--	19.10
16' to 20'	--	SF	--	--	--	21.20
21' to 25'	--	SF	--	--	--	28.90
26' to 35'	--	SF	--	--	--	32.50
36' to 45' with 3 walers	--	SF	--	--	--	34.60
Tie backs, drilled and plugged, including tie	--	LF	--	--	--	28.40
Rakers, steel	--	Lb	--	--	--	1.00

Underpinning foundations, normal conditions

	Craft@Hrs	Unit	Material	Labor	Equipment	Total
Hand mining, hard pack	--	CF	--	--	--	3.90
Form work for concrete, SF one side	--	SF	--	--	--	4.65
3,500 PSI low slump concrete, in place	--	CY	--	--	--	88.80

Site Work 2

	Craft@Hrs	Unit	Material	Labor	Equipment	Total
Temporary shoring, 4' OC	--	SF	--	--	--	14.00
Permanent shoring, 4' OC	--	SF	--	--	--	26.90
Dry pack, concrete	--	CF	--	--	--	41.80

Wellpoint Dewatering. Based on header pipe connecting 2" diameter jetted wellpoints 5' on center. Includes header pipe, wellpoints, filter sand, pumps and swing joints. The header pipe length is usually equal to the perimeter of the area excavated. These costs include the subcontractor's overhead and profit.

Header pipe and accessories, per LF installed, rented 1 month and removed

	Craft@Hrs	Unit	Material	Labor	Equipment	Total
6" pipe	C5@.208	LF	.72	7.74	3.64	12.10
8" pipe	C5@.253	LF	.87	9.41	4.42	14.70

Wellpoints, 2", with sand fill, per point jetted, rented 1 month and removed

	Craft@Hrs	Unit	Material	Labor	Equipment	Total
14' deep	C5@1.00	Ea	2.20	37.20	23.60	63.00
18' deep	C5@1.28	Ea	2.77	47.60	30.13	80.50

Setting and removing pumps, valves, swing joints, outflow, etc., per foot

	Craft@Hrs	Unit	Material	Labor	Equipment	Total
of header pipe installed	C5@.127	LF	--	4.72	--	--

Wellpoint diesel pump

	Craft@Hrs	Unit	Material	Labor	Equipment	Total
6" pump rented 1 month	--	Mo	--	--	1,650.00	--
8" pump rented 1 month	--	Mo	--	--	2,000.00	--
Jetting pump, rented 1 month	--	LS	--	--	1,800.00	--

Typical cost for installation and removal of wellpoint system with wellpoints 14' deep and placed 5' on center along 6" header pipe. Based on equipment rented for 1 month. Add (if required) for additional months of rental, fuel for the pumps, pump operators, a standby pump, water truck for jetting, outflow pipe, permits and consultant costs.

6" header pipe and accessories, system per LF installed, rented 1 month and removed

	Craft@Hrs	Unit	Material	Labor	Equipment	Total
100 LF system	C5@.546	LF	1.20	20.30	42.00	63.50
200 LF system	C5@.546	LF	1.20	20.30	25.00	46.50
500 LF system	C5@.546	LF	1.20	20.30	15.00	36.50
1,000 LF system	C5@.546	LF	1.20	20.30	12.00	33.50
Add for 8" header pipe	--	%	5.0	2.0	6.0	--
Add for 18' wellpoint depth	--	%	9.0	18.0	17.0	--
Add for second month	--	%	--	--	50.0	--
Add for each additional month	--	%	--	--	30.0	--
Add for operator, per hour	TO@1.00	Hr	--	43.00	--	--

Pipe Jacking. Typical costs for jacking .50" thick wall pipe casing under an existing roadway. Costs include leaving casing in place. Add 15% when ground water is present. Add 100% for light rock conditions. Includes jacking pits on both sides. Costs shown are for casing diameter. These costs include the subcontractor's overhead and profit.

	Craft@Hrs	Unit	Material	Labor	Equipment	Total
2"	C5@.289	LF	4.80	10.70	3.83	19.33
3"	C5@.351	LF	5.75	13.10	4.66	23.51
4"	C5@.477	LF	7.95	17.70	6.42	32.07
6"	C5@.623	LF	10.20	23.20	8.38	41.78
8"	C5@.847	LF	13.90	31.50	11.40	56.80
10"	C5@1.24	LF	20.40	46.10	16.60	83.10
12"	C5@1.49	LF	24.50	55.40	20.20	100.10
16"	C5@1.78	LF	29.30	66.20	23.80	119.30
17"	C5@2.09	LF	34.30	77.70	27.90	139.90
24"	C5@7.27	LF	119.00	270.00	97.30	486.30
30"	C5@7.88	LF	130.00	293.00	106.00	529.00
36"	C5@8.61	LF	142.00	320.00	115.00	577.00
42"	C5@9.29	LF	152.00	345.00	124.00	621.00
48"	C5@10.2	LF	167.00	379.00	137.00	683.00

Site Work 2

	Craft@Hrs	Unit	Material	Labor	Equipment	Total

Asbestos Cement Pipe (Transite). Class 2400 sewer pipe or Class 3000 storm drain pipe, standard 13' lengths (Certain Teed Products). Costs include couplers, inspection and test and are based on truck load quantities. These costs include the subcontractor's overhead and profit. Excavation, bedding material or backfill are not included.

	Craft@Hrs	Unit	Material	Labor	Equipment	Total
8" pipe	P5@.077	LF	3.00	2.82	.33	6.15
10" pipe	P5@.080	LF	4.60	2.93	.34	7.87
12" pipe	P5@.083	LF	6.50	3.04	.35	9.89
18" pipe	P5@.116	LF	15.00	4.24	.72	19.96
21" pipe	P5@.123	LF	18.50	4.50	.78	23.78
24" pipe	P5@.134	LF	26.50	4.90	.84	32.24
Add for class 3300 pipe	--	%	15.0	--	--	--
Add for class 4000 pipe	--	%	20.0	--	--	--

Polyethylene Rigid Drainage Pipe. Standard 10' lengths, 1500 lb crunch rating, plain or perforated and high impact ABS glued-on fittings. These costs include the subcontractor's overhead and profit but no excavation or backfill.

	Craft@Hrs	Unit	Material	Labor	Equipment	Total
3" pipe	CL@.070	LF	.55	2.39	--	2.94
4" pipe	CL@.073	LF	.75	2.49	--	3.24
6" pipe	CL@.079	LF	1.55	2.69	--	4.24
Elbows fittings (1/4 bend)						
3"	CL@.168	Ea	1.70	5.73	--	7.43
4"	CL@.250	Ea	2.45	8.53	--	10.98
6"	CL@.332	Ea	7.40	11.30	--	18.70
Tee fittings						
3"	CL@.251	Ea	1.25	8.56	--	9.81
4"	CL@.375	Ea	1.75	12.80	--	14.55
6"	CL@.500	Ea	11.30	17.10	--	28.40

Polyethylene Flexible Drainage Tubing. Corrugated drainage tubing, plain or perforated and snap-on ABS fittings. These costs include the subcontractor's overhead and profit but no excavation or backfill.

	Craft@Hrs	Unit	Material	Labor	Equipment	Total
Tubing						
3" (10' length or 100' coil)	CL@.009	LF	.45	.31	--	.76
4" (10' length or 250' coil)	CL@.011	LF	.50	.38	--	.88
5" (180' coil)	CL@.013	LF	.80	.44	--	1.24
6" (100' coil)	CL@.014	LF	1.20	.48	--	1.68
8" (20' or 40' length)	CL@.017	LF	2.00	.58	--	2.58
10" (20' length)	CL@.021	LF	3.85	.72	--	4.57
12" (20' length)	CL@.022	LF	5.40	.75	--	6.15
Snap fittings, elbows or tees						
3"	CL@.025	Ea	3.00	.85	--	3.85
4"	CL@.025	Ea	3.65	.85	--	4.50
5"	CL@.025	Ea	5.50	.85	--	6.35
6"	CL@.025	Ea	6.30	.85	--	7.15

Cast Iron Soil Pipe. Service weight single hub type, 5' lengths, with rubber ring gaskets, installed in an open trench. These costs include the subcontractor's overhead and profit but no excavation, dewatering, shoring, bedding or backfill.

	Craft@Hrs	Unit	Material	Labor	Equipment	Total
2" pipe	P6@.079	LF	6.50	3.23	.95	10.68
3" pipe	P6@.082	LF	8.90	3.35	1.00	13.25
4" pipe	P6@.085	LF	11.60	3.47	1.05	16.12
5" pipe	P6@.088	LF	16.10	3.60	1.10	20.80
6" pipe	P6@.093	LF	19.90	3.80	1.20	24.90
8" pipe	P6@.103	LF	31.90	4.21	1.40	37.51
10" pipe	P6@.119	LF	52.80	4.86	1.70	59.36
12" pipe	P6@.132	LF	65.00	5.40	1.80	72.20

Site Work 2

	Craft@Hrs	Unit	Material	Labor	Equipment	Total

Non-Reinforced Concrete Pipe. Smooth wall, standard strength, 3' lengths with tongue and groove mortar joint ends. These costs include the subcontractor's overhead and profit but no excavation, sand bed, backfill, shoring or dewatering.

	Craft@Hrs	Unit	Material	Labor	Equipment	Total
6" pipe	C5@.076	LF	2.60	2.83	.47	5.90
8" pipe	C5@.088	LF	4.00	3.27	.62	7.89
10" pipe	C5@.088	LF	4.80	3.27	.62	8.69
12" pipe	C5@.093	LF	6.80	3.46	.63	10.89
15" pipe	C5@.105	LF	8.30	3.91	.64	12.85
18" pipe	C5@.117	LF	10.40	4.35	.71	15.46
21" pipe	C5@.117	LF	13.50	4.35	.71	18.56
24" pipe	C5@.130	LF	19.70	4.84	.82	25.36
30" pipe	C5@.143	LF	34.20	5.32	.92	40.44
36" pipe	C5@.158	LF	37.30	5.88	.99	44.17

Non-Reinforced Perforated Concrete Underdrain Pipe. Perforated smooth wall, standard strength, 3' lengths with tongue and groove mortar joint ends. These costs include the subcontractor's overhead and profit but no excavation, rock fill, backfill, shoring or dewatering.

	Craft@Hrs	Unit	Material	Labor	Equipment	Total
6" pipe	C5@.076	LF	3.00	2.83	.45	6.28
8" pipe	C5@.104	LF	4.60	3.87	.60	9.07

Reinforced Concrete Pipe. Class III, 1350 "D" load. Tongue and groove mortar joint ends. These costs include the subcontractor's overhead and profit but no sand bed, trenching or backfill. Based on 500' minimum job.

	Craft@Hrs	Unit	Material	Labor	Equipment	Total
12" pipe	C5@.134	LF	14.00	4.98	.88	19.86
15" pipe	C5@.193	LF	17.50	7.18	1.24	25.92
18" pipe	C5@.272	LF	30.90	10.10	1.71	42.71
24" pipe	C5@.365	LF	41.20	13.60	2.38	57.18
27" pipe	C5@.396	LF	46.40	14.70	2.54	63.64
30" pipe	C5@.426	LF	51.40	15.80	2.74	69.94
36" pipe	C5@.511	LF	61.80	19.00	3.31	84.11
42" pipe	C5@.567	LF	72.00	21.10	3.67	96.77
48" pipe	C5@.623	LF	82.40	23.20	3.98	109.58
54" pipe	C5@.707	LF	92.60	26.30	4.50	123.40
60" pipe	C5@.785	LF	103.00	29.20	5.07	137.27
66" pipe	C5@.914	LF	113.00	34.00	5.85	152.85
72" pipe	C5@.914	LF	122.00	34.00	5.85	161.85
84" pipe	C5@1.04	LF	144.00	38.70	6.62	189.32
90" pipe	C5@1.04	LF	154.00	38.70	6.62	199.32
96" pipe	C5@1.18	LF	166.00	43.90	7.56	217.46
Deduct for Class II reinforced concrete pipe	--	%	-4.0	--	--	--

Reinforced Elliptical Concrete Pipe. Class III, 1350 "D", C-502-72, 8' lengths, tongue and groove mortar joint ends. These costs include the subcontractor's overhead and profit but no sand bed, excavation or backfill.

	Craft@Hrs	Unit	Material	Labor	Equipment	Total
19" x 30" pipe, 24" pipe equivalent	C5@.356	LF	26.90	13.20	3.88	43.98
24" x 38" pipe, 30" pipe equivalent	C5@.424	LF	35.70	15.80	4.50	56.00
29" x 45" pipe, 36" pipe equivalent	C5@.516	LF	58.00	19.20	5.59	82.79
34" x 53" pipe, 42" pipe equivalent	C5@.581	LF	68.30	21.60	6.31	96.21
38" x 60" pipe, 48" pipe equivalent	C5@.637	LF	85.40	23.70	6.73	115.83
48" x 76" pipe, 60" pipe equivalent	C5@.797	LF	141.00	29.60	8.59	179.19
53" x 83" pipe, 66" pipe equivalent	C5@.923	LF	174.00	34.30	9.78	218.08
58" x 91" pipe, 72" pipe equivalent	C5@.923	LF	209.00	34.30	9.78	253.08

Site Work 2

	Craft@Hrs	Unit	Material	Labor	Equipment	Total

Flared Concrete End Sections. For round concrete drain pipe, precast. These costs include the subcontractor's overhead and profit.

	Craft@Hrs	Unit	Material	Labor	Equipment	Total
12" opening, 530 lbs	C5@2.41	Ea	258.00	89.60	24.20	371.80
15" opening, 740 lbs	C5@2.57	Ea	303.00	95.60	26.50	425.10
18" opening, 990 lbs	C5@3.45	Ea	335.00	128.00	36.40	499.40
24" opening, 1,520 lbs	C5@3.45	Ea	415.00	128.00	36.40	579.40
30" opening, 2,190 lbs	C5@3.96	Ea	562.00	147.00	41.70	750.70
36" opening, 4,100 lbs	C5@3.96	Ea	712.00	147.00	41.70	900.70
42" opening, 5,380 lbs	C5@4.80	Ea	1,000.00	179.00	48.60	1,227.60
48" opening, 6,550 lbs	C5@5.75	Ea	1,190.00	214.00	58.00	1,462.00
54" opening, 8,000 lbs	C5@7.10	Ea	1,530.00	264.00	116.00	1,910.00

Precast Reinforced Concrete Box Culvert Pipe. ASTM C-850 with tongue and groove mortar joint ends. These costs include the subcontractor's overhead and profit but no excavation, backfill, shoring or dewatering. Laying length is 8' for smaller cross sections and 6' or 4' for larger cross sections.

	Craft@Hrs	Unit	Material	Labor	Equipment	Total
4' x 3'	P5@.273	LF	96.50	9.98	5.43	111.91
5' x 5'	P5@.336	LF	144.00	12.30	6.62	162.92
6' x 4'	P5@.365	LF	167.00	13.30	7.35	187.65
6' x 6'	P5@.425	LF	194.00	15.50	8.54	218.04
7' x 4'	P5@.442	LF	202.00	16.20	8.90	227.10
7' x 7'	P5@.547	LF	248.00	20.00	11.00	279.00
8' x 4'	P5@.479	LF	217.00	17.50	9.06	243.56
8' x 8'	P5@.613	LF	280.00	22.40	11.70	314.10
9' x 6'	P5@.658	LF	300.00	24.10	12.50	336.60
9' x 9'	P5@.773	LF	354.00	28.30	14.80	397.10
10' x 6'	P5@.787	LF	360.00	28.80	12.80	401.60
10' x 10'	P5@.961	LF	436.00	35.10	18.20	489.30
Deduct for ASTM C-789 precast reinforced concrete box culvert pipe	--	%	-8.0	--	--	--

Corrugated Metal Pipe Galvanized, 20' lengths, installed in an open trench. Costs include couplers. These costs include the subcontractor's overhead and profit but no trenching, shoring, dewatering or backfill. Truckload quantities.

Round pipe

	Craft@Hrs	Unit	Material	Labor	Equipment	Total
8", 16 gauge (.064)	C5@.102	LF	7.15	3.79	.67	11.61
10", 16 gauge (.064)	C5@.102	LF	7.45	3.79	.67	11.91
12", 16 gauge (.064)	C5@.120	LF	7.75	4.46	.78	12.99
15", 16 gauge (.064)	C5@.120	LF	9.65	4.46	.78	14.89
18", 16 gauge (.064)	C5@.152	LF	11.50	5.65	.98	18.13
24", 14 gauge (.079)	C5@.196	LF	17.90	7.29	1.29	26.48
30", 14 gauge (.079)	C5@.196	LF	22.30	7.29	1.29	30.88
36", 14 gauge (.079)	C5@.231	LF	26.90	8.59	1.50	36.99
42", 14 gauge (.079)	C5@.231	LF	31.40	8.59	1.50	41.49
48", 14 gauge (.079)	C5@.272	LF	35.60	10.10	1.81	47.51
60", 12 gauge (.109)	C5@.311	LF	60.90	11.60	2.07	74.57

Oval pipe

	Craft@Hrs	Unit	Material	Labor	Equipment	Total
18" x 11", 16 gauge (.0640)	C5@.133	LF	11.40	4.95	.88	17.23
22" x 13", 16 gauge (.0640)	C5@.167	LF	13.70	6.21	1.09	21.00
29" x 18", 14 gauge (.079)	C5@.222	LF	21.20	8.26	1.45	30.91
36" x 22", 14 gauge (.079)	C5@.228	LF	26.40	8.48	1.50	36.38
43" x 27", 14 gauge (.079)	C5@.284	LF	31.90	10.60	1.86	44.36

Flared end sections for oval pipe

	Craft@Hrs	Unit	Material	Labor	Equipment	Total
18" x 11"	C5@1.59	Ea	62.30	59.10	10.25	131.65
22" x 13"	C5@1.75	Ea	73.40	65.10	11.40	149.90

Site Work 2

	Craft@Hrs	Unit	Material	Labor	Equipment	Total
29" x 18"	C5@1.85	Ea	107.00	68.80	12.20	188.00
36" x 22"	C5@2.38	Ea	176.00	88.50	15.60	280.10
43" x 27"	C5@2.85	Ea	299.00	106.00	18.80	423.80
Add for bituminous coating, with paved invert						
On round pipe	--	%	20.0	--	--	--
On oval pipe or flared ends	--	%	24.0	--	--	--
Deduct for aluminum pipe, round						
oval or flared ends	--	%	-12.0	--	--	--

Corrugated Metal Nestable Pipe, split-in-half, type I, 16 gauge, no excavation or grading included. These costs include the subcontractor's overhead and profit.

	Craft@Hrs	Unit	Material	Labor	Equipment	Total
12" diameter, galvanized	C5@.056	LF	11.60	2.08	.38	14.06
18" diameter, galvanized	C5@.064	LF	17.20	2.38	.43	20.01
Add for bituminous coating,						
with paved invert	--	%	20.0	--	--	--

Drainage Tile. Vitrified clay drainage tile pipe. Standard weight, 6' joint length with 1 coupler per joint. Couplers consist of a rubber sleeve with stainless steel compression bands. These costs include the subcontractor's overhead and profit but no excavation or backfill.

	Craft@Hrs	Unit	Material	Labor	Equipment	Total
4" tile	CL@.095	LF	2.00	3.24	--	5.24
6" tile	CL@.119	LF	5.50	4.06	--	9.56
8" tile	CL@.143	LF	7.75	4.88	--	12.63
10" tile	CL@.165	LF	12.70	5.63	--	18.33
Add for covering 4" to 10" tile,						
with 30 lb felt building paper	CL@.048	LF	.10	1.64	--	1.74
Elbows (1/4 bends), costs include one coupler						
4"	CL@.286	Ea	9.80	9.75	--	19.55
6"	CL@.360	Ea	25.90	12.30	--	38.20
8"	CL@.430	Ea	34.50	14.70	--	49.20
10"	CL@.494	Ea	59.80	16.80	--	76.60
Tees, costs include two couplers						
4"	CL@.430	Ea	14.40	14.70	--	29.10
6"	CL@.540	Ea	32.80	18.40	--	51.20
8"	CL@.643	Ea	48.30	21.90	--	70.20
10"	CL@.744	Ea	104.00	25.40	--	129.40

Corrugated Polyethylene Culvert Pipe. Heavy duty, 20' lengths. These costs include the subcontractor's overhead and profit but no excavation or backfill. Based on 500' minimum job. Various sizes may be combined for total footage. Couplings are split type that snap onto the pipe.

	Craft@Hrs	Unit	Material	Labor	Equipment	Total
8" diameter pipe	C5@.055	LF	2.10	2.05	.21	4.36
10" diameter pipe	C5@.038	LF	4.00	1.41	.26	5.67
12" diameter pipe	C5@.053	LF	5.70	1.97	.31	7.98
18" diameter pipe	C5@.111	LF	10.70	4.13	.78	15.61
8" coupling	C5@.105	Ea	5.00	3.91	.72	9.63
10" coupling	C5@.114	Ea	6.40	4.24	.80	11.44
12" coupling	C5@.158	Ea	7.50	5.88	1.04	14.42
18" coupling	C5@.331	Ea	13.80	12.30	2.17	28.27

PVC Sewer Pipe. Bell and spigot ends with rubber ring gasket in bell. Installed in open trench, no backfill, shoring or trenching included, 13' lengths. These costs include the subcontractor's overhead and profit.

	Craft@Hrs	Unit	Material	Labor	Equipment	Total
4" pipe	C5@.047	LF	1.00	1.75	.31	3.06
6" pipe	C5@.056	LF	2.10	2.08	.41	4.59
8" pipe	C5@.061	LF	3.60	2.27	.47	6.34

Site Work 2

	Craft@Hrs	Unit	Material	Labor	Equipment	Total
PVC Sewer Pipe, continued						
10" pipe	C5@.079	LF	5.90	2.94	.52	9.36
12" pipe	C5@.114	LF	7.75	4.24	.78	12.77

Vitrified Clay Pipe. C-200, extra strength, installed in open trench, no backfill, shoring or trenching included. 4" through 12" are 6'0" joint length, over 12" are 7'6" joint length. Costs shown include one coupler per joint. Couplers consist of a rubber sleeve with stainless steel compression bands. These costs include the subcontractor's overhead and profit.

	Craft@Hrs	Unit	Material	Labor	Equipment	Total
4" pipe	CL@.081	LF	2.75	2.76	--	5.51
6" pipe	CL@.175	LF	6.95	5.97	--	12.92
8" pipe	CL@.183	LF	10.00	6.24	--	16.24
10" pipe	CL@.203	LF	16.50	6.92	--	23.42
12" pipe	CL@.225	LF	22.10	7.67	--	29.77
15" pipe	C5@.233	LF	23.00	8.67	1.55	33.22
18" pipe	C5@.254	LF	32.30	9.45	1.66	43.41
21" pipe	C5@.283	LF	42.90	10.50	1.86	55.26
24" pipe	C5@.334	LF	56.90	12.40	2.17	71.47
27" pipe	C5@.396	LF	68.00	14.70	2.59	85.29
30" pipe	C5@.452	LF	84.00	16.80	3.00	103.80
1/4 bends including couplers						
4"	CL@.494	Ea	13.00	16.80	--	29.80
6"	CL@.524	Ea	35.00	17.90	--	52.90
8"	CL@.549	Ea	71.80	18.70	--	90.50
10"	CL@.610	Ea	83.00	20.80	--	103.80
12"	CL@.677	Ea	107.00	23.10	--	130.10
15"	C5@.715	Ea	288.00	26.60	4.71	319.31
18"	C5@.844	Ea	395.00	31.40	5.49	431.89
21"	C5@.993	Ea	540.00	36.90	6.52	583.42
24"	C5@1.17	Ea	700.00	43.50	7.66	751.16
30"	C5@1.37	Ea	1,050.00	51.00	9.06	1,110.06
Wyes and tees including couplers						
4"	CL@.744	Ea	19.20	25.40	--	44.60
6"	CL@.787	Ea	44.10	26.80	--	70.90
8"	CL@.823	Ea	67.50	28.10	--	95.60
10"	CL@.915	Ea	145.00	31.20	--	176.20
12"	CL@1.02	Ea	160.00	34.80	--	194.80
15"	C5@1.10	Ea	238.00	40.90	7.25	286.15
18"	C5@1.30	Ea	332.00	48.30	8.54	388.84
21"	C5@1.53	Ea	446.00	56.90	10.04	512.94
24"	C5@1.80	Ea	586.00	66.90	11.90	664.80
27"	C5@2.13	Ea	730.00	79.20	13.90	823.10
30"	C5@2.50	Ea	875.00	93.00	16.40	984.40

Sewer Main Cleaning. Ball method, typical costs for cleaning new sewer mains prior to inspection, includes operator and equipment. Add $175 to total costs for setting up and removing equipment. Minimum cost will be $400. These costs include the subcontractor's overhead and profit. Costs per 1,000 LF of main diameter listed.

	Craft@Hrs	Unit	Material	Labor	Equipment	Total
6" diameter	--	MLF	--	--	--	62.50
8" diameter	--	MLF	--	--	--	82.50
10" diameter	--	MLF	--	--	--	100.00
12" diameter	--	MLF	--	--	--	125.00
14" diameter	--	MLF	--	--	--	142.00
16" diameter	--	MLF	--	--	--	165.00
18" diameter	--	MLF	--	--	--	185.00

Site Work 2

	Craft@Hrs	Unit	Material	Labor	Equipment	Total
24" diameter	--	MLF	--	--	--	245.00
36" diameter	--	MLF	--	--	--	375.00
48" diameter	--	MLF	--	--	--	500.00

Water Main Sterilization. Gas chlorination method, typical costs for sterilizing new water mains prior to inspection, includes operator and equipment. Add $175 to total costs for setting up and removing equipment. Minimum cost will be $400. These costs include the subcontractor's overhead and profit. Cost per 1,000 LF of main diameter listed.

	Craft@Hrs	Unit	Material	Labor	Equipment	Total
3" diameter	--	MLF	--	--	--	30.00
4" diameter	--	MLF	--	--	--	41.50
6" diameter	--	MLF	--	--	--	62.50
8" diameter	--	MLF	--	--	--	82.50
10" diameter	--	MLF	--	--	--	100.00
12" diameter	--	MLF	--	--	--	125.00
14" diameter	--	MLF	--	--	--	145.00
16" diameter	--	MLF	--	--	--	165.00
18" diameter	--	MLF	--	--	--	185.00
24" diameter	--	MLF	--	--	--	245.00
36" diameter	--	MLF	--	--	--	345.00
48" diameter	--	MLF	--	--	--	500.00

Soil Covers. These costs include the subcontractor's overhead and profit.

	Craft@Hrs	Unit	Material	Labor	Equipment	Total
Using staked burlap and tar over straw	CL@.019	SY	.60	.65	--	1.25
Using copolymer-base sprayed-on liquid soil sealant						
On slopes for erosion control	CL@.012	SY	.30	.41	--	.71
On level ground for dust abatement	CL@.012	SY	.25	.41	--	.66

Catch Basins. These costs include the subcontractor's overhead and profit. Catch basin including 6" concrete top and base but no excavation or backfill. Add for frames and grates from the cost below

	Craft@Hrs	Unit	Material	Labor	Equipment	Total
4' diameter precast concrete basin						
4' deep	S6@19.0	Ea	900.00	704.00	128.00	1,732.00
6' deep	S6@23.7	Ea	1,120.00	878.00	160.00	2,158.00
8' deep	S6@28.9	Ea	1,360.00	1,070.00	197.00	2,627.00
4' diameter concrete block radial basin						
4' deep	S6@20.2	Ea	960.00	749.00	133.00	1,842.00
6' deep	S6@25.7	Ea	1,220.00	952.00	172.00	2,344.00
8' deep	S6@32.6	Ea	1,530.00	1,210.00	219.00	2,959.00
Light duty grate, gray iron, asphalt coated						
Grate on pipe bell						
6" diameter, 13 lb	S6@.205	Ea	8.75	7.60	1.40	17.75
8" diameter, 25 lb	S6@.247	Ea	23.50	9.15	1.60	34.25
Frame and grate						
8" diameter, 55 lb	S6@.424	Ea	50.00	15.70	2.90	68.60
17" diameter, 135 lb	S6@.823	Ea	120.00	30.50	5.60	156.10
Medium duty, frame and grate						
11" diameter, 70 lb	S6@.441	Ea	60.00	16.30	2.95	79.25
15" diameter, 120 lb	S6@.770	Ea	110.00	28.50	5.20	143.70
Radial grate, 20" diameter, 140 lb	S6@1.68	Ea	125.00	62.30	11.30	198.60
Heavy duty frame and grate						
Flat grate						
11-1/2" diameter, 85 lb	S6@.537	Ea	80.00	19.90	3.60	103.50
20" diameter, 235 lb	S6@2.01	Ea	210.00	74.50	13.50	298.00
21" diameter, 315 lb	S6@2.51	Ea	260.00	93.00	17.10	370.10

Site Work 2

	Craft@Hrs	Unit	Material	Labor	Equipment	Total
Heavy duty frame and grate, continued						
24" diameter, 350 lb	S6@2.51	Ea	290.00	93.00	17.10	400.10
30" diameter, 555 lb	S6@4.02	Ea	425.00	149.00	27.20	601.20
Convex or concave grate						
20" diameter, 200 lb	S6@1.83	Ea	150.00	67.80	11.90	229.70
20" diameter, 325 lb	S6@2.51	Ea	245.00	93.00	17.10	355.10
Beehive grate and frame						
11" diameter, 80 lb	S6@.495	Ea	110.00	18.30	3.30	131.60
15" diameter, 120 lb	S6@1.83	Ea	105.00	67.80	12.30	185.10
21" diameter, 285 lb	S6@2.51	Ea	215.00	93.00	17.10	325.10
24" diameter, 375 lb	S6@2.87	Ea	300.00	106.00	19.30	425.30

Manholes. These costs include the subcontractor's overhead and profit.
Precast concrete manholes, including concrete top and base. No excavation or backfill included. Add for steps, frames and lids from the costs listed below

	Craft@Hrs	Unit	Material	Labor	Equipment	Total
3' x 6' to 8' deep	S6@17.5	Ea	1,040.00	649.00	118.00	1,807.00
3' x 9' to 12' deep	S6@19.4	Ea	1,250.00	719.00	131.00	2,100.00
3' x 13' to 16' deep	S6@23.1	Ea	1,550.00	856.00	155.00	2,561.00
3' diameter, 4' deep	S6@12.1	Ea	620.00	448.00	81.70	1,149.70
4' diameter, 5' deep	S6@22.8	Ea	750.00	845.00	154.00	1,749.00
4' diameter, 6' deep	S6@26.3	Ea	825.00	975.00	178.00	1,978.00
4' diameter, 7' deep	S6@30.3	Ea	925.00	1,120.00	206.00	2,251.00
4' diameter, 8' deep	S6@33.7	Ea	1,020.00	1,250.00	227.00	2,497.00
4' diameter, 9' deep	S6@38.2	Ea	1,090.00	1,420.00	257.00	2,767.00
4' diameter, 10' deep	S6@43.3	Ea	1,180.00	1,600.00	295.00	3,075.00
Concrete block radial manholes, 4' inside diameter, no excavation or backfill included						
4' deep	M1@8.01	Ea	255.00	300.00	52.90	607.90
6' deep	M1@13.4	Ea	380.00	502.00	87.80	969.80
8' deep	M1@20.0	Ea	560.00	749.00	131.00	1,440.00
10' deep	M1@26.7	Ea	645.00	1,000.00	177.00	1,822.00
Depth over 10'	M1@2.29	LF	72.00	85.80	15.00	172.80
2' depth cone block for 30" grate	M1@5.34	LF	110.00	200.00	34.20	344.20
2'6" depth cone block for 24" grate	M1@6.43	LF	143.00	241.00	42.40	426.40
Manhole steps, cast iron, heavy type, asphalt coated						
10" x 14-1/2" in job-built manhole	M1@.254	Ea	12.50	9.51	--	22.01
12" x 10" in precast manhole	--	Ea	11.20	--	--	11.20
Gray iron manhole frames, asphalt coated, standard sizes						
Light duty frame and lid						
15" diameter, 65 lb	S6@.660	Ea	57.20	24.50	4.50	86.20
22" diameter, 140 lb	S6@1.42	Ea	124.00	52.60	9.63	186.23
Medium duty frame and lid						
11" diameter, 75 lb	S6@1.51	Ea	67.00	56.00	10.20	133.20
20" diameter, 185 lb	S6@3.01	Ea	155.00	112.00	20.40	287.40
Heavy duty frame and lid						
17" diameter, 135 lb	S6@2.51	Ea	118.00	93.00	17.10	228.10
21" diameter, 315 lb	S6@3.77	Ea	263.00	140.00	25.50	428.50
24" diameter, 375 lb	S6@4.30	Ea	304.00	159.00	29.20	492.20
Connect new drain line to existing manhole, no excavation or backfill included	S6@6.83	Ea	81.50	253.00	46.30	380.80
Connect existing drain line to new manhole, no excavation or backfill included	S6@3.54	Ea	50.00	131.00	24.00	205.00
Manhole repairs and alterations, typical costs						
Repair manhole leak with grout	S6@15.8	Ea	285.00	586.00	108.00	979.00
Repair inlet leak with grout	S6@15.8	Ea	300.00	586.00	108.00	994.00

Site Work 2

	Craft@Hrs	Unit	Material	Labor	Equipment	Total
Grout under manhole frame	S6@1.29	Ea	5.00	47.80	8.80	61.60
Drill and grout pump setup cost	S6@17.4	LS	300.00	645.00	118.00	1,063.00
Grout for pressure grouting	--	Gal	5.00	--	--	--
Replace brick in manhole wall	M1@8.79	SF	8.00	329.00	--	337.00
Replace brick under manhole frame	M1@8.79	LS	25.00	329.00	--	354.00
Raise existing frame and cover 2"	CL@4.76	LS	150.00	162.00	--	312.00
Raise existing frame and cover more than 2", per each 1" added	CL@.250	LS	50.00	8.53	--	58.53
T.V. inspection of pipe interior	--	LF	--	--	--	1.80
Grouting concrete pipe joints						
6" to 30" diameter	S6@.315	Ea	14.00	11.70	2.17	27.87
33" to 60" diameter	S6@1.12	Ea	27.50	41.50	7.50	76.50
66" to 72" diameter	S6@2.05	Ea	55.50	76.00	14.00	145.50

Accessories for Site Utilities. These costs include the subcontractor's overhead and profit.

	Craft@Hrs	Unit	Material	Labor	Equipment	Total
Curb inlets, gray iron, asphalt coated, heavy duty frame, grate and curb box						
20" x 11", 260 lb	S6@2.75	Ea	226.00	102.00	18.70	346.70
20" x 16.5", 300 lb	S6@3.04	Ea	248.00	113.00	20.60	381.60
20" x 17", 400 lb	S6@4.30	Ea	316.00	159.00	29.20	504.20
19" x 18", 500 lb	S6@5.06	Ea	375.00	188.00	34.30	597.30
30" x 17", 600 lb	S6@6.04	Ea	411.00	224.00	40.90	675.90
Gutter inlets, gray iron, asphalt coated, heavy duty frame and grate						
8" x 11.5", 85 lb	S6@.444	Ea	75.80	16.50	3.00	95.30
22" x 17", 260 lb, concave	S6@1.84	Ea	216.00	68.20	12.30	296.50
22.3" x 22.3", 475 lb	S6@2.87	Ea	238.00	106.00	19.30	363.30
Two 29.8" x 17.8", 750 lb	S6@3.37	Ea	359.00	125.00	22.80	506.80
Trench inlets, ductile iron, light duty frame and grate for pedestrian traffic, 53.5" x 8.3", 100 lb	S6@.495	Ea	82.20	18.30	3.42	103.92
Trench inlets, gray iron, asphalt coated frame and grate or solid cover						
Light duty, 1-1/4"						
8" wide grate	S6@.247	LF	31.40	9.15	1.76	42.31
12" wide grate	S6@.275	LF	44.30	10.20	1.86	56.36
8" wide solid cover	S6@.247	LF	34.60	9.15	1.76	45.51
12" wide solid cover	S6@.275	LF	49.80	10.20	1.86	61.86
Heavy duty, 1-3/4						
8" wide grate	S6@.275	LF	34.60	10.20	1.86	46.66
12" wide grate	S6@.309	LF	50.80	11.50	2.07	64.37
8" wide solid grate	S6@.275	LF	38.90	10.20	1.86	50.96
12" wide solid cover	S6@.309	LF	54.10	11.50	2.07	67.67
Curb inlet catch basins, complete, including excavation						
4' throat, 8' deep, for 12" to 30" pipe	S6@61.0	Ea	578.00	2,260.00	412.00	3,250.00
4' throat, 8' deep, for 36" to 48" pipe	S6@63.2	Ea	704.00	2,340.00	428.00	3,472.00

Concrete headwalls, by pipe diameter, assuming height overall is 18" more than pipe diameter. Width is 3 times height, 8" thick. Includes formwork and reinforcing but no excavation, typical prices

	Craft@Hrs	Unit	Material	Labor	Equipment	Total
24" pipe, straight headwall	M2@22.5	Ea	373.00	833.00	145.00	1,351.00
30" pipe, straight headwall	M2@31.0	Ea	525.00	1,150.00	201.00	1,876.00
36" pipe, straight headwall	M2@41.3	Ea	700.00	1,530.00	264.00	2,494.00
42" pipe, straight headwall	M2@51.6	Ea	865.00	1,910.00	26.70	2,801.70
12" pipe, two 45 degree wings	M2@10.6	Ea	143.00	392.00	68.20	603.20
18" pipe, two 45 degree wings	M2@17.0	Ea	239.00	629.00	109.00	977.00
24" pipe, two 45 degree wings	M2@27.0	Ea	414.00	999.00	175.00	1,588.00
30" pipe, two 45 degree wings	M2@37.0	Ea	594.00	1,370.00	239.00	2,203.00

Site Work 2

	Craft@Hrs	Unit	Material	Labor	Equipment	Total

Accessories for Site Utilities, continued
Curb valve box, with concrete cover

	Craft@Hrs	Unit	Material	Labor	Equipment	Total
19" x 14" x 12" box	CL@1.13	Ea	24.20	38.50	--	62.70
20" x 14" x 12" box	CL@1.13	Ea	27.50	38.50	--	66.00

Oil and water separators, concrete, 4" inlet and outlet, complete, typical prices

	Craft@Hrs	Unit	Material	Labor	Equipment	Total
5' x 10' x 5'	S5@41.6	Ea	1,550.00	1,500.00	273.00	3,323.00
7' x 13' x 5'	S5@187.	Ea	2,900.00	6,730.00	1,220.00	10,850.00

Yard drains, excavating and backfill included

	Craft@Hrs	Unit	Material	Labor	Equipment	Total
18" x 18"	C5@2.06	Ea	179.00	76.60	13.50	269.10
Depth extensions, 18" x 18" x 12" deep	C5@1.07	Ea	38.30	39.80	6.99	85.09

Gas meter box, rectangular concrete with cast iron cover, includes excavating and backfill

	Craft@Hrs	Unit	Material	Labor	Equipment	Total
	C5@2.00	Ea	65.90	74.40	13.10	153.40

Asphalt Paving. These costs include the subcontractor's overhead and profit. For lime stabilization or soil cement treatment, refer to the section on Soil Stabilization

	Craft@Hrs	Unit	Material	Labor	Equipment	Total
Fine grading existing subbase to within 1/10'	P5@.007	SY	--	.26	.46	.72
Watering and restaking existing subbase	P5@.004	SY	.01	.15	.24	.40

Stone aggregate base, Class 2, (at $11.00 per CY or $8.00 ton). Includes grading and compacting of base material but no grading or compacting of the existing subbase

	Craft@Hrs	Unit	Material	Labor	Equipment	Total
1" thick base (36 SY per CY)	P5@.008	SY	.31	.29	.41	1.01
4" thick base (9 SY per CY)	P5@.020	SY	1.22	.73	1.00	2.95
6" thick base (6 SY per CY)	P5@.033	SY	1.83	1.21	1.52	4.56
8" thick base (4.5 SY per CY)	P5@.039	SY	2.44	1.43	1.82	5.69
10" thick base (3.6 SY per CY)	P5@.042	SY	3.06	1.54	2.01	6.61
12" thick base (3 SY per CY)	P5@.048	SY	3.66	1.76	2.28	7.70
Bituminous prime coat, MC70 (at $1.50 per gallon and .25 gallon per SY), 269 gallons per ton	C4@.001	SY	.38	.04	.02	.44
Bituminous tack coat, MC70 (at $1.50 per gallon and .15 gallon per SY), 269 gallons per ton	C4@.001	SY	.23	.04	.02	.29

Asphalt binder course for flexible pavement (at $34.00 per ton), spread and rolled

	Craft@Hrs	Unit	Material	Labor	Equipment	Total
1" thick (18 SY per ton)	P5@.028	SY	1.88	1.02	.50	3.40
1-1/2" thick (12 SY per ton)	P5@.036	SY	2.83	1.32	.61	4.76
2" thick (9 SY per ton)	P5@.039	SY	3.78	1.43	.66	5.87
3" thick (6 SY per ton)	P5@.050	SY	5.66	1.83	.90	8.39
Add for wear course (at $36.00 per ton)	--	%	6.0	--	--	--
Tar emulsion protective seal coat	C4@.034	SY	1.20	1.20	.17	2.57
Asphalt slurry seal	C4@.043	SY	1.50	1.52	.26	3.28
Rubberized asphalt interlayer	C4@.052	SY	2.00	1.83	.88	4.71

Cold milling of asphalt surface to 1" depth. Equipment cost is $171 per hour and includes a motorized street broom (at $35 per day), a pavement planer (at $950 per day), 2 dump trucks (at $160 per day each) and a pickup truck (at $60 per day).

	Craft@Hrs	Unit	Material	Labor	Equipment	Total
Productivity is 600 SY per hour.	C6@.017	SY	.02	.62	.29	.93

Asphalt pavement repairs to conform to existing street, strips of 100 to 500 SF

	Craft@Hrs	Unit	Material	Labor	Equipment	Total
Asphalt to 1" thick	P5@.342	SY	1.95	12.50	.60	15.05
Asphalt 2" thick	P5@.425	SY	3.90	15.50	.66	20.06
Asphalt 2" thick and 12" stone base	P5@.527	SY	7.60	19.30	6.50	33.40
30" fill compacted to 90%, 6" aggregate base, tack coat, 2" asphalt	P5@.701	SY	9.00	25.60	8.40	43.00
Clean and sweep pavement, by hand	CL@.013	SY	--	.44	--	.44

Pavement sweeping, using a self-propelled vacuum street sweeper, 3 CY capacity at $20.00 per hr (5,000 SF per hr)

	Craft@Hrs	Unit	Material	Labor	Equipment	Total
	T0@.200	MSF	--	8.60	4.00	12.60

Site Work 2

	Craft@Hrs	Unit	Material	Labor	Equipment	Total

Asphalt Paving Composite Costs. These costs include the subcontractor's overhead and profit. Parking lot, typical 1,300 SY job: Strip 4" of topsoil (60 CY) and pile on site; excavate 300 CY of soil and dispose to 5 miles; import, spread and compact 250 CY of stone aggregate base (6"); apply prime coat, 2" binder and wear course; form and pour 250 LF of 6" x 18" curb and 15 SY of concrete valley gutter; paint 4" parking stripes, spread topsoil, fertilize and seed 600 SY of adjacent lawn area.

	Craft@Hrs	Unit	Material	Labor	Equipment	Total
Cost per SY of parking area	P7@.493	SY	7.85	19.00	3.15	30.00
Cost per SF of parking area	P7@.055	SF	.87	2.11	.35	3.33

Access road, typical 500' job: Strip 250 CY of topsoil and pile on site; excavate 400 CY of soil and dispose to 5 miles; import, spread and compact 400 CY of stone aggregate base (8"); apply prime coat, 2" binder and wear course; paint 4" traffic stripes; spread topsoil, fertilize and seed 800 SY of adjacent lawn area.

	Craft@Hrs	Unit	Material	Labor	Equipment	Total
Cost per SY of roadway	P7@.249	SY	7.83	9.72	3.95	21.50
Cost per SF of roadway	P7@.028	SF	.87	1.08	.44	2.39

Concrete Paving. These costs include the subcontractor's overhead and profit. No grading or base preparation included.

	Craft@Hrs	Unit	Material	Labor	Equipment	Total
Steel edge forms to 15", rented	C8@.028	LF	--	1.08	.50	1.58

2,000 PSI concrete (at $46.90 per CY before markup) placed directly from the chute of a ready-mix truck, vibrated and broom finished. Costs include 2% allowance for waste but no forms, reinforcing, or joints are included

	Craft@Hrs	Unit	Material	Labor	Equipment	Total
4" thick (80 SF per CY)	P8@.012	SF	.69	.45	.09	1.23
6" thick (54 SF per CY)	P8@.016	SF	1.02	.59	.14	1.75
8" thick (40 SF per CY)	P8@.023	SF	1.36	.85	.19	2.40
10" thick (32 SF per CY)	P8@.028	SF	1.70	1.04	.24	2.98
12" thick (27 SF per CY)	P8@.032	SF	2.04	1.19	.27	3.50
15" thick (21 SF per CY)	P8@.041	SF	2.59	1.52	.35	4.46

Fibrous concrete (at $85.00 per CY before markup) placed directly from the chute of a ready-mix truck, vibrated and broom finished. Costs include 2% allowance for waste but no forms, reinforcing, or joints are included

	Craft@Hrs	Unit	Material	Labor	Equipment	Total
5" thick (65 SF per CY)	P8@.016	SF	1.54	.59	.10	2.23
8" thick (40 SF per CY)	P8@.026	SF	2.47	.97	.12	3.56

Joint Treatment for Concrete Pavement. These costs include the subcontractor's overhead and profit.

Seal joint in concrete pavement with ASTM D-3569 PVC coal tar (at $12.00 per gallon). By joint size

	Craft@Hrs	Unit	Material	Labor	Equipment	Total
3/8" x 1/2" (128 LF per gallon)	PA@.002	LF	.09	.08	--	.17
1/2" x 1/2" (96 LF per gallon)	PA@.002	LF	.13	.08	--	.21
1/2" x 3/4" (68 LF per gallon)	PA@.003	LF	.18	.12	--	.30
1/2" x 1" (50 LF per gallon)	PA@.003	LF	.24	.12	--	.36
1/2" x 1-1/4" (40 LF per gallon)	PA@.004	LF	.30	.16	--	.46
1/2" x 1-1/2" (33 LF per gallon)	PA@.006	LF	.36	.24	--	.60
3/4" x 3/4" (46 LF per gallon)	PA@.004	LF	.26	.16	--	.42
3/4" x 1-1/4" (33 LF per gallon)	PA@.006	LF	.36	.24	--	.60
3/4" x 1-1/2" (27 LF per gallon)	PA@.007	LF	.44	.28	--	.72
3/4" x 2" (22 LF per gallon)	PA@.008	LF	.55	.33	--	.88
1" x 1" (25 LF per gallon)	PA@.008	LF	.48	.33	--	.81

Backer rod for joints in concrete pavement,

	Craft@Hrs	Unit	Material	Labor	Equipment	Total
1/2"	PA@.002	LF	.02	.08	--	.10

Clean joint or crack in concrete pavement by sandblasting, no salvage of sand

	Craft@Hrs	Unit	Material	Labor	Equipment	Total
1/2" x 1/2"	PA@.004	LF	.03	.16	.02	.21
1/2" x 3/4"	PA@.007	LF	.04	.28	.03	.35
1/2" x 1"	PA@.008	LF	.05	.33	.04	.42
1/2" x 1-1/4"	PA@.009	LF	.06	.37	.04	.47
1/2" x 1-1/2"	PA@.011	LF	.07	.45	.05	.57

Site Work 2

	Craft@Hrs	Unit	Material	Labor	Equipment	Total

Joint Treatment for Concrete Pavement, continued
Clean out old joint sealer in concrete pavement by sandblasting and reseal with PVC coal tar sealer, including new baker rod

	Craft@Hrs	Unit	Material	Labor	Equipment	Total
1/2" x 1/2"	PA@.009	LF	.18	.37	.06	.61
1/2" x 3/4"	PA@.013	LF	.24	.53	.16	.93
1/2" x 1"	PA@.013	LF	.31	.53	.16	1.00
1/2" x 1-1/4"	PA@.018	LF	.38	.73	.22	1.33
1/2" x 1-1/2"	PA@.022	LF	.45	.89	.26	1.60

Pavement Striping and Markers. These costs include the subcontractor's overhead and profit, small quantities.

Pavement line markings, cost per coat

	Craft@Hrs	Unit	Material	Labor	Equipment	Total
4" wide, white	PA@.005	LF	.05	.20	.02	.27
6" wide, white	PA@.005	LF	.07	.20	.02	.29
Reflective white	PA@.016	SF	.22	.65	.08	.95
Parking lot stall markings, single line per stall, including layout	PA@.072	Ea	.95	2.92	.21	4.08

Thermoplastic line markings, cost per coat

	Craft@Hrs	Unit	Material	Labor	Equipment	Total
4" wide, white	PA@.005	LF	.26	.20	.02	.48
4" wide, yellow	PA@.005	LF	.24	.20	.02	.46
12" wide, white	PA@.013	LF	.74	.53	.08	1.35

Airfield pavement line markings, reflective, TT-P-1952B

	Craft@Hrs	Unit	Material	Labor	Equipment	Total
White	PA@.010	SF	.25	.41	.03	.69
Yellow	PA@.010	SF	.22	.41	.03	.66

Traffic lines and symbols, preformed reflective tape

	Craft@Hrs	Unit	Material	Labor	Equipment	Total
4" wide lines	PA@.004	LF	.96	.16	--	1.12
Arrows	PA@2.75	Ea	68.50	112.00	--	180.50
Letters, per letter	PA@.928	Ea	32.50	37.70	--	70.20
Handicapped symbol	PA@2.58	Ea	57.50	105.00	--	162.50

Buttons, reflective, large quantities, including epoxy

	Craft@Hrs	Unit	Material	Labor	Equipment	Total
Type A, round, non-reflective	PA@.039	Ea	1.15	1.58	--	2.73
Type B, and D, 2 way reflective	PA@.043	Ea	2.25	1.75	--	4.00
Type C, 2 way reflective	PA@.032	Ea	2.50	1.30	--	3.80
Type G, and H, 1 way reflective	PA@.032	Ea	2.25	1.30	--	3.55

Reflective stakes, hand driven

	Craft@Hrs	Unit	Material	Labor	Equipment	Total
Steel to 6', Type F, 1 way reflective	CL@.117	Ea	17.50	3.99	--	21.49
Amber, reverse amber, Type I, 2 face	CL@.119	Ea	18.00	4.06	--	22.06

	Craft@Hrs	Unit	Material	Labor	Total

Traffic signs, reflectorized, with a 2" galvanized steel pipe post 10' long set 2' into the ground. Includes digging of hole with a manual auger and backfill

	Craft@Hrs	Unit	Material	Labor	Total
Stop, 24" x 24"	CL@1.25	Ea	90.20	42.60	132.80
Yield, 30" triangle	CL@1.25	Ea	82.30	42.60	124.90
Speed limit, Exit, etc, 18" x 24"	CL@1.25	Ea	78.30	42.60	120.90
Warning, 24" x 24"	CL@1.25	Ea	82.85	42.60	125.45

Brick Paving. These costs include the subcontractor's overhead and profit but no site preparation. Extruded hard red brick laid flat and without mortar. (Brick cost shown per M is cost per thousand before markup.)

Running bond

	Craft@Hrs	Unit	Material	Labor	Total
4" x 8" x 2-1/4" (4.5 per SF at $400 per M)	M1@.114	SF	2.07	4.27	6.34
4" x 8" x 1-5/8" (4.5 per SF at $386 per M)	M1@.112	SF	2.00	4.19	6.19
3-5/8" x 7-5/8" x 2-1/4" (5.2 per SF at $450 per M)	M1@.130	SF	2.70	4.87	7.57
3-5/8" x 7-5/8" x 1-5/8" (5.2 per SF at $368 per M)	M1@.127	SF	2.20	4.76	6.96
Add for herringbone pattern	--	%	--	30.0	--

Site Work 2

	Craft@Hrs	Unit	Material	Labor	Total
Basketweave					
4" x 8" x 2-1/4" (4.5 per SF at $400 per M)	M1@.133	SF	2.07	4.98	7.05
4" x 8" x 1-5/8" (4.5 per SF at $386 per M)	M1@.130	SF	2.00	4.87	6.87
Brick laid on edge and without mortar, running bond					
4" x 8" x 2-1/4" (8 per SF at $400 per M)	M1@.251	SF	3.68	9.40	13.08
3-5/8" x 7-5/8" x 2-1/4" (8.4 per SF at $450 per M)	M1@.265	SF	4.35	9.92	14.27
Brick laid flat with 3/8" mortar joints and mortar setting bed					
Running bond					
4" x 8" x 2-1/4" (3.9 per SF at $400 per M)	M1@.127	SF	2.24	4.76	7.00
4" x 8" x 1-5/8" (3.9 per SF at $386 per M)	M1@.124	SF	2.16	4.64	6.80
3-5/8" x 7-5/8" x 2-1/4" (4.5 per SF at $450 per M)	M1@.144	SF	2.59	5.39	7.98
3-5/8" x 7-5/8" x 1-5/8" (4.5 per SF at $368 per M)	M1@.140	SF	2.38	5.24	7.62
Add for herringbone pattern	--	%	--	20.0	--
Basketweave					
4" x 8" x 2-1/4" (3.9 per SF at $400 per M)	M1@.147	SF	2.24	5.51	7.75
4" x 8" x 1-5/8" (3.9 per SF at $386 per M)	M1@.144	SF	2.16	5.39	7.55
3-5/8" x 7-5/8" x 2-1/4" (4.5 per SF at $450 per M)	M1@.147	SF	2.59	5.51	8.10
3-5/8" x 7-5/8" x 1-5/8" (4.5 per SF at $368 per M)	M1@.144	SF	2.38	5.39	7.77
Brick laid on edge with 3/8" mortar joints and setting bed, running bond					
4" x 8" x 2-1/4" (6.5 per SF at $400 per M)	M1@.280	SF	3.74	10.50	14.24
3-5/8" x 7-5/8" x 2-1/4" (6.9 per SF at $450 per M)	M1@.314	SF	4.46	11.80	16.26
Mortar color for brick laid in mortar, add per SF of paved area					
Red or yellow	--	SF	.06	--	--
Black or brown	--	SF	.10	--	--
Green	--	SF	.27	--	--
Base underlayment for paving brick					
Add for 2" sand base (124 SF at $11 per ton)	M1@.009	SF	.09	.34	.43
Add for 15 lb felt underlayment	M1@.002	SF	.10	.08	.18
Add for asphalt tack coat (40 SF @ $5 per gallon)	M1@.005	SF	.13	.19	.32
Add for 2% neoprene tack coat (40 SF @ $7 per gal.)	M1@.005	SF	.18	.19	.37
Brick edging, 8" deep, set in concrete with dry joints					
Headers 4" x 8" x 2-1/4" (3/LF at $400/M)	M1@.147	LF	1.86	5.51	7.37
Headers 3-5/8" x 7-5/8" x 2-1/4" (3.3/LF at $450/M)	M1@.159	LF	2.30	5.96	8.26
Rowlocks 3-5/8" x 7-5/8" x 2-1/4" (4.5/LF at $450/M)	M1@.251	LF	3.14	9.40	12.54
Rowlocks 4" x 8" x 2-1/4" (4.7/LF at $400/M)	M1@.265	LF	2.90	9.92	12.82
Brick edging, 8" deep, set in concrete with 3/8" mortar joints					
Headers 4" x 8" x 2-1/4" (2.8/LF at $400/M)	M1@.159	LF	1.73	5.96	7.69
Headers 3-5/8" x 7-5/8" x 2-1/4" (3/LF at $450/M)	M1@.173	LF	2.10	6.48	8.58
Rowlocks 3-5/8" x 7-5/8" x 2-1/4" (4.2/LF at $450/M)	M1@.298	LF	2.93	11.20	14.13
Rowlocks 4" x 8" x 2-1/4" (4.5/LF at $400/M)	M1@.314	LF	2.79	11.80	14.59

Masonry, Stone and Slate Paving. These costs include the subcontractor's overhead and profit but no site preparation. Costs for stone and slate vary widely. (Costs shown in parentheses are before markup.)

	Craft@Hrs	Unit	Material	Labor	Total
Interlocking 9" x 4-1/2" concrete pavers, dry joints, natural gray					
2-3/8" thick (3.5 per SF at $.40 each)	M1@.058	SF	1.61	2.17	3.78
3-1/8" thick (3.5 per SF at $.45 each)	M1@.063	SF	1.81	2.36	4.17
Add for standard colors (3.5 per SF at $.05 each)	--	SF	.20	--	--
Add for custom colors (3.5 per SF at $.10 each)	--	SF	.40	--	--
Patio blocks, 8" x 16" x 2", dry joints					
Natural gray (1.125 per SF at $.90 each)	M1@.052	SF	1.16	1.95	3.11
Standard colors (1.125 per SF at $1.00 each)	M1@.052	SF	1.29	1.95	3.24

Site Work 2

	Craft@Hrs	Unit	Material	Labor	Total
Masonry, Stone and Slate Paving, continued					
Flagstone pavers in sand bed with dry joints					
Random ashlar, 1-1/4"	M1@.193	SF	2.30	7.23	9.53
Random ashlar, 2-1/2"	M1@.228	SF	3.50	8.54	12.04
Irregular fitted, 1-1/4"	M1@.208	SF	1.95	7.79	9.74
Irregular fitted, 2-1/2"	M1@.251	SF	3.00	9.40	12.40
Flagstone pavers in mortar bed with mortar joints					
Random ashlar, 1-1/4"	M1@.208	SF	2.40	7.79	10.19
Random ashlar, 2-1/2"	M1@.251	SF	3.65	9.40	13.05
Irregular fitted, 1-1/4"	M1@.228	SF	2.05	8.54	10.59
Irregular fitted, 2-1/2"	M1@.277	SF	3.10	10.40	13.50
Granite paving, modular					
Sand bed, dry jts, 4" x 4" x 3" (9/SF at $.85)	M1@.179	SF	8.79	6.70	15.49
Sand bed, dry jts, 12" x 12" x 3" (1/SF at $7.95)	M1@.208	SF	9.14	7.79	16.93
Mortar bed and 1" jts, 4" x 4" x 3" (5.8/SF at $.85)	M1@.193	SF	9.75	7.23	16.98
Mortar bed and 1" jts, 12" x 12" x 3" (.85/SF at $7.95)	M1@.228	SF	10.00	8.54	18.54
Limestone flag paving					
Sand bed, dry joints					
Random ashlar, 1-1/4"	M1@.193	SF	2.40	7.23	9.63
Random ashlar, 2-1/2"	M1@.228	SF	4.60	8.54	13.14
Irregular fitted, 1-1/2"	M1@.208	SF	2.10	7.79	9.89
Irregular fitted, 2-1/2"	M1@.251	SF	3.60	9.40	13.00
Mortar bed and joints					
Random ashlar, 1"	M1@.208	SF	2.15	7.79	9.94
Random ashlar, 2-1/2"	M1@.251	SF	4.75	9.40	14.15
Irregular fitted, 1"	M1@.228	SF	1.90	8.54	10.44
Irregular fitted, 2-1/2"	M1@.277	SF	3.60	10.40	14.00
Slate flag paving, natural cleft					
Sand bed, dry joints					
Random ashlar, 1-1/4"	M1@.173	SF	7.70	6.48	14.18
Irregular fitted, 1-1/4"	M1@.193	SF	7.15	7.23	14.38
Mortar bed and joints					
Random ashlar, 3/4"	M1@.173	SF	6.10	6.48	12.58
Random ashlar, 1"	M1@.208	SF	7.15	7.79	14.94
Irregular fitted, 3/4"	M1@.193	SF	5.65	7.23	12.88
Irregular fitted, 1"	M1@.228	SF	6.60	8.54	15.14
Add for sand rubbed slate	--	SF	1.50	--	--

Curbs and Gutters. These costs include the subcontractor's overhead and profit. Costs assume 3 uses of the forms and 2,000 PSI concrete (at $46.90 per CY before markup) placed directly from the chute of a ready-mix truck)

Concrete cast-in-place curb, including 2% waste, forms and finishing but no excavation or backfill

	Craft@Hrs	Unit	Material	Labor	Equipment	Total
Vertical curb						
6" x 12" straight curb	P9@.109	LF	2.00	4.25	.93	7.18
6" x 12" curved curb	P9@.150	LF	2.05	5.86	1.29	9.20
6" x 18" straight curb	P9@.119	LF	3.00	4.65	1.04	8.69
6" x 18" curved curb	P9@.166	LF	3.00	6.48	1.45	10.93
6" x 24" straight curb	P9@.123	LF	4.00	4.80	1.09	9.89
6" x 24" curved curb	P9@.179	LF	4.00	6.99	1.55	12.54
Rolled curb and gutter, 6" roll						
18" x 6" base, straight curb	P9@.143	LF	4.00	5.58	1.24	10.82
18" x 6" base, curved curb	P9@.199	LF	4.00	7.77	1.71	13.48
24" x 6" base, straight curb	P9@.168	LF	5.00	6.56	1.45	13.01
24" x 6" base, curved curb	P9@.237	LF	5.00	9.25	2.07	16.32

Site Work 2

	Craft@Hrs	Unit	Material	Labor	Equipment	Total
Concrete cast-in-place driveway apron						
4" x 2' x 8' inclined	P9@.198	SF	3.95	7.73	1.71	13.39
5" x 2' x 8' inclined	P9@.263	SF	4.50	10.30	2.28	17.08
Concrete valley gutter, 6" thick	P9@.111	SF	1.45	4.33	.98	6.76
Asphaltic concrete curb, straight, placed during paving						
8" x 6", placed by hand	P5@.141	LF	.95	5.16	.93	7.04
8" x 6", machine extruded	P5@.067	LF	.75	2.45	.52	3.72
8" x 8", placed by hand	P5@.141	LF	1.25	5.16	.93	7.34
8" x 8", machine extruded	P5@.068	LF	1.00	2.49	.58	4.07
Add for curved sections	--	%	--	25.0	25.0	--
Asphaltic concrete speed bumps, 3" high, 24" wide (135 LF and 2 tons per CY), placed on parking lot surface and rolled, 100 LF job	P5@.198	LF	.45	7.24	1.55	9.24
Painting speed bumps, 24" wide, white	P5@.022	LF	.25	.81	.05	1.11
Granite, split face, small quantities, typical prices						
6" x 18" straight	P5@.247	LF	25.00	9.03	1.97	36.00
6" x 24" tapered 12" base	P5@.296	LF	36.00	10.80	2.43	49.23
6" x 18" radial flat face, with radius over 5'	P5@.345	LF	45.00	12.60	2.69	60.29

Walkways. These costs include the subcontractor's overhead and profit.

Asphalt, temporary, including installation and breakout, but no hauling

	Craft@Hrs	Unit	Material	Labor	Equipment	Total
2" to 2-1/2" thick	P5@.015	SF	.75	.55	.16	1.46

Concrete, 2,000 PSI (at $46.90 per CY before markup) plain scored, placed directly from the chute of a ready-mix truck, vibrated and broom finished. Costs include 2% allowance for waste but no forms, reinforcing, dividers or joints

	Craft@Hrs	Unit	Material	Labor	Equipment	Total
4" thick (80 SF per CY)	P8@.012	SF	.68	.45	.22	1.35
Add for mesh reinforcing	RB@.003	SF	.12	.14	.03	.29
Add for integral colors, most pastels	--	SF	.90	--	--	--
Add for 1/2" color top course	P8@.012	SF	1.00	.45	.09	1.54
Add for steel trowel finish	CM@.011	SF	.02	.44	.10	.56
Add for seeded aggregate finish	CM@.011	SF	.10	.44	.10	.64
Add for exposed aggregate wash process	CM@.006	SF	.05	.24	.07	.36
Headers and dividers, complete						
2" x 4", treated pine	C8@.027	LF	.45	1.04	--	1.49
2" x 6", treated pine	C8@.028	LF	.70	1.08	--	1.78
2" x 6", redwood, construction heart	C8@.028	LF	.94	1.08	--	2.02
2" x 4", patio type dividers, untreated	C8@.030	LF	.30	1.15	--	1.45

Post Holes. Drilling 2' deep by 10" to 12" diameter fence post holes. Costs will be higher on slopes or where access is limited. These costs include the subcontractor's overhead and profit but no layout, setting of fence posts or disposal of excavated material.

	Craft@Hrs	Unit	Material	Labor	Equipment	Total
Light to medium soil, laborer working with a hand auger, 4 holes per hour	CL@.250	Ea	--	8.53	.17	8.70
Medium to heavy soil, laborer working with a breaking bar and hand auger, 2 holes per hour	CL@.500	Ea	--	17.10	.33	17.43
Medium to heavy soil, 2 laborers working with a gas powered auger (at $55 per day), 15 holes per hour	CL@.133	Ea	--	4.54	.46	5.00
Broken rock, laborer and equipment operator working with a compressor and jackhammer (at $100 per day), 2 holes per hour	S1@1.00	Ea	--	38.50	6.25	44.75
Rock, laborer and an equipment operator working with a truck-mounted drill (at $675 per day), 12 holes per hour	S1@.166	Ea	--	6.36	6.96	13.32

Site Work 2

	Craft@Hrs	Unit	Material	Labor	Equipment	Total

Fencing, Chain Link. These costs include the subcontractor's overhead and profit.
Fence, industrial grade 9 gauge 2" x 2" galvanized steel chain link fabric and framework, including 2" line posts at 10' set in concrete in augured post holes and 1-5/8" top rail, tension panels at corners and abrupt grade changes. Add for gates, gate posts and corner posts as required. Equipment is a gasoline powered 12" auger at $5.50 an hour.
For scheduling purposes, estimate that a 2-man crew will install 125 to 130 LF of fence per 8-hour day

	Craft@Hrs	Unit	Material	Labor	Equipment	Total
4' high galvanized fence	CL@.125	LF	3.50	4.26	.34	8.10
5' high galvanized fence	CL@.125	LF	3.90	4.26	.34	8.50
6' high galvanized fence	CL@.125	LF	4.30	4.26	.34	8.90
7' high galvanized fence	CL@.125	LF	4.75	4.26	.34	9.35
8' high galvanized fence	CL@.125	LF	5.15	4.26	.34	9.75
10' high galvanized fence	CL@.156	LF	5.50	5.32	.43	11.25
12' high galvanized fence, with middle rail	CL@.195	LF	5.80	6.65	.54	12.99
Add for vertical aluminum privacy slats	CL@.016	SF	.65	.55	--	1.20
Add for vinyl coated fabric and posts	--	%	10.0	--	--	--
Add for under 200 LF quantities	--	%	5.0	5.0	5.0	--
Add for 6 gauge fabric	--	%	45.0	10.0	10.0	--
Deduct for highway quantities	--	%	-10.0	--	--	--
Deduct for 11 gauge chain link fabric	--	%	-10.0	--	--	--
Deduct for "C" section line posts	--	%	-15.0	--	--	--

Gates, driveway or walkway. 9 gauge 2" x 2" galvanized steel chain link with frame and tension bars. Costs shown are per square foot of frame area. Add for hardware and gate posts, from below.

	Craft@Hrs	Unit	Material	Labor	Equipment	Total
With 1-3/8" diameter pipe frame	CL@.025	SF	3.10	.85	--	3.95
With 1-5/8" diameter pipe frame	CL@.025	SF	3.60	.85	--	4.45

Gate hardware, per gate

	Craft@Hrs	Unit	Material	Labor	Equipment	Total
Walkway gate	CL@.250	Ea	14.50	8.50	--	23.00
Driveway gate	CL@.250	Ea	35.50	8.50	--	44.00
Add for vinyl coated fabric and frame	--	%	10.0	--	--	--

Corner, end or gate posts, complete, heavyweight, 2-1/2" outside diameter galvanized, with fittings and brace. Costs include 2/3 CF sack of concrete per post

	Craft@Hrs	Unit	Material	Labor	Equipment	Total
4' high fence	CL@.612	Ea	17.30	20.90	1.68	39.88
5' high fence	CL@.668	Ea	19.50	22.80	1.84	44.14
6' high fence	CL@.735	Ea	21.80	25.00	2.02	48.82
7' high fence	CL@.800	Ea	24.10	27.30	2.20	53.60
8" high fence	CL@.882	Ea	26.40	30.10	2.43	58.93
10' high fence	CL@.962	Ea	29.75	32.80	2.64	65.19
12' high fence, with middle rail	CL@1.00	Ea	32.70	34.10	2.75	69.55
Add for vinyl coated posts	--	%	10.0	--	--	--
Deduct for 2" outside diameter	--	%	-15.0	--	--	--

Barbed wire topper for galvanized chain link fence

	Craft@Hrs	Unit	Material	Labor	Equipment	Total
Single strand barbed wire	CL@.004	LF	.10	.14	--	.24
3 strands on one side, with support arm	CL@.008	LF	.70	.27	--	.97
3 strand on each side, with support arm	CL@.013	LF	1.40	.44	--	1.84
Double coil of 24" to 30" barbed tape	CL@.038	LF	5.50	1.30	--	6.80

Fencing, Wood. Costs shown include posts set in concrete. Gates have 2" x 4" frame and are 3' wide. Add for concrete footings, gate hardware or painting as required from costs shown at the end of this section. These costs include the subcontractor's overhead and profit.

	Craft@Hrs	Unit	Material	Labor	Total
Barbed wire fence, 4' high, wood posts	CL@.121	LF	1.50	4.13	5.63

Basketweave fence, redwood, "B" grade, 1" x 6" boards, 2" x 4" stringers or spreaders, 4" x 4" posts

	Craft@Hrs	Unit	Material	Labor	Total
Tight weave, 4' high	C8@.090	LF	13.40	3.46	16.86
Tight weave, 6' high	C8@.095	LF	21.10	3.66	24.76
Wide span, 4' to 6' high	C8@.095	LF	20.00	3.66	23.66

Site Work 2

	Craft@Hrs	Unit	Material	Labor	Total
Board fence, 1" x 6" boards, 2" x 4" rails, 4" x 4" posts					
Fir or larch					
3' high, 2 rail	C8@.110	LF	3.27	4.23	7.50
4' high, 3 rail	C8@.110	LF	4.34	4.23	8.57
6' high, 3 rail	C8@.119	LF	5.76	4.58	10.34
3' high gate	C8@.808	Ea	21.60	31.10	52.70
4' high gate	C8@.846	Ea	28.70	32.50	61.20
6' high gate	C8@1.05	Ea	38.00	40.40	78.40
Red cedar, #2 & better					
3' high, 2 rail	C8@.110	LF	4.18	4.23	8.41
4' high, 3 rail	C8@.112	LF	5.62	4.31	9.93
6' high, 3 rail	C8@.119	LF	7.52	4.58	12.10
3' high gate	C8@.808	Ea	27.60	31.10	58.70
4' high gate	C8@.846	Ea	37.10	32.50	69.60
6' high gate	C8@1.05	Ea	49.60	40.40	90.00
Redwood, B grade					
3' high, 2 rail	C8@.110	LF	11.70	4.23	15.93
4' high, 3 rail	C8@.110	LF	17.10	4.23	21.33
6' high, 3 rail	C8@.119	LF	22.30	4.58	26.88
3' high gate	C8@.808	Ea	51.00	31.10	82.10
4' high gate	C8@.846	Ea	61.70	32.50	94.20
6' high gate	C8@1.05	Ea	88.30	40.40	128.70
Board and batten fence, 1" x 6" boards, 1" x 2" battens, 2" x 4" rails, 4" x 4" posts					
Fir or larch					
3' high, 2 rail	C8@.119	LF	3.78	4.58	8.36
4' high, 3 rail	C8@.124	LF	5.02	4.77	9.79
6' high, 3 rail	C8@.130	LF	6.77	5.00	11.77
3' high gate	C8@.808	Ea	24.90	31.10	56.00
4' high gate	C8@.846	Ea	33.10	32.50	65.60
6' high gate	C8@1.05	Ea	44.70	40.40	85.10
Red cedar, #2 grade					
3' high, 2 rail	C8@.119	LF	4.88	4.58	9.46
4' high, 3 rail	C8@.124	LF	6.62	4.77	11.39
6' high, 3 rail	C8@.130	LF	10.50	5.00	15.50
3' high gate	C8@.808	Ea	32.20	31.10	63.30
4' high gate	C8@.846	Ea	43.70	32.50	76.20
6' high gate	C8@1.05	Ea	69.40	40.40	109.80
Redwood, B grade					
3' high, 2 rail	C8@.119	LF	14.20	4.58	18.78
4' high, 3 rail	C8@.124	LF	19.20	4.77	23.97
6' high, 3 rail	C8@.130	LF	25.80	5.00	30.80
3' high gate	C8@.808	Ea	93.40	31.10	124.50
4' high gate	C8@.846	Ea	126.00	32.50	158.50
6' high gate	C8@1.05	Ea	170.00	40.40	210.40
Open rail fence, red cedar					
Split 2" x 4" rails, 3' high, 2 rail	C8@.095	LF	2.42	3.66	6.08
Split 2" x 4" rails, 4' high, 3 rail	C8@.102	LF	2.77	3.92	6.69
Rustic 4" round, 3' high, 2 rail	C8@.095	LF	2.97	3.66	6.63
Rustic 4" round, 4' high, 3 rail	C8@.102	LF	3.31	3.92	7.23
Picket fence, 1" x 2" pickets, 2" x 4" rails, 4" x 4" posts					
Fir or larch					
3' high, 2 rail	C8@.105	LF	3.55	4.04	7.59
5' high, 3 rail	C8@.113	LF	5.55	4.35	9.90

Site Work 2

	Craft@Hrs	Unit	Material	Labor	Total
Picket fence, Fir or larch, continued					
3' high gate	C8@.770	Ea	24.10	29.60	53.70
5' high gate	C8@.887	Ea	36.60	34.10	70.70
Red cedar, #2 and better					
3' high, 2 rail	C8@.105	LF	4.43	4.04	8.47
5' high, 3 rail	C8@.113	LF	6.90	4.35	11.25
3' high gate	C8@.770	Ea	29.20	29.60	58.80
5' high gate	C8@.887	Ea	45.50	34.10	79.60
Redwood, B grade					
3' high, 2 rail	C8@.105	LF	10.10	4.04	14.14
5' high, 3 rail	C8@.105	LF	15.80	4.04	19.84
3' high gate	C8@.770	Ea	35.70	29.60	65.30
5' high gate	C8@.887	Ea	53.30	34.10	87.40
Stockade fence, red cedar, #2 and better					
Halved poles, 6' high	C8@.090	LF	6.28	3.46	9.74
Halved poles, 8' high	C8@.098	LF	8.15	3.77	11.92
Whole round posts, 6' high	C8@.090	LF	9.75	3.46	13.21
Whole round posts, 8' high	C8@.098	LF	12.80	3.77	16.57
Concrete fence post footing,					
8" diameter, 2' deep, typical	C8@.136	Ea	3.50	5.23	8.73
Gate hardware					
Latch, standard duty	C8@.281	Ea	3.25	10.80	14.05
Latch, heavy duty	C8@.281	Ea	7.50	10.80	18.30
Hinges, standard, per pair	C8@.374	Pr	6.00	14.40	20.40
Hinges, heavy duty, per pair	C8@.374	Pr	14.50	14.40	28.90
No-sag cable kit	C8@.281	Ea	8.50	10.80	19.30
Self closing spring	C8@.281	Ea	7.50	10.80	18.30
Paint or stain fence or gate with a roller, 2 coats, 2 sides					
3' high fence	PA@.093	LF	.65	3.78	4.43
4' high fence	PA@.115	LF	.80	4.67	5.47
5' high fence	PA@.128	LF	1.00	5.20	6.20
6' high fence	PA@.171	LF	1.20	6.94	8.14
3' or 4' high gate (includes brushwork)	PA@.043	LF	1.70	1.75	3.45
5' or 6' high gate (includes brushwork)	PA@.147	LF	2.80	5.97	8.77

Guardrails and Bumpers. These costs include the subcontractor's overhead and profit. Equipment is a forklift at $10.50 per hour.

	Craft@Hrs	Unit	Material	Labor	Equipment	Total
Steel guardrail, standard beam	H1@.250	LF	12.50	10.60	.65	23.75
Posts, wood, treated, 4" x 4" x 36" high	H1@1.00	Ea	7.35	42.70	2.63	52.68
Guardrail two sides, channel rail & side blocks	H1@1.00	LF	23.00	42.70	2.63	68.33
Sight screen	--	LF	--	--	--	2.50
Highway median barricade, precast concrete						
2'8" high x 10' long	H1@.500	Ea	225.00	21.40	1.31	247.71
Rental per month	--	Ea	5.50	--	--	--
Add for reflectors, 2 per 10' section	--	LS	11.00	--	--	--
Parking bumper, precast concrete, including dowels						
3' wide	CL@.381	Ea	15.00	13.00	--	28.00
6' wide	CL@.454	Ea	28.00	15.50	--	43.50

Bollards. Posts mounted in walkways to limit vehicle entry. Equipment is a forklift at $10.50 per hour. Add for concrete footings, if required, from costs shown at the end of this section. These costs include the subcontractor's overhead and profit.

Cast iron bollards, ornamental, surface mounted

	Craft@Hrs	Unit	Material	Labor	Equipment	Total
12" diameter base, 42" high	S1@1.00	Ea	645.00	38.50	5.25	688.75

Site Work 2

	Craft@Hrs	Unit	Material	Labor	Equipment	Total
17" diameter base, 42" high	S1@1.00	Ea	1,100.00	38.50	5.25	1,143.75
Lighted, electrical work not included,						
17" diameter base, 42" high	S1@1.00	Ea	1,200.00	38.50	5.25	1,243.75
Concrete bollards, precast, exposed aggregate, surface mounted						
Square, 12" x 12" x 30" high	S1@1.00	Ea	220.00	38.50	5.25	263.75
Round, 12" diameter x 30" high	S1@1.00	Ea	225.00	38.50	5.25	268.75
Granite bollards, dowelled to concrete foundation or slab (foundation not included)						
Square, smooth matt finish, flat top						
16" x 16" x 30", 756 lbs	S1@1.00	Ea	780.00	38.50	5.25	823.75
16" x 16" x 54", 1,320 lbs	S1@1.50	Ea	975.00	57.80	7.88	1,040.68
Square, rough finish, pyramid top						
12" x 12" x 24", 330 lbs	S1@1.00	Ea	350.00	38.50	5.25	393.75
Round, smooth matt finish, flat top, 12" diameter, 18" high	S1@1.00	Ea	750.00	38.50	5.25	793.75
Octagonal, smooth matt finish, flat top, 24" x 24" x 20", 740 lbs	S1@1.00	Ea	1,150.00	38.50	5.25	1,193.75
Pipe bollards, concrete filled steel pipe, painted yellow, 8' long, set 4' in concrete, includes digging hole						
6" diameter pipe	S1@.750	Ea	75.00	28.90	3.94	107.84
8" diameter pipe	S1@.750	Ea	115.00	28.90	3.94	147.84
12" diameter pipe	S1@.750	Ea	195.00	28.90	3.94	227.84
Wood bollards, pressure treated timber, includes hand-digging and backfill of hole but no disposal of excess soil. Costs per MBF, shown in parentheses, are before markup. Length shown is portion exposed, costs include 3' bury depth.						
6" x 6" x 36" high (at $1,150 MBF)	S1@.750	Ea	23.80	28.90	3.94	56.64
8" x 8" x 30" high (at $1,100 MBF)	S1@.750	Ea	37.00	28.90	3.94	69.84
8" x 8" x 36" high (at $1,100 MBF)	S1@.750	Ea	40.40	28.90	3.94	73.24
8" x 8" x 42" high (at $1,100 MBF)	S1@.750	Ea	43.80	28.90	3.94	76.64
12" x 12" x 24" high (at $1,000 MBF)	S1@.750	Ea	69.00	28.90	3.94	101.84
12" x 12" x 36" high (at $1,000 MBF)	S1@.750	Ea	82.80	28.90	3.94	115.64
12" x 12" x 42" high (at $1,000 MBF)	S1@.750	Ea	89.70	28.90	3.94	122.54
Footings for bollards, add if required						
Permanent concrete footing	C8@.500	Ea	15.00	19.20	--	34.20
Footing for removable bollards	C8@.500	Ea	110.00	19.20	--	129.20

Softball Backstops. Regulation size galvanized chain link backstops with 4" OD galvanized posts each 8', top, bottom and 2 middle brace rails, truss rod supported 4' high overhang panel with 9 gauge chain link mesh and 1-1/8" OD frame. Includes posts set 3'6" in concrete, 16 LF of 6 gauge mesh behind home plate and 16 LF of 9 gauge mesh on both first and third base lines (48 LF overall). These costs include the subcontractor's overhead and profit. Equipment is a forklift at $10.50 per hour.

	Craft@Hrs	Unit	Material	Labor	Equipment	Total
14' high	S6@24.0	Ea	3,100.00	890.00	84.00	4,074.00
18' high	S6@24.0	Ea	3,750.00	890.00	84.00	4,724.00
20' high	S6@24.0	Ea	4,450.00	890.00	84.00	5,424.00
22' high	S6@24.0	Ea	5,350.00	890.00	84.00	6,324.00
24' high	S6@24.0	Ea	6,650.00	890.00	84.00	7,624.00
Add for vinyl coated backstops	--	%	60.0	--	--	--

Playing Fields. Typical costs. These costs include a concrete base and the subcontractor's overhead and profit.

	Craft@Hrs	Unit	Material	Labor	Equipment	Total
Baseball field, includes rough and fine grading and backstop						
Seeded	--	Ea	--	--	--	16,700.00
Sodded	--	Ea	--	--	--	35,500.00
Football field, includes rough and fine grading and goal posts						
Seeded	--	Ea	--	--	--	12,800.00
Sodded	--	Ea	--	--	--	33,000.00

Site Work 2

	Craft@Hrs	Unit	Material	Labor	Equipment	Total

Playing Fields, continued

Softball field, includes rough and fine grading and backstop

Seeded	--	Ea	--	--	--	11,700.00
Sodded	--	Ea	--	--	--	24,200.00

Basketball court, asphaltic concrete, includes rough and fine grading and goals

Plain surface	--	Ea	--	--	--	14,000.00
Color finished surface	--	Ea	--	--	--	18,600.00

18 hole golf course
  Construction of 18 hole course

High	--	Ea	--	--	--	2,400,000.00
Low	--	Ea	--	--	--	927,000.00

  Irrigation system (including pump station)

High	--	Ea	--	--	--	618,000.00
Low	--	Ea	--	--	--	206,000.00

  Golf course maintenance equipment

High	--	Ea	--	--	--	412,000.00
Low	--	Ea	--	--	--	154,500.00

  Maintenance building

High	--	Ea	--	--	--	257,500.00
Low	--	Ea	--	--	--	103,000.00

  Total golf course development costs

High	--	Ea	--	--	--	4,000,000.00
Low	--	Ea	--	--	--	1,500,000.00

Playing fields, synthetic surface

Turf, including base and turf	--	SF	--	--	--	7.40
Turf, 3 layer, base foam, turf	--	SF	--	--	--	8.25
Uniturf, embossed running surface, 3/8"	--	SF	--	--	--	5.46

Running track

Volcanic cinder, 7"	--	SF	--	--	--	1.25
Bituminous and cork, 2"	--	SF	--	--	--	1.96
Regulation 1/4 mile synthetic track, 55,000 SF	--	LS	--	--	--	438,000.00

Playground Equipment. These costs include the subcontractor's overhead and profit. Equipment is a forklift (at $10.50 per hour)

	Craft@Hrs	Unit	Material	Labor	Equipment	Total
Slide tower with climber ladder, dimensional wood, galvanized pipe ladder rungs, 12' x 18", stainless steel slide	S6@16.0	Ea	2,300.00	593.00	56.00	2,949.00
Fire pole tower with tire swing, dimensional wood, galvanized fire pole, chain ladder	S6@18.0	Ea	2,700.00	667.00	63.00	3,430.00
All component structure, two 4' high and two 6' platforms, beam and suspension bridges, 12' x 18" slide, horizontal ladder, fire pole, tire and belt swings, ladders, balance beam, chain ladder	S6@72.0	Ea	11,500.00	2,670.00	252.00	14,422.00

Balance beams
  12' beam, 12" high

Metal	S6@1.50	Ea	253.00	55.60	5.25	313.85
Wood	S6@1.50	Ea	213.00	55.60	5.25	273.85
Multi-level wood beam, three 12' sections, 0' to 2' high, concrete footings	S6@3.00	Ea	626.00	112.00	10.50	748.50

Geodesic dome, galvanized pipe

8' diameter, 4' high	S6@24.0	Ea	453.00	889.00	84.00	1,426.00
13' diameter, 5' high	S6@36.0	Ea	1,110.00	1,330.00	126.00	2,566.00
16' diameter, 6' high	S6@48.0	Ea	1,060.00	1,780.00	168.00	3,008.00
17' diameter, 7' high	S6@48.0	Ea	1,090.00	1,780.00	168.00	3,038.00

Rope net, galvanized steel mast, nylon-coated steel cables

8' high, 30' diameter	S6@12.0	Ea	3,920.00	445.00	42.00	4,407.00

Site Work 2

	Craft@Hrs	Unit	Material	Labor	Equipment	Total
13' high, 17' x 17'	S6@18.0	Ea	10,300.00	667.00	63.00	11,030.00
20' high, 26' x 26'	S6@24.0	Ea	13,800.00	889.00	84.00	14,773.00
26' high, 43' x 43'	S6@30.0	Ea	23,100.00	1,110.00	105.00	24,315.00
Double mast, 20' high, 61' x 20'	S6@30.0	Ea	24,400.00	1,110.00	105.00	25,615.00

Horizontal ladder
  Galvanized pipe

	Craft@Hrs	Unit	Material	Labor	Equipment	Total
12' long, 6.5' high	S6@6.00	Ea	487.00	223.00	21.00	731.00
16' long, 7.5' high	S6@9.00	Ea	489.00	333.00	31.50	853.50

  Dimensional wood rails, galvanized pipe rungs

	Craft@Hrs	Unit	Material	Labor	Equipment	Total
8' long	S6@6.00	Ea	785.00	223.00	21.00	1,029.00
10' long	S6@6.00	Ea	902.00	223.00	21.00	1,146.00

Galvanized pipe pull-up bars

	Craft@Hrs	Unit	Material	Labor	Equipment	Total
2 bars at 2 heights	S6@6.00	Ea	211.00	223.00	21.00	455.00
3 bars at 3 heights	S6@9.00	Ea	244.00	333.00	31.50	608.50

Wood post pull-up bars, galvanized pipe rungs

	Craft@Hrs	Unit	Material	Labor	Equipment	Total
2 bars at 2 heights	S6@6.00	Ea	350.00	223.00	21.00	594.00
3 bars at 3 heights	S6@9.00	Ea	496.00	333.00	31.50	860.50

Sandboxes, prefabricated
  Dimensional wood, 10" high, includes sand

	Craft@Hrs	Unit	Material	Labor	Equipment	Total
6' x 6'	S6@6.00	Ea	405.00	223.00	21.00	649.00
8' x 8'	S6@9.00	Ea	460.00	333.00	31.50	824.50

  Painted galvanized steel, 12" high, pine seats, includes sand

	Craft@Hrs	Unit	Material	Labor	Equipment	Total
10' x 10'	S6@9.00	Ea	384.00	333.00	31.50	748.50
12' x 12'	S6@12.0	Ea	492.00	445.00	42.00	979.00

Seesaws
  Galvanized pipe frame, wood, metal or plastic seats

	Craft@Hrs	Unit	Material	Labor	Equipment	Total
1 unit, 2 seater	S6@9.00	Ea	254.00	333.00	31.50	618.50
2 units, 4 seater	S6@12.0	Ea	541.00	445.00	42.00	1,028.00
4 units, 8 seater	S6@18.0	Ea	1,029.00	667.00	63.00	1,759.00
6 units, 12 seater	S6@24.0	Ea	1,170.00	889.00	84.00	2,143.00
Wood frame, 2 wood seats	S6@12.0	Ea	529.00	445.00	42.00	1,016.00

Slides, 18" stainless steel bed, galvanized pipe frame

	Craft@Hrs	Unit	Material	Labor	Equipment	Total
8' long	S6@9.00	Ea	679.00	333.00	31.50	1,043.50
10' long	S6@12.0	Ea	785.00	445.00	42.00	1,272.00
12' long	S6@15.0	Ea	870.00	556.00	52.50	1,478.50
16' long	S6@18.0	Ea	965.00	667.00	63.00	1,695.00
20' long	S6@24.0	Ea	1,180.00	889.00	84.00	2,153.00

Swings, two leg ends, belt seats, galvanized pipe frame
  8' high

	Craft@Hrs	Unit	Material	Labor	Equipment	Total
2 seats	S6@12.0	Ea	331.00	445.00	42.00	818.00
4 seats	S6@18.0	Ea	562.00	667.00	63.00	1,292.00
6 seats	S6@24.0	Ea	785.00	889.00	84.00	1,758.00
8 seats	S6@30.0	Ea	902.00	1,110.00	105.00	2,117.00

  10' high

	Craft@Hrs	Unit	Material	Labor	Equipment	Total
2 seats	S6@18.0	Ea	436.00	667.00	63.00	1,166.00
4 seats	S6@24.0	Ea	647.00	889.00	84.00	1,620.00
6 seats	S6@30.0	Ea	1,008.00	1,110.00	105.00	2,223.00
8 seats	S6@36.0	Ea	1,230.00	1,330.00	126.00	2,686.00

Swings, three leg ends, belt seats, galvanized pipe frame
  10' high

	Craft@Hrs	Unit	Material	Labor	Equipment	Total
3 seats	S6@24.0	Ea	647.00	889.00	84.00	1,620.00
6 seats	S6@30.0	Ea	997.00	1,110.00	105.00	2,212.00
9 seats	S6@36.0	Ea	1,130.00	1,330.00	126.00	2,586.00

Site Work 2

	Craft@Hrs	Unit	Material	Labor	Equipment	Total
Swings, continued						
12' high						
3 seats	S6@30.0	Ea	785.00	1,110.00	105.00	2,000.00
6 seats	S6@36.0	Ea	1,220.00	1,330.00	126.00	2,676.00
9 seats	S6@42.0	Ea	1,370.00	1,560.00	147.00	3,077.00
Wheelchair swing platform	S6@12.0	Ea	483.00	445.00	42.00	970.00
Swings, T-bar galvanized pipe frame, belt seats						
2 seats	S6@18.0	Ea	530.00	667.00	63.00	1,260.00
3 seats	S6@21.0	Ea	817.00	778.00	73.50	1,668.50
4 seats	S6@24.0	Ea	913.00	889.00	84.00	1,886.00
Tetherball set, galvanized pipe post, nylon rope, ball, concrete footing, ground sleeve	S6@3.00	Ea	239.00	112.00	10.50	361.50
Whirls, painted galvanized steel						
6' diameter	S6@12.0	Ea	809.00	445.00	42.00	1,296.00
8' diameter	S6@18.0	Ea	1,220.00	667.00	63.00	1,950.00
10' diameter	S6@24.0	Ea	1,680.00	889.00	84.00	2,653.00
Fitness trails						
10 stations, 5 structures, 11 signs	S6@36.0	Ea	4,640.00	1,330.00	126.00	6,096.00
15 stations, 7 structures, 16 signs	S6@54.0	Ea	6,960.00	2,000.00	189.00	9,149.00
20 stations, 10 structures, 21 signs	S6@72.0	Ea	9,290.00	2,660.00	252.00	12,202.00

Athletic Equipment. These costs include the subcontractor's overhead and profit. Equipment is a forklift (at $10.50 per hour).

	Craft@Hrs	Unit	Material	Labor	Equipment	Total
Football goals						
Regulation goal, galvanized steel pipe						
Single support	S6@18.0	Pr	1,910.00	667.00	63.00	2,640.00
Double support	S6@24.0	Pr	1,990.00	889.00	84.00	2,963.00
Combination football/soccer goal, regulation size, galvanized						
steel pipe, double support	S6@24.0	Pr	1,320.00	889.00	84.00	2,293.00
Soccer goals, regulation size, galvanized steel frame, with net	S6@24.0	Pr	1,760.00	889.00	84.00	2,733.00
Basketball goals, aluminum fan backboard, hoop and net						
Single support, 5' backboard extension	S6@12.0	Pr	650.00	445.00	42.00	1,137.00
Double goal back to back	S6@15.0	Ea	925.00	556.00	52.50	1,533.50
Volleyball nets, galvanized posts, net, and ground sleeves						
Per net	S6@18.0	Ea	390.00	667.00	63.00	1,120.00

Tennis Courts. Typical costs for championship size single playing court installation including normal fine grading. These costs include the subcontractor's overhead and profit but no site preparation, excavation, drainage, or retaining walls.

	Craft@Hrs	Unit	Material	Labor	Equipment	Total
Court, 60' x 120' asphaltic concrete or portland cement slab (4" thickness built up to 6" thickness at perimeter). Includes compacted base material and wire mesh reinforcing in concrete	--	LS	--	--	--	18,000.00
Fence, 10' high galvanized chain link (360 LF) at court perimeter, including gates	--	LS	--	--	--	4,400.00
Court surface seal coat (7,200 SF) one color plus 2" white boundary lines	--	LS	--	--	--	1,900.00
Lighting (typical installation)	--	LS	--	--	--	7,000.00
Net posts, with tension reel. Equipment is a forklift at $10.50 per hour.						
Galvanized steel, 3-1/2" diameter	S6@6.00	Pr	275.00	223.00	21.00	519.00
Galvanized steel, 4-1/2" diameter	S6@6.00	Pr	380.00	223.00	21.00	624.00
Tennis nets						
Nylon	--	Ea	--	--	--	200.00

Site Work 2

	Craft@Hrs	Unit	Material	Labor	Equipment	Total
Steel	--	Ea	--	--	--	400.00
Ball wall, fiberglass, 8' x 5', attached to wall	S6@1.50	Ea	500.00	55.60	5.25	560.85
Windbreak, 9' high, closed mesh polypropylene, lashed to chain link fence	S6@.003	SF	.08	.11	.02	.21

Athletic Benches These costs include the subcontractor's overhead and profit.
Backed athletic bench, galvanized pipe frame, embedded

	Craft@Hrs	Unit	Material	Labor	Total
Aluminum seat and back					
6' long	C8@.930	Ea	212.00	35.80	247.80
8' long	C8@1.16	Ea	245.00	44.60	289.60
15' long	C8@1.31	Ea	392.00	50.40	442.40
Fiberglass seat and back					
10' long	C8@.930	Ea	301.00	35.80	336.80
12' long	C8@1.16	Ea	351.00	44.60	395.60
16' long	C8@1.31	Ea	469.00	50.40	519.40
Galvanized steel seat and back, 15' long	C8@1.31	Ea	285.00	50.40	335.40
Backless athletic bench, galvanized pipe frame, embedded					
Aluminum seat, 6' long	C8@.776	Ea	137.00	29.90	166.90
Aluminum seat, 8' long	C8@.776	Ea	156.00	29.90	185.90
Aluminum seat, 15' long	C8@.846	Ea	237.00	32.50	269.50
Wood seat, 6' long	C8@.776	Ea	111.00	29.90	140.90
Wood seat, 8' long	C8@.776	Ea	117.00	29.90	146.90
Wood seat, 15' long	C8@.846	Ea	176.00	32.50	208.50

Striping for Athletic Courts
Marked with white or yellow striping paint

	Craft@Hrs	Unit	Material	Labor	Total
Tennis court or volleyball court	PA@1.60	Ea	12.40	65.00	77.40
Lines 3" wide	PA@.002	LF	.04	.08	.12
Lines 4" wide	PA@.002	LF	.06	.08	.14
Letters, symbols, shapes	PA@.014	LF	.28	.57	.85
Lines 4" wide, reflectorized	PA@.002	LF	.06	.08	.14
Lines 8" wide, reflectorized	PA@.004	LF	.08	.16	.24
Letters, symbols, shapes, reflectorized	PA@.014	LF	.39	.57	.96
Marked with reflectorized thermoplastic tape					
Lines 4" wide	PA@.007	LF	.28	.28	.56
Lines 6" wide	PA@.008	LF	.43	.33	.76
Letters, symbols, shapes	PA@.035	SF	1.21	1.42	2.63

Surface Preparation for Athletic Courts

	Craft@Hrs	Unit	Material	Labor	Total
Acid etch concrete surface	PA@.013	SY	.12	.53	.65
Primer on asphalt	PA@.006	SY	.15	.24	.39
Primer on concrete	PA@.006	SY	.25	.24	.49
Asphalt emulsion filler, 1 coat	PA@.008	SY	.30	.33	.63
Acrylic filler, 1 coat	PA@.008	SY	.55	.33	.88
Rubber/acrylic compound, per coat	PA@.012	SY	1.25	.49	1.74
Acrylic emulsion texture, per coat					
Sand filled	PA@.006	SY	.80	.24	1.04
Rubber filled	PA@.006	SY	.80	.24	1.04
Acrylic emulsion color coat, per coat					
Tan, brown, or red	PA@.006	SY	.25	.24	.49
Greens	PA@.006	SY	.27	.24	.51
Blue	PA@.006	SY	.31	.24	.55
Rubber cushioned concrete pavers, 18" x 18", dry joints	FL@.039	SF	9.50	1.64	11.14

Site Work 2

	Craft@Hrs	Unit	Material	Labor	Total

Bicycle, Moped and Motorcycle Security Racks. Installation kit includes tamper proof bolts and quickset cement. Locks supplied by users. Labor is to install units on concrete surface. Cost per bicycle or motorcycle of capacity. Add cost for drilling holes in concrete as shown. These costs include the subcontractor's overhead and profit, based on minimum purchase of six racks.

	Craft@Hrs	Unit	Material	Labor	Total
Basic bicycle rack for locking rear wheels (RR-100)	C8@.575	Ea	63.00	22.10	85.10
Basic bicycle rack for locking rear wheels and cable for front wheels (RR-200)	C8@.575	Ea	68.30	22.10	90.40
Basic bicycle rack for locking rear wheels and chassis housing for front wheels (RR-300)	C8@.862	Ea	168.00	33.20	201.20
Moped or motorcycle rack with 24" security chain (RR-700)	C8@.575	Ea	99.80	22.10	121.90

Add for mounting track for use on asphalt or non-concrete surfaces. No foundation work included. Each 10' track section holds 6 bicycle racks or 4 moped or motorcycle racks.

	Craft@Hrs	Unit	Material	Labor	Total
Per 10' track section, including drilling holes	C8@2.50	Ea	89.30	96.20	185.50
Add for purchasing less than six mounting racks	--	%	15.0	--	--
Add for drilling mounting holes in concrete, 2 required per rack, cost per rack	C8@.125	Ea	--	4.81	--

Bicycle Racks. These costs include the subcontractor's overhead and profit.
Galvanized pipe rack, embedded or surface mounted to concrete, cost per rack

	Craft@Hrs	Unit	Material	Labor	Total
Double entry					
5' long, 7 bikes	C8@3.31	Ea	235.00	127.00	362.00
10' long, 14 bikes	C8@3.31	Ea	334.00	127.00	461.00
20' long, 28 bikes	C8@4.88	Ea	617.00	188.00	805.00
30' long, 42 bikes	C8@6.46	Ea	900.00	249.00	1,149.00
Single side entry					
5' long, 4 bikes	C8@2.41	Ea	224.00	92.70	316.70
10' long, 8 bikes	C8@2.41	Ea	253.00	92.70	345.70
20' long, 16 bikes	C8@3.55	Ea	461.00	137.00	598.00
Wood and metal rack, embedded or surface mounted					
Double entry					
10' long, 14 bikes	C8@3.31	Ea	384.00	127.00	511.00
20' long, 28 bikes	C8@4.88	Ea	691.00	188.00	879.00
Single entry					
10' long, 8 bikes	C8@2.41	Ea	308.00	92.70	400.70
20' long, 10 bikes	C8@3.55	Ea	550.00	137.00	687.00
Bike post bollard, dimensional wood, concrete footing	C8@.223	Ea	146.00	8.58	154.58
Single tube galvanized steel, looped wave pattern, 30" high, surface mounted					
3' long, 5 bikes	C8@3.45	Ea	603.00	133.00	736.00
5' long, 7 bikes	C8@3.45	Ea	691.00	133.00	824.00
7' long, 9 bikes	C8@4.61	Ea	900.00	177.00	1,077.00
9' long, 11 bikes	C8@4.61	Ea	1,040.00	177.00	1,217.00
Bonded color, add	--	Ea	92.60	--	--
Add for embedded mounting	C8@1.77	Ea	19.80	68.10	87.90

Sprinkler Irrigation Systems. These costs include the subcontractor's overhead and profit.
Typical complete system costs, including PVC pipe, heads, valves, fittings, trenching and backfill. Per SF watered. Add 10% for irrigation systems installed in areas subject to freezing hazard.

	Craft@Hrs	Unit	Material	Labor	Equipment	Total
Large areas using pop-up impact heads, 200 to 300 SF per head, manual system						
To 5,000 SF job	CL@.454	CSF	14.60	15.50	3.90	34.00
Over 5,000 SF to 10,000 SF	CL@.415	CSF	13.00	14.20	3.80	31.00

Site Work 2

	Craft@Hrs	Unit	Material	Labor	Equipment	Total
Over 10,000 SF	CL@.381	CSF	11.80	13.00	3.20	28.00
Add for automatic system	--	%	--	--	--	20.0
Deduct for hose bibs only (no sprinkler heads)	--	%	--	--	--	-10.0
Golf courses, parks, per acre	CL@72.0	LS	2,000.00	2,455.00	545.00	5,000.00
Small areas, 5,000 SF and under, spray heads						
Strip, automatic, shrub type	CL@.826	CSF	26.30	28.20	7.00	61.50
Commercial type, manual	CL@.759	CSF	19.30	25.90	6.80	52.00
Residential type, manual	CL@.759	CSF	26.30	25.90	6.80	59.00
Add for automatic system	--	%	--	--	--	20.0
Trenching for sprinkler systems, per 100 LF						
Main lines with a 45 HP riding type trencher, 18" wide (at $27 per hour)						
12" deep, 150 LF per hour	CL@.666	CLF	--	22.70	18.00	40.70
18" deep, 125 LF per hour	CL@.800	CLF	--	27.30	21.60	48.90
24" deep, 100 LF per hour	CL@1.00	CLF	--	34.10	27.00	61.10
Lateral lines with a 20 HP riding type trencher, 12" wide (at $13 per hour)						
8" deep, 200 LF per hour	CL@.500	CLF	--	17.00	6.50	23.50
12" deep, 125 LF per hour	CL@.800	CLF	--	27.30	21.60	48.90
Add for hard soil (shelf, rock field, hardpan)	--	%		20.0	20.0	--
Add for removing and replacing existing turf	--	%		5.0	5.0	--

	Craft@Hrs	Unit	Material	Labor	Total
Boring under walkways and pavement					
To 8' wide, by hand, 8 LF per hour	CL@.125	LF	--	4.11	--
Backfill and compact trench, by hand					
Mains to 24" deep, 125 LF per hour	CL@.799	CLF	--	27.20	--
Laterals to 12" deep, 190 LF per hour	CL@.527	CLF	--	18.00	--
Restore turf above pipe after sprinkler installation					
Replace and roll sod	CL@.019	SY	.04	.65	.69
Reseed by hand, cover, water and mulch	CL@.010	SY	.10	.34	.44
Connection to existing water line					
Residential or small commercial tap	CL@.939	Ea	17.50	32.00	49.50
Residential or small commercial stub	CL@.683	Ea	12.50	23.30	35.80
Medium commercial stub	CL@1.25	Ea	40.00	42.60	82.60
Add in freezing area	--	%	10.0	30.0	--
Backflow preventers					
Atmospheric vacuum breaker, brass, with shutoff					
3/4" to 1"	CL@.470	Ea	20.00	16.00	36.00
1-1/4" to 2"	CL@.625	Ea	43.00	21.30	64.30
Pressure type vacuum breaker, with two gate valves					
1" to 1-1/4"	CL@.750	Ea	165.00	25.60	190.60
1-1/2" to 2"	CL@.939	Ea	227.00	32.00	259.00
Double check valve assembly					
1-1/2", with two gate valves	CL@.750	Ea	300.00	25.60	325.60
2", with two gate valves	CL@.939	Ea	346.00	32.00	378.00
3", with two gate valves	CL@1.25	Ea	1,140.00	42.60	1,182.60
4", with two gate valves	CL@1.50	Ea	1,650.00	51.20	1,701.20
Reduced pressure backflow preventer					
1"	CL@.750	Ea	286.00	25.60	311.60
1-1/2"	CL@.939	Ea	476.00	32.00	508.00
2" with gate valves	CL@.939	Ea	559.00	32.00	591.00
3" flanged	CL@1.25	Ea	1,660.00	42.60	1,702.60
4" flanged	CL@1.50	Ea	2,330.00	51.20	2,381.20
6" flanged	CL@1.88	Ea	3,570.00	64.10	3,634.10

Site Work 2

	Craft@Hrs	Unit	Material	Labor	Total

Sprinkler Irrigation Systems, continued
Automatic control valves, brass with electric solenoid, no vacuum breaker or feeder wire included. Add valve box below

	Craft@Hrs	Unit	Material	Labor	Total
3/4" or 1"	CL@.341	Ea	66.70	11.60	78.30
1-1/4" or 1-1/2"	CL@.418	Ea	91.60	14.30	105.90
2"	CL@.470	Ea	114.00	16.00	130.00
2-1/2"	CL@.625	Ea	223.00	21.30	244.30
3"	CL@.750	Ea	244.00	25.60	269.60
Deduct for brass valves, hydraulic solenoid	--	%	-10.0	--	--
Deduct for plastic valves, electric solenoid	--	%	-50.0	--	--
Deduct for plastic valves, hydraulic solenoid	--	%	-60.0	--	--
Add for flow control, 3/4" and 1" valves	--	Ea	2.50	--	--

Valve boxes, plastic box with plastic lid

	Craft@Hrs	Unit	Material	Labor	Total
8" round	CL@.314	Ea	7.14	10.70	17.84
10" round	CL@.375	Ea	11.90	12.80	24.70
12" rectangular	CL@.537	Ea	16.70	18.30	35.00
6" rectangular extension	CL@.375	Ea	14.30	12.80	27.10
Jumbo rectangular	CL@.625	Ea	25.00	21.30	46.30
Add for security bolt	CL@.038	Ea	.60	1.30	1.90

Sprinkler controllers, solid state electronics, add electrical connection and feeder wire to valves

	Craft@Hrs	Unit	Material	Labor	Total
Indoor mount, 4 station	CL@.960	Ea	90.50	32.70	123.20
Indoor mount, 6 station	CL@.960	Ea	106.00	32.70	138.70
Indoor mount, 8 station	CL@1.10	Ea	143.00	37.50	180.50

Outdoor mount, add pedestal cost below

	Craft@Hrs	Unit	Material	Labor	Total
4 station	CL@2.77	Ea	106.00	94.50	200.50
6 station	CL@2.77	Ea	142.00	94.50	236.50
8 station	CL@2.77	Ea	167.00	94.50	261.50
10-12 station	CL@3.72	Ea	217.00	127.00	344.00
16-18 station	CL@3.72	Ea	619.00	127.00	746.00
20-24 station	CL@4.45	Ea	833.00	152.00	985.00
Add for pedestal for outdoor units	CL@5.73	Ea	47.10	195.00	242.10
Add for dual program controllers	--	%	5.0	--	--
Deduct for mechanical-electrical controller	--	%	-20.0	--	--

Underground control wire for automatic valves, includes waterproof connectors, wire laid in an open trench, two conductors per valve, costs are per pair of conductors

	Craft@Hrs	Unit	Material	Labor	Total
14 gauge	CL@.149	CLF	9.52	5.03	14.60
12 gauge	CL@.149	CLF	17.90	5.08	22.98
10 gauge	CL@.200	CLF	29.80	6.82	36.62
8 gauge	CL@.302	CLF	46.50	10.30	56.80

Manual control valves, with vacuum breaker, brass

	Craft@Hrs	Unit	Material	Labor	Total
3/4"	CL@.518	Ea	17.60	17.70	35.30
1"	CL@.610	Ea	24.50	20.80	45.30
1-1/4"	CL@.677	Ea	40.10	23.10	63.20
1-1/2"	CL@.796	Ea	50.00	27.10	77.10
2"	CL@.866	Ea	69.20	29.50	98.70

Pipe, PVC, solvent welded, including typical fittings but no excavation
Class 200 PVC pipe

	Craft@Hrs	Unit	Material	Labor	Total
3/4" diameter	CL@.005	LF	.19	.17	.36
1" diameter	CL@.005	LF	.25	.17	.42
1-1/4" diameter	CL@.006	LF	.45	.21	.66
1-1/2" diameter	CL@.006	LF	.62	.21	.83
2" diameter	CL@.008	LF	.90	.27	1.17
Deduct for Class 160 PVC pipe	--	%	-30.0	--	--

Site Work 2

	Craft@Hrs	Unit	Material	Labor	Total
Class 315 PVC pipe					
1/2" diameter	CL@.005	LF	.18	.17	.35
3/4" diameter	CL@.005	LF	.33	.17	.50
1" diameter	CL@.005	LF	.48	.17	.65
1-1/4" diameter	CL@.006	LF	.71	.21	.92
1-1/2" diameter	CL@.006	LF	.89	.21	1.10
2" diameter	CL@.008	LF	1.55	.27	1.82
Schedule 40 PVC pipe					
1/2" diameter	CL@.005	LF	.30	.17	.47
3/4" diameter	CL@.005	LF	.38	.17	.55
1" diameter	CL@.006	LF	.53	.21	.74
1-1/4" diameter	CL@.006	LF	.71	.21	.92
1-1/2" diameter	CL@.007	LF	.95	.24	1.19
2" diameter	CL@.008	LF	1.19	.27	1.46
PVC bell and spigot pipe, Class 160					
1-1/2" diameter	CL@.013	LF	.50	.44	.94
2" diameter	CL@.015	LF	.71	.51	1.22
2-1/2" diameter	CL@.018	LF	1.01	.61	1.62
3" diameter	CL@.021	LF	1.46	.72	2.18
4" diameter	CL@.030	LF	2.35	1.02	3.37
6" diameter	CL@.076	LF	5.24	2.59	7.83
8" diameter	CL@.111	LF	9.52	3.79	13.31
10" diameter	CL@.225	LF	13.90	7.67	21.57
Polyethylene pipe, 80 PSI					
3/4" diameter	CL@.006	LF	.15	.21	.36
1" diameter	CL@.006	LF	.20	.21	.41
1-1/4" diameter	CL@.007	LF	.34	.24	.58
1-1/2" diameter	CL@.008	LF	.41	.27	.68
2" diameter	CL@.009	LF	.50	.31	.81
Polyethylene pipe, 100 PSI					
3/4" diameter	CL@.006	LF	.17	.21	.38
1" diameter	CL@.006	LF	.24	.21	.45
1-1/4" diameter	CL@.007	LF	.42	.24	.66
1-1/2" diameter	CL@.008	LF	.57	.27	.84
2" diameter	CL@.009	LF	.93	.31	1.24
Add for 125 PSI polyethylene pipe	--	%	30.0	--	--

Shrub Sprinklers and Bubblers. These costs include a plastic pipe riser as noted at each head and the subcontractor's overhead and profit.

	Craft@Hrs	Unit	Material	Labor	Total
Bubblers (5' spacing, 12" riser), plastic	CL@.032	Ea	1.57	1.09	2.66
Flex riser	CL@.035	Ea	1.99	1.19	3.18
Double swing joint	CL@.035	Ea	2.40	1.19	3.59
Triple swing joint	CL@.035	Ea	3.66	1.19	4.85
Add for brass head and metal riser	--	Ea	4.60	--	--
Shrub spray (8' to 18' spacing, 12" riser)	CL@.038	Ea	1.67	1.30	2.97
Flex riser	CL@.038	Ea	2.09	1.30	3.39
Double swing joint	CL@.038	Ea	2.51	1.30	3.81
Triple swing joint	CL@.041	Ea	2.93	1.40	4.33
Add for brass head and metal riser	--	Ea	4.60	--	--
Stream spray (12' to 22' spacing, 12" riser)					
Plastic riser	CL@.041	Ea	2.82	1.40	4.22
Flex riser	CL@.041	Ea	3.29	1.40	4.69
Double swing joint	CL@.045	Ea	3.71	1.54	5.25
Triple swing joint	CL@.045	Ea	4.15	1.54	5.69

Site Work 2

	Craft@Hrs	Unit	Material	Labor	Total
Shrub Sprinklers and Bubblers, Stream spray, continued					
Add for brass head and metal riser	--	Ea	4.60	--	--
Rotary head, plastic (20' to 40' spacing, 12" riser)					
Plastic riser	CL@.063	Ea	10.35	2.15	12.50
Flex riser	CL@.063	Ea	11.00	2.15	13.15
Double swing joint	CL@.067	Ea	11.40	2.29	13.69
Triple swing joint	CL@.067	Ea	12.00	2.29	14.29
Rotary head, metal (20' to 40' spacing, 12" riser)					
Plastic riser	CL@.063	Ea	15.70	2.15	17.85
Flex riser	CL@.063	Ea	16.10	2.15	18.25
Double swing joint	CL@.067	Ea	16.50	2.29	18.79
Triple swing joint	CL@.067	Ea	16.90	2.29	19.19
Hi-pop spray, 6" pop-up (6" riser)	CL@.041	Ea	6.48	1.40	7.88
Flex riser	CL@.043	Ea	6.95	1.47	8.42
Double swing joint	CL@.045	Ea	7.32	1.54	8.86
Triple swing joint	CL@.045	Ea	7.84	1.54	9.38
Rotary, hi-pop, 10" to 12" pop-up	CL@.076	Ea	27.70	2.59	30.29
Flex riser	CL@.076	Ea	28.20	2.59	30.79
Double swing joint	CL@.079	Ea	29.10	2.69	31.79
Triple swing joint	CL@.079	Ea	29.40	2.69	32.09
Add when shrub sprinklers and bubblers are installed in existing shrub and flower beds	--	%	--	25.0	--

Lawn Sprinklers. These costs include a plastic pipe riser as noted at each head and the subcontractor's overhead and profit.

	Craft@Hrs	Unit	Material	Labor	Total
1" pop-up spray, plastic head (10' to 12' spacing)					
Single riser	CL@.045	Ea	3.29	1.54	4.83
Double swing joint	CL@.048	Ea	4.29	1.64	5.93
Triple swing joint	CL@.048	Ea	4.76	1.64	6.40
Add for brass head and metal riser	--	Ea	6.80	--	--
2" pop-up spray, plastic head (10' to 12' spacing)					
Single riser	CL@.045	Ea	4.49	1.54	6.03
Double swing joint	CL@.048	Ea	5.12	1.64	6.76
Triple swing joint	CL@.048	Ea	5.96	1.64	7.60
Add for brass head and metal riser	--	Ea	9.70	--	--
Rotary pop-up spray (20' to 30' spacing)					
Regular	CL@.094	Ea	15.90	3.21	19.11
Double swing joint	CL@.098	Ea	17.70	3.34	21.04
Triple swing joint	CL@.102	Ea	18.90	3.48	22.38
Rotary pop-up spray (30' to 50' spacing)					
Regular	CL@.124	Ea	18.70	4.23	22.93
Double swing joint	CL@.132	Ea	20.60	4.50	25.10
Triple swing joint	CL@.133	Ea	21.70	4.54	26.24
Rotary head (50' to 70' spacing), full or part circle					
Without valve	CL@.130	Ea	74.50	4.43	78.93
With integral check valve	CL@.130	Ea	95.00	4.43	99.43
With integral automatic electric valve	CL@.137	Ea	98.50	4.67	103.17
Rotary head (70' to 80' spacing), full or part circle					
Without valve	CL@.149	Ea	95.00	5.08	100.08
With integral check valve	CL@.149	Ea	115.00	5.08	120.08
With integral automatic electric valve	CL@.159	Ea	137.00	5.42	142.42
Rotary head metal cover	--	Ea	6.30	--	--

Site Work 2

	Craft@Hrs	Unit	Material	Labor	Total
Quick coupling valves, add swing joint costs below					
3/4" regular	CL@.089	Ea	19.30	3.04	22.34
3/4" 2-piece	CL@.089	Ea	22.60	3.04	25.64
1" regular	CL@.094	Ea	27.70	3.21	30.91
1" 2-piece	CL@.094	Ea	33.10	3.21	36.31
1-1/4" regular	CL@.108	Ea	34.90	3.68	38.58
1-1/2" regular	CL@.124	Ea	42.10	4.23	46.33
Add for double swing joint	CL@.132	Ea	3.61	4.50	8.11
Add for triple swing joint	CL@.133	Ea	6.01	4.54	10.55
Locking vinyl cover for quick coupling valves	--	Ea	9.00	--	--
Hose bib, 3/4" on 12" galvanized riser	CL@.041	Ea	9.61	1.40	11.01

Planting Bed Preparation. These costs include the subcontractor's overhead and profit.

Spreading topsoil from piles on site, based on topsoil delivered to the site at $16 per CY

	Craft@Hrs	Unit	Material	Labor	Equipment	Total
Hand work, 10 CY job, level site, 10' throw	CL@.500	CY	16.00	17.00	--	33.00
Hand work, 10 CY job, move 25' in wheelbarrow	CL@1.00	CY	16.00	34.10	--	50.10
With 1/2 CY loader, (at $20.00 per hour) 100 CY job, level site	TO@.080	CY	16.00	3.44	1.60	21.04

Spreading granular or powdered soil conditioners, based on fertilizer at $.15 per pound spread 20 pounds per 1,000 square feet (MSF). Add mixing cost below

	Craft@Hrs	Unit	Material	Labor	Equipment	Total
By hand, 2,250 SF per hour	CL@.445	MSF	3.00	15.20	--	18.20
Hand broadcast spreader, 9,000 SF per hour	CL@.111	MSF	3.00	3.79	.07	6.86
Push gravity spreader, 3,000 SF per hour	CL@.333	MSF	3.00	11.30	.42	14.72
Push broadcast spreader, 20,000 SF per hour	CL@.050	MSF	3.00	1.74	.09	4.83
Tractor drawn broadcast spreader, 87 MSF per hour	CL@.011	MSF	3.00	.38	.22	3.60

Spreading organic soil conditioners, based on peat humus at $.15 per pound spread. 200 pounds per 1,000 square feet (MSF). Add mixing cost below

	Craft@Hrs	Unit	Material	Labor	Equipment	Total
By hand, 1,000 SF per hour	CL@1.00	MSF	30.00	34.10	--	64.10
Push gravity spreader, 2,750 SF per hour	CL@.363	MSF	30.00	12.40	.45	42.85
Manure spreader, 34,000 SF per hour	CL@.029	MSF	30.00	1.02	.36	31.38

Fertilizer and soil conditioners

	Craft@Hrs	Unit	Material	Labor	Equipment	Total
Most nitrate fertilizers, ammonium sulfate	--	Lb	.15	--	--	--
Lawn and garden fertilizer, calcium nitrate	--	Lb	.30	--	--	--
Hydrated lime, sodium nitrate, superphosphate	--	Lb	.20	--	--	--
Ground limestone	--	Ton	110.00	--	--	--
Ground dolomitic limestone	--	Ton	125.00	--	--	--
Composted manure	--	Lb	.06	--	--	--
Peat humus	--	Lb	.15	--	--	--
Vermiculite, perlite	--	CF	2.50	--	--	--

Spreading lime at 70 pounds per 1,000 SF. Based on an 8 acre job using a tractor-drawn spreader at $20.00 per hour and 1 acre per hour. Add mixing cost below. Note: 1 acre equals 43,560 square feet.

	Craft@Hrs	Unit	Material	Labor	Equipment	Total
Lime at $5.75 per 100 pounds	TO@1.00	Acre	175.00	43.00	20.00	238.00
Lime at $5.75 per 100 pounds	TO@.656	Ton	115.00	28.20	13.10	156.30

Site Work 2

	Craft@Hrs	Unit	Material	Labor	Equipment	Total

Planting Bed Preparation, continued
Mixing soil and fertilizer in place. Medium soil. See adjustments for light or heavy soil below. Add soil conditioner cost above

	Craft@Hrs	Unit	Material	Labor	Equipment	Total
Mixing by hand						
2" deep, 9 SY per hour	CL@.111	SY	--	3.79	--	--
4" deep, 7.5 SY per hour	CL@.133	SY	--	4.54	--	--
6" deep, 5.5 SY per hour	CL@.181	SY	--	6.17	--	--
Mixing with a 26" tiller rented for $32 per day						
2" deep, 110 SY per hour	CL@.009	SY	--	.31	.03	.34
4" deep, 90 SY per hour	CL@.011	SY	--	.38	.04	.42
6" deep, 65 SY per hour	CL@.015	SY	--	.51	.06	.57
8" deep, 45 SY per hour	CL@.022	SY	--	.75	.09	.84
Mixing with a 6' disk and 20 HP garden tractor rented for $100 per day						
2" deep, 50,000 SF per hour	TO@.020	MSF	--	.86	.25	1.11
4" deep, 40,000 SF per hour	TO@.025	MSF	--	1.08	.31	1.39
6" deep, 30,000 SF per hour	TO@.033	MSF	--	1.42	.41	1.83
8" deep, 20,000 SF per hour	TO@.050	MSF	--	2.19	.63	2.82
Mixing with a 12' disk and 40 HP utility tractor rented for $130 per day						
2" deep, 90,000 SF per hour	TO@.011	MSF	--	.47	.18	.65
4" deep, 70,000 SF per hour	TO@.014	MSF	--	.60	.23	.83
6" deep, 53,000 SF per hour	TO@.019	MSF	--	.82	.31	1.13
8" deep, 35,000 SF per hour	TO@.028	MSF	--	1.20	.46	1.66
Add for heavy soils (clay, wet or rocky soil)						
Mixing by hand	--	%	--	30.0	--	--
Mixing with a 26" tiller	--	%	--	60.0	60.0	--
Mixing with a 6' or 12' disk	--	%	--	20.0	20.0	--
Deduct for light soils (sand or soft loam)						
Mixing by hand	--	%	--	-20.0	--	--
Mixing with a 26" tiller	--	%	--	-30.0	-30.0	--
Mixing with a 6' or 12' disk	--	%	--	-15.0	-15.0	--
Leveling the surface for lawn or planting bed						
By hand, 110 SY per hour	CL@.009	SY	--	.31	--	--
With 6' drag harrow, at $100 per day,						
60,000 SF per hour	TO@.017	MSF	--	.73	.21	.94
With 12' drag harrow, at $130 per day,						
115,000 SF per hour	TO@.009	MSF	--	.39	.15	.54
Plant bed preparation						
With a 1/2 CY wheel loader, at $160 per day,						
12.5 CY per hour	TO@.080	CY	--	3.43	1.60	5.03
By hand to 18" deep, 8.5 SY/hr	CL@.118	SY	--	4.02	--	--
Mixing planting soil						
By hand, .9 CY per hour	CL@1.11	CY	--	37.90	--	--
With 1/2 CY wheel loader, at $160 per day,						
7 CY per hour	TO@.143	CY	--	6.15	2.86	9.01
With a soil shredder, at $170 per day,						
11 CY per hour	TO@.090	CY	--	3.86	1.91	5.77

Seeding and Planting. These costs include the subcontractor's overhead and profit.
Seeding with a hand broadcast spreader, 5 pounds per 1,000 SF, 4,000 SF per hour, seed at $3.20 per pound

	Craft@Hrs	Unit	Material	Labor	Equipment	Total
	CL@.250	MSF	16.00	8.56	.16	24.72

Seeding with push broadcast spreader, 5 pounds per 1,000 SF, 10,000 SF per hour, seed at $3.20 per pound

	Craft@Hrs	Unit	Material	Labor	Equipment	Total
	CL@.100	MSF	16.00	3.41	.19	19.60

Seeding with a mechanical seeder at 175 pounds per acre, 5 acre job

	Craft@Hrs	Unit	Material	Labor	Equipment	Total
Seed at $3.20 per pound	TO@13.1	Acre	560.00	563.00	262.00	1,385.00
Seed at $3.20 per pound	TO@.300	MSF	12.90	12.90	6.00	31.80

Site Work 2

	Craft@Hrs	Unit	Material	Labor	Equipment	Total

Hydroseeding (spray application of seed, binder and fertilizer slurry), 3 pounds of Alta Fescue seed applied in a 12 pound mix per 1,000 SF

	Craft@Hrs	Unit	Material	Labor	Equipment	Total
10,000 SF job	C4@1.12	MSF	15.40	39.50	17.00	71.90
25,000 SF job	C4@.645	MSF	15.40	22.70	9.85	47.95
50,000 SF job	C4@.399	MSF	15.40	14.10	6.05	35.55
Add for work in congested areas	--	%	--	--	--	50.0

Sealing soil, application by a hydroseeding unit of a copolymer based liquid for erosion control

	Craft@Hrs	Unit	Material	Labor	Equipment	Total
	C4@.020	SY	.18	.71	.02	.91

Grass seed

	Craft@Hrs	Unit	Material	Labor	Equipment	Total
Most Kentucky bluegrass	--	Lb	4.30	--	--	--
Most fescue	--	Lb	2.10	--	--	--
Most ryegrass, timothy, orchard grass	--	Lb	1.60	--	--	
Annual ryegrass	--	Lb	.55	--	--	
Bentgrass, certified	--	Lb	10.50	--	--	--
Native grass PLS (bluestream, blue grama)	--	Lb	16.00	--	--	--

Sodding, placed on level ground at 25 SY per hour. Add delivery cost below

	Craft@Hrs	Unit	Material	Labor	Equipment	Total
Northeast, Midwest, bluegrass blend	CL@.040	SY	1.30	1.36	--	2.66
Great Plains, South, bluegrass	CL@.040	SY	1.60	1.36	--	2.96
West, bluegrass, tall fescue	CL@.040	SY	2.90	1.36	--	4.26
South, Bermuda grass, centipede grass	CL@.040	SY	2.65	1.36	--	4.01
South, Zoysia grass, St. Augustine	CL@.040	SY	4.25	1.36	--	5.61
Typical delivery cost, to 90 miles	--	SY	.43	--	--	--
Add for staking sod on slopes	CL@.010	SY	.09	.34	--	.43

Work grass seed into soil, no seed included

	Craft@Hrs	Unit	Material	Labor	Equipment	Total
By hand, 220 SY per hour	CL@.005	SY	--	.17	--	--
6' harrow and 20 HP tractor, at $100 per day, 60 MSF per hour	TO@.017	MSF	--	.73	.21	.94
12' harrow and 40 HP tractor, at $130 per day, 100 MSF per hour	TO@.010	MSF	--	.43	.16	.59

Apply top dressing over seed or stolons, 300 pounds per 1,000 SF (2.7 pounds per SY), top dressing at $.15 per pound

	Craft@Hrs	Unit	Material	Labor	Equipment	Total
By hand, 65 SY per hour	CL@.015	SY	.41	.51	--	.92
With manure spreader, 10,000 SF per hour	CL@.125	MSF	45.00	4.26	--	49.26

Roll sod or soil surface with a roller

	Craft@Hrs	Unit	Material	Labor	Equipment	Total
With a push roller, 400 SY per hour	CL@.002	SY	--	.07	--	--
With a 20 HP tractor, at $100 per day, 25,000 SF per hour	TO@.040	MSF	--	1.76	.50	2.26

Sprigging, by hand

	Craft@Hrs	Unit	Material	Labor	Equipment	Total
6" spacing, 50 SY per hour	CL@.020	SY	.18	.68	--	.86
9" spacing, 100 SY per hour	CL@.010	SY	.12	.34	--	.46
12" spacing, 150 SY per hour	CL@.007	SY	.09	.24	--	.33
Hybrid Bermuda grass stolons, per bushel	--	Ea	3.00	--	--	--

Ground cover, large areas, no surface preparation

	Craft@Hrs	Unit	Material	Labor	Equipment	Total
Ice plant (80 SF per flat)	CL@.008	SF	.15	.27	--	.42
Strawberry (75 SF per flat)	CL@.008	SF	.18	.27	--	.45
Ivy and similar (70 SF per flat)	CL@.008	SF	.19	.27	--	.46
Gravel bed, pea gravel, at 12.00 a ton, spread by hand	CL@.555	CY	16.80	18.90	--	35.70

Trees and shrubs, complete. See also Landscaping in the Residential Division

Shrubs, most varieties

	Craft@Hrs	Unit	Material	Labor	Equipment	Total
1 gallon, 100 units	S1@.085	Ea	7.00	3.28	.17	10.45

Site Work 2

	Craft@Hrs	Unit	Material	Labor	Equipment	Total
Seeding and Planting, Shrubs, continued						
1 gallon, over 100 units	S1@.076	Ea	4.50	2.93	.15	7.58
5 gallons, 100 units	S1@.443	Ea	17.00	17.10	.89	34.99
Trees, most varieties, complete, staked, typical costs						
5 gallon, 4' to 6' high	S1@.716	Ea	21.50	27.60	1.43	50.53
15 gallon, 8' to 10' high	S1@1.37	Ea	70.00	52.80	2.74	125.54
20" x 24" box, 10' to 12' high	S1@4.54	Ea	265.00	175.00	9.08	449.08
Specimen size 36" box, 14' to 20' high	S1@11.6	Ea	700.00	447.00	23.20	1,170.20
Guaranteed establishment of plants, % of contract price	--	%	--	--	--	12.0
Edging						
Redwood benderboard, 6" x 3/8", staked	CL@.013	LF	.43	.44	--	.87
Redwood benderboard, 5/16" x 4", staked	CL@.013	LF	.26	.44	--	.70
Redwood headerboard, 2" x 4", staked, specification grade	CL@.127	LF	2.75	4.33	--	7.08
Decomposed granite edging, saturated and rolled, 3" deep	CL@.011	SF	.25	.38	--	.63
Landscape stepping stones (concrete pavers), laid on level ground	CL@.024	SF	3.25	.82	--	4.07
Creosote piles in 2' to 5' sections, 9" to 12" diameter	CL@.045	LF	2.50	1.54	--	4.04

Marine Work. Typical costs for jobs with a contract value of $25,000 and up. These costs include the subcontractor's overhead and profit but no mobilization costs. Mobilization costs can be expected to vary widely, from $5,000 to $50,000. Add for same as required by job location and job site conditions.
Dredging. No hauling included

	Craft@Hrs	Unit	Material	Labor	Equipment	Total
General dredging based on using a 45 ton crawler mounted dragline at $80 per hour and a 2-man crew,						
Spoil dumped on bank, or loaded in trucks	H2@.040	CY	--	1.62	1.60	3.22
Levee embankment, no imports	H2@.044	CY	--	1.79	1.76	3.55
Riprap, random size stones						
Dumped from truck, based on using a 10 ton dump truck at $20 per hour,						
50 to 150 pounds per stone,	TD@.143	Ton	25.00	5.38	2.86	33.24
Sacked and hand placed, 100 to 500 CY job	CL@2.98	CY	30.00	102.00	--	132.00
Filter material, 140 mil nonwoven fabric, 1,000 SY job	CL@.032	SY	2.00	1.09	--	3.09
Wharfs, including fenders and utilities. Based on using equipment costing $100 per hour and a 3-man crew,						
Timber with 4" wood deck	M2@1.33	SF	25.00	49.20	44.00	118.50
Timber with concrete deck	P9@1.50	SF	26.00	58.50	50.00	134.50
Concrete piles. Based on using equipment costing $140 per hour and a 7-man crew,						
with 4" precast concrete deck lifted into place	S8@1.00	SF	40.00	40.70	20.00	100.70
Fender system (dolphin) consisting of 13 piles (up to 50' in height), 12" x 12" chocks and wales, and all hardware. Based on using equipment costing $140 per hour and a 7-man crew						
Per LF of length overall	S8@1.75	LF	75.00	71.20	35.00	181.20

Railroad Sidings. Costs are based on single track siding quantities using new materials. Equipment cost is $30 per hour and includes a flat bed truck with an A-frame hoist (at $20 per hour), an air compressor with pneumatic tools suitable for this type work (at $6 per hour) and welding & cutting equipment (at $4 per hour). For scheduling purposes estimate that a 5-man crew will install 20 to 25 LF of two-rail siding in an 8-hour day. These costs include the subcontractor's overhead and profit, but no excavation or site preparation.

Site Work 2

	Craft@Hrs	Unit	Material	Labor	Equipment	Total

Siding, with 2 rails, including cross ties, plates, splice bars, track spikes, bolts, ballast, alignment & surfacing of track and date nails in ties. (See component costs listed on next page.)

	Craft@Hrs	Unit	Material	Labor	Equipment	Total
Railroad siding complete, per LF	P5@1.77	LF	70.26	65.01	10.61	145.88
Components for railroad siding costs listed above, per LF of track:						
Rail, 115 lb (77 lbs per LF of track)	P5@.720	LF	20.21	26.50	4.32	51.03
Cross ties, treated, 7" x 9" x 8'6", type 1						
(25 each 40' of track)	P5@.231	LF	16.25	8.70	1.39	26.34
Tie plates (1.25 per LF of track)	--	LF	8.44	--	--	8.44
Splice bars w/bolts (1 per 20 LF of track)	--	LF	2.33	--	--	2.33
Track spikes, 9/16" x 5-1/2"						
(1 keg per 115 LF of track)	--	LF	.68	--	--	.68
Track bolts, 1" diameter, 5-1/2" long						
(22 per 100 LF of track)	--	LF	.38	--	--	.38
Ballast 3/4" to 1-1/2" quarry run gravel						
(1.25 tons per LF of track, $ 17.50 per ton)	P5@.482	LF	21.90	17.60	2.89	42.39
Align and surface rail (per LF of track)	P5@.331	LF	--	12.10	1.99	14.09
Date nails in ties, 3/16" x 2", aluminum, $.12 each						
(6 per 10 LF of track)	P5@.003	LF	.07	.11	.02	.20
Individual component costs:						
Rail, by American Railroad Association Class						
90 lb, class A, new	P5@18.7	Ton	580.00	684.00	112.00	1,376.00
90 lb, class B, relayer	P5@18.7	Ton	330.00	684.00	112.00	1,126.00
115 lb, class A, new	P5@18.7	Ton	525.00	684.00	112.00	1,321.00
Remove existing track, 2 rails, for salvage	P5@.217	LF	2.15	7.94	1.30	11.39
Cross ties, treated						
7" x 9" x 8'6", type 1	P5@.376	Ea	26.00	13.80	2.26	42.06
7" x 9" x 8'6", type 1 (25% 7" x 8")	P5@.376	Ea	22.00	13.80	2.26	38.06
Remove cross ties for disposal	P5@.238	Ea	--	8.70	1.43	10.13
Tie plates						
For 90 lb rail	--	Ea	5.50	--	--	--
For 115 lb rail	--	Ea	6.75	--	--	--
For 90 lb rail, used	--	Ea	1.50	--	--	--
For 90 lb rail, special plates for turnouts	--	Ea	10.90	--	--	--
Splice bars, with bolts						
For 90 lb rail	--	Ea	41.00	--	--	--
For 115 lb rail	--	Ea	46.60	--	--	--
For 90 lb rail, used	--	Ea	20.50	--	--	--
Compromise splice bar	--	Ea	381.60	--	--	--
Compromise angle splice bar	--	Ea	1,248.00	--	--	--
Track spikes						
9/16" square, 5-1/2" long, new, per keg	--	Ea	78.20	--	--	--
5/8" square, 6" long, new, per keg	--	Ea	77.30	--	--	--
Spring rail anchor, for 100 lb rail, new	--	Ea	1.57	--	--	--
5/8" or 9/16" used spikes, per keg	--	Ea	52.80	--	--	--
Swage locking pins, 1" diameter						
1-1/2" grip range	--	Ea	2.98	--	--	--
2-1/4" grip range	--	Ea	2.90	--	--	--
3-1/2" grip range	--	Ea	2.40	--	--	--
Swage locking collar	--	Ea	3.45	--	--	--
Hardened swage locking spacer	--	Ea	.28	--	--	--
Rail clip, 2-3/4" x 2-1/4", 15/16" thick	--	Ea	2.80	--	--	--
Gage rod, 1-1/4", adjustable 2 ends, new	--	Ea	21.10	--	--	--
Rail brace, for 100 lb class A rail	--	Ea	42.00	--	--	--
Track bolts, 1" diameter, 5-1/2" long	--	Ea	1.72	--	--	--

Site Work 2

	Craft@Hrs	Unit	Material	Labor	Equipment	Total

Railroad Sidings, continued

	Craft@Hrs	Unit	Material	Labor	Equipment	Total
Wheel stops, per pair	P5@2.79	Pr	380.00	102.00	16.70	498.70
Bumping post, 705 lb	P5@33.1	Ea	1,200.00	1,210.00	200.00	2,610.00
Bumping post, 1,100 lb	P5@33.1	Ea	2,800.00	1,210.00	200.00	4,210.00
Bromicil weed control, manually sprayed (15 lbs/acre)	CL@9.67	Acre	180.00	330.00	--	510.00

Railroad Siding Turnouts and Switches. Costs are based on single track siding installations using new materials. Equipment cost is $30 per hour and includes a flat bed truck with an A-frame hoist (at $20 per hour), an air compressor with pneumatic tools suitable for this type work (at $6 per hour) and welding & cutting equipment (at $4 per hour). For scheduling purposes estimate that a 5-man crew will install one railroad siding turnout in 6 to 7 working days (8-hour days). These costs include the subcontractor's overhead and profit, but no excavation or site preparation.

Railroad siding turnout with #8 straight split switch, including switch stand, ties, spikes, ballast, alignment & track surfacing and date nails in ties. (See individual component costs below.)

	Craft@Hrs	Unit	Material	Labor	Equipment	Total
Railroad siding turnout as described above	P5@269.	Ea	13,131.72	9,810.20	1,611.68	24,553.60

Component costs per turnout for #8 turnout as listed above:

	Craft@Hrs	Unit	Material	Labor	Equipment	Total
Straight split #8 switch with solid frog, 115 lb rail	P5@80.0	Ea	7,800.00	2,930.00	480.00	11,210.00
Switch ties, treated, 7" x 9" (139 per switch)	P5@51.0	Ea	2,350.00	1,870.00	306.00	4,526.00
Switch stand, adjustable, with reflectorized target	--	Ea	640.00	--	--	640.00
Track spikes, 9/16" square, 5-1/2" long, (3 kegs per switch)	--	Ea	235.00	--	--	235.00
Ballast, 3/4" to 1-1/2" quarry run gravel (120 tons per switch at $17.50 per ton)	P5@46.3	Ea	2,100.00	1,690.00	278.00	4,068.00
Align and surface rail (275 LF of track)	P5@91.0	Ea	--	3,330.00	546.00	3,876.00
Date nails in ties, 3/16" x 2", aluminum, (56 per switch)	P5@.280	Set	6.72	10.20	1.68	18.60

Other turnout and switch costs are listed below:

	Craft@Hrs	Unit	Material	Labor	Equipment	Total
Number 8 straight split switch, 90 lb class A rail with solid frog	P5@70.0	Ea	3,800.00	2,560.00	420.00	6,780.00
Switch points for #8 switch, 90 lb rail, 16'6" long, new	P5@16.0	Ea	2,900.00	585.00	96.00	3,581.00
Bolted frog for #8 switch, 90 lb class A rail	P5@24.0	Ea	5,300.00	878.00	144.00	6,322.00

Number 8 turnouts, 100 lb class A rail

	Craft@Hrs	Unit	Material	Labor	Equipment	Total
Straight split switch, solid frog	P5@80.0	Ea	12,400.00	2,930.00	480.00	15,810.00
Straight split switch with bolted frog	P5@80.0	Ea	11,700.00	2,930.00	480.00	15,110.00
Tongue and mate switch with solid frog	P5@80.0	Ea	25,600.00	2,930.00	480.00	29,010.00
Tongue and mate switch with bolted frog	P5@71.0	Ea	17,300.00	2,600.00	426.00	20,326.00
Straight double tongue switch and solid frog	P5@60.0	Ea	40,300.00	2,190.00	360.00	42,850.00
Number 6 turnout, straight split switch, 90 lb class A rail with solid frog	P5@80.0	Ea	24,500.00	2,930.00	480.00	27,910.00
Number 4 turnout, straight split switch, 90 lb class A rail with solid frog	P5@80.0	Ea	24,000.00	2,930.00	480.00	27,410.00
Crossing angle, 14 degree, 15', straight, 100 lb class A track, built-up	P5@80.0	Ea	14,800.00	2,930.00	480.00	18,210.00
Number 8 straight crossover with 2 number 8 straight split switches, track spaced at 13'	P5@160.	Ea	46,300.00	5,850.00	960.00	53,110.00
Expansion joint with two 90 lb class A rails, with bolts and plates	P5@16.0	Ea	1,130.00	585.00	96.00	1,811.00
Rebuild frog	P5@16.0	Ea	930.00	585.00	96.00	1,611.00
Remove turnout in salvage condition	P5@40.0	Ea	--	1,460.00	240.00	--
Remove switch ties for disposal	P5@.376	Ea	--	13.80	2.26	--

Site Work 2

	Craft@Hrs	Unit	Material	Labor	Equipment	Total

Railroad Grade Crossings. Costs are based on single crossing quantity installations using new materials. Equipment cost is $30 per hour and includes a flat bed truck with an A-frame hoist (at $20 per hour), an air compressor with pneumatic tools suitable for this type work (at $6 per hour) and welding & cutting equipment (at $4 per hour). These costs include the subcontractor's overhead and profit, but no excavation or site preparation.

Grade crossing, single track, 2" asphalt on 8" stone base, 25' x 15' overall (42 SY), including flangeway and prime coat, but no grading. (See individual component costs listed below.)

	Craft@Hrs	Unit	Material	Labor	Equipment	Total
Cost per grade crossing as described above	P5@25.6	Ea	859.50	935.50	153.60	1,948.60
Cost per square yard for grade crossing described above, based on 42 SY	P5@.601	SY	20.46	22.27	3.66	46.39
Components for grade crossing listed above, costs per 42 SY grade crossing						
Flat bar type flangeway, 50 LF	P5@20.0	Ea	550.00	731.00	120.00	1,401.00
Stone aggregate base, 8" thick, 10 CY	P5@4.00	Ea	175.00	146.00	24.00	345.00
Bituminous binder and wear course, 2", 42 SY	P5@1.25	Ea	115.00	45.70	7.50	168.20
Bituminous prime coat, 42 SY	P5@.350	Ea	19.50	12.80	2.10	34.40
Flangeway (for placement adjacent to track), per linear foot of rail						
6" x 6" timbers, treated, 3/4" x 16" screws	P5@.211	LF	9.00	7.71	1.27	17.98
Flat bar type	P5@.400	LF	11.00	14.60	2.40	28.00
Rubber grade crossing, with shims, rubber pads and header board, 8'4" wide, per LF of track	P5@.371	LF	300.00	13.57	2.23	315.80

Concrete 3

	Craft@Hrs	Unit	Material	Labor	Equipment	Total

Excavation for Concrete Work. These costs include the subcontractor's overhead and profit, but no shoring or disposal. Typical soil conditions.

	Craft@Hrs	Unit	Material	Labor	Equipment	Total
Batterboards, lay out for footings, from known markers, per corner	C8@1.24	Ea	5.00	47.70	--	52.70
Trenching with a 1/2 CY utility backhoe, small jobs, good soil conditions, no backfill included. Equipment cost is $20.00 per hour at 15 CY per hour						
18" x 24" depth, .111 CY per LF (135 LF/Hr)	S1@.015	LF	--	.58	.15	.73
24" x 36" depth, .222 CY per LF (68 LF/Hr)	S1@.029	LF	--	1.12	.29	1.41
36" x 48" depth, .444 CY per LF (34 LF/Hr)	S1@.059	LF	--	2.27	.59	2.86
48" x 60" depth, .741 CY per LF (20 LF/Hr)	S1@.099	LF	--	3.82	.99	4.81
Per cubic yard, 15 CY per hour	S1@.133	CY	--	5.13	1.33	6.46
Hand labor						
Clearing trench shoulders of obstructions	CL@.017	LF	--	.58	--	--
Breaking hard pan or rock outcropping and stack	CL@3.23	CY	--	110.00	--	--
Column pads and small piers, average soil conditions	CL@1.32	CY	--	45.00	--	--
Backfilling against foundations, no import or export of materials						
Hand, including hand tamp	CL@.860	CY	--	29.30	--	--
Backfill with small tractor (at $20 per hour), minimum compaction, wheel-rolled	S1@.055	CY	--	2.12	.55	2.67
Backfill with small tractor & pneumatic tamper, spec grade compaction (at $28 per hour)	S1@.141	CY	--	5.43	1.97	7.40
Disposal of excess backfill material with 3/4 CY crawler loader (at $20 per hour), no compaction						
Spot spread, 100' haul	S1@.034	CY	--	1.31	.34	1.65
Area spread, 4" deep (covers 81 SF per CY)	S1@.054	CY	--	2.08	.54	2.62

Concrete 3

	Craft@Hrs	Unit	Material	Labor	Equipment	Total

Excavation for Concrete Work, continued
Grading for slabs
  Rough grading by crawler tractor with dozer blade
    (1000 SF and $40 per hour) — S1@.001 SF -- .04 .02 .06
  Fine grading, by hand, light to medium soil
    (165 to 170 SF per hour) — CL@.006 SF -- .21 -- --
  Fine grading, by crawler tractor with dozer blade
    (1000 SF and $40 per hour) — S1@.001 SF -- .04 .02 .06
Capillary fill, at $17.50 per CY, one CY covers 81 SF 4" deep
  4", hand graded and rolled, small job — CL@1.41 SF .22 .46 .08 .76
  4", machine graded and rolled, larger job — S1@.066 SF .22 .02 .13 .37
  Add for each extra 1" hand graded — CL@.354 SF .05 .12 .02 .19
  Add for each extra 1" machine graded — S1@.017 SF .05 .06 .03 .14
Sand fill, at $15 per CY, 1 CY covers 162 SF at 2" deep
  2" sand cushion, hand spread, 1 throw — CL@.341 SF .09 .11 -- .20
  Add for each additional 1" — CL@.171 SF .02 .06 -- .08
Waterproof membrane, polyethylene, over sand bed, including 20% lap and waste
  .004" (4 mil) clear or black — CL@.119 SF .02 .04 -- .06
  .006" (6 mil) clear or black — CL@.119 SF .04 .04 -- .08

Formwork for Concrete. Labor costs include the time needed to prepare formwork sketches at the job site, measure for the forms, fabricate, erect, align, brace, strip, clean and stack the forms. Costs for reinforcing and concrete are not included. Multiple use of forms shows the cost per use when a form is used several times on the same job without being totally disassembled or discarded. Normal cleaning and repairs are included in the labor cost on lines that show multiple use of forms. Cost will be higher if plans are not detailed enough to identify the work to be performed. No salvage value is assumed except as noted. Costs listed are per square foot of form in contact with the concrete (SFCA). These costs include the subcontractor's overhead and profit and assume a number 2 and better lumber price of $500 per MBF and $710 per MSF for 3/4" plyform before markup and waste allowance.

Wall footing, grade beam or tie beam forms These figures assume nails, stakes and form oil costing $.30 per square foot and 2.5 board feet of lumber are used for each square foot of form in contact with the concrete (SFCA). To calculate the quantity of formwork required, multiply the depth of footing in feet by the length of footing in feet. Then double the results if two sides will be formed. For scheduling purposes, estimate that a crew of 5 can lay out, fabricate and erect 600 to 700 SF of footing, grade beam or tie beam forms in an 8-hour day.

	Craft@Hrs	Unit	Material	Labor	Total
1 use	F5@.070	SFCA	1.81	2.75	4.56
3 uses	F5@.050	SFCA	1.00	1.97	2.97
5 uses	F5@.040	SFCA	.76	1.57	2.33
Add for stepped footings	--	%	25.0	40.0	--
Add for keyed joint, 1 use	F5@.020	LF	.73	.79	1.52

Reinforcing bar supports for footing, grade beam and tie beam forms. Bars suspended from 2" x 4" lumber, including stakes. Based on .86 BF of lumber per LF

	Craft@Hrs	Unit	Material	Labor	Total
1 use	F5@.060	LF	.52	2.36	2.88
3 uses	F5@.050	LF	.29	1.97	2.26
5 uses	F5@.040	LF	.22	1.57	1.79

Integral starter wall forms (stem walls) formed monolithic with footings, heights up to 4'0", with 3 BF of lumber per SFCA plus an allowance of $.36 per SFCA for nails, stakes and form oil

	Craft@Hrs	Unit	Material	Labor	Total
1 use	F5@.100	SFCA	2.18	3.93	6.11
3 uses	F5@.075	SFCA	1.20	2.95	4.15
5 uses	F5@.067	SFCA	.92	2.64	3.56

Concrete 3

	Craft@Hrs	Unit	Material	Labor	Total

Bulkheads or pour-stops. When integral starter walls are formed, forms will usually be required at the ends of each wall or footing. These are usually called bulkheads or pour-stops. These figures assume the use of 1.1 SF of 3/4" plyform per SFCA plus an allowance of $.08 per SFCA for nails, stakes and bracing

	Craft@Hrs	Unit	Material	Labor	Total
1 use	F5@.150	SFCA	1.02	5.90	6.92
3 uses	F5@.120	SFCA	.56	4.72	5.28
5 uses	F5@.100	SFCA	.43	3.93	4.36

Column footing or pile cap forms These figures assume nails, stakes and form oil costing $.42 per square foot and 3.5 board feet of lumber are used for each square foot of form in contact with the concrete. For scheduling purposes, estimate that a crew of 5 can lay out, fabricate and erect 500 to 600 SF of square or rectangular footing forms in an 8-hour day or 350 to 450 SF of octagonal, hexagonal or triangular footing forms.

Square or rectangular column forms

	Craft@Hrs	Unit	Material	Labor	Total
1 use	F5@.090	SFCA	2.53	3.54	6.07
3 uses	F5@.070	SFCA	1.39	2.75	4.14
5 uses	F5@.060	SFCA	1.06	2.36	3.42

Octagonal, hexagonal or triangular forms

	Craft@Hrs	Unit	Material	Labor	Total
1 use	F5@.117	SFCA	2.53	4.60	7.13
3 uses	F5@.091	SFCA	1.39	3.58	4.97
5 uses	F5@.078	SFCA	1.06	3.07	4.13

Reinforcing bar supports for column forms. Bars suspended from 2" x 4" lumber, including stakes. Based on .86 BF of lumber per LF

	Craft@Hrs	Unit	Material	Labor	Total
1 use	F5@.060	LF	.52	2.36	2.88

Anchor bolt templates or dowel supports for column forms. Based on using nails, stakes, bracing, and form oil costing $.08 per SF and 1.1 SF of 3/4" plyform for each SF of column base or dowel support. No anchor bolts included

	Craft@Hrs	Unit	Material	Labor	Total
1 use	F5@.150	SF	1.02	5.90	6.92
3 uses	F5@.100	SF	.56	3.93	4.49
5 uses	F5@.090	SF	.43	3.54	3.97

Slab-on-grade forms These figures assume nails, stakes, form oil, and accessories costing $.06 are used for each board foot of lumber. Note that costs listed below are per linear foot of form. To convert a linear foot cost to a cost per square foot of form, first multiply the BF per LF figure by .333. Then divide the costs shown by that factor. For example, 12" to 24" high edge forms average 4.5 board feet of lumber per linear foot (4.5 BF per LF). Multiply 4.5 by .333 to get 1.5 square feet (SF) per linear foot (LF) of form. Then divide the costs shown by 1.5. For example, the cost per SF for 3 uses would be $1.10 ($1.65 divided by 1.5) for material plus $3.12 ($4.68 divided by 1.5) for labor.

For scheduling purposes, estimate that a crew of 5 will lay out, fabricate and erect the following quantities of slab-on-grade edge forms in an 8-hour day:

Slabs to 6" high	= 615 to 720 LF	Over 6" to 12" high	= 450 to 500 LF
Over 12" to 24" high	= 320 to 350 LF	Over 24" to 36" high	= 240 to 260 LF

Edge forms to 6" high, 1.5 BF per LF of form

	Craft@Hrs	Unit	Material	Labor	Total
1 use	F5@.065	LF	1.00	2.56	3.56
3 uses	F5@.061	LF	.55	2.40	2.95
5 uses	F5@.055	LF	.42	2.16	2.58

Edge forms over 6" high to 12" high, 2.25 BF per LF of form

	Craft@Hrs	Unit	Material	Labor	Total
1 use	F5@.090	LF	1.50	3.54	5.04
3 uses	F5@.086	LF	1.24	3.38	4.62
5 uses	F5@.080	LF	.63	3.15	3.78

Concrete 3

	Craft@Hrs	Unit	Material	Labor	Total

Slab-on-grade forms, continued
Edge forms over 12" high to 24" high, 4.5 BF per LF of form

	Craft@Hrs	Unit	Material	Labor	Total
1 use	F5@.124	LF	3.00	4.88	7.88
3 uses	F5@.119	LF	1.65	4.68	6.33
5 uses	F5@.115	LF	1.26	4.52	5.78

Edge forms over 24" high to 36" high, 7.5 BF per LF of forms

	Craft@Hrs	Unit	Material	Labor	Total
1 use	F5@.166	LF	5.00	6.53	11.53
3 uses	F5@.160	LF	2.75	6.29	9.04
5 uses	F5@.155	LF	2.10	6.10	8.20

Add for 2" x 2" tapered keyed joint, one-piece, .38 BF per LF

	Craft@Hrs	Unit	Material	Labor	Total
1 use	F5@.044	LF	.23	1.73	1.96
3 uses	F5@.040	LF	.13	1.57	1.70
5 uses	F5@.035	LF	.10	1.38	1.48

Blockout and slab depression forms. Figure the linear feet required.
Use the linear foot costs for slab edge forms for the appropriate height
and add to the material and labor cost   --   %   20.0   25.0   --

Wall forms. Formwork over 6' high includes an allowance for a work platform and handrail built on one side of the form for use by the concrete placing crew.

Heights to 4', includes 1.1 SF of plyform, 1.5 BF of lumber and $.33 for nails, ties and oil per SFCA

	Craft@Hrs	Unit	Material	Labor	Total
1 use	F5@.119	SFCA	2.17	4.68	6.85
3 uses	F5@.080	SFCA	1.20	3.15	4.35
5 uses	F5@.069	SFCA	.91	2.71	3.62

Heights over 4' to 6', includes 1.1 SF of plyform, 2.0 BF of lumber and $.49 for nails, ties and oil per SFCA

	Craft@Hrs	Unit	Material	Labor	Total
1 use	F5@.140	SFCA	2.64	5.51	8.15
3 uses	F5@.100	SFCA	1.45	3.93	5.38
5 uses	F5@.080	SFCA	1.11	3.15	4.26

Heights over 6' to 12', includes 1.2 SF of plyform, 2.5 BF of lumber and $.58 for nails, ties and oil per SFCA

	Craft@Hrs	Unit	Material	Labor	Total
1 use	F5@.160	SFCA	3.11	6.29	9.40
3 uses	F5@.110	SFCA	1.71	4.33	6.04
5 uses	F5@.100	SFCA	1.30	3.93	5.23

Heights over 12' to 16', includes 1.2 SF of plyform, 3.0 BF of lumber and $.65 for nails, ties and oil per SFCA

	Craft@Hrs	Unit	Material	Labor	Total
1 use	F5@.180	SFCA	3.49	7.08	10.57
3 uses	F5@.130	SFCA	1.92	5.11	7.03
5 uses	F5@.110	SFCA	1.46	4.33	5.79

Heights over 16', includes 1.3 SF of plyform, 3.5 BF of lumber and $.74 for nails, ties and oil per SFCA

	Craft@Hrs	Unit	Material	Labor	Total
1 use	F5@.199	SFCA	4.00	7.83	11.83
3 uses	F5@.140	SFCA	2.20	5.51	7.71
5 uses	F5@.119	SFCA	1.68	4.68	6.36

Architectural form liner for wall forms

	Craft@Hrs	Unit	Material	Labor	Total
Low cost liners	F5@.020	SFCA	1.00	.79	1.79
Average cost liners	F5@.100	SFCA	1.50	3.93	5.43

Reveal strips 1" deep by 2" wide, one-piece, wood

	Craft@Hrs	Unit	Material	Labor	Total
1 use	F5@.069	LF	.15	2.71	2.86
3 uses	F5@.050	LF	.08	1.97	2.05
5 uses	F5@.020	LF	.06	.79	.85

Battered wall forms (wall form is inclined from the vertical)

	Craft@Hrs	Unit	Material	Labor	Total
Add for 1 side battered	--	%	5.0	10.0	--
Add for 2 sides battered	--	%	10.0	15.0	--

Concrete 3

	Craft@Hrs	Unit	Material	Labor	Total

Blockouts for openings. Form area is the opening perimeter times the depth. These figures assume that nails and form oil costing $.07 per square foot, 1.2 SF of plyform and .5 board feet of lumber are used per SF of form. Blockouts usually can be used only once

	Craft@Hrs	Unit	Material	Labor	Total
1 use	F5@.250	SF	1.39	9.84	11.23

Bulkheads or pour-stops. These costs assume that nails, form oil and accessories costing $.15 per square foot, 1.1 SF of plyform and 1 board foot of lumber are used per SF of form

1 use	F5@.220	SF	1.68	8.66	10.34
3 uses	F5@.149	SF	.92	5.86	6.78
5 uses	F5@.130	SF	.71	5.11	5.82

Add for keyed wall joint, two-piece, tapered, 1.2 BF per SF

1 use	F5@.100	SF	.73	3.93	4.66
3 uses	F5@.069	SF	.40	2.71	3.11
5 uses	F5@.061	SF	.31	2.40	2.71

Curved wall forms

Smooth radius, add to straight wall cost	--	%	25.0	20.0	--
8' chord sections, add to straight wall cost	--	%	15.0	20.0	--

Gang-formed modular sized plyform wood wall forms made from the materials described in this section will cost less

To 16' high by 16' wide, deduct	--	%	-15.0	-15.0	--
To 16' high by 24' wide, deduct	--	%	-20.0	-20.0	--

Haunches or ledges. Area is the width of the ledge times the length. These costs assume nails, form oil and accessories costing $.10, 2 square feet of plyform, and 1/2 board foot of lumber are used per SF of ledge

1 use	F5@.300	SF	2.10	11.80	13.90
3 uses	F5@.210	SF	1.15	8.26	9.41
5 uses	F5@.180	SF	.88	7.08	7.96

Steel framed plywood forms Rented steel framed plywood forms can reduce forming costs on many jobs. Where rented forms are used 3 times a month, use the wall forming costs shown for 3 uses at the appropriate height but deduct 25% from the material cost and 50% from the labor manhours and labor costs. Savings will be smaller where layouts change from one use to the next, when form penetrations must be made and repaired before returning a form, where form delivery costs are high, and where non-standard form sizes are needed.

Column forms for square or rectangular columns Quantities shown in parenthesis give cross section area of a square column in square inches. When estimating a rectangular column, use the costs for the square column with closest cross section area. For scheduling purposes estimate that a crew of 5 can lay out, fabricate and erect about 550 to 600 SFCA of square or rectangular column forms in an 8-hour day.

Up to 12" x 12" (144 square inches), using nails, snap ties, oil and column clamps costing $1.09, 1.15 SF of plyform and 2 BF of lumber per SFCA

1 use	F5@.130	SFCA	3.28	5.11	8.39
3 uses	F5@.090	SFCA	1.80	3.54	5.34
5 uses	F5@.080	SFCA	1.38	3.15	4.53

Over 12" x 12" to 16" x 16" (256 square inches), using nails, snap ties, oil and column clamps costing $.89, 1.125 SF of plyform and 2.1 BF of lumber per SFCA

1 use	F5@.100	SFCA	3.12	3.93	7.05
3 uses	F5@.069	SFCA	1.71	2.71	4.42
5 uses	F5@.061	SFCA	1.31	2.40	3.71

Over 16" x 16" to 20" x 20" (400 square inches), using nails, snap ties, oil and column clamps costing $.79, 1.1 SF of plyform and 2.2 BF of lumber per SFCA

1 use	F5@.080	SFCA	3.06	3.15	6.21
3 uses	F5@.061	SFCA	1.68	2.40	4.08
5 uses	F5@.050	SFCA	1.28	1.97	3.25

Concrete 3

	Craft@Hrs	Unit	Material	Labor	Total

Column forms for square or rectangular columns, continued
Over 20" x 20" to 24" x 24" (576 square inches), using nails, snap ties, oil and column clamps costing $.64, 1.1 SF of plyform and 2.4 BF of lumber per SFCA

1 use	F5@.069	SFCA	3.03	2.71	5.74
3 uses	F5@.050	SFCA	1.67	1.97	3.64
5 uses	F5@.040	SFCA	1.27	1.57	2.84

Over 24" x 24" to 30" x 30" (900 square inches), using nails, snap ties, oil and column clamps costing $.60, 1.1 SF of plyform and 2.4 BF of lumber per SFCA

1 use	F5@.061	SFCA	2.98	2.40	5.38
3 uses	F5@.040	SFCA	1.64	1.57	3.21
5 uses	F5@.031	SFCA	1.25	1.22	2.47

Over 30" x 30" to 36" x 36" (1,296 square inches), using nails, snap ties, oil and column clamps costing $.48, 1.1 SF of plyform and 2.4 BF of lumber per SFCA

1 use	F5@.050	SFCA	2.86	1.97	4.83
3 uses	F5@.040	SFCA	1.57	1.57	3.14
5 uses	F5@.031	SFCA	1.20	1.22	2.42

Over 36" x 36" to 48" x 48" (2,304 square inches), using nails, snap ties, oil and column clamps costing $.37, 1.1 SF of plyform and 2.5 BF of lumber per SFCA

1 use	F5@.040	SFCA	2.81	1.57	4.38
3 uses	F5@.031	SFCA	1.55	1.22	2.77
5 uses	F5@.020	SFCA	1.18	.79	1.97

Column capitals for square columns Capital forms for square columns usually have four symmetrical sides. Length and width at the top of the capital are usually twice the length and width at the bottom of the capital. Height is usually the same as the width at the capital base. These costs assume capitals installed not over 12' above floor level, use of nails, form oil, shores and accessories costing $3.18, 1.5 SF of plyform and 2 BF of lumber per SF of contact area (SFCA). Complexity of these forms usually makes more than 3 uses impractical. For scheduling purposes estimate that a crew of 5 can lay out, fabricate and erect the following quantities of capital formwork in an 8-hour day:

       70 to 80 SF for 12" to 16" columns   90 to 100 SF for 20" to 24" columns
       110 to 120 SF for 30" to 36" columns   160 to 200 SF for 48" columns

Up to 12" x 12" column, 6.0 SFCA

1 use of forms	F5@4.01	Ea	18.10	158.00	176.10
3 uses of forms	F5@3.00	Ea	9.96	118.00	127.96

Over 12" x 12" to 16" x 16" column, 10.7 SFCA

1 use of forms	F5@6.00	Ea	29.80	236.00	265.80
3 uses of forms	F5@4.01	Ea	16.40	158.00	174.40

Over 16" x 16" to 20" x 20" column, 16.6 SFCA

1 use of forms	F5@8.31	Ea	44.50	327.00	371.50
3 uses of forms	F5@6.00	Ea	24.50	236.00	260.50

Over 20" x 20" to 24" x 24" column, 24.0 SFCA

1 use of forms	F5@12.0	Ea	63.00	472.00	535.00
3 uses of forms	F5@9.00	Ea	34.60	354.00	388.60

Over 24" x 24" to 30" x 30" column, 37.5 SFCA

1 use of forms	F5@16.0	Ea	96.60	629.00	725.60
3 uses of forms	F5@12.0	Ea	53.10	472.00	525.10

Over 30" x 30" to 36" x 36" column, 54.0 SFCA

1 use of forms	F5@20.0	Ea	138.00	787.00	925.00
3 uses of forms	F5@16.0	Ea	76.00	629.00	705.00

Over 36" x 36" to 48" x 48", 96.0 SFCA

1 use of forms	F5@24.0	Ea	242.00	944.00	1,186.00
3 uses of forms	F5@18.0	Ea	133.00	708.00	841.00

Concrete 3

	Craft@Hrs	Unit	Material	Labor	Total

Column forms for round columns Use the figures below to estimate the cost of round fiber tube forms that are peeled off when the concrete has cured. Costs include setting, aligning, bracing and stripping and assume that bracing and collars at top and bottom can be used 5 times. Column forms over 12'0" long will cost more per linear foot. Costs are per linear foot for light weight column form. Plastic lined forms have one vertical seam. For scheduling purposes, estimate that a crew of 5 can lay out, erect and brace the following quantities of 10' to 12' high round fiber tube forms in an 8-hour day: twenty 8" to 12" diameter columns; sixteen 14" to 20" columns and twelve 24" to 48" columns.

	Craft@Hrs	Unit	Material	Labor	Total
8" spiral type	F5@.166	LF	5.50	6.53	12.03
8" plastic lined	F5@.166	LF	7.45	6.53	13.98
10" spiral type	F5@.182	LF	5.80	7.16	12.96
10" plastic lined	F5@.177	LF	7.70	6.96	14.66
12" spiral type	F5@.207	LF	6.10	8.14	14.24
12" plastic lined	F5@.207	LF	8.00	8.14	16.14
14" spiral type	F5@.224	LF	6.50	8.81	15.31
14" plastic lined	F5@.224	LF	8.45	8.81	17.26
16" spiral type	F5@.232	LF	7.00	9.13	16.13
16" plastic lined	F5@.232	LF	10.20	9.13	19.33
20" spiral type	F5@.250	LF	8.90	9.84	18.74
20" plastic lined	F5@.250	LF	12.70	9.84	22.54
24" spiral type	F5@.275	LF	10.50	10.80	21.30
24" plastic lined	F5@.275	LF	15.40	10.80	26.20
30" spiral type	F5@.289	LF	13.70	11.40	25.10
30" plastic lined	F5@.289	LF	19.60	11.40	31.00
36" spiral type	F5@.308	LF	15.80	12.10	27.90
36" plastic lined	F5@.308	LF	22.10	12.10	34.20
48" spiral type	F5@.334	LF	39.20	13.10	52.30
48" plastic lined	F5@.334	LF	49.10	13.10	62.20

Column capitals for round columns Use these figures to estimate costs for rented prefabricated steel conical shaped capital forms. Costs assume a minimum rental period of 1 month and include an allowance for nails and form oil. Labor shown assumes the capital will be supported from the deck formwork above the capital and includes laying out and cutting the hole in the deck to receive the capital. Capital bottom diameter is sized to fit into the column tube form. The column form size (diameter) does not significantly affect the capital form rental cost. Dimensions shown below are the top diameter. Form weights are shown in parentheses. Only 3 uses of these forms are usually possible in a 30-day period. Setting forms and installing reinforcing steel takes one day, concrete placing takes another day, curing requires 7 to 10 days, removing and cleaning forms takes another day. For scheduling purposes estimate that a crew of 5 can lay out, cut openings in floor deck formwork and install an average of 8 or 9 round column capital forms in an 8-hour day.

	Craft@Hrs	Unit	Material	Labor	Total
3'6", (100 pounds) 1 use	F5@4.01	Ea	142.00	158.00	300.00
3 uses	F5@4.01	Ea	47.30	158.00	205.30
4'0" (125 pounds) 1 use	F5@4.01	Ea	151.00	158.00	309.00
3 uses	F5@4.01	Ea	50.40	158.00	208.40
4'6" (150 pounds) 1 use	F5@4.30	Ea	158.00	169.00	327.00
3 uses	F5@4.30	Ea	52.50	169.00	221.50
5'0" (175 pounds) 1 use	F5@4.51	Ea	173.00	177.00	350.00
3 uses	F5@4.51	Ea	57.80	177.00	234.80
5'6" (200 pounds) 1 use	F5@4.51	Ea	189.00	177.00	366.00
3 uses	F5@4.51	Ea	63.00	177.00	240.00
6'0" (225 pounds) 1 use	F5@4.70	Ea	205.00	185.00	390.00
3 uses	F5@4.70	Ea	68.30	185.00	253.30

Concrete 3

	Craft@Hrs	Unit	Material	Labor	Total

Beam and girder forms Using nails, snap ties and oil costing $.52, 1.3 SF of plyform and 2.1 BF of lumber per SFCA. For scheduling purposes, estimate that a crew of 5 can lay out, fabricate and erect 250 to 300 SF of beam and girder forms in an 8-hour day.

	Craft@Hrs	Unit	Material	Labor	Total
1 use	F5@.149	SFCA	3.00	5.62	8.62
3 uses	F5@.141	SFCA	1.65	5.55	7.20
5 uses	F5@.120	SFCA	1.25	4.72	5.97

Shores for beams and girders Using 4" x 4" wooden posts, nails, adjustable post clamps and accessories costing $3.00 and 1.25 BF of miscellaneous lumber per shore. These figures assume an average shore height of 12'0" with a 2'0" long horizontal 4" x 4" head and 2" x 6" diagonal brace. Estimate that a crew of 5 can lay out, fabricate and install 30 to 35 adjustable wood shores in an 8-hour day when shores have an average height of 12'0". Quantity and spacing of shores is dictated by the size of the beams and girders. Generally, allow one shore for each 6 LF of beam or girder.

	Craft@Hrs	Unit	Material	Labor	Total
1 use	F5@1.50	Ea	20.70	59.00	79.70
3 uses	F5@1.00	Ea	11.40	39.30	50.70
5 uses	F5@.748	Ea	8.70	29.40	38.10

Costs for heavy duty shoring are listed under General Requirements, Heavy Duty Shoring.

Elevated slab forms These are costs for forms per square foot of finished slab area and include shoring and stripping at heights to 12'. Bays are assumed to be 20'0" x 20'0" bays.

Waffle slab-joist pans, monolithic. Using rented forms (30-day rental period minimum). For scheduling purposes estimate that a crew of 5 can lay out, fabricate and install 500 to 600 SF of waffle slab-joist pan forms and shoring in an 8-hour day. Two uses of the forms in a 30-day period is usually the maximum.

Metal pans, 10,000 to 20,000 SF

	Craft@Hrs	Unit	Material	Labor	Total
1 use	F5@.100	SF	3.55	3.93	7.48
2 uses	F5@.066	SF	1.95	2.60	4.55

Fiberglass pans, 10,000 to 20,000 SF

	Craft@Hrs	Unit	Material	Labor	Total
1 use	F5@.077	SF	3.55	3.03	6.58
2 uses	F5@.061	SF	1.95	2.40	4.35

Flat slab with beams, monolithic. Using nails, braces, stiffeners and oil costing $.65, 1.2 SF of plyform and 3.2 BF of lumber per SFCA for one-way beam systems. Two-way beam systems will increase the forming material by approximately 40%. For scheduling purposes, estimate that a crew of 5 can lay out, fabricate and erect 350 to 400 SF of one-way beam systems and 250 to 300 SF of two-way beam systems in an 8-hour day, including shoring.

	Craft@Hrs	Unit	Material	Labor	Total
One-way beam	F5@.110	SFCA	3.60	4.33	7.93
One-way beam	F5@.080	SFCA	2.00	3.15	5.15
Two-way beam	F5@.152	SFCA	5.00	5.98	10.98
Two-way beam	F5@.100	SFCA	2.75	3.93	6.68

Structural flat slab. Using nails, braces, stiffeners and oil costing $.48, 1.2 SF of plyform and 2.25 BF of lumber per SFCA. For scheduling purposes estimate that a crew of 5 can lay out, fabricate and erect 500 to 600 SFCA of structural flat slab forms in an 8-hour day, including shoring.

	Craft@Hrs	Unit	Material	Labor	Total
Edge forms	F5@.072	SFCA	1.80	2.83	4.63
Slab forms	F5@.069	SFCA	2.85	2.71	5.56
Slab forms	F5@.048	SFCA	1.60	1.89	3.49
Slab forms	F5@.043	SFCA	1.20	1.69	2.89

Additional forming costs for elevated slabs, if required. Based on single use of plyform and lumber. For scheduling purposes estimate that a 5 man crew can install 500 SFCA per 8-hour day.

	Craft@Hrs	Unit	Material	Labor	Total
Control joints	F5@.031	LF	1.95	1.22	3.17
Curbs & pads	F5@.083	SFCA	1.80	3.27	5.07
Depressions	F5@.083	SFCA	.86	3.27	4.13
Keyed joints	F5@.031	LF	.82	1.22	2.04

Concrete 3

	Craft@Hrs	Unit	Material	Labor	Total

Miscellaneous formwork Based on single use of plyform and lumber. For scheduling purposes, estimate that a crew of 5 can lay out, fabricate and erect 225 to 250 SFCA of these types of forms in an 8-hour day.

	Craft@Hrs	Unit	Material	Labor	Total
Flat soffits & landings	F5@.166	SFCA	4.40	6.53	10.93
Sloping soffits	F5@.166	SFCA	4.60	6.53	11.13
Stair risers (steps)	F5@.120	SFCA	1.70	4.72	6.42

Driveway, curb and sidewalk forms See Site Work section.

Reinforcing for Cast-in-Place Concrete. These costs include the subcontractor's overhead and profit. Reinforcing bars, ASTM A615 Grade A60 deformed bars, in place, including tieing, dowels, supports, accessories and 10% for waste.

	Craft@Hrs	Unit	Material	Labor	Total
Beams and girders, #3 and #4	RB@.007	Lb	.31	.33	.64
Beams and girders, #5 and #6	RB@.006	Lb	.29	.28	.57
Bond beams, #3 and #4	RB@.007	Lb	.31	.33	.64
Bond beams, #5 and #6	RB@.006	Lb	.29	.28	.57
Columns, #3 and #4	RB@.007	Lb	.31	.33	.64
Columns, #5 and #6	RB@.006	Lb	.29	.28	.57
Foundations, #3 and #4	RB@.006	Lb	.31	.28	.59
Foundations, #5 and #6	RB@.005	Lb	.29	.23	.52
Slabs, #3 and #4	RB@.006	Lb	.31	.28	.59
Slabs, #5 and #6	RB@.005	Lb	.29	.23	.52
Stairs, #3 and #4	RB@.007	Lb	.31	.33	.64
Stairs, #5 and #6	RB@.006	Lb	.29	.28	.57
Walls, #3 and #4	RB@.006	Lb	.31	.28	.59
Walls, #5 and #6	RB@.005	Lb	.29	.23	.52
Any of above					
#7 and #8 bars	RB@.004	Lb	.29	.19	.48
#9 bars	RB@.004	Lb	.28	.19	.47
#10 bars	RB@.004	Lb	.27	.19	.46
Deduct for lots of 10 tons or more	--	Lb	-.01	--	--
Add for galvanized bars	RB@.001	Lb	.52	.05	.57
Add for epoxy coated bars	RB@.002	Lb	.62	.09	.71

Reinforcing bar weld splicing

	Craft@Hrs	Unit	Material	Labor	Total
Number 8 bars, 1"	RB@.356	Ea	1.50	16.60	18.10
Number 9 bars, 1-1/8"	RB@.401	Ea	1.50	18.70	20.20
Number 10 bars, 1-1/4"	RB@.445	Ea	1.50	20.80	22.30

Reinforcing bar clip splicing, sleeve and wedge, with hand-held hydraulic ram

	Craft@Hrs	Unit	Material	Labor	Total
Number 3 bars, 3/8"	RB@.194	Ea	2.63	9.07	11.70
Number 4 bars, 1/2"	RB@.212	Ea	3.36	9.91	13.27
Number 5 bars, 5/8"	RB@.236	Ea	7.88	11.00	18.88
Number 6 bars, 3/4"	RB@.258	Ea	14.70	12.10	26.80
Gun and wedge driver rental, per day	--	Ea	47.30	--	--

Spiral caisson and round column reinforcing, 3/8" diameter hot rolled steel spirals with main vertical bars as shown, shop fabricated delivered ready to install, with typical engineering and drawings, per vertical foot

	Craft@Hrs	Unit	Material	Labor	Total
16" diameter with 6 #6 bars, 16.8 lbs per LF	RB@.041	LF	15.20	1.92	17.12
24" diameter with 6 #6 bars, 28.5 lbs per LF	RB@.085	LF	26.00	3.97	29.97
36" diameter with 8 #10 bars, 51.2 lbs per LF	RB@.216	LF	47.20	10.10	57.30

Welded wire mesh, electric weld, including 10% waste and overlap.

	Craft@Hrs	Unit	Material	Labor	Total
2" x 2" W.9 x W.9 (#12 x #12), slabs	RB@.004	SF	.39	.19	.58
4" x 4" W1.4 x W1.4 (#10 x #10), slabs	RB@.003	SF	.17	.14	.31
4" x 4" W2.0 x W2.0 (#8 x #8), slabs	RB@.004	SF	.21	.19	.40
4" x 4" W2.9 x W2.9 (#6 x #6), slabs	RB@.005	SF	.30	.23	.53

Concrete 3

	Craft@Hrs	Unit	Material	Labor	Total
Reinforcing for Cast-in-Place Concrete, Welded wire mesh, continued					
4" x 4" W4.0 x W4.0 (#4 x #4), slabs	RB@.006	SF	.33	.28	.61
6" x 6" W1.4 x W1.4 (#10 x #10), slabs	RB@.004	SF	.12	.19	.31
6" x 6" W2.0 x W2.0 (#8 x #8), slabs	RB@.004	SF	.17	.19	.36
6" x 6" W2.9 x W2.9 (#6 x #6), slabs	RB@.004	SF	.24	.19	.43
6" x 6" W4.0 x W4.0 (#4 x #4), slabs	RB@.005	SF	.27	.23	.50
2" x 2" W.9 x W.9 (#12 x #12), beams and columns	RB@.020	SF	.39	.94	1.33
Add for lengthwise cut, LF of cut	RB@.002	LF	--	.09	--
Add for galvanized mesh	--	%	100.0	--	--
Add for epoxy coated mesh	--	SF	1.00	--	--

Ready-Mix Concrete. Ready-mix delivered by truck. Typical prices for most cities. Includes delivery up to 20 miles for 9 CY or more, 3" to 4" slump. Material cost only, no placing or pumping included. All material costs which include concrete are based on these figures. Costs shown are before markup.

	Craft@Hrs	Unit	Material	Labor	Total
5.0 sack mix, 2,000 PSI	--	CY	46.90	--	--
6.0 sack mix, 3,000 PSI	--	CY	50.60	--	--
6.6 sack mix, 3,500 PSI	--	CY	54.70	--	--
7.1 sack mix, 4,000 PSI	--	CY	56.70	--	--
8.5 sack mix, 5,000 PSI	--	CY	67.50	--	--
Extra costs for ready-mix concrete					
Add for less than 9 CY per load	--	CY	10.00	--	--
Add for delivery over 20 miles	--	Mile	.60	--	--
Add for standby charge in excess of 5 minutes per CY delivered, per minute of extra time	--	Ea	1.00	--	--
Add for super plasticized mix, 7"-8" slump	--	%	8.0	--	--
Add for high early strength concrete					
5 sack mix	--	CY	7.00	--	--
6 sack mix	--	CY	8.80	--	--
Add for lightweight aggregate, typical	--	CY	29.00	--	--
Add for pump mix (pea-gravel aggregate)	--	CY	7.00	--	--
Add for granite aggregate, typical	--	CY	2.90	--	--
Add for white cement (architectural)	--	CY	35.00	--	--
Add for 1% calcium chloride	--	CY	1.10	--	--
Add for chemical compensated shrinkage	--	CY	11.50	--	--
Add for colored concrete, typical prices					
Adobe	--	CY	10.00	--	--
Black	--	CY	19.00	--	--
Blended red	--	CY	9.50	--	--
Brown	--	CY	13.50	--	--
Green	--	CY	15.50	--	--
Yellow	--	CY	12.00	--	--

Concrete Placing. No forms, finishing or reinforcing included. These costs include the subcontractor's overhead and profit. Material cost is based on 3,000 PSI concrete at $50.60 per CY before markup. Pump mix is $57.60 before markup.

	Craft@Hrs	Unit	Material	Labor	Equipment	Total
Columns						
By crane	H3@.873	CY	58.20	33.00	21.20	112.40
By pump	M2@.738	CY	66.20	27.30	13.90	107.40
Slabs-on-grade						
Direct from chute	CL@.430	CY	58.20	14.70	.89	73.79
By crane	H3@.594	CY	58.20	22.40	9.56	90.16
By pump	M2@.456	CY	66.20	16.90	6.06	89.16
With buggy	CL@.543	CY	58.20	18.50	2.73	79.43

Concrete 3

	Craft@Hrs	Unit	Material	Labor	Equipment	Total
Elevated slabs						
By crane	H3@.868	CY	58.20	32.80	12.60	103.60
By pump	M2@.690	CY	66.20	25.50	7.77	99.47
Footings, pile caps, foundations						
Direct from chute	CL@.564	CY	58.20	19.20	.97	78.37
By pump	M2@.546	CY	66.20	20.20	9.02	95.42
With buggy	CL@.701	CY	58.20	23.90	3.03	85.13
Beams and girders						
By pump	M2@.608	CY	66.20	22.50	21.80	110.50
By crane	H3@.807	CY	58.20	30.50	17.10	105.80
Stairs						
Direct from chute	CL@.814	CY	58.20	27.80	1.27	87.27
By crane	H3@1.05	CY	58.20	39.60	13.50	111.30
By pump	M2@.817	CY	66.20	30.20	8.61	105.01
With buggy	CL@.948	CY	58.20	32.30	3.88	94.38
Walls and grade beams to 4' high						
Direct from chute	CL@.479	CY	58.20	16.30	.76	75.26
By pump	M2@.476	CY	66.20	17.60	6.52	90.32
With buggy	CL@.582	CY	58.20	19.80	2.54	80.54
Walls over 4' to 8' high						
By crane	H3@.821	CY	58.20	31.00	15.00	104.20
By pump	M2@.608	CY	66.20	22.50	10.01	98.71
Walls over 8' to 16' high						
By crane	H3@.895	CY	58.20	33.80	18.10	110.10
By pump	M2@.667	CY	66.20	24.70	12.60	103.50
Walls over 16' high						
By crane	H3@.986	CY	58.20	37.20	20.90	116.30
By pump	M2@.704	CY	66.20	26.10	14.10	106.40

Concrete Slabs-on-Grade. Complete costs including fine grading, edge forms, placing 3,000 PSI concrete, wire mesh, finishing and curing. These costs include the subcontractor's overhead and profit.

	Craft@Hrs	Unit	Material	Labor	Equipment	Total
Cost breakdown for 4" x 4' x 6' equipment pad						
Hand trimming and shaping, 24 SF	CL@.512	LS	--	17.50	--	17.50
Lay out, set and strip edge forms,						
20 LF, 1 use	C8@1.30	LS	20.00	50.00	.88	70.88
Place W2.9 x W2.9 x 6" x 6" mesh, 24 SF	RB@.097	LS	5.76	4.53	--	10.29
Place ready-mix concrete, from chute						
.3 CY at $58.20 per CY (see note below)	CL@.127	LS	17.50	4.33	.24	22.07
Finish concrete, broom finish	CM@.591	LS	--	23.70	2.12	25.82
Cure concrete, curing paper	CL@.124	LS	1.15	4.23	--	5.38
Total job cost for 24 SF pad	--@2.75	LS	**44.41**	**104.29**	**3.24**	**151.94**
Cost per SF for 24 SF job	--@.115	SF	**1.85**	**4.34**	**.14**	**6.33**
Cost per CY of concrete, .3 CY job	--@9.17	CY	**148.03**	**347.63**	**10.80**	**506.46**
Cost breakdown for 8" x 10' x 10' equipment pad						
Hand trimming and shaping, 100 SF	CL@2.79	LS	--	95.10	--	95.10
Lay out, set and strip edge forms,						
40 LF, 5 uses	C8@3.28	LS	25.20	126.00	3.79	154.99
Place W2.9 x W2.9 x 6" x 6" mesh,						
100 SF	RB@.405	LS	24.00	18.90	--	42.90
Place ready-mix concrete, from chute						
2.5 CY at $58.20 per CY (see note below)	CL@1.08	LS	145.50	36.80	2.25	184.55
Finish concrete, broom finish	CM@2.11	LS	--	84.70	9.12	93.82

Concrete 3

	Craft@Hrs	Unit	Material	Labor	Equipment	Total
Concrete Slabs-on-Grade, continued						
Cure concrete, curing paper	CL@.515	LS	4.80	17.60	--	22.40
Total job cost for 100 SF pad	--@10.2	LS	199.50	379.10	15.16	593.76
Cost per SF for 100 SF job	--@.102	SF	2.00	3.79	.15	5.94
Cost per CY of concrete, 2.5 CY job	--@4.08	CY	79.80	151.64	6.06	237.50

Note: Ready-mix concrete unit price at $ 58.20 per CY, used both places above, assumes a minimum of 9 cubic yards (CY) will be delivered (per load) with the excess used elsewhere on the same job.

	Craft@Hrs	Unit	Material	Labor	Equipment	Total
Add for each CY less than 9 delivered	--	CY	11.50	--	--	--

Floor slabs, slab-on-grade reinforced concrete including typical excavation, gravel fill, forms, vapor barrier, wire mesh, 3000 PSI concrete, finishing and curing. Based on 100' x 75' slab (7,500 SF) See detailed cost breakdown below

	Craft@Hrs	Unit	Material	Labor	Equipment	Total
4" slab	--@.024	SF	1.38	.94	.14	2.46
5" slab	--@.025	SF	1.56	.99	.14	2.69
6" slab	--@.026	SF	1.74	1.04	.14	2.92
Cost breakdown for 4" x 100' x 75' slab as described above:						
Grade sandy loam site using a D-4 tractor (at $28 per hour), 140 CY, balanced job, (no import or export)	S1@2.66	LS	--	103.00	37.20	140.20
Buy and spread 6" crushed rock base using a D-4 tractor, 140 CY	S1@15.8	LS	2,100.00	609.00	221.00	2,930.00
Lay out, set and strip edge forms, 350 LF, 5 uses	C8@19.6	LS	147.00	754.00	27.30	928.30
Place .006" polyethylene vapor barrier 7,500 SF	CL@8.84	LS	375.00	301.00	--	676.00
Place W2.9 x W2.9 x 6" x 6" mesh, 7,500 SF	RB@30.5	LS	1,800.00	1,430.00	--	3,230.00
Place and remove 2" x 2" keyway, 200 LF, 1 use	C8@4.61	LS	146.00	177.00	--	323.00
Place 4" concrete, 93 CY, from chute	CL@39.9	LS	5,413.00	1,360.00	80.50	6,853.50
Float finish 7,500 SF	CM@52.4	LS	--	2,100.00	710.00	2,810.00
Cure with slab curing paper	CM@5.58	LS	360.00	224.00	--	584.00
Total job cost	--@180.	LS	10,341.00	7,058.00	1,076.00	18,475.00
Cost per SF for 7,500 SF job	--@.024	SF	1.38	.94	.14	2.46
Cost per CY for 93 CY job	--@1.94	CY	111.19	75.89	11.57	198.65
Cost per each additional 1" of concrete	--@.001	SF	.18	.05	--	.23

Cast-in-Place Concrete. Typical costs including forms, concrete, reinforcing, finishing and curing. These costs include the subcontractor's overhead and profit. Use these figures for preliminary estimates on jobs that require a minimum of 1,000 CY of concrete.

	Craft@Hrs	Unit	Material	Labor	Total
Foundations					
Institutional or office buildings	--	CY	--	--	285.00
Heavy engineered structures	--	CY	--	-	310.00
Structural concrete walls					
Single story, 8" wall	--	CY	--	-	380.00
Single story, 10" wall	--	CY	--	-	435.00
Multi-story, 8" wall	--	CY	--	-	530.00
Multi-story, 10"	--	CY	--	-	560.00
Slip formed 8" wall	--	CY	--	-	465.00
Structural slabs, including shoring					
6", 1 way beams, 100 pounds reinforcing	--	CY	--	-	470.00
6", 2 way beams, 225 pounds reinforcing	--	CY	--	-	455.00
8", flat, with double mat reinforcing over steel frame by others, 100 pounds reinforcing	--	CY	--	-	415.00
12" flat, 250 pounds reinforcing bars per CY	--	CY	--	-	375.00

Concrete 3

	Craft@Hrs	Unit	Material	Labor	Total
7", post-tensioned, to 1 pound tempered bars	--	CY	--	-	460.00
6" to 8" including beam jacketing	--	CY	--	-	440.00
6" to 8" on permanent metal form	--	CY	--	-	395.00
8" concurrent with slipform construction	--	CY	--	-	425.00
Lift slab construction, 6" to 8"	--	SF	--	-	9.00
Pan and joist slabs 3" thick, 30" pan, including shoring					
6" x 16" joist	--	CY	--	--	470.00
6" x 12" joist	--	CY	--	--	410.00
6" x 10" joist	--	CY	--	--	360.00
Pan and joist slabs 4-1/2" thick, 20" pan, including shoring					
6" x 20" joist	--	CY	--	--	365.00
6" x 16" joist	--	CY	--	--	360.00
6" x 12" joist	--	CY	--	--	345.00
Parapet and fascia, 6" thick	--	CY	--	--	555.00
Loading docks, based on 8" walls and 8" slab	--	CY	--	--	245.00
Beams and girders, not slab integrated					
12" x 24", typical	--	CY	--	--	760.00
18" x 24", typical	--	CY	--	--	620.00
Columns, average reinforcing					
Square, wood formed, with chamfer					
12"	--	CY	--	--	1,025.00
16"	--	CY	--	--	940.00
18"	--	CY	--	--	845.00
20"	--	CY	--	--	760.00
24"	--	CY	--	--	650.00
Round, fiber tube formed, 12" to 18", light reinforcing	--	CY	--	--	675.00

Cast-in-Place Concrete Steps on Grade. Costs include layout, fabrication and placing forms, setting and tieing steel reinforcing, installation of steel nosing, placing 2,000 PSI concrete directly from the chute of a ready-mix truck, finishing, stripping forms and curing. Costs assume excavation, back filling and compaction have been completed by others. Crew is a carpenter, laborer and finisher. Formwork is based on using number 2 and better lumber at $500 per MBF before markup or waste allowance, and concrete at $46.90 per CY before markup. Costs shown below include the subcontractor's overhead and profit. For scheduling purposes, estimate that a crew of 3 can set forms and place steel for and pour 4 to 5 CY of concrete cast-in-place steps in an 8-hour day.

Costs shown are for concrete steps with a 6" riser height and 12" tread depth supported by a 6" thick monolithic slab. The lines below explain how to use these unit prices to estimate stairways of various widths and heights.

	Craft@Hrs	Unit	Material	Labor	Total
Fabricate and erect forms, 2 uses of forms, includes stakes, braces, form oil and nails					
Side forms, 1.7 BF per LF of tread	P9@.080	LF	.74	3.12	3.86
Riser forms, 1.5 BF per LF of tread	P9@.040	LF	.65	1.56	2.21
Rebars, #3, A60, 1 pound per LF of tread	P9@.009	LF	.31	.35	.66
Embedded steel step nosing, 2-1/2" x 2-1/2" x 1/4" black mild steel angle iron, 4.5 pounds per LF of tread	P9@.004	LF	1.80	.16	1.96
Concrete, .03 CY per LF of tread	P9@.030	LF	1.62	1.17	2.79
Total cost per LF of risers	P9@.163	**LF**	**5.12**	**6.36**	**11.48**
Total cost per CY of concrete	P9@5.47	**CY**	**170.67**	**212.00**	**382.67**

To quickly estimate the cost of cast-in-place concrete steps on grade, do this:
Multiply the width of the steps (in feet) times their height in feet and multiply the result times 2.
This will give the total liner feet (LF) of risers in the steps. Multiply the total LF by the costs per LF.

Concrete 3

	Craft@Hrs	Unit	Material	Labor	Total

Sample estimates for cast-in-place concrete steps on grade using the procedure and costs on preceding page:
  2'6" wide by 3' high   (6 steps) would be: 2.5' x 3' x 2 = 15 LF x $11.48 = $172
  3' wide by 3' high   (6 steps) would be: 3' x 3' x 2 = 18 LF x $11.48 = $207
  4' wide by 4' high   (8 steps) would be: 4' x 4' x 2 = 32 LF x $11.48 = $367
  7' wide by 4'6" high   (9 steps) would be: 7' x 4.5' x 2 = 63 LF x $11.48 = $723

Tip for using the above with the Estimate Writer computer estimating program:
  Calculate the LF quantity as shown, select the line "Total Cost per LF of risers" and input the quantity.

Concrete Estimates by Building Type. Typical costs including forms, concrete, reinforcing, finishing, curing and the subcontractor's overhead and profit. Use these figures for preliminary estimates on jobs that require a minimum of 1,000 CY of concrete.

	Craft@Hrs	Unit	Material	Labor	Total
Highway structures, overpasses	--	CY	--	--	425.00
Hospitals, concrete walls and slabs on steel frame	--	CY	--	--	495.00
Hospitals, multi-story, concrete frame and walls	--	CY	--	--	440.00
Office buildings, concrete frame, walls and slabs	--	CY	--	--	410.00
Military structures, all concrete	--	CY	--	--	445.00
Parking structures	--	CY	--	--	420.00
Stadiums, auditoriums, typical	--	CY	--	--	450.00
College and university buildings	--	CY	--	--	515.00

Concrete Accessories

	Craft@Hrs	Unit	Material	Labor	Total
Bentonite granules, 50 pound bag	--	Ea	14.50	--	--
Bond breaker and form release at $6.00 per gallon, Thompson's CBA, spray on					
Lumber & plywood form release, 200 SF/gal	CL@.006	SF	.03	.17	.20
Metal form release, 600 SF per gallon	CL@.003	SF	.01	.10	.11
Casting bed form breaker, 300 SF per gallon	CL@.005	SF	.02	.10	.12
Cure coat, 300 SF per gallon	CL@.005	SF	.02	.10	.12
Chamfer strip, reusable polyethylene					
1/2" radius	CC@.015	LF	.27	.64	.91
3/4" radius	CC@.015	LF	.43	.64	1.07
1" radius	CC@.015	CLF	.49	.64	1.13
Column clamps for plywood forms, per set, by column size, Econ-o-clamp					
8" x 8" to 24" x 24", purchase	--	Set	75.00	--	--
8" x 8" to 30" x 30", purchase	--	Set	86.00	--	--
10" x 10" to 36" x 36", purchase	--	Set	98.00	--	--
Monthly rental cost as % of purchase price	--	%	5.0	--	--
Column clamps for lumber forms, per set, by column size, Symons					
10" x 10" to 28" x 28", purchase	--	Set	61.00	--	--
16" x 16" to 40" x 40", purchase	--	Set	75.00	--	--
21" x 21" to 50" x 50", purchase	--	Set	98.00	--	--
10" x 10" to 28" x 28", rent per month, per set	--	Mo	3.50	--	--
16" x 16" to 40" x 40", rent per month, per set	--	Mo	4.00	--	--
21" x 21" to 50" x 50", rent per month, per set	--	Mo	6.90	--	--
Concrete sleeves, circular polyethylene liner, any slab thickness					
1-1/2" to 3" diameter	--	Ea	1.50	--	--
4" to 6" diameter	--	Ea	3.00	--	--
Embedded iron, installation only	C8@.045	Lb	--	1.73	--
Expansion joints, cane fiber, asphalt impregnated, premoulded					
1/2" x 4", slabs	C8@.028	LF	.28	1.08	1.36
1/2" x 6", in walls	C8@.035	LF	.39	1.35	1.74
1/2" x 8", in walls	C8@.052	LF	.50	2.00	2.50
Footing ties, 6" to 12" wide form	--	Ea	.36	--	--
Form spreaders, 4" to 10"	--	Ea	.35	--	--

Concrete 3

	Craft@Hrs	Unit	Material	Labor	Total
Form oil, 800 to 1,200 SF per gallon	--	Gal	5.00	--	--
Grout for machine bases, mixed and placed (add for edge forms if required, from below)					
Five Star 3-part epoxy (at $135 per CF)					
per SF for each 1" of thickness	C8@.170	SF	11.10	6.54	17.64
NBEC non-shrink (at $30 per CF)					
per SF for each 1" of thickness	C8@.170	SF	2.50	6.54	9.04
Embeco 636, metallic aggregate (at $65 per CF)					
per SF for each 1" of thickness	C8@.163	SF	5.40	6.27	11.67
Forms for grout, based on using 2" x 4" lumber, 1 use	C8@.195	LF	1.00	7.50	8.50
Inserts, Unistrut, average	C8@.068	LF	6.50	2.62	9.12
She bolt form clamps, with cathead and bolt, purchase					
17" rod	--	Ea	6.00	--	--
21" rod	--	Ea	6.40	--	--
24" rod	--	Ea	6.85	--	--
Monthly rental as % of purchase price	--	%	15.0	--	--
Snap ties, average, 4,000 pound, 6" to 10"	--	Ea	.40	--	--
Water stop, 3/8"					
Rubber, 6"	C8@.059	LF	4.40	2.27	6.67
Rubber, 9"	C8@.059	LF	7.00	2.27	9.27
PVC, 9"	C8@.059	LF	4.00	2.27	6.27
Waler brackets, 2" x 4" or 2" x 6"	--	Ea	2.25	--	--
Waler jacks (scaffold support on form), purchase	--	Ea	35.00	--	--
Rental per month, each	--	Mo	5.00	--	--

Concrete Slab Finishes. These costs include the subcontractor's overhead and profit.

	Craft@Hrs	Unit	Material	Labor	Total
Float finish	CM@.007	SF	.07	.28	.35
Trowel finishing					
Steel, machine work	CM@.010	SF	.08	.40	.48
Steel, hand work	CM@.013	SF	--	.52	--
Broom finish	CM@.008	SF	.06	.32	.38
Scoring concrete surface, hand work	CM@.005	LF	--	.20	--
Exposed aggregate (washed, including finishing), no disposal of slurry	CM@.015	SF	.10	.60	.70
Metallic color and hardener, dry shake application					
Natural gray	--	Lb	.32	--	--
Charcoal, russet, brown red, walnut	--	Lb	.46	--	--
Green	--	Lb	.53	--	--
Light duty floors (40 pounds per 100 SF)	CM@.008	SF	--	.32	--
Pastel shades (60 pounds per 100 SF)	CM@.012	SF	--	.48	--
Heavy duty (90 pounds per 100 SF)	CM@.015	SF	--	.60	--
Add for colored wax, 800 SF and $25 per gal	CM@.260	CSF	3.19	10.40	13.59
Sidewalk grits, abrasive, 25 pounds per 100 SF					
Aluminum oxide, $65 per 100 pound bag	CM@.008	SF	.16	.32	.48
Silicon carbide, $125 per 100 pound bag	CM@.008	SF	.31	.32	.63
Metallic surface hardener, dry shake, Masterplate 200, at $.60 per pound					
Moderate duty, 1 pound per SF	CM@.008	SF	.60	.32	.92
Heavy duty, 1.5 pounds per SF	CM@.009	SF	.90	.36	1.26
Concentrated traffic, 2 pounds per SF	CM@.010	SF	1.20	.40	1.60
Add for light-reflective hardener	--	%	15.0	--	--
Non-metallic color and hardener at $.60 per pound, troweled on, 2 layer applications; red, gray or black					
60 pounds per 100 SF, non-traffic areas	CM@.020	SF	.36	.80	1.16
100 pounds per 100 SF, traffic area	CM@.024	SF	.60	.96	1.56
Add for other standard colors	--	%	50.0	--	--

Concrete 3

	Craft@Hrs	Unit	Material	Labor	Total

Concrete Slab Finishes, continued
Liquid curing and sealing compound, sprayed-on,

	Craft@Hrs	Unit	Material	Labor	Total
Mastercure, 400 SF and $20 per gallon	CL@.004	SF	.05	.14	.19
Sweep, scrub and wash down	CL@.006	SF	.02	.20	.22
Finish treads and risers					
No abrasives, no plastering, per SF of tread	CM@.038	SF	--	1.53	--
With abrasives, plastered, per SF of tread	CM@.058	SF	.60	2.33	2.93
Wax	CL@.002	SF	.04	.07	.11

Concrete Wall Finishes. These costs include the subcontractor's overhead and profit.

	Craft@Hrs	Unit	Material	Labor	Total
Cut back ties and patch	CM@.010	SF	.11	.40	.51
Remove fins	CM@.007	LF	.04	.28	.32
Grind smooth	CM@.020	SF	.08	.80	.88
Sack, simple	CM@.012	SF	.05	.48	.53
Bush hammer light, green concrete	CM@.020	SF	.12	.80	.92
Bush hammer standard, cured concrete	CM@.029	SF	.12	1.16	1.28
Bush hammer, heavy, cured concrete	CM@.077	SF	.12	3.09	3.21
Needle gun treatment, large areas	CM@.058	SF	.12	2.33	2.45
Wire brush, green	CM@.014	SF	.04	.56	.60
Wash with acid and rinse	CM@.004	SF	.26	.16	.42
Break fins, patch voids, Carborundum rub	CM@.032	SF	.06	1.28	1.34
Break fins, patch voids, burlap grout rub	CM@.024	SF	.06	.96	1.02
Sandblasting, no scaffolding included, flat vertical surfaces					
Light	--	SF	--	--	1.00
Medium, to expose aggregate	--	SF	--	--	1.25
Heavy, with dust control requirements	--	SF	--	--	2.00
Note: Minimum charge for sandblasting	--	LS	--	--	300.00

Miscellaneous Concrete Finishes. These costs include the subcontractor's overhead and profit.

	Craft@Hrs	Unit	Material	Labor	Total
Monolithic natural aggregate topping					
1/16"	P9@.006	SF	.07	.23	.30
3/16"	P9@.018	SF	.21	.70	.91
1/2"	P9@.021	SF	.16	.82	.98
Integral colors, typical, figure 8 pounds per sack of cement	--	Lb	--	--	1.75
Mono rock, 3/8" wear course only	--	SF	--	--	1.00
Kalman 3/4", wear course only	--	SF	--	--	1.50
Acid etching, 5% muriatic acid	--	SF	--	--	.30
Felton sand, for use with white cement	--	CY	30.00	--	--

Specially Placed Concrete. These costs include the subcontractor's overhead and profit.

	Craft@Hrs	Unit	Material	Labor	Total
Gunite, no forms and no reinforcing included					
Flat plane, vertical areas, 1" thick	--	SF	--	--	2.25
Curved arch, barrel, per 1" thickness	--	SF	--	--	3.00
Pool or retaining walls, per 1" thickness	--	SF	--	--	2.30
Add for architectural finish	--	SF	--	--	1.00
Wire mesh for gunite 2" x 2", W.9 x W.9	--	SF	--	--	1.40
Pressure grouting, large quantity, 50/50 mix	--	CY	--	--	480.00

Precast Concrete. Panel costs are based on 3,000 PSI natural gray concrete with a smooth or board finish on one side. Placing (lifting) costs include handling, hoisting, alignment, bracing, permanent connections and the subcontractor's overhead and profit. Erection costs vary with the type of crane. Heavier lifts over a longer radius (reach) require larger cranes. Equipment costs include the crane and a

Concrete 3

	Craft@Hrs	Unit	Material	Labor	Equipment	Total

300 amp gas powered welding machine. Add the cost of grouting or caulking joints (about $2 per LF). Costs will be higher on small jobs (less than 35,000 SF).

Precast wall panels Costs include delivery to 40 miles, typical reinforcing steel and embedded items but no lifting. See placing costs below.

	Craft@Hrs	Unit	Material	Labor	Equipment	Total
4" thick wall panels (52 lbs per SF)	--	SF	6.55	--	--	--
5" thick wall panels (65 lbs per SF)	--	SF	7.15	--	--	--
6" thick wall panels (78 lbs per SF)	--	SF	7.75	--	--	--
8" thick wall panels (105 lbs per SF)	--	SF	8.35	--	--	--
Add for insulated sandwich wall panels 2-1/2" of cladding and 2" of fiberboard	--	SF	2.50	--	--	--
Add for high strength concrete						
3,500 lb concrete mix	--	%	2.0	--	--	--
4,000 lb concrete mix	--	%	3.0	--	--	--
4,500 lb concrete mix	--	%	4.0	--	--	--
5,000 lb concrete mix	--	%	6.0	--	--	--
Add for white facing or integral white mix	--	%	20.0	--	--	--
Add for broken rib or fluted surface design	--	%	20.0	--	--	--
Add for exposed aggregate finish	--	%	17.0	--	--	--
Add for sandblasted surface finish	--	%	12.0	--	--	--

Placing precast structural wall panels Figures in parentheses show the panels placed per day with a 100 ton crawler-mounted crane and welding machine at $720 per day.

	Craft@Hrs	Unit	Material	Labor	Equipment	Total
1 ton panels to 180' reach (23 per day)	H4@2.79	Ea	--	135.00	31.40	166.40
1 to 3 tons, to 121' reach (21 per day)	H4@3.05	Ea	--	148.00	34.30	182.30
3 to 5 tons, to 89' reach (18 per day)	H4@3.54	Ea	--	171.00	39.80	210.80
5 to 7 tons, to 75' reach (17 per day)	H4@3.76	Ea	--	182.00	42.30	224.30
7 to 9 tons, to 65' reach (16 per day)	H4@4.00	Ea	--	194.00	45.00	239.00
9 to 11 tons, to 55' reach (15 per day)	H4@4.28	Ea	--	207.00	48.20	255.20
11 to 13 tons, to 50' reach (14 per day)	H4@4.56	Ea	--	221.00	51.30	272.30
13 to 15 tons, to 43' reach (13 per day)	H4@4.93	Ea	--	239.00	55.50	294.50
15 to 20 tons, to 36' reach (11 per day)	H4@5.81	Ea	--	281.00	65.40	346.40
20 to 25 tons to 31' reach (10 per day)	H4@6.40	Ea	--	310.00	72.00	382.00
For placing panels with cladding, add	--	%	--	5.0	5.0	--
For placing sandwich wall panels, add	--	%	--	10.0	10.0	--
If a 50 ton crane can be used, deduct	--	%	--	--	-30.0	--

Placing precast partition wall panels Figures in parentheses show the panels placed per day with a 50 ton truck-mounted crane and welding machine at $496 per day.

	Craft@Hrs	Unit	Material	Labor	Equipment	Total
1 ton panels to 100' reach (23 per day)	H4@2.79	Ea	--	135.00	21.60	156.60
1 to 2 tons, to 82' reach (21 per day)	H4@3.05	Ea	--	148.00	23.60	171.60
2 to 3 tons, to 65' reach (18 per day)	H4@3.54	Ea	--	171.00	27.40	198.40
3 to 4 tons, to 57' reach (16 per day)	H4@4.00	Ea	--	194.00	31.00	225.00
4 to 5 tons, to 50' reach (15 per day)	H4@4.28	Ea	--	207.00	33.20	240.20
5 to 6 tons, to 44' reach (14 per day)	H4@4.56	Ea	--	221.00	35.30	256.30
6 to 7 tons, to 40' reach (13 per day)	H4@4.93	Ea	--	239.00	38.20	277.20
7 to 8 tons, to 36' reach (11 per day)	H4@5.81	Ea	--	281.00	45.00	326.00
8 to 10 tons, to 31' reach (10 per day)	H4@6.40	Ea	--	310.00	49.60	359.60
10 to 12 tons to 26' reach (9 per day)	H4@7.11	Ea	--	344.00	55.10	399.10
12 to 14 tons, to 25' reach (8 per day)	H4@8.02	Ea	--	388.00	62.20	450.20
14 to 16 tons, to 21' reach (7 per day)	H4@9.14	Ea	--	443.00	70.80	513.80
If a 35 ton crane can be used, deduct	--	%	--	--	-10.0	--

Concrete 3

	Craft@Hrs	Unit	Material	Labor	Equipment	Total

Precast flat floor slab or floor plank Costs include delivery to 40 miles, 3,000 PSI concrete, typical reinforcing steel and embedded items. See placing costs below.

	Craft@Hrs	Unit	Material	Labor	Equipment	Total
4" thick floor panels (52 lbs per SF)	--	SF	4.60	--	--	--
5" thick floor panels (65 lbs per SF)	--	SF	5.55	--	--	--
6" thick floor panels (78 lbs per SF)	--	SF	6.45	--	--	--
8" thick floor panels (105 lbs per SF)	--	SF	7.00	--	--	--

Placing panels with a 100 ton crawler-mounted crane and welding machine at $720 per day. Figures in parentheses show the number of panels placed per day

	Craft@Hrs	Unit	Material	Labor	Equipment	Total
2 ton panels to 150' reach (23 per day)	H4@2.79	Ea	--	135.00	31.40	166.40
2 to 3 tons, to 121' reach (21 per day)	H4@3.05	Ea	--	148.00	34.30	182.30
3 to 4 tons, to 105' reach (19 per day)	H4@3.37	Ea	--	163.00	37.90	200.90
4 to 5 tons, to 88' reach (18 per day)	H4@3.56	Ea	--	172.00	40.00	212.00
5 to 6 tons, to 82' reach (17 per day)	H4@3.76	Ea	--	182.00	42.30	224.30
6 to 7 tons, to 75' reach (16 per day)	H4@4.00	Ea	--	194.00	45.00	239.00
7 to 8 tons, to 70' reach (15 per day)	H4@4.28	Ea	--	207.00	48.20	255.20
8 to 9 tons, to 65' reach (14 per day)	H4@4.56	Ea	--	221.00	51.30	272.30
9 to 10 tons, to 60' reach (11 per day)	H4@5.82	Ea	--	282.00	65.50	347.50
20 to 25 tons to 31' reach (10 per day)	H4@6.40	Ea	--	310.00	72.00	382.00

Precast combination beam and slab units Beam and slab units with beam on edge of slab. Weight of beams and slabs will be about 4,120 lbs per CY. See placing costs below.

	Craft@Hrs	Unit	Material	Labor	Equipment	Total
Slabs with no intermediate beam	--	CY	400.00	--	--	--
Slabs with one intermediate beam	--	CY	430.00	--	--	--
Slabs with two intermediate beams	--	CY	460.00	--	--	--

Placing deck units with a 100 ton crawler-mounted crane and a welding machine at $720 per day. Figures in parentheses show the number of units placed per day.

	Craft@Hrs	Unit	Material	Labor	Equipment	Total
5 ton units to 88' reach (21 per day)	H4@3.05	Ea	--	148.00	34.30	182.30
5 to 10 tons, to 60' reach (19 per day)	H4@3.37	Ea	--	163.00	37.90	200.90
10 to 15 tons, to 43' reach (17 per day)	H4@3.76	Ea	--	182.00	42.30	224.30
15 to 20 tons, to 36' reach (15 per day)	H4@4.28	Ea	--	207.00	48.20	255.20
20 to 25 tons, to 31' reach (14 per day)	H4@4.57	Ea	--	221.00	51.40	272.40
If a 150 ton crane is required, add	--	%	--	--	18.0	--

Precast beams, girders and joists Beams, girders and joists, costs per linear foot of span, including 3,000 PSI concrete, reinforcing steel, embedded items and delivery to 40 miles. See placing costs below.

	Craft@Hrs	Unit	Material	Labor	Equipment	Total
1,000 lb. load per foot, 15' span	--	LF	42.90	--	--	--
1,000 lb. load per foot, 20' span	--	LF	49.00	--	--	--
1,000 lb. load per foot, 30' span	--	LF	55.10	--	--	--
3,000 lb. load per foot, 10' span	--	LF	42.60	--	--	--
3,000 lb. load per foot, 20' span	--	LF	55.50	--	--	--
3,000 lb. load per foot, 30' span	--	LF	67.40	--	--	--
5,000 lb. load per foot, 10' span	--	LF	42.60	--	--	--
5,000 lb. load per foot, 20' span	--	LF	61.30	--	--	--
5,000 lb. load per foot, 30' span	--	LF	73.60	--	--	--

Placing beams, girders and joists with a 100 ton crawler-mounted crane and a welding machine at $720 per day. Figures in parentheses show the number of units placed per day

	Craft@Hrs	Unit	Material	Labor	Equipment	Total
5 ton units to 88' reach (18 per day)	H4@3.56	Ea	--	172.00	40.00	212.00
5 to 10 tons, to 60' reach (17 per day)	H4@3.76	Ea	--	182.00	42.30	224.30
10 to 15 tons, to 43' reach (16 per day)	H4@4.00	Ea	--	194.00	45.00	239.00
15 to 20 tons, to 36' reach (15 per day)	H4@4.27	Ea	--	207.00	48.00	255.00
If a 150 ton crane is required, add	--	%	--	--	18.0	--

Concrete 3

	Craft@Hrs	Unit	Material	Labor	Equipment	Total

Precast concrete columns Precast concrete columns, 12" x 12" to 36" x 36" including concrete, reinforcing steel, embedded items and delivery to 40 miles. See placing costs below.

	Craft@Hrs	Unit	Material	Labor	Equipment	Total
Columns with 500 lbs of reinforcement per CY	--	CY	416.00	--	--	--
Columns with 350 lbs of reinforcement per CY	--	CY	406.00	--	--	--
Columns with 200 lbs of reinforcement per CY	--	CY	393.00	--	--	--

Placing concrete columns with a 100 ton crawler-mounted crane and a welding machine at $720 per day. Figures in parentheses show the number of units placed per day

	Craft@Hrs	Unit	Material	Labor	Equipment	Total
3 ton columns to 121' reach (29 per day)	H4@2.21	Ea	--	107.00	24.90	131.90
3 to 5 tons, to 89' reach (27 per day)	H4@2.37	Ea	--	115.00	26.70	141.70
5 to 7 tons, to 75' reach (25 per day)	H4@2.55	Ea	--	124.00	28.70	152.70
7 to 9 tons, to 65' reach (23 per day)	H4@2.79	Ea	--	135.00	31.40	166.40
9 to 11 tons, to 55' reach (22 per day)	H4@2.92	Ea	--	141.00	32.90	173.90
11 to 13 tons, to 50' reach (20 per day)	H4@3.20	Ea	--	155.00	36.00	191.00
13 to 15 tons, to 43' reach (18 per day)	H4@3.56	Ea	--	172.00	40.00	212.00

Precast concrete stairs Costs include concrete, reinforcing, embedded steel, nosing, and delivery to 40 miles. See placing costs below.

	Craft@Hrs	Unit	Material	Labor	Equipment	Total
Stairs, 44" to 48" wide, 10' to 12' rise, "U" or "L" shape, including typical landings, cost per each step (7" rise)	--	Ea	20.00	--	--	--

Placing precast concrete stairs with a 50 ton truck-mounted crane and a welding machine at $496 per day. Figures in parentheses show the number of stair flights placed per day.

	Craft@Hrs	Unit	Material	Labor	Equipment	Total
2 ton stairs to 82' reach (21 per day)	H4@3.05	Ea	--	148.00	23.60	171.60
2 to 3 tons, to 65' reach (19 per day)	H4@3.37	Ea	--	163.00	26.10	189.10
3 to 4 tons, to 57' reach (18 per day)	H4@3.54	Ea	--	171.00	27.40	198.40
4 to 5 tons, to 50' reach (17 per day)	H4@3.76	Ea	--	182.00	29.60	211.60
5 to 6 tons, to 44' reach (16 per day)	H4@4.00	Ea	--	194.00	31.00	225.00
6 to 7 tons, to 40' reach (15 per day)	H4@4.28	Ea	--	207.00	33.20	240.20
7 to 8 tons, to 36' reach (14 per day)	H4@4.56	Ea	--	221.00	35.30	256.30
8 to 9 tons, to 33' reach (10 per day)	H4@6.40	Ea	--	310.00	49.60	359.60
If a 35 ton crane can be used, deduct	--	%	--	--	-10.0	--

Precast double tees 8' wide, 24" deep. Design load is shown in pounds per square foot (PSF). Costs include lifting and placing but no slab topping. Equipment cost is $448 per day for a 50 ton truck crane

	Craft@Hrs	Unit	Material	Labor	Equipment	Total
30' to 35' span, 115 PSF	H4@.019	SF	4.80	.92	.13	5.85
30' to 35' span, 141 PSF	H4@.019	SF	5.30	.92	.13	6.35
35' to 40' span, 78 PSF	H4@.018	SF	4.80	.87	.13	5.80
35' to 40' span, 98 to 143 PSF	H4@.018	SF	5.30	.87	.13	6.30
40' to 45' span, 53 PSF	H4@.017	SF	4.80	.82	.12	5.74
40' to 45' span, 69 to 104 PSF	H4@.017	SF	5.30	.82	.12	6.24
40' to 45' span, 134 PSF	H4@.017	SF	5.50	.82	.12	6.44
45' to 50' span, 77 PSF	H4@.016	SF	5.30	.78	.11	6.19
45' to 50' span, 101 PSF	H4@.016	SF	5.50	.78	.11	6.39

Concrete topping, based on 350 CY job 3,000 PSI design mix with lightweight aggregate, no forms or finishing included.

	Craft@Hrs	Unit	Material	Labor	Equipment	Total
By crane	H3@.868	CY	58.20	32.80	12.50	103.50
By pump	M2@.690	CY	66.20	25.50	9.00	100.70

Hollow core roof planks, not including topping, 35 PSF live load, 10' to 15' span

	Craft@Hrs	Unit	Material	Labor	Equipment	Total
4" thick	H4@.018	SF	2.95	.87	.13	3.95
6" thick	H4@.018	SF	3.60	.87	.13	4.60
8" thick	H4@.020	SF	4.22	.97	.14	5.33
10" thick	H4@.026	SF	43.70	1.26	.18	45.14

Concrete 3

	Craft@Hrs	Unit	Material	Labor	Equipment	Total

Prestressed concrete support poles Costs shown are per lineal foot for sizes and lengths shown. Equipment cost is $448 per day for a 50 ton truck crane. These costs include the subcontractor's overhead and profit but no base or excavation. Lengths other than as shown will cost more per foot.
10" x 10" poles with four 7/16" 270 KSI ultimate tendon strand strength

	Craft@Hrs	Unit	Material	Labor	Equipment	Total
10', 14', 20', 25' or 30' long	H4@.028	LF	7.70	1.36	.20	9.26

12" x 12" poles with six 7/16" 270 KSI ultimate tendon strand strength

	Craft@Hrs	Unit	Material	Labor	Equipment	Total
10', 14' or 20' long	H4@.028	LF	9.40	1.36	.20	10.96
25' or 30' long	H4@.028	LF	10.00	1.36	.20	11.56

Tilt-Up Concrete Construction. These costs include the subcontractor's overhead and profit. Costs will be higher if access is not available from all sides or obstructions delay the work. Except as noted, no grading, compacting, engineering, design, permit or inspection fees are included.

Foundations and footings for tilt-up Costs assume normal soil conditions with no forming or shoring required and are based on concrete poured directly from the chute of a ready-mix truck. Soil excavated and not needed for backfill can be stored on site and used by others. For scheduling purposes, estimate that a crew of 5 can set forms for and pour 40 to 50 CY of concrete footings and foundations in an 8-hour day.

Lay out foundation from known point on site, set stakes and mark for excavation.

	Craft@Hrs	Unit	Material	Labor	Equipment	Total
Per SF of floor slab	C8@.001	SF	--	.04	--	--
Excavation with a utility backhoe	S1@.075	CY	--	2.89	.80	3.69

Set and tie grade A60 reinforcing bars, typical foundation has 50 to 60 pounds of

	Craft@Hrs	Unit	Material	Labor	Equipment	Total
reinforcing bar per CY of concrete	RB@.008	Lb	.30	.37	--	.67

Place stringer supports for reinforcing bars (at 1.1 LF of support per LF of rebar), using 2" x 4" or 2" x 6" lumber at $525 per MBF before markup or waste allowance. Includes typical stakes and hardware.

	Craft@Hrs	Unit	Material	Labor	Equipment	Total
Based on 3 uses of supports	C8@.017	LF	.30	.65	--	.95

(Inspection of reinforcing may be required before concrete is placed.)
Make and place steel column anchor bolt templates (using 1.1 SF of plywood per SF of column base). Based on 2 uses of 3/4" plyform at $24 per 4' x 8' sheet ($.75 per SF) before markup or allowance for waste. If templates are furnished by the steel column fabricator, include only the labor cost to place the templates as shown

	Craft@Hrs	Unit	Material	Labor	Equipment	Total
Make and place templates	C8@.160	SF	.47	6.16	--	6.63
Place templates furnished by others	C8@.004	SF	--	4.04	--	--

Blockout form at foundation, step footings, columns, etc., per SF of contact area, using 2" x 6" lumber at $525 per MBF before markup or allowance for waste. Includes typical stakes and hardware.

	Craft@Hrs	Unit	Material	Labor	Equipment	Total
Based on 4 uses of forms	C8@.018	SF	.18	.69	--	.87

Column anchor bolts, usually 4 per column. J-hook bolt includes flat washer and two nuts

	Craft@Hrs	Unit	Material	Labor	Equipment	Total
1/2" x 6"	C8@.037	Ea	.38	1.42	--	1.80
3/4" x 12"	C8@.037	Ea	1.45	1.42	--	2.87

Concrete, 2,000 PSI design mix, at $46.90 per CY before markup or 3% allowance for waste,

	Craft@Hrs	Unit	Material	Labor	Equipment	Total
placed directly from chute	CL@.552	CY	55.60	18.80	.95	75.35
Remove, clean and stack reinforcing supports	CL@.020	LF	--	.68	--	--

Dry pack non-shrink grout under column bases, at $17.50 per CF before markup. Typical grout area is

	Craft@Hrs	Unit	Material	Labor	Equipment	Total
2' x 2' x 4" (1.33 CF of grout)	CM@.625	Ea	20.70	25.10	--	45.80

Panel support pads. (Supports panel when grout is placed.) One needed per panel. Based on

	Craft@Hrs	Unit	Material	Labor	Equipment	Total
3 CF of 2,000 PSI concrete per pad	C8@.141	Ea	6.16	5.42	--	11.58

Backfill at foundation with utility backhoe at $22 per hour

	Craft@Hrs	Unit	Material	Labor	Equipment	Total
90 CY per hour	S1@.022	CY	--	.85	.24	1.09
Dispose of excess soil on site, using backhoe	S1@.011	CY	--	.42	.12	.54

Concrete test cylinders, at $10 each,

	Craft@Hrs	Unit	Material	Labor	Equipment	Total
typically 5 required per 100 CY of concrete	--	CY	.50	--	--	--

Concrete 3

	Craft@Hrs	Unit	Material	Labor	Equipment	Total

Floor slabs for tilt-up These costs assume that the site has been graded and compacted by others before slab work begins. The costs below include fine grading the site, placing a 2" layer of sand, a vapor barrier and a second 2" layer of sand over the barrier. Each layer of sand should be screeded level. Reinforcing may be either wire mesh or bars. When sand and vapor barrier are used, concrete may have to be pumped in place. If inspection of the reinforcing is required, the cost will usually be included in the building permit fee. Floor slabs are usually poured 6" deep in 14' to 18' wide strips with edge forms on both sides and ends. Edge forms are used as supports for the mechanical screed which consolidates the concrete with pneumatic vibrators as the surface is leveled. The screed is pulled forward by a winch and cable. Edge forms should have an offset edge to provide a groove which is filled by the adjacent slab. The slab perimeter should end 2' to 3' inside the finished building walls. This "pour strip" is filled in with concrete when the walls are up and braces have been removed. About 2 days after pouring, the slab should be saw-cut 1-1/2" to 2" deep and 1/4" wide each 20' both ways. This controls cracking as the slab cures. For scheduling purposes, estimate that a finishing crew of 10 workers can place and finish 10,000 to 12,000 square feet of slab in 8 hours.
These costs include the subcontractor's overhead and profit.

	Craft@Hrs	Unit	Material	Labor	Equipment	Total
Fine grade the building pad with a utility tractor with blade and bucket, at $22 per hour						
2,500 SF per hour	S6@.001	SF	--	.04	.01	.05
Spread sand 4" deep in 2" layers, (1.3 tons is 1 cubic yard and covers 81 SF at 4" depth), using a utility tractor at $22 per hour						
with sand at $11.00 per ton, 60 SF per hour	S6@.006	SF	.14	.22	.04	.40
Vapor barrier, 6 mil polyethylene	CL@.001	SF	.05	.03	--	.08
Welded wire mesh, 6" x 6", W1.4 x W1.4, including 10% for overlap and waste	RB@.004	SF	.12	.19	--	.31
Reinforcing bars, grade A60, typically #4 bars on 2' centers each way, with 1.05 LF bars per SF at $.18 per LF	RB@.004	SF	.19	.19	--	.38
Dowel out for #4 bars at construction joints	RB@.004	LF	.18	.19	--	.37
Edge forms with keyed joint, rented	F5@.013	LF	1.85	.51	--	2.36
Ready-mix concrete for the slab, 2,000 PSI, chute mix at $46.90 per CY and pump mix at $57.60 per CY before markup or 3% waste						
Placed from the chute of the ready-mix truck	CL@.421	CY	55.60	14.40	.86	70.86
Pump mix, including pumping cost	M2@.442	CY	68.20	16.40	6.00	90.60
Finish the floor slab, using power trowels	CM@.003	SF	--	.12	.04	.16
Vibrating screed and accessories, (includes screeds, fuel, air compressor and hoses) rented, per SF of slab	--	SF	--	--	.10	--
Curing compound, spray applied, at $5.00 and 250 SF per gallon	CL@.001	SF	.02	.03	--	.05
Saw cut green concrete, 1-1/2" to 2" deep	C8@.010	LF	--	.39	.25	.64
Strip, clean and stack edge forms	CL@.011	LF	--	.38	--	--
Ready-mix concrete for the pour strip, 2,000 PSI, chute mix at $46.90 per CY and pump mix at $57.60 per CY before markup or 3% waste						
Placed from the chute of the ready-mix truck	CL@.421	CY	55.60	14.40	.86	70.86
Pump mix, including pumping cost	M2@.442	CY	68.20	16.40	6.00	90.60
Finish the pour strip	CM@.004	SF	--	.16	.06	.22
Special inspection of the concrete pour (if required) at $40.00 per hour, based on 10,000 SF in 8 hours	--	SF	.03	--	--	--
Clean and caulk the saw joints (caulk at $20.00 per gallon), based on 50 LF per hour and 1,500 LF per gallon	CL@.020	LF	.01	.68	--	.69
Clean slab (broom clean)	CL@.001	SF	--	.03	--	--

Concrete 3

	Craft@Hrs	Unit	Material	Labor	Equipment	Total

Tilt-up wall panels Wall thickness is usually a nominal 6" (5-1/2" actual) or a nominal 8" (7-1/2"). Panel heights over 40' are uncommon. Typical panel width is 20'. Construction period will be longer and the costs higher than the figures shown below when panels must be poured and cured in multiple layers. When the floor area isn't large enough to form all wall panels on the slab, stacking will be required. When calculating the volume of concrete required, the usual practice is to multiply overall panel dimensions (width, length and thickness) and deduct only for panel openings that exceed 25 square feet. When pilasters (thickened wall sections) are formed as part of the panel, add the cost of extra formwork and the additional concrete. Use the production rates that follow for scheduling a job.

Production rates for tilt-up

Lay out and form panels, place reinforcing bars and install embedded lifting devices: 2,000 to 3,000 SF of panel face (measured one side) per 8-hour day for a crew of 5. Note that reinforcing may have to be inspected before concrete is poured.

Place and finish concrete: 4,000 to 5,000 SF of panel per 8-hour day for a crew of 5. Continuous inspection of the concrete pour by a licensed inspector may be required. The minimum charge will usually be 4 hours at $40.00 per hour plus travel cost.

Install ledgers with embedded anchor bolts on panels before the concrete hardens: 150 to 200 LF of ledger per 8-hour day for a crew of 2.

Sack and patch exposed panel face before panel is erected: 2,000 to 3,000 SF of panel face per 8-hour day for a crew of 2.

Install panel braces before lifting panels: 30 to 40 braces per 8-hour day for a crew of 2. Panels usually need 3 or 4 braces. Braces are usually rented for 30 days.

Tilt up and set panels: 20 to 24 panels per 8-hour day for a crew of 11 (6 workers, the crane operator, oiler and 3 riggers). If the crew sets 22 panels averaging 400 SF each in an 8-hour day, the crew is setting 1,100 SF per hour. These figures include time for setting and aligning panels, drilling the floor slab for brace bolts, and securing the panel braces. Panel tilt-up will go slower if the crane can't work from the floor slab when lifting panels. Good planning of the panel lifting sequence and easy exit of the crane from the building interior will increase productivity. Size and weight of the panel has little influence on the setting time if the crane is capable of making the lift and is equipped with the right rigging equipment, spreader bars and strongbacks. Panels will usually have to cure for 14 days before lifting unless high early strength concrete is used.

Weld or bolt steel embedded in the panels: 16 to 18 connections per 8-hour day for a crew of 2 working from ladders. Work more than 30 feet above the floor may take more time.

Sack and patch panel faces after erection: 400 SF of face per 8-hour day for a crew of 2.

Tilt-up panel costs

	Craft@Hrs	Unit	Material	Labor	Equipment	Total
Engineering fee, per panel, typical	--	Ea	--	--	--	25.00

Fabricate and lay out forms for panel edges and blockouts, assumes 2 uses and includes typical hardware and chamfer

	Craft@Hrs	Unit	Material	Labor	Equipment	Total
2" x 6" at $525 per MBF before markup	F5@.040	LF	.34	1.57	--	1.91
2" x 8" at $515 per MBF before markup	F5@.040	LF	.45	1.57	--	2.02

Reveals (1" x 4" accent strips on exterior panel face), based on 1 use of lumber, nailed to slab,

	Craft@Hrs	Unit	Material	Labor	Equipment	Total
at $520 per MBF before markup	C8@.018	LF	.21	.69	--	.90

Clean and prepare the slab prior to placing the concrete for the panels,

	Craft@Hrs	Unit	Material	Labor	Equipment	Total
1,250 SF per hour	CL@.001	SF	--	.03	--	--

Releasing agent (bond breaker), sprayed on,

	Craft@Hrs	Unit	Material	Labor	Equipment	Total
250 SF and $5.00 per gallon	CL@.001	SF	.02	.03	--	.05

Place and tie grade A60 reinforcing bars and supports. Based on using #3 and #4 bars, including waste and laps. A typical panel has 80 to 100 pounds of reinforcing steel per CY of concrete in the panel

	Craft@Hrs	Unit	Material	Labor	Equipment	Total
Reinforcing per CY of concrete, at $.31 per lb	RB@.500	CY	27.90	23.40	--	51.30
Reinforcing per SF of 6" thick panel	RB@.009	SF	.52	.43	--	.95
Reinforcing per SF of 8" thick panel	RB@.012	SF	.69	.58	--	1.27

Concrete 3

	Craft@Hrs	Unit	Material	Labor	Equipment	Total

Ledger boards with anchor bolts. Based on 4" x 12" treated lumber at $865 per MBF before markup and 3/4" x 12" bolts at $1.45 each placed each 2 feet, installed prior to lifting panel

Typical ledgers and anchor bolts as described	C8@.353	LF	4.70	13.60	--	18.30

Embedded steel for pickup and brace point hardware.
Usually 10 to 12 points per panel, quantity and type depend on panel size

Embedded pickup & brace hardware, typical	C8@.001	SF	.20	.04	--	.24

Note: Inspection of reinforcing may be required before concrete is placed.)

Concrete, 3,000 PSI, placed direct from chute (at $50.60 per CY before markup or 3% waste)

6" thick panels (54 SF per CY)	CL@.008	SF	1.11	.27	.02	1.40
8" thick panels (40.5 SF per CY)	CL@.010	SF	1.48	.34	.02	1.84

Concrete, 3,000 PSI, pumped in place (at $57.60 per CY before markup or 3% waste)

6" thick panels (54 SF per CY)	M2@.008	SF	1.26	.30	.12	1.68
8" thick panels (40.5 SF per CY)	M2@.011	SF	1.68	.41	.15	2.24
Finish concrete panels, one face before lifting	P8@.004	SF	--	.15	.05	.20

Continuous inspection of pour (at $40.00 per hour), if required, based on 5,000 to 6,000 SF being placed per 8-hour day

Typical cost for continuous inspection of pour	--	SF	--	--	--	.06

Curing wall panels with spray-on curing compound, per SF of face,

based on 250 SF and $5.00 per gallon	CL@.001	SF	.02	.03	--	.05
Strip, clean and stack 2" edge forms	CL@.030	LF	--	1.02	--	--

Braces for panels.
Install rented panel braces, based on an average of three braces per panel for panels averaging 400 to 500 SF each and rental cost of $10 per month each. See also, Remove panel braces.

Typical cost for braces, per SF of panel	F5@.005	SF	--	.20	.08	.28

Miscellaneous equipment rental, typical costs per SF of wall panel

Power trowels and vibrators	--	SF	--	--	.02	--
Compressors, fuel, hoses, rotohammers	--	SF	--	--	.03	--

Lifting panels into place, with crane and riggers, using a 140 ton truck crane

Move crane on and off site, typical	--	LS	--	--	1,590.00	--
Hourly cost (4 hr minimum applies), per hour	T2@11.0	Hr	--	465.00	220.00	685.00
Set, brace and align panels, per SF of panel measured on one face, based on 1,100 SF per hour	T2@.010	SF	--	.42	.22	.64

Allowance for shims, bolts for panel braces and miscellaneous hardware, (labor cost is included with associated items)

Typical cost per SF of panel	--	SF	.15	--	--	--

Weld or bolt panels together, typically 2 connections are required per panel. Equipment is a welding machine at $24.00 per day

Typical cost per connection	T3@.667	Ea	60.00	27.00	1.00	88.00

Continuous inspection of welding, if required, at $40 per hour. ($160 minimum applies)

Cost per connection (3 welds per hour)	--	Ea	13.30	--	--	--

Set grout forms at exterior wall perimeter. Based on using 2" x 12" boards at $515 per MBF before markup or allowance for waste. Costs shown include stakes and assume 4 uses of the forms

Forms per linear foot (LF)	F5@.019	LF	.65	.75	--	1.40

Pour grout under panel bottoms. Place 2,000 PSI concrete (grout) directly from the chute of ready-mix truck (at $52.00 per CY before markup or allowance for waste),

typically 1.85 CY per 100 LF of wall	M2@.005	LF	1.16	.19	--	1.35

Sack and patch wall, per SF of area (measure area to be sacked and patched both inside and outside),

per SF of wall face(s)	P8@.007	SF	.01	.26	--	.27

Caulking and backing strip for panel joints. Cost per LF for caulking both inside & outside face of joint,

per linear foot (LF) of joint	PA@.055	LF	.10	2.23	--	2.33
Remove panel braces, stack & load onto trucks	CL@.168	Ea	--	5.73	--	--
Final cleanup & patch floor, per SF of floor	P8@.003	SF	.01	.11	--	.12

Concrete 3

	Craft@Hrs	Unit	Material	Labor	Equipment	Total

Tilt-up panel costs, continued
Additional costs for tilt-up panels, if required

	Craft@Hrs	Unit	Material	Labor	Equipment	Total
Architectural form liner, minimum cost	F5@.001	SF	.85	.04	--	.89
Architectural form liner, typical cost	F5@.003	SF	1.25	.12	--	1.37
Sandblast, light and medium	--	SF	--	--	--	1.25
Sandblast, heavy	--	SF	--	--	--	1.65
Truck dock bumpers, installed only	C8@2.00	Ea	--	76.90	--	--
Dock levelers, installed only	C8@7.46	Ea	--	287.00	--	--
Angle iron guard around door opening, 2" x 2" x 1/4", mild steel						
(4' lengths)	C8@.120	LF	1.75	4.62	--	6.37
Pipe guard at door opening, 6" black steel, concrete						
filled	C8@.281	LF	17.00	10.80	--	27.80
Embedded reglet at parapet wall,						
aluminum	F5@.019	LF	1.00	.75	--	1.75
Integral color in concrete, light colors, add to concrete						
price per CY	--	CY	65.00	--	--	--
Backfill outside perimeter wall						
with utility tractor at $22 and 90 CY per hour	S1@.022	CY	--	.85	.24	1.09

Cast-in-place columns for tilt-up Concrete columns are often designed as part of the entryway of a tilt-up building or to support lintel panels over a glass storefront. Both round and square freestanding columns are used. They must be formed and poured before the lintel panels are erected. Round columns are formed with plastic-treated fiber tubes (Sonotube is one manufacturer) that are peeled off when the concrete has cured. Square columns are usually formed with plywood and 2" x 4" wood braces or adjustable steel column clamps.

Round plastic-lined fiber tube forms, set and aligned, based on 12' long tubes.
Longer tubes will cost more per LF

	Craft@Hrs	Unit	Material	Labor	Equipment	Total
12" diameter tubes	M2@.222	LF	8.50	8.22	--	16.72
18" diameter tubes	M2@.258	LF	12.50	9.55	--	22.05
24" diameter tubes	M2@.292	LF	16.00	10.80	--	26.80

Rectangular plywood forms, using 1.1 SF of 3/4" plyform at $24 per 4' x 8' sheet ($.75 per SF) before markup, and 2" x 4" bracing 3' OC,

	Craft@Hrs	Unit	Material	Labor	Equipment	Total
per SF of contact area based on 4 uses	F5@.083	SF	5.95	3.27	--	9.22

Column braces, 2" x 4" at $460 per MBF before markup, with 5 BF per LF of column height,

	Craft@Hrs	Unit	Material	Labor	Equipment	Total
per LF of column height based on 2 uses	F5@.005	LF	1.35	.20	--	1.55

Column clamps rented at $4 for 30 days, 1 per LF of column height, cost per LF of column height,

	Craft@Hrs	Unit	Material	Labor	Equipment	Total
per LF of column based on 2 uses a month	F5@.040	LF	2.00	1.57	--	3.57

Reinforcing bars, grade A60, placed and tied, typically 150 to 200 pounds per CY of concrete

	Craft@Hrs	Unit	Material	Labor	Equipment	Total
Cages, round or rectangular (at $.34/lb)	RB@1.50	CY	59.50	70.10	--	129.60
Vertical bars attached to cages (at $.29/lb)	RB@1.50	CY	50.80	70.10	--	120.90

Ready-mix concrete, 3,000 PSI pump mix, at $57.60 before markup or 3% waste allowance,

	Craft@Hrs	Unit	Material	Labor	Equipment	Total
place by pump	M2@.670	CY	68.20	24.80	7.00	100.00

Strip off fiber forms and dismantle bracing, per LF of height

	Craft@Hrs	Unit	Material	Labor	Equipment	Total
12" or 18" diameter	CL@.075	LF	--	2.56	--	--
24" diameter	CL@.095	LF	--	3.24	--	--

Strip wood forms and dismantle bracing,

	Craft@Hrs	Unit	Material	Labor	Equipment	Total
per SF of contact area, no salvage	CL@.011	SF	--	.38	--	--
Clean and stack column clamps	CL@.098	Ea	--	3.34	--	--
Sack and patch concrete column face	P8@.011	SF	.01	.41	--	.42

Allowance for miscellaneous equipment rental (vibrators, compressors, hoses, etc.),

	Craft@Hrs	Unit	Material	Labor	Equipment	Total
per CY of concrete in column	--	CY	--	--	2.30	--

Concrete 3

Trash enclosures for tilt-up These usually consist of a 6" concrete slab measuring 5' x 10' and precast concrete tilt-up panels 6' high. A 6" ledger usually runs along the wall interior. Overall wall length is usually 20 feet. Cost for the trash enclosure foundation, floor slab and wall will be approximately the same as for the tilt-up foundation, floor slab and wall. The installed cost of a two-leaf steel gate measuring 6' high and 8' long overall will be about $300 including hardware and the subcontractor's overhead and profit.

Equipment Foundations. Except as noted, no grading, compacting, engineering, design, permit or inspection fees are included. Costs assume normal soil conditions with no shoring or dewatering required and are based on concrete poured directly from the chute of a ready-mix truck. Costs will be higher if access is not available from all sides or if obstructions delay the work. These costs include the subcontractor's overhead and profit.

Labor costs include the time needed to prepare formwork sketches at the job site, measure for the forms, fabricate, erect, align and brace the forms, place and tie the reinforcing steel and embedded steel items, pour and finish the concrete and, strip, clean and stack the forms.

Reinforcing steel costs assume an average of #3 and #4 bars at 12" on center each way. Weights shown include an allowance for waste and laps.

Form costs assume 3 uses of the forms and show the cost per use when a form is used on the same job without being totally disassembled. Normal cleaning and repairs are included. No salvage value is assumed. Costs listed are per square foot of form in contact with the concrete (SFCA) and assume a number 2 and better lumber price of $525 per MBF before markup and a price of $740 per MSF for 3/4" plyform before markup.

For scheduling purposes, estimate that a crew of 5 can set the forms and place and tie the reinforcing steel for and pour 8 to 10 CY of concrete equipment foundations in an 8-hour day.

Rule-of-thumb estimates. The costs below are based on the volume of concrete in the foundation. If plans for the foundation have not been prepared, estimate the quantity of concrete required from the weight of the equipment using the rule-of-thumb estimating data below. Design considerations may alter the actual quantities required. Based on the type foundation required, use the cost per CY given in the detailed cost breakdown section which follows the table below.

Estimating concrete quantity from the weight of equipment:
1. Rectangular pad type foundations --
    Boilers and similar equipment mounted at grade:
      Allow 1 CY per ton
    Centrifugal pumps and compressors, including weight of motor:
      Up to 2 HP, allow .5 CY per ton
      Over 2 to 5 HP, allow .75 CY per ton
      Over 5 to 10 HP, allow 1 CY per ton
      Over 10 to 20 HP, allow 1.5 CY per ton
      Over 20 to 30 HP, allow 2 CY per ton
      Over 30 HP, allow 3 CY per ton
    Reciprocating pumps and compressors, including weight of motor:
      Up to 2 HP, allow .75 CY per ton
      Over 2 to 5 HP, allow 1.2 CY per ton
      Over 5 to 10 HP, allow 1.5 CY per ton
      Over 10 to 20 HP, allow 2.2 CY per ton
      Over 20 to 30 HP, allow 3 CY per ton
      Over 30 HP, allow 4.5 CY per ton
2. Rectangular pad and stem wall type foundations --
    Horizontal tanks or shell and tube heat exchangers, allow 6 CY per ton
3. Square pad and pedestal type foundations --
    Vertical tanks not over 3' in diameter, allow 6 CY per ton
4. Octagonal pad and pedestal type foundations --
    Vertical tanks over 3' in diameter, allow 4 CY per ton of empty weight

Concrete 3

	Craft@Hrs	Unit	Material	Labor	Equipment	Total

Rectangular pad type foundations For pumps, compressors, boilers and similar equipment mounted at grade. Form costs assume nails, stakes and form oil costing $.35 per square foot and 1.1 SF of 3/4" plyform plus 2.2 BF of lumber per square foot of contact surface. Costs shown are per cubic yard (CY) of concrete within the foundation. These costs include the subcontractor's overhead and profit.

Lay out foundation from known point on site, set stakes and mark for excavation,
	Craft@Hrs	Unit	Material	Labor	Equipment	Total
with 1 CY per CY of concrete	C8@.147	CY	.05	5.66	--	5.71

Excavation with utility backhoe, at 28 CY and $22 per hour,
with 1.5 CY per CY of concrete	S1@.108	CY	--	4.16	1.19	5.35

Formwork, material at $1.37 SF and labor at .150 manhours per SF
with 15 SF per CY of concrete	F5@2.25	CY	20.60	88.50	--	109.10

Reinforcing bars, Grade A60, set and tied, material at $.31 per lb and labor .006 manhours per lb
with 25 lb per CY of concrete	RB@.150	CY	7.75	7.00	--	14.75

Anchor bolts, sleeves, embedded steel, material at $1.60 per lb and labor at .040 manhours per lb
with 5 lb per CY of concrete	C8@.200	CY	8.00	7.69	--	15.69

Ready-mix concrete, 3,000 PSI mix at $50.60 per CY before markup or allowance for 10% waste
with 1 CY per CY of concrete	C8@.385	CY	64.00	14.80	.88	79.68

Concrete test cylinders including test reports, at $10 each
with 5 per 100 CY of concrete	--	CY	.50	--	--	.50

Finish, sack and patch concrete, material at $.02 SF and labor at .010 manhours per SF
with 15 SF per CY of concrete	P8@.150	CY	.30	5.56	--	5.86

Backfill around foundation with utility backhoe at 40 CY and $22 per hour,
with .5 CY per CY of concrete	S1@.025	CY	--	.96	.28	1.24

Dispose of excess soil on site with utility backhoe at 80 CY and $22 per hour,
with 1 CY per CY of concrete	S1@.025	CY	--	.96	.28	1.24
Rectangular pad foundations, total cost	--@3.44	CY	101.20	135.29	2.63	239.12

Rectangular pad and stem wall type foundations For horizontal cylindrical tanks and shell and tube heat exchangers mounted above grade. Form costs assume nails, stakes and form oil costing $.50 per square foot and 1.1 SF of 3/4" plyform plus 4.5 BF of lumber per square foot of contact surface. Costs shown are per cubic yard (CY) of concrete within the foundation. These costs include the subcontractor's overhead and profit.

Lay out foundation from known point on site, set stakes and mark for excavation,
	Craft@Hrs	Unit	Material	Labor	Equipment	Total
with 1 CY per CY of concrete	C8@.147	CY	.05	5.66	--	5.71

Excavation with utility backhoe, at 28 CY and $22 per hour,
with 1 CY per CY of concrete	S1@.072	CY	--	2.77	.79	3.56

Formwork, material at $2.15 SF and labor at .200 manhours per SF
with 60 SF per CY of concrete	F5@12.0	CY	129.00	472.00	--	601.00

Reinforcing bars, Grade A60, set and tied, material at $.31 per lb and labor at .006 manhours per lb
with 45 lb per CY of concrete	RB@.270	CY	14.00	12.60	--	26.60

Anchor bolts, sleeves, embedded steel, material at $1.60 per lb and labor at .040 manhours per lb
with 5 lb per CY of concrete	C8@.200	CY	8.00	7.69	--	15.69

Ready-mix concrete, 3,000 PSI mix, at $50.60 before markup or 10% waste allowance
with 1 CY per CY of concrete	C8@.385	CY	64.00	14.80	.88	79.68

Concrete test cylinders including test reports, at $10 each
with 5 per 100 CY of concrete	--	CY	.50	--	--	.50

Finish, sack and patch concrete, material at $.02 SF and labor at .010 manhours per SF
with 60 SF per CY of concrete	P8@.600	CY	1.20	22.30	--	23.50

Backfill around foundation with utility backhoe, at 40 CY and $22 per hour,
with .5 CY per CY of concrete	S1@.025	CY	--	.96	.28	1.24

Dispose of excess soil on site with utility backhoe, at 80 CY and $22 per hour,
with .5 CY per CY of concrete	S1@.013	CY	--	.50	.14	.64
Rectangular pad and stem wall type foundations,						
total cost per CY of concrete	--@13.7	CY	216.75	539.28	2.09	758.12

Concrete 3

	Craft@Hrs	Unit	Material	Labor	Equipment	Total

Square pad and pedestal type foundations For vertical cylindrical tanks not over 3' in diameter. Form costs assume nails, stakes and form oil costing $.35 per square foot and 1.1 SF of 3/4" plyform plus 3 BF of lumber per square foot of contact surface. Costs shown are per cubic yard (CY) of concrete within the foundations. These costs include the subcontractor's overhead and profit.

Lay out foundation from known point on site, set stakes and mark for excavation,
  with 1 CY per CY of concrete   C8@.147   CY   .05   5.66   --   5.71
Excavation with utility backhoe, at 28 CY and $22 per hour,
  with 1 CY per CY of concrete   S1@.072   CY   --   2.77   .79   3.56
Formwork, material at $1.60 SF and labor at .192 manhours per SF
  with 13 SF per CY of concrete   F5@2.50   CY   20.80   98.40   --   119.20
Reinforcing bars, Grade A60, set and tied, material at $.31 per lb and labor at .006 manhours per lb
  with 20 lb per CY of concrete   RB@.120   CY   6.20   5.60   --   11.80
Anchor bolts, sleeves, embedded steel, material at $1.60 per lb and labor at .040 manhours per lb
  with 2 lb per CY of concrete   C8@.080   CY   3.20   3.08   --   6.28
Ready-mix concrete, 3,000 PSI mix, at $50.60 before markup or 10% waste allowance
  with 1 CY per CY   C8@.385   CY   64.00   14.80   .88   79.68
Concrete test cylinders including test reports, at $10 each
  with 5 per 100 CY   --   CY   .50   --   --   .50
Finish, sack and patch concrete, material at $.02 per SF and labor at .010 manhours per SF
  with 60 SF per CY of concrete   P8@.600   CY   1.20   22.30   --   23.50
Backfill around foundation with utility backhoe, at 40 CY and $22 per hour,
  with .5 CY per CY of concrete   S1@.025   CY   --   .96   .28   1.24
Dispose of excess soil on site with utility backhoe, at 80 CY and $22 per hour,
  with .5 CY per CY of concrete   S1@.013   CY   --   .50   .14   .64
Square pad and pedestal type foundations,
  total cost per CY of concrete   --@3.94   CY   95.95   154.07   2.09   252.11

Octagonal pad and pedestal type foundations For vertical cylindrical tanks over 3' in diameter. Form costs assume nails, stakes and form oil costing $.35 per square foot and 1.1 SF of 3/4" plyform plus 4.5 BF of lumber per square foot of contact surface. Costs shown are per cubic yard (CY) of concrete within the foundation. These costs include the subcontractor's overhead and profit.

Lay out foundation from known point on site, set stakes and mark for excavation,
  with 1 CY per CY of concrete   C8@.147   CY   .05   5.66   --   5.71
Excavation with utility backhoe, at 28 CY and $22 per hour,
  with 1 CY per CY of concrete   S1@.072   CY   --   2.77   .79   3.56
Formwork, material at $2.00 SF and labor at .250 manhours per SF
  with 10 SF per CY of concrete   F5@2.50   CY   20.00   98.40   --   118.40
Reinforcing bars, Grade A60, set and tied, material at $.31 per lb and labor at .006 manhours per lb
  with 20 lb per CY of concrete   RB@.120   CY   6.20   5.60   --   11.80
Anchor bolts, sleeves, embedded steel, material at $1.60 per lb and labor at .040 manhours per lb
  with 2 lb per CY of concrete   C8@.080   CY   3.20   3.08   --   6.28
Ready-mix concrete, 3,000 PSI mix, at $50.60 before markup or 10% waste allowance
  with 1 CY per CY of concrete   C8@.385   CY   64.00   14.80   .88   79.68
Concrete test cylinders including test reports, at $10 ea
  with 5 per 100 CY of concrete   --   CY   .50   --   --   .50
Finish, sack and patch concrete, material at $.02 SF and labor at .010 manhours per SF
  with 10 SF per CY of concrete   P8@.100   CY   .20   3.71   --   3.91
Backfill around foundation with utility backhoe, at 40 CY and $22 per hour,
  with .5 CY per CY of concrete   S1@.025   CY   --   .96   .28   1.24
Dispose of excess soil on site with utility backhoe, at 80 CY and $22 per hour,
  with .5 CY per CY of concrete   S1@.013   CY   --   .50   .14   .64
Octagonal pad and pedestal type foundations,
  total cost per CY of concrete   --@3.44   CY   94.15   135.48   2.09   231.72

Concrete 3

Cementitious Decks. Typical costs including the subcontractor's overhead and profit.

Insulating concrete deck, poured	Craft@Hrs	Unit	Material	Labor	Total
Residential quality, 1-1/2" thick					
Under 3,000 SF	--	SF	--	--	.92
Over 3,000 SF to 10,000 SF	--	SF	--	--	.78
Over 10,000 SF	--	SF	--	--	.66
Poured on metal deck, 2-5/8" thick, lightweight					
Concrete fill, trowel finish	--	SF	--	--	1.21
Vermiculite, screed and float, field mixed	--	SF	--	--	1.37
Gypsum deck, screed and float	--	SF	--	--	1.06
Exterior concrete deck for walking service, troweled					
Lightweight, 2-5/8", 110 lbs per CF	--	SF	--	--	1.63
Add for 4,000 PSI mix	--	SF	--	--	.11
Add for 5,000 PSI mix	--	SF	--	--	.18
Cool deck at pool side, 1/8" to 3/16" thick, colors, finished	--	SF	--	--	1.52
Insulating unit deck, 24 gauge metal deck, key deck reinforcement, zonolite fill, 2-1/2" thick	--	SF	--	--	2.90
Add for 3-1/2" insulperm, for R-19, 2 hour system rating	--	SF	--	--	3.50
Truss tee system, 2-1/2" poured gypsum, key deck reinforcing with ceiling boards					
1/2" sheetrock boards	--	SF	--	--	2.40
1" wood fiber boards	--	SF	--	--	2.90
1" acoustical boards	--	SF	--	--	3.40
Fiber deck, tongue and groove, cementitious planks, complete					
2" thickness	--	SF	--	--	1.56
2-1/2"	--	SF	--	--	1.82
3"	--	SF	--	--	2.07

Masonry 4

	Craft@Hrs	Unit	Material	Labor	Total

Clay Brick. These costs include the subcontractor's overhead and profit.

	Craft@Hrs	Unit	Material	Labor	Total
Brick veneer, single withe, running bond, 3/8" joint					
Face brick, 4" (at $270.00 per 1,000) up to 8' high.					
Commercial	M1@.140	SF	2.63	5.24	7.87
School or institutional	M1@.144	SF	2.73	5.39	8.12
Norman brick	M1@.146	SF	2.84	5.47	8.31
Roman brick	M1@.200	SF	3.58	7.49	11.07
Glazed brick, standard size	M1@.289	SF	5.25	10.80	16.05
Brick walls, common brick, running bond					
8" wall	M1@.203	SF	3.97	7.60	11.57
12" wall	M1@.303	SF	6.00	11.30	17.30
Add for:					
American bond	--	%	--	--	15.0
Basketweave bond	--	%	--	--	20.0
Flemish bond	--	%	--	--	25.0
Herringbone pattern	--	%	--	--	30.0
Soldier bond	--	%	--	--	20.0
Stacked bond	--	%	--	--	15.0
Institutional inspection	--	%	--	--	20.0

Masonry 4

	Craft@Hrs	Unit	Material	Labor	Total
Short runs or cut-up work	--	%	--	--	20.0
Pilasters to 15 SF	--	%	--	--	100.0
Walls over 1 story, per story	--	%	--	--	10.0
Modular brick	--	%	--	--	15.0
Colored mortar, red, tan, brown	--	SF	.26	--	--
Clay brick specialties					
Arches	--	LF	--	--	18.50
Caps and coping	M1@.144	LF	2.70	5.39	8.09
Cuts	M1@.158	LF	--	5.92	--
Heads, anchored	M1@1.37	SF	26.40	51.30	77.70
Jambs	M1@.112	LF	2.24	4.19	6.43
Dove tail anchors	M1@.013	Ea	.30	.49	.79
Mortar filling for 2" wall cavity, type S mortar at $2.50 per CF, per SF of wall face	M1@.013	SF	.50	.49	.99
Zonolite fill for 2" cavity wall	M1@.008	SF	.25	.30	.55

Acid-Proof Brick. Typical prices for red shale, ASTM C-279 modified type II brick with 1/8" mortar joints installed over a membrane of either elastomeric (floating floor) bonded to the floor or wall, sheet membrane with a special seam treatment, hot mop asphalt, cold-applied adhesive troweled on, or special spray-applied material. The setting bed and joint material will be either 3 part epoxy bonded to concrete, carbon-filled furan mortar, potassium silicate mortar, or other special mortars. Expansion joints are required to prevent cracking of the installed brickwork. Brick will be either 3-3/4" x 8" face, 1-1/8" thick (splits); 3-3/4" x 8" face, 2-1/4" thick (singles); 3-3/4" x 8" face, 4-1/2" thick (doubles); or 4" x 8" face, 1-3/8" thick (pavers). Costs are based on a 2,000 SF job. For smaller projects, increase the cost 5% for each 500 SF or portion of 500 SF less than 2,000 SF. For scheduling purposes estimate that a crew of 4 can install 75 to 100 SF of membrane and brick in an 8-hour day. These costs assume that membrane is laid on a suitable surface prepared by others and include the subcontractor's overhead and profit.

	Craft@Hrs	Unit	Material	Labor	Total
Installed costs for acid-proof brick					
Less complex jobs	--	SF	--	--	20.00
More complex jobs	--	SF	--	--	25.00
Minimum job cost	--	LS	--	--	8,000.00

Quarry Tile. Unglazed natural red tile set in a portland cement bed with mortar joints. These costs include the subcontractor's overhead and profit.

	Craft@Hrs	Unit	Material	Labor	Total
Quarry floor tile					
4" x 4" x 1/2", 1/8" straight joints	M1@.112	SF	2.80	4.19	6.99
6" x 6" x 1/2", 1/4" straight joints	M1@.101	SF	2.60	3.78	6.38
6" x 6" x 3/4", 1/4" straight joints	M1@.105	SF	3.45	3.93	7.38
8" x 8" x 3/4", 3/8" straight joints	M1@.095	SF	3.25	3.56	6.81
8" x 8" x 1/2", 1/4" hexagon joints	M1@.112	SF	3.65	4.19	7.84
Quarry wall tile					
4" x 4" x 1/2", 1/8" straight joints	M1@.125	SF	2.80	4.68	7.48
6" x 6" x 3/4", 1/4" straight joints	M1@.122	SF	3.45	4.57	8.02
Quarry tile stair treads					
6" x 6" x 3/4", 12" wide tread	M1@.122	LF	3.75	4.57	8.32
Quarry tile window sill					
6" x 6" x 3/4" tile on 6" wide sill	M1@.078	LF	3.62	2.92	6.54

Masonry 4

	Craft@Hrs	Unit	Material	Labor	Total
Quarry Tile, continued					
Quarry tile trim or cove base					
5" x 6" x 1/2" straight top	M1@.083	LF	2.80	3.11	5.91
6" x 6" x 3/4" round top	M1@.087	LF	3.25	3.26	6.51
Deduct for tile set in epoxy bed with grout joints	--	%	--	-33.0	--
Add for tile set in epoxy bed with grout joints	--	SF	1.05	--	--

Pavers and Floor Tile. These costs include the subcontractor's overhead and profit.

	Craft@Hrs	Unit	Material	Labor	Total
Brick, excluding platform cost					
Plate, glazed	M1@.114	SF	2.35	4.27	6.62
Laid in sand	M1@.089	SF	1.80	3.33	5.13
Pavers, on concrete, grouted	M1@.107	SF	3.15	4.01	7.16
Add for special patterns	--	%	--	--	60.0
Slate	M1@.099	SF	2.75	3.71	6.46
Terrazzo tiles, 1/4" thick					
Standard	M1@.078	SF	4.35	2.92	7.27
Granite	M1@.085	SF	7.10	3.18	10.28
Brick steps	M1@.186	SF	2.80	6.97	9.77

Interlocking Paving Stones. These costs include subcontractor's overhead and profit, fine sand to fill joints and machine vibration, but no excavation.

	Craft@Hrs	Unit	Material	Labor	Total
Interlocking pavers, rectangular, 60mm thick	M1@.083	SF	1.15	3.11	4.26
Interlocking pavers, hexagonal, 80mm thick	M1@.085	SF	1.35	3.18	4.53
Interlocking pavers, multi-angle, 80mm thick	M1@.081	SF	1.17	3.03	4.20
Concrete masonry grid pavers (erosion control)	M1@.061	SF	2.03	2.28	4.31
Add for a 1" sand cushion	M1@.001	SF	.08	.04	.12
Add for a 2" sand cushion	M1@.002	SF	.10	.08	.18
Deduct for over 5,000 SF	--	%	--	--	-10.0

Load Bearing Concrete Block. Prices shown in parentheses are per block before markup. Costs per square foot include allowance for mortar, waste, and subcontractors overhead and profit.

Hollow block, 8" high x 16" long standard units, 3/8" joints. No foundations, reinforcing or grout included

	Craft@Hrs	Unit	Material	Labor	Total
Regular weight units (1.05 blocks per SF)					
4" wide ($.70 per block)	M1@.081	SF	.85	3.03	3.88
6" wide ($.80 per block)	M1@.089	SF	.98	3.33	4.31
8" wide ($1.00 per block)	M1@.095	SF	1.23	3.56	4.79
10" wide ($1.70 per block)	M1@.101	SF	2.08	3.78	5.86
12" wide ($1.65 per block)	M1@.117	SF	2.03	4.38	6.41
Lightweight units (1.05 blocks per SF)					
4" wide ($.83 per block)	M1@.081	SF	1.02	3.03	4.05
6" wide ($1.11 per block)	M1@.089	SF	1.36	3.33	4.69
8" wide ($1.42 per block)	M1@.094	SF	1.74	3.52	5.26
10" wide ($1.97 per block)	M1@.098	SF	2.42	3.67	6.09
12" wide ($2.06 per block)	M1@.114	SF	2.53	4.27	6.80

Glazed Concrete Block. Prices shown in parentheses are per block before markup. Costs per square foot include allowance for mortar, waste, and subcontractors overhead and profit. These costs include the subcontractor's overhead and profit.

8" high x 16" long, 3/8" joints, no reinforcing, grout or foundations included. Based on 1.05 blocks per square foot (SF).

	Craft@Hrs	Unit	Material	Labor	Total
Glazed 1 side					
2" wide ($6.50 per block)	M1@.102	SF	7.55	3.82	11.37
4" wide ($6.60 per block)	M1@.107	SF	7.65	4.01	11.66

Masonry 4

	Craft@Hrs	Unit	Material	Labor	Total
6" wide ($6.80 per block)	M1@.111	SF	7.90	4.16	12.06
8" wide ($7.00 per block)	M1@.118	SF	8.05	4.42	12.47
Glazed 2 sides					
4" wide ($9.50 per block)	M1@.122	SF	11.00	4.57	15.57
6" wide ($9.70 per block)	M1@.127	SF	11.20	4.76	15.96
8" wide ($9.90 per block)	M1@.136	SF	11.40	5.09	16.49
Glazed cove base, glazed 1 side, 8" high					
4" wide ($5.70 per block and 3 per LF)	M1@.127	LF	18.80	4.76	23.56
6" wide ($5.80 per block and 2 per LF)	M1@.137	LF	12.80	5.13	17.93
8" wide ($5.85 per block and 1.5 per LF)	M1@.147	LF	9.65	5.51	15.16

Decorative Concrete Block. Prices shown in parentheses are per block before markup. Costs per square foot include allowance for mortar, waste, and subcontractor's overhead and profit.

Regular weight units, 1.05 blocks per SF.

Split face hollow block, 8" high x 16" long, 3/8" joints. No foundations, grout or reinforcing included

	Craft@Hrs	Unit	Material	Labor	Total
4" wide ($1.10 per block)	M1@.110	SF	1.40	4.12	5.52
6" wide ($1.20 per block)	M1@.121	SF	1.55	4.53	6.08
8" wide ($1.80 per block)	M1@.133	SF	2.30	4.98	7.28
10" wide ($2.05 per block)	M1@.140	SF	2.65	5.24	7.89
12" wide ($2.90 per block)	M1@.153	SF	3.75	5.73	9.48

Split rib hollow block, 8" high x 16" long, 3/8" joints. No foundations, grout or reinforcing included

	Craft@Hrs	Unit	Material	Labor	Total
4" wide ($1.00 per block)	M1@.107	SF	1.30	4.01	5.31
6" wide ($1.05 per block)	M1@.115	SF	1.35	4.31	5.66
8" wide ($1.75 per block)	M1@.121	SF	2.25	4.53	6.78
10" wide ($2.20 per block)	M1@.130	SF	2.85	4.87	7.72
12" wide ($2.55 per block)	M1@.146	SF	3.30	5.47	8.77

Screen block, 3/8" joints, no foundations or reinforcing included

	Craft@Hrs	Unit	Material	Labor	Total
6" x 6" x 4" wide (4 per SF at $.50 each)	M1@.210	SF	2.50	7.87	10.37
8" x 8" x 4" wide (2.25 per SF at $1.10)	M1@.140	SF	3.00	5.24	8.24
12" x 12" x 4" wide (1 per SF at $2.95 each)	M1@.127	SF	3.60	4.76	8.36
8" x 16" x 8" wide (1.11 per SF at $2.80 each)	M1@.115	SF	3.80	4.31	8.11

Acoustical slotted block, by rated Noise Reduction Coefficient (NRC), no foundation or reinforcing included

	Craft@Hrs	Unit	Material	Labor	Total
4" wide, with metal baffle, NRC, .45 to .55	M1@.121	SF	2.93	4.53	7.46
6" wide, with metal baffle, NRC, .45 to .55	M1@.130	SF	3.43	4.87	8.30
8" wide, with metal baffle, NRC, .45 to .55	M1@.121	SF	4.24	4.53	8.77
8" wide, with fiber filler, NRC, .65 to .75	M1@.143	SF	5.59	5.36	10.95

Concrete Block Bond Beams. For door and window openings, no grout or reinforcing included. Cost per linear foot of bond beam block

	Craft@Hrs	Unit	Material	Labor	Total
4" wide, 8" high, 16" long	M1@.065	LF	1.00	2.43	3.43
6" wide, 8" high, 16" long	M1@.068	LF	1.20	2.55	3.75
8" wide, 8" high, 16" long	M1@.072	LF	1.50	2.70	4.20
10" wide, 8" high, 16" long	M1@.079	LF	1.55	2.96	4.51
12" wide, 8" high, 16" long	M1@.089	LF	1.60	3.33	4.93
12" wide, 8" high double bond beam, 16" long	M1@.102	LF	1.60	3.82	5.42
8" wide, 16" high, 16" long	M1@.127	LF	2.40	4.76	7.16
10" wide, 16" high, 16" long	M1@.143	LF	2.60	5.36	7.96
12" wide, 16" high, 16" long	M1@.159	LF	3.05	5.96	9.01

Grout for Concrete Block. These costs include the subcontractor's overhead and profit.

Grout for block cores, type "S" at $2.00 per cubic foot, before markup or 3% allowance for waste. Costs shown are based on 2 cores per 16" block.

Masonry 4

	Craft@Hrs	Unit	Material	Labor	Total
Grout for Concrete Block, continued					
Grout per SF of wall, all cells filled, pumped in place and rodded					
4" wide block, .11 CF per SF	M1@.018	SF	.26	.67	.93
6" wide block, .22 CF per SF	M1@.025	SF	.52	.94	1.46
8" wide block, .32 CF per SF	M1@.035	SF	.76	1.31	2.07
10" wide block, .45 CF per SF	M1@.035	SF	1.07	1.31	2.38
12" wide block, .59 CF per SF	M1@.039	SF	1.40	1.46	2.86
Grout per LF of cell filled measured vertically					
4" wide block, .06 CF per SF	M1@.010	LF	.14	.38	.52
6" wide block, .11 CF per SF	M1@.015	LF	.26	.56	.82
8" wide block, .16 CF per SF	M1@.021	LF	.38	.79	1.17
10" wide block, .23 CF per SF	M1@.022	LF	.55	.82	1.37
12" wide block, .30 CF per SF	M1@.024	LF	.71	.90	1.61
Add for type M mortar (2,500 PSI)	--	%	10.0	--	--
Grout for bond beams, 3,000 PSI concrete, by block size and linear foot of beam, no block or reinforcing included, based on block cavity at 50% of block volume					
8" high x 4" wide, .10 CF per LF	M1@.005	LF	.24	.19	.43
8" high x 6" wide, .15 CF per LF	M1@.010	LF	.36	.38	.74
8" high x 8" wide, .22 CF per LF	M1@.017	LF	.52	.64	1.16
8" high x 10" wide, .30 CF per LF	M1@.023	LF	.71	.86	1.57
8" high x 12" wide, .40 CF per LF	M1@.032	LF	.95	1.20	2.15
16" high x 8" wide, .44 CF per LF	M1@.061	LF	1.04	2.28	3.32
16" high x 10" wide, .60 CF per LF	M1@.058	LF	1.42	2.17	3.59
16" high x 12" wide, .80 CF per LF	M1@.079	LF	1.90	2.96	4.86
Additional Costs for Concrete Block. Add these costs to the figures above.					
Stack bond	--	%	--	--	16.0
Sill blocks	M1@.087	LF	1.21	3.26	4.47
Cutting blocks	M1@.137	LF	1.15	5.13	6.28
Add for cut-up jobs	--	%	--	--	10.0
Add for institutional grade work	--	%	--	--	20.0
Add for colored block	--	%	15.0	--	--
Grout hollow metal door frame in a masonry wall					
Single door	M1@.668	Ea	5.36	25.00	30.36
Double door	M1@.851	Ea	7.21	31.90	39.11
Masonry Reinforcing and Flashing. These costs include the subcontractor's overhead and profit. Reinforcing bars for concrete block, ASTM A615 grade A60 bars, cost per linear foot including 10% overlap and waste.					
#3 bars (3/8", 5,319 LF per ton), horizontal	M1@.005	LF	.33	.19	.52
#3 bars placed vertically	M1@.006	LF	.33	.23	.56
#3 galvanized bars, horizontal	M1@.005	LF	.54	.19	.73
#3 galvanized bars placed vertically	M1@.006	LF	.54	.23	.77
#4 bars (1/2", 2,994 LF per ton), horizontal	M1@.008	LF	.29	.30	.59
#4 bars placed vertically	M1@.010	LF	.29	.38	.67
#4 galvanized bars, horizontal	M1@.008	LF	.53	.30	.83
#4 galvanized bars placed vertically	M1@.010	LF	.53	.38	.91
#5 bars (5/8", 1,918 LF per ton), horizontal	M1@.009	LF	.29	.34	.63
#5 bars placed vertically	M1@.012	LF	.29	.45	.74
#5 galvanized bars, horizontal	M1@.009	LF	.52	.34	.86
#5 galvanized bars placed vertically	M1@.012	LF	.52	.45	.97
#6 bars (3/4", 1,332 LF per ton), horizontal	M1@.010	LF	.29	.38	.67
#6 bars placed vertically	M1@.013	LF	.29	.49	.78
#6 galvanized bars, horizontal	M1@.014	LF	.48	.52	1.00

Masonry 4

	Craft@Hrs	Unit	Material	Labor	Total
#6 galvanized bars placed vertically	M1@.014	LF	.48	.52	1.00
Wall ties					
Rectangular, galvanized	M1@.005	Ea	.15	.19	.34
Rectangular, copper coated	M1@.005	Ea	.17	.19	.36
Cavity "Z" type, galvanized	M1@.005	Ea	.14	.19	.33
Cavity "Z" type, copper coated	M1@.005	Ea	.15	.19	.34
Masonry anchors, steel, including bolts					
3/16" x 18" long	M1@.173	Ea	1.67	6.48	8.15
3/16" x 24" long	M1@.192	Ea	2.12	7.19	9.31
1/4" x 18" long	M1@.216	Ea	2.15	8.09	10.24
1/4" x 24" long	M1@.249	Ea	2.52	9.33	11.85
Wall reinforcing					
Truss type, plain					
4"	M1@.002	LF	.14	.08	.22
6"	M1@.002	LF	.15	.08	.23
8"	M1@.003	LF	.19	.11	.30
10"	M1@.003	LF	.21	.11	.32
12"	M1@.004	LF	.22	.15	.37
Truss type, galvanized					
4"	M1@.002	LF	.18	.08	.26
6"	M1@.002	LF	.20	.08	.28
8"	M1@.003	LF	.24	.11	.35
10"	M1@.003	LF	.26	.11	.37
12"	M1@.004	LF	.29	.15	.44
Ladder type, galvanized, 8" wide	M1@.003	LF	.18	.11	.29
Control joints					
Cross-shaped PVC	M1@.024	LF	2.10	.90	3.00
8" wide PVC	M1@.047	LF	2.60	1.76	4.36
Closed cell 1/2" joint filler	M1@.012	LF	.30	.45	.75
Closed cell 3/4" joint filler	M1@.012	LF	.60	.45	1.05
Through the wall flashing					
5 ounce copper	M1@.023	SF	2.10	.86	2.96
.030" elastomeric sheeting	M1@.038	SF	.45	1.42	1.87

Clay Backing Tile. These costs include the subcontractor's overhead and profit.

	Craft@Hrs	Unit	Material	Labor	Total
Load bearing, 12" x 12"					
4" thick	M1@.066	SF	3.09	2.47	5.56
6" thick	M1@.079	SF	3.50	2.96	6.46
8" thick	M1@.091	SF	4.17	3.41	7.58
Deduct for non-load bearing tile	--	%	-8.0	-10.0	--

Structural Glazed Tile. Heavy-duty, fire-safe, structural glazed facing tile for interior surfaces. These costs include the subcontractor's overhead and profit. No foundations or reinforcing included.

	Craft@Hrs	Unit	Material	Labor	Total
Glazed one side, all colors, 5-1/3" x 12"					
2" thick	M1@.125	SF	4.45	4.68	9.13
4" thick	M1@.136	SF	5.20	5.09	10.29
6" thick	M1@.141	SF	7.30	5.28	12.58
8" thick	M1@.172	SF	8.10	6.44	14.54
Glazed one side, all colors, 8" x 16" block					
2" thick	M1@.072	SF	3.75	2.70	6.45
4" thick	M1@.072	SF	4.10	2.70	6.80
6" thick	M1@.079	SF	5.75	2.96	8.71
8" thick	M1@.085	SF	6.65	3.18	9.83

Masonry 4

	Craft@Hrs	Unit	Material	Labor	Total

Structural Glazed Tile, continued
Acoustical glazed facing tile, with small holes on a smooth ceramic glazed facing, and one mineral fiberglass sound batt located in the hollow core behind perforated face

	Craft@Hrs	Unit	Material	Labor	Total
8" x 8" face size unit, 4" thick	M1@.136	SF	4.00	5.09	9.09
8" x 16" face size unit, 4" thick	M1@.141	SF	6.20	5.28	11.48

Reinforced glazed security tile, with vertical cores and knockouts for horizontal reinforcing, 8" x 16" block

	Craft@Hrs	Unit	Material	Labor	Total
4" thick	M1@.092	SF	5.70	3.45	9.15
6" thick	M1@.095	SF	7.90	3.56	11.46
8" thick	M1@.101	SF	9.40	3.78	13.18
Add to 8" security tile for horizontal and vertical reinforcing and grout cores to meet high security standards	--	SF	--	--	3.36
Add for glazed 2 sides	--	%	55.0	10.0	--
Base, glazed 1 side	M1@.190	LF	3.95	7.12	11.07
Cap, glazed 2 sides	M1@.193	LF	4.50	7.23	11.73
Deduct for large areas	--	%	--	--	-10.0
Add for institutional inspection	--	%	--	--	10.0

Ceramic Veneer Facing. These costs include the subcontractor's overhead and profit.

	Craft@Hrs	Unit	Material	Labor	Total
Applied vertically	M1@1.74	SF	7.00	65.20	72.20
Applied horizontally in molds for attachment to precast panels	M1@.221	SF	6.70	8.28	14.98
Brick plate applied to walls	M1@.196	SF	3.20	7.34	10.54

Glass Block. Clear 3-7/8" thick block installed in walls to 8' high. Costs shown include 1/4" mortar joints, caulking, wall ties and 6% waste. Heights over 8' to 10' require additional scaffolding and handling and may increase labor costs up to 40%. Mortar is 1 part portland cement, 1/2 part lime, and 4 parts sand, plus integral waterproofer. Costs for continuous panel reinforcing in horizontal joints are listed below.

Flat glass block, under 1,000 SF job

	Craft@Hrs	Unit	Material	Labor	Total
6" x 6", 4 per SF, $3.00 each	M1@.323	SF	16.70	12.10	28.80
8" x 8", 2.25 per SF, $4.00 each	M1@.245	SF	12.50	9.18	21.68
12" x 12", 1 per SF, $11.00 each	M1@.202	SF	15.30	7.57	22.87
Deduct for jobs over 1,000 SF to 5,000 SF	--	%	--	--	-8.0
Deduct for jobs over 5,000 SF	--	%	--	--	-20.0
Deduct for Thinline interior glass block	--	%	-10.0	--	--
Add for thermal control fiberglass inserts	--	%	--	--	10.0
Replacing 8" x 8" x 4" clear glass block, 10 unit minimum, per block replaced	M1@.851	Ea	13.50	31.90	45.40

Additional costs for glass block

	Craft@Hrs	Unit	Material	Labor	Total
Solar reflective blocks, add	--	%	50.0	--	--
Fiberglass or polymeric expansion joints	--	LF	.65	--	--
Reinforcing steel mesh	--	LF	.45	--	--
Wall anchors, 2' long (used where blocks aren't set into a wall)	--	Ea	2.00	--	--
Asphalt emulsion, 600 LF per gallon	--	Gal	5.25	--	--
Waterproof sealant, 180 LF per gallon	--	Gal	7.50	--	--
Cleaning block after installation, using sponge and water only, per SF cleaned	CL@.023	SF	--	.78	--

Stone Work. The delivered price of stone can be expected to vary widely. These costs include mortar, based on 1/2" joints, and the subcontractor's overhead and profit but no scaffolding. Figures in parentheses show typical quantities installed per 8-hour day by a crew of 2.

Masonry 4

	Craft@Hrs	Unit	Material	Labor	Total
Rough stone veneer, 4", placed over stud wall					
Lava stone (79 SF)	M4@.202	SF	6.25	7.77	14.02
Rubble stone (74 SF)	M4@.216	SF	7.19	8.30	15.49
Most common veneer (76 SF)	M4@.210	SF	6.77	8.07	14.84
Cut stone,					
Granite, 7/8" thick, interior (48 SF)	M4@.333	SF	19.00	12.80	31.80
Granite, 1-1/4" thick, exterior (48 SF)	M4@.333	SF	21.00	12.80	33.80
Limestone, 2" thick (48 SF)	M4@.333	SF	10.00	12.80	22.80
Limestone, 3" thick (48 SF)	M4@.333	SF	13.00	12.80	25.80
Marble, 7/8" thick, interior (48 SF)	M4@.333	SF	18.00	12.80	30.80
Marble, 1-1/4" thick, interior (48 SF)	M4@.333	SF	21.00	12.80	33.80
Sandstone, 2" thick (36 SF)	M4@.444	SF	11.00	17.05	28.05
Sandstone, 3" thick (36 SF)	M4@.444	SF	12.00	17.05	29.05
Add for honed finish	--	SF	--	--	2.85
Add for polished finish	--	SF	--	--	10.00
Marble specialties, typical prices					
Floors, 7/8" thick, exterior (48 SF)	M1@.333	SF	18.00	12.80	30.80
Thresholds, 1-1/4" thick (67 LF)	M1@.239	LF	11.70	8.95	20.65
Bases, 7/8" x 6" (51 LF)	M1@.317	LF	10.32	11.90	22.22
Columns, plain (15 CF)	M1@1.10	CF	106.00	41.20	147.20
Columns, fluted (13 CF)	M1@1.26	CF	150.00	47.20	197.20
Window stools, 5" x 1" (60 LF)	M1@.268	LF	12.10	10.00	22.10
Toilet partitions (1.5 Ea)	M1@10.9	Ea	573.00	408.00	981.00
Stair treads, 1-1/4" x 11" (50 LF)	M1@.323	LF	26.10	12.10	38.20
Limestone, rough cut large blocks (39 CF)	M1@.409	CF	30.20	15.30	45.50
Manufactured stone veneer, including mortar, and mastic, Cultured Stone					
Rough or smooth terrazzo type (50 SF)	M1@.317	SF	6.51	11.90	18.41
Lava rock type (121 SF)	M1@.139	SF	3.50	5.21	8.71
Brick type (121 SF)	M1@.139	SF	3.39	5.21	8.60
Random cast type (121 SF)	M1@.139	SF	3.75	5.21	8.96

Masonry Lintels and Coping. These costs include the subcontractor's overhead and profit.

	Craft@Hrs	Unit	Material	Labor	Total
Reinforced precast concrete lintels					
4" wide x 8" high, to 6'6" long	M1@.073	LF	4.00	2.73	6.73
6" wide x 8" high, to 6'6" long	M1@.078	LF	7.25	2.92	10.17
8" wide x 8" high, to 6'6" long	M1@.081	LF	7.50	3.03	10.53
10" wide x 8" high, to 6'6" long	M1@.091	LF	10.00	3.41	13.41
4" wide x 8" high, over 6'6" to 10' long	M1@.073	LF	7.50	2.73	10.23
6" wide x 8" high, over 6'6" to 10' long	M1@.077	LF	8.95	2.88	11.83
8" wide x 8" high, over 6'6" to 10' long	M1@.081	LF	10.40	3.03	13.43
10" wide x 8" high, over 6'6" to 10' long	M1@.091	LF	13.00	3.41	16.41
Precast concrete coping, 5" average thickness, 4' to 8' long					
10" wide, gray concrete	M1@.190	LF	6.75	7.12	13.87
12" wide, gray concrete	M1@.200	LF	7.45	7.49	14.94
12" wide, white concrete	M1@.226	LF	18.85	8.46	27.31

Cleaning and Pointing Masonry. These costs assume the masonry surface is in fair to good condition with no unusual damage. Add the cost of protecting adjacent surfaces such as trim or the base of the wall and the cost of scaffolding, if required. Labor required to presoak or saturate the area cleaned is included in the labor cost. Work more than 12' above floor level will increase costs. These costs include the subcontractor's overhead and profit.

	Craft@Hrs	Unit	Material	Labor	Total
Brushing (hand cleaning) masonry, includes the cost of detergent or chemical solution					
Light cleanup (100 SF per manhour)	M1@.010	SF	.02	.38	.40

Masonry 4

	Craft@Hrs	Unit	Material	Labor	Total

Cleaning and Pointing Masonry, Brushing masonry, continued

	Craft@Hrs	Unit	Material	Labor	Total
Medium scrub (75 SF per manhour)	M1@.013	SF	.03	.49	.52
Heavy (50 SF per manhour)	M1@.020	SF	.04	.75	.79

Water blasting masonry, using rented 400 to 700 PSI power washer with 3 to 8 gallon per minute flow rate, includes blaster rental at $50.00 per day

	Craft@Hrs	Unit	Material	Labor	Total
Smooth face (250 SF per manhour)	M1@.004	SF	.03	.15	.18
Rough face (200 SF per manhour)	M1@.005	SF	.04	.19	.23

Sandblasting masonry, using 250 PSI compressor (with accessories and sand) at $142.50 per day

	Craft@Hrs	Unit	Material	Labor	Total
Smooth face (50 SF per manhour)	M1@.020	SF	.38	.75	1.13
Rough face (40 SF per manhour)	M1@.025	SF	.46	.94	1.40

Steam cleaning masonry, using rented steam cleaning rig (with accessories) at $52.50 per day

	Craft@Hrs	Unit	Material	Labor	Total
Smooth face (75 SF per manhour)	M1@.013	SF	.10	.49	.59
Rough face (55 SF per manhour)	M1@.018	SF	.13	.67	.80
Add for masking adjacent surfaces	--	%	--	5.0	--
Add for difficult stain removal	--	%	50.0	50.0	--
Add for working from scaffold	--	%	--	20.0	--
Repointing brick, cut out joint, mask, and regrout, 30 SF per manhour	M1@.033	SF	.03	1.24	1.27

Masonry Cleaning. Costs to clean masonry surfaces using commercial cleaning agents, including washing and scaffolding equipment. These costs include the subcontractor's overhead and profit.

	Craft@Hrs	Unit	Material	Labor	Total
Typical cleaning of granite, sandstone, terra cotta and brick surfaces	CL@.015	SF	.08	.51	.59
Cleaning of heavily carbonated limestone or cast stone surfaces	CL@.045	SF	.16	1.54	1.70

Fireplaces. Common brick, including the chimney foundation, damper, flue lining, stack to 15' and the subcontractor's overhead and profit.

	Craft@Hrs	Unit	Material	Labor	Total
30" box, brick to 5', large quantity	M1@30.3	Ea	610.00	1,130.00	1,740.00
36" box, brick to 5', custom	M1@33.7	Ea	730.00	1,260.00	1,990.00
42" box, brick to 8', custom	M1@48.1	Ea	865.00	1,800.00	2,665.00
48" box, brick to 8', custom	M1@53.1	Ea	1,010.00	1,990.00	3,000.00

Flue lining, 2' lengths

	Craft@Hrs	Unit	Material	Labor	Total
8" x 8"	M1@.133	LF	3.65	4.98	8.63
8" x 10"	M1@.153	LF	4.30	5.73	10.03
12" x 12"	M1@.176	LF	4.80	6.59	11.39
Stone veneer to 9', up to 48 SF, typical	--	LS	--	--	985.00
Raised hearth, up to 16 SF, typical	--	LS	--	--	410.00
Extra stack, additional 10'	--	LS	--	--	510.00
Prefabricated cast concrete 30" box and 15' chimney, large quantity, installed	M1@5.43	Ea	1,030.00	203.00	1,233.00
Face and hearth for prefabricated fireplace	--	LS	--	--	426.00
Face brick, mantle to ceiling	M1@.451	SF	6.25	16.90	23.15

Pargeting and Waterproofing. These costs include the subcontractor's overhead and profit.

	Craft@Hrs	Unit	Material	Labor	Total
Pargeting, 2 coats, 1/2" thick	M3@.070	SF	.15	2.64	2.79
Pargeting, 2 coats, waterproof, 3/4" thick	M3@.085	SF	.16	3.20	3.36
Waterproofing, per coat of asphalt	CL@.007	SF	.12	.24	.36
Dampproofing, asphalt primer at 1 gallon per 100 SF, dampproof coat at 30 pounds per 100 SF	CL@.029	SF	.60	.99	1.59

Metals 5

	Craft@Hrs	Unit	Material	Labor	Equipment	Total

Pre-Engineered Steel Buildings. 26 gauge colored galvanized steel roof and siding with 4 in 12 (20 pound live load) roof. Cost per SF of floor area. These costs include the subcontractor's overhead and profit but no foundation or floor slab. Add delivery cost to the site.

	Craft@Hrs	Unit	Material	Labor	Equipment	Total
40' x 100' (4,000 SF)						
14' eave height	H5@.074	SF	6.00	3.17	.80	9.97
16' eave height	H5@.081	SF	6.70	3.47	.85	11.02
20' eave height	H5@.093	SF	7.60	3.98	.95	12.53
60' x 100' (6,000 SF)						
14' eave height	H5@.069	SF	5.75	2.95	.75	9.45
16' eave height	H5@.071	SF	5.90	3.04	.75	9.69
20' eave height	H5@.081	SF	6.80	3.47	.85	11.12
80' x 100' (8,000 SF)						
14' eave height	H5@.069	SF	5.65	2.95	.75	9.35
16' eave height	H5@.071	SF	5.90	3.04	.75	9.69
20' eave height	H5@.080	SF	6.80	3.43	.85	11.08
100' x 100' (10,000 SF)						
14' eave height	H5@.067	SF	5.60	2.87	.69	9.16
16' eave height	H5@.071	SF	5.90	3.04	.75	9.69
20' eave height	H5@.080	SF	6.60	3.43	.85	10.88
100' x 150' (15,000 SF)						
14' eave height	H5@.065	SF	5.30	2.78	.69	8.77
16' eave height	H5@.069	SF	5.65	2.95	.75	9.35
20' eave height	H5@.074	SF	6.30	3.17	.80	10.27
100' x 200' (20,000 SF)						
14' eave height	H5@.063	SF	5.20	2.70	.64	8.54
16' eave height	H5@.066	SF	5.45	2.83	.69	8.97
20' eave height	H5@.071	SF	5.90	3.04	.75	9.69
140' x 150' (21,000 SF)						
14' eave height	H5@.058	SF	4.80	2.48	.64	7.92
16' eave height	H5@.059	SF	4.95	2.53	.64	8.12
20' eave height	H5@.066	SF	5.55	2.83	.69	9.07
140' x 175' (24,500 SF)						
14' eave height	H5@.057	SF	4.70	2.44	.64	7.78
16' eave height	H5@.058	SF	4.80	2.48	.64	7.92
20' eave height	H5@.065	SF	5.30	2.78	.69	8.77
160' x 200' (32,000 SF)						
14' eave height	H5@.055	SF	4.55	2.36	.58	7.49
16' eave height	H5@.056	SF	4.70	2.40	.58	7.68
20' eave height	H5@.061	SF	5.10	2.61	.64	8.35
200' x 200' (40,000 SF)						
14' eave height	H5@.052	SF	4.40	2.23	.53	7.16
16' eave height	H5@.053	SF	4.50	2.27	.58	7.35
20' eave height	H5@.058	SF	4.80	2.48	.64	7.92
Add for personnel door						
3'0" x 6'8"	H5@2.83	Ea	360.00	121.00	29.80	510.80
Add for overhead door						
8'0" x 10'0"	H5@8.02	Ea	870.00	343.00	84.20	1,297.20
Add for interior liner panel, 26 gauge colored galvanized steel,						
per SF of wall	H5@.012	SF	1.15	.51	.12	1.78
Add for foam insulated sandwich panel roofing and siding,						
per SF of roof and sidewall	H5@.005	SF	3.50	.21	.07	3.78
Add for plastic skylight roof panel,						
per SF of panel	H5@.013	SF	3.00	.56	.16	3.72

Metals 5

	Craft@Hrs	Unit	Material	Labor	Equipment	Total
Pre-Engineered Steel Buildings, continued						
Add for 3-1/2", R-11 blanket insulation, per SF of wall or floor	H5@.005	SF	.60	.21	.09	.90
Add for 9" throat ridge ventilator	H5@.193	LF	30.00	8.26	2.03	40.29
Add for 24 gauge colored eave gutter	H5@.028	LF	4.60	1.20	.32	6.12
Add for 4" x 4" colored downspouts	H5@.014	LF	4.10	.60	.12	4.82
Add for 30 pound live load roof design	--	%	--	--	--	12.0

Railings and Kickplates. These costs include the subcontractor's overhead and profit.

Aluminum railings. 3'6" high with 1-1/2" square horizontal rails top and bottom, 5/8" square vertical bars at 4" OC, and 1-1/2" square posts spaced 4' OC. Welded construction with standard mill finish.

	Craft@Hrs	Unit	Material	Labor	Equipment	Total
Installed on balconies	H6@.092	LF	29.60	3.92	--	33.52
Installed on stairways	H6@.115	LF	33.00	4.89	--	37.89
Add for anodized finish, bronze or black	--	%	30.0	--	--	--

Brass railings. Round tubular type, bright finish, standard 12' sections, .050" wall commercial quality with supports at 3' OC attached to existing structure using brass screws. Labor includes layout, measure, cut and end preparation

	Craft@Hrs	Unit	Material	Labor	Equipment	Total
1-1/2" tubing	H6@.058	LF	4.14	2.47	--	6.61
2" tubing	H6@.058	LF	5.28	2.47	--	7.75
Fittings, slip-on type, no soldering or brazing required						
1-1/2" elbows	H6@.092	Ea	11.80	3.92	--	15.72
1-1/2" tees	H6@.137	Ea	13.80	5.83	--	19.63
1-1/2" crosses	H6@.277	Ea	15.80	11.80	--	27.60
2" elbows	H6@.115	Ea	13.30	4.89	--	18.19
2" tees	H6@.171	Ea	17.90	7.28	--	25.18
2" crosses	H6@.346	Ea	20.20	14.70	--	34.90

Kickplates, 1/4" thick flat steel plate, shop prime painted, attached to steel railings, stairways or landings

	Craft@Hrs	Unit	Material	Labor	Equipment	Total
4" high	H6@.058	LF	3.81	2.47	1.71	7.99
6" high	H6@.069	LF	4.16	2.94	2.07	9.17
Add for galvanized finish	--	%	18.0	--	--	--

Pipe and chain railings. Welded steel pipe posts with U-brackets attached to accept 1/4" chain, shop prime painted. Installed in existing embedded sleeves. Posts are 5'0" long (1'6" in the sleeve and 3'6" above grade). Usual spacing is 4' OC. Add chain per lineal foot

	Craft@Hrs	Unit	Material	Labor	Equipment	Total
1-1/4" diameter post	H6@.287	Ea	10.60	12.20	--	22.80
1-1/2" diameter post	H6@.287	Ea	12.40	12.20	--	24.60
2" diameter post	H6@.343	Ea	16.90	14.60	--	31.50
2-1/2" diameter post	H6@.343	Ea	23.10	14.60	--	37.70
Add for galvanized finish	--	%	10.0	--	--	--
Chain, 1/4" diameter galvanized	H6@.058	LF	1.53	2.47	--	4.00

Steel handrails. Wall mounted, with brackets 5' OC, based on 12' lengths, welded steel, shop prime painted

	Craft@Hrs	Unit	Material	Labor	Equipment	Total
1-1/4" diameter rail	H6@.092	LF	4.76	3.92	2.77	11.45
1-1/2" diameter rail	H6@.102	LF	5.54	4.34	3.09	12.97
Add for galvanized finish	--	%	20.0	--	--	--

Steel railings. Floor mounted, 3'6" high with posts 5' OC, based on 12' length, welded steel, shop prime painted 1-1/4" diameter rails and posts,

	Craft@Hrs	Unit	Material	Labor	Equipment	Total
2-rail type	H6@.115	LF	19.40	4.89	3.47	27.76
3-rail type	H6@.125	LF	25.40	5.32	3.79	34.51
1-1/2" diameter rails and posts,						
2-rail type	H6@.125	LF	23.70	5.32	3.79	32.81
3-rail type	H6@.137	LF	31.00	5.83	4.16	40.99
Add for galvanized finish	--	%	20.0	--	--	--

Metals 5

	Craft@Hrs	Unit	Material	Labor	Equipment	Total

Wrought iron railings. Standard 3'6" high, welded steel construction, shop prime painted, attached to existing structure

	Craft@Hrs	Unit	Material	Labor	Equipment	Total
Stock patterns	H6@.171	LF	13.70	7.28	5.18	26.16
Custom patterns	H6@.287	LF	20.40	12.20	8.63	41.23
Stairway handrails	H6@.343	LF	22.20	14.60	10.30	47.10

Space Frame System. 10,00 SF or more. These costs include the subcontractor's overhead and profit.
5' module

	Craft@Hrs	Unit	Material	Labor	Equipment	Total
4.5 lb live load	H5@.146	SF	14.65	6.10	1.40	22.15
4' module, add	H5@.025	SF	2.10	1.10	.26	3.46

Steel Floor and Roof Decking. These costs assume a 40,000 SF job not over six stories high and include the subcontractor's overhead and profit. Weight and cost include 10% for waste and connections.

Floor and roof decking, 1-1/2" standard rib, simple non-composite section, non-cellular, shop prime painted finish

	Craft@Hrs	Unit	Material	Labor	Equipment	Total
22 gauge, 2.1 lbs per SF	H4@.013	SF	1.20	.63	.16	1.99
20 gauge, 2.5 lbs per SF	H4@.013	SF	1.25	.63	.16	2.04
18 gauge, 3.3 lbs per SF	H4@.013	SF	1.50	.63	.16	2.29
16 gauge, 4.1 lbs per SF	H4@.013	SF	1.70	.63	.16	2.49
Add for galvanized finish	--	%	12.5	--	--	--

Steel floor deck system. Consists of 18 gauge, 1-1/2" standard rib, shop prime painted steel decking, installed over a structural steel beam and girder support system (cost of support structure not included) covered with 3" thick lightweight ready-mix concrete at $83.90 per CY including 3% for waste, for 6.0 sack mix, 3,000 PSI placed by pump, reinforced with 6" x 6", W1.4 x W1.4. welded wire mesh. Costs shown assume shoring is not required and include curing and finishing the concrete.

	Craft@Hrs	Unit	Material	Labor	Equipment	Total
18 gauge decking	H7@.013	SF	1.45	.63	.16	2.24
Welded wire mesh	RB@.004	SF	.10	.19	--	.29
Lightweight concrete	M2@.004	SF	.80	.15	.06	1.01
Cure and finish concrete	P8@.011	SF	.10	.41	--	.51
Total floor deck system	--@.032	SF	2.45	1.38	.22	4.05
Add for 16 gauge decking	--	SF	.20	--	--	.20

Steel roof deck system. Consists of 18 gauge, 1-1/2" standard rib, shop prime painted steel decking, installed over open web steel joists 4' OC (costs of joists not included) covered with 1-1/4" R-8.3 urethane insulation board overlaid with .045" loose laid membrane EPDM (ethylene propylene diene monomer) elastomeric roofing.

	Craft@Hrs	Unit	Material	Labor	Equipment	Total
18 gauge decking	H7@.013	SF	1.50	.63	.16	2.29
Insulation board	A1@.011	SF	1.05	.45	--	1.50
EPDM roofing	R3@.008	SF	.85	.34	--	1.19
Total roof deck system	--@.032	SF	3.40	1.42	.16	4.98
Add for 16 gauge decking	--	SF	.20	--	--	.20
Add open web steel joists	H7@.005	SF	.50	.24	.05	.79

Joist pricing assumes H or J series @ 9.45 lbs per LF spanning 20'0", spaced 4' OC.

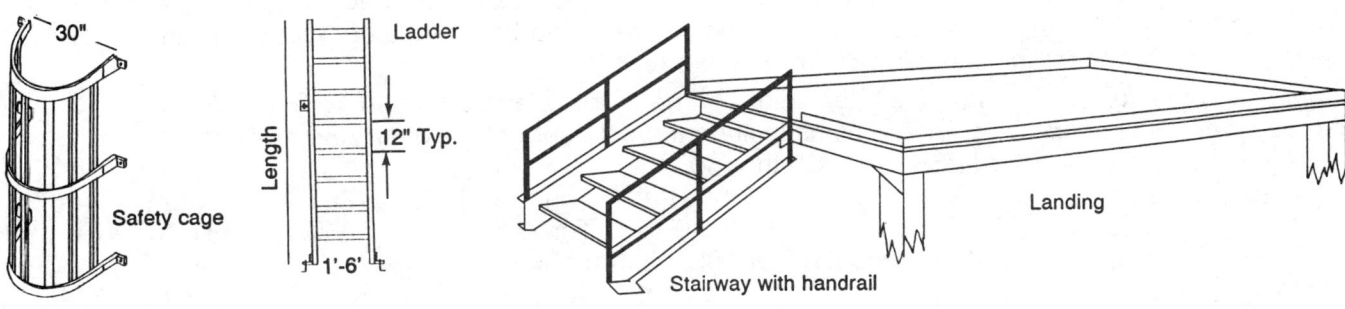

See costs on following page

Metals 5

	Craft@Hrs	Unit	Material	Labor	Equipment	Total

Steel Ladders, Stairways, and Landings. Material costs assume shop fabrication and assembly (as permitted by job conditions), shop prime and delivery complete with erection bolts. Labor includes time to unload and handle the material, erect, install permanent bolts, welds and field touch-up painting. Equipment cost is $32.00 per hour and includes a 10 ton hydraulic truck-mounted crane at $22.00 per hour plus $10.00 per hour for an electric powered welding machine, an oxygen/acetylene welding and cutting torch, a diesel powered 100 CFM air compressor with pneumatic grinders, ratchet wrenches, hoses, and other tools usually associated with work of this type.
These costs include the subcontractor's overhead and profit. (See sketches on preceding page)

Ladders, vertical type. 1'6" wide, attached to existing structure

	Craft@Hrs	Unit	Material	Labor	Equipment	Total
Without safety cage	H7@.100	LF	25.00	4.86	3.35	33.21
With steel safety cage	H7@.150	LF	30.00	7.29	4.85	42.14
Add for galvanized finish	--	%	10.0	--	--	--

Stairways. Cost per linear foot based on sloping length. Rule of thumb: the sloping length of stairways is the vertical rise times 1.6. Costs assume standard C shape structural steel channels for stringers, 12 gauge steel pan 10" treads with 7-1/2" closed risers. Posts are 2" diameter tubular steel, 5' OC. Handrails are 3'6" high, 1-1/2" diameter 2-rail type, on both sides of the stairs. Installation assumes stairs and landings are attached to an existing structure on two sides.

	Craft@Hrs	Unit	Material	Labor	Equipment	Total
3'0" wide	H7@.469	LF	170.00	22.80	15.70	208.50
4'0" wide	H7@.630	LF	225.00	30.60	21.00	276.60
5'0" wide	H7@.781	LF	285.00	38.00	26.00	349.00

Landings. Costs assume standard C shape structural steel channels for joists and frame, 10 gauge steel pan deck areas, shop prime painted. Add for support legs from Structural Steel Section.

	Craft@Hrs	Unit	Material	Labor	Equipment	Total
Cost per square foot	H7@.250	SF	90.80	12.20	8.34	111.34
Add for galvanized finish	--	%	10.0	--	--	--

Spiral stairways. Cost per LF of vertical rise, floor to floor. Costs assume a 6" diameter round tubular steel center column, 12 gauge steel pan type 10" treads with 7-1/2" open risers, and 2" x 2" tubular steel handrail with 1" x 1" tubular steel posts at each riser, shop prime painted. Add for railings at top of stairways as required.

	Craft@Hrs	Unit	Material	Labor	Equipment	Total
4'0" diameter	H7@.151	LF	200.00	7.34	4.22	211.56
6'0" diameter	H7@.151	LF	275.00	7.34	4.22	286.56
8'0" diameter	H7@.151	LF	320.00	7.34	4.22	331.56

Add railings and kickplates from Railings and Kickplates Section.

Structural Steel. Costs shown below are for 60 to 100 ton jobs up to six stories high where the fabricated structural steel is purchased from a steel fabricator located within 50 miles of the job site. Material costs are for ASTM A36 steel, fabricated according to plans, specifications and AISC recommendations and delivered to the job site with connections attached, piece marked and ready for erection. Labor includes time to unload and handle the steel, erect and plumb, install permanent bolts, weld and all other work usually associated with erecting the steel including field touch-up painting. Equipment cost is $68.00 per hour including markup. Equipment includes a 100 ton crawler mounted crane at $56.00 per hour, plus $12.00 per hour for an electric powered welding machine, an oxygen/acetylene welding and cutting torch, a diesel powered 100 CFM air compressor with pneumatic grinders, ratchet wrenches, hoses, and other tools usually associated with work of this type. Structural steel is usually estimated by the tons of the steel in the project. Labor for erection will vary with the weight of steel per linear foot. For example a member weighing 25 pounds per foot will require more manhours per ton than one weighing 60 pounds per foot.

Open web steel joists. Estimate the total weight of "bare" joists and then add 10% to allow for connections and accessories. These costs include the subcontractor's overhead and profit.

"H" and "J" series, 8H3 thru 26H10 or 8J3 thru 30J8

	Craft@Hrs	Unit	Material	Labor	Equipment	Total
Under 10 lbs per LF	H8@10.0	Ton	915.00	476.00	476.00	1,867.00
Over 10 to 20 lbs per LF	H8@8.00	Ton	880.00	381.00	381.00	1,642.00

"LH" and "LJ" series, 18LH02 thru 48LH10 or 18LJ02 thru 48LJ11

	Craft@Hrs	Unit	Material	Labor	Equipment	Total
Under 20 lbs per LF	H8@8.00	Ton	840.00	381.00	381.00	1,602.00

Metals 5

	Craft@Hrs	Unit	Material	Labor	Equipment	Total
Over 20 to 50 lbs per LF	H8@7.00	Ton	810.00	333.00	333.00	1,476.00
Over 50 lbs per LF	H8@6.00	Ton	775.00	285.00	285.00	1,345.00
"DLH" or "DLJ" series, 52DLH10 thru 72DLH18 or 52DLJ12 thru 52DLJ20						
20 to 50 lbs per LF	H8@8.00	Ton	890.00	381.00	90.60	1,361.60
Over 50 lbs per LF	H8@6.00	Ton	850.00	285.00	67.00	1,202.00

Structural shapes. Estimate the total weight of "bare"" fabricated structural shapes and then add 15% to allow for the weight of connections. These costs include the subcontractor's overhead and profit.

	Craft@Hrs	Unit	Material	Labor	Equipment	Total
Beams, purlins, and girts						
Under 20 pounds per LF	H8@14.0	Ton	1,720.00	666.00	158.00	2,544.00
From 20 to 50 lbs per LF	H8@11.0	Ton	1,110.00	523.00	124.00	1,757.00
Over 50 to 75 lbs per LF	H8@9.40	Ton	1,040.00	447.00	106.00	1,593.00
Over 75 to 100 lbs per LF	H8@8.00	Ton	1,010.00	381.00	89.90	1,480.90
Columns						
Under 20 pounds per LF	H8@12.0	Ton	1,610.00	571.00	135.00	2,316.00
From 20 to 50 lbs per LF	H8@10.0	Ton	1,035.00	476.00	112.00	1,623.00
Over 50 to 75 lbs per LF	H8@9.00	Ton	965.00	428.00	101.10	1,494.10
Over 75 to 100 lbs per LF	H8@7.00	Ton	990.00	333.00	78.70	1,401.70
Sag rods and X type bracing						
Under 1 pound per LF	H8@82.1	Ton	2,620.00	3,910.00	922.00	7,452.00
From 1 to 3 lbs per LF	H8@48.0	Ton	2,410.00	2,280.00	540.00	5,230.00
Over 3 to 6 lbs per LF	H8@34.8	Ton	2,250.00	1,660.00	393.00	4,303.00
Over 6 to 10 lbs per LF	H8@28.0	Ton	2,070.00	1,330.00	314.00	3,714.00
Trusses and girders						
Under 20 pounds per LF	H8@10.0	Ton	1,600.00	476.00	112.00	2,188.00
From 20 to 50 lbs per LF	H8@7.50	Ton	1,030.00	357.00	84.30	1,471.30
Over 50 to 75 lbs per LF	H8@6.00	Ton	990.00	285.00	67.50	1,342.50
Over 75 to 100 lbs per LF	H8@5.00	Ton	960.00	238.00	56.20	1,254.20

Structural Steel by Square Foot of Floor. Use these figures for preliminary estimates on buildings with 20,000 to 80,000 SF per story, 100 PSF live load and 14' to 16' story height. A "bay" is the center to center distance between columns. Costs assume steel at $1,600 to $1,700 per ton including installation labor and material. Include all area under the roof when calculating the total square footage. These costs include the subcontractor's overhead and profit.

	Craft@Hrs	Unit	Material	Labor	Equipment	Total
Single story						
20' x 20' bays	--	SF	--	--	--	7.00
30' x 30' bays	--	SF	--	--	--	8.50
40' x 40' bays	--	SF	--	--	--	10.30
Two story to six story						
20' x 20' bays	--	SF	--	--	--	7.80
30' x 30' bays	--	SF	--	--	--	9.50
40' x 40' bays	--	SF	--	--	--	11.20

Rectangular Tubular Column with top bracket and base plate

Round Tubular Column with top bracket and base plate

Tubular Columns. Rectangular, round, and square structural columns. When estimating total weight of columns, figure the column weight and add 30% to allow for a shop attached U-bracket on the top, a square base plate on the bottom and four anchor bolts. This should be sufficient for columns 15' to 20' in length. These costs include the subcontractor's overhead and profit.

Metals 5

	Craft@Hrs	Unit	Material	Labor	Equipment	Total

Tubular Columns, continued (See illustrations on preceding page)
Rectangular tube columns, in pounds per LF

	Craft@Hrs	Unit	Material	Labor	Equipment	Total
3" x 2" to 12" x 4", to 20 lbs per LF	H8@9.00	Ton	1,900.00	428.00	100.90	2,428.90
6" x 4" to 16" x 8", 21 to 50 lbs per LF	H8@8.00	Ton	1,530.00	381.00	89.60	2,000.60
12" x 6" to 20" x 12", 51 to 75 lbs per LF	H8@6.00	Ton	1,490.00	285.00	67.00	1,842.00
14" x 10" to 20" x 12", 76 to 100 lbs/LF	H8@5.00	Ton	1,450.00	238.00	55.60	1,743.60

Round columns, in pounds per LF

	Craft@Hrs	Unit	Material	Labor	Equipment	Total
3" to 6" diameter pipe, to 20 lbs per LF	H8@9.00	Ton	2,010.00	428.00	100.90	2,538.90
4" to 12" pipe, 20 to 50 lbs per LF	H8@8.00	Ton	1,610.00	381.00	92.30	2,083.30
6" to 12" pipe, 51 to 75 lbs per LF	H8@6.00	Ton	1,570.00	285.00	67.00	1,922.00

Square tube columns, in pounds per LF

	Craft@Hrs	Unit	Material	Labor	Equipment	Total
2" x 2" to 8" x 8", to 20 lbs per LF	H8@9.00	Ton	1,810.00	428.00	100.90	2,338.90
4" x 4" to 12" x 12", 21 to 50 lbs per LF	H8@8.00	Ton	1,490.00	381.00	92.30	1,963.30
8" x 8" to 16" x 16", 51 to 75 lbs per LF	H8@6.00	Ton	1,460.00	285.00	67.00	1,812.00
10" x 10" to 16" x 16", 76 to 100 lbs/LF	H8@5.00	Ton	1,430.00	238.00	55.60	1,723.60
Add for concrete fill in columns, 3,000 PSI design mix, placed by pump	M2@.464	CY	72.40	16.30	17.20	105.90

Fabricated Metals. These costs include the subcontractor's overhead and profit.

	Craft@Hrs	Unit	Material	Labor	Total
Catch basin grating and frame, 24" x 24", standard duty	CL@8.68	Ea	240.00	296.00	536.00

Manhole rings with cover

	Craft@Hrs	Unit	Material	Labor	Total
24", 330 lbs	CL@6.80	Ea	230.00	232.00	462.00
30", 400 lbs	CL@6.80	Ea	240.00	232.00	472.00
38", 730 lbs	CL@6.80	Ea	660.00	232.00	892.00

Welded steel grating. Based on Klemp Corporation. Weight shown is approximate. Costs shown are per square foot of grating, shop prime painted or galvanized after fabrication as shown.

Grating bars as shown, 1/2" cross bars at 4" OC

	Craft@Hrs	Unit	Material	Labor	Total
1-1/4" x 3/16" at 1-1/4" OC, 9 lbs per SF, painted	IW@.019	SF	4.90	.97	5.87
1-1/4" x 3/16" at 1-1/4" OC, 9 lbs per SF, galvanized	IW@.019	SF	7.25	.97	8.22
Add for banding on edges	--	LF	3.30	--	--
2" x 1/4" at 1-1/4" OC, 18.7 lbs per SF, painted	IW@.026	SF	10.25	1.33	11.58
2" x 1/4" at 1-1/4" OC, 18.7 lbs per SF, galvanized	IW@.026	SF	15.00	1.33	16.33
Add for banding on edges	--	LF	4.20	--	--
Add for toe-plates, steel, shop prime painted, attached after grating is in place, 1/4" thick x 4" high	IW@.048	LF	5.55	2.45	8.00

Checkered-steel floor plate, shop prime painted or galvanized after fabrication as shown

	Craft@Hrs	Unit	Material	Labor	Total
1/4", 11-1/4 lbs per SF, painted	IW@.025	SF	10.30	1.28	11.58
3/8", 16-1/2 lbs per SF, painted	IW@.025	SF	14.70	1.28	15.98
1/4", 11-1/4 lbs per SF, galvanized	IW@.025	SF	11.10	1.28	12.38
3/8", 16-1/2 lbs per SF, galvanized	IW@.025	SF	16.30	1.28	17.58

Checkered-steel trench cover plate, shop prime painted or galvanized after fabrication as shown

	Craft@Hrs	Unit	Material	Labor	Total
1/4", 11-1/4 lbs per SF, primed	IW@.045	SF	13.40	2.30	15.70
1/4", 11-1/4 lbs per SF, galvanized	IW@.045	SF	15.50	2.30	17.80
3/8", 16-1/2 lbs per SF, primed	IW@.045	SF	19.30	2.30	21.60
3/8", 16-1/2 lbs per SF, galvanized	IW@.045	SF	22.20	2.30	24.50
Canopy framing, steel	IW@.009	Lb	.85	.46	1.31
Miscellaneous supports, steel	IW@.019	Lb	1.35	.97	2.32
Miscellaneous supports, aluminum	IW@.061	Lb	9.25	3.11	12.36

Embedded steel, in concrete

	Craft@Hrs	Unit	Material	Labor	Total
Light, to 20 lbs per LF	T3@.042	Lb	1.65	1.70	3.35
Medium, over 20 to 50 lbs per LF	T3@.035	Lb	1.50	1.41	2.91
Heavy, over 50 lbs per LF	T3@.025	Lb	1.25	1.01	2.26

Metals 5

	Craft@Hrs	Unit	Material	Labor	Total
Carpenters iron, general	T3@.038	Lb	1.60	1.54	3.14
Channel sill, 1/8" steel, 1-1/2" x 8" wide	T3@.131	LF	12.90	5.29	18.19
Angle sill, 1/8" steel, 1-1/2" x 1-1/2"	T3@.095	LF	4.15	3.84	7.99
Fire escapes, ladder and balcony, per floor	IW@16.5	Ea	1,550.00	842.00	2,392.00
Grey iron foundry items	T3@.014	Lb	.75	.57	1.32
Stair nosing, 2-1/2" x 2-1/2", set in concrete	T3@.139	LF	6.45	5.62	12.07
Cast iron					
Lightweight sections	T3@.012	Lb	.80	.49	1.29
Heavy section	T3@.010	Lb	.65	.40	1.05
Cast column bases					
Iron, 16" x 16", custom	T3@.759	Ea	71.90	30.70	102.60
Iron, 32" x 32", custom	T3@1.54	Ea	300.00	62.20	362.20
Aluminum, 8" column, stock	T3@.245	Ea	16.80	9.90	26.70
Aluminum, 10" column, stock	T3@.245	Ea	21.60	9.90	31.50
Wheel type corner guards, per LF of height	T3@.502	LF	48.00	20.30	68.30
Angle corner guards, with anchors	T3@.018	Lb	1.45	.73	2.18
Channel door frames, with anchors	T3@.012	Lb	1.35	.49	1.84
Steel lintels, painted	T3@.013	Lb	.80	.53	1.33
Steel lintels, galvanized	T3@.013	Lb	1.05	.53	1.58
Anodizing, add to cost of fabrication					
Aluminum	--	%	--	--	15.0
Bronze	--	%	--	--	30.0
Black	--	%	--	--	50.0
Galvanizing, add to cost of fabrication	--	Lb	--	--	.35

Expansion Shields. Masonry anchors set in block or concrete including drilling. With nut and washer. Length indicates drilled depth. These costs include the subcontractor's overhead and profit.

	Craft@Hrs	Unit	Material	Labor	Total
One wedge anchors, 1/4" diameter					
1-3/4" long	CL@.079	Ea	.50	2.69	3.19
2-1/4" long	CL@.114	Ea	.59	3.89	4.48
3-1/4" long	CL@.171	Ea	.76	5.83	6.59
One wedge anchors, 1/2" diameter					
4-1/4" long	CL@.227	Ea	1.38	7.74	9.12
5-1/2" long	CL@.295	Ea	.83	10.10	10.93
7" long	CL@.409	Ea	2.25	13.90	16.15
One wedge anchors, 3/4" diameter					
5-1/2" long	CL@.286	Ea	3.69	9.75	13.44
8-1/2" long	CL@.497	Ea	5.20	16.90	22.10
10" long	CL@.692	Ea	6.14	23.60	29.74
One wedge anchors, 1" diameter					
6" long	CL@.295	Ea	9.20	10.10	19.30
9" long	CL@.515	Ea	11.30	17.60	28.90
12" long	CL@.689	Ea	13.10	23.50	36.60
Flush self-drilling anchors					
1/4" x 1-1/4"	CL@.094	Ea	.52	3.21	3.73
5/16" x 1-1/4"	CL@.108	Ea	.60	3.68	4.28
3/8" x 1-1/2"	CL@.121	Ea	.80	4.13	4.93
1/2" x 1-1/2"	CL@.159	Ea	1.20	5.42	6.62
5/8" x 2-3/8"	CL@.205	Ea	1.87	6.99	8.86
3/4" x 2-3/8"	CL@.257	Ea	3.02	8.76	11.78

Ornamental Metal. These costs include the subcontractor's overhead and profit.

	Craft@Hrs	Unit	Material	Labor	Total
Ornamental screens, aluminum	H6@.135	SF	24.80	5.75	30.55
Ornamental screens, extruded metal	H6@.151	SF	21.30	6.43	27.73

Metals 5

	Craft@Hrs	Unit	Material	Labor	Total
Ornamental Metal, continued					
Gates, wrought iron, 6' x 7', typical	--	Ea	--	--	1,240.00
Sun screens, aluminum stock design, manual	H6@.075	SF	26.80	3.19	29.99
Sun screens, aluminum stock design, motorized	H6@.102	SF	49.60	4.15	53.75
Brass and bronze work, general fabrication	H6@.079	Lb	18.20	3.36	21.56
Aluminum work, general fabrication	H6@.160	Lb	13.60	6.81	20.41
Stainless steel, general fabrication	H6@.152	Lb	14.10	6.47	20.57

Security Gratings. These costs include the subcontractor's overhead and profit.
Door grating, 1" tool steel vertical bars at 4" OC, 3/8" x 2-1/4" steel frame and horizontal bars at 18" OC,

	Craft@Hrs	Unit	Material	Labor	Total
3' x 7' overall	H6@10.5	Ea	1,080.00	427.00	1,507.00
Add for butt hinges and locks, per door	H6@2.67	Ea	324.00	114.00	438.00
Add for field painting, per door	H6@1.15	Ea	5.75	48.90	54.65

Window grating, 3/4" tool steel vertical bars at 4" OC, 3/8" x 2-1/4" steel frame and horizontal bars 6" OC, 20" x 30" overall,

	Craft@Hrs	Unit	Material	Labor	Total
set with 1/2" bolts at 6" OC	H6@3.21	Ea	141.00	137.00	278.00

Louver vent grating, 1" x 3/16", "I" bearing bars 1-3/16" OC with 3/16" spaced bars 4" OC and banded ends. Set with 1/2" bolts

	Craft@Hrs	Unit	Material	Labor	Total
at 1" OC, primed	H6@.114	SF	17.20	4.85	22.05

Steel plate window covers, hinged, 3/8" plate steel, 32" W x 48" H overall, 10 lbs per SF,

	Craft@Hrs	Unit	Material	Labor	Total
including hardware	H6@.150	SF	8.05	6.38	14.43

Wood and Plastic 6

	Craft@Hrs	Unit	Material	Labor	Total

Rough Carpentry. Industrial and commercial work. Price shown in parentheses is before markup or allowance for waste. Costs shown include typical waste and the subcontractor's overhead and profit.
Stud walls, including plates, blocks and bracing, 16" OC, 8' heights, parallel plates, per MBF

	Craft@Hrs	Unit	Material	Labor	Total
2" x 4" (at $450 per MBF)	C8@18.1	MBF	535.00	696.00	1,231.00
2" x 6" (at $460 per MBF)	C8@19.4	MBF	545.00	746.00	1,291.00
Stub (jack) studs, 2' to 6' long (at $450 per MBF)	C8@30.1	MBF	535.00	1,110.00	1,645.00

Stud walls, including plates, blocks and bracing, 16" OC, 10' to 16' heights, parallel plates, per MBF

	Craft@Hrs	Unit	Material	Labor	Total
2" x 4" (at $450 per MBF)	C8@19.9	MBF	535.00	766.00	1,301.00
2" x 6" (at $460 per MBF)	C8@19.4	MBF	545.00	746.00	1,291.00

Stud walls, including plates, blocks, bracing, flat bottom plate, sloping top plate, 8' to 16' high, 16" OC, per MBF

	Craft@Hrs	Unit	Material	Labor	Total
2" x 4" (at $450 per MBF)	C8@28.0	MBF	535.00	1,034.00	1,569.00
2" x 6" (at $460 per MBF)	C8@28.0	MBF	545.00	1,080.00	1,625.00

Stud walls, including plates, blocking, headers and diagonal bracing, per SF of wall, 16" OC

	Craft@Hrs	Unit	Material	Labor	Total
2" x 4", 8' high, 1.1 BF per SF (at $450 per MBF)	C8@.021	SF	.59	.78	1.37
2" x 4", 10' high, 1.0 BF per SF (at $450 per MBF)	C8@.021	SF	.54	.81	1.35
2" x 6", 8' high, 1.6 BF per SF (at $460 per MBF)	C8@.031	SF	.88	1.15	2.03
2" x 6", 10' high, 1.5 BF per SF (at $460 per MBF)	C8@.030	SF	.83	1.15	1.98
2" x 6", 8' high, 12" OC, 2.0 BF per SF (at $460)	C8@.037	SF	1.10	1.36	2.46
2" x 4", on 2" x 6" plates, staggered studs, 1.8 BF per SF (at $450 per MBF)	C8@.034	SF	.96	1.31	2.27

Underpinning, treated material, .60 ACZA treated

	Craft@Hrs	Unit	Material	Labor	Total
Pier caps, 2" x 6" (at $460 + $150 per MBF)	C8@61.8	MBF	725.00	2,280.00	3,005.00

Mud sills, treated material, add bolts or fastener

	Craft@Hrs	Unit	Material	Labor	Total
Bolted, 2" x 4" (at $450 + $150 per MBF)	C8@43.4	MBF	710.00	1,670.00	2,380.00
Bolted, 2" x 6" (at $460 + $150 per MBF)	C8@43.4	MBF	725.00	1,600.00	2,325.00
Shot, 2" x 4" (at $450 + $150 per MBF)	C8@35.5	MBF	710.00	1,370.00	2,080.00
Shot, 2" x 6" (at $460 + $150 per MBF)	C8@35.5	MBF	725.00	1,370.00	2,095.00

Wood and Plastic 6

	Craft@Hrs	Unit	Material	Labor	Total
Anchor bolts, with nut and washer, set and aligned in concrete					
3/8" x 8"	P8@.131	Ea	.33	4.86	5.19
3/8" x 10"	P8@.139	Ea	.40	5.16	5.56
3/8" x 12"	P8@.148	Ea	.46	5.49	5.95
1/2" x 8"	P8@.139	Ea	.45	5.16	5.61
1/2" x 10"	P8@.154	Ea	.56	5.72	6.28
1/2" x 12"	P8@.157	Ea	.63	5.83	6.46
5/8" x 8"	P8@.162	Ea	.98	6.01	6.99
5/8" x 10"	P8@.168	Ea	1.13	6.24	7.37
5/8" x 12"	P8@.174	Ea	1.32	6.46	7.78
5/8" x 18"	P8@.190	Ea	2.06	7.05	9.11
3/4" x 8"	P8@.209	Ea	1.44	7.76	9.20
3/4" x 12"	P8@.209	Ea	2.01	7.76	9.77
3/4" x 18"	P8@.260	Ea	2.58	9.65	12.23
7/8" x 8"	P8@.244	Ea	2.07	9.06	11.13
7/8" x 12"	P8@.260	Ea	2.30	9.65	11.95
7/8" x 18"	P8@.296	Ea	6.90	11.00	17.90
7/8" x 24"	P8@.341	Ea	9.78	12.70	22.48
Add for galvanized anchor bolts	--	%	30.0	--	--
Shot fasteners	CC@.023	Ea	.50	.99	1.49
Plates, (add for .60 ACZA treatment at $176 per MBF including markup, if required)					
2" x 4" (at $450 per MBF)	C8@18.4	MBF	535.00	708.00	1,243.00
2" x 6" (at $460 per MBF)	C8@15.7	MBF	550.00	604.00	1,154.00
Posts, #1 structural					
4" x 4" (at $825 per MBF)	C8@33.6	MBF	977.00	1,240.00	2,217.00
6" x 6" (at $1,020 per MBF)	C8@32.8	MBF	1,210.00	1,260.00	2,470.00
Headers, 4" x 12" (at $655 per MBF)	C8@19.0	MBF	758.00	731.00	1,489.00
Diagonal bracing, 1" x 4", let in (at $400 per MBF)	C8@23.3	MBF	480.00	896.00	1,376.00
Blocking and nailers, 2" x 6", # 3 (at $360 per MBF)	C8@24.4	MBF	437.00	939.00	1,376.00
Blocking, solid diagonal bracing, 2" x 4" (at $450/MBF)	C8@38.2	MBF	535.00	1,410.00	1,945.00
Beams and girders					
4" x 8" to 4" x 12" (at $600 per MBF)	C8@32.8	MBF	715.00	1,260.00	1,975.00
6" x 8" to 6" x 12" (at $1,040 per MBF)	C8@29.3	MBF	1,230.00	1,080.00	2,310.00
Joists and blocking, including rim joists, #2 & better					
2" x 6" (at $460 per MBF)	C8@19.3	MBF	550.00	742.00	1,292.00
2" x 8" (at $440 per MBF)	C8@19.0	MBF	520.00	731.00	1,251.00
2" x 10" (at $530 per MBF)	C8@17.2	MBF	630.00	662.00	1,292.00
2" x 12" (at $530 per MBF)	C8@16.6	MBF	630.00	639.00	1,269.00
Floor joists and blocking including rim joists and 6% waste, by SF of floor, 16" OC, #2 & Btr					
2" x 4", .75 BF per SF (at $450 per MBF)	C8@.012	SF	.40	.46	.86
2" x 6", 1.02 BF per SF (at $460 per MBF)	C8@.017	SF	.56	.65	1.21
2" x 8", 1.36 BF per SF (at $440 per MBF)	C8@.018	SF	.71	.69	1.40
2" x 10", 1.71 BF per SF (at $530 per MBF)	C8@.019	SF	1.08	.73	1.81
2" x 12", 2.05 BF per SF (at $530 per MBF)	C8@.021	SF	1.30	.81	2.11
Posts and mullions, solid timbers, select ($1,260/MBF)	C8@32.0	MBF	1,490.00	1,180.00	2,670.00
Purlins, struts, ridge boards, #2 and better					
4" x 10" (at $755 per MBF)	C8@30.6	MBF	895.00	1,180.00	2,075.00
4" x 12" (at $765 per MBF)	C8@30.6	MBF	906.00	1,130.00	2,036.00
6" x 10" (at $1,136 per MBF)	C8@24.4	MBF	1,345.00	939.00	2,284.00
6" x 12" (at $1,225 per MBF)	C8@24.4	MBF	1,451.00	939.00	2,390.00

Wood and Plastic 6

	Craft@Hrs	Unit	Material	Labor	Total

Rough Carpentry, continued
Rafters and outriggers, to 4 in 12 pitch, #2 & Btr

	Craft@Hrs	Unit	Material	Labor	Total
2" x 4" to 2" x 8" (at $485 per MBF)	C8@32.5	MBF	575.00	1,200.00	1,775.00
2" x 10" to 2" x 12" (at $540 per MBF)	C8@29.8	MBF	638.00	1,150.00	1,788.00
Add for over 4 in 12 pitch roof	--	%	--	30.0	--

Rafters, including frieze blocks, ribbon, strongbacks, 24" OC, per SF of roof plan area, #2 & Btr

	Craft@Hrs	Unit	Material	Labor	Total
2" x 4", .44 BF per SF (at $450 per MBF)	C8@.015	SF	.22	.58	.80
2" x 6", .65 BF per SF (at $460 per MBF)	C8@.021	SF	.36	.81	1.17
2" x 8", .86 BF per SF (at $440 per MBF)	C8@.027	SF	.45	1.04	1.49
2" x 10", 1.10 BF per SF (at $530 per MBF)	C8@.031	SF	.70	1.19	1.89
2" x 12", 1.40 BF per SF (at $530 per MBF)	C8@.041	SF	.88	1.58	2.46

Plywood and Lumber Sheathing. Price shown in parentheses is before markup or waste allowance. Costs shown include nails, normal waste and the subcontractor's overhead and profit.
Floors, hand nailed, underlayment C-D grade, B cross core

	Craft@Hrs	Unit	Material	Labor	Total
1/2" (at $.32 per SF)	C8@8.60	MSF	393.00	331.00	724.00
5/8" (at $.37 per SF)	C8@9.19	MSF	454.00	354.00	808.00
3/4" (at $.50 per SF)	C8@9.79	MSF	605.00	377.00	982.00
5/8" T&G (at $.40 per SF)	C8@9.79	MSF	490.00	377.00	867.00
3/4" T&G (at $.47 per SF)	C8@10.7	MSF	563.00	412.00	975.00
1-1/8" T&G, 2-4-1 (at $.85 per SF)	C8@14.1	MSF	1,030.00	542.00	1,572.00
Deduct for machine nailing	--	MSF	--	-101.00	--

Walls, B-D, rough, standard, interior grade

	Craft@Hrs	Unit	Material	Labor	Total
5/16" (at $.27 per SF)	C8@15.3	MSF	332.00	589.00	921.00
3/8" (at $.30 per SF)	C8@15.3	MSF	365.00	589.00	954.00
1/2" (3 ply) (at $.35 per SF)	C8@15.9	MSF	418.00	612.00	1,030.00
5/8" (at $.46 per SF)	C8@18.1	MSF	556.00	696.00	1,252.00
3/4" (at $.55 per SF)	C8@18.1	MSF	660.00	696.00	1,356.00
5/8" fire retardant, A-B, scr'd to metal frame ($.67)	C8@26.1	MSF	810.00	963.00	1,773.00

Roofs, C-D, rough, structural #1

	Craft@Hrs	Unit	Material	Labor	Total
3/8" (at $.26 per SF)	C8@9.47	MSF	315.00	364.00	679.00
1/2" (at $.32 per SF)	C8@9.47	MSF	393.00	364.00	757.00
5/8" (at $.38 per SF)	C8@9.47	MSF	454.00	364.00	818.00
3/4" (at $.50 per SF)	C8@11.3	MSF	605.00	435.00	1,040.00
Add for T&G edge (at $.017 per SF)	C8@.542	MSF	20.00	20.90	40.90
Deduct for non-structural	--	%	-15.0	--	--
Add for nailing shear walls	--	%	--	20.0	--
Deduct for machine nailing	--	%	--	-25.0	--
Blocking for plywood joints, 2" x 4" (at $450 per MBF)	C8@15.3	MBF	535.00	589.00	1,124.00

Lumber floor sheathing

	Craft@Hrs	Unit	Material	Labor	Total
Diagonal, 1" x 6" Std & Btr, 1.14 BF/SF ($500/MBF)	C8@17.7	MSF	700.00	654.00	1,354.00
Straight 2" x 6" commercial grade T&G fir, 2.28 BF per SF (at $450 per MBF)	C8@32.0	MSF	1,245.00	1,230.00	2,475.00
Straight 2" x 6" commercial grade T&G cedar, 2.28 BF per SF (at $590 per MBF)	C8@32.0	MSF	1,626.00	1,180.00	2,806.00
Timber deck, SYP, 5/4" x 6", premium grade, 1.25 BF per SF (at $660 per MBF)	C8@12.0	MBF	1,000.00	462.00	1,462.00

Lumber wall sheathing, diagonal, 1" x 6"

	Craft@Hrs	Unit	Material	Labor	Total
Std & Btr, 1.14 BF per SF (at $615 per MBF)	C8@24.8	MSF	852.00	954.00	1,806.00

Lumber roof sheathing, diagonal, 1" x 6"

	Craft@Hrs	Unit	Material	Labor	Total
Std & Btr, 1.14 BF per SF (at $615 per MBF)	C8@16.0	MSF	852.00	616.00	1,468.00

Wood and Plastic 6

	Craft@Hrs	Unit	Material	Labor	Total

Panelized Wood Roof Systems. These costs include the subcontractor's overhead and profit and will apply on 20,000 to 30,000 SF jobs where roof panels are fabricated near the point of installation and moved into place with a forklift. Costs will be higher on smaller jobs and jobs with poor access. 4' x 8' and 4' x 10' roof panels are made of plywood nailed to 2" x 4" or 2" x 6" stiffeners spaced 24" on centers and running the long dimension of the panel. Based on fabricated panels lifted by forklift (with metal hangers attached) and nailed to 4" x 12" or 4" x 14" roof purlins spaced 8' on center. Equipment cost is $14 per hour for a truck with hand tools, a portable air compressor with air hoses and two pneumatic-operated nailing guns. Costs are per 1,000 square feet (MSF) of roof surface and do not include roof supports except as noted. Lumber costs assume a lumber price of $430 per MBF for stiffeners and $620 per MBF for purlins before markup and a price of $320 per MSF for 1/2" plywood before markup.

For scheduling purposes, estimate that a crew of 5 can fabricate 6,000 to 6,400 SF of 4' x 8' or 4' x 10' panels in an 8-hour day. Estimate that the same crew of 5 can install and nail down 7,600 to 8,000 SF of panels in an 8-hour day.

4' x 8' roof panels with 1/2" CDX structural #1 plywood attached
to 2" x 4" const. grade fir stiffeners

	Craft@Hrs	Unit	Material	Labor	Total
Panels fabricated, ready to install	F6@6.60	MSF	655.00	259.00	914.00
Equipment & labor to install and nail panels	F6@5.21	MSF	74.00	191.00	265.00
Total installed cost, 4' x 8' panels	**F6@11.8**	**MSF**	**729.00**	**450.00**	**1,179.00**

4' x 10' roof panels with 1/2" CDX structural #1 plywood attached
to 2" x 6" const. grade fir stiffeners

	Craft@Hrs	Unit	Material	Labor	Total
Panels fabricated, ready to install	F6@6.22	MSF	800.00	244.00	1,044.00
Equipment & labor to install and nail panels	F6@4.91	MSF	70.00	180.00	250.00
Total installed cost, 4' x 10' panels	**F6@11.1**	**MSF**	**870.00**	**424.00**	**1,294.00**

8' x 20' and 8' x 32' roof panels can be fabricated on site using the same crew and equipment. However, the stiffeners run across the short dimension of the panel and one purlin is attached to the outside edge of the long dimension of each panel before lifting into place. Figures below include the cost of purlins but not supporting beams. For scheduling purposes, estimate that a crew of 5 can fabricate 6,400 to 7,000 SF of 8' x 20' or 8' x 32' panels in an 8-hour day. Estimate that the same crew of 5 can install and nail down 10,000 to 11,000 SF of panels in an 8-hour day.

8' x 20' panels with purlins attached
 4' x 10' x 1/2" CDX structural #1 plywood attached
  to 2" x 4" const. grade fir stiffeners and 4" x 12" #1 grade fir purlins

	Craft@Hrs	Unit	Material	Labor	Total
Panels fabricated, ready to install	F6@6.22	MSF	1,420.00	244.00	1,664.00
Equipment & labor to install and nail panels	F6@3.96	MSF	56.00	145.00	201.00
Total installed cost, 8' x 20' panels	**F6@10.1**	**MSF**	**1,476.00**	**389.00**	**1,865.00**

8' x 32' panel with purlins attached
 4' x 8' x 1/2" CDX structural #1 plywood attached
  to 2" x 4" const. grade fir stiffeners and 4" x 12" #1 grade fir purlins

	Craft@Hrs	Unit	Material	Labor	Total
Panels fabricated, ready to install	F6@5.81	MSF	1,390.00	228.00	1,618.00
Equipment & labor to install and nail panels	F6@3.55	MSF	50.00	130.00	180.00
Total installed cost, 8' x 32' panels	**F6@9.36**	**MSF**	**1,440.00**	**358.00**	**1,798.00**

Wood and Plastic 6

	Craft@Hrs	Unit	Material	Labor	Total

Panelized Wood Roof Systems, continued
Add to any of the panel costs shown on preceding page as may be required for:

	Craft@Hrs	Unit	Material	Labor	Total
5/8" CDX #1 structural plywood, instead of 1/2" attached before panels are erected	--	MSF	65.00	--	65.00
Roof hatches, includes reinforced opening, curb & hatch, installed before panel is erected	F6@.224	SF	15.00	8.80	23.80

Panelized Wood Wall Panels. These costs include the subcontractor's overhead and profit and will apply on jobs with 20,000 to 30,000 SF of wall panels. Panel material costs are for prefabricated panels purchased from a modular plant located within 50 miles of the job site. Costs will be higher on smaller jobs and on jobs with poor access. Costs assume wall panels are fabricated according to job plans and specifications, delivered piece-marked, ready to install and finish. Labor costs for installation of panels are listed at the end of this section and include unloading and handling, sorting, moving the panels into position using a forklift, installing and nailing. Maximum panel height is usually 12'. Equipment cost is $14.00 per hour and includes a forklift, hand tools, a portable air compressor with air hoses and two pneumatic-operated nailing guns. Costs are per square foot of wall, measured on one face. For scheduling purposes, estimate that a crew of 4 can place and fasten 3,500 to 4,500 SF of panels in an 8-hour day.

	Craft@Hrs	Unit	Material	Labor	Equipment	Total
2" x 4" studs 16" OC with gypsum wallboard on both sides						
1/2" wallboard	--	SF	1.24	--	--	--
5/8" wallboard	--	SF	1.32	--	--	--
2" x 4" studs 16" OC with gypsum wallboard inside and plywood outside						
1/2" wallboard and 1/2" CDX plywood	--	SF	1.40	--	--	--
1/2" wallboard and 5/8" CDX plywood	--	SF	1.50	--	--	--
5/8" wallboard and 1/2" CDX plywood	--	SF	1.40	--	--	--
5/8" wallboard and 5/8" CDX plywood	--	SF	1.55	--	--	--
2" x 6" studs 12" OC with gypsum wallboard on both sides						
1/2" wallboard	--	SF	1.40	--	--	--
5/8" wallboard	--	SF	1.50	--	--	--
2" x 6" studs 12" OC with gypsum wallboard inside and plywood outside						
1/2" wallboard and 1/2" CDX plywood	--	SF	1.58	--	--	--
1/2" wallboard and 5/8" CDX plywood	--	SF	1.66	--	--	--
5/8" wallboard and 1/2" CDX plywood	--	SF	1.58	--	--	--
5/8" wallboard and 5/8" CDX plywood	--	SF	1.72	--	--	--
2" x 6" studs 16" OC with gypsum wallboard on both sides						
1/2" wallboard	--	SF	1.36	--	--	--
5/8" wallboard	--	SF	1.45	--	--	--
2" x 6" studs 16" OC with gypsum wallboard inside and plywood outside						
1/2" wallboard and 1/2" CDX plywood	--	SF	1.54	--	--	--
1/2" wallboard and 5/8" CDX plywood	--	SF	1.62	--	--	--
5/8" wallboard and 1/2" CDX plywood	--	SF	1.54	--	--	--
5/8" wallboard and 5/8" CDX plywood	--	SF	1.68	--	--	--
Add to any of the above for the following, installed by panel fabricator:						
R-19 insulation	--	SF	.40	--	--	--
R-11 insulation	--	SF	.37	--	--	--
Vapor barrier	--	SF	.05	--	--	--
Door or window openings to 16 SF each	--	SF	.60	--	--	--
Door or window openings over 16 SF each	--	SF	.54	--	--	--
Electrical receptacles	--	Ea	30.00	--	--	--
Electrical switches	--	Ea	34.25	--	--	--
Labor and equipment cost to install panels						
Up to 8' high	F7@.003	SF	--	.12	.27	.39
Over 8' to 10' high	F7@.002	SF	--	.08	.24	.32
Over 10' to 12' high	F7@.002	SF	--	.08	.22	.30

Wood and Plastic 6

	Craft@Hrs	Unit	Material	Labor	Total

Sills and Ledgers Bolted to Concrete or Masonry Walls. Costs are per linear foot of sill or ledger using standard and better pressure treated fir with anchor bolt holes drilled each 2'6" center. Labor costs include handling, measuring, cutting to length and drilling holes at floor level and installation on a scaffold 10' to 20' above floor level. Figures in parentheses show the board feet per linear foot of sill or ledger including 5% waste. For scheduling purposes, estimate that a crew of 2 can install 150 to 200 LF of sills with a single row of anchor bolts in an 8-hour day when working from scaffolds. Estimate that a crew of 2 can install 150 to 175 LF of ledgers with a single row of anchor bolt holes or 50 to 60 LF of ledgers with a double row of anchor bolt holes in an 8-hour day working from scaffolds. Add the cost of anchor bolts and scaffold, if required. These costs include the subcontractor's overhead and profit.

	Craft@Hrs	Unit	Material	Labor	Total
Sills with a single row of anchor bolt holes					
2" x 4" at $685 per MBF (0.70 BF per LF)	C8@.040	LF	.48	1.54	2.02
2" x 6" at $700 per MBF (1.05 BF per LF)	C8@.040	LF	.74	1.54	2.28
2" x 8" at $670 per MBF (1.40 BF per LF)	C8@.040	LF	.94	1.54	2.48
2" x 10" at $780 per MBF (1.75 BF per LF)	C8@.040	LF	1.36	1.54	2.90
2" x 12" at $780 per MBF(2.10 BF per LF)	C8@.040	LF	1.64	1.54	3.18
4" x 6" at $1,020 per MBF (2.10 BF per LF)	C8@.051	LF	2.15	1.96	4.11
4" x 8" at $1,020 per MBF (2.80 BF per LF)	C8@.051	LF	2.85	1.96	4.81
4" x 10" at $1,020 per MBF (3.33 BF per LF)	C8@.059	LF	3.40	2.27	5.67
4" x 12" at $1,020 per MBF (4.20 BF per LF)	C8@.059	LF	4.28	2.27	6.55
Ledgers with a single row of anchor bolt holes					
2" x 4" at $670 per MBF (0.70 BF per LF)	C8@.059	LF	.48	2.27	2.75
4" x 6" at $1,020 per MBF (2.10 BF per LF)	C8@.079	LF	2.15	3.04	5.19
4" x 8" at $1,020 per MBF (2.80 BF per LF)	C8@.079	LF	2.85	3.04	5.89
Ledgers with a double row of anchor bolt holes					
4" x 4" at $1,020 per MBF (1.40 BF per LF)	C8@.120	LF	1.42	4.62	6.04
4" x 6" at $1,020 per MBF (2.10 BF per LF))	C8@.160	LF	2.15	6.16	8.31
4" x 8" at $1,020 per MBF (2.80 BF per LF)	C8@.160	LF	2.85	6.16	9.01
4" x 12" at $1,020 per MBF (4.20 BF per LF)	C8@.250	LF	4.28	9.62	13.90

Miscellaneous Rough Carpentry. These costs include the subcontractor's overhead and profit.

	Craft@Hrs	Unit	Material	Labor	Total
Furring at $495 per MBF including 5% for waste & nails					
2" x 4" on masonry walls	C8@.024	LF	.42	.92	1.34
1" x 4" machine nailed to ceiling	C8@.017	LF	.20	.65	.85
1" x 4" nailed on concrete	C8@.016	LF	.20	.62	.82
Skip sheathing, on roof rafters, machine nailed, at $750 per MBF including 7% for waste & nails					
1" x 6" at 6" centers (1 BF per SF)					
Per MBF lumber	C8@27.0	MBF	620.00	1,040.00	1,660.00
Per SF of roof area	C8@.027	SF	.60	1.04	1.64
1" x 4" at 6" centers (.67 BF per SF)					
Per MBF lumber	C8@14.6	MBF	620.00	562.00	1,182.00
Per SF of roof area	C8@.027	SF	.42	1.04	1.46
1" x 4" at 4" centers (1 BF per SF)					
Per MBF lumber	C8@27.0	MBF	620.00	1,040.00	1,660.00
Per SF of roof area	C8@.027	SF	.60	1.04	1.64
Ribbons and strongbacks	C8@26.7	MBF	535.00	1,030.00	1,565.00
Backing for other trades	C8@31.7	MBF	535.00	1,170.00	1,705.00
Beams, solid timbers, 6" and larger, #1 structural, specified lengths	C8@23.9	MBF	1,210.00	919.00	2,129.00
Hardboard, 4' x 8', used as underlayment					
1/8" tempered	C8@.011	SF	.40	.42	.82
1/4" tempered	C8@.011	SF	.57	.42	.99
Add for attaching to metal frame	--	%	--	50.0	--

Wood and Plastic 6

	Craft@Hrs	Unit	Material	Labor	Total
Miscellaneous Rough Carpentry, continued					
Glassweld panels					
1/8", finish 1 side	C8@.040	SF	3.72	1.54	5.26
1/4", finish 2 sides	C8@.055	SF	4.96	2.12	7.08
15 pound felt on walls	C8@3.31	MSF	45.00	127.00	172.00
Kraft flashing with foil 2 sides, 30 pound	C8@4.04	MSF	81.00	155.00	236.00
Insulation board on walls, 4' x 8'					
Sound control board, 1/2"	C8@.010	SF	.30	.39	.69
Building board, 1/2"	C8@.010	SF	.40	.39	.79
Asphalt impregnated 1/2" Firtex	C8@.012	SF	.30	.46	.76

Wood Trusses. Material costs are for prefabricated wood trusses shipped to job site fully assembled ready for installation. Equipment is a forklift at $16 per hour. Labor includes unloading and handling and installing the trusses not over 20' above grade. These costs include the subcontractor's overhead and profit.

	Craft@Hrs	Unit	Material	Labor	Equipment	Total
Truss joists, at 2' OC, complete, including blocking						
"I" series	F6@.012	SF	1.45	.47	.19	2.11
Simple	F6@.011	SF	1.76	.43	.18	2.37
Compound	F6@.010	SF	2.02	.39	.16	2.57
Hardware, shoes (at $1.00 per pound)	--	SF	.10	--	--	--
Add for job under 3,000 SF	--	%	--	--	--	20.0
Roof trusses, 2" x 4" top and bottom chord						
Fink truss "W"						
24' span, 3 in 12 slope	F6@.712	Ea	43.30	28.00	11.10	82.40
24' span, 4 in 12 slope	F6@.676	Ea	41.80	26.60	10.60	79.00
28' span, 5 in 12 slope	F6@.840	Ea	51.60	33.00	13.00	97.60
32' span, 5 in 12 slope	F6@1.11	Ea	67.10	43.60	17.30	128.00
40' span, 5 in 12 slope	F6@1.49	Ea	90.80	58.60	23.20	172.60
Gable truss						
28' span, 5 in 12 slope	F6@.840	Ea	64.50	33.00	13.00	110.50
32' span, 5 in 12 slope	F6@1.11	Ea	87.10	43.60	17.30	148.00
40' span, 5 in 12 slope	F6@1.49	Ea	126.00	58.60	23.20	207.80

Laminated Glued Beams (Glu-lam beams). The following gives typical costs for horizontal laminated glu-lam beams. Figures in parentheses give approximate weight per linear foot of beam. Equipment cost is based on using a 4,000 pound capacity forklift, with 20' maximum lift, at $21.50 per hour. These costs include the subcontractor's overhead and profit.

	Craft@Hrs	Unit	Material	Labor	Equipment	Total
Steel connectors and shoes for glu-lam beams, including bolts,						
typical per end or splice connection	--	Ea	50.00	--	--	--
Drill holes and attach connectors at grade before installing beam,						
based on number of holes in the connector, with work performed at job site,						
labor per each hole	F6@.236	Ea	--	9.27	--	--
Install beams with connections attached	F6@.133	LF	--	5.22	.57	5.79
Stock glu-lam beams, typical costs						
West coast Douglas fir/hemlock, 24F-V5, H=155 PSI, exterior glue, camber 5000' R						
Industrial grade, as is, no wrap. For other grades see additive adjustment data that follows below.						
2-1/2" x 6" (3.6 pounds per LF)	--	LF	2.60	--	--	--
2-1/2" x 9" (5.4 pounds per LF)	--	LF	3.90	--	--	--
2-1/2" x 12" (7.1 pounds per LF)	--	LF	5.15	--	--	--
West coast Douglas fir, 24F-V4, H=165 PSI, exterior glue, camber is 5000' R for beams through 15" deep and 2000' R for beams over 15" deep.						
Industrial grade, as is, no wrap. For other grades see additive adjustment data that follows.						
3-1/8" x 6" (4.4 pounds per LF)	--	LF	3.70	--	--	--

Wood and Plastic 6

	Craft@Hrs	Unit	Material	Labor	Equipment	Total
3-1/8" x 9" (6.6 pounds per LF)	--	LF	5.50	--	--	--
3-1/8" x 10-1/2" or 12" (8.0 pounds per LF)	--	LF	6.50	--	--	--
3-1/8" x 15" (10.0 pounds per LF)	--	LF	8.10	--	--	--
3-1/8" x 18" (13.2 pounds per LF)	--	LF	9.70	--	--	--
3-1/2" x 6" (5.0 pounds per LF)	--	LF	3.60	--	--	--
3-1/2" x 9" (7.5 pounds per LF)	--	LF	5.40	--	--	--
3-1/2" x 12" (9.0 pounds per LF)	--	LF	7.20	--	--	--
3-1/2" x 15" (10.0 pounds per LF)	--	LF	9.00	--	--	--
3-1/2" x 19-1/2" (16.2 pounds per LF)	--	LF	11.70	--	--	--
3-1/2" x 21" (17.5 pounds per LF)	--	LF	12.60	--	--	--
5-1/8" x 6" (7.3 pounds per LF)	--	LF	5.10	--	--	--
5-1/8" x 9" (10.7 pounds per LF)	--	LF	7.60	--	--	--
5-1/8" x 12" (14.5 pounds per LF)	--	LF	10.00	--	--	--
5-1/8" x 15" (18.1 pounds per LF)	--	LF	12.50	--	--	--
5-1/8" x 18" (21.7 pounds per LF)	--	LF	14.90	--	--	--
5-1/8" x 21" (25.4 pounds per LF)	--	LF	17.40	--	--	--
5-1/8" x 24" (29.0 pounds per LF)	--	LF	19.90	--	--	--
6-3/4" x 12" (19.1 pounds per LF)	--	LF	13.80	--	--	--
6-3/4" x 15" (23.8 pounds per LF)	--	LF	17.10	--	--	--
6-3/4" x 18" (28.6 pounds per LF)	--	LF	20.40	--	--	--
6-3/4" x 21" (33.3 pounds per LF)	--	LF	23.70	--	--	--
6-3/4" x 24" (38.1 pounds per LF)	--	LF	27.00	--	--	--
Additive adjustments, add to any of the beam costs shown in this section for:						
Industrial grade, clean, wrapped, add	--	%	12.5	--	--	--
Architectural grade, wrapped, add	--	%	20.0	--	--	--
One-coat rez-latex stain, add	--	%	10.0	--	--	--
Transportation to job site, per delivery, maximum 50 miles one way, add	--	LS	200.00	--	--	--

Wood Stairs. These costs include the subcontractor's overhead and profit. Cost per riser.

	Craft@Hrs	Unit	Material	Labor	Total
Hardwood treads, closed riser, straight run, 36" riser, precut materials	C8@.749	Ea	78.00	28.80	106.80
Hardwood treads, switchback, closed riser, 36" riser, precut material	C8@.876	Ea	95.00	33.70	128.70
Dadoed open riser softwood industrial stairway, 36" wide, two 2" x 12" stringers, 2" x 12" treads, 2" x 4" handrail each side, 2" x 2" balusters at 8" on one side, 4" x 4" support post, 9' rise, (250 BF of lumber), cost per 7" riser	C8@.721	Ea	14.70	27.70	42.40
Stair landings, finished hardwood, framed and supported	C8@.184	SF	12.30	7.08	19.38
Hardwood, four-way, closed, 36" wide riser, per riser	C8@1.16	Ea	103.80	44.60	148.40
Circular, hardwood, closed riser, mill made, 4' diameter, cost per 7-1/2" riser	C8@.740	Ea	247.00	28.50	275.50
Circular, hardwood, open riser, mill made, 4, diameter, cost per 7-1/2" riser	C8@.740	Ea	236.00	28.50	264.50
Built on job, tract type, 36" wide riser, semicircular, hardwood, carriage mill made, per riser	C8@.819	Ea	85.00	31.50	116.50
Stair railing, not including hardware, hand rail only					
Single piece, 1-5/8" x 1-3/4", fir or pine	C8@.117	LF	2.65	4.50	7.15
Single piece, 1-5/8" x 2-5/8", fir or pine, detailed	C8@.117	LF	3.30	4.32	7.62
Custom design, hardwood, typical	C8@.188	LF	20.30	7.23	27.53

Thermal and Moisture Protection 7

	Craft@Hrs	Unit	Material	Labor	Total

Dampproofing. These costs include the subcontractor's overhead and profit.

	Craft@Hrs	Unit	Material	Labor	Total
Asphalt wall primer, per coat, gallon at $7.50 covers 250 SF	CL@.010	SF	.03	.34	.37
Asphalt emulsion, wall, per coat, gallon at $6.00 covers 50 SF, brush on	CL@.013	SF	.13	.44	.57
Hot mop concrete wall, 2 coats and glass fabric	CL@.033	SF	.25	1.08	1.33
Hot mop deck, 2 ply, felt, typical price	--	Sq	--	--	65.00
Hot mop deck, 3 ply, felt, typical price	--	Sq	--	--	90.00
Add for 1/2" asphalt fiberboard	--	Sq	--	--	45.00
Bituthene waterproofing membrane, 1/16" plain surface	CL@.025	SF	2.00	.85	2.85
Waterproof baths in high rise	--	SF	--	--	2.00
Iron compound, internal coating, 2 coats	--	SF	--	--	1.50
Roof vapor barrier, .004" polyethylene	CL@.002	SF	.05	.07	.12

Exterior Wall and Finish System (EIFS). Exterior non-structural wall finish applied to concrete, masonry, stucco or exterior grade gypsum sheathing. System consists of a plaster finish coat approximately 1/4" thick, with integral color, troweled on over a glass fiber mesh that is embedded in an adhesive base coat. The base coat is applied over polystyrene insulation boards that are bonded to an existing flat vertical wall surface with adhesive. Costs shown are per SF of finished wall surface on structures not over three stories in height. These costs include the subcontractor's overhead and profit but no surface preparation or wall costs. Cost for scaffolding is not included: add for scaffolding if required. For scheduling purposes, estimate that a crew of three can field apply 400 SF of exterior wall insulation and finish systems in an 8-hour day.

	Craft@Hrs	Unit	Material	Labor	Total
Adhesive mixture, 2 coats 2.1 SF per SF of wall area including waste	F8@.013	SF	.45	.51	.96
Glass fiber mesh 1.1 SF per SF of wall area including waste	F8@.006	SF	.25	.23	.48
Insulation board, 1" thick 1.05 SF per SF of wall area including waste	F8@.013	SF	.15	.51	.66
Textured finish coat 1.05 SF per SF of wall area including waste	F8@.024	SF	.50	.94	1.44
Total with 1" thick insulation	**F8@.056**	**SF**	**1.35**	**2.19**	**3.54**
Add for 2" thick insulation board	--	SF	.15	--	--
Add for 3" thick insulation board	--	SF	.25	--	--
Add for 4" thick insulation board	--	SF	.40	--	--

Insulation. These costs include the subcontractor's overhead and profit.
Roof Insulation. Due to regulations that may be imposed on manufacturers by government agencies, estimators should verify with material suppliers the cost of urethanes, isocyanurates, phenolics and polystyrene roof insulation board before preparing estimates on major projects.

	Craft@Hrs	Unit	Material	Labor	Total
Fiberglass board roof insulation					
3/4", R-2.80, C-0.36	R3@.570	Sq	37.80	24.00	61.80
1", R-4.20, C-0.24	R3@.636	Sq	45.40	26.80	72.20
1-3/8", R-5.30, C-0.19	R3@.684	Sq	58.35	28.80	87.15
1-5/8", R-6.70, C-0.15	R3@.706	Sq	66.00	29.80	95.80
2-1/4", R-8.30, C-0.12	R3@.797	Sq	80.00	33.60	113.60
Perlite-urethane composition board roof insulation					
1-1/2", R-6.70, C-0.15	R3@.478	Sq	75.00	20.20	95.20
1-5/8", R-7.70, C-0.13	R3@.478	Sq	80.00	20.20	100.20
2", R-10.0, C-0.10	R3@.684	Sq	94.20	28.80	123.00
2-1/4", R-12.5, C-0.08	R3@.728	Sq	114.00	30.70	144.70
2-1/2", R-14.3, C-0.07	R3@.728	Sq	128.75	30.70	159.45

Thermal and Moisture Protection 7

	Craft@Hrs	Unit	Material	Labor	Total
2-3/4", R-16.7, C-0.06	R3@.820	Sq	152.50	34.60	187.10
3-1/4", R-20.0, C-0.05	R3@.889	Sq	184.40	37.50	221.90
Phenolic board roof insulation					
1.2", R-10.0, C-0.10	R3@.500	Sq	50.00	21.10	71.10
1.5", R-12.5, C-0.08	R3@.592	Sq	56.20	25.00	81.20
1.75", R-14.6, C-0.07	R3@.592	Sq	61.00	25.00	86.00
2", R-16.7, C-0.06	R3@.684	Sq	70.40	28.80	99.20
2.4", R-20.0, C-0.05	R3@.728	Sq	76.80	30.70	107.50
3.0", R-25.0, C-0.04	R3@.889	Sq	102.70	37.50	140.20
3.3", R-28.0, C-0.035	R3@1.05	Sq	132.50	44.30	176.80
Urethane board roof insulation					
3/4", R-5.30, C-0.19	R3@.478	Sq	42.40	20.20	62.60
1", R-6.70, C-0.15	R3@.478	Sq	55.20	20.20	75.40
1-1/4", R-8.30, C-0.12	R3@.478	Sq	61.00	20.20	81.20
1-1/2", R-11.1, C-0.09	R3@.570	Sq	73.60	24.00	97.60
2", R-14.3, C-0.07	R3@.684	Sq	96.25	28.80	125.05
2-1/4", R-16.7, C-0.06	R3@.684	Sq	103.80	28.80	132.60
Perlite board roof insulation					
1", R-2.80, C-0.36	R3@.639	Sq	42.40	26.90	69.30
1-1/2", R-4.20, C-0.24	R3@.728	Sq	62.00	30.70	92.70
2", R-5.30, C-0.19	R3@.820	Sq	76.20	34.60	110.80
2-1/2", R-6.70, C-0.15	R3@.956	Sq	94.70	40.30	135.00
3", R-8.30, C-0.12	R3@1.03	Sq	115.00	43.40	158.40
4", R-10.0, C-0.10	R3@1.23	Sq	153.00	51.90	204.90
5-1/4", R-14.3, C-0.07	R3@1.57	Sq	202.00	66.20	268.20
Tapered 3/16" per foot	R3@.495	Sq	58.00	20.90	78.90
Cellular Foamglass board insulation, on roofs					
1-1/2"	R3@.775	Sq	153.20	32.70	185.90
Tapered 3/16" per foot	R3@.775	Sq	140.00	32.70	172.70
Polystyrene board roof insulation					
1", R-4.20, C-0.24	R3@.545	Sq	32.45	23.00	55.45
1-1/2", R-6.30, C-0.16	R3@.639	Sq	43.30	26.90	70.20
2", R-8.30, C-0.12	R3@.797	Sq	56.20	33.60	89.80
Fasteners for board type insulation on roofs					
Typical 2-1/2" to 3-1/2" screws with 3" metal disks					
I-60	R3@.053	Sq	2.45	2.23	4.68
I-90	R3@.080	Sq	3.70	3.37	7.07
Add for plastic disks	--	%	20.0	--	--
Fiberglass batts, placed between or over framing members,					
3-1/2" unfaced, R-11, between ceiling joists	A1@.006	SF	.27	.24	.51
3-1/2" unfaced, R-11, on suspended ceiling,					
working off ladders from below	A1@.018	SF	.27	.73	1.00
3-1/2" unfaced, R-11, in crawl space	A1@.020	SF	.27	.81	1.08
3-1/2" unfaced, R-19, between studs	A1@.006	SF	.27	.24	.51
6-1/4" unfaced, R-19, between ceiling joists	A1@.006	SF	.42	.24	.66
6-1/4" unfaced, R-19, on suspended ceiling,					
working off ladders from below	A1@.018	SF	.42	.73	1.15
6-1/4" unfaced, R-19, in crawl space	A1@.020	SF	.42	.81	1.23
6-1/4" unfaced, R-19, between studs	A1@.006	SF	.42	.24	.66
Add for supporting batts on wire rods, 1 per SF	A1@.001	SF	.03	.04	.07
Add for foil one side	--	SF	.03	--	--
Add for foil two sides	--	SF	.05	--	--
Add for Kraft paper one side	--	SF	.02	--	--
Add for Kraft paper two sides	--	SF	.03	--	--

Thermal and Moisture Protection 7

	Craft@Hrs	Unit	Material	Labor	Total

Insulation, continued
Fasteners for batts placed under a ceiling deck, including flat washers

	Craft@Hrs	Unit	Material	Labor	Total
2-1/2" long	A1@.025	Ea	.16	1.02	1.18
4-1/2" long	A1@.025	Ea	.18	1.02	1.20
6-1/2" long	A1@.025	Ea	.25	1.02	1.27
Domed self-locking washers	A1@.005	Ea	.14	.20	.34

Blown fiberglass or mineral wool, over ceiling joists. Add blowing equipment cost at $264.00 per day

	Craft@Hrs	Unit	Material	Labor	Total
R-11, 5"	A1@.008	SF	.15	.33	.48
R-13, 6"	A1@.010	SF	.20	.41	.61
R-19, 9"	A1@.014	SF	.27	.57	.84

Urethane board installed on perimeter walls

	Craft@Hrs	Unit	Material	Labor	Total
1", R-6.70, C-0.15	A1@.010	SF	.55	.41	.96
1-1/2", R-11.0, C-0.09	A1@.010	SF	.72	.41	1.13
2", R-14.3, C-0.07	A1@.011	SF	.97	.45	1.42

Polystyrene board installed on perimeter walls

	Craft@Hrs	Unit	Material	Labor	Total
1", R-4.20, C-0.24	A1@.010	SF	.34	.41	.75
1-1/2", R-6.30, C-0.16	A1@.010	SF	.45	.41	.86
2", R-8.30, C-0.12	A1@.010	SF	.57	.41	.98

Perlite (at $1.80 per CF) or vermiculite poured in concrete block cores

	Craft@Hrs	Unit	Material	Labor	Total
4" wall, 8.1 SF per CF	M1@.004	SF	.25	.15	.40
6" wall, 5.4 SF per CF	M1@.007	SF	.36	.26	.62
8" wall, 3.6 SF per CF	M1@.009	SF	.55	.34	.89
10" wall, 3.0 SF per CF	M1@.011	SF	.65	.41	1.06
12" wall, 2.1 SF per CF	M1@.016	SF	.97	.60	1.57
Poured perlite or vermiculite in wall cavities	M1@.032	CF	1.95	1.20	3.15
Sound board, 1/2", on walls	C8@.010	SF	.34	.39	.73
Sound board, 1/2", on floors	C8@.009	SF	.34	.35	.69

Cold box insulation

	Craft@Hrs	Unit	Material	Labor	Total
2" polystyrene	A1@.012	SF	1.06	.49	1.55
1" cork	A1@.013	SF	.83	.53	1.36
Add for overhead work	--	%	--	--	10.0
Add for poor access or enclosed areas	--	%	--	--	15.0
Add for 1 hour fire rating	--	%	--	--	15.0

Aluminum Wall Cladding System, Exterior. These costs assume installation by a manufacturer-trained contractor, on a framed and insulated exterior building surface, including track system, panels and trim, and subcontractor's overhead and profit. Conspec Systems
Techwall solid aluminum panel, 1/8" thick, joints sealed with silicon sealer

	Craft@Hrs	Unit	Material	Labor	Total
Kynar 500 paint finish	SM@.184	SF	19.80	8.46	28.26
Anodized aluminum finish	SM@.184	SF	18.80	8.46	27.26
Add for radius panels	--	%	20.0	--	--

Cladding, Preformed Roofing and Siding. Applied on metal framing. These costs include the subcontractor's overhead and profit.
Corrugated or ribbed roofing, colored galvanized steel, 9/16" deep, uninsulated

	Craft@Hrs	Unit	Material	Labor	Total
18 gauge	SM@.026	SF	2.34	1.20	3.54
20 gauge	SM@.026	SF	2.06	1.20	3.26
22 gauge	SM@.026	SF	1.83	1.20	3.03
24 gauge	SM@.026	SF	1.58	1.20	2.78

Corrugated siding, colored galvanized steel, 9/16" deep, uninsulated

	Craft@Hrs	Unit	Material	Labor	Total
18 gauge	SM@.034	SF	2.34	1.56	3.90
20 gauge	SM@.034	SF	2.06	1.56	3.62
22 gauge	SM@.034	SF	1.83	1.56	3.39

Thermal and Moisture Protection 7

	Craft@Hrs	Unit	Material	Labor	Total
Ribbed siding, colored galvanized steel, 1-3/4" deep, box rib					
18 gauge	SM@.034	SF	2.47	1.56	4.03
20 gauge	SM@.034	SF	2.19	1.56	3.75
22 gauge	SM@.034	SF	1.93	1.56	3.49
Corrugated aluminum roofing, natural finish					
.020" aluminum	SM@.026	SF	1.41	1.20	2.61
.032" aluminum	SM@.026	SF	1.75	1.20	2.95
Add for factory painted finish	--	%	40.0	--	--
Corrugated aluminum siding, natural finish					
.020" aluminum	SM@.034	SF	1.42	1.56	2.98
.032" aluminum	SM@.034	SF	1.76	1.56	3.32
Add for factory painted finish	--	%	10.0	--	--
Solid vinyl, .040", horizontal siding	SM@.026	SF	.76	1.20	1.96
Add for insulated siding backer panel	SM@.001	SF	.16	.05	.21
Vinyl window and door trim	SM@.040	LF	.27	1.84	2.11
Vinyl siding and fascia system	SM@.032	SF	.78	1.47	2.25
Laminated sandwich panel siding, enameled 22 gauge aluminum face one side, polystyrene core					
1" core	SM@.069	SF	3.91	3.17	7.08
1-1/2" core	SM@.069	SF	4.38	3.17	7.55

Membrane Roofing. These costs include the subcontractor's overhead and profit.

	Craft@Hrs	Unit	Material	Labor	Total
Built-up roofing, asphalt felt					
3 ply, smooth surface top sheet	R3@1.61	Sq	52.00	67.90	119.90
4 ply, base sheet, 3-ply felt, smooth surface top sheet	R3@1.85	Sq	60.70	78.00	138.70
5 ply, 20 year	R3@2.09	Sq	74.25	88.10	162.35
Built-up fiberglass felt roof, base and					
3-ply felt	R3@1.93	Sq	70.60	81.40	152.00
Add for each additional ply	R3@.188	Sq	10.50	7.93	18.43
Add for light rock dress off (250 lbs per Sq)	R3@.448	Sq	9.90	18.90	28.80
Add for heavy rock dress off (400 lbs per Sq)	R3@.718	Sq	14.75	30.30	45.05
Remove and replace 400 lb per sq rock, including new flood coat	R3@1.32	Sq	23.80	55.70	79.50
Aluminized coating for built-up roofing	R3@.599	Sq	17.40	25.30	42.70
Fire rated type	R3@.609	Sq	21.60	25.70	47.30
Modified APP, SBS flashing membrane	R3@.033	SF	.65	1.39	2.04
Cap sheet	R3@.277	Sq	18.00	11.70	29.70
Modified asphalt, APP, SBS base sheet and membrane	R3@1.97	Sq	68.50	83.10	151.60
Roll roofing					
90 lb mineral surface	R3@.399	Sq	20.70	16.80	37.50
19" wide selvage edge mineral surface (110 lb)	R3@.622	Sq	39.20	26.20	65.40
Double coverage selvage edge roll (140 lb)	R3@.703	Sq	39.10	29.60	68.70
Asphalt impregnated walkway for built-up roofing					
1/2"	R3@.016	SF	1.50	.68	2.18
3/4"	R3@.017	SF	1.92	.72	2.64
1"	R3@.020	SF	2.20	.84	3.04
Strip off existing 4-ply roof, no disposal included	R3@1.32	Sq	--	55.70	--

Elastomeric Roofing. These costs include the subcontractor's overhead and profit.

	Craft@Hrs	Unit	Material	Labor	Total
Butyl					
1/16"	R3@1.93	Sq	120.50	81.40	201.90
1/32"	R3@1.93	Sq	104.00	81.40	185.40
Neoprene, 1/16"	R3@2.23	Sq	180.00	94.00	274.00

Thermal and Moisture Protection 7

	Craft@Hrs	Unit	Material	Labor	Total
Elastomeric Roofing, continued					
GRM membrane, 1/16"	R3@2.23	Sq	190.00	94.00	284.00
Silicone-urethane foam roof system					
Remove gravel and prepare existing built-up roof	R3@1.26	Sq	4.50	53.10	57.60
Apply base and top coats of silicone rubber (.025")	R3@1.03	Sq	98.20	43.40	141.60
Urethane foam 1" thick (R-7.1)	R3@.446	Sq	86.60	18.80	105.40
Urethane foam 2" thick (R-14.3)	R3@.795	Sq	147.00	33.50	180.50
Spray-on mineral granules, 50 lbs per CSF	R3@.411	Sq	14.10	17.30	31.40
EPDM roofing system					
.045" loose laid membrane	R3@.743	Sq	80.65	31.30	111.95
.060" adhered membrane (excluding adhesive)	R3@1.05	Sq	92.50	44.30	136.80
Bonding adhesive	R3@.305	Sq	37.40	12.90	50.30
Lap splice cement, per 100 LF	R3@.223	CLF	22.00	9.40	31.40
Ballast rock, 3/4" to 1-1/2" gravel	R3@.520	Sq	15.15	21.90	37.05
Neoprene sheet flashing, .060"	R3@.021	SF	2.00	.89	2.89
CSPE (Hypalon) .045" single ply membrane	R3@.743	Sq	127.00	31.30	158.30
Add for mechanical fasteners and washers					
6" long, 12" OC at laps	--	Ea	.58	--	--
12" long, 12" OC at laps	--	Ea	1.20	--	--
Cant Strips					
3" fiber	R3@.017	LF	.35	.72	1.07
4" fiber or wood	R3@.017	LF	.44	.72	1.16

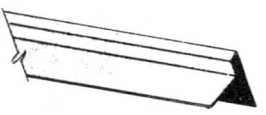 

Flashing and Sheet Metal. These costs include the subcontractor's overhead and profit.

	Craft@Hrs	Unit	Material	Labor	Total
Prefinished metal fascia and mansards, standing beam and batten					
Straight or simple	SM@.054	SF	3.49	2.48	5.97
Curved or complex	SM@.082	SF	5.26	3.77	9.03
Coping, gravel stop and flashing	SM@.037	SF	2.53	1.70	4.23
Sheet metal wainscot, galvanized	SM@.017	SF	.70	.78	1.48
Aluminum flashing, .032"					
Coping and wall cap, 16" girth	SM@.069	LF	2.86	3.17	6.03
Counter flash, 8"	SM@.054	LF	2.50	2.48	4.98
Reglet flashing, 8"	SM@.058	LF	1.03	2.67	3.70
Neoprene gasket for flashing	SM@.015	LF	.78	.69	1.47
Gravel stop and fascia, 10"	SM@.084	LF	2.97	3.86	6.83
Side wall flashing, 9"	SM@.027	LF	1.55	1.24	2.79
Gravel stop					
4"	SM@.058	LF	3.38	2.67	6.05
6"	SM@.063	LF	3.85	2.90	6.75
8"	SM@.068	LF	4.65	3.13	7.78
12"	SM@.075	LF	7.72	3.45	11.17
Valley, 24"	SM@.070	LF	1.99	3.22	5.21
Copper flashing, 16 oz					
Coping and wall cap, 16" girth	SM@.074	LF	6.30	3.40	9.70
Counter flash, 8"	SM@.057	LF	3.92	2.62	6.54
Gravel stop					
4"	SM@.058	LF	4.88	2.67	7.55
6"	SM@.063	LF	5.65	2.90	8.55

Thermal and Moisture Protection 7

	Craft@Hrs	Unit	Material	Labor	Total
8"	SM@.068	LF	6.64	3.13	9.77
10"	SM@.075	LF	8.15	3.45	11.60
Reglet flashing, 8"	SM@.054	LF	3.05	2.48	5.53
Neoprene gasket for flashing	SM@.015	LF	.70	.69	1.39
Side wall flashing, 6"	SM@.027	LF	2.15	1.24	3.39
Base, 20 oz	SM@.072	LF	4.43	3.31	7.74
Valley, 24"	SM@.069	LF	3.95	3.17	7.12
Sheet lead flashing					
Plumbing vents, soil stacks, 4 lbs/SF material					
4" pipe size	P6@.709	Ea	19.00	29.00	48.00
6" pipe size	P6@.709	Ea	22.20	29.00	51.20
8" pipe size	P6@.709	Ea	27.20	29.00	56.20
Roof drains, flat pan type, 36" x 36"					
2.5 lbs/SF material	P6@1.49	Ea	23.20	60.90	84.10
4 lbs/SF material	P6@1.49	Ea	38.90	60.90	99.80
6 lbs/SF material	P6@1.49	Ea	55.60	60.90	116.50
Stainless steel flashing					
Fascia roof edge, .018"	SM@.054	LF	3.72	2.48	6.20
Base, .018"	SM@.054	LF	4.43	2.48	6.91
Counter flash, .015"	SM@.054	LF	3.51	2.48	5.99
Valley, .015", 24"	SM@.069	LF	3.49	3.17	6.66
Reglets, .020", 8"	SM@.035	LF	3.03	1.61	4.64
Galvanized sheet metal flashing					
Cap and counter flash, 8"	SM@.046	LF	1.15	2.12	3.27
Coping and wall cap, 16" girth	SM@.069	LF	1.49	3.17	4.66
Gravel stop, 6"	SM@.060	LF	.44	2.76	3.20
Neoprene gasket	SM@.005	LF	.65	.23	.88
Pitch pockets, 24 gauge					
4" x 4"	SM@.584	Ea	17.10	26.80	43.90
6" x 6"	SM@.584	Ea	21.50	26.80	48.30
8" x 8"	SM@.584	Ea	25.40	26.80	52.20
8" x 10"	SM@.777	Ea	29.70	35.70	65.40
8" x 12"	SM@.777	Ea	35.40	35.70	71.10
Reglet flashing, 8"	SM@.054	LF	1.14	2.48	3.62
Shingles (chimney flash)	RF@.064	LF	.87	2.96	3.83
Side wall flashing, 9"	SM@.027	LF	.85	1.24	2.09
"W" valley, 24"	SM@.069	LF	1.66	3.17	4.83
Plumbers counter flash cone, 4" diameter	P8@.204	Ea	5.62	7.57	13.19
Roof safes and caps					
4" galvanized sheet metal	SM@.277	Ea	10.90	12.70	23.60
4" aluminum	SM@.277	Ea	12.30	12.70	25.00
Pitch pockets, 16 oz copper, filled					
4" x 4"	SM@.547	Ea	29.50	25.10	54.60
6" x 6"	SM@.547	Ea	37.50	25.10	62.60
8" x 8"	SM@.547	Ea	44.00	25.10	69.10
8" x 10"	SM@.679	Ea	51.50	31.20	82.70
8" x 12"	SM@.679	Ea	61.70	31.20	92.90
Corner guards, 4" x 4"	SM@.039	LF	11.40	1.79	13.19
Copper sheet metal roofing, 16 oz					
Batten seam	SM@.065	SF	3.88	2.99	6.87
Standing seam	SM@.056	SF	3.78	2.57	6.35

Thermal and Moisture Protection 7

	Craft@Hrs	Unit	Material	Labor	Total
Gutters and Downspouts. These costs include the subcontractor's overhead and profit.					
Aluminum, .032", including hangers					
Fascia gutter, 5"	SM@.050	LF	1.60	2.30	3.90
Box gutter, 4"	SM@.050	LF	1.40	2.30	3.70
Dropouts, elbows, for either of above	SM@.101	Ea	3.12	4.64	7.76
Downspouts, to 24' height					
2" x 3"	SM@.037	LF	1.15	1.70	2.85
3" x 4"	SM@.047	LF	1.80	2.16	3.96
4" diameter, round, 12' to 24' high	SM@.047	LF	1.45	2.16	3.61
Add for height over 24'	SM@.017	LF	--	.78	--
Copper, 16 oz, including hangers					
Box gutter, 4"	SM@.054	LF	4.00	2.48	6.48
Half round gutter					
4"	SM@.039	LF	3.47	1.79	5.26
5"	SM@.051	LF	4.35	2.34	6.69
6"	SM@.056	LF	5.38	2.57	7.95
Dropouts, elbows, for either of above	SM@.154	Ea	8.69	7.08	15.77
Downspouts, to 24' height					
2" x 3"	SM@.036	LF	3.83	1.66	5.49
3" x 4"	SM@.047	LF	5.18	2.16	7.34
3" diameter	SM@.037	LF	2.54	1.70	4.24
4" diameter	SM@.047	LF	3.88	2.16	6.04
5" diameter	SM@.049	LF	4.71	2.25	6.96
Add for heights over 24'	SM@.017	LF	--	.78	--
Scuppers					
8" x 8"	SM@.972	Ea	99.10	44.70	143.80
10" x 10"	SM@.972	Ea	117.00	44.70	161.70
Galvanized steel, 26 gauge, including hangers					
Box gutter 4"	SM@.054	LF	.98	2.48	3.46
Dropouts, elbows	SM@.082	Ea	1.55	3.77	5.32
Fascia gutter					
5" face	SM@.054	LF	1.20	2.48	3.68
7" face	SM@.062	LF	1.60	2.85	4.45
Dropouts, elbows, for either of above	SM@.088	Ea	2.35	4.05	6.40
Downspouts, with all fittings, height to 24'					
2" x 3"	SM@.036	LF	.85	1.66	2.51
3" x 4"	SM@.047	LF	1.30	2.16	3.46
3" round	SM@.037	LF	1.30	1.70	3.00
6" round	SM@.055	LF	1.20	2.53	3.73
Add for heights over 24'	SM@.017	LF	--	.78	--
Roof sump, 18" x 18" x 5"	SM@.779	Ea	6.70	35.80	42.50
Scupper, 6" x 6" x 8"	SM@.591	Ea	19.80	27.20	47.00
Stainless steel, .015" thick, including hangers					
Box gutter, 5" wide, 4-1/2" high	SM@.054	LF	6.25	2.48	8.73
Dropouts, elbows	SM@.076	Ea	9.20	3.49	12.69
Downspout, 4" x 5", to 24' high	SM@.051	LF	5.60	2.34	7.94
Add for heights over 24'	SM@.017	LF	--	.78	--
Vents, Louvers and Screens. These costs include the subcontractor's overhead and profit.					
Fixed louvers, with screen, typical	SM@.402	SF	9.57	18.50	28.07
Door louvers, typical	SM@.875	Ea	20.70	40.20	60.90
Manual operating louvers, typical	SM@.187	SF	15.50	8.60	24.10
Cooling tower screens	SM@.168	SF	8.54	7.72	16.26
Bird screens with frame	SM@.028	SF	1.29	1.29	2.58

Thermal and Moisture Protection 7

	Craft@Hrs	Unit	Material	Labor	Total
Concrete block vents, aluminum	M1@.370	SF	11.40	13.90	25.30
Frieze vents, with screen, 14" x 4"	C8@.198	Ea	2.33	7.62	9.95
Foundation vents, 6" x 14", ornamental	C8@.317	Ea	6.47	12.20	18.67
Attic vents, with louvers, 14" x 24"	C8@.377	Ea	18.60	14.50	33.10
Architectural facade screen, aluminum	--	SF	--	--	17.80
Add for enamel or light anodized	--	SF	--	--	2.00
Add for porcelain or heavy anodized	--	SF	--	--	5.00

Roof Accessories. These costs include the subcontractor's overhead and profit.

	Craft@Hrs	Unit	Material	Labor	Total
Roof hatches, not including ladder, steel					
30" x 30"	C8@3.09	Ea	336.00	119.00	455.00
30" x 72"	C8@4.69	Ea	812.00	180.00	992.00
Ceiling access hatches, 30" x 30"	C8@1.59	Ea	199.00	61.20	260.20
Smoke vents					
Aluminum, 48" x 48"	C8@3.09	Ea	1,020.00	119.00	1,139.00
Galvanized, 48" x 48"	C8@3.09	Ea	895.00	119.00	1,014.00
Fusible "shrink-out" heat and smoke vent (PVC dome in aluminum frame),					
4' x 8'	C8@5.15	Ea	812.00	198.00	1,010.00
Roof scuttle, aluminum, 2'6" x 3'0"	C8@3.50	Ea	414.00	135.00	549.00
Ventilators					
Rotary, wind driven					
6" diameter	SM@.913	Ea	55.90	42.00	97.90
12" diameter	SM@.913	Ea	69.30	42.00	111.30
24" diameter	SM@1.36	Ea	197.00	62.50	259.50
Mushroom type, motorized, single speed, including damper and bird screen but no electrical work					
8", 180 CFM	SM@2.93	Ea	274.00	135.00	409.00
12", 1,360 CFM	SM@2.93	Ea	455.00	135.00	590.00
18", 2,000 CFM	SM@4.00	Ea	652.00	184.00	836.00
24", 4,000 CFM	SM@4.86	Ea	1,004.00	223.00	1,227.00
Add for two-speed motor	--	%	35.0	--	--
Add for explosive proof units	--	Ea	336.00	--	--

Sealants and Caulking. By joint size. Costs are based on C.R. Laurence Co. products in 5 gallon quantities and assume 10% waste. Twelve 10.6 ounce cartridges equals one gallon and yields 231 cubic inches of caulk or sealant. Productivity shown assumes one roofer using a bulk dispenser. Add the cost of joint cleaning, if required. These costs include the subcontractor's overhead and profit.

Oil base caulk, $16.50 per gallon	Craft@Hrs	Unit	Material	Labor	Equipment	Total
1/4" x 1/4", 277 LF/gallon, 63 LF per hour	RF@.016	LF	.06	.74	.02	.82
1/4" x 3/8", 185 LF/gallon, 60 LF per hour	RF@.017	LF	.09	.79	.02	.90
1/4" x 1/2", 139 LF/gallon, 58 LF per hour	RF@.017	LF	.12	.79	.02	.93
3/8" x 3/8", 123 LF/gallon, 58 LF per hour	RF@.017	LF	.14	.79	.02	.95
3/8" x 1/2", 92 LF/gallon, 57 LF per hour	RF@.017	LF	.18	.79	.02	.99
3/8" x 5/8", 74 LF/gallon, 56 LF per hour	RF@.018	LF	.23	.83	.03	1.09
3/8" x 3/4", 60 LF/gallon, 52 LF per hour	RF@.019	LF	.28	.88	.03	1.19
1/2" x 1/2", 69 LF/gallon, 56 LF per hour	RF@.018	LF	.24	.83	.03	1.10
1/2" x 5/8", 55 LF/gallon, 50 LF per hour	RF@.020	LF	.30	.92	.03	1.25
1/2" x 3/4", 46 LF/gallon, 48 LF per hour	RF@.021	LF	.36	.97	.03	1.36
1/2" x 1", 35 LF/gallon, 45 LF per hour	RF@.022	LF	.48	1.02	.03	1.53
1" x 1", 17 LF/gallon, 30 LF per hour	RF@.033	LF	.98	1.53	.03	2.54
Add for cartridge application	--	%	25.0	25.0	--	--
Butyl rubber sealant, $20 per gallon						
1/4" x 1/4", 277 LF/gallon, 50 LF per hour	RF@.020	LF	.08	.92	.02	1.02

Thermal and Moisture Protection 7

	Craft@Hrs	Unit	Material	Labor	Equipment	Total
Sealants and Caulking, Butyl rubber sealant, continued						
1/2" x 1/2", 69 LF/gallon, 45 LF per hour	RF@.022	LF	.29	1.02	.02	1.33
1/2" x 3/4", 46 LF/gallon, 40 LF per hour	RF@.025	LF	.44	1.16	.02	1.62
3/4" x 3/4", 31 LF/gallon, 35 LF per hour	RF@.029	LF	.65	1.34	.02	2.01
1" x 1", 17 LF/gallon, 24 LF per hour	RF@.042	LF	1.18	1.94	.02	3.14
Add for cartridge application	--	%	25.0	25.0	--	--
Add for clear butyl sealant	--	%	25.0	--	--	--
Deduct for acrylic latex caulk	--	%	-5.0	--	--	--
Polyurethane, 1 or 2 component sealant, $34 per gallon						
1/4" x 1/4", 277 LF/gallon, 63 LF per hour	RF@.016	LF	.13	.74	.02	.89
1/4" x 3/8", 185 LF/gallon, 60 LF per hour	RF@.017	LF	.19	.79	.02	1.00
1/4" x 1/2", 139 LF/gallon, 58 LF per hour	RF@.017	LF	.25	.79	.02	1.06
3/8" x 3/8", 123 LF/gallon, 58 LF per hour	RF@.017	LF	.28	.79	.02	1.09
3/8" x 1/2", 92 LF/gallon, 57 LF per hour	RF@.017	LF	.37	.79	.02	1.18
3/8" x 5/8", 74 LF/gallon, 56 LF per hour	RF@.018	LF	.46	.83	.03	1.32
3/8" x 3/4", 60 LF/gallon, 52 LF per hour	RF@.019	LF	.57	.88	.03	1.48
1/2" x 1/2", 69 LF/gallon, 56 LF per hour	RF@.018	LF	.49	.83	.03	1.35
1/2" x 5/8", 55 LF/gallon, 50 LF per hour	RF@.020	LF	.62	.92	.03	1.57
1/2" x 3/4", 46 LF/gallon, 48 LF per hour	RF@.021	LF	.74	.97	.03	1.74
1/2" x 7/8", 40 LF/gallon, 46 LF per hour	RF@.022	LF	.85	1.02	.03	1.90
1/2" x 1", 35 LF/gallon, 45 LF per hour	RF@.022	LF	.97	1.02	.03	2.02
3/4" x 3/4", 31 LF/gallon, 44 LF per hour	RF@.023	LF	1.10	1.06	.03	2.19
1" x 1", 17 LF/gallon, 30 LF per hour	RF@.033	LF	2.00	1.53	.03	3.56
Add for cartridge application	--	%	25.0	25.0	--	--
Silicone RTV 1 component sealant, $43 per gallon						
1/4" x 1/4", 277 LF/gallon, 92 LF per hour	RF@.011	LF	.15	.51	.02	.68
1/2" x 1/2", 69 LF/gallon, 82 LF per hour	RF@.012	LF	.60	.55	.02	1.17
1/2" x 3/4", 46 LF/gallon, 55 LF per hour	RF@.018	LF	.90	.83	.02	1.75
3/4" x 3/4", 31 LF/gallon, 41 LF per hour	RF@.024	LF	1.30	1.11	.02	2.43
1/8" x 1", 138 LF/gallon, 87 LF per hour	RF@.012	LF	.30	.55	.02	.87
1/8" x 3", 46 LF/gallon, 55 LF per hour	RF@.019	LF	.90	.88	.02	1.80
1/4" x 3", 23 LF/gallon, 37 LF per hour	RF@.027	LF	1.80	1.25	.02	3.07
1/4" x 6", 11.6 LF/gallon, 19 LF per hour	RF@.054	LF	3.50	2.50	.02	6.02
1/2" x 6", 5.8 LF/gallon, 10 LF per hour	RF@.100	LF	7.05	4.62	.04	11.71
1/2" x 9", 3.9 LF/gallon, 7 LF per hour	RF@.143	LF	10.50	6.61	.05	17.16
1/2" x 12", 2.9 LF/gallon, 4 LF per hour	RF@.251	LF	14.10	11.60	.08	25.78
Add for cartridge application	--	%	25.0	25.0	--	--
Polysulfide or polyurethane sealant, 2 component, $35 per gallon						
1/4" x 1/4", 277 LF/gallon, 63 LF per hour	RF@.016	LF	.14	.74	.03	.91
1/4" x 1/2", 139 LF/gallon, 58 LF per hour	RF@.017	LF	.26	.79	.02	1.07
3/8" x 1/2", 92 LF/gallon, 57 LF per hour	RF@.017	LF	.40	.79	.02	1.21
1/2" x 1/2", 69 LF/gallon, 69 LF per hour	RF@.014	LF	.50	.65	.02	1.17
3/8" x 3/4", 60 LF/gallon, 52 LF per hour	RF@.019	LF	.60	.88	.03	1.51
Backing rods for sealant						
Open cell foam, 3/8" rod, 1/4" joint	RF@.010	LF	.05	.46	--	.51
Open cell foam, 5/8" rod, 1/2" joint	RF@.010	LF	.08	.46	--	.54
Open cell foam, 1" rod, 3/4" joint	RF@.010	LF	.15	.46	--	.61
Closed cell, 3/8" rod	RF@.011	LF	.10	.51	--	.61
Closed cell, 5/8" rod	RF@.011	LF	.12	.51	--	.63
Closed cell, 1-1/8" rod	RF@.012	LF	.21	.55	--	.76

Expansion Joints, Preformed. These costs include the subcontractor's overhead and profit.

Wall and ceiling, aluminum cover	Craft@Hrs	Unit	Material	Labor	Total
Drywall or panel type	CC@.067	LF	14.10	2.87	16.97

Thermal and Moisture Protection 7

	Craft@Hrs	Unit	Material	Labor	Total
Plaster type	CC@.067	LF	14.10	2.87	16.97
Floor to wall, 3" x 3"	CC@.038	LF	7.75	1.63	9.38
Floor, 3"	CC@.038	LF	9.65	1.63	11.28
Neoprene joint with aluminum cover	CC@.079	LF	20.60	3.38	23.98
Wall to roof joint	CC@.094	LF	8.55	4.03	12.58
Bellows type expansion joints, butyl with neoprene backer					
16 oz copper, 4" wide	CC@.046	LF	12.90	1.97	14.87
28 gauge stainless steel, 4" wide	CC@.046	LF	12.90	1.97	14.87
26 gauge galvanized sheet metal, 4" wide	CC@.046	LF	7.60	1.97	9.57
16 oz copper, 6" wide	CC@.046	LF	14.80	1.97	16.77
28 gauge stainless steel, 6" wide	CC@.046	LF	14.80	1.97	16.77
26 gauge galvanized sheet metal, 6" wide	CC@.046	LF	9.20	1.97	11.17
Neoprene gaskets, closed cell					
1/8" x 2"	CC@.019	LF	.66	.81	1.47
1/8" x 6"	CC@.023	LF	1.55	.99	2.54
1/4" x 2"	CC@.022	LF	.78	.94	1.72
1/4" x 6"	CC@.024	LF	1.65	1.03	2.68
1/2" x 6"	CC@.025	LF	2.45	1.07	3.52
1/2" x 8"	CC@.028	LF	3.60	1.20	4.80
Acoustical caulking, drywall or plaster type	CC@.038	LF	.28	1.63	1.91
Polyisobutylene tapes, non-drying					
Polybutene	CC@.028	LF	.22	1.20	1.42
Polyisobutylene/butyl, preformed	CC@.020	LF	.10	.86	.96

Expansion Joint Covers, Subcontract. Surface type, aluminum

	Craft@Hrs	Unit	Material	Labor	Total
Floor, 1-1/2"	CC@.194	LF	21.85	8.31	30.16
Wall and ceiling, 1-1/2"	CC@.185	LF	12.25	7.93	20.18
Gymnasium base	CC@.099	LF	9.85	4.24	14.09
Roof, typical	CC@.145	LF	20.90	6.21	27.11

Cast-in-Place Concrete Fireproofing. Costs per linear foot (LF) of steel beam, column or girder. These costs include the subcontractor's overhead and profit. Use the figures below to estimate the average cubic feet (CF) of concrete per linear foot and square feet (SF) of form required per linear foot of each type of beam, column or girder. Quantities include concrete required to fill the void between the web and flange and provide 2" protection at the flanges. A small change in concrete thickness at flanges has very little effect on cost.

	Craft@Hrs	Unit	Material	Labor	Equipment	Total
W36 x 135 to W36 x 300						
5.13 CF and 10.07 SF per LF	P9@2.00	LF	48.00	78.10	3.00	129.10
W33 x 118 to W30 x 241						
4.59 CF and 9.35 SF per LF	P9@1.84	LF	43.80	71.80	2.70	118.30
W30 x 99 to W30 x 211						
3.78 CF and 8.40 SF per LF	P9@1.63	LF	37.80	63.60	2.25	103.65
W27 x 84 to W27 x 178						
3.51 CF and 8.19 SF per LF	P9@1.57	LF	36.00	61.30	2.05	99.35
W24 x 55 to W24 x 162						
2.70 CF and 6.90 SF per LF	P9@1.31	LF	29.17	51.10	1.55	81.82
W21 x 44 to W21 x 147						
2.43 CF and 6.75 SF per LF	P9@1.26	LF	27.60	49.20	1.45	78.25
W18 x 37 to W18 x 71						
1.62 CF and 5.34 SF per LF	P9@.974	LF	20.40	38.00	1.00	59.40
W16 x 26 to W16 x 100						
1.62 CF and 5.40 SF per LF	P9@.987	LF	20.60	38.50	1.00	60.10
W14 x 90 to W14 x 730						
2.97 CF and 7.15 SF per LF	P9@1.37	LF	31.00	53.50	1.70	86.20

Thermal and Moisture Protection 7

	Craft@Hrs	Unit	Material	Labor	Equipment	Total
Cast-in-Place Concrete Fireproofing, continued						
W14 x 22 to W14 x 82						
1.35 CF and 4.70 SF per LF	P9@.854	LF	17.60	33.30	.75	51.65
W12 x 65 to W12 x 336						
2.16 CF and 6.08 SF per LF	P9@1.13	LF	24.70	44.10	1.30	70.10
W12 x 40 to W12 x 58						
1.62 CF and 5.40 SF per LF	P9@.987	LF	20.60	38.50	.95	60.05
W12 x 14 to W12 x 35						
1.08 CF and 4.44 SF per LF	P9@.790	LF	15.80	30.80	.60	47.20
W10 x 33 to W10 x 112						
1.35 CF and 4.75 SF per LF	P9@.862	LF	17.70	33.60	.75	52.05
W10 x 12 to W10 x 30						
1.08 CF and 4.56 SF per LF	P9@.808	LF	16.00	31.50	.60	48.10
W8 x 24 to W8 x 67						
1.08 CF & 4.40 SF per LF	P9@.784	LF	15.70	30.60	.60	46.90
W8 x 10 to W8 x 21						
.81 CF & 3.87 SF per LF	P9@.678	LF	13.10	26.50	.45	40.05
W6 x 15 to W6 x 25						
.81 CF & 3.84 SF per LF	P9@.672	LF	13.00	26.20	.45	39.65
W6 x 9 to W6 x 16						
.54 CF & 2.92 SF per LF	P9@.507	LF	9.60	19.80	.30	29.70
W5 x 16 to W5 x 19						
.54 CF & 2.86 SF per LF	P9@.499	LF	9.45	19.50	.30	29.25
W4 x 13						
.54 CF and 3.20 SF per LF	P9@.552	LF	10.30	21.50	.30	32.10
M4 x 13 to M14 x 18						
.54 CF & 2.90 SF per LF	P9@.504	LF	9.50	19.70	.30	29.50
S24 x 80 to S24 x 121						
2.43 CF and 7.11 SF per LF	P9@1.32	LF	28.40	51.50	1.45	81.35
S20 x 66 to S20 x 96						
1.89 CF and 6.09 SF per LF	P9@1.12	LF	23.55	43.70	1.15	68.40
S15 x 43 to S18 x 70						
1.62 CF and 5.82 SF per LF	P9@1.05	LF	21.65	41.00	.95	63.60
S12 x 35 to S12 x 50						
1.08 CF and 4.44 SF per LF	P9@.790	LF	15.70	30.80	.60	47.10
S8 x 23 to S10 x 35						
0.81 CF &3.81 SF per LF	P9@.670	LF	13.00	26.20	.45	39.65
S6 x 13 to S7 x 20						
0.54 CF & 2.91 SF per LF	P9@.504	LF	9.55	19.70	.30	29.55

Job size. Use $7,500.00 as the minimum subcontract price for work of this type.
Concrete thickness for fireproofing will usually be:
  Members at least 6" x 6" but less than 8" x 8", 3" for a 4 hour rating, 2" for a 3 hour rating,
  1-1/2" for a 2 hour rating, and 1" for a 1 hour rating.
  Members at least 8" x 8" but less than 12" x 12", 2-1/2" for a 4 hour rating,
  2" for a 3 hour rating, 1" for a 1 or 2 hour rating.
  Members 12" x 12" or greater, 2" for a 4 hour rating and 1" for a 1, 2, or 3 hour rating.
Labor costs listed here include the time needed to prepare formwork sketches at the job site,
  measure for the forms, fabricate, erect, align and brace the forms, cut, bend, place and tie
  the reinforcing steel, install embedded steel items, place and finish the concrete and strip, clean
  and stack the forms.
Labor costs assume a concrete forming, placing and finishing crew of 1 carpenter, 1 laborer and 1
  finisher working at an average cost of $39.04 per manhour including markup. Concrete placing
  assumes the use of a concrete pump. For scheduling purposes, estimate that a crew of 3 will
  fireproof the following quantities of steel in an 8-hour day: 15 LF of W21 through W36 members,

Thermal and Moisture Protection 7

	Craft@Hrs	Unit	Material	Labor	Equipment	Total

25 LF of W8 through W18 members, 22 LF of S12 through S24 members, and 42 LF of M4 through W6 or S6 through S8 members.

Material costs assume reinforcing steel is #3 and #4 bars at 12" on center each way and includes waste and laps. Form cost assumes 3 uses without completely disassembling the form. Normal cleaning and repairs are included. No salvage value is assumed. Costs for forms assume use of number 2 and better lumber at $605 per MBF and $850 per MSF for 3/4" plyform.

Concrete and steel for fireproofing

	Craft@Hrs	Unit	Material	Labor	Equipment	Total
Concrete, 3,000 PSI pump mix, at $78.50 CY including 5% waste and pump cost at $15.00 per CY	P9@.600	CY	78.50	23.40	15.00	116.90
Grade A60 reinforcing bars, set and tied, material at $.30 lb with 50 lbs per CY of concrete	P9@.790	CY	15.00	30.80	--	45.80
Hangers, snap-ties, misc. embedded steel material at $3.00 lb with 10 lb per CY of concrete	P9@.664	CY	30.00	25.90	--	55.90
Concrete test cylinders including test reports, at $10.00 each with 5 per 100 CY	--	CY	.50	--	--	.50
Total cost for concrete and steel	P9@2.05	CY	124.00	80.10	15.00	219.10

Forms for fireproofing These costs assume the following materials are used per square foot of contact area (SFCA): Nails, clamps and form oil costing $.90 per square foot, 1.2 SF of 3/4" plyform and 3.7 BF of lumber per SFCA

	Craft@Hrs	Unit	Material	Labor	Equipment	Total
Make, erect, align & strip forms, 3 uses	P9@.150	SF	2.38	5.86	--	8.24
Sack and patch concrete SF	P9@.010	SF	.05	.39	--	.44
Total cost for forms and finishing, per use	P9@.160	SF	2.43	6.25	--	8.68

Sample fireproofing estimate for ten W24 x 55 beams 20' long. Note from the preceding table that this beam requires 2.7 cubic feet of concrete per linear foot and 6.9 square feet of form per linear foot. Using labor and material prices from above to estimate 200 linear feet of beam:

	Craft@Hrs	Unit	Material	Labor	Equipment	Total
Concrete (200 x 2.7 = 540 CF or 20 CY)	P9@41.0	CY	2,480.00	1,600.00	300.00	4,380.00
Form & finish (200 x 6.9 = 1,380 SF)	P9@221.	SF	3,353.40	8,630.00	--	11,983.40
Total job cost	P9@262.	LS	5,833.40	10,200.00	300.00	16,333.40
Cost per linear foot of beam	P9@1.31	LF	29.17	51.00	1.50	81.67

Spray-Applied Fireproofing. Fire endurance coating made from inorganic vermiculite and portland cement. Costs assume a 10,000 board foot job. (One BF is one square foot covered 1" thick.) For smaller jobs, increase the cost by 5% for each 500 BF less than 10,000 BF. Use $2,000.00 as a minimum subcontract price. For thicknesses other than 1", adjust these costs proportionately. For scheduling purposes, estimate that a crew of 2 plasterers and 1 helper can apply 200 to 250 board feet per hour. These costs include the subcontractor's overhead and profit.

	Craft@Hrs	Unit	Material	Labor	Total
Structural steel columns	--	BF	--	--	1.15
Structural steel beams	--	BF	--	--	1.00
Purlins, girts, and miscellaneous members	--	BF	--	--	.95
Decks, ceilings or walls	--	BF	--	--	.90
Add for 18 gauge 2" hex mesh reinforcing	--	%	--	--	20.0
Add for key coat bonder on primed surfaces	--	SF	--	--	.20

Rule-of-thumb method for estimating spray-applied fireproofing on bare structural steel by member size:

	Beams	Cost LF	Columns	Cost LF
W36 x 135 to 300	10 BF/LF	10.00	11 BF/LF	12.65
W33 x 118 to 241	9 BF/LF	9.00	10 BF/LF	11.50
W30 x 99 to 211	8 BF/LF	8.00	9 BF/LF	10.35
W27 x 84 to 178	7 BF/LF	7.00	8 BF/LF	9.20
W24 x 55 to 182	6.5 BF/LF	6.50	7.5 BF/LF	8.62

Thermal and Moisture Protection 7

Spray-Applied Fireproofing, continued

	Beams	Cost LF	Columns	Cost LF
W21 x 44 to 147	6 BF/LF	6.00	7 BF/LF	8.05
W18 x 35 to 118	5 BF/LF	5.00	6 BF/LF	6.90
W16 x 28 to 57	3.5 BF/LF	3.50	4.5 BF/LF	5.18
W14 x 61 to 132	5.5 BF/LF	5.50	6.5 BF/LF	7.48
W14 x 22 to 63	4 BF/LF	4.00	5 BF/LF	5.75
W12 x 65 to 190	5 BF/LF	5.00	6 BF/LF	6.90
W12 x 40 to 58	4 BF/LF	4.00	5 BF/LF	5.75
W12 x 14 to 35	3 BF/LF	3.00	4 BF/LF	4.60
W10 x 49 to 112	4 BF/LF	4.00	5 BF/LF	5.75
W10 x 22 to 45	3 BF/LF	3.00	4 BF/LF	4.60
W10 x 12 to 19	2 BF/LF	2.00	3 BF/LF	3.45
W8 x 24 to 67	3 BF/LF	3.00	4 BF/LF	4.60
W6 x 9 to 25	2 BF/LF	2.00	3 BF/LF	3.45
W5 x 16 to 19	2 BF/LF	2.00	2.5 BF/LF	2.82
W4 x 13	1 BF/LF	1.00	2 BF/LF	2.30

Purlins and girts. Cost per LF

			Purlins or girts	
MC 18 members			4 BF/LF	3.80
MC 10 to MC 13 members			3 BF/LF	2.85
MC 8 members			2 BF/LF	1.90
C 10 to C 15 members			3 BF/LF	2.85
C 7 to C 9 members			2 BF/LF	1.90

Doors and Windows 8

	Craft@Hrs	Unit	Material	Labor	Total

Complete Hollow Metal Door Assembly. These figures show the costs normally associated with installing an exterior hollow core steel door and include the subcontractor's overhead and profit.

Hollow metal exterior 3' x 7' flush door, with frame, hardware and trim, complete

	Craft@Hrs	Unit	Material	Labor	Total
Total cost door, frame & trim as described below	C8@7.62	Ea	793.45	292.90	1,086.35
Stock hollow metal flush door, 16 gauge, 3' x 7' x 1-3/4", non-rated	C8@.721	Ea	277.00	27.70	304.70
Closer plate reinforcing on door	--	Ea	11.10	--	11.10
Stock 18 gauge frame, 6" jamb, non-rated	C8@.944	Ea	88.40	36.30	124.70
Closer plate on frame	--	Ea	5.00	--	5.00
Three hinges, 4-1/2" x 4-1/2"	C8@.578	LS	55.35	22.20	77.55
Lockset, mortise type	C8@.949	Ea	208.00	36.50	244.50
Saddle type threshold, aluminum, 3'	C8@.276	Ea	15.50	10.60	26.10
Standard duty closer	C8@.787	Ea	83.00	30.30	113.30
Bronze weatherstripping, 20 LF	C8@2.58	LS	40.60	99.30	139.90
Paint with primer and 2 coats enamel	C8@.781	LS	9.50	30.00	39.50

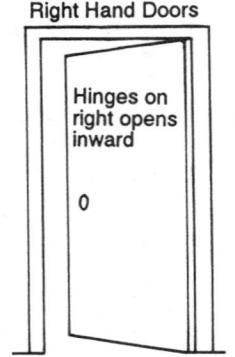
Right Hand Doors
Hinges on right opens inward

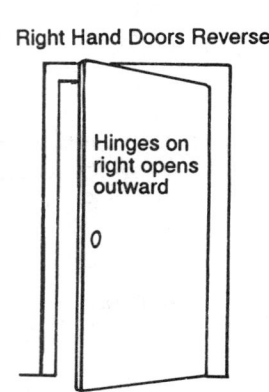
Right Hand Doors Reverse
Hinges on right opens outward

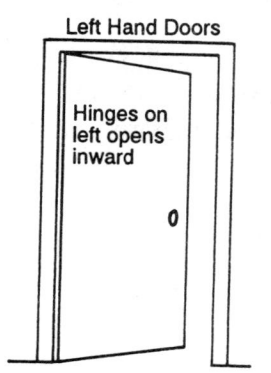
Left Hand Doors
Hinges on left opens inward

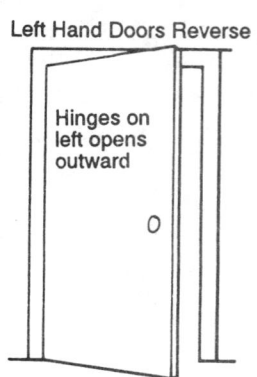
Left Hand Doors Reverse
Hinges on left opens outward

Doors and Windows 8

	Craft@Hrs	Unit	Material	Labor	Total

Hollow Metal Exterior Doors. Commercial quality doors. See illustrations on preceding page. These costs include the subcontractor's overhead and profit but no frame, hinges, lockset, trim or finishing.

	Craft@Hrs	Unit	Material	Labor	Total
6'8" high hollow flush doors, 1-3/8" thick, non-rated					
2'6" wide, 20 gauge	C8@.681	Ea	157.00	26.20	183.20
2'8" wide, 20 gauge	C8@.683	Ea	161.00	26.30	187.30
3'0" wide, 20 gauge	C8@.721	Ea	173.00	27.70	200.70
6'8" high hollow flush doors, 1-3/4" thick, non-rated					
2'6" wide, 20 gauge	C8@.681	Ea	190.00	26.20	216.20
2'8" wide, 20 gauge	C8@.697	Ea	196.00	26.80	222.80
2'8" wide, 18 gauge	C8@.697	Ea	227.00	26.80	253.80
2'8" wide, 16 gauge	C8@.697	Ea	266.00	26.80	292.80
3'0" wide, 20 gauge	C8@.721	Ea	206.00	27.70	233.70
3'0" wide, 18 gauge	C8@.721	Ea	237.00	27.70	264.70
3'0" wide, 16 gauge	C8@.721	Ea	277.00	27.70	304.70
7'0" high hollow flush doors, 1-3/4" thick, non-rated					
2'6" wide, 20 gauge	C8@.757	Ea	201.00	29.10	230.10
2'8" wide, 20 gauge	C8@.735	Ea	206.00	28.30	234.30
2'8" wide, 18 gauge	C8@.757	Ea	237.00	29.10	266.10
2'8" wide, 16 gauge	C8@.757	Ea	275.00	29.10	304.10
3'0" wide, 20 gauge	C8@.816	Ea	215.00	31.40	246.40
3'0" wide, 18 gauge	C8@.816	Ea	247.00	31.40	278.40
3'0" wide, 16 gauge	C8@.816	Ea	287.00	31.40	318.40
Additional costs for hollow metal doors, cost per door					
Add for factory applied steel astragal	--	Ea	30.00	--	--
Add for steel astragal set on site	C8@.496	Ea	30.00	19.10	49.10
Add for 90 minute "B" fire label rating	--	Ea	33.20	--	--
Add for R-7 polyurethane foam core	--	Ea	80.30	--	--
Add for 10" x 10" wired glass panel	--	Ea	58.00	--	--
Add for closer reinforcing plate	--	Ea	11.10	--	--
Add for chain or bolt reinforcing plate	--	Ea	85.70	--	--
Add for rim exit latch reinforcing	--	Ea	24.60	--	--
Add for vertical exit latch reinforcing	--	Ea	44.40	--	--
Add for pull plate reinforcing	--	Ea	17.90	--	--
Add for galvanizing 1-3/8" doors	--	Ea	22.70	--	--
Add for galvanizing 1-3/4" door	--	Ea	27.70	--	--
Add for cutouts to 4 SF	--	Ea	46.40	--	--
Add for cutouts over 4 SF	--	SF	9.80	--	--
Add for stainless steel doors	--	Ea	782.00	--	--
Add for baked enamel finish	--	Ea	58.90	--	--
Add for porcelain enamel finish	--	Ea	141.00	--	--
Add for larger sizes	--	SF	3.30	--	--
Add for special dapping	--	Ea	25.10	--	--
Kalamein doors					
90 minute rating	C8@.164	SF	15.50	6.31	21.81
2 hour rating	C8@.164	SF	25.00	6.31	31.31

Hollow Metal Door Frames. 18 gauge prefinished hollow metal door frames, non-rated, stock sizes, for 1-3/4" or 1-3/8" doors. These costs include the subcontractor's overhead and profit but no door, hinges, lockset, trim or finishing.

	Craft@Hrs	Unit	Material	Labor	Total
6'8" high					
To 3'6" wide, to 4-1/2" jamb	C8@.944	Ea	76.60	36.30	112.90
To 3'6" wide, 4-3/4" to 6" jamb	C8@.944	Ea	80.70	36.30	117.00
To 3'6" wide, over 6" jamb	C8@.944	Ea	108.00	36.30	144.30

Doors and Windows 8

	Craft@Hrs	Unit	Material	Labor	Total
Hollow Metal Door Frames, 6'8" high, continued					
Over 4' wide, 4-1/2" jamb	C8@.944	Ea	92.30	36.30	128.60
Over 4' wide, 4-3/4" to 6" jamb	C8@.944	Ea	97.20	36.30	133.50
Over 4' wide, over 6" jamb	C8@.944	Ea	125.50	36.30	161.80
7' high					
To 3'6" wide, to 4-1/2" jamb	C8@.944	Ea	83.40	36.30	119.70
To 3'6" wide, 4-3/4" to 6" jamb	C8@.944	Ea	88.40	36.30	124.70
To 3'6" wide, over 6" jamb	C8@.944	Ea	114.00	36.30	150.30
Over 4' wide, 4-1/2" jamb	C8@.944	Ea	98.80	36.30	135.10
Over 4' wide, 4-3/4" to 6" jamb	C8@.944	Ea	103.20	36.30	139.50
Over 4' wide, over 6" jamb	C8@.944	Ea	130.60	36.30	166.90
8' high					
To 3'6" wide, to 4-1/2" jamb	C8@1.05	Ea	96.60	40.40	137.00
To 3'6" wide, 4-3/4" to 6" jamb	C8@1.05	Ea	101.10	40.40	141.50
To 3'6" wide, over 6" jamb	C8@1.05	Ea	130.00	40.40	170.40
Over 4' wide, 4-1/2" jamb	C8@1.05	Ea	112.00	40.40	152.40
Over 4' wide, 4-3/4" to 6" jamb	C8@1.05	Ea	115.00	40.40	155.40
Over 4' wide, over 6" jamb	C8@1.05	Ea	147.00	40.40	187.40
Cost additions or deductions for hollow metal door frames, per frame					
Add for 90 minute UL label	--	Ea	24.40	--	--
Add for aluminum casing	--	Ea	20.50	--	--
Add for communicator frame (back-to-back doors)	--	Ea	15.00	--	--
Add for galvanized finish	--	%	12.0	--	--
Add for lengthening, 7' to 8'10"	--	Ea	13.20	--	--
Add for extra hinge reinforcing	--	Ea	4.50	--	--
Add for stainless steel frames	--	Ea	432.00	--	--
Add for porcelain enamel finish	--	Ea	140.00	--	--
Add for reinforcing for chain and bolt	--	Ea	5.75	--	--
Add for closer plate reinforcing	--	Ea	5.00	--	--
Add for exit latch reinforcing	--	Ea	9.30	--	--
Add for concrete filled frames	C8@.382	Ea	6.45	14.70	21.15
Add for borrowed lite	C8@.086	SF	6.15	3.31	9.46
Add for fixed transom lite	C8@.099	SF	8.10	3.81	11.91
Add for movable transom lite	C8@.107	SF	9.30	4.12	13.42
Add for window frame sections	C8@.071	LF	5.00	2.73	7.73
Add for wall frame sections	C8@.071	LF	5.70	2.73	8.43
Deduct for 22 gauge frames	--	Ea	-12.50	--	--

Prehung Steel Doors. 18 gauge primed insulated 1-3/4" thick entry doors with sweep and 18 gauge steel frame. Includes three 4" x 4" x 1/4" hinges but no lockset

	Craft@Hrs	Unit	Material	Labor	Total
Flush doors					
Jamb to 5-1/2" wide, 2'8" x 6'8"	C8@1.05	Ea	330.00	40.40	370.40
Jamb to 5-1/2" wide, 3'0" x 6'8"	C8@1.05	Ea	332.00	40.40	372.40
Jamb over 5-1/2" wide, 2'8" x 6'8"	C8@1.05	Ea	335.00	40.40	375.40
Jamb over 5-1/2" wide, 3'0" x 6'8"	C8@1.05	Ea	340.00	40.40	380.40
6 or 8 panel doors					
Jamb to 5-1/2" wide, 2'8" x 6'8"	C8@1.05	Ea	340.00	40.40	380.40
Jamb to 5-1/2" wide, 3'0" x 6'8"	C8@1.05	Ea	340.00	40.40	380.40
Jamb over 5-1/2" wide, 2'8" x 6'8"	C8@1.05	Ea	345.00	40.40	385.40
Jamb over 5-1/2" wide, 3'0" x 6'8"	C8@1.05	Ea	346.00	40.40	386.40
Add for Factory Mutual fire rating					
90 minute	--	Ea	18.00	--	--
60 minute	--	Ea	14.00	--	--
Add for 16 gauge door and frame	--	Ea	11.00	--	--

Doors and Windows 8

	Craft@Hrs	Unit	Material	Labor	Total
Deduct for 22 gauge and door	--	Ea	-3.50	--	--
Add for galvanized steel finish	--	Ea	9.50	--	--
Deduct for unfinished door	--	Ea	-13.00	--	--
Add for installed weatherstripping	--	Ea	28.00	--	--
Add for aluminum threshold	--	Ea	27.50	--	--
Deduct for wood door replacing steel door	--	Ea	-32.50	--	--

Wood Doors and Frames. Commercial and institutional quality. Unfinished, no hardware included. These costs include the subcontractor's overhead and profit.

	Craft@Hrs	Unit	Material	Labor	Total
Hollow core 1-3/8" thick flush interior doors					
2'6" x 6'8", hardboard face	C8@.936	Ea	29.40	36.00	65.40
2'8" x 6'8", hardboard face	C8@.936	Ea	39.40	36.00	75.40
3'0" x 6'8", hardboard face	C8@.982	Ea	41.70	37.80	79.50
2'6" x 6'8", hardwood face	C8@.936	Ea	41.20	36.00	77.20
2'8" x 6'8", hardwood face	C8@.936	Ea	48.80	36.00	84.80
3'0" x 6'8", hardwood face	C8@.982	Ea	51.00	37.80	88.80
Hollow core 1-3/4" thick flush exterior doors					
2'6" x 6'8", hardwood face	C8@.982	Ea	48.70	37.80	86.50
2'8" x 6'8", hardwood face	C8@.982	Ea	51.40	37.80	89.20
3'0" x 6'8", hardwood face	C8@1.01	Ea	55.80	38.90	94.70
2'6" x 7'0", hardwood face	C8@.982	Ea	55.10	37.80	92.90
2'8" x 7'0", hardwood face	C8@.982	Ea	58.00	37.80	95.80
3'0" x 7'0", hardwood face	C8@1.01	Ea	61.80	38.90	100.70
Solid core 1-3/4" thick flush exterior doors, "B" label					
2'6" x 6'8", hardwood face	C8@1.05	Ea	203.00	40.40	243.40
2'8" x 6'8", hardwood face	C8@1.05	Ea	219.00	40.40	259.40
3'0" x 6'8", hardwood face	C8@1.10	Ea	224.00	42.30	266.30
2'0" x 7'0", hardwood face	C8@1.05	Ea	193.00	40.40	233.40
2'4" x 7'0", hardwood face	C8@1.05	Ea	203.00	40.40	243.40
2'6" x 7'0", hardwood face	C8@1.05	Ea	235.00	40.40	275.40
2'8" x 7'0", hardwood face	C8@1.05	Ea	243.00	40.40	283.40
3'0" x 7'0", hardwood face	C8@1.10	Ea	257.00	42.30	299.30
Exterior door frames, fir, with oak sill					
2'6" x 6'8"	C8@1.48	Ea	103.00	56.90	159.90
2'8" x 6'8"	C8@1.50	Ea	104.00	57.70	161.70
3'0" x 6'8"	C8@1.57	Ea	107.00	60.40	167.40
5'0" x 6'8", for two doors	C8@1.85	Ea	119.00	71.20	190.20
6'0" x 6'8", for two doors	C8@1.99	Ea	125.00	76.60	201.60
2'6" x 7'0"	C8@1.48	Ea	109.00	56.90	165.90
2'8" x 7'0"	C8@1.50	Ea	110.00	57.70	167.70
3'0" x 7'0"	C8@1.57	Ea	112.00	60.40	172.40
5'0" x 7'0", for two doors	C8@1.85	Ea	125.00	71.20	196.20
6'0" x 7'0", for two doors	C8@1.99	Ea	132.00	76.60	208.60
Interior door jambs, fir, without stops					
2'6" x 6'8"	C8@.871	Ea	20.60	33.50	54.10
2'8" x 6'8"	C8@.879	Ea	20.90	33.80	54.70
3'0" x 6'8"	C8@.922	Ea	21.60	35.50	57.10
5'0" x 6'8", for two doors	C8@1.10	Ea	23.80	42.30	66.10
6,0" x 6'8", for two doors	C8@1.16	Ea	25.10	44.60	69.70
2'6" x 7'0"	C8@.871	Ea	21.60	33.50	55.10
2'8" x 7'0"	C8@.879	Ea	21.90	33.80	55.70
3'0" x 7,0"	C8@.922	Ea	22.60	35.50	58.10
5'0" x 7'0", for two doors	C8@1.10	Ea	25.40	42.30	67.70
6'0" x 7'0", for two doors	C8@1.16	Ea	26.30	44.60	70.90

Doors and Windows 8

	Craft@Hrs	Unit	Material	Labor	Total

Wood Doors and Frames, continued
Prehung flush doors with hinge set, casing and jambs, but no lockset, 1-3/4", solid core, hardwood veneer, prefinished

	Craft@Hrs	Unit	Material	Labor	Total
To 2'8" x 6'8"	C8@.971	Ea	203.00	37.40	240.40
3'0" x 6'8"	C8@.971	Ea	210.00	37.40	247.40
3'6" x 6'8"	C8@1.06	Ea	217.00	40.80	257.80
To 2'8" x 7'0"	C8@1.06	Ea	230.00	40.80	270.80
3'0" x 7'0"	C8@1.10	Ea	217.00	42.30	259.30
3'6" x 7'0"	C8@1.29	Ea	237.00	49.60	286.60
To 3'0" x 8'0"	C8@1.63	Ea	278.00	62.70	340.70
3'6" x 8'0"	C8@1.72	Ea	292.00	66.20	358.20
Deduct for hollow core 1-3/8" prehung doors	--	%	-25.0	-10.0	--

Special Doors. These costs include the subcontractor's overhead and profit but no electrical work. Equipment is a forklift at $16.00 per hour.

	Craft@Hrs	Unit	Material	Labor	Equipment	Total
Fire doors, overhead roll-up type, sectional steel, fusible link operated, UL label						
6' x 7', manual operation	F7@13.4	Ea	1,075.00	545.00	53.70	1,673.70
5' x 8', manual operation	F7@13.4	Ea	1,020.00	545.00	53.70	1,618.70
14' x 14', chain hoist operated	F7@26.2	Ea	2,975.00	1,070.00	105.00	4,150.00
18' x 14', chain hoist operated	F7@32.9	Ea	4,120.00	1,340.00	132.00	5,592.00
Fire doors, sliding type, one-piece steel, including hardware, fusible link, motor operated, UL label						
4' x 7'	F7@13.3	Ea	1,055.00	541.00	53.20	1,649.20
6' x 7'	F7@14.8	Ea	1,420.00	602.00	59.20	2,081.20
10' x 10'	F7@32.1	Ea	2,740.00	1,310.00	128.00	4,178.00
Grilles, aluminum, overhead roll-up type, horizontal rods at 2" OC, chain hoist operated						
8' x 8', clear anodized aluminum	F7@18.5	Ea	1,800.00	753.00	74.00	2,627.00
8' x 8', medium bronze aluminum	F7@18.5	Ea	2,570.00	753.00	74.00	3,397.00
10' x 10', anodized aluminum	F7@21.8	Ea	2,450.00	887.00	87.00	3,424.00
18' x 8', anodized aluminum	F7@26.3	Ea	3,740.00	1,070.00	105.00	4,915.00
18' x 18', anodized, motor operated	F7@28.7	Ea	4,390.00	1,170.00	115.00	5,675.00
Refrigerator doors, manually operated, with hardware and frame						
Galvanized, cooler, 3' x 7', hinged	F7@6.60	Ea	1,100.00	269.00	26.40	1,395.40
Stainless, cooler, 3' x 7', hinged	F7@6.60	Ea	2,000.00	269.00	26.40	2,295.40
Stainless, freezer, 3' x 7', hinged	F7@6.60	Ea	2,250.00	269.00	26.40	2,545.40
Stainless, cooler, 4' x 7', hinged	F7@6.60	Ea	2,000.00	269.00	26.40	2,295.40
Galvanized, cooler, 5' x 7', sliding,	F7@7.93	Ea	2,200.00	323.00	31.70	2,554.70
Galvanized, freezer, 5' x 7' sliding	F7@7.93	Ea	2,750.00	323.00	31.70	3,104.70
Revolving doors, 7' diameter, complete						
Aluminum	F7@50.3	Ea	26,000.00	2,050.00	201.00	28,251.00
Stainless steel	F7@50.3	Ea	29,500.00	2,050.00	201.00	31,751.00
Bronze	F7@69.0	Ea	37,900.00	2,810.00	276.00	40,986.00
Service doors, aluminum, overhead roll-up type						
4' x 4', manual counter shutter	F7@13.6	Ea	712.00	553.00	54.40	1,319.40
8' x 4', manual operation	F7@12.7	Ea	1,030.00	517.00	50.70	1,597.70
10' x 10', chain hoist operated	F7@12.7	Ea	1,595.00	517.00	50.70	2,162.70
Service doors, steel, overhead roll-up, chain hoist operated						
8' x 8'	F7@11.7	Ea	1,020.00	476.00	47.00	1,543.00
10' x 10'	F7@11.7	Ea	1,270.00	476.00	47.00	1,793.00
12' x 12'	F7@14.7	Ea	1,700.00	598.00	58.80	2,356.80
14' x 14'	F7@18.2	Ea	1,850.00	741.00	72.80	2,663.80
18' x 18'	F7@22.0	Ea	2,520.00	895.00	88.00	3,503.00
18' x 14', 240 volt motor operated	F7@25.0	Ea	3,300.00	1,020.00	100.00	4,420.00

Doors and Windows 8

	Craft@Hrs	Unit	Material	Labor	Equipment	Total

Service doors, weatherstripped, steel, overhead roll-up type, chain hoist operated, bottom and side astragal, with hood baffles but no insulation

	Craft@Hrs	Unit	Material	Labor	Equipment	Total
10' x 10'	F7@12.8	Ea	1,250.00	521.00	51.20	1,822.20
14' x 14'	F7@26.9	Ea	1,705.00	1,090.00	108.00	2,903.00
Add for insulated weatherstripped doors	--	%	150.0	--	--	--

Stock room doors, double-action .063" aluminum, with hardware and bumper strips both sides, two 7' high doors per opening, plastic laminate finish, 14" x 9" lite in each door

	Craft@Hrs	Unit	Material	Labor	Equipment	Total
4' wide opening	F7@3.36	Ea	635.00	137.00	13.40	785.40
5' wide opening	F7@3.36	Ea	725.00	137.00	13.40	875.40
6' wide opening	F7@3.98	Ea	785.00	162.00	15.90	962.90
7' wide opening	F7@3.98	Ea	865.00	162.00	15.90	1,042.90
Add for dual bumpers on both sides	--	LS	225.00	--	--	--
Deduct for no bumpers, aluminum finish	--	%	-50.0	--	--	--
Deduct for single door hung in opening	--	%	-50.0	-50.0	--	--
Add for heavy duty doors, spring bumpers	--	%	50.0	--	--	--

Vault doors, minimum security

	Craft@Hrs	Unit	Material	Labor	Equipment	Total
3' x 7', 2 hour fire rating	F7@22.0	Ea	2,200.00	895.00	88.00	3,183.00
4' x 7', 2 hour fire rating	F7@23.4	Ea	3,150.00	952.00	93.60	4,195.60
3' x 7', 4 hour fire rating	F7@27.2	Ea	2,415.00	1,110.00	119.00	3,644.00
4' x 7', 4 hour fire rating	F7@28.7	Ea	3,270.00	1,170.00	115.00	4,555.00

Storefronts and Entrances. These figures include the subcontractor's overhead and profit. Commercial grade, 4-1/2" section, anodized aluminum with clear glass, includes tempered safety glass in lower sections.

	Craft@Hrs	Unit	Material	Labor	Total
Stub wall to 8' high	G1@.089	SF	13.20	3.46	16.66
Floor to 8' high	G1@.150	SF	14.00	5.84	19.84
Floor to 10' high	G1@.150	SF	14.50	5.84	20.34
Add for solar grey glass	--	SF	1.78	--	--
Add for medium bronze anodized finish	--	SF	1.78	--	--
Add for black anodized finish	--	SF	3.29	--	--
Add for extra aluminum sections	G1@.096	LF	10.29	3.74	14.03

Entrance doors, center pivot concealed closer, add to costs above without deducting for SF door area

	Craft@Hrs	Unit	Material	Labor	Total
To 3' x 7', narrow stile	G1@7.88	Ea	800.00	307.00	1,107.00
To 3' x 7', heavy section	G1@8.42	Ea	950.00	328.00	1,278.00
To 3' x 7', tempered glass	G1@10.4	Ea	1,740.00	405.00	2,145.00
Add for center stop	G1@.329	Ea	31.30	12.80	44.10
Add for floor check	G1@4.44	Ea	286.00	173.00	459.00
Add for medium bronze finish	--	Ea	220.00	--	--
Add for black anodized finish	--	Ea	340.00	--	--
Add for larger sizes	G1@.161	SF	22.45	6.27	28.72
Add for automatic opener	G1@24.0	Ea	3,420.00	934.00	4,354.00

Institutional storefronts, 7" section, anodized aluminum, clear polished plate glass

	Craft@Hrs	Unit	Material	Labor	Total
Stub wall to 9' high	G1@.136	SF	16.10	5.29	21.39
Stub wall to 13' high	G1@.167	SF	17.10	6.50	23.60
Floor to 9' high	G1@.161	SF	17.50	6.27	23.77
Floor to 10' high	G1@.181	SF	18.75	7.04	25.79
Add for solar grey glass	--	SF	1.75	--	--
Add for medium bronze anodized finish	--	SF	1.75	--	--
Add for black anodized finish	--	SF	3.35	--	--
Add for extra aluminum sections	G1@.108	LF	12.15	4.20	16.35

Entrances, center pivot concealed closer, add to costs above without deducting SF of door area

	Craft@Hrs	Unit	Material	Labor	Total
3' x 7', aluminum and glass	G1@8.58	Ea	1,080.00	334.00	1,414.00

Doors and Windows 8

	Craft@Hrs	Unit	Material	Labor	Total
Storefronts and Entrances, Entrances, continued					
3' x 7', 5/8", tempered glass	G1@11.4	Ea	2,150.00	444.00	2,594.00
Add for black anodized finish	--	SF	3.40	--	--
Add for larger sizes	G1@.242	SF	26.90	9.42	36.32
Add for automatic opener	G1@35.0	Ea	3,820.00	1,360.00	5,180.00
Add for floor check	G1@1.88	Ea	325.00	73.20	398.20
Monumental storefronts, 9" section, aluminum section and clear plate glass					
Floor to 9' high	G1@.164	SF	24.40	6.38	30.78
Floor to 11' high	G1@.164	SF	24.75	6.38	31.13
Floor to 13' high	G1@.179	SF	25.35	6.97	32.32
Add for solar grey glass	--	SF	1.85	--	--
Add for medium bronze finish	--	SF	1.85	--	--
Add for black anodized finish	--	SF	3.35	--	--
Add for extra aluminum sections	G1@.147	LF	17.10	5.72	22.82
Add for lacquered bronze finish	G1@.294	SF	12.60	11.40	24.00
Add for automatic opener	G1@31.3	Ea	4,225.00	1,220.00	5,445.00
Steel storefronts, glazed					
Primed steel framing	G1@.079	SF	13.80	3.07	16.87
Stainless steel framing	G1@.164	SF	36.90	6.38	43.28
Stainless steel door, 1 lite	G1@8.31	Ea	1,690.00	323.00	2,013.00
Stainless steel door, 1 lite, with frame	G1@11.4	Ea	2,625.00	444.00	3,069.00
Stainless steel doors, 1 lite each, pair, with frame	G1@15.1	LS	5,390.00	588.00	5,978.00
Double hung sash in conjunction with wall	--	SF	12.65	--	--
Panels for storefront framing, subtract $4.00 per SF for glass from the framing costs above and add the following:					
1", glassweld 1 side	--	SF	12.50	--	--
1", porcelain enamel 1 side	--	SF	11.75	--	--
Aluminum anodized bronze	--	SF	11.80	--	--
Aluminum anodized black	--	SF	12.60	--	--
Stainless steel, 1" panels	--	SF	16.60	--	--
Glassweld, 1/8", 4" x 10"	G1@.045	SF	3.70	1.75	5.45
Aluminum trim for glassweld, 12" length	G1@.017	LF	.65	.66	1.31
Aluminum corners for glassweld	G1@.028	LF	.92	1.09	2.01
Commercial and Industrial Grade Steel Windows. Glazed, including the subcontractor's overhead and profit.					
Industrial grade, fixed 100%	G1@.086	SF	14.85	3.35	18.20
Industrial grade, vented 50%	G1@.086	SF	18.75	3.35	22.10
Projected, vented 50%	G1@.086	SF	21.30	3.35	24.65
Add for screen, SF of screen	--	SF	2.50	--	--
Hardware. These costs include the subcontractor's overhead and profit. Commercial and industrial quality. Installation costs assume doors manufactured to receive hardware.					
Typical finish hardware costs, (doors, cabinets, toilet rooms), in place					
Commercial structure, economy	CC@.002	SF	.14	.09	.23
Commercial structure, standard	CC@.004	SF	.35	.17	.52
School or institution, not including panic hardware, per door	CC@1.94	Ea	294.00	83.10	377.10
Hospital, not including panic hardware, per door	CC@2.46	Ea	459.00	105.00	564.00
Office, not including panic hardware, per door	CC@1.57	Ea	206.00	67.30	273.30
Locksets, heavy residential or light duty commercial, "A" series, chrome finish					
Passage latch, no lock	CC@.689	Ea	36.90	29.50	66.40
Privacy latch, button lock	CC@.689	Ea	42.70	29.50	72.20

Doors and Windows 8

	Craft@Hrs	Unit	Material	Labor	Total
Entrance lockset, key lock	CC@.745	Ea	75.00	31.90	106.90
Classroom or storeroom, key lock	CC@.745	Ea	82.40	31.90	114.30
Dummy knob	CC@.247	Ea	15.80	10.60	26.40
Add for lever handle (handicapped)	--	%	30.0	--	--
Add for brass or bronze finish	--	%	5.0	--	--
Locksets, heavy duty commercial, chrome finish					
"D" series, bored entrance lockset	CC@.745	Ea	190.00	31.90	221.90
"E" series grip handle, with trim	CC@.855	Ea	208.00	36.60	244.60
"H" series single deadbolt-lockset	CC@.745	Ea	92.30	31.90	124.20
"H" series double deadbolt-lockset	CC@.745	Ea	121.00	31.90	152.90
"L" series, mortise deadbolt-lockset	CC@.855	Ea	208.00	36.60	244.60
Add for lever handle (handicapped)	--	%	20.0	--	--
Add for brass or bronze finish	--	%	8.0	--	--
Deadbolts and deadlatches, "B" series, chrome finish					
Standard deadbolt, single cylinder	CC@.745	Ea	27.70	31.90	59.60
Standard deadbolt, double cylinder	CC@.745	Ea	38.10	31.90	70.00
Standard turnbolt, no cylinder	CC@.745	Ea	21.30	31.90	53.20
Nightlatch, single cylinder	CC@.745	Ea	67.00	31.90	98.90
Nightlatch, double cylinder	CC@.745	Ea	73.90	31.90	105.80
Nightlatch, no cylinder, turnbolt	CC@.745	Ea	57.80	31.90	89.70
Heavy duty deadbolt, single cylinder	CC@.745	Ea	51.90	31.90	83.80
Heavy duty deadbolt, double cylinder	CC@.745	Ea	64.60	31.90	96.50
Heavy duty turnbolt, no cylinder	CC@.745	Ea	51.90	31.90	83.80
Extra heavy deadbolt, single cylinder	CC@.745	Ea	82.00	31.90	113.90
Extra heavy deadbolt, double cylinder	CC@.745	Ea	97.00	31.90	128.90
Add for brass or bronze finish	--	Ea	2.32	--	--
Hinges, butt type					
3-1/2" x 3-1/2"	CC@.357	Pr	8.62	15.30	23.92
5" x 5", hospital swing clear	CC@.547	Pr	112.00	23.40	135.40
4", spring, single acting	CC@.243	Pr	28.10	10.40	38.50
4-1/2", heavy duty	CC@.347	Pr	52.70	14.90	67.60
7", spring, double acting	CC@.491	Pr	69.30	21.00	90.30
Floor-mounted, interior, per door	CC@.665	Ea	231.00	28.50	259.50
Floor-mounted, exterior, per 3' door	CC@.665	Ea	248.00	28.50	276.50
Floor-mounted, exterior, over 3' wide door	CC@.799	Ea	525.00	34.20	559.20
Invisible, blind doors, interior, per door	CC@.574	Ea	27.20	24.60	51.80
Invisible, for blind doors, exterior	CC@.574	Pr	34.90	24.60	59.50
Closer, surface mounted					
Interior doors	CC@.616	Ea	64.80	26.40	91.20
Exterior doors	CC@.709	Ea	83.00	30.40	113.40
Exterior, heavy duty	CC@.894	Ea	105.00	38.30	143.30
Floor mounted	CC@1.16	Ea	179.00	49.70	228.70
Rim lock panic door exit hardware, with trim, for aluminum, steel or wood doors					
Satin aluminum finish	CC@1.50	Ea	279.00	64.30	343.30
Dark bronze finish	CC@1.50	Ea	292.00	64.30	356.30
Polished chrome finish	CC@1.50	Ea	359.00	64.30	423.30
Add for external lockset	CC@.401	Ea	39.20	17.20	56.40
Vertical rod panic door exit hardware, satin aluminum finish, with trim					
For metal or wood doors, external rod	CC@1.79	Ea	386.00	76.70	462.70
For wood doors, concealed rod	CC@2.14	Ea	453.00	91.70	544.70
For aluminum doors, concealed rod, satin	CC@2.14	Ea	371.00	91.70	462.70
Add for external lockset	CC@.401	Ea	49.70	17.20	66.90
Add for bronze or black finish	--	%	42.0	--	--
Add for chrome finish	--	%	80.0	--	--

Doors and Windows 8

	Craft@Hrs	Unit	Material	Labor	Total

Hardware, continued
Mortise panic door exit hardware, with trim, for aluminum, steel or wood doors

	Craft@Hrs	Unit	Material	Labor	Total
Satin aluminum finish	CC@2.41	Ea	381.00	103.00	484.00
Dark bronze finish	CC@2.41	Ea	392.00	103.00	495.00
Polished chrome finish	CC@2.41	Ea	464.00	103.00	567.00
Add for external lockset	CC@.401	Ea	51.90	17.20	69.10
Threshold, aluminum, 36"	CC@.241	Ea	15.50	10.30	25.80
Threshold, bronze, 36"	CC@.241	Ea	63.50	10.30	73.80
Kick plates, 10" x 34"					
16 gauge, bronze	CC@.452	Ea	37.80	19.40	57.20
18 gauge, stainless	CC@.452	Ea	31.40	19.40	50.80
Push plates, bronze, 4" x 16"	CC@.190	Ea	24.00	8.14	32.14
Pull plates, bronze, 4" x 16"	CC@.190	Ea	31.30	8.14	39.44
Knob dummy, single	CC@.369	Ea	26.40	15.80	42.20
Flush bolt, automatic	CC@.740	Ea	127.00	31.70	158.70
Flush bolt, extension	CC@.357	Ea	15.90	15.30	31.20
Surface bolts, 4"	CC@.308	Ea	4.83	13.20	18.03
Deadbolt, specification grade	CC@.196	Ea	44.70	8.40	53.10
Surface bolts, 6"	CC@.310	Ea	6.62	13.30	19.92
Dustproof strike	CC@.276	Ea	14.10	11.80	25.90
Automatic door bottom, aluminum, 36"	CC@.494	Ea	27.60	21.20	48.80
Letter drop plate	CC@.369	Ea	26.30	15.80	42.10
Door stop					
Rubber tip, screw base	CC@.060	Ea	3.78	2.57	6.35
Floor mounted, with holder	CC@.279	Ea	15.60	12.00	27.60
Wall mounted, with holder	CC@.279	Ea	15.30	12.00	27.30
Overhead mounted	CC@.528	Ea	90.00	22.60	112.60
Door viewer	CC@.183	Ea	10.90	7.84	18.74
Interview panel, grille	CC@.308	Ea	23.10	13.20	36.30
Cabinet hardware, per LF of face, typical					
Commercial grade	CC@.145	LF	9.82	6.21	16.03
Institutional grade	CC@.181	LF	12.40	7.75	20.15

Weatherstripping. These costs include the subcontractor's overhead and profit.
Bronze and neoprene, 3' x 7' door

	Craft@Hrs	Unit	Material	Labor	Total
Wood door	CC@1.53	Ea	16.40	65.50	81.90
Steel door, adjustable	CC@2.32	Ea	40.50	99.40	139.90
Astragal, mortise mounted, bronze, adjustable	CC@.064	LF	10.03	2.74	12.77
Glue-back foam for 3' x 7' door	CC@.006	LF	.26	.26	.52

Glazing. These costs include the subcontractor's overhead and profit.
1/8" sheet (window) glass
Single strength "B"

	Craft@Hrs	Unit	Material	Labor	Total
To 60" width plus length	G1@.057	SF	2.15	2.22	4.37
Over 60" to 70" width plus length	G1@.057	SF	2.20	2.22	4.42
Over 70" to 80" width plus length	G1@.057	SF	2.40	2.22	4.62
Over 80" width plus length	G1@.057	SF	2.45	2.22	4.67
Double strength "B"					
To 60" width plus length	G1@.057	SF	2.80	2.22	5.02
Over 60" to 70" width plus length	G1@.057	SF	2.90	2.22	5.12
Over 70" to 80" width plus length	G1@.057	SF	3.00	2.22	5.22
Over 80" width plus length	G1@.057	SF	3.10	2.22	5.32
Tempered "B"	G1@.057	SF	4.55	2.22	6.77

Doors and Windows 8

	Craft@Hrs	Unit	Material	Labor	Total
Obscure	G1@.057	SF	4.30	2.22	6.52
3/16"					
Crystal (clear)	G1@.068	SF	3.85	2.65	6.50
Tempered	G1@.068	SF	6.15	2.65	8.80
1/4"					
Float, clear, quality 3	G1@.102	SF	3.45	3.97	7.42
Float, bronze or gray, quality 3	G1@.102	SF	3.75	3.97	7.72
Float, obscure	G1@.102	SF	5.40	3.97	9.37
Float, heat absorbing	G1@.102	SF	5.90	3.97	9.87
Float, safety, laminated, clear	G1@.102	SF	6.50	3.97	10.47
Float, tempered	G1@.102	SF	5.90	3.97	9.87
Spandrel, plain	G1@.102	SF	7.95	3.97	11.92
Spandrel, tinted gray or bronze	G1@.102	SF	8.75	3.97	12.72
Tempered, reflective	G1@.102	SF	9.10	3.97	13.07
3/8"					
Float, clear	G1@.128	SF	6.05	4.98	11.03
Float, tinted	G1@.128	SF	8.50	4.98	13.48
Obscure, tempered	G1@.128	SF	9.10	4.98	14.08
Corrugated glass	G1@.128	SF	7.80	4.98	12.78
1/2" float, tempered	G1@.157	SF	22.20	6.11	28.31
1" bullet resistant, 12" x 12" panels	G1@1.32	SF	86.80	51.40	138.20
2" bullet resistant, 15 to 20 SF	G1@1.62	SF	51.80	63.00	114.80
Insulating glass, 2 layers of 1/8", 1/2" overall, 10 to 15 SF lites					
"B" quality sheet glass	G1@.131	SF	6.90	5.10	12.00
Float, polished bronze or gray	G1@.131	SF	9.60	5.10	14.70
Insulating glass, 2 layers of 1/4" float, 1" overall, 30 to 40 SF lites					
Clear float glass	G1@.136	SF	10.60	5.29	15.89
Plexiglass					
Clear 1/8"	G1@.086	SF	2.60	3.35	5.95
Clear 1/4"	G1@.086	SF	3.55	3.35	6.90
Colors 1/4"	G1@.086	SF	4.90	3.35	8.25
Shatterproof 1/4"	G1@.086	SF	5.65	3.35	9.00
Wired glass, 1/4", type 3					
Clear	G1@.103	SF	6.70	4.01	10.71
Hammered	G1@.103	SF	5.35	4.01	9.36
Obscure	G1@.103	SF	5.10	4.01	9.11
Mirrors, unframed					
Sheet glass, 3/16"	G1@.103	SF	5.45	4.01	9.46
Float glass, 1/4"	G1@.106	SF	8.45	4.12	12.57
Reflective, 1 way, in wood stops	G1@.118	SF	15.50	4.59	20.09
Small lites, adhesive mount	G1@.057	SF	3.55	2.22	5.77

Curtain Walls. Typical costs. These costs include the subcontractor's overhead and profit.

Structural sealant curtain wall systems with aluminum or steel frame

	Craft@Hrs	Unit	Material	Labor	Total
Regular weight, float glass	--	SF	--	--	27.00
Heavy duty, float glass	--	SF	--	--	29.20
Corning glasswall glass	--	SF	--	--	31.50
Add for spandrel glass	--	SF	--	--	3.80
Add for heat absorbing glass	--	SF	--	--	6.75
Add for low transmission glass	--	SF	--	--	4.40
Add for bronze anodizing	--	%	--	--	15.0
Add for black anodizing	--	%	--	--	18.0
Add for 1" insulating glass	--	SF	--	--	6.80

Finishes 9

	Craft@Hrs	Unit	Material	Labor	Total

Lathing and Furring. These costs include the subcontractor's overhead and profit but no plaster.
Furring, no lath or plaster included, galvanized

Walls					
3/4" x 5/8" hat channel, 16" OC	C8@.243	SY	3.40	9.35	12.75
3/4" x 3/4" at 16" OC	C8@.243	SY	3.05	9.35	12.40
1-1/2" x 1-1/2" at 16" OC	C8@.269	SY	4.10	10.30	14.40
Beams and columns					
3/4" x 3/4" at 12" OC	C8@.423	SY	3.80	16.30	20.10
3/4" x 3/4" at 16" OC	C8@.382	SY	3.05	14.70	17.75

Ceilings, main runners at 4' OC, furring channels at 16" OC, 8 gauge wire at 4' OC

1-1/2" x 3/4", standard centering	C8@.344	SY	4.40	13.20	17.60
1-1/2" x 7/8", hat channel, standard centering	C8@.363	SY	4.25	14.00	18.25
3-1/4" x 1-1/2" x 3/4", triple hung	C8@.683	SY	8.05	26.30	34.35
1-1/2" x 3/4", coffered	C8@.887	SY	7.60	34.10	41.70
Add for resilient spring system	--	SY	--	--	10.60

Lathing, walls and ceilings, nailed or wired in place

2.5 lb, standard diamond galvanized	F8@.099	SY	2.40	3.87	6.27
2.5 lb, painted	F8@.099	SY	2.20	3.87	6.07
3.4 lb, standard diamond galvanized	F8@.099	SY	2.60	3.87	6.47
3.4 lb, painted copper alloy	F8@.099	SY	2.40	3.87	6.27
1/8" flat rib, 3.4 lb, galvanized	F8@.099	SY	2.60	3.87	6.47
1/8" flat rib, 2.75 lb, painted	F8@.099	SY	2.50	3.87	6.37
3/8" rib lath, galvanized, 3.4 lb	F8@.099	SY	2.90	3.87	6.77
3/8" painted copper alloy, 3.4 lb	F8@.099	SY	2.60	3.87	6.47
Wire mesh, 15 lb felt and line wire					
1" mesh, 18 gauge	F8@.079	SY	1.75	3.09	4.84
1-1/2" mesh, 17 gauge	F8@.079	SY	1.85	3.09	4.94
Paper back lath	F8@.063	SY	2.40	2.46	4.86
Stucco rite 2" corner bead, 16 gauge	F8@.028	CLF	1.85	1.09	2.94
Aqua lath, 2" x 16 gauge	F8@.076	SY	2.50	2.97	5.47
Gypsum lath, plain or perforated, 1/2"					
Clipped in place	F8@.094	SY	3.30	3.67	6.97
Nailed in place	F8@.099	SY	3.15	3.87	7.02
Asphalt coated 1/2" gypsum lath (nailed)	F8@.059	SY	2.60	2.30	4.90
Lead lined gypsum lath, 4 lb	F8@.348	SY	38.80	13.60	52.40
Add for screwed to 16 gauge metal studs	--	SY	--	--	1.55
Add for screwed to 25 gauge metal studs	--	SY	--	--	1.35

Plastering. These costs include the subcontractor's overhead and profit but no lath.
Gypsum interior plaster, lime putty trowel finish

Two coat application, on ceilings	F8@.381	SY	3.80	14.90	18.70
Two coat application, on walls	F8@.330	SY	3.80	12.90	16.70
Three coat application, on ceilings	F8@.448	SY	5.00	17.50	22.50
Three coat application, on walls	F8@.405	SY	5.00	15.80	20.80

Gypsum interior vermiculite plaster, trowel finish

Two coat application, on ceilings	F8@.381	SY	4.00	14.90	18.90
Two coat application, on walls	F8@.330	SY	4.00	12.90	16.90
Three coat application, on ceilings	F8@.448	SY	5.20	17.50	22.70
Three coat application, on walls	F8@.405	SY	5.20	15.80	21.00

Keene's cement plaster, troweled lime putty medium hard finish

Two coat application, on ceilings	F8@.435	SY	4.30	17.00	21.30
Two coat application, on walls	F8@.381	SY	4.30	14.90	19.20
Three coat application, on ceilings	F8@.521	SY	5.30	20.30	25.60

Finishes 9

	Craft@Hrs	Unit	Material	Labor	Total
Three coat application, on walls	F8@.459	SY	5.30	17.90	23.20
Portland cement stucco on exterior walls, 3 coats totaling 1" thick					
Natural gray, sand float finish	F8@.567	SY	3.80	22.10	25.90
Natural gray, trowel finish	F8@.640	SY	3.80	25.00	28.80
White cement, sand float finish	F8@.675	SY	4.50	26.40	30.90
White cement, trowel finish	F8@.735	SY	4.50	28.70	33.20
Portland cement stucco on soffits, 3 coats totaling 1" thick					
Natural gray, sand float finish	F8@.678	SY	3.80	26.50	30.30
Natural gray, trowel finish	F8@.829	SY	3.80	32.40	36.20
White cement, sand float finish	F8@.878	SY	4.45	34.30	38.75
White cement, trowel finish	F8@1.17	SY	4.45	45.70	50.15
Brown and scratch coat base for tile	F8@.308	SY	3.00	12.00	15.00
Scratch coat only for tile	F8@.152	SY	1.40	5.94	7.34
Thin coat plaster, with board but no studs					
1/2" board on wood studs	F8@.265	SY	3.40	10.30	13.70
5/8" board on wood studs	F8@.265	SY	5.85	10.30	16.15
1/2" board on metal studs	F8@.277	SY	4.80	10.80	15.60
5/8" board on metal studs	F8@.262	SY	6.10	10.20	16.30
Simulated acoustic texture on ceilings	F8@.052	SY	.81	2.03	2.84
Texture coat on exterior walls and soffits	F8@.045	SY	.81	1.76	2.57
Patching gypsum plaster, including lath repair but no studding					
To 5 SF repairs	P5@.282	SF	.75	10.30	11.05
Over 5 SF repairs	P5@.229	SF	.75	8.38	9.13
Repair cracks only ($150.00 per job minimum typical)	P5@.055	LF	.13	2.01	2.14
Plastering bead and expansion joint, galvanized					
Stop and casing, square nose	F8@.028	LF	.38	1.09	1.47
Corner bead, 3/4" radius	F8@.023	LF	.41	.90	1.31
Corner bead, expanded, zinc nose	F8@.024	LF	.82	.94	1.76
Expansion joint, 3/4", 26 gauge, 1 piece	F8@.025	LF	.82	.98	1.80
Expansion joint, 1-1/2", 2 piece	F8@.028	LF	1.40	1.09	2.49
Plastering accessories					
Base screed, 1/2", 26 gauge	F8@.032	LF	.50	1.25	1.75
Vents 1-1/2", galvanized	F8@.042	LF	1.40	1.64	3.04
Vents 4", galvanized	F8@.061	LF	1.65	2.38	4.03
Archbead, plastic nose	F8@.028	LF	.85	1.09	1.94
Steel access doors, with casing and ground					
12" x 12"	CC@.391	Ea	32.20	16.80	49.00
18" x 18"	CC@.391	Ea	46.00	16.80	62.80
24" x 24"	CC@.559	Ea	69.00	23.90	92.90
Plaster mouldings, ornate designs					
2"	CC@.090	LF	3.60	3.86	7.46
4"	CC@.115	LF	7.40	4.93	12.33
6"	CC@.132	LF	10.20	5.66	15.86

Hollow Metal Stud Partitions for Plaster or Gypsum Wallboard. Galvanized metal studs, including studs, runners and shoes. These costs include the subcontractor's overhead and profit but no lath or plaster or gypsum wallboard (drywall).

	Craft@Hrs	Unit	Material	Labor	Total
Non-load bearing partitions, 25 gauge, 16" on center spacing					
1-5/8" wide studs	C8@.026	SF	.38	1.00	1.38
2-1/2" wide studs	C8@.028	SF	.50	1.08	1.58
3-5/8" wide studs	C8@.030	SF	.62	1.15	1.77
4" wide studs	C8@.032	SF	.70	1.23	1.93
6" wide studs	C8@.034	SF	.85	1.31	2.16

Finishes 9

	Craft@Hrs	Unit	Material	Labor	Total
Hollow Metal Stud Partitions, Non-load bearing, continued					
Add for 20 gauge, 16" OC spacing	--	%	25.0	--	--
Deduct for 25 gauge, 24" OC spacing	--	%	-20.0	-15.0	--
Load bearing partitions, 16 gauge "C" section, 16" on center spacing					
2-1/2" wide studs	C8@.037	SF	1.35	1.42	2.77
3-3/5" wide studs	C8@.042	SF	1.50	1.62	3.12
4" wide studs	C8@.045	SF	1.60	1.73	3.33
6" wide studs	C8@.048	SF	2.05	1.85	3.90
Deduct for 20 gauge, 16" OC spacing	--	%	-30.0	-15.0	--
Deduct for 16 gauge, 24" OC spacing	--	%	-20.0	-10.0	--
Add for 16 gauge, 12" OC spacing	--	%	50.0	20.0	--
Add for jobs under 4,500 square feet	--	%	--	--	10.0

Gypsum Wallboard. Commercial grade work. These costs include the subcontractor's overhead and profit. Material costs include 6% for waste. Costs per square foot of area covered.

	Craft@Hrs	Unit	Material	Labor	Total
Gypsum wallboard nailed or screwed to wood framing or wood furring, no taping or finishing included					
3/8" on walls	CD@.007	SF	.18	.30	.48
3/8" on ceilings	CD@.009	SF	.18	.38	.56
3/8" on furred columns and beams	CD@.013	SF	.18	.55	.73
1/2" on walls	CD@.007	SF	.18	.30	.48
1/2" on ceilings	CD@.009	SF	.18	.38	.56
1/2" on furred columns and beams	CD@.015	SF	.18	.64	.82
5/8" on walls	CD@.007	SF	.20	.30	.50
5/8" on ceilings	CD@.010	SF	.20	.43	.63
5/8" on furred columns and beams	CD@.015	SF	.20	.64	.84
Add for taping and finishing wall joints	CD@.007	SF	.04	.30	.34
Add for taping and finishing ceiling joints	CD@.009	SF	.04	.38	.42
Gypsum wallboard clipped to metal furring, no taping or finishing included					
3/8" on wall furring	CD@.008	SF	.20	.34	.54
3/8" on ceiling furring	CD@.011	SF	.20	.47	.67
1/2" on wall furring	CD@.009	SF	.20	.38	.58
1/2" on ceiling furring	CD@.011	SF	.20	.47	.67
1/2" on column or beam furring	CD@.018	SF	.20	.77	.97
5/8" on wall furring	CD@.009	SF	.22	.38	.60
5/8" on ceiling furring	CD@.011	SF	.22	.47	.69
5/8" on column or beam furring	CD@.018	SF	.22	.77	.99
1/2", two layers on ceiling furring	CD@.022	SF	.39	.94	1.33
1/2", two layers on furring	CD@.017	SF	.39	.72	1.11
1/4" sound board	CD@.008	SF	.22	.34	.56
Add for taping and finishing wall joints	CD@.007	SF	.04	.30	.34
Add for taping and finishing ceiling joints	CD@.009	SF	.04	.38	.42
Add for taping only, no joint finishing	CD@.003	SF	.03	.13	.16
Vinyl clad gypsum board, adhesive or clip application on walls					
1/2", no mouldings included	CD@.010	SF	.69	.43	1.12
5/8", no mouldings included	CD@.011	SF	.72	.47	1.19
Additional costs for gypsum wallboard					
Add for foil-backed board	--	SF	.06	--	--
Add for fire resistant board	--	SF	.03	--	--
Add for water resistant board	--	SF	.12	--	--
Add for 10' or 12' wall heights	CD@.001	SF	.02	.04	.06
Add for school jobs	CD@.002	SF	--	.09	--
Add for trowel textured finish	CD@.009	SF	.06	.38	.44
Add for adhesive application, 1/4" bead					

Finishes 9

	Craft@Hrs	Unit	Material	Labor	Total
Studs 16" on center	CD@.002	SF	.06	.09	.15
Joists 16" on center	CD@.002	SF	.09	.09	.18
Bead and casing					
Corner bead, 1-1/4" x 1-1/4"	CD@.017	LF	.18	.72	.90
Stop or casing	CD@.021	LF	.22	.89	1.11
Jamb casing	CD@.023	LF	.22	.98	1.20

Metal Framed Shaft Walls. Shaft walls are used to enclose vertical shafts that surround pipe chases, electrical conduit, elevators or stairwells. A two-hour wall can be made from 1" gypsum wallboard screwed to a 2-1/2" metal stud partition on the shaft side with two layers of 5/8" gypsum wallboard screwed to the partition exterior face. The wall cavity is filled with fiberglass batt insulation and the drywall is taped and finished on both sides. This wall will be 4-3/4" thick. Labor includes installing metal studs, installing the insulation, hanging, taping and finishing the drywall, and cleanup. Metal studs and insulation include a 10% allowance for waste. Wallboard costs include corner and edge trim and a 15% allowance for waste. Costs shown are per square foot of wall measured on one side. These costs include the subcontractor's overhead and profit. For scheduling purposes, estimate that a crew of 4 can install metal studs and insulation, hang, tape and finish 250 SF of shaft wall per 8-hour day.

	Craft@Hrs	Unit	Material	Labor	Total
Metal studs, "C-T" section, 2-1/2" wide, 16 gauge, 24" on center, complete with top runner and bottom plate	C8@.036	SF	1.25	1.39	2.64
1" type X gypsum shaftboard	D2@.046	SF	.45	1.76	2.21
2 layers 5/8" type X gypsum wallboard	D2@.038	SF	.80	1.46	2.26
Fiberglass insulation, foil faced,	A1@.006	SF	.25	.24	.49
Total for metal framed shaft wall as described	—@.126	SF	2.75	4.85	7.60

Ceramic Tile. Set in adhesive and grouted. Add the cost of adhesive below. No scratch or brown coat included. Based on standard U.S. grades and stock colors. Custom colors, designs and imported tile will cost more. These costs include the subcontractor's overhead and profit.

	Craft@Hrs	Unit	Material	Labor	Total
4-1/4" x 4-1/4" glazed wall tile					
Smooth gloss glaze, minimum grade	T4@.061	SF	1.04	2.25	3.29
Smooth gloss glaze, standard grade	T4@.061	SF	1.35	2.25	3.60
Matte finish, better grade	T4@.061	SF	1.66	2.25	3.91
High gloss finish	T4@.061	SF	2.07	2.25	4.32
Commercial grades					
Group 1 (matte glaze)	T4@.061	SF	2.00	2.25	4.25
Group 2 (bright glaze)	T4@.061	SF	2.50	2.25	4.75
Group 3 (crystal glaze)	T4@.061	SF	2.85	2.25	5.10
6" wall tile, smooth glaze					
6" x 6", gray or brown	T4@.061	SF	1.35	2.25	3.60
6" x 8", quilted look	T4@.061	SF	2.75	2.25	5.00
6" x 8", fume look	T4@.061	SF	2.43	2.25	4.68
Trim pieces for glazed wall tile					
Surface bullnose	T4@.077	LF	1.45	2.84	4.29
Surface bullnose corner	T4@.039	Ea	.94	1.44	2.38
Sink rail or cap	T4@.077	LF	4.14	2.84	6.98
Radius bullnose	T4@.077	LF	1.91	2.84	4.75
Quarter round or bead	T4@.077	LF	2.28	2.84	5.12
Outside corner or bead	T4@.039	Ea	.72	1.44	2.16
Base	T4@.077	LF	4.57	2.84	7.41
Soap holder, soap dish	T4@.265	Ea	11.10	9.78	20.88
Tissue holder, towel bar	T4@.268	Ea	15.20	9.90	25.10
Glazed floor tile, 1/4" thick					
Minimum grade, standard colors	T4@.060	SF	2.23	2.22	4.45
Better grades, patterns	T4@.060	SF	2.75	2.22	4.97

Finishes 9

	Craft@Hrs	Unit	Material	Labor	Total
Ceramic Tile, Glazed floor tile, continued					
Commercial grade, 6" x 6" to 6" x 9"					
Reds and browns	T4@.060	SF	2.40	2.22	4.62
Most colors	T4@.060	SF	2.50	2.22	4.72
Blues and greens	T4@.060	SF	2.70	2.22	4.92
Add for abrasive surface	--	SF	.27	--	--
1" x 1" mosaic tile, back-mounted, natural clays					
Group I colors (browns, reds)	T4@.055	SF	2.15	2.03	4.18
Group II colors (tans, grays)	T4@.055	SF	2.40	2.03	4.43
Group III colors (charcoal, blues)	T4@.055	SF	2.60	2.03	4.63
Add for extruded porcelain mosaic tile	--	%	20.0	--	20.0
Adhesive for ceramic tile					
Thinset mortar on concrete (25 lbs at $6.00 covers 100 SF)	--	SF	.06	--	.06
Latex floor and wall adhesive (gallon at $5.00 covers 50 SF)	--	SF	.10	--	--
Epoxy adhesive on concrete or wood (12.5 lbs at $19.50 covers 80 SF)	--	SF	.24	--	--
Acrylic adhesive on counter tops or wood (25 lbs at $15.50 covers 80 SF)	--	SF	.19	--	--
Decorator tile panels, institutional grade, 6" x 9" x 3/4", typical costs					
Unglazed domestic	T4@.102	SF	5.28	3.77	9.05
Wash-glazed domestic	T4@.102	SF	9.05	3.77	12.82
Full glazed domestic	T4@.102	SF	9.56	3.77	13.33
Glazed imported, decorative	T4@.102	SF	24.10	3.77	27.87
Deduct for imported tile	--	%	-20.0	--	--
Plastic tile					
4-1/4" x 4-1/4" x .110"	T4@.052	SF	1.51	1.92	3.43
4-1/4" x 4-1/4" x .050"	T4@.052	SF	1.16	1.92	3.08
Aluminum tile, 4-1/4"	T4@.073	SF	2.72	2.70	5.42
Copper on aluminum, 4-1/4" x 4-1/4"	T4@.076	SF	3.02	2.81	5.83
Stainless steel tile, 4-1/4" x 4-1/4"	T4@.076	SF	4.83	2.81	7.64
Quarry tile set in portland cement 5" high base tile	T4@.109	LF	2.79	4.02	6.81
4" x 4" x 1/2" floor tile	T4@.096	SF	3.10	3.54	6.64
6" x 6" x 1/2" floor tile	T4@.080	SF	2.69	2.95	5.64
Quarry tile set in furan resin					
5" high base tile	T4@.096	LF	1.97	3.54	5.51
6" x 6" x 3/4" floor tile	T4@.093	SF	4.71	3.43	8.14
Add for abrasive finish	--	SF	.40	--	--
Terrazzo. These costs include the subcontractor's overhead and profit.					
Floors, 1/2" terrazzo topping on an underbed bonded to an existing concrete slab, #1 and #2 chips in gray portland cement. Add divider strips below					
2", epoxy or polyester	T4@.128	SF	2.15	4.73	6.88
2" conductive	T4@.134	SF	2.80	4.95	7.75
2" with #3 and larger chips	T4@.134	SF	3.80	4.95	8.75
3" with 15 lb felt and sand cushion	T4@.137	SF	2.80	5.06	7.86
2-3/4" with mesh, felt and sand cushion	T4@.137	SF	2.70	5.06	7.76
Add for:					
White portland cement	--	SF	.32	--	--
Non-slip abrasive, light	--	SF	--	--	1.00
Non-slip abrasive, heavy	--	SF	--	--	1.35
Countertops, mud set LF of front	--	LF	--	--	30.40
Wainscot, precast, mud set	T4@.198	SF	6.65	7.31	13.96

Finishes 9

	Craft@Hrs	Unit	Material	Labor	Total
Cove base, mud set	--	LF	--	--	7.85
Add for 2,000 SF job	--	%	--	--	25.0
Add for 1,000 SF job	--	%	--	--	100.0
Add for waterproof membrane	--	SF	--	--	1.15
Divider strips					
Brass, 12 gauge	T4@.009	LF	2.25	.33	2.58
White metal, 12 gauge	T4@.009	LF	1.00	.33	1.33
Brass 4' OC each way	T4@.004	SF	.90	.15	1.05
Brass 2' OC each way	T4@.007	SF	2.15	.26	2.41

Acoustical Treatment. These costs include the subcontractor's overhead and profit.

Tile board and panels, no suspension grid included, applied to ceilings with staples, not including furring

Decorative (non-acoustic) Armstrong ceiling tile, 12" x 12" x 1/2", T&G

	Craft@Hrs	Unit	Material	Labor	Total
Washable white or Grenoble	C8@.017	SF	.50	.65	1.15
Chaperone	C8@.017	SF	.55	.65	1.20
Conestoga	C8@.017	SF	.60	.65	1.25
Windstone	C8@.017	SF	.66	.65	1.31

Acoustic (Cushiontone) Armstrong ceiling tile, 12" x 12" x 1/2", T&G

	Craft@Hrs	Unit	Material	Labor	Total
Classic random perf or Verona	C8@.017	SF	.73	.65	1.38
Textured random perf	C8@.017	SF	.88	.65	1.53
Fissured or Plank n' Plaster	C8@.017	SF	1.15	.65	1.80
Rush Square	C8@.017	SF	1.25	.65	1.90
Colonial Sampler	C8@.017	SF	1.41	.65	2.06
Constitution	C8@.017	SF	1.77	.65	2.42

Fire-rated Armstrong acoustic tile, mineral fiber 12" x 12" x 5/8", beveled edge

	Craft@Hrs	Unit	Material	Labor	Total
Fissured Minitone	C8@.017	SF	1.05	.65	1.70
Cortega Minitone	C8@.017	SF	1.05	.65	1.70
Add for adhesive applications	C8@.001	SF	.16	.04	.20
Deduct for wall applications	--	%	--	-10.0	--

Sound-absorbing fabric covered wall panels with splines for attaching to structural walls, Armstrong Soundsoak, 9' or 10' high

	Craft@Hrs	Unit	Material	Labor	Total
Embossed 60, with internal splines, 30" wide, 3/4" thick	C8@.038	SF	6.01	1.46	7.47
Soundsoak 80, with internal splines, 24" wide, 1" thick	C8@.038	SF	8.15	1.46	9.61
Vinyl Soundsoak, 24" wide, 3/4" thick	C8@.038	SF	5.23	1.46	6.69
Encore, 30" wide, 3/4" thick	C8@.038	SF	8.26	1.46	9.72
Aluminum starter spline for panels	C8@.020	LF	.94	.77	1.71
Aluminum internal spline for panels	C8@.021	LF	.99	.81	1.80

Suspended Ceiling Grid Systems. 5,000 SF job. These costs include the subcontractor's overhead and profit. See Ceiling Tile below.

"T" bar suspension system, no tile included, suspended from ceiling joists with wires

	Craft@Hrs	Unit	Material	Labor	Total
1' x 1' grid	C8@.016	SF	.48	.62	1.10
2' x 2' grid	C8@.011	SF	.44	.42	.86
2' x 4' grid	C8@.010	SF	.42	.39	.81
4' x 4' grid	C8@.009	SF	.33	.35	.68
Add for jobs under 5,000 SF					
Less than 2,500 SF	--	%	--	--	65.0
2,500 SF to 4,500 SF	--	%	--	--	50.0

Finishes 9

	Craft@Hrs	Unit	Material	Labor	Total

Ceiling Tile. Tile laid in suspended ceiling grid. No ceiling grid included. Labor column shows the cost of laying tile in a suspended ceiling grid. These costs include the subcontractor's overhead and profit.

1/2" thick non-acoustic (Armstrong Temlock) panels, 2' x 4'

	Craft@Hrs	Unit	Material	Labor	Total
Plain white	C8@.004	SF	.50	.15	.65
Conestoga or Chaperone	C8@.004	SF	.65	.15	.80
Grenoble or Windstone vinyl	C8@.004	SF	.70	.15	.85
Glenwood vinyl	C8@.004	SF	.73	.15	.88

1/2" thick acoustic (Armstrong Cushiontone) panels, 2' x 4'

	Craft@Hrs	Unit	Material	Labor	Total
Kingsley	C8@.004	SF	.51	.15	.66
Classic	C8@.004	SF	.63	.15	.78
Verona	C8@.004	SF	.67	.15	.82
Plaza	C8@.004	SF	.80	.15	.95

5/8" thick (Armstrong Fashiontone) panels, 2' x 4' mineral fiber panels

	Craft@Hrs	Unit	Material	Labor	Total
Random perf or Cortega	C8@.004	SF	.55	.15	.70
Textured or fissured	C8@.004	SF	.64	.15	.79
Rock Castle	C8@.004	SF	.62	.15	.77
Royal Oak	C8@.004	SF	1.19	.15	1.34
Victoria	C8@.004	SF	1.27	.15	1.42
Cumberland	C8@.004	SF	1.09	.15	1.24

5/8" special purpose panels, 2' x 4'

	Craft@Hrs	Unit	Material	Labor	Total
Panels with embossed aluminum foil face	C8@.004	SF	.91	.15	1.06
Clean room rated panels	C8@.004	SF	1.12	.15	1.27

3/4" thick panels

	Craft@Hrs	Unit	Material	Labor	Total
2' x 2' reveal edge panels	C8@.005	SF	.85	.19	1.04
12" x 12" concealed edge panels	C8@.007	SF	1.04	.27	1.31
2' x 2' non-woven fabric covered with radius reveal	C8@.012	SF	5.40	.46	5.86

7/8" sound rated mineral fiber panels NRC (75-85), 2' x 4'

	Craft@Hrs	Unit	Material	Labor	Total
White panels	C8@.004	SF	.98	.15	1.13
Fire-rated white panels	C8@.004	SF	1.06	.15	1.21
Rabbeted reveal edge white panels	C8@.004	SF	1.14	.15	1.29

Translucent panels, 2' x 4'

	Craft@Hrs	Unit	Material	Labor	Total
Flat mist white	C8@.004	SF	.75	.15	.90
Clear or frosted prismatic	C8@.004	SF	.70	.15	.85
Arctic opal or cracked ice	C8@.004	SF	.70	.15	.85
Eggcrate louvers	C8@.004	SF	1.70	.15	1.85

Wood Flooring. These costs include the subcontractor's overhead and profit.

Oak, 3/4" x 2-1/4", unfinished

	Craft@Hrs	Unit	Material	Labor	Total
Number 1 common, red and white	F9@.047	SF	1.85	1.79	3.64
Select, plain red and white	F9@.048	SF	2.10	1.83	3.93
Plank natural oak, 3/4", 3" to 5" wide	F9@.048	SF	3.60	1.83	5.43

Parquet, prefinished, 5/16"

	Craft@Hrs	Unit	Material	Labor	Total
Red oak, 9" x 9"	F9@.053	SF	3.15	2.02	5.17
Tropical walnut and cherry, 9" x 9"	F9@.053	SF	4.40	2.02	6.42
Imported teak, neutral finish, 12" x 12"	F9@.053	SF	3.40	2.02	5.42

Gym floors, 3/4", maple, including finishing

	Craft@Hrs	Unit	Material	Labor	Total
On sleepers and membrane	F9@.086	SF	3.20	3.28	6.48
On steel spring system	F9@.114	SF	4.00	4.34	8.34
Connected with steel channels	F9@.097	SF	3.65	3.70	7.35

Softwood floors, fir, unfinished

	Craft@Hrs	Unit	Material	Labor	Total
"B" grade	F9@.031	SF	3.10	1.18	4.28
"C" grade	F9@.031	SF	2.50	1.18	3.68

Finishes 9

	Craft@Hrs	Unit	Material	Labor	Total
Wood block flooring					
Natural, 1-1/2" thick	F9@.023	SF	3.00	.88	3.88
Creosoted, 2" thick	F9@.025	SF	2.20	.95	3.15
Creosoted, 2-1/2" thick	F9@.027	SF	2.55	1.03	3.58
Creosoted, 3" thick	F9@.027	SF	3.05	1.03	4.08
Sand, scrape and edge hardwood floor					
Large room or hall, using 12" drum sander and 7" disc sander					
Three cuts	F9@.993	CSF	4.40	37.80	42.20
Four cuts	F9@1.14	CSF	5.50	43.40	48.90
Closet or 4' x 10' area, using 7" disc sander					
Three cuts	F9@1.02	CSF	3.30	38.90	42.20
Four cuts	F9@1.25	CSF	4.40	47.60	52.00
Finishing wood flooring					
Fill and stain	F9@.567	CSF	4.40	21.60	26.00
Two coats shellac	F9@.794	CSF	7.70	30.20	37.90
Wax	F9@.114	CSF	3.30	4.34	7.64

Resilient Flooring. These costs include the subcontractor's overhead and profit.

	Craft@Hrs	Unit	Material	Labor	Total
Asphalt tile, 1/8"					
"B" grade	F9@.018	SF	1.16	.69	1.85
"C" grade	F9@.018	SF	1.31	.69	2.00
"D" grade	F9@.018	SF	1.26	.69	1.95
Vinyl composition and vinyl tile					
Minimum grade 1/16" vinyl composition tile	F9@.016	SF	.68	.61	1.29
Most .045" no wax, self-stick tile	F9@.016	SF	.95	.61	1.56
Most .050" no wax, self-stick tile	F9@.016	SF	1.21	.61	1.82
Most .080" no wax, self-stick tile	F9@.016	SF	1.58	.61	2.19
Most 1/8" no wax vinyl tile	F9@.016	SF	2.31	.61	2.92
Imperial texture, 3/32", fire rated	F9@.016	SF	.95	.61	1.56
Imperial texture, 1/8", fire rated	F9@.016	SF	1.16	.61	1.77
Add for latex adhesive (1 gallon at $11.00 covers 150 SF)	--	SF	.077	--	--
Solid vinyl tile, Azrock					
1/16" standard	F9@.022	SF	2.84	.84	3.68
1/8" heavy duty	F9@.021	SF	3.94	.80	4.74
Vinyl sheet flooring, Tarkett					
Multiflor granite	F9@.023	SF	2.10	.88	2.98
Acoustiflor granite					
Heat welded seams	F9@.028	SF	2.94	1.07	4.01
Chemical welded seams	F9@.024	SF	2.73	.91	3.64
Salute	F9@.023	SF	3.31	.88	4.19
Conductiflor	F9@.036	SF	5.62	1.37	6.99
Gymflor, including game lines	F9@.036	SF	5.25	1.37	6.62
Optima	F9@.021	SF	2.52	.80	3.32
Sheet vinyl flooring, Armstrong					
Royelle series	F9@.018	SF	.84	.69	1.53
Imperial Accotone series, no wax	F9@.018	SF	1.00	.69	1.69
Sundial Solarian, .077', no wax	F9@.018	SF	1.84	.69	2.53
Classic Corlon, .085"	F9@.021	SF	1.68	.80	2.48
Crown Corlon series	F9@.021	SF	2.31	.80	3.11
Crosswalk, .100", for entryways	F9@.024	SF	3.62	.91	4.53
Flat lay, labor only when not coved	F9@.021	SF	--	.80	--
Self-cove only, labor for small quantities	F9@.142	LF	--	5.41	5.41

Finishes 9

	Craft@Hrs	Unit	Material	Labor	Total
Resilient Flooring, continued					
Add for latex adhesive (1 gallon at $11.00 covers 150 SF)	--	SF	.073	--	--
Plastic Interlocking Safety Floor Tiles, Duragrid. Based on a job with 1,000 to 5,000 SF of floor tiles.					
Firm, smooth tiles, for athletic courts, gym floors, industrial areas					
12" x 12" x 1/2"	F9@.003	SF	4.57	.11	4.68
2" x 12" x 1/2" line strips	F9@.003	LF	1.21	.11	1.32
Firm, non-slip (nippled) tile, for tennis courts, wet-weather surface					
12" x 12" x 1/2"	F9@.003	SF	3.31	.11	3.42
2" x 12" x 1/2" line strips	F9@.003	LF	1.21	.11	1.32
Soft tiles, for anti-fatigue matting, household or deck areas					
12" x 12" x 1/2"	F9@.003	SF	3.57	.11	3.68
2" x 12" x 1/2" line strips	F9@.003	LF	1.21	.11	1.32
Mitered edging strip, 2" x 12" x 1/2"	F9@.003	LF	1.21	.11	1.32
Monogramming/graphics pegs	F9@.085	C	2.42	3.24	5.66
Add for jobs under 1,000 SF	--	%	10.0	20.0	--
Deduct for jobs over 5,000 SF	--	%	-5.0	-5.0	--
Rubber Flooring and Accessories					
Rubber tile flooring					
17-13/16" x 17-13/16" x 1/8" circular pattern	F9@.021	SF	4.20	.80	5.00
Base, top set rubber, 1/8" thick					
2-1/2" high	F9@.018	LF	.63	.69	1.32
4" high	F9@.018	LF	.74	.69	1.43
6" high	F9@.018	LF	1.00	.69	1.69
Base corners, top set rubber, 1/8" thick					
2-1/2" high	F9@.028	Ea	1.00	1.07	2.07
4" high	F9@.028	Ea	1.05	1.07	2.12
6" high	F9@.028	Ea	1.37	1.07	2.44
Stair treads, molded rubber, 12-1/2" tread width, per LF of tread length					
1/4" circle design, colors	F9@.067	LF	9.19	2.55	11.74
Rib design, light duty	F9@.067	LF	6.30	2.55	8.85
Tread with abrasive strip, light duty	F9@.071	LF	9.19	2.70	11.89
Tread with abrasive strip, heavy duty	F9@.071	LF	11.00	2.70	13.70
Stair risers, molded rubber, 7" x 1/8"	F9@.038	LF	2.21	1.45	3.66
Stringer cover, .100"					
10" high	F9@.106	LF	2.26	4.04	6.30
Add for project under 2,000 SF	--	%	--	--	25.0
Underlayment, particleboard, 3/8"	C8@.009	SF	.38	.35	.73
Composition Flooring, 1/4". These costs include the subcontractor's overhead and profit.					
Acrylic	F9@.075	SF	3.25	2.86	6.11
Epoxy terrazzo	F9@.105	SF	3.78	4.00	7.78
Conductive epoxy terrazzo	F9@.132	SF	5.20	5.03	10.23
Troweled neoprene	F9@.082	SF	4.00	3.12	7.12
Polyester	F9@.065	SF	2.90	2.48	5.38

Carpeting. These costs include the subcontractor's overhead and profit. Glue-down carpet installation includes sweeping the floor and adhesive. Carpet over pad installation includes laying the pad and stapling or spot adhesive.

	Craft@Hrs	Unit	Material	Labor	Total
Acrylic, with 50 oz pad					
28 oz	F9@.114	SY	15.20	4.34	19.54
32 oz	F9@.114	SY	16.30	4.34	20.64

Finishes 9

	Craft@Hrs	Unit	Material	Labor	Total
40 oz	F9@.114	SY	18.90	4.34	23.24
Nylon, continuous filament, with 50 oz pad					
20 oz	F9@.114	SY	7.87	4.34	12.21
24 oz	F9@.114	SY	9.21	4.34	13.55
Nylon, level loop, with 50 oz pad					
18 oz	F9@.114	SY	9.21	4.34	13.55
24 oz	F9@.114	SY	10.90	4.34	15.24
28 oz	F9@.114	SY	13.00	4.34	17.34
Antron nylon, 20 oz, anti-static, no pad	F9@.114	SY	8.44	4.34	12.78
Wool, commercial, 34 oz, with 50 oz pad	F9@.114	SY	21.70	4.34	26.04
Pads, labor cost is included with carpet					
40 oz jute	--	SY	2.48	--	--
50 oz jute and hair	--	SY	2.17	--	--
50 oz hair	--	SY	2.74	--	--
76 oz rubber waffle	--	SY	2.69	--	--
100 oz rubber waffle	--	SY	3.57	--	--
58 oz rubber slab	--	SY	2.54	--	--
68 oz rubber slab	--	SY	2.90	--	--
3/8" urethane	--	SY	1.81	--	--
1/2" urethane	--	SY	2.12	--	--
9/16" urethane	--	SY	2.28	--	--
Urethane rebound 5.0 lb, 5/8"	--	SY	3.42	--	--

Carpet Tile
Modular carpet tiles

	Craft@Hrs	Unit	Material	Labor	Total
"Faculty 11" tiles, 26 oz face weight	F9@.170	SY	22.40	6.48	28.88
"Surfaces" tiles, 38 oz face weight	F9@.170	SY	29.00	6.48	35.48
"Facilities 10" tiles, 32 oz face weight	F9@.170	SY	27.60	6.48	34.08

Velvet Tile. Recycled tire material for entrances, stairs, wet areas, etc. Installation includes adhesive. Add for border strips below. El-Do Velvet Tile

	Craft@Hrs	Unit	Material	Labor	Total
12" x 12" tiles	F9@.044	SF	5.75	1.68	7.43
10" x 10" tiles	F9@.044	SF	5.98	1.68	7.66
6" x 6" tiles	F9@.044	SF	6.90	1.68	8.58
5" x 5" tiles	F9@.051	SF	6.33	1.94	8.27
4" x 4" tiles	F9@.051	SF	8.05	1.94	9.99
Aluminum border strips for velvet tile installations	F9@.030	LF	4.00	1.14	5.14

Painting. These costs include the subcontractor's overhead and profit but no surface preparation, scaffolding or masking unless noted.
In place costs, by SF of floor area, typical costs
  Commercial or industrial, 2 coats flat

	Craft@Hrs	Unit	Material	Labor	Total
and enamel	--	SF	--	--	.90
Institutional, 2 coats flat, 3 coats enamel	--	SF	--	--	1.10
Hospitals, public buildings	--	SF	--	--	1.15

Surface preparation for painting, typical costs, SF of area prepared for painting, including normal preparation on walls and ceilings. Hand work, no unusual conditions

	Craft@Hrs	Unit	Material	Labor	Total
Light cleaning (wipe down or hose down)	PA@.001	SF	.02	.04	.06
New cement asbestos siding	PA@.002	SF	.02	.08	.10
Painted cement asbestos siding	PA@.003	SF	.02	.12	.14
New masonry or concrete	PA@.002	SF	.02	.08	.10
Painted masonry or concrete	PA@.003	SF	.03	.12	.15
New plaster or wallboard	PA@.001	SF	.02	.04	.06

Finishes 9

	Craft@Hrs	Unit	Material	Labor	Total
Painting, continued					
Painted plaster or wallboard	PA@.003	SF	.02	.12	.14
New wood	PA@.001	SF	.02	.04	.06
Painted wood	PA@.003	SF	.02	.12	.14
Painted steel (light brushing)	PA@.007	SF	.03	.28	.31
Cabinets and shelves, per square foot of surface painted, brush work					
Wood, 2 coats sealer, 2 coats enamel	PA@.026	SF	.27	1.06	1.33
Wood, 2 coats enamel	PA@.014	SF	.14	.57	.71
Wood, seal and varnish, 3 coats	PA@.016	SF	.17	.65	.82
Wood, back priming	PA@.006	SF	.05	.24	.29
Metal, primer and 2 coats enamel	PA@.014	SF	.17	.57	.74
Cement asbestos siding, 2 coats acrylic latex					
Brush	PA@.016	SF	.09	.65	.74
Roller	PA@.009	SF	.10	.37	.47
Spray	PA@.001	SF	.12	.04	.16
Concrete walls, 2 coats acrylic latex					
Brush	PA@.012	SF	.09	.49	.58
Roller	PA@.006	SF	.12	.24	.36
Spray	PA@.003	SF	.13	.12	.25
Doors, brush application, per square foot painted					
Plywood, 2 coats primer, 1 coat alkyd	PA@.016	SF	.17	.65	.82
Wood, 2 coats alkyd oil base paint	PA@.010	SF	.11	.41	.52
Metal, 1 coat primer, 2 coats enamel	PA@.016	SF	.17	.65	.82
Metal, 2 coats enamel	PA@.012	SF	.12	.49	.61
Deduct for roller application	--	%	--	-25.0	--
Deduct for spray application	--	%	--	-65.0	--
Add for spray application	--	%	25.0	--	--
Add for panel or carved doors	--	%	10.0	35.0	--
Masonry walls, 1 coat primer, 2 coats acrylic latex					
Brush	PA@.019	SF	.19	.77	.96
Roller	PA@.010	SF	.20	.41	.61
Spray	PA@.003	SF	.23	.12	.35
Masonry walls, 2 coats acrylic latex					
Brush	PA@.014	SF	.15	.57	.72
Roller	PA@.006	SF	.16	.24	.40
Spray	PA@.002	SF	.17	.08	.25
Metal ladders and cages, per linear foot of height, widths to 4', brush work					
Wire brush ladder	PA@.027	LF	--	1.10	--
Wire brush ladder and cage	PA@.054	LF	--	2.19	--
Primer, 2 coats enamel, ladder or cage	PA@.019	LF	.62	.77	1.39
Metal trim with brush, figure trim less than 6" wide as being 1 square foot for each linear foot					
1 coat primer, 1 coat enamel	PA@.012	SF	.16	.49	.65
Metal walls, 1 coat primer, 2 coats enamel					
Brush	PA@.013	SF	.13	.53	.66
Roller	PA@.008	SF	.14	.33	.47
Spray	PA@.003	SF	.17	.12	.29
Metal walls, 2 coats enamel					
Brush	PA@.009	SF	.09	.37	.46
Roller	PA@.006	SF	.10	.24	.34
Spray	PA@.001	SF	.11	.04	.15
Most metals, primer and 2 coats enamel					
Brush	PA@.023	SF	.14	.93	1.07
Roller	PA@.010	SF	.15	.41	.56

Finishes 9

	Craft@Hrs	Unit	Material	Labor	Total
Spray	PA@.004	SF	.18	.16	.34
Miscellaneous metals, 2 coats enamel					
Brush	PA@.015	SF	.10	.61	.71
Roller	PA@.007	SF	.11	.28	.39
Spray	PA@.003	SF	.13	.12	.25
Pipe, figure 1 SF per LF for pipe 4" diameter or less					
Per coat, brush or glove	PA@.008	SF	.07	.33	.40
Per coat, spray	PA@.004	SF	.11	.16	.27
Pipe insulation, glue size, primer, enamel	PA@.013	SF	.13	.53	.66
Plaster or wallboard ceilings, 2 coats acrylic latex					
Brush	PA@.011	SF	.10	.45	.55
Roller	PA@.007	SF	.13	.28	.41
Spray	PA@.003	SF	.17	.12	.29
Plaster or wallboard walls, enamel coats, per coat					
Brush	PA@.005	SF	.06	.20	.26
Roller	PA@.003	SF	.07	.12	.19
Spray	PA@.001	SF	.08	.04	.12
Plaster or wallboard walls, primer coat					
Brush	PA@.005	SF	.06	.20	.26
Roller	PA@.003	SF	.06	.12	.18
Spray	PA@.001	SF	.07	.04	.11
Roofing, 2 coats aluminum paint					
Roller	PA@.005	SF	.12	.20	.32
Spray	PA@.001	SF	.16	.04	.20
Steel tank exteriors, 2 coats red lead and 1 coat enamel					
Brush	PA@.013	SF	.31	.53	.84
Roller	PA@.008	SF	.40	.33	.73
Spray	PA@.002	SF	.41	.08	.49
Structural steel, brush, medium or heavy members					
Primer and 2 coats enamel	PA@.012	SF	.16	.49	.65
2 coats enamel	PA@.008	SF	.10	.33	.43
2 coats primer	PA@.008	SF	.11	.33	.44
2 coats enamel, spray	PA@.004	SF	.17	.16	.33
Epoxy touch-up and 1 coat enamel	PA@.008	SF	.20	.33	.53
Structural steel, brush, light members					
Primer and 2 coats enamel	PA@.024	SF	.16	.97	1.13
2 coats enamel	PA@.018	SF	.10	.73	.83
2 coats primer	PA@.018	SF	.11	.73	.84
Stucco walls, 2 coats acrylic latex					
Brush	PA@.010	SF	.12	.41	.53
Roller	PA@.006	SF	.15	.24	.39
Spray	PA@.003	SF	.16	.12	.28
Window sash and trim, per linear foot of sash or trim painted each side, brush					
Wood, 1 coat primer, 2 coats oil base	PA@.022	LF	.16	.89	1.05
Wood, 2 coats oil base	PA@.014	LF	.11	.57	.68
Metal, 1 coat primer, 2 coats enamel	PA@.018	LF	.15	.73	.88
Metal, 2 coats enamel	PA@.012	LF	.09	.49	.58
Wood or plywood ceilings, brush application					
2 coats sealer, 2 coats enamel	PA@.023	SF	.27	.93	1.20
2 coats enamel	PA@.010	SF	.13	.41	.54
Wood or plywood walls, 2 coats primer, 1 coat alkyd					
Brush	PA@.013	SF	.11	.53	.64
Roller	PA@.007	SF	.13	.28	.41
Spray	PA@.005	SF	.16	.20	.36

Finishes 9

	Craft@Hrs	Unit	Material	Labor	Total
Painting, continued					
Wood or plywood walls, 2 coats primer					
Brush	PA@.009	SF	.10	.37	.47
Roller	PA@.005	SF	.11	.20	.31
Spray	PA@.003	SF	.15	.12	.27
Wood or plywood walls, brush application					
2 coats sealer and varnish, 2 coats enamel	PA@.019	SF	.25	.77	1.02
2 coats enamel	PA@.008	SF	.13	.33	.46
Sealer and varnish, 3 coats	PA@.010	SF	.23	.41	.64
Wood trim, brush, figure trim less than 6" wide at 1 SF per LF					
1 coat primer, 1 coat enamel	PA@.012	SF	.16	.49	.65
Sealer and varnish, 3 coats	PA@.013	SF	.22	.53	.75

Paper Wallcoverings. Based on 2,000 SF job. These costs include the subcontractor's overhead and profit. For estimating purposes figure an average roll of wallpaper will cover 40 SF.

	Craft@Hrs	Unit	Material	Labor	Total
Paperhanging, average quality, labor only	PA@.011	SF	--	.45	--
Paperhanging, good quality, labor only	PA@.013	SF	--	.53	--
Vinyl wallcovering, typical costs					
7 oz, light grade	PA@.012	SF	.38	.49	.87
14 oz, medium grade	PA@.013	SF	.65	.53	1.18
22 oz, heavy grade, premium quality	PA@.017	SF	.91	.69	1.60
Vinyl, 14 oz, on aluminum backer	PA@.035	SF	.69	1.42	2.11
Flexwood, many veneers, typical job	PA@.053	SF	2.17	2.15	4.32
Cloth wallcovering					
Linen, acrylic backer, scotch guarded	PA@.021	SF	.92	.85	1.77
Grass cloth	PA@.021	SF	1.42	.85	2.27
Felt	PA@.027	SF	1.80	1.10	2.90
Cork sheathing, 1/8", typical	PA@.027	SF	1.16	1.10	2.26
Blank stock (underliner)	PA@.011	SF	.38	.45	.83
Paper backed vinyl, prepasted, typical	PA@.012	SF	.59	.49	1.08
Untrimmed paper, hand prints, foils, custom	PA@.029	SF	2.58	1.18	3.76
Gypsum impregnated jute fabric (Flexi-wall), with #500 adhesive					
Heavy duty wall liner #609	PA@.017	SF	.53	.69	1.22
Scotland or Indian jute weave	PA@.017	SF	.68	.69	1.37
French weave	PA@.017	SF	.88	.69	1.57
Trimmed paper, hand prints, foils	PA@.012	SF	1.03	.49	1.52
Add for job less than 200 SF (5 rolls)	--	LS	--	--	20.00
Add for small rooms (kitchen, bath)	--	%	10.0	20.0	--
Add for patterns	--	%	--	--	15.0
Deduct for job over 2,000 SF (50 rolls)	--	%	--	--	-5.0
Deduct for job over 4,000 SF (100 rolls)	--	%	--	--	-10.0

Laminated Plastic Wallcovering. Adhesive application, standard patterns and colors. These costs include the subcontractor's overhead and profit.

	Craft@Hrs	Unit	Material	Labor	Total
1/16", most colors and textures	PA@.028	SF	1.45	1.14	2.59
Vertical grade, .032", most finishes	PA@.028	SF	1.25	1.14	2.39
1/16" acid resisting	PA@.035	SF	1.80	1.42	3.22
.020" backing sheet	PA@.017	SF	.32	.69	1.01
1/16" suede, leather, wood grain, solid colors	PA@.028	SF	2.00	1.14	3.14
Black, #840-7	PA@.028	SF	4.40	1.14	5.54
Metallic slate	PA@.028	SF	7.60	1.14	8.74
Add for post forming grade	PA@.004	SF	.12	.16	.28
Add for adhesive	--	SF	.21	--	--

Finishes 9

	Craft@Hrs	Unit	Material	Labor	Total
Hardboard Wallcovering. Vinyl clad, printed. Add metal trim at $.30 per LF. These costs include the subcontractor's overhead and profit.					
3/16"	C8@.025	SF	.70	.96	1.66
1/4"	C8@.025	SF	.91	.96	1.87
3/16" pegboard	C8@.025	SF	1.08	.96	2.04
1/4" pegboard	C8@.025	SF	1.18	.96	2.14
Add for patterns and deep colors	--	%	20.0	--	--
Add for adhesive, 1/4" bead, 16" OC	--	SF	.11	--	--

Specialties 10

	Craft@Hrs	Unit	Material	Labor	Total
Chalkboard and Tackboards. These costs include the subcontractor's overhead and profit.					
Typical in-place cost					
School and institutional grade chalkboards					
With trim, map and chalk rail	--	SF	--	--	10.50
Without trim, average	--	SF	--	--	5.50
Tackboards with trim, cork or vinyl	--	SF	--	--	9.50
Vertical sliding boards	--	SF	--	--	46.50
Horizontal sliding boards	--	SF	--	--	30.50
Swing leaf panels	--	SF	--	--	27.00
Chalkboards, no frame included					
Cement asbestos, 1/4"	CC@.038	SF	4.12	1.63	5.75
Tempered hardboard, 1/4"	CC@.034	SF	2.43	1.46	3.89
Tempered hardboard, 1/2"	CC@.036	SF	3.88	1.54	5.42
Metal back, 24 gauge, 1/4", porcelain enamel	CC@.041	SF	6.96	1.76	8.72
Metal back, 24 gauge, 1/2", porcelain enamel	CC@.048	SF	7.21	2.06	9.27
Slate, 3/8"	CC@.055	SF	10.30	2.36	12.66
Tackboards, no frame or furring included					
Cork sheets					
Unbacked, 1/8"	CC@.032	SF	2.84	1.37	4.21
Unbacked, 1/4"	CC@.032	SF	3.02	1.37	4.39
Burlap backed, 1/8"	CC@.032	SF	4.12	1.37	5.49
Burlap backed, 1/4"	CC@.032	SF	4.58	1.37	5.95
Vinyl covered fiberboard, 1/2"	CC@.032	SF	4.17	1.37	5.54
1/4" vinyl cork on 1/4" hardboard	CC@.032	SF	5.68	1.37	7.05
Frames, trim, aluminum	CC@.018	LF	1.57	.77	2.34
Chalk tray, aluminum	CC@.040	LF	4.17	1.71	5.88
Map and display rail, aluminum	CC@.032	LF	2.84	1.37	4.21

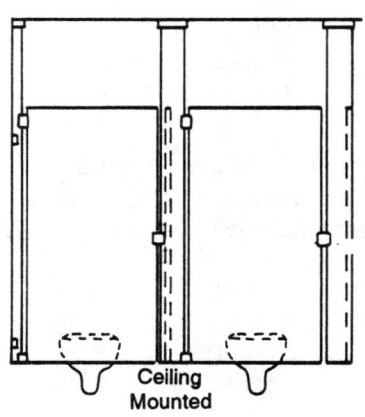
Ceiling Mounted

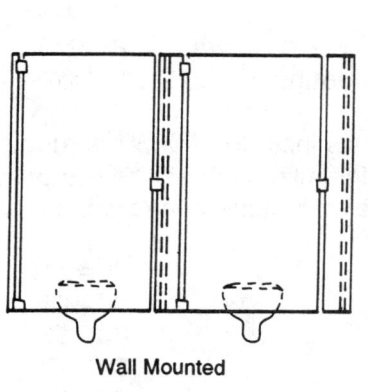
Wall Mounted

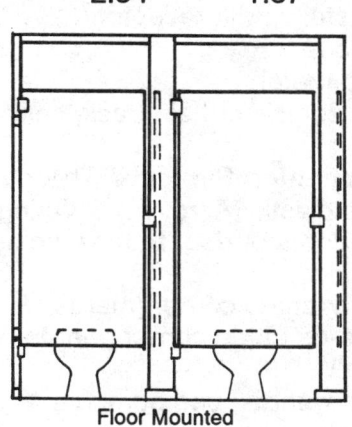
Floor Mounted

See costs on the following page

Specialties 10

	Craft@Hrs	Unit	Material	Labor	Total

Compartments and Cubicles. See illustrations on the preceding page. These costs include the subcontractor's overhead and profit.

Toilet partitions, floor or ceiling mounted, standard sizes, includes 1 wall and 1 door

	Craft@Hrs	Unit	Material	Labor	Total
Baked enamel	T5@2.66	Ea	330.00	107.00	437.00
Porcelain enamel	T5@2.66	Ea	680.00	107.00	787.00
Laminated plastic	T5@2.66	Ea	440.00	107.00	547.00
Stainless steel	T5@2.66	Ea	790.00	107.00	897.00
Add for top quality or wheel chair units	--	Ea	--	--	55.00

Urinal screens, wall mounted with brackets, 18" wide

	Craft@Hrs	Unit	Material	Labor	Total
Standard quality, baked enamel	T5@1.95	Ea	130.00	78.10	208.10
Porcelain enamel	T5@1.03	Ea	265.00	41.20	306.20
Laminated plastic	T5@1.03	Ea	190.00	41.20	231.20
Stainless steel	T5@1.03	Ea	280.00	41.20	321.20
Add for floor mounted, 24" wide	T5@.476	Ea	30.40	19.10	49.50
Add for top quality units	--	Ea	--	--	31.00

Sight screens, 3' x 7'

	Craft@Hrs	Unit	Material	Labor	Total
Baked enamel, floor mounted	T5@1.28	Ea	175.00	51.30	226.30
Porcelain enamel, floor mounted	T5@1.28	Ea	290.00	51.30	341.30
Laminated plastic, floor mounted	T5@1.28	Ea	210.00	51.30	261.30
Stainless steel, floor mounted	T5@1.28	Ea	360.00	51.30	411.30
Add for top quality or ceiling-mounted units	--	Ea	--	--	38.50

Accessories mounted on partitions

	Craft@Hrs	Unit	Material	Labor	Total
Coat hooks and door stop, chrome	--	Ea	--	--	8.00
Purse shelf, chrome, 5" x 14"	--	Ea	--	--	21.00

Dressing cubicles, 80" high, with curtain, floor standing, wall hung

	Craft@Hrs	Unit	Material	Labor	Total
Baked enamel	T5@3.41	Ea	330.00	137.00	467.00
Porcelain enamel	T5@3.41	Ea	590.00	137.00	727.00
Laminated plastic	T5@3.41	Ea	420.00	137.00	557.00
Stainless steel	T5@3.41	Ea	700.00	137.00	837.00

Shower compartments, industrial type, with receptor but no door or plumbing

	Craft@Hrs	Unit	Material	Labor	Total
Baked enamel	T5@2.68	Ea	890.00	107.00	997.00
Porcelain enamel	T5@2.68	Ea	1,590.00	107.00	1,697.00
Laminated plastic	T5@2.68	Ea	1,050.00	107.00	1,157.00
Stainless steel	T5@2.68	Ea	1,760.00	107.00	1,867.00
Add for soap dish	--	Ea	--	--	11.00
Add for curtain rod	--	Ea	--	--	7.50

Doors for industrial shower compartments

	Craft@Hrs	Unit	Material	Labor	Total
Wire or tempered glass, 24" x 72"	G1@1.03	Ea	115.00	40.10	155.10
Plastic, 24" x 72"	G1@1.03	Ea	78.00	40.10	118.30
Plastic, 1 panel door and 1 side panel	G1@1.03	Ea	165.00	40.10	205.10

Shower stalls, with receptor, 32" x 32", with door but no plumbing

	Craft@Hrs	Unit	Material	Labor	Total
Fiberglass	P6@2.65	Ea	305.00	108.00	413.00
Painted metal	P6@2.65	Ea	320.00	108.00	428.00
Hospital cubicle curtain track, ceiling mount	MW@.060	LF	5.45	2.60	8.05

Wall Protection Systems. These costs include a .063" extruded aluminum retainer, with a .100" extruded plastic (Acrovyn) covering, installed with adhesives or mechanical fasteners over existing drywall, doors or door frames; costs also include contractor's overhead and profit. No substrate costs are included.

	Craft@Hrs	Unit	Material	Labor	Total
Flush-mounted corner guards	CC@.194	LF	15.40	8.31	23.71
Surface-mounted corner guards	CC@.069	LF	8.90	2.96	11.86
Handrail	CC@.111	LF	15.40	4.76	20.16
Bumper guards, bed bumpers	CC@.138	LF	18.40	5.91	24.31
Door frame protectors	CC@.056	LF	13.00	2.40	15.40

Specialties 10

	Craft@Hrs	Unit	Material	Labor	Total
Door protectors	CC@.083	SF	3.55	3.56	7.11
Protective wallcovering (vinyl only)	PA@.087	SF	3.55	3.53	7.08

Louvers and Vents, Fixed. These costs include the subcontractor's overhead and profit.
Door louver vents, 1" thick, inverted "V", 18 gauge cold rolled steel with baked enamel finish. Labor costs include installation in pre-cut opening

	Craft@Hrs	Unit	Material	Labor	Total
12" x 8"	SM@.311	Ea	24.70	14.30	39.00
12" x 12"	SM@.311	Ea	32.40	14.30	46.70
16" x 12"	SM@.311	Ea	39.10	14.30	53.40
18" x 12"	SM@.311	Ea	46.90	14.30	61.20
20" x 8"	SM@.311	Ea	52.50	14.30	66.80
20" x 12"	SM@.311	Ea	55.10	14.30	69.40
20" x 16"	SM@.311	Ea	59.20	14.30	73.50
20" x 20"	SM@.311	Ea	63.90	14.30	78.20
24" x 12"	SM@.311	Ea	53.60	14.30	67.90
24" x 16"	SM@.311	Ea	58.70	14.30	73.00
24" x 18"	SM@.311	Ea	64.90	14.30	79.20
24" x 24"	SM@.311	Ea	68.00	14.30	82.30
26" x 26"	SM@.311	Ea	78.30	14.30	92.60

Wall louvers, aluminum mill finish or galvanized steel, fixed louvers, overall sizes, cost per SF of overall dimension

	Craft@Hrs	Unit	Material	Labor	Total
Under 2 SF	SM@.140	SF	14.60	6.44	21.04
Over 2 SF to 5 SF	SM@.107	SF	13.30	4.92	18.22
Over 5 SF to 10 SF	SM@.090	SF	12.60	4.14	16.74
Screen vents, galvanized, 8" x 5" x 4" deep	CC@.293	Ea	21.60	12.60	34.20
Gable louvers, fixed, aluminum mill finish	CC@.171	SF	10.30	7.33	17.63
Penthouse louvers, anodized aluminum finish, fixed horizontal louvers	CC@.247	SF	28.30	10.60	38.90
Continuous aluminum screen vent, 4" wide	CC@.040	LF	1.38	1.71	3.09
Soffit vent, 12" x 10"	CC@.201	Ea	3.19	8.61	11.80
Block vents, 8" x 16" x 4", with screen	CC@.391	Ea	37.60	16.80	54.40

Access Flooring. Standard 24" x 24" steel flooring panels with 1/8" high pressure laminate surface. Costs shown include the floor panels and the supporting understructures listed. Add for accessories below. These costs include the subcontractor's overhead and profit. Raised Floor Installation, Inc.

Stringerless system, up to 10" finish floor height

	Craft@Hrs	Unit	Material	Labor	Total
Under 1,000 SF job	D3@.058	SF	10.50	2.48	12.98
1,000 to 5,000 SF job	D3@.046	SF	9.00	1.97	10.97
Over 5,000 SF job	D3@.039	SF	8.00	1.67	9.67

Snap-on stringer system, up to 10" finish floor height

	Craft@Hrs	Unit	Material	Labor	Total
Under 1,000 SF job	D3@.058	SF	10.80	2.48	13.28
1,000 to 5,000 SF job	D3@.046	SF	9.80	1.97	11.77
Over 5,000 SF job	D3@.039	SF	8.80	1.67	10.47

Rigid bolted stringer system, up to 18" finish floor height

	Craft@Hrs	Unit	Material	Labor	Total
Under 1,000 SF job	D3@.058	SF	11.10	2.48	13.58
1,000 to 5,000 SF job	D3@.046	SF	9.80	1.97	11.77
Over 5,000 SF job	D3@.039	SF	8.30	1.67	9.97
Corner lock system, under 2,500 SF office areas, 6" to 10" finish floor height, blank panels	D3@.036	SF	5.50	1.54	7.04

Used flooring, most floor system types

	Craft@Hrs	Unit	Material	Labor	Total
Unrefurbished, typical	D3@.058	SF	7.72	2.48	10.20
Refurbished, new panel surface and trim	D3@.058	SF	9.98	2.48	12.46

Specialties 10

	Craft@Hrs	Unit	Material	Labor	Total
Access Flooring, continued					
Extras for all floor systems, add to the costs on previous page,					
Carpeted panels	--	SF	1.05	--	--
Concrete-filled steel panels	--	SF	.55	--	--
Ramps, per SF of actual ramp area	D3@.104	SF	7.14	4.45	11.59
Non-skid tape on ramp surface	D3@.033	SF	4.41	1.41	5.82
Portable ramp for temporary use	--	SF	--	--	50.00
Steps, per SF of actual tread area	D3@.208	SF	11.60	8.91	20.51
Fascia at steps and ramps, to 18" high	D3@.117	LF	7.50	5.01	12.51
Cove base, 4" high, black or brown	D3@.026	LF	1.25	1.11	2.36
Guardrail	D3@.573	LF	44.00	24.50	68.50
Handrail	D3@.573	LF	20.00	24.50	44.50
Cable cutouts					
Standard "L" type trim	D3@.407	Ea	7.72	17.40	25.12
Type "F" trim, under floor plenum systems	D3@.407	Ea	8.82	17.40	26.22
Perforated air distribution panels, 24" x 24", deduct area of one standard					
floor panel and add	--	Ea	66.50	--	--
Air distribution grilles, 6" x 18", with adjustable damper,					
including cutout	D3@.673	Ea	52.00	28.80	80.80
Fire extinguishing systems, smoke detectors					
with halon gas suppression	--	SF	--	--	4.95
Automatic fire alarm	--	SF	--	--	.75
Deduct for over $5,000 job	--	SF	-.37	--	--

Flagpoles. Including erection cost but no foundation. See foundation costs below. By pole height above ground, except as noted. Buried dimension is usually 10% of pole height. These costs include the subcontractor's overhead and profit.

	Craft@Hrs	Unit	Material	Labor	Total
Fiberglass, semi-gloss white, tapered, external halyard, commercial grade					
25' with 5.90" butt diameter	D4@8.19	Ea	1,010.00	320.00	1,330.00
30' with 6.50" butt diameter	D4@9.21	Ea	1,240.00	360.00	1,600.00
35' with 6.90" butt diameter	D4@9.96	Ea	1,575.00	389.00	1,964.00
39' with 7.10" butt diameter	D4@10.2	Ea	1,820.00	398.00	2,218.00
50' with 9.90" butt diameter	D4@15.1	Ea	4,110.00	590.00	4,700.00
60' with 10.50" butt diameter	D4@18.5	Ea	5,320.00	722.00	6,042.00
70' with 11.40" butt diameter	D4@21.7	Ea	6,805.00	847.00	7,652.00
80' with 16.00" butt diameter	D4@22.8	Ea	11,100.00	890.00	11,990.00
Add for nautical yardarm mast, based on height of pole,					
per foot of pole height	--	LF	28.00	--	--
Add for internal halyard, jam cleat type					
Heights to 40'	--	Ea	--	--	180.00
Heights over 40'	--	Ea	--	--	1,100.00
Aluminum, satin finish, tapered, external halyard, heavy duty type					
20' x .188" wall, with 5" butt diam.	D4@8.19	Ea	950.00	320.00	1,270.00
25' x .188" wall, with 5" butt diam.	D4@8.19	Ea	1,180.00	320.00	1,500.00
30' x .188" wall, with 6" butt diam.	D4@9.21	Ea	1,340.00	360.00	1,700.00
35' x .188" wall, with 7" butt diam.	D4@9.96	Ea	1,760.00	389.00	2,149.00
40' x .188" wall, with 8" butt diam.	D4@10.2	Ea	2,410.00	398.00	2,808.00
45' x .188" wall, with 8" butt diam.	D4@10.5	Ea	2,760.00	410.00	3,170.00
50' x .188" wall, with 10" butt diam.	D4@15.1	Ea	3,780.00	590.00	4,370.00
60' x .250" wall, with 12" butt diam.	D4@18.5	Ea	8,140.00	722.00	8,862.00
70' x .250" wall, with 12" butt diam.	D4@21.7	Ea	8,800.00	847.00	9,647.00
80' x .375" wall, with 12" butt diam.	D4@25.2	Ea	11,090.00	984.00	12,074.00

Specialties 10

			Clear	Bronze	Black
Add for anodized finish					
Heights to 50'	--	%	15.0	20.0	25.0
Heights over 50'	--	%	10.0	15.0	20.0

Wall-mounted aluminum poles, including wall mount on concrete or masonry structure

	Craft@Hrs	Unit	Material	Labor	Total
8'	D4@7.95	Ea	710.00	310.00	1,020.00
10'	D4@9.69	Ea	825.00	378.00	1,203.00
15' with wall support	D4@11.1	Ea	1,290.00	433.00	1,723.00
20' with wall support	D4@14.4	Ea	1,700.00	562.00	2,262.00
Steel poles, white enamel finish, lightweight					
12'	D4@6.41	Ea	60.00	250.00	310.00
18'	D4@6.41	Ea	80.00	250.00	330.00
22'	D4@7.13	Ea	165.00	278.00	443.00
30'	D4@8.97	Ea	345.00	350.00	695.00
Concrete poles, off-white or brown finish, with internal halyard, overall heights including buried portion					
20'	D4@6.41	Ea	1,390.00	250.00	1,640.00
30'	D4@8.83	Ea	1,855.00	345.00	2,200.00
40'	D4@10.4	Ea	2,370.00	406.00	2,776.00
50'	D4@14.4	Ea	5,360.00	562.00	5,922.00
60'	D4@16.0	Ea	7,470.00	625.00	8,095.00
70'	D4@19.2	Ea	9,580.00	750.00	10,330.00
80'	D4@22.3	Ea	11,740.00	871.00	12,611.00
90'	D4@27.2	Ea	14,200.00	1,060.00	15,260.00
100'	D4@31.8	Ea	15,760.00	1,240.00	17,000.00
Add for yardarm nautical mast					
11' aluminum yardarm	--	Ea	840.00	--	--
13'6" aluminum yardarm	--	Ea	900.00	--	--
12' fiberglass yardarm	--	Ea	900.00	--	--
Add for locking cleat box	D4@2.42	Ea	110.00	94.50	204.50
Add for internal halyard					
20'	--	Ea	370.00	--	--
50'	--	Ea	4,120.00	--	--
70'	--	Ea	6,800.00	--	--
Add for eagle mounted on spindle rod					
12" bronze	--	Ea	67.00	--	--
12" natural	--	Ea	82.90	--	--
16" bronze	--	Ea	113.30	--	--
16" natural	--	Ea	140.00	--	--

Poured concrete flagpole foundations, including excavation, metal collar, steel casing, base plate, sand fill and ground spike, by pole height above ground, typical prices

	Craft@Hrs	Unit	Material	Labor	Total
30' or less	P9@10.5	Ea	53.50	410.00	463.50
35'	P9@10.8	Ea	53.50	422.00	475.50
40'	P9@12.5	Ea	65.00	488.00	553.00
45'	P9@13.2	Ea	67.00	515.00	582.00
50'	P9@15.8	Ea	82.50	617.00	699.50
60'	P9@19.2	Ea	98.00	749.00	847.00
70'	P9@22.5	Ea	123.50	878.00	1,001.50
80'	P9@26.0	Ea	155.00	1,010.00	1,165.00
90' to 100'	P9@30.1	Ea	195.00	1,170.00	1,365.00

Direct embedded pole base, augered in place, no concrete required, by pole height

	Craft@Hrs	Unit	Material	Labor	Total
20' to 25' pole	D4@1.88	Ea	280.00	73.40	353.40
25' to 40' pole	D4@2.71	Ea	400.00	106.00	506.00
45' to 55' pole	D4@2.99	Ea	450.00	117.00	567.00
60' to 70' pole	D4@3.42	Ea	500.00	134.00	634.00

Specialties 10

	Craft@Hrs	Unit	Material	Labor	Total

Identifying Devices. These costs include the subcontractor's overhead and profit.
Architectural signage, self-adhesive plastic signs with high contrast white lettering

	Craft@Hrs	Unit	Material	Labor	Total
4" x 12", stock wording	PA@.191	Ea	17.70	7.76	25.46
3" x 8", stock wording	PA@.191	Ea	15.60	7.76	23.36
Add per sign for custom wording	--	Ea	6.00	--	--
Add for screw-mounted aluminum sign frame	PA@.096	Ea	27.00	3.90	30.90

Directory boards, felt-covered letter boards in an aluminum and glass case, indoor or outdoor wall mount with 3/4" Gothic letters

	Craft@Hrs	Unit	Material	Labor	Total
18" x 2', 1 door	C8@.635	Ea	170.00	24.40	194.40
2' x 3', 1 door	C8@1.31	Ea	210.00	50.40	260.40
4' x 3', 2 doors	C8@1.91	Ea	560.00	73.50	633.50
5' x 3', 2 doors	C8@2.24	Ea	640.00	86.20	726.20
6' x 4', 2 doors	C8@2.82	Ea	740.00	108.00	848.00
18" x 2', open face	C8@.635	Ea	90.00	24.40	114.40
2' x 3', open face	C8@1.31	Ea	122.00	50.40	172.40
4' x 3', open face	C8@1.91	Ea	220.00	73.50	293.50
5' x 3', open face	C8@2.23	Ea	305.00	85.80	390.80
Add for bronze or stainless steel frame	--	%	90.0	--	--
Add for illuminated panel	--	%	50.0	--	--
Deduct for boards without glass face	--	%	-50.0	--	--
Add for custom lettered header	--	%	10.0	--	--
Add for recessed directory boards	--	%	10.0	--	--
Deduct for cork boards	--	%	-10.0	--	--
Hand-painted lettering, per SF overall	--	SF	--	--	10.50
Hand-painted gold leaf, per square foot	--	SF	--	--	22.50

Illuminated letters, porcelain enamel finish, stock letters, aluminum with Plexiglass face, add electrical work

	Craft@Hrs	Unit	Material	Labor	Total
12" high x 2-1/2" wide, 1 tube per letter	C8@.849	Ea	195.00	32.70	227.70
24" high x 5" wide, 2 tubes per letter	C8@1.61	Ea	395.00	61.90	456.90
36" high x 6-1/2" wide, 2 tubes per letter	C8@3.28	Ea	620.00	126.00	746.00

Fabricated back-lighted letters, aluminum with neon illumination, baked enamel finish, script, add electrical work

	Craft@Hrs	Unit	Material	Labor	Total
12" high x 3" deep, 1 tube	G1@.834	Ea	105.00	32.50	137.50
24" high x 6" deep, 2 tubes	G1@1.41	Ea	230.00	54.90	284.90
36" high x 8" deep, 2 tubes	G1@2.53	Ea	430.00	98.40	528.40

Stock cast aluminum letters, various styles, on masonry wall

	Craft@Hrs	Unit	Material	Labor	Total
2" high	C8@.215	Ea	9.71	8.27	17.98
6"	C8@.320	Ea	42.00	12.30	54.30
12"	C8@.320	Ea	86.60	12.30	98.90
14"	C8@.456	Ea	131.00	17.50	148.50
Add for baked enamel finish	--	%	--	--	2.0
Add for satin bronze finish	--	%	--	--	75.0
Add for duranodic color finish	--	%	--	--	20.0

Stock moulded plastic letters, various styles, on masonry wall

	Craft@Hrs	Unit	Material	Labor	Total
2" high	C8@.215	Ea	2.00	8.27	10.27
3"	C8@.215	Ea	3.05	8.27	11.32
4"	C8@.215	Ea	3.90	8.27	12.17
5"	C8@.215	Ea	7.50	8.27	15.77
7"	C8@.320	Ea	12.70	12.30	25.00
10"	C8@.320	Ea	23.20	12.30	35.50
Stainless steel letters, 6"	C8@.320	Ea	80.00	12.30	92.30

Plexiglass signs, plastic letters, aluminum frame, interior illumination, add electrical connection and supports.

	Craft@Hrs	Unit	Material	Labor	Total
Exterior type, single side	G1@.120	SF	57.80	4.67	62.47

Specialties 10

	Craft@Hrs	Unit	Material	Labor	Total
Exterior type, double side	G1@.158	SF	68.80	6.15	74.95
Interior type, single side	G1@.120	SF	53.60	4.67	58.27
Interior type, double side	G1@.158	SF	63.50	6.15	69.65

Plaques, including standard lettering, screw applied

	Craft@Hrs	Unit	Material	Labor	Total
Aluminum, 24" x 18"	G1@2.70	Ea	630.00	105.00	735.00
Aluminum, 24" x 30'	G1@2.91	Ea	1,160.00	113.00	1,273.00
Bronze, 24" x 18"	G1@2.70	Ea	919.00	105.00	1,024.00
Bronze, 24" x 36"	G1@2.91	Ea	1,700.00	113.00	1,813.00
Bronze or aluminum, 12" x 3"	G1@.900	Ea	131.00	35.00	166.00

Safety and warning signs, stock signs ("High Voltage," "Fire Exit," etc.)
  Indoor, 30 gauge baked enamel

	Craft@Hrs	Unit	Material	Labor	Total
10" x 7"	PA@.099	Ea	15.10	4.02	19.12
14" x 10"	PA@.099	Ea	18.70	4.02	22.72
20" x 14"	PA@.099	Ea	26.90	4.02	30.92
Add for outdoor baked enamel signs attached with mechanical fasteners	--	%	10.0	100.0	--

Indoor, color-printed vinyl film, pressure sensitive back.

	Craft@Hrs	Unit	Material	Labor	Total
10" x 7"	PA@.099	Ea	10.10	4.02	14.12
14" x 10"	PA@.099	Ea	11.80	4.02	15.82

Turnstiles. These costs include the subcontractor's overhead and profit.

	Craft@Hrs	Unit	Material	Labor	Total
Non-register type, manual	D4@2.78	Ea	577.00	109.00	686.00
Register type, manual	D4@2.78	Ea	824.00	109.00	933.00
Register type, portable	--	Ea	1,290.00	--	--
Attached ticket collection box, register type	--	Ea	196.00	--	--

Lockers, Metal. Baked enamel finish, key lock. These costs include the subcontractor's overhead and profit.
12" wide x 60" high x 15" deep overall dimensions

	Craft@Hrs	Unit	Material	Labor	Total
Single tier, cost per locker	D4@.571	Ea	131.00	22.30	153.30
Double tier, cost per 2 lockers	D4@.571	Ea	147.00	22.30	169.30
Triple tier, cost per 3 lockers	D4@.571	Ea	152.00	22.30	174.30
Quadruple tier, cost per 4 lockers	D4@.571	Ea	175.00	22.30	197.30
Add for 72" height	--	%	10.0	--	--
Add for 15" width	--	%	25.0	--	--
Heavy duty 18" x 18" x 72" galvanized steel lockers, 1 tier, 3 shelves	D4@.608	Ea	316.00	23.70	339.70
Add for combination lock, per lock	--	Ea	11.40	--	--

Benches, Locker Room. Wooden, floor-mounted. These costs include the subcontractor's overhead and profit.

	Craft@Hrs	Unit	Material	Labor	Total
Maple, 6' long	D4@.753	Ea	175.00	29.40	204.40

Postal Specialties. These costs include the subcontractor's overhead and profit.
Mail chutes, cost per floor, typical
  Aluminum and glass

	Craft@Hrs	Unit	Material	Labor	Total
8-3/4" x 3-1/2"	D4@11.1	Ea	500.00	433.00	933.00
14-1/4" x 4-5/8"	D4@11.1	Ea	600.00	433.00	1,033.00
14-1/4" x 8-5/8"	D4@11.1	Ea	700.00	433.00	1,133.00

  Stainless or bronze

	Craft@Hrs	Unit	Material	Labor	Total
8-3/4" x 3-1/2"	D4@11.1	Ea	600.00	433.00	1,033.00
14-1/4" x 4-5/8"	D4@11.1	Ea	650.00	433.00	1,083.00
14-1/4" x 8-5/8"	D4@13.9	Ea	800.00	543.00	1,343.00

Specialties 10

	Craft@Hrs	Unit	Material	Labor	Total
Postal Specialties, continued					
Collection boxes					
Single box, 36" H x 20" W x 12" deep					
Aluminum	D4@4.16	Ea	1,500.00	162.00	1,662.00
Stainless or bronze	D4@4.16	Ea	1,650.00	162.00	1,812.00
Double box, 36" H x 40" W x 12" deep					
Add to cost of single box	--	%	100.0	--	--

Movable Office Partitions. Prefabricated units. Cost per SF of partition wall measured one side. These costs include the subcontractor's overhead and profit. Use 100 SF as minimum job size.

	Craft@Hrs	Unit	Material	Labor	Total
Gypsum board on 2-1/2" metal studs (includes finish on both sides)					
1/2" board, unpainted, STC 38	C8@.028	SF	2.95	1.08	4.03
5/8" board, unpainted, STC 40	C8@.028	SF	3.01	1.08	4.09
1/2" or 5/8", unpainted, STC 45	C8@.028	SF	3.38	1.08	4.46
1/2" vinyl-covered board, STC 38	C8@.028	SF	3.69	1.08	4.77
5/8" vinyl-covered board, STC 40	C8@.028	SF	3.80	1.08	4.88
Metal-covered, baked enamel finish	C8@.028	SF	8.76	1.08	9.84
Add for factory vinyl wrap, 15 to 22 oz	--	SF	.32	--	--
Windows, including frame and glass					
3'6" x 2'0"	--	Ea	--	--	76.00
3'6" x 4'0"	--	Ea	--	--	110.00
Wood doors, hollow core, prefinished,					
1-3/4", 3'0" x 7'0"	--	Ea	--	--	200.00
Metal door jambs, 3'6" x 7'0"	--	Ea	--	--	135.00
Metal door jambs, 6'0" x 7'0"	--	Ea	--	--	220.00
Passage hardware set	--	Ea	--	--	51.50
Lockset hardware	--	LS	--	--	69.50
Corner for gypsum partitions	--	Ea	--	--	85.00
Corner for metal-covered partitions	--	Ea	--	--	167.00
Starter, gypsum partitions	--	Ea	--	--	29.00
Starter, metal-covered partitions	--	Ea	--	--	61.00
Metal base	--	LF	--	--	1.70
Cubicles and welded booths, 5' x 5' x 5'	C8@.028	SF	10.02	1.08	11.10
Banker type divider partition, 5'6" high including 18" glass top, 2" thick, vinyl finish, not including door or gates	--	SF	--	--	8.50

Demountable Partitions. Prefinished units. Cost per SF of partition measured one side. These costs include the subcontractor's overhead and profit. Use 100 SF as minimum job size.

	Craft@Hrs	Unit	Material	Labor	Total
Air wall, non-rated	--	SF	--	--	20.00
Modular panels, spring pressure, mounted to floor and ceiling with removable tracks	--	SF	--	--	10.00
Accordion folding partitions, vinyl-covered particleboard core, top hung, not including structural supports or placing track in concrete					
Economy, to 8' high, STC 36	C8@.090	SF	9.10	3.46	12.56
Economy, to 30' wide x 17' high, STC 41	C8@.124	SF	12.05	4.77	16.82
Good quality, to 30' W x 17' H, STC 43	C8@.124	SF	13.40	4.77	18.17
Better quality, to 30' W x 17' H, STC 44	C8@.124	SF	13.90	4.77	18.67
Better quality, large openings, STC 45	C8@.124	SF	13.30	4.77	18.07
Better quality, extra large openings, STC 47	C8@.124	SF	14.00	4.77	18.77
Add for prefinished birch or ash wood slats accordion folding partitions with vinyl hinges, 15' W x 8' H maximum	--	%	45.0	--	--

Specialties 10

	Craft@Hrs	Unit	Material	Labor	Total

Folding fire partitions, accordion, with automatic closing system, baked enamel finish, with standard hardware and motor, cost per 5' x 7' module

	Craft@Hrs	Unit	Material	Labor	Total
20 minute rating	C8@22.6	Ea	3,875.00	869.00	4,744.00
60 minute rating	C8@22.6	Ea	4,950.00	869.00	5,819.00
90 minute rating	C8@22.6	Ea	5,320.00	869.00	6,189.00
2 hour rating	C8@22.6	Ea	6,100.00	869.00	6,969.00

Folding leaf partitions, not including header or floor track recess, vinyl covered

	Craft@Hrs	Unit	Material	Labor	Total
Metal panels, 7.5 lbs per SF, top hung, hinged, (STC 52), 16' high x 60' wide maximum	--	SF	--	--	39.50
Metal panels, 7.5 lbs per SF, top hung, single panel pivot, (STC 48), 18' high x 60' wide maximum	--	SF	--	--	40.50
Wood panels, hinged, floor or ceiling hung, 6 lbs per SF, (STC 40), 12' high x 36' wide maximum	--	SF	--	--	32.50
Add for laminated plastic	--	SF	--	--	2.00
Add for unfinished wood veneer	--	SF	--	--	2.50
Add for chalkboard, large area	--	SF	--	--	2.00

Woven Wire Partitions. 10 gauge painted, 1-1/2" steel mesh. These costs include the subcontractor's overhead and profit.

Wall panels, 8' high, installed on a concrete slab

	Craft@Hrs	Unit	Material	Labor	Total
1' wide	H6@.437	Ea	74.90	18.60	93.50
2' wide	H6@.491	Ea	87.80	20.90	108.70
3' wide	H6@.545	Ea	99.70	23.20	122.90
4' wide	H6@.545	Ea	112.00	23.20	135.20
5' wide	H6@.598	Ea	116.00	25.50	141.50
Add for galvanized mesh	--	SF	.45	--	--
Corner posts for 8' high walls	H6@.270	Ea	22.50	11.50	34.00
Stiffener channel posts for 8' walls	H6@.277	Ea	45.60	11.80	57.40
Service window panel, 8' H x 5' W panel	H6@.836	Ea	259.00	35.60	294.60

Sliding doors, 8' high

	Craft@Hrs	Unit	Material	Labor	Total
3' wide	H6@2.94	Ea	383.00	125.00	508.00
4' wide	H6@3.29	Ea	405.00	140.00	545.00
5' wide	H6@3.29	Ea	433.00	140.00	573.00
6' wide	H6@3.58	Ea	507.00	152.00	659.00

Hinged doors, 8' high

	Craft@Hrs	Unit	Material	Labor	Total
3' wide	H6@2.85	Ea	304.00	121.00	425.00
4' wide	H6@3.14	Ea	338.00	134.00	472.00

Dutch doors, 8' high

	Craft@Hrs	Unit	Material	Labor	Total
3' wide	H6@2.85	Ea	468.00	121.00	589.00
4' wide	H6@3.14	Ea	552.00	134.00	686.00

Complete wire partition installation including posts and typical doors, per SF of wall and ceiling panel

	Craft@Hrs	Unit	Material	Labor	Total
Painted wall panels	H6@.029	SF	3.49	1.23	4.72
Galvanized wall panels	H6@.029	SF	4.00	1.23	5.23
Painted ceiling panels	H6@.059	SF	3.15	2.51	5.66
Galvanized ceiling panels	H6@.059	SF	3.61	2.51	6.12

Truck Scales. With dial and readout. Steel platform. These costs include the subcontractor's overhead and profit but no concrete work or excavation.

	Craft@Hrs	Unit	Material	Labor	Total
5 ton, 6' x 8' platform	D4@45.9	Ea	7,250.00	1,790.00	9,040.00
20 ton, 10' x 24' platform	D4@70.2	Ea	10,600.00	2,740.00	13,340.00
50 ton, 10' x 60' platform	D4@112.	Ea	18,400.00	4,370.00	22,770.00
80 ton, 10' x 60' platform	D4@112.	Ea	21,500.00	4,370.00	25,870.00

Specialties 10

	Craft@Hrs	Unit	Material	Labor	Total

Recessed Wall and Floor Safes. Safes with 1" thick steel door, 1/4" hardplate protecting locking mechanism, welded deadbolts, and key or combination lock, internal relock device. These costs include the subcontractor's overhead and profit.

In-floor safes

	Craft@Hrs	Unit	Material	Labor	Total
15" L x 14" W x 15" D, 12" x 12" door	D4@2.33	Ea	509.00	91.00	600.00
15" L x 14" W x 20" D, 12" x 12" door	D4@2.33	Ea	525.00	91.00	616.00
24" L x 24" W x 17" D, 12" x 12" door	D4@2.33	Ea	730.00	91.00	821.00
Add for installation in existing building	--	%	--	100.0	--

Storage Shelving. These costs include the subcontractor's overhead and profit.
Steel industrial shelving, enamel finish, 2 sides & back, 12" deep x 30" wide x 75" high

	Craft@Hrs	Unit	Material	Labor	Total
5 shelves high	D4@.758	Ea	87.40	29.60	117.00
8 shelves high	D4@1.21	Ea	123.00	47.20	170.20

Steel industrial shelving, enamel finish, open on sides, 12" deep x 36" wide x 75" high

	Craft@Hrs	Unit	Material	Labor	Total
5 shelves high	D4@.686	Ea	52.70	26.80	79.50
8 shelves high	D4@1.06	Ea	79.50	41.40	120.90

Stainless steel modular shelving, 20" deep x 36" wide x 63" high,

	Craft@Hrs	Unit	Material	Labor	Total
3 shelves, 16 gauge	D4@1.19	Ea	885.00	46.50	931.50
Stainless steel wall shelf, 12" deep, bracket mount	D4@.675	LF	40.40	26.40	66.80

Stainless steel work counters, with undercounter storage shelf, 16 gauge

	Craft@Hrs	Unit	Material	Labor	Total
4' wide x 2' deep x 3' high	D4@1.36	Ea	971.00	53.10	1,024.10
6' wide x 2' deep x 3' high	D4@2.01	Ea	1,470.00	78.50	1,548.50

Library type metal shelving, 90" high, 8" deep, 5 shelves

	Craft@Hrs	Unit	Material	Labor	Total
30" long units	D4@1.17	Ea	156.00	45.70	201.70
36" long units	D4@1.30	Ea	168.00	50.80	218.80

Telephone Enclosures. Typical costs including the subcontractor's overhead and profit.

	Craft@Hrs	Unit	Material	Labor	Total
Shelf type, wall mounted, 15" D x 28" W x 30" H	--	Ea	--	--	535.80
Desk type	--	Ea	--	--	515.00
Full height booth	--	Ea	--	--	2,650.00
Wall mounted stainless steel directory shelf, 3 directory	--	Ea	--	--	560.00

Toilet and Bath Accessories, material cost only. See labor below. These costs are based on subcontractor-supplied units.

	Unit	Commercial or Industrial Grade	Institutional Grade
Toilet paper dispensers			
Single roll, surface mounted	Ea	12.90	22.55
Double roll, surface mounted	Ea	15.25	25.00
Single roll, recessed	Ea	17.20	22.55
Cabinet, double roll, recessed	Ea	--	144.00
Toilet seat cover dispensers			
Recessed	Ea	61.40	88.00
Surfaced mounted	Ea	37.00	43.00
With sanitary napkin disposal	Ea	205.00	295.00
Feminine napkin dispenser, recessed	Ea	270.00	400.00
Feminine napkin dispenser, surface mounted	Ea	40.40	55.40
Feminine napkin disposer, recessed	Ea	85.60	119.50
Soap dish, chrome, surface mounted	Ea	6.80	12.05
Soap dish, chrome, recessed	Ea	7.40	12.25
Soap dispenser			
Liquid, 20 oz, deck mounted	Ea	31.40	38.70
Liquid, 40 oz, deck mounted	Ea	38.50	46.75

Specialties 10

		Commercial or Industrial Grade	Institutional Grade
Powder	Ea	27.70	40.00
System, liquid, with tank and 3 stations	Ea	--	770.00
Add for additional stations	Ea	--	66.20
Soap dishes			
With grab bar, recessed	Ea	15.65	19.80
With grab bar, surface mounted	Ea	24.60	31.20
Grab bars, stainless steel, wall mounted			
1" x 12"	Ea	24.10	28.00
1" x 18"	Ea	26.00	30.00
1" x 24"	Ea	29.40	32.00
1" x 30"	Ea	32.30	34.80
1" x 36"	Ea	34.50	36.70
1" x 48"	Ea	38.50	40.00
1-1/4" x 18"	Ea	34.20	45.40
1-1/4" x 24"	Ea	37.30	48.00
1-1/4" x 36"	Ea	45.50	57.50
1", 90 degree angle, 30"	Ea	54.00	61.40
1-1/4", 90 degree angle, 30"	Ea	66.85	91.20
Towel bars			
18" chrome	Ea	12.25	20.00
24" chrome	Ea	13.30	20.40
18" stainless steel	Ea	--	23.70
24" stainless steel	Ea	--	26.50
Paper towel dispensers			
Recessed	Ea	112.30	160.70
Semi-recessed or wall mounted	Ea	140.00	235.00
Recessed, with waste receptacle	Ea	306.00	432.00
Semi-recessed with waste receptacle	Ea	375.00	540.00
Waste receptacles, recessed	Ea	211.00	305.00
Waste receptacles, semi-recessed	Ea	290.00	414.00
Medicine cabinets, with mirror			
Swing door, 16" x 22", wall hung	Ea	50.00	75.00
Sliding doors, recessed	Ea	88.00	112.30
Swing door, 16" x 22", recessed	Ea	96.30	125.00
Mirrors, unframed, 1/4" float glass, polished edges			
To 5 SF	SF	5.80	--
Over 5 SF	SF	7.00	--
Mirrors, stainless steel frame			
18" x 24"	Ea	--	78.00
18" x 32"	Ea	--	86.60
18" x 36"	Ea	--	90.75
24" x 30"	Ea	--	111.25
24" x 48"	Ea	--	215.00
24" x 60"	Ea	--	265.00
30" x 72"	Ea	--	315.00
Shelves, stainless steel, 6" deep			
18" long	Ea	34.10	37.40
24" long	Ea	36.15	42.00
36" long	Ea	47.00	57.40
72" long	Ea	76.60	105.00
Shower rod, end flanges, vinyl curtain, 6' long, straight	Ea	40.00	50.00

Specialties 10

		Commercial or Industrial Grade	Institutional Grade
Toilet and Bath Accessories, continued			
Robe hooks			
Single	Ea	6.80	10.00
Double	Ea	8.10	10.90
Single with door stop	Ea	7.50	12.05
Straddle bar, 24" x 24" x 20", stainless steel	Ea	77.00	92.00
Urinal bar, 24" long, stainless steel	Ea	62.60	78.00
Ash urns, wall mounted, aluminum	Ea	90.30	115.40
Hot air hand dryer, 110 volt, add electric connection	Ea	320.00	353.00

	Craft@Hrs	Unit	Material	Labor	Total
Labor to install toilet and bath accessories					
Toilet paper dispensers					
Surface mounted	CC@.182	Ea	--	7.80	--
Recessed	CC@.310	Ea	--	13.30	--
Toilet seat cover dispensers					
Surface mounted	CC@.413	Ea	--	17.70	--
Recessed	CC@.662	Ea	--	28.40	--
Feminine napkin dispenser, surface mounted	CC@.760	Ea	--	32.60	--
Feminine napkin disposer, recessed	CC@.882	Ea	--	37.80	--
Soap dishes	CC@.230	Ea	--	9.85	--
Soap dispensers, surface mounted	CC@.254	Ea	--	10.90	--
Soap dispensers, recessed	CC@.628	Ea	--	26.90	--
Soap dispenser system	CC@2.91	Ea	--	125.00	--
Paper towel dispensers and waste receptacles					
Semi-recessed	CC@1.96	Ea	--	84.00	--
Recessed	CC@2.57	Ea	--	110.00	--
Medicine cabinets with mirror, swing door					
wall hung	CC@.943	Ea	--	40.40	--
Mirrors					
To 5 SF	CG@.566	Ea	--	24.80	--
Over 5 to 10 SF	CG@.999	Ea	--	43.70	--
Over 10 SF	CG@.125	SF	--	5.47	--
Towel bars and grab bars	CC@.286	Ea	--	12.30	--
Shelves, stainless steel, to 72"	CC@.569	Ea	--	24.40	--
Robe hooks	CC@.182	Ea	--	7.80	--
Shower rod	CC@.557	Ea	--	23.90	--
Straddle and urinal bars	CC@.484	Ea	--	20.70	--
Ash urns	CC@.564	Ea	--	24.20	--
Hot air hand dryer, add for electric circuit wiring	CE@.908	Ea	--	42.10	--

Coat and Hat Racks. Assembly and installation of prefabricated coat and hat racks. These costs include the subcontractor's overhead and profit. Bevco

	Craft@Hrs	Unit	Material	Labor	Total
Wall-mounted coat and hat rack with pilfer-resistant hangers, chrome (#200-SHB)					
24" length, 9 hangers	CC@.665	Ea	70.25	28.50	98.75
36" length, 13 hangers	CC@.748	Ea	76.75	32.00	108.75
48" length, 17 hangers	CC@.833	Ea	90.70	35.70	126.40
60" length, 21 hangers	CC@.916	Ea	100.00	39.20	139.20
72" length, 25 hangers	CC@.997	Ea	112.00	42.70	154.70
Deduct for no hangers	--	%	-30.0	--	--
Wall-mounted hook bar, chrome (#H-2)					
24" length, 5 hooks	CC@.252	Ea	17.20	10.80	28.00

Specialties 10

	Craft@Hrs	Unit	Material	Labor	Total
36" length, 7 hooks	CC@.303	Ea	21.75	13.00	34.75
48" length, 9 hooks	CC@.352	Ea	28.50	15.10	43.60
60" length, 11 hooks	CC@.391	Ea	33.20	16.80	50.00
72" length, 13 hooks	CC@.457	Ea	37.70	19.60	57.30

Church Glass. Material cost includes fabrication of the lite. Labor cost includes setting in a prepared opening only. Minimum labor cost is usually $150 per job. Add the cost of scaffold or a hoist, if needed. These costs include the subcontractor's overhead and profit.

	Craft@Hrs	Unit	Material	Labor	Total
Simple artwork stained glass	G1@.184	SF	63.80	7.16	70.96
Moderate artwork stained glass	G1@.318	SF	161.00	12.40	173.40
Elaborate artwork stained glass	G1@.633	SF	513.00	24.60	537.60
Faceted glass					
Simple artwork	G1@.184	SF	57.10	7.16	64.26
Moderate artwork	G1@.318	SF	116.00	12.40	128.40
Elaborate artwork	G1@.633	SF	156.00	24.60	180.60
Colored glass, no artwork					
Single pane	G1@.056	SF	6.38	2.18	8.56
Patterned	G1@.184	SF	13.30	7.16	20.46
Small pieces	G1@.257	SF	34.30	10.00	44.30

Equipment 11

Bank Equipment. These costs include the subcontractor's overhead and profit.

	Craft@Hrs	Unit	Material	Labor	Total
Vault door and frame, 78" x 32", 10", single					
2 hour, 10 bolt	D4@11.9	Ea	2,400.00	465.00	2,865.00
4 hour, 10 bolt	D4@11.9	Ea	2,780.00	465.00	3,245.00
6 hour, 10 bolt	D4@11.9	Ea	3,520.00	465.00	3,985.00
Vault door and frame, 78" x 40", 10", double					
4 hour, 18 bolt	D4@20.8	Ea	4,315.00	812.00	5,127.00
6 hour, 18 bolt	D4@20.8	Ea	4,670.00	812.00	5,482.00
Teller windows					
Drive-up, motorized drawer, projected	D4@26.8	Ea	9,900.00	1,050.00	10,950.00
Walk-up, one teller	D4@17.6	Ea	5,100.00	687.00	5,787.00
Walk-up, two teller	D4@17.6	Ea	6,075.00	687.00	6,762.00
After hours depository doors					
Envelope and bag, whole chest, complete	D4@37.1	Ea	8,600.00	1,450.00	10,050.00
Envelope only	D4@4.72	Ea	995.00	184.00	1,179.00
Flush mounted, bag only	D4@5.44	Ea	1,540.00	212.00	1,752.00
Teller counter, modular components	C8@.532	LF	92.70	20.50	113.20
Square check desks, 48" x 48", 4 person	C8@2.00	Ea	1,170.00	76.90	1,246.90
Round check desks, 48" diameter, 4 person	C8@2.00	Ea	1,250.00	76.90	1,326.90
Rectangular check desks, 72" x 24", 4 person	C8@2.00	Ea	1,200.00	76.90	1,276.90
Rectangular check desks, 36" x 72", 8 person	C8@2.00	Ea	2,300.00	76.90	2,376.90
Safe deposit boxes, modular units					
42 openings, 2" x 5"	D4@3.42	Ea	1,820.00	134.00	1,954.00
30 openings, 2" x 5"	D4@3.42	Ea	1,435.00	134.00	1,569.00
18 openings, 2" x 5"	D4@3.42	Ea	1,145.00	134.00	1,279.00
Base, 32" x 24" x 3"	D4@1.11	Ea	107.00	43.30	150.30
Canopy top	D4@.857	Ea	57.40	33.50	90.90
Cash dispensing automatic teller, with phone	CE@84.3	Ea	42,000.00	3,910.00	45,910.00
Film camera surveillance system, 3 cameras	CE@15.6	Ea	2,640.00	723.00	3,363.00
Perimeter alarm system, typical	CE@57.7	Ea	6,650.00	2,670.00	9,320.00

Equipment 11

	Craft@Hrs	Unit	Material	Labor	Total

Ecclesiastical Equipment. Labor includes unpacking, layout and complete installation of prefinished furniture and equipment. These costs include the subcontractor's overhead and profit.

	Craft@Hrs	Unit	Material	Labor	Total
Lecterns, 16" x 24"					
Plain	C8@4.48	Ea	566.00	172.00	738.00
Deluxe	C8@4.48	Ea	1,060.00	172.00	1,232.00
Premium	C8@4.48	Ea	2,020.00	172.00	2,192.00
Pulpits					
Plain	C8@4.77	Ea	642.00	184.00	826.00
Deluxe	C8@4.77	Ea	1,020.00	184.00	1,204.00
Premium	C8@4.77	Ea	2,550.00	184.00	2,734.00
Arks, with curtain					
Plain	C8@5.97	Ea	666.00	230.00	896.00
Deluxe	C8@5.97	Ea	870.00	230.00	1,100.00
Premium	C8@5.97	Ea	1,470.00	230.00	1,700.00
Arks, with door					
Plain	C8@8.95	Ea	754.00	344.00	1,098.00
Deluxe	C8@8.95	Ea	1,300.00	344.00	1,644.00
Premium	C8@8.95	Ea	2,020.00	344.00	2,364.00
Pews, bench type					
Plain	C8@.382	LF	41.80	14.70	56.50
Deluxe	C8@.382	LF	45.50	14.70	60.20
Premium	C8@.382	LF	51.80	14.70	66.50
Pews, seat type					
Plain	C8@.464	LF	52.40	17.90	70.30
Deluxe	C8@.491	LF	68.00	18.90	86.90
Premium	C8@.523	LF	81.10	20.10	101.20
Kneelers					
Plain	C8@.185	LF	10.40	7.12	17.52
Deluxe	C8@.185	LF	13.10	7.12	20.22
Premium	C8@.185	LF	20.40	7.12	27.52
Cathedral chairs, shaped wood					
Without accessories	C8@.235	Ea	133.00	9.04	142.04
With book rack	C8@.235	Ea	145.00	9.04	154.04
With book rack and kneeler	C8@.235	Ea	161.00	9.04	170.04
Cathedral chairs, upholstered					
Without accessories	C8@.285	Ea	141.00	11.00	152.00
With book rack	C8@.285	Ea	150.00	11.00	161.00
With book rack and kneeler	C8@.285	Ea	179.00	11.00	190.00
Adaptable fabric upholstered chair seating (Sauder)					
Hardwood laminated frame, interlocking legs, book rack, stackable, "Modlok"					
Standard	C8@.066	Ea	114.00	2.54	116.54
Deluxe	C8@.066	Ea	120.00	2.54	122.54
Solid oak frame, stackable, "Oaklok"	C8@.066	Ea	93.00	2.54	95.54
Laminated beech frame, stackable, "Plylok"	C8@.066	Ea	101.00	2.54	103.54
Laminated beech frame, stacks & folds, "Plyfold"	C8@.066	Ea	87.80	2.54	90.34
Confessionals, single, with curtain					
Plain	C8@9.11	Ea	2,460.00	350.00	2,810.00
Deluxe	C8@9.11	Ea	2,930.00	350.00	3,280.00
Premium	C8@9.11	Ea	3,350.00	350.00	3,700.00
Add for door in place of curtain	--	Ea	387.00	--	--
Confessionals, double, with curtain					
Plain	C8@13.7	Ea	3,870.00	527.00	4,397.00
Deluxe	C8@13.7	Ea	4,500.00	527.00	5,027.00
Premium	C8@13.7	Ea	5,540.00	527.00	6,067.00

Equipment 11

	Craft@Hrs	Unit	Material	Labor	Total
Communion rail, hardwood, with standards					
Plain	C8@.315	LF	44.00	12.10	56.10
Deluxe	C8@.315	LF	55.40	12.10	67.50
Premium	C8@.315	LF	65.90	12.10	78.00
Communion rail, carved oak, with standards					
Embellished	C8@.315	LF	71.70	12.10	83.80
Ornate	C8@.315	LF	95.70	12.10	107.80
Premium	C8@.315	LF	157.00	12.10	169.10
Bronze or stainless steel communion rail, with standards	H6@.285	LF	104.00	12.10	116.10
Altars, hardwood, sawn					
Plain	C8@5.97	Ea	773.00	230.00	1,003.00
Deluxe	C8@5.97	Ea	911.00	230.00	1,141.00
Premium	C8@5.97	Ea	1,260.00	230.00	1,490.00
Altars, carved hardwood					
Plain	C8@5.97	Ea	2,350.00	230.00	2,580.00
Deluxe	C8@6.73	Ea	5,440.00	259.00	5,699.00
Premium	C8@7.30	Ea	6,170.00	281.00	6,451.00
Altars, with Nimmerillium fonts, marble	D4@27.8	Ea	4,030.00	1,090.00	5,120.00
Altars, marble or granite					
Plain	D4@29.1	Ea	4,180.00	1,140.00	5,320.00
Deluxe	D4@29.1	Ea	6,070.00	1,140.00	7,210.00
Premium	D4@29.1	Ea	9,000.00	1,140.00	10,140.00
Altars, marble legs and base					
Deluxe	D4@29.1	Ea	2,510.00	1,140.00	3,650.00
Premium	D4@29.1	Ea	3,610.00	1,140.00	4,750.00

Theater & Stage Curtains. These costs include the subcontractor's overhead and profit.

Straight track for manual curtains, see lifting and drawing equipment following

	Craft@Hrs	Unit	Material	Labor	Total
Standard duty					
24' wide stage	D4@.486	LF	14.25	19.00	33.25
40' wide stage	D4@.315	LF	14.25	12.30	26.55
60' wide stage	D4@.233	LF	14.25	9.10	23.35
Medium duty					
24' wide stage	D4@.486	LF	18.00	19.00	37.00
40' wide stage	D4@.315	LF	18.00	12.30	30.30
60' wide stage	D4@.233	LF	18.00	9.10	27.10
Heavy duty					
24' wide stage	D4@.486	LF	20.50	19.00	39.50
40' wide stage	D4@.315	LF	20.50	12.30	32.80
60' wide stage	D4@.233	LF	20.50	9.10	29.60

Curved track, manual operation, see lifting and drawing equipment following

	Craft@Hrs	Unit	Material	Labor	Total
Standard duty					
24' wide stage	D4@.486	LF	33.00	19.00	52.00
40' wide stage	D4@.371	LF	27.00	14.50	41.50
60' wide stage	D4@.315	LF	25.00	12.30	37.30
Heavy duty					
24' wide stage	D4@.584	LF	78.00	22.80	100.80
40' wide stage	D4@.427	LF	68.00	16.70	84.70
60' wide stage	D4@.342	LF	65.00	13.40	78.40

Cyclorama track, manual, see lifting and drawing equipment following

	Craft@Hrs	Unit	Material	Labor	Total
24' wide stage	D4@.486	LF	21.00	19.00	40.00
40' wide stage	D4@.315	LF	21.00	12.30	33.30
60' wide stage	D4@.233	LF	21.00	9.10	30.10

Equipment 11

	Craft@Hrs	Unit	Material	Labor	Total

Theater & Stage Curtains, continued
Stage curtains. Complete costs for custom fabricated, flameproof stage curtains including labor to hang. No track or curtain lifting or drawing equipment included. By percent of added fullness

	Craft@Hrs	Unit	Material	Labor	Total
Heavyweight velour					
0% fullness	--	SF	--	--	2.30
50% fullness	--	SF	--	--	2.60
75% fullness	--	SF	--	--	2.90
100% fullness	--	SF	--	--	3.10
Lightweight velour					
0% fullness	--	SF	--	--	2.00
50% fullness	--	SF	--	--	2.30
75% fullness	--	SF	--	--	2.55
100% fullness	--	SF	--	--	2.75
Muslin, 0% fullness	--	SF	--	--	1.00
Repp cloth					
0% fullness	--	SF	--	--	1.90
50% fullness	--	SF	--	--	2.30
75% fullness	--	SF	--	--	2.55
100% fullness	--	SF	--	--	2.75

Stage curtain lifting and drawing equipment. Labor cost is for installation of equipment only. Add for electrical connection as shown below

	Craft@Hrs	Unit	Material	Labor	Total
Lift curtain equipment					
1/8 HP	D4@7.50	Ea	1,500.00	293.00	1,793.00
1/2 HP	D4@7.50	Ea	2,500.00	293.00	2,793.00
3/4 HP	D4@7.50	Ea	2,650.00	293.00	2,943.00
Draw curtain equipment					
1/8 HP	D4@7.50	Ea	1,500.00	293.00	1,793.00
1/2 HP	D4@7.50	Ea	2,050.00	293.00	2,343.00
3/4 HP	D4@7.50	Ea	2,175.00	293.00	2,468.00
Add for electrical connection, 40' wire run with conduit, circuit breaker, junction box and starter switch	--	LS	--	--	750.00

Educational Equipment. Labor includes unpacking, layout and complete installation of prefinished furniture and equipment. These costs include the subcontractor's overhead and profit.

	Craft@Hrs	Unit	Material	Labor	Total
Wardrobes, 40" x 78" x 26"					
Teacher type	C8@2.00	Ea	1,100.00	76.90	1,176.90
Student type	C8@2.00	Ea	550.00	76.90	626.90
Seating, per seat					
Lecture room, pedestal, with folding arm	C8@2.00	Ea	140.00	76.90	216.90
Horizontal sections	C8@2.00	Ea	67.00	76.90	143.90
Tables, fixed pedestal					
48" x 16"	C8@2.00	Ea	310.00	76.90	386.90
With chairs, 48" x 16"	C8@2.00	Ea	440.00	76.90	516.90
Projection screen, ceiling mounted, pull down	D4@.019	SF	3.55	.74	4.29
Videotape recorder	CE@5.32	Ea	800.00	246.00	1,046.00
T.V. camera, color	CE@5.43	Ea	646.00	252.00	898.00
T.V. monitor, wall mounted, color	CE@4.08	Ea	486.00	189.00	675.00
Language laboratory study carrels, plastic laminated wood					
Individual, 48" x 30" x 54"	C8@1.43	Ea	199.00	55.00	254.00
Two position, 73" x 30" x 47"	C8@1.93	Ea	538.00	74.20	612.20
Island, 4 position, 66" x 66" x 47"	C8@3.17	Ea	608.00	122.00	730.00

Equipment 11

	Craft@Hrs	Unit	Material	Labor	Total

Checkroom Equipment. These costs include the subcontractor's overhead and profit.
Automatic electric checkroom conveyor, 350 coat capacity.

	Craft@Hrs	Unit	Material	Labor	Total
Conveyor and controls	D4@15.2	Ea	2,860.00	593.00	3,453.00
Add for electrical connection for checkroom conveyor, 40' wire run with conduit, circuit breaker, and j-box	--	LS	--	--	325.00

Food Service Equipment. These costs include the subcontractor's overhead and profit but no electrical work or plumbing. Add the cost of equipment or fixture hookup when needed.

Work tables and counters, stainless steel top, with 6" splash

	Craft@Hrs	Unit	Material	Labor	Total
With straight edge	C8@1.10	LF	180.70	42.30	223.00
With rolled edge	C8@1.10	LF	175.40	42.30	217.70
Serving fixtures					
Economy, stainless top, galvanized frame	C8@1.49	LF	164.20	57.30	221.50
Best grade, all stainless	C8@1.49	LF	234.80	57.30	292.10
Additional costs for work tables and serving fixtures					
Base and intermediate shelf, stainless	--	LF	--	--	52.15
Base and intermediate shelf, galvanized	--	LF	--	--	27.40
Angle or pipe stretchers	--	LF	--	--	17.05
Tray slide rack, stainless steel	--	LF	--	--	59.40
Display shelf, sneeze guard and bracket	--	LF	--	--	73.10
Add for each additional shelf	--	LF	--	--	38.50
Plastic laminate facing on plywood	--	LF	--	--	32.00
18 gauge stainless steel panel facing	--	LF	--	--	51.20
22 gauge stainless facing on plywood	--	LF	--	--	30.00
Marine front coping	--	LF	--	--	12.00
Soiled dish table, stainless steel, 12" splash, 4" raised front	--	LF	--	--	23.50
Slop gutter, 4" x 4", with sump and screen	--	LF	--	--	47.00
Weld and polish 30" miter joint	--	Ea	--	--	147.70
Doors, up to 24", stainless steel	--	Ea	--	--	109.50
Drawers, stainless steel	--	Ea	--	--	182.75
Drawers, galvanized pan, stainless steel	--	Ea	--	--	143.50
Punchouts, (waste holes, cans, etc.)	--	Ea	--	--	50.50
Ventilating grilles, 24" x 12"	--	Ea	--	--	76.50
Maple cutting boards, 2" laminated	--	Ea	--	--	21.20
Richlite cutting boards, 2" laminated	--	Ea	--	--	22.00
Pot wash sink, 24" x 24" x 24"	--	Ea	--	--	625.00
Vegetable sink	--	Ea	--	--	547.00
Mixer valve	--	Ea	--	--	90.00
Lever operated drain valve	--	Ea	--	--	68.90
Bain Marie, gas or steam, 6'	C8@3.47	Ea	2,260.00	133.00	2,393.00
Bain Marie, electric	C8@3.47	Ea	2,160.00	133.00	2,293.00
Broiler, infrared, Fostoria Model 1000A	C8@105.	Ea	66,600.00	4,040.00	70,640.00
Carbonated beverage dispenser, 4 station	C8@3.88	Ea	1,995.00	149.00	2,144.00
Charbroiler, General Electric (#CB-51)	C8@5.07	Ea	3,770.00	195.00	3,965.00
Cup and plate dispenser, heated	C8@1.54	Ea	420.00	59.20	479.20
Deep fat fryer, 125 lb	C8@7.46	Ea	1,675.00	287.00	1,962.00
Dishwashers, by rack capacity per hour					
50 (#431C-2)	C8@5.07	Ea	4,575.00	195.00	4,770.00
115 (#1390)	C8@20.7	Ea	11,875.00	796.00	12,671.00
275 (#1390)	C8@20.7	Ea	16,520.00	796.00	17,316.00
Display case fluorescent fixture, 6' long	C8@2.00	Ea	162.10	76.90	239.00

Equipment 11

	Craft@Hrs	Unit	Material	Labor	Total

Food Service Equipment, continued

Electric sub panel with prewired case, 100 amp connected load

	Craft@Hrs	Unit	Material	Labor	Total
typical price, per case	--	Ea	--	--	380.00
Doughnut machine, (#582, C), size 100	C8@12.2	Ea	8,610.00	469.00	9,079.00
Equipment base, 4" channel, (assembly length)	C8@.407	Ea	9.20	15.70	24.90

Food mixing machines, (#38-G)

	Craft@Hrs	Unit	Material	Labor	Total
80 quart	C8@17.6	Ea	5,225.00	677.00	5,902.00
140 quart	C8@29.3	Ea	10,740.00	1,130.00	11,870.00
Fry pan, tilting type 12,000 Btu	C8@5.07	Ea	4,790.00	195.00	4,985.00

Garbage disposers, stainless steel cone, commercial

	Craft@Hrs	Unit	Material	Labor	Total
1 HP	P6@4.46	Ea	645.00	182.00	827.00
1-1/2 HP	P6@4.46	Ea	1,075.00	182.00	1,257.00
3 HP	P6@4.46	Ea	1,610.00	182.00	1,792.00
7-1/2 HP	P6@19.5	Ea	4,200.00	797.00	4,997.00
Glass rack dispenser, shelf level	C8@2.47	Ea	540.00	95.00	635.00
Griddle, including stand, 24" x 36"	C8@7.46	Ea	1,725.00	287.00	2,012.00
Hot chocolate dispenser	C8@.963	Ea	810.00	37.00	847.00
Hot food wells, 12" x 20", electric	CE@1.85	Ea	320.00	85.70	405.70
Ice-maker, 500 lbs per day	P6@9.23	Ea	8,290.00	377.00	8,667.00
Ice tea dispenser	C8@2.93	Ea	215.00	113.00	328.00

Kettles, steel jacketed

	Craft@Hrs	Unit	Material	Labor	Total
5 gallon	C8@3.88	Ea	1,075.00	149.00	1,224.00
20 gallon	C8@7.78	Ea	1,990.00	299.00	2,289.00
40 gallon	C8@7.78	Ea	2,365.00	299.00	2,664.00
60 gallon	C8@7.78	Ea	3,015.00	299.00	3,314.00
80 gallon	C8@8.68	Ea	3,230.00	334.00	3,564.00
40 gallon, leg mounted, free standing	C8@7.78	Ea	2,365.00	299.00	2,664.00

Ovens

	Craft@Hrs	Unit	Material	Labor	Total
Roll thru dispatch oven (#1869)	D4@47.8	Ea	29,115.00	1,870.00	30,985.00
Convection oven (#MIL-0-43679)	D4@9.55	Ea	6,350.00	373.00	6,723.00
Revolving tray oven, size 12	D4@26.2	Ea	16,730.00	1,020.00	17,750.00
Revolving tray oven, size 18	D4@26.2	Ea	19,930.00	1,020.00	20,950.00
Revolving tray oven, size 24	D4@28.8	Ea	24,780.00	1,120.00	25,900.00
Pot and pan rack, table mounted, 6'	C8@2.00	Ea	350.00	76.90	426.90
Pot washers, sink mounted	P6@4.89	Ea	1,075.00	200.00	1,275.00

Range hoods

	Craft@Hrs	Unit	Material	Labor	Total
Filterless, Gaylord Model "A"	T5@23.1	Ea	17,240.00	925.00	18,165.00
Filterless, Gaylord Model "BD"	T5@23.1	Ea	11,875.00	925.00	12,800.00
Filter, Gaylord Model "BD"	T5@26.3	Ea	15,075.00	1,050.00	16,125.00

Refrigerator, under counter or wall, 5', built-in compressor

	Craft@Hrs	Unit	Material	Labor	Total
	C8@3.07	Ea	1,405.00	118.00	1,523.00

Refrigerator-freezer, under counter

	Craft@Hrs	Unit	Material	Labor	Total
Box only, LF of front	C8@.223	LF	215.00	8.58	223.58
Add for each door	C8@1.00	Ea	235.00	38.50	273.50
Add for each drawer	C8@.499	Ea	430.00	19.20	449.20
Add for refrigerator coil	P6@2.98	Ea	180.00	122.00	302.00
Remote refrigerator compressor, 1/4 HP	P6@2.98	Ea	1,135.00	122.00	1,257.00
Shelf heater, infrared	CE@2.57	Ea	215.00	119.00	334.00

Shelves

	Craft@Hrs	Unit	Material	Labor	Total
Table mounted shelf rack, 12' long, 12" wide	C8@2.00	Ea	54.00	76.90	130.90
Wall mounted shelf (shelf area), stainless	C8@.223	SF	25.50	8.58	34.08

Sinks, counter mounted, steel

	Craft@Hrs	Unit	Material	Labor	Total
Utility sink, 24" x 14" x 14"	P6@2.46	Ea	565.00	101.00	666.00

Equipment 11

	Craft@Hrs	Unit	Material	Labor	Total
Utility sink, 24" x 36" x 20"	P6@3.90	Ea	960.00	159.00	1,119.00
Soiled dish sink, with pre-rinse unit	P6@4.74	Ea	5,060.00	194.00	5,254.00
Soiled dish conveyor, 16" wide to 15' long	D4@63.0	Ea	5,930.00	2,460.00	8,390.00
Tables, mobile, hot food, size 5	C8@2.93	Ea	4,090.00	113.00	4,203.00
Toaster, conveyor type, 300 slices per hour	D4@2.88	Ea	755.00	112.00	867.00
Vegetable peeling machines					
30 lb capacity	D4@4.38	Ea	1,725.00	171.00	1,896.00
50 lb capacity	D4@4.38	Ea	2,045.00	171.00	2,216.00
Water cooler, drop-in 7.5 gallons per hour	P6@1.40	Ea	900.00	57.20	957.20
Water heater, under-sink mount, electric	P6@4.89	Ea	600.00	200.00	800.00

Gymnasium and Athletic Equipment. These costs include the subcontractor's overhead and profit.

	Craft@Hrs	Unit	Material	Labor	Total
Basketball backstops					
Fixed, wall mounted	D4@8.00	Ea	300.00	312.00	612.00
Swing-up, wall mounted	D4@10.0	Ea	800.00	390.00	1,190.00
Ceiling suspended, to 20', swing-up manual	D4@16.0	Ea	1,800.00	625.00	2,425.00
Add for glass backstop, fan shape	--	Ea	--	--	585.00
Add for glass backstop, rectangular	--	Ea	--	--	645.00
Add for power operation	--	Ea	--	--	570.00
Gym floor, not including subfloor or base					
Synthetic gym floor, 3/16"	--	SF	--	--	5.20
Synthetic gym floor, 3/8"	--	SF	--	--	6.75
Built-up wood "spring system" maple floor	--	SF	--	--	7.75
Rubber cushioned maple floor	--	SF	--	--	6.70
Maple floor on sleepers and membrane, unfinished	--	SF	--	--	5.45
Apparatus inserts	--	Ea	--	--	35.80
Bleachers, telescoping, manual, per seat	D4@.187	Ea	40.00	7.30	47.30
Bleachers, hydraulic operated, per seat	D4@.235	Ea	54.50	9.17	63.67
Scoreboard, basketball, single face					
Economy	CE@7.01	Ea	1,200.00	325.00	1,525.00
Good	CE@13.0	Ea	1,650.00	602.00	2,252.00
Premium	CE@21.6	Ea	2,600.00	1,000.00	3,600.00

Rolling Ladders. Rolling ladders for mercantile, library or industrial use. These costs include the subcontractor's overhead and profit. Putnam Rolling Ladder

Oak rolling ladder with top and bottom rolling fixtures, non-slip treads, pre-finished. By floor to track height. Painted aluminum hardware. Add track below

	Craft@Hrs	Unit	Material	Labor	Total
7'6"	D4@.819	Ea	280.00	32.00	312.00
9'0"	D4@.819	Ea	296.00	32.00	328.00
11'8"	D4@1.09	Ea	328.00	42.60	370.60
12'0"	D4@1.09	Ea	344.00	42.60	386.60
13'9"	D4@1.09	Ea	360.00	42.60	402.60
14'0"	D4@1.09	Ea	376.00	42.60	418.60
Add for bend at ladder top or bottom	--	Ea	70.00	--	--
Add for brass or chrome hardware	--	Ea	85.00	--	--
Add for ladder work shelf	--	Ea	40.00	--	--
Add for handrail	--	Ea	37.50	--	--
Rolling ladder track, 7/8" slotted steel, per foot of length, 9' to 14' high					
Painted aluminum finish	D4@.089	LF	3.50	3.48	6.98
Painted black finish	D4@.089	LF	4.50	3.48	7.98
Brass or chrome plated finish	D4@.089	LF	11.50	3.48	14.98
Corner bend track, 90 degree, 30" radius					

Equipment 11

	Craft@Hrs	Unit	Material	Labor	Total
Rolling Ladders, continued					
Painted finish	D4@.363	Ea	37.50	14.20	51.70
Chrome or brass finish	D4@.363	Ea	60.00	14.20	74.20

Service Station Equipment. These costs include the subcontractor's overhead and profit but no excavation, concrete work, piping or electrical work except as noted.

	Craft@Hrs	Unit	Material	Labor	Total
Air compressor, 1-1/2 HP with receiver	PF@5.54	Ea	1,730.00	255.00	1,985.00
Air compressor, 3 HP with receiver	PF@5.54	Ea	1,980.00	255.00	2,235.00
Hose reels with hose					
Air hose, 50', heavy duty	PF@8.99	Ea	495.00	414.00	909.00
Water hose, 50' and valve	PF@8.01	Ea	495.00	369.00	864.00
Lube rack, 3 hose, remote pump	PF@48.4	Ea	4,910.00	2,230.00	7,140.00
Lube rack, 5 hose, remote pump	PF@48.4	Ea	5,930.00	2,230.00	8,160.00
Air-operated pumps for hose reels, fit 55 gallon drum					
Motor oil or gear oil	PF@1.00	Ea	805.00	46.00	851.00
Lube oil	PF@1.00	Ea	1,720.00	46.00	1,766.00
Gasoline pump, with cabinet, industrial type, 15 gallons per minute	PF@5.93	Ea	1,630.00	273.00	1,903.00
Add for totalizer	--	Ea	235.00	--	--
Submerged turbine, 1/3 HP, serves 4 dispensers	PF@22.5	Ea	975.00	1,040.00	2,015.00
Submerged turbine, 3/4 HP, serves 8 dispensers	PF@22.5	Ea	1,200.00	1,040.00	2,240.00
Gasoline dispenser, computing, single hose, single product not including rough-in of pit box	PF@9.94	Ea	3,300.00	458.00	3,758.00
Gasoline dispenser, computing, dual hose, dual product, not including rough-in of pit box	PF@9.94	Ea	5,800.00	458.00	6,258.00
Pit box for submerged pump, 22" x 22", installed in existing pit	PF@1.26	Ea	125.00	58.00	183.00
Fill boxes, 12" diameter, cast iron	PF@.949	Ea	60.60	43.70	104.30
Air and water combination underground reels and box, stainless steel cover, with hose	PF@9.58	Ea	650.00	441.00	1,091.00
Add for electric thermal unit	--	Ea	90.70	--	--
Hoist, single post frame contact, 8,000 lb semi hydraulic	D4@71.3	Ea	3,450.00	2,780.00	6,230.00
Hoist, single post frame contact, 8,000 lb fully hydraulic	D4@71.3	Ea	3,700.00	2,780.00	6,480.00
Hoist, two post, pneumatic, 11,000 lb	D4@114.	Ea	5,250.00	4,450.00	9,700.00
Hoist, two post, pneumatic, 24,000 lb	D4@188.	Ea	9,580.00	7,340.00	16,920.00
Two-post fully hydraulic hoists					
10,000 pound capacity	D4@188.	Ea	4,310.00	7,340.00	11,650.00
13,000 pound capacity	D4@188.	Ea	5,480.00	7,340.00	12,820.00
18,500 pound capacity	D4@188.	Ea	7,250.00	7,340.00	14,590.00
24,000 pound capacity	D4@188.	Ea	9,770.00	7,340.00	17,110.00
26,000 pound capacity	D4@188.	Ea	11,400.00	7,340.00	18,740.00
Cash box and pedestal stand	D4@2.55	Ea	160.00	99.60	259.60
Tire changer, 120 PSI pneumatic, auto tire	D4@10.6	Ea	2,440.00	414.00	2,854.00
Tire changer, hydraulic, truck tire	D4@21.4	Ea	7,570.00	835.00	8,405.00
Exhaust fume system, underground, complete, per station	D5@9.52	Ea	525.00	381.00	906.00

Equipment 11

	Craft@Hrs	Unit	Material	Labor	Total

Parking Control Equipment. Costs are for parking equipment only, no concrete work, paving, or electrical work included. Labor costs include installation and testing. These costs include the subcontractor's overhead and profit.

Parking gates, controllable by attendant, vehicle detector, ticket dispenser, card reader, or coin/token machine

	Craft@Hrs	Unit	Material	Labor	Total
Gate with 10' arm	D4@9.13	Ea	2,655.00	356.00	3,011.00
Folding arm gate (low headroom), 10'	D4@9.13	Ea	3,005.00	356.00	3,361.00
Vehicle detector, for installation in approach lane, including cutting and grouting	D4@4.96	Ea	365.00	194.00	559.00
Ticket dispenser (spitter), with date/time imprinter, actuated by vehicle detector or hand button	D4@9.13	Ea	5,045.00	356.00	5,401.00

Card readers

	Craft@Hrs	Unit	Material	Labor	Total
Non-programmable, parking areas with long term parking privileges such as condominiums	D4@3.39	Ea	690.00	132.00	822.00
Programmable, parking areas for users paying a periodic fee to renew parking privileges, per reader	D4@3.39	Ea	4,120.00	132.00	4,252.00
Lane spikes, dragon teeth, 6' wide, installation including cutting and grouting	D4@9.69	Ea	1,075.00	378.00	1,453.00
Lighted warning sign for lane spikes	D4@5.71	Ea	510.00	223.00	733.00

Loading Dock Equipment. Costs are based on new construction and assume that equipment is set in prepared locations. No excavation, concrete work or electrical work included. These costs include the subcontractor's overhead and profit.

	Craft@Hrs	Unit	Material	Labor	Equipment	Total
Dock levelers, Systems, Inc.						
Edge-of-dock leveler, pit type	D6@2.13	Ea	600.00	92.10	35.00	727.10
Mechanical platform dock leveler, recessed						
6' wide x 6' long	D6@3.24	Ea	1,900.00	140.00	48.90	2,088.90
6' wide x 8' long	D6@3.24	Ea	2,230.00	140.00	50.00	2,420.00
7' wide x 6' long	D6@3.24	Ea	2,210.00	140.00	50.00	2,400.00
7' wide x 8' long	D6@3.24	Ea	2,380.00	140.00	48.90	2,568.90
Hydraulic, pit type						
6' wide x 6' long	D6@4.03	Ea	2,935.00	174.00	41.70	3,150.70
6' wide x 8' long	D6@4.03	Ea	3,150.00	174.00	41.70	3,365.70

Dock seals, foam pad type, 12" projection, with anchor brackets (bolts not included), Systems, Inc. door opening size as shown

	Craft@Hrs	Unit	Material	Labor	Equipment	Total
8' wide x 8' high	D6@3.77	Ea	500.00	163.00	18.00	681.00
8' wide x 10' high	D6@3.77	Ea	600.00	163.00	18.00	781.00
Dock lifts, portable, electro-hydraulic scissor lifts, Systems, Inc.						
4,000 lb capacity, 6' x 6' dock	D6@2.04	Ea	5,220.00	88.20	--	5,308.20
Dock lifts, pit recessed, electro-hydraulic scissor lifts, Systems, Inc.						
5,000 lb capacity, 6' x 8' platform	D6@11.6	Ea	5,745.00	502.00	54.50	6,301.50
6,000 lb capacity, 6' x 8' platform	D6@11.6	Ea	5,990.00	502.00	54.50	6,546.50
8,000 lb capacity						
6' x 8' platform	D6@11.6	Ea	8,700.00	502.00	54.50	9,256.50
8' x 10' platform	D6@11.6	Ea	9,800.00	502.00	54.50	10,356.50
10,000 lb capacity						
6' x 8' platform	D6@11.6	Ea	10,270.00	502.00	54.60	10,826.60
8' x 10' platform	D6@11.6	Ea	11,375.00	502.00	54.60	11,931.60
12,000 lb capacity						
6' x 10' platform	D6@15.5	Ea	12,750.00	670.00	67.00	13,487.00
8' x 12' platform	D6@15.5	Ea	13,900.00	670.00	67.00	14,637.00
15,000 lb capacity						
6' x 10' platform	D6@15.5	Ea	14,200.00	670.00	67.00	14,937.00

Equipment 11

	Craft@Hrs	Unit	Material	Labor	Equipment	Total
Loading Dock Equipment, 15,000 lb capacity, continued						
8' x 12' platform	D6@15.5	Ea	15,500.00	670.00	67.00	16,237.00
18,000 lb capacity						
6' x 10' platform	D6@15.5	Ea	14,650.00	670.00	81.40	15,401.40
8' x 12' platform	D6@15.5	Ea	16,100.00	670.00	81.40	16,851.40
20,000 lb capacity						
6' x 12' platform	D6@15.5	Ea	16,700.00	670.00	81.40	17,451.40
8' x 12' platform	D6@15.5	Ea	16,800.00	670.00	81.40	17,551.40
Combination dock lift and dock leveler, electro-hydraulic scissor lifts, pit recessed, 5,000 lb capacity, Advance						
6' x 7'2" platform	D6@11.6	Ea	7,900.00	502.00	54.60	8,456.60
6' x 8' platform	D6@11.6	Ea	8,140.00	502.00	54.60	8,696.60
Dock bumpers, laminated rubber, not including masonry anchors						
10" high x 14" wide x 4-1/2" thick	D4@.248	Ea	47.00	9.68	--	56.68
10" high x 14" wide x 6" thick	D4@.299	Ea	64.30	11.70	--	76.00
10" high x 24" wide x 4-1/2" thick	D4@.269	Ea	55.30	10.50	--	65.80
10" high x 24" wide x 6" thick	D4@.358	Ea	77.30	14.00	--	91.30
10" high x 36" wide x 4-1/2" thick	D4@.416	Ea	67.60	16.20	--	83.80
10" high x 36" wide x 6" thick	D4@.499	Ea	94.30	19.50	--	113.80
12" high x 14" wide x 4-1/2" thick	D4@.259	Ea	56.50	10.10	--	66.60
12" high x 24" wide x 4-1/2" thick	D4@.328	Ea	65.70	12.80	--	78.50
12" high x 36" wide x 4-1/2" thick	D4@.432	Ea	81.40	16.90	--	98.30
22" high x 22" wide x 3" thick	D4@.251	Ea	71.00	9.80	--	80.80
Add for 4 masonry anchors	D4@1.19	LS	2.00	46.50	--	48.50
Barrier posts, 8' long concrete-filled steel posts set 4' underground in concrete						
3" diameter	P8@.498	Ea	67.30	18.50	--	85.80
4" diameter	P8@.498	Ea	97.10	18.50	--	115.60
6" diameter	P8@.752	Ea	119.00	27.90	--	146.90
8" diameter	P8@1.00	Ea	145.00	37.10	--	182.10
Loading dock shelter, fabric covered, steel frame, 24" extension for						
10' x 10' door	D4@9.91	Ea	800.00	387.00	--	1,187.00
Perimeter door seal, 12" x 12", vinyl cover	D4@4.00	LF	325.00	156.00	--	481.00

Medical Refrigeration Equipment. Labor costs include unpacking, assembly and installation. These costs include the subcontractor's overhead and profit. Jewett Refrigerator
Blood bank refrigerators, upright enameled steel cabinets with locking glass doors, 7 day recording thermometer, LED temperature display, failure alarm, interior lights, and removable drawers. 2 to 4 degree C operating range

	Craft@Hrs	Unit	Material	Labor	Total
5 drawer, 240 bag, 16.9 CF, 29" wide x 30" deep x 76" high, #BBR17	D4@2.25	Ea	6,985.00	87.80	7,072.80
6 drawer, 360 bag, 24.8 CF, 29" wide x 36" deep x 80" high, #BBR25	D4@2.25	Ea	7,500.00	87.80	7,587.80
10 drawer, 480 bag, 37.4 CF, 59" wide x 30" deep x 77" high, #BBR37	D4@3.42	Ea	10,065.00	134.00	10,199.00
12 drawer, 720 bag, 55.0 CF, 59" wide x 36" deep x 80" high, #BBR55	D4@3.42	Ea	10,400.00	134.00	10,534.00

Blood plasma storage freezers, upright enameled steel cabinets with locking steel doors, 7 day recording thermometer, LED temperature display, failure alarm, and removable drawers. -35 degree C operating temperature

	Craft@Hrs	Unit	Material	Labor	Total
6 drawer, 265 pack, 13.2 CF, 34" wide x 30" deep x 76" high, #BPL13	D4@2.25	Ea	10,370.00	87.80	10,457.80
7 drawer, 415 pack, 20.7 CF, 34" wide x 36" deep x 80" high, #BPL21	D4@2.25	Ea	13,350.00	87.80	13,437.80

Equipment 11

	Craft@Hrs	Unit	Material	Labor	Total
14 drawer, 830 pack, 40.6 CF, 59" wide x 36" deep x 81" high, #BPL41	D4@3.42	Ea	18,810.00	134.00	18,944.00

Laboratory refrigerators, upright enameled steel cabinets with locking steel doors, enameled steel interior, and adjustable stainless steel shelves. -2 degree C operating temperature

	Craft@Hrs	Unit	Material	Labor	Total
4 shelf, 16.9 CF, 29" wide x 30" deep x 77" high #LR17B	D4@2.25	Ea	4,025.00	87.80	4,112.80
8 shelf, 37.4 CF, 59" wide x 30" deep x 77" high #LR37B	D4@3.42	Ea	6,000.00	134.00	6,134.00

Pharmacy refrigerators, upright enameled steel cabinets with locking steel doors, enameled steel interior, and adjustable stainless steel drawers and shelves. -2 degree C operating temperature

	Craft@Hrs	Unit	Material	Labor	Total
6 drawer, 1 shelf, 24.8 CF, 29" wide x 36" deep x 80" high, #PR25B	D4@2.25	Ea	4,850.00	87.80	4,937.80
9 drawer, 5 shelf, 55.0 CF, 59" wide x 36" deep x 80" high, #PR55B	D4@3.42	Ea	7,500.00	134.00	7,634.00

Hospital/Lab/Pharmacy undercounter refrigerators, stainless steel interior and exterior, locking door. 24" wide x 24" deep x 34-1/2" high. 2 degree C operating temperature

	Craft@Hrs	Unit	Material	Labor	Total
5.4 CF, with ice cube compartment cooling system and two stainless steel racks, #UC5B	D4@1.69	Ea	1,930.00	66.00	1,996.00
5.4 CF, with blower coil cooling system, stainless steel racks, and auto-defrost, #UC5B	D4@1.67	Ea	1,930.00	65.20	1,995.20
5.4 CF with cold wall cooling system, stainless steel racks, and auto-defrost, #UC5C	D4@1.69	Ea	2,200.00	66.00	2,266.00

Hospital/Lab/Pharmacy undercounter freezers, stainless steel interior and exterior, locking door. 24" wide x 24" deep x 34-1/2" high. -20 degree C operating temperature

	Craft@Hrs	Unit	Material	Labor	Total
4.6 CF, stainless steel racks, auto-defrost, #UCF5B	D4@1.69	Ea	3,070.00	66.00	3,136.00
4.6 CF, stainless steel racks, manual hot gas defrost system, #UCF5C	D4@1.69	Ea	3,435.00	66.00	3,501.00

Morgue refrigerators

	Craft@Hrs	Unit	Material	Labor	Total
1 or 2 body roll-in type, 39" wide x 96" deep x 75" high, with 1/2 HP, 4380 Btu/hour condenser, #1SPEC	D4@15.8	Ea	7,480.00	617.00	8,097.00
2 body sliding tray type, 39" wide x 96" deep x 75" high, with 1/2 HP, 4380 Btu/hour condenser, #2EC	D4@18.0	Ea	9,900.00	703.00	10,603.00
2 body side opening type, 96" wide x 39" deep x 75" high, with 1/2 HP, 4380 Btu/hour condenser, #2SC	D4@18.0	Ea	9,950.00	703.00	10,653.00
4 or 6 body with roll-in type lower compartments and sliding tray type upper compartments. 73" wide x 96" deep x 102" high with 3/4 HP, 6060 Btu/hour condenser, #4SPEC2W	D4@22.5	Ea	15,950.00	878.00	16,828.00

Furnishings 12

	Craft@Hrs	Unit	Material	Labor	Total

Cabinets. Cost per linear foot of face or back dimension. Base cabinets are 34" high x 24" deep, wall cabinets are 42" high x 12" deep and full height cabinets are 94" high x 24" deep. Labor shown is for installing shop fabricated units. These costs include the subcontractor's overhead and profit. Add for countertops at the end of this section.

Metal cabinets, shop and commercial type, with hardware but no countertop

	Craft@Hrs	Unit	Material	Labor	Total
Base cabinet, drawer, door and shelf	D4@.379	LF	95.00	14.80	109.80
Wall cabinet with door and 2 shelves	D4@.462	LF	87.00	18.00	105.00
Full height cabinet, 5 shelves	D4@.595	LF	178.00	23.20	201.20
Library shelving, 48" high x 8" deep modules	D4@.336	LF	63.20	13.10	76.30
Wardrobe units, 72" high x 42" deep	D4@.398	LF	156.00	15.50	171.50

Furnishings 12

	Craft@Hrs	Unit	Material	Labor	Total
Cabinets, continued					
Wood cabinets, laminated plastic face, classroom type, with hardware but no countertop					
Base cabinet, drawer, door and shelf	C8@.391	LF	133.15	15.00	148.15
Wall cabinet with door and 2 shelves	C8@.475	LF	116.50	18.30	134.80
Full height cabinet with doors	C8@.608	LF	187.20	23.40	210.60
Wood cabinets, hospital type, laminated plastic face, with hardware but no countertop					
Base cabinet, drawer, door and shelf	C8@.982	LF	142.00	37.80	179.80
Wall, with 2 shelves and door	C8@1.15	LF	129.00	44.20	173.20
Full height with doors	C8@1.35	LF	191.00	51.90	242.90
Laboratory cabinets, metal, with hardware but no countertops					
Base cabinets, doors	D4@.785	LF	88.80	30.60	119.40
Base cabinets, drawers	D4@.785	LF	166.00	30.60	196.60
Base cabinets, island base	D4@.785	LF	166.00	30.60	196.60
Wall cabinets, doors	D4@.785	LF	144.10	30.60	174.70
Wardrobe or storage cabinets, 80" high	D4@.939	LF	166.00	36.70	202.70
Fume hoods, steel, without duct, typical price	T5@3.88	Ea	664.00	155.00	819.00
Add for duct, without electric work, typical price	T5@16.6	LF	995.00	665.00	1,660.00
Custom made wood cabinets, no hardware or countertops included, unfinished, based on 36" wide modules					
Standard grade, mahogany or laminated plastic face					
Base	C8@.256	LF	54.50	9.85	64.35
Wall	C8@.380	LF	41.50	14.60	56.10
Full height	C8@.426	LF	95.00	16.40	111.40
Custom grade, birch					
Base	C8@.361	LF	83.60	13.90	97.50
Wall	C8@.529	LF	71.10	20.40	91.50
Full height	C8@.629	LF	131.00	24.20	155.20
Sink fronts	C8@.361	LF	69.00	13.90	82.90
Drawer units, 4 drawers	C8@.361	LF	137.00	13.90	150.90
Apron (knee space)	C8@.148	LF	17.60	5.69	23.29
Additional costs for custom cabinets					
Ash face	--	LF	3.11	--	--
Walnut face	--	LF	46.70	--	--
Edge banding	--	LF	5.19	--	--
Prefinished exterior	--	LF	5.19	--	--
Cabinet widths averaging 24"	--	%	30.0	10.0	--
Prefinished interior	--	LF	8.82	--	--
Standard hardware, installed	--	LF	4.15	--	--
Institutional grade hardware, installed	--	LF	6.29	--	--
Drawer roller guides, per drawer	--	Ea	6.75	--	--
Drawer roller guides, suspension	--	Ea	21.80	--	--
Extra drawers					
12" wide	--	Ea	40.50	--	--
18" wide	--	Ea	46.70	--	--
24" wide	--	Ea	50.90	--	--
Countertops, shop fabricated, delivered ready for installation					
Laminated plastic tops 2'0" wide over plywood					
Custom work, square edge front, 4" splash	C8@.187	LF	22.75	7.19	29.94
Add for square edge 9" backsplash	--	LF	7.70	--	--
Add for solid colors	--	%	10.0	--	--
Add for acid proof tops	--	%	100.0	--	--
Komar or simulated marble tops					
Standard vanity unit	C8@.192	LF	24.20	7.39	31.59
Custom moulded top with splash	C8@.049	SF	13.60	1.89	15.49

Furnishings 12

	Craft@Hrs	Unit	Material	Labor	Total
Stainless steel, 2'0" wide with backsplash	C8@.366	LF	67.00	14.10	81.10

Window Treatment. These costs include the subcontractor's overhead and profit.
Shades, custom made, cost per SF of glass covered

	Craft@Hrs	Unit	Material	Labor	Total
Standard roll-up unit	D7@.013	SF	2.35	.49	2.84
Better quality roll-up unit	D7@.013	SF	3.30	.49	3.79
Decorator quality unit	D7@.013	SF	8.25	.49	8.74

Mini-blinds, custom made, cost per SF of glass covered.

	Craft@Hrs	Unit	Material	Labor	Total
Aluminum, horizontal, 1"	D7@.013	SF	3.10	.49	3.59
Cloth type, vertical	D7@.013	SF	4.90	.49	5.39
Steel, horizontal, 2"	D7@.013	SF	2.15	.49	2.64

Fabrics. These costs include the subcontractor's overhead and profit. Measure glass size only. Includes allowance for overlap, pleating and all hardware.
Curtains

	Craft@Hrs	Unit	Material	Labor	Total
Lead mesh (soundproofing)	--	SY	--	--	19.90
Lead mesh (x-ray) 2.5 lbs per SF	--	SY	--	--	22.45

Draperies, 8' height, fabric quality will cause major variation in prices, per LF of opening, typical prices.

	Craft@Hrs	Unit	Material	Labor	Total
Standard quality, window or sliding door	--	LF	--	--	16.25
Better quality, window	--	LF	--	--	43.45
Better quality, sliding door	--	LF	--	--	37.20
Drapery lining	--	LF	--	--	2.80
Fiberglass fabric	--	LF	--	--	44.00
Filter light control type	--	LF	--	--	40.00
Flameproofed	--	LF	--	--	45.00
Velour, grand	--	LF	--	--	48.15

Recessed Open Link Entry Mats. These costs assume that mats are mounted in an aluminum frame recessed into a concrete walkway and include the subcontractor's overhead and profit but no concrete work, special designs or lettering.

	Craft@Hrs	Unit	Material	Labor	Total
Aluminum link, 3/8" thick	D4@.560	SF	12.20	21.90	34.10
Steel link, 3/8" thick	D4@.560	SF	5.50	21.90	27.40
Aluminum hinged strips, carpet inserts, 1/2"	D4@.560	SF	27.75	21.90	49.65
Black Koroseal, 1/2" thick, open link	D4@.560	SF	10.75	21.90	32.65
Aluminum frame for entry mats	D4@.086	LF	4.50	3.36	7.86

Entrance Mats and Recessed Foot Grilles. These costs assume installation of an extruded aluminum tread foot grille on a level surface or over a catch basin with or without a drain, and include subcontractor's overhead and profit. Conspec Systems
Pedigrid/Pedimat extruded aluminum entrance mat

	Craft@Hrs	Unit	Material	Labor	Total
Grid mat with carpet treads	D4@.150	SF	28.80	5.86	34.66
Grid mat with vinyl treads	D4@.150	SF	30.95	5.86	36.81
Grid mat with pool or shower vinyl treads	D4@.150	SF	34.70	5.86	40.56
Grid mat with abrasive treads	D4@.150	SF	34.70	5.86	40.56
Grid mat with serrated aluminum treads	D4@.150	SF	34.70	5.86	40.56
Recessed catch basin with 2" drain	D4@.302	SF	40.00	11.80	51.80
Recessed catch basin without drain	D4@.242	SF	38.00	9.45	47.45
Level base recessed mount	D4@.150	SF	34.70	5.86	40.56
Surface-mount frame, vinyl	--	LF	3.40	--	--
Recessed-mount frame, bronze aluminum	D4@.092	LF	8.00	3.59	11.59

Furnishings 12

	Craft@Hrs	Unit	Material	Labor	Total
Seating. These costs include the subcontractor's overhead and profit but no concrete work.					
Theater style					
Economy	D4@.424	Ea	97.30	16.60	113.90
Lodge, rocking type	D4@.470	Ea	182.50	18.30	200.80
Bleachers, open riser, extruded aluminum, cost per 18" wide seat					
Portable bleachers	D4@.090	Ea	25.20	3.51	28.71
Portable bleachers with guardrails	D4@.136	Ea	32.00	5.31	37.31
Team bleachers, stationary	D4@.090	Ea	35.50	3.51	39.01
Stadium seating, on concrete foundation by others, extruded aluminum, cost per 18" seat width					
10" flat bench seating, seatboard only	D4@.090	Ea	11.80	3.51	15.31
12" contour bench seating, seatboard only	D4@.090	Ea	13.10	3.51	16.61
Chair seating, molded plastic	D4@.310	Ea	62.60	12.10	74.70
Replacement bench seating, stadium or bleacher, extruded aluminum, cost per 18" seat width					
10" wide flat seatboard only	D4@.069	Ea	7.68	2.69	10.37
10" wide flat seatboard and footboard	D4@.136	Ea	15.90	5.31	21.21
12" contour seatboard only	D4@.069	Ea	9.03	2.69	11.72

Special Construction 13

	Craft@Hrs	Unit	Material	Labor	Total
Audiometric Rooms. These costs include the subcontractor's overhead and profit, but no electrical work or audio equipment. Typical price based on room 500 SF and over,					
per SF of wall and ceiling surface area	--	SF	--	--	38.50
Incinerators. Typical subcontract prices based on pounds per hour capacity, electrically operated, no scrubber or stack included.					
50 pounds	--	Ea	--	--	11,000.00
100 pounds	--	Ea	--	--	12,000.00
200 pounds	--	Ea	--	--	15,000.00
500 pounds	--	Ea	--	--	29,500.00
1,000 pounds	--	Ea	--	--	85,000.00
Deduct for gas units	--	%	--	--	-8.0
Integrated Ceilings. These costs include the subcontractor's overhead and profit.					
Integrated suspended ceiling and lighting system, typical costs, "T" bar grid, fixtures and mineral fiber tile, no ventilating included, with electric connection only					
55 foot candles	--	SF	--	--	3.80
70 foot candles	--	SF	--	--	4.40
85 foot candles	--	SF	--	--	4.85
100 foot candles	--	SF	--	--	5.10
Luminous ceiling systems, fixtures and ceiling, no ventilating included					
80% plastic, 100 foot candles	--	SF	--	--	7.10
70% plastic, 80 foot candles	--	SF	--	--	6.75
60% plastic, 70 foot candles	--	SF	--	--	6.10
50% plastic, 55 foot candles	--	SF	--	--	5.65
Add for jobs under 3,000 SF	--	%	--	--	33.0

Modular Building Construction (Manufactured housing)
Relocatable structures and institutional type housing for remote areas. Costs are per square foot covered by the roof and include the subcontractor's overhead and profit. Based on 2" x 4" wood studs 16" OC, 1/2" gypsum wallboard inside and 1/2" CDX plywood outside, with minimum plumbing, heating

Special Construction 13

	Craft@Hrs	Unit	Material	Labor	Equipment	Total

and electrical systems for the intended use. Costs include factory assembly, delivery by truck to 50 miles and set up on site. Add 1.5% for each additional 50 miles of delivery. No site preparation, earthwork, foundations, or furnishings included. Cost for smaller, more complex structures will be higher.

Portable structures (temporary offices, school classrooms, etc.)

Medium quality	--	SF	--	--	--	30.50
Better quality	--	SF	--	--	--	36.50

Institutional housing, four-plex modules

Medium quality	--	SF	--	--	--	32.00
Better quality	--	SF	--	--	--	33.50

Construction or mining camp barracks, mess halls, kitchens, etc.

Single story structures	--	SF	--	--	--	29.00
Add for 2nd story (per SF of floor)	--	%	--	--	--	10.0

Bus Stop Shelters. Prefabricated structure systems, shipped knocked-down complete with all necessary hardware and foundation anchors and installation instructions, assembled at installation site. These costs include the subcontractor's overhead and profit but no foundation or floor slab. Dimensions are approximate outside measurements. Based on Columbia Equipment Company.

Three-sided open front type. Clear satin silver anodized aluminum structure and fascia, 1/4" clear acrylic sheet glazing in panels and white baked enamel finish aluminum V-beam roof.

9'2" L x 5'3" W x 7'5" H

	Craft@Hrs	Unit	Material	Labor	Total
Two rear panels, single panel each side	C8@8.00	Ea	2,200.00	308.00	2,508.00
Four rear panels, two panels each side	C8@12.0	Ea	2,300.00	462.00	2,762.00

11'3" L x 5'3" W x 7'5" H

Four rear panels, two panels each side	C8@12.0	Ea	2,700.00	462.00	3,162.00

Add to three-sided open front shelters for the following:
Wind break front entrance panel

5'3" W x 7'5" H, with two glazed panels	C8@2.00	Ea	260.00	76.90	336.90

Bench, all aluminum vandal-resistant seat & backrest

8'0" long	C8@1.00	Ea	215.00	38.50	253.50
10'0" long	C8@1.50	Ea	275.00	57.70	332.70

Light fixture with unbreakable Lexan diffuser, wiring not included

100 watt incandescent, w/photoelectric cell control	C8@2.00	Ea	115.00	76.90	191.90

Radiant heater, with vandal-resistant electric heating element in heavy duty enclosure and metal mesh guard over opening, wiring not included

2500 watt unit (no controls)	C8@1.50	Ea	470.00	57.70	527.70
5000 watt unit (no controls)	C8@1.50	Ea	550.00	57.70	607.70
Electronic timer control, add to above	C8@1.00	Ea	250.00	38.50	288.50

Integrated map/schedule display panel, anodized aluminum frame tamper-proof fasteners and "spanner-head" tool for opening, full panel width x 30" high

3/16" clear Plexiglass	C8@1.00	Ea	75.00	38.50	113.50
3/16" polycarbonate with mar-resistant coating	C8@1.00	Ea	105.00	38.50	143.50

Graphics, words "Bus Stop" in white letters, on fascia

First side	--	LS	65.00	--	--
Each additional side	--	Ea	22.50	--	--

Special Purpose Rooms and Buildings

Air-supported pool or tennis court enclosures, polyester reinforced vinyl dome, with zippered entry air lock doors, cable tiedowns and blower, cost per square foot of

area covered. Air Structures	D4@.030	SF	3.00	1.17	4.17
Add for 40" revolving door	D4@2.14	Ea	1,500.00	83.50	1,583.50

Special Construction 13

	Craft@Hrs	Unit	Material	Labor	Total
Special Purpose Rooms and Buildings, continued					
Add for 66" air lock door	D4@3.28	Ea	1,500.00	128.00	1,628.00
Domes, net area covered, complete, including roofing, typical costs					
Astronomy dome, fiberglass skin, revolving, 10' to 40' diameter base	--	SF	--	--	123.00
Belem truss, corrugated metal top & bottom	--	SF	--	--	13.50
Cable suspended with air-floated fabric cover	--	SF	--	--	15.50
Cable suspended with fabric cover	--	SF	--	--	16.50
Cable suspended with steel deck	--	SF	--	--	17.50
Fink dome, steel truss ribs with corrugated cover	--	SF	--	--	18.00
Geodesic, aluminum	--	SF	--	--	21.90
Glulam beams and decking, ribbed	--	SF	--	--	16.20
Hyperbolic paraboloid, thin shell concrete	--	SF	--	--	21.90
Steel rib frame with skylight cover	--	SF	--	--	31.80
Steel rib frame with skylight cover, partial opening	--	SF	--	--	61.80
Triodesic, glulam beams with decking	--	SF	--	--	16.90
Greenhouses, single wall fiberglass panels in aluminum frame, with foundation but no floor or equipment					
To 1,000 SF	--	SF	--	--	20.30
Over 1,000 SF to 4,000 SF	--	SF	--	--	14.30
Over 4,000 SF to 10,000 SF	--	SF	--	--	13.20
Radiation Protection. These costs include the subcontractor's overhead and profit.					
Lead-lined lath, cost per SF of lath					
2 pound	LA@.031	SF	4.26	1.27	5.53
4 pound	LA@.039	SF	5.65	1.60	7.25
6 pound	LA@.054	SF	7.05	2.21	9.26
8 pound	LA@.071	SF	11.25	2.91	14.16
Lead-lined glass windows with lead frames (to 2' x 3')	G1@.572	SF	165.00	22.30	187.30
Lead-lined doors to 4 pounds per SF	C4@.167	SF	22.70	5.89	28.59
Lead-lined door frames	C4@2.38	Ea	285.00	83.90	368.90

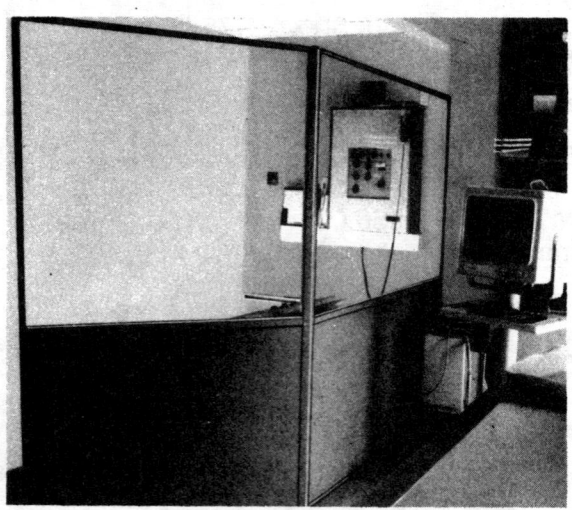

X-Ray Viewing Panels

X-Ray Viewing Panels, Clear Lead-plastic. Based on CLEAR-Pb, by Nuclear Associates. Based on SF of panel (rounded up to the next higher whole square foot). For panels 12 square feet or larger, add a crating charge of $60.00. Panels larger than 72" x 96" are special order items. Weights shown are approximate.

	Craft@Hrs	Unit	Material	Labor	Total
7mm thick, 0.3mm lead equivalence, 2.3 lbs/SF	G1@.115	SF	91.10	4.48	95.58

Special Construction 13

	Craft@Hrs	Unit	Material	Labor	Total
12mm thick, 0.5mm lead equivalence, 3.9 lbs/SF	G1@.195	SF	124.00	7.59	131.59
18mm thick, 0.8mm lead equivalence, 5.9 lbs/SF	G1@.294	SF	134.00	11.40	145.40
22mm thick, 1.0mm lead equivalence, 7.2 lbs/SF	G1@.361	SF	139.00	14.00	153.00
35mm thick, 1.5mm lead equivalence, 11.5 lbs/SF	G1@.574	SF	155.00	22.30	177.30
46mm thick, 2.0mm lead equivalence, 15.0 lbs/SF	G1@.751	SF	205.00	29.20	234.20

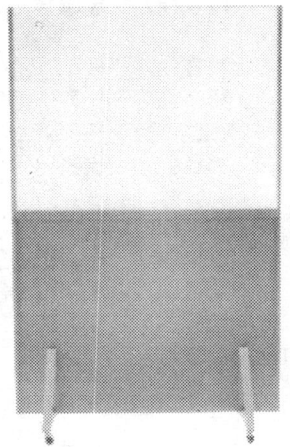
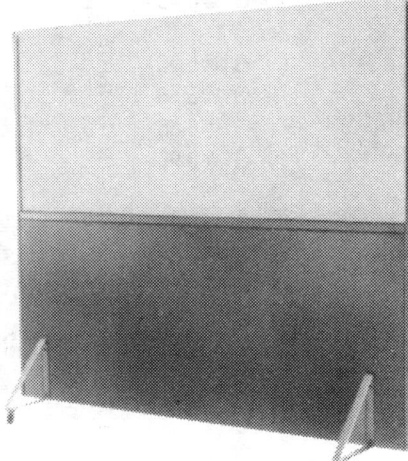

Mobile X-Ray Barriers

Mobile X-Ray Barriers. Clear lead-plastic window panels on the upper portion and opaque panels on the lower portion, mounted within a framework with casters on the bottom. Based on Nuclear Associates. Labor shown is to uncrate factory assembled barrier and attach casters.

	Craft@Hrs	Unit	Material	Labor	Total
30" W x 75" H overall 0.5mm lead equiv. window panel 30" x 24" and 0.8mm lead equiv. opaque panel 30" x 48"	MW@.501	Ea	1,550.00	21.70	1,571.70
48" W x 75" H overall					
0.5mm lead equiv. window panel 48" x 36" and 0.8mm lead equiv. opaque panel 48" x 36"	MW@.501	Ea	2,550.00	21.70	2,571.70
1.0mm lead equiv. window panel 48" x 36" and 1.5mm lead equiv. opaque panel 48" x 36"	MW@.501	Ea	3,310.00	21.70	3,331.70
72" W x 75" H, overall					
0.5mm lead equiv. window panel 72" x 36" and 0.8mm lead equiv. opaque panel 72" x 36"	MW@.751	Ea	2,880.00	32.60	2,912.60
1.0mm lead equiv. window panel 72" x 36" and 1.5mm lead equiv. opaque panel 72" x 36"	MW@.751	Ea	3,600.00	32.60	3,632.60

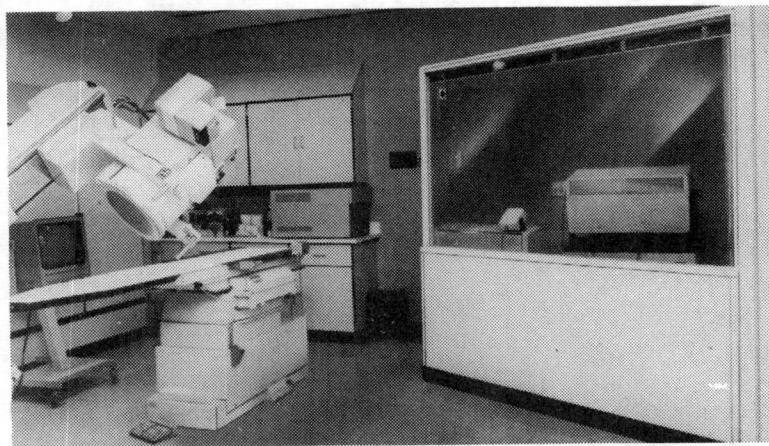

Modular X-Ray Barriers

Modular X-Ray Barriers. Panels are mounted within a framework for attaching to floor, wall or ceiling. Shipped unassembled. Based on Nuclear Associates. Structural supports not included. Costs shown are based on typical 36" wide x 84" high panel sections. Clear lead-plastic window panels 48" high are the upper portion and opaque leaded steel panels 36" high are the bottom portion.

Special Construction 13

	Craft@Hrs	Unit	Material	Labor	Total
Modular X-Ray Barriers, continued					
1-section barrier, 36" W x 84" H overall					
0.5mm lead equiv. panels	G1@2.78	Ea	2,440.00	108.00	2,548.00
0.8mm lead equiv. panels	G1@2.78	Ea	2,775.00	108.00	2,883.00
1.0mm lead equiv. panels	G1@3.34	Ea	2,635.50	130.00	2,765.50
1.5mm lead equiv. panels	G1@4.46	Ea	2,830.00	174.00	3,004.00
2-section barrier, 72" W x 84" H overall					
0.5mm lead equiv. panels	G1@5.56	Ea	5,255.00	216.00	5,471.00
0.8mm lead equiv. panels	G1@5.56	Ea	5,520.00	216.00	5,736.00
1.0mm lead equiv. panels	G1@6.68	Ea	5,635.00	260.00	5,895.00
1.5mm lead equiv. panels	G1@8.90	Ea	6,050.00	346.00	6,396.00
3-section barrier, 108" W x 84" H overall					
0.5mm lead equiv. panels	G1@8.34	Ea	8,015.00	325.00	8,340.00
0.8mm lead equiv. panels	G1@8.34	Ea	8,420.00	325.00	8,745.00
1.0mm lead equiv. panels	G1@10.0	Ea	8,600.00	389.00	8,989.00
1.5mm lead equiv. panels	G1@13.4	Ea	9,210.00	521.00	9,731.00
Larger than 3-section barriers, add to the cost of 1-section barriers for each section over 3					
Add per 18" W x 84" H section	--	%	50.0	--	--
Add per 36" W x 84" H section	--	%	100.0	--	--

Sound Isolation Rooms. Cost per SF of wall and ceiling. These costs include the subcontractor's overhead and profit.

Blast absorbing 4" perforated metal	--	SF	--	--	37.90
Sound deadening enclosure, 4"	--	SF	--	--	25.40

Swimming Pools. Complete gunite pools including filter, chlorinator, ladder, diving board, basic electrical system, but excluding deck beyond edge of coping. Typical cost per SF of water surface. These costs include the subcontractor's overhead and profit.

Apartment building	--	SF	--	--	32.00
Community	--	SF	--	--	41.50
Hotel or resort	--	SF	--	--	38.50
School, training (42' x 75')	--	SF	--	--	43.50
School, Olympic (42' x 165')	--	SF	--	--	64.50

Water Tanks, Elevated. Typical subcontract prices for 100' high steel tanks, including design, fabrication and erection. Complete tank includes 36" pipe riser with access manhole, 8" overflow to ground, ladder with protective cage, vent, balcony catwalk, concrete foundation, painting and test. No well, pump, fencing, drainage or water distribution piping included. Costs are per tank with gallon capacity as shown. Tanks built in areas without earthquake or high wind risk may cost 15% to 25% less.

80,000 gallon	--	Ea	--	--	260,000.00
100,000 gallon	--	Ea	--	--	270,000.00
150,000 gallon	--	Ea	--	--	323,000.00
200,000 gallon	--	Ea	--	--	446,000.00
300,000 gallon	--	Ea	--	--	611,000.00
400,000 gallon	--	Ea	--	--	690,000.00
Add or subtract for each foot of height more or less than 100, per 100 gallons of capacity	--	Ea	--	--	.57
Cathodic protection system, per tank	--	LS	--	--	12,500.00
Electric service, including float switches, obstruction marker lighting system, and hookup of cathodic protection system, per tank	--	LS	--	--	9,500.00

Special Construction 13

	Craft@Hrs	Unit	Material	Labor	Total

Solar Water Heating Systems. These costs include the subcontractor's overhead and profit.
Solar collector panels, no piping included

Typical 24 SF panel	P6@.747	Ea	291.00	30.50	321.50
Supports for solar panels, per panel	P6@2.25	Ea	30.40	92.00	122.40
Control panel for solar heating system	P6@1.56	Ea	103.00	63.80	166.80

Solar collector system pumps, no wiring or piping included

1.4 GPM, 1/20 HP	P6@1.50	Ea	193.00	61.30	254.30
5.4 GPM, 7.9' head	P6@1.89	Ea	199.00	77.20	276.20
10.5 GPM, 11.6' head	P6@2.25	Ea	358.00	92.00	450.00
21.6 GPM, 8.5' head	P6@2.25	Ea	358.00	92.00	450.00
12.6 GPM, 20.5' head	P6@3.26	Ea	552.00	133.00	685.00
18.6 GPM, 17.3' head	P6@3.26	Ea	552.00	133.00	685.00
22.5 GPM, 13.3' head	P6@3.26	Ea	552.00	133.00	685.00
46.5 GPM, 13.7' head	P6@4.59	Ea	827.00	188.00	1,015.00
52.8 GPM, 11.9' head	P6@4.10	Ea	717.00	168.00	885.00
53.4 GPM, 18.8' head	P6@4.10	Ea	728.00	168.00	896.00
60.9 GPM, 21.0' head	P6@5.25	Ea	965.00	215.00	1,180.00

Solar hot water storage tanks, no piping and no pad included

40 gallon, horizontal	P6@1.12	Ea	212.00	45.80	257.80
110 gallon, horizontal	P6@3.54	Ea	859.00	145.00	1,004.00
180 gallon, vertical	P6@5.43	Ea	1,440.00	222.00	1,662.00
650 gallon, horizontal	P6@5.84	Ea	2,120.00	239.00	2,359.00
1,250 gallon, horizontal	P6@5.84	Ea	3,080.00	239.00	3,319.00
1,500 gallon, vertical	P6@11.4	Ea	4,880.00	466.00	5,346.00
2,200 gallon, horizontal	P6@17.1	Ea	4,460.00	699.00	5,159.00
2,700 gallon, vertical	P6@14.9	Ea	5,090.00	609.00	5,699.00

Conveying Systems 14

	Craft@Hrs	Unit	Material	Labor	Total

Dumbwaiters. These costs include the subcontractor's overhead and profit but no allowance for constructing the hoistway walls or electrical work.
Manual, 2 stop, 25 feet per minute, no doors, by rated capacity, 24" x 24" x 36" high car

25 pounds	D8@39.8	Ea	1,370.00	1,670.00	3,040.00
50 pounds	D8@39.8	Ea	1,420.00	1,670.00	3,090.00
75 pounds to 200 pounds	D8@42.7	Ea	1,930.00	1,800.00	3,730.00
Add for each additional stop	--	Ea	--	--	1,280.00

Electric, with machinery mounted above, floor loading, no security gates included

25 lbs, 25 FPM, 2 stop, no doors	D8@37.7	Ea	1,550.00	1,590.00	3,140.00
25 lbs, 25 FPM, 2 stop, manual doors	D8@106.	Ea	3,710.00	4,460.00	8,170.00
25 lbs, 50 FPM, 2 stop, manual doors	D8@107.	Ea	3,820.00	4,500.00	8,320.00
50 lbs, 25 FPM, 2 stop, no doors	D8@42.2	Ea	1,700.00	1,770.00	3,470.00
50 lbs, 25 FPM, 2 stop, manual doors	D8@106.	Ea	4,220.00	4,460.00	8,680.00
50 lbs, 50 FPM, 2 stop, manual doors	D8@107.	Ea	4,220.00	4,500.00	8,720.00
75 lbs, 25 FPM, 2 stop, no doors	D8@42.2	Ea	1,700.00	1,770.00	3,470.00
75 lbs, 25 FPM, 2 stop, manual doors	D8@107.	Ea	5,090.00	4,500.00	9,590.00
75 lbs, 50 FPM, 2 stop, manual doors	D8@111.	Ea	5,090.00	4,670.00	9,760.00
100 lbs, 25 FPM, 2 stop, no doors	D8@50.6	Ea	2,470.00	2,130.00	4,600.00
100 lbs, 25 FPM, 2 stop, manual doors	D8@107.	Ea	5,240.00	4,500.00	9,740.00
100 lbs, 50 FPM, 2 stop, manual doors	D8@111.	Ea	5,350.00	4,670.00	10,020.00
100 lbs, 100 FPM, 5 stop, manual doors	D8@162.	Ea	5,600.00	6,810.00	12,410.00
200 lbs, 25 FPM, 2 stop, no doors	D8@52.3	Ea	2,540.00	2,200.00	4,740.00

Conveying Systems 14

	Craft@Hrs	Unit	Material	Labor	Total
Dumbwaiters, Electric, continued					
200 lbs, 25 FPM, 2 stop, manual doors	D8@111.	Ea	5,350.00	4,670.00	10,020.00
200 lbs, 100 FPM, 5 stop, manual doors	D8@162.	Ea	6,540.00	6,810.00	13,350.00
300 lbs, 50 FPM, 2 stop, manual doors	D8@113.	Ea	5,560.00	4,750.00	10,310.00
300 lbs, 100 FPM, 5 stop, manual doors	D8@171.	Ea	7,390.00	7,190.00	14,580.00
400 lbs, 50 FPM, 2 stop, manual doors	D8@114.	Ea	5,620.00	4,790.00	10,410.00
400 lbs, 100 FPM, 5 stop, manual doors	D8@174.	Ea	8,200.00	7,320.00	15,520.00
500 lbs, 50 FPM, 2 stop, manual doors	D8@117.	Ea	5,830.00	4,920.00	10,750.00
500 lbs, 100 FPM, 5 stop, manual doors	D8@178.	Ea	8,620.00	7,480.00	16,100.00

Elevators, Passenger. Typical costs including the subcontractor's overhead and profit but no allowance for the shaftwall.

	Craft@Hrs	Unit	Material	Labor	Total
Hydraulic, office type					
100 FPM, 10 to 13 passenger, automatic exit door, 7' x 5' cab, 2,500 lb capacity					
3 stop	--	LS	--	--	41,000.00
4 stop	--	LS	--	--	44,500.00
5 stop	--	LS	--	--	47,800.00
Each additional stop over 5	--	Ea	--	--	4,500.00
150 FPM, 10 to 13 passenger, center opening automatic door, 7' x 5' cab, 2,500 lb capacity					
3 stop	--	LS	--	--	43,600.00
4 stop	--	LS	--	--	46,200.00
5 stop	--	LS	--	--	50,300.00
6 stop	--	LS	--	--	55,000.00
Each additional stop over 6	--	Ea	--	--	5,550.00
Geared, 350 FPM, 3,500 lb automatic, selective collective, 5' x 8' cab					
5 stop	--	LS	--	--	74,600.00
6 stop	--	LS	--	--	78,800.00
10 stop	--	LS	--	--	95,800.00
15 stop	--	LS	--	--	127,900.00
Each additional stop	--	Ea	--	--	4,900.00
Add for 5,000 lb capacity	--	LS	--	--	2,600.00
Gearless, 500 FPM, 3,500 lb, 6' x 9' cab					
10 stop	--	LS	--	--	134,000.00
15 stop	--	LS	--	--	154,000.00
Add for each additional stop	--	Ea	--	--	4,900.00
Gearless, 700 FPM, 4,500 lb, 6' x 9' cab, passenger					
10 stop	--	LS	--	--	145,000.00
15 stop	--	LS	--	--	166,000.00
20 stop	--	LS	--	--	197,000.00
Add for each additional stop	--	Ea	--	--	5,100.00
Gearless, 1,000 FPM, 4,500 lb, 6' x 9' cab, passenger					
10 stop	--	LS	--	--	174,000.00
Add for each additional stop	--	Ea	--	--	6,700.00
Sidewalk elevators, 2 stops, 2,500 pounds	--	LS	--	--	31,200.00

Elevators, Freight. These costs include the subcontractor's overhead and profit but no allowance for the shaftwall.

	Craft@Hrs	Unit	Material	Labor	Total
Hydraulic, 2,500 pound capacity, 100 FPM, single entry cab, manual vertical doors, 2 stops	--	LS	--	--	33,000.00
Hydraulic, 6,000 pound capacity, 100 FPM, single entry cab, manual vertical doors, 5 stops	--	LS	--	--	48,400.00
Hydraulic, 6,000 pound capacity, 100 FPM, single entry cab, powered vertical doors, 5 stops	--	LS	--	--	57,100.00

Conveying Systems 14

	Craft@Hrs	Unit	Material	Labor	Total
Hydraulic, 10,000 pound capacity, 100 FPM, single entry cab					
Manual vertical doors, 5 stops	--	LS	--	--	54,800.00
Powered vertical doors, 5 stops	--	LS	--	--	61,300.00
Geared, 2,500 pound capacity, 100 FPM, single entry					
cab, manual doors, 2 stops	--	LS	--	--	53,000.00

Hoists and Cranes. These costs include the subcontractor's overhead and profit but no electrical work.
Electric hoists, manual trolley, 20 FPM lift. Add monorail costs below

	Craft@Hrs	Unit	Material	Labor	Total
1/2 ton, swivel mount, 25' lift	D4@9.34	Ea	3,580.00	365.00	3,945.00
1/2 ton, swivel mount, 45' lift	D4@13.7	Ea	3,780.00	535.00	4,315.00
1/2 ton, swivel mount, 75' lift	D4@18.0	Ea	4,200.00	703.00	4,903.00
1 ton, geared trolley, 20' lift	D4@11.9	Ea	4,250.00	465.00	4,715.00
1 ton, geared trolley, 30' lift	D4@15.8	Ea	4,500.00	617.00	5,117.00
1 ton, geared trolley, 55' lift	D4@19.2	Ea	4,920.00	750.00	5,670.00
2 ton, geared trolley, 15' lift	D4@15.8	Ea	5,220.00	617.00	5,837.00
2 ton, geared trolley, 35' lift	D4@15.8	Ea	5,530.00	617.00	6,147.00
2 ton, geared trolley, 55' lift	D4@15.8	Ea	6,040.00	617.00	6,657.00
2 ton, geared trolley, 70' lift	D4@15.8	Ea	6,350.00	617.00	6,967.00

Electric hoists, power trolley, 20 FPM trolley speed. Add monorail costs below

	Craft@Hrs	Unit	Material	Labor	Total
1/2 ton, 50 FPM lift to 25'	D4@8.94	Ea	4,160.00	349.00	4,509.00
1/2 ton, 100 FPM lift to 25'	D4@15.8	Ea	6,140.00	617.00	6,757.00
1/2 ton, 50 FPM lift to 45'	D4@8.94	Ea	4,400.00	349.00	4,749.00
1/2 ton, 100 FPM lift to 45'	D4@15.8	Ea	6,420.00	617.00	7,037.00
1/2 ton, 50 FPM lift to 75'	D4@11.9	Ea	4,820.00	465.00	5,285.00
1/2 ton, 100 FPM lift to 75'	D4@18.0	Ea	6,760.00	703.00	7,463.00
1 ton, 50 FPM lift to 20'	D4@15.8	Ea	4,250.00	617.00	4,867.00
1 ton, 50 FPM lift to 30'	D4@15.8	Ea	4,500.00	617.00	5,117.00
1 ton, 50 FPM lift to 55'	D4@18.0	Ea	4,920.00	703.00	5,623.00
2 ton, 50 FPM lift to 15'	D4@17.9	Ea	5,120.00	699.00	5,819.00
2 ton, 50 FPM lift to 35'	D4@19.9	Ea	5,640.00	777.00	6,417.00
2 ton, 50 FPM lift to 55'	D4@19.9	Ea	6,040.00	777.00	6,817.00
2 ton, 50 FPM lift to 70'	D4@19.9	Ea	6,350.00	777.00	7,127.00

Monorail for electric hoists, channel type

	Craft@Hrs	Unit	Material	Labor	Total
100 pounds per LF	D4@.443	LF	7.50	17.30	24.80
200 pounds per LF	D4@.555	LF	12.00	21.70	33.70
300 pounds per LF	D4@.761	LF	18.00	29.70	47.70

Jib cranes, self-supporting, swinging 8' boom, 220 degree rotation

	Craft@Hrs	Unit	Material	Labor	Total
1,000 pounds	D4@7.90	Ea	1,310.00	308.00	1,618.00
2,000 pounds	D4@11.9	Ea	1,390.00	465.00	1,855.00
3,000 pounds	D4@13.5	Ea	1,580.00	527.00	2,107.00
4,000 pounds	D4@13.5	Ea	1,860.00	527.00	2,387.00
6,000 pounds	D4@15.8	Ea	1,990.00	617.00	2,607.00
10,000 pounds	D4@19.7	Ea	2,690.00	769.00	3,459.00

Jib cranes, wall mounted, swinging 8' boom, 180 degree rotation

	Craft@Hrs	Unit	Material	Labor	Total
1,000 pounds	D4@8.19	Ea	690.00	320.00	1,010.00
2,000 pounds	D4@13.5	Ea	740.00	527.00	1,267.00
4,000 pounds	D4@13.5	Ea	1,140.00	527.00	1,667.00
6,000 pounds	D4@15.8	Ea	1,270.00	617.00	1,887.00
10,000 pounds	D4@19.7	Ea	2,430.00	769.00	3,199.00

Material Handling Systems. These costs include the subcontractor's overhead and profit.
Conveyors, typical subcontract price. Foundations, support structures or electrical work not included.

	Craft@Hrs	Unit	Material	Labor	Total
Belt type, 24" width	--	LF	--	--	136.00
Mail conveyors, automatic, electronic					

Conveying Systems 14

	Craft@Hrs	Unit	Material	Labor	Total
Material Handling Systems, continued					
Horizontal	--	LF	--	--	1,210.00
Vertical, per 12' floor	--	Ea	--	--	15,400.00

Chutes, linen or solid waste handling, prefabricated unit including roof vent, 1-1/2 hour "B" rated doors, discharge and sprinkler system, gravity feed. Costs shown are for each floor based on 8' to 10' floor to floor height.

	Craft@Hrs	Unit	Material	Labor	Total
Light duty, aluminum, 20" diameter	D4@5.85	Ea	585.00	228.00	813.00
Standard, 18 gauge steel, 24" diameter	D4@5.85	Ea	700.00	228.00	928.00
Standard, 18 gauge steel, 30" diameter	D4@5.85	Ea	840.00	228.00	1,068.00
Heavy duty, 18 gauge stainless steel					
24" diameter	D4@5.85	Ea	1,010.00	228.00	1,238.00
30" diameter	D4@5.85	Ea	1,090.00	228.00	1,318.00
Manual door with stainless steel rim	D4@2.17	Ea	295.00	84.70	379.70
Disinfecting and sanitizing unit	D4@2.17	Ea	140.00	84.70	224.70
Discharge storage unit					
Aluminum	D4@1.00	Ea	510.00	39.00	549.00
Stainless steel	D4@1.00	Ea	915.00	39.00	954.00

Turntables. These costs include the subcontractor's overhead and profit but no foundations or electrical work.

Baggage handling carousels

	Craft@Hrs	Unit	Material	Labor	Total
Round, 20' diameter	D4@208.	Ea	26,600.00	8,120.00	34,720.00
Round, 25' diameter	D4@271.	Ea	35,900.00	10,600.00	46,500.00

Moving Stairs and Walks. Typical subcontract prices including the subcontractor's overhead and profit.

Escalators, 90 FPM, steel trim, 32" step width, opaque balustrade

	Craft@Hrs	Unit	Material	Labor	Total
13' rise	--	Ea	--	--	91,400.00
14' rise	--	Ea	--	--	93,800.00
15' rise	--	Ea	--	--	95,400.00
16' rise	--	Ea	--	--	99,400.00
17' rise	--	Ea	--	--	101,500.00
18' rise	--	Ea	--	--	103,500.00
19' rise	--	Ea	--	--	105,600.00
20' rise	--	Ea	--	--	108,600.00
21' rise	--	Ea	--	--	121,000.00
Add for 48" width	--	%	--	--	12.0
Add for glass balustrade	--	%	--	--	15.0

Pneumatic Tube Systems, Twin Tubes. These costs include the subcontractor's overhead and profit.

	Craft@Hrs	Unit	Material	Labor	Total
3" diameter, two station,					
single connecting tube 100' long	--	LS	--	--	3,500.00
3" diameter, two station,					
twin tube, 100' between stations	--	LS	--	--	4,800.00

Mechanical 15

Threaded Schedule 40 Black Steel Pipe with Threaded Fittings and Hangers. These costs include the subcontractor's overhead and profit. Threaded black steel pipe with fittings and hangers as noted below installed in a building to 10' above floor level. Schedule 40 A53 or A120 pipe and 150 lb malleable iron fittings. Use these figures for preliminary estimates.

Mechanical 15

	Craft@Hrs	Unit	Material	Labor	Total
1/2" threaded pipe installed with a clevis hanger every 7' and					
With a coupling each 20'	M5@.082	LF	1.19	3.29	4.48
With a tee each 20	M5@.097	LF	1.24	3.89	5.13
With an ell each 5'	M5@.159	LF	1.34	6.37	7.71
3/4" threaded pipe installed on a clevis hanger every 7' and					
With a coupling each 20'	M5@.086	LF	1.34	3.45	4.79
With a tee each 20'	M5@.101	LF	1.41	4.05	5.46
With an ell each 5'	M5@.167	LF	1.50	6.69	8.19
1" threaded pipe installed on a clevis hanger every 7' and					
With a coupling each 20'	M5@.094	LF	1.88	3.77	5.65
With a tee each 20'	M5@.111	LF	2.00	4.45	6.45
With an ell each 5'	M5@.186	LF	2.20	7.46	9.66
1-1/2" threaded pipe installed on a clevis hanger every 9' and					
With a coupling each 20'	M5@.098	LF	2.47	3.92	6.39
With a tee each 20'	M5@.114	LF	2.71	4.57	7.28
With an ell each 5'	M5@.194	LF	3.15	7.77	10.92
2" threaded pipe installed on a clevis hanger every 10' and					
With a coupling each 20'	M5@.111	LF	3.18	4.45	7.63
With a tee each 20'	M5@.136	LF	3.54	5.45	8.99
With an ell each 5'	M5@.240	LF	4.18	9.61	13.79
Add for installation heights over 10' to 20'	--	%	--	25.0	--

Black Steel Pipe with Welded Fittings and Hangers. These costs include the subcontractor's overhead and profit. Butt welded black steel pipe with welded fittings and hangers as noted below, installed in a building to 10' above floor level. Schedule 40 pipe and carbon steel fittings. Use these figures for preliminary estimates.

	Craft@Hrs	Unit	Material	Labor	Total
2-1/2" A120 pipe installed with a clevis hanger each 10' and					
With a weld each 20'	M7@.138	LF	4.60	5.80	10.40
With a tee each 20'	M7@.253	LF	5.37	10.60	15.97
With an ell each 5'	M7@.546	LF	6.02	23.00	29.02
3" A120 pipe installed with a clevis hanger each 12' and					
With a weld each 20'	M7@.158	LF	10.20	6.64	16.84
With a tee each 20'	M7@.278	LF	11.40	11.70	23.10
With an ell each 5'	M7@.620	LF	10.60	26.10	36.70
4" A120 pipe installed with a clevis hanger each 14' and					
With a weld each 20'	M7@.191	LF	10.50	8.03	18.53
With a tee each 20'	M7@.355	LF	13.00	14.90	27.90
With an ell each 5'	M7@.811	LF	15.60	34.10	49.70
6" A53 electric resistance welded pipe installed with a clevis hanger each 17' and					
With a weld each 20'	M7@.266	LF	20.50	11.20	31.70
With a tee each 20'	M7@.501	LF	23.80	21.10	44.90
With an ell and a hanger each 10'	M7@1.15	LF	25.90	48.40	74.30
8" A53 electric resistance welded pipe installed with a clevis hanger each 19' and					
With a weld each 20'	M7@.330	LF	30.30	13.90	44.20
With a tee each 20'	M7@.638	LF	36.30	26.80	63.10
With an ell each 5'	M7@1.46	LF	51.40	61.40	112.80
Add for installation heights over 10' to 20'	--	%	--	25.0	--

Black Steel Pipe with Fittings and Supports Priced Separately. Use these figures when final plans are available. Pipe 1-1/2" diameter and smaller is welded seam. Pipe 2" diameter and larger is welded seam or seamless. Installation heights indicate height above the working floor. These costs include the subcontractor's overhead and profit.

Mechanical 15

	Craft@Hrs	Unit	Material	Labor	Total

1/2" black steel (A53 or A120) threaded pipe and fittings, installed in a building
Pipe, no fittings or supports included, 1/2", threaded ends

	Craft@Hrs	Unit	Material	Labor	Total
Schedule 40 to 10' high	M5@.038	LF	.77	1.52	2.29
Schedule 40 over 10' to 20'	M5@.049	LF	.77	1.96	2.73
Schedule 80 to 10' high	M5@.038	LF	1.18	1.52	2.70
Schedule 80 over 10' to 20'	M5@.049	LF	1.18	1.96	3.14

90 degree ells, 1/2", threaded except as noted

	Craft@Hrs	Unit	Material	Labor	Total
150 lb, malleable iron	M5@.489	Ea	.75	19.60	20.35
300 lb, malleable iron	M5@.580	Ea	4.30	23.20	27.50
2,000 lb, forged steel	M5@.494	Ea	6.60	19.80	26.40
3,000 lb, forged steel	M5@.520	Ea	7.15	20.80	27.95
3,000 lb, socket welded, forged steel	M5@.557	Ea	7.55	22.30	29.85
6,000 lb, forged steel	M5@.526	Ea	12.45	21.10	33.55

45 degree ells, 1/2", threaded

	Craft@Hrs	Unit	Material	Labor	Total
150 lb, malleable iron	M5@.489	Ea	1.15	19.60	20.75
300 lb, malleable iron	M5@.580	Ea	5.75	23.20	28.95
2,000 lb, forged steel	M5@.494	Ea	7.80	19.80	27.60
3,000 lb, forged steel	M5@.520	Ea	9.20	20.80	30.00
3,000 lb, socket welded, forged steel	M5@.557	Ea	7.55	22.30	29.85
6,000 lb, forged steel	M5@.526	Ea	16.10	21.10	37.20

Tees, 1/2"

	Craft@Hrs	Unit	Material	Labor	Total
150 lb, malleable iron	M5@.729	Ea	.95	29.20	30.15
300 lb, malleable iron	M5@.840	Ea	6.20	33.70	39.90
2,000 lb, forged steel	M5@.729	Ea	8.50	29.20	37.70
3,000 lb, forged steel	M5@.729	Ea	9.35	29.20	38.55
3,000 lb, socket welded, forged steel	M5@.812	Ea	8.75	32.50	41.25
6,000 lb, forged steel	M5@.755	Ea	15.70	30.20	45.90
150 lb, reducing, malleable iron	M5@.575	Ea	1.30	23.00	24.30
300 lb, reducing, malleable iron	M5@.575	Ea	6.95	23.00	29.95

Caps, 1/2"

	Craft@Hrs	Unit	Material	Labor	Total
150 lb, malleable iron	M5@.393	Ea	.75	15.70	16.45
300 lb, malleable iron	M5@.468	Ea	3.00	18.70	21.70

Couplings, 1/2"

	Craft@Hrs	Unit	Material	Labor	Total
150 lb, malleable iron	M5@.432	Ea	1.00	17.30	18.30
300 lb, malleable iron	M5@.526	Ea	3.40	21.10	24.50
3,000 lb, forged steel	M5@.494	Ea	2.65	19.80	22.45
6,000 lb, forged steel	M5@.520	Ea	5.10	20.80	25.90

Unions, 1/2"

	Craft@Hrs	Unit	Material	Labor	Total
150 lb, malleable iron	M5@.393	Ea	4.90	15.70	20.60
300 lb, malleable iron	M5@.440	Ea	8.00	17.60	25.60
Threadolets, 3,000 lb forged steel, 1/2"	M5@1.46	Ea	3.95	58.50	62.45

Welding flanges, forged steel, 1/2"

	Craft@Hrs	Unit	Material	Labor	Total
150 lb, slip-on flange	M5@.731	Ea	6.40	29.30	35.70
300 lb, slip-on flange	M5@.973	Ea	7.80	39.00	46.80
600 lb, slip-on flange	M5@1.08	Ea	10.10	43.30	53.40
150 lb, neck flange	M5@.731	Ea	8.90	29.30	38.20
300 lb, neck flange	M5@.731	Ea	9.40	29.30	38.70
600 lb, neck flange	M5@.973	Ea	13.10	39.00	52.10
150 lb, threaded flange	M5@.489	Ea	7.85	19.60	27.45
300 lb, threaded flange	M5@.781	Ea	9.50	31.30	40.80
600 lb, threaded flange	M5@.859	Ea	13.50	34.40	47.90
For galvanized pipe, add	--	%	32.0	--	--
Galvanized 150 lb. malleable fittings, add	--	%	15.0	--	--
Galvanized 300 lb. malleable fittings, add	--	%	55.0	--	--

Mechanical 15

	Craft@Hrs	Unit	Material	Labor	Total

3/4" black steel (A53 or A120) threaded pipe and fittings, installed in a building
Pipe, no fittings or supports included, 3/4", threaded ends

	Craft@Hrs	Unit	Material	Labor	Total
Schedule 40 to 10' high	M5@.038	LF	.92	1.52	2.44
Schedule 40 over 10' to 20'	M5@.049	LF	.92	1.96	2.88
Schedule 80 to 10' high	M5@.038	LF	1.50	1.52	3.02
Schedule 80 over 10' to 20'	M5@.049	LF	1.50	1.96	3.46
90 degree ells, 3/4"					
150 lb, malleable iron	M5@.533	Ea	.81	21.40	22.21
300 lb, malleable iron	M5@.630	Ea	4.60	25.20	29.80
2,000 lb, forged steel	M5@.690	Ea	7.10	27.60	34.70
3,000 lb, forged steel	M5@.690	Ea	9.00	27.60	36.60
3,000 lb, socket welded, forged steel	M5@.742	Ea	7.70	29.70	37.40
6,000 lb, forged steel	M5@.723	Ea	15.60	29.00	44.60
45 degree ells, 3/4"					
150 lb, malleable iron	M5@.541	Ea	1.45	21.70	23.15
300 lb, malleable iron	M5@.630	Ea	6.35	25.20	31.55
2,000 lb, forged steel	M5@.690	Ea	8.90	27.60	36.50
3,000 lb, forged steel	M5@.690	Ea	10.50	27.60	38.10
3,000 lb, socket welded, forged steel	M5@.744	Ea	9.10	29.80	38.90
6,000 lb, forged steel	M5@.723	Ea	22.50	29.00	51.50
Tees, 3/4"					
150 lb, malleable iron	M5@.799	Ea	1.40	32.00	33.40
300 lb, malleable iron	M5@.898	Ea	6.80	36.00	42.80
2,000 lb, forged steel	M5@.999	Ea	11.60	40.00	51.60
3,000 lb, forged steel	M5@.986	Ea	14.40	39.50	53.90
3,000 lb, socket welded, forged steel	M5@1.21	Ea	18.20	48.50	66.70
6,000 lb, forged steel	M5@1.04	Ea	21.30	41.70	63.00
150 lb, reducing, malleable iron	M5@.643	Ea	2.30	25.80	28.10
300 lb, reducing, malleable iron	M5@.643	Ea	7.80	25.80	33.60
Caps, 3/4"					
150 lb, malleable iron	M5@.440	Ea	1.00	17.60	18.60
300 lb, malleable iron	M5@.513	Ea	3.90	20.60	24.50
Couplings, 3/4"					
150 lb, malleable iron	M5@.489	Ea	1.20	19.60	20.80
300 lb, malleable iron	M5@.583	Ea	4.00	23.40	27.40
3,000 lb, forged steel	M5@.671	Ea	3.65	26.90	30.55
6,000 lb, forged steel	M5@.690	Ea	5.50	27.60	33.10
Unions, 3/4"					
150 lb, malleable iron	M5@.435	Ea	5.58	17.40	22.98
300 lb, malleable iron	M5@.474	Ea	8.60	19.00	27.60
Threadolets, carbon steel, 3/4"	M5@1.66	Ea	4.95	66.50	71.45
Welding flanges, forged steel, 3/4"					
150 lb, slip-on flange	M5@.731	Ea	6.40	29.30	35.70
300 lb, slip-on flange	M5@.973	Ea	7.80	39.00	46.80
600 lb, slip-on flange	M5@1.08	Ea	10.10	43.30	53.40
150 lb, neck flange	M5@.731	Ea	9.15	29.30	38.45
300 lb, neck flange	M5@.731	Ea	9.75	29.30	39.05
600 lb, neck flange	M5@.973	Ea	13.10	39.00	52.10
150 lb, threaded flange	M5@.585	Ea	8.20	23.40	31.60
300 lb, threaded flange	M5@.781	Ea	10.75	31.30	42.05
600 lb, threaded flange	M5@.859	Ea	13.50	34.40	47.90
For galvanized pipe, add	--	%	32.0	--	--
Galvanized 150 lb. malleable fittings, add	--	%	25.0	--	--
Galvanized 300 lb. malleable fittings, add	--	%	55.0	--	--

Mechanical 15

	Craft@Hrs	Unit	Material	Labor	Total

1" black steel (A53 or A120) threaded pipe and fittings, installed in a building
Pipe, no fittings or supports included, 1", threaded ends

	Craft@Hrs	Unit	Material	Labor	Total
Schedule 40 to 10' high	M5@.042	LF	1.30	1.68	2.98
Schedule 40 over 10' to 20'	M5@.053	LF	1.30	2.12	3.42
Schedule 80 to 10' high	M5@.040	LF	2.00	1.60	3.60
Schedule 80 over 10' to 20'	M5@.057	LF	2.00	2.28	4.28
90 degree ells, 1"					
150 lb, malleable iron	M5@.679	Ea	1.60	27.20	28.80
300 lb, malleable iron	M5@.734	Ea	6.30	29.40	35.70
2,000 lb, forged steel	M5@.903	Ea	9.20	36.20	45.40
3,000 lb, forged steel	M5@.911	Ea	13.30	36.50	49.80
3,000 lb, socket welded, forged steel	M5@1.15	Ea	10.10	46.10	56.20
6,000 lb, forged steel	M5@.952	Ea	20.30	38.10	58.40
Schedule 40, butt welded long turn	M5@.679	Ea	6.25	27.20	33.45
Schedule 80, butt welded long turn	M5@1.57	Ea	7.60	62.90	70.50
45 degree ells, 1"					
150 lb, malleable iron	M5@.679	Ea	1.80	27.20	29.00
300 lb, malleable iron	M5@.734	Ea	7.10	29.40	36.50
2,000 lb, forged steel	M5@.903	Ea	10.40	36.20	46.60
3,000 lb, forged steel	M5@.903	Ea	14.60	36.20	50.80
3,000 lb, socket welded, forged steel	M5@1.15	Ea	11.90	46.10	58.00
6,000 lb, forged steel	M5@.952	Ea	22.80	38.10	60.90
Schedule 40, carbon steel, butt welded	M5@1.05	Ea	4.40	42.10	46.50
Schedule 80, carbon steel, butt welded	M5@1.57	Ea	5.30	62.90	68.20
Tees, 1"					
150 lb, malleable iron	M5@1.02	Ea	2.40	40.90	43.30
300 lb, malleable iron	M5@1.15	Ea	7.40	46.10	53.50
2,000 lb, forged steel	M5@1.32	Ea	15.85	52.90	68.75
3,000 lb, forged steel	M5@1.35	Ea	18.30	54.10	72.40
3,000 lb, socket welded, forged steel	M5@1.71	Ea	15.00	68.50	83.50
6,000 lb, forged steel	M5@1.39	Ea	35.20	55.70	90.90
Schedule 40, carbon steel, butt welded	M5@1.57	Ea	11.70	62.90	74.60
Schedule 80, carbon steel, butt welded	M5@2.62	Ea	15.10	105.00	120.10
150 lb, reducing, malleable iron	M5@.765	Ea	2.70	30.60	33.30
300 lb, reducing, malleable iron	M5@.783	Ea	10.90	31.40	42.30
Caps, 1"					
150 lb, malleable iron	M5@.544	Ea	1.20	21.80	23.00
300 lb, malleable iron	M5@.645	Ea	5.25	25.80	31.05
Schedule 40, carbon steel, butt welded	M5@.687	Ea	2.95	27.50	30.45
Schedule 80, carbon steel, butt welded	M5@1.05	Ea	3.20	42.10	45.30
Couplings, 1"					
150 lb, malleable iron	M5@.604	Ea	1.80	24.20	26.00
300 lb, malleable iron	M5@.716	Ea	4.60	28.70	33.30
3,000 lb, forged steel	M5@.903	Ea	5.50	36.20	41.70
6,000 lb, forged steel	M5@.911	Ea	8.00	36.50	44.50
Unions, 1"					
150 lb, malleable iron	M5@.544	Ea	7.25	21.80	29.05
300 lb, malleable iron	M5@.598	Ea	10.55	24.00	34.55
Threadolets, carbon steel, 1"	M5@1.99	Ea	5.20	79.70	84.90
Welding flanges, forged steel, 1"					
150 lb, slip-on flange	M5@.788	Ea	6.40	31.60	38.00
300 lb, slip-on flange	M5@.919	Ea	8.50	36.80	45.30
600 lb, slip-on flange	M5@1.23	Ea	10.40	49.30	59.70
150 lb, neck flange	M5@.788	Ea	9.15	31.60	40.75

Mechanical 15

	Craft@Hrs	Unit	Material	Labor	Total
300 lb, neck flange	M5@1.05	Ea	10.00	42.10	52.10
600 lb, neck flange	M5@1.31	Ea	13.70	52.50	66.20
150 lb, threaded flange	M5@.731	Ea	8.80	29.30	38.10
300 lb; threaded flange	M5@1.05	Ea	11.30	42.10	53.40
600 lb, threaded flange	M5@1.15	Ea	14.10	46.10	60.20
For galvanized pipe, add	--	%	32.0	--	--
Galvanized 150 lb. malleable fittings, add	--	%	20.0	--	--
Galvanized 300 lb. malleable fittings, add	--	%	60.0	--	--

1-1/4" black steel (A53 or A120) threaded pipe and fittings, installed in a building

Pipe, no fittings or supports included, 1-1/4", threaded ends

	Craft@Hrs	Unit	Material	Labor	Total
Schedule 40 to 10' high	M5@.046	LF	1.67	1.84	3.51
Schedule 40 over 10' to 20'	M5@.057	LF	1.67	2.28	3.95
Schedule 80 to 10' high	M5@.049	LF	2.74	1.96	4.70
Schedule 80 over 10' to 20'	M5@.062	LF	2.74	2.48	5.22

90 degree ells, 1-1/4"

	Craft@Hrs	Unit	Material	Labor	Total
150 lb, malleable iron	M5@.734	Ea	2.60	29.40	32.00
300 lb, malleable iron	M5@.843	Ea	9.00	33.80	42.80
2,000 lb, forged steel	M5@1.04	Ea	15.25	41.70	56.95
3,000 lb, forged steel	M5@1.04	Ea	22.50	41.70	64.20
3,000 lb, socket welded, forged steel	M5@1.44	Ea	15.90	57.70	73.60
6,000 lb, forged steel	M5@1.10	Ea	34.10	44.10	78.20
Schedule 40 carbon steel, butt weld, long turn	M5@1.57	Ea	4.40	62.90	67.30
Schedule 80 carbon steel, but weld, long turn	M5@2.10	Ea	6.10	84.10	90.20

45 degree ells, 1-1/4"

	Craft@Hrs	Unit	Material	Labor	Total
150 lb, malleable iron	M5@.734	Ea	3.30	29.40	32.70
300 lb, malleable iron	M5@.843	Ea	11.30	33.80	45.10
2,000 lb, forged steel	M5@1.04	Ea	17.00	41.70	58.70
3,000 lb, forged steel	M5@1.04	Ea	20.95	41.70	62.65
3,000 lb, socket welded, forged steel	M5@1.44	Ea	16.65	57.70	74.35
6,000 lb, forged steel	M5@1.10	Ea	67.30	44.10	111.40
Schedule 40, carbon steel, butt welded	M5@1.57	Ea	4.40	62.90	67.30
Schedule 80, carbon steel, butt welded	M5@2.10	Ea	6.10	84.10	90.20

Tees, 1-1/4"

	Craft@Hrs	Unit	Material	Labor	Total
150 lb, malleable iron	M5@1.11	Ea	3.95	44.50	48.45
300 lb, malleable iron	M5@1.24	Ea	10.90	49.70	60.60
2,000 lb, forged steel	M5@1.54	Ea	23.45	61.70	85.15
3,000 lb, forged steel	M5@1.57	Ea	30.40	62.90	93.30
3,000 lb, socket welded, forged steel	M5@2.13	Ea	20.40	85.30	105.70
6,000 lb, forged steel	M5@1.63	Ea	44.94	65.30	110.24
Schedule 40, carbon steel, butt welded	M5@2.10	Ea	14.00	84.10	98.10
Schedule 80, carbon steel, butt welded	M5@2.76	Ea	15.10	111.00	126.10
150 lb, reducing, malleable iron	M5@.864	Ea	4.30	34.60	38.90
300 lb, reducing, malleable iron	M5@.911	Ea	14.30	36.50	50.80

Caps, 1-1/4"

	Craft@Hrs	Unit	Material	Labor	Total
150 lb, malleable iron	M5@.598	Ea	1.60	24.00	25.60
300 lb, malleable iron	M5@.697	Ea	5.50	27.90	33.40
Schedule 40, carbon steel, butt welded	M5@1.05	Ea	4.20	42.10	46.30
Schedule 80, carbon steel, butt welded	M5@1.57	Ea	4.50	62.90	67.40

Couplings, 1-1/4"

	Craft@Hrs	Unit	Material	Labor	Total
150 lb, malleable iron	M5@.664	Ea	2.30	26.60	28.90
300 lb, malleable iron	M5@.770	Ea	5.45	30.80	36.25
3,000 lb, forged steel	M5@1.04	Ea	10.25	41.70	51.95
6,000 lb, forged steel	M5@1.04	Ea	15.80	41.70	57.50

Mechanical 15

	Craft@Hrs	Unit	Material	Labor	Total

1-1/4" black steel (A53 or A120) threaded pipe and fittings, continued

Unions, 1-1/4"
150 lb, malleable iron	M5@.588	Ea	10.00	23.60	33.60
300 lb, malleable iron	M5@.643	Ea	16.20	25.80	42.00
Threadolets, carbon steel, 1-1/4"	M5@2.19	Ea	7.30	87.70	95.00

Welding flanges, forged steel, 1-1/4"
150 lb, slip-on flange	M5@.788	Ea	6.40	31.60	38.00
300 lb, slip-on flange	M5@1.15	Ea	9.40	46.10	55.50
150 lb, neck flange	M5@1.04	Ea	9.50	41.70	51.20
300 lb, neck flange	M5@1.04	Ea	11.20	41.70	52.90
150 lb, threaded flange	M5@.788	Ea	8.80	31.60	40.40
300 lb, threaded flange	M5@1.05	Ea	13.50	42.10	55.60
For galvanized pipe, add	--	%	32.0	--	--
Galvanized 150 lb. malleable fittings, add	--	%	20.0	--	--
Galvanized 300 lb. malleable fittings, add	--	%	65.0	--	--

1-1/2" black steel (A53 or A120) threaded pipe and fittings, installed in a building

Pipe, no fittings or supports included, 1-1/2", threaded ends
Schedule 40 to 10' high	M5@.047	LF	1.98	1.88	3.86
Schedule 40 over 10' to 20'	M5@.063	LF	1.98	2.52	4.50
Schedule 80 to 10' high	M5@.051	LF	3.30	2.04	5.34
Schedule 80 over 10' to 20'	M5@.066	LF	3.30	2.64	5.94

90 degree ells, 1-1/2"
150 lb, malleable iron	M5@.788	Ea	3.40	31.60	35.00
300 lb, malleable iron	M5@.900	Ea	10.60	36.10	46.70
2,000 lb, forged steel	M5@1.13	Ea	23.70	45.30	69.00
3,000 lb, forged steel	M5@1.17	Ea	33.45	46.90	80.35
3,000 lb, socket welded, forged steel	M5@1.72	Ea	21.10	68.90	90.00
6,000 lb, forged steel	M5@1.22	Ea	53.20	48.90	102.10
Sch. 40 carbon steel, butt weld, long turn	M5@2.10	Ea	4.40	84.10	88.50
Sch. 80 carbon steel, butt weld, long turn	M5@2.63	Ea	6.70	105.00	111.70

45 degree ells, 1-1/2"
150 lb, malleable iron	M5@.788	Ea	4.10	31.60	35.70
300 lb, malleable iron	M5@.903	Ea	14.70	36.20	50.90
2,000 lb, forged steel	M5@1.17	Ea	19.80	46.90	66.70
3,000 lb, forged steel	M5@1.71	Ea	32.00	68.50	100.50
3,000 lb, socket welded, forged steel	M5@1.72	Ea	20.60	68.90	89.50
6,000 lb, forged steel	M5@1.21	Ea	67.30	48.50	115.80
Schedule 40, carbon steel, butt welded	M5@2.10	Ea	4.40	84.10	88.50
Schedule 80, carbon steel, butt welded	M5@2.63	Ea	6.70	105.00	111.70

Tees, 1-1/2"
150 lb, malleable iron	M5@1.17	Ea	4.90	46.90	51.80
300 lb, malleable iron	M5@1.32	Ea	13.40	52.90	66.30
2,000 lb, forged steel	M5@1.69	Ea	33.30	67.70	101.00
3,000 lb, forged steel	M5@1.75	Ea	41.70	70.10	111.80
3,000 lb, socket welded, forged steel	M5@2.52	Ea	31.80	101.00	132.80
6,000 lb, forged steel	M5@1.78	Ea	61.00	71.30	132.30
Schedule 40, carbon steel, butt welded	M5@2.63	Ea	17.40	105.00	122.40
Schedule 80, carbon steel, butt welded	M5@4.16	Ea	17.20	167.00	184.20
150 lb, reducing, malleable iron	M5@.960	Ea	5.45	38.50	43.95
300 lb, reducing, malleable iron	M5@.997	Ea	17.35	39.90	57.25

Caps, 1-1/2"
150 lb, malleable iron	M5@.643	Ea	2.20	25.80	28.00
300 lb, malleable iron	M5@.752	Ea	8.50	30.10	38.60

Mechanical 15

	Craft@Hrs	Unit	Material	Labor	Total
Schedule 40, carbon steel, butt welded	M5@1.05	Ea	4.20	42.10	46.30
Schedule 80, carbon steel, butt welded	M5@1.57	Ea	4.50	62.90	67.40
Couplings, 1-1/2"					
150 lb, malleable iron	M5@.721	Ea	3.10	28.90	32.00
300 lb, malleable iron	M5@.822	Ea	8.40	32.90	41.30
3,000 lb, forged steel	M5@1.13	Ea	13.80	45.30	59.10
6,000 lb, forged steel	M5@1.17	Ea	17.60	46.90	64.50
Unions, 1-1/2"					
150 lb, malleable iron	M5@.627	Ea	12.20	25.10	37.30
300 lb, malleable iron	M5@.679	Ea	19.90	27.20	47.10
Threadolets, carbon steel, 1-1/2"	M5@2.43	Ea	8.00	97.30	105.30
Welding flanges, forged steel, 1-1/2"					
150 lb, slip-on flange	M5@.960	Ea	6.90	38.50	45.40
300 lb, slip-on flange	M5@1.15	Ea	9.75	46.10	55.85
600 lb, slip-on flange	M5@1.26	Ea	12.80	50.50	63.30
150 lb, neck flange	M5@1.40	Ea	9.50	56.10	65.60
300 lb, neck flange	M5@1.04	Ea	11.60	41.70	53.30
600 lb, neck flange	M5@1.31	Ea	16.20	52.50	68.70
150 lb, threaded flange	M5@.840	Ea	11.90	33.70	45.60
300 lb, threaded flange	M5@1.15	Ea	13.50	46.10	59.60
600 lb, threaded flange	M5@1.26	Ea	16.30	50.50	66.80
For galvanized pipe, add	--	%	32.0	--	--
Galvanized 150 lb. malleable fittings, add	--	%	20.0	--	--
Galvanized 300 lb. malleable fittings, add	--	%	55.0	--	--

2" black steel (A53 or A120) threaded pipe and fittings, installed in a building
Pipe, no fittings or supports included, 2", threaded ends

	Craft@Hrs	Unit	Material	Labor	Total
Schedule 40 to 10' high	M7@.053	LF	2.70	2.23	4.93
Schedule 40 over 10' to 20'	M7@.069	LF	2.70	2.90	5.60
Schedule 80 to 10' high	M7@.057	LF	4.00	2.40	6.40
Schedule 80 over 10' to 20'	M7@.073	LF	4.00	3.07	7.07
90 degree ells, 2"					
150 lb, malleable iron	M7@.801	Ea	5.00	33.70	38.70
300 lb, malleable iron	M7@.891	Ea	14.80	37.50	52.30
2,000 lb, forged steel	M7@1.32	Ea	29.90	55.50	85.40
3,000 lb, forged steel	M7@1.34	Ea	40.70	56.30	97.00
3,000 lb, socket welded, forged steel	M7@2.20	Ea	32.10	92.50	124.60
6,000 lb, forged steel	M7@1.44	Ea	93.15	60.60	153.75
Schedule 40, carbon steel, butt weld, long turn	M7@2.00	Ea	4.60	84.10	88.70
Schedule 80, carbon steel, butt weld, long turn	M7@3.23	Ea	6.00	136.00	142.00
45 degree ells, 2"					
150 lb, malleable iron	M7@.767	Ea	5.30	32.30	37.60
300 lb, malleable iron	M7@.911	Ea	21.75	38.30	60.05
2,000 lb, forged steel	M7@1.32	Ea	35.90	55.50	91.40
3,000 lb, forged steel	M7@1.34	Ea	44.05	56.30	100.35
3,000 lb, socket welded, forged steel	M7@2.20	Ea	35.90	92.50	128.40
6,000 lb, forged steel	M7@1.44	Ea	94.10	60.60	154.70
Schedule 40, carbon steel, butt welded	M7@2.00	Ea	4.40	84.10	88.50
Schedule 80, carbon steel, butt welded	M7@3.23	Ea	5.70	136.00	141.70
Tees, 2"					
150 lb, malleable iron	M7@1.20	Ea	7.15	50.50	57.65
300 lb, malleable iron	M7@1.35	Ea	19.80	56.80	76.60
2,000 lb, forged steel	M7@1.94	Ea	41.30	81.60	122.90

Mechanical 15

	Craft@Hrs	Unit	Material	Labor	Total
2" black steel (A53 or A120) threaded pipe and fittings, Tees, continued					
3,000 lb, forged steel	M7@2.00	Ea	55.20	84.10	139.30
3,000 lb, socket welded, forged steel	M7@3.23	Ea	44.60	136.00	180.60
6,000 lb, forged steel	M7@2.17	Ea	116.60	91.20	207.80
Schedule 40, carbon steel, butt welded	M7@2.95	Ea	12.30	124.00	136.30
Schedule 80, carbon steel, butt welded	M7@4.79	Ea	14.60	201.00	215.60
150 lb, reducing, malleable iron	M7@1.03	Ea	8.50	43.30	51.80
300 lb, reducing, malleable iron	M7@1.07	Ea	26.00	45.00	71.00
Caps, 2"					
150 lb, malleable iron	M7@.648	Ea	2.85	27.20	30.05
300 lb, malleable iron	M7@.757	Ea	11.80	31.80	43.60
Schedule 40, carbon steel, butt welded	M7@.995	Ea	3.80	41.80	45.60
Schedule 80, carbon steel, butt welded	M7@1.60	Ea	4.20	67.30	71.50
Couplings, 2"					
150 lb, malleable iron	M7@.722	Ea	4.50	30.40	34.90
300 lb, malleable iron	M7@.821	Ea	11.70	34.50	46.20
3,000 lb, forged steel	M7@1.32	Ea	17.50	55.50	73.00
6,000 lb, forged steel	M7@.846	Ea	31.30	35.60	66.90
Unions, 2"					
150 lb, malleable iron	M7@.685	Ea	15.35	28.80	44.15
300 lb, malleable iron	M7@.730	Ea	25.00	30.70	55.70
Threadolets, carbon steel, 2"	M7@2.80	Ea	9.20	118.00	127.20
Welding flanges, forged steel, 2"					
150 lb, slip-on flange	M7@.903	Ea	6.40	38.00	44.40
300 lb, slip-on flange	M7@1.40	Ea	9.70	58.90	68.60
600 lb, slip-on flange	M7@1.77	Ea	12.80	74.40	87.20
150 lb, neck flange	M7@.995	Ea	9.50	41.80	51.30
300 lb, neck flange	M7@.993	Ea	11.60	41.80	53.40
600 lb, neck flange	M7@1.60	Ea	15.10	67.30	82.40
150 lb, threaded flange	M7@.846	Ea	12.30	35.60	47.90
300 lb, threaded flange	M7@1.25	Ea	13.50	52.60	66.10
600 lb, threaded flange	M7@1.43	Ea	17.10	60.10	77.20
Butt welded joints, 2" pipe					
Schedule 40 pipe	M7@.990	Ea	--	41.60	--
Schedule 80 pipe	M7@1.60	Ea	--	67.30	--
For galvanized pipe, add	--	%	32.0	--	--
Galvanized 150 lb. malleable fittings, add	--	%	15.0	--	--
Galvanized 300 lb. malleable fittings, add	--	%	60.0	--	--
2-1/2" black steel pipe (A53 or A120) and welded fittings, installed in a building					
Pipe, no fittings or supports included, 2-1/2", plain ends					
Schedule 40 to 10' high	M7@.076	LF	3.70	3.20	6.90
Schedule 40 over 10' to 20' high	M7@.090	LF	3.70	3.79	7.49
Schedule 80 to 10' high	M7@.086	LF	5.10	3.62	8.72
Schedule 80 over 10' to 20' high	M7@.103	LF	5.10	4.33	9.43
90 degree ells, 2-1/2"					
Schedule 40, carbon steel, butt weld, long turn	M7@3.65	Ea	7.15	153.00	160.15
Schedule 80, carbon steel, butt weld, long turn	M7@3.95	Ea	9.70	166.00	175.70
45 degree ells, 2-1/2"					
Schedule 40, carbon steel, butt welded	M7@3.65	Ea	6.80	153.00	159.80
Schedule 80, carbon steel, butt welded	M7@3.95	Ea	9.20	166.00	175.20
Tees, 2-1/2"					
Schedule 40, carbon steel, butt welded	M7@3.65	Ea	15.40	153.00	168.40
Schedule 80, carbon steel, butt welded	M7@3.95	Ea	19.50	166.00	185.50

Mechanical 15

	Craft@Hrs	Unit	Material	Labor	Total
Reducing tee, Sch. 40, carbon steel, welded	M7@2.25	Ea	20.80	94.60	115.40
Caps, 2-1/2"					
Schedule 40, carbon steel, butt welded	M7@1.20	Ea	4.10	50.50	54.60
Schedule 80, carbon steel, butt welded	M7@1.20	Ea	4.50	50.50	55.00
Reducers, Schedule 40, carbon steel, butt weld	M7@1.44	Ea	8.75	60.60	69.35
Bolt-ups, 2-1/2"	M7@.469	Ea	4.40	19.70	24.10
Weldolets, Schedule 40, carbon steel, 2-1/2"	M7@3.52	Ea	25.50	148.00	173.50
Welding flanges, forged steel, 2-1/2"					
150 lb, slip-on flange	M7@1.10	Ea	9.00	46.30	55.30
300 lb, slip-on flange	M7@1.80	Ea	11.30	75.70	87.00
600 lb, slip-on flange	M7@2.00	Ea	18.00	84.10	102.10
150 lb, neck flange	M7@1.20	Ea	10.70	50.50	61.20
300 lb, neck flange	M7@1.20	Ea	14.40	50.50	64.90
600 lb, neck flange	M7@2.00	Ea	20.60	84.10	104.70
150 lb, threaded flange	M7@.990	Ea	12.60	41.60	54.20
300 lb, threaded flange	M7@1.25	Ea	16.50	52.60	69.10
Butt welded joints, 2-1/2" pipe					
Schedule 40 pipe	M7@1.20	Ea	--	50.50	--
Schedule 80 pipe	M7@2.03	Ea	--	85.40	--

3" black steel pipe (A53 or A120) and welded fittings, installed in a building

	Craft@Hrs	Unit	Material	Labor	Total
Pipe, no fittings or supports included, 3", plain end					
Schedule 40 to 10' high	M7@.081	LF	4.88	3.41	8.29
Schedule 40 over 10' to 20' high	M7@.098	LF	4.88	4.12	9.00
Schedule 80 to 10' high	M7@.092	LF	6.90	3.87	10.77
Schedule 80 over 10' to 20' high	M7@.110	LF	6.90	4.63	11.53
90 degree ells, 3"					
Schedule 40, carbon steel, butt weld, long turn	M7@2.80	Ea	7.00	118.00	125.00
Schedule 80, carbon steel, butt weld, long turn	M7@4.49	Ea	10.40	189.00	199.40
45 degree ells, 3"					
Schedule 40, carbon steel, butt weld	M7@2.80	Ea	6.00	118.00	124.00
Schedule 80, carbon steel, butt weld	M7@4.49	Ea	8.80	189.00	197.80
Tees, 3"					
Schedule 40, carbon steel, butt welded	M7@4.24	Ea	15.60	178.00	193.60
Schedule 80, carbon steel, butt welded	M7@6.90	Ea	21.30	290.00	311.30
Reducing tee, Schedule 40, carbon steel, butt weld	M7@2.73	Ea	20.00	115.00	135.00
Caps, 3"					
Schedule 40, carbon steel, butt welded	M7@1.40	Ea	3.65	58.90	62.55
Schedule 80, carbon steel, butt welded	M7@2.31	Ea	4.40	97.10	101.50
Reducers, Schedule 40, carbon steel, butt weld	M7@1.66	Ea	6.50	69.80	76.30
Bolt-ups, 3"	M7@.469	Ea	4.40	19.70	24.10
Weldolets, Schedule 40, carbon steel, 3"	M7@4.24	Ea	26.40	178.00	204.40
Welding flanges, forged steel, 3"					
150 lb, slip-on flange	M7@1.29	Ea	9.20	54.20	63.40
300 lb, slip-on flange	M7@2.11	Ea	13.40	88.70	102.10
600 lb, slip-on flange	M7@2.51	Ea	18.00	106.00	124.00
150 lb, neck flange	M7@1.40	Ea	11.60	58.90	70.50
300 lb, neck flange	M7@1.40	Ea	16.60	58.90	75.50
600 lb, neck flange	M7@2.31	Ea	21.70	97.10	118.80
150 lb, threaded flange	M7@1.25	Ea	14.60	52.60	67.20
300 lb, threaded flange	M7@1.40	Ea	19.50	58.90	78.40
600 lb, threaded flange	M7@2.00	Ea	28.60	84.10	112.70
Butt welded joints, 3" pipe					
Schedule 40 pipe	M7@1.40	Ea	--	58.90	--

Mechanical 15

	Craft@Hrs	Unit	Material	Labor	Total
3" black steel pipe (A53 or A120) and welded fittings, Butt welded joints, continued					
Schedule 80 pipe	M7@2.31	Ea	--	97.10	--
4" black steel pipe (A53 or A120) and welded fittings, installed in a building					
Pipe, no fittings or supports included, 4", plain end					
Schedule 40 to 10' high	M7@.100	LF	7.10	4.21	11.31
Schedule 40 over 10' to 20' high	M7@.121	LF	7.10	5.09	12.19
Schedule 80 to 10' high	M7@.121	LF	10.20	5.09	15.29
Schedule 80 over 10' to 20' high	M7@.144	LF	10.20	6.06	16.26
90 degree ells, 4"					
Schedule 40 carbon steel, butt weld, long turn	M7@3.80	Ea	11.70	160.00	171.70
Schedule 80 carbon steel, butt weld, long turn	M7@6.06	Ea	17.20	255.00	272.20
45 degree ells, 4"					
Schedule 40, carbon steel, butt welded	M7@3.80	Ea	9.95	160.00	169.95
Schedule 80, carbon steel, butt welded	M7@6.06	Ea	14.60	255.00	269.60
Tees, 4"					
Schedule 40, carbon steel, butt welded	M7@5.61	Ea	21.60	236.00	257.60
Schedule 80, carbon steel, butt welded	M7@9.28	Ea	34.80	390.00	424.80
Reducing, Schedule 40 carbon steel, butt weld	M7@3.35	Ea	24.70	141.00	165.70
Caps, 4"					
Schedule 40, carbon steel, butt welded	M7@1.89	Ea	4.70	79.50	84.20
Schedule 80, carbon steel, butt welded	M7@3.10	Ea	6.20	130.00	136.20
Reducers, Schedule 40, carbon steel, butt weld	M7@2.11	Ea	8.60	88.70	97.30
Bolt-ups, 4"	M7@1.69	Ea	11.70	71.10	82.80
Weldolets, Schedule 40, carbon steel, 4"	M7@5.46	Ea	33.75	230.00	263.75
Welding flanges, forged steel, 4"					
150 lb, slip-on flange	M7@1.69	Ea	11.70	71.10	82.80
300 lb, slip-on flange	M7@2.80	Ea	19.50	118.00	137.50
600 lb, slip-on flange	M7@3.23	Ea	32.60	136.00	168.60
150 lb, neck flange	M7@1.89	Ea	14.60	79.50	94.10
300 lb, neck flange	M7@1.89	Ea	23.40	79.50	102.90
600 lb, neck flange	M7@3.10	Ea	40.90	130.00	170.90
150 lb, threaded flange	M7@1.74	Ea	16.20	73.20	89.40
300 lb, threaded flange	M7@1.80	Ea	28.20	75.70	103.90
600 lb, threaded flange	M7@2.20	Ea	50.90	92.50	143.40
Butt welded joints, 4" pipe					
Schedule 40 pipe	M7@1.89	Ea	--	79.50	--
Schedule 80 pipe	M7@3.10	Ea	--	130.00	--
6" black steel pipe (A53 or A106) and welded fittings, installed in a building					
Pipe, no fittings or supports included, 6", plain end					
Schedule 40 to 10' high	M8@.138	LF	13.30	5.94	19.24
Schedule 40 over 10' to 20'	M8@.165	LF	13.30	7.10	20.40
Schedule 80 to 10' high	M8@.182	LF	20.20	7.84	28.04
Schedule 80 over 10' to 20'	M8@.217	LF	20.20	9.34	29.54
90 degree ells, 6"					
Schedule 40 carbon steel, butt weld, long turn	M8@5.48	Ea	28.60	236.00	264.60
Schedule 80 carbon steel, butt weld, long turn	M8@8.95	Ea	42.60	385.00	427.60
45 degree ells, 6"					
Schedule 40, carbon steel, butt welded	M8@5.48	Ea	21.40	236.00	257.40
Schedule 80, carbon steel, butt welded	M8@8.95	Ea	31.90	385.00	416.90
Tees, 6"					
Schedule 40, carbon steel, butt welded	M8@8.24	Ea	39.20	355.00	394.20
Schedule 80, carbon steel, butt welded	M8@13.3	Ea	55.30	573.00	628.30

Mechanical 15

	Craft@Hrs	Unit	Material	Labor	Total
Reducing, Schedule 40 carbon steel, butt weld	M8@4.68	Ea	50.00	201.00	251.00
Caps, 6"					
Schedule 40, carbon steel, butt welded	M8@2.74	Ea	9.10	118.00	127.10
Schedule 80, carbon steel, butt welded	M8@4.39	Ea	13.40	189.00	202.40
Reducers, Schedule 40, carbon steel, butt weld	M8@3.03	Ea	17.70	130.00	147.70
Bolt-ups, 6"	M8@1.95	Ea	13.70	83.90	97.60
Welding flanges, forged steel, 6"					
150 lb, slip-on flange	M8@2.45	Ea	19.15	105.00	124.15
300 lb, slip-on flange	M8@4.00	Ea	32.40	172.00	204.40
150 lb, neck flange	M8@2.72	Ea	22.70	117.00	139.70
300 lb, neck flange	M8@2.72	Ea	38.80	117.00	155.80
Butt welded joints, 6" pipe					
Schedule 40 pipe	M8@2.72	Ea	--	117.00	--
Schedule 80 pipe	M8@4.39	Ea	--	189.00	--

8" black steel pipe (A53 or A106) and welded fittings, installed in a building

	Craft@Hrs	Unit	Material	Labor	Total
Pipe, no fittings or supports included, 8" plain end					
Schedule 40 to 10' high	M8@.178	LF	19.80	7.66	27.46
Schedule 40 over 10' to 20'	M8@.216	LF	19.80	9.30	29.10
Schedule 80 to 10' high	M8@.238	LF	30.50	10.20	40.70
Schedule 80 over 10' to 20' high	M8@.284	LF	30.50	12.20	42.70
90 degree ells, 8"					
Schedule 40 carbon steel, butt weld, long turn	M8@7.03	Ea	53.30	303.00	356.30
Schedule 80 carbon steel, butt weld, long turn	M8@10.7	Ea	80.00	461.00	541.00
45 degree ells, 8"					
Schedule 40, carbon steel, butt welded	M8@7.03	Ea	37.30	303.00	340.30
Schedule 80, carbon steel, butt welded	M8@10.7	Ea	55.70	461.00	516.70
Tees, 8"					
Schedule 40, carbon steel, butt welded	M8@10.6	Ea	73.00	456.00	529.00
Schedule 80, carbon steel, butt welded	M8@16.1	Ea	112.00	693.00	805.00
Reducing, Schedule 40, carbon steel, butt weld	M8@6.33	Ea	95.00	273.00	368.00
Caps, 8"					
Schedule 40, carbon steel, butt welded	M8@3.57	Ea	13.10	154.00	167.10
Schedule 80, carbon steel, butt welded	M8@5.34	Ea	18.00	230.00	248.00
Reducers, Schedule 40, carbon steel, butt weld	M8@3.86	Ea	26.50	166.00	192.50
Bolt-ups, 8"	M8@1.95	Ea	15.20	83.90	99.10
Welding flanges, forged steel, 8"					
150 lb, slip-on flange	M8@3.15	Ea	29.10	136.00	165.10
300 lb, slip-on flange	M8@4.95	Ea	54.00	213.00	267.00
150 lb, neck flange	M8@3.57	Ea	42.40	154.00	196.40
300 lb, neck flange	M8@3.57	Ea	67.00	154.00	221.00
Butt welded joints, 8" pipe					
Schedule 40 pipe	M8@3.57	Ea	--	154.00	--
Schedule 80 pipe	M8@5.34	Ea	--	230.00	--

10" black steel pipe (A53 or A106) and welded fittings, installed in a building

	Craft@Hrs	Unit	Material	Labor	Total
Pipe, no fittings or supports included, 10", plain end					
Schedule 40 to 20' high	M8@.500	LF	28.50	21.50	50.00
Carbon steel, butt welded fittings					
90 degree ells, long turn	M8@6.45	Ea	96.00	278.00	374.00
45 degree ells	M8@6.45	Ea	67.30	278.00	345.30
Tees	M8@9.77	Ea	122.10	421.00	543.10
Reducing tees	M8@8.10	Ea	161.00	349.00	510.00
Caps	M8@3.27	Ea	20.80	141.00	161.80

Mechanical 15

	Craft@Hrs	Unit	Material	Labor	Total
10" black steel pipe (A53 or A106) and welded fittings, continued					
Reducers	M8@5.09	Ea	39.20	219.00	258.20
Bolt-ups, 10"	M8@3.44	Ea	39.20	148.00	187.20
Welding flanges, forged steel, 10"					
150 lb, slip-on flange	M8@3.15	Ea	52.65	136.00	188.65
300 lb, slip-on flange	M8@3.15	Ea	103.00	136.00	239.00
150 lb, weld neck flange	M8@3.27	Ea	69.10	141.00	210.10
300 lb, weld neck flange	M8@3.57	Ea	135.10	154.00	289.10
12" black steel pipe (A53 or A106) and welded fittings, installed in a building					
Pipe, no fittings or supports included, 12", plain end					
Schedule 40 to 20' high	M8@.597	LF	37.50	25.70	63.20
Carbon steel, butt welded fittings					
90 degree ell, long turn	M8@7.54	Ea	137.00	325.00	462.00
45 degree ells	M8@7.54	Ea	96.00	325.00	421.00
Tees	M8@11.4	Ea	177.10	491.00	668.10
Reducing tees	M8@10.1	Ea	236.00	435.00	671.00
Caps	M8@3.86	Ea	29.00	166.00	195.00
Reducers	M8@6.18	Ea	59.20	266.00	325.20
Bolt-ups, 12"	M8@3.44	Ea	38.40	148.00	186.40
Welding flanges, forged steel, 12"					
150 lb, slip-on flange	M8@3.71	Ea	78.00	160.00	238.00
300 lb, slip-on flange	M8@3.71	Ea	131.00	160.00	291.00
150 lb, weld neck flange	M8@3.86	Ea	101.00	166.00	267.00
300 lb, weld neck flange	M8@3.86	Ea	169.50	166.00	335.50

Galvanized Steel Pipe (A53 or A120) and Threaded Fittings, Schedule 40, installed in a building. See Black Steel Pipe on preceding pages for additive costs for galvanized.

Cast Iron Hub and Spigot Soil Pipe. Standard weight, installed as noted either in a building horizontally or vertically to 10' above floor level or in an open trench with no excavation or backfill included. These costs include the subcontractor's overhead and profit.

	Craft@Hrs	Unit	Material	Labor	Total
2" pipe and fittings					
Gaskets	P6@.066	Ea	2.00	2.70	4.70
Lead (1.5 lbs) and oakum (.15 lbs) joints	P6@.160	Ea	1.80	6.54	8.34
Regular bends, cleanouts, reducers	P6@.431	Ea	4.75	17.60	22.35
Sweep bends, traps	P6@.431	Ea	10.80	17.60	28.40
combination wye and bend	P6@.854	Ea	9.50	34.90	44.40
Wyes, tees	P6@.854	Ea	9.45	34.90	44.35
Double wyes, crosses	P6@1.28	Ea	14.70	52.30	67.00
Pipe installed in a building	P6@.135	LF	3.30	5.52	8.82
Pipe installed in an open trench	P6@.046	LF	3.00	1.88	4.88
3" pipe and fittings					
Gaskets	P6@.077	Ea	2.65	3.15	5.80
Lead (2.25 lbs) and oakum (.23 lbs) joints	P6@.214	Ea	2.75	8.75	11.50
Regular bends, cleanouts, reducers	P6@.482	Ea	7.95	19.70	27.65
Sweep bends, traps	P6@.482	Ea	16.15	19.70	35.85
Combination wye and bend	P6@.956	Ea	15.20	39.10	54.30
Wyes, tees	P6@.956	Ea	17.60	39.10	56.70
Double wyes, crosses	P6@1.44	Ea	25.10	58.90	84.00
Pipe installed in a building	P6@.197	LF	4.60	8.05	12.65
Pipe installed in an open trench	P6@.051	LF	4.20	2.08	6.28

Mechanical 15

	Craft@Hrs	Unit	Material	Labor	Total
4" pipe and fittings					
Gaskets	P6@.088	Ea	3.40	3.60	7.00
Lead (3 lbs) and oakum (.3 lbs) joints	P6@.268	Ea	3.65	11.00	14.65
Regular bends, cleanouts, reducers	P6@.642	Ea	10.85	26.20	37.05
Sweep bends, traps	P6@.642	Ea	23.55	26.20	49.75
Combination wye and bend	P6@1.28	Ea	21.50	52.30	73.80
Wyes, tees	P6@1.28	Ea	22.50	52.30	74.80
Double wyes, crosses	P6@1.93	Ea	34.80	78.90	113.70
Pipe installed in a building	P6@.219	LF	6.00	8.95	14.95
Pipe installed in an open trench	P6@.071	LF	5.50	2.90	8.40
5" pipe and fittings					
Gaskets	P6@.098	Ea	5.25	4.01	9.26
Lead (3.75 lbs) and oakum (.38 lbs) joints	P6@.372	Ea	4.60	15.20	19.80
Regular bends, cleanouts, reducers	P6@.798	Ea	14.40	32.60	47.00
Sweep bends, traps	P6@.798	Ea	42.10	32.60	74.70
Combination wye and bend	P6@1.60	Ea	35.80	65.40	101.20
Wyes, tees	P6@1.60	Ea	38.65	65.40	104.05
Double wyes, crosses	P6@2.39	Ea	53.35	97.70	151.05
Pipe installed in a building	P6@.249	LF	6.30	10.20	16.50
Pipe installed in an open trench	P6@.090	LF	7.70	3.68	11.38
6" pipe and fitting					
Gaskets	P6@.111	Ea	5.45	4.54	9.99
Lead (4.5 lbs) and oakum (.45 lbs) joints	P6@.428	Ea	5.50	17.50	23.00
Regular bends, cleanouts, reducers	P6@.958	Ea	19.70	39.20	58.90
Sweep bends, traps	P6@.958	Ea	56.50	39.20	95.70
Combination wye and bend	P6@1.93	Ea	45.20	78.90	124.10
Wyes and tees	P6@1.93	Ea	49.40	78.90	128.30
Double wyes, crosses	P6@2.86	Ea	69.25	117.00	186.25
Pipe installed in a building	P6@.268	LF	10.00	11.00	21.00
Pipe installed in an open trench	P6@.108	LF	9.40	4.41	13.81
8" pipe and fittings					
Gaskets	P6@.129	Ea	12.10	5.27	17.37
Lead (6.8 lbs) and oakum (.68 lbs) joints	P6@.642	Ea	8.25	26.20	34.45
Regular bends, cleanouts, reducers	P6@1.23	Ea	47.05	50.30	97.35
Sweep bends, traps	P6@1.23	Ea	151.20	50.30	201.50
Combination wye and bend	P6@2.45	Ea	118.00	100.00	218.00
Wyes and tees	P6@2.45	Ea	125.30	100.00	225.30
Double wyes, crosses	P6@3.65	Ea	146.85	149.00	295.85
Pipe installed in a building	P6@.344	LF	24.00	14.10	38.10
Pipe installed in an open trench	P6@.140	LF	22.00	5.72	27.72
Add for extra heavy cast iron soil pipe	--	%	50.0	10.0	--
Add for double hub pipe, 5' lengths	--	%	40.0	--	--
Add for pipe in 5' rather than 10' lengths	--	%	25.0	--	--

Cast Iron No-Hub Soil Pipe. Standard weight, installed vertically or horizontally in a building to 10' above floor level. No hangers included. Deduct 30% from the labor cost for pipe installed in an open trench. Deduct 20% from the labor cost for fittings installed in an open trench. These costs include the subcontractor's overhead and profit.

	Craft@Hrs	Unit	Material	Labor	Total
1-1/2" pipe and fittings					
Pipe	P6@.058	LF	3.40	2.37	5.77
Neoprene sleeve, stainless steel coupling	P6@.206	Ea	6.75	8.42	15.17
Quarter bend	P6@.477	Ea	4.50	19.50	24.00
Sweep quarter bend	P6@.477	Ea	9.80	19.50	29.30
Eighth bend	P6@.477	Ea	3.55	19.50	23.05

Mechanical 15

	Craft@Hrs	Unit	Material	Labor	Total
Cast Iron No-Hub Soil Pipe, continued					
Sanitary tee	P6@.714	Ea	6.00	29.20	35.20
Sanitary cross	P6@.714	Ea	8.10	29.20	37.30
Plug	P6@.238	Ea	1.95	9.73	11.68
Wye	P6@.714	Ea	6.30	29.20	35.50
Wye and eighth bend "combo"	P6@.714	Ea	6.40	29.20	35.60
Tapped tee	P6@.477	Ea	6.90	19.50	26.40
P trap	P6@.477	Ea	6.90	19.50	26.40
Tapped cross	P6@.477	Ea	8.40	19.50	27.90
2" pipe and fittings					
Pipe	P6@.085	LF	3.35	3.47	6.82
Neoprene sleeve, stainless steel coupling	P6@.206	Ea	6.75	8.42	15.17
Quarter bend	P6@.568	Ea	4.80	23.20	28.00
Sweep quarter bend	P6@.568	Ea	9.95	23.20	33.15
Eighth bend	P6@.568	Ea	3.65	23.20	26.85
Sanitary tee	P6@.856	Ea	6.70	35.00	41.70
Sanitary cross	P6@.856	Ea	10.15	35.00	45.15
Plug	P6@.285	Ea	1.85	11.60	13.45
Wye	P6@.856	Ea	6.40	35.00	41.40
Double wye	P6@1.14	Ea	8.90	46.60	55.50
Wye and eighth bend "combo"	P6@.856	Ea	6.65	35.00	41.65
Double wye and eighth bend "combo"	P6@1.14	Ea	19.30	46.60	65.90
Test tee with plug	P6@.579	Ea	7.95	23.70	31.65
Tapped tee	P6@.568	Ea	7.95	23.20	31.15
P trap	P6@.568	Ea	6.90	23.20	30.10
Tapped cross	P6@.568	Ea	10.20	23.20	33.40
3" pipe and fittings					
Pipe	P6@.124	LF	4.65	5.07	9.72
Neoprene sleeve, stainless steel coupling	P6@.206	Ea	7.40	8.42	15.82
Quarter bend	P6@.663	Ea	6.30	27.10	33.40
Sweep quarter bend	P6@.663	Ea	11.25	27.10	38.35
Eighth bend	P6@.663	Ea	5.20	27.10	32.30
Sanitary tee	P6@.999	Ea	8.00	40.80	48.80
Sanitary cross	P6@.999	Ea	15.95	40.80	56.75
Plug	P6@.334	Ea	3.30	13.70	17.00
Wye	P6@.999	Ea	8.75	40.80	49.55
Double wye	P6@1.33	Ea	16.45	54.40	70.85
Wye and eighth bend "combo"	P6@.999	Ea	10.70	40.80	51.50
Double wye and eighth bend "combo"	P6@1.33	Ea	24.20	54.40	78.60
Reducers	P6@.619	Ea	3.40	25.30	28.70
Test tee with plug	P6@.663	Ea	12.95	27.10	40.05
Tapped tee	P6@.663	Ea	18.00	27.10	45.10
P trap	P6@.663	Ea	15.70	27.10	42.80
Tapped cross	P6@.663	Ea	12.80	27.10	39.90
Closet flange	P6@.334	Ea	8.85	13.70	22.55
4" pipe and fittings					
Pipe	P6@.161	LF	6.00	6.58	12.58
Neoprene sleeve, stainless steel coupling	P6@.206	Ea	8.35	8.42	16.77
Quarter bend	P6@.762	Ea	9.50	31.10	40.60
Sweep quarter bend	P6@.790	Ea	18.60	32.30	50.90
Eighth bend	P6@.762	Ea	6.80	31.10	37.90
Sanitary tee	P6@1.14	Ea	12.55	46.60	59.15
Sanitary cross, 4" x 4"	P6@1.51	Ea	30.00	61.70	91.70
Sanitary cross, 4" x 3"	P6@1.51	Ea	24.45	61.70	86.15

Mechanical 15

	Craft@Hrs	Unit	Material	Labor	Total
Sanitary cross, 4" x 2"	P6@1.51	Ea	20.25	61.70	81.95
Plug	P6@.380	Ea	4.45	15.50	19.95
Wye	P6@1.14	Ea	14.20	46.60	60.80
Double wye	P6@1.51	Ea	33.85	61.70	95.55
Wye and eighth bend "combo"	P6@1.14	Ea	19.70	46.60	66.30
Double wye and eighth bend, 4" x 4" "combo"	P6@1.51	Ea	46.25	61.70	107.95
Double wye and eighth bend, 4" x 3" "combo"	P6@1.51	Ea	30.40	61.70	92.10
Double wye and eighth bend, 4" x 2" "combo"	P6@1.21	Ea	20.90	49.50	70.40
Reducers	P6@.714	Ea	5.35	29.20	34.55
Test tee with plug	P6@.762	Ea	20.49	31.10	51.59
Tapped tee	P6@.762	Ea	9.90	31.10	41.00
P trap	P6@.790	Ea	27.30	32.30	59.60
Tapped cross	P6@.790	Ea	17.10	32.30	49.40
Closet flange, 3" deep	P6@.380	Ea	8.85	15.50	24.35
Closet flange, 6" deep	P6@.380	Ea	19.75	15.50	35.25
6" pipe and fittings					
Pipe	P6@.238	LF	10.35	9.73	20.08
Neoprene sleeve, stainless steel coupling	P6@.206	Ea	19.20	8.42	27.62
Quarter bend	P6@.948	Ea	23.15	38.70	61.85
Sweep quarter bend	P6@.948	Ea	40.40	38.70	79.10
Eighth bend	P6@.948	Ea	15.55	38.70	54.25
Sanitary tee	P6@1.43	Ea	36.10	58.40	94.50
Plug	P6@.477	Ea	8.35	19.50	27.85
Wye, 6" x 6"	P6@1.43	Ea	36.35	58.40	94.75
Wye, 6" x 4"	P6@1.43	Ea	24.20	58.40	82.60
Wye, 6" x 2"	P6@1.43	Ea	19.60	58.40	78.00
Double wye, 6" x 6"	P6@1.89	Ea	58.30	77.20	135.50
Double wye, 6" x 4"	P6@1.89	Ea	47.30	77.20	124.50
Wye and eighth bend, 6" x 6" "combo"	P6@1.43	Ea	49.60	58.40	108.00
Wye and eighth bend, 6" x 4" "combo"	P6@1.43	Ea	32.90	58.40	91.30
Wye and eighth bend, 6" x 3" "combo"	P6@1.43	Ea	32.00	58.40	90.40
Wye and eighth bend, 6" x 2" "combo"	P6@1.43	Ea	25.30	58.40	83.70
Reducers	P6@.856	Ea	13.30	35.00	48.30
Test tee with plug	P6@.948	Ea	44.70	38.70	83.40
P trap	P6@.948	Ea	62.45	38.70	101.15
8" pipe and fittings					
Pipe	P6@.380	LF	21.35	15.50	36.85
Neoprene sleeve, stainless steel coupling	P6@.206	Ea	30.80	8.42	39.22
Quarter bend	P6@1.23	Ea	61.70	50.30	112.00
Sweep quarter bend	P6@1.23	Ea	123.95	50.30	174.25
Eighth bend	P6@1.23	Ea	41.40	50.30	91.70
Sanitary tee, 8" x 8"	P6@1.87	Ea	105.55	76.40	181.95
Sanitary tee, 8" x 6"	P6@1.87	Ea	78.65	76.40	155.05
Plug	P6@.619	Ea	26.30	25.30	51.60
Wye, 8" x 8"	P6@1.87	Ea	85.45	76.40	161.85
Wye, 8" x 6"	P6@1.87	Ea	60.45	76.40	136.85
Wye, 8" x 4"	P6@1.87	Ea	48.90	76.40	125.30
Double wye, 8" x 8"	P6@2.45	Ea	157.95	100.00	257.95
Wye and eighth bend, 8" x 8" "combo"	P6@1.87	Ea	124.35	76.40	200.75
Wye and eighth bend, 8" x 6" "combo"	P6@1.87	Ea	97.45	76.40	173.85
Wye and eighth bend, 8" x 4" "combo"	P6@1.87	Ea	65.10	76.40	141.50
Reducers	P6@1.05	Ea	22.90	42.90	65.80
Test tee with plug	P6@1.23	Ea	132.30	50.30	182.60

Mechanical 15

	Craft@Hrs	Unit	Material	Labor	Total
Cast Iron No-Hub Soil Pipe, continued					
10" pipe and fittings					
Pipe	P6@.446	LF	27.55	18.20	45.75
Neoprene sleeve, stainless steel coupling	P6@.206	Ea	40.55	8.42	48.97
Eighth bend	P6@1.51	Ea	78.10	61.70	139.80
Plug	P6@.762	Ea	37.80	31.10	68.90
Wye, 10" x 10"	P6@2.27	Ea	183.05	92.80	275.85
Wye, 10" x 8"	P6@2.27	Ea	160.80	92.80	253.60
Wye, 10" x 6"	P6@2.27	Ea	133.90	92.80	226.70
Wye, 10" x 4"	P6@2.27	Ea	128.00	92.80	220.80
Reducer, 10" x 8"	P6@1.23	Ea	50.45	50.30	100.75
Reducer, 10" x 6"	P6@1.23	Ea	42.10	50.30	92.40
Reducer, 10" x 4"	P6@1.23	Ea	37.60	50.30	87.90

Ductile Iron Pressure Pipe. These costs include the subcontractor's overhead and profit.
Slip joint 150 lb ductile iron pipe, including trenching to 6', pipe, backfill, compaction, regrading and seeding but no dewatering or shoring

	Craft@Hrs	Unit	Material	Labor	Equipment	Total
4"	U1@.436	LF	7.70	17.30	3.61	28.61
6"	U1@.457	LF	10.70	18.10	3.76	32.56
8"	U1@.478	LF	16.20	19.00	4.07	39.27
10"	U1@.530	LF	19.30	21.00	4.33	44.63
12"	U1@.525	LF	22.90	20.80	4.53	48.23
14"	U1@.541	LF	24.90	21.50	4.64	51.04
16"	U1@.567	LF	26.75	22.50	5.00	54.25
18"	U1@.627	LF	28.90	24.90	5.25	59.05

Mechanical joint 150 lb ductile iron pipe, including trenching to 6', pipe, backfill, compaction, regrading and seeding, no dewatering or shoring

	Craft@Hrs	Unit	Material	Labor	Equipment	Total
3"	U1@.496	LF	8.20	19.70	3.81	31.71
4"	U1@.496	LF	8.50	19.70	3.91	32.11
6"	U1@.527	LF	12.90	20.90	3.91	37.71
8"	U1@.538	LF	16.40	21.40	4.12	41.92

Slip joint 150 lb ductile iron pipe, including trenching to 6', typical dewatering and sheeting, pipe, backfill, compaction, regrading and seeding

	Craft@Hrs	Unit	Material	Labor	Equipment	Total
4"	U1@.876	LF	9.80	34.80	45.60	90.20
6"	U1@.895	LF	13.65	35.50	45.60	94.75
8"	U1@.903	LF	17.00	35.90	45.60	98.50
10"	U1@.947	LF	20.90	37.60	45.80	104.30
12"	U1@.968	LF	24.60	38.40	46.40	109.40
14"	U1@.981	LF	26.75	39.00	47.40	113.15
16"	U1@1.01	LF	28.45	40.10	47.70	116.25
18"	U1@1.07	LF	30.55	42.50	47.90	120.95
Add for excavation over 6' to 10'	--	%	--	40.0	10.0	--
Add for sheeting and shoring 10' trench	S6@2.01	LF	12.30	74.50	12.10	98.90

Mechanical joint 150 lb ductile iron pipe, including trenching to 6', typical dewatering, pipe, backfill, compaction, regrading and seeding

	Craft@Hrs	Unit	Material	Labor	Equipment	Total
3"	U1@.905	LF	9.90	35.90	47.40	93.20
4"	U1@.908	LF	10.15	36.10	47.60	93.85
6"	U1@.945	LF	14.60	37.50	48.10	100.20
8"	U1@.976	LF	18.15	38.80	48.70	105.65
Add for excavation over 6' to 10'	--	%	--	40.0	10.0	--
Add for sheeting and shoring 10' trench	U1@.924	LF	12.30	36.70	3.74	52.74

Slip joint 150 lb cement-lined Class 52 ductile iron pipe, including cost of lifting equipment, no excavation, sheeting, backfill or dewatering

	Craft@Hrs	Unit	Material	Labor	Equipment	Total
3"	U1@.183	LF	7.25	7.27	.65	15.17

Mechanical 15

	Craft@Hrs	Unit	Material	Labor	Equipment	Total
4"	U1@.186	LF	7.60	7.39	.65	15.64
6"	U1@.188	LF	11.15	7.47	.65	19.27
8"	U1@.200	LF	14.70	7.94	.65	23.29
10"	U1@.224	LF	18.40	8.90	.77	28.07
12"	U1@.239	LF	22.15	9.49	.77	32.41
14"	U1@.254	LF	23.60	10.10	.83	34.53
16"	U1@.270	LF	25.50	10.70	1.00	37.20
18"	U1@.286	LF	27.70	11.40	1.14	40.24
20"	U1@.378	LF	33.20	15.00	1.25	49.45
24"	U1@.454	LF	39.20	18.00	1.45	58.65
30"	U1@.506	LF	60.00	20.10	1.71	81.81
36"	U1@.564	LF	86.10	22.40	1.95	110.45
42"	U1@.645	LF	109.20	25.60	2.27	137.07
48"	U1@.659	LF	140.70	26.20	2.48	169.38
Add for mechanical joint 150 lb cement-lined Class 52 ductile iron pipe	--	%	12.0	50.0	20.0	--

Asbestos Cement Water Pipe. Prebelled Class 150 fluid tight pressure pipe with couplings, including lifting equipment but no excavation or backfill. Standard 13' lengths. These costs include the subcontractor's overhead and profit.

	Craft@Hrs	Unit	Material	Labor	Equipment	Total
6" pipe	U1@.112	LF	4.90	4.45	.52	9.87
8" pipe	U1@.125	LF	7.75	4.96	.62	13.33
10" pipe	U1@.142	LF	12.05	5.64	.70	18.39
12" pipe	U1@.157	LF	15.90	6.23	.78	22.91
14" pipe	U1@.177	LF	21.05	7.03	.89	28.97
16" pipe	U1@.177	LF	26.60	7.03	.89	34.52
Add for Class 200 pipe	--	%	20.0	--	--	--
Deduct for Class 100 pipe	--	%	-10.0	--	--	--
Add for 3'3" lengths, fully machined	--	%	40.0	100.0	100.0	--

Gas Distribution Lines. These costs assume pipe is laid and connected in an open trench and include the contractor's overhead and profit but no excavation or backfill. The crew is one plumber and a laborer.

Polyethylene pipe, 60 PSI

	Craft@Hrs	Unit	Material	Labor	Equipment	Total
1-1/4" diameter coils	P6@.064	LF	.56	2.62	.21	3.39
1-1/2" diameter coils	P6@.064	LF	.76	2.62	.21	3.59
2" diameter coils	P6@.071	LF	.96	2.90	.23	4.09
3" diameter coils	P6@.086	LF	2.00	3.52	.27	5.79
3" diameter in 38' lengths with couplings	P6@.133	LF	1.65	5.44	.66	7.75
4" diameter in 38' lengths with couplings	P6@.168	LF	3.20	6.87	.83	10.90
6" diameter in 38' lengths with couplings	P6@.182	LF	6.90	7.44	.93	15.27
8" diameter in 38' lengths with couplings	P6@.218	LF	12.80	8.91	1.11	22.82

Natural gas diaphragm meters, direct digital reading, 5 to 100 PSI, not including temperature and pressure compensation

	Craft@Hrs	Unit	Material	Labor	Equipment	Total
250 CFH at 5 PSI	P6@2.88	Ea	96.40	118.00	1.43	215.83
425 CFH at 10 PSI	P6@4.38	Ea	235.00	179.00	2.20	416.20
425 CFH at 25 PSI	P6@4.38	Ea	310.00	179.00	2.20	491.20
800 CFH at 20 PSI	P6@5.28	Ea	886.00	216.00	2.60	1,104.60

Natural gas pressure regulators with screwed connections

	Craft@Hrs	Unit	Material	Labor	Equipment	Total
3/4"	P6@.729	Ea	24.70	29.80	.33	54.83
1"	P6@.772	Ea	24.70	31.60	.35	56.65
1-1/4"	P6@.872	Ea	132.00	35.60	.39	167.99
1-1/2"	P6@1.04	Ea	132.00	42.50	.47	174.97
2"	P6@1.55	Ea	1,065.00	63.30	.76	1,129.06

Mechanical 15

	Craft@Hrs	Unit	Material	Labor	Equipment	Total

PVC Water Pipe. No excavation or backfill included. These costs include the subcontractor's overhead and profit.

Class 200 Size to Diameter Ratio (SDR) 21 white bell-end pipe

	Craft@Hrs	Unit	Material	Labor	Equipment	Total
3/4" pipe	P6@.029	LF	.28	1.19	.04	.37
1" pipe	P6@.029	LF	.37	1.19	.04	1.60
1-1/4" pipe	P6@.029	LF	.57	1.19	.04	1.80
1-1/2" pipe	P6@.029	LF	.75	1.19	.04	1.98
2" pipe	P6@.036	LF	1.20	1.47	.04	2.71
2-1/2" pipe	P6@.037	LF	1.70	1.51	.05	3.26
3" pipe	P6@.065	LF	2.50	2.56	.08	5.14
4" pipe	P6@.068	LF	4.15	2.78	.09	7.02
6" pipe	P6@.082	LF	9.05	3.35	1.10	13.50

90 degree ells schedule 40

	Craft@Hrs	Unit	Material	Labor	Equipment	Total
3/4" ells	P6@.241	LF	.68	9.85	.16	10.69
1" ells	P6@.241	LF	1.20	9.85	.16	11.21
1-1/4" ells	P6@.326	LF	2.15	13.30	.21	15.66
1-1/2" ells	P6@.385	LF	2.30	15.70	.24	18.24
2" ells	P6@.489	LF	3.60	20.00	.31	23.91
2-1/2" ells	P6@.724	LF	11.00	29.60	.45	41.05
3" ells	P6@1.32	LF	13.15	53.90	.82	67.87
4" ells	P6@1.57	LF	23.55	64.20	.97	88.72
6" ells	P6@1.79	LF	74.90	73.20	1.13	149.23
Deduct for 45 degree ells	--	%	-5.0	--	--	--

Tees

	Craft@Hrs	Unit	Material	Labor	Equipment	Total
3/4" tees	P6@.385	LF	.86	15.70	.24	16.80
1" tees	P6@.385	LF	1.60	15.70	.24	17.54
1-1/4" tees	P6@.451	LF	2.55	18.40	.28	21.23
1-1/2" tees	P6@.581	LF	3.10	23.70	.36	27.16
2" tees	P6@.724	LF	4.45	29.60	.45	34.50
2-1/2" tees	P6@.951	LF	14.70	38.90	.60	54.20
3" tees	P6@1.57	LF	19.30	64.20	.98	84.48
4" tees	P6@2.09	LF	34.90	85.40	1.30	121.60
6" tees	P6@2.70	LF	117.70	110.00	1.70	229.40

Class 150 AWWA C900 Size to Diameter Ratio (SDR) 18 white bell-end pipe

	Craft@Hrs	Unit	Material	Labor	Equipment	Total
4" pipe	U1@.249	LF	4.05	9.89	1.10	15.04
6" pipe	U1@.278	LF	8.00	11.00	1.23	20.23
8" pipe	U1@.312	LF	13.65	12.40	1.37	27.42
10" pipe	U1@.357	LF	22.50	14.20	1.58	38.28
12" pipe	U1@.499	LF	26.80	19.80	2.20	48.80

Cast Iron Flanged Pipe. AWWA C151 pipe with cast iron flanges on both ends. Labor shown includes inspection and testing. Add for thrust block from below. These costs include subcontractors overhead and profit but no excavation, shoring, dewatering, fittings, valves or backfill.

	Craft@Hrs	Unit	Material	Labor	Equipment	Total
4" pipe, 260 pounds per 18' section	U1@.057	LF	7.90	2.26	.59	10.75
6" pipe, 405 pounds per 18' section	U1@.075	LF	9.15	2.98	.59	12.72
8" pipe, 570 pounds per 18' section	U1@.075	LF	12.90	2.98	.78	16.66
10" pipe, 740 pounds per 18' section	U1@.094	LF	17.85	3.73	.98	22.56
12" pipe, 930 pounds per 18' section	U1@.094	LF	24.15	3.73	.98	28.86
Adjust for 15' sections	--	%	-15.0	+20.0	+20.0	--
Adjust for 12' sections	--	%	-25.0	+50.0	+50.0	--
Adjust for 10' sections	--	%	-20.0	+80.0	+80.0	--
Adjust for 8' sections	--	%	-12.0	+125.0	+125.0	--
Adjust for 6' sections	--	%	+5.0	+200.0	+200.0	--

Mechanical 15

	Craft@Hrs	Unit	Material	Labor	Equipment	Total
Adjust for 4' sections	--	%	+35.0	+350.0	+350.0	--
Adjust for 3' sections	--	%	+70.0	+500.0	+500.0	--
Deduct for push-in joint cast iron pipe	--	%	-5.0	-10.0	-10.0	--

Cast Iron Fittings, Valves and Accessories. C110-64, C111-64 and ASA A21.10-A21.11-64. No excavation, shoring, backfill or dewatering included. These costs include the subcontractor's overhead and profit.

Cast iron mechanical joint 90 degree ells

	Craft@Hrs	Unit	Material	Labor	Equipment	Total
4"	U1@.808	Ea	94.80	32.10	6.70	133.60
6"	U1@1.06	Ea	128.80	42.10	8.76	179.66
8"	U1@1.06	Ea	187.10	42.10	8.76	237.96
10"	U1@1.32	Ea	270.55	52.40	10.90	333.85
12"	U1@1.32	Ea	358.85	52.40	10.90	422.15
14"	U1@1.32	Ea	730.00	52.40	10.90	793.30
16"	U1@1.65	Ea	928.25	65.50	13.70	1,007.45
Deduct for 1/8 or 1/16 bends	--	%	-20.0	--	--	--

Cast iron mechanical joint tees

	Craft@Hrs	Unit	Material	Labor	Equipment	Total
4" x 4"	U1@1.22	Ea	133.65	48.40	10.10	192.15
6" x 4"	U1@1.59	Ea	175.75	63.10	13.10	251.95
6" x 6"	U1@1.59	Ea	190.35	63.10	13.10	266.55
8" x 8"	U1@1.59	Ea	286.75	63.10	13.10	362.95
10" x 10"	U1@1.59	Ea	432.55	63.10	13.10	508.75
12" x 12"	U1@2.48	Ea	571.05	98.50	20.50	690.05
Add for wyes	--	%	40.0	--	--	--
Add for crosses	--	%	30.0	33.0	33.0	--

Cast iron mechanical joint reducers

	Craft@Hrs	Unit	Material	Labor	Equipment	Total
6" x 4"	U1@1.06	Ea	102.85	42.10	8.76	153.71
8" x 6"	U1@1.06	Ea	153.10	42.10	8.76	203.96
10" x 8"	U1@1.32	Ea	210.60	52.40	10.90	273.90
12" x 10"	U1@1.32	Ea	288.35	52.40	10.90	351.65

AWWA mechanical joint gate valves, with extension valve box

	Craft@Hrs	Unit	Material	Labor	Equipment	Total
3"	U1@3.10	Ea	228.15	123.00	25.40	376.55
4"	U1@4.04	Ea	271.60	160.00	33.40	465.00
6"	U1@4.80	Ea	351.55	191.00	39.70	582.25
8"	U1@5.14	Ea	546.90	204.00	42.20	793.10
10"	U1@5.67	Ea	850.00	225.00	46.90	1,121.90
12"	U1@6.19	Ea	1,074.25	246.00	51.20	1,371.45
Indicator post, upright, adjustable	U1@2.10	Ea	567.00	83.40	17.30	667.70

Backflow preventers

	Craft@Hrs	Unit	Material	Labor	Equipment	Total
3"	P6@11.3	Ea	1,139.00	462.00	--	1,601.00
4"	P6@15.3	Ea	1,644.00	625.00	--	2,269.00
6"	P6@22.7	Ea	2,765.00	928.00	--	3,693.00
8"	P6@28.0	Ea	5,150.00	1,140.00	--	6,290.00
Block structure for backflow preventer, 5' x 7' x 3'4"	M1@31.7	Ea	661.00	1,190.00	--	1,851.00
Sheer gate, 8", with adjustable release handle	U1@2.44	Ea	1,130.00	96.90	420.00	1,646.90
Adjustable length valve boxes, for up to 20" valves, 5' depth	U1@1.75	Ea	66.15	69.50	14.40	150.05

Concrete thrust blocks, minimum formwork

	Craft@Hrs	Unit	Material	Labor	Equipment	Total
1/4 CY	C8@1.20	Ea	43.25	46.20	--	89.45
1/2 CY	C8@2.47	Ea	87.40	95.00	--	182.40
3/4 CY	C8@3.80	Ea	132.00	146.00	--	278.00
1 CY	C8@4.86	Ea	178.50	187.00	--	365.50

Mechanical 15

	Craft@Hrs	Unit	Material	Labor	Equipment	Total

Cast Iron Fittings, Valves and Accessories, continued
Tapping saddles, double strap, iron body, tap size to 2"

	Craft@Hrs	Unit	Material	Labor	Equipment	Total
Pipe to 4"	U1@2.38	Ea	24.50	94.50	19.70	138.70
8" pipe	U1@2.89	Ea	34.30	115.00	23.90	173.20
10" pipe	U1@3.15	Ea	43.20	125.00	26.00	194.20
12" pipe	U1@3.52	Ea	54.25	140.00	29.00	223.25
14" pipe, bronze body	U1@3.78	Ea	284.00	150.00	31.10	465.10

Tapping valves, mechanical joint

	Craft@Hrs	Unit	Material	Labor	Equipment	Total
4"	U1@6.93	Ea	275.00	275.00	57.30	607.30
6"	U1@8.66	Ea	375.00	344.00	71.40	790.40
8"	U1@9.32	Ea	550.00	370.00	76.80	996.80
10"	U1@10.4	Ea	830.00	413.00	86.30	1,329.30
12"	U1@11.5	Ea	1,265.00	457.00	95.10	1,817.10

Labor and equipment tapping pipe, by hole size

	Craft@Hrs	Unit	Material	Labor	Equipment	Total
4"	U1@5.90	Ea	--	234.00	48.70	282.70
6"	U1@9.32	Ea	--	370.00	76.80	446.80
8"	U1@11.7	Ea	--	465.00	96.20	561.20
10"	U1@15.5	Ea	--	616.00	128.00	744.00
12"	U1@23.4	Ea	--	929.00	194.00	1,123.00

Copper Tube and Soldered Copper Fittings. Note that copper tube and fitting prices change very quickly as the price of copper changes. During 1991 prices increased about 5% and then declined about 20%. More price changes can be expected in 1992. The costs for copper tube and fittings in the following sections include the subcontractor's overhead and profit.

Type L hard copper tube with typical fittings and hangers Use these figures for preliminary estimates when the exact number of fittings and hangers have not been calculated. Solder is 50% lead and 50% tin. Installed exposed either horizontally or vertically in a building and up to 10' above floor level

	Craft@Hrs	Unit	Material	Labor	Total
1/2" type L hard copper tube with clevis hangers and fittings as noted					
Coupling each 20' and a hanger each 6'	P6@.091	LF	1.10	3.72	4.82
Tee each 20' and a hanger each 6'	P6@.098	LF	1.15	4.01	5.16
Two ells each 10' and a hanger each 6'	P6@.131	LF	1.20	5.35	6.55
3/4" type L hard copper tube with clevis hangers and fittings as noted					
Coupling each 20' and a hanger each 6'	P6@.093	LF	1.60	3.80	5.40
Tee each 20' and hanger each 6'	P6@.098	LF	1.70	4.01	5.71
Two ells each 10' and a hanger each 6'	P6@.136	LF	1.85	5.56	7.41
1" type L hard copper tube with clevis hangers and fittings as noted					
Coupling each 20' and a hanger each 6'	P6@.098	LF	2.20	4.01	6.21
Tee each 20' and a hanger each 6'	P6@.103	LF	2.50	4.21	6.71
Two ells each 10' and a hanger each 6'	P6@.148	LF	2.75	6.05	8.80
1-1/2" type L hard copper tube with clevis hangers and fittings as noted					
Coupling each 20' and a hanger each 8'	P6@.116	LF	3.65	4.74	8.39
Tee each 20' and a hanger each 8'	P6@.119	LF	4.35	4.86	9.21
Two ells each 10' and a hanger each 8'	P6@.173	LF	5.00	7.07	12.07
2" type L hard copper tube with clevis hangers and fittings as noted					
Coupling each 20' and a hanger each 8'	P6@.120	LF	5.50	4.90	10.40
Tee each 20' and hanger each 8'	P6@.122	LF	6.60	4.99	11.59
Two ells each 10' and a hanger each 8'	P6@.178	LF	8.00	7.28	15.28
Add for installations over 10' to 20' high	--	%	--	25.0	--
Add for soft copper type L tube	--	%	10.0	--	--

Mechanical 15

	Craft@Hrs	Unit	Material	Labor	Total

Type L and M copper tube No fittings or hangers included. See fittings and hangers in the sections that follow. Tube installed exposed either horizontally or vertically in a building and up to 10 above floor level. Add 25% to the labor cost for heights over 10'

Type L hard copper tube

	Craft@Hrs	Unit	Material	Labor	Total
1/2" tube	P6@.050	LF	.85	2.04	2.89
3/4" tube	P6@.050	LF	1.35	2.04	3.39
1" tube	P6@.050	LF	1.85	2.04	3.89
1-1/4" tube	P6@.066	LF	2.55	2.70	5.25
1-1/2" tube	P6@.066	LF	3.25	2.70	5.95
2" tube	P6@.066	LF	4.90	2.70	7.60
2-1/2" tube	P6@.075	LF	7.10	3.07	10.17
3" tube	P6@.103	LF	9.60	4.21	13.81
4" tube	P6@.128	LF	16.00	5.23	21.23
Add for soft type L copper tube	--	%	10.0	--	--

Type M hard copper tube

	Craft@Hrs	Unit	Material	Labor	Total
1/2" tube	P6@.048	LF	.60	1.96	2.56
3/4" tube	P6@.048	LF	1.00	1.96	2.96
1" tube	P6@.048	LF	1.40	1.96	3.36
1-1/4" tube	P6@.063	LF	2.05	2.58	4.63
1-1/2" tube	P6@.063	LF	2.85	2.58	5.43
2" tube	P6@.063	LF	4.50	2.58	7.08
2-1/2" tube	P6@.071	LF	6.15	2.90	9.05
3" tube	P6@.098	LF	7.80	4.01	11.81
4" tube	P6@.122	LF	14.00	4.99	18.99

Type K hard copper tube Installed in an open trench to 5' deep. No fittings, excavation or backfill included. See the cost of fittings in the section that follows

	Craft@Hrs	Unit	Material	Labor	Total
3/4" tube	P6@.024	LF	1.85	.98	2.83
1" tube	P6@.024	LF	2.40	.98	3.38
1-1/4" tube	P6@.028	LF	3.05	1.14	4.19
1-1/2" tube	P6@.029	LF	4.00	1.19	5.19
2" tube	P6@.032	LF	5.90	1.31	7.21
2-1/2" tube	P6@.030	LF	8.60	1.23	9.83
Add for soft copper type K tube	--	%	5.0	--	--

Copper pressure fittings. Soldered wrought copper fittings. Solder is 50% lead and 50% tin. Installed exposed either horizontally or vertically in a building up to 10' above floor level or in an open trench. Add 25% to the labor cost when fittings are installed in a building at heights over 10'. Use these figures when final plans are available

1/2" copper fittings

	Craft@Hrs	Unit	Material	Labor	Total
1/2" 90 degree ell	P6@.280	Ea	.50	11.40	11.90
1/2" 45 degree ell	P6@.280	Ea	.95	11.40	12.35
1/2" tee	P6@.426	Ea	.90	17.40	18.30
1/2" reducing tee (reduced branch only)	P6@.449	Ea	4.40	18.40	22.80
1/2" cap	P6@.147	Ea	.35	6.01	6.36
1/2" coupling	P6@.329	Ea	.40	13.40	13.80
1/2" tube to male thread adapter	P6@.147	Ea	1.15	6.01	7.16

3/4" copper fittings

	Craft@Hrs	Unit	Material	Labor	Total
3/4" 90 degree ell	P6@.303	Ea	1.15	12.40	13.55
3/4" 45 degree ell	P6@.303	Ea	1.60	12.40	14.00
3/4" tee	P6@.456	Ea	2.15	18.60	20.75
3/4" reducing tee (reduced branch only)	P6@.479	Ea	2.00	19.60	21.60
3/4" cap	P6@.166	Ea	.60	6.78	7.38
3/4" coupling	P6@.365	Ea	.80	14.90	15.70

Mechanical 15

	Craft@Hrs	Unit	Material	Labor	Total
Copper pressure fittings, continued					
3/4" tube to male thread adapter	P6@.166	Ea	1.80	6.78	8.58
1" copper fittings					
1" 90 degree ell	P6@.367	Ea	2.75	15.00	17.75
1" 45 degree ell	P6@.367	Ea	4.10	15.00	19.10
1" tee	P6@.551	Ea	6.30	22.50	28.80
1" reducing tee (reduced branch only)	P6@.576	Ea	6.60	23.50	30.10
1" cap	P6@.201	Ea	1.55	8.22	9.77
1" coupling	P6@.449	Ea	1.60	18.40	20.00
1" tube to male thread adapter	P6@.201	Ea	4.65	8.22	12.87
1-1/4" copper fittings					
1-1/4" 90 degree ell	P6@.385	Ea	4.30	15.70	20.00
1-1/4" 45 degree ell	P6@.385	Ea	5.70	15.70	21.40
1-1/4" tee	P6@.604	Ea	10.00	24.70	34.70
1-1/4" reducing tee (reduced branch only)	P6@.645	Ea	9.60	26.40	36.00
1-1/4" cap	P6@.222	Ea	2.25	9.07	11.32
1-1/4" coupling	P6@.507	Ea	2.80	20.70	23.50
1-1/4" tube to male thread adapter	P6@.222	Ea	6.90	9.07	15.97
1-1/2" copper fittings					
1-1/2" 90 degree ell	P6@.426	Ea	6.70	17.40	24.10
1-1/2" 45 degree ell	P6@.426	Ea	6.75	17.40	24.15
1-1/2" tee	P6@.645	Ea	13.85	26.40	40.25
1-1/2" reducing tee (reduced branch only)	P6@.737	Ea	10.90	30.10	41.00
1-1/2" cap	P6@.256	Ea	3.25	10.50	13.75
1-1/2" coupling	P6@.566	Ea	3.70	23.10	26.80
1-1/2" tube to male thread adapter	P6@.256	Ea	7.90	10.50	18.40
2" copper fittings					
2" 90 degree ell	P6@.454	Ea	12.25	18.60	30.85
2" 45 degree ell	P6@.454	Ea	11.25	18.60	29.85
2" tee	P6@.683	Ea	21.60	27.90	49.50
2" reducing tee (reduced branch only)	P6@.849	Ea	16.30	34.70	51.00
2" cap	P6@.293	Ea	5.90	12.00	17.90
2" coupling	P6@.655	Ea	6.10	26.80	32.90
2" tube to male thread adapter	P6@.293	Ea	13.40	12.00	25.40
2-1/2" copper fittings					
2-1/2" 90 degree ell	P6@.558	Ea	23.20	22.80	46.00
2-1/2" 45 degree ell	P6@.558	Ea	24.15	22.80	46.95
2-1/2" tee	P6@.839	Ea	43.35	34.30	77.65
2-1/2" reducing tee (reduced branch only)	P6@1.01	Ea	49.40	41.30	90.70
2-1/2" cap	P6@.357	Ea	18.40	14.60	33.00
2-1/2" coupling	P6@.719	Ea	10.90	29.40	40.30
2-1/2" tube to male thread adapter	P6@.357	Ea	41.30	14.60	55.90
3" copper fittings					
3" 90 degree ell	P6@.918	Ea	32.55	37.50	70.05
3" 45 degree ell	P6@.918	Ea	35.80	37.50	73.30
3" tee	P6@1.03	Ea	86.80	42.10	128.90
3" reducing tee (reduced branch only)	P6@1.28	Ea	54.70	52.30	107.00
3" cap	P6@.456	Ea	25.20	18.60	43.80
3" coupling	P6@.918	Ea	20.00	37.50	57.50
3" tube to male thread adapter	P6@.456	Ea	56.30	18.60	74.90
4" copper fittings					
4" 90 degree ell	P6@1.38	Ea	75.30	56.40	131.70
4" 45 degree ell	P6@1.38	Ea	75.30	56.40	131.70
4" tee	P6@1.62	Ea	143.00	66.20	209.20

Mechanical 15

	Craft@Hrs	Unit	Material	Labor	Total
4" reducing tee (reduced branch only)	P6@1.86	Ea	97.00	76.00	173.00
4" cap	P6@.691	Ea	49.40	28.20	77.60
4" coupling	P6@1.38	Ea	39.00	56.40	95.40
4" tube to male thread adapter	P6@.691	Ea	85.50	28.20	113.70
Add to the costs of similar fittings above					
For *reducing* ells and tees, add	--	%	100.0	--	--
For *long turn* ells, add	--	%	105.0	--	--
For *reducing* couplings, add	--	%	40.0	--	--
For tube to *female* thread adapter, add	--	%	10.0	--	--

Copper "DWV" Drainage Tube. Installed in a building including tube and miscellaneous materials (one fitting each 10' and typical supports). These costs include the subcontractor's overhead and profit.

	Craft@Hrs	Unit	Material	Labor	Total
1-1/4" tube at $1.96/LF, misc. materials at $1.77/LF	P6@.249	LF	3.73	10.20	13.93
1-1/2" tube at $2.46/LF, misc. materials at $1.74/LF	P6@.263	LF	4.20	10.70	14.90
2" tube at $3.27/LF, misc. materials at $3.18/LF	P6@.331	LF	6.45	13.50	19.95
3" tube at $5.15/LF, misc. materials at $6.35/LF	P6@.357	LF	11.50	14.60	26.10
4" tube at $9.30/LF, misc. materials at $18.30/LF	P6@.415	LF	27.60	17.00	44.60

Schedule 40 PVC Pipe with Typical Fittings and Supports. Pipe installed in a building up to 10' above floor level. Add 25% to the labor cost for pipe installed over 10' and up to 20' above floor level. Use these figures for preliminary estimating. These costs include the subcontractor's overhead and profit. Based on one coupling, one elbow and one hanger every 10'.

	Craft@Hrs	Unit	Material	Labor	Total
1/2" pipe at $.20/LF, misc. materials at $.40/LF	P6@.127	LF	.60	5.19	5.79
3/4" pipe at $.25/LF, misc. materials at $.45/LF	P6@.141	LF	.70	5.76	6.46
1" pipe at $.40/LF, misc. materials at $.70/LF	P6@.166	LF	1.10	6.78	7.88
1-1/4" pipe at $.50/LF, misc. materials at $.90/LF	P6@.193	LF	1.40	7.89	9.29
1-1/2" pipe at $.60/LF, misc. materials at $.95/LF	P6@.213	LF	1.55	8.71	10.26
2" pipe at $.82/LF, misc. materials at $1.58/LF	P6@.235	LF	2.40	9.60	12.00
2-1/2" pipe at $1.30/LF, misc. materials at $3.45/LF	P6@.249	LF	4.75	10.20	14.95
3" pipe at $1.65/LF, misc. materials at $4.70/LF	P6@.316	LF	6.35	12.90	19.25
4" pipe at $2.40/LF, misc. materials at $6.90/LF	P6@.477	LF	9.30	19.50	28.80

Schedule 40 PVC Pressure Pipe and Socket Type Fittings. Pipe and fittings priced separately, installed in a building vertically or horizontally up to 10' above floor level. These costs include the subcontractor's overhead and profit.

	Craft@Hrs	Unit	Material	Labor	Total
1/2" Schedule 40 pipe and fittings					
1/2" pipe	P6@.036	LF	.20	1.47	1.67
1/2" 90 degree ell	P6@.164	Ea	.60	6.70	7.30
1/2" tee	P6@.254	Ea	.75	10.40	11.15
3/4" Schedule 40 pipe and fittings					
3/4" pipe	P6@.036	LF	.25	1.47	1.72
3/4" 90 degree ell	P6@.219	Ea	.70	8.95	9.65
3/4" tee	P6@.331	Ea	.85	13.50	14.35
1" Schedule 40 pipe and fittings					
1" pipe	P6@.034	LF	.40	1.39	1.79
1" 90 degree ell	P6@.273	Ea	1.20	11.20	12.40
1" tee	P6@.410	Ea	1.60	16.80	18.40
1-1/4" Schedule 40 pipe and fittings					
1-1/4" pipe	P6@.038	LF	.50	1.55	2.05
1-1/4" 90 degree ell	P6@.331	Ea	2.15	13.50	15.65
1-1/4" tee	P6@.497	Ea	2.55	20.30	22.85
1-1/2" Schedule 40 pipe and fittings					
1-1/2" pipe	P6@.040	LF	.60	1.64	2.24
1-1/2" 90 degree ell	P6@.387	Ea	2.30	15.80	18.10

Mechanical 15

	Craft@Hrs	Unit	Material	Labor	Total
Schedule 40 PVC Pressure Pipe and Socket Type Fittings, 1-1/2", continued					
1-1/2" tee	P6@.586	Ea	3.10	24.00	27.10
2" Schedule 40 pipe and fittings					
2" pipe	P6@.046	LF	.80	1.88	2.68
2" 90 degree ell	P6@.444	Ea	3.60	18.10	21.70
2" tee	P6@.663	Ea	4.45	27.10	31.55
2-1/2" Schedule 40 pipe and fittings					
2-1/2" pipe	P6@.053	LF	1.30	2.17	3.47
2-1/2" 90 degree ell	P6@.505	Ea	11.00	20.60	31.60
2-1/2" tee	P6@.757	Ea	14.70	30.90	45.60
3" Schedule 40 pipe and fittings					
3" pipe	P6@.061	LF	1.65	2.49	4.14
3" 90 degree ell	P6@.568	Ea	13.15	23.20	36.35
3" tee	P6@.859	Ea	19.30	35.10	54.40
4" Schedule 40 pipe and fittings					
4" pipe	P6@.093	LF	2.40	3.80	6.20
4" 90 degree ell	P6@.890	Ea	23.55	36.40	59.95
4" tee	P6@1.33	Ea	35.00	54.40	89.40
Add for heights over 10' to 20'	--	%	--	25.0	--
Add for Schedule 80 (gray) PVC pipe	--	%	40.0	--	--

Schedule 80 CPVC Hot Water Pipe. Pipe and Schedule 80 CPVC socket type fittings installed in a building vertically or horizontally up to 10' above floor level. These costs include the subcontractor's overhead and profit.

	Craft@Hrs	Unit	Material	Labor	Total
1/2" Schedule 80 pipe and fittings					
1/2" pipe	P6@.037	LF	.60	1.51	2.11
1/2" 90 degree ell	P6@.172	Ea	3.90	7.03	10.93
1/2" tee	P6@.268	Ea	5.80	11.00	16.80
3/4" Schedule 80 pipe and fittings					
3/4" pipe	P6@.037	LF	.80	1.51	2.31
3/4" 90 degree ell	P6@.233	Ea	5.00	9.52	14.52
3/4" tee	P6@.347	Ea	9.30	14.20	23.50
1" Schedule 80 pipe and fittings					
1" pipe	P6@.037	LF	1.20	1.51	2.71
1" 90 degree ell	P6@.291	Ea	7.85	11.90	19.75
1" tee	P6@.438	Ea	11.40	17.90	29.30
1-1/4" Schedule 80 pipe and fittings					
1-1/4" pipe	P6@.040	LF	1.70	1.64	3.34
1-1/4" 90 degree ell	P6@.347	Ea	17.00	14.20	31.20
1-1/4" tee	P6@.515	Ea	24.00	21.00	45.00
1-1/2" Schedule 80 pipe and fittings					
1-1/2" pipe	P6@.042	LF	2.05	1.72	3.77
1-1/2" 90 degree ell	P6@.367	Ea	19.00	15.00	34.00
1-1/2" tee	P6@.617	Ea	27.50	25.20	52.70
2" Schedule 80 pipe and fittings					
2" pipe	P6@.048	LF	2.80	1.96	4.76
2" 90 degree ell	P6@.461	Ea	23.00	18.80	41.80
2" tee	P6@.696	Ea	30.60	28.40	59.00
Add for heights over 10' to 20'	--	%	--	25.0	--
Deduct for Schedule 40 CPVC pipe	--	%	-12.0	--	--
Add for threaded CPVC fittings	--	%	60.0	25.0	--

Mechanical 15

	Craft@Hrs	Unit	Material	Labor	Total

PVC "DWV" Drain Pipe. Installed in a building including typical fittings and supports. (One fitting each 10' and typical supports) These costs include the subcontractor's overhead and profit.

	Craft@Hrs	Unit	Material	Labor	Total
1-1/2" pipe at $.60/LF, materials at $.40/LF	P6@.201	LF	1.00	8.22	9.22
2" pipe at $.82/LF, materials at $.53/LF	P6@.256	LF	1.35	10.50	11.85
3" pipe at $1.65/LF, materials at $1.25/LF	P6@.280	LF	2.90	11.40	14.30
4" pipe at $2.40/LF, materials at $2.70/LF	P6@.319	LF	5.10	13.00	18.10
6" pipe at $4.05/LF, materials at $4.10/LF	P6@.395	LF	8.15	16.10	24.25

ABS "DWV" Drain Pipe. Installed in a building including typical fittings and supports. (One fitting each 10' and typical supports) These costs include the subcontractor's overhead and profit.

	Craft@Hrs	Unit	Material	Labor	Total
1-1/2" pipe at $.60/LF, materials at $.50/LF	P6@.201	LF	1.10	8.22	9.32
2" pipe at $.82/LF, materials at $.58/LF	P6@.256	LF	1.40	10.50	11.90
3" pipe at $1.70/LF, materials at $1.45/LF	P6@.280	LF	3.15	11.40	14.55
4" pipe at $2.40/LF, materials at $2.40/LF	P6@.319	LF	4.80	13.00	17.80
6" pipe at $6.30/LF, materials at $4.00/LF	P6@.395	LF	10.30	16.10	26.40

Polyethylene Pipe. Installed in an open trench. No excavation, shoring, backfill or dewatering included. These costs include the subcontractor's overhead and profit.

160 PSI polyethylene pipe, size to diameter ratio (SDR) 11 to 1

	Craft@Hrs	Unit	Material	Labor	Total
3"	P6@.108	LF	1.65	4.41	6.06
4"	P6@.108	LF	2.75	4.41	7.16
5"	P6@.108	LF	4.25	4.41	8.66
6"	P6@.108	LF	5.85	4.41	10.26
8"	P6@.145	LF	9.95	5.93	15.88
10"	P6@.145	LF	15.50	5.93	21.43
12"	P6@.145	LF	21.75	5.93	27.68

100 PSI polyethylene pipe, size to diameter ratio (SDR) 17 to 1

	Craft@Hrs	Unit	Material	Labor	Total
4"	P6@.108	LF	1.85	4.41	6.26
6"	P6@.108	LF	3.95	4.41	8.36
8"	P6@.145	LF	6.65	5.93	12.58
10"	P6@.145	LF	10.35	5.93	16.28
12"	P6@.145	LF	14.55	5.93	20.48

Low pressure polyethylene pipe, size to diameter ratio (SDR) 21 to 1

	Craft@Hrs	Unit	Material	Labor	Total
4"	P6@.108	LF	1.50	4.41	5.91
5"	P6@.108	LF	2.27	4.41	6.68
6"	P6@.108	LF	3.32	4.41	7.73
7"	P6@.145	LF	3.75	5.93	9.68
8"	P6@.145	LF	5.50	5.93	11.43
10"	P6@.145	LF	8.50	5.93	14.43
12"	P6@.145	LF	11.95	5.93	17.88
14"	P6@.169	LF	14.40	6.91	21.31
16"	P6@.169	LF	18.80	6.91	25.71
18"	P6@.169	LF	23.75	6.91	30.66
20"	P6@.212	LF	29.35	8.66	38.01
22"	P6@.212	LF	35.50	8.66	44.16
24"	P6@.212	LF	42.25	8.66	50.91

Fittings, molded, butt fusion joint

	Craft@Hrs	Unit	Material	Labor	Total
4" 90 degree ell	P6@3.72	Ea	35.25	152.00	187.25
4" 45 degree ell	P6@5.58	Ea	35.25	228.00	263.25
6" 90 degree ell	P6@3.75	Ea	73.60	153.00	226.60
6" 45 degree ell	P6@5.58	Ea	73.60	228.00	301.60
4" to 3" reducer	P6@3.75	Ea	21.50	153.00	174.50
6" to 4" reducer	P6@5.58	Ea	51.25	228.00	279.25
3" transition section	P6@3.29	Ea	45.00	134.00	179.00

Mechanical 15

	Craft@Hrs	Unit	Material	Labor	Total
Polyethylene Pipe, Fittings, continued					
6" transition section	P6@5.58	Ea	200.00	228.00	428.00
10" by 4" branch saddle	P6@8.39	Ea	65.00	343.00	408.00
3" flanged adapter, low pressure	P6@3.29	Ea	33.60	134.00	167.60
4" flanged adapter, low pressure	P6@3.75	Ea	48.15	153.00	201.15
6" flanged adapter, low pressure	P6@5.58	Ea	63.15	228.00	291.15
8" flanged adapter, low pressure	P6@6.55	Ea	90.50	268.00	358.50
10" flanged adapter, low pressure	P6@6.55	Ea	130.50	268.00	398.50
4" flanged adapter, 150 PSI	P6@3.75	Ea	52.50	153.00	205.50
6" flanged adapter, 150 PSI	P6@5.58	Ea	74.29	228.00	302.29
Fittings, fabricated, 100 PSI					
10" 90 degree ell	P6@8.39	Ea	370.00	343.00	713.00
10" 45 degree ell	P6@8.39	Ea	205.00	343.00	548.00
4" 30 degree ell	P6@3.75	Ea	60.00	153.00	213.00
8" to 6" reducer	P6@6.55	Ea	110.00	268.00	378.00
10" to 8" reducer	P6@8.39	Ea	130.00	343.00	473.00
4" 45 degree wye	P6@3.75	Ea	158.00	153.00	311.00
10" 45 degree wye	P6@8.26	Ea	500.00	338.00	838.00

Fiberglass Reinforced Plastic Pipe. Installed with adhesive bonded bell and spigot fittings. No trenching, backfill or supports included. Uninsulated pipe. These costs include the subcontractor's overhead and profit.

	Craft@Hrs	Unit	Material	Labor	Total
2" pipe and fittings					
Pipe	P6@.087	LF	4.31	3.56	7.87
90 degree ell	P6@.744	Ea	26.00	30.40	56.40
45 degree ell	P6@.744	Ea	29.10	30.40	59.50
Tee	P6@1.10	Ea	53.80	45.00	98.80
3" pipe and fittings					
Pipe	P6@.091	LF	5.59	3.72	9.31
90 degree ell	P6@1.27	Ea	29.10	51.90	81.00
45 degree ell	P6@1.27	Ea	25.80	51.90	77.70
Tee	P6@1.93	Ea	64.60	78.90	143.50
4" pipe and fittings					
Pipe	P6@.100	LF	7.53	4.09	11.62
90 degree ell	P6@1.50	Ea	34.40	61.30	95.70
45 degree ell	P6@1.50	Ea	36.10	61.30	97.40
Tee	P6@2.22	Ea	75.30	90.70	166.00
6" pipe and fittings					
Pipe	P6@.194	LF	11.40	7.93	19.33
90 degree ell	P6@3.01	Ea	67.80	123.00	190.80
45 degree ell	P6@3.01	Ea	73.40	123.00	196.40
Tee	P6@4.54	Ea	149.00	186.00	335.00
8" pipe and fittings					
Pipe	P6@.268	LF	20.20	11.00	31.20
90 degree ell	P6@3.31	Ea	184.00	135.00	319.00
45 degree ell	P6@3.31	Ea	189.00	135.00	324.00
Tee	P6@5.38	Ea	245.00	220.00	465.00
10" pipe and fittings					
Pipe	P6@.344	LF	30.30	14.10	44.40
90 degree ell	P6@3.31	Ea	325.00	135.00	460.00
45 degree ell	P6@3.31	Ea	337.00	135.00	472.00
Tee	P6@5.02	Ea	426.00	205.00	631.00

Mechanical 15

	Craft@Hrs	Unit	Material	Labor	Total

Stainless Steel Pipe. These costs include the subcontractor's overhead and profit.
1/2" stainless steel Schedule 40 pipe and butt welded 3,000 lb fittings installed in a building

Pipe, butt welded, no fittings or supports included, 1/2"

	Craft@Hrs	Unit	Material	Labor	Total
Type 304L pipe installed to 10' high	M5@.094	LF	3.25	3.77	7.02
Type 304L pipe over 10' to 20' high	M5@.105	LF	3.25	4.21	7.46
Type 316L pipe installed to 10' high	M5@.094	LF	4.30	3.77	8.07
Type 316L pipe installed over 10' to 20' high	M5@.105	LF	4.30	4.21	8.51

90 degree ells, butt welded, long radius, 1/2"

	Craft@Hrs	Unit	Material	Labor	Total
Type 304	M5@.703	Ea	5.70	28.20	33.90
Type 316	M5@.703	Ea	6.80	28.20	35.00

45 degree ells, butt welded, long radius, 1/2"

	Craft@Hrs	Unit	Material	Labor	Total
Type 304	M5@.703	Ea	7.50	28.20	35.70
Type 316	M5@.703	Ea	8.55	28.20	36.75

Tees, butt welded, 1/2"

	Craft@Hrs	Unit	Material	Labor	Total
Type 304	M5@1.05	Ea	15.30	42.10	57.40
Type 316	M5@1.05	Ea	17.40	42.10	59.50

Caps, butt welded, 1/2"

	Craft@Hrs	Unit	Material	Labor	Total
Type 304	M5@.354	Ea	14.40	14.20	28.60
Type 316	M5@.354	Ea	16.30	14.20	30.50

3/4" stainless steel Schedule 40 pipe and butt welded, 3,000 lb fittings installed in a building

Pipe, butt welded, no fittings or supports included, 3/4"

	Craft@Hrs	Unit	Material	Labor	Total
Type 304L pipe to 10' high	M5@.096	LF	4.00	3.85	7.85
Type 304L pipe over 10' to 20' high	M5@.105	LF	4.00	4.21	8.21
Type 316L pipe to 10' high	M5@.094	LF	4.85	3.77	8.62
Type 316L pipe over 10' to 20' high	M5@.105	LF	4.85	4.21	9.06

90 degree ells, butt welded, long radius, 3/4"

	Craft@Hrs	Unit	Material	Labor	Total
Type 304	M5@.703	Ea	5.70	28.20	33.90
Type 316	M5@.703	Ea	6.85	28.20	35.05

45 degree ells, butt welded, long radius, 3/4"

	Craft@Hrs	Unit	Material	Labor	Total
Type 304	M5@.703	Ea	7.50	28.20	35.70
Type 316	M5@.703	Ea	8.65	28.20	36.85

Tees, butt welded, 3/4"

	Craft@Hrs	Unit	Material	Labor	Total
Type 304	M5@1.05	Ea	15.75	42.10	57.85
Type 316	M5@1.05	Ea	18.10	42.10	60.20

Caps, butt welded, 3/4"

	Craft@Hrs	Unit	Material	Labor	Total
Type 304	M5@.354	Ea	16.75	14.20	30.95
Type 316	M5@.354	Ea	17.30	14.20	31.50

1" stainless steel Schedule 40 pipe and butt welded, 3,000 lb fittings installed in a building

Pipe, butt welded, no fittings or supports included, 1"

	Craft@Hrs	Unit	Material	Labor	Total
Type 304L pipe to 10' high	M5@.103	LF	5.05	4.13	9.18
Type 304L pipe over 10' to 20' high	M5@.123	LF	5.05	4.93	9.98
Type 316L pipe to 10' high	M5@.103	LF	6.45	4.13	10.58
Type 316L pipe over 10' to 20' high	M5@.123	LF	6.45	4.93	11.38

90 degree ells, butt welded, long radius, 1"

	Craft@Hrs	Unit	Material	Labor	Total
Type 304	M5@.846	Ea	5.70	33.90	39.60
Type 316	M5@.846	Ea	6.85	33.90	40.75

45 degree ells, butt welded, long radius, 1"

	Craft@Hrs	Unit	Material	Labor	Total
Type 304	M5@.846	Ea	7.50	33.90	41.40
Type 316	M5@.846	Ea	8.50	33.90	42.40

Tees, butt welded, 1"

	Craft@Hrs	Unit	Material	Labor	Total
Type 304	M5@1.26	Ea	23.60	50.50	74.10
Type 316	M5@1.26	Ea	25.00	50.50	75.50

Caps, butt welded, 1"

	Craft@Hrs	Unit	Material	Labor	Total
Type 304	M5@.422	Ea	16.90	16.90	33.80

Mechanical 15

	Craft@Hrs	Unit	Material	Labor	Total
Stainless Steel Pipe, continued					
Caps, butt welded, 1", Type 316	M5@.422	Ea	17.40	16.90	34.30
1-1/4" stainless steel Schedule 40 pipe and butt welded, 3,000 lb fittings installed in a building					
Pipe, butt welded, no fittings or supports included, 1-1/4"					
Type 304L pipe to 10', high	M5@.111	LF	6.35	4.45	10.80
Type 304L pipe over 10' to 20' high	M5@.131	LF	6.35	5.25	11.60
Type 316L pipe to 10' high	M5@.111	LF	8.15	4.45	12.60
Type 316L pipe over 10' to 20' high	M5@.131	LF	8.15	5.25	13.40
90 degree ells, butt welded, long radius, 1-1/4"					
Type 304	M5@.947	Ea	12.60	37.90	50.50
Type 316	M5@.947	Ea	15.30	37.90	53.20
45 degree ells, butt welded, long radius, 1-1/4"					
Type 304	M5@.947	Ea	14.60	37.90	52.50
Type 316	M5@.880	Ea	17.30	35.30	52.60
Tees, butt welded, 1-1/4"					
Type 304	M5@1.41	Ea	42.20	56.50	98.70
Type 316	M5@1.41	Ea	48.70	56.50	105.20
Caps, butt welded, 1-1/4"					
Type 304	M5@.466	Ea	26.60	18.70	45.30
Type 316	M5@.466	Ea	34.50	18.70	53.20
1-1/2" stainless steel Schedule 40 pipe and butt welded, 3,000 lb fittings installed in a building					
Pipe, butt welded, no fittings or supports included, 1-1/2"					
Type 304L pipe to 10' high	M5@.116	LF	7.45	4.65	12.10
Type 304L pipe over 10' to 20' high	M5@.140	LF	7.45	5.61	13.06
Type 316L pipe to 10' high	M5@.116	LF	9.70	4.65	14.35
Type 316L pipe over 10' to 20' high	M5@.140	LF	9.70	5.61	15.31
90 degree ells, butt welded, long radius, 1-1/2"					
Type 304	M5@1.02	Ea	8.70	40.90	49.60
Type 316	M5@1.02	Ea	10.40	40.90	51.30
45 degree ells, butt welded, long radius, 1-1/2"					
Type 304	M5@1.02	Ea	9.45	40.90	50.35
Type 316	M5@1.02	Ea	10.40	40.90	51.30
Tees, butt welded, 1-1/2"					
Type 304	M5@1.54	Ea	41.70	61.70	103.40
Type 316	M5@1.54	Ea	48.00	61.70	109.70
Caps, butt welded, 1-1/2"					
Type 304	M5@.515	Ea	13.80	20.60	34.40
Type 316	M5@.515	Ea	15.70	20.60	36.30
2" stainless steel Schedule 40 pipe and butt welded, 3,000 lb fittings installed in a building					
Pipe, butt welded, no fittings or supports included, 2"					
Type 304L pipe to 10' high	M7@.118	LF	9.70	4.96	14.66
Type 304L pipe over 10' to 20' high	M7@.141	LF	9.70	5.93	15.63
Type 316L pipe to 10' high	M7@.118	LF	12.70	4.96	17.66
Type 316L pipe over 10' to 20' high	M7@.141	LF	12.70	5.93	18.63
90 degree ells, butt welded, long radius, 2"					
Type 304	M7@1.25	Ea	15.30	52.60	67.90
Type 316	M7@1.25	Ea	17.50	52.60	70.10
45 degree ells, butt welded, long radius, 2"					
Type 304	M7@1.25	Ea	13.80	52.60	66.40
Type 316	M7@1.25	Ea	15.70	52.60	68.30
Tees, butt welded, 2"					
Type 304	M7@1.91	Ea	42.60	80.30	122.90
Type 316	M7@1.91	Ea	47.30	80.30	127.60

Mechanical 15

	Craft@Hrs	Unit	Material	Labor	Total
Caps, butt welded, 2"					
Type 304	M7@.625	Ea	20.60	26.30	46.90
Type 316	M7@.625	Ea	15.80	26.30	42.10

Hangers and Supports. By intended pipe size. These costs include the subcontractor's overhead and profit.

	Craft@Hrs	Unit	Material	Labor	Total
Angle brackets, steel, by rod size					
3/8"	M5@.070	Ea	.69	2.80	3.49
1/2"	M5@.070	Ea	1.18	2.80	3.98
5/8"	M5@.070	Ea	1.39	2.80	4.19
3/4"	M5@.070	Ea	1.79	2.80	4.59
7/8"	M5@.070	Ea	2.72	2.80	5.52
Angle supports, wall mounted, welded steel					
12" x 18", medium weight	M5@.546	Ea	57.00	21.90	78.90
18" x 24", medium weight	M5@.546	Ea	70.25	21.90	92.15
24" x 30", medium weight	M5@.546	Ea	88.80	21.90	110.70
12" x 30", heavy weight	M5@.546	Ea	80.80	21.90	102.70
18" x 24", heavy weight	M5@.546	Ea	113.60	21.90	135.50
24" x 30", heavy weight	M5@.546	Ea	130.40	21.90	152.30
C-clamps, steel, by rod size					
3/8"	M5@.070	Ea	2.24	2.80	5.04
1/2"	M5@.070	Ea	2.66	2.80	5.46
5/8"	M5@.070	Ea	3.90	2.80	6.70
3/4"	M5@.070	Ea	5.60	2.80	8.40
7/8"	M5@.070	Ea	8.40	2.80	11.20
Clevis or swivel ring hangers, adjustable, black steel, no hanger rod or angle bracket included					
1/2"	M5@.156	Ea	1.36	6.25	7.61
3/4"	M5@.156	Ea	1.36	6.25	7.61
1"	M5@.156	Ea	1.47	6.25	7.72
1-1/4"	M5@.173	Ea	1.68	6.93	8.61
1-1/2"	M5@.173	Ea	1.82	6.93	8.75
2"	M5@.173	Ea	2.20	6.93	9.13
2-1/2"	M5@.189	Ea	3.36	7.57	10.93
3"	M5@.189	Ea	4.13	7.57	11.70
4"	M5@.189	Ea	5.07	7.57	12.64
6"	M5@.202	Ea	8.12	8.09	16.21
8"	M5@.202	Ea	13.34	8.09	21.43
Add for galvanized	--	%	35.0	--	--
Add for copper	--	%	35.0	--	--
Copper tube riser clamps					
1/2"	M5@.216	Ea	7.71	8.65	16.36
3/4"	M5@.216	Ea	7.71	8.65	16.36
1"	M5@.216	Ea	7.71	8.65	16.36
1-1/4"	M5@.216	Ea	7.71	8.65	16.36
1-1/2"	M5@.216	Ea	9.80	8.65	18.45
2"	M5@.216	Ea	10.20	8.65	18.85
2-1/2"	M5@.237	Ea	13.25	9.49	22.74
3"	M5@.237	Ea	14.75	9.49	24.24
3-1/2" and 4"	M5@.237	Ea	14.75	9.49	24.24
Pipe clamps, medium duty					
1/2" or 3/4"	M5@.156	Ea	3.05	6.25	9.30
1"	M5@.156	Ea	3.20	6.25	9.45
1-1/4" or 1-1/2"	M5@.173	Ea	4.50	6.93	11.43
2"	M5@.173	Ea	4.80	6.93	11.73

Mechanical 15

	Craft@Hrs	Unit	Material	Labor	Total
Hangers and Supports, Pipe clamps, continued					
3"	M5@.189	Ea	5.70	7.57	13.27
4"	M5@.189	Ea	7.84	7.57	15.41
6"	M5@.202	Ea	19.67	8.09	27.76
Pipe hooks, 10 gauge wire, nail-on					
1/2" x 4" long	M5@.072	Ea	.05	2.88	2.93
1/2" x 6" long	M5@.072	Ea	.06	2.88	2.94
3/4" x 4" long	M5@.072	Ea	.05	2.88	2.93
3/4" x 6" long	M5@.072	Ea	.06	2.88	2.94
1" x 6" long	M5@.072	Ea	.08	2.88	2.96
Pipe rolls					
1" pipe roll only	M5@.144	Ea	2.98	5.77	8.75
2"	M5@.144	Ea	3.92	5.77	9.69
3"	M5@.268	Ea	4.27	10.70	14.97
4"	M5@.346	Ea	8.75	13.90	22.65
6"	M5@.346	Ea	13.16	13.90	27.06
8"	M5@.385	Ea	21.90	15.40	37.30
1" pipe roll, complete	M5@.216	Ea	15.12	8.65	23.77
2"	M5@.216	Ea	16.05	8.65	24.70
3"	M5@.403	Ea	18.27	16.10	34.37
4"	M5@.518	Ea	23.27	20.80	44.07
6"	M5@.518	Ea	34.45	20.80	55.25
8"	M5@.578	Ea	52.80	23.20	76.00
2" pipe roll stand, complete	M5@.323	Ea	40.00	12.90	52.90
3"	M5@.323	Ea	40.00	12.90	52.90
4"	M5@.604	Ea	54.50	24.20	78.70
6"	M5@.775	Ea	54.50	31.00	85.50
8"	M5@.775	Ea	95.20	31.00	126.20
Pipe straps, galvanized, two hole, lightweight					
3/8"	M5@.069	Ea	.25	2.76	3.01
1/2"	M5@.072	Ea	.15	2.88	3.03
3/4"	M5@.080	Ea	.15	3.21	3.36
1"	M5@.088	Ea	.21	3.53	3.74
1-1/4"	M5@.096	Ea	.26	3.85	4.11
1-1/2"	M5@.104	Ea	.30	4.17	4.47
2"	M5@.112	Ea	.40	4.49	4.89
2-1/2"	M5@.119	Ea	1.85	4.77	6.62
3"	M5@.128	Ea	2.40	5.13	7.53
3-1/2"	M5@.135	Ea	3.00	5.41	8.41
4"	M5@.144	Ea	3.40	5.77	9.17
Plumber's tape, galvanized, 3/4" x 10' roll					
26 gauge	--	LF	.10	--	--
22 gauge	--	LF	.15	--	--
Saddles, 1" covering					
3/4"	M5@.232	Ea	12.30	9.29	21.59
1"	M5@.232	Ea	12.30	9.29	21.59
2"	M5@.297	Ea	19.15	11.90	31.05
4"	M5@.354	Ea	20.50	14.20	34.70
6"	M5@.354	Ea	24.40	14.20	38.60
Sliding pipe guides, welded					
2"	M5@.354	Ea	13.65	14.20	27.85
3"	M5@.354	Ea	14.87	14.20	29.07
4"	M5@.354	Ea	15.70	14.20	29.90
6"	M5@.354	Ea	18.35	14.20	32.55

Mechanical 15

	Craft@Hrs	Unit	Material	Labor	Total
8"	M5@.471	Ea	32.20	18.90	51.10
10"	M5@.471	Ea	42.70	18.90	61.60
Threaded hanger rod, by rod size, in 6' to 10' lengths, per LF					
3/8"	--	LF	.88	--	--
1/2"	--	LF	1.40	--	--
5/8"	--	LF	2.15	--	--
3/4"	--	LF	3.75	--	--
U bolts, with hex nuts, light duty					
1/2"	M5@.089	Ea	.82	3.57	4.39
1"	M5@.089	Ea	.88	3.57	4.45
2"	M5@.089	Ea	1.15	3.57	4.72
4"	M5@.089	Ea	2.20	3.57	5.77
6"	M5@.135	Ea	4.30	5.41	9.71
Wall sleeves, cast iron					
4"	M5@.241	Ea	4.90	9.65	14.55
6"	M5@.284	Ea	6.10	11.40	17.50
8"	M5@.377	Ea	8.10	15.10	23.20

Gate Valves. These costs include the subcontractor's overhead and profit.

	Craft@Hrs	Unit	Material	Labor	Total
1/2" bronze gate valves					
125 lb, threaded	M5@.552	Ea	12.05	22.10	34.15
125 lb, soldered	M5@.312	Ea	8.80	12.50	21.30
150 lb, threaded	M5@.570	Ea	15.20	22.80	38.00
200 lb, threaded	M5@.570	Ea	29.00	22.80	51.80
3/4" bronze gate valves					
125 lb, threaded	M5@.617	Ea	14.70	24.70	39.40
125 lb, soldered	M5@.424	Ea	10.40	17.00	27.40
150 lb, threaded	M5@.617	Ea	18.10	24.70	42.80
200 lb, threaded	M5@.617	Ea	36.00	24.70	60.70
1" bronze gate valves					
125 lb, threaded	M5@.666	Ea	19.70	26.70	46.40
125 lb, soldered	M5@.502	Ea	14.40	20.10	34.50
150 lb, threaded	M5@.666	Ea	23.60	26.70	50.30
200 lb, threaded	M5@.666	Ea	51.70	26.70	78.40
1-1/2" bronze gate valves					
125 lb, threaded	M5@.783	Ea	34.30	31.40	65.70
125 lb, soldered	M5@.614	Ea	24.40	24.60	49.00
150 lb, threaded	M5@.783	Ea	39.50	31.40	70.90
200 lb, threaded	M5@.783	Ea	80.60	31.40	112.00
2" bronze gate valves					
125 lb, threaded	M5@.833	Ea	43.60	33.40	77.00
125 lb, soldered	M5@.666	Ea	34.75	26.70	61.45
150 lb, threaded	M5@.833	Ea	54.30	33.40	87.70
200 lb, threaded	M5@.833	Ea	116.80	33.40	150.20
2-1/2" gate valves					
125 lb, bronze, soldered	M5@.757	Ea	81.10	30.30	111.40
125 lb, cast iron, flanged	M7@.950	Ea	162.00	39.90	201.90
250 lb, cast iron, flanged	M7@.950	Ea	443.50	39.90	483.40
3" cast iron gate valves, 125 lb, flanged	M7@1.09	Ea	180.80	45.80	226.60
4" cast iron gate valves, 125 lb, flanged	M7@2.17	Ea	263.35	91.20	354.55
4" cast iron gate valves, 250 lb, flanged	M7@2.16	Ea	728.30	90.80	819.10

Mechanical 15

	Craft@Hrs	Unit	Material	Labor	Total

Steel Gate, Globe and Check Valves. These costs include the subcontractor's overhead and profit.
Forged steel Class 800 bolted bonnet OS&Y gate valves, screwed ends

	Craft@Hrs	Unit	Material	Labor	Total
1/2"	M5@.497	Ea	50.00	19.90	69.90
3/4"	M5@.531	Ea	56.10	21.30	77.40
1"	M5@.604	Ea	68.05	24.20	92.25
1-1/2"	M5@.757	Ea	130.40	30.30	160.70
2"	M5@.893	Ea	166.30	35.80	202.10

Cast steel Class 150 gate valves, OS&Y, flanged ends

	Craft@Hrs	Unit	Material	Labor	Total
2"	M7@1.00	Ea	450.00	42.00	492.00
2-1/2"	M7@1.11	Ea	610.00	46.70	656.70
3"	M7@1.39	Ea	610.00	58.50	668.50
4"	M7@1.92	Ea	780.00	80.70	860.70

Forged steel Class 800 bolted bonnet OS&Y globe valves, screwed or socket weld ends

	Craft@Hrs	Unit	Material	Labor	Total
1/2"	M5@.497	Ea	75.60	19.90	95.50
3/4"	M5@.531	Ea	85.70	21.30	107.00
1"	M5@.604	Ea	107.10	24.20	131.30
1-1/2"	M5@.757	Ea	221.15	30.30	251.45
2"	M7@.849	Ea	285.00	35.70	320.70

Forged steel Class 800 bolted cap horizontal piston check valves, screwed or socket weld ends

	Craft@Hrs	Unit	Material	Labor	Total
1/2"	M5@.497	Ea	66.80	19.90	86.70
3/4"	M5@.531	Ea	77.00	21.30	98.30
1"	M5@.604	Ea	90.10	24.20	114.30
1-1/2"	M5@.757	Ea	175.80	30.30	206.10
2"	M5@.893	Ea	245.10	35.80	280.90

Outside Screw and Yoke (OS&Y) All Iron Gate Valves. These costs include the subcontractor's overhead and profit.
2" flanged iron OS&Y gate valves

	Craft@Hrs	Unit	Material	Labor	Total
125 lb	M7@.814	Ea	156.90	34.20	191.10
250 lb	M7@.814	Ea	356.40	34.20	390.60

2-1/2" flanged iron OS&Y gate valves

	Craft@Hrs	Unit	Material	Labor	Total
125 lb	M7@.908	Ea	162.00	38.20	200.20
250 lb	M7@.908	Ea	443.50	38.20	481.70

3" flanged iron OS&Y gate valves

	Craft@Hrs	Unit	Material	Labor	Total
125 lb	M7@1.14	Ea	180.80	47.90	228.70
250 lb	M7@1.14	Ea	494.00	47.90	541.90

4" flanged iron OS&Y gate valves

	Craft@Hrs	Unit	Material	Labor	Total
125 lb	M7@1.97	Ea	263.35	82.80	346.15
250 lb	M7@1.97	Ea	728.30	82.80	811.10

6" flanged iron OS&Y gate valves

	Craft@Hrs	Unit	Material	Labor	Total
125 lb	M7@2.50	Ea	443.45	105.00	548.45
250 lb	M7@2.50	Ea	1,232.90	105.00	1,337.90

Butterfly Valves, Ductile Iron. Wafer body, flanged, with lever lock. 150 lb valves have aluminum bronze disk, stainless steel stem and EPDM seals. 200 lb valves have nickle plated ductile iron disk and Buna-N seals. These costs include the subcontractor's overhead and profit.
2" butterfly valves

	Craft@Hrs	Unit	Material	Labor	Total
150 lb	M7@1.02	Ea	55.90	42.90	98.80
200 lb	M7@1.02	Ea	55.10	42.90	98.00

2-1/2" butterfly valves

	Craft@Hrs	Unit	Material	Labor	Total
150 lb	M7@1.97	Ea	58.30	82.80	141.10
200 lb	M7@1.97	Ea	57.50	82.80	140.30

3" butterfly valves

	Craft@Hrs	Unit	Material	Labor	Total
150 lb	M7@1.88	Ea	63.20	79.10	142.30

Mechanical 15

	Craft@Hrs	Unit	Material	Labor	Total
200 lb	M7@2.05	Ea	62.40	86.20	148.60
4" butterfly valves					
150 lb	M7@2.33	Ea	77.75	98.00	175.75
200 lb	M7@2.56	Ea	77.00	108.00	185.00
6" butterfly valves					
150 lb	M7@2.98	Ea	136.10	125.00	261.10
200 lb	M7@3.23	Ea	133.65	136.00	269.65
8" butterfly valves, with gear operator					
150 lb	M7@3.77	Ea	300.00	159.00	459.00
200 lb	M7@4.24	Ea	284.30	178.00	462.30
10" butterfly valves, with gear operator					
150 lb	M7@4.76	Ea	380.70	200.00	580.70
200 lb	M7@5.34	Ea	354.80	225.00	579.80

Globe Valves. These costs include the subcontractor's overhead and profit. 125 lb and 150 lb ratings are saturated steam pressure. 200 lb rating is working steam pressure, union bonnet. 125 lb valves have bronze disks. 150 lb valves have composition disks. 300 lb valves have regrinding disks.

	Craft@Hrs	Unit	Material	Labor	Total
1/2" threaded bronze globe valves					
125 lb	M5@.533	Ea	15.40	21.40	36.80
150 lb	M5@.533	Ea	20.40	21.40	41.80
300 lb	M5@.533	Ea	32.80	21.40	54.20
3/4" threaded bronze globe valves					
125 lb	M5@.632	Ea	19.50	25.30	44.80
150 lb	M5@.632	Ea	28.10	25.30	53.40
300 lb	M5@.632	Ea	45.40	25.30	70.70
1" threaded bronze globe valves					
125 lb	M5@.687	Ea	26.10	27.50	53.60
150 lb	M5@.687	Ea	43.35	27.50	70.85
300 lb	M5@.687	Ea	65.50	27.50	93.00
1-1/2" threaded bronze globe valves					
125 lb	M5@.760	Ea	45.00	30.40	75.40
150 lb.	M5@.760	Ea	81.30	30.40	111.70
300 lb.	M5@.760	Ea	113.40	30.40	143.80
2" threaded bronze globe valves					
125 lb	M5@1.04	Ea	67.40	41.70	109.10
150 lb	M5@1.04	Ea	125.40	41.70	167.10
300 lb	M5@1.19	Ea	173.25	47.70	220.95
2-1/2" threaded bronze globe valves					
150 lb	M5@1.37	Ea	250.00	54.90	304.90
300 lb	M5@1.58	Ea	550.00	63.30	613.30
Flanged iron body globe valves, OS&Y					
3" 125 lb	M7@1.60	Ea	386.80	67.30	454.10
3" 250 lb	M7@1.85	Ea	648.00	77.80	725.80
4" 125 lb	M7@2.25	Ea	553.15	94.60	647.75
4" 250 lb	M7@2.58	Ea	905.30	108.00	1,013.30
6" 125 lb	M7@2.93	Ea	990.40	123.00	1,113.40
6" 250 lb	M7@3.33	Ea	1,627.30	140.00	1,767.30
8" 125 lb	M7@3.45	Ea	1,940.00	145.00	2,085.00
8" 250 lb	M7@3.95	Ea	2,701.00	166.00	2,867.00

Swing Check Valves. These costs include the subcontractor's overhead and profit.

	Craft@Hrs	Unit	Material	Labor	Total
1/2" threaded bronze check valves					
125 lb	M5@.578	Ea	13.75	23.20	36.95
125 lb, soldered	M5@.375	Ea	10.80	15.00	25.80

Mechanical 15

	Craft@Hrs	Unit	Material	Labor	Total
Swing Check Valves, 1/2" threaded bronze, continued					
125 lb, vertical lift check	M5@.578	Ea	26.20	23.20	49.40
150 lb, bronze disk	M5@.578	Ea	21.25	23.20	44.45
200 lb, swing, regrinding bronze disk	M5@.632	Ea	22.35	25.30	47.65
3/4" threaded bronze check valves					
125 lb	M5@.617	Ea	16.90	24.70	41.60
125 lb, soldered	M5@.432	Ea	13.10	17.30	30.40
125 lb, vertical check	M5@.617	Ea	35.70	24.70	60.40
150 lb, bronze disk	M5@.617	Ea	26.60	24.70	51.30
200 lb, swing, regrinding bronze disk	M5@.682	Ea	27.85	27.30	55.15
1" threaded bronze check valves					
125 lb	M5@.700	Ea	23.00	28.00	51.00
125 lb, soldered	M5@.515	Ea	18.00	20.60	38.60
125 lb, vertical check	M5@.700	Ea	44.85	28.00	72.85
150 lb, bronze disk	M5@.700	Ea	35.85	28.00	63.85
200 lb, swing, regrinding bronze disk	M5@.768	Ea	37.50	30.80	68.30
1-1/2" threaded bronze check valves					
125 lb	M5@.880	Ea	38.05	35.30	73.35
125 lb, soldered	M5@.687	Ea	29.55	27.50	57.05
125 lb, vertical check	M5@.880	Ea	75.85	35.30	111.15
150 lb, bronze disk	M5@.880	Ea	61.80	35.30	97.10
200 lb, swing, regrinding bronze disk	M5@.968	Ea	64.75	38.80	103.55
2" threaded bronze check valves					
125 lb	M5@1.04	Ea	55.50	41.70	97.20
125 lb, soldered	M5@.799	Ea	43.20	32.00	75.20
125 lb, vertical check	M5@1.04	Ea	112.55	41.70	154.25
150 lb, bronze disk	M5@1.04	Ea	91.40	41.70	133.10
200 lb, swing, regrinding bronze disk	M5@1.14	Ea	95.65	45.70	141.35
2-1/2" flanged iron bolted cap swing check valves					
125 lb	M7@1.29	Ea	135.15	54.20	189.35
125 lb, all iron	M7@1.29	Ea	164.40	54.20	218.60
250 lb	M7@1.47	Ea	442.25	61.80	504.05
3" flanged iron bolted cap swing check valves					
125 lb	M7@1.60	Ea	146.70	67.30	214.00
125 lb, all iron	M7@1.60	Ea	178.35	67.30	245.65
250 lb	M7@1.86	Ea	555.75	78.20	633.95
4" flanged iron bolted cap swing check valves					
125 lb	M7@2.26	Ea	232.30	95.00	327.30
125 lb, all iron	M7@2.26	Ea	282.10	95.00	377.10
250 lb	M7@2.61	Ea	714.30	110.00	824.30
6" flanged iron bolted cap swing check valves					
125 lb	M7@2.93	Ea	395.00	123.00	518.00
125 lb, all iron	M7@2.93	Ea	480.00	123.00	603.00
250 lb	M7@3.33	Ea	1,313.00	140.00	1,453.00
8" flanged iron check valves, 125 lb	M7@3.45	Ea	745.40	145.00	890.40
10" flanged iron check valves, 125 lb	M7@4.42	Ea	1,272.40	186.00	1,458.40
12" flanged iron check valves, 125 lb	M7@6.45	Ea	1,977.00	271.00	2,248.00

Miscellaneous Valves and Regulators. These costs include the subcontractor's overhead and profit.
Angle valves, bronze, 150 lb, threaded, Teflon disk

	Craft@Hrs	Unit	Material	Labor	Total
1/2"	M5@.580	Ea	35.30	23.20	58.50
3/4"	M5@.617	Ea	48.30	24.70	73.00
1"	M5@.700	Ea	70.75	28.00	98.75
1-1/4"	M5@.833	Ea	90.80	33.40	124.20

Mechanical 15

	Craft@Hrs	Unit	Material	Labor	Total
1-1/2"	M5@.877	Ea	121.10	35.10	156.20
Backflow preventers					
3/4", screwed, bronze, with 2 gate valves	M5@.684	Ea	200.00	27.40	227.40
1", screwed, bronze, with 2 gate valves	M5@.827	Ea	250.00	33.10	283.10
2", screwed, with 2 gate valves	M5@3.96	Ea	505.00	159.00	664.00
6", without gate valves, flanged, iron body	M7@12.4	Ea	3,120.00	521.00	3,641.00
6", with 2 gate valves, flanged, iron body	M7@15.3	Ea	3,865.00	643.00	4,508.00
8", without gate valves, flanged, iron body	M7@16.0	Ea	7,850.00	673.00	8,523.00
8", with 2 gate valves, flanged, iron body	M7@16.5	Ea	9,420.00	694.00	10,114.00
Ball valves, bronze, 150 lb steam working pressure, threaded, Teflon seat					
1/2"	M5@.575	Ea	5.40	23.00	28.40
3/4"	M5@.617	Ea	8.15	24.70	32.85
1"	M5@.695	Ea	11.05	27.80	38.85
1-1/4"	M5@.830	Ea	19.40	33.30	52.70
1-1/2"	M5@.874	Ea	25.00	35.00	60.00
2"	M5@1.19	Ea	30.50	47.70	78.20
2-1/2"	M7@1.28	Ea	107.90	53.80	161.70
3"	M7@1.60	Ea	164.40	67.30	231.70
4" flanged, 150 lb, carbon steel	M7@1.43	Ea	710.00	60.10	770.10

Fixture wall valves (stops) with handwheel shutoff, 3/8" flexible compression connection tube and chrome escutcheon plate

	Craft@Hrs	Unit	Material	Labor	Total
Threaded angle stops					
12" tube	PL@.294	Ea	14.40	14.00	28.40
20" tube	PL@.294	Ea	15.30	14.00	29.30
36" tube	PL@.473	Ea	16.60	22.50	39.10
Soldered angle stops					
12" tube	PL@.344	Ea	14.50	16.40	30.90
20" tube	PL@.344	Ea	15.40	16.40	31.80
36" tube	PL@.550	Ea	16.60	26.20	42.80
Threaded straight stops					
12" tube	PL@.294	Ea	14.80	14.00	28.80
36" tube	PL@.473	Ea	17.00	22.50	39.50
Add for loose key or screwdriver shutoff rather than handwheel shutoff	--	Ea	3.30	--	--
Subtract for straight or elbow connection without valve	--	Ea	5.00	--	--
Gas regulator valves, threaded, with automatic shutoff and relief					
1"	P6@.670	Ea	85.00	27.40	112.40
1-1/4"	P6@.918	Ea	85.00	37.50	122.50
1-1/2"	P6@1.51	Ea	90.00	61.70	151.70
Gas stops, ground key, lever handle, screwed					
1/2"	P6@.619	Ea	6.00	25.30	31.30
3/4"	P6@.619	Ea	7.45	25.30	32.75
1"	P6@.943	Ea	12.65	38.50	51.15
1-1/4"	P6@.943	Ea	17.40	38.50	55.90
1-1/2"	P6@1.20	Ea	27.30	49.00	76.30
2"	P6@1.71	Ea	45.80	69.90	115.70
Hose bib, 3/4", brass	P6@.433	Ea	3.67	17.70	21.37
Hose bib with wall box, with door and frame, 3/4", brass, Lumaloy	P6@1.05	Ea	105.00	42.90	147.90
Hose gate valve, with cap, 175 lb, 2-1/2",					
Non-rising stem	P6@1.05	Ea	290.00	42.90	332.90
Rising stem	P6@1.05	Ea	112.00	42.90	154.90

Mechanical 15

	Craft@Hrs	Unit	Material	Labor	Total
Miscellaneous Valves and Regulators, continued					
Pressure regulator valves, air and water, 200 degree, threaded					
1/2", iron body, low pressure	P6@.836	Ea	80.00	34.20	114.20
3/4", iron body, low pressure	P6@.936	Ea	96.40	38.30	134.70
1", iron body, low pressure	P6@1.01	Ea	125.80	41.30	167.10
1-1/2", iron body, low pressure	P6@1.29	Ea	194.10	52.70	246.80
2", iron body, low pressure	P6@1.57	Ea	281.70	64.20	345.90
1/2", bronze, low pressure	P6@.836	Ea	111.55	34.20	145.75
3/4", bronze, low pressure	P6@.936	Ea	132.95	38.30	171.25
1", bronze, low pressure	P6@1.01	Ea	202.50	41.30	243.80
1-1/2", bronze, low pressure	P6@1.27	Ea	293.70	51.90	345.60
2", bronze, low pressure	P6@1.57	Ea	533.15	64.20	597.35
Pressure regulator valves, bronze, Class 300, 25 to 75 PSI, with "Y" strainer					
3/4"	P6@.936	Ea	91.20	38.30	129.50
1"	P6@1.01	Ea	136.65	41.30	177.95
1-1/4"	P6@1.27	Ea	208.75	51.90	260.65
1-1/2"	P6@1.27	Ea	270.00	51.90	321.90
2"	P6@1.57	Ea	392.00	64.20	456.20
2-1/2"	P6@1.88	Ea	587.00	76.80	663.80
3"	P6@2.29	Ea	734.00	93.60	827.60
Square head or flat head cock, 125 lb, brass body					
1/2"	P6@.617	Ea	9.80	25.20	35.00
3/4"	P6@.658	Ea	12.00	26.90	38.90
1"	P6@.744	Ea	17.60	30.40	48.00
1-1/4"	P6@.775	Ea	25.45	31.70	57.15
Water control valves, threaded, 3-way with actuator					
1/2", CV 2.2	M5@1.41	Ea	215.00	56.50	271.50
3/4", CV 8.6	M5@1.75	Ea	215.00	70.10	285.10
1", CV 13.9	M5@1.95	Ea	250.00	78.10	328.10
1-1/2", CV 21	M5@2.45	Ea	300.00	98.10	398.10
2", CV 30	M5@2.97	Ea	380.00	119.00	499.00
2-1/2", CV 54	M7@3.52	Ea	1,070.00	148.00	1,218.00
3", CV 80	M7@4.49	Ea	1,290.00	189.00	1,479.00
4", CV 157	M7@7.35	Ea	2,440.00	309.00	2,749.00
5", CV 238	M7@7.89	Ea	2,595.00	332.00	2,927.00
6", CV 347	M7@8.29	Ea	3,025.00	349.00	3,374.00
Piping Specialties. These costs include the subcontractor's overhead and profit.					
Stainless steel bellows type expansion joints, 150 lb. Flanged units include bolt and gasket sets					
2" pipe, 3" travel, welded	M7@1.77	Ea	575.00	74.40	649.40
2" pipe, 8" travel, welded	M7@1.77	Ea	705.00	74.40	779.40
3" pipe, 4" travel, welded	M7@2.40	Ea	645.00	101.00	746.00
3" pipe, 6" travel, welded	M7@2.40	Ea	705.00	101.00	806.00
4" pipe, 3" travel, welded	M7@3.35	Ea	675.00	141.00	816.00
4" pipe, 6" travel, welded	M7@3.35	Ea	840.00	141.00	981.00
5" pipe, 4" travel, flanged	M7@3.52	Ea	1,785.00	148.00	1,933.00
5" pipe, 7" travel, flanged	M7@4.09	Ea	1,840.00	172.00	2,012.00
6" pipe, 4" travel, flanged	M7@4.24	Ea	1,900.00	178.00	2,078.00
8" pipe, 8" travel, flanged	M7@5.46	Ea	2,815.00	230.00	3,045.00
Ball joints, welded ends, 150 lb					
3"	M7@2.40	Ea	285.00	101.00	386.00
4"	M7@3.35	Ea	420.00	141.00	561.00
5"	M7@4.09	Ea	710.00	172.00	882.00
6"	M7@4.24	Ea	760.00	178.00	938.00

Mechanical 15

	Craft@Hrs	Unit	Material	Labor	Total
8"	M7@4.79	Ea	1,330.00	201.00	1,531.00
10"	M7@5.81	Ea	1,640.00	244.00	1,884.00
14"	M7@7.74	Ea	2,290.00	325.00	2,615.00

Victaulic couplings, painted, labor includes grooving two ends and installing gasket and coupling

	Craft@Hrs	Unit	Material	Labor	Total
2"	M7@1.08	Ea	15.90	45.40	61.30
3"	M7@1.22	Ea	20.20	51.30	71.50
4"	M7@1.32	Ea	28.90	55.50	84.40
6"	M7@1.49	Ea	51.20	62.70	113.90
8"	M7@2.22	Ea	85.30	93.40	178.70
10"	M7@2.78	Ea	120.00	117.00	237.00
12"	M7@3.23	Ea	138.00	136.00	274.00
14"	M7@3.80	Ea	207.00	160.00	367.00
16"	M7@4.24	Ea	272.00	178.00	450.00
Add for galvanized couplings	--	%	20.0	--	--

In-Line Circulating Pumps. Flanged, for hot or cold water circulation. By flange size and pump horsepower rating. See flange costs below. No electrical work included. These costs include the subcontractor's overhead and profit.

Horizontal drive, iron body

	Craft@Hrs	Unit	Material	Labor	Total
3/4" to 1-1/2", 1/12 HP	D4@2.62	Ea	151.00	102.00	253.00
2", 1/4 HP	D4@3.23	Ea	261.00	126.00	387.00
2-1/2" or 3", 1/4 HP	D4@3.79	Ea	373.00	148.00	521.00
3/4" to 1-1/2", 1/6 HP	D4@2.62	Ea	220.00	102.00	322.00
3/4" to 1-1/2", 1/4 HP	D4@2.62	Ea	292.00	102.00	394.00
3", 1/3 HP	D9@4.87	Ea	506.00	196.00	702.00
3", 1/2 HP	D9@4.87	Ea	530.00	196.00	726.00
3", 3/4 HP	D9@4.87	Ea	612.00	196.00	808.00

Horizontal drive, all bronze

	Craft@Hrs	Unit	Material	Labor	Total
3/4" to 1-1/2", 1/12 HP	D4@2.62	Ea	231.00	102.00	333.00
3/4" to 1-1/2", 1/6 HP	D4@2.62	Ea	346.00	102.00	448.00
3/4" to 1-1/2", 1/4 HP	D4@2.62	Ea	408.00	102.00	510.00
2", 1/4 HP	D9@3.13	Ea	428.00	126.00	554.00
2-1/2", 1/4 HP	D9@3.68	Ea	668.00	148.00	816.00
3", 1/4 HP	D9@4.87	Ea	668.00	196.00	864.00
3", 1/3 HP	D9@4.87	Ea	898.00	196.00	1,094.00
3", 1/2 HP	D9@4.87	Ea	925.00	196.00	1,121.00
3", 3/4 HP	D9@4.87	Ea	963.00	196.00	1,159.00

Vertical drive, iron body

	Craft@Hrs	Unit	Material	Labor	Total
3/4" to 1/2", 1/6 HP	D4@2.62	Ea	304.00	102.00	406.00
2", 1/4 HP	D9@3.13	Ea	361.00	126.00	487.00
2-1/2", 1/4 HP	D9@3.68	Ea	427.00	148.00	575.00
3", 1/4 HP	D9@4.87	Ea	427.00	196.00	623.00

Pump connectors, vibration isolators, in line, stainless steel braided connections, threaded

	Craft@Hrs	Unit	Material	Labor	Total
3/4" x 11"	M5@.804	Ea	23.80	32.20	56.00
1" x 12"	M5@.903	Ea	28.90	36.20	65.10
1-1/4" x 13"	M5@.955	Ea	35.10	38.30	73.40
1-1/2" x 14"	M5@1.21	Ea	38.60	48.50	87.10
2" x 15"	M7@1.24	Ea	53.80	52.10	105.90
3" x 17"	M7@1.24	Ea	90.00	52.10	142.10

Flanges for circulating pumps, per set of two, including bolts and gaskets

	Craft@Hrs	Unit	Material	Labor	Total
3/4" to 1-1/2", iron	--	Ea	4.40	--	--
2", iron	--	Ea	11.50	--	--
2-1/2" to 3", iron	--	Ea	29.90	--	--

Mechanical 15

	Craft@Hrs	Unit	Material	Labor	Total
In-Line Circulating Pumps, Flanges for circulating pumps, continued					
3/4" to 1", bronze	--	Ea	15.00	--	--
2", bronze	--	Ea	34.60	--	--
2-1/2" to 3", iron	--	Ea	86.30	--	--
3/4" copper sweat flanges	--	Ea	11.40	--	--
1" copper sweat flanges	--	Ea	16.80	--	--
1-1/2" copper sweat flanges	--	Ea	18.10	--	--

Base Mounted Centrifugal Pumps. Single stage cast iron pumps with flanges but no electrical connection. These costs include the subcontractor's overhead and profit.

	Craft@Hrs	Unit	Material	Labor	Total
50 GPM at 100' head	D9@4.50	Ea	1,570.00	181.00	1,751.00
50 GPM at 200' head	D9@5.95	Ea	1,790.00	240.00	2,030.00
50 GPM at 300' head	D9@12.2	Ea	2,000.00	491.00	2,491.00
200 GPM at 100' head	D9@9.60	Ea	2,020.00	387.00	2,407.00
200 GPM at 200' head	D9@10.2	Ea	2,080.00	411.00	2,491.00
200 GPM at 300' head	D9@14.6	Ea	2,270.00	588.00	2,858.00
240 GPM, 7-1/2 horsepower	D9@8.85	Ea	1,340.00	356.00	1,696.00
300 GPM, 5 horsepower	D9@5.59	Ea	1,500.00	225.00	1,725.00
340 GPM, 7-1/2 horsepower	D9@9.60	Ea	1,650.00	387.00	2,037.00
370 GPM, 5 horsepower	D9@5.75	Ea	1,530.00	232.00	1,762.00
375 GPM, 7-1/2 horsepower	D9@10.2	Ea	1,730.00	411.00	2,141.00
465 GPM, 10 horsepower	D9@10.9	Ea	1,800.00	439.00	2,239.00

Vertical Turbine Pumps 3,550 RPM. These costs include the subcontractor's overhead and profit but no electrical work.

	Craft@Hrs	Unit	Material	Labor	Total
Cast iron single stage vertical turbine pumps					
50 GPM at 50' head	D9@14.1	Ea	2,700.00	568.00	3,268.00
50 GPM at 100' head	D9@14.1	Ea	3,110.00	568.00	3,678.00
100 GPM at 300' head	D9@20.5	Ea	4,020.00	826.00	4,846.00
200 GPM at 50' head	D9@15.4	Ea	3,380.00	620.00	4,000.00
Cast iron multi-stage vertical turbine pumps					
50 GPM at 100' head	D9@15.6	Ea	3,070.00	628.00	3,698.00
50 GPM at 200' head	D9@15.6	Ea	3,520.00	628.00	4,148.00
50 GPM at 300' head	D9@15.6	Ea	4,010.00	628.00	4,638.00
100 GPM at 100' head	D9@15.6	Ea	3,300.00	628.00	3,928.00
100 GPM at 200' head	D9@15.6	Ea	3,700.00	628.00	4,328.00
100 GPM at 300' head	D9@15.6	Ea	4,260.00	628.00	4,888.00
200 GPM at 100' head	D9@15.6	Ea	3,600.00	628.00	4,228.00
200 GPM at 200' head	D9@20.5	Ea	4,260.00	826.00	5,086.00
200 GPM at 300' head	D9@20.5	Ea	4,560.00	826.00	5,386.00
Bronze single stage vertical turbine pumps					
50 GPM at 50' head	D9@15.6	Ea	2,680.00	628.00	3,308.00
100 GPM at 50' head	D9@15.6	Ea	2,870.00	628.00	3,498.00
200 GPM at 50' head	D9@15.6	Ea	3,110.00	628.00	3,738.00
Bronze multi-stage vertical turbine pumps					
50 GPM at 50' head	D9@15.6	Ea	3,070.00	628.00	3,698.00
50 GPM at 100' head	D9@15.6	Ea	3,300.00	628.00	3,928.00
50 GPM at 150' head	D9@15.6	Ea	3,390.00	628.00	4,018.00
100 GPM at 50' head	D9@14.5	Ea	3,070.00	584.00	3,654.00
100 GPM at 100' head	D9@15.6	Ea	3,300.00	628.00	3,928.00
100 GPM at 150' head	D9@15.6	Ea	3,480.00	628.00	4,108.00
200 GPM at 50' head	D9@15.6	Ea	3,300.00	628.00	3,928.00
200 GPM at 100' head	D9@15.6	Ea	3,600.00	628.00	4,228.00
200 GPM at 150' head	D9@15.6	Ea	3,970.00	628.00	4,598.00

Mechanical 15

	Craft@Hrs	Unit	Material	Labor	Total

Sump Pumps. Bronze 1,750 RPM sump pumps. These costs include the subcontractor's overhead and profit but no electrical work.

	Craft@Hrs	Unit	Material	Labor	Total
25 GPM at 25' head	M7@28.5	Ea	3,190.00	1,200.00	4,390.00
25 GPM at 50' head	M7@28.5	Ea	3,350.00	1,200.00	4,550.00
25 GPM at 100' head	M7@28.5	Ea	3,490.00	1,200.00	4,690.00
25 GPM at 150' head	M7@37.7	Ea	4,800.00	1,590.00	6,390.00
50 GPM at 25' head	M7@28.5	Ea	3,190.00	1,200.00	4,390.00
50 GPM at 50' head	M7@28.5	Ea	3,350.00	1,200.00	4,550.00
50 GPM at 100' head	M7@28.5	Ea	3,490.00	1,200.00	4,690.00
50 GPM at 150' head	M7@37.7	Ea	4,900.00	1,590.00	6,490.00
100 GPM at 25' head	M7@28.5	Ea	3,400.00	1,200.00	4,600.00
100 GPM at 50' head	M7@28.5	Ea	3,350.00	1,200.00	4,550.00
100 GPM at 100' head	M7@28.5	Ea	4,290.00	1,200.00	5,490.00
100 GPM at 150' head	M7@37.7	Ea	4,890.00	1,590.00	6,480.00

Miscellaneous Pumps 1,750 RPM. No electrical work included. These costs include the subcontractor's overhead and profit.

	Craft@Hrs	Unit	Material	Labor	Total
Salt water pump, 135 GPM at 300' head	D9@19.9	Ea	8,860.00	802.00	9,662.00
Sewage pumps, vertical mount, centrifugal					
1-1/2 HP, 100 GPM at 17' head	P6@14.6	Ea	3,000.00	597.00	3,597.00
10 HP, 300 GPM at 70' head	P6@18.3	Ea	4,950.00	748.00	5,698.00
Medical vacuum pump, duplex, tank mount with 30 gallon tank,					
1/2 HP, 20" vacuum	D4@20.6	Ea	5,070.00	804.00	5,874.00
Oil pump, base mounted, 1-1/2 GPM at 15' head,					
1/3 HP, 120 volt	D4@.923	Ea	146.00	36.00	182.00

Fiberglass Pipe Insulation with AP (All Purpose) Jacket. By nominal pipe diameter. Also see fittings, flanges and valves at the end of this section. These costs include the subcontractor's overhead and profit.

	Craft@Hrs	Unit	Material	Labor	Total
1/2" diameter pipe					
1/2" thick insulation	A1@.060	LF	.87	2.44	3.31
1" thick insulation	A1@.060	LF	1.02	2.44	3.46
1-1/2" thick insulation	A1@.072	LF	2.11	2.93	5.04
3/4" diameter pipe					
1/2" thick insulation	A1@.060	LF	.96	2.44	3.40
1" thick insulation	A1@.060	LF	1.17	2.44	3.61
1-1/2" thick insulation	A1@.072	LF	2.16	2.93	5.09
1" diameter pipe					
1/2" thick insulation	A1@.060	LF	1.02	2.44	3.46
1" thick insulation	A1@.060	LF	1.22	2.44	3.66
1-1/2" thick insulation	A1@.072	LF	2.32	2.93	5.25
1-1/4" diameter pipe					
1/2" thick insulation	A1@.060	LF	1.13	2.44	3.57
1" thick insulation	A1@.072	LF	1.36	2.93	4.29
1-1/2" thick insulation	A1@.077	LF	2.45	3.13	5.58
1-1/2" diameter pipe					
1/2" thick insulation	A1@.060	LF	1.22	2.44	3.66
1" thick insulation	A1@.072	LF	1.50	2.93	4.43
1-1/2" thick insulation	A1@.077	LF	2.62	3.13	5.75
2" thick insulation	A1@.082	LF	4.05	3.34	7.39
2" diameter pipe					
1/2" thick insulation	A1@.060	LF	1.30	2.44	3.74

Mechanical 15

	Craft@Hrs	Unit	Material	Labor	Total
Fiberglass Pipe Insulation with AP (All Purpose) Jacket, 2" diameter, continued					
1" thick insulation	A1@.072	LF	1.60	2.93	4.53
1-1/2" thick insulation	A1@.077	LF	2.85	3.13	5.98
2" thick insulation	A1@.082	LF	4.20	3.34	7.54
2-1/2" diameter pipe					
1/2" thick insulation	A1@.060	LF	1.50	2.44	3.94
1" thick insulation	A1@.072	LF	1.75	2.93	4.68
1-1/2" thick insulation	A1@.082	LF	3.10	3.34	6.44
2" thick insulation	A1@.082	LF	4.50	3.34	7.84
3" diameter pipe					
1/2" thick insulation	A1@.072	LF	1.65	2.93	4.58
1" thick insulation	A1@.082	LF	2.00	3.34	5.34
1-1/2" thick insulation	A1@.089	LF	3.20	3.62	6.82
2" thick insulation	A1@.089	LF	4.80	3.62	8.42
4" diameter pipe					
1/2" thick insulation	A1@.072	LF	2.10	2.93	5.03
1" thick insulation	A1@.082	LF	2.55	3.34	5.89
1-1/2" thick insulation	A1@.089	LF	3.65	3.62	7.27
2" thick insulation	A1@.089	LF	5.60	3.62	9.22
5" diameter pipe					
1/2" thick insulation	A1@.072	LF	2.45	2.93	5.38
1" thick insulation	A1@.082	LF	3.00	3.34	6.34
1-1/2" thick insulation	A1@.089	LF	4.10	3.62	7.72
2" thick insulation	A1@.095	LF	6.30	3.87	10.17
6" diameter pipe					
1/2" thick insulation	A1@.082	LF	2.55	3.34	5.89
1" thick insulation	A1@.095	LF	3.10	3.87	6.97
1-1/2" thick insulation	A1@.095	LF	4.20	3.87	8.07
2" thick insulation	A1@.095	LF	6.50	3.87	10.37
8" diameter pipe					
1" thick insulation	A1@.095	LF	4.50	3.87	8.37
1-1/2" thick insulation	A1@.101	LF	5.30	4.11	9.41
2" thick insulation	A1@.101	LF	8.00	4.11	12.11
10" diameter pipe					
1" thick insulation	A1@.095	LF	5.25	3.87	9.12
1-1/2" thick insulation	A1@.101	LF	6.65	4.11	10.76
2" thick insulation	A1@.267	LF	9.55	10.90	20.45
12" diameter pipe					
1" thick insulation	A1@.101	LF	5.95	4.11	10.06
1-1/2" thick insulation	A1@.101	LF	7.30	4.11	11.41
2" thick insulation	A1@.118	LF	10.55	4.80	15.35
Add for .016" aluminum jacket, SF of surface	A1@.018	SF	.40	.73	1.13

Pipe fittings and flanges (for each fitting or flange use the cost for 3 LF of pipe of the same pipe size)
Valves
  Body only (for each valve body use the cost for 5 LF of pipe of the same pipe size)
  Body and bonnet or yoke (for each valve use the cost for 10 LF of pipe of the same pipe size)
  Flanged valves (add the cost for insulation of flanges per above)

Calcium Silicate Pipe Insulation. Without cover. By nominal pipe diameter. Also see fittings, flanges and valves at the end of this section. These costs include the subcontractor's overhead and profit.

	Craft@Hrs	Unit	Material	Labor	Total
6" diameter pipe					
2" thick insulation	A1@.109	LF	8.25	4.44	12.69
4" thick insulation	A1@.133	LF	17.80	5.41	23.21
6" thick insulation	A1@.178	LF	29.70	7.25	36.95

Mechanical 15

	Craft@Hrs	Unit	Material	Labor	Total
8" diameter pipe					
2" thick insulation	A1@.115	LF	9.55	4.68	14.23
4" thick insulation	A1@.143	LF	17.75	5.82	23.57
6" thick insulation	A1@.200	LF	29.70	8.14	37.84
10" diameter pipe					
2" thick insulation	A1@.122	LF	12.00	4.97	16.97
4" thick insulation	A1@.143	LF	25.20	5.82	31.02
6" thick insulation	A1@.230	LF	41.00	9.36	50.36
12" diameter pipe					
2" thick insulation	A1@.127	LF	13.30	5.17	18.47
4" thick insulation	A1@.158	LF	29.00	6.43	35.43

Pipe fittings and flanges (for each fitting or flange use the cost for 3 LF of pipe of the same pipe size)
Valves
  Body only (for each valve body use the cost for 5 LF of pipe of the same pipe size)
  Body and bonnet or yoke (for each valve use the cost for 10 LF of pipe of the same pipe size)
  Flanged valves (add the cost for insulation of flanges per above)

Calcium Silicate Pipe Insulation with .016" Aluminum Jacket. By nominal pipe diameter. Also see fittings, flanges and valves at the end of this section. These costs include the subcontractor's overhead and profit.

	Craft@Hrs	Unit	Material	Labor	Total
6" diameter pipe					
2" thick insulation	A1@.183	LF	8.25	7.45	15.70
4" thick insulation	A1@.218	LF	17.75	8.87	26.62
6" thick insulation	A1@.288	LF	29.70	11.70	41.40
8" diameter pipe					
2" thick insulation	A1@.194	LF	9.55	7.90	17.45
4" thick insulation	A1@.228	LF	21.40	9.28	30.68
6" thick insulation	A1@.408	LF	34.70	16.60	51.30
10" diameter pipe					
2" thick insulation	A1@.202	LF	12.00	8.22	20.22
4" thick insulation	A1@.239	LF	25.20	9.73	34.93
6" thick insulation	A1@.337	LF	41.00	13.70	54.70
12" diameter pipe					
2" thick insulation	A1@.208	LF	13.30	8.47	21.77
4" thick insulation	A1@.261	LF	29.05	10.60	39.65

Pipe fittings and flanges (for each fitting or flange use the cost for 3 LF of pipe of the same pipe size)
Valves
  Body only (for each valve body use the cost for 5 LF of pipe of the same pipe size)
  Body and bonnet or yoke (for each valve use the cost for 10 LF of pipe of the same pipe size)
  Flanged valves (add the cost for insulation of flanges per above)

Closed Cell Polyethylene Pipe and Tubing Insulation, semi split, 1/2" wall thickness, no cover, by nominal pipe or tube diameter. Also see fittings, flanges and valves at the end of this section. These costs include the subcontractor's overhead and profit.

	Craft@Hrs	Unit	Material	Labor	Total
1/4"	A1@.039	LF	.26	1.59	1.85
3/8"	A1@.039	LF	.26	1.59	1.85
1/2"	A1@.039	LF	.26	1.59	1.85
3/4"	A1@.039	LF	.31	1.59	1.90
1"	A1@.042	LF	.34	1.71	2.05
1-1/4"	A1@.042	LF	.44	1.71	2.15
1-1/2"	A1@.042	LF	.53	1.71	2.24
2"	A1@.046	LF	.68	1.87	2.55
2-1/2"	A1@.056	LF	.90	2.28	3.18
3"	A1@.082	LF	1.15	3.34	4.49

Mechanical 15

	Craft@Hrs	Unit	Material	Labor	Total

Closed Cell Polyethylene Pipe and Tubing Insulation, continued
4"	A1@.095	LF	1.63	3.87	5.50
Deduct for 3/8" insulation wall thickness	--	%	-50.0	--	--
Add for 3/4" insulation wall thickness	--	%	75.0	--	--

Pipe fittings and flanges (for each fitting or flange use the cost for 3 LF of pipe or tube of the same size)
Valves
  Body only (for each valve body use the cost for 5 LF of pipe or tube of the same size)
  Body and bonnet or yoke (for each valve use the cost for 10 LF of pipe or tube of the same size)
  Flanged valves (add the cost for insulation of flanges per above)

Meters, Gauges and Indicators. These costs include the subcontractor's overhead and profit.

	Craft@Hrs	Unit	Material	Labor	Total
Water meters					
Disc type, 1-1/2"	P6@1.00	Ea	115.00	40.90	155.90
Turbine type, 1-1/2"	P6@1.00	Ea	173.00	40.90	213.90
Displacement type, AWWA C7000, 1"	P6@.941	Ea	178.00	38.50	216.50
Displacement type, AWWA C7000, 1-1/2"	P6@1.23	Ea	614.00	50.30	664.30
Displacement type, AWWA C7000, 2"	P6@1.57	Ea	917.00	64.20	981.20
Curb stop and waste valve, 1"	P6@1.00	Ea	64.00	40.90	104.90
Fuel meter, direct reading	P6@1.00	Ea	105.00	40.90	145.90
Thermoflow indicators, soldered					
1-1/4"	M5@2.01	Ea	263.00	80.50	343.50
2"	M5@2.01	Ea	300.00	80.50	380.50
2-1/2"	M7@1.91	Ea	442.00	80.30	522.30
Cast steel steam meters, in-line, flanged, 300 pound					
1" pipe, threaded	M5@1.20	Ea	2,480.00	48.10	2,528.10
2" pipe	M7@1.14	Ea	2,800.00	47.90	2,847.90
3" pipe	M7@2.31	Ea	3,100.00	97.10	3,197.10
4" pipe	M7@2.71	Ea	3,460.00	114.00	3,574.00
Cast steel by-pass steam meters, 2", by line pipe size					
5" line	M7@15.4	Ea	5,900.00	648.00	6,548.00
6" line	M7@16.7	Ea	6,100.00	702.00	6,802.00
8" line	M7@17.7	Ea	6,250.00	744.00	6,994.00
10" line	M7@18.5	Ea	6,500.00	778.00	7,278.00
12" line	M7@19.7	Ea	6,750.00	828.00	7,578.00
14" line	M7@21.4	Ea	7,150.00	900.00	8,050.00
16" line	M7@22.2	Ea	7,370.00	934.00	8,304.00
Add for steam meter pressure-compensated counter	--	Ea	1,080.00	--	--
Add for contactor used with dial counter	--	Ea	505.00	--	--
Add for wall or panel-mounted remote totalizer with contactor	--	Ea	805.00	--	--
Add for direct reading pressure gauge	M5@.859	Ea	63.40	34.40	97.80
Add for direct reading thermometer, stainless steel case, with trim, 2% accuracy					
3" dial	M5@.674	Ea	50.30	27.00	77.30
5" dial	M5@.674	Ea	76.10	27.00	103.10

Plumbing Specialties. These costs include the subcontractor's overhead and profit.

	Craft@Hrs	Unit	Material	Labor	Total
Access doors, painted steel, with screwdriver catch, Elmco					
8" x 8"	SM@.536	Ea	30.00	24.60	54.60
12" x 12"	SM@.536	Ea	34.40	24.60	59.00
18" x 18"	SM@.788	Ea	48.50	36.20	84.70
24" x 24"	SM@.918	Ea	73.50	42.20	115.70
36" x 36"	SM@1.21	Ea	136.00	55.60	191.60
Add for fire rating	--	%	400.0	--	--

Mechanical 15

	Craft@Hrs	Unit	Material	Labor	Total
Add for stainless steel	--	%	300.0	--	--
Add for cylinder lock	--	Ea	11.00	--	--
Cleanouts, floor, with polished bronze top, Zurn					
2" pipe, 4-1/8" round top	P6@.930	Ea	67.60	38.00	105.60
3" pipe, 5-1/8" round top	P6@1.09	Ea	72.30	44.50	116.80
4" pipe, 6-1/8" round top	P6@1.25	Ea	98.00	51.10	149.10
6" pipe, 8" round top	P6@1.50	Ea	166.20	61.30	227.50
Cleanouts, to grade, cast iron, round flanged housing, bronze top					
2" pipe	P6@1.25	Ea	50.00	51.10	101.10
3" pipe	P6@1.32	Ea	53.60	53.90	107.50
4" pipe	P6@1.83	Ea	69.50	74.80	144.30
Wall cleanout tees, chrome cover with bronze plug					
2" pipe	P6@1.21	Ea	17.80	49.50	67.30
4" pipe	P6@1.42	Ea	31.60	58.00	89.60
6" pipe	P6@1.47	Ea	73.70	60.10	133.80
Roof (jack) flashing, round					
4"	SM@.311	Ea	5.45	14.30	19.75
6"	SM@.416	Ea	12.00	19.10	31.10
8"	SM@.516	Ea	16.00	23.70	39.70
Drain, general use, medium duty, PVC plastic, 3" or 4" pipe					
3" x 4", stainless steel strainer	P6@1.16	Ea	7.10	47.40	54.50
3" x 4", brass strainer	P6@1.32	Ea	14.00	53.90	67.90
4" with 5" brass strainer & spigot outlet	P6@1.55	Ea	30.00	63.30	93.30
Drain, floor, cast iron, bronze top, with flashing ring					
2" to 4" pipe, 9" diameter, standard sump	P6@1.17	Ea	75.70	47.80	123.50
6" pipe, 12" diameter top, medium sump	P6@1.55	Ea	125.50	63.30	188.80
Drains, roof, cast iron mushroom strainer, bottom outlet					
2" to 4"	P6@1.75	Ea	84.40	71.50	155.90
5" or 6"	P6@1.98	Ea	120.50	80.90	201.40
8"	P6@2.36	Ea	157.00	96.50	253.50
Deduct for PVC body and dome	--	%	-40.0	--	--
Drains, shower, with bronze top, no hub					
2" to 4" pipe, 5" strainer	P6@1.47	Ea	54.30	60.10	114.40
2" to 4" pipe, 7" strainer	P6@1.65	Ea	76.30	67.40	143.70
2" to 4" pipe, 10" strainer	P6@2.07	Ea	132.00	84.60	216.60
Add for threaded drains	--	%	30.0	--	--

Plumbing Equipment. These costs include the subcontractor's overhead and profit. Valves, supports, vents, gas or electric hookup and related equipment are not included except as noted

	Craft@Hrs	Unit	Material	Labor	Total
Garbage disposers, commercial, stainless steel, with water flow interlock switch, Insinkerator					
1/2 HP, 1-1/2" drain connection	PL@1.47	Ea	660.00	70.00	730.00
3/4 HP, 1-1/2" drain connection	PL@2.09	Ea	927.00	99.60	1,026.60
1 HP, 1-1/2" drain connection	PL@3.92	Ea	990.00	187.00	1,177.00
1-1/2 HP, 2" drain connection	PL@5.24	Ea	1,450.00	250.00	1,700.00
Water heaters, commercial, gas fired, 3 year warranty, Rheem Glasslined Energy Miser					
60 MBtu, 55 GPH, 50 gallon	P6@3.98	Ea	742.00	163.00	905.00
75 MBtu, 75 GPH, 68 gallon	P6@5.63	Ea	791.00	230.00	1,021.00
98 MBtu, 89 GPH, 50 gallon	P6@10.4	Ea	1,800.00	425.00	2,225.00
180 MBtu, 164 GPH, 98 gallon	P6@15.1	Ea	2,520.00	617.00	3,137.00
250 MBtu, 228 GPH, 76 gallon	P6@19.1	Ea	2,700.00	781.00	3,481.00
500 MBtu, 455 GPH, 66 gallon, ASME	P6@24.1	Ea	5,080.00	985.00	6,065.00
Add for propane fired units, any of above	--	Ea	64.80	--	--

Mechanical 15

	Craft@Hrs	Unit	Material	Labor	Total
Plumbing Equipment, continued					
Water heaters, commercial, electric, 208/240 volts, with immersion thermostat, Rheem Glasslined Energy Miser,					
50 gallons, 9 KW	P6@4.72	Ea	1,500.00	193.00	1,693.00
50 gallons, 27 KW	P6@5.84	Ea	2,140.00	239.00	2,379.00
85 gallons, 18 KW	P6@6.42	Ea	1,940.00	262.00	2,202.00
120 gallons, 45 KW	P6@8.41	Ea	3,380.00	344.00	3,724.00
Water heater stands, 16 gauge steel					
18" x 18" square x 18" high	--	Ea	32.40	--	--
24" x 24" square x 18" high	--	Ea	45.40	--	--
Water heater safety pans with side drain					
20" diameter	--	Ea	8.55	--	--
26" diameter	--	Ea	14.25	--	--
Grease interceptors					
2" cast iron, 4 GPM, 8 lb capacity	P6@3.24	Ea	309.00	132.00	441.00
2" cast iron, 10 GPM, 20 lb capacity	P6@4.13	Ea	505.00	169.00	674.00
2" steel, 15 GPM, 30 lb capacity	P6@5.63	Ea	748.00	230.00	978.00
3" steel, 20 GPM, 40 lb capacity	P6@16.8	Ea	915.00	687.00	1,602.00
Hair interceptors, 1-1/2" cast iron					
Small	P6@3.24	Ea	116.00	132.00	248.00
Large	P6@4.72	Ea	163.00	193.00	356.00
Add for nickel bronze construction	--	Ea	100.00	--	--
Sewage ejector pumps, packaged systems, with vertical pumps, float switch controls, poly tank, cover and fittings, no electrical work included					
Single pump systems					
1/2 HP, 2" line, 20" x 30" tank	P6@12.7	Ea	496.00	519.00	1,015.00
3/4 HP, 2" line, 24" x 36" tank	P6@15.1	Ea	1,185.00	617.00	1,802.00
1 HP, 3" line, 24" x 36" tank	P6@24.0	Ea	1,400.00	981.00	2,381.00
Duplex pump systems					
1/2 HP, 2" line, 30" x 36" tank	P6@24.3	Ea	1,815.00	993.00	2,808.00
3/4 HP, 2" line, 30" x 36" tank	P6@29.1	Ea	2,360.00	1,190.00	3,550.00
1 HP, 3" line, 30" x 36" tank	P6@45.6	Ea	2,725.00	1,860.00	4,585.00
Sump pumps, with float switch, no electrical work included					
Submersible type					
1/4 HP, 1-1/2" outlet, ABS plastic	P6@2.58	Ea	106.00	105.00	211.00
1/3 HP, 1-1/2" outlet, ABS plastic	P6@2.93	Ea	112.00	120.00	232.00
1/2 HP, 1-1/2" outlet, ABS plastic	P6@3.24	Ea	135.00	132.00	267.00
Upright type					
1/3 HP, 1-1/4" outlet, cast iron	P6@3.80	Ea	150.00	155.00	305.00
1/3 HP, 1-1/4" outlet, bronze	P6@3.80	Ea	220.00	155.00	375.00
1/3 HP, 1-1/4" outlet, ABS plastic	P6@3.80	Ea	113.00	155.00	268.00
1/2 HP, 1-1/4" outlet, cast iron heavy duty	P6@4.21	Ea	170.00	172.00	342.00
1/2 HP, 1-1/4" outlet, bronze heavy duty	P6@4.21	Ea	248.00	172.00	420.00
Tanks, See also ASME tanks					
Expansion tanks, galvanized steel, ASME code, 125 PSI					
15 gallons	P6@1.00	Ea	335.00	40.90	375.90
30 gallons	P6@1.00	Ea	375.00	40.90	415.90
60 gallons	P6@1.50	Ea	460.00	61.30	521.30
80 gallons	P6@1.50	Ea	495.00	61.30	556.30
Hot water storage tanks, glass lined					
62 gallons	P6@5.97	Ea	833.00	244.00	1,077.00
190 gallons	P6@7.93	Ea	1,622.00	324.00	1,946.00
980 gallons	P6@19.9	Ea	5,738.00	813.00	6,551.00
1,230 gallons	P6@19.9	Ea	6,533.00	813.00	7,346.00

Mechanical 15

	Craft@Hrs	Unit	Material	Labor	Total
1,615 gallons	P6@23.8	Ea	8,610.00	973.00	9,583.00

Septic tanks, including typical excavation and piping, see also water softening systems

	Craft@Hrs	Unit	Material	Labor	Total
500 gallons, polyethylene	U2@15.2	Ea	382.00	587.00	969.00
1,000 gallons, fiberglass	U2@16.1	Ea	546.00	622.00	1,168.00
2,000 gallons, concrete	U2@17.0	Ea	822.00	656.00	1,478.00
5,000 gallons, concrete	U2@22.3	Ea	3,060.00	861.00	3,921.00

Water softeners, automatic two tank unit, 3/4" valve

	Craft@Hrs	Unit	Material	Labor	Total
32,000 grain capacity	P6@3.59	Ea	570.00	147.00	717.00
60,000 grain capacity	P6@3.59	Ea	720.00	147.00	867.00

Plumbing Rough-in and Fixtures for commercial and industrial buildings. This section shows costs for typical rough-in of drain, waste and vent (DWV) and water supply piping using either plastic or metal pipe and fittings. Plastic DWV pipe is usually used with plastic supply pipe and cast iron DWV is usually used with copper supply pipe. Fixture costs are based on good to better quality fixtures and include all trim and valves. Detailed estimates of the piping requirements for each type plumbing fixture were made using the Uniform Plumbing Code as a guideline. For scheduling purposes, estimate that a 2-man crew can install rough-in piping for 3 to 4 plumbing fixtures in an 8-hour day. The same crew will install 3 to 4 fixtures a day. When installing less than 4 fixtures in a building, increase the rough-in cost 25% for each fixture less than 4. Add the cost of the appropriate rough-in to the fixture as shown. These costs include the subcontractor's overhead and profit.

Bathtubs Rough-in includes 50' of DWV pipe and 60' of supply pipe with typical couplings, fittings, flashing and hangers.

Recessed white enameled steel tub, 5' x 30" x 15", with mixing valve, shower head and pop-up waste

	Craft@Hrs	Unit	Material	Labor	Total
Plastic DWV and supply rough-in	P6@3.72	Ea	148.00	152.00	300.00
Cast iron DWV and copper supply rough-in	P6@4.69	Ea	545.00	192.00	737.00
Tub, trim and valves	P6@3.72	Ea	280.00	152.00	432.00

Recessed white enameled cast iron tub, 5' x 32" x 15", with mixing valve, shower head and pop-up waste

	Craft@Hrs	Unit	Material	Labor	Total
Plastic DWV and supply rough-in	P6@3.72	Ea	148.00	152.00	300.00
Cast iron DWV and copper supply rough-in	P6@4.69	Ea	545.00	192.00	737.00
Tub, trim and valves	P6@3.72	Ea	755.00	152.00	907.00

Tub and shower combination, fiberglass, with 3 integral walls and grab bars, 5' x 32" x 75" H

	Craft@Hrs	Unit	Material	Labor	Total
Plastic DWV and supply rough-in	P6@3.72	Ea	148.00	152.00	300.00
Cast iron DWV and copper supply rough-in	P6@4.69	Ea	545.00	192.00	737.00
Fiberglass tub and shower, trim and valves	P6@3.59	Ea	635.00	147.00	782.00

Cleanouts for waste piping. Usually one cleanout is required at the end of each waste line. Rough-in includes 30' of pipe, typical fittings and hangers and a cast iron body and countersunk cadmium plated plug

	Craft@Hrs	Unit	Material	Labor	Total
2" plastic DWV pipe, 2" cast iron cleanout	P6@1.86	Ea	50.90	76.00	126.90
2" cast iron DWV pipe, 2" cast iron cleanout	P6@2.80	Ea	156.00	114.00	270.00
3" plastic DWV pipe, 3" cast iron cleanout	P6@1.86	Ea	77.45	76.00	153.45
3" cast iron DWV pipe, 3" cast iron cleanout	P6@2.80	Ea	221.00	114.00	335.00
4" plastic DWV pipe, 4" cast iron cleanout	P6@1.86	Ea	116.00	76.00	192.00
4" cast iron DWV pipe, 4" cast iron cleanout	P6@2.80	Ea	339.00	114.00	453.00
6" plastic DWV pipe, 6" cast iron cleanout	P6@2.34	Ea	323.00	95.60	418.60
6" cast iron DWV pipe, 6" cast iron cleanout	P6@3.72	Ea	592.00	152.00	744.00

Additional costs for cleanout access covers

	Craft@Hrs	Unit	Material	Labor	Total
Round polished bronze cover	--	Ea	--	--	12.00
Square chrome plated cover	--	Ea	--	--	135.00
Square bronze hinged cover box	--	Ea	--	--	95.50

Mechanical 15

	Craft@Hrs	Unit	Material	Labor	Total

Drinking fountains Rough-in includes 80' of DWV pipe and 40' of supply pipe with typical couplings, fittings, flashing and hangers.
Wall hung drinking fountain, 14" wide, 13" high, non-electric, white vitreous china, with chrome plated spout, self-closing hand operated valve and automatic volume control

	Craft@Hrs	Unit	Material	Labor	Total
Plastic DWV and supply rough-in	P6@3.72	Ea	118.00	152.00	270.00
Cast iron DWV and copper supply rough-in	P6@4.69	Ea	478.00	192.00	670.00
Drinking fountain, trim and valves	P6@2.80	Ea	354.00	114.00	468.00

Floor mount or wall mount type, electric water cooler, 8 gallons per hour capacity, cold service only (no electrical work included)

	Craft@Hrs	Unit	Material	Labor	Total
Plastic DWV and supply rough-in	P6@3.72	Ea	118.00	152.00	270.00
Cast iron DWV and copper supply rough-in	P6@4.69	Ea	478.00	192.00	670.00
Water cooler, trim and valves	P6@2.80	Ea	584.00	114.00	698.00
Add for wall-mounted cooler for handicapped	--	Ea	--	--	185.00

Emergency drench shower and eye wash safety stations with ball valve and safety sign. Includes valved water supply with 30' of pipe, typical valves, fittings and hangers. Add the cost of a floor drain from below if required.
Drench shower with chain operator

	Craft@Hrs	Unit	Material	Labor	Total
Plastic pipe, plastic head, no pipe standard	P6@7.44	Ea	205.00	304.00	509.00
Copper pipe, chrome head, galvanized standard	P6@9.30	Ea	403.00	380.00	783.00
Add for eye wash fountain combination with foot operated ball valve	--	Ea	506.00	--	--

Drench shower and eye wash fountain combination, galvanized pipe construction, walk-thru type safety station, with 18 spray heads, 1 dual spray head and eye wash, and push operated ball valves

	Craft@Hrs	Unit	Material	Labor	Total
Shower with plastic pipe rough-in	P6@63.0	Ea	2,420.00	2,570.00	4,990.00
Shower with copper pipe rough-in	P6@14.9	Ea	2,490.00	609.00	3,099.00

Floor drains Includes 30' of pipe, typical fittings and hangers and a cast iron drain body with 6" diameter nickle bronze grate.

	Craft@Hrs	Unit	Material	Labor	Total
2" plastic DWV rough-in, 2" cast iron floor drain	P6@2.80	Ea	94.00	114.00	208.00
2" cast iron DWV rough-in, 2" cast iron floor drain	P6@3.72	Ea	204.00	152.00	356.00
3" plastic DWV rough-in, 3" cast iron floor drain	P6@2.80	Ea	124.00	114.00	238.00
3" cast iron DWV rough-in, 3" cast iron floor drain	P6@3.72	Ea	274.00	152.00	426.00
4" plastic DWV rough-in, 4" cast iron floor drain	P6@2.80	Ea	171.00	114.00	285.00
4" cast iron DWV rough-in, 4" cast iron floor drain	P6@3.72	Ea	378.00	152.00	530.00
Add for a sediment bucket, any of above	--	Ea	--	--	26.50
Add for a backwater valve, any of above	--	Ea	--	--	48.50

Floor sinks Includes 30' of pipe, typical fittings and hangers and an 8" square sink with cast iron top and body.

	Craft@Hrs	Unit	Material	Labor	Total
2" plastic DWV rough-in, 2" cast iron floor drain	P6@2.80	Ea	243.00	114.00	357.00
2" cast iron DWV rough-in, 2" cast iron floor drain	P6@3.72	Ea	352.00	152.00	504.00
3" plastic DWV rough-in, 3" cast iron floor drain	P6@2.80	Ea	273.00	114.00	387.00
3" cast iron DWV rough-in, 3" cast iron floor drain	P6@3.72	Ea	422.00	152.00	574.00
4" plastic DWV rough-in, 4" cast iron floor drain	P6@2.80	Ea	310.00	114.00	424.00
4" cast iron DWV rough-in, 4" cast iron floor drain	P6@3.72	Ea	541.00	152.00	693.00
Add for sink with acid-resisting finish, any of above	--	Ea	--	--	23.00

Kitchen sinks Rough-in includes 80' of DWV pipe and 60' of supply pipe with typical couplings, fittings, flashing and hangers.
Single compartment kitchen sink, 24" long, 21" wide, 7-5/8" deep, self rimming, countertop mounted, white enameled steel with single handle swing spout, chrome plated faucet and plastic handled spray

	Craft@Hrs	Unit	Material	Labor	Total
Plastic DWV and supply rough-in	P6@2.80	Ea	147.00	114.00	261.00
Cast iron DWV and copper supply rough-in	P6@3.72	Ea	566.00	152.00	718.00

Mechanical 15

	Craft@Hrs	Unit	Material	Labor	Total
Sink, trim and valves	P6@2.80	Ea	301.00	114.00	415.00

Double compartment kitchen sink, 32" long, 21" wide, 6-1/2" deep, self rimming, countertop mounted, white enameled steel with single handle swing spout, chrome plated faucet and plastic handled spray

	Craft@Hrs	Unit	Material	Labor	Total
Plastic DWV and supply rough-in	P6@2.80	Ea	147.00	114.00	261.00
Cast iron DWV and copper supply rough-in	P6@3.72	Ea	526.00	152.00	678.00
Sink, trim and valves	P6@2.80	Ea	315.00	114.00	429.00

Lavatories Rough-in includes 50' of DWV pipe and 60' of supply pipe with typical couplings, fittings, flashing and hangers.

Wall hung lavatory, 20" long, 18" deep, white vitreous china with single handle water faucet and pop-up waste

	Craft@Hrs	Unit	Material	Labor	Total
Plastic DWV and supply rough-in	P6@2.80	Ea	148.00	114.00	262.00
Cast iron DWV and copper supply rough-in	P6@3.72	Ea	526.00	152.00	678.00
Lavatory, trim and valves	P6@1.86	Ea	330.00	76.00	406.00
Add for wheelchair type	--	Ea	--	--	420.00

Countertop mounted 22" long, 19" deep, self-rimming type, white vitreous china with single handle water faucet and pop-up waste

	Craft@Hrs	Unit	Material	Labor	Total
Plastic DWV and supply rough-in	P6@2.80	Ea	148.00	114.00	262.00
Cast iron DWV and copper supply rough-in	P6@3.72	Ea	526.00	152.00	678.00
Lavatory, trim and valves	P6@1.86	Ea	389.00	76.00	465.00

Service sinks Rough-in includes 30' of DWV pipe and 70' of supply pipe with typical couplings, fittings, flashing and hangers.

Wall hung service sink, 22" x 18", with rough chrome-plated double lever handle faucet

	Craft@Hrs	Unit	Material	Labor	Total
Plastic DWV and supply rough-in	P6@3.72	Ea	95.00	152.00	247.00
Cast iron DWV and copper supply rough-in	P6@5.61	Ea	225.00	229.00	454.00
Service sink, trim and valves	P6@2.80	Ea	549.00	114.00	663.00

Urinals Rough-in includes 50' of DWV pipe and 30' of supply pipe with typical couplings, fittings, flashing and hangers.

Wall hung urinal, white vitreous china, washout flush action, 3/4" top spud with hand operated flush valve/vacuum breaker

	Craft@Hrs	Unit	Material	Labor	Total
Plastic DWV and supply rough-in	P6@1.86	Ea	107.00	76.00	183.00
Cast iron DWV and copper supply rough-in	P6@2.80	Ea	372.00	114.00	486.00
Urinal, trim and valves	P6@1.86	Ea	489.00	76.00	565.00

Floor mounted urinal, 18" wide, sloping front white vitreous china, washout flush action, 3/4" top spud with hand operated flush valve/vacuum breaker

	Craft@Hrs	Unit	Material	Labor	Total
Plastic DWV and supply rough-in	P6@1.86	Ea	107.00	76.00	183.00
Cast iron DWV and copper supply rough-in	P6@2.80	Ea	372.00	114.00	486.00
Urinal, trim and valves	P6@1.86	Ea	617.00	76.00	693.00

Wash fountains Rough-in includes 60' of DWV pipe and 60' of supply pipe with typical couplings, fittings, flashing and hangers.

Circular floor mounted wash fountain, 54" diameter, precast stone, standard height, with foot operated water spray control

	Craft@Hrs	Unit	Material	Labor	Total
Plastic DWV and supply rough-in	P6@4.21	Ea	164.00	172.00	336.00
Cast iron DWV and copper supply rough-in	P6@9.33	Ea	513.00	381.00	894.00
Wash fountain, trim and valves	P6@7.47	Ea	1,910.00	305.00	2,215.00
Add for stainless steel in lieu of precast stone	--	Ea	--	--	450.00

Semi-circular wall mounted, wash fountain, 54" diameter, precast stone, standard height, with foot operated water spray control

	Craft@Hrs	Unit	Material	Labor	Total
Plastic DWV and supply rough-in	P6@4.21	Ea	164.00	172.00	336.00
Cast iron DWV and copper supply rough-in	P6@9.33	Ea	513.00	381.00	894.00
Wash fountain, trim and valves	P6@7.47	Ea	1,707.00	305.00	2,012.00

Mechanical 15

	Craft@Hrs	Unit	Material	Labor	Total

Wash fountains, Semi-circular, continued
Add for stainless steel in lieu of precast stone -- Ea -- -- 300.00

Wash sinks Rough-in includes 50' of DWV pipe and 50' of supply pipe with typical couplings, fittings, flashing and hangers.
Wall hung wash sink, white enameled cast iron, 48" long, 18" deep, trough-type with 2 twin-handled faucets and 2 soap dishes

Plastic DWV and supply rough-in	P6@3.72	Ea	142.00	152.00	294.00
Cast iron DWV and copper supply rough-in	P6@4.69	Ea	337.00	192.00	529.00
Wash sink, trim and valves	P6@4.69	Ea	1,103.00	192.00	1,295.00

Water closets Rough-in includes 30' of DWV pipe and 30' of supply pipe with typical couplings, fittings, flashing and hangers.
Floor mounted, tank type white vitreous china closet with elongated bowl and toilet seat with cover

Plastic DWV and supply rough-in	P6@3.72	Ea	89.65	152.00	241.65
Cast iron DWV and copper supply rough-in	P6@4.69	Ea	290.00	192.00	482.00
Water closet, trim and valves	P6@1.87	Ea	267.00	76.40	343.40

Wall hung, tank type white vitreous china closet with elongated bowl, carrier and toilet seat with cover

Plastic DWV and supply rough-in	P6@3.72	Ea	89.65	152.00	241.65
Cast iron DWV and copper supply rough-in	P6@4.69	Ea	290.00	192.00	482.00
Water closet, carrier, trim and valves	P6@2.80	Ea	563.00	114.00	677.00

Water heaters Rough-in includes 40' of water pipe (gas models include 20' of gas supply pipe and 10' of double wall vent pipe) with typical couplings, fittings, and hangers. All are commercial grade with a 3 year limited warranty.
Gas fired, with glass lined insulated storage tank, immersion thermostat, vent hood, pressure and temperature relief valve and tank drain valve

Plastic water piping rough-in	P6@3.72	Ea	161.00	152.00	313.00
Copper water piping rough-in	P6@4.00	Ea	211.00	163.00	374.00
Black steel gas piping rough-in	P6@1.40	Ea	56.20	57.20	113.40
Double wall vent piping rough-in	P6@1.40	Ea	585.00	57.20	642.20
80 gallon heater, trim and valves	P6@1.87	Ea	1,006.00	76.40	1,082.40
100 gallon heater, trim and valves	P6@1.87	Ea	1,200.00	76.40	1,276.40

Electric models, 240 volt AC factory fused 18,000 watt heating elements, with glass lined insulated storage tank, immersion thermostat, pressure and temperature relief valve and tank drain valve

Plastic water piping rough-in	P6@3.72	Ea	161.00	152.00	313.00
Copper water piping rough-in	P6@4.21	Ea	211.00	172.00	383.00
82 gallon heater, trim and valves	P6@1.87	Ea	1,985.00	76.40	2,061.40
120 gallon heater, trim and valves	P6@1.87	Ea	2,385.00	76.40	2,461.40

Fire Protection Equipment. These costs include the subcontractor's overhead and profit.
Sprinkler systems, including valves and connection. Costs will be higher where room sizes are smaller or where head coverage averages less than 110 SF
Exposed systems, wet, complete, cost per SF of floor protected

5,000 SF	--	SF	--	--	2.00
Over 5,000 to 15,000 SF	--	SF	--	--	1.70
15,000 SF or more	--	SF	--	--	1.50

Concealed systems, wet, complete, cost per SF of floor protected

5,000 SF	--	SF	--	--	2.25
Over 5,000 to 15,000 SF	--	SF	--	--	1.80
15,000 SF or more	--	SF	--	--	1.70
Add for dry systems	--	%	--	--	15.0

Mechanical 15

	Craft@Hrs	Unit	Material	Labor	Total
Fire sprinkler system components					
Sprinkler heads only, 212 degree, brass, for exposed piping system	SP@.450	Ea	5.50	21.60	27.10
Sprinkler heads only, 212 degree, chrome, for concealed piping system	SP@.601	Ea	6.80	28.90	35.70
Water powered gong (local alarm)	SP@1.94	Ea	141.00	93.20	234.20
Fire department inlet connection	SP@2.41	Ea	151.00	116.00	267.00
Wall plate (sign) for inlet connection	SP@.827	Ea	29.70	39.70	69.40
4" swing check valve, flanged, iron body	SP@3.57	Ea	276.00	172.00	448.00
6" swing check valve, flanged, iron body	SP@5.40	Ea	474.00	259.00	733.00
1/2" ball drip at each check valve and fire department connection	SP@.193	Ea	12.00	9.27	21.27
Inspector's test connection (excluding concrete cutting)	SP@.973	Ea	46.90	46.80	93.70
Cabinet with 6 spare heads and wrench	SP@.291	LS	53.70	14.00	67.70
Field testing and flushing	--	LS	--	--	140.00
Disinfection of distribution system	--	LS	--	--	140.00
Hydraulic design (sprinklers in a high density or high risk area), typical	--	LS	--	--	2,200.00
Escutcheon plate collar for exposed sprinkler heads, chrome	SP@.071	Ea	.60	3.41	4.01
Wet system components (where freezing is not a hazard)					
4" valve, flanged to grooved pipe	SP@1.24	Ea	287.00	59.60	346.60
6" valve, flanged to flanged pipe	SP@1.60	Ea	359.00	76.90	435.90
8" valve, flange to flange	SP@4.16	Ea	479.00	200.00	679.00
Wet valve trim (retard chamber and gauges)	SP@.825	Ea	255.00	39.60	294.60
4" alarm valve	SP@2.41	Ea	328.00	116.00	444.00
8" alarm valve	SP@9.67	Ea	891.00	465.00	1,356.00
Retard pressure switch	SP@2.89	Ea	469.00	139.00	608.00
Dry system components (distribution piping holds no water)					
4" dry pipe valve, flanged	SP@2.24	Ea	688.00	108.00	796.00
6" dry pipe valve, flanged	SP@2.89	Ea	834.00	139.00	973.00
Dry valve trim and gauges	SP@.825	Ea	255.00	39.60	294.60
Air maintenance device	SP@.781	Ea	135.00	37.50	172.50
Low air supervisory unit	SP@1.11	Ea	427.00	53.30	480.30
Low air pressure switch	SP@.784	Ea	135.00	37.70	172.70
Pressure switch (double circuit, open and close)	SP@1.92	Ea	130.00	92.30	222.30
Connection fee (typical)	--	LS	--	--	650.00
Flow detector check valve (if required)	--	LS	--	--	3,450.00
Underground valve vault (if required)	--	LS	--	--	3,000.00
Wall hydrants for sprinkler systems, brass					
Single outlet, 2-1/2" x 2-1/2"	SP@.825	Ea	234.00	39.60	273.60
2-way outlet, 2-1/2" x 2-1/2" x 4"	SP@.825	Ea	662.00	39.60	701.60
3-way outlet, 2-1/2" x 2-1/2" x 3-1/3" x 4"	SP@.825	Ea	979.00	39.60	1,018.60
Pumper connection, 6"	SP@2.89	Ea	745.00	139.00	884.00
Roof manifold, with valves	SP@3.90	Ea	443.00	187.00	630.00
Fire hose cabinets, recessed, 24" x 30" x 5" with rack, full glass door, 1-1/2" hose, valve and nozzle, Standard					
Painted steel, 75' hose, 24" x 30"	SP@1.55	Ea	397.00	74.50	471.50
Painted steel, 100' hose, 24" x 36"	SP@1.55	Ea	443.00	74.50	517.50
Aluminum, 75' hose	SP@1.55	Ea	391.00	74.50	465.50
Stainless steel, 75' hose	SP@1.55	Ea	464.00	74.50	538.50
Add for hose and extinguisher cabinet	--	%	10.0	--	--

Mechanical 15

	Craft@Hrs	Unit	Material	Labor	Total

Fire Protection Equipment, continued
Fire extinguishers, factory charged, complete with hose and horn and wall mounting bracket.

	Craft@Hrs	Unit	Material	Labor	Total
5 lb carbon dioxide	C8@.488	Ea	172.00	18.80	190.80
10 lb carbon dioxide	C8@.488	Ea	258.00	18.80	276.80
15 lb carbon dioxide	C8@.488	Ea	300.00	18.80	318.80
5 lb dry chemical	C8@.488	Ea	50.20	18.80	69.00
20 lb dry chemical	C8@.488	Ea	135.00	18.80	153.80

Extinguisher cabinets, no extinguishers included, painted steel, recessed

	Craft@Hrs	Unit	Material	Labor	Total
9" x 24" x 5" full glass door	C8@1.67	Ea	77.00	64.20	141.20
9" x 24" x 5" break glass door	C8@1.67	Ea	85.00	64.20	149.20
12" x 27" x 8" full glass door	C8@1.67	Ea	102.00	64.20	166.20
12" x 27" x 8" break glass door	C8@1.67	Ea	110.00	64.20	174.20
Add for semi-recessed or surface mount	--	%	10.0	--	--
Add for chrome cabinets	--	%	15.0	--	--
Add for stainless steel cabinets	--	Ea	150.00	--	--

Tanks. These costs include the subcontractor's overhead and profit.
Underground oil storage, steel, including fill sand and pavement fittings but no excavation, piping, paving, or concrete

	Craft@Hrs	Unit	Material	Labor	Total
550 gallon	M7@4.86	Ea	800.00	204.00	1,004.00
2,500 gallon	M8@18.5	Ea	1,860.00	796.00	2,656.00
5,000 gallon	M8@24.0	Ea	3,350.00	1,030.00	4,380.00
10,000 gallon	M8@37.3	Ea	5,740.00	1,610.00	7,350.00

Above ground steel oil storage tanks, including tank, fittings and supports but no pipe, concrete or excavation

	Craft@Hrs	Unit	Material	Labor	Total
275 gallon	M7@2.98	Ea	284.00	125.00	409.00
500 gallon	M7@3.80	Ea	605.00	160.00	765.00
1,000 gallon	M8@7.64	Ea	990.00	329.00	1,319.00
1,500 gallon	M8@8.29	Ea	1,100.00	357.00	1,457.00
2,000 gallon	M8@10.5	Ea	1,425.00	452.00	1,877.00
5,000 gallon	M8@29.6	Ea	3,960.00	1,270.00	5,230.00
Add for vent caps	--	Ea	.90	7.45	8.35
Add for filler cap	--	Ea	20.80	7.45	28.25

Underground fiberglass oil storage tanks, including fittings and supports but no pipe, concrete or excavation

	Craft@Hrs	Unit	Material	Labor	Total
550 gallon	M7@8.36	Ea	1,835.00	352.00	2,187.00
1,000 gallon	M8@16.8	Ea	3,305.00	723.00	4,028.00
2,000 gallon	M8@16.8	Ea	3,740.00	723.00	4,463.00
4,000 gallon	M8@25.9	Ea	4,550.00	1,110.00	5,660.00
6,000 gallon	M8@33.5	Ea	5,070.00	1,440.00	6,510.00
8,000 gallon	M8@33.5	Ea	5,700.00	1,440.00	7,140.00
10,000 gallon	M8@33.5	Ea	6,240.00	1,440.00	7,680.00
12,000 gallon	M8@84.4	Ea	7,700.00	3,630.00	11,330.00
15,000 gallon	M8@89.0	Ea	8,950.00	3,830.00	12,780.00
20,000 gallon	M8@89.0	Ea	10,660.00	3,830.00	14,490.00

Propane tanks with valves, underground, including fill sand and fittings, 1,000 gallon

	Craft@Hrs	Unit	Material	Labor	Total
	M8@10.0	Ea	1,630.00	431.00	2,061.00

Heating, Ventilating and Air Conditioning. These costs include the subcontractor's overhead and profit.
Typical in-place cost per square foot of floor. Complete heating, ventilating and cooling system except as noted. Costs will vary widely with climate and design considerations

	Craft@Hrs	Unit	Material	Labor	Total
Auditoriums and theaters	--	SF	--	--	12.00
Banks	--	SF	--	--	8.00

Mechanical 15

	Craft@Hrs	Unit	Material	Labor	Total
Colleges	--	SF	--	--	11.00
Commercial stores, small markets	--	SF	--	--	4.00
Dormitories	--	SF	--	--	4.50
Hospitals	--	SF	--	--	14.50
Institutional structures	--	SF	--	--	9.00
Manufacturing plants, heating only	--	SF	--	--	2.00
Medical clinics	--	SF	--	--	7.00
Office buildings, small	--	SF	--	--	5.50
Office buildings, multi-story	--	SF	--	--	6.00
Schools	--	SF	--	--	7.50

Heat Generators. No pipe or electric runs included. These costs include the subcontractor's overhead and profit.

Unit heaters, suspended, with flue (at $110) and valve, gas fired, blower style with aluminized heat exchanger, Modine

	Craft@Hrs	Unit	Material	Labor	Total
50 MBtu input	M5@5.33	Ea	1,006.00	214.00	1,220.00
75 MBtu	M5@5.91	Ea	1,065.00	237.00	1,302.00
125 MBtu	M5@7.00	Ea	1,285.00	280.00	1,565.00
170 MBtu	M5@9.45	Ea	1,470.00	379.00	1,849.00
225 MBtu	M5@10.1	Ea	1,760.00	405.00	2,165.00
300 MBtu	M5@11.5	Ea	2,270.00	461.00	2,731.00
400 MBtu	M5@13.5	Ea	2,775.00	541.00	3,316.00
Add for intermittent pilot	--	LS	160.00	--	--
Add for high efficiency units	--	Ea	500.00	--	--
Deduct for air propeller style	--	%	-30.0	--	--
Add for switch and thermostat	CE@1.09	Ea	50.00	50.50	100.50
Add for typical electric run, conduit	E4@5.75	Ea	60.00	231.00	291.00

Duct heaters, indoor, with flue (at $110) and valve, gas fired, intermittent pilot, with aluminized heat exchanger

	Craft@Hrs	Unit	Material	Labor	Total
100 MBtu input	M5@5.20	Ea	807.00	208.00	1,015.00
175 MBtu	M5@5.91	Ea	980.00	237.00	1,217.00
225 MBtu	M5@6.32	Ea	1,135.00	253.00	1,388.00
300 MBtu	M5@8.02	Ea	1,350.00	321.00	1,671.00
350 MBtu	M5@8.98	Ea	1,510.00	360.00	1,870.00
400 MBtu	M5@11.0	Ea	1,635.00	441.00	2,076.00
Deduct for standard ignition	--	LS	-150.00	--	--
Add for switch and thermostat	CE@2.14	Ea	50.00	99.10	149.10
Add for typical electric run, conduit	E4@5.75	Ea	60.00	231.00	291.00

Duct heaters, roof mounted, and valve, gas fired, intermittent pilot, with aluminized heat exchanger

	Craft@Hrs	Unit	Material	Labor	Total
100 MBtu	M5@4.53	Ea	1,170.00	181.00	1,351.00
175 MBtu	M5@6.19	Ea	1,460.00	248.00	1,708.00
225 MBtu	M5@8.02	Ea	1,610.00	321.00	1,931.00
300 MBtu	M5@9.45	Ea	1,960.00	379.00	2,339.00
350 MBtu	M5@9.84	Ea	2,100.00	394.00	2,494.00
400 MBtu	M5@12.3	Ea	2,400.00	493.00	2,893.00

Furnaces, upflow, gas, wall type, with valve, vent (at $110) and electronic ignition

	Craft@Hrs	Unit	Material	Labor	Total
35 MBtu	M5@4.24	Ea	640.00	170.00	810.00
55 MBtu	M5@4.24	Ea	705.00	170.00	875.00
65 MBtu	M5@4.53	Ea	710.00	181.00	891.00
Add for unit with cooling coils	--	Ea	100.00	--	--
Add for thermostat	CE@1.00	Ea	42.70	50.00	92.70

Furnaces, horizontal, gas, floor type, with valve, vent (at $110) and electronic ignition

	Craft@Hrs	Unit	Material	Labor	Total
30 MBtu	M5@4.42	Ea	650.00	177.00	827.00
45 MBtu	M5@5.15	Ea	685.00	206.00	891.00

Mechanical 15

	Craft@Hrs	Unit	Material	Labor	Total
Heat Generators, continued					
65 MBtu	M5@5.33	Ea	780.00	214.00	994.00
Add for unit with cooling coils	--	Ea	97.90	--	--
Add for thermostat	CE@1.08	Ea	42.70	50.00	92.70
Duct heaters, with thermostat and relay, no enclosure, electric					
3.4 MBtu, 1 kw	H9@1.08	Ea	251.00	49.80	300.80
10 MBtu, 3 kw	H9@1.27	Ea	263.00	58.60	321.60
18 MBtu, 5 kw	H9@1.37	Ea	314.00	63.20	377.20
25 MBtu, 7.5 kw	H9@1.93	Ea	389.00	89.10	478.10
34 MBtu, 10 kw	H9@2.29	Ea	475.00	106.00	581.00
40 MBtu, 12 kw	H9@2.81	Ea	593.00	130.00	723.00
Air curtain electric heaters, wall mounted, includes electrical connection only					
37" long, 9.5 kw	H9@6.33	Ea	467.00	292.00	759.00
49" long, 12.5 kw	H9@8.05	Ea	656.00	372.00	1,028.00
61" long, 16 kw	H9@11.2	Ea	788.00	517.00	1,305.00
Air curtain gas heaters, wall mounted, includes electrical connection only and no venting					
36" long, 256 MBtu, 2,840 CFM blower	M5@11.1	Ea	1,120.00	445.00	1,565.00
60" long, 290 MBtu, 4,740 CFM blower	M5@11.1	Ea	1,210.00	445.00	1,655.00
72" long, 400 MBtu, 5,660 CFM blower	M5@13.2	Ea	1,360.00	529.00	1,889.00
96" long, 525 MBtu, 7,580 CFM blower	M1@14.1	Ea	1,440.00	528.00	1,968.00
120" long, 630 MBtu, 9,480 CFM blower	M1@16.5	Ea	1,570.00	618.00	2,188.00
144" long, 850 MBtu, 11,400 CFM blower	M1@23.4	Ea	2,250.00	876.00	3,126.00
Baseboard electric heaters, with remote thermostat,					
185 watts per LF	CE@.175	LF	16.00	8.11	24.11

Boilers Shipped skid mounted. Costs shown include the boiler shell and structure, I-beam skids, combustion chamber, firetubes, crown sheet, tube sheets (for firetube), water tubes, mud drum(s), steam drum (for watertube boiler), finned watertubes and check valves (for hot water generators), refractory brick lining, explosion doors, combustion air inlets and stack outlet. Add the cost of the combustion train, boiler trim, fuel train piping, electrical wiring, refractory, stack, feedwater system, combustion controls, water treatment, expansion tanks (if hot water) or condensate return (if steam) and system start-up from the following pages. Costs shown are by boiler horsepower (BHP) rating. These costs include the subcontractor's overhead and profit. Equipment cost, for boilers up to 100 boiler horsepower, is $70 per hour and includes a 35 ton truck crane (at $50) and a 2 ton capacity forklift (at $20). Equipment cost, for boilers over 100 boiler horsepower, is $110 per hour and includes a 90 ton truck crane (at $75) and a 4 ton capacity forklift (at $35).

	Craft@Hrs	Unit	Material	Labor	Equipment	Total
Package Scotch marine fire tube boilers 15 PSI steam, 30 PSI water, oil or gas fired						
20 BHP, 670 MBtu/Hr	H8@8.00	Ea	14,500.00	380.00	93.30	14,973.30
30 BHP, 1,004 MBtu/Hr	H8@8.00	Ea	16,250.00	380.00	93.30	16,723.30
40 BHP, 1,339 MBtu/Hr	H8@8.00	Ea	17,150.00	380.00	93.30	17,623.30
50 BHP, 1,674 MBtu/Hr	H8@8.00	Ea	18,000.00	380.00	93.30	18,473.30
60 BHP, 2,009 MBtu/Hr	H8@8.00	Ea	20,000.00	380.00	93.30	20,473.30
70 BHP, 2,343 MBtu/Hr	H8@8.00	Ea	21,500.00	380.00	93.30	21,973.30
80 BHP, 2,678 MBtu/Hr	H8@8.00	Ea	23,000.00	380.00	93.30	23,473.30
100 BHP, 3,348 MBtu/Hr	H8@8.00	Ea	27,400.00	380.00	93.30	27,873.30
125 BHP, 4,184 MBtu/Hr	H8@16.0	Ea	29,300.00	761.00	293.00	30,354.00

Mechanical 15

	Craft@Hrs	Unit	Material	Labor	Equipment	Total
150 BHP, 5,021 MBtu/Hr	H8@16.0	Ea	31,400.00	761.00	293.00	32,454.00
200 BHP, 6,695 MBtu/Hr	H8@16.0	Ea	37,200.00	761.00	293.00	38,254.00
250 BHP, 8,369 MBtu/Hr	H8@32.0	Ea	41,000.00	1,520.00	587.00	43,107.00
300 BHP, 10,043 MBtu/Hr	H8@32.0	Ea	44,100.00	1,520.00	587.00	46,207.00
350 BHP, 11,761 MBtu/Hr	H8@32.0	Ea	49,800.00	1,520.00	587.00	51,907.00
400 BHP, 13,390 MBtu/Hr	H8@32.0	Ea	51,300.00	1,520.00	587.00	53,407.00
500 BHP, 16,738 MBtu/Hr	H8@32.0	Ea	61,100.00	1,520.00	587.00	63,207.00
600 BHP, 20,085 MBtu/Hr	H8@32.0	Ea	68,300.00	1,520.00	587.00	70,407.00
750 BHP, 25,106 MBtu/Hr	H8@64.0	Ea	72,100.00	3,050.00	1,173.00	76,323.00
800 BHP, 26,780 MBtu/Hr	H8@64.0	Ea	84,000.00	3,050.00	1,173.00	88,223.00

Package Scotch marine fire tube boilers 150 PSI steam, oil or gas fired

	Craft@Hrs	Unit	Material	Labor	Equipment	Total
60 BHP, 2,009 MBtu/Hr	H8@8.00	Ea	23,000.00	380.00	93.30	23,473.30
70 BHP, 2,343 MBtu/Hr	H8@8.00	Ea	24,700.00	380.00	93.30	25,173.30
80 BHP, 2,678 MBtu/Hr	H8@8.00	Ea	26,500.00	380.00	93.30	26,973.30
100 BHP, 3,348 MBtu/Hr	H8@8.00	Ea	31,500.00	380.00	93.30	31,973.30
125 BHP, 4,184 MBtu/Hr	H8@16.0	Ea	33,600.00	761.00	293.00	34,654.00
150 BHP, 5,021 MBtu/Hr	H8@16.0	Ea	36,100.00	761.00	293.00	37,154.00
200 BHP, 6,695 MBtu/Hr	H8@16.0	Ea	47,200.00	761.00	293.00	48,254.00
250 BHP, 8,369 MBtu/Hr	H8@32.0	Ea	50,700.00	1,520.00	587.00	52,807.00
300 BHP, 10,043 MBtu/Hr	H8@32.0	Ea	57,200.00	1,520.00	587.00	59,307.00
350 BHP, 11,716 MBtu/Hr	H8@32.0	Ea	59,000.00	1,520.00	587.00	61,107.00
400 BHP, 13,390 MBtu/Hr	H8@32.0	Ea	70,300.00	1,520.00	587.00	72,407.00
500 BHP, 16,738 MBtu/Hr	H8@32.0	Ea	78,500.00	1,520.00	587.00	80,607.00
600 BHP, 20,085 MBtu/Hr	H8@32.0	Ea	79,800.00	1,520.00	587.00	81,907.00
750 BHP, 25,106 MBtu/Hr	H8@64.0	Ea	83,000.00	3,050.00	1,173.00	87,223.00
800 BHP, 26,780 MBtu/Hr	H8@64.0	Ea	96,600.00	3,050.00	1,173.00	100,823.00

Package firebox boilers 30 PSI hot water-rated, oil or gas fired

	Craft@Hrs	Unit	Material	Labor	Equipment	Total
13.4 BHP, 450 MBtu/Hr	H8@8.00	Ea	8,630.00	380.00	93.30	9,103.30
16.4 BHP, 550 MBtu/Hr	H8@8.00	Ea	8,800.00	380.00	93.30	9,273.30
19.4 BHP, 650 MBtu/Hr	H8@8.00	Ea	9,100.00	380.00	93.30	9,573.30
22.4 BHP, 750 MBtu/Hr	H8@8.00	Ea	9,350.00	380.00	93.30	9,823.30
28.4 BHP, 950 MBtu/Hr	H8@8.00	Ea	9,700.00	380.00	93.30	10,173.30
34.4 BHP, 1,150 MBtu/Hr	H8@8.00	Ea	11,200.00	380.00	93.30	11,673.30
40.3 BHP, 1,350 MBtu/Hr	H8@8.00	Ea	12,300.00	380.00	93.30	12,773.30
46.3 BHP, 1,550 MBtu/Hr	H8@8.00	Ea	14,000.00	380.00	93.30	14,473.30
52.3 BHP, 1,750 MBtu/Hr	H8@8.00	Ea	16,400.00	380.00	93.30	16,873.30
61.1 BHP, 2,050 MBtu/Hr	H8@8.00	Ea	18,100.00	380.00	93.30	18,573.30
70.0 BHP, 2,350 MBtu/Hr	H8@8.00	Ea	19,400.00	380.00	93.30	19,873.30
80.0 BHP, 2,650 MBtu/Hr	H8@8.00	Ea	20,000.00	380.00	93.30	20,473.30
100 BHP, 3,350 MBtu/Hr	H8@8.00	Ea	22,500.00	380.00	93.30	22,973.30
126.9 BHP, 4,250 MBtu/Hr	H8@16.0	Ea	26,700.00	761.00	293.00	27,754.00
150.8 BHP, 5,050 MBtu/Hr	H8@16.0	Ea	28,900.00	761.00	293.00	29,954.00

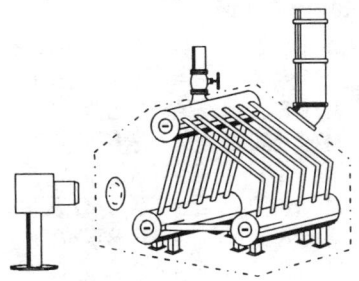

Mechanical 15

	Craft@Hrs	Unit	Material	Labor	Equipment	Total

Watertube industrial high pressure boilers. See illustration on preceding page. Add trim, burner and fuel train costs following this section.

	Craft@Hrs	Unit	Material	Labor	Equipment	Total
480 BHP, 16,200 MBtu/Hr	H8@32.0	Ea	170,000.00	1,520.00	587.00	172,107.00
526 BHP, 17,800 MBtu/Hr	H8@32.0	Ea	177,000.00	1,520.00	587.00	179,107.00
572 BHP, 19,300 MBtu/Hr	H8@32.0	Ea	181,000.00	1,520.00	587.00	183,107.00
618 BHP, 20,800 MBtu/Hr	H8@32.0	Ea	184,000.00	1,520.00	587.00	186,107.00
664 BHP, 22,400 MBtu/Hr	H8@32.0	Ea	193,000.00	1,520.00	587.00	195,107.00
710 BHP, 23,800 MBtu/Hr	H8@64.0	Ea	196,000.00	3,050.00	1,170.00	200,220.00
747 BHP, 25,000 MBtu/Hr	H8@64.0	Ea	199,000.00	3,050.00	1,170.00	203,220.00
792 BHP, 26,100 MBtu/Hr	H8@64.0	Ea	205,000.00	3,050.00	1,170.00	209,220.00
838 BHP, 28,100 MBtu/Hr	H8@64.0	Ea	211,000.00	3,050.00	1,170.00	215,220.00
884 BHP, 29,600 MBtu/Hr	H8@64.0	Ea	217,000.00	3,050.00	1,170.00	221,220.00
930 BHP, 31,100 MBtu/Hr	H8@64.0	Ea	221,000.00	3,050.00	1,170.00	225,220.00
1,000 BHP, 33,400 MBtu/Hr	H8@64.0	Ea	225,000.00	3,050.00	1,170.00	229,220.00

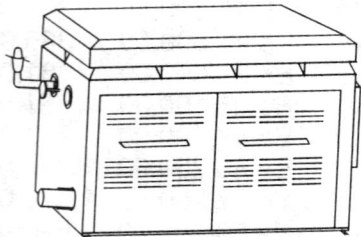

Hydronic hot water generators, 30 PSI, with cast iron burners, factory installed boiler trim and circulating pump. For commercial and light industrial applications. Equipment cost is $20.00 per hour for a 2 ton capacity forklift. These costs include the subcontractor's overhead and profit.

	Craft@Hrs	Unit	Material	Labor	Equipment	Total
4 BHP, 133 MBtu/Hr	M5@4.00	Ea	1,380.00	160.00	40.00	1,580.00
8 BHP, 263 MBtu/Hr	M5@4.00	Ea	1,830.00	160.00	40.00	2,030.00
12 BHP, 404 MBtu/Hr	M5@4.00	Ea	2,360.00	160.00	40.00	2,560.00
15 BHP, 514 MBtu/Hr	M5@5.00	Ea	3,580.00	200.00	50.00	3,830.00
19 BHP, 624 MBtu/Hr	M5@5.00	Ea	4,750.00	200.00	50.00	5,000.00
22 BHP, 724 MBtu/Hr	M5@6.00	Ea	4,940.00	240.00	60.00	5,240.00
29 BHP, 962 MBtu/Hr	M5@6.00	Ea	5,890.00	240.00	60.00	6,190.00
34 BHP, 1,125 MBtu/Hr	M7@8.00	Ea	6,570.00	337.00	53.40	6,960.40
40 BHP, 1,336 MBtu/Hr	M7@8.00	Ea	7,580.00	337.00	53.40	7,970.40
47 BHP, 1,571 MBtu/Hr	M7@9.00	Ea	7,850.00	379.00	60.00	8,289.00
53 BHP, 1,758 MBtu/Hr	M7@10.0	Ea	8,560.00	421.00	66.70	9,047.70
63 BHP, 2,100 MBtu/Hr	M7@12.0	Ea	8,870.00	505.00	80.00	9,455.00
89 BHP, 3,001 MBtu/Hr	M7@18.0	Ea	11,408.00	757.00	120.00	12,285.00
118 BHP, 4,001 MBtu/Hr	M7@22.0	Ea	11,800.00	925.00	147.00	12,872.00

Combustion Train for Boilers Costs shown include the burner, burner head, burner mount flange, associated refractory and gasketing, main fuel valve, draft damper and fan (for a power burner), or burner face plate (for an atmospheric or induced-draft burner). Equipment cost is $20.00 per hour for a 2 ton capacity forklift. These costs include the subcontractor's overhead and profit.
Costs shown are based on boiler horsepower rating (BHP)

Combustion train for package Scotch marine firetube boilers

	Craft@Hrs	Unit	Material	Labor	Equipment	Total
20 BHP, 670 MBtu/Hr	M7@8.00	Ea	1,390.00	337.00	53.40	1,780.40
30 BHP, 1,004 MBtu/Hr	M7@8.00	Ea	1,880.00	337.00	53.40	2,270.40
40 BHP, 1,339 MBtu/Hr	M7@8.00	Ea	2,570.00	337.00	53.40	2,960.40
50 BHP, 1,674 MBtu/Hr	M7@8.00	Ea	3,220.00	337.00	53.40	3,610.40
60 BHP, 2,009 MBtu/Hr	M7@8.00	Ea	3,560.00	337.00	53.40	3,950.40
70 BHP, 2,343 MBtu/Hr	M7@8.00	Ea	3,890.00	337.00	53.40	4,280.40
80 BHP, 2,678 MBtu/Hr	M7@8.00	Ea	3,950.00	337.00	53.40	4,340.40
100 BHP, 3,348 MBtu/Hr	M7@8.00	Ea	4,320.00	337.00	53.40	4,710.40

Mechanical 15

	Craft@Hrs	Unit	Material	Labor	Equipment	Total
125 BHP, 4,184 MBtu/Hr	M7@8.00	Ea	4,500.00	337.00	53.40	4,890.40
150 BHP, 5,021 MBtu/Hr	M7@8.00	Ea	4,980.00	337.00	53.40	5,370.40
200 BHP, 6,695 MBtu/Hr	M7@8.00	Ea	5,190.00	337.00	53.40	5,580.40
250 BHP, 8,369 MBtu/Hr	M7@8.00	Ea	5,720.00	337.00	53.40	6,110.40
300 BHP, 10,043 MBtu/Hr	M7@8.00	Ea	6,390.00	337.00	53.40	6,780.40
350 BHP, 11,761 MBtu/Hr	M7@8.00	Ea	6,550.00	337.00	53.40	6,940.40
400 BHP, 13,390 MBtu/Hr	M7@8.00	Ea	6,750.00	337.00	53.40	7,140.40
500 BHP, 16,738 MBtu/Hr	M7@16.0	Ea	7,320.00	673.00	107.00	8,100.00
600 BHP, 20,085 MBtu/Hr	M7@16.0	Ea	7,830.00	673.00	107.00	8,610.00
750 BHP, 25,106 MBtu/Hr	M7@32.0	Ea	8,710.00	1,350.00	214.00	10,274.00
800 BHP, 26,780 MBtu/Hr	M7@32.0	Ea	8,680.00	1,350.00	214.00	10,244.00
Combustion train for package firebox boilers						
13.4 BHP, 450 MBtu/Hr	M7@8.00	Ea	725.00	337.00	53.40	1,115.40
16.4 BHP, 550 MBtu/Hr	M7@8.00	Ea	950.00	337.00	53.40	1,340.40
17.4 BHP, 650 MBtu/Hr	M7@8.00	Ea	1,130.00	337.00	53.40	1,520.40
22.4 BHP, 750 MBtu/Hr	M7@8.00	Ea	1,340.00	337.00	53.40	1,730.40
28.4 BHP, 950 MBtu/Hr	M7@8.00	Ea	1,440.00	337.00	53.40	1,830.40
34.4 BHP, 1,150 MBtu/Hr	M7@8.00	Ea	1,590.00	337.00	53.40	1,980.40
40.3 BHP, 1,350 MBtu/Hr	M7@8.00	Ea	1,710.00	337.00	53.40	2,100.40
46.3 BHP, 1,550 MBtu/Hr	M7@8.00	Ea	1,890.00	337.00	53.40	2,280.40
52.3 BHP, 1,750 MBtu/Hr	M7@8.00	Ea	2,100.00	337.00	53.40	2,490.40
61.1 BHP, 2,050 MBtu/Hr	M7@8.00	Ea	2,280.00	337.00	53.40	2,670.40
70.0 BHP, 2,350 MBtu/Hr	M7@8.00	Ea	2,340.00	337.00	53.40	2,730.40
80.0 BHP, 2,650 MBtu/Hr	M7@8.00	Ea	2,450.00	337.00	53.40	2,840.40
100.0 BHP, 3,350 MBtu/Hr	M7@8.00	Ea	2,580.00	337.00	53.40	2,970.40
126.9 BHP, 4,250 MBtu/Hr	M7@8.00	Ea	2,740.00	337.00	53.40	3,130.40
150.0 BHP, 5,050 MBtu/Hr	M7@8.00	Ea	3,000.00	337.00	53.40	3,390.40

Natural Gas Fuel Train Piping Meets requirements of Underwriters' Laboratories. Supplied pre-assembled. Add the cost of venting and piping to a gas meter. These costs include the subcontractor's overhead and profit.
Costs shown are based on boiler horsepower rating (BHP)

Natural gas fuel train piping for Scotch marine firetube and industrial watertube boilers

	Craft@Hrs	Unit	Material	Labor	Equipment	Total
20 BHP, 670 MBtu/Hr	M5@8.00	Ea	310.00	321.00	--	631.00
30 BHP, 1,004 MBtu/Hr	M5@8.00	Ea	425.00	321.00	--	746.00
40 BHP, 1,339 MBtu/Hr	M5@8.00	Ea	650.00	321.00	--	971.00
50 BHP, 1,674 MBtu/Hr	M5@8.00	Ea	775.00	321.00	--	1,096.00
60 BHP, 2,009 MBtu/Hr	M5@8.00	Ea	849.00	321.00	--	1,170.00
70 BHP, 2,343 MBtu/Hr	M5@8.00	Ea	900.00	321.00	--	1,221.00
80 BHP, 2,678 MBtu/Hr	M5@8.00	Ea	975.00	321.00	--	1,296.00
100 BHP, 3,348 MBtu/Hr	M5@8.00	Ea	1,130.00	321.00	--	1,451.00
125 BHP, 4,184 MBtu/Hr	M5@8.00	Ea	1,350.00	321.00	--	1,671.00
150 BHP, 5,021 MBtu/Hr	M5@8.00	Ea	1,590.00	321.00	--	1,911.00
200 BHP, 6,695 MBtu/Hr	M5@8.00	Ea	1,900.00	321.00	--	2,221.00
250 BHP, 8,369 MBtu/Hr	M5@8.00	Ea	2,350.00	321.00	--	2,671.00
300 BHP, 10,043 MBtu/Hr	M5@8.00	Ea	2,480.00	321.00	--	2,801.00
350 BHP, 11,761 MBtu/Hr	M5@8.00	Ea	2,660.00	321.00	--	2,981.00
400 BHP, 13,390 MBtu/Hr	M5@8.00	Ea	2,850.00	321.00	--	3,171.00
500 BHP, 16,738 MBtu/Hr	M5@16.0	Ea	3,210.00	641.00	--	3,851.00
600 BHP, 20,085 MBtu/Hr	M5@16.0	Ea	3,700.00	641.00	--	4,341.00
750 BHP, 25,106 MBtu/Hr	M5@16.0	Ea	4,000.00	641.00	--	4,641.00
800 BHP, 26,780 MBtu/Hr	M5@16.0	Ea	4,180.00	641.00	--	4,821.00
Natural gas fuel train piping for package firebox boilers						
13.4 BHP, 450 MBtu/Hr	M5@8.00	Ea	415.00	321.00	--	736.00

Mechanical 15

	Craft@Hrs	Unit	Material	Labor	Equipment	Total
Natural Gas Fuel Train Piping, continued						
16.4 BHP, 550 MBtu/Hr	M5@8.00	Ea	475.00	321.00	--	796.00
19.4 BHP, 650 MBtu/Hr	M5@8.00	Ea	510.00	321.00	--	831.00
22.4 BHP, 750 MBtu/Hr	M5@8.00	Ea	560.00	321.00	--	881.00
28.4 BHP, 950 MBtu/Hr	M5@8.00	Ea	590.00	321.00	--	911.00
34.4 BHP, 1,150 MBtu/Hr	M5@8.00	Ea	650.00	321.00	--	971.00
40.3 BHP, 1,350 MBtu/Hr	M5@8.00	Ea	775.00	321.00	--	1,096.00
46.3 BHP, 1,550 MBtu/Hr	M5@8.00	Ea	815.00	321.00	--	1,136.00
52.3 BHP, 1,750 MBtu/Hr	M5@8.00	Ea	975.00	321.00	--	1,296.00
61.0 BHP, 2,050 MBtu/Hr	M5@8.00	Ea	1,100.00	321.00	--	1,421.00
70.0 BHP, 2,350 MBtu/Hr	M5@8.00	Ea	1,250.00	321.00	--	1,571.00
80.0 BHP, 2,650 MBtu/Hr	M5@8.00	Ea	1,310.00	321.00	--	1,631.00
100.0 BHP, 3,350 MBtu/Hr	M5@8.00	Ea	1,550.00	321.00	--	1,871.00
126.9 BHP, 4,250 MBtu/Hr	M5@8.00	Ea	1,610.00	321.00	--	1,931.00
150.0 BHP, 5,050 MBtu/Hr	M5@8.00	Ea	1,750.00	321.00	--	2,071.00
Add for gas train piping approval						
Factory Mutual approval	--	%	7.0	--	--	--
Industrial Risk Insurers approval	--	%	10.0	--	--	--

Oil Fuel Train Piping. Number 2 oil. Costs shown include recirculating pump, electric oil preheater, regulating valve, check valves, isolating pump stop cocks, oil line pressure relief valve and pressure gauges. Add the cost of concrete pump pad, electrical connections, relief line piping back to the oil tank and tank connecting piping. These costs include the subcontractor's overhead and profit.
Costs shown are based on boiler horsepower rating (BHP)

	Craft@Hrs	Unit	Material	Labor	Equipment	Total
20 BHP, 670 MBtu/Hr	M5@4.00	Ea	790.00	160.00	--	950.00
30 BHP, 1,004 MBtu/Hr	M5@4.00	Ea	790.00	160.00	--	950.00
40 BHP, 1,339 MBtu/Hr	M5@4.00	Ea	875.00	160.00	--	1,035.00
50 BHP, 1,674 MBtu/Hr	M5@4.00	Ea	875.00	160.00	--	1,035.00
60 BHP, 2,009 MBtu/Hr	M5@4.00	Ea	875.00	160.00	--	1,035.00
70 BHP, 2,343 MBtu/Hr	M5@4.00	Ea	955.00	160.00	--	1,115.00
80 BHP, 2,678 MBtu/Hr	M5@4.00	Ea	955.00	160.00	--	1,115.00
100 BHP, 3,348 MBtu/Hr	M5@8.00	Ea	1,240.00	321.00	--	1,561.00
125 BHP, 4,184 MBtu/Hr	M5@8.00	Ea	1,600.00	321.00	--	1,921.00
150 BHP, 5,021 MBtu/Hr	M5@8.00	Ea	1,600.00	321.00	--	1,921.00
200 BHP, 6,695 MBtu/Hr	M5@8.00	Ea	2,100.00	321.00	--	2,421.00
250 BHP, 8,369 MBtu/Hr	M5@8.00	Ea	2,500.00	321.00	--	2,821.00
300 BHP, 10,043 MBtu/Hr	M5@16.0	Ea	2,680.00	641.00	--	3,321.00
350 BHP, 11,761 MBtu/Hr	M5@16.0	Ea	2,990.00	641.00	--	3,631.00
400 BHP, 13,390 MBtu/Hr	M5@16.0	Ea	3,560.00	641.00	--	4,201.00
500 BHP, 16,738 MBtu/Hr	M5@32.0	Ea	3,630.00	1,280.00	--	4,910.00
600 BHP, 20,085 MBtu/Hr	M5@32.0	Ea	3,800.00	1,280.00	--	5,080.00
750 BHP, 25,106 MBtu/Hr	M5@32.0	Ea	4,150.00	1,280.00	--	5,430.00
800 BHP, 26,780 MBtu/Hr	M5@32.0	Ea	4,150.00	1,280.00	--	5,430.00
900 BHP, 30,127 MBtu/Hr	M5@32.0	Ea	4,890.00	1,280.00	--	6,170.00
1,000 BHP, 33,475 MBtu/Hr	M5@32.0	Ea	4,890.00	1,280.00	--	6,170.00
Add for #4-6 (heavy) oil	--	%	10.0	--	--	--
Add for tie-in with dual-fuel burner	--	%	5.0	--	--	--

Electrical Service for Boilers Costs shown include typical main power tie-in and fusing, circuit breaker panel, and main bus. Assumes controls, fan, blower switchgear and control console are pre-wired. Electrical service for package Scotch marine firetube boilers and Industrial watertube boilers. These costs include the subcontractor's overhead and profit.

Mechanical 15

	Craft@Hrs	Unit	Material	Labor	Equipment	Total
Costs shown are based on boiler horsepower rating (BHP)						
20 to 100 BHP, 670 to 3,348 MBtu/Hr	CE@4.00	Ea	385.00	185.00	--	570.00
125 to 200 BHP, 4,184 to 6,695 MBtu/Hr	CE@4.00	Ea	475.00	185.00	--	660.00
250 BHP, 8,369 MBtu/Hr	CE@8.00	Ea	475.00	371.00	--	846.00
300 BHP, 10,043 MBtu/Hr	CE@8.00	Ea	560.00	371.00	--	931.00
350 BHP, 11,761 MBtu/Hr	CE@8.00	Ea	560.00	371.00	--	931.00
400 BHP, 13,390 MBtu/Hr	CE@8.00	Ea	560.00	371.00	--	931.00
500 BHP, 16,738 MBtu/Hr	CE@16.0	Ea	725.00	741.00	--	1,466.00
600 BHP, 20,085 MBtu/Hr	CE@16.0	Ea	725.00	741.00	--	1,466.00
750 BHP, 25,106 MBtu/Hr	CE@32.0	Ea	1,240.00	1,480.00	--	2,720.00
800 BHP, 26,780 MBtu/Hr	CE@32.0	Ea	1,240.00	1,480.00	--	2,720.00
Electrical service for package firebox boilers						
13.4 to 70.0 BHP, 450 to 2,350 MBtu/Hr	CE@4.00	Ea	360.00	185.00	--	545.00
80.0 BHP, 2,650 MBtu/Hr	CE@4.00	Ea	390.00	185.00	--	575.00
100.0 BHP, 3,350 MBtu/Hr	CE@4.00	Ea	390.00	185.00	--	575.00
126.9 BHP, 4,250 MBtu/Hr	CE@4.00	Ea	411.00	185.00	--	596.00
150.0 BHP, 5,050 MBtu/Hr	CE@4.00	Ea	483.00	185.00	--	668.00

Refractory for package Scotch marine firetube, firebox and watertube boilers. Most new boilers are shipped from the factory with brick installed at the combustion chamber and in turnarounds for the combustion gases. If refractory is installed on the job site to repair shipping damage or to create firedoor seals, use the figures below. These costs include the subcontractor's overhead and profit. Costs shown are per 100 pounds of castable refractory used.

	Craft@Hrs	Unit	Material	Labor	Equipment	Total
Refractory brick, repairs and seals	M1@4.00	100lb	35.00	150.00	--	185.00

Boiler Feedwater Pumps Including a duplex pump connected to the water main service connection on the inlet side and the feedwater inlet on the boiler on the pump discharge side. With larger units, the feedwater system draws from either a deaerator/condenser unit, a zeolite-based water softener, or from a chemical feedwater treatment tank. These costs include the subcontractor's overhead and profit. Add the cost of piping and pump actuating valves. Equipment cost is $20.00 per hour for a 2 ton forklift. Costs shown are based on boiler horsepower rating (BHP)

	Craft@Hrs	Unit	Material	Labor	Equipment	Total
Feedwater pumps for package Scotch marine firetube and industrial watertube boilers						
20 to 40 BHP, 670 to 1,339 MBtu/Hr	M7@8.00	Ea	4,600.00	337.00	53.40	4,990.40
50 to 80 BHP, 1,674 to 2,678 MBtu/Hr	M7@8.00	Ea	5,500.00	337.00	53.40	5,890.40
100 BHP, 3,348 MBtu/Hr	M7@8.00	Ea	6,100.00	337.00	53.40	6,490.40
125 to 200 BHP, 4,184 to 6,695 MBtu/Hr	M7@16.0	Ea	6,300.00	673.00	107.00	7,080.00
250 to 400 BHP, 8,369 to 13,390 MBtu/Hr	M7@32.0	Ea	6,900.00	1,350.00	214.00	8,464.00
500 BHP, 16,738 MBtu/Hr	M7@32.0	Ea	7,300.00	1,350.00	214.00	8,864.00
600 BHP, 20,085 MBtu/Hr	M7@32.0	Ea	7,600.00	1,350.00	214.00	9,164.00
750 BHP, 25,106 MBtu/Hr	M7@32.0	Ea	7,800.00	1,350.00	214.00	9,364.00
800 BHP, 26,780 MBtu/Hr	M7@32.0	Ea	10,600.00	1,350.00	214.00	12,164.00
Feedwater pumps for package firebox boilers						
13.4 to 34.4 BHP, 450 to 1,150 MBtu/Hr	M7@8.00	Ea	2,500.00	337.00	53.40	2,890.40
40.3 BHP, 1,350 MBtu/Hr	M7@8.00	Ea	2,870.00	337.00	53.40	3,260.40
46.3 BHP, 1,550 MBtu/Hr	M7@8.00	Ea	2,870.00	337.00	53.40	3,260.40
52.3 BHP, 1,750 MBtu/Hr	M7@8.00	Ea	3,590.00	337.00	53.40	3,980.40
61.0 BHP, 2,050 MBtu/Hr	M7@8.00	Ea	3,590.00	337.00	53.40	3,980.40
70.0 BHP, 2,350 MBtu/Hr	M7@8.00	Ea	4,300.00	337.00	53.40	4,690.40
80.0 BHP, 2,650 MBtu/Hr	M7@8.00	Ea	4,300.00	337.00	53.40	4,690.40
100.0 BHP, 3,350 MBtu/Hr	M7@8.00	Ea	5,500.00	337.00	53.40	5,890.40
126.9 BHP, 4,250 MBtu/Hr	M7@16.0	Ea	5,500.00	673.00	107.00	6,280.00
150.0 BHP, 5,050 MBtu/Hr	M7@16.0	Ea	6,100.00	673.00	107.00	6,880.00

Mechanical 15

	Craft@Hrs	Unit	Material	Labor	Equipment	Total

Natural Gas Combustion Controls Costs shown include automatic recycling or continuous pilot, spark igniter and coil (for electric ignition), timer to control the purge cycle, pilot, main burner (hi-lo-off) or main burner (proportional control), flame safeguard and sensor and combustion control panel.
Costs shown are based on boiler horsepower rating (BHP).
Natural gas combustion controls for Scotch marine firetube boilers and watertube boilers. These costs include the subcontractor's overhead and profit.

	Craft@Hrs	Unit	Material	Labor	Equipment	Total
20 to 80 BHP, 670 to 2,678 MBtu/Hr	M9@4.00	Ea	1,850.00	185.00	--	2,035.00
100 BHP, 3,348 MBtu/Hr	M9@4.00	Ea	2,135.00	185.00	--	2,320.00
125 BHP, 4,184 MBtu/Hr	M9@8.00	Ea	2,135.00	369.00	--	2,504.00
150 BHP, 5,021 MBtu/Hr	M9@8.00	Ea	2,135.00	369.00	--	2,504.00
200 BHP, 6,695 MBtu/Hr	M9@8.00	Ea	2,135.00	369.00	--	2,504.00
250 BHP, 8,369 MBtu/Hr	M9@16.0	Ea	2,135.00	739.00	--	2,874.00
300 BHP, 10,043 MBtu/Hr	M9@16.0	Ea	2,565.00	739.00	--	3,304.00
350 BHP, 11,761 MBtu/Hr	M9@16.0	Ea	2,980.00	739.00	--	3,719.00
400 BHP, 13,390 MBtu/Hr	M9@16.0	Ea	2,980.00	739.00	--	3,719.00
500 BHP, 16,738 MBtu/Hr	M9@32.0	Ea	2,980.00	1,480.00	--	4,460.00
600 BHP, 20,085 MBtu/Hr	M9@32.0	Ea	3,590.00	1,480.00	--	5,070.00
750 BHP, 25,106 MBtu/Hr	M9@64.0	Ea	4,150.00	2,960.00	--	7,110.00
800 BHP, 26,780 MBtu/Hr	M9@64.0	Ea	4,790.00	2,960.00	--	7,750.00
Natural gas combustion controls for package firebox boilers						
13.4 to 80.0 BHP, 450 to 2,650 MBtu/Hr	M9@4.00	Ea	1,850.00	185.00	--	2,035.00
100.0 BHP, 3,350 MBtu/Hr	M9@4.00	Ea	2,135.00	185.00	--	2,320.00
126.9 BHP, 4,250 MBtu/Hr	M9@8.00	Ea	2,135.00	369.00	--	2,504.00
150.0 BHP, 5,050 MBtu/Hr	M9@8.00	Ea	2,135.00	369.00	--	2,504.00
Add for #2 oil-fired burner controls	--	%	--	--	--	10.0
Add for heavy oil-fired burner controls	--	%	--	--	--	20.0
Add for dual fuel burner controls	--	%	--	--	--	70.0

Water Softening Systems Costs shown include the metering valve, tank, pump, and zeolite media. Add the cost of piping to connect the water softening system to the main feedwater tank and pump. These costs include the subcontractor's overhead and profit. Equipment cost is $20.00 per hour for a 2 ton capacity forklift.
Costs shown are based on boiler horsepower rating (BHP)
Water softening for package Scotch marine firetube boilers and Industrial watertube boilers

	Craft@Hrs	Unit	Material	Labor	Equipment	Total
20 to 80 BHP, 670 to 2,678 MBtu/Hr	M7@4.00	Ea	3,100.00	168.00	26.70	3,294.70
100 BHP, 3,348 MBtu/Hr	M7@4.00	Ea	3,600.00	168.00	26.70	3,794.70
125 BHP, 4,184 MBtu/Hr	M7@8.00	Ea	4,200.00	337.00	53.40	4,590.40
150 BHP, 5,021 MBtu/Hr	M7@8.00	Ea	5,300.00	337.00	53.40	5,690.40
200 BHP, 6,695 MBtu/Hr	M7@8.00	Ea	5,300.00	337.00	53.40	5,690.40
250 BHP, 8,369 MBtu/Hr	M7@8.00	Ea	6,300.00	337.00	53.40	6,690.40
300 BHP, 10,043 MBtu/Hr	M7@16.0	Ea	8,000.00	673.00	107.00	8,780.00
350 BHP, 11,761 MBtu/Hr	M7@16.0	Ea	8,000.00	673.00	107.00	8,780.00
400 BHP, 13,390 MBtu/Hr	M7@16.0	Ea	10,000.00	673.00	107.00	10,780.00
500 BHP, 16,738 MBtu/Hr	M7@32.0	Ea	10,000.00	1,350.00	214.00	11,564.00
600 BHP, 20,085 MBtu/Hr	M7@32.0	Ea	12,000.00	1,350.00	214.00	13,564.00
750 BHP, 25,106 MBtu/Hr	M7@64.0	Ea	14,300.00	2,700.00	427.00	17,427.00
800 BHP, 26,780 MBtu/Hr	M7@64.0	Ea	14,300.00	2,700.00	427.00	17,427.00
Water softening for package firebox boilers						
13.4 to 80 BHP, 450 to 2,650 MBtu/Hr	M7@4.00	Ea	3,100.00	168.00	26.70	3,294.70
100.0 BHP, 3,350 MBtu/Hr	M7@4.00	Ea	3,100.00	168.00	26.70	3,294.70
126.9 BHP, 4,250 MBtu/Hr	M7@8.00	Ea	3,600.00	337.00	53.40	3,990.40
150.0 BHP, 5,050 MBtu/Hr	M7@8.00	Ea	4,200.00	337.00	53.40	4,590.40

Mechanical 15

	Craft@Hrs	Unit	Material	Labor	Equipment	Total

Start-up, Shakedown, and Calibration. Costs shown include minor changes and adjustments to meet requirements of the boiler inspector, adjustment of combustion efficiency at the burner by a control technician, pump and metering calibration testing and steam or hot water metering on start-up. These costs include the subcontractor's overhead and profit.
Costs shown are based on boiler horsepower rating (BHP)

Start-up of package Scotch marine firetube boilers and industrial watertube boilers. If a factory technician performs the start-up, add this cost at a minimum of $400 (8 hours at $50 per hour).

	Craft@Hrs	Unit	Material	Labor	Equipment	Total
20 to 100 BHP, 670 to 3,348 MBtu/Hr	M9@4.00	Ea	--	185.00	--	--
125 to 200 BHP, 4,184 to 6,695 MBtu/Hr	M9@8.00	Ea	--	369.00	--	--
250 to 400 BHP, 8,369 to 13,390 MBtu/Hr	M9@16.0	Ea	--	739.00	--	--
500 BHP, 16,738 MBtu/Hr	M9@32.0	Ea	--	1,480.00	--	--
600 BHP, 20,085 MBtu/Hr	M9@32.0	Ea	--	1,480.00	--	--
750 BHP, 25,106 MBtu/Hr	M9@64.0	Ea	--	2,960.00	--	--
800 BHP, 26,780 MBtu/Hr	M9@64.0	Ea	--	2,960.00	--	--

Start-up on package firebox boilers

	Craft@Hrs	Unit	Material	Labor	Equipment	Total
13.4 to 100 BHP, 450 to 3,350 MBtu/Hr	M9@4.00	Ea	--	185.00	--	--
126.9 BHP, 4,250 MBtu/Hr	M9@8.00	Ea	--	369.00	--	--
150.0 BHP, 5,050 MBtu/Hr	M9@8.00	Ea	--	369.00	--	--

Stack economizer unit for boilers Stack waste heat recovery system, based on an operating pressure of 150 PSI steam. Costs shown include piping connection to existing boiler, economizer fin tube coils, feedwater pump, check valves, thermostatic regulating valve, stack dampers, pressure relief valve, stack mount flanges and insulating enclosure, but no architectural modifications. These costs include the subcontractor's overhead and profit. Equipment cost is $20.00 per hour for a 2 ton forklift.
Costs shown are based on boiler horsepower rating (BHP)

	Craft@Hrs	Unit	Material	Labor	Equipment	Total
500 BHP boiler	M7@40.0	Ea	21,000.00	1,680.00	267.00	22,947.00
750 BHP boiler	M7@48.0	Ea	26,000.00	2,020.00	320.00	28,340.00
800 BHP boiler	M7@48.0	Ea	28,000.00	2,220.00	320.00	30,540.00
1,000 BHP boiler	M7@56.0	Ea	35,000.00	2,360.00	374.00	37,734.00

Condensate receiver and pumping unit, stainless steel, Costs shown include a factory assembled tank, pump with motor, float switch, condensate inlet, feedwater tank, connection, level gauge, thermometer, solenoid valve for direct boiler feed. These costs include the subcontractor's overhead and profit. No concrete mounting pad required. Equipment cost is $20.00 per hour for a 2 ton forklift. Unit sizing and costs are based on storage for one minute at the designed steam return rate.

	Craft@Hrs	Unit	Material	Labor	Equipment	Total
750 lbs per hour	M7@4.00	Ea	2,250.00	168.00	26.70	2,444.70
1,500 lbs per hour	M7@4.00	Ea	2,250.00	168.00	26.70	2,444.70
3,000 lbs per hour	M7@4.00	Ea	2,890.00	168.00	26.70	3,084.70
5,000 lbs per hour	M7@6.00	Ea	3,150.00	252.00	40.00	3,442.00
6,250 lbs per hour	M7@6.00	Ea	3,510.00	252.00	40.00	3,802.00

Deaerator/condenser units for boilers, Pressurized jet spray. Costs shown include an ASME storage receiver, stand, jet spray deaerator head, makeup control valve, makeup controller, pressure control valve, level controls, pressure and temperature controls, pumped condensate inlet, pumps and motor, steam inlet and system accessories. Add for a concrete mounting pad and piping to the condensate return line (inlet) and piping to the boiler feedwater system (outlet). These costs include the subcontractor's overhead and profit. Equipment cost is $20.00 per hour for a 2 ton forklift.
Costs shown are based on boiler horsepower rating (BHP)

	Craft@Hrs	Unit	Material	Labor	Equipment	Total
100 BHP	M7@8.00	Ea	23,600.00	337.00	53.40	23,990.40
200 BHP	M7@12.0	Ea	24,600.00	505.00	80.00	25,185.00
300 BHP	M7@14.0	Ea	26,000.00	589.00	94.00	26,683.00
400 BHP	M7@16.0	Ea	29,600.00	673.00	107.00	30,380.00
600 BHP	M7@24.0	Ea	31,600.00	1,010.00	160.00	32,770.00

Mechanical 15

	Craft@Hrs	Unit	Material	Labor	Equipment	Total

Deaerator/condenser units for boilers, continued

	Craft@Hrs	Unit	Material	Labor	Equipment	Total
800 BHP	M7@36.0	Ea	34,800.00	1,520.00	240.00	36,560.00
1,000 BHP	M7@48.0	Ea	40,600.00	2,020.00	320.00	42,940.00

Chemical feed duplex pump, packaged unit. Costs shown include the chemical tanks, electronic metering, valve and stand. These costs include the subcontractor's overhead and profit. Add the cost of piping to the feedwater tank. Equipment cost is $20.00 per hour for a 2 ton forklift. Costs shown are based on boiler horsepower rating (BHP)

	Craft@Hrs	Unit	Material	Labor	Equipment	Total
20 to 100 BHP	M7@3.00	Ea	850.00	126.00	20.00	996.00
125 to 200 BHP	M7@3.00	Ea	1,100.00	126.00	20.00	1,246.00
200 to 400 BHP	M7@6.00	Ea	1,100.00	252.00	40.00	1,392.00
500 to 600 BHP	M7@6.00	Ea	2,500.00	252.00	40.00	2,792.00
700 BHP	M7@7.00	Ea	2,800.00	295.00	46.70	3,141.70
800 BHP	M7@8.00	Ea	2,800.00	337.00	53.40	3,190.40
1,000 BHP	M7@10.0	Ea	3,780.00	421.00	66.70	4,267.70

Package boiler feedwater systems Costs shown include duplex pumps with heater assembly, thermal lining, pressure gauges, makeup feeder valve, ASME feedwater tank and stand, single phase electric motor and level gauge. These costs include the subcontractor's overhead and profit. Add the costs for a concrete mounting pad and piping to the boilers. Equipment cost is $20.00 per hour for a 2 ton forklift. Costs shown are based on boiler horsepower rating (BHP)

	Craft@Hrs	Unit	Material	Labor	Equipment	Total
15 to 50 BHP	M7@5.00	Ea	6,070.00	210.00	33.40	6,313.40
60 to 80 BHP	M7@5.00	Ea	6,670.00	210.00	33.40	6,913.40
100 BHP	M7@6.00	Ea	6,785.00	253.00	40.00	7,078.00
125 to 150 BHP	M7@6.00	Ea	7,237.00	253.00	40.00	7,530.00
200 BHP	M7@8.00	Ea	7,481.00	337.00	53.40	7,871.40
250 BHP	M7@8.00	Ea	7,797.00	337.00	53.40	8,187.40
300 BHP	M7@8.00	Ea	8,207.00	337.00	53.40	8,597.40
400 BHP	M7@8.00	Ea	8,510.00	337.00	53.40	8,900.40
500 BHP	M7@8.00	Ea	10,382.00	337.00	53.40	10,772.40
600 BHP	M7@12.0	Ea	10,860.00	505.00	80.00	11,445.00
750 BHP	M7@12.0	Ea	12,135.00	505.00	80.00	12,720.00

ASME Tanks 125 PSI rating, vertical cylindrical, epoxy-lined, with pipe legs, two 1/2" openings, a 1" temperature and pressure relief valve opening and a 1" drain opening. These costs include the subcontractor's overhead and profit. Add the cost of a concrete pad and an insulation jacket. Equipment cost is $20.00 per hour for a 2 ton forklift.
Costs shown are by tank capacity, diameter, length and overall height.

	Craft@Hrs	Unit	Material	Labor	Equipment	Total
115 gallons, 24" dia., 48" long, 63" high	M7@4.00	Ea	1,410.00	168.00	26.70	1,604.70
261 gallons, 30" dia., 72" long, 90" high	M7@4.00	Ea	1,930.00	168.00	26.70	2,124.70
327 gallons, 36" dia., 60" long, 81" high	M7@4.00	Ea	2,250.00	168.00	26.70	2,444.70
462 gallons, 42" dia., 60" long, 84" high	M7@4.00	Ea	3,400.00	168.00	26.70	3,594.70
534 gallons, 42" dia., 72" long, 96" high	M7@4.00	Ea	3,590.00	168.00	26.70	3,784.70
606 gallons, 42" dia., 84" long, 108" high	M7@5.00	Ea	3,790.00	210.00	33.40	4,033.40
712 gallons, 48" dia., 72" long, 101" high	M7@5.00	Ea	4,630.00	210.00	33.40	4,873.40
803 gallons, 54" dia., 60" long, 90" high	M7@5.00	Ea	5,540.00	210.00	33.40	5,783.40
922 gallons, 54" dia., 72" long, 102" high	M7@6.00	Ea	5,800.00	253.00	40.00	6,093.00
1,017 gallons, 60" dia., 60" long, 93" high	M7@6.00	Ea	6,520.00	253.00	40.00	6,813.00
1,164 gallons, 60" dia., 72" long, 105" high	M7@6.00	Ea	6,830.00	253.00	40.00	7,123.00
1,311 gallons, 60" dia., 84" long, 117" high	M7@8.00	Ea	7,170.00	337.00	53.40	7,560.40
1,486 gallons, 60" dia., 96" long, 129" high	M7@8.00	Ea	7,470.00	337.00	53.40	7,860.40
1,616 gallons, 66" dia., 84" long, 120" high	M7@8.00	Ea	8,550.00	337.00	53.40	8,940.40
1,953 gallons, 72" dia., 84" long, 123" high	M7@8.00	Ea	9,750.00	337.00	53.40	10,140.40
2,378 gallons, 72" dia., 108" long, 147" high	M7@8.00	Ea	11,400.00	337.00	53.40	11,790.40

Mechanical 15

	Craft@Hrs	Unit	Material	Labor	Equipment	Total
2,590 gallons, 72" dia., 120" long, 159" high	M7@8.00	Ea	11,800.00	337.00	53.40	12,190.40
3,042 gallons, 84" dia., 96" long, 131" high	M7@8.00	Ea	14,000.00	337.00	53.40	14,390.40
4,500 gallons, 96" dia., 108" long, 159" high	M7@8.00	Ea	19,000.00	337.00	53.40	19,390.40
5,252 gallons, 96" dia., 132" long, 183" high	M7@8.00	Ea	25,300.00	337.00	53.40	25,690.40
Add for each 2-1/2" 150 lb. flanged opening	--	Ea	104.00	--	--	--
Add for each 11" x 15" manhole	--	Ea	350.00	--	--	--

Continuous blowdown heat recovery systems Costs shown include a flash receiver (50 PSI ASME rated tank), plate and frame heat exchanger, pneumatic level controller and valve, air filter regulator, water gauge, pressure gauge, backflush piping, pressure relief valve connected to mud drum and drain. Add for concrete pad mount, piping and electrical wiring. These costs include the subcontractor's overhead and profit. Equipment cost is $20.00 per hour for a 2 ton forklift.
Costs shown by pounds of rated makeup capacity.

	Craft@Hrs	Unit	Material	Labor	Equipment	Total
5,000 pounds per hour	M7@8.00	Ea	21,700.00	337.00	53.40	22,090.40
10,000 pounds per hour	M7@8.00	Ea	42,600.00	337.00	53.40	42,990.40
25,000 pounds per hour	M7@8.00	Ea	54,000.00	337.00	53.40	54,390.40
50,000 pounds per hour	M7@8.00	Ea	86,000.00	337.00	53.40	86,390.40

Boiler stack Costs shown assume a 20-foot stack height and include rigging, breeching, connection of stack sections, guy wires where required, vendor-supplied stack supports and insulation. These costs include the subcontractor's overhead and profit. Add the cost of a concrete base. Equipment cost is $110 per hour and includes a 90 ton truck crane (at $75) and a 4 ton capacity forklift (at $35).

	Craft@Hrs	Unit	Material	Labor	Equipment	Total
10" diameter	H8@8.00	Ea	1,500.00	381.00	147.00	2,028.00
12" diameter	H8@8.00	Ea	1,800.00	381.00	147.00	2,328.00
16" diameter	H8@10.0	Ea	2,200.00	476.00	184.00	2,860.00
20" diameter	H8@12.0	Ea	2,800.00	571.00	220.00	3,591.00
24" diameter	H8@16.0	Ea	3,600.00	761.00	294.00	4,655.00
30" diameter	H8@32.0	Ea	4,300.00	1,520.00	587.00	6,407.00
56" diameter	H8@40.0	Ea	8,500.00	1,910.00	734.00	11,144.00
Add for each 10' section beyond 20' high	--	%	--	--	--	15.0

Tankless indirect water heater (provides domestic hot water from boiler coil insert). Costs shown include heat exchanger, coil, circulating pump, thermostatically controlled valve, bypass piping and air bleeder valve. These costs include the subcontractor's overhead and profit. Add the cost of an equipment base, electrical wiring, piping and an insulated hot water storage tank.

	Craft@Hrs	Unit	Material	Labor	Equipment	Total
240 gallons per hour	M5@8.00	Ea	8,500.00	321.00	--	8,821.00
360 gallons per hour	M5@8.00	Ea	10,750.00	321.00	--	11,071.00
480 gallons per hour	M5@10.0	Ea	12,300.00	401.00	--	12,701.00
840 gallons per hour	M5@12.0	Ea	24,800.00	481.00	--	25,281.00
920 gallons per hour	M5@16.0	Ea	27,900.00	641.00	--	28,541.00
1,240 gallons per hour	M5@20.0	Ea	31,000.00	801.00	--	31,801.00

Boiler trim Costs shown include the pressure relief valve, level gauge (for steam systems), level controller & low water cutoff, pressure/temperature gauge, feedwater valve & actuator, main valve, blowdown valve, cleanout manholes and gaskets. These costs include the subcontractor's overhead and profit.
Costs shown by pounds of rated makeup capacity.

	Craft@Hrs	Unit	Material	Labor	Equipment	Total
20 to 100 BHP	M5@4.00	Ea	1,275.00	160.00	--	1,435.00
100 to 200 BHP	M5@6.00	Ea	1,400.00	240.00	--	1,640.00
200 to 400 BHP	M5@8.00	Ea	1,800.00	321.00	--	2,121.00
400 to 600 BHP	M5@8.00	Ea	2,700.00	321.00	--	3,021.00
600 to 800 BHP	M5@8.00	Ea	3,200.00	321.00	--	3,521.00
800 to 1000 BHP	M5@10.0	Ea	3,800.00	401.00	--	4,201.00

Mechanical 15

	Craft@Hrs	Unit	Material	Labor	Equipment	Total

Compressors Costs shown include the concrete mounting pad, vibration isolation and anchor bolting, steel stand and structural supports, rigging, bolts and gasket sets for flanges, integrated intercooling and lubrication systems, typical vendor engineering and calibration, and inlet air cleaners. These costs include the subcontractor's overhead and profit. Add the cost of electrical connections, control valving, service bypass/isolation piping and the cost of piping to load and control valving. Equipment cost is $20.00 per hour for a 2 ton forklift. See prices on following page.

	Craft@Hrs	Unit	Material	Labor	Equipment	Total
16.5 CFM at 175 PSI, 5 HP, 80 gallon tank	M7@2.00	Ea	1,860.00	84.10	13.40	1,957.50
34.4 CFM at 175 PSI, 10 HP, 120 gallon tank	M7@2.00	Ea	3,180.00	84.10	13.40	3,277.50
53.7 CFM at 175 PSI, 15 HP, 120 gallon tank	M7@2.00	Ea	4,500.00	84.10	13.40	4,597.50
91.0 CFM at 175 PSI, 25 HP, 120 gallon tank	M7@4.00	Ea	5,600.00	168.00	26.70	5,794.70

Shell and tube heat exchangers Based on 200 GPM maximum flow rate, 50 PSIG maximum for steam. Costs shown include the concrete mounting pad, anchor bolting, steel structural supports, holes through the structure to accommodate piping, insulation, flange bolt and gasket sets where appropriate, rigging, typical vendor engineering and calibration. Add the cost of temperature, pressure and flowrate controls, pumps and pump ancillaries for liquid heat transfer media supply. These costs include the subcontractor's overhead and profit. Equipment cost is $20.00 per hour for a 2 ton forklift. Costs shown by square feet of rated heat transfer area.

	Craft@Hrs	Unit	Material	Labor	Equipment	Total
9.36 SF	M7@4.00	Ea	1,540.00	168.00	26.70	1,734.70
15.64 SF	M7@4.00	Ea	1,660.00	168.00	26.70	1,854.70
22.98 SF	M7@4.00	Ea	2,460.00	168.00	26.70	2,654.70
30.82 SF	M7@4.00	Ea	2,710.00	168.00	26.70	2,904.70
38.66 SF	M7@5.00	Ea	2,966.00	211.00	33.40	3,210.40
60.80 SF	M7@7.00	Ea	4,080.00	295.00	46.70	4,421.70

Tubeaxial fans Costs shown include vibration isolation and anchor bolting, mating of flanges at both end to duct, steel structural supports and electric motor with v-belt drive. Add the cost of electrical switchgear, wiring and a filter box. These costs include the subcontractor's overhead and profit.

	Craft@Hrs	Unit	Material	Labor	Equipment	Total
24" diameter, 10,310 CFM	M9@2.00	Ea	849.00	92.40	--	941.40
30" diameter, 16,495 CFM	M9@2.00	Ea	1,130.00	92.40	--	1,222.40
34" diameter, 19,290 CFM	M9@4.00	Ea	1,230.00	185.00	--	1,415.00
36" diameter, 20,650 CFM	M9@4.00	Ea	1,250.00	185.00	--	1,435.00
42" diameter, 33,005 CFM	M9@6.00	Ea	1,720.00	277.00	--	1,997.00
48" diameter, 38,100 CFM	M9@8.00	Ea	2,100.00	369.00	--	2,469.00

	Craft@Hrs	Unit	Material	Labor	Total

Steam Heating Specialties. These costs include the subcontractor's overhead and profit.

Bucket traps, screwed connections

	Craft@Hrs	Unit	Material	Labor	Total
3/4"	PF@.706	Ea	74.20	32.50	106.70
1"	PF@.826	Ea	295.00	38.00	333.00
1-1/4"	PF@.944	Ea	402.00	43.50	445.50
1-1/2"	PF@1.06	Ea	499.00	48.80	547.80

Inverted bucket steam traps, semi-steel, screwed connections

	Craft@Hrs	Unit	Material	Labor	Total
3/4" side inlet and outlet	PF@.706	Ea	64.60	32.50	97.10
1" side inlet and outlet	PF@.826	Ea	228.00	38.00	266.00
1-1/4" side inlet and outlet	PF@.944	Ea	346.00	43.50	389.50
1-1/2" bottom inlet and top outlet	PF@1.06	Ea	384.00	48.80	432.80

Mechanical 15

	Craft@Hrs	Unit	Material	Labor	Total
Inverted bucket traps with stainless steel interior					
1/2"	PF@.620	Ea	68.90	28.50	97.40
3/4"	PF@.706	Ea	68.90	32.50	101.40
1"	PF@.826	Ea	226.00	38.00	264.00
1-1/4"	PF@.944	Ea	345.00	43.50	388.50
Inverted bucket traps with brass interior					
3/4"	PF@.540	Ea	121.00	24.90	145.90
1"	PF@.826	Ea	331.00	38.00	369.00
1-1/4"	PF@.944	Ea	456.00	43.50	499.50
Inverted bucket traps, cast steel body, high temperature, screw connected					
3/4"	PF@.706	Ea	387.00	32.50	419.50
1"	PF@.826	Ea	497.00	38.00	535.00
1-1/4"	PF@.944	Ea	864.00	43.50	907.50
1-1/2"	PF@1.06	Ea	1,690.00	48.80	1,738.80
2"	PF@1.18	Ea	1,740.00	54.30	1,794.30
Pressure and temperature relief valves, maximum 250 PSI, maximum temperature 450 degrees F, screwed connections					
1/2"	PF@1.11	Ea	260.00	51.10	311.10
3/4"	PF@1.11	Ea	277.00	51.10	328.10
1"	PF@1.30	Ea	305.00	59.80	364.80
1-1/4"	PF@1.49	Ea	392.00	68.60	460.60
1-1/2"	PF@1.67	Ea	473.00	76.90	549.90
2"	PF@1.85	Ea	629.00	85.20	714.20
Float traps, use Float and thermostatic traps					
Float and thermostatic traps, 15 PSI, cast iron, screwed connections					
3/4"	PF@.733	Ea	79.40	33.70	113.10
1"	PF@.820	Ea	95.60	37.70	133.30
1-1/4"	PF@.976	Ea	133.00	44.90	177.90
1-1/2"	PF@1.04	Ea	190.00	47.90	237.90
2"	PF@1.22	Ea	305.00	56.20	361.20
Float and thermostatic traps, 30 PSI, cast iron, screwed connections					
3/4"	PF@.733	Ea	92.40	33.70	126.10
1"	PF@.820	Ea	138.00	37.70	175.70
1-1/4"	PF@.976	Ea	138.00	44.90	182.90
1-1/2"	PF@1.04	Ea	287.00	47.90	334.90
Float and thermostatic traps, 75 PSI or 125 PSI, cast iron, screwed connections					
3/4"	PF@.733	Ea	117.00	33.70	150.70
1"	PF@.820	Ea	200.00	37.70	237.70
1-1/4"	PF@.976	Ea	200.00	44.90	244.90
1-1/2"	PF@1.04	Ea	328.00	47.90	375.90
Thermostatic traps, 25 PSI, angle type, cast iron, screwed connections					
1/2"	PF@.733	Ea	30.00	33.70	63.70
3/4"	PF@.733	Ea	50.20	33.70	83.90
1"	PF@.820	Ea	71.30	37.70	109.00
1/2" swivel type	PF@.733	Ea	40.50	33.70	74.20
3/4" straight type	PF@.733	Ea	54.30	33.70	88.00
1/2" vertical type	PF@.733	Ea	39.70	33.70	73.40
Thermostatic traps, 50 PSI, angle type, bronze, screwed connections					
1/2"	PF@.733	Ea	54.30	33.70	88.00
3/4"	PF@.733	Ea	73.00	33.70	106.70
1"	PF@.820	Ea	88.30	37.70	126.00

Mechanical 15

	Craft@Hrs	Unit	Material	Labor	Total

Steam Heating Specialties, continued
Thermostatic traps, 125 PSI, cast iron, screwed connections

3/4"	PF@.733	Ea	95.60	33.70	129.30
1"	PF@.820	Ea	143.00	37.70	180.70

Thermostatic disc traps, 10 PSI to 250 PSI differential, screwed connections

3/4"	PF@.733	Ea	130.00	33.70	163.70
1"	PF@.820	Ea	169.00	37.70	206.70
Air vent for hot water, 1/4" MPT	PF@.454	Ea	6.75	20.90	27.65

Boiler low water cutoff

20 PSI steam	PF@1.30	Ea	98.00	59.80	157.80
50 PSI water	PF@1.30	Ea	122.00	59.80	181.80

Boiler feeder and cutoff combination,

75 PSI maximum	PF@1.30	Ea	320.00	59.80	379.80
Add for alarm switch and auto reset	PF@1.30	Ea	35.90	59.80	95.70

Condensate pumps, bronze, with 10 & 20 gal. steel receiver and controls

Simplex, 20 PSI, 18 GPM, 1/3 HP	M5@9.00	Ea	895.00	361.00	1,256.00
Simplex, 30 PSI, 18 GPM, 1/2 HP	M5@9.00	Ea	1,090.00	361.00	1,451.00
Simplex, 40 PSI, 18 GPM, 3/4 HP	M5@13.5	Ea	1,140.00	541.00	1,681.00
Duplex, 20 PSI, 18 GPM, 1/3 HP	M5@13.5	Ea	1,740.00	541.00	2,281.00
Duplex, 30 PSI, 18 GPM, 1/2 HP	M5@20.2	Ea	2,130.00	809.00	2,939.00
Duplex, 40 PSI, 18 GPM, 3/4 HP	M5@22.2	Ea	2,200.00	889.00	3,089.00

Hydronic Heating Specialties. These costs include the subcontractor's overhead and profit.
Air eliminator-purger, screwed connections

3/4", cast iron, 15 PSI steam	PF@.558	Ea	96.10	25.70	121.80
3/4" or 1", cast iron, 150 PSI water	PF@.558	Ea	78.00	25.70	103.70
3/4", brass, 150 PSI water	PF@.558	Ea	148.00	25.70	173.70
3/4", cast iron, 300 PSI water	PF@.558	Ea	127.00	25.70	152.70
Airtrol fitting, 3/4"	PF@.493	Ea	29.20	22.70	51.90
Air eliminator vents, 1/4"	PF@.511	Ea	16.50	23.50	40.00

Atmospheric vacuum breakers, screwed

3/4"	PF@.493	Ea	112.00	22.70	134.70
1"	PF@.561	Ea	296.00	25.80	321.80

Anti-siphon vacuum breakers, brass

1/2"	PF@.459	Ea	12.80	21.10	33.90
3/4"	PF@.493	Ea	15.30	22.70	38.00
1"	PF@.561	Ea	24.00	25.80	49.80
1-1/4	PF@.667	Ea	39.50	30.70	70.20
1-1/2"	PF@.704	Ea	46.50	32.40	78.90
2"	PF@.829	Ea	72.40	38.20	110.60

Balancing valves, screwed connections

1/2"	PF@.459	Ea	7.05	21.10	28.15
3/4"	PF@.493	Ea	8.30	22.70	31.00
1"	PF@.561	Ea	12.00	25.80	37.80
1-1/2"	PF@.701	Ea	15.20	32.30	47.50

Flow check valves, screwed connections, horizontal type

3/4"	PF@.493	Ea	28.90	22.70	51.60
1"	PF@.561	Ea	41.30	25.80	67.10

Pressure reducing valves

1/2", low pressure, brass body	PF@.459	Ea	18.80	21.10	39.90
1/2", dual unit, low pressure, cast iron body	PF@.459	Ea	35.80	21.10	56.90
3/4", low pressure, brass body	PF@.493	Ea	27.50	22.70	50.20
1/2", high pressure, brass body	PF@.493	Ea	22.00	22.70	44.70

Mechanical 15

	Craft@Hrs	Unit	Material	Labor	Total
3/4", high pressure, brass body	PF@.493	Ea	30.40	22.70	53.10
Pressure relief valves, standard heating boiler valves, 15 PSI, cast iron, screwed connections					
3/4" x 1", 2,010 lbs/hr	PF@.561	Ea	137.00	25.80	162.80
1" x 1-1/4", 4,455 lbs/hr	PF@.667	Ea	156.00	30.70	186.70
1-1/4" x 1-1/2", 5,620 lbs/hr	PF@.704	Ea	200.00	32.40	232.40
1-1/2" x 2", 5,260 lbs/hr	PF@.829	Ea	240.00	38.20	278.20
3" x 4", 785 lbs/hr	PF@1.64	Ea	1,570.00	75.50	1,645.50
4" x 4", 1,225 lbs/hr	PF@2.29	Ea	1,720.00	105.00	1,825.00
Combination pressure relief and reducing valves					
1/2", brass body	PF@.459	Ea	58.00	21.10	79.10
3/4", cast iron body	PF@.493	Ea	90.10	22.70	112.80
Wye pattern strainers, screwed connections, 250 PSI, cast iron body					
3/4"	PF@.497	Ea	15.20	22.90	38.10
1"	PF@.558	Ea	22.30	25.70	48.00
1-1/4"	PF@.667	Ea	27.60	30.70	58.30
1-1/2"	PF@.701	Ea	35.30	32.30	67.60
2"	PF@.829	Ea	50.70	38.20	88.90
Wye pattern strainers, screwed connections, 300 PSI, cast bronze body					
3/4"	PF@.493	Ea	52.20	22.70	74.90
1"	PF@.561	Ea	79.50	25.80	105.30
1-1/4"	PF@.667	Ea	112.00	30.70	142.70
1-1/2"	PF@.701	Ea	156.00	32.30	188.30
2"	PF@.829	Ea	216.00	38.20	254.20
Wye pattern strainers, screwed connections, 900 PSI, cast steel body					
3/4"	PF@.497	Ea	85.00	22.90	107.90
1"	PF@.558	Ea	109.00	25.70	134.70
1-1/4"	PF@.667	Ea	158.00	30.70	188.70
1-1/2"	PF@.701	Ea	200.00	32.30	232.30
2"	PF@.829	Ea	275.00	38.20	313.20
Wye pattern strainers, flanged connections, 125 PSI, cast iron body					
2-1/2" with 1" drain	M5@1.52	Ea	173.00	60.90	233.90
3" with 1-1/4" drain	M5@1.88	Ea	211.00	75.30	286.30
4" with 1-1/2" drain	M5@2.64	Ea	340.00	106.00	446.00
6" with 2" drain	M7@3.23	Ea	655.00	136.00	791.00
8" with 2" drain	M7@3.80	Ea	1,127.00	160.00	1,287.00
10" with 2" drain	M7@5.09	Ea	2,320.00	214.00	2,534.00
Tempering valves, screwed connections, high temperature					
3/4", screwed	PF@.493	Ea	88.50	22.70	111.20
1", screwed	PF@.558	Ea	97.50	25.70	123.20
1-1/4", screwed	PF@.667	Ea	162.00	30.70	192.70
1-1/2", screwed	PF@.701	Ea	177.00	32.30	208.30
2", screwed	PF@.829	Ea	265.00	38.20	303.20
Thermostatic mixing valves, screwed connection					
3/4"	PF@.493	Ea	68.10	22.70	90.80
1"	PF@.561	Ea	240.00	25.80	265.80
1-1/2"	PF@.704	Ea	263.00	32.40	295.40
2"	PF@.829	Ea	441.00	38.20	479.20
1/2", sweat connection	PF@.459	Ea	64.90	21.10	86.00
3/4", sweat connection	PF@.493	Ea	87.40	22.70	110.10
Liquid level gauges					
3/4", aluminum	PF@.493	Ea	211.00	22.70	233.70
3/4", 125 PSI PVC	PF@.493	Ea	220.00	22.70	242.70
1/2", 175 PSI bronze	PF@.459	Ea	41.30	21.10	62.40

Mechanical 15

	Craft@Hrs	Unit	Material	Labor	Total
Hydronic Heating Specialties, Liquid level gauges, continued					
3/4", 150 PSI stainless steel	PF@.493	Ea	202.50	22.70	225.20
1", 150 PSI stainless steel	PF@.561	Ea	226.80	25.80	252.60

Steam and Hydronic Radiation Units. Includes cost of setting and connecting the unit, fittings and trim but no electrical work. These costs include the subcontractor's overhead and profit.

	Craft@Hrs	Unit	Material	Labor	Total
Baseboard radiators, per linear foot, wall mounted					
1/2" tube and cover	PF@.456	LF	6.05	21.00	27.05
3/4" tube and cover	PF@.493	LF	6.50	22.70	29.20
3/4" tube and cover, high capacity	PF@.493	LF	7.70	22.70	30.40
1/2" tube alone	PF@.350	LF	3.30	16.10	19.40
3/4" tube alone	PF@.377	LF	3.75	17.40	21.15
Cover only	PF@.137	LF	3.95	6.31	10.26
Finned tube wall type commercial radiation units, 18 gauge, flat top cover, with accessories					
1-1/4", copper element, aluminum fins, one tier	PF@.504	LF	21.00	23.20	44.20
1-1/4", steel element and fins, one tier	PF@.509	LF	18.70	23.40	42.10
2", steel element and fins, one tier	PF@.509	LF	21.15	23.40	44.55
1-1/4", copper element, aluminum fins, two tier	PF@.792	LF	38.00	36.50	74.50
1-1/4", steel element and fins, two tier	PF@.795	LF	33.70	36.60	70.30
2", steel element and fins, two tier	PF@.795	LF	38.25	36.60	74.85
1-1/4", copper element, aluminum fins, three tier	PF@.967	LF	55.00	44.50	99.50
1-1/4", steel element and fins, three tier	PF@.967	LF	48.50	44.50	93.00
2", steel element and fins, three tier	PF@.967	LF	55.00	44.50	99.50
Finned tube wall type commercial radiation units, 16 gauge, sloping top cover, with accessories					
1-1/4", copper element, aluminum fins, one tier	PF@.504	LF	28.90	23.20	52.10
1-1/4", steel element and fins, one tier	PF@.509	LF	27.00	23.40	50.40
2", steel element and fins, one tier	PF@.509	LF	29.10	23.40	52.50
1-1/4", copper element, aluminum fins, two tier	PF@.792	LF	46.00	36.50	82.50
1-1/4", steel element and fins, two tier	PF@.792	LF	45.20	36.50	81.70
2", steel element and fins, two tier	PF@.792	LF	46.00	36.50	82.50
1-1/4", copper element, aluminum fins, three tier	PF@.967	LF	62.10	44.50	106.60
1-1/4", steel element and fins, three tier	PF@.967	LF	56.00	44.50	100.50
2", steel element and fins, three tier	PF@.967	LF	62.10	44.50	106.60
Horizontal fan driven steam unit heaters					
18 MBtu, 300 CFM	M5@1.93	Ea	193.80	77.30	271.10
45 MBtu, 500 CFM	M5@1.93	Ea	319.50	77.30	396.80
60 MBtu, 700 CFM	M5@1.93	Ea	360.00	77.30	437.30
85 MBtu, 1,000 CFM	M5@1.93	Ea	380.00	77.30	457.30
Vertical fan driven steam unit heaters					
12.5 MBtu, 200 CFM	M5@1.93	Ea	175.00	77.30	252.30
17 MBtu, 300 CFM	M5@1.93	Ea	178.00	77.30	255.30
40 MBtu, 500 CFM	M5@1.93	Ea	345.00	77.30	422.30
60 MBtu, 700 CFM	M5@1.93	Ea	400.00	77.30	477.30
70 MBtu, 1,000 CFM	M5@1.93	Ea	505.00	77.30	582.30
Horizontal fan driven hydronic unit heaters					
12.5 MBtu, 200 CFM	M5@1.93	Ea	210.00	77.30	287.30
17 MBtu, 300 CFM	M5@1.93	Ea	232.00	77.30	309.30
25 MBtu, 500 CFM	M5@1.93	Ea	265.00	77.30	342.30
30 MBtu, 700 CFM	M5@1.93	Ea	280.00	77.30	357.30
50 MBtu, 1,000 CFM	M5@1.93	Ea	324.00	77.30	401.30
60 MBtu, 1,300 CFM	M5@1.93	Ea	385.00	77.30	462.30
Vertical fan driven hydronic unit heaters					
12.5 MBtu, 200 CFM	M5@1.93	Ea	235.00	77.30	312.30
17 MBtu, 300 CFM	M5@1.93	Ea	240.00	77.30	317.30

Mechanical 15

	Craft@Hrs	Unit	Material	Labor	Total
25 MBtu, 500 CFM	M5@1.93	Ea	310.00	77.30	387.30
30 MBtu, 700 CFM	M5@1.93	Ea	325.00	77.30	402.30
50 MBtu, 1,000 CFM	M5@1.93	Ea	390.00	77.30	467.30
60 MBtu, 1,300 CFM	M5@1.93	Ea	430.00	77.30	507.30
Air curtain blower units, steam					
104 MBtu, 36" long, 1,275 CFM	M5@12.5	Ea	605.00	501.00	1,106.00
133 MBtu, 48" long, 1,600 CFM	M5@16.9	Ea	605.00	677.00	1,282.00
149 MBtu, 36" long, 2,840 CFM	M5@16.9	Ea	605.00	677.00	1,282.00
172 MBtu, 60" long, 2,100 CFM	M5@20.8	Ea	675.00	833.00	1,508.00
190 MBtu, 48" long, 3,790 CFM	M5@23.3	Ea	605.00	933.00	1,538.00
225 MBtu, 60" long, 4,740 CFM	M5@23.3	Ea	710.00	933.00	1,643.00
265 MBtu, 84" long, 3,000 CFM	M7@28.0	Ea	930.00	1,180.00	2,110.00
290 MBtu, 96" long, 3,400 CFM	M7@31.8	Ea	930.00	1,340.00	2,270.00
350 MBtu, 120" long, 4,200 CFM	M7@39.7	Ea	1,100.00	1,670.00	2,770.00
400 MBtu, 144" long, 5,000 CFM	M7@44.9	Ea	1,230.00	1,890.00	3,120.00
525 MBtu, 120" long, 9,480 CFM	M7@40.7	Ea	1,080.00	1,710.00	2,790.00
630 MBtu, 144" long, 11,375 CFM	M7@49.4	Ea	1,230.00	2,080.00	3,310.00
Air curtain blower units, hydronic					
56 MBtu, 36" long, 1,275 CFM	M5@9.58	Ea	605.00	384.00	989.00
70 MBtu, 48" long, 1,600 CFM	M5@9.58	Ea	605.00	384.00	989.00
92 MBtu, 60" long, 2,100 CFM	M5@9.58	Ea	675.00	384.00	1,059.00
96 MBtu, 36" long, 2,840 CFM	M5@9.58	Ea	605.00	384.00	989.00
103 MBtu, 72" long, 2,500 CFM	M7@11.3	Ea	755.00	475.00	1,230.00
125 MBtu, 84" long, 3,000 CFM	M7@11.9	Ea	930.00	500.00	1,430.00
149 MBtu, 96" long, 3,400 CFM	M7@11.9	Ea	930.00	500.00	1,430.00
190 MBtu, 120" long, 4,200 CFM	M7@17.9	Ea	1,080.00	753.00	1,833.00
225 MBtu, 144" long, 5,000 CFM	M7@17.9	Ea	1,245.00	753.00	1,998.00
256 MBtu, 36" long, 7,580 CFM	M5@14.2	Ea	930.00	569.00	1,499.00
290 MBtu, 120" long, 9,480 CFM	M7@15.8	Ea	1,080.00	664.00	1,744.00
350 MBtu, 144" long, 11,375 CFM	M7@18.0	Ea	1,230.00	757.00	1,987.00
Fan coil units, 200 CFM, two pipe, with electric controls, single coil					
Ceiling hung, with cabinet	M5@4.24	Ea	671.00	170.00	841.00
Ceiling hung, without cabinet	M5@3.80	Ea	549.00	152.00	701.00
Floor mounted, with cabinet	M5@4.24	Ea	628.00	170.00	798.00
Add for 4 pipe unit with 2 coils	--	%	--	--	12.0
Fan coil units, 300 CFM, two pipe, with electric controls, single coil					
Ceiling hung, with cabinet	M5@4.24	Ea	720.00	170.00	890.00
Ceiling hung, without cabinet	M5@3.80	Ea	585.00	152.00	737.00
Floor mounted, with cabinet	M5@4.24	Ea	671.00	170.00	841.00
Add for 4 pipe unit with 2 coils	--	%	--	--	12.0
Fan coil units, 400 CFM, two pipe, with electric controls, single coil					
Ceiling hung, with cabinet	M5@5.15	Ea	774.00	206.00	980.00
Ceiling hung, without cabinet	M5@4.76	Ea	616.00	191.00	807.00
Floor mounted, with cabinet	M5@5.15	Ea	847.00	206.00	1,053.00
Add for 4 pipe unit with 2 coils	--	%	--	--	12.0
Fan coil units, 600 CFM, two pipe, with electric controls, single coil					
Ceiling hung, with cabinet	M5@6.19	Ea	847.00	248.00	1,095.00
Ceiling hung, without cabinet	M5@5.46	Ea	671.00	219.00	890.00
Floor mounted, with cabinet	M5@6.19	Ea	792.00	248.00	1,040.00
Add for 4 pipe unit with 2 coils	--	%	--	--	12.0
Fan coil units, 800 CFM, two pipe, with electric controls, single coil					
Ceiling hung, with cabinet	M5@7.00	Ea	1,250.00	280.00	1,530.00
Ceiling hung, without cabinet	M5@6.32	Ea	994.00	253.00	1,247.00
Floor mounted, with cabinet	M5@7.00	Ea	1,350.00	280.00	1,630.00

Mechanical 15

	Craft@Hrs	Unit	Material	Labor	Total

Steam and Hydronic Radiation Units, Fan coil units, continued

Add for 4 pipe unit with 2 coils	--	%	--	--	12.0

Fan coil units, 1,000 CFM, two pipe, with electric controls, single coil

Ceiling hung, with cabinet	M5@7.60	Ea	1,370.00	304.00	1,674.00
Ceiling hung, without cabinet	M5@6.74	Ea	1,080.00	270.00	1,350.00
Floor mounted, with cabinet	M5@7.60	Ea	1,480.00	304.00	1,784.00
Add for 4 pipe unit with 2 coils	--	%	--	--	12.0

Fan coil units, 1,200 CFM, two pipe, with electric controls, single coil

Ceiling hung, with cabinet	M5@7.60	Ea	1,450.00	304.00	1,754.00
Ceiling hung, without cabinet	M5@6.74	Ea	1,160.00	270.00	1,430.00
Floor mounted, with cabinet	M5@7.60	Ea	1,570.00	304.00	1,874.00
Add for 4 pipe unit with 2 coils	--	%	--	--	12.0

Cabinet unit heaters, exposed, ceiling hung hot water units

200 CFM	M5@4.96	Ea	568.00	199.00	767.00
300 CFM	M5@5.02	Ea	622.00	201.00	823.00
400 CFM	M5@5.96	Ea	678.00	239.00	917.00
600 CFM	M5@7.23	Ea	814.00	290.00	1,104.00
800 CFM	M5@8.46	Ea	984.00	339.00	1,323.00
1,000 CFM	M5@9.58	Ea	1,100.00	384.00	1,484.00
1,200 CFM	M5@11.5	Ea	1,180.00	461.00	1,641.00
2,000 CFM	M5@13.4	Ea	1,450.00	537.00	1,987.00

Heating and Cooling Combinations. These costs include the subcontractor's overhead and profit but no ducting.

Gas heat and direct expansion cooling, roof mounted, single zone with disposable filters, isolation dampers, curbs, control and valving

2 ton cooling, 80 MBtu heating	M5@23.4	Ea	2,860.00	937.00	3,797.00
3 ton cooling, 100 MBtu heating	M5@26.3	Ea	3,630.00	1,050.00	4,680.00
4 ton cooling, 140 MBtu heating	M5@30.2	Ea	4,630.00	1,210.00	5,840.00
5 ton cooling, 140 MBtu heating	M5@33.3	Ea	5,520.00	1,330.00	6,850.00
8 ton cooling, 200 MBtu heating	M7@40.0	Ea	8,610.00	1,680.00	10,290.00
10 ton cooling, 260 MBtu heating	M7@44.4	Ea	10,800.00	1,870.00	12,670.00
15 ton cooling, 300 MBtu heating	M7@57.6	Ea	14,300.00	2,420.00	16,720.00
20 ton cooling, 400 MBtu heating	M7@69.0	Ea	19,400.00	2,900.00	22,300.00
30 ton cooling, 500 MBtu heating	M7@114.	Ea	24,500.00	4,790.00	29,290.00
40 ton cooling, 760 MBtu heating	M7@149.	Ea	30,900.00	6,270.00	37,170.00
60 ton cooling 1,000 MBtu heating	M7@216.	Ea	45,300.00	9,080.00	54,380.00
Add for multi-zone units, 2nd zone	--	%	6.0	6.0	--
Add for multi-zone units, 3rd to 8th zone, per zone	--	%	3.0	3.0	--

Heat pumps, direct expansion type, complete packaged units

Thru-wall units

2 ton cooling, 26,000 Btu heating	M5@9.84	Ea	1,450.00	394.00	1,844.00
3 ton cooling, 37,000 Btu heating	M5@11.5	Ea	2,110.00	461.00	2,571.00
4 ton cooling, 52,000 Btu heating	M5@16.6	Ea	2,790.00	665.00	3,455.00
5 ton cooling, 61,000 Btu heating	M5@18.8	Ea	3,510.00	753.00	4,263.00

Roof mounted, ducted

8 ton cooling, 90,000 Btu heating	M7@24.6	Ea	6,920.00	1,030.00	7,950.00
10 ton cooling, 120,000 Btu heating	M7@29.3	Ea	8,770.00	1,230.00	10,000.00
15 ton cooling, 180,000 Btu heating	M7@34.7	Ea	12,600.00	1,460.00	14,060.00
20 ton cooling, 240,000 Btu heating	M7@46.7	Ea	16,600.00	1,960.00	18,560.00

Plenum mounted

1 ton, 13,000 Btu heating	M5@9.58	Ea	1,850.00	384.00	2,234.00
2 ton, 26,000 Btu heating	M5@12.8	Ea	2,310.00	513.00	2,823.00
3 ton, 37,000 Btu heating	M5@15.3	Ea	2,820.00	613.00	3,433.00

Mechanical 15

	Craft@Hrs	Unit	Material	Labor	Total
4 ton, 52,000 Btu heating	M5@19.3	Ea	3,220.00	773.00	3,993.00
Vibration isolators, per set					
15 ton unit	M7@1.83	Ea	185.00	77.00	262.00
20 ton unit	M7@2.14	Ea	211.00	90.00	301.00
30 ton unit	M7@2.53	Ea	248.00	106.00	354.00
40 ton unit	M7@2.95	Ea	285.00	124.00	409.00

Computer Room Heating and Cooling. These systems have high efficiency filtration equipment, alarms and high reliability controls which make them suitable for computer room applications. Includes remote condenser and compressor, heating and cooling coils, valving and trim but no boilers, cooling towers, ducting, electrical wiring, equipment base, water pumps or piping. These costs include the subcontractor's overhead and profit.

	Craft@Hrs	Unit	Material	Labor	Total
Direct expansion cooling, electric heating, air cooled condenser					
3 tons	M5@16.9	Ea	9,200.00	677.00	9,877.00
5 tons	M5@18.0	Ea	11,000.00	721.00	11,721.00
7.5 tons	M5@25.5	Ea	14,100.00	1,020.00	15,120.00
10 tons	M7@26.3	Ea	15,300.00	1,110.00	16,410.00
15 tons	M7@29.5	Ea	17,900.00	1,240.00	19,140.00
Direct expansion cooling, steam heating, air cooled condenser					
3 tons	M5@18.0	Ea	10,700.00	721.00	11,421.00
5 tons	M5@19.1	Ea	12,900.00	765.00	13,665.00
7.5 tons	M5@25.3	Ea	16,100.00	1,010.00	17,110.00
10 tons	M7@27.3	Ea	16,200.00	1,150.00	17,350.00
15 tons	M7@30.8	Ea	18,900.00	1,300.00	20,200.00
Direct expansion cooling, hydronic heating, air cooled condenser					
3 tons	M5@18.0	Ea	10,340.00	721.00	11,061.00
5 tons	M5@19.1	Ea	11,500.00	765.00	12,265.00
7.5 tons	M5@25.4	Ea	14,900.00	1,020.00	15,920.00
10 tons	M7@27.3	Ea	16,000.00	1,150.00	17,150.00
15 tons	M7@35.5	Ea	18,500.00	1,490.00	19,990.00
Direct expansion cooling, water cooled condenser, electric heat					
3 tons	M5@17.7	Ea	9,570.00	709.00	10,279.00
5 tons	M5@19.1	Ea	11,000.00	765.00	11,765.00
7.5 tons	M5@27.1	Ea	14,300.00	1,090.00	15,390.00
10 tons	M7@29.0	Ea	15,300.00	1,220.00	16,520.00
15 tons	M7@32.0	Ea	18,300.00	1,350.00	19,650.00
Direct expansion cooling, water cooled condenser, steam heat					
3 tons	M5@18.8	Ea	10,700.00	753.00	11,453.00
5 tons	M5@20.0	Ea	12,000.00	801.00	12,801.00
7.5 tons	M5@28.1	Ea	15,600.00	1,130.00	16,730.00
10 tons	M7@29.5	Ea	16,200.00	1,240.00	17,440.00
15 tons	M7@32.8	Ea	19,400.00	1,380.00	20,780.00
Direct expansion cooling, water cooled condenser, hydronic heat					
3 tons	M5@18.8	Ea	10,370.00	753.00	11,123.00
5 tons	M5@20.0	Ea	11,700.00	801.00	12,501.00
7.5 tons	M5@28.1	Ea	15,400.00	1,130.00	16,530.00
10 tons	M7@29.8	Ea	16,000.00	1,250.00	17,250.00
15 tons	M7@32.8	Ea	19,300.00	1,380.00	20,680.00
Chilled water cooling, electric heating, not including chiller					
7.5 tons	M5@20.7	Ea	9,950.00	829.00	10,779.00
10 tons	M7@22.3	Ea	10,510.00	938.00	11,448.00
15 tons	M7@25.3	Ea	12,100.00	1,060.00	13,160.00
Chilled water cooling, steam heating, not including chiller or boiler					
7.5 tons	M5@21.5	Ea	10,900.00	861.00	11,761.00

Mechanical 15

	Craft@Hrs	Unit	Material	Labor	Total

Computer Room Heating and Cooling, Chilled water cooling, continued

	Craft@Hrs	Unit	Material	Labor	Total
10 tons	M7@23.2	Ea	11,500.00	976.00	12,476.00
15 tons	M7@26.6	Ea	13,000.00	1,120.00	14,120.00

Chilled water cooling, hydronic heating, not including chiller or boiler

	Craft@Hrs	Unit	Material	Labor	Total
7.5 tons	M5@21.7	Ea	10,600.00	869.00	11,469.00
10 tons	M7@23.3	Ea	11,300.00	980.00	12,280.00
15 tons	M7@26.8	Ea	12,700.00	1,130.00	13,830.00
Desiccant dehumidifier, 700 CFM, 20 lbs per hour, 15 kw	M5@39.3	Ea	11,400.00	1,570.00	12,970.00
Humidifier package for heating system	--	Ea	1,320.00	--	--

Air Cooled Remote Condensers. Includes motor, condenser, subcooler, casing, fan guard, control panel, thermostat and holding charge. No piping, cooling coils, electrical connection, carpentry, concrete, roofing or lifting included. By tons of rated capacity. These costs include the subcontractor's overhead and profit.

	Craft@Hrs	Unit	Material	Labor	Total
1.5 ton, single circuit	M5@2.71	Ea	836.00	109.00	945.00
2 ton, single circuit	M5@2.71	Ea	1,050.00	109.00	1,159.00
2.5 ton, single circuit	M5@2.71	Ea	1,200.00	109.00	1,309.00
3 ton, single circuit	M5@3.77	Ea	1,250.00	151.00	1,401.00
5 ton, single circuit	M5@4.32	Ea	1,410.00	173.00	1,583.00
7.5 ton, single circuit	M5@6.19	Ea	1,930.00	248.00	2,178.00
10 ton, dual circuit	M7@7.44	Ea	2,720.00	313.00	3,033.00
15 ton, dual circuit	M7@8.19	Ea	3,400.00	344.00	3,744.00
20 ton, single circuit	M7@8.69	Ea	3,870.00	365.00	4,235.00
20 ton, dual circuit	M7@8.69	Ea	4,490.00	365.00	4,855.00
25 ton, single circuit	M7@9.68	Ea	4,600.00	407.00	5,007.00
25 ton, dual circuit	M7@9.68	Ea	5,230.00	407.00	5,637.00
30 ton, single circuit	M7@10.4	Ea	5,330.00	437.00	5,767.00
30 ton, dual circuit	M7@10.4	Ea	5,960.00	437.00	6,397.00
40 ton, single circuit	M7@10.9	Ea	6,790.00	458.00	7,248.00
40 ton, dual circuit	M7@10.9	Ea	7,580.00	458.00	8,038.00
50 ton, single circuit	M7@11.4	Ea	8,200.00	479.00	8,679.00
50 ton, dual circuit	M7@11.4	Ea	9,090.00	479.00	9,569.00
60 ton, single circuit	M7@12.4	Ea	9,610.00	521.00	10,131.00
60 ton, dual circuit	M7@12.4	Ea	10,240.00	521.00	10,761.00
80 ton, dual circuit	M7@13.9	Ea	12,200.00	584.00	12,784.00
100 ton, dual circuit	M7@14.9	Ea	15,200.00	627.00	15,827.00
120 ton, dual circuit	M7@15.6	Ea	18,100.00	656.00	18,756.00
Add for low ambient air dampers	--	%	10.0	8.0	--

Chillers. Includes starters, connection, electric drive and trim but no cooling tower, electrical wiring, piping, lifting or equipment base. These costs include the subcontractor's overhead and profit.
Air cooled reciprocating packaged chillers, requires remote condenser

	Craft@Hrs	Unit	Material	Labor	Total
10 ton	M7@51.4	Ea	12,500.00	2,160.00	14,660.00
15 ton	M7@51.4	Ea	13,100.00	2,160.00	15,260.00
20 ton	M7@56.3	Ea	14,900.00	2,370.00	17,270.00
25 ton	M7@56.3	Ea	16,500.00	2,370.00	18,870.00
30 ton	M7@59.1	Ea	17,900.00	2,490.00	20,390.00
40 ton	M7@66.7	Ea	21,400.00	2,800.00	24,200.00
50 ton	M7@84.6	Ea	25,400.00	3,560.00	28,960.00
60 ton	M7@97.5	Ea	30,200.00	4,100.00	34,300.00
75 ton	M7@126.	Ea	35,700.00	5,300.00	41,000.00
80 ton	M7@138.	Ea	39,200.00	5,800.00	45,000.00
100 ton	M8@150.	Ea	43,700.00	6,460.00	50,160.00

Mechanical 15

	Craft@Hrs	Unit	Material	Labor	Total
120 ton	M8@158.	Ea	56,600.00	6,800.00	63,400.00
150 ton	M8@170.	Ea	72,100.00	7,320.00	79,420.00
180 ton	M8@180.	Ea	79,600.00	7,750.00	87,350.00
200 ton	M8@191.	Ea	86,000.00	8,220.00	94,220.00

Packaged air cooled reciprocating chillers, with condenser

	Craft@Hrs	Unit	Material	Labor	Total
20 ton	M7@59.1	Ea	17,700.00	2,490.00	20,190.00
25 ton	M7@61.5	Ea	20,100.00	2,590.00	22,690.00
30 ton	M7@64.3	Ea	22,700.00	2,700.00	25,400.00
40 ton	M7@72.0	Ea	26,200.00	3,030.00	29,230.00
50 ton	M7@95.0	Ea	27,600.00	3,990.00	31,590.00
60 ton	M7@108.	Ea	35,400.00	4,540.00	39,940.00
75 ton	M7@138.	Ea	42,700.00	5,800.00	48,500.00
80 ton	M7@151.	Ea	45,800.00	6,350.00	52,150.00
100 ton	M8@165.	Ea	53,100.00	7,100.00	60,200.00
120 ton	M8@176.	Ea	66,600.00	7,580.00	74,180.00

Heat recovery air cooled reciprocating chillers, with starter and remote condenser

	Craft@Hrs	Unit	Material	Labor	Total
40 ton	M7@79.7	Ea	31,800.00	3,350.00	35,150.00
50 ton	M7@103.	Ea	36,800.00	4,330.00	41,130.00
60 ton	M7@115.	Ea	40,400.00	4,840.00	45,240.00
75 ton	M7@151.	Ea	46,100.00	6,350.00	52,450.00
100 ton	M8@180.	Ea	52,500.00	7,750.00	60,250.00

Heat recovery water cooled reciprocating chillers, with starter and remote condenser

	Craft@Hrs	Unit	Material	Labor	Total
40 ton	M7@79.7	Ea	31,400.00	3,350.00	34,750.00
50 ton	M7@103.	Ea	37,300.00	4,330.00	41,630.00
60 ton	M7@115.	Ea	41,300.00	4,840.00	46,140.00
75 ton	M7@157.	Ea	49,500.00	6,600.00	56,100.00
100 ton	M8@180.	Ea	57,800.00	7,750.00	65,550.00

Water cooled reciprocating chillers, with starter and condenser

	Craft@Hrs	Unit	Material	Labor	Total
10 ton	M7@53.8	Ea	11,300.00	2,260.00	13,560.00
15 ton	M7@53.8	Ea	11,400.00	2,260.00	13,660.00
20 ton	M7@59.1	Ea	13,100.00	2,490.00	15,590.00
25 ton	M7@61.5	Ea	14,900.00	2,590.00	17,490.00
30 ton	M7@64.3	Ea	16,000.00	2,700.00	18,700.00
40 ton	M7@74.4	Ea	20,200.00	3,130.00	23,330.00
50 ton	M7@95.0	Ea	24,200.00	3,990.00	28,190.00
60 ton	M7@105.	Ea	27,000.00	4,420.00	31,420.00
75 ton	M7@141.	Ea	32,500.00	5,930.00	38,430.00
80 ton	M7@154.	Ea	35,000.00	6,480.00	41,480.00
100 ton	M8@168.	Ea	37,700.00	7,230.00	44,930.00
120 ton	M8@183.	Ea	44,500.00	7,880.00	52,380.00
150 ton	M8@196.	Ea	59,000.00	8,440.00	67,440.00
180 ton	M8@206.	Ea	62,500.00	8,870.00	71,370.00
200 ton	M8@216.	Ea	69,000.00	9,300.00	78,300.00

Centrifugal chillers, water cooled, with single bundle condenser but no cooling tower

	Craft@Hrs	Unit	Material	Labor	Total
80 ton	M7@203.	Ea	66,000.00	8,540.00	74,540.00
130 ton	M8@208.	Ea	69,400.00	8,950.00	78,350.00
160 ton	M8@218.	Ea	71,200.00	9,380.00	80,580.00
180 ton	M8@231.	Ea	75,600.00	9,940.00	85,540.00
230 ton	M8@264.	Ea	79,100.00	11,400.00	90,500.00
280 ton	M8@339.	Ea	87,000.00	14,600.00	101,600.00
360 ton	M8@339.	Ea	97,200.00	14,600.00	111,800.00
460 ton	M8@376.	Ea	119,000.00	16,200.00	135,200.00
530 ton	M8@376.	Ea	131,000.00	16,200.00	147,200.00
670 ton	M8@500.	Ea	159,000.00	21,500.00	180,500.00

Mechanical 15

	Craft@Hrs	Unit	Material	Labor	Total

Chillers, Centrifugal chillers, continued
Add for absorption type steam driven chillers   --   %   9.0   --   --

Cooling Towers for Chiller Compressors. Packaged units including electric drive and connection but no wiring, equipment base, pumps or lifting. These costs include the subcontractor's overhead and profit.
  Crossflow induced draft units with propeller drive
    100 tons   M8@82.7   Ea   5,330.00   3,560.00   8,890.00
    200 tons   M8@158.   Ea   10,240.00   6,800.00   17,040.00
    300 tons   M8@201.   Ea   12,900.00   8,650.00   21,550.00
    400 tons   M8@231.   Ea   14,600.00   9,940.00   24,540.00
    600 tons   M8@293.   Ea   19,500.00   12,600.00   32,100.00
    800 tons   M8@364.   Ea   24,200.00   15,700.00   39,900.00
    1,000 tons   M8@417.   Ea   27,600.00   18,000.00   45,600.00
  Counterflow forced draft units with centrifugal blower drive
    100 tons   M8@32.5   Ea   7,110.00   1,400.00   8,510.00
    200 tons   M8@60.1   Ea   14,000.00   2,590.00   16,590.00
    300 tons   M8@77.8   Ea   16,300.00   3,350.00   19,650.00
    400 tons   M8@92.9   Ea   21,600.00   4,000.00   25,600.00
    600 tons   M8@186.   Ea   31,100.00   8,010.00   39,110.00
    800 tons   M8@264.   Ea   41,400.00   11,400.00   52,800.00
    1,000 tons   M8@298.   Ea   51,800.00   12,800.00   64,600.00
  Chemical feeder, 5 gallon capacity   M5@7.55   Ea   272.00   302.00   574.00

Auxiliary Heating and Cooling Equipment. These costs include the subcontractor's overhead and profit but no electric work.
  Humidifiers, duct type, with regulator and trim, automatic, steam
    50 lbs per hour   M5@5.39   Ea   3,010.00   216.00   3,226.00
    75 lbs per hour   M5@6.32   Ea   3,270.00   253.00   3,523.00
    100 lbs per hour   M5@7.83   Ea   3,530.00   314.00   3,844.00
    150 lbs per hour   M5@9.16   Ea   3,690.00   367.00   4,057.00
    200 lbs per hour   M5@9.45   Ea   4,750.00   379.00   5,129.00
    300 lbs per hour   M5@11.7   Ea   5,280.00   469.00   5,749.00
  Pumps for chilled water cooling systems, 50 ft nominal head, with motor
    7-1/2 GPM   M5@6.19   Ea   438.00   248.00   686.00
    20 GPM   M5@7.55   Ea   464.00   302.00   766.00
    30 GPM   M5@10.4   Ea   707.00   417.00   1,124.00
    40 GPM   M5@11.6   Ea   797.00   465.00   1,262.00
    60 GPM   M5@16.4   Ea   1,100.00   657.00   1,757.00
    90 GPM   M7@15.9   Ea   1,220.00   669.00   1,889.00
    120 GPM   M7@20.0   Ea   1,500.00   841.00   2,341.00
    150 GPM   M7@21.7   Ea   1,610.00   912.00   2,522.00
    200 GPM   M8@26.2   Ea   2,040.00   1,130.00   3,170.00
    250 GPM   M8@29.3   Ea   2,300.00   1,260.00   3,560.00
    300 GPM   M8@34.7   Ea   2,810.00   1,490.00   4,300.00
    400 GPM   M8@37.6   Ea   3,010.00   1,620.00   4,630.00
    600 GPM   M8@43.6   Ea   3,450.00   1,880.00   5,330.00
    800 GPM   M8@49.0   Ea   4,010.00   2,110.00   6,120.00
    1,000 GPM   M8@56.0   Ea   4,570.00   2,410.00   6,980.00
    1,500 GPM   M8@57.0   Ea   4,810.00   2,450.00   7,260.00
  Prefabricated chimney, insulating galvanized metal, 1,400 degree F, double wall, UL listed, 28 gauge aluminum jacket, with typical supports. Figure the cost of each ell or tee at four times the linear foot cost for chimney
    12"   M7@.288   LF   55.40   12.10   67.50

Mechanical 15

	Craft@Hrs	Unit	Material	Labor	Total
18"	M7@.429	LF	91.40	18.00	109.40
24"	M7@.658	LF	116.00	27.70	143.70
30"	M7@.876	LF	156.00	36.80	192.80
36"	M7@1.10	LF	209.00	46.30	255.30

Air Handling Units. Modular medium pressure variable air velocity units with an air mixing box, hot and cold water coils as indicated, dampers, coil casing, vibration isolators and disposable filters but no plumbing or electrical work. These costs include the subcontractor's overhead and profit.

	Craft@Hrs	Unit	Material	Labor	Total
1,500 CFM unit, 2 HP					
One zone, 4 rows of cooling coils	M7@7.02	Ea	2,270.00	295.00	2,565.00
One zone, 1 row of heating coils	M7@7.02	Ea	2,060.00	295.00	2,355.00
One zone, 4 rows cooling, 1 row heating coils	M7@8.41	Ea	2,760.00	354.00	3,114.00
3,000 CFM unit, 3 HP					
One zone, 4 rows of cooling coils	M7@12.6	Ea	2,930.00	530.00	3,460.00
Multi-zone, 4 rows of cooling coils	M7@16.5	Ea	4,550.00	694.00	5,244.00
One zone, 1 row of heating coils	M7@12.6	Ea	2,530.00	530.00	3,060.00
Multi-zone, 1 row of heating coils	M7@16.5	Ea	4,220.00	694.00	4,914.00
One zone, 4 rows cooling, 1 row heating coils	M7@15.4	Ea	3,540.00	648.00	4,188.00
Multi-zone, 4 rows cooling, 1 row heating coils	M7@19.8	Ea	5,200.00	833.00	6,033.00
4,000 CFM unit, 5 HP					
One zone, 4 rows of cooling coils	M7@14.0	Ea	3,410.00	589.00	3,999.00
Multi-zone, 4 rows of cooling coils	M7@18.3	Ea	5,160.00	770.00	5,930.00
One zone, 1 row of heating coils	M7@14.0	Ea	2,980.00	589.00	3,569.00
Multi-zone, 1 row of heating coils	M7@18.3	Ea	4,810.00	770.00	5,580.00
One zone, 4 rows cooling, 1 row heating coils	M7@17.0	Ea	4,220.00	715.00	4,935.00
Multi-zone, 4 rows cooling, 1 row heating coils	M7@22.1	Ea	5,960.00	929.00	6,889.00
5,000 CFM unit, 5 HP					
One zone, 4 rows of cooling coils	M7@15.5	Ea	3,960.00	652.00	4,612.00
Multi-zone, 4 rows of cooling coils	M7@20.1	Ea	5,750.00	845.00	6,595.00
One zone, 1 row of heating coils	M7@15.5	Ea	3,400.00	652.00	4,052.00
Multi-zone, 1 row of heating coils	M7@20.1	Ea	5,170.00	845.00	6,015.00
One zone, 4 rows cooling, 1 row heating coils	M7@18.6	Ea	4,830.00	782.00	5,612.00
Multi-zone, 4 rows cooling, 1 row heating coils	M7@24.3	Ea	6,630.00	1,020.00	7,650.00
6,000 CFM unit, 7.5 HP					
One zone, 4 rows of cooling coils	M7@16.7	Ea	4,530.00	702.00	5,232.00
Multi-zone, 4 rows of cooling coils	M7@21.9	Ea	6,360.00	921.00	7,281.00
One zone, 1 row of heating coils	M7@16.7	Ea	3,890.00	702.00	4,592.00
Multi-zone, 1 row of heating coils	M7@21.9	Ea	5,720.00	921.00	6,641.00
One zone, 4 rows cooling, 1 row heating coils	M7@20.4	Ea	5,540.00	858.00	6,398.00
Multi-zone, 4 rows cooling, 1 row heating coils	M7@26.8	Ea	7,340.00	1,130.00	8,470.00
7,000 CFM unit, 7.5 HP					
One zone, 4 rows of cooling coils	M7@18.3	Ea	5,070.00	770.00	5,840.00
Multi-zone, 4 rows of cooling coils	M7@23.7	Ea	6,910.00	997.00	7,907.00
One zone, 1 row of heating coils	M7@18.3	Ea	4,310.00	770.00	5,080.00
Multi-zone, 1 row of heating coils	M7@23.7	Ea	6,160.00	997.00	7,157.00
One zone, 4 rows cooling, 1 row heating coils	M7@22.2	Ea	6,180.00	934.00	7,114.00
Multi-zone, 4 rows cooling, 1 row heating coils	M7@28.8	Ea	8,050.00	1,210.00	9,260.00
8,500 CFM unit, 10 HP					
One zone, 4 rows of cooling coils	M7@21.1	Ea	5,830.00	887.00	6,717.00
Multi-zone, 4 rows of cooling coils	M7@27.0	Ea	7,760.00	1,140.00	8,900.00
One zone, 1 row of heating coils	M7@21.8	Ea	4,810.00	917.00	5,727.00
Multi-zone, 1 row of heating coils	M7@27.0	Ea	6,670.00	1,140.00	7,810.00
One zone, 4 rows cooling, 1 row heating coils	M7@25.4	Ea	7,060.00	1,070.00	8,130.00
Multi-zone, 4 rows cooling, 1 row heating coils	M7@32.8	Ea	8,990.00	1,380.00	10,370.00

Mechanical 15

	Craft@Hrs	Unit	Material	Labor	Total

Air Handling Units, continued

10,500 CFM unit, 10 HP

	Craft@Hrs	Unit	Material	Labor	Total
One zone, 4 rows of cooling coils	M7@25.1	Ea	6,610.00	1,060.00	7,670.00
Multi-zone, 4 rows of cooling coils	M7@32.8	Ea	7,300.00	1,380.00	8,680.00
One zone, 1 row of heating coils	M7@25.1	Ea	5,350.00	1,060.00	6,410.00
Multi-zone, 1 row of heating coils	M7@32.8	Ea	7,430.00	1,380.00	8,810.00
One zone, 4 rows cooling, 1 row heating coils	M7@30.8	Ea	8,000.00	1,300.00	9,300.00
Multi-zone, 4 rows cooling, 1 row heating coils	M7@40.2	Ea	10,080.00	1,690.00	11,770.00

12,500 CFM unit, 15 HP

	Craft@Hrs	Unit	Material	Labor	Total
One zone, 4 rows of cooling coils	M8@31.3	Ea	7,800.00	1,350.00	9,150.00
Multi-zone, 4 rows of cooling coils	M8@41.2	Ea	9,870.00	1,770.00	11,640.00
One zone, 1 row of heating coils	M8@31.3	Ea	6,400.00	1,350.00	7,750.00
Multi-zone, 1 row of heating coils	M8@41.2	Ea	8,590.00	1,770.00	10,360.00
One zone, 4 rows cooling, 1 row heating coils	M8@38.8	Ea	9,360.00	1,670.00	11,030.00
Multi-zone, 4 rows cooling, 1 row heating coils	M8@49.7	Ea	10,800.00	2,140.00	12,940.00

15,500 CFM unit, 20 HP

	Craft@Hrs	Unit	Material	Labor	Total
One zone, 4 rows of cooling coils	M8@38.8	Ea	9,690.00	1,670.00	11,360.00
Multi-zone, 4 rows of cooling coils	M8@49.7	Ea	11,400.00	2,140.00	13,540.00
One zone, 1 row of heating coils	M8@38.8	Ea	7,900.00	1,670.00	9,570.00
Multi-zone, 1 row of heating coils	M8@49.7	Ea	10,080.00	2,140.00	12,220.00
One zone, 4 rows cooling, 1 row heating coils	M8@46.6	Ea	11,900.00	2,010.00	13,910.00
Multi-zone, 4 rows cooling, 1 row heating coils	M8@60.6	Ea	14,100.00	2,610.00	16,710.00

17,500 CFM unit, 20 HP

	Craft@Hrs	Unit	Material	Labor	Total
One zone, 4 rows of cooling coils	M8@42.7	Ea	13,900.00	1,840.00	15,740.00
Multi-zone, 4 rows of cooling coils	M8@55.5	Ea	16,300.00	2,390.00	18,690.00
One zone, 1 row of heating coils	M8@42.7	Ea	11,900.00	1,840.00	13,740.00
Multi-zone, 1 row of heating coils	M8@55.5	Ea	14,300.00	2,390.00	16,690.00
One zone, 4 rows cooling, 1 row heating coils	M8@51.4	Ea	16,400.00	2,210.00	18,610.00
Multi-zone, 4 rows cooling, 1 row heating coils	M8@66.9	Ea	18,700.00	2,880.00	21,580.00

20,000 CFM unit 25 HP

	Craft@Hrs	Unit	Material	Labor	Total
One zone, 4 rows of cooling coils	M8@50.4	Ea	16,300.00	2,170.00	18,470.00
Multi-zone, 4 rows of cooling coils	M8@65.2	Ea	18,600.00	2,810.00	21,410.00
One zone, 1 row of heating coils	M8@50.4	Ea	13,900.00	2,170.00	16,070.00
Multi-zone, 1 row of heating coils	M8@65.7	Ea	16,600.00	2,830.00	19,430.00
One zone, 4 rows cooling, 1 row heating coils	M8@61.6	Ea	19,000.00	2,650.00	21,650.00
Multi-zone, 4 rows cooling, 1 row heating coils	M8@79.5	Ea	21,400.00	3,420.00	24,820.00

25,000 CFM unit, 25 HP

	Craft@Hrs	Unit	Material	Labor	Total
One zone, 4 rows of cooling coils	M8@63.0	Ea	19,200.00	2,710.00	21,910.00
Multi-zone, 4 rows of cooling coils	M8@81.7	Ea	21,400.00	3,520.00	24,920.00
One zone, 1 row of heating coils	M8@63.0	Ea	16,100.00	2,710.00	18,810.00
Multi-zone, 1 row of heating coils	M8@81.7	Ea	18,400.00	3,520.00	21,920.00
One zone, 4 rows cooling, 1 row heating coils	M8@76.4	Ea	22,200.00	3,290.00	25,490.00
Multi-zone, 4 rows cooling, 1 row heating coils	M8@99.4	Ea	25,400.00	4,280.00	29,680.00

31,500 CFM unit, 30 HP

	Craft@Hrs	Unit	Material	Labor	Total
One zone, 4 rows of cooling coils	M8@76.6	Ea	23,200.00	3,300.00	26,500.00
Multi-zone, 4 rows of cooling coils	M8@99.4	Ea	25,900.00	4,280.00	30,180.00
One zone, 1 row of heating coils	M8@76.6	Ea	19,700.00	3,300.00	23,000.00
Multi-zone, 1 row of heating coils	M8@99.4	Ea	22,200.00	4,280.00	26,480.00
One zone, 4 rows cooling, 1 row heating coils	M8@93.1	Ea	26,000.00	4,010.00	30,010.00
Multi-zone, 4 rows cooling, 1 row heating coils	M8@121.	Ea	29,800.00	5,210.00	35,010.00
Add for direct expansion refrigerant coils rather than cold water coils	--	%	8.0	--	--
Add for steam coil rather than hot water coil	--	%	10.0	--	--

Mechanical 15

	Craft@Hrs	Unit	Material	Labor	Total

Roof Mounted Exhaust Fans. Belt driven centrifugal aluminum blade roof exhaust sets with hood and bird screen. Includes drip-proof 3 phase motor. No dampers, curb or starter. CFM rating at 1/4" static pressure. FPM rating at blade tip. These costs include the subcontractor's overhead and profit but no roofing, electrical work, lifting or carpentry.

	Craft@Hrs	Unit	Material	Labor	Total
1/4 HP units, 2 phase					
640 CFM, 2,618 FPM	T5@1.80	Ea	568.00	72.10	640.10
940 CFM, 2,604 FPM	T5@1.80	Ea	658.00	72.10	730.10
1,050 CFM, 3,325 FPM	T5@1.80	Ea	568.00	72.10	640.10
1,170 CFM, 2,373 FPM	T5@1.80	Ea	943.00	72.10	1,015.10
2,380 CFM, 3,382 FPM	T5@1.80	Ea	774.00	72.10	846.10
2,440 CFM, 4,501 FPM	T5@1.80	Ea	659.00	72.10	731.10
2,880 CFM, 3,859 FPM	T5@1.77	Ea	774.00	70.90	844.90
3,660 CFM, 3,437 FPM	T5@2.24	Ea	952.00	89.70	1,041.70
1/3 HP units					
1,280 CFM, 3,774 FPM	T5@1.80	Ea	722.00	72.10	794.10
2,760 CFM, 4,950 FPM	T5@2.24	Ea	814.00	89.70	903.70
3,200 CFM, 4,173 FPM	T5@2.24	Ea	980.00	89.70	1,069.70
4,070 CFM, 3,694 FPM	T5@2.24	Ea	1,120.00	89.70	1,209.70
5,030 CFM, 3,251 FPM	T5@2.24	Ea	1,530.00	89.70	1,619.70
1/2 HP unit, 7,000 CFM, 3,449 FPM	T5@3.52	Ea	2,060.00	141.00	2,201.00
3/4 HP unit, 11,300 CFM, 3,232 FPM	T5@5.47	Ea	3,470.00	219.00	3,689.00
1 HP units					
3,890 CFM, 6,769 FPM	T5@2.97	Ea	825.00	119.00	944.00
6,120 CFM, 2,400 FPM	T5@7.24	Ea	3,700.00	290.00	3,990.00
1-1/2 HP units					
5,830 CFM, 6,932 FPM	T5@2.97	Ea	1,009.00	119.00	1,128.00
6,380 CFM, 3,817 FPM	T5@2.97	Ea	1,610.00	119.00	1,729.00
2 HP units					
8,460 CFM, 6,721 FPM	T5@2.97	Ea	1,200.00	119.00	1,319.00
10,970 CFM, 5,906 FPM	T5@3.52	Ea	1,620.00	141.00	1,761.00
13,000 CFM, 5,456 FPM	T5@3.52	Ea	2,170.00	141.00	2,311.00
18,330 CFM, 4,488 FPM	T5@5.47	Ea	3,440.00	219.00	3,659.00
3 HP units					
11,250 CFM, 4,854 FPM	T5@3.52	Ea	2,320.00	141.00	2,461.00
12,470 CFM, 6,620 FPM	T5@4.56	Ea	1,820.00	183.00	2,003.00
21,720 CFM, 5,131 FPM	T5@5.47	Ea	3,660.00	219.00	3,879.00
5 HP units					
18,490 CFM, 7,405 FPM	T5@6.35	Ea	2,500.00	254.00	2,754.00
30,130 CFM, 5,681 FPM	T5@7.24	Ea	3,780.00	290.00	4,070.00
7-1/2 HP units					
31,110 CFM, 6,965 FPM	T5@7.24	Ea	4,000.00	290.00	4,290.00
35,470 CFM, 6,533 FPM	T5@7.24	Ea	4,210.00	290.00	4,500.00
10 HP unit, 39,400 CFM, 7,172 FPM	T5@8.12	Ea	4,350.00	325.00	4,675.00
Add for explosion-proof motors	--	%	35.0	--	--

Direct drive centrifugal aluminum blade roof mounted 115 volt exhaust sets with hood and bird screen. Includes drip-proof motor but no dampers, ducting, curb, starter, roofing, electrical work, lifting or carpentry. CFM rating at 1/4" static pressure. These costs include the subcontractor's overhead and profit.

	Craft@Hrs	Unit	Material	Labor	Total
60 to 390 CFM, 1/40 HP	T5@1.80	Ea	378.00	72.10	450.10
145 to 590 CFM, 1/22 HP	T5@1.80	Ea	391.00	72.10	463.10
295 to 860 CFM, 1/11 HP	T5@1.80	Ea	378.00	72.10	450.10
235 to 1,300 CFM, 1/6 HP	T5@1.80	Ea	589.00	72.10	661.10
415 to 1,630 CFM, 1/4 HP	T5@1.80	Ea	601.00	72.10	673.10

Mechanical 15

	Craft@Hrs	Unit	Material	Labor	Total

Roof Mounted Exhaust Fans, continued

	Craft@Hrs	Unit	Material	Labor	Total
590 to 2,045 CFM, 1/3 HP	T5@1.80	Ea	625.00	72.10	697.10
805 CFM, 3,235 FPM tip speed, 1/20 HP	T5@1.80	Ea	631.00	72.10	703.10
1,455 CFM, 4,360 FPM tip speed, 1/6 HP	T5@1.80	Ea	691.00	72.10	763.10
1,720 CFM, 3,870 FPM tip speed, 1/6 HP	T5@1.80	Ea	715.00	72.10	787.10
1,385 CFM, 3,655 FPM tip speed, 1/6 HP	T5@1.80	Ea	721.00	72.10	793.10
2,260 CFM, 4,930 FPM tip speed, 1/3 HP	T5@1.80	Ea	788.00	72.10	860.10
2,700 CFM, 5,220 FPM tip speed, 1/3 HP	T5@1.80	Ea	854.00	72.10	926.10

Air Handling Equipment. These costs include the subcontractor's overhead and profit but no roofing, lifting, electrical work or carpentry.

Exhaust utility sets, with vibration mounts, rating at 3/4" static pressure

	Craft@Hrs	Unit	Material	Labor	Total
600 CFM	T5@3.23	Ea	647.00	129.00	776.00
1,000 CFM	T5@5.10	Ea	922.00	204.00	1,126.00
2,000 CFM	T5@7.00	Ea	1,300.00	280.00	1,580.00
4,000 CFM	T5@7.81	Ea	1,470.00	313.00	1,783.00
10,000 CFM	T5@12.9	Ea	2,460.00	517.00	2,977.00
16,000 CFM	T5@14.0	Ea	2,680.00	561.00	3,241.00
20,000 CFM	T5@16.9	Ea	3,330.00	677.00	4,007.00
30,000 CFM	T5@24.8	Ea	5,250.00	993.00	6,243.00
40,000 CFM	T6@31.0	Ea	6,790.00	1,300.00	8,090.00
60,000 CFM	T6@43.7	Ea	10,150.00	1,840.00	11,990.00
80,000 CFM	T6@54.9	Ea	13,100.00	2,310.00	15,410.00

Supply fans, low pressure, 1-1/2" static pressure rating, utility sets

	Craft@Hrs	Unit	Material	Labor	Total
600 CFM	T5@5.10	Ea	753.00	204.00	957.00
1,000 CFM	T5@6.46	Ea	1,120.00	259.00	1,379.00
2,000 CFM	T5@9.16	Ea	1,550.00	367.00	1,917.00
4,000 CFM	T5@9.97	Ea	1,730.00	399.00	2,129.00
6,000 CFM	T5@11.6	Ea	2,620.00	464.00	3,084.00
10,000 CFM	T5@15.9	Ea	2,770.00	637.00	3,407.00
16,000 CFM	T5@19.7	Ea	3,360.00	789.00	4,149.00
20,000 CFM	T5@21.8	Ea	3,810.00	873.00	4,683.00
30,000 CFM	T5@30.2	Ea	5,810.00	1,210.00	7,020.00
40,000 CFM	T6@34.3	Ea	7,710.00	1,440.00	9,150.00
60,000 CFM	T6@48.7	Ea	11,700.00	2,050.00	13,750.00
80,000 CFM	T6@60.3	Ea	15,300.00	2,530.00	17,830.00

Supply fans, high pressure utility sets with vibration mounts, 3-1/2" static pressure

	Craft@Hrs	Unit	Material	Labor	Total
2,000 CFM	T5@9.16	Ea	1,550.00	367.00	1,917.00
4,000 CFM	T5@9.43	Ea	2,170.00	378.00	2,548.00
10,000 CFM	T5@14.3	Ea	3,250.00	573.00	3,823.00
16,000 CFM	T5@20.5	Ea	4,600.00	821.00	5,421.00
20,000 CFM	T5@24.5	Ea	5,400.00	981.00	6,381.00
30,000 CFM	T5@27.3	Ea	6,550.00	1,090.00	7,640.00
40,000 CFM	T6@37.7	Ea	9,360.00	1,580.00	10,940.00
60,000 CFM	T6@51.1	Ea	14,800.00	2,150.00	16,950.00
80,000 CFM	T6@58.8	Ea	24,100.00	2,470.00	26,570.00

Relief and intake ventilators. Aluminum ventilators with screen. Labor assumes installation on curb. No curb included. These costs include the subcontractor's overhead and profit.

	Craft@Hrs	Unit	Material	Labor	Total
12" x 12"	T5@1.54	Ea	338.00	61.70	399.70
16" x 16"	T5@1.64	Ea	395.00	65.70	460.70
20" x 20"	T5@1.83	Ea	467.00	73.30	540.30
24" x 24"	T5@1.92	Ea	541.00	76.90	617.90
30" x 30"	T5@2.44	Ea	659.00	97.70	756.70
36" x 36"	T5@2.68	Ea	936.00	107.00	1,043.00

Mechanical 15

	Craft@Hrs	Unit	Material	Labor	Total
42" x 42"	T5@2.94	Ea	1,250.00	118.00	1,368.00
48" x 48"	T5@3.28	Ea	1,840.00	131.00	1,971.00
Filters and frames					
Pre-filters, throwaway, 2' x 2' x 2"	T5@.051	Ea	10.49	2.04	12.53
Pre-filter frames, 16 gauge, 2' x 2' x 2"	T5@.063	Ea	14.40	2.52	16.92
Filter and frame, per 1,000 CFM, typical	T5@.058	Ea	13.00	2.32	15.32
HP 80% filter, 24" x 24" x 24"	T5@.302	SF	65.90	12.10	78.00
HP 80% filter frames, 24" x 24" x 24", 16 gauge	T5@.513	Ea	313.00	20.50	333.50
Activated carbon plastic panel filter and housing, complete					
24" x 24" x 29"	T5@.867	Ea	532.00	34.70	566.70
Washable plastic permanent filters, no frame					
1/2" x 14" x 25"	T5@.133	Ea	2.65	5.33	7.98
1" x 20" x 25"	T5@.133	Ea	4.72	5.33	10.05
2" x 20" x 25"	T5@.133	Ea	6.15	5.33	11.48
Roll stock washable plastic permanent filters					
1/2" thick	T5@.050	SF	1.17	2.00	3.17
1" thick	T5@.050	SF	1.43	2.00	3.43
2" thick	T5@.050	SF	1.75	2.00	3.75
Disposable fiberglass filters, with plastic frame					
1" x 24" x 24"	T5@.135	Ea	2.49	5.41	7.90
2" x 24" x 24"	T5@.135	Ea	2.97	5.41	8.38

Air Distribution Equipment. These costs include the subcontractor's overhead and profit but no electrical work or controls.

	Craft@Hrs	Unit	Material	Labor	Total
Mixing boxes, constant volume, with pneumatic regulator					
200 CFM	T5@2.84	Ea	408.00	114.00	522.00
350 CFM	T5@3.36	Ea	440.00	135.00	575.00
550 CFM	T5@3.65	Ea	504.00	146.00	650.00
850 CFM	T5@4.04	Ea	604.00	162.00	766.00
1,400 CFM	T5@4.71	Ea	710.00	189.00	899.00
2,400 CFM	T5@5.39	Ea	922.00	216.00	1,138.00
3,200 CFM	T5@5.78	Ea	1,110.00	231.00	1,341.00
5,000 CFM	T5@6.46	Ea	1,270.00	259.00	1,529.00
Add for variable volume mixing boxes	--	%	--	--	14.0
Air terminal units, with re-heat coils, CFM at 1" static pressure					
Size #7, 400 to 650 CFM with 1.1 SF coil	T5@3.36	Ea	371.00	135.00	506.00
Size #8, 600 to 850 CFM with 1.6 SF coil	T5@4.04	Ea	413.00	162.00	575.00
Size #9, 800 to 1,050 CFM with 2.8 SF coil	T5@4.45	Ea	514.00	178.00	692.00
Size #10, 600 to 1,000 CFM with 2.0 SF coil	T5@4.45	Ea	541.00	178.00	719.00
Size #12, 1,200 to 1,800 CFM with 2.1 SF coil	T5@4.71	Ea	763.00	189.00	952.00
Size #14, 2,000 to 2,800 CFM with 3.4 SF coil	T5@5.78	Ea	943.00	231.00	1,174.00
Air terminal units, constant volume, single duct, without coils					
200 CFM	T5@1.99	Ea	164.00	79.70	243.70
350 CFM	T5@2.40	Ea	228.00	96.10	324.10
500 CFM	T5@2.68	Ea	254.00	107.00	361.00
800 CFM	T5@3.23	Ea	297.00	129.00	426.00
1,600 CFM	T5@3.78	Ea	376.00	151.00	527.00
2,400 CFM	T5@4.45	Ea	451.00	178.00	629.00
3,000 CFM	T5@5.52	Ea	546.00	221.00	767.00
4,000 CFM	T5@6.07	Ea	615.00	243.00	858.00
Add for double duct system without coils	--	%	30.0	15.0	--
Air terminal units, constant volume, single duct with re-heat coils and trim					
200 CFM	T5@2.84	Ea	191.00	114.00	305.00

Mechanical 15

	Craft@Hrs	Unit	Material	Labor	Total

Air Distribution Equipment, Air terminal units, continued

	Craft@Hrs	Unit	Material	Labor	Total
350 CFM	T5@3.36	Ea	270.00	135.00	405.00
500 CFM	T5@3.78	Ea	302.00	151.00	453.00
800 CFM	T5@4.30	Ea	382.00	172.00	554.00
1,600 CFM	T5@5.10	Ea	456.00	204.00	660.00
2,400 CFM	T5@5.52	Ea	519.00	221.00	740.00
3,000 CFM	T5@6.87	Ea	578.00	275.00	853.00
4,000 CFM	T5@7.55	Ea	657.00	302.00	959.00

Re-heat coils for air terminal units, double row

	Craft@Hrs	Unit	Material	Labor	Total
1/2 SF	T5@1.19	Ea	170.00	47.60	217.60
1 SF	T5@1.27	Ea	201.00	50.90	251.90
1-1/2 SF	T5@1.32	Ea	239.00	52.90	291.90
2 SF	T5@1.37	Ea	270.00	54.90	324.90
2-1/2 SF	T5@1.43	Ea	286.00	57.30	343.30
3 SF	T5@1.51	Ea	307.00	60.50	367.50
3-1/2 SF	T5@1.56	Ea	339.00	62.50	401.50
4 SF	T5@1.67	Ea	371.00	66.90	437.90

Duct Work. Low pressure, to 3.5" static pressure with duct velocity up to 2,000 feet per minute. Material cost shown assumes shop fabricated duct. Labor shown is for installing the duct. These costs include the subcontractor's overhead and profit.

For estimating purposes calculate the weight per square foot of metal with table below. Sheet metal gauge is determined by the longest side dimension of duct. Once the gauge is known, multiply the square feet of sheet metal by pounds per square foot. The values given here include an allowance for waste, supports, and joints. Consider each fitting to be 3' long and reducing fitting to be 5' long regardless of the actual length

(Al = aluminum, Stl = galvanized or stainless steel)

To 12", 26 gauge minimum, Al 0.9 lbs per SF, Stl 1.3 lbs per SF
   13" to 30", 24 gauge minimum, Al 1.0 lbs per SF, Stl 1.5 lbs per SF
   31" to 54", 22 gauge minimum, Al 1.2 lbs per SF, Stl 1.8 lbs per SF
   55" to 84", 20 gauge minimum, Al 1.4 lbs per SF, Stl 2.1 lbs per SF
   85" and over, 18 gauge min., Al 1.9 lbs per SF, Stl 2.8 lbs per SF
   16 gauge, Al 2.3 lbs per SF, Stl 3.5 lbs per SF

Aluminum ductwork with supports and accessories, low pressure

Rectangular

	Craft@Hrs	Unit	Material	Labor	Total
To 2,500 lb job	T6@.100	Lb	2.45	4.20	6.65
Over 2,500 to 8,000 lb job	T6@.090	Lb	2.24	3.78	6.02
Over 8,000 lb job	T6@.080	Lb	2.08	3.36	5.44

Round

	Craft@Hrs	Unit	Material	Labor	Total
To 2,500 lb job	T6@.082	Lb	2.18	3.44	5.62
Over 2,500 lb job	T6@.072	Lb	2.00	3.03	5.03

Galvanized steel

Rectangular, shop fabricated, including typical waste, supports, joints and accessories

	Craft@Hrs	Unit	Material	Labor	Total
Under 1,000 lb job	T6@.078	Lb	.98	3.28	4.26
Over 1,000 to 5,000 lb job	T6@.069	Lb	.81	2.90	3.71
Over 5,000 lb job	T6@.060	Lb	.69	2.52	3.21

Round, shop fabricated, including typical waste, supports, joints and accessories

	Craft@Hrs	Unit	Material	Labor	Total
Under 1,000 lb job	T6@.073	Lb	.84	3.07	3.91
Over 1,000 lb to 5,000 lb job	T6@.064	Lb	.68	2.69	3.37
Over 5,000 lb job	T6@.056	Lb	.59	2.35	2.94

Stainless steel duct, typical prices with supports and accessories

	Craft@Hrs	Unit	Material	Labor	Total
Rectangular	T6@.077	Lb	2.64	3.24	5.88
Round	T6@.064	Lb	2.13	2.69	4.82

Mechanical 15

	Craft@Hrs	Unit	Material	Labor	Total
Vinyl wrapped galvanized steel or aluminum round duct, not insulated, with supports and accessories					
3", 26 gauge	T5@.024	LF	.53	.96	1.49
4", 26 gauge	T5@.028	LF	.64	1.12	1.76
6", 26 gauge	T5@.038	LF	.91	1.52	2.43
8", 26 gauge	T5@.051	LF	1.25	2.04	3.29
10", 24 gauge	T5@.081	LF	1.90	3.24	5.14
12", 24 gauge	T5@.102	LF	2.15	4.08	6.23
16", 24 gauge	T5@.128	LF	2.95	5.13	8.08
20", 24 gauge	T5@.156	LF	3.80	6.25	10.05
24", 22 gauge	T5@.186	LF	5.05	7.45	12.50
30", 22 gauge	T5@.250	LF	6.50	10.00	16.50
36", 22 gauge	T5@.336	LF	7.90	13.50	21.40
42", 20 gauge	T5@.518	LF	10.00	20.70	30.70
48", 20 gauge	T5@.586	LF	11.80	23.50	35.30
Duct lining, mineral fiber, nonflexible,					
1" thick, 1-1/2 pound density	--	SF	.46	--	--
Sound trap fill	--	CF	15.00	--	--
Flexible duct, 1.25" fiberglass insulation with vinyl cover (NFPA 90A)					
5" diameter	T5@.059	LF	1.80	2.36	4.16
6" diameter	T5@.059	LF	1.90	2.36	4.26
7" diameter	T5@.059	LF	2.15	2.36	4.51
8" diameter	T5@.067	LF	2.40	2.68	5.08
10" diameter	T5@.067	LF	2.95	2.68	5.63
12" diameter	T5@.101	LF	3.60	4.04	7.64
14" diameter	T5@.101	LF	4.35	4.04	8.39
16" diameter	T5@.101	LF	4.95	4.04	8.99
Add for 3" wide fabric connector strips for flexible duct	T5@.213	Ea	1.80	8.53	10.33
Fiberglass duct insulation, 1.00 pound density, flexible					
1-1/2" thick, no vapor barrier	A1@.022	SF	.40	.90	1.30
2" thick, no vapor barrier	A1@.028	SF	.46	1.14	1.60
1-1/2" thick, with vapor barrier	A1@.023	SF	.41	.94	1.35
2" thick, with vapor barrier	A1@.029	SF	.47	1.18	1.65
Rigid fiberglass duct insulation, 3 pound density, with vapor barrier					
2" thick	A1@.053	SF	1.61	2.16	3.77
3" thick	A1@.077	SF	2.87	3.13	6.00
4" thick	A1@.088	SF	3.78	3.58	7.36
6" thick	A1@.114	SF	5.10	4.64	9.74
Weatherproof duct insulation, with vapor barrier and mastic, 3" thick					
Polystyrene board	A1@.129	SF	2.48	5.25	7.73
Urethane board	A1@.172	SF	3.44	7.00	10.44

Air Distribution Diffusers, Grilles and Registers. Baked enamel finish, commercial quality, installed at heights to 10'. These costs include the subcontractor's overhead and profit.

	Craft@Hrs	Unit	Material	Labor	Total
Round ceiling diffusers, with adjustable dampers					
6" diameter	SM@.391	Ea	27.60	18.00	45.60
8" diameter	SM@.391	Ea	35.50	18.00	53.50
10" diameter	SM@.391	Ea	42.40	18.00	60.40
12" diameter	SM@.486	Ea	49.20	22.30	71.50
14" diameter	SM@.538	Ea	64.20	24.70	88.90
16" diameter	SM@.538	Ea	72.20	24.70	96.90
18" diameter	SM@.584	Ea	120.00	26.80	146.80
20" diameter	SM@.682	Ea	144.00	31.40	175.40

Mechanical 15

	Craft@Hrs	Unit	Material	Labor	Total

Air Distribution Diffusers, Grilles and Registers, continued
Square ceiling diffusers, adjustable flow, with sponge gasket

	Craft@Hrs	Unit	Material	Labor	Total
6" x 6"	SM@.411	Ea	32.90	18.90	51.80
9" x 9"	SM@.411	Ea	40.10	18.90	59.00
12" x 12"	SM@.486	Ea	50.00	22.30	72.30
15" x 15"	SM@.541	Ea	81.20	24.90	106.10
18" x 18"	SM@.584	Ea	106.00	26.80	132.80
21" x 21"	SM@.682	Ea	129.00	31.40	160.40
24" x 24"	SM@.670	Ea	149.00	30.80	179.80

Ceiling grid lay-in perforated metal supply diffusers, flush mount in 24" x 24" grid

	Craft@Hrs	Unit	Material	Labor	Total
6" x 6"	SM@.468	Ea	29.40	21.50	50.90
8" x 8"	SM@.468	Ea	29.60	21.50	51.10
9" x 9"	SM@.468	Ea	30.50	21.50	52.00
10" x 10"	SM@.468	Ea	31.00	21.50	52.50
12" x 12"	SM@.468	Ea	31.60	21.50	53.10
15" x 15"	SM@.468	Ea	32.90	21.50	54.40
18" x 18"	SM@.468	Ea	33.30	21.50	54.80
22" x 22"	SM@.468	Ea	34.90	21.50	56.40

Ceiling grid lay-in perforated metal return air registers with opposed blade dampers, flush mount in 24" x 24" grid

	Craft@Hrs	Unit	Material	Labor	Total
6" x 6"	SM@.468	Ea	25.40	21.50	46.90
8" x 8"	SM@.468	Ea	25.90	21.50	47.40
9" x 9"	SM@.468	Ea	26.30	21.50	47.80
10" x 10"	SM@.468	Ea	26.90	21.50	48.40
12" x 12"	SM@.468	Ea	27.60	21.50	49.10
15" x 15"	SM@.468	Ea	29.60	21.50	51.10
18" x 18"	SM@.468	Ea	31.60	21.50	53.10
22" x 22"	SM@.468	Ea	33.30	21.50	54.80

Double deflection wall registers

	Craft@Hrs	Unit	Material	Labor	Total
8" x 4"	SM@.273	Ea	12.40	12.60	25.00
8" x 6"	SM@.273	Ea	15.50	12.60	28.10
10" x 6"	SM@.273	Ea	17.00	12.60	29.60
12" x 6"	SM@.273	Ea	18.20	12.60	30.80
16" x 12"	SM@.273	Ea	27.00	12.60	39.60
18" x 12"	SM@.273	Ea	31.00	12.60	43.60
20" x 12"	SM@.273	Ea	32.00	12.60	44.60

Rectangular ceiling return air registers, single deflection, with opposed blade dampers

	Craft@Hrs	Unit	Material	Labor	Total
6" x 6"	SM@.368	Ea	11.70	16.90	28.60
8" x 8"	SM@.477	Ea	12.70	21.90	34.60
10" x 8"	SM@.477	Ea	17.60	21.90	39.50
10" x 10"	SM@.477	Ea	20.00	21.90	41.90
12" x 12"	SM@.554	Ea	23.60	25.50	49.10
14" x 14"	SM@.554	Ea	29.40	25.50	54.90
16" x 8"	SM@.627	Ea	22.30	28.80	51.10
16" x 16"	SM@.627	Ea	33.60	28.80	62.40
18" x 8"	SM@.627	Ea	23.60	28.80	52.40
20" x 20"	SM@.627	Ea	45.40	28.80	74.20
48" x 24"	SM@1.84	Ea	142.00	84.60	226.60
24" x 12"	SM@.918	Ea	32.10	42.20	74.30
24" x 18"	SM@.918	Ea	46.90	42.20	89.10
36" x 24"	SM@1.18	Ea	103.20	54.20	157.40
36" x 30"	SM@1.18	Ea	138.00	54.20	192.20

Wall return air registers

	Craft@Hrs	Unit	Material	Labor	Total
12" x 12"	SM@.261	Ea	19.20	12.00	31.20

Mechanical 15

	Craft@Hrs	Unit	Material	Labor	Total
16" x 16"	SM@.261	Ea	27.10	12.00	39.10
18" x 18"	SM@.261	Ea	33.30	12.00	45.30
20" x 20"	SM@.261	Ea	36.80	12.00	48.80
24" x 24"	SM@.261	Ea	48.70	12.00	60.70
Face hinged filter grilles					
16" x 20" vertical	SM@.529	Ea	37.80	24.30	62.10
16" x 25" vertical	SM@.529	Ea	42.50	24.30	66.80
20" x 25" vertical	SM@.529	Ea	52.50	24.30	76.80
20" x 30" vertical	SM@.529	Ea	66.50	24.30	90.80
20" x 16" horizontal	SM@.529	Ea	38.00	24.30	62.30
20" x 20" horizontal	SM@.529	Ea	44.00	24.30	68.30
25" x 16" horizontal	SM@.529	Ea	42.50	24.30	66.80
25" x 20" horizontal	SM@.529	Ea	52.50	24.30	76.80
Ceiling exhaust grilles, egg crate style					
6" x 6"	SM@.391	Ea	13.30	18.00	31.30
8" x 8"	SM@.391	Ea	15.10	18.00	33.10
10" x 10"	SM@.391	Ea	17.30	18.00	35.30
12" x 12"	SM@.391	Ea	20.60	18.00	38.60
14" x 14"	SM@.541	Ea	23.70	24.90	48.60
16" x 16"	SM@.541	Ea	31.10	24.90	56.00
18" x 18"	SM@.541	Ea	37.10	24.90	62.00
Floor diffusers with parallel valves					
2-1/2" x 10"	SM@.236	Ea	13.20	10.80	24.00
4" x 10"	SM@.236	Ea	15.10	10.80	25.90
4" x 12"	SM@.259	Ea	16.90	11.90	28.80
4" x 14"	SM@.259	Ea	19.30	11.90	31.20
6" x 14"	SM@.277	Ea	25.40	12.70	38.10
Ceiling return air grilles					
6" x 6"	SM@.391	Ea	8.32	18.00	26.32
8" x 8"	SM@.391	Ea	9.18	18.00	27.18
10" x 10"	SM@.391	Ea	10.70	18.00	28.70
12" x 12"	SM@.391	Ea	13.00	18.00	31.00
14" x 14"	SM@.541	Ea	17.00	24.90	41.90
16" x 16"	SM@.541	Ea	18.90	24.90	43.80
18" x 18"	SM@.541	Ea	24.60	24.90	49.50
20" x 20"	SM@.541	Ea	30.90	24.90	55.80
Round ceiling return air registers					
6" diameter	SM@.391	Ea	13.10	18.00	31.10
8" diameter	SM@.391	Ea	14.80	18.00	32.80
10" diameter	SM@.391	Ea	17.10	18.00	35.10
12" diameter	SM@.486	Ea	27.30	22.30	49.60
14" diameter	SM@.541	Ea	34.20	24.90	59.10
16" diameter	SM@.584	Ea	38.90	26.80	65.70

Mechanical System Controls. These costs include the subcontractor's overhead and profit.

Controls for built-up systems, percent of equipment cost					
Under $25M	--	%	--	--	25.0
$25M to $50M	--	%	--	--	21.0
$50M to $100M	--	%	--	--	18.0
$100M to $250M	--	%	--	--	16.0
$250M to $500M	--	%	--	--	13.0
Add for computerized control	--	%	--	--	10.0
Controls for package systems, percent of equipment cost					
Under $25M	--	%	--	--	8.0

Mechanical 15

	Craft@Hrs	Unit	Material	Labor	Total
Mechanical System Controls, Controls for package systems, continued					
$25M to $100M	--	%	--	--	6.0
$100M to $500M	--	%	--	--	5.0
Primary control devices, including normal wiring					
Thermostats					
Heat only	CE@1.05	Ea	27.75	48.60	76.35
Heating and cooling, setback	CE@1.19	Ea	73.10	55.10	128.20
Fan relays					
Single unit	CE@.635	Ea	44.00	29.40	73.40
Multi-unit	CE@1.32	Ea	136.60	61.20	197.80
Humidistats	CE@.815	Ea	50.00	37.80	87.80
Aquastats	CE@1.07	Ea	77.30	49.60	126.90
Timers, 24 hour	CE@2.06	Ea	96.00	95.40	191.40
Backdraft dampers. Horizontal-parallel type aluminum blade backdraft dampers installed in duct, 1,500 FPM maximum velocity					
12" x 12"	T5@.305	Ea	67.80	12.20	80.00
16" x 16"	T5@.344	Ea	86.50	13.80	100.30
20" x 20"	T5@.440	Ea	104.00	17.60	121.60
24" x 24"	T5@.609	Ea	125.00	24.40	149.40
28" x 28"	T6@.832	Ea	142.80	35.00	177.80
32" x 32"	T6@.899	Ea	160.00	37.80	197.80
36" x 36"	T6@.966	Ea	202.80	40.60	243.40
40" x 40"	T6@1.03	Ea	225.00	43.30	268.30
44" x 44"	T6@1.10	Ea	270.00	46.20	316.20
48" x 48"	T6@1.15	Ea	290.00	48.30	338.30
Add for up-parallel type	--	%	10.0	--	--
Add for down-parallel type	--	%	35.0	--	--
Control dampers, round duct mounted, not including motor operators, galvanized steel					
6" diameter	T5@1.18	Ea	15.50	47.20	62.70
8" diameter	T5@1.18	Ea	16.70	47.20	63.90
10" diameter	T5@1.18	Ea	18.00	47.20	65.20
12" diameter	T5@1.18	Ea	18.75	47.20	65.95
14" diameter	T6@1.32	Ea	21.60	55.50	77.10
16" diameter	T6@1.50	Ea	23.80	63.00	86.80
18" diameter	T6@1.67	Ea	27.75	70.20	97.95
20" diameter	T6@2.27	Ea	30.00	95.40	125.40
Control dampers, parallel horizontal blade, duct mounted, not including motor operators, galvanized steel. Dimensions are Width x Height					
Standard leakage units					
12" x 12"	T5@.429	Ea	27.20	17.20	44.40
16" x 16"	T5@.429	Ea	33.30	17.20	50.50
20" x 20"	T5@.429	Ea	44.70	17.20	61.90
24" x 24"	T5@.429	Ea	51.65	17.20	68.85
28" x 28"	T6@.650	Ea	63.45	27.30	90.75
32" x 32"	T6@.747	Ea	74.50	31.40	105.90
36" x 36"	T6@.794	Ea	86.50	33.40	119.90
40" x 40"	T6@.941	Ea	100.00	39.50	139.50
44" x 44"	T6@1.04	Ea	110.00	43.70	153.70
48" x 48"	T6@1.20	Ea	121.00	50.40	171.40
48" x 52"	T6@1.20	Ea	126.00	50.40	176.40
48" x 56"	T6@1.35	Ea	136.00	56.70	192.70
48" x 60"	T6@1.35	Ea	150.00	56.70	206.70
48" x 64"	T6@1.40	Ea	154.00	58.80	212.80
48" x 68"	T6@1.44	Ea	162.50	60.50	223.00

Mechanical 15

	Craft@Hrs	Unit	Material	Labor	Total
48" x 72"	T6@1.50	Ea	167.70	63.00	230.70
Add for opposed horizontal blade units	--	%	8.0	--	--
Add for parallel blade low leakage units	--	%	75.0	--	--
Control damper operating motors, low voltage					
Actuator for standard leakage up to 36 SF	SM@.370	Ea	265.00	17.00	282.00
Actuator for low leakage up to 24 SF	SM@.382	Ea	280.00	17.60	297.60
Fire dampers, 1.5 hour rating, curtain type, duct mounted					
12" x 12"	T5@.497	Ea	27.00	19.90	46.90
16" x 16"	T5@.497	Ea	38.00	19.90	57.90
20" x 20"	T5@.497	Ea	45.15	19.90	65.05
24" x 24"	T5@.497	Ea	54.55	19.90	74.45
29" x 28"	T6@.675	Ea	66.25	28.40	94.65
32" x 32"	T6@.700	Ea	80.00	29.40	109.40
36" x 30"	T6@1.58	Ea	94.60	66.40	161.00
40" x 40"	T6@.963	Ea	117.00	40.50	157.50
44" x 44"	T6@1.15	Ea	136.60	48.30	184.90
48" x 48"	T6@1.35	Ea	152.15	56.70	208.85
Electric pressure or temperature receivers or transmitters	CE@1.24	Ea	28.70	57.40	86.10
Fan coil unit controls					
Pneumatic room thermostats, three-way	CE@8.65	Ea	190.00	401.00	591.00
Pneumatic room thermostats, two-way	CE@8.65	Ea	208.00	401.00	609.00
Heater/cooler changeover control	CE@22.6	Ea	518.00	1,050.00	1,568.00
Water cooled chiller temperature controls					
100 tons, 2 flow switches	CE@26.6	Ea	475.00	1,230.00	1,705.00
155 tons, 2 flow switches, one 3-way valve	CE@26.6	Ea	1,585.00	1,230.00	2,815.00
150 tons, 1 temperature controller, one 3-way valve, two remote switches	CE@30.4	Ea	1,565.00	1,410.00	2,975.00
Air handling unit control systems					
Room thermostat, outside air thermostat, room air/supply air high limit switch, room air/outside air motor damper, 2-way valve	H9@47.5	Ea	1,090.00	2,190.00	3,280.00
Room air thermostat, room air/outside air min-max damper, room air/supply air high limit switch, enthalpy sensor, 2-way valve	H9@42.8	Ea	1,250.00	1,980.00	3,230.00
Room air thermostat, room air/outside air min-max damper, room air/outside air limit switch, 2-way valve	H9@50.0	Ea	1,335.00	2,310.00	3,645.00
Room thermostat, outside air damper, 2 high limit switches, two 2-way valves	H9@29.2	Ea	700.00	1,350.00	2,050.00
Room thermostat, two smoke detectors, enthalpy sensor and switch, 2-way valve	H9@52.5	Ea	2,370.00	2,420.00	4,790.00
Heating/cooling changeover control	H9@34.8	Ea	1,400.00	1,610.00	3,010.00
Unit heater controls, room thermostat with electro-pneumatic switch	M9@17.4	Ea	210.00	804.00	1,014.00
Analyzers and detectors					
Oxygen analyzers	M9@3.26	Ea	5,700.00	151.00	5,851.00
Nitrogen analyzers	M9@19.4	Ea	6,110.00	896.00	7,006.00
Boiler smoke detectors	M9@19.4	Ea	14,600.00	896.00	15,496.00
Viscosity recorder and control system	M9@19.4	Ea	5,080.00	896.00	5,976.00
Flow meter for oxygen	M9@2.19	Ea	775.00	101.00	876.00
Energy monitoring control system hardware					
Fan coil units	M9@12.9	Ea	550.00	596.00	1,146.00
Chillers and cooling towers	M9@152.	Ea	6,440.00	7,020.00	13,460.00
Air handling units, temperature only	M9@37.8	Ea	1,600.00	1,750.00	3,350.00
Air handling units, temperature and humidity	M9@38.2	Ea	1,675.00	1,760.00	3,435.00
Steam or hydronic heating system	M9@12.8	Ea	550.00	591.00	1,141.00

Mechanical 15

	Craft@Hrs	Unit	Material	Labor	Total
Mechanical System Controls, continued					
Unit heaters	M9@13.4	Ea	570.00	619.00	1,189.00
Steam/hot water converters	M9@13.4	Ea	570.00	619.00	1,189.00

Roof Curbs. Prefabricated metal roof curb. 8" high with 3.5" thick insulation, liner and raised cant strip. No roofing or carpentry work included.

	Craft@Hrs	Unit	Material	Labor	Total
15" x 15"	D4@.582	Ea	97.00	22.70	119.70
17" x 17"	D4@.582	Ea	106.00	22.70	128.70
19" x 19"	D4@.616	Ea	111.00	24.00	135.00
21" x 21"	D4@.649	Ea	115.00	25.30	140.30
23" x 23"	D4@.787	Ea	129.00	30.70	159.70
25" x 25"	D4@.819	Ea	137.00	32.00	169.00
28" x 28"	D4@.857	Ea	149.00	33.50	182.50
32" x 32"	D4@.921	Ea	160.00	36.00	196.00
36" x 36"	D4@.990	Ea	172.00	38.70	210.70
40" x 40"	D4@1.03	Ea	190.00	40.20	230.20
44" x 44"	D4@1.11	Ea	209.00	43.30	252.30
48" x 48"	D4@1.16	Ea	225.00	45.30	270.30
52" x 52"	D4@1.20	Ea	248.00	46.80	294.80
56" x 56"	D4@1.23	Ea	260.00	48.00	308.00
60" x 60"	D4@1.30	Ea	270.00	50.80	320.80
64" x 64"	D4@1.34	Ea	295.00	52.30	347.30
68" x 68"	D4@1.40	Ea	305.00	54.70	359.70
72" x 72"	D4@1.48	Ea	320.00	57.80	377.80

Electrical 16

	Craft@Hrs	Unit	Material	Labor	Total

Typical In-Place Costs. Preliminary estimates per square foot of floor area for all electrical work, including service entrance, distribution and lighting fixtures, by building type.

	Craft@Hrs	Unit	Material	Labor	Total
Commercial stores	--	SF	--	--	5.20
Market buildings	--	SF	--	--	5.90
Recreation facilities	--	SF	--	--	6.30
Schools	--	SF	--	--	7.30
Colleges	--	SF	--	--	9.90
Hospitals	--	SF	--	--	16.10
Office buildings	--	SF	--	--	6.30
Warehouses	--	SF	--	--	2.90
Parking lots	--	SF	--	--	1.00
Garages	--	SF	--	--	4.70

Underground Power and Communication Duct. Type EB (encased burial) PVC duct including duct spacers, fittings, 3" concrete envelope around duct, excavation deep enough to allow 24" cover over the duct, backfill and tamp. Cost per LF of duct bank. No wire, shoring, dewatering, pavement removal or paving included. Direct burial (DB) power and communication duct will cost about 25% more. These costs include the subcontractor's overhead and profit.

Single duct run	Craft@Hrs	Unit	Material	Labor	Equipment	Total
2" (duct at $ .40 per linear foot)	E1@.060	LF	1.19	2.45	.21	3.85
3" (duct at $ .59 per linear foot)	E1@.067	LF	1.62	2.73	.24	4.59
4" (duct at $ .96 per linear foot)	E1@.075	LF	2.19	3.06	.27	5.52
5" (duct at $1.45 per linear foot)	E1@.079	LF	2.95	3.22	.33	6.50
6" (duct at $2.08 per linear foot)	E1@.088	LF	3.67	3.59	.38	7.64

Electrical 16

	Craft@Hrs	Unit	Material	Labor	Equipment	Total
Two duct run						
2"	E1@.114	LF	2.12	4.65	.35	7.12
3"	E1@.123	LF	2.85	5.02	.41	8.28
4"	E1@.140	LF	3.93	5.71	.51	10.15
5"	E1@.150	LF	5.26	6.12	.57	11.95
6"	E1@.166	LF	6.85	6.77	.66	14.28
Three duct run						
2"	E1@.174	LF	3.15	7.09	.39	10.63
3"	E1@.174	LF	4.30	7.09	.48	11.87
4"	E1@.198	LF	5.90	8.07	.56	14.53
5"	E1@.211	LF	7.90	8.60	.62	17.12
6"	E1@.235	LF	10.60	9.58	.77	20.95
Four duct run						
2"	E1@.209	LF	3.51	8.52	.43	12.46
3"	E1@.227	LF	5.03	9.26	.47	14.76
4"	E1@.258	LF	7.10	10.50	.61	18.21
5"	E1@.268	LF	9.70	10.90	.76	21.36
6"	E1@.304	LF	12.80	12.40	.85	26.05
Six duct run						
2"	E1@.304	LF	5.21	12.40	.51	18.12
3"	E1@.332	LF	7.22	13.50	.63	21.35
4"	E1@.373	LF	10.80	15.20	.78	26.78
5"	E1@.399	LF	14.20	16.30	.94	31.44
6"	E1@.445	LF	18.80	18.10	1.10	38.00
Eight duct run						
2"	E1@.399	LF	6.73	16.30	.57	23.60
3"	E1@.422	LF	9.41	17.20	.69	27.30
4"	E1@.488	LF	14.20	19.90	.90	35.00
5"	E1@.527	LF	18.60	21.50	1.09	41.19
6"	E1@.527	LF	24.60	21.50	1.26	47.36
Twelve duct run						
2"	E1@.245	LF	9.62	9.99	.79	20.40
3"	E1@.644	LF	13.50	26.30	1.04	40.84
4"	E1@.729	LF	19.70	29.70	1.26	50.66
5"	E1@.777	LF	27.00	31.70	1.57	60.27
6"	E1@.867	LF	36.30	35.30	1.86	73.46

Underground Electric Duct, Steel. Includes excavation deep enough to allow 24" cover over the duct, galvanized rigid steel conduit as noted below, PVC duct spacers as required, fittings, 3" concrete envelope around duct, backfill and tamp. Cost per LF of duct bank. No wire, shoring, dewatering, pavement removal or paving included. These costs include the subcontractor's overhead and profit.

	Craft@Hrs	Unit	Material	Labor	Equipment	Total
Single duct run						
2" (duct at $2.70 per linear foot)	E1@.094	LF	4.46	3.83	.24	8.53
3" (duct at $5.74 per linear foot)	E1@.131	LF	7.77	5.34	.24	13.35
4" (duct at $8.53 per linear foot)	E1@.168	LF	11.00	6.85	.27	18.12
5" (duct at $17.05 per linear foot)	E1@.209	LF	20.60	8.52	.33	29.45
6" (duct at $24.08 per linear foot)	E1@.286	LF	28.80	11.70	.38	40.88
Two duct run						
2"	E1@.182	LF	7.73	7.42	.35	15.50
3"	E1@.266	LF	14.80	10.80	.41	26.01
4"	E1@.348	LF	21.00	14.20	.51	35.71
5"	E1@.432	LF	40.40	17.60	.57	58.57
6"	E1@.591	LF	56.10	24.10	.66	80.86

Electrical 16

	Craft@Hrs	Unit	Material	Labor	Equipment	Total
Underground Electric Duct, Steel, continued						
Three duct run						
2"	E1@.279	LF	11.30	11.40	.39	23.09
3"	E1@.391	LF	21.90	15.90	.48	38.28
4"	E1@.516	LF	31.30	21.00	.56	52.86
5"	E1@.644	LF	59.80	26.30	.62	86.72
6"	E1@.683	LF	84.00	27.80	.76	112.56
Four duct run						
2"	E1@.358	LF	14.50	14.60	.43	29.53
3"	E1@.496	LF	25.00	20.20	.46	45.66
4"	E1@.670	LF	40.70	27.30	.61	68.61
5"	E1@.834	LF	80.10	34.00	.76	114.86
6"	E1@1.16	LF	111.00	47.30	.84	159.14

Concrete Envelopes for Duct. No wire, duct, excavation, pavement removal or replacement included. These costs include the subcontractor's overhead and profit. Based on ready-mix concrete at $45.00 per CY before markup or allowance for waste

	Craft@Hrs	Unit	Material	Labor	Total
9" x 9" (.56 CF per linear foot)	E2@.020	LF	1.10	.76	1.86
12" x 12" (1.00 CF per linear foot)	E2@.036	LF	1.96	1.37	3.33
12" x 24" (2.00 CF per linear foot)	E2@.072	LF	3.73	2.75	6.48
Per CY	E2@1.21	CY	52.80	46.20	99.00

Concrete Pull and Junction Boxes. Complete installations including excavation, backfill, reinforced concrete with appropriate inserts, cast iron frame, cover and pulling hooks. Overall dimensions. Includes local delivery cost and the subcontractor's overhead and profit.

	Craft@Hrs	Unit	Material	Labor	Total
Precast handholes, 4' deep					
2' wide, 3' long	E1@4.24	Ea	552.00	173.00	725.00
3' wide, 3' long	E1@5.17	Ea	793.00	211.00	1,004.00
4' wide, 4' long	E1@13.2	Ea	1,120.00	538.00	1,658.00
Cast-in-place handholes, including formwork, 4' deep					
2' wide, 3' long	E3@39.8	Ea	610.00	1,630.00	2,240.00
3' wide, 3' long	E3@59.2	Ea	857.00	2,420.00	3,277.00
4' wide, 4' long	E3@93.5	Ea	1,480.00	3,820.00	5,300.00
Precast power manholes, 7' deep					
4' wide, 6' long	E1@24.6	Ea	1,650.00	1,000.00	2,650.00
6' wide, 8' long	E1@26.6	Ea	1,810.00	1,080.00	2,890.00
8' wide, 10' long	E1@26.6	Ea	1,940.00	1,080.00	3,020.00
Cast-in-place power manholes, including formwork 7' deep					
4' wide, 6' long	E3@81.2	Ea	1,290.00	3,320.00	4,610.00
6' wide, 8' long	E3@103.	Ea	1,630.00	4,210.00	5,840.00
8' wide, 10' long	E3@118.	Ea	2,170.00	4,820.00	6,990.00

Light and Power Circuits. Costs per circuit including stranded THW copper wire (except where noted), conduit, junction box, cover, straps, fasteners and connectors, but without fixture, receptacle or switch, except as noted.
Exit light or fire alarm bell circuit, 20 amp, with 4" octagonal junction box and 30' of 1/2" EMT conduit and copper THW wire

	Craft@Hrs	Unit	Material	Labor	Total
3 #12 solid wire	E4@2.68	Ea	25.50	108.00	133.50
4 #12 solid wire	E4@2.90	Ea	27.30	117.00	144.30
5 #12 solid wire	E4@3.14	Ea	29.10	126.00	155.10

Suspended ceiling lighting circuit, 20 amp, with 4" square box, copper THW wire and 30 feet of 1/2" EMT

	Craft@Hrs	Unit	Material	Labor	Total
3 #12 solid wire	E4@2.68	Ea	25.60	108.00	133.60
4 #12 solid wire	E4@2.90	Ea	27.40	117.00	144.40

Electrical 16

	Craft@Hrs	Unit	Material	Labor	Total
5 #12 solid wire	E4@3.14	Ea	29.20	126.00	155.20

Under slab circuit, with 4" square junction box, 10' of 1/2" RSC and 20' of 1/2" PVC conduit

	Craft@Hrs	Unit	Material	Labor	Total
3 #12 wire, 20 amp	E4@1.75	Ea	26.50	70.40	96.90
4 #12 wire, 20 amp	E4@1.97	Ea	28.30	79.20	107.50
5 #12 wire, 20 amp	E4@2.21	Ea	30.10	88.90	119.00
3 #10 wire, 30 amp	E4@2.00	Ea	29.20	80.40	109.60
4 #10 wire, 30 amp	E4@2.31	Ea	31.90	92.90	124.80

Under slab circuit, with 4" square junction box, 10' of 1/2" RSC conduit, 20' of 1/2" PVC conduit and 6' of flex conduit

	Craft@Hrs	Unit	Material	Labor	Total
3 #12 wire, 20 amp	E4@2.41	Ea	45.10	96.90	142.00
4 #12 wire, 20 amp	E4@2.69	Ea	48.40	108.00	156.40
3 #10 wire, 30 amp	E4@2.75	Ea	50.40	111.00	161.40
4 #10 wire, 30 amp	E4@3.81	Ea	52.70	153.00	205.70

20 amp switch circuit, 277 volt, with 30' of 1/2" EMT conduit, switch, cover and 2 indenter connectors

	Craft@Hrs	Unit	Material	Labor	Total
One gang box, single pole switch	E4@2.82	Ea	41.30	113.00	154.30
One gang box, three-way switch	E4@2.93	Ea	42.20	118.00	160.20
One gang box, four-way switch	E4@3.14	Ea	61.30	126.00	187.30
Two gang box, single pole switch	E4@2.95	Ea	46.20	119.00	165.20
Two gang box, three-way switch	E4@3.08	Ea	47.10	124.00	171.10
Two gang box, four-way switch	E4@3.24	Ea	66.20	130.00	196.20

15 amp switch circuit, 125 volt, with 30' of #14 two conductor Romex non-metallic cable (NM) with ground, switch, cover and switch connection

	Craft@Hrs	Unit	Material	Labor	Total
One gang box, single pole switch	E4@1.05	Ea	16.80	42.20	59.00
One gang box, three-way switch	E4@1.18	Ea	20.40	47.50	67.90
Two gang box, single pole switch	E4@1.17	Ea	21.70	47.10	68.80
Two gang box, single pole switch and receptacle	E4@1.36	Ea	34.50	54.70	89.20
Two gang box, single pole and 3-way switch	E4@1.43	Ea	33.60	57.50	91.10
Two gang box, single two pole and one 3-way switch	E4@1.66	Ea	33.60	66.80	100.40

20 amp 2 pole duplex receptacle outlet circuit, 125 volt, 30' of 1/2" EMT conduit, wire and receptacle, connectors and box

	Craft@Hrs	Unit	Material	Labor	Total
3 #12 wire	E4@2.80	Ea	43.70	113.00	156.70
4 #12 wire	E4@3.06	Ea	45.50	123.00	168.50
5 #12 wire	E4@3.26	Ea	47.30	131.00	178.30

Duplex receptacle outlet circuit with 30' of Romex non-metallic cable (NM) with ground, receptacle, cover and connection

	Craft@Hrs	Unit	Material	Labor	Total
15 amp, 125 volt, #14/2 wire	E4@1.11	Ea	20.70	44.60	65.30
15 amp, 125 volt, #14/2 wire with ground fault circuit interrupter (GFCI)	E4@1.30	Ea	25.20	52.30	77.50
20 amp, 125 volt, #12/2 wire	E4@1.36	Ea	24.50	54.70	79.20
30 amp, 250 volt, #10/3 wire, dryer circuit	E4@1.77	Ea	30.80	71.20	102.00
50 amp, 250 volt, #8/3 wire, range circuit	E4@2.06	Ea	45.80	82.90	128.70

Circuit breaker panels, 4 wire 120/208 volts, flush-mounted, with 225 amp (maximum) main breaker, 14 bolt-in 15 to 50 amp breakers, grounding bus, 50' of aluminum entrance cable as noted, with straps and connectors

	Craft@Hrs	Unit	Material	Labor	Total
Panel and 1/0 SEU cable, 100 amp service	E4@14.3	Ea	807.00	575.00	1,382.00
Panel and 1/0 SER cable, 100 amp service	E4@15.5	Ea	822.00	623.00	1,445.00
Panel and 2/0 SEU cable, 150 amp service	E4@14.9	Ea	800.00	599.00	1,399.00
Panel and 2/0 SER cable, 150 amp service	E4@16.7	Ea	847.00	672.00	1,519.00

Rigid Galvanized Steel Conduit. Installed exposed in a building either vertically or horizontally up to 10' above floor level. No wire, fittings or supports included except as noted. These costs include the subcontractor's overhead, and profit.

Rigid galvanized steel conduit with one coupling each 10'

	Craft@Hrs	Unit	Material	Labor	Total
1/2"	E4@.043	LF	1.24	1.73	2.97

Electrical 16

	Craft@Hrs	Unit	Material	Labor	Total
Rigid Galvanized Steel Conduit, continued					
3/4"	E4@.054	LF	1.54	2.17	3.71
1"	E4@.066	LF	2.23	2.66	4.89
1-1/4"	E4@.077	LF	2.84	3.10	5.94
1-1/2"	E4@.100	LF	3.56	4.02	7.58
2"	E4@.116	LF	4.62	4.67	9.29
2-1/2"	E4@.143	LF	7.93	5.75	13.68
3"	E4@.164	LF	9.82	6.60	16.42
3-1/2"	E4@.187	LF	12.10	7.52	19.62
4"	E4@.221	LF	14.30	8.89	23.19
5"	E4@.275	LF	29.10	11.10	40.20
6"	E4@.383	LF	41.20	15.40	56.60
Add for red plastic coated rigid steel conduit	--	%	80.0	--	--
Deduct for rigid steel conduit installed in a concrete slab or open trench	--	%	-5.0	-40.0	--
90 or 45 degree ells, standard rigid galvanized steel, threaded both ends					
1/2"	E4@.170	Ea	5.64	6.84	12.48
3/4"	E4@.190	Ea	6.82	7.64	14.46
1"	E4@.210	Ea	10.10	8.45	18.55
1-1/4"	E4@.311	Ea	14.60	12.50	27.10
1-1/2"	E4@.378	Ea	17.40	15.20	32.60
2"	E4@.448	Ea	25.60	18.00	43.60
2-1/2"	E4@.756	Ea	42.20	30.40	72.60
3"	E4@1.10	Ea	63.20	44.20	107.40
3-1/2"	E4@1.37	Ea	109.00	55.10	164.10
4"	E4@1.73	Ea	127.00	69.60	196.60
5"	E4@2.42	Ea	286.00	97.30	383.30
Add for PVC coated ells	--	%	40.0	--	--
Rigid steel threaded couplings					
1/2"	E4@.062	Ea	1.68	2.49	4.17
3/4"	E4@.080	Ea	2.07	3.22	5.29
1"	E4@.092	Ea	2.92	3.70	6.62
1-1/4"	E4@.102	Ea	3.64	4.10	7.74
1-1/2"	E4@.140	Ea	4.60	5.63	10.23
2"	E4@.170	Ea	6.08	6.84	12.92
2-1/2"	E4@.202	Ea	13.50	8.12	21.62
3"	E4@.230	Ea	18.40	9.25	27.65
3-1/2"	E4@.275	Ea	24.60	11.10	35.70
4"	E4@.425	Ea	25.90	17.10	43.00
Three-piece "Erickson" unions, by pipe size					
1/2"	E4@.285	Ea	1.99	11.50	13.49
3/4"	E4@.324	Ea	2.57	13.00	15.57
1"	E4@.394	Ea	4.80	15.80	20.60
1-1/4"	E4@.541	Ea	9.76	21.80	31.56
1-1/2"	E4@.645	Ea	12.30	25.90	38.20
2"	E4@.860	Ea	24.60	34.60	59.20
2-1/2"	E4@1.08	Ea	59.40	43.40	102.80
3"	E4@1.15	Ea	90.10	46.30	136.40
3-1/2"	E4@1.40	Ea	145.00	56.30	201.30
4"	E4@1.62	Ea	171.00	65.20	236.20
5"	E4@1.65	Ea	363.00	66.40	429.40
Type A, C, LB, E, LL or LR two hub malleable threaded conduit bodies, with covers					
1/2"	E4@.191	Ea	4.42	7.68	12.10
3/4"	E4@.238	Ea	5.30	9.57	14.87

Electrical 16

	Craft@Hrs	Unit	Material	Labor	Total
1"	E4@.285	Ea	7.96	11.50	19.46
1-1/4"	E4@.358	Ea	13.80	14.40	28.20
1-1/2"	E4@.456	Ea	18.00	18.30	36.30
2"	E4@.598	Ea	29.60	24.10	53.70
2-1/2"	E4@1.43	Ea	62.00	57.50	119.50
3"	E4@1.77	Ea	82.60	71.20	153.80
3-1/2"	E4@2.38	Ea	135.00	95.70	230.70
4"	E4@2.80	Ea	153.00	113.00	266.00
Add for explosion-proof 2 hub junction boxes	--	%	300.0	100.0	--

Type T malleable threaded conduit boxes, with covers

	Craft@Hrs	Unit	Material	Labor	Total
1/2"	E4@.285	Ea	5.53	11.50	17.03
3/4"	E4@.358	Ea	6.64	14.40	21.04
1"	E4@.433	Ea	9.96	17.40	27.36
1-1/4"	E4@.534	Ea	14.60	21.50	36.10
1-1/2"	E4@.679	Ea	19.50	27.30	46.80
2"	E4@.889	Ea	30.10	35.80	65.90
2-1/2"	E4@2.15	Ea	65.10	86.50	151.60
3"	E4@2.72	Ea	85.70	109.00	194.70
3-1/2"	E4@3.63	Ea	159.00	146.00	305.00
4"	E4@4.22	Ea	178.00	170.00	348.00
Add for explosion-proof 3 hub junction boxes	--	%	230.0	100.0	--

Conduit seal fittings, explosion proof, malleable iron or ferrous alloy, horizontal or vertical, male or female

	Craft@Hrs	Unit	Material	Labor	Total
1/2"	E4@.689	Ea	8.18	27.70	35.88
3/4"	E4@.798	Ea	9.73	32.10	41.83
1"	E4@.990	Ea	12.20	39.80	52.00
1-1/4"	E4@1.19	Ea	15.90	47.90	63.80
1-1/2"	E4@1.43	Ea	23.40	57.50	80.90
2"	E4@1.76	Ea	30.50	70.80	101.30
2-1/2"	E4@1.85	Ea	52.30	74.40	126.70
3"	E4@1.97	Ea	59.80	79.20	139.00
3-1/2"	E4@2.15	Ea	150.00	86.50	236.50
4"	E4@2.33	Ea	233.00	93.70	326.70
5"	E4@2.51	Ea	516.00	101.00	617.00

Entrance caps, threaded

	Craft@Hrs	Unit	Material	Labor	Total
1/2"	E4@.319	Ea	.09	12.80	12.89
3/4"	E4@.373	Ea	.11	15.00	15.11
1"	E4@.376	Ea	.15	15.10	15.25
1-1/4"	E4@.482	Ea	.18	19.40	19.58
1-1/2"	E4@.536	Ea	.27	21.60	21.87
2"	E4@.591	Ea	.49	23.80	24.29
2-1/2"	E4@.671	Ea	1.63	27.00	28.63
3"	E4@.749	Ea	2.37	30.10	32.47
3-1/2"	E4@.834	Ea	3.14	33.50	36.64
4"	E4@.912	Ea	3.91	36.70	40.61
5"	E4@1.01	Ea	9.84	40.60	50.44
Add for heights over 10' to 20' above ground level	--	%	--	20.0	--

Electric Metallic Tubing (EMT). Installed exposed in a building either vertically or horizontally up to 10' above floor level. No wire, fittings or supports included except as noted. These costs include the subcontractor's overhead and profit.

Electrical 16

	Craft@Hrs	Unit	Material	Labor	Total
EMT with fittings and hangers. These figures assume a typical installation with tube bent once each 10', one tap-on coupling each 10', one set screw connector each 50', and one stamped steel strap with lag screw each 6'. Use these costs for preliminary estimates					
1/2"	E4@.059	LF	.61	2.37	2.98
3/4"	E4@.067	LF	.84	2.70	3.54
1"	E4@.080	LF	1.35	3.22	4.57
1-1/4"	E4@.092	LF	2.05	3.70	5.75
1-1/2"	E4@.100	LF	2.78	4.02	6.80
2"	E4@.119	LF	3.67	4.79	8.46
2-1/2"	E4@.143	LF	9.72	5.75	15.47
EMT tubing with one bend and one set screw coupling each 10' (no hangers included)					
1/2"	E4@.036	LF	.46	1.45	1.91
3/4"	E4@.043	LF	.68	1.73	2.41
1"	E4@.055	LF	1.10	2.21	3.31
1-1/4"	E4@.067	LF	1.72	2.70	4.42
1-1/2"	E4@.070	LF	2.32	2.82	5.14
2"	E4@.086	LF	2.99	3.46	6.45
2-1/2"	E4@.106	LF	8.26	4.26	12.52
3"	E4@.127	LF	9.58	5.11	14.69
3-1/2"	E4@.144	LF	12.10	5.79	17.89
4"	E4@.170	LF	13.60	6.84	20.44
EMT 90 or 45 degree ells					
1/2"	E4@.143	Ea	3.68	5.75	9.43
3/4"	E4@.170	Ea	3.71	6.84	10.55
1"	E4@.190	Ea	4.13	7.64	11.77
1-1/4"	E4@.216	Ea	5.65	8.69	14.34
1-1/2"	E4@.288	Ea	7.15	11.60	18.75
2"	E4@.358	Ea	11.40	14.40	25.80
2-1/2"	E4@.648	Ea	30.40	26.10	56.50
3"	E4@.858	Ea	47.40	34.50	81.90
4"	E4@1.33	Ea	74.70	53.50	128.20
EMT set screw steel connectors (regular grade)					
1/2"	E4@.085	Ea	.71	3.42	4.13
3/4"	E4@.096	Ea	1.16	3.86	5.02
1"	E4@.124	Ea	2.19	4.99	7.18
1-1/4"	E4@.143	Ea	4.26	5.75	10.01
1-1/2"	E4@.162	Ea	7.87	6.52	14.39
2"	E4@.237	Ea	10.90	9.53	20.43
2-1/2"	E4@.363	Ea	37.60	14.60	52.20
3"	E4@.479	Ea	44.90	19.30	64.20
3-1/2"	E4@.567	Ea	63.00	22.80	85.80
4"	E4@.655	Ea	70.20	26.30	96.50
Deduct for similar sizes of connectors					
Competitive grade set screw connectors	--	%	-45.0	--	--
Indenter connectors	--	%	-35.0	--	--
Competitive grade compression connectors	--	%	-25.0	--	--
Competitive grade insulated throat set screw connectors	--	%	-20.0	--	--
Insulated throat indenter connectors	--	%	-15.0	--	--
Tap-on connectors	--	%	-15.0	--	--
Competitive grade insulated throat compression connectors	--	%	-5.0	--	--

Electrical 16

	Craft@Hrs	Unit	Material	Labor	Total
Add for similar sizes of connectors					
Regular grade set screw insulated throat connectors	--	%	30.0	--	--
Regular grade compression connectors	--	%	50.0	--	--
Regular grade insulated throat compression connectors	--	%	80.0	--	--
Add for heights over 10' to 20' above floor level	--	%	--	20.0	--

EMT Conduit with Wire. THW wire pulled in EMT conduit bent each 10', with one tap-on coupling each 10', one set screw connector each 50' and a strap and lag screw each 6'. Installed exposed on walls or ceilings. Add 15% to the labor cost for heights over 10' to 20'. Use these figures for preliminary estimates. These costs include the subcontractor's overhead and profit.

	Craft@Hrs	Unit	Material	Labor	Total
1-1/2" EMT conduit and wire					
2 #1 aluminum, 1 #8 copper wire	E4@.140	LF	3.99	5.63	9.62
3 #1 aluminum, 1 #8 copper wire	E4@.156	LF	4.51	6.27	10.78
4 #1 aluminum, 1 #8 copper wire	E4@.173	LF	5.03	6.96	11.99
4 #4 copper, 1 #8 copper wire	E4@.162	LF	3.91	6.52	10.43
2 #3 copper, 1 #8 copper wire	E4@.140	LF	3.91	5.63	9.54
3 #3 copper, 1 #8 copper wire	E4@.155	LF	4.39	6.23	10.62
4 #3 copper, 1 #8 copper wire	E4@.170	LF	4.87	6.84	11.71
2 #1 copper, 1 #8 copper wire	E4@.146	LF	4.57	5.87	10.44
3 #1 copper, 1 #8 copper wire	E4@.164	LF	5.38	6.60	11.98
4 #1 copper, 1 #8 copper wire	E4@.183	LF	6.19	7.36	13.55
2" EMT conduit and wire					
2 #2/0 aluminum, 1 #6 copper wire	E4@.170	LF	5.35	6.84	12.19
3 #2/0 aluminum, 1 #6 copper wire	E4@.190	LF	6.07	7.64	13.71
4 #2/0 aluminum, 1 #6 copper wire	E4@.210	LF	6.79	8.45	15.24
2 #3/0 aluminum, 1 #6 copper wire	E4@.174	LF	5.69	7.00	12.69
3 #3/0 aluminum, 1 #6 copper wire	E4@.195	LF	6.58	7.84	14.42
4 #3/0 aluminum, 1 #6 copper wire	E4@.218	LF	7.47	8.77	16.24
2 #1/0 copper, 1 #6 copper wire	E4@.171	LF	5.81	6.88	12.69
3 #1/0 copper, 1 #6 copper wire	E4@.193	LF	6.76	7.76	14.52
4 #1/0 copper, 1 #6 copper wire	E4@.213	LF	7.71	8.57	16.28
2 #3/0 copper, 1 #4 copper wire	E4@.175	LF	6.83	7.04	13.87
3 #3/0 copper, 1 #4 copper wire	E4@.197	LF	8.29	7.92	16.21
2-1/2" EMT conduit and wire					
2 250 MCM aluminum, 1 #4 copper wire	E4@.210	LF	12.30	8.45	20.75
3 250 MCM aluminum, 1 #4 copper wire	E4@.237	LF	13.50	9.53	23.03
4 #3 copper, 1 #4 copper wire	E4@.216	LF	11.90	8.69	20.59
2 #4 copper, 1 #2 copper wire	E4@.186	LF	10.80	7.48	18.28
3 #4 copper, 1 #2 copper wire	E4@.198	LF	11.00	7.96	18.96
4 #4 copper, 1 #2 copper wire	E4@.210	LF	11.30	8.45	19.75
3 250 MCM copper, 1 #2 copper wire	E4@.253	LF	16.70	10.20	26.90
3" EMT conduit and wire					
4 250 MCM aluminum, 1 #4 copper wire	E4@.288	LF	16.30	11.60	27.90
4 250 MCM copper, 1 #2 copper wire	E4@.290	LF	20.40	11.70	32.10
5 250 MCM copper, 1 #2 copper wire	E4@.319	LF	22.50	12.80	35.30

EMT Conduit Circuits. Cost per circuit based on 30' run from the panel. Includes THW copper wire pulled in conduit, 2 compression connectors, 2 couplings, conduit bending, straps, bolts and washers. Three-wire circuits are 120, 240 or 277 volt 1 phase, with 2 conductors and ground. Four-wire circuits are 120/208 or 480 volt 3 phase, with 3 conductors and ground. Five-wire circuits are 120/208 or 480 volt 3 phase, with 3 conductors, neutral and ground. Installed exposed in vertical or horizontal runs to 10' above floor level in a building. Use the figures on the following page for preliminary estimates. These costs include the subcontractor's overhead and profit.

Electrical 16

	Craft@Hrs	Unit	Material	Labor	Total
EMT Conduit Circuits, continued					
15 amp circuits					
3 #14 wire, 1/2" conduit	E4@2.39	Ea	21.90	96.10	118.00
4 #14 wire, 1/2" conduit	E4@2.58	Ea	23.10	104.00	127.10
20 amp circuits					
3 #12 wire, 1/2" conduit	E4@2.49	Ea	23.70	100.00	123.70
4 #12 wire, 1/2" conduit	E4@2.72	Ea	25.50	109.00	134.50
5 #12 wire, 1/2" conduit	E4@2.95	Ea	27.30	119.00	146.30
30 amp circuits					
3 #10 wire, 1/2" conduit	E4@2.75	Ea	26.40	111.00	137.40
4 #10 wire, 1/2" conduit	E4@3.08	Ea	29.10	124.00	153.10
5 #10 wire, 3/4" conduit	E4@3.58	Ea	38.70	144.00	182.70
50 amp circuits					
3 #8 wire, 3/4" conduit	E4@2.98	Ea	40.50	120.00	160.50
4 #8 wire, 3/4" conduit	E4@3.32	Ea	45.60	134.00	179.60
5 #8 wire, 1" conduit	E4@4.02	Ea	66.00	162.00	228.00
65 amp circuits					
2 #6, 1 #8 wire, 1" conduit	E4@3.42	Ea	60.00	138.00	198.00
3 #6, 1 #8 wire, 1" conduit	E4@3.76	Ea	67.20	151.00	218.20
4 #6, 1 #8 wire, 1-1/4" conduit	E4@4.48	Ea	95.40	180.00	275.40
85 amp circuits					
2 #4, 1 #8 wire, 1-1/4" conduit	E4@3.89	Ea	81.00	156.00	237.00
3 #4, 1 #8 wire, 1-1/4" conduit	E4@4.25	Ea	88.20	171.00	259.20
4 #4, 1 #8 wire, 1-1/2" conduit	E4@4.48	Ea	95.40	180.00	275.40
100 amp circuits					
2 #3, 1 #8 wire, 1-1/2" conduit	E4@4.20	Ea	117.00	169.00	286.00
3 #3, 1 #8 wire, 1-1/2" conduit	E4@4.66	Ea	132.00	187.00	319.00
4 #3, 1 #8 wire, 1-1/2" conduit	E4@5.10	Ea	146.00	205.00	351.00
125 amp circuits					
2 #1, 1 #8 wire, 1-1/2" conduit	E4@4.40	Ea	137.00	177.00	314.00
3 #1, 1 #8 wire, 1-1/2" conduit	E4@4.92	Ea	161.00	198.00	359.00
4 #1, 1 #8 wire, 1-1/2" conduit	E4@5.47	Ea	186.00	220.00	406.00
150 amp circuits					
2 #1/0, 1 #6 wire, 2" conduit	E4@5.16	Ea	174.00	208.00	382.00
3 #1/0, 1 #6 wire, 2" conduit	E4@5.78	Ea	203.00	232.00	435.00
4 #1/0, 1 #6 wire, 2" conduit	E4@6.43	Ea	231.00	259.00	490.00
200 amp circuits					
2 #3/0, 1 #4 wire, 2" conduit	E4@5.49	Ea	205.00	221.00	426.00
3 #3/0, 1 #4 wire, 2" conduit	E4@6.27	Ea	248.00	252.00	500.00
4 #3/0, 1 #4 wire, 2 1/2" conduit	E4@7.80	Ea	474.00	314.00	788.00
225 amp circuits					
2 #4/0, 1 #2 wire, 2-1/2" conduit	E4@6.48	Ea	418.00	261.00	679.00
3 #4/0, 1 #2 wire, 2-1/2" conduit	E4@7.33	Ea	472.00	295.00	767.00
4 #4/0, 1 #2 wire, 2-1/2" conduit	E4@8.19	Ea	527.00	329.00	856.00
250 amp circuits					
2 250 MCM, 1 #2 wire, 2-1/2" conduit	E4@6.63	Ea	437.00	267.00	704.00
3 250 MCM, 1 #2 wire, 3" conduit	E4@8.24	Ea	548.00	331.00	879.00
4 250 MCM, 1 #2 wire, 3" conduit	E4@9.56	Ea	612.00	385.00	997.00
Add for heights over 10' to 20'	--	%	--	15.0	--

Intermediate Metal Conduit (IMC). Installed exposed in a building either vertically or horizontally up to 10' above floor level. No wire, fittings or supports included except as noted. These costs include the subcontractor's overhead and profit.

Electrical 16

	Craft@Hrs	Unit	Material	Labor	Total
IMC conduit and one coupling each 10'					
1/2"	E4@.038	LF	.96	1.53	2.49
3/4"	E4@.048	LF	1.11	1.93	3.04
1"	E4@.056	LF	1.65	2.25	3.90
1-1/4"	E4@.073	LF	2.11	2.94	5.05
1-1/2"	E4@.086	LF	2.47	3.46	5.93
2"	E4@.101	LF	3.37	4.06	7.43
2-1/2"	E4@.127	LF	6.81	5.11	11.92
3"	E4@.146	LF	8.78	5.87	14.65
3-1/2"	E4@.164	LF	10.30	6.60	16.90
4"	E4@.194	LF	12.10	7.80	19.90
90 degree ells					
1/2"	E4@.160	Ea	4.40	6.44	10.84
3/4"	E4@.191	Ea	5.77	7.68	13.45
1"	E4@.225	Ea	8.31	9.05	17.36
1-1/4"	E4@.282	Ea	11.60	11.30	22.90
1-1/2"	E4@.347	Ea	14.70	14.00	28.70
2"	E4@.407	Ea	21.20	16.40	37.60
2-1/2"	E4@.692	Ea	36.00	27.80	63.80
3"	E4@.995	Ea	55.10	40.00	95.10
3-1/2"	E4@1.25	Ea	91.90	50.30	142.20
4"	E4@1.58	Ea	107.00	63.50	170.50
Add for heights over 10' to 20' above the floor	--	%	--	20.0	--
Deduct for IMC conduit installed in a concrete slab or open trench	--	%	-5.0	-40.0	--

PVC Schedule 40 Conduit. Installed in or under a building slab. No wire, fittings or supports included except as noted. These costs include the subcontractor's overhead and profit.

	Craft@Hrs	Unit	Material	Labor	Total
PVC conduit with one coupling each 10', regular type 40 heavy wall conduit					
1/2"	E4@.030	LF	.32	1.21	1.53
3/4"	E4@.036	LF	.41	1.45	1.86
1"	E4@.046	LF	.62	1.85	2.47
1-1/4"	E4@.056	LF	.84	2.25	3.09
1-1/2"	E4@.061	LF	1.01	2.45	3.46
2"	E4@.072	LF	1.31	2.90	4.21
2-1/2"	E4@.089	LF	2.10	3.58	5.68
3"	E4@.105	LF	2.51	4.22	6.73
3-1/2"	E4@.121	LF	3.23	4.87	8.10
4"	E4@.151	LF	3.52	6.07	9.59
5"	E4@.176	LF	5.09	7.08	12.17
6"	E4@.208	LF	6.75	8.37	15.12
Add for type 80 extra heavy wall PVC conduit	--	%	30.0	10.0	--
Deduct for type A thin wall PVC conduit	--	%	-35.0	--	--
Rigid straight couplings					
1/2"	E4@.040	Ea	.26	1.61	1.87
3/4"	E4@.051	Ea	.32	2.05	2.37
1"	E4@.062	Ea	.50	2.49	2.99
1-1/4"	E4@.078	Ea	.67	3.14	3.81
1-1/2"	E4@.092	Ea	.89	3.70	4.59
2"	E4@.102	Ea	1.24	4.10	5.34
2-1/2"	E4@.113	Ea	2.19	4.55	6.74
3"	E4@.129	Ea	3.49	5.19	8.68
3-1/2"	E4@.144	Ea	3.94	5.79	9.73
4"	E4@.156	Ea	5.39	6.27	11.66

Electrical 16

	Craft@Hrs	Unit	Material	Labor	Total
PVC Schedule 40 Conduit, Rigid straight couplings, continued					
5"	E4@.191	Ea	13.30	7.68	20.98
6"	E4@.226	Ea	17.10	9.09	26.19
90 degree regular weight ells					
1/2"	E4@.082	Ea	.83	3.30	4.13
3/4"	E4@.102	Ea	.91	4.10	5.01
1"	E4@.124	Ea	1.43	4.99	6.42
1-1/4"	E4@.154	Ea	2.02	6.19	8.21
1-1/2"	E4@.195	Ea	2.73	7.84	10.57
2"	E4@.205	Ea	3.97	8.25	12.22
2-1/2"	E4@.226	Ea	7.21	9.09	16.30
3"	E4@.259	Ea	12.70	10.40	23.10
3-1/2"	E4@.288	Ea	17.50	11.60	29.10
4"	E4@.313	Ea	21.90	12.60	34.50
5"	E4@.381	Ea	38.60	15.30	53.90
6"	E4@.453	Ea	65.10	18.20	83.30
Deduct for 45 or 30 degree regular ells	--	%	-30.0	--	--
Add for extra heavy weight ells	--	%	70.0	--	--
Add for 24" radius sweep ells	E4@.187	Ea	5.00	7.52	12.52
Add for 36" radius sweep ells	E4@.187	Ea	10.00	7.52	17.52
Male or female terminal adapters					
1/2"	E4@.065	Ea	.33	2.61	2.94
3/4"	E4@.081	Ea	.59	3.26	3.85
1"	E4@.097	Ea	.74	3.90	4.64
1-1/4"	E4@.121	Ea	.93	4.87	5.80
1-1/2"	E4@.144	Ea	1.14	5.79	6.93
2"	E4@.162	Ea	1.66	6.52	8.18
2-1/2"	E4@.178	Ea	2.79	7.16	9.95
3"	E4@.203	Ea	4.05	8.17	12.22
3-1/2"	E4@.226	Ea	5.32	9.09	14.41
4"	E4@.244	Ea	6.98	9.81	16.79
5"	E4@.301	Ea	13.70	12.10	25.80
6"	E4@.358	Ea	16.60	14.40	31.00
Bell ends					
1"	E4@.105	Ea	1.80	4.22	6.02
1-1/4"	E4@.115	Ea	2.16	4.63	6.79
1-1/2"	E4@.124	Ea	2.16	4.99	7.15
2"	E4@.143	Ea	3.36	5.75	9.11
2-1/2"	E4@.221	Ea	3.53	8.89	12.42
3"	E4@.285	Ea	3.94	11.50	15.44
3-1/2"	E4@.381	Ea	4.24	15.30	19.54
4"	E4@.479	Ea	4.69	19.30	23.99
5"	E4@.578	Ea	7.24	23.20	30.44
6"	E4@.741	Ea	7.94	29.80	37.74
PVC conduit access fittings, type LL, LB, LR, C or E					
1/2"	E4@.100	Ea	2.12	4.02	6.14
3/4"	E4@.132	Ea	2.98	5.31	8.29
1"	E4@.164	Ea	3.29	6.60	9.89
1-1/4"	E4@.197	Ea	4.75	7.92	12.67
1-1/2"	E4@.234	Ea	5.40	9.41	14.81
2"	E4@.263	Ea	9.48	10.60	20.08
Add for type T conduit access fittings	--	%	25.0	50.0	--
Add for PVC conduit installed exposed either vertically or horizontally in building walls to 10' above floor level	--	%	5.0	35.0	--

Electrical 16

	Craft@Hrs	Unit	Material	Labor	Total
Add for heights over 10' to 20' above the floor	--	%	--	20.0	--

PVC Conduit with Wire. Includes THW wire pulled in type 40 rigid PVC heavy wall conduit placed in a slab. Use these figures for preliminary estimates. These costs include the subcontractor's overhead and profit. Add for fittings as required.

	Craft@Hrs	Unit	Material	Labor	Total
1/2" PVC conduit with wire					
3 wire, #12 solid copper	E4@.054	LF	.50	2.17	2.67
4 wire, #12 solid copper	E4@.061	LF	.56	2.45	3.01
5 wire, #12 solid copper	E4@.069	LF	.62	2.78	3.40
3 wire, #10 solid copper	E4@.062	LF	.59	2.49	3.08
4 wire, #10 solid copper	E4@.073	LF	.68	2.94	3.62
3/4" PVC conduit with wire					
5 wire, #10 solid copper	E4@.089	LF	.86	3.58	4.44
3 wire, #8 copper	E4@.069	LF	.92	2.78	3.70
4 wire, #8 copper	E4@.080	LF	1.09	3.22	4.31
1" PVC conduit with wire					
5 wire, #8 copper	E4@.100	LF	1.47	4.02	5.49
2 #6 copper, 1 #8 copper wire	E4@.081	LF	1.27	3.26	4.53
3 #6 copper, 1 #8 copper wire	E4@.092	LF	1.51	3.70	5.21
1-1/2" PVC conduit with wire					
4 #6 copper, 1 #8 copper wire	E4@.117	LF	2.14	4.71	6.85
2 #4 copper, 1 #8 copper wire	E4@.097	LF	1.66	3.90	5.56
2 #1 aluminum, 1 #8 copper wire	E4@.102	LF	2.22	4.10	6.32
3 #1 aluminum, 1 #8 copper wire	E4@.119	LF	2.74	4.79	7.53
4 #4 copper, 1 #8 copper wire	E4@.124	LF	2.14	4.99	7.13
3 #3 copper, 1 #8 copper wire	E4@.116	LF	2.62	4.67	7.29
4 #3 copper, 1 #8 copper wire	E4@.132	LF	3.10	5.31	8.41
2 #1 copper, 1 #8 copper wire	E4@.108	LF	2.80	4.34	7.14
3 #1 copper, 1 #8 copper wire	E4@.125	LF	3.61	5.03	8.64
4 #1 copper, 1 #8 copper wire	E4@.143	LF	4.42	5.75	10.17
2" PVC conduit with wire					
4 #1 aluminum, 1 #8 copper wire	E4@.154	LF	3.92	6.19	10.11
2 2/0 aluminum, 1 #6 copper wire	E4@.123	LF	2.99	4.95	7.94
3 2/0 aluminum, 1 #6 copper wire	E4@.143	LF	3.71	5.75	9.46
4 2/0 aluminum, 1 #6 copper wire	E4@.163	LF	4.43	6.56	10.99
2 3/0 aluminum, 1 #6 copper wire	E4@.127	LF	3.33	5.11	8.44
3 3/0 aluminum, 1 #6 copper wire	E4@.148	LF	4.22	5.95	10.17
4 3/0 aluminum, 1 #6 copper wire	E4@.171	LF	5.11	6.88	11.99
2 1/0 copper, 1 #6 copper wire	E4@.124	LF	3.45	4.99	8.44
3 1/0 copper, 1 #6 copper wire	E4@.146	LF	4.40	5.87	10.27
4 1/0 copper, 1 #6 copper wire	E4@.167	LF	5.35	6.72	12.07
2 3/0 copper, 1 #4 copper wire	E4@.136	LF	4.47	5.47	9.94
3 3/0 copper, 1 #4 copper wire	E4@.162	LF	5.93	6.52	12.45
2-1/2" PVC conduit with wire					
2 250 MCM aluminum, 1 #4 copper wire	E4@.156	LF	4.72	6.27	10.99
3 250 MCM aluminum, 1 #4 copper wire	E4@.183	LF	5.91	7.36	13.27
4 3/0 copper, 1 #4 copper wire	E4@.206	LF	8.53	8.29	16.82
2 4/0 copper, 1 #2 copper wire	E4@.164	LF	6.31	6.60	12.91
3 4/0 copper, 1 #2 copper wire	E4@.193	LF	8.12	7.76	15.88
4 4/0 copper, 1 #2 copper wire	E4@.221	LF	9.93	8.89	18.82
2 250 MCM copper, 1 #2 copper wire	E4@.170	LF	6.95	6.84	13.79
3" PVC conduit with wire					
4 250 MCM aluminum, 1 #4 copper wire	E4@.226	LF	7.51	9.09	16.60
2 350 MCM aluminum, 1 #2 copper wire	E4@.186	LF	6.48	7.48	13.96

Electrical 16

	Craft@Hrs	Unit	Material	Labor	Total
PVC Conduit with Wire, 3", continued					
3 350 MCM aluminum, 1 #2 copper wire	E4@.217	LF	8.17	8.73	16.90
3 250 MCM copper, 1 #2 copper wire	E4@.217	LF	9.49	8.73	18.22
4 250 MCM copper, 1 #2 copper wire	E4@.248	LF	11.60	9.98	21.58
3-1/2" PVC conduit with wire					
4 350 MCM aluminum, 1 #2 copper wire	E4@.264	LF	10.60	10.60	21.20
4" PVC conduit with wire					
6 350 MCM aluminum, 1 #2 copper wire	E4@.293	LF	10.90	11.80	22.70
7 350 MCM aluminum, 1 #2 copper wire	E4@.391	LF	15.80	15.70	31.50
3 500 MCM copper, 1 1/0 copper wire	E4@.311	LF	16.80	12.50	29.30
4 500 MCM copper, 1 1/0 copper wire	E4@.355	LF	21.00	14.30	35.30

PVC-RSC Conduit Circuits. Cost per circuit for a 30' run from the panel. Based on THW wire pulled in 20' of type 40 rigid PVC heavy wall conduit placed in a slab and 10' of galvanized rigid steel conduit installed exposed in a building. Includes two 90 degree RSC ells, 2 PVC terminal adapters, hangers and bolts. Use these figures for preliminary estimates. These costs include the subcontractor's overhead and profit.

	Craft@Hrs	Unit	Material	Labor	Total
20 amp circuits					
3 #12 copper wire, 1/2" conduit	E4@2.23	Ea	36.50	89.70	126.20
4 #12 copper wire, 1/2" conduit	E4@2.47	Ea	38.40	99.30	137.70
5 #12 copper wire, 1/2" conduit	E4@2.56	Ea	40.20	103.00	143.20
30 amp circuits					
3 #10 copper wire, 1/2" conduit	E4@2.50	Ea	39.30	101.00	140.30
4 #10 copper wire, 1/2" conduit	E4@2.82	Ea	42.00	113.00	155.00
5 #10 copper wire, 3/4" conduit	E4@3.47	Ea	52.40	140.00	192.40
50 amp circuits					
3 #8 copper wire, 3/4" conduit	E4@2.85	Ea	54.20	115.00	169.20
4 #8 copper wire, 3/4" conduit	E4@3.21	Ea	59.40	129.00	188.40
5 #8 copper wire, 1" conduit	E4@3.96	Ea	81.80	159.00	240.80
65 amp circuits					
2 #6 copper, 1 #8 copper, 1" conduit	E4@3.26	Ea	74.20	131.00	205.20
3 #6 copper, 1 #8 copper, 1" conduit	E4@3.60	Ea	81.50	145.00	226.50
4 #6 copper, 1 #8 copper, 1-1/2" conduit	E4@4.97	Ea	112.00	200.00	312.00
85 amp circuits					
2 #4 copper, 1 #8 copper, 1-1/2" conduit	E4@4.35	Ea	112.00	175.00	287.00
3 #4 copper, 1 #8 copper, 1-1/2" conduit	E4@4.77	Ea	119.00	192.00	311.00
4 #4 copper, 1 #8 copper, 1-1/2" conduit	E4@5.18	Ea	127.00	208.00	335.00
100 amp circuits					
2 #3 copper, 1 #8 copper, 1-1/2" conduit	E4@4.40	Ea	127.00	177.00	304.00
3 #3 copper, 1 #8 copper, 1-1/2" conduit	E4@5.10	Ea	141.00	205.00	346.00
4 #3 copper, 1 #8 copper, 1-1/2" conduit	E4@5.42	Ea	156.00	218.00	374.00
2 #1 aluminum, 1 #8 copper, 1-1/2" conduit	E4@4.53	Ea	129.00	182.00	311.00
3 #1 aluminum, 1 #8 copper, 1-1/2" conduit	E4@5.00	Ea	145.00	201.00	346.00
4 #1 aluminum, 1 #8 copper, 1-1/2" conduit	E4@5.13	Ea	155.00	206.00	361.00
125 amp circuits					
2 #1 copper, 1 #8 copper, 1-1/2" conduit	E4@4.69	Ea	147.00	189.00	336.00
3 #1 copper, 1 #8 copper, 1-1/2" conduit	E4@5.23	Ea	172.00	210.00	382.00
4 #1 copper, 1 #8 copper, 1-1/2" conduit	E4@5.80	Ea	197.00	233.00	430.00
2 2/0 aluminum, 1 #6 copper, 2" conduit	E4@5.54	Ea	160.00	223.00	383.00
3 2/0 aluminum, 1 #6 copper, 2" conduit	E4@6.14	Ea	199.00	247.00	446.00
4 2/0 aluminum, 1 #6 copper, 2" conduit	E4@6.76	Ea	221.00	272.00	493.00
150 amp circuits					
2 1/0 copper, 1 #6 copper, 2" conduit	E4@5.21	Ea	191.00	210.00	401.00
3 1/0 copper, 1 #6 copper, 2" conduit	E4@6.30	Ea	220.00	253.00	473.00

Electrical 16

	Craft@Hrs	Unit	Material	Labor	Total
4 1/0 copper, 1 #6 copper, 2" conduit	E4@7.44	Ea	249.00	299.00	548.00
2 3/0 aluminum, 1 #6 copper, 2" conduit	E4@5.86	Ea	187.00	236.00	423.00
3 3/0 aluminum, 1 #6 copper, 2" conduit	E4@6.61	Ea	215.00	266.00	481.00
4 3/0 aluminum, 1 #6 copper, 2" conduit	E4@10.1	Ea	311.00	406.00	717.00
200 amp circuits					
2 3/0 copper, 1 #4 copper, 2" conduit	E4@6.11	Ea	279.00	246.00	525.00
3 3/0 copper, 1 #4 copper, 2" conduit	E4@6.71	Ea	273.00	270.00	543.00
4 3/0 copper, 1 #4 copper, 2-1/2" conduit	E4@10.3	Ea	404.00	414.00	818.00
2 250 MCM alum., 1 #4 copper, 2-1/2" conduit	E4@8.76	Ea	299.00	352.00	651.00
3 250 MCM alum., 1 #4 copper, 2-1/2" conduit	E4@9.59	Ea	335.00	386.00	721.00
4 250 MCM alum., 1 #4 copper, 3" conduit	E4@11.3	Ea	435.00	454.00	889.00
225 amp circuits					
2 4/0 copper, 1 #2 copper, 2-1/2" conduit	E4@8.94	Ea	336.00	360.00	696.00
3 4/0 copper, 1 #2 copper, 2-1/2" conduit	E4@9.85	Ea	391.00	396.00	787.00
4 4/0 copper, 1 #2 copper, 2-1/2" conduit	E4@9.64	Ea	447.00	388.00	835.00
250 amp circuits					
2 250 MCM copper, 1 #2 copper, 2-1/2" conduit	E4@9.15	Ea	364.00	368.00	732.00
3 250 MCM copper, 1 #2 copper, 3" conduit	E4@11.0	Ea	493.00	442.00	935.00
4 250 MCM copper, 1 #2 copper, 3" conduit	E4@12.0	Ea	557.00	483.00	1,040.00
2 350 MCM aluminum, 1 #2 copper, 3" conduit	E4@10.6	Ea	400.00	426.00	826.00
3 350 MCM aluminum, 1 #2 copper, 3" conduit	E4@11.9	Ea	452.00	479.00	931.00
4 350 MCM aluminum, 1 #2 copper, 3-1/2" conduit	E4@14.9	Ea	614.00	599.00	1,213.00
400 amp circuits					
3 500 MCM copper, 1 1/0 copper, 4" conduit	E4@16.7	Ea	882.00	672.00	1,554.00
4 500 MCM copper, 1 1/0 copper, 4" conduit	E4@18.1	Ea	1,010.00	728.00	1,738.00
6 350 MCM aluminum, 1 2/0 copper, 4" conduit	E4@18.6	Ea	739.00	748.00	1,487.00
7 350 MCM aluminum, 1 2/0 copper, 4" conduit	E4@20.5	Ea	856.00	825.00	1,681.00

Rigid Aluminum Conduit. Installed exposed in a building either vertically or horizontally up to 10' above floor level. No wire, fittings or supports included except as noted. These costs include the subcontractor's overhead and profit.

Rigid aluminum conduit with one coupling and one bend each 10'

	Craft@Hrs	Unit	Material	Labor	Total
1/2"	E4@.040	LF	1.05	1.61	2.66
3/4"	E4@.048	LF	1.40	1.93	3.33
1"	E4@.059	LF	2.00	2.37	4.37
1-1/4"	E4@.075	LF	2.623	3.02	5.64
1-1/2"	E4@.093	LF	3.26	3.74	7.00
2"	E4@.108	LF	4.35	4.34	8.69

Rigid aluminum conduit with one coupling each 10'

	Craft@Hrs	Unit	Material	Labor	Total
2-1/2"	E4@.081	LF	6.88	3.26	10.14
3"	E4@.097	LF	9.04	3.90	12.94
3-1/2"	E4@.129	LF	10.80	5.19	15.99
4"	E4@.162	LF	12.90	6.52	19.42
5"	E4@.203	LF	18.40	8.17	26.57
6"	E4@.242	LF	24.20	9.73	33.93

Type C, LL, LR, LB or LC threaded aluminum conduit bodies, with aluminum covers and gaskets

	Craft@Hrs	Unit	Material	Labor	Total
1/2"	E4@.313	Ea	7.07	12.60	19.67
3/4"	E4@.345	Ea	8.61	13.90	22.51
1"	E4@.464	Ea	12.80	18.70	31.50
1-1/4"	E4@.648	Ea	16.30	26.10	42.40
1-1/2"	E4@.777	Ea	17.80	31.30	49.10
2"	E4@.902	Ea	35.30	36.30	71.60
2-1/2"	E4@1.42	Ea	61.60	57.10	118.70
3"	E4@1.97	Ea	84.20	79.20	163.40

Electrical 16

	Craft@Hrs	Unit	Material	Labor	Total

Rigid Aluminum Conduit, Type C, LL, LR, LB, or LC, continued

	Craft@Hrs	Unit	Material	Labor	Total
3-1/2"	E4@2.60	Ea	146.00	105.00	251.00
4"	E4@3.06	Ea	167.00	123.00	290.00
Add for type T aluminum conduit bodies	--	%	25.0	30.0	--
Add for heights over 10' to 20' above the floor	--	%	--	20.0	--
For fittings, use rigid galvanized steel conduit fitting costs and deduct	--	%	--	-15.0	--

Lightweight Flexible Metal Conduit. Installed exposed in a building either vertically or horizontally up to 10' above floor level. No wire, fittings or supports included except as noted. Typical flexible conduit connection for electrical equipment consisting of a 4" square junction box, 6' of aluminum reduced wall conduit, two 1-screw connectors and solid copper wire pulled in conduit as noted. Per 6' connection

	Craft@Hrs	Unit	Material	Labor	Total
3/8" conduit, 3 #14 THHN copper wire	E4@.630	Ea	6.21	25.30	31.51
1/2" conduit, 4 #12 THHN copper wire	E4@.733	Ea	9.05	29.40	38.45
1/2" conduit, 4 #10 THHN copper wire	E4@.785	Ea	9.77	31.60	41.37
3/4" conduit, 4 #8 THHN copper wire	E4@.883	Ea	13.80	35.50	49.30
1" conduit, 4 #4 THHN copper wire	E4@1.12	Ea	21.40	45.00	66.40

Flexible aluminum reduced wall conduit installed in a building to 10' above floor level. No wire, fittings or supports included

	Craft@Hrs	Unit	Material	Labor	Total
3/8"	E4@.030	LF	.23	1.21	1.44
1/2"	E4@.039	LF	.33	1.57	1.90
3/4"	E4@.051	LF	.46	2.05	2.51
1"	E4@.078	LF	.87	3.14	4.01
1-1/4"	E4@.113	LF	1.10	4.55	5.65
1-1/2"	E4@.147	LF	1.40	5.91	7.31
2"	E4@.191	LF	2.06	7.68	9.74
2-1/2"	E4@.282	LF	2.62	11.30	13.92
3"	E4@.394	LF	3.03	15.80	18.83
Add for standard wall flex steel conduit	--	%	7.0	10.0	--
Deduct for reduced wall flex steel conduit	--	%	-14.0	--	--

Flexible metal one screw tite-bite straight connectors

	Craft@Hrs	Unit	Material	Labor	Total
3/8"	E4@.070	Ea	.90	2.82	3.72
1/2"	E4@.086	Ea	1.66	3.46	5.12
3/4"	E4@.100	Ea	2.33	4.02	6.35
1"	E4@.113	Ea	4.04	4.55	8.59
1-1/4"	E4@.148	Ea	6.21	5.95	12.16
1-1/2"	E4@.191	Ea	9.86	7.68	17.54
2"	E4@.260	Ea	13.60	10.50	24.10
2-1/2"	E4@.326	Ea	25.50	13.10	38.60
3"	E4@.435	Ea	35.60	17.50	53.10
Add for 90 or 45 degree tite-bite connectors	--	%	100.0	--	--
Add for height over 10' to 20' above the floor	--	%	--	20.0	--

Liquid-Tight Flexible Metal Conduit (Sealtite). Installed exposed in a building either vertically or horizontally up to 10' above floor level. No wire, fittings or supports included except as noted. These costs include the subcontractor's overhead and profit.

Typical liquid-tight flexible conduit connection for electrical equipment consisting of 6' of conduit, 2 connectors and solid copper wire pulled in conduit as noted. Per 6' connection

	Craft@Hrs	Unit	Material	Labor	Total
3/8" conduit, 3 #12 THW copper wire	E4@.622	Ea	18.60	25.00	43.60
1/2" conduit, 4 #12 THW copper wire	E4@.746	Ea	20.10	30.00	50.10
1/2" conduit, 4 #10 THW copper wire	E4@.813	Ea	20.80	32.70	53.50
3/4" conduit, 3 #8, 1 #10 THW copper wire	E4@.891	Ea	29.20	35.80	65.00
1" conduit, 3 #6, 1 #8 THW copper wire	E4@1.23	Ea	48.70	49.50	98.20

Electrical 16

	Craft@Hrs	Unit	Material	Labor	Total

Liquid-tight flexible metal conduit installed in a building to 10' above floor level. No wire, supports or fittings included (Type E.F. or L.T. extra flex)

	Craft@Hrs	Unit	Material	Labor	Total
3/8"	E4@.045	LF	2.11	1.81	3.92
1/2"	E4@.051	LF	2.30	2.05	4.35
3/4"	E4@.067	LF	3.11	2.70	5.81
1"	E4@.094	LF	4.75	3.78	8.53
1-1/4"	E4@.116	LF	6.48	4.67	11.15
1-1/2"	E4@.132	LF	8.78	5.31	14.09
2"	E4@.156	LF	11.10	6.27	17.37
2-1/2"	E4@.234	LF	20.30	9.41	29.71
3"	E4@.313	LF	28.10	12.60	40.70
4"	E4@.624	LF	40.90	25.10	66.00

Add for other types of liquid-tight flexible metal conduit

Type U.A. or L.A.	--	%	50.0	--	--
Type O.R. or L.O.R.	--	%	20.0	--	--
Type H.C. or A.T.	--	%	28.0	--	--

Straight sealtite connectors for flexible metal conduit

	Craft@Hrs	Unit	Material	Labor	Total
3/8"	E4@.110	Ea	2.44	4.42	6.86
1/2"	E4@.129	Ea	2.44	5.19	7.63
3/4"	E4@.140	Ea	3.48	5.63	9.11
1"	E4@.221	Ea	7.44	8.89	16.33
1-1/4"	E4@.244	Ea	12.80	9.81	22.61
1-1/2"	E4@.293	Ea	18.20	11.80	30.00
2"	E4@.492	Ea	25.30	19.80	45.10
2-1/2"	E4@.741	Ea	112.00	29.80	141.80
3"	E4@.982	Ea	126.00	39.50	165.50
4"	E4@1.47	Ea	155.00	59.10	214.10
Add for height over 10' to 20' above the floor	--	%	--	20.0	--
Add for insulated throat sealtite connectors	--	%	20.0	--	--
Add for 45 or 90 degree sealtite connectors	--	%	30.0	--	--

Conduit Hangers and Supports. By conduit size. Installed exposed in a building to 10' above floor level. Hangers and supports installed at heights over 10' will add about 20% to the labor cost. These costs include the subcontractor's overhead and profit.

Right angle conduit supports

	Craft@Hrs	Unit	Material	Labor	Total
3/8"	E4@.242	Ea	2.08	9.73	11.81
1/2"	E4@.242	Ea	1.83	9.73	11.56
3/4"	E4@.242	Ea	1.91	9.73	11.64
1"	E4@.269	Ea	2.11	10.80	12.91
1-1/4"	E4@.269	Ea	2.52	10.80	13.32
1-1/2"	E4@.269	Ea	2.66	10.80	13.46
2"	E4@.295	Ea	3.88	11.90	15.78
2-1/2"	E4@.295	Ea	5.49	11.90	17.39
3"	E4@.295	Ea	5.52	11.90	17.42
3-1/2"	E4@.324	Ea	7.69	13.00	20.69
4"	E4@.324	Ea	8.29	13.00	21.29
Add for parallel conduit supports	--	%	50.0	--	--
Add for edge conduit supports	--	%	120.0	--	--

Conduit hangers, with bolt, labor includes the cost of cutting threaded rod to length and attaching the rod but not the rod itself. Add the material cost of threaded rod from below

	Craft@Hrs	Unit	Material	Labor	Total
1/2"	E4@.097	Ea	.58	3.90	4.48
3/4"	E4@.097	Ea	.64	3.90	4.54
1"	E4@.108	Ea	.86	4.34	5.20
1-1/4"	E4@.108	Ea	1.02	4.34	5.36

Electrical 16

	Craft@Hrs	Unit	Material	Labor	Total
Conduit Hangers and Supports, Conduit hangers, with bolt, continued					
1-1/2"	E4@.108	Ea	1.28	4.34	5.62
2"	E4@.119	Ea	1.44	4.79	6.23
2-1/2"	E4@.119	Ea	1.55	4.79	6.34
3"	E4@.119	Ea	1.88	4.79	6.67
3-1/2"	E4@.129	Ea	2.17	5.19	7.36
4"	E4@.129	Ea	4.18	5.19	9.37
All threaded rod, plated steel, per linear foot of rod					
1/4", 20 thread	--	LF	.78	--	--
5/16", 18 thread	--	LF	1.09	--	--
3/8", 16 thread	--	LF	1.29	--	--
1/2", 13 thread	--	LF	2.06	--	--
5/8", 11 thread	--	LF	3.07	--	--
Deduct for plain steel rod	--	%	-10.0	--	--
Rod beam clamps, for 1/4" or 3/8" rod					
1" flange	E4@.234	Ea	1.44	9.41	10.85
1-1/2" flange	E4@.234	Ea	3.74	9.41	13.15
2" flange	E4@.234	Ea	5.27	9.41	14.68
2-1/2" flange	E4@.234	Ea	7.79	9.41	17.20
Conduit clips, for EMT conduit					
1/2" or 3/4"	E4@.062	Ea	.16	2.49	2.65
1"	E4@.069	Ea	.21	2.78	2.99
1-1/4"	E4@.069	Ea	.30	2.78	3.08
1-1/2"	E4@.069	Ea	.32	2.78	3.10
2"	E4@.075	Ea	.37	3.02	3.39
One hole heavy duty stamped steel conduit straps					
1/2"	E4@.039	Ea	.28	1.57	1.85
3/4"	E4@.039	Ea	.32	1.57	1.89
1"	E4@.039	Ea	.70	1.57	2.27
1-1/4"	E4@.039	Ea	.96	1.57	2.53
1-1/2"	E4@.039	Ea	1.24	1.57	2.81
2"	E4@.039	Ea	2.15	1.57	3.72
2-1/2"	E4@.053	Ea	3.65	2.13	5.78
3"	E4@.053	Ea	4.08	2.13	6.21
3-1/2"	E4@.059	Ea	6.33	2.37	8.70
4"	E4@.059	Ea	7.97	2.37	10.34
Add for malleable iron one hole straps	--	%	30.0	--	--
Deduct for nail drive straps	--	%	-20.0	-30.0	--
Deduct for two hole stamped steel EMT straps	--	%	-70.0	--	--
Hanger channel, 1-1/2" x 1-1/2", based on 12" length No holes, solid back,					
12 gauge steel	E4@.119	Ea	2.88	4.79	7.67
Holes 1-1/2" on center, 12 gauge aluminum	E4@.119	Ea	4.05	4.79	8.84
Channel strap for rigid steel conduit or EMT conduit, by conduit size, with bolts					
1/2"	E4@.020	Ea	.80	.80	1.60
3/4"	E4@.020	Ea	.90	.80	1.70
1"	E4@.020	Ea	.97	.80	1.77
1-1/4"	E4@.023	Ea	1.12	.93	2.05
1-1/2"	E4@.023	Ea	1.23	.93	2.16
2"	E4@.023	Ea	1.34	.93	2.27
2-1/2"	E4@.039	Ea	1.67	1.57	3.24
3"	E4@.039	Ea	1.81	1.57	3.38
3-1/2"	E4@.039	Ea	2.26	1.57	3.83
4"	E4@.062	Ea	2.48	2.49	4.97
5"	E4@.062	Ea	3.64	2.49	6.13

Electrical 16

	Craft@Hrs	Unit	Material	Labor	Total
6"	E4@.062	Ea	4.64	2.49	7.13
Lag screws, flattened end, with bolt					
1/4" x 3"	E4@.092	Ea	.44	3.70	4.14
1/4" x 4"	E4@.092	Ea	.49	3.70	4.19
5/16" x 3-1/2"	E4@.092	Ea	.52	3.70	4.22
3/8" x 4"	E4@.119	Ea	.60	4.79	5.39
1/2" x 5"	E4@.129	Ea	.67	5.19	5.86
Lag screw short expansion shields, without screws					
1/4"	E4@.175	Ea	.22	7.04	7.26
5/16"	E4@.175	Ea	.24	7.04	7.28
3/8"	E4@.272	Ea	.38	10.90	11.28
1/2"	E4@.321	Ea	.59	12.90	13.49
5/8"	E4@.321	Ea	.83	12.90	13.73
3/4"	E4@.345	Ea	1.14	13.90	15.04
Add for long lag screw expansion shields	--	%	25.0	25.0	--
Self drilling masonry anchors					
1/4"	E4@.127	Ea	.72	5.11	5.83
5/16"	E4@.127	Ea	.96	5.11	6.07
3/8"	E4@.191	Ea	1.07	7.68	8.75
1/2"	E4@.191	Ea	1.60	7.68	9.28
5/8"	E4@.259	Ea	2.89	10.40	13.29
3/4"	E4@.259	Ea	5.04	10.40	15.44
7/8"	E4@.332	Ea	7.60	13.40	21.00

Electrical Wireway (Surface Duct). Wall mounted flangeless hinged cover or screw cover enameled steel wiring raceway. No wire included. Section prices include 1 connector or end plate with each section. These figures include the subcontractor's overhead and profit.

	Craft@Hrs	Unit	Material	Labor	Total
2-1/2" x 2-1/2" wireway					
1' sections	E4@.265	Ea	7.07	10.70	17.77
2' sections	E4@.391	Ea	10.10	15.70	25.80
3' sections	E4@.671	Ea	13.60	27.00	40.60
4' sections	E4@.702	Ea	18.90	28.20	47.10
5' sections	E4@.865	Ea	22.60	34.80	57.40
10' sections	E4@1.76	Ea	46.90	70.80	117.70
Hangers and brackets	E4@.222	Ea	4.15	8.93	13.08
Tee or cross pull boxes	E4@1.17	Ea	19.50	47.10	66.60
Internal or external corners	E4@.894	Ea	16.60	36.00	52.60
4" x 4" wireway					
1' sections	E4@.272	Ea	7.68	10.90	18.58
2' sections	E4@.407	Ea	11.20	16.40	27.60
3' sections	E4@.609	Ea	16.60	24.50	41.10
4' sections	E4@.811	Ea	22.60	32.60	55.20
5' sections	E4@.902	Ea	24.90	36.30	61.20
10' sections	E4@1.85	Ea	51.00	74.40	125.40
Hangers and brackets	E4@.334	Ea	5.30	13.40	18.70
Tee or cross pull boxes	E4@1.79	Ea	22.60	72.00	94.60
Internal or external corners	E4@1.35	Ea	18.90	54.30	73.20
6" x 6" wireway					
1' sections	E4@.326	Ea	14.80	13.10	27.90
2' sections	E4@.490	Ea	18.40	19.70	38.10
3' sections	E4@.733	Ea	25.70	29.50	55.20
4' sections	E4@.972	Ea	34.00	39.10	73.10
5' sections	E4@1.08	Ea	36.70	43.40	80.10
10' sections	E4@2.21	Ea	87.00	88.90	175.90

Electrical 16

	Craft@Hrs	Unit	Material	Labor	Total
Electrical Wireway (Surface Duct), 6" x 6", continued					
Hangers and brackets	E4@.391	Ea	9.45	15.70	25.15
Tee or cross pull boxes	E4@2.06	Ea	25.70	82.90	108.60
Internal or external corners	E4@1.58	Ea	21.00	63.50	84.50
8" x 8" wireway					
1' sections	E4@.389	Ea	23.30	15.60	38.90
2' sections	E4@.591	Ea	36.00	23.80	59.80
3' sections	E4@.878	Ea	54.40	35.30	89.70
4' sections	E4@1.17	Ea	66.00	47.10	113.10
5' sections	E4@1.30	Ea	74.10	52.30	126.40
10' sections	E4@2.66	Ea	138.00	107.00	245.00
Hangers and brackets	E4@.451	Ea	12.40	18.10	30.50
Tee or cross pull boxes	E4@2.39	Ea	47.60	96.10	143.70
Internal or external corners	E4@1.82	Ea	34.00	73.20	107.20
12" x 12" wireway					
1' sections	E4@.547	Ea	32.70	22.00	54.70
2' sections	E4@.819	Ea	63.20	32.90	96.10
3' sections	E4@1.23	Ea	94.50	49.50	144.00
4' sections	E4@1.64	Ea	114.00	66.00	180.00
5' sections	E4@1.82	Ea	131.00	73.20	204.20
Hangers and brackets	E4@.541	Ea	35.40	21.80	57.20
Tee or cross pull boxes	E4@2.85	Ea	95.20	115.00	210.20
Internal or external corners	E4@2.15	Ea	61.90	86.50	148.40
Add for raintight (exterior) wireway	--	%	75.0	--	--
Add for flanged latch cover wireway	--	%	125.0	--	--

Wiremold Raceway. Installed on a finished wall in a building. These costs include the subcontractor's overhead and profit.

	Craft@Hrs	Unit	Material	Labor	Total
Raceway including one coupling each 10 feet					
#200 2-piece midget raceway	E4@.046	LF	.51	1.85	2.36
#500 2 wire surface mounted raceway	E4@.059	LF	.59	2.37	2.96
#700 4 wire surface mounted raceway	E4@.067	LF	.68	2.70	3.38
#1500 "pancake" surface mounted raceway	E4@.080	LF	.97	3.22	4.19
Wiremold fittings					
#200 90 degree flat ell	E4@.100	Ea	1.17	4.02	5.19
#200 internal ell	E4@.100	Ea	1.93	4.02	5.95
#200 extension adapter	E4@.199	Ea	3.66	8.00	11.66
#200, #500, #700 1-pole switch and box	E4@.253	Ea	6.49	10.20	16.69
#200, #500, #700 1-pole switch box only	E4@.360	Ea	4.35	14.50	18.85
#200, #500, #700 duplex receptacle and box	E4@.360	Ea	7.67	14.50	22.17
#500, #700 90 degree flat ell	E4@.100	Ea	.81	4.02	4.83
#500, #700 internal twisted ell	E4@.100	Ea	2.37	4.02	6.39
#500, #700 utility box	E4@.360	Ea	4.62	14.50	19.12
#500, #700 corner box	E4@.360	Ea	7.08	14.50	21.58
#500, #700 fixture box	E4@.360	Ea	5.85	14.50	20.35
#1500 90 degree flat ell	E4@.151	Ea	2.70	6.07	8.77
#1500 internal ell	E4@.151	Ea	2.79	6.07	8.86
#1500 junction box	E4@.360	Ea	4.98	14.50	19.48

Cable Tray and Ducts. These costs include the subcontractor's overhead and profit.

	Craft@Hrs	Unit	Material	Labor	Total
Cable tray, steel, with typical fittings and supports, ladder type					
6" wide	E4@.166	LF	7.07	6.68	13.75
9" wide	E4@.189	LF	7.68	7.60	15.28
12" wide	E4@.211	LF	14.80	8.49	23.29

Electrical 16

	Craft@Hrs	Unit	Material	Labor	Total
18" wide	E4@.280	LF	23.30	11.30	34.60
24" wide	E4@.383	LF	25.60	15.40	41.00

Steel underfloor duct, including typical supports, fittings and accessories

	Craft@Hrs	Unit	Material	Labor	Total
3-1/4" wide, 1 cell	E4@.097	LF	3.91	3.90	7.81
3-1/4" wide, 2 cell	E4@.104	LF	7.93	4.18	12.11
7-1/4" wide, 1 cell	E4@.148	LF	9.16	5.95	15.11
7-1/4" wide, 2 cell	E4@.148	LF	14.00	5.95	19.95

Copper Service Entrance Wire. Type USE-RHH-RHW (XLPE) crosslinked polyethylene 600 volt stranded copper service entrance wire pulled in conduit. No excavation or conduit included. Labor cost assumes three bundled conductors are pulled at one time. By American Wire Gauge. These costs include the subcontractor's overhead and profit. Note that wire prices can change very quickly.

	Craft@Hrs	Unit	Material	Labor	Total
#12	E4@.008	LF	.11	.32	.43
#10	E4@.009	LF	.15	.36	.51
#8	E4@.011	LF	.23	.44	.67
#6	E4@.013	LF	.27	.52	.79
#4	E4@.016	LF	.41	.64	1.05
#2	E4@.018	LF	.64	.72	1.36
#1	E4@.019	LF	.84	.76	1.60
#1/0	E4@.024	LF	1.01	.97	1.98
#2/0	E4@.028	LF	1.23	1.13	2.36
#3/0	E4@.032	LF	1.53	1.29	2.82
#4/0	E4@.036	LF	1.85	1.45	3.30
250 MCM	E4@.039	LF	2.32	1.57	3.89
300 MCM	E4@.040	LF	2.75	1.61	4.36
350 MCM	E4@.045	LF	3.10	1.81	4.91
400 MCM	E4@.046	LF	3.61	1.85	5.46
500 MCM	E4@.047	LF	4.20	1.89	6.09
600 MCM	E4@.054	LF	5.88	2.17	8.05
750 MCM	E4@.066	LF	7.09	2.66	9.75
For aerial distribution, deduct	--	%	--	-25.0	--
Add for each additional conductor pulled	--	%	100.0	80.0	--

Bare Copper Groundwire. Stranded, pulled in conduit with conductor wires. No excavation or duct included. By American Wire Gauge. These costs include the subcontractor's overhead and profit. Note that wire prices can change very quickly.

	Craft@Hrs	Unit	Material	Labor	Total
#8	E4@.008	LF	.16	.32	.48
#6	E4@.009	LF	.23	.36	.59
#4	E4@.012	LF	.37	.48	.85
#2	E4@.014	LF	.58	.56	1.14
#1	E4@.015	LF	.79	.60	1.39
#1/0	E4@.018	LF	.93	.72	1.65
#2/0	E4@.021	LF	1.11	.85	1.96
#3/0	E4@.025	LF	1.43	1.01	2.44
#4/0	E4@.028	LF	1.78	1.13	2.91

Aluminum Service Entrance Wire. Type USE-RHH-RHW (XLPE) crosslinked polyethylene 600 volt stranded aluminum service entrance wire pulled in conduit. No excavation or conduit included. Labor cost assumes three bundled conductors are pulled at one time. By American Wire Gauge. These costs include the subcontractor's overhead and profit. Note that wire prices can change very quickly.

	Craft@Hrs	Unit	Material	Labor	Total
#6	E4@.012	LF	.24	.48	.72
#4	E4@.014	LF	.29	.56	.85
#2	E4@.017	LF	.39	.68	1.07
#1	E4@.017	LF	.56	.68	1.24

Electrical 16

	Craft@Hrs	Unit	Material	Labor	Total
Aluminum Service Entrance Wire, continued					
#1/0	E4@.022	LF	.65	.89	1.54
#2/0	E4@.026	LF	.75	1.05	1.80
#3/0	E4@.030	LF	.92	1.21	2.13
#4/0	E4@.034	LF	1.05	1.37	2.42
250 MCM	E4@.035	LF	1.31	1.41	2.72
300 MCM	E4@.038	LF	1.76	1.53	3.29
350 MCM	E4@.040	LF	1.82	1.61	3.43
400 MCM	E4@.042	LF	2.05	1.69	3.74
500 MCM	E4@.043	LF	2.36	1.73	4.09
600 MCM	E4@.050	LF	2.89	2.01	4.90
700 MCM	E4@.054	LF	3.38	2.17	5.55
750 MCM	E4@.059	LF	3.45	2.37	5.82
1000 MCM	E4@.074	LF	5.06	2.98	8.04
For aerial distribution, deduct	--	%	--	-25.0	--
Add for each additional conductor pulled	--	%	100.0	80.0	--

Bare Aluminum Groundwire. Stranded, pulled in conduit with conductor wires. No excavation or conduit included. By American Wire Gauge. These costs include the subcontractor's overhead and profit. Note that wire prices can change very quickly.

	Craft@Hrs	Unit	Material	Labor	Total
#4, .0392 pounds per foot, $2.27 per pound	E4@.011	LF	.17	.44	.61
#2, .0623 pounds per foot, $2.15 per pound	E4@.012	LF	.29	.48	.77
#1/0, .0991 pounds per foot, $2.07 per pound	E4@.016	LF	.36	.64	1.00
#2/0, .1249 pounds per foot, $2.02 per pound	E4@.019	LF	.62	.76	1.38
#3/0, .1575 pounds per foot, $2.04 per pound	E4@.022	LF	.77	.89	1.66
#4/0, .1986 pounds per foot, $2.03 per pound	E4@.025	LF	.98	1.01	1.99

Copper Building Wire. Type THW and single conductor flame-retardant, moisture and heat resistant thermoplastic insulated 600 volt stranded (except as noted) copper building wire pulled in conduit. No conduit included. Labor cost assumes three bundled conductors are pulled at one time. Listed by American Wire Gauge. These costs include the subcontractor's overhead and profit. Sizes smaller than 8 gauge are type THHN flame-retardant, moisture and heat resistant thermoplastic insulated copper wire with extruded nylon jacket. Note that wire prices can change very quickly.

	Craft@Hrs	Unit	Material	Labor	Total
#14, stranded THHN-THWN	E4@.006	LF	.05	.24	.29
#14, solid THHN-THWN	E4@.006	LF	.04	.24	.28
#12, stranded THHN-THWN	E4@.007	LF	.07	.28	.35
#12 solid THHN-THWN	E4@.008	LF	.06	.32	.38
#10, stranded THHN-THWN	E4@.010	LF	.11	.40	.51
#10 solid THHN-THWN	E4@.011	LF	.09	.44	.53
#8	E4@.011	LF	.17	.44	.61
#6	E4@.012	LF	.24	.48	.72
#4	E4@.013	LF	.24	.52	.76
#3	E4@.015	LF	.48	.60	1.08
#2	E4@.017	LF	.59	.68	1.27
#1	E4@.018	LF	.81	.72	1.53
#1/0	E4@.021	LF	.95	.85	1.80
#2/0	E4@.023	LF	1.13	.93	2.06
#3/0	E4@.026	LF	1.46	1.05	2.51
#4/0	E4@.029	LF	1.81	1.17	2.98
250 MCM	E4@.032	LF	2.13	1.29	3.42
300 MCM	E4@.035	LF	2.56	1.41	3.97
350 MCM	E4@.038	LF	2.90	1.53	4.43
400 MCM	E4@.042	LF	3.42	1.69	5.11
500 MCM	E4@.046	LF	4.12	1.85	5.97

Electrical 16

	Craft@Hrs	Unit	Material	Labor	Total
600 MCM	E4@.051	LF	5.42	2.05	7.47
750 MCM	E4@.055	LF	6.65	2.21	8.86
1000 MCM	E4@.066	LF	10.20	2.66	12.86
Add for each additional conductor pulled	--	%	100.0	80.0	--
Add for type XHHW (XLP) crosslinked polyethylene thermosetting moisture and heat-resistant insulated copper building wire	--	%	5.0	--	--
Add for type THHN flame-retardant, moisture and heat resistant thermoplastic insulated copper wire with extruded nylon jacket, 8 gauge and over	--	%	10.0	--	--

Aluminum Building Wire. Type THW single conductor flame-retardant, moisture and heat resistant thermoplastic insulated 600 volt stranded aluminum building wire pulled in conduit. No conduit included. Labor cost assumes three bundled conductors are pulled at one time. Listed by American Wire Gauge. These costs include the subcontractor's overhead and profit. Note that wire prices can change very quickly.

	Craft@Hrs	Unit	Material	Labor	Total
#6	E4@.009	LF	.21	.36	.57
#4	E4@.011	LF	.26	.44	.70
#2	E4@.014	LF	.36	.56	.92
#1	E4@.016	LF	.52	.64	1.16
#1/0	E4@.018	LF	.61	.72	1.33
#2/0	E4@.020	LF	.72	.80	1.52
#3/0	E4@.022	LF	.89	.89	1.78
#4/0	E4@.025	LF	1.01	1.01	2.02
250 MCM	E4@.027	LF	1.19	1.09	2.28
300 MCM	E4@.030	LF	1.57	1.21	2.78
350 MCM	E4@.032	LF	1.69	1.29	2.98
400 MCM	E4@.035	LF	1.89	1.41	3.30
500 MCM	E4@.039	LF	2.18	1.57	3.75
600 MCM	E4@.043	LF	2.67	1.73	4.40
700 MCM	E4@.046	LF	3.10	1.85	4.95
750 MCM	E4@.047	LF	3.19	1.89	5.08
1000 MCM	E4@.055	LF	4.70	2.21	6.91
Add for each additional conductor pulled	--	%	100.0	80.0	--
Add for XHHW crosslinked polyethylene thermosetting moisture and heat resistant insulated aluminum building wire	--	%	10.0	--	--

UF Direct Burial Cable. Copper type UF and UF-NMC thermoplastic jacketed "A-Z" cable, 600 volt. No excavation or backfill included. These costs include the subcontractor's overhead and profit. Note that wire prices can change very quickly.

	Craft@Hrs	Unit	Material	Labor	Total
# 6, 1 conductor, no ground wire	E4@.005	LF	.34	.20	.54
# 4, 1 conductor, no ground wire	E4@.006	LF	.54	.24	.78
# 2, 1 conductor, no ground wire	E4@.007	LF	.76	.28	1.04
#14, 2 conductors, no ground wire	E4@.005	LF	.15	.20	.35
#12, 2 conductors, no ground wire	E4@.005	LF	.19	.20	.39
#10, 2 conductors, no ground wire	E4@.005	LF	.29	.20	.49
#14, 2 conductors, with ground wire	E4@.006	LF	.17	.24	.41
#12, 2 conductors, with ground wire	E4@.006	LF	.23	.24	.47
#10, 2 conductors, with ground wire	E4@.006	LF	.34	.24	.58
# 8, 2 conductors, with ground wire	E4@.007	LF	1.01	.28	1.29
# 6, 2 conductors, with ground wire	E4@.009	LF	1.51	.36	1.87
#14, 3 conductors, no ground wire	E4@.007	LF	.21	.28	.49
#12, 3 conductors, no ground wire	E4@.008	LF	.30	.32	.62
#10, 3 conductors, no ground wire	E4@.009	LF	.44	.36	.80
# 8, 3 conductors, no ground wire	E4@.010	LF	1.20	.40	1.60
# 6, 3 conductors, no ground wire	E4@.011	LF	1.75	.44	2.19

Electrical 16

	Craft@Hrs	Unit	Material	Labor	Total
UF Direct Burial Cable, continued					
#14, 3 conductors, with ground wire	E4@.010	LF	.24	.40	.64
#12, 3 conductors, with ground wire	E4@.010	LF	.34	.40	.74
#10, 3 conductors, with ground wire	E4@.011	LF	.49	.44	.93
# 8, 3 conductors, with ground wire	E4@.012	LF	1.32	.48	1.80
# 6, 3 conductors, with ground wire	E4@.013	LF	1.91	.52	2.43

Romex Cable. Non-metallic (NM) sheathed copper cable, 600 volt, with full size ground wire, installed in frame building, including boring out and pulling cable. These costs include the subcontractor's overhead and profit. Note that wire prices can change very quickly.

	Craft@Hrs	Unit	Material	Labor	Total
#14 wire, 2 conductor	E4@.027	LF	.12	1.09	1.21
#12 wire, 2 conductor	E4@.030	LF	.17	1.21	1.38
#10 wire, 2 conductor	E4@.031	LF	.29	1.25	1.54
# 8 wire, 2 conductor	E4@.040	LF	.62	1.61	2.23
# 6 wire, 2 conductor	E4@.045	LF	.90	1.81	2.71
#14 wire, 3 conductor	E4@.030	LF	.20	1.21	1.41
#12 wire, 3 conductor	E4@.031	LF	.28	1.25	1.53
#10 wire, 3 conductor	E4@.035	LF	.43	1.41	1.84
# 8 wire, 3 conductor	E4@.043	LF	.93	1.73	2.66
# 6 wire, 3 conductor	E4@.059	LF	1.29	2.37	3.66
# 4 wire, 3 conductor	E4@.067	LF	1.94	2.70	4.64
Deduct for no ground wire					
#14 and #12 wire	--	%	-5.0	-5.0	--
#10, #8 and #6 wire	--	%	-15.0	-5.0	--

Armored Cable. BX (type AC) copper conductors with moisture-resistant and flame-retardant cover wrapped in flexible steel cover, 600 volt, with bonding strip, installed with clamps or staples in a frame building or embedded in masonry, including approved bushings at terminations. These costs include the subcontractor's overhead and profit. Note that wire prices can change very quickly.

	Craft@Hrs	Unit	Material	Labor	Total
#14 wire, 2 conductor, solid	E4@.030	LF	.38	1.21	1.59
#12 wire, 2 conductor, solid	E4@.030	LF	.42	1.21	1.63
#10 wire, 2 conductor, solid	E4@.031	LF	.72	1.25	1.97
# 8 wire, 2 conductor, stranded	E4@.035	LF	1.25	1.41	2.66
# 6 wire, 2 conductor, stranded	E4@.039	LF	1.60	1.57	3.17
# 4 wire, 2 conductor, stranded	E4@.043	LF	2.73	1.73	4.46
#14 wire, 3 conductor, solid	E4@.031	LF	.48	1.25	1.73
#12 wire, 3 conductor, solid	E4@.035	LF	.60	1.41	2.01
#10 wire, 3 conductor, solid	E4@.039	LF	.94	1.57	2.51
# 8 wire, 3 conductor, stranded	E4@.043	LF	1.44	1.73	3.17
# 6 wire, 3 conductor, stranded	E4@.047	LF	1.57	1.89	3.46
# 4 wire, 3 conductor, stranded	E4@.051	LF	2.38	2.05	4.43
# 2 wire, 3 conductor, stranded	E4@.056	LF	3.29	2.25	5.54
#14 wire, 4 conductor, solid	E4@.035	LF	.67	1.41	2.08
#12 wire, 4 conductor, solid	E4@.038	LF	.86	1.53	2.39
#10 wire, 4 conductor, solid	E4@.043	LF	1.42	1.73	3.15
# 8 wire, 4 conductor, stranded	E4@.048	LF	2.28	1.93	4.21
# 6 wire, 4 conductor, stranded	E4@.054	LF	2.87	2.17	5.04
# 4 wire, 4 conductor, stranded	E4@.059	LF	4.11	2.37	6.48

Power Cable. Single conductor medium voltage ozone resistant stranded copper cable pulled in conduit. No conduit included. Labor cost assumes three bundled conductors are pulled at one time on runs up to 100 feet. Labor cost will be higher on longer cable pulls and about 50% lower when table is laid in an open trench. No splicing included. Listed by American Wire Gauge. These costs include the subcontractor's overhead and profit. Note that wire prices can change very quickly.

Electrical 16

	Craft@Hrs	Unit	Material	Labor	Total
5,000 volt, tape shielded, crosslinked polyethylene (XLP) insulated, with PVC jacket					
#6	E4@.014	LF	1.73	.56	2.29
#4	E4@.016	LF	1.87	.64	2.51
#2	E4@.021	LF	2.25	.85	3.10
#1/0	E4@.026	LF	3.10	1.05	4.15
#2/0	E4@.028	LF	3.54	1.13	4.67
#4/0	E4@.038	LF	4.39	1.53	5.92
250 MCM	E4@.039	LF	4.94	1.57	6.51
350 MCM	E4@.047	LF	6.85	1.89	8.74
500 MCM	E4@.053	LF	9.55	2.13	11.68
750 MCM	E4@.065	LF	12.90	2.61	15.51
Deduct for unshielded 5,000 volt power cable	--	%	-40.0	--	--
15,000 volt, tape shielded, ethylene propylene rubber (EPR) insulated, with PVC jacket					
#2	E4@.024	LF	3.17	.97	4.14
#1	E4@.025	LF	3.51	1.01	4.52
#1/0	E4@.030	LF	3.84	1.21	5.05
#2/0	E4@.032	LF	4.48	1.29	5.77
#4/0	E4@.043	LF	5.87	1.73	7.60
250 MCM	E4@.046	LF	6.64	1.85	8.49
350 MCM	E4@.055	LF	8.19	2.21	10.40
500 MCM	E4@.062	LF	9.92	2.49	12.41
750 MCM	E4@.075	LF	14.70	3.02	17.72

Snow Melting Cable. For concrete pavement. Self-regulating heater cable 208-277 volt AC, encased in concrete pavement for walkways, steps, loading ramps or parking garages. Based on Raychem ElectroMelt™ System for concrete pavement. The heating cable is cut to length at the installation site and typically laid in a serpentine pattern using 12-inch center to center spacing fastened to the top of the reinforcing using nylon cable ties before the concrete is placed. For scheduling purposes estimate that one man can lay out, cut to length, install and tie 600 LF of cable in an 8-hour day. Based on 1.1 LF of cable per SF of pavement including ties, "return bends" and waste, this yields 545 SF of pavement. These costs include the subcontractor's overhead and profit but no reinforcing, concrete or concrete placing. A rule of thumb for sizing circuit breakers required for start-up at 0 °F with heating cable voltage at 220 volts AC is to allow 0.20 amps per LF of cable.

	Craft@Hrs	Unit	Material	Labor	Total
Heater cable (1.1 LF per SF of pavement)	CE@.015	SF	5.75	.70	6.45
Power connection kit or end seal kit	CE@.500	Ea	16.70	23.20	39.90
Cable splice kit	CE@.250	Ea	21.10	11.60	32.70
Cable expansion joint kit	CE@.250	Ea	16.80	11.60	28.40
ElectroMelt™ junction box	CE@.500	Ea	91.80	23.20	115.00
System controller, automatic	CE@1.50	Ea	670.00	69.50	739.50
Thermostat	CE@.500	Ea	164.00	23.20	187.20
Ground fault protection device	CE@.969	Ea	684.00	44.90	728.90
Cable markers	CE@.250	Ea	64.10	11.60	75.70

Ice Melting Cable. For roofs, gutters and downspouts. Self-regulating heater cable 120 or 208-277 volt AC, run exposed on the surface of the roof or within gutters. Based on Raychem IceMelt™ System for roofs and gutters. The heating cable is cut to length at the installation site and typically laid in a serpentine pattern using 24-inch center to center spacing fastened to the roof using clips supplied by Raychem. For scheduling purposes estimate that one man can lay out, cut to length and install 400 LF of cable in an 8-hour day. Based on 1.6 LF of cable per SF of protected roof area including clips, "return bends" and waste, this yields 250 SF of protected area. For each LF of gutter or downspout add costs equal to one SF of roof. These costs include the subcontractor's overhead and profit but no roofing, gutters, or downspouts. A rule of thumb for sizing circuit breakers required for start-up at 0 °F with heating cable voltage at 120 volts AC is to allow 0.17 amps per LF of cable, at 220 volts AC allow 0.11 amps per LF of cable.

Electrical 16

	Craft@Hrs	Unit	Material	Labor	Total
Ice Melting Cable, continued					
Heater cable (1.5 LF per SF of protected area)	CE@.033	SF	4.28	1.53	5.81
Power connection kit	CE@.500	Ea	27.50	23.20	50.70
Cable splice kit	CE@.374	Ea	15.30	17.30	32.60
End seal kit	CE@.374	Ea	7.88	17.30	25.18
Downspout hanger	CE@.374	Ea	13.80	17.30	31.10

Electrical Outlet Boxes. Steel boxes installed on an exposed wall or ceiling. The material cost for phenolic, PVC and fiberglass boxes will be 50% to 70% less. These costs include the subcontractor's overhead, profit and fasteners but no switch or receptacle.

	Craft@Hrs	Unit	Material	Labor	Total
Square outlet boxes, 4" x 4"					
1-1/2" deep, 3/4" and 1/2" knockouts	E4@.187	Ea	2.31	7.52	9.83
2-1/8" deep, 3/4" and 1/2" knockouts	E4@.245	Ea	4.04	9.85	13.89
1-1/2" deep, with side mounting bracket	E4@.187	Ea	3.20	7.52	10.72
2-1/8" deep, with side mounting bracket	E4@.245	Ea	5.55	9.85	15.40
1-1/2" deep extension rings	E4@.120	Ea	3.11	4.83	7.94
Add for boxes with Romex or BX clamps	--	Ea	1.94	--	--
Add for 4" x 4" steel flush cover blanks	E4@.039	Ea	.58	1.57	2.15
Add for plaster rings to 3/4" deep	E4@.080	Ea	1.54	3.22	4.76
Octagon outlet boxes, 4"					
1-1/2" deep, 1/2" and 3/4" knockouts	E4@.187	Ea	1.78	7.52	9.30
1-1/2" with Romex clamps	E4@.187	Ea	2.25	7.52	9.77
1-1/2" with mounting bracket	E4@.187	Ea	2.85	7.52	10.37
2-1/8" deep, 1/2" and 3/4" knockouts	E4@.245	Ea	2.83	9.85	12.68
2-1/8" deep with Romex clamps	E4@.245	Ea	3.75	9.85	13.60
1-1/2" deep extension rings	E4@.120	Ea	2.73	4.83	7.56
1-1/2" with 21" bar set	E4@.282	Ea	3.62	11.30	14.92
3" deep concrete ring	E4@.516	Ea	4.40	20.80	25.20
Add for 4" steel flush cover blanks	E4@.039	Ea	.55	1.57	2.12
Add for plaster rings to 3/4" deep	E4@.080	Ea	2.43	3.22	5.65
Handy boxes, 4" x 2-1/8"					
1-1/2" deep, 1/2" knockouts	E4@.187	Ea	1.45	7.52	8.97
1-7/8" deep, 3/4" knockouts	E4@.187	Ea	2.06	7.52	9.58
2-1/8" deep, 3/4" knockouts	E4@.245	Ea	2.44	9.85	12.29
2-1/8" deep with side mounting bracket	E4@.245	Ea	2.81	9.85	12.66
1-1/2" deep extension rings	E4@.120	Ea	2.05	4.83	6.88
Blank or switch cover	E4@.039	Ea	.64	1.57	2.21
Weatherproof box and cover	E4@.059	Ea	3.88	2.37	6.25
Switch boxes, 3" x 2", square corner gangable boxes with mounting ears					
2" deep, 1/2" knockouts	E4@.245	Ea	1.86	9.85	11.71
2-1/2" deep, 3/4" knockouts	E4@.245	Ea	2.55	9.85	12.40
3-1/2" deep, 3/4' knockouts	E4@.324	Ea	2.54	13.00	15.54
Add for boxes with Romex or BX clamps	--	Ea	.62	--	--
Gang switch boxes, 2" x 3" x 1-13/16" deep, with cover					
2 gang	E4@.242	Ea	9.75	9.73	19.48
3 gang	E4@.342	Ea	12.10	13.80	25.90
4 gang	E4@.342	Ea	16.60	13.80	30.40
5 gang	E4@.482	Ea	22.90	19.40	42.30
6 gang	E4@.482	Ea	42.60	19.40	62.00
7 gang	E4@.604	Ea	70.00	24.30	94.30
8 gang	E4@.604	Ea	78.20	24.30	102.50
Floor boxes, watertight, cast iron, round					
4" x 3-3/4" deep, non-adjustable	E4@1.54	Ea	33.60	61.90	95.50
3-3/4" x 2" deep, semi-adjustable	E4@1.54	Ea	48.70	61.90	110.60

Electrical 16

	Craft@Hrs	Unit	Material	Labor	Total
4" x 3-3/4" deep, adjustable	E4@1.54	Ea	57.90	61.90	119.80
Round floor cover plates	E4@.156	Ea	24.10	6.27	30.37

Galvanized or grey enamel NEMA class 1 (indoor) pull boxes, with screw cover

	Craft@Hrs	Unit	Material	Labor	Total
4" x 4" x 4" deep	E4@.373	Ea	4.49	15.00	19.49
4" x 6" x 4" deep	E4@.373	Ea	5.34	15.00	20.34
6" x 6" x 4" deep	E4@.373	Ea	6.39	15.00	21.39
6" x 8" x 4" deep	E4@.391	Ea	7.53	15.70	23.23
6" x 12" x 4" deep	E4@.443	Ea	9.68	17.80	27.48
8" x 8" x 4" deep	E4@.391	Ea	8.79	15.70	24.49
8" x 10" x 4" deep	E4@.484	Ea	10.10	19.50	29.60
10" x 10" x 4" deep	E4@.492	Ea	11.60	19.80	31.40
10" x 12" x 4" deep	E4@.539	Ea	13.10	21.70	34.80
12" x 12" x 4" deep	E4@.593	Ea	14.90	23.90	38.80
12" x 15" x 4" deep	E4@.663	Ea	17.50	26.70	44.20
15" x 18" x 4" deep	E4@.785	Ea	23.60	31.60	55.20
Add for 6" deep boxes	--	%	15.0	15.0	--

Cast aluminum NEMA class 3R (weatherproof) screw cover junction boxes, flanged, surface mounted, with cover

	Craft@Hrs	Unit	Material	Labor	Total
6" x 6" x 4" deep	E4@.930	Ea	71.70	37.40	109.10
6" x 8" x 4" deep	E4@.930	Ea	90.80	37.40	128.20
6" x 12" x 6" deep	E4@1.42	Ea	171.00	57.10	228.10
8" x 8" x 4" deep	E4@.982	Ea	123.00	39.50	162.50
8" x 12" x 6" deep	E4@1.30	Ea	195.00	52.30	247.30
12" x 12" x 6" deep	E4@1.35	Ea	282.00	54.30	336.30
12" x 12" x 8" deep	E4@1.73	Ea	323.00	69.60	392.60
12" x 24" x 6" deep	E4@1.97	Ea	715.00	79.20	794.20
18" x 36" x 8" deep	E4@2.56	Ea	1,870.00	103.00	1,973.00
Add for recessed cover, flush mount boxes	--	%	5.0	--	--

Hinged cover panel enclosures, NEMA 1, enamel, with cover

	Craft@Hrs	Unit	Material	Labor	Total
16" x 12" x 6" deep	E4@1.30	Ea	65.60	52.30	117.90
20" x 20" x 6" deep	E4@1.30	Ea	93.00	52.30	145.30
30" x 20" x 6" deep	E4@1.77	Ea	118.00	71.20	189.20
24" x 20" x 8" deep	E4@1.30	Ea	112.00	52.30	164.30
30" x 24" x 8" deep	E4@1.77	Ea	143.00	71.20	214.20
36" x 30" x 8" deep	E4@1.77	Ea	188.00	71.20	259.20

Electrical Receptacles. Standard commercial grade ivory or brown receptacles with cover and screws. White or gray receptacles will cost about 10% more. No outlet boxes included. Labor includes connecting wire, securing device in the box, and attaching cover. These costs include the subcontractor's overhead and profit.

15 amp, 125 volt self-grounding, two pole, three wire duplex receptacles

	Craft@Hrs	Unit	Material	Labor	Total
Minimum grade, screwless, not self-grounding	CE@.162	Ea	5.28	7.51	12.79
Side terminals	CE@.162	Ea	5.58	7.51	13.09
Back and side terminals	CE@.162	Ea	8.06	7.51	15.57
Feed thru wiring, back and side terminals	CE@.162	Ea	12.80	7.51	20.31
Safety ground, side terminals	CE@.162	Ea	17.30	7.51	24.81
Hospital grade	CE@.162	Ea	12.30	7.51	19.81
Ground fault circuit interrupter receptacle	CE@.324	Ea	21.30	15.00	36.30
Add for NEMA 5 single receptacles	--	%	15.0	--	--

20 amp, 125 volt self-grounding, two pole, three wire duplex receptacles

	Craft@Hrs	Unit	Material	Labor	Total
Side terminals	CE@.162	Ea	8.06	7.51	15.57
Back and side terminals	CE@.162	Ea	11.20	7.51	18.71
Feed thru wiring, back and side terminals	CE@.162	Ea	15.10	7.51	22.61

Electrical 16

	Craft@Hrs	Unit	Material	Labor	Total
Electrical Receptacles, continued					
Hospital grade	CE@.162	Ea	13.80	7.51	21.31
Ground fault circuit interrupter receptacle	CE@.324	Ea	37.60	15.00	52.60
Add for NEMA 5 single receptacles	--	%	15.0	--	--
250 volt receptacles, self-grounding, 2 pole, 3 wire, back & side terminals					
15 amp, duplex	CE@.187	Ea	8.78	8.66	17.44
15 amp, duplex, feed thru wiring	CE@.187	Ea	14.60	8.66	23.26
15 amp, single	CE@.187	Ea	17.30	8.66	25.96
20 amp, duplex	CE@.187	Ea	11.60	8.66	20.26
20 amp, duplex, feed thru wiring	CE@.187	Ea	17.90	8.66	26.56
20 amp, single	CE@.187	Ea	13.60	8.66	22.26
Clock receptacle, 2 pole, 15 amp, 125 volt	CE@.329	Ea	15.20	15.20	30.40
120/208 volt 20 amp 4 pole duplex receptacle	CE@.795	Ea	21.00	36.80	57.80
125/250 volt 3 pole receptacles, flush mount					
15 amp/10 amp	CE@.714	Ea	14.90	33.10	48.00
20 amp	CE@.820	Ea	15.20	38.00	53.20
277 volt 50 amp 2 pole receptacle	CE@.268	Ea	21.50	12.40	33.90
Dryer receptacle, 250 volt, 30/50 amp, 3 wire	CE@.536	Ea	24.00	24.80	48.80
Accessories for 50 amp dryer receptacles					
Straight arm grip plug	CE@.288	Ea	30.50	13.30	43.80
Plastic plug with cap, angle	CE@.288	Ea	43.90	13.30	57.20
Cord sets for dryer receptacles					
36 inch, three #10 wires	CE@.288	Ea	11.70	13.30	25.00
48 inch, three #10 wires	CE@.288	Ea	13.20	13.30	26.50
60 inch, three #10 wires	CE@.288	Ea	22.70	13.30	36.00

Electrical Switches. Commercial grade, 120 to 277 volt rating, ivory or brown. Add 10% for white or gray. Includes cover plate but no fixture boxes except as noted. Labor includes connecting wire to switch, securing switch in box and attaching cover plate. These costs include the subcontractor's overhead and profit.

	Craft@Hrs	Unit	Material	Labor	Total
15 amp switches, back and side wired					
One pole switch	CE@.112	Ea	8.30	5.19	13.49
Two pole switch	CE@.309	Ea	13.60	14.30	27.90
Three-way switch	CE@.227	Ea	11.90	10.50	22.40
Four-way switch	CE@.309	Ea	28.40	14.30	42.70
20 amp switches, back and side wired					
One pole switch	CE@.187	Ea	12.70	8.66	21.36
Two pole switch	CE@.433	Ea	14.50	20.10	34.60
Three-way switch	CE@.291	Ea	13.60	13.50	27.10
Four-way switch	CE@.435	Ea	32.70	20.20	52.90
30 amp switches, side wired					
One pole switch	CE@.246	Ea	18.40	11.40	29.80
Two pole switch	CE@.475	Ea	23.50	22.00	45.50
Three-way switch	CE@.372	Ea	22.70	17.20	39.90
Four-way switch	CE@.475	Ea	33.10	22.00	55.10
20 amp weatherproof switches, lever handle, with cover					
One pole switch	CE@.187	Ea	9.59	8.66	18.25
Two pole switch	CE@.433	Ea	14.10	20.10	34.20
Three-way switch	CE@.291	Ea	9.41	13.50	22.91
Single pole, two gang, 10 amp	CE@.358	Ea	14.50	16.60	31.10
Dimmer switches, push for off					
600 watt, one pole	CE@.417	Ea	7.60	19.30	26.90
600 watt, three way	CE@.626	Ea	12.30	29.00	41.30
Fluorescent dimmer, 10 lamp load	CE@.626	Ea	44.80	29.00	73.80

Electrical 16

	Craft@Hrs	Unit	Material	Labor	Total
Astro dial time switch, 40 amp, with box	CE@3.00	Ea	139.00	139.00	278.00
15 minute timer switch, wall box mounted	CE@.426	Ea	12.80	19.70	32.50
Single pole, 1 throw time switch, 277 volt	CE@.890	Ea	62.90	41.20	104.10
Float switches					
Low level float switch	CE@1.03	Ea	43.70	47.70	91.40
120 volt, mercury contacts, cords, mounts, breaker and rods	CE@1.68	Ea	146.00	77.80	223.80
460 volt, stainless steel floats, with mounts	CE@4.60	Ea	907.00	213.00	1,120.00
Lighting contactors, three pole					
20 amp	CE@1.50	Ea	114.00	69.50	183.50
30 amp	CE@1.71	Ea	121.00	79.20	200.20
60 amp	CE@2.03	Ea	242.00	94.10	336.10
100 amp	CE@2.31	Ea	403.00	107.00	510.00
One way 15 amp toggle switch with neon pilot	CE@.327	Ea	8.88	15.20	24.08
Three way 15 amp toggle switch with neon pilot	CE@.372	Ea	14.30	17.20	31.50
Photoelectric switches, flush, with wall plate					
120 volt, 1,000 watt	CE@.426	Ea	15.30	19.70	35.00
208 volt, 1,800 watt	CE@.626	Ea	19.60	29.00	48.60
480 volt, 3,000 watt	CE@.795	Ea	27.60	36.80	64.40
210-250 volt, 3,000 watt, 2 pole, 2 throw	CE@.809	Ea	132.00	37.50	169.50
Button control stations, surface mounted, NEMA class 1, standard duty, 120/240 volt, with enclosure					
Start-stop switch, 2 button	CE@.624	Ea	22.90	28.90	51.80
Start-stop switch with lockout, 2 button	CE@.624	Ea	36.60	28.90	65.50
Forward, reverse and stop buttons, 3 button	CE@.624	Ea	48.70	28.90	77.60
Forward, reverse, stop and lockout, 3 button	CE@.624	Ea	51.80	28.90	80.70
Manual toggle starter switches, surface mounted, NEMA class 1, 120/240 volt, non-reversing, with enclosure					
Size 0 motors, 2 pole	CE@.613	Ea	124.00	28.40	152.40
Size 1 motors, 3 pole	CE@.698	Ea	158.00	32.30	190.30
Size 1P motors, 2 pole	CE@.901	Ea	185.00	41.70	226.70
Manual button starter switches, surface mounted, NEMA class 1, 110 to 240 volt, 2 pole, 1 phase, with enclosure, start, stop, reset, with relay					
Size 00 motors	CE@.712	Ea	141.00	33.00	174.00
Size 0 motors	CE@.712	Ea	155.00	33.00	188.00
Size 1 motors	CE@.820	Ea	174.00	38.00	212.00
Size 1-1/2 motors	CE@.975	Ea	216.00	45.20	261.20

Grounding Devices. No excavation or concrete included. These costs include the subcontractor's overhead and profit.

	Craft@Hrs	Unit	Material	Labor	Total
Copper clad grounding rods, driven					
5/8" x 8'	E4@.365	Ea	12.80	14.70	27.50
5/8" x 10'	E4@.420	Ea	16.60	16.90	33.50
3/4" x 8'	E4@.365	Ea	20.70	14.70	35.40
3/4" x 10'	E4@.420	Ea	25.60	16.90	42.50
Ground rod clamps					
5/8"	E4@.249	Ea	3.08	10.00	13.08
3/4"	E4@.249	Ea	4.40	10.00	14.40
Coupling for threaded ground rod, 5/8"	E4@.249	Ea	7.20	10.00	17.20
Static discharge ground reel with 50' of nylon covered cable	E4@1.48	Ea	128.00	59.50	187.50
Copper bus bar, 2" x 1/4"	E4@.225	LF	13.70	9.05	22.75
Copper braid (1" x 1/8" x 4") used as door ground	E4@.295	Ea	1.86	11.90	13.76
Copper bonding connector strap, 3/4" x 1/8" x 10"	E4@.738	Ea	3.55	29.70	33.25

Electrical 16

	Craft@Hrs	Unit	Material	Labor	Total

Grounding Devices, continued
Brazed connections for wire

#6 wire	E4@.194	Ea	.57	7.80	8.37
#2 wire	E4@.194	Ea	.83	7.80	8.63
#2/0 wire	E4@.295	Ea	1.36	11.90	13.26
#4/0 wire	E4@.391	Ea	2.26	15.70	17.96
Fusion welded connection to ground	E4@.982	Ea	5.46	39.50	44.96

Grounding at a cold water main and with a 5/8" x 8' copper clad ground rod, includes 50' grounding conductor, clamps, connectors, EMT conduit, bushing, and rod

100 amp service, #8 copper wire	E4@1.13	Ea	35.80	45.40	81.20
150 amp service, #6 copper wire	E4@1.16	Ea	39.00	46.70	85.70

Electric Motors. General purpose, 3 phase, open drip-proof, industrial duty, 1,740 RPM, 208-230/460 volt, rigid welded or solid base, furnished and placed, no hookup included. These costs include the subcontractor's overhead and profit. Dayton

1/2 HP	E4@1.99	Ea	187.00	80.00	267.00
3/4 HP	E4@1.99	Ea	201.00	80.00	281.00
1 HP	E4@1.99	Ea	166.00	80.00	246.00
1-1/2 HP	E4@1.99	Ea	187.00	80.00	267.00
2 HP	E4@1.99	Ea	202.00	80.00	282.00
3 HP	E4@1.99	Ea	219.00	80.00	299.00
5 HP	E4@2.22	Ea	243.00	89.30	332.30
7-1/2 HP	E4@2.40	Ea	344.00	96.50	440.50
10 HP	E4@2.51	Ea	423.00	101.00	524.00
15 HP	E4@2.57	Ea	527.00	103.00	630.00
20 HP	E4@2.85	Ea	656.00	115.00	771.00
25 HP	E4@3.19	Ea	830.00	128.00	958.00

Electric Motor Connection. Includes a fusible indoor heavy duty safety switch, junction box and cover, connectors, fittings, ells, adapters, supports, anchors and conduit with wire appropriate for the load connected but no starter or motor. These costs include the subcontractor's overhead and profit.

Single phase, using 10' RSC, 20' PVC and 6' flex conduit, NEMA class 1 (indoor) switch, 230 volts

For motor to 2 HP, 20 amps	E4@6.35	Ea	79.40	255.00	334.40
For 2.5 to 3 HP motor, 30 amps	E4@6.63	Ea	90.70	267.00	357.70
For 3 to 7.5 HP motor, 45 amps	E4@6.68	Ea	134.00	269.00	403.00
For 8 to 15 HP motor, 90 amps	E4@13.4	Ea	302.00	539.00	841.00

Three phase wiring using 10' RSC, 20' PVC and 6' flex conduit, NEMA class 1 (indoor) switch, 600 volts

For motor to 2 HP, 20 amps	E4@7.18	Ea	196.00	289.00	485.00
For 2 to 5 HP motor, 30 amps	E4@7.90	Ea	200.00	318.00	518.00
For 7 to 25 HP motor, 45 amps	E4@8.30	Ea	253.00	334.00	587.00
For 25 to 50 HP motor, 90 amps	E4@9.30	Ea	517.00	374.00	891.00
For 60 to 100 HP motor, 135 amps	E4@14.1	Ea	784.00	567.00	1,351.00
For 125 to 200 HP motor, 270 amps	E4@21.9	Ea	2,000.00	881.00	2,881.00

Single phase, using 10' RSC, 20' PVC and 6' flex conduit, NEMA class 3R (weatherproof) switch, 230 volts

For motor to 2 HP, 20 amps	E4@6.63	Ea	130.00	267.00	397.00
For 2 to 5 HP motor, 30 amps	E4@6.97	Ea	134.00	280.00	414.00
For 5 to 7.5 HP motor, 45 amps	E4@6.94	Ea	194.00	279.00	473.00

Three phase wiring using 10' RSC, 20' PVC and 6' flex conduit, NEMA class 3R (weatherproof) switch, 600 volts

For motor to 1 HP motor, 20 amps	E4@7.10	Ea	219.00	286.00	505.00
For 1 to 10 HP motor, 30 amps	E4@7.25	Ea	223.00	292.00	515.00
For 10 to 25 HP motor, 45 amps	E4@9.51	Ea	320.00	382.00	702.00

Electrical 16

	Craft@Hrs	Unit	Material	Labor	Total
For 25 to 50 HP motor, 90 amps	E4@9.51	Ea	532.00	382.00	914.00
100 HP					
For 60 to 100 HP motor, 135 amps	E4@12.1	Ea	790.00	487.00	1,277.00
For 125 to 200 HP motor, 270 amps	E4@21.6	Ea	1,770.00	869.00	2,639.00

Complete installation including magnetic motor starter, overload relay, all wiring, a heavy duty fusible safety switch, junction box and cover, connectors, fittings, adapters, supports, anchors and conduit appropriate for the load connected. No motor included. These costs include the subcontractor's overhead and profit.

Single phase, using 10' RSC, 20' PVC and 6' flex conduit, NEMA class 1 (indoor) starter, 230 volts

	Craft@Hrs	Unit	Material	Labor	Total
For motor to 2 HP, 20 amps	E4@8.11	Ea	298.00	326.00	624.00
For 2 to 3 HP motor, 30 amps	E4@8.32	Ea	322.00	335.00	657.00
For 3 to 8 HP motor, 45 amps	E4@8.45	Ea	518.00	340.00	858.00
For 8 to 15 HP motor, 90 amps	E4@15.5	Ea	925.00	623.00	1,548.00

Three phase wiring using 10' RSC, 20' PVC and 6' flex conduit, NEMA class 1 (indoor) starter, 600 volts

	Craft@Hrs	Unit	Material	Labor	Total
For motor to 2 HP, 20 amps	E4@9.90	Ea	570.00	398.00	968.00
For 2 to 5 HP motor, 30 amps	E4@10.6	Ea	595.00	426.00	1,021.00
For 7 to 25 HP motor, 45 amps	E4@20.5	Ea	647.00	825.00	1,472.00
For 25 to 50 HP motor, 90 amps	E4@23.1	Ea	1,150.00	929.00	2,079.00
For 60 to 100 HP motor, 135 amps	E4@26.9	Ea	1,780.00	1,080.00	2,860.00
For 125 to 200 HP motor, 270 amps	E4@34.5	Ea	3,900.00	1,390.00	5,290.00

Single phase, using 10' RSC, 20' PVC and 6' flex conduit, NEMA class 3R (weatherproof) starter, 230 volts

	Craft@Hrs	Unit	Material	Labor	Total
For motor to 2 HP, 20 amps	E4@8.91	Ea	370.00	358.00	728.00
For 2 to 5 HP motor, 30 amps	E4@9.15	Ea	404.00	368.00	772.00
For 5 to 7.5 HP motor, 45 amps	E4@9.25	Ea	705.00	372.00	1,077.00

Three phase wiring using 10' RSC, 20' PVC and 6' flex conduit, NEMA class 3R (weatherproof) starter, 600 volts

	Craft@Hrs	Unit	Material	Labor	Total
For motor to 5 HP, 20 amps	E4@10.9	Ea	751.00	438.00	1,189.00
For 5 to 10 HP motor, 30 amps	E4@11.7	Ea	784.00	471.00	1,255.00
For 10 to 25 HP motor, 45 amps	E4@22.5	Ea	884.00	905.00	1,789.00
For 25 to 50 HP motor, 90 amps	E4@25.4	Ea	1,460.00	1,020.00	2,480.00
For 60 to 100 HP motor, 135 amps	E4@29.0	Ea	2,380.00	1,170.00	3,550.00
For 125 to 200 HP motor, 270 amps	E4@37.3	Ea	5,360.00	1,500.00	6,860.00

Motor Starters. Magnetic operated full voltage non-reversing motor controllers with thermal overload relays and enclosure. These costs include the subcontractor's overhead and profit.

Two pole, 1 phase contactors, NEMA class 1 (indoor), electrically held with holding interlock, one reset only

	Craft@Hrs	Unit	Material	Labor	Total
Size 00, 1 HP, 9 amp	E4@1.47	Ea	98.00	59.10	157.10
Size 0, 2 HP, 18 amp	E4@1.62	Ea	124.00	65.20	189.20
Size 1, 3 HP, 27 amp	E4@1.73	Ea	146.00	69.60	215.60
Size 2, 7-1/2 HP, 45 amp	E4@1.73	Ea	300.00	69.60	369.60
Size 3, 15 HP, 90 amp	E4@2.02	Ea	493.00	81.20	574.20
Size 4, 20 HP, 135 amp	E4@2.17	Ea	1,160.00	87.30	1,247.30

Three pole polyphase NEMA class 1 (indoor) AC magnetic combination starters with fusible disconnect and overload relays but no heaters, 208 to 240 volts

	Craft@Hrs	Unit	Material	Labor	Total
NEMA size 0, 3 HP, 30 amps	E4@2.75	Ea	395.00	111.00	506.00
NEMA size 1, 5 HP, 30 amps	E4@3.89	Ea	416.00	156.00	572.00
NEMA size 1, 5 HP, 60 amps	E4@3.89	Ea	424.00	156.00	580.00
NEMA size 2, 10 HP, 60 amps	E4@5.34	Ea	655.00	215.00	870.00
NEMA size 2, 20 HP, 100 amps	E4@5.34	Ea	705.00	215.00	920.00
NEMA size 3, 20 HP, 100 amps	E4@6.76	Ea	1,100.00	272.00	1,372.00
NEMA size 3, 40 HP, 200 amps	E4@6.76	Ea	1,200.00	272.00	1,472.00

Electrical 16

	Craft@Hrs	Unit	Material	Labor	Total

Motor Starters, Three pole polyphase NEMA class 1, continued

	Craft@Hrs	Unit	Material	Labor	Total
NEMA size 4, 30 HP, 200 amps	E4@12.0	Ea	2,120.00	483.00	2,603.00
NEMA size 4, 75 HP, 400 amps	E4@12.0	Ea	2,330.00	483.00	2,813.00

Accessories for any motor starter on preceding page

	Craft@Hrs	Unit	Material	Labor	Total
Add for 440 to 600 volt starters	--	%	2.0	--	--
Add for NEMA type 3R (rainproof) enclosure	--	%	40.0	10.0	--
Add for NEMA type 4 (waterproof) enclosure	--	%	40.0	20.0	--
Add for NEMA type 12 (dust-tight) enclosure	--	%	15.0	15.0	--
Add for starters in an oversize enclosure	--	%	21.0	--	--
Deduct for starters with circuit breakers	--	%	-4.0	--	--

On and off switches for starter enclosures

	Craft@Hrs	Unit	Material	Labor	Total
Switch kit without pilot light	E4@.744	Ea	22.90	29.90	52.80
Switch kit with pilot light	E4@.744	Ea	34.30	29.90	64.20

Safety Switches. Wall mounted switches with enclosures as noted. No fuses or hubs included. These costs include the subcontractor's overhead and profit.

Heavy duty (NEMA-1) 600 volt 2, 3 or 4 pole fusible safety switches

	Craft@Hrs	Unit	Material	Labor	Total
30 amp	E4@3.06	Ea	137.00	123.00	260.00
60 amp	E4@3.89	Ea	164.00	156.00	320.00
100 amp	E4@4.20	Ea	305.00	169.00	474.00
200 amp	E4@6.92	Ea	443.00	278.00	721.00
400 amp	E4@11.4	Ea	1,320.00	459.00	1,779.00
600 amp	E4@14.3	Ea	1,940.00	575.00	2,515.00
800 amp	E4@19.6	Ea	3,350.00	788.00	4,138.00
1200 amp	E4@22.7	Ea	4,410.00	913.00	5,323.00

Heavy duty rainproof (NEMA-3R) 600 volt 2, 3 or 4 pole fusible safety switches

	Craft@Hrs	Unit	Material	Labor	Total
30 amp	E4@3.32	Ea	160.00	134.00	294.00
60 amp	E4@4.61	Ea	231.00	185.00	416.00
100 amp	E4@4.77	Ea	334.00	192.00	526.00
200 amp	E4@7.51	Ea	449.00	302.00	751.00
400 amp	E4@12.4	Ea	1,090.00	499.00	1,589.00
600 amp	E4@15.6	Ea	2,400.00	627.00	3,027.00
800 amp	E4@21.4	Ea	4,310.00	861.00	5,171.00
1200 amp	E4@25.8	Ea	5,810.00	1,040.00	6,850.00

Heavy duty watertight (NEMA-4) 600 volt 3 pole, 4 wire fusible safety switches

	Craft@Hrs	Unit	Material	Labor	Total
30 amp	E4@3.76	Ea	577.00	151.00	728.00
60 amp	E4@5.21	Ea	635.00	210.00	845.00
100 amp	E4@5.34	Ea	1,270.00	215.00	1,485.00
200 amp	E4@8.52	Ea	1,780.00	343.00	2,123.00
400 amp	E4@13.7	Ea	3,380.00	551.00	3,931.00
600 amp	E4@17.0	Ea	4,840.00	684.00	5,524.00

Heavy duty dust-tight (NEMA-12) 600 volt 2 pole or 3 pole solid neutral fusible safety switches

	Craft@Hrs	Unit	Material	Labor	Total
30 amp	E4@3.32	Ea	228.00	134.00	362.00
60 amp	E4@4.20	Ea	237.00	169.00	406.00
100 amp	E4@4.48	Ea	391.00	180.00	571.00
200 amp	E4@7.51	Ea	552.00	302.00	854.00
400 amp	E4@12.7	Ea	1,240.00	511.00	1,751.00
600 amp	E4@15.9	Ea	1,940.00	639.00	2,579.00

General duty (NEMA-1) 240 volt 3 pole, 4 wire non-fusible safety switches

	Craft@Hrs	Unit	Material	Labor	Total
30 amp	E4@2.58	Ea	29.70	104.00	133.70
60 amp	E4@3.45	Ea	50.70	139.00	189.70
100 amp	E4@3.60	Ea	104.00	145.00	249.00
200 amp	E4@6.50	Ea	216.00	261.00	477.00
400 amp	E4@10.8	Ea	642.00	434.00	1,076.00

Electrical 16

	Craft@Hrs	Unit	Material	Labor	Total
600 amp	E4@12.7	Ea	1,230.00	511.00	1,741.00
General duty rainproof (NEMA-3R) 240 volt 3 pole, 4 wire non-fusible safety switches					
30 amp	E4@2.88	Ea	73.40	116.00	189.40
60 amp	E4@3.76	Ea	111.00	151.00	262.00
100 amp	E4@3.89	Ea	204.00	156.00	360.00
200 amp	E4@7.23	Ea	368.00	291.00	659.00
400 amp	E4@11.8	Ea	915.00	475.00	1,390.00
600 amp	E4@14.0	Ea	1,820.00	563.00	2,383.00
Heavy duty watertight (NEMA-4) 240 volt 3 pole, 4 wire non-fusible safety switches					
30 amp	E4@3.19	Ea	510.00	128.00	638.00
60 amp	E4@3.76	Ea	606.00	151.00	757.00
100 amp	E4@4.33	Ea	1,240.00	174.00	1,414.00
200 amp	E4@7.80	Ea	1,680.00	314.00	1,994.00
400 amp	E4@13.0	Ea	3,510.00	523.00	4,033.00
600 amp	E4@15.4	Ea	4,960.00	619.00	5,579.00
Heavy duty dust-tight (NEMA-12) 240 volt 3 pole, 4 wire non-fusible safety switches					
30 amp	E4@2.88	Ea	149.00	116.00	265.00
60 amp	E4@3.76	Ea	183.00	151.00	334.00
100 amp	E4@4.04	Ea	264.00	162.00	426.00
200 amp	E4@7.23	Ea	354.00	291.00	645.00
400 amp	E4@11.8	Ea	912.00	475.00	1,387.00
600 amp	E4@14.0	Ea	1,460.00	563.00	2,023.00
Add for conduit hubs					
3/4" to 1-1/2", to 100 amp	--	Ea	7.11	--	--
2", 200 amp	--	Ea	12.50	--	--
2-1/2", 200 amp	--	Ea	20.80	--	--

Service Entrance and Distribution Switchboards By amps of rated capacity. Use these figures to estimate the cost of service entrance and distribution switchgear when complete plans are not available. These costs include the subcontractor's overhead and profit.

Light commercial switchgear, costs are per amp of rated capacity

		Unit			Total
Service side enclosure with meter set	--	Amp	--	--	1.80
Pull section	--	Amp	--	--	.70
Fire alarm breaker	--	Amp	--	--	.45
Main disconnect	--	Amp	--	--	1.60
Distribution side, including breakers equal to 150% of the main breaker capacity	--	Amp	--	--	5.30
Total cost for light commercial switchgear	--	Amp	--	--	9.85
Commercial and industrial switchgear, costs are per amp of rated capacity					
Service side enclosure with meter set	--	Amp	--	--	5.75
Pull section	--	Amp	--	--	1.50
Fire alarm breaker	--	Amp	--	--	.50
Main disconnect	--	Amp	--	--	2.40
Distribution side, including breakers equal to 150% of the main breaker capacity	--	Amp	--	--	5.80
Total cost, commercial and industrial switchgear	--	Amp	--	--	15.90

Service Entrance and Distribution Switchboards. NEMA Class 1 indoor, 600 volt, 3 phase, 4 wire, for 240/480 volt insulated case main breakers. Basic structure is 90" high by 21" deep. Width varies with equipment capacity. Labor cost includes setting and leveling on a prepared pad but excludes the pad cost. Add breaker, instrumentation and accessory costs for a complete installation. Based on Westinghouse Pow-R-Gear. These switchboards are custom designed for each installation. Costs can vary widely. Multiple units ordered at the same time may reduce costs per unit by 25% or more. These costs include the subcontractor's overhead and profit. See prices on the following page.

Electrical 16

	Craft@Hrs	Unit	Material	Labor	Total

Service Entrance and Distribution Switchboards, continued

	Craft@Hrs	Unit	Material	Labor	Total
600 amp bus	E4@22.9	Ea	2,750.00	921.00	3,671.00
1,000 amp bus	E4@22.9	Ea	3,260.00	921.00	4,181.00
1,200 amp bus	E4@22.9	Ea	3,480.00	921.00	4,401.00
1,600 amp bus	E4@22.9	Ea	3,990.00	921.00	4,911.00
2,000 amp bus	E4@22.9	Ea	4,440.00	921.00	5,361.00
2,500 amp bus	E4@22.9	Ea	5,330.00	921.00	6,251.00
3,000 amp bus	E4@22.9	Ea	6,110.00	921.00	7,031.00
4,000 amp bus	E4@22.9	Ea	7,860.00	921.00	8,781.00

240/480 volt draw-out main breakers, 100K amp interrupt capacity. Includes connecting and testing breakers only. Based on Pow-R-Trip

	Craft@Hrs	Unit	Material	Labor	Total
100 to 250 amp, manual operation	E4@10.2	Ea	8,110.00	410.00	8,520.00
300 to 800 amp, manual operation	E4@10.2	Ea	8,110.00	410.00	8,520.00
1,000 to 1,600 amp, manual operation	E4@17.5	Ea	17,100.00	704.00	17,804.00
2,000 amp, manual operation	E4@23.2	Ea	21,000.00	933.00	21,933.00
2,500 amp, manual operation	E4@23.2	Ea	35,400.00	933.00	36,333.00
3,000 amp, manual operation	E4@23.2	Ea	39,600.00	933.00	40,533.00
4,000 amp, manual operation	E4@23.2	Ea	47,500.00	933.00	48,433.00
100 to 250 amp electric operation	E4@10.2	Ea	11,100.00	410.00	11,510.00
300 to 800 amp electric operation	E4@10.2	Ea	11,100.00	410.00	11,510.00
1,000 to 1,600 amp electric operation	E4@17.5	Ea	18,700.00	704.00	19,404.00
2,000 amp electric operation	E4@23.2	Ea	24,700.00	933.00	25,633.00
2,500 amp electric operation	E4@23.2	Ea	40,400.00	933.00	41,333.00
3,000 amp electric operation	E4@23.2	Ea	43,500.00	933.00	44,433.00
4,000 amp electric operation	E4@23.2	Ea	62,000.00	933.00	62,933.00

Additional space for future draw-out breakers

	Craft@Hrs	Unit	Material	Labor	Total
100 to 250 amp	--	Ea	2,180.00	--	--
300 to 800 amp	--	Ea	2,180.00	--	--
1,000 to 1,600 amp	--	Ea	3,170.00	--	--
2,000 amp	--	Ea	3,560.00	--	--
2,500 amp	--	Ea	3,800.00	--	--
3,000 amp	--	Ea	6,930.00	--	--
4,000 amp	--	Ea	9,590.00	--	--

Service Entrance and Distribution Switchboards, Weatherproof. 600 volt, 3 phase, 4 wire, for 480 volt insulated case main breakers. Basic structure is 90" high by 21" deep. Width varies with equipment capacity. Labor cost includes setting and leveling on a prepared pad but excludes the pad cost. Add breaker, instrumentation and accessory costs for a complete installation. These switchboards are custom designed for each installation. Costs can vary widely. Multiple units ordered at the same time can reduce costs per unit by 25% or more. These costs include the subcontractor's overhead and profit.

	Craft@Hrs	Unit	Material	Labor	Total
600 amp bus	E4@23.2	Ea	3,310.00	933.00	4,243.00
1,000 amp bus	E4@23.2	Ea	3,870.00	933.00	4,803.00
1,200 amp bus	E4@23.2	Ea	4,120.00	933.00	5,053.00
1,600 amp bus	E4@23.2	Ea	4,770.00	933.00	5,703.00
2,000 amp bus	E4@23.2	Ea	5,290.00	933.00	6,223.00
2,500 amp bus	E4@23.2	Ea	6,370.00	933.00	7,303.00
3,000 amp bus	E4@23.2	Ea	6,470.00	933.00	7,403.00
4,000 amp bus	E4@23.2	Ea	9,430.00	933.00	10,363.00

480 volt manual operated draw-out breakers. By amp interrupt capacity (AIC) as noted. Includes connecting and testing factory installed breakers only

	Craft@Hrs	Unit	Material	Labor	Total
50 to 800 amp, 30K AIC	E4@10.2	Ea	11,200.00	410.00	11,610.00
50 to 1,600 amp, 50K AIC	E4@17.5	Ea	23,500.00	704.00	24,204.00
2,000 amp	E4@23.2	Ea	29,400.00	933.00	30,333.00
1,200 to 3,200 amp	E4@23.2	Ea	51,700.00	933.00	52,633.00

Electrical 16

	Craft@Hrs	Unit	Material	Labor	Total
4,000 amp	E4@23.2	Ea	82,300.00	933.00	83,233.00

480 volt manual operated draw-out breakers. 200K amp interrupt capacity (AIC). Includes connecting and testing factory installed breakers only

	Craft@Hrs	Unit	Material	Labor	Total
50 to 800 amp	E4@10.2	Ea	16,500.00	410.00	16,910.00
50 to 1,600 amp	E4@17.5	Ea	29,600.00	704.00	30,304.00
1,200 to 3,200 amp	E4@23.2	Ea	75,000.00	933.00	75,933.00
4,000 amp	E4@23.2	Ea	121,000.00	933.00	121,933.00

480 volt electrically operated draw-out breakers. By amp interrupt capacity (AIC) as noted. Includes connecting and testing factory installed breakers only

	Craft@Hrs	Unit	Material	Labor	Total
50 to 800 amp, 30K AIC	E4@10.2	Ea	14,800.00	410.00	15,210.00
50 to 1,600 amp, 50K AIC	E4@17.5	Ea	30,500.00	704.00	31,204.00
2,000 amp	E4@23.2	Ea	38,900.00	933.00	39,833.00
1,200 to 3,200 amp	E4@23.2	Ea	58,200.00	933.00	59,133.00
4,000 amp	E4@23.2	Ea	89,800.00	933.00	90,733.00

480 volt electrically operated draw-out breakers. 200K amp interrupt capacity (AIC). Includes connecting and testing factory installed breakers only

	Craft@Hrs	Unit	Material	Labor	Total
50 to 800 amp	E4@10.2	Ea	20,100.00	410.00	20,510.00
50 to 1,600 amp	E4@17.5	Ea	40,500.00	704.00	41,204.00
1,200 to 3,200 amp	E4@23.2	Ea	83,800.00	933.00	84,733.00
4,000 amp	E4@23.2	Ea	121,000.00	933.00	121,933.00

Additional space for future draw-out breakers

	Craft@Hrs	Unit	Material	Labor	Total
50 to 800 amp	--	Ea	2,560.00	--	--
50 to 1,600 amp	--	Ea	4,120.00	--	--
2,000 amp	--	Ea	5,660.00	--	--
1,200 to 3,200 amp	--	Ea	8,810.00	--	--
4,000 amp	--	Ea	16,200.00	--	--

Switchboard Instrumentation and Accessories. Add these costs to the cost of the basic enclosure and breakers to find the complete switchboard cost. These figures are for factory-installed instrumentation and accessories and include the cost of connecting and testing only.

Bus duct connection, 3 phase, 4 wire

	Craft@Hrs	Unit	Material	Labor	Total
225 amp	CE@.516	Ea	483.00	23.90	506.90
400 amp	CE@.516	Ea	561.00	23.90	584.90
600 amp	CE@.516	Ea	684.00	23.90	707.90
800 amp	CE@.516	Ea	1,090.00	23.90	1,113.90
1,000 amp	CE@.516	Ea	1,230.00	23.90	1,253.90
1,350 amp	CE@1.01	Ea	1,610.00	46.80	1,656.80
1,600 amp	CE@1.01	Ea	1,870.00	46.80	1,916.80
2,000 amp	CE@1.01	Ea	2,290.00	46.80	2,336.80
2,500 amp	CE@1.01	Ea	2,740.00	46.80	2,786.80
3,000 amp	CE@1.01	Ea	3,110.00	46.80	3,156.80
4,000 amp	CE@1.01	Ea	3,470.00	46.80	3,516.80

Recording meters

	Craft@Hrs	Unit	Material	Labor	Total
Voltmeter	CE@1.01	Ea	7,010.00	46.80	7,056.80
Ammeter	CE@1.01	Ea	7,160.00	46.80	7,206.80
Wattmeter	CE@1.01	Ea	8,950.00	46.80	8,996.80
Varmeter	CE@1.01	Ea	9,570.00	46.80	9,616.80
Power factor meter	CE@1.01	Ea	9,370.00	46.80	9,416.80
Frequency meter	CE@1.01	Ea	9,370.00	46.80	9,416.80
Watt-hour meter, 3 element	CE@1.01	Ea	11,900.00	46.80	11,946.80

Non-recording meters

	Craft@Hrs	Unit	Material	Labor	Total
Voltmeter	CE@.516	Ea	1,180.00	23.90	1,203.90
Ammeter for incoming line	CE@.516	Ea	1,190.00	23.90	1,213.90
Ammeter for feeder circuits	CE@.516	Ea	506.00	23.90	529.90

Electrical 16

	Craft@Hrs	Unit	Material	Labor	Total

Switchboard Instrumentation and Accessories, Non-recording meters, continued

	Craft@Hrs	Unit	Material	Labor	Total
Wattmeter	CE@1.01	Ea	3,930.00	46.80	3,976.80
Varmeter	CE@1.01	Ea	3,540.00	46.80	3,586.80
Power factor meter	CE@1.01	Ea	2,880.00	46.80	2,926.80
Frequency meter	CE@1.01	Ea	3,390.00	46.80	3,436.80
Watt-hour meter, 3 element	CE@1.01	Ea	3,540.00	46.80	3,586.80
Instrument phase select switch	CE@.516	Ea	611.00	23.90	634.90
Adjustable short time pickup and delay	CE@.516	Ea	921.00	23.90	944.90
Ground fault trip pickup and delay, 4 wire	CE@.525	Ea	973.00	24.30	997.30
Ground fault test panel	CE@.516	Ea	1,060.00	23.90	1,083.90
Shunt trip for breakers	CE@.516	Ea	707.00	23.90	730.90
Key interlock for breakers	CE@.516	Ea	707.00	23.90	730.90
Breaker lifting and transport truck	--	Ea	4,800.00	--	--
Breaker lifting device	--	Ea	4,160.00	--	--
Current transformers, by primary capacity					
800 amps and less	CE@1.01	Ea	853.00	46.80	899.80
1,000 to 1,500 amps	CE@1.01	Ea	1,240.00	46.80	1,286.80
2,000 to 6,000 amps	CE@1.01	Ea	1,490.00	46.80	1,536.80
Current transformer mount	--	Ea	415.00	--	--
Potential transformer	CE@1.01	Ea	1,180.00	46.80	1,226.80
Potential transformer mount	--	Ea	264.00	--	--

Feeder Section Breaker Panels. Flush or surface mounted 277/480 volt, 3 phase, 4 wire, FA frame panels. These costs include the subcontractor's overhead and profit.

	Craft@Hrs	Unit	Material	Labor	Total
225 amp, with 22K amp interrupt capacity breaker	E4@6.76	Ea	1,130.00	272.00	1,402.00
225 amp, with 25K amp interrupt capacity breaker	E4@6.76	Ea	2,070.00	272.00	2,342.00
400 amp, with 30K amp interrupt capacity breaker	E4@10.8	Ea	2,470.00	434.00	2,904.00
400 amp, with 35K amp interrupt capacity breaker	E4@10.8	Ea	2,940.00	434.00	3,374.00
225 amp, with main lugs only	E4@6.76	Ea	376.00	272.00	648.00
400 amp, with main lugs only	E4@10.8	Ea	528.00	434.00	962.00
600 amp, with main lugs only	E4@19.3	Ea	697.00	776.00	1,473.00

Transformers. Indoor dry type transformer, floor mounted, including connections. These costs include the subcontractor's overhead and profit.

Single phase light and power circuit transformer, 240/480 volt primary, 120/240 secondary, no taps for voltage change

	Craft@Hrs	Unit	Material	Labor	Total
1 KVA	E4@3.76	Ea	160.00	151.00	311.00
2 KVA	E4@4.20	Ea	236.00	169.00	405.00
3 KVA	E4@4.33	Ea	302.00	174.00	476.00
5 KVA	E4@5.47	Ea	411.00	220.00	631.00
7.5 KVA	E4@8.34	Ea	565.00	335.00	900.00
10 KVA	E4@9.38	Ea	655.00	377.00	1,032.00
15 KVA	E4@12.6	Ea	984.00	507.00	1,491.00
25 KVA	E4@12.6	Ea	1,470.00	507.00	1,977.00
37.5 KVA	E4@14.8	Ea	1,700.00	595.00	2,295.00
50 KVA	E4@16.7	Ea	2,070.00	672.00	2,742.00
75 KVA	E4@17.6	Ea	3,300.00	708.00	4,008.00
100 KVA	E4@21.2	Ea	3,810.00	853.00	4,663.00

Three phase light and power circuit transformers, 480 volt primary, 120/208 volt secondary, with 6 taps for voltage change

	Craft@Hrs	Unit	Material	Labor	Total
15 KVA	E4@8.34	Ea	699.00	335.00	1,034.00
30 KVA	E4@10.5	Ea	1,510.00	422.00	1,932.00
45 KVA	E4@12.6	Ea	1,820.00	507.00	2,327.00

Electrical 16

	Craft@Hrs	Unit	Material	Labor	Total
75 KVA	E4@14.8	Ea	2,730.00	595.00	3,325.00
112.5 KVA	E4@15.7	Ea	3,630.00	631.00	4,261.00
150 KVA	E4@16.7	Ea	4,730.00	672.00	5,402.00
225 KVA	E4@18.9	Ea	6,420.00	760.00	7,180.00
300 KVA	E4@21.2	Ea	8,110.00	853.00	8,963.00
500 KVA	E4@23.1	Ea	9,320.00	929.00	10,249.00

Three phase 5000 volt load center transformers, 4160 volt primary, 480/277 volt secondary, with voltage change taps, insulation class H, 80 degree system

	Craft@Hrs	Unit	Material	Labor	Total
225 KVA	E4@29.5	Ea	9,510.00	1,190.00	10,700.00
300 KVA	E4@33.7	Ea	11,300.00	1,360.00	12,660.00
500 KVA	E4@42.5	Ea	17,100.00	1,710.00	18,810.00
750 KVA	E4@53.1	Ea	22,500.00	2,140.00	24,640.00
1,000 KVA	E4@63.2	Ea	26,500.00	2,540.00	29,040.00

Panelboards. Surface mounted lighting and appliance distribution panelboards for bolt-on circuit breakers. These costs include the subcontractor's overhead and profit but no breakers.

4 wire 120/240 volt 3 phase AC panelboards with main lugs only for bolt-on breakers

	Craft@Hrs	Unit	Material	Labor	Total
8 circuits	E4@6.19	Ea	392.00	249.00	641.00
10 circuits	E4@6.92	Ea	428.00	278.00	706.00
12 circuits	E4@7.51	Ea	467.00	302.00	769.00
14 circuits	E4@8.21	Ea	503.00	330.00	833.00
16 circuits	E4@8.94	Ea	540.00	360.00	900.00
18 circuits	E4@10.7	Ea	577.00	430.00	1,007.00
20 circuits	E4@11.3	Ea	614.00	454.00	1,068.00
22 circuits	E4@12.1	Ea	650.00	487.00	1,137.00
24 circuits	E4@13.1	Ea	687.00	527.00	1,214.00
26 circuits	E4@14.1	Ea	725.00	567.00	1,292.00
28 circuits	E4@14.7	Ea	761.00	591.00	1,352.00
30 circuits	E4@15.3	Ea	798.00	615.00	1,413.00
32 circuits	E4@16.2	Ea	856.00	652.00	1,508.00
34 circuits	E4@16.9	Ea	893.00	680.00	1,573.00
36 circuits	E4@18.0	Ea	930.00	724.00	1,654.00
38 circuits	E4@19.0	Ea	968.00	764.00	1,732.00
40 circuits	E4@19.9	Ea	1,000.00	800.00	1,800.00
42 circuits	E4@20.9	Ea	1,040.00	841.00	1,881.00
Deduct for 3 wire 120/240 panelboards with main lugs only	--	%	-10.0	-15.0	--

4 wire 120/208 volt 3 phase AC panelboards for bolt-on breakers, with main circuit breaker only

	Craft@Hrs	Unit	Material	Labor	Total
8 circuits	E4@5.62	Ea	515.00	226.00	741.00
10 circuits	E4@7.05	Ea	553.00	284.00	837.00
12 circuits	E4@8.52	Ea	589.00	343.00	932.00
14 circuits	E4@10.1	Ea	703.00	406.00	1,109.00
16 circuits	E4@11.4	Ea	740.00	459.00	1,199.00
18 circuits	E4@12.9	Ea	777.00	519.00	1,296.00
20 circuits	E4@14.1	Ea	813.00	567.00	1,380.00
22 circuits	E4@15.7	Ea	850.00	631.00	1,481.00
24 circuits	E4@16.9	Ea	887.00	680.00	1,567.00
26 circuits	E4@18.6	Ea	924.00	748.00	1,672.00
28 circuits	E4@19.7	Ea	961.00	792.00	1,753.00
30 circuits	E4@21.5	Ea	998.00	865.00	1,863.00
32 circuits	E4@22.7	Ea	1,500.00	913.00	2,413.00
34 circuits	E4@24.2	Ea	1,550.00	973.00	2,523.00
36 circuits	E4@25.5	Ea	1,580.00	1,030.00	2,610.00
38 circuits	E4@27.2	Ea	1,630.00	1,090.00	2,720.00

Electrical 16

	Craft@Hrs	Unit	Material	Labor	Total
Panelboards, 4 wire 120/208 volt, continued					
40 circuits	E4@28.2	Ea	1,660.00	1,130.00	2,790.00
42 circuits	E4@30.1	Ea	1,690.00	1,210.00	2,900.00
Deduct for 3 wire panelboards with circuit breaker mains only	--	%	-10.0	-15.0	--

Loadcenters. Surface mounted 240 volt NEMA 1 (indoor) distribution panels for plug-in branch breakers. Add the cost of distribution breakers and conduit hubs. These costs include the subcontractor's overhead and profit.

	Craft@Hrs	Unit	Material	Labor	Total
Single phase 3 wire 120/240 volt loadcenters with main circuit breaker only					
100 amps, 14 circuits	E4@1.20	Ea	77.90	48.30	126.20
100 amps, 22 circuits	E4@1.52	Ea	98.40	61.10	159.50
150 amps, 30 circuits	E4@2.30	Ea	185.00	92.50	277.50
225 amps, 42 circuits	E4@2.93	Ea	269.00	118.00	387.00
Add for NEMA-4 raintight loadcenters	--	%	25.0	--	--
Three phase 4 wire 120/208 volt loadcenters with main circuit breaker only					
100 amps, 12 circuits	E4@1.21	Ea	180.00	48.70	228.70
100 amps, 18 circuits	E4@1.60	Ea	222.00	64.40	286.40
125 amps, 30 circuits	E4@1.91	Ea	479.00	76.80	555.80
150 amps, 24 circuits	E4@2.18	Ea	425.00	87.70	512.70
150 amps, 42 circuits	E4@2.82	Ea	550.00	113.00	663.00
200 amps, 42 circuits	E4@3.63	Ea	507.00	146.00	653.00
225 amps, 42 circuits	E4@3.78	Ea	577.00	152.00	729.00
400 amps, 24 circuits	E4@3.91	Ea	1,320.00	157.00	1,477.00
400 amps, 42 circuits	E4@4.51	Ea	1,580.00	181.00	1,761.00
Add for NEMA-4 raintight loadcenters	--	%	25.0	--	--
Single phase 3 wire, 120/240 volt loadcenters with main lugs only					
150 amps, 12 circuits	E4@1.20	Ea	69.00	48.30	117.30
150 amps, 16 circuits	E4@1.20	Ea	80.50	48.30	128.80
150 amps, 20 circuits	E4@1.52	Ea	88.50	61.10	149.60
200 amps, 30 circuits	E4@1.70	Ea	121.00	68.40	189.40
200 amps, 40 circuits	E4@2.03	Ea	169.00	81.60	250.60
225 amps, 42 circuits	E4@2.77	Ea	201.00	111.00	312.00
Three phase 4 wire 120/208 volt loadcenters with main lugs only					
125 amps, 12 circuits	E4@1.19	Ea	77.00	47.90	124.90
150 amps, 18 circuits	E4@1.47	Ea	109.00	59.10	168.10
200 amps, 30 circuits	E4@1.90	Ea	149.00	76.40	225.40
200 amps, 42 circuits	E4@2.18	Ea	198.00	87.70	285.70
225 amps, 42 circuits	E4@2.75	Ea	227.00	111.00	338.00
400 amps, 24 circuits	E4@3.63	Ea	436.00	146.00	582.00
400 amps, 42 circuits	E4@3.78	Ea	484.00	152.00	636.00
600 amps, 24 circuits	E4@3.91	Ea	555.00	157.00	712.00
600 amps, 42 circuits	E4@6.53	Ea	602.00	263.00	865.00

Circuit Breakers. No enclosures included. 10,000 amp interrupt capacity except as noted. These costs include the subcontractor's overhead and profit.

	Craft@Hrs	Unit	Material	Labor	Total
Plug-in molded case 120 volt, 100 amp frame circuit breakers, 1" or 1/2" module					
15 to 50 amps, single pole	CE@.270	Ea	7.30	12.50	19.80
15 to 60 amps, two pole	CE@.374	Ea	16.80	17.30	34.10
70 amps, two pole	CE@.563	Ea	34.10	26.10	60.20
80 to 100 amps, two pole	CE@.680	Ea	47.30	31.50	78.80
Plug-in molded case 240 volt, 100 amp frame circuit breakers					
15 thru 60 amps, single pole	CE@.417	Ea	27.80	19.30	47.10
15 thru 60 amps, two pole	CE@.577	Ea	53.00	26.70	79.70

Electrical 16

	Craft@Hrs	Unit	Material	Labor	Total
15 thru 60 amps, three pole	CE@.863	Ea	65.00	40.00	105.00
Plug-in molded case 480 volt, 100 amp frame circuit breakers					
15 thru 60 amps, single pole	CE@.577	Ea	55.60	26.70	82.30
15 thru 60 amps, two pole	CE@.577	Ea	133.00	26.70	159.70
15 thru 60 amps, three pole	CE@.897	Ea	171.00	41.60	212.60
Plug-in molded case 600 volt, 100 amp frame circuit breakers					
15 thru 60 amps, two pole	CE@.577	Ea	153.00	26.70	179.70
15 thru 60 amps, three pole	CE@.897	Ea	198.00	41.60	239.60
Plug-in molded case 240 volt, 100 amp frame circuit breakers					
70 thru 100 amps, single pole	CE@.863	Ea	35.80	40.00	75.80
70 thru 100 amps, two pole	CE@1.22	Ea	87.50	56.50	144.00
70 thru 100 amps, three pole	CE@1.71	Ea	97.50	79.20	176.70
Plug-in molded case 480 volt, 100 amp frame circuit breakers					
70 thru 100 amps, single pole	CE@1.22	Ea	68.50	56.50	125.00
70 thru 100 amps, two pole	CE@1.22	Ea	172.00	56.50	228.50
70 thru 100 amps, three pole	CE@1.71	Ea	202.00	79.20	281.20
Plug-in molded case 600 volt, 100 amp frame circuit breakers					
70 thru 100 amps, two pole	CE@1.22	Ea	194.00	56.50	250.50
70 thru 100 amps, three pole	CE@1.71	Ea	243.00	79.20	322.20
Plug-in molded case 250/600 volt, 225 amp frame circuit breakers					
110 thru 225 amps, two pole	CE@2.06	Ea	487.00	95.40	582.40
110 thru 225 amps, three pole	CE@2.57	Ea	611.00	119.00	730.00
Plug-in molded case 250/600 volt, 400 amp frame circuit breakers					
250 thru 400 amps, two pole	CE@2.88	Ea	834.00	133.00	967.00
250 thru 400 amps, three pole	CE@4.33	Ea	1,010.00	201.00	1,211.00
450 thru 600 amps, two pole	CE@4.33	Ea	1,180.00	201.00	1,381.00
Plug-in molded case 250/600 volt, 1,000 amp frame circuit breakers					
450 thru 600 amps, three pole	CE@6.04	Ea	2,130.00	280.00	2,410.00
700 thru 800 amps, two pole	CE@4.73	Ea	1,670.00	219.00	1,889.00
700 thru 800 amps, three pole	CE@6.71	Ea	2,140.00	311.00	2,451.00
900 thru 1,000 amps, two pole	CE@6.98	Ea	2,360.00	323.00	2,683.00
900 thru 1,000 amps, three pole	CE@9.35	Ea	2,770.00	433.00	3,203.00
Plug-in molded case 600 volt, 2,000 amp frame circuit breakers					
600 thru 2,000 amps, two pole	CE@9.60	Ea	2,330.00	445.00	2,775.00
600 thru 2,000 amps, three pole	CE@11.9	Ea	2,670.00	551.00	3,221.00
Space only for factory assembled panels					
Single pole space	--	Ea	15.20	--	--
Two pole space	--	Ea	18.90	--	--
Three pole space	--	Ea	25.40	--	--
Bolt-on molded case 240 volt circuit breakers					
15 thru 60 amps, one pole	CE@.320	Ea	8.90	14.80	23.70
15 thru 60 amps, two pole	CE@.484	Ea	19.50	22.40	41.90
15 thru 60 amps, three pole	CE@.638	Ea	62.50	29.60	92.10
Bolt-on molded case 240 volt circuit breakers					
70 thru 100 amps, two pole	CE@.712	Ea	51.00	33.00	84.00
70 thru 100 amps, three pole	CE@.946	Ea	90.00	43.80	133.80
Bolt-on molded case 480 volt circuit breakers, 14,000 amp interrupt capacity					
15 thru 60 amps, two pole	CE@.484	Ea	152.00	22.40	174.40
15 thru 60 amps, three pole	CE@.712	Ea	193.00	33.00	226.00
70 thru 100 amps, three pole	CE@.804	Ea	193.00	37.20	230.20
Bolt-on molded case 480 volt circuit breakers, 22,000 amp interrupt capacity					
70 thru 225 amps, three pole	CE@1.23	Ea	543.00	57.00	600.00

Electrical 16

	Craft@Hrs	Unit	Material	Labor	Total

Fuses. No enclosures included. These costs include the subcontractor's overhead and profit.

250 volt cartridge fuses, one time non-renewable

	Craft@Hrs	Unit	Material	Labor	Total
To 30 amps	CE@.077	Ea	.94	3.57	4.51
35 to 60 amps	CE@.077	Ea	1.18	3.57	4.75
70 to 100 amps	CE@.077	Ea	5.79	3.57	9.36
110 to 200 amps	CE@.077	Ea	14.00	3.57	17.57
225 to 400 amps	CE@.077	Ea	21.10	3.57	24.67
450 to 600 amps	CE@.077	Ea	36.90	3.57	40.47

600 volt cartridge fuses, one time non-renewable

To 30 amps	CE@.077	Ea	4.34	3.57	7.91
35 to 60 amps	CE@.077	Ea	6.26	3.57	9.83
70 to 100 amps	CE@.077	Ea	13.20	3.57	16.77
110 to 200 amps	CE@.077	Ea	26.30	3.57	29.87
225 to 400 amps	CE@.087	Ea	52.50	4.03	56.53
450 to 600 amps	CE@.124	Ea	75.60	5.75	81.35

250 volt class H renewable fuses

To 10 amps	CE@.077	Ea	6.07	3.57	9.64
12 to 30 amps	CE@.077	Ea	5.59	3.57	9.16
35 to 60 amps	CE@.077	Ea	11.00	3.57	14.57
70 to 100 amps	CE@.077	Ea	24.60	3.57	28.17
110 to 200 amps	CE@.077	Ea	55.40	3.57	58.97
225 to 400 amps	CE@.080	Ea	100.00	3.71	103.71
450 to 600 amps	CE@.122	Ea	153.00	5.65	158.65

600 volt class H renewable fuses

To 10 amps	CE@.077	Ea	15.30	3.57	18.87
12 to 30 amps	CE@.077	Ea	13.90	3.57	17.47
35 to 60 amps	CE@.077	Ea	21.80	3.57	25.37
70 to 100 amps	CE@.077	Ea	49.50	3.57	53.07
110 to 200 amps	CE@.077	Ea	96.10	3.57	99.67
225 to 400 amps	CE@.153	Ea	145.00	7.09	152.09
450 to 600 amps	CE@.223	Ea	198.00	10.30	208.30

250 volt class K5 non-renewable cartridge fuses

10 to 30 amps	CE@.077	Ea	3.12	3.57	6.69
35 to 60 amps	CE@.077	Ea	4.66	3.57	8.23
70 to 100 amps	CE@.077	Ea	10.50	3.57	14.07
110 to 200 amps	CE@.077	Ea	27.30	3.57	30.87
225 to 400 amps	CE@.230	Ea	41.70	10.70	52.40
450 to 600 amps	CE@.297	Ea	88.20	13.80	102.00
650 to 1,200 amps	CE@.372	Ea	114.00	17.20	131.20
1,400 to 1,600 amps	CE@.804	Ea	156.00	37.20	193.20
1,800 to 2,000 amps	CE@1.00	Ea	198.00	46.30	244.30

600 volt class K5 non-renewable cartridge fuses

10 to 30 amps	CE@.077	Ea	6.38	3.57	9.95
35 to 60 amps	CE@.077	Ea	9.70	3.57	13.27
70 to 100 amps	CE@.077	Ea	20.10	3.57	23.67
110 to 200 amps	CE@.129	Ea	40.20	5.98	46.18
225 to 400 amps	CE@.257	Ea	80.50	11.90	92.40
450 to 600 amps	CE@.354	Ea	137.00	16.40	153.40
650 to 1,200 amps	CE@.372	Ea	194.00	17.20	211.20
1,400 to 1,600 amps	CE@.797	Ea	268.00	36.90	304.90
1,800 to 2,000 amps	CE@1.00	Ea	338.00	46.30	384.30
Add for type K5 600 volt fuses	--	%	20.0	--	--
Deduct for type J 600 volt fuses	--	%	-50.0	--	--

Electrical 16

	Craft@Hrs	Unit	Material	Labor	Total
600 volt class L fuses, bolt-on					
600 to 1,200 amps	CE@.333	Ea	304.00	15.40	319.40
1,500 to 1,600 amps	CE@.615	Ea	395.00	28.50	423.50
1,800 to 2,000 amps	CE@.809	Ea	522.00	37.50	559.50
2,500 amps	CE@1.10	Ea	591.00	51.00	642.00
3,000 amps	CE@1.25	Ea	682.00	57.90	739.90
3,500 to 4,000 amps	CE@1.74	Ea	1,080.00	80.60	1,160.60
5,000 amps	CE@2.33	Ea	1,320.00	108.00	1,428.00

Overhead Electrical Distribution. These costs include the subcontractor's overhead and profit. Poles, pressure treated wood, class 4, type C, set with crane or boom, including augered holes. Equipment is a flatbed truck with a boom and auger at $20.00 per hour.

	Craft@Hrs	Unit	Material	Labor	Equipment	Total
25'	E1@7.62	Ea	230.00	311.00	31.60	572.60
30'	E1@7.62	Ea	295.00	311.00	31.60	637.60
35'	E1@9.72	Ea	384.00	396.00	40.30	820.30
40'	E1@11.5	Ea	496.00	469.00	47.60	1,012.60
Cross arms, wood, with typical hardware						
4'	E1@2.86	Ea	54.30	117.00	11.90	183.20
5'	E1@2.86	Ea	68.10	117.00	11.90	197.00
6'	E1@4.19	Ea	99.50	171.00	17.40	287.90
8'	E1@4.91	Ea	130.00	200.00	20.30	350.30
Transformers, pole mounted, single phase, 7620/13200 Y primary, 120/240 secondary, with two 2-1/2% taps above and below						
10 KVA	E1@4.14	Ea	781.00	169.00	17.20	967.20
15 KVA	E1@4.14	Ea	900.00	169.00	17.20	1,086.20
25 KVA	E1@4.14	Ea	1,110.00	169.00	17.20	1,296.20
36.5 KVA	E1@4.14	Ea	1,460.00	169.00	17.20	1,646.20
50 KVA	E1@4.14	Ea	1,690.00	169.00	17.20	1,876.20
75 KVA	E1@6.16	Ea	2,480.00	251.00	25.50	2,756.50
100 KVA	E1@6.16	Ea	2,830.00	251.00	25.50	3,106.50
167 KVA	E1@6.16	Ea	4,370.00	251.00	25.50	4,646.50
250 KVA	E1@6.16	Ea	4,450.00	251.00	25.50	4,726.50
333 KVA	E1@6.16	Ea	4,810.00	251.00	25.50	5,086.50
500 KVA	E1@6.16	Ea	6,710.00	251.00	25.50	6,986.50
Fused cutouts, pole mounted, 5 KV						
50 amp	E1@2.56	Ea	91.70	104.00	10.60	206.30
100 amp	E1@2.56	Ea	91.70	104.00	10.60	206.30
250 amp	E1@2.56	Ea	104.00	104.00	10.60	218.60
Switches, disconnect, pole mounted, pole arm throw, 5 KV, set of 3 with throw and lock						
400 amp	E1@45.8	Ea	1,550.00	1,870.00	189.00	3,609.00
600 amp	E1@45.8	Ea	2,960.00	1,870.00	189.00	5,019.00
1,200 amp	E1@51.4	Ea	5,850.00	2,100.00	213.00	8,163.00

Fluorescent Lighting. Commercial, industrial and architectural grade. Rapid start. With ballasts but no lamps, wire or conduit. Includes setting and connecting prewired fixtures only. These costs include the subcontractor's overhead and profit.
Surface mounted fluorescent fixtures, better quality, baked enamel finish, with hinged wraparound acrylic prismatic lens suitable for office or classroom use, 4' long, Lithonia VC series

	Craft@Hrs	Unit	Material	Labor	Total
Two 40 watt lamp fixture (#240-A)	CE@.773	Ea	85.30	35.80	121.10
Four 40 watt lamp fixture (#440-A)	CE@.773	Ea	131.00	35.80	166.80
Eight 40 watt lamp fixture (#8T440-A)	CE@.960	Ea	261.00	44.50	305.50
Add for 40 watt lamps	--	Ea	4.10	--	--

Electrical 16

	Craft@Hrs	Unit	Material	Labor	Total

Fluorescent Lighting, continued
Surface mounted fluorescent fixtures, baked enamel finish, with wraparound acrylic prismatic lens suitable for industrial use, 4' long, Lithonia series

	Craft@Hrs	Unit	Material	Labor	Total
One 40 watt lamp fixture (CB140A)	CE@.806	Ea	60.20	37.30	97.50
Two 40 watt lamp fixture (LB240A)	CE@.858	Ea	51.60	39.80	91.40
Four 40 watt lamp fixture (LB440A)	CE@1.07	Ea	87.90	49.60	137.50
Add for 40 watt lamps	--	Ea	4.10	--	--

Surface mounted steel sided fluorescent fixtures with acrylic prismatic diffuser, Lithonia M/2M series

	Craft@Hrs	Unit	Material	Labor	Total
Two 40 watt "U" lamps, 2' x 2' (2M2U40A12120)	CE@.955	Ea	58.30	44.20	102.50
Two 40 watt lamps, 1' x 4' (M240A12120)	CE@.955	Ea	59.30	44.20	103.50
Four 40 watt lamps, 2' x 4' (2M240A12120)	CE@.955	Ea	93.00	44.20	137.20
Six 40 watt lamps, 2' x 4' (4M666FWA12120)	CE@1.05	Ea	224.00	48.60	272.60
Add for 40 watt "U" lamps	--	Ea	13.00	--	--
Add for 40 watt straight lamps	--	Ea	4.26	--	--

Wall surface mounted fluorescent fixtures, baked enamel finish, 2 lamps, 8" wide 5" high, hinged acrylic prismatic lens, Day-Brite

	Craft@Hrs	Unit	Material	Labor	Total
26" long fixture (A2220W)	CE@.737	Ea	132.00	34.10	166.10
38" long fixture (A2230W)	CE@.795	Ea	148.00	36.80	184.80
50" long fixture (A2240W)	CE@.858	Ea	148.00	39.80	187.80
Add for stainless steel finish	--	%	75.0	--	--
Add for 24" 20 watt lamps	--	Ea	4.83	--	--
Add for 36" 30 watt lamps	--	Ea	7.04	--	--
Add for 48" 40 watt lamps	--	Ea	3.26	--	--

Wall surface mounted valance fluorescent fixtures, walnut finished metal reflector, Markstone

	Craft@Hrs	Unit	Material	Labor	Total
One lamp, 4' long fixture (M2220TS)	CE@.721	Ea	57.80	33.40	91.20
Two lamp, 4' long fixture (M2240RS)	CE@1.08	Ea	81.40	50.00	131.40
Add for 48" 40 watt lamps	--	Ea	3.31	--	--

Industrial pendant or surface mounted open striplights, baked enamel finish, Lithonia C series

	Craft@Hrs	Unit	Material	Labor	Total
One or two 40 watt lamps, 4' long (C240)	CE@.865	Ea	29.90	40.10	70.00
One or two 75 watt lamps, 8' long (C296)	CE@1.02	Ea	42.70	47.30	90.00
One or two 55 watt lamps, 6' long (C272)	CE@1.02	Ea	42.70	47.30	90.00
Add for 40 watt lamps	--	Ea	3.26	--	--
Add for 75 or 55 watt lamps	--	Ea	7.56	--	--
Add for 24" stem and canopy set, per fixture	--	Ea	8.72	--	--

Industrial pendant or surface mounted fluorescent fixtures, 12" wide baked enamel, steel reflector-shield, Lithonia L/LA series

	Craft@Hrs	Unit	Material	Labor	Total
Two 40 watt lamps, 4' long (L-240-120)	CE@.903	Ea	37.10	41.80	78.90
Two 75 watt lamps, 8' long (L-296-120)	CE@1.21	Ea	60.80	56.10	116.90
Add for 40 watt lamps	--	Ea	3.26	--	--
Add for 75 watt lamps	--	Ea	7.67	--	--

Ceiling grid mounted fluorescent troffer fixtures, baked enamel finish, with hinged acrylic prismatic lens with energy saving ballast, Metalux

	Craft@Hrs	Unit	Material	Labor	Total
Two 40 watt "U" lamp fixture, 2' x 2'	CE@.908	Ea	48.00	42.10	90.10
Two 40 watt lamp fixture, 2' x 4'	CE@.998	Ea	48.50	46.20	94.70
Four 40 watt lamp fixture, 2' x 4'	CE@1.05	Ea	59.00	48.60	107.60
Add for plaster frames, per fixture	--	Ea	13.20	--	--

Air-handling grid mounted floating door, heat transfer troffer fixtures with acrylic lens, no ductwork included, Miller

	Craft@Hrs	Unit	Material	Labor	Total
1' x 4', two 40 watt lamps	CE@.908	Ea	98.70	42.10	140.80
1' x 4', three 40 watt lamps	CE@.908	Ea	116.00	42.10	158.10
2' x 2', one U-shape 40 watt lamp	CE@.809	Ea	106.00	37.50	143.50
2' x 4', two 40 watt lamps	CE@1.00	Ea	110.00	46.30	156.30
2' x 4', three 40 watt lamps	CE@1.05	Ea	121.00	48.60	169.60

Electrical 16

	Craft@Hrs	Unit	Material	Labor	Total
2' x 4', four 40 watt lamps	CE@1.05	Ea	121.00	48.60	169.60
4' x 4', four 40 watt lamps	CE@1.15	Ea	287.00	53.30	340.30
4' x 4', five 40 watt lamps	CE@1.15	Ea	299.00	53.30	352.30
4' x 4', six 40 watt lamps	CE@1.15	Ea	313.00	53.30	366.30
Add for 40 watt 4' lamps	--	Ea	4.22	--	--
Add for 40 watt "U" lamps	--	Ea	12.80	--	--

Surface mounted vandal resistant luminaire with tamper-resistant screw-type latching, Day-Brite

	Craft@Hrs	Unit	Material	Labor	Total
2 lamps, 12" x 48"	CE@.761	Ea	146.00	35.30	181.30
2 lamps, 24" x 48"	CE@.761	Ea	200.00	35.30	235.30
4 lamps, 24" x 48"	CE@.761	Ea	235.00	35.30	270.30

Parabolic fluorescent luminaire, 6" x 6" x 4' with semi-specular anodized aluminum reflector and parabolic louvers, I.C.E. AR series

	Craft@Hrs	Unit	Material	Labor	Total
One 40 watt lamp, surface mount	CE@.955	Ea	100.00	44.20	144.20
Two 40 watt lamps, surface mount	CE@.955	Ea	100.00	44.20	144.20
One 40 watt lamp, bracket or pendant mount	CE@.955	Ea	142.00	44.20	186.20
Two 40 watt lamps, bracket or pendant mount	CE@.955	Ea	142.00	44.20	186.20

Round surface mounted fluorescent fixtures, aluminum housing with matte black finish, polycarbonate or acrylic opal globe, Dayton

	Craft@Hrs	Unit	Material	Labor	Total
22 to 33 watt, 14" diameter	CE@.478	Ea	54.60	22.10	76.70
32 to 40 watt, 20" diameter	CE@.478	Ea	61.60	22.10	83.70
Add for circline tubes	--	Ea	13.00	--	--

Outdoor sign fluorescent fixtures, anodized aluminum housing with parabolic specular reflector and acrylic lens, with high output lamps and remote ballast

	Craft@Hrs	Unit	Material	Labor	Total
One 60 watt lamp, 48" long fixture	CE@2.39	Ea	232.00	111.00	343.00
One 85 watt lamp, 72" long fixture	CE@2.39	Ea	248.00	111.00	359.00
One 110 watt lamp, 96" long fixture	CE@2.39	Ea	270.00	111.00	381.00
Two 85 watt lamps, 144" long fixture	CE@2.39	Ea	370.00	111.00	481.00
Add for high output T12 800 MA lamps, sign white					
60 watt	--	Ea	13.80	--	--
85 watt	--	Ea	15.40	--	--
110 watt	--	Ea	22.60	--	--

Round continuous cable-supported fluorescent fixtures, 6" diameter extruded aluminum with baked enamel finish and clear prismatic lens, no cable support included

	Craft@Hrs	Unit	Material	Labor	Total
Cost per linear foot of fixture	CE@.179	LF	64.50	8.29	72.79
Add for 90 degree elbows	CE@.475	Ea	166.00	22.00	188.00
Add for in-line connectors	CE@.475	Ea	83.30	22.00	105.30

Wet or damp location surface mounted fluorescent fixtures, hinge mounted acrylic diffuse lens, fiberglass or plastic housing with gaskets, Day-Brite

	Craft@Hrs	Unit	Material	Labor	Total
Two 40 watt lamps, 4' long fixture (R41241)	CE@1.21	Ea	105.00	56.10	161.10
Two 40 watt lamps, 8' long fixture (R81441)	CE@1.86	Ea	279.00	86.20	365.20
Add for 40 watt lamps	--	Ea	3.26	--	--

High Intensity Discharge Lighting. Commercial and architectural grade fixtures. With ballasts but no lamps, wire, conduit or concrete except as noted. See lamp costs below. Labor includes setting and connecting prewired fixtures only. These costs include the contractor's overhead and profit.

Recess mounted commercial luminaire, 2' x 2' x 13", aluminum reflector glass lens, Lithonia TC series

	Craft@Hrs	Unit	Material	Labor	Total
150 watt, high pressure sodium	CE@1.15	Ea	207.00	53.30	260.30
175 watt, metal halide	CE@1.15	Ea	197.00	53.30	250.30
250 watt, high pressure sodium	CE@1.15	Ea	239.00	53.30	292.30
400 watt, metal halide	CE@1.15	Ea	207.00	53.30	260.30

Surface mounted commercial luminaire, 2' x 2' x 15", baked enamel finish with tempered glass lens, Art Metal MET series

	Craft@Hrs	Unit	Material	Labor	Total
100 watt, high pressure sodium	CE@1.42	Ea	400.00	65.80	465.80
150 watt, high pressure sodium	CE@1.42	Ea	410.00	65.80	475.80

Electrical 16

	Craft@Hrs	Unit	Material	Labor	Total

High Intensity Discharge Lighting, Surface mounted commercial luminaire, continued

	Craft@Hrs	Unit	Material	Labor	Total
250 watt, high pressure sodium	CE@1.42	Ea	437.00	65.80	502.80
175 watt, metal halide	CE@1.42	Ea	365.00	65.80	430.80
400 watt, metal halide	CE@1.42	Ea	406.00	65.80	471.80
Add for quartz restrike	--	Ea	61.20	--	--

Pendant mounted indirect metal halide luminaire, aluminum reflector with glass lens

	Craft@Hrs	Unit	Material	Labor	Total
175 watt, square	CE@1.71	Ea	332.00	79.20	411.20
250 watt, square	CE@1.71	Ea	404.00	79.20	483.20
400 watt, square	CE@1.71	Ea	501.00	79.20	580.20
175 watt, round	CE@1.71	Ea	361.00	79.20	440.20
250 watt, round	CE@1.71	Ea	413.00	79.20	492.20
400 watt, round	CE@1.71	Ea	530.00	79.20	609.20

Pendant mounted indirect high pressure sodium luminaire, aluminum reflector with glass lens

	Craft@Hrs	Unit	Material	Labor	Total
150 watt, square	CE@1.71	Ea	409.00	79.20	488.20
250 watt, square	CE@1.71	Ea	419.00	79.20	498.20
400 watt, square	CE@1.71	Ea	562.00	79.20	641.20
150 watt, round	CE@1.71	Ea	419.00	79.20	498.20
250 watt, round	CE@1.71	Ea	446.00	79.20	525.20
400 watt, round	CE@1.71	Ea	594.00	79.20	673.20

"High Tek" recessed round open reflector fixtures, Art Metal 10VAG series

	Craft@Hrs	Unit	Material	Labor	Total
70 watt, high pressure sodium	CE@.969	Ea	266.00	44.90	310.90
150 watt, high pressure sodium	CE@.969	Ea	297.00	44.90	341.90
175 watt, metal halide	CE@1.12	Ea	260.00	51.90	311.90
250 watt, metal halide	CE@1.12	Ea	276.00	51.90	327.90
400 watt, metal halide	CE@1.12	Ea	366.00	51.90	417.90

Multiple-groove baffle recessed round luminaire, Art Metal 12VAC series

	Craft@Hrs	Unit	Material	Labor	Total
70 watt, high pressure sodium	CE@.969	Ea	222.00	44.90	266.90
150 watt, high pressure sodium	CE@.969	Ea	266.00	44.90	310.90
175 watt, metal halide	CE@1.12	Ea	291.00	51.90	342.90
250 watt, metal halide	CE@1.12	Ea	266.00	51.90	317.90
400 watt, metal halide	CE@1.12	Ea	364.00	51.90	415.90

Regressed lens recessed round fixtures, Art Metal 18VBC series

	Craft@Hrs	Unit	Material	Labor	Total
70 watt, high pressure sodium	CE@.969	Ea	291.00	44.90	335.90
150 watt, high pressure sodium	CE@.969	Ea	316.00	44.90	360.90
175 watt, metal halide	CE@1.12	Ea	263.00	51.90	314.90
250 watt, metal halide	CE@1.12	Ea	247.00	51.90	298.90

Handball and racquetball court luminaire, 2' x 2' x 12", with impact-resistant prismatic glass lens

	Craft@Hrs	Unit	Material	Labor	Total
250 watt, metal halide	CE@1.12	Ea	261.00	51.90	312.90
400 watt, metal halide	CE@1.12	Ea	274.00	51.90	325.90
150 watt, high pressure sodium	CE@1.12	Ea	204.00	51.90	255.90
250 watt, high pressure sodium	CE@1.12	Ea	214.00	51.90	265.90
400 watt, high pressure sodium	CE@1.12	Ea	441.00	51.90	492.90

Bollard light, 42" high and 8" diameter extruded aluminum column with anodized finish, acrylic lens with alzac reflecting cone, 70 watt high pressure sodium lamp set with anchor bolts in manufacturer's 12" x 12" x 24" concrete base

	Craft@Hrs	Unit	Material	Labor	Total
	E4@6.61	Ea	557.00	266.00	823.00

Outdoor wall pack, aluminum housing with hinged polycarbonate access door and lens, Hubbell

	Craft@Hrs	Unit	Material	Labor	Total
100 watt, high pressure sodium	CE@.955	Ea	222.00	44.20	266.20
150 watt, high pressure sodium	CE@.955	Ea	234.00	44.20	278.20
175 watt, mercury vapor	CE@.955	Ea	165.00	44.20	209.20
175 watt, metal halide	CE@.955	Ea	202.00	44.20	246.20

Utility wall pack, aluminum housing with hinged polycarbonate or tempered glass lens, Holophane Wallpack 2

	Craft@Hrs	Unit	Material	Labor	Total
70 watt high pressure sodium	CE@.955	Ea	233.00	44.20	277.20
100 watt high pressure sodium	CE@.955	Ea	250.00	44.20	294.20

Electrical 16

	Craft@Hrs	Unit	Material	Labor	Total
150 watt high pressure sodium	CE@.955	Ea	256.00	44.20	300.20
175 watt metal halide	CE@.955	Ea	295.00	44.20	339.20
250 watt mercury vapor	CE@.955	Ea	232.00	44.20	276.20
Add for photo cell automatic control	--	Ea	33.10	--	--

High bay mercury vapor industrial fixture, flange type, cast aluminum housing, Abolite

	Craft@Hrs	Unit	Material	Labor	Total
400 watt fixture	CE@2.61	Ea	289.00	121.00	410.00
Add for power hook and cord	CE@.250	Ea	16.30	11.60	27.90
Add for thru wire power hook receptacle	CE@.187	Ea	9.08	8.66	17.74
Add for wire guard on lens	CE@.196	Ea	16.30	9.08	25.38

High bay metal halide industrial fixture, cast aluminum housing, spun aluminum reflector, lens is tempered glass

	Craft@Hrs	Unit	Material	Labor	Total
400 watt, porcelain enamel reflector	CE@2.61	Ea	170.00	121.00	291.00
1,000 watt, alzac aluminum	CE@2.61	Ea	245.00	121.00	366.00
Add for power hook and cord	CE@.250	Ea	16.30	11.60	27.90
Add for thru wire power hook receptacle	CE@.199	Ea	28.70	9.22	37.92
Add for wire guard on lens	CE@.199	Ea	15.10	9.22	24.32

Low bay high pressure sodium industrial fixture, cast aluminum with epoxy finish, acrylic lens

	Craft@Hrs	Unit	Material	Labor	Total
100 watt fixture	CE@1.42	Ea	270.00	65.80	335.80
150 watt fixture	CE@1.39	Ea	270.00	64.40	334.40
250 watt fixture	CE@1.39	Ea	308.00	64.40	372.40
Add for power hook and cord	CE@.250	Ea	16.00	11.60	27.60
Add for thru wire power hook receptacle	CE@.199	Ea	28.10	9.22	37.32
Add for wire guard on lens	CE@.199	Ea	17.00	9.22	26.22

Low bay metal halide industrial downlight, cast aluminum with epoxy finish, shock resistant glass lens

	Craft@Hrs	Unit	Material	Labor	Total
175 watt fixture	CE@1.07	Ea	212.00	49.60	261.60
250 watt fixture	CE@1.39	Ea	308.00	64.40	372.40
Add for power hook and cord	CE@.250	Ea	16.00	11.60	27.60
Add for thru wire power hook receptacle	CE@.199	Ea	32.00	9.22	41.22
Add for wire guard on lens	CE@.199	Ea	19.30	9.22	28.52

Lamps for high intensity discharge luminaires

	Craft@Hrs	Unit	Material	Labor	Total
400 watt mercury vapor	--	Ea	31.80	--	--
1,000 watt mercury vapor	--	Ea	68.40	--	--
35 watt high pressure sodium	--	Ea	49.90	--	--
70 watt high pressure sodium	--	Ea	49.90	--	--
100 watt high pressure sodium	--	Ea	60.50	--	--
150 watt high pressure sodium	--	Ea	60.50	--	--
250 watt high pressure sodium	--	Ea	64.70	--	--
400 watt high pressure sodium	--	Ea	69.00	--	--
1,000 watt high pressure sodium	--	Ea	202.00	--	--
175 watt metal halide	--	Ea	54.60	--	--
250 watt metal halide	--	Ea	57.30	--	--
400 watt metal halide	--	Ea	53.00	--	--
1,000 watt metal halide	--	Ea	133.00	--	--
Add for diffuse coated lamps	--	Ea	12.80	--	--

Incandescent Lighting. Commercial and architectural grade. No wire, conduit or concrete included. Includes setting and connecting prewired fixtures only. These costs include the subcontractor's overhead and profit.

Step illumination light fixtures, 16 gauge steel housing with white enamel finish, 5" high, 4" deep, 11" long

	Craft@Hrs	Unit	Material	Labor	Total
Polycarbonate or tempered glass lens fixture	CE@1.07	Ea	38.70	49.60	88.30
Louvered sheet metal front fixture	CE@1.07	Ea	32.90	49.60	82.50
Add for brushed aluminum trim, 2 lamps	--	Ea	68.40	--	--

Electrical 16

	Craft@Hrs	Unit	Material	Labor	Total

Incandescent Lighting, continued
Adjustable recessed spotlight, 20 gauge steel housing with matt black finish, low gloss white enamel trim ring, specular aluminum reflector with heat-resistant spread lens, 30 degree adjustable with 350 degree rotation

150 watt fixture	CE@1.07	Ea	135.00	49.60	184.60

Semi-recessed baffle downlight, 20 gauge steel housing with aluminum plaster ring, 7" baffle with brushed aluminum exterior finish, Prescolite

75 watt fixture (1252-916)	CE@1.07	Ea	81.20	49.60	130.80
150 watt fixture (1260-926)	CE@1.07	Ea	99.20	49.60	148.80

Recessed round, 6" diameter by 8" high housing, white finish, drop opel or fresnel lens, 100 watt, Markstone A1-F40

	CE@1.07	Ea	20.70	49.60	70.30

Adjustable semi-recessed spotlight, 20 gauge galvanized steel with die cast aluminum ball movable 45 degrees vertically and 350 degrees in rotation, matt white painted trim ring, Lightolier 1122

75 watt fixture	CE@1.07	Ea	50.40	49.60	100.00

Exterior wall fixtures, cast anodized aluminum with white polycarbonate globe, gaskets between globe and housing and between housing and base, Marco B11 Series

Cylinder globe fixture	CE@.320	Ea	21.70	14.80	36.50
Round globe fixture	CE@.430	Ea	24.40	19.90	44.30
Add for 100 watt lamp	--	Ea	1.01	--	--

Exterior ceiling fixtures, cast anodized aluminum with white polycarbonate globe, gaskets between globe and housing and between housing and base, Marco B11 Series

Cylinder globe fixture	CE@.320	Ea	21.70	14.80	36.50
Round globe fixture	CE@.430	Ea	25.40	19.90	45.30
Add for 150 watt lamp	--	Ea	.91	--	--

Obstruction light, meets Federal Aviation Administration Specification L-810, cast aluminum housing, two one-piece red, heat-resistant fresnel globes, with photoelectric control, mounted on 1" rigid steel conduit, including junction box and mounting plate

Two 100 watt lamp fixture	CE@4.91	Ea	67.10	227.00	294.10

Vandal Resistant Lighting. Commercial and architectural grade. Includes setting and connecting prewired fixtures but no wire, conduit or lamps. These costs include the subcontractor's overhead and profit.
Ceiling mounted 12" x 12" x 6" fixture, polycarbonate prismatic lens

Two 100 watt incandescent lamps	CE@.764	Ea	78.00	35.40	113.40
35 watt high pressure sodium lamp	CE@.764	Ea	228.00	35.40	263.40
50 watt high pressure sodium lamp	CE@.764	Ea	237.00	35.40	272.40
70 watt high pressure sodium lamp	CE@.764	Ea	237.00	35.40	272.40
22 watt fluorescent circline lamp	CE@.955	Ea	52.00	44.20	96.20

Wall mounted 6" wide, 9" high, 7" deep fixture, polycarbonate prismatic diffuser

One 100 watt incandescent lamp	CE@.572	Ea	55.20	26.50	81.70
13 watt fluorescent lamp	CE@.764	Ea	86.40	35.40	121.80
35 watt high pressure sodium lamp	CE@.475	Ea	234.00	22.00	256.00
50 watt high pressure sodium lamp	CE@.475	Ea	237.00	22.00	259.00
70 watt high pressure sodium lamp	CE@.475	Ea	224.00	22.00	246.00

Lamps for vandal resistant fixtures

100 watt incandescent lamp	--	Ea	.91	--	--
13 watt fluorescent lamp	--	Ea	9.55	--	--
35 to 70 watt high pressure sodium lamp	--	Ea	47.80	--	--

Lighted Exit Signs. Includes setting and connecting prewired fixtures but no wire or conduit. These costs include the subcontractor's overhead and profit.
Universal mount (wall or ceiling), 6" letters on one side, stencil face, painted steel housing, two 15 watt lamps

	CE@.764	Ea	78.20	35.40	113.60

Electrical 16

	Craft@Hrs	Unit	Material	Labor	Total
Universal mount, 6" letters on two sides of polycarbonate housing, two 15 watt lamps	CE@.764	Ea	126.00	35.40	161.40
Add for battery backup for exit signs	CE@.381	Ea	158.00	17.70	175.70
Add for two circuit supply	--	Ea	12.10	--	--
Add for 24" stem hanger	--	Ea	16.50	--	--

Explosion Proof Lighting. No wire, conduit or concrete included. Includes setting and connecting prewired fixtures only. These costs include the subcontractor's overhead and profit.

Explosion proof incandescent fixtures, cast aluminum housing, heat-resistant prestressed globe, with fiberglass reinforced polyester reflector

	Craft@Hrs	Unit	Material	Labor	Total
60 watt fixture	CE@1.74	Ea	59.40	80.60	140.00
200 watt fixture	CE@1.74	Ea	177.00	80.60	257.60

Explosion proof (Class 1 Division 2) fixtures, cast aluminum housing, heat-resistant prestressed globe, with fiberglass reinforced polyester reflector

	Craft@Hrs	Unit	Material	Labor	Total
250 watt high pressure sodium fixture	CE@2.57	Ea	741.00	119.00	860.00
175 watt metal halide fixture	CE@2.57	Ea	570.00	119.00	689.00
400 watt metal halide fixture	CE@2.57	Ea	712.00	119.00	831.00

Explosion proof (Class 1 Division 2) fixtures, cast aluminum housing, heat-resistant prestressed globe, with fiberglass reinforced polyester reflector

	Craft@Hrs	Unit	Material	Labor	Total
70 watt high pressure sodium fixture	CE@2.57	Ea	387.00	119.00	506.00
250 watt high pressure sodium fixture	CE@2.57	Ea	524.00	119.00	643.00
175 watt metal halide fixture	CE@2.57	Ea	356.00	119.00	475.00
400 watt metal halide fixture	CE@2.57	Ea	503.00	119.00	622.00

Emergency Lighting. Includes setting and connecting prewired fixtures but no wire or conduit. These costs include the subcontractor's overhead and profit.

Shelf-mounted solid state battery pack with three-rate charger, lead-acid battery, baked enamel finish, test switch, pilot light and automatic overload breaker, cost per battery pack

	Craft@Hrs	Unit	Material	Labor	Total
Six volt, with 2 halogen lamps	CE@1.74	Ea	187.00	80.60	267.60
Twelve volt, with 2 sealed beam lamps	CE@1.74	Ea	224.00	80.60	304.60
Emergency battery pack for one 80 watt tube	CE@.955	Ea	287.00	44.20	331.20
Emergency battery pack for one 40 watt tube	CE@.955	Ea	261.00	44.20	305.20

Ceiling or wall surface mounted 9" x 9" fixture with two 8 watt tungsten halogen lamps, acrylic prismatic lens, 6 volt NiCad battery with two rate charger, automatic transfer switch and pilot light — CE@1.62 Ea 201.00 75.10 276.10

Remote emergency lighting fixtures, 9 watt tungsten halogen lamps, 12 volt, no battery included

	Craft@Hrs	Unit	Material	Labor	Total
Two pivoting heads, wall mount 	CE@.825	Ea	83.30	38.20	121.50
One pivoting head, wall mount	CE@.478	Ea	37.20	22.10	59.30
Round, flush ceiling mount	CE@.825	Ea	39.20	38.20	77.40
Square, surface wall mount	CE@.737	Ea	52.50	34.10	86.60

Pivot head channel mount 9 watt tungsten halogen fixture. NiCad battery with test switch, pilot light, two-rate charger, automatic transfer, and low-voltage cutoff

	Craft@Hrs	Unit	Material	Labor	Total
2 head fixture	CE@.825	Ea	267.00	38.20	305.20
1 head fixture	CE@.825	Ea	178.00	38.20	216.20

Yard and Street Lighting. Includes setting and connecting prewired fixtures with ballast but no pole, concrete work, excavation, wire, lamps or conduit. See pole costs below. Equipment is a flatbed truck with a boom & auger at $20.00 per hour. These costs include the subcontractor's overhead and profit. Mast arm mounted rectangular high pressure sodium flood, aluminum housing with baked enamel finish, anodized aluminum reflector, tempered glass lens, slip-fitting for mast mount, photoelectric cell

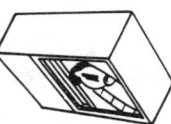

	Craft@Hrs	Unit	Material	Labor	Equipment	Total
70 watt	E1@1.55	Ea	300.00	63.20	6.42	369.62
100 watt	E1@1.55	Ea	296.00	63.20	6.42	365.62

Electrical 16

	Craft@Hrs	Unit	Material	Labor	Equipment	Total

Yard and Street Lighting, continued

	Craft@Hrs	Unit	Material	Labor	Equipment	Total
150 watt	E1@1.55	Ea	309.00	63.20	6.42	378.62
250 watt	E1@1.55	Ea	332.00	63.20	6.42	401.62
400 watt	E1@1.55	Ea	393.00	63.20	6.42	462.62
Add for 6' x 2" mounting arm	E1@.547	Ea	64.40	22.30	2.30	89.00

Yoke mounted high intensity discharge floor, die-cast anodized aluminum housing with tempered glass lens

	Craft@Hrs	Unit	Material	Labor	Equipment	Total
400 watt high pressure sodium	E1@2.71	Ea	298.00	110.00	11.30	419.30
1,000 watt high pressure sodium	E1@2.71	Ea	452.00	110.00	11.30	573.30
400 watt metal halide	E1@2.71	Ea	283.00	110.00	11.30	404.30
1,000 watt metal halide	E1@2.71	Ea	380.00	110.00	11.30	501.30
400 watt mercury vapor	E1@2.71	Ea	283.00	110.00	11.30	404.30
1,000 watt mercury vapor	E1@2.71	Ea	380.00	110.00	11.30	501.30

Round flood, aluminum reflector, impact-resistant glass lens

	Craft@Hrs	Unit	Material	Labor	Equipment	Total
400 watt, high pressure sodium	E1@2.71	Ea	344.00	110.00	11.30	465.30
1,000 watt, high pressure sodium	E1@2.71	Ea	552.00	110.00	11.30	673.30
400 watt, metal halide	E1@2.71	Ea	318.00	110.00	11.30	439.30
1,000 watt, metal halide	E1@2.71	Ea	400.00	110.00	11.30	521.30
1,500 watt, high pressure sodium	E1@2.71	Ea	477.00	110.00	11.30	598.30

Cobra head high intensity flood, anodized aluminum reflector, glass lens, die cast aluminum housing, with photoelectric control but no bulb or mounting arm

	Craft@Hrs	Unit	Material	Labor	Equipment	Total
150 watt high pressure sodium	E1@2.71	Ea	184.00	110.00	11.30	305.30
200 watt high pressure sodium	E1@2.71	Ea	207.00	110.00	11.30	328.30
250 watt mercury vapor	E1@2.71	Ea	161.00	110.00	11.30	282.30
400 watt high pressure sodium	E1@2.71	Ea	314.00	110.00	11.30	435.30
Add for 6' x 2" mounting arm	E1@.568	Ea	67.60	23.20	2.37	93.17

Square metal halide flood, anodized aluminum housing, tempered glass door, trunnion or yoke mount

	Craft@Hrs	Unit	Material	Labor	Equipment	Total
400 watt	E1@2.71	Ea	298.00	110.00	11.30	419.30
1,000 watt	E1@2.71	Ea	354.00	110.00	11.30	475.30

Yard and Street Lighting Poles. These costs include the subcontractor's overhead and profit. Hole for a pole foundation dug with a truck-mounted auger, no soil disposal included, per CF of undisturbed soil

	Craft@Hrs	Unit	Material	Labor	Equipment	Total
	TO@.064	CF	--	2.75	1.23	3.98

Concrete pole foundations, formed, poured and finished, with anchor bolts. Material costs shown assume concrete, forms and anchor bolts at $150.00 per CY including subcontractor's markup

	Craft@Hrs	Unit	Material	Labor	Equipment	Total
12" diameter, 36" deep, for 12' pole	E3@.746	Ea	18.00	30.50	3.09	51.59
24" diameter, 72" deep, for 30' pole	E3@1.14	Ea	144.00	46.60	4.64	195.24
30" diameter, 76" deep, for 40' pole	E3@1.54	Ea	240.00	63.00	6.39	309.39

Yard and street light poles, including base plate but no top brackets. See labor costs below. Crouse-Hinds

		Unit	Round steel	Square steel	Aluminum
12' high		Ea	502.00	391.00	940.00
16' high		Ea	582.00	465.00	1,100.00
20' high, 6" wide		Ea	851.00	543.00	1,470.00
25' high, 6" wide		Ea	1,020.00	734.00	1,840.00
30' high, 8" wide		Ea	1,160.00	1,290.00	2,560.00
35' high, 8" wide		Ea	1,320.00	1,990.00	3,440.00
40' high, 9" wide		Ea	1,790.00	2,100.00	4,070.00
50' high, 10" wide		Ea	3,020.00	3,170.00	--

Labor setting steel light poles. Equipment is a flatbed truck with a boom & auger at $20.00 per hour.

	Craft@Hrs	Unit	Material	Labor	Equipment	Total
To 20' high	E1@2.89	Ea	--	118.00	11.90	129.90
25' to 30' high	E1@4.35	Ea	--	177.00	18.40	195.40
35' to 40' high	E1@5.73	Ea	--	234.00	23.60	257.60

Electrical 16

	Craft@Hrs	Unit	Material	Labor	Equipment	Total
50' high	E1@8.57	Ea	--	349.00	35.30	384.30
Deduct for aluminum poles	--	%	--	-20.0	-20.0	--

Top brackets for steel and aluminum light poles, installed prior to setting pole

	Craft@Hrs	Unit	Material	Labor	Equipment	Total
Single arm	CE@.500	Ea	79.00	23.20	2.37	104.57
Two arm	CE@.500	Ea	92.30	23.20	2.37	117.87
Three arm	CE@.991	Ea	119.00	45.90	4.64	169.54
Four arm	CE@.991	Ea	142.00	45.90	4.64	192.54

Area Lighting, Single Pole Installation. Includes excavation for pole and 150' conduit run, pole foundation, 30' square steel pole and arm for each fixture, 150' of 2" PVC conduit with RSC terminations, 450' of #8 copper wire, and 480 volt high pressure sodium 400 watt luminaire with lamp. Equipment is a flatbed truck with a boom & auger at $20.00 per hour. These costs include the subcontractor's overhead and profit. Cost per pole.

	Craft@Hrs	Unit	Material	Labor	Equipment	Total
1 fixture on pole	E1@30.4	Ea	2,540.00	1,240.00	126.00	3,906.00
2 fixtures on pole	E1@32.5	Ea	3,280.00	1,330.00	134.00	4,744.00
3 fixtures on pole	E1@34.8	Ea	4,050.00	1,420.00	143.00	5,613.00
4 fixtures on pole	E1@36.8	Ea	4,790.00	1,500.00	151.00	6,441.00
Deduct for aluminum wire per pole	--	LS	--	--	--	-34.50

Area Lighting, Multiple Pole Installation. Includes excavation for poles and conduit, pole foundations, 30' square steel poles with one arm and one fixture per pole. Each pole after the first includes 150' of 2" PVC conduit with RSC terminations and 450' of wire. Based on one 480 volt high pressure sodium luminaire with lamp on each pole. Equipment is a flatbed truck with a boom & auger at $20.00 per hour. These costs include the subcontractor's overhead and profit. Cost per pole.

Aluminum wire, 250 watt lamps

	Craft@Hrs	Unit	Material	Labor	Equipment	Total
First 4 poles, #6 wire	E1@25.4	Ea	2,410.00	1,040.00	105.00	3,555.00
Next 8 poles, #4 wire	E1@30.7	Ea	2,520.00	1,250.00	127.00	3,897.00
Next 12 poles, #2 wire	E1@32.2	Ea	2,570.00	1,310.00	133.00	4,013.00
Next 8 poles, 1/0 wire	E1@33.5	Ea	2,620.00	1,370.00	138.00	4,128.00
Next 12 poles, 2/0 wire	E1@34.8	Ea	2,680.00	1,420.00	143.00	4,243.00
Next 12 poles, 3/0 wire	E1@33.5	Ea	2,730.00	1,370.00	138.00	4,238.00

Copper wire, 400 watt lamps

	Craft@Hrs	Unit	Material	Labor	Equipment	Total
First 2 poles, #8 wire	E1@20.1	Ea	2,360.00	819.00	82.70	3,261.70
Next 5 poles, #6 wire	E1@28.1	Ea	2,570.00	1,150.00	116.00	3,836.00
Next 5 poles, #4 wire	E1@29.4	Ea	2,620.00	1,200.00	122.00	3,942.00
Next 3 poles, #3 wire	E1@30.7	Ea	2,680.00	1,250.00	127.00	4,057.00
Next 5 poles, #2 wire	E1@32.2	Ea	2,730.00	1,310.00	133.00	4,173.00
Next 7 poles, #1 wire	E1@33.5	Ea	2,780.00	1,370.00	138.00	4,288.00
Next 8 poles, 1/0 wire	E1@34.8	Ea	2,840.00	1,420.00	143.00	4,403.00
Next 7 poles, 2/0 wire	E1@36.1	Ea	2,950.00	1,470.00	148.00	4,568.00
Next 13 poles, 3/0 wire	E1@37.3	Ea	3,050.00	1,520.00	153.00	4,723.00
Next 7 poles, 4/0 wire	E1@38.9	Ea	3,160.00	1,590.00	161.00	4,911.00

Standby Power Generators. Diesel generators including exhaust system, fuel tank and pump, automatic transfer switch, equipment pad, vibration isolators, associated piping, conduit, and lifting into position. Typical prices including the subcontractor's overhead and profit. Use these figures for preliminary estimates.

	Craft@Hrs	Unit	Material	Labor	Total
10 kw	--	Ea	--	--	10,750.00
30 kw	--	Ea	--	--	19,150.00
45 kw	--	Ea	--	--	23,080.00
50 kw	--	Ea	--	--	24,080.00
60 kw	--	Ea	--	--	28,920.00
75 kw	--	Ea	--	--	30,130.00

Electrical 16

	Craft@Hrs	Unit	Material	Labor	Total
Standby Power Generators, continued					
90 kw	--	Ea	--	--	35,060.00
100 kw	--	Ea	--	--	38,890.00
125 kw	--	Ea	--	--	39,710.00
150 kw	--	Ea	--	--	48,870.00
175 kw	--	Ea	--	--	52,790.00
200 kw	--	Ea	--	--	53,110.00
300 kw	--	Ea	--	--	77,280.00
350 kw	--	Ea	--	--	91,280.00
400 kw	--	Ea	--	--	92,590.00
450 kw	--	Ea	--	--	128,400.00
500 kw	--	Ea	--	--	131,700.00
600 kw	--	Ea	--	--	162,200.00
750 kw	--	Ea	--	--	222,600.00
900 kw	--	Ea	--	--	269,900.00
1,000 kw	--	Ea	--	--	273,500.00
1,100 kw	--	Ea	--	--	340,700.00

Automatic Transfer Switches for standby power systems. Single phase, 3 wire, solid neutral, 600 volt, 50 to 60 cycle. NEMA Class 1 (indoor). These costs include the subcontractor's overhead and profit.

	Craft@Hrs	Unit	Material	Labor	Total
30 amp	E4@4.85	Ea	1,830.00	195.00	2,025.00
40 amp	E4@4.97	Ea	2,040.00	200.00	2,240.00
80 amp	E4@4.97	Ea	2,850.00	200.00	3,050.00
100 amp	E4@4.97	Ea	3,220.00	200.00	3,420.00
150 amp	E4@4.97	Ea	4,400.00	200.00	4,600.00
225 amp	E4@8.76	Ea	6,870.00	352.00	7,222.00
260 amp	E4@10.8	Ea	7,620.00	434.00	8,054.00
400 amp	E4@15.1	Ea	9,400.00	607.00	10,007.00
600 amp	E4@16.2	Ea	11,300.00	652.00	11,952.00
800 amp	E4@18.9	Ea	12,900.00	760.00	13,660.00
1,000 amp	E4@21.4	Ea	16,400.00	861.00	17,261.00
1,200 amp	E4@25.9	Ea	17,400.00	1,040.00	18,440.00

Security and Alarm Systems. Sensor costs include connection but no wire runs. These costs include the subcontractor's overhead and profit.

Monitor panels, including accessory section and connection to signal and power supply wiring, cost per panel.

	Craft@Hrs	Unit	Material	Labor	Total
1 zone wall-mounted cabinet with 1 monitor panel, tone, standard line supervision, and 115 volt power supply	E4@2.02	Ea	946.00	81.20	1,027.20
5 zone wall-mounted cabinet with 5 monitor panels, 10 zone monitor rack, 5 monitor panel blanks, tone, standard line supervision and 115 volt power supply	E4@8.21	Ea	5,360.00	330.00	5,690.00
10 zone wall-mounted cabinet with 10 monitor panels, 10 zone monitor rack, tone, standard line supervision and 115 volt power supply	E4@12.1	Ea	6,150.00	487.00	6,637.00
10 zone monitor rack and 10 monitor panels with tone and standard line supervision but no power supply and no cabinet	E4@11.1	Ea	5,100.00	446.00	5,546.00
1 zone wall-mounted cabinet with 1 monitor panel, tone, high security line supervision and 115 volt power supply	E4@2.02	Ea	1,260.00	81.20	1,341.20
5 zone wall-mounted cabinet with 10 zone monitor rack, 5 monitor panel blanks, tone, high security line supervision but no power supply	E4@7.82	Ea	6,720.00	315.00	7,035.00
10 zone wall-mounted cabinet with 10 monitor panels, 10 zone monitor rack, tone, high security line supervision and 115 volt power supply	E4@12.0	Ea	7,880.00	483.00	8,363.00
Emergency power indicator for monitor panel	E4@1.14	Ea	252.00	45.90	297.90

Electrical 16

	Craft@Hrs	Unit	Material	Labor	Total
Monitor panels					
Panel with accessory section, tone and standard line supervision	E4@.990	Ea	415.00	39.80	454.80
Panel with accessory section, tone and high security line supervision	E4@.904	Ea	625.00	36.40	661.40
Monitor racks					
1 zone with 115 volt power supply	E4@.904	Ea	342.00	36.40	378.40
10 zone with 115 volt power supply	E4@.904	Ea	1,760.00	36.40	1,796.40
10 zone with no power supply	E4@.904	Ea	1,350.00	36.40	1,386.40
Monitor cabinets					
1 zone monitor for wall mounted cabinet	E4@.990	Ea	536.00	39.80	575.80
5 zone monitor for wall mounted cabinet	E4@.990	Ea	689.00	39.80	728.80
10 zone monitor for wall mounted cabinet	E4@.990	Ea	840.00	39.80	879.80
20 zone monitor for wall mounted cabinet	E4@.990	Ea	1,130.00	39.80	1,169.80
50 zone monitor for wall mounted cabinet	E4@.990	Ea	2,540.00	39.80	2,579.80
50 zone monitor for floor mounted cabinet	E4@1.04	Ea	2,760.00	41.80	2,801.80
Balanced magnetic door switches					
Surface mounted	E4@2.02	Ea	105.00	81.20	186.20
Surface mounted with remote test	E4@2.02	Ea	147.00	81.20	228.20
Flush mounted	E4@2.02	Ea	101.00	81.20	182.20
Mounted bracket or spacer	E4@.249	Ea	6.04	10.00	16.04
Photoelectric sensors					
500' range, 12 volt DC	E4@2.02	Ea	294.00	81.20	375.20
800' range, 12 volt DC	E4@2.02	Ea	330.00	81.20	411.20
Fence type, 6 beam, without wire or trench	E4@3.89	Ea	10,700.00	156.00	10,856.00
Fence type, 9 beam, without wire or trench	E4@3.89	Ea	13,400.00	156.00	13,556.00
Capacitance wire grid systems					
Surface type	E4@.990	Ea	73.50	39.80	113.30
Duct type	E4@.990	Ea	59.90	39.80	99.70
Tube grid kit	E4@.990	Ea	109.00	39.80	148.80
Vibration sensors, to 30 per zone	E4@2.02	Ea	130.00	81.20	211.20
Audio sensors, to 30 per zone	E4@2.02	Ea	153.00	81.20	234.20
Inertia sensors, outdoor, without trenching	E4@2.02	Ea	105.00	81.20	186.20
Inertia sensors, indoor	E4@2.02	Ea	66.70	81.20	147.90
Electric sensor cable, 300 meters, no trenching	E4@3.89	Ea	1,560.00	156.00	1,716.00
Ultrasonic transmitters, to 20 per zone					
Omni-directional	E4@2.02	Ea	65.70	81.20	146.90
Directional	E4@2.02	Ea	73.50	81.20	154.70
Ultrasonic transceivers, to 20 per zone					
Omni-directional	E4@2.02	Ea	73.50	81.20	154.70
Directional	E4@2.02	Ea	78.80	81.20	160.00
High security	E4@2.02	Ea	268.00	81.20	349.20
Standard security	E4@2.02	Ea	109.00	81.20	190.20
Microwave perimeter sensor, to 4 per zone	E4@3.89	Ea	4,360.00	156.00	4,516.00
Passive interior infrared sensors, to 20 per zone					
Wide pattern	E4@2.02	Ea	599.00	81.20	680.20
Narrow pattern	E4@2.02	Ea	567.00	81.20	648.20
Access and secure control units					
For balanced magnetic door switches	E4@3.06	Ea	330.00	123.00	453.00
For photoelectric sensors	E4@3.06	Ea	599.00	123.00	722.00
For capacitance sensors	E4@3.06	Ea	604.00	123.00	727.00
For audio and vibration sensors	E4@3.06	Ea	699.00	123.00	822.00
For inertia sensors	E4@3.06	Ea	625.00	123.00	748.00

Electrical 16

	Craft@Hrs	Unit	Material	Labor	Total
Security and Alarm Systems, Access and secure control units, continued					
For nimmer (m.b.) detector	E4@3.06	Ea	515.00	123.00	638.00
For electric cable sensors	E4@3.06	Ea	3,130.00	123.00	3,253.00
For ultrasonic sensors	E4@3.06	Ea	961.00	123.00	1,084.00
For microwave sensors	E4@3.06	Ea	589.00	123.00	712.00
Accessories					
Tamper assembly for monitor cabinet	E4@.990	Ea	65.70	39.80	105.50
Monitor panel blank	E4@.249	Ea	7.35	10.00	17.35
Audible alarm	E4@.990	Ea	76.70	39.80	116.50
Audible alarm control	E4@.990	Ea	305.00	39.80	344.80
Termination screw terminal cabinet, 25 pair	E4@.990	Ea	220.00	39.80	259.80
Termination screw terminal cabinet, 50 pair	E4@1.36	Ea	357.00	54.70	411.70
Termination screw terminal cabinet, 150 pair	E4@3.45	Ea	589.00	139.00	728.00
Universal termination with remote test (for cabinets and control panels)	E4@1.47	Ea	54.10	59.10	113.20
Universal termination without remote test	E4@.990	Ea	31.50	39.80	71.30
High security line supervision termination with tone and remote test	E4@1.47	Ea	273.00	59.10	332.10
High security line supervision termination with tone only	E4@1.47	Ea	252.00	59.10	311.10
12" door cord for capacitance sensor	E4@.249	Ea	8.40	10.00	18.40
Insulation block kit for capacitance sensor	E4@.990	Ea	42.00	39.80	81.80
Termination block for capacitance sensor	E4@.249	Ea	8.94	10.00	18.94
200 zone graphic display	E4@3.06	Ea	4,200.00	123.00	4,323.00
210 zone multiplexer event recorder unit	E4@3.89	Ea	4,100.00	156.00	4,256.00
Guard alert display	E4@3.06	Ea	940.00	123.00	1,063.00
Uninterruptable power supply (battery)	E4@2.02	Ea	1,920.00	81.20	2,001.20
12 volt, 40VA plug-in transformer	E4@.495	Ea	36.20	19.90	56.10
18 volt, 40VA plug-in transformer	E4@.495	Ea	24.70	19.90	44.60
24 volt, 40VA plug-in transformer	E4@.495	Ea	17.90	19.90	37.80
Test relay	E4@.332	Ea	55.70	13.40	69.10
Sensor test control (average for sensors)	E4@.990	Ea	73.00	39.80	112.80
Cable, 2 pair 22 gauge, twisted, 1 pair shielded, run in exposed walls or pulled in raceway and connected	E4@.008	LF	.53	.32	.85
Fire Alarm and Detection Systems. No wiring included except as noted. These costs include the subcontractor's overhead and profit.					
Pedestal mounted master fire alarm box	E4@3.89	Ea	1,410.00	156.00	1,566.00
Automatic coded signal transmitter	E4@3.89	Ea	746.00	156.00	902.00
Fire alarm control panel, simplex	E4@3.89	Ea	699.00	156.00	855.00
Fire detection annunciator panel					
8 zone drop	E4@3.89	Ea	462.00	156.00	618.00
12 zone drop	E4@5.03	Ea	657.00	202.00	859.00
16 zone drop	E4@5.93	Ea	946.00	239.00	1,185.00
8 zone lamp panel only	E4@4.33	Ea	152.00	174.00	326.00
12 zone lamp panel only	E4@5.93	Ea	225.00	239.00	464.00
16 zone lamp panel only	E4@7.23	Ea	300.00	291.00	591.00
Battery charger and cabinet, simplex	E4@6.19	Ea	405.00	249.00	654.00
Add for nickel cadmium batteries	E4@3.89	LS	436.00	156.00	592.00
Fire alarm pull stations, manual operation	E4@.497	Ea	22.50	20.00	42.50
Fire alarm 10" bell with outlet box	E4@.497	Ea	73.50	20.00	93.50
Fire alarm 4" trouble bell with outlet box	E4@.642	Ea	44.20	25.80	70.00
Magnetic door holder	E4@1.33	Ea	68.90	53.50	122.40
Combination door holder and closer	E4@2.48	Ea	278.00	99.70	377.70

Electrical 16

	Craft@Hrs	Unit	Material	Labor	Total
Detex door lock	E4@2.72	Ea	33.00	109.00	142.00
Thermodetector	E4@.497	Ea	9.99	20.00	29.99
Ionization AC smoke detector, with wiring	E4@.710	Ea	171.00	28.60	199.60
Fixed temp. and rate of rise smoke detector	E4@.746	Ea	15.20	30.00	45.20
Carbon dioxide pressure switch	E4@.746	Ea	5.25	30.00	35.25

Telephone Wiring. These costs include the subcontractor's overhead and profit.

Plywood backboard mount for circuit wiring, set on masonry wall

	Craft@Hrs	Unit	Material	Labor	Total
4' x 4' x 3/4"	E4@.938	Ea	25.40	37.70	63.10
4' x 8' x 3/4"	E4@1.10	Ea	39.20	44.20	83.40

Cable tap in a manhole or junction box

	Craft@Hrs	Unit	Material	Labor	Total
25 to 50 pair cable	E4@7.80	Ea	103.00	314.00	417.00
100 to 200 pair cable	E4@10.1	Ea	117.00	406.00	523.00
300 pair cable	E4@12.0	Ea	117.00	483.00	600.00
400 pair cable	E4@13.0	Ea	130.00	523.00	653.00

Cable terminations in a manhole or junction box

	Craft@Hrs	Unit	Material	Labor	Total
25 pair cable	E4@1.36	Ea	9.31	54.70	64.01
50 pair cable	E4@1.36	Ea	10.60	54.70	65.30
100 pair cable	E4@2.72	Ea	21.20	109.00	130.20
150 pair cable	E4@3.45	Ea	32.10	139.00	171.10
200 pair cable	E4@4.61	Ea	48.20	185.00	233.20
300 pair cable	E4@5.03	Ea	63.90	202.00	265.90
400 pair cable	E4@5.34	Ea	84.90	215.00	299.90

Communications cable, pulled in conduit or walls and connected

	Craft@Hrs	Unit	Material	Labor	Total
2 pair cable	E4@.008	LF	.06	.32	.38
25 pair cable	E4@.030	LF	.62	1.21	1.83
50 pair cable	E4@.043	LF	1.32	1.73	3.05
75 pair cable	E4@.043	LF	1.66	1.73	3.39
100 pair cable	E4@.043	LF	1.93	1.73	3.66
150 pair cable	E4@.045	LF	2.70	1.81	4.51
200 pair cable	E4@.051	LF	3.53	2.05	5.58
300 pair cable	E4@.051	LF	5.19	2.05	7.24

Telephone outlets (no junction box included)

	Craft@Hrs	Unit	Material	Labor	Total
Wall outlet	E4@.396	Ea	5.63	15.90	21.53
Floor outlet, flush mounted	E4@.495	Ea	7.00	19.90	26.90

Closed Circuit Television Systems. These costs include the subcontractor's overhead and profit.

	Craft@Hrs	Unit	Material	Labor	Total
Outdoor television camera (16 mm f11.6 lens, black and white) in weatherproof housing, wall or roof mount	E4@1.67	Ea	2,120.00	67.20	2,187.20
Self-terminating outlet box with cover plate and male connector	E4@.689	Ea	5.30	27.70	33.00
Coaxial cable (RG59/J), 75 ohms, no raceway included	E4@.006	LF	.13	.25	.37
In-line cable taps (PTU) for 36 TV system	E4@.689	Ea	7.88	28.90	36.78
Cable blocks for in-line taps	--	Ea	2.97	--	--
Desk type black and white TV monitor, 12" diagonal screen, 75 ohms, 0.5 to 2.0 Vp-p external sync input	E4@2.77	Ea	398.00	116.00	514.00

Public Address Paging Systems. These costs include the subcontractor's overhead and profit.

	Craft@Hrs	Unit	Material	Labor	Total
Signal cable, #22 solid copper polyethylene insulated, PVC jacketed wire, pulled in conduit, no conduit included	E4@.008	LF	.21	.33	0.54
Microphone, desk type, push to talk, 60 to 10,000 Hertz, 150 ohms, with 7' shielded cable	E4@.839	Ea	106.00	35.10	141.10

Electrical 16

	Craft@Hrs	Unit	Material	Labor	Total

Public Address Paging Systems, continued

Wall mounted enameled steel amplifier cabinet (18" high, 24" wide, 6" deep) with electrical receptacle and lockable door

| | E4@1.67 | Ea | 197.00 | 70.00 | 267.00 |

Amplifier, 50 watts, 40 to 20,000 Hertz, 8 ohms, output, with treble and bass control and magnetic tape input, includes transformer

| | E4@.801 | Ea | 377.00 | 33.60 | 410.60 |

Reflex wall mounted paging horns, 250 to 13,000 Hertz, 8 ohms, 30 watts, 14" wide, 6" high, 13" deep, gray baked enamel

	Craft@Hrs	Unit	Material	Labor	Total
Interior horns	E4@1.67	Ea	66.30	70.00	136.30
Exterior horns, weatherproof	E4@2.54	Ea	144.00	106.00	250.00

Testing Electrical Installations. These costs include the subcontractor's overhead and profit.

	Craft@Hrs	Unit	Material	Labor	Total
Devices	CE@.125	Ea	--	6.04	--
Fixtures	CE@.167	Ea	--	8.08	--

How to Use Estimate Writer

Estimate Writer is the *1992 National Construction Estimator* on computer disk. If you use this book to compile cost estimates, the disk in the envelope inside the back cover can save time and help prevent many of the most common estimating mistakes. To use Estimate Writer, you'll need a computer running DOS 3.2 or higher with at least 5 Mb of free space on a hard disk and at least 384K of memory.

Estimate Writer comes in highly compressed form on a single 5¼" (1.2 Mb, high density) floppy disk. If your computer can't use 5¼" high density disks, instructions on page 563 explain how to order Estimate Writer on 5¼" (360K) or 3½" (720K) disks, or transfer Estimate Writer from the 1.2 Mb disk you have to 360K or 720K disks.

Estimate Writer has 3 parts:
>The Estimate Window
>The N.C.E. Window
>The Split Window.

Estimate Window

The Estimate Window is like a simple word processing program with special capabilities estimators need. For example, the Estimate Window totals labor, material and equipment columns automatically. It also multiplies the cost per unit by the number of units to find the extended cost or *extension*. And you can add the overhead and profit percentages of your choice at the end of every estimate. In short, the Estimate Window is a handy tool most estimators could use, even without the other two parts of Estimate Writer.

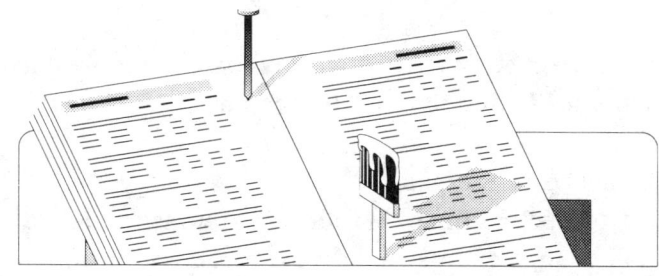

Illustrations by Bob Lee

N.C.E. Window

The N.C.E. Window has the entire *1992 National Construction Estimator* in electronic form. You can page through the whole book screen by screen, go quickly to any page you choose, or search the index by keyword to find the item you want (in seconds). In some ways, using the N.C.E. Window is easier than using the book itself.

Split Window

The Split Window is where you'll compile most estimates. It combines the N.C.E. Window (at the top of the screen) and the Estimate Window (at the bottom of the screen). Press F8 to select the line you want to copy from the N.C.E. Window. Then type the estimated material quantity. Estimate Writer copies that line (with costs totaled) to your estimate (the Estimate Window). That's the real power of Estimate Writer.

Getting Around in Estimate Writer
Use the Option Menu to jump between the three parts of Estimate Writer (the Estimate Window, N.C.E. Window and Split Window).

How to Use Estimate Writer

Any time the upper right corner of your screen shows:

> <Esc> for Option Menu

press Escape (*Esc* on your keyboard) to make the Option Menu pop onto the screen.

Other Advantages
Of course, Estimate Writer lets you insert, delete and change estimates, add overhead and profit and then print the estimate on paper. Maybe best of all, Estimate Writer can create a text file (sometimes called *ASCII*) that most of the popular word processing programs can use. That lets you dress up your estimates: center headings, put words in big, bold type, include section subtotals, check spelling automatically — or anything else your word processor can handle.

Need More Estimating Power?
Estimate Writer was created by SDSI Business Systems as an estimate *writing* program. SDSI also offers The Advanced Estimator, a comprehensive estimating system that records costs, lets you create a cost estimating database and interfaces with scheduling programs. If you need features not found in Estimate Writer, contact SDSI Business Systems at 27475 Ynez Road, Suite 385, Temecula, CA 92591, (714) 677-6785.

Getting Started
The pages that follow explain how to use Estimate Writer. If you have trouble loading the program, call Craftsman Book Company at **619-438-7828**. If you have trouble using the program, please read the instructions on disk before calling. From the Main Menu, select *Help*. Then select *Display Instructions*. You can also print the instructions on paper. Select *Print Instructions*.

Installing Estimate Writer
Don't try to install Estimate Writer on your hard disk with the DOS copy command. Instead, turn your computer on and go to the DOS prompt (such as C:>). Put the Estimate Writer disk in a floppy disk drive (such as A) and change control to that drive (such as by typing A: and pressing Enter). At the command prompt (such as A:>), type INSTALL and press Enter.

> A:>INSTALL

Then follow instructions on the screen:

> **Welcome to Estimate Writer**
>
> The first step is to load all programs and cost estimates. Estimate Writer assumes you're loading from the A disk drive:
>
> Load from diskette drive A
>
> If you're not loading from the A drive, change diskette drive by typing over the letter A above. Press Enter.
>
> Estimate Writer assumes you're loading to the C hard drive and the ESTWRITE directory:
>
> Load to hard drive C:\ESTWRITE
>
> If you want to load to a different drive or directory, change the drive letter or directory name above. Estimate Writer will create the directory (if needed).
>
> Press <Enter> to Accept or <Esc> to Change

When installation is complete, the Main Menu will be on the screen. Instructions for using Estimate Writer begin on the next page.

You'll find that Estimate Writer usually offers clues about what you should do next. When in doubt, look for a hint at the bottom of the screen.

When you finish using Estimate Writer, you'll be in the directory where Estimate Writer is installed. The next time you want to use Estimate Writer, change control to that same directory (such as by typing CD \ESTWRITE). Type ESTIMATE and press Enter. For example:

> C:\ESTWRITE>ESTIMATE

That restarts Estimate Writer at the Main Menu.

How to Use Estimate Writer

The Main Menu

This is the starting point. Move the highlight up or down with the ↑ or ↓ key (with Num Lock off) to make your choice and press Enter. Or, you can select by typing the first letter of the line you want.

- *Revise an Estimate* to revise an estimate you've already made.
- *Start an Estimate* to begin a new estimate.
- *Print an Estimate* you've already created.
- *Delete an Estimate* to remove an estimate from your hard disk.
- *Transfer an Estimate* to make a copy of an estimate your word processor can use.
- *Help* to rebuild or repaginate an estimate or to see instructions for Estimate Writer.
- *Quit* to exit Estimate Writer.

Start an Estimate

Begin a new estimate by moving the highlight down to *Start an Estimate*. Then press Enter. Or, you can simply type S. Next, type the name you'll give to this new estimate.

- The name can be up to 8 characters long.
- The name can be any combination of letters and numbers. But avoid using symbols.
- The name can't be one you've used already. To see names already in use, press the key labeled F5.

When you've typed the new estimate name (such as SAMPLE in the example at the right), press Enter to begin estimating.

If you decide not to start a new estimate, press Esc and type *M* to return to the Main Menu.

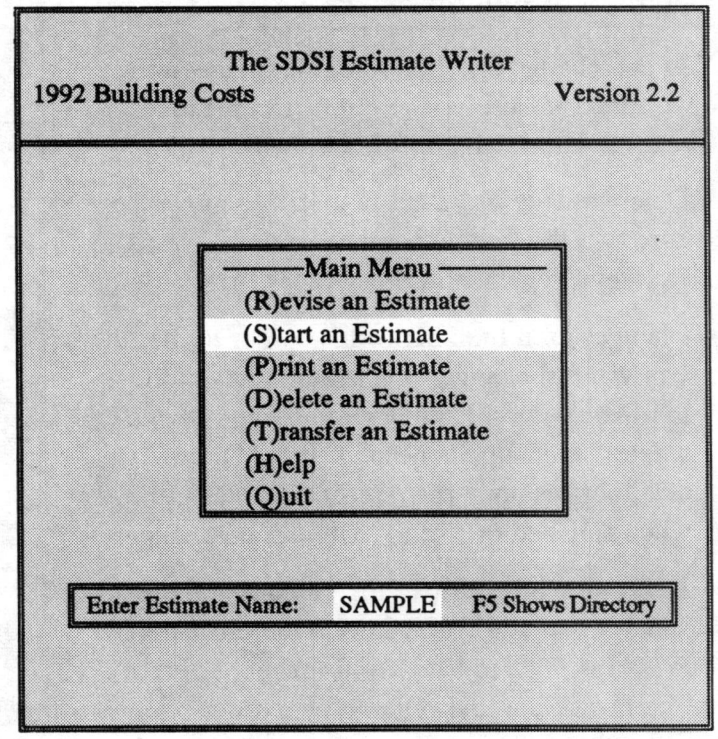

Typing in the Estimate Window

This is where you type text like titles and material descriptions (not dollar and cents costs). Try typing a job address like "1356 Tigertail Road." When you finish a line, or reach the right margin, press Enter to begin the next line.

- If you notice a mistake on a line *before* pressing Enter, erase with the backspace key or back up with the ← key (with Num Lock off).
- Until you press Enter at the end of a line, you can insert or delete anywhere on the line. Move the cursor to where you want to make a change. Then press Del to erase, or press Ins and type the addition. When the line is perfect, press Enter.
- If you notice a mistake after a line is finished, don't worry. You can change any line later.
- You type in the Estimating Window in the *insert mode*. Press Esc to leave insert mode.

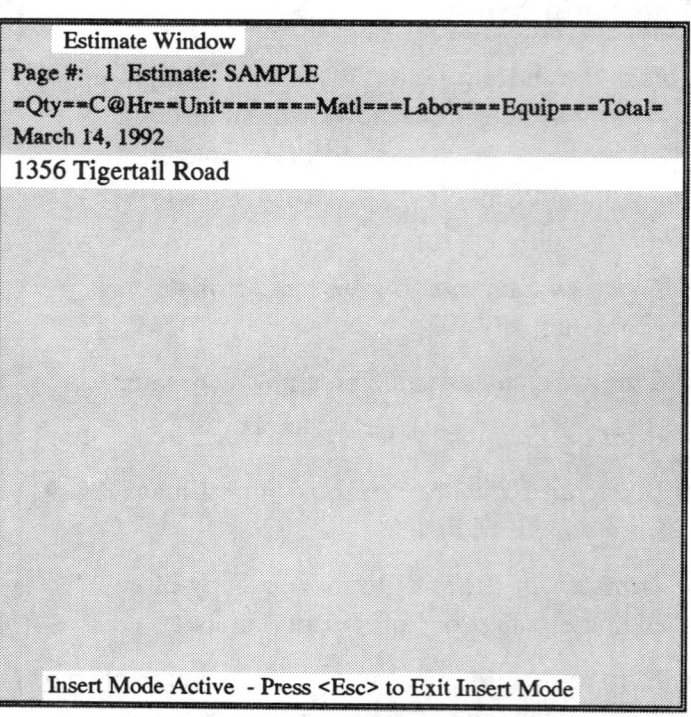

Insert Mode is on. Note the bottom line.

Delete, Change or Insert Text

You can change any line in the Estimate Window. First, be sure you're out of insert mode (press Esc) and that Num Lock is off.

- Delete a line: Use the ↑ or ↓ key to move the cursor to the line you want to delete. Press the Del key. Type Y and press Enter.
- Changes on a line: Move the cursor to the line you want to change. Press the F3 key. With the ← or → key, Home or End key, move to where the change is needed. Press Del to delete or Ins and type an insert. Press Enter when finished or Esc to exit with no changes.
- Insert new lines: Move the cursor to where you want to add a new line to the estimate. Press Ins. You're back in insert mode. Press Enter to skip a line. Then type a description (like "Concrete slab on grade" as in the example). Press Enter to accept the insert. Press Esc to leave insert mode.

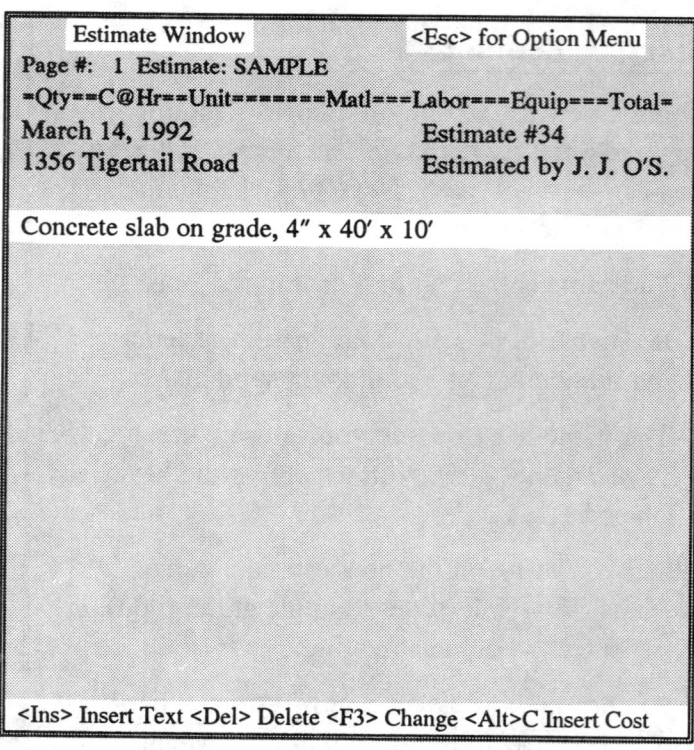

Delete, change, or insert in the Estimate Window

How to Use Estimate Writer

Costs in the Estimate Window

Next, move the cursor to a line where you want to insert a cost estimate. Hold the Alt key down and press C to open a cost estimating box.

- Type the quantity (400 in the sample). Press Enter. Type the manhours per unit (.034). Press Enter. Type the units (SF). Press Enter.

- Type the material cost per unit. Press Enter. Type labor and equipment unit costs the same way. The total cost and cost extensions (quantity times cost) appear automatically.

- If you're estimating only the total cost (as for a subcontracted item), skip over the hours, material, labor, and equipment columns by pressing Enter. Type only a total cost.

- If you want to change any figure, type Q, H, U, M, L, E, or T and press Enter. Type A and Enter to accept the insert or Esc to cancel.

When you accept the cost insert, unit costs (not extensions) appear in the Estimate Window.

Making Choices in the Option Menu

Use the Option Menu to jump between parts of Estimate Writer. To see the options, press Esc in either the Estimate Window, the N.C.E. Window or the Split Window.

- First, move the highlight to *Turn to N.C.E. Page*. Press Enter. Type any page number you want to see. Press Enter. Use the arrow keys or PgUp and PgDn to go from page to page. Press Esc to return to the Option Menu.

- Next, select *Estimate Full Window*. Press Enter. Estimate Writer turns to the page of the estimate you worked on last. To go to any other estimate page, select *Go To Estimate Page*. Press Esc to return to the Option Menu.

- Select *N.C.E. Full Window*. Press Enter and you're back at the N.C.E. page you saw last.

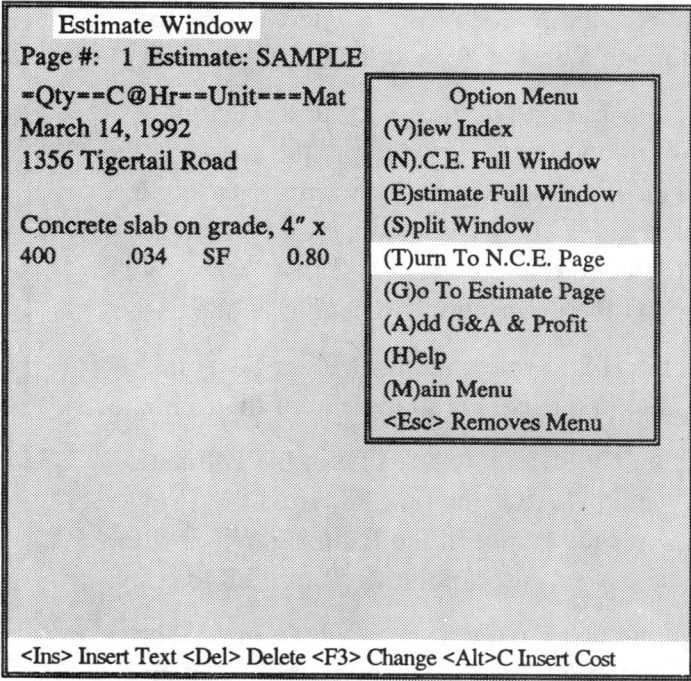

The Estimate Window shows costs per unit (such as per square foot). Your printed estimates show *extended* costs (unit price times the number of units).

Press Esc to bring the Option Menu onto the screen.

557

A Quick Way to Find Costs

To find cost estimates quickly, Press Esc and Enter. That selects *View Index* from the Option Menu. Press Enter again to *Search by Keyword*.

- Type the name of what you're looking for. In this case, type *driveway* and Press Enter.

- Estimate Writer will find the word in the index closest to the name you typed.

- Press the ↓ key three times to put the cursor on 47 in the left column opposite *forms* and under the index entry *Driveway*. Press Enter. Estimate Writer goes to the top of page 47.

- Press the PgDn key once. At the bottom of the screen you'll see cost estimates for driveway and walkway edge forms. Let's copy some of this description to our estimate.

Copying Text in the Split Window

Press Esc to see the Option Menu. Select *Split Window*. Press Enter. Press PgDn. We're going to copy a line of descriptions — not dollar costs — from the N.C.E. Window (screen top) to the Estimate Window (screen bottom).

- Press End to put the cursor in the Estimate Window. Press Home to put the cursor in the N.C.E. Window. Move the cursor to the line in the N.C.E. Window that you want to copy. (Select the line that starts *2" x 4" form*. We'll copy the description first and then the costs.)

- Press F8 to select the line. Press F8 again. That line is copied to the end of the estimate.

- You could *copy to* any line in the Estimate Window. After the first F8, press the I key. Move the cursor in the Estimate Window to the correct insert point and press F8 to copy.

- You don't have to copy a whole line. Shorten the right end of the selection with ← after the first F8. Shorten the left end by pressing → before the first F8.

```
  Index Window                              Index Menu
A                                     (S)earch by Keyword
Abbreviations................220     (R)esume Estimating
Abrasion........................226   (H)elp
Abrasive                              (M)ain Menu
   concrete finish.........311-312    <Esc> Removes Menu
   stair treads............141, 378
   terrazzo...................374     Keyword
Absorption....................226     block.....................327
Absorption chillers.........488       caulk......................37
Acacia veneer................124      designs....................74
Access                                engineering................74
   control units..........549-550     partitions................390
   doors.......................181    plaster...................371
   flooring..........385-386, 390     rooms.....................412
   stairs, folding...........202      tile, ceiling..........40-41
Access hatches                        tile, demolition....54, 244
   ceiling.....................355    wallboard texture.......375
   plumbing..................458      Acoustical adhesive......13
   sheet metal...............181      Acrovyn wall cover..384-385
Accordion folding                     Acrylic
   doors........................64       blank panels............39
   partitions................390        diffusers............39, 540
Acid cleaning, concrete....312         primer.............151-152
```

```
 N.C.E. Window                        <Esc> for Option Menu
 47 Edge Forms         Craft@Hrs  Unit   Matl  Labor  Total
                       B2@.027    LF     .41    .59   1.00
 Driveway and walkway edge forms. Material costs include stakes, nails,
 and form oil (@ $.16 per LF) and 5% waste. No stripping included.
 Per LF of edge form.
 2" x 4" form, Std & Btr (@$430 per MBF, .7 BF per LF)
   1 use                B2@.050   LF     .46    1.10   1.56
   3 use                B2@.050   LF     .25    1.10   1.35
   5 use                B2@.050   LF     .19    1.10   1.29
 2" x 6" form, Std & Btr (@$450 per MBF, 1.05 BF per LF)
 =Qty==C@Hr===Unit========Matl====Labor===Equip==Total
 March 14, 1992                    Estimate #34
 1356 Tigertail Road               Estimated by J. J. O'S.

 Concrete slab on grade, 4" x 40' x 10'
 400       .034     SF       0.80    0.30   0.00   1.10

 2" x 4" form, Std & Btr (@$430 per MBF, .7 BF per LF)

 Estimate Window    <F8> to Select - PgUp, PgDn ↑ ↓ <End> for Est Window
```

Use the F8 key to copy a line of text from the NCE Window to the Estimate Window.

How to Use Estimate Writer

Copying Costs in the Split Window

With the cursor on the line you want to copy (in this case, at *1 use*), press F8. That opens an estimating box. Type the quantity (100 in the example at the right) and press Enter. The total and extensions appear automatically.

- To change any cost in the estimating box, type Q, H, U, M, L, E or T. Then type the new number. Press Enter. Press F8 to copy those costs to the last line of your estimate.

- If you want to copy costs to the middle of the estimate, press I after entering the quantity. Move the cursor in the Estimate Window to the correct insert point. Then press F8 to copy.

- You can't make changes in the estimate when the screen is split. But it's easy to switch to the full Estimate Window where you can make any change. Just type Esc and E.

```
  N.C.E. Window
47 Edge Forms              Craft@Hrs   Unit   Matl   Labor   Total
                           B2@.027     LF     .41    .59     1.00

Driveway and walkway edge forms  Material costs include stakes, nails,
and form oil (@ $.16 per LF) and 5% waste. No stripping included.
Per LF of edge form.
2" x 4" form, Std & Btr (@$430 per MBF, .7 BF per LF)
  1 use                    B2@.050    LF     .46    1.10    1.56
  3 use                    B2@.050    LF     .25    1.10    1.35
  5 use                    B2@.050    LF     .19    1.10    1.29
2" x 6" form, Std & Btr (@$450 per MBF, 1.05 BF per LF)
=Qty==C@Hr==Unit========Matl===Labor===Equip===Total=
March 14, 1992                    Estimate #34
  Q    H      U       M       L        E        T
  Qty  C@Hr   Unit    Matl    Labor    Equip    Total
       B2@
  100  .050   LF      0.46    1.10     0.00     1.56
       Extensions     46.00   110.00   0.00     156.00

Q,H,U,M,L,E,T to Edit, <F8> to Add, <I> to Insert, <Esc> to Abort:
Estimate Window
```

Making Percentage Adjustments

Some lines in the N.C.E. show a percentage adjustment to make if certain conditions apply. To see an example, call the Option Menu with Esc. Select *Turn to N.C.E. Page*. Type 49 and press Enter. Be sure the window is split (press Esc S). Press PgUp. Press the ↓ key until the cursor is at the line *Add for light bending*. Then press F8 to open an estimating box as shown at the right.

- The highlight in the box is on the line below the adjustment (10.00% in this case).

- If the estimated cost for placing rebars on this job will be $150, type 150.00 and press Enter. Estimate Writer shows 15.00 (10% of $150) in the total column of the cost adjustment line.

- Press F8 and the adjustment is copied to the last line of the Estimate Window.

```
  N.C.E. Window
48 Steel Reinf. Bars           Craft@Hrs   Unit   Matl   Labor   Total
                               RI@.008     Lb     .25    .20     .45
1-1/8" diameter, #9 bar (3.40 lb per LF)
                               RI@.008     Lb     .25    .20     .45
1-1/4" diameter, #10 bar (4.30 lb per LF)
                               RI@.007     Lb     .25    .17     .42
1-3/8" diameter, #11 bar (5.31 lb per LF)
                               RI@.007     Lb     .25    .17     .42
Add for less than 5,000 lb job  --         Lb     .05    .10     .15
Add for light bending           --         %      --     10.0    --
Deduct for structural slabs     --         %      --     -10.0   --
=Qty==C@Hr==Unit========Matl===Labor===Equip===Total=
March 14, 1992                    Estimate #34
                        M        L        E        T
  Qty  C@Hr   Unit      Matl     Labor    Equip    Total
       --               0.00%    10.00%   0.00%    0.00%
  1    LS               0.00     150.00   0.00     150.00
  Cost Adjustment       0.00     15.00    0.00     15.00

M,L,E,T to Edit, <F8> to Add, <I> to Insert, <Esc> to Abort:
Estimate Window
```

Copying From Multiple Cost Lines

Some sections of the N.C.E. have all material, all labor or all subcontract costs in each row. Adhesives, near the top of page 13, is an example. Suppose you're estimating a ¼" bead of subfloor adhesive on 250 SF of joists spaced 12" center to center. Turn to N.C.E. page 13 (Esc T 13). Split the window (Esc S). Press PgDn. Here's how to use these cost estimates.

- Press ↓ three times so the cursor is on the line *12" OC members*. Then press F8 to open the estimating box.

- Press → to move the highlight to 12.20. Press Enter. Type the estimated quantity (2.5 for 250 SF). Press Enter. The total cost per unit and cost extensions appear automatically. Note the example at the right. Press F8 to copy unit costs to the Estimate Window.

```
N.C.E. Window
13 Panel Adhesives                    1/8"    1/4"    3/8"    1/2"
including 6% waste.                          Bead diameter
                                      1/8"    1/4"    3/8"    1/2"
Subfloor adhesive (at $.13 per oz), on floors
   12" OC members        CSF   5.40   12.20   13.75   21.90
   16" OC members        CSF   4.20    9.90   10.90   17.45
   20" OC members        CSF   3.80    8.45    9.75   14.95
Wall sheathing or shear panel adhesive (at $.14 per oz), on walls
   12" OC members        CSF   3.12   12.60   28.85   51.50
   16" OC members        CSF   2.90   11.60   26.30   46.75
   20" OC members        CSF   2.80   11.20   25.65   44.60
=Qty==C@Hr==Unit========Matl===Labor===Equip===Total=
March 14, 1992                        Estimate #34

   Q      H     U        M      L       E       T
  Qty    C@Hr  Unit     Matl   Labor  Equip   Total

   2.5         CSF     12.20    0.00    0.00   12.20
         Extensions    30.50    0.00    0.00   30.50

Q,H,U,M,L,E,T to Edit, <F8> to Add, <I> to Insert, <Esc> to Abort:

Estimate Window
```

Finishing and Printing Your Estimate

Adding General & Administrative expense and Profit to your estimate is easy. Just select the line *Add G&A & Profit* in the Option Menu. Then type the percentages you want to use. For example, if you wanted to add 25% to the estimate for G&A expense and 10% for Profit:

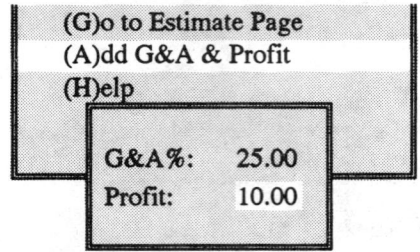

You can enter or change G&A expense and profit at any time. These items won't appear in the Estimate Window. But they'll be at the end of your printed estimate.

Saving your estimate to disk is automatic. Selecting *Main Menu* from the Options Menu creates three files on your hard disk. The names of all are the same as the name you gave the estimate. But the file suffixes (the three letters after the period) are different. The estimate suffix is .TXT. The index suffix is .NDX. The version of the estimate saved before repagination has the suffix .BAK. Estimates are repaginated each time they're saved to disk.

Print your estimate by selecting *Print an Estimate* from the Main Menu. You'll need 8½" by 11" continuous form paper.

Select *Transfer an Estimate* on the Main Menu to turn an estimate into a text or ASCII file (suffix .ASC). Most word processing programs (such as WordPerfect, Microsoft Word and Wordstar) can read ASCII files. Transfer the estimate to the directory where your word processing program is stored. With your favorite word processor, load the estimate name (with the suffix .ASC) from that directory. Set narrow margins so the estimate can be at least 7-7/8" wide. Then center headings, change type to bold or italic, check spelling automatically, create subtotals or multiply all labor costs by a constant. The only limit is your ingenuity and the features in your word processing program. Then print the estimate. But be sure to select monospaced type such as courier. Otherwise cost columns may not line up correctly.

How to Use Estimate Writer

```
┌─────────────────────────────────────────────┐
│ Estimate Window         <Esc> for Option Menu│
│ Page #:  1  Estimate: SAMPLE                │
│ Qty--C@Hr---Unit-------Matl---Labor--Equip====Total│
│ March 14, 1992              Estimate #34    │
│ 1356 Tigertail Road         Estimated by J. J. O'S.│
│                                             │
│ Concrete slab on grade, 4" x 40' x 10'      │
│ 400    .034    SF    0.80   0.30   0.00  1.10│
│                                             │
│ 2" x 4" form, Std & Btr (@$430 per MBF, .7BF per LF)│
│   1 use                                     │
│ 100    B2@.049 LF    .46    1.03   0.00  1.49│
│                                             │
│ Finishing 400 SF (subcontract)              │
│ 1              LS    0.00   0.00   0.00  50.00│
│                                             │
│ Form stripping, 2" x 4"                     │
│ 100    BL@.012 LF    --     .21    .00   0.21│
│                                             │
│ <Ins> Insert Text <Del> Delete <F3> Change <Alt>C Insert Cost│
└─────────────────────────────────────────────┘
```

```
┌─────────────────────────────────────────────┐
│                Estimate Report      Page: 1 │
│ Estimate Name: SAMPLE                       │
│                                             │
│ Description                                 │
│ Qty   Craft@Hrs  Unit   Material  Labor  Equip  Total│
│ ..........................................│
│ March 14, 1992              Estimate #34    │
│ 1356 Tigertail Road         Estimated by J. J. O'S.│
│                                             │
│ Concrete slab on grade, 4" x 40' x 10'      │
│ 400   13.6      SF      320.00  120.00  0.00  440.00│
│                                             │
│ 2" x 4" form, Std & Btr (@$430 per MBF, .7BF per LF)│
│   1 use                                     │
│ 100   B2@ 4.9   LF       46.00  103.00  0.00  149.00│
│                                             │
│ Finishing 400 SF (subcontract)              │
│ 1               LS        0.00    0.00  0.00   50.00│
│                                             │
│ Form stripping, 2" x 4"                     │
│ 100   BL@1.2    LF       --      21.00  0.00   21.00│
│ ..........................................│
│ Total Material, Labor, and Equipment:       │
│       19                366.00  244.00  0.00  610.00*│
│ Items Not Included in Material, Labor, and Equipment:  50.00│
│                                             │
│                              Subtotal:  660.00*│
│                                             │
│                       G&A %:   25.00    165.00│
│                       Profit %: 10.00    82.50│
│                                             │
│                       Estimate Total:   907.50**│
└─────────────────────────────────────────────┘
```

If your Estimate Window looks like the screen above, your printed (and transferred) estimate should look like the Estimate Report at the right.

Notice the differences between what you see in the Estimate Window (above) and the Estimate Report (at the right). The Estimate Window shows unit costs — material, labor, equipment and total cost per unit, such as per square foot or per linear foot. The Estimate Report shows only extended costs (quantity times unit cost).

For example, in the Estimate Window the cost in the total column for concrete slab on grade is $1.10 per square foot. The Estimate Report shows the total cost as $440.00. That's the quantity (400 square feet) times the cost per square foot ($1.10).

The column *Craft@Hrs* in the Estimate Report shows extended manhours (units times the hours per unit). This column also shows the crew codes (B2 and BL in the example at the right above). For more information on these crews, see pages 5 to 7 of the *1992 National Construction Estimator*. There's no crew code in the *Craft@Hrs* column of the Estimate Report for subcontract costs or estimates you create using the Alt-C key in the Estimate Window.

Below the line *Total Material, Labor, and Equipment* in the Estimate Report you see totals for material, labor and equipment columns and manhours (rounded to the nearest hour). The star (*) after 610.00 marks this figure as a subtotal.

The line *Items Not Included in Material, Labor, and Equipment* shows the total of subcontract or lump sum costs (where you entered only a total cost without any material, labor, or equipment cost).

The line *Subtotal* shows the sum of all costs above (including subcontract costs) and has a star at the right margin indicating it's a subtotal.

General and Administrative cost (G&A) and *Profit* don't appear in the Estimate Window. But they'll be in the Estimate Report if you've entered percentages for G&A and profit in the

Option Menu. Notice in the sample Estimate Report that G&A ($165.00) is 25% of the subtotal ($660.00) but Profit is 10% of the sum of the subtotal ($660.00) and G&A ($165.00).

Estimate Total is the sum of all costs and has two stars to show that it's the grand total.

Transferring Files

Estimate Writer suggests that you transfer a file to the directory C:\WP. That may be a good choice if you use WordPerfect and if WordPerfect is on the directory WP of drive C. But you can transfer an estimate to any directory on any drive. Just type the drive letter, a colon, a backslash and the name of the directory (for example, C:\WORD). Then use the Del key to delete anything that remains of Estimate Writer's suggestion of C:\WP.

You can also transfer a file to a floppy disk. For example, to transfer to disk drive A, type an A: over the C: suggested by Estimate Writer. Then press Del three times to delete \WP. Press Enter and your estimate will be transferred to drive A with the same estimate name and the file suffix .ASC.

Changing Columns of Costs

Sometimes you may want to revise all costs in an estimate up or down a few percent. Of course, you can revise costs one at a time in the Estimate Window with F3. But there's a quicker way to change entire columns of costs.

Suppose you want to make the following revisions to an estimate you're ready to print:

- Material costs up 10%.
- Labor costs down 15%.
- Equipment costs up 5%.
- Total (subcontract) costs up 20%.

Before printing or transferring an estimate, Estimate Writer asks, *"Do you want to adjust a column of costs in the estimate by a percentage? (Y/N)."* Answer "Y" and you'll open the Revise by a Percent box. Then type the changes you want in each estimate column:

```
         Revise an Estimate Column
              by a Percentage
         -----------------------------
     Labor hours column:          0.00%
     Material costs column:      10.00%
     Labor costs column:        -15.00%
     Equipment costs column:      5.00%
     Total (only) cost column:   20.00%
```
Any Changes? (Y/N): N

To reduce costs in a column, type a minus sign and then the percentage to deduct. Percent changes to costs in the *Total (only) cost column* will apply only on lines where material, labor and equipment costs aren't shown (such as for a quote by a subcontractor).

Cost adjustments in the Revise by a Percent box don't change your estimate in the Estimate Window. They affect only a single printing or a single transfer of the estimate.

Making a Copy of an Estimate

Every time you go back to the Main Menu (press Esc and M), Estimate Writer replaces the previous version of that estimate (if there was one) with the current version. So how do you revise an estimate without changing the estimate already on disk?

Here's an example. Suppose you have the estimate JOB1 on disk. You want to leave JOB1 alone but create a new estimate (JOB2) that's like JOB1 but with a few changes. You could use Transfer an Estimate on the Main Menu to transfer JOB1 to your word processing program. Most word processors make it easy to create multiple copies of a document. But there's another way to duplicate an entire estimate.

Using the DOS copy command (as explained in *Microsoft MS-DOS User's Guide),* copy JOB1.TXT to a new file JOB2.TXT in the same directory. Then copy JOB1.NDX to a new file JOB2.NDX in the same directory. That creates a new estimate JOB2 which is just like JOB1.

If You Can't Use 5¼" Disks

The Estimate Writer disk in the envelope at the back of this book is a 5¼" (1.2 Mb) high density disk. It won't work in a computer with a double density disk drive ("General Failure reading drive A").

If your computer can't use the high density disk, you can order Estimate Writer on either four 5¼" (360K) double density or two 3½" (720K) disks. Send your check for $10 to Craftsman Book Company, 6058 Corte del Cedro, Carlsbad, CA 92009 and request the disk size you need.

Before ordering other disks, try this. Get the help of a friend or associate who has a computer that can read 5¼" 1.2 Mb disks. Copy the entire Estimate Writer disk to the hard disk of that computer using the DOS copy command (as explained in *Microsoft MS-DOS User's Guide*). Then, using four empty floppy disks that your computer can read, copy Estimate Writer files from the hard disk to the four floppy disks:

1. To the first disk, copy files INSTALL.EXE, PKUNZIP.EXE and PROGRAMS.ZIP.
2. To the second disk, copy NCEFILE1.ZIP.
3. To the third disk, copy NCEFILE2.ZIP.
4. To the fourth disk, copy NCEINDEX.ZIP.

Then follow the installation instructions. The installation program will request the second, third and fourth disks when needed.

Customizing the N.C.E. Data Base

If you find costs in the N.C.E. Window that aren't right for your jobs, change the N.C.E. data base so it meets your needs.

Suppose you usually install 6" x 6' 9" door jambs at a material cost of $28 and at a labor cost of $24. There aren't any 6" door jambs in the *1992 National Construction Estimator*. But near the bottom of page 58 you'll see 5¼" finger-joint jambs. Let's change 5¼" jambs to 6" jambs and change the costs in the data base.

- In the N.C.E. Window (Esc N), turn to page 58 (Esc, T, 58, Enter). Press PgDn twice. Press ↓ 6 times to move to *5-1/4" set*. Press F3. Your screen should look like this:

N.C.E. Window 58 Door Frames	Craft@Hrs	Unit	Matl	Labor	Total
Door Jambs, Exterior					
Select finger-joint, kiln dried pine, 6'9-1/2" high					
4-1/8" set	BC@.800	Ea	22.50	18.90	41.40
4-1/2" set	BC@.800	Ea	23.70	18.90	42.60
4-3/4" set	BC@.800	Ea	24.60	18.90	43.50
5-1/4" set	BC@.800	Ea	24.60	20.10	44.70
Solid stock kiln dried fir, 6'9-1/2" high					
4-1/8" set	BC@.800	Ea	31.00	18.90	49.90
4-1/2" set	BC@.800	Ea	34.40	18.90	53.30
4-3/4" set	BC@.850	Ea	36.00	18.90	54.90
5-1/4" set	BC@.800	Ea	39.30	20.10	59.40
Door Jambs, Interior. Flat jamb with square cut heads and rabbeted sides, pine, 6'9-1/2" high					
3-9/16" set	BC@.800	Ea	9.70	18.90	28.60
4-9/16" set	BC@.800	Ea	11.90	18.90	30.80
4-11/16" set	BC@.800	Ea	12.90	18.90	31.80
5-1/4" set	BC@.800	Ea	14.80	20.10	34.90
Add for solid jambs	--	Ea	17.50	--	--

Move Cursor and Make Changes, <Enter> to Review, <Esc> to Exit

- With Num Lock off, press → twice so the highlight is on 5 in *5-1/4" set*. Type *6" set* and press Del 4 times to delete 4 extra letters.

- Press Ins. (Note the blinking insert highlight). Tap the spacebar 4 times. Press Ins again to turn the blinking insert highlight off.

- Notice that decimals in the cost columns at the right are restored to vertical alignment by using a combination of the Ins (insert) and Del (delete) keys.

- Press → until the highlight reaches *24.60*. Type the new material cost (28.00) over the old cost. Change 20.10 in the labor column to 24.00 using the same procedure.

- Press Enter to review your work. If there's a mistake, press F3 to try again or Esc to exit without making any change in the data base.
- Press Enter again and the change becomes a permanent part of your data base.

Notes on Customizing the Data Base

- The F3 key lets you be creative. But don't get too creative. Change a line to gibberish and your permanent data base includes gibberish. Rather than write nonsense to disk, use the Esc key to exit without any change.
- There's an easy way to avoid problems: Type changes over a line that's like the line you want to create. For example, type over a line that includes material, labor and total costs when you're creating a line of material, labor and total cost estimates. Type over a line that's all text (without figures in cost columns) when you're creating a new line that's all text.
- Estimate Writer lets you change lines, not add lines to the data base. If you want to add new lines of data, call us at 619-438-7828 for instructions.
- Making changes to the data base works only in the N.C.E. Window. F3 doesn't do anything in the Split Window.
- Try to keep decimal points in cost columns in vertical alignment. If you're off by a little, Estimate Writer can make small adjustments so the costs stay aligned in columns.
- There's no need to change figures in the Total column. Totals are adjusted automatically.
- If your data base gets polluted with random garbage, no problem. You can always start fresh by installing Estimate Writer again.

Using TSR Programs

Terminate and Stay Resident (TSR) programs may cause unpredictable errors in Estimate Writer. We recommend that you begin using Estimate Writer without any TSR programs loaded. Once you've found that Estimate Writer works reliably on your computer, experiment with TSR programs.

Be Sure Num Lock is Off

Usually you'll want Num Lock off when using Estimate Writer. The numeric keypad on most computer keyboards works two ways. With the Num Lock light on, the numeric keypad produces numbers. With the light off, the numeric keypad moves the cursor: left, right, up, down, Home, End, PgUp and PgDn. If you want to move the cursor (rather than type numbers), be sure the Num Lock light is off. Most computers have the Num Lock light on when they're first turned on. Press the Num Lock key to turn the light off.

Back Up Your Files

It's easy to erase files and directories accidentally. And hard disks don't work forever. That's why it's wise to make regular backups of your most important estimates. Estimate Writer can be re-installed again at any time from the installation disk. So there's probably no need to make backup copies of the Estimate Writer program files (suffix .EXE) or data files (suffix .NCE). But either copy your estimates (suffix .TXT) and estimate indexes (suffix .NDX) to a floppy disk or make a backup of the ESTWRITE directory.

Errors and the TEMPEST File

We recommend that you return to the Main Menu every 10 or 15 minutes when creating estimates. That saves your estimate to disk so a

How to Use Estimate Writer

power failure (or some other problem) doesn't cause a loss of your valuable work.

If something unexpected (like a power failure) happens when Estimate Writer is writing to your hard disk, a file may be corrupted and become unusable. If that happens, try running *Rebuild Estimate* from the Help option on the Estimate Writer Main Menu.

Every time you transfer, print, or repaginate an estimate, Estimate Writer creates a temporary estimate file (TEMPEST.TXT) and a temporary estimate index file (TEMPEST.NDX). These files are a safety measure. If the transfer, print, or repaginate you've requested isn't completed for any reason, the TEMPEST files have a good copy of your estimate and the estimate index files.

When a transfer, print, or repaginate is completed successfully, the TEMPEST files are erased from the disk. If the TEMPEST files aren't erased for any reason, you'll see the following message on the screen.

```
            The SDSI Estimate Writer
1992 Building Costs                Version 2.2




         A temporary work file exists. The last Print
         Repaginate, or Transfer did not execute properly.
         Press <Enter> to return to the Menu. Print or display
         Instructions from the Help Menu and refer to the
         Error Messages section for more information.
```

To restore your estimate, quit Estimate Writer at the Main Menu but stay in the Estimate Writer directory (such as ESTWRITE). Delete the .TXT and .NDX files you were working on. For example, if your estimate name was SAMPLE, erase the files SAMPLE.TXT and SAMPLE.NDX. Then rename TEMPEST.TXT as SAMPLE.TXT and rename TEMPEST.NDX as SAMPLE.NDX. You should be able to start Estimate Writer (type ESTIMATE) and revise the estimate you were working on. The *Microsoft MS-DOS User's Guide* explains how to delete and rename files. In any case, Estimate Writer won't operate as long as a TEMPEST file is in the same directory with Estimate Writer.

Estimate Writer is New

Estimate Writer is both a new program and a new concept in construction estimating programs: It's a disk that comes with a book, works like a book and has all the same information as the book it comes with.

Like most new programs, there are lots of ways Estimate Writer could be improved. We're planning to make improvements in future editions. For example, we plan to make Estimate Writer Version 3.0 look and work much more like a Windows program.

You probably have some suggestions and we'd like to have them. If you want to offer suggestions (or report a program bug), call us at 619-438-7828, fax us at 619-438-0398 or mail us a note at The Estimating Department, Craftsman Book Company, 6058 Corte del Cedro, Carlsbad, CA 92009. We appreciate your interest in Estimate Writer.

If You're Having Trouble . . .

We know how frustrating it can be to learn a new program. If you need help with Estimate Writer, call 619-438-7828 between 8 A.M. and 5 P.M. Pacific time Monday through Friday. Before calling, please re-read these instructions to be sure your question isn't answered here.

How to Use Estimate Writer

The Most Frequently Asked Questions

Since Estimate Writer was introduced in December of 1990, hundreds of users have called us to get assistance. If the instructions aren't clear or don't answer your questions, we invite you to call us at 619-438-7828. To save you (and us) the trouble of a phone call, we've listed below answers to the most common questions from Estimate Writer users.

What's "General Failure reading drive A"

If your computer reports a "general failure" and asks "Abort, Retry, Fail?" when you try to install Estimate Writer, it's probably because your computer can't use high density floppy disks.

Many computers made before 1989 have a double-density (360K) disk drive. The disk attached inside the back cover of the *1992 National Construction Estimator* is high density (1.2Mb). It won't work in a computer with a double density disk drive. So what do you do? There are three choices:

- First, you can copy the high density disk onto four double density disks. Instructions on page 563 explain how to use a computer with a high density (1.2 Mb) disk drive to transfer Estimate Writer to four double density 5¼" (360K) disks or two 3½" (720K) disks.

- Second, you can fill out and mail the white card in the envelope attached to the back cover of this book. The cost for Estimate Writer on either two 3½" (720K) disks or four 5¼" (360K) disks is $10 plus the actual shipping cost (and 73¢ tax in California only). If the card is missing, mail your request to Craftsman Book Company, 6058 Corte del Cedro, Carlsbad, CA 92009. Enclose a check for $10 (plus 73¢ tax in CA only) or use your charge card.

- Third, you can call us (619-438-7828) or Fax us (619-438-0398) and order Estimate Writer on either two 3½" (720K) disks or four 5¼" (360K) disks. The cost is $10 plus the actual shipping cost (and 73¢ tax in California only). We can ship Estimate Writer by air anywhere in the U.S. for $4.50. You'll usually receive air shipments on the next business day.

Why Can't I See the Cursor?

Estimate Writer determines whether you have a monochrome or color video display board and selects screen colors correct for the type of board present. That works fine if you have a monitor appropriate for the type of display card installed. But if you have a *color display board* and a *monochrome monitor,* you may find that highlighted lines of text aren't visible (disappear) until the highlight is removed (by typing Enter). Switching Estimate Writer from color to monochrome mode may solve the problem. So, how do you do that?

Estimate Writer Release 2.1 and later gives users the option of selecting the display type of their choice, either monochrome or color. Here's how to switch between monochrome and color displays:

- Go to the Estimate Writer Main Menu. Select either *Revise an Estimate* or *Start an Estimate.* Instead of typing an estimate name, press the F9 key. Estimate Writer asks, *Monochrome or Color? (M/C):*

- Type M to select mono. Type C if you want to return to color. Then press Enter.

- Estimate Writer then asks for the default word processing directory: *Enter Default W/P Directory:* Type the name of the word processing directory where you expect to transfer most estimates. For example, type *C:\WORD* and press Enter. You're always free to select a different drive and directory when actually transferring estimates.

- Switching from color to mono or from mono to color doesn't take effect immediately and completely. Quit Estimate Writer and start again (by typing *ESTIMATE* in the ESTWRITE directory). That puts the change into effect.

Can I Use Your Costs in my Spreadsheet?

Yes. The database files in Estimate Writer (those with a .NCE extension) are ASCII files and can be read by any spreadsheet or database program that accepts text files. You may want to edit the files to remove the columns and lines used by Estimate Writer code. You'll probably want to insert column delimiters (usually tab characters) to replace the space bar spaces in these ASCII files.

Can I Use Estimate Writer on my Mac?

Yes. We understand that Universal SoftPC offered by Insignia Solutions (800-848-7677) will let Macintosh users run programs written for computers that use MS-DOS. Of course, Estimate Writer running on the Mac is still Estimate Writer. It won't have the usual Macintosh user interface.

You should be able to use Estimate Writer on any computer that uses MS-DOS and its variations (such as PC DOS or Compaq DOS).

How do I See Extended Costs on Screen?

Estimate Writer shows only unit costs in the N.C.E. window. To see the cost extensions (quantity times the price), print the estimate. If that takes too long, transfer the estimate to your word processing directory. Then switch to your word processing program and open the file with the same name as your estimate and with the file name extension of .ASC. You'll see all extended costs, overhead, profit and the estimate total.

Does Estimate Writer Run Under Windows?

Yes, if you can run Windows in enhanced mode. But we recommend that you use Estimate Writer outside of Windows until you're satisfied that Estimate Writer is working perfectly. Then try running it under Windows.

My Columns Aren't Adding Up Right

Material, labor, equipment and total columns will add up correctly unless you're using the insert mode (Ins key) to enter cost estimates. Use Alt-C to add costs to your estimate and the Insert key to add text (descriptions).

Why Can't I do This (or That)?

Estimate Writer is an estimate writing program. It isn't intended to make labor and material take-offs, track job costs, generate a lumber list or schedule work. Other programs are available that handle tasks like these (and many others).

If you need an estimate writing program with additional features, you'll be pleased to know that Estimate Writer Version 3.0 will be available in early 1992. It has most of what users of earlier version have asked for (mouse scrolling, drop-down menus, save as, automatic subtotals, etc.). To get your free copy, complete and return the Free Spring Update card in the envelope in the back of this book.

Can I Cut Out What I Don't Need?

Estimate Writer requires 5,200,000 bytes free on your hard disk. That's a lot of space. Suppose you use Estimate Writer only for commercial electrical estimates. Everything you need is on pages 500 to 552 of the *National Construction Estimator*. Is it safe to erase everything but those pages from your disk?

No. It's not safe. Calling on a page that's been erased from the disk can have unpredictable results, including loss of data. If you're short of disk space, a better choice is to use a program like Cubit (1-800-272-9900) or Stacker (1-800-225-1128). Either will compress Estimate Writer into about one-half the normal space (2½ Mb instead of 5 Mb).

My Install Program Won't Work

Computers using versions of DOS earlier than 3.2 and computers with unusual hardware or software configurations may not be able to use the automatic installation program on the Estimate Writer disk.

If your computer reports, "DOS error 5," or "Error Reading Drive A" during installation, try manual installation. Here's how.

- Create a directory for Estimate Writer on your hard disk. For example, at the C prompt, type MD ESTWRITE and press Enter.
- Change control to that new directory. For example, type CD \ESTWRITE and press Enter.
- Put the first (or only) Estimate Writer disk in your floppy drive and change control to that disk drive. For example, type A:
- Copy all Estimate Writer files from the floppy disk (or disks) to your hard disk. For example, COPY A:*.* C:\ESTWRITE and press Enter.
- Change control to the hard disk directory where Estimate Writer is stored. For example, type C:
- Type PKUNZIP *.ZIP and press Enter to explode (decompress) the files.
- When decompression is complete, type ESTIMATE and press Enter to begin using Estimate Writer.
- When you're satisfied that Estimate Writer is working correctly, quit the program and erase the files with a .ZIP extension from your disk. This removes files used during installation. For example, you might type ERASE C:\ESTWRITE\ *.ZIP and press Enter.

Your Costs are Too High (or Too Low)

It's easy to adjust costs in the database to your job site when using Estimate Writer. See page 562.

Can My Computer use Estimate Writer?

To use Estimate Writer, you need a computer that uses DOS 3.2 or higher, 5 Mb of free space on the hard disk, 384K of memory and a high density 5¼" disk drive. Here's how to figure out if your computer qualifies.

1. Most personal computers use DOS. To find out the version of DOS you're using, go to the DOS prompt (such as C:>). Type VER and press Enter. Any DOS numbered 3.2 or higher (such as 3.2, 3.3, or 4.01) will run Estimate Writer. The vast majority of computers that use DOS are running DOS version 3.2 or higher.

2. To find the amount of free space on your hard disk, go to the DOS prompt for the hard disk (such as C:>). Type DIR in any directory or sub-directory. DOS will display the names of files. The last line of the display will show the number of files in that directory and the free space available on that disk, such as: 45 file(s) 5242880 bytes free. If the number of bytes free is more than 5400000, you have enough disk space to use Estimate Writer.

Index

A

Abbreviations 220
Abrasion 226
Abrasive
   concrete finish 311-312
   stair treads 141, 378
   terrazzo 374
Absorption 226
Absorption chillers 488
Acacia veneer 124
Access
   control units 549-550
   doors 181
   flooring 385-386, 390
   stairs, folding 202
Access hatches
   ceiling 355
   plumbing 458
   sheet metal 181
Accordion folding
   doors 64
   partitions 390
Acid cleaning, concrete 312
Acid etching 285, 312
Acid resistant countertops ...406
Acid wash 52
Acid-proof brick 325
Acknowledgments 12
Acoustic ceiling texture 96
Acoustical
   block 327
   caulk 37
   designs 74
   engineering 74
   partitions 390
   plaster 371
   rooms 412
   tile, ceiling 40-41
   tile, demolition 54, 244
   wallboard texture 375
Acoustical adhesive 13
Acrovyn wall cover 384-385
Acrylic
   blank panels 39
   diffusers 39, 540
   primer 151-152
   skylights 197
   waterproofing 105
Activated carbon filters 493
Actuating devices
   electrical 527
   garage doors 91
   mechanical controls 499
ACZA treatment, lumber 128
Additives
   coloring 46, 277
   concrete 46, 306, 311-312
Address systems, public ..551-552
Adhesives 13-14
   ceramic tile 14, 374
   flooring 14, 86, 89
   hardboard 96
   roofing 171
   tile 13-14, 205
   wallpaper 211
Administrative overhead 130
Admixtures
   coloring 46, 138, 311-312
   concrete 277, 311-312
Adobe block 137
Aerial equipment 236
Aerial mapping 225
Aerial platforms, rental 236
Afromosia veneer 124
Aggregate 311-312
   base 51, 272
   concrete 14
   exposed, concrete 277
   finish 311-312
   paving 272
   roofing 174
   seeded 160
   testing 226
Air cleaners, filters 493
Air compressors
   rentals 78, 232
   service station 402
Air conditioning
   equipment 484-488
   residential 103
   wiring 70
Air curtain heaters 468, 483
Air distribution equipment 493-494
Air exhausters
   fans 78, 491-492
   roof ventilators 355
Air filtration equipment 492-493
Air hammer 252
Air handling control 497-500
Air handling equipment ..492-493
Air hose
   reel 402
   rental 233
Air mixing boxes 493
Air moving equipment 493-494
Air purger 480
Air supported enclosure .409-410
Air switch 203
Air terminal units 493
Air tool operator, wage 221
Air tools 54-56, 231
   earthwork 78
   pneumatic 54-56
Air wall 390
Airless paint sprayer 155
Airlock doors 246
Airtrol fitting 480
Alarm
   circuit 502
   computer floor 386
   systems 548-550
   wiring 70, 179
Alarms
   bank 395
   fire 179, 550-551
   security 179, 548-550
   smoke 179
Alder lumber 123
Aldrin 199
Aliphatic resin glue 13
Alkali, resistant primer ...151
Alkyd primers 151
Altars 397
Aluminum
   anodized 365
   cladding 350
   coating 172
   conduit 513-514
   cupola roof 53
   curtain wall 369
   demolition 56
   doors 60, 65, 67-68
   downspouts 354
   flagpoles 386
   flashing 183, 352-353
   foil 21
   foil insulation 104
   gutters 354
   letters 388
   louvers 39, 354-355
   metal work 339-340
   nails 150
   ornamental 339-340
   oxide 311
   panels, preformed 351
   pet doors 65
   railing 334
   roof paint 152
   roofing 175, 351
   screen doors 67
   sheet 352-353
   siding 194-195, 351
   stairs 201
   supports 338
   vents 198, 355
   wiring 520
American walnut 125
Ammeter 533
Ammonium sulfate 291
Amplifier 552
Analyzers, air 499
Anchor bolts
   foundation 91, 339
   templates 299
   tilt-up 316, 318-319
Anchorage, boat 294
Anchors
   chimney 83
   concrete 91, 339
   conduit hanger 515-517
   embedded steel 339
   glass block 138
   masonry ...139, 316, 329, 341, 517
   timber connectors 206-209
Angle
   clips 206
   iron 139
   irons, fireplace 83
   supports, pipe 445
   valves 450
Angles, steel 339
Annunciator panel 550
Anodizing 339
Anzac siding 126
Appalachian oak 123
Appliances
   cabinets 23
   dishwashers 164
   garbage disposers 163-164
   ranges 170-171
   solar water heaters 164-165
   water heaters 164-165
   water softeners 165
Appraisal fees 14-15
Apprentice costs included 3
Apron, concrete driveway ...277
Aquabar 21
Aquastats 498
Arbitration costs 15
Archaid 112
Archbead 112, 371
Arches
   masonry 140, 325
Architectural
   drawings 15-16, 68-69
   facade screen 354-355
   form lining 300
   screens 339-340
   signage 388
   veneers 124-125
Area modification
   factors 10-11
   how to use 4
Area walls, sheet metal 182
Arks 396
Armed guard 180
Armored locks 99
Armstrong flooring 85-86
Artificial stone 331
As-built drawings 68-69
Asbestos
   breaching 248
   cement water pipe 433
   containment 246
   decontamination 247
   disposal bags 249
   duct insulation 248-249
   dump fees 249
   encapsulation 249
   fiber drums 249
   pipe insulation removal ..246
   removal 246-249
   testing 21
   worker, wage 221
Ash
   doors 42, 44, 61-64
   dumps 83
   flooring 87-89
   lumber 123
   paneling 157-159
   urns 394
   veneer 124
Ashlar veneer 140
ASME tanks 476-477
Asphalt
   access road 273
   built-up roofing 171-172, 351
   coatings 152, 172
   cold milling 272
   curb 277
   dampproofing 332, 348
   emulsion 138, 171-172, 330
   felt 21, 174, 351
   felt roofing 351
   mix 42
   parking lot 273
   paving 159-160, 272
   paving, demolition 56, 242
   planing 272
   primer 285
   roll sheathing 21
   roof coating 152
   roofing 173-174
   sawing 251
   scarify 258
   shingles, roofing 173-174
   speed bumps 277
   tack coat 272
   tile 85, 91, 377
   walks 277
   wall sheathing ..105, 130, 346
   waterproofing 332, 348
Asphaltic concrete testing ..227
Assemblies
   ceiling 27-28
   floor 30
   hollow metal door 360
   openings, doors and windows 27
   wall 36
Astragal moulding 145, 361, 368
Astro dial switch 527
Astronomy domes 410
Athletic courts, painting ..285
Athletic equipment 284
   gymnasium 401
   swimming pools 412
Athletic fields 281-282
Attic
   access doors 181
   fans 71, 78
   stairs, folding 202
   vents 184
Audio
   detectors 179
   sensors 549
Audiometric rooms 408
Auditorium
   acoustic design fees 74
   seating 408
Auger holes
   caissons 261
   post holes 277
Auto hoist 402
Automatic openers
   entrance doors 365
   garage door 91
   storefront doors 366
Automatic transfer switches .548
Avodire veneer 124
Awning windows 213-214
Awnings 16-17

B

Backer board, gypsum drywall 95-96
Backer board, tile 206
Backfill
   earthwork 75-76
   tilt-up 316
   trench 287
Backflow
   preventers 287, 435, 450
   valves 224
Backhoes 76
   excavation 255
   rental 77
   trenching 253-254, 297
Backing
   carpentry 25, 345
   carpet 379
   clay tile 329
Backing rods 273, 356
Backsplash 53

Index

Backstops
  baseball 281
  basketball 284, 401
Baffle downlight 544
Baffles, audiometric rooms 408
Baggage handlers 416
Bain Marie 399
Balancing valves 480
Ball cleaning 268-269
Ball joint, pipe 452-453
Ball valves 450
Ballast, railroad 295
Balusters 200-202
  painting 156
Band moulding 144
Bank equipment 395
  alarm system 395
  camera surveillance 395
Bar
  doors 59-60
  sinks 162
  tops 53
Barbed wire 238
Barbed wire fence 81, 278
Barricade
  lights 17
  roadway 17, 280
  temporary enclosure 238
Barrier posts 404
Barrier, foil 21
Bars
  bath accessories 18-19, 393
  masonry 305, 328-329
  playground 283
  reinforcing steel 48-49, 305
  welding 305
Base
  asphalt 159, 272
  ceramic tile 205, 330, 373
  column 206-207, 339
  corners 378
  marble 331
  moulding 144, 148
  resilient flooring 86, 378
  terrazzo 375
  top set 378
  wood 144
Base course, paving 159-160, 272
Base screed 112, 371
Base shoe 144-145
Base wages
  commercial & industrial 221
  residential 8
Base, floor 86
Baseball
  backstops 281
  fields 281-282
Baseboard heating
  electric 468
  hot water 100
  radiators 482
Basement
  columns 31
  doors 18
  excavation 255
Basic wage cost 3
Basketball
  backstops 284, 401
  courts 282
Basketweave fence 81, 278
Basswood 123
Bath accessories 18-19, 392-394
  ceramic tile 204-205, 373-374
  compartments 384, 461
  doors 186
  enclosures 187
  fans 78
  shower and tub doors 186-187
  shower cabinets 461
  vanities 210
Bath fixtures 160, 461
Bath, waterproof 348
Bathroom
  cabinets 210
  fan wiring 70
  heaters 101
  sinks 162, 463
  vanities 210
  wallboard 96

Baths, steam 202
Bathtub
  caulk 38
  doors 186-187
  rough-in 461
Bathtubs 161, 461
Batten fence 279
Battens, lumber 144
Batter walls 300
Batterboards 114, 297
Battery, electric fence 80-81
Batts, insulation 104, 349-350
Bay windows 214-215
Bead
  moulding 145-146
  wallboard 373
Beadex 96
Beads, lath & plaster 112, 370-371
Beam, shoring 304
Beams 304
  carpentry 24, 28, 341
  ceiling 31
  clamps, conduit 516
  collar 28
  concrete 298
  fireproofing 359
  formwork 298, 304
  grade 49-50, 298
  laminated timbers 346-347
  lumber 119
  precast 314
  steel 337
Bearing capacity test 199
Bed bumpers 384
Bed moulding 146
Beds, wall 19
Beech
  lumber 123
  veneer 124
Beehive grate 270
Belem truss 410
Bell footings 261
Bell wire, door bell 57
Bell wiring, electrical work .. 71
Bellows, pipe 452
Bells
  door 57
  fire alarm 550
Belt conveyors 415-416
Benches
  athletic 285
  church 396
  locker room 389
  stadium 408
Benderboard 111
Benefits, fringe 221
Benin veneer 124
Bentonite
  granules 310
  waterproofing 211
Berber carpet 37
Berths, boat facilities 294
Bestile hardboard 96
Bibs, hose 167, 451
Bicycle racks 286
Bidet 162
Bike post bollard 286
Bin, trash 240
Birch
  doors 44, 61-64
  lumber 123
  paneling 158
  veneer 124
Bird screen 354-355
Birdstop 177
Bituminous
  paving 159, 272
  paving, removal 241
  roofing 173-174, 351
  shingles 173-174
Bituthene roofing 348
Black steel pipe 416-428
Blackboard, chalkboard 383
Blanket insulation 104, 105
Blast absorbing rooms 412
Blasting, excavation 251-252
Bleachers 401, 408
Blind stop moulding 145

Blinds
  shades 407
  shutters 187
  Venetian 213, 407
Block
  adobe 137
  concrete 134-137, 326-328
  demolition 244
  detailed concrete 135
  flooring 88-89, 377
  frieze 342
  glass building 137-138
  masonry ... 134-137, 139, 326-328
  painting 155
  primer, dry mix 151
  sawing 250
  vents 385
  walls, concrete 134-135
Blocking
  carpentry 27, 341
  fire 27
  floor joists 29
  sheathing 342
Blockout forms
  cast in place slabs 300
  tilt-up 316
Blocks, pier 32
Blood storage refrigerator ... 404
Blowers 489-494
  fans 78-79
  fireplace 84-85
  heating 483
Blown insulation 105-106, 350
Bluegrass seeding 293
Blueprinting 19-20
Board
  fencing 81-82, 278-279
  forming 47
  insulation ... 104-105, 346, 348-350
Boards, treatment 128
Boat facilities 294
Boiler
  chemical feed systems 476
  combustion controls 474
  combustion train 470-471
  condensate receiver pumps . 475
  deaearator/condenser units . 475-476
  economizer units 475
  electric service 472-473
  feedwater pumps 473
  feedwater systems 476
  foundations 321
  fuel train piping 470-472
  refractory 473
  smoke detector 499
  stacks 477
  startup 475
  trim 477
  water softening systems ... 474
Boilermaker, wage 221
Boilers 468-477
Bollard light 542
Bollards
  bicycle 286
  cast iron 280-281
  granite 281
Bolts
  door 367
  foundation 91, 339
  tilt-up 316
Bond beams, concrete block ... 327
Bond breaker 310
  tilt-up 318
Bonding agents
  adhesives 14
  flooring 86
  roofing 171-172, 352
Bonds
  insurance 106
  masonry 140
Bookcases
  library 392
  painting 157
Boomlifts 236
Booths, phone 392
Boring
  caissons 261
  soil testing 199, 229
  under walkways 287

Bottom plates, carpentry 32
Boxes
  junction 524-525
  outlet 72
  pull 525
  switch 72
  valve 288
Braces
  tilt-up 318
  timber 206-207, 209
  trench, rental 235
  wall 209
Bracing
  bulkheads 262-263
  carpentry 25, 341
  temporary 25
  trenches 235, 262
Brackets
  pipe 445
  waler 311
Branch circuits 535-536
Brass
  fabrications 340
  railing 334
Breaker
  ball 231-232
  bond 310
  form 310
  panels 531-534
  vacuum 287
Breakers
  circuit 72, 536-537
  paving, rental 232
Breaking, demolition .. 54-55, 241-245
Brick
  acid-proof 325
  arches 140
  artificial 159
  cleaning 331-332
  common 140, 324-325
  demolition 54, 244
  face brick 131, 332
  firebrick 133
  flooring 133-134, 141
  jumbo 132
  manhole 271
  masonry 131, 324, 332
  modular 132
  mortar 132-133
  moulding 144
  Norman 131, 140
  padre 132
  painting 155
  paving . 133-134, 160, 274-275, 326
  plate 326, 330
  pointing 331-332
  primer 151
  reinforcing 140
  Roman 140
  sandblasting 177
  sawing 250
  steps 326
  testing 228
  ties 206
  veneer 140
  walls 140, 331-332
Bricklayer's tender, wage
  commercial & industrial ... 221
  residential 8
Bricklayer, wage
  commercial & industrial ... 221
  residential 8
Bridging
  carpentry 25
  joist 108, 206, 341
Broiler, food 399
Bromicil weed control 296
Bronze fabrications 340
Broom finish, concrete 311
Brown coat plaster 160, 371
Brush chipper 237
Brush removal 76, 240
Brush, wire 52
Brushing, painting 380-382
Bubblers, sprinkler 289
Bubinga veneer 124
Bucket traps 478-479
Buckets, rental 233

Index

Buggies, rental 233
Builder's adhesive 14
Builder's hardware 98, 366-368
  nails 148-151
  rough carpentry 108-109, 206
Building
  appraisal 14-15
  demolition 54-57, 243
  inspection service 20-21
  moving 240-241
  paper 21-22
  permits 22
  posts, steel 108
  sewer 180-181
  temporary, rental 239-240
  wiring 70-72, 519-521
Building board 210, 344
Building laborer, wage
  commercial & industrial 221
  residential 8
Built-up girders
  carpentry 31
  steel 337
Built-up roofing 173, 351
  demolition 56
Bulkhead forms 299
Bulkheads, shoring 262-263
Bulldozer, rental 77
Bulldozing 76, 256-257
Bullet-resistant glass 369
Bulletin boards 383
  directories 388
Bumper guards 384
Bumpers, dock 320, 404
Bungalow siding, resawn 125-126
Burlap soil stabilization 269
Burners
  incinerators 408
  ranges 170-171
Bus bar, grounding 527
Bus duct connection 533
Bus stop shelters 409
Business overhead 130
Butt hinges 99, 367
Butterfly valves 448-449
Butternut veneer 124
Button switch 527
Buttons, reflective 274
Butyl caulking 37, 355-356
Butyl roofing 351
BX cable 522
By-passing doors 44

C

C-clamps, pipe 445
Cabinet
  hardware 368
  locks 98-99
  shelving 186
  unit heaters 484
Cabinets 405-406, 484
  bath 141-142, 210, 393-394
  china cases 24
  classroom 406
  electrical 524-525
  finish carpentry 406
  fire 465-466
  hospital .406
  kitchen 22-24
  laboratory 406
  lavatory .210
  lighting fixtures 116, 141-142
  linen, shelving 186
  medicine 141-142, 393-394
  metal 405
  painting 154, 157, 380
  shop 405
  shower 187, 461
  utility 24
  vanity 210
  wardrobe 398
Cable
  armored 522
  BX 522
  coaxial .551
  communications 551
  electrical 521-524
  ice melting 523-524
  phone 551
  power 521-524
  Romex 70, 522
  snow melting 523
  UF 521-522
Cable cranes 231
Cable suspended roofs 410
Cable tap 551
Cable tray 518-519
Cabs, elevator 73, 414
Cafe doors 59-60
Caissons 261
  reinforcing 305
Calcium chloride 46, 306
Calcium nitrate 291
Calcium pipe insulation 456-457
Calcium silicate insulation . . . 456-457
California pavers 133
Cane fiber joint 310
Canopy
  door 16
  range hoods 170-171
  steel framing 338
Cant strips
  carpentry 26
  roofing 105, 352
Canvas, tarpaulins 203
Cap
  flashing .353
  masonry 136, 140, 325
  moulding 144
  tile 330
Capillary fill 299
  aggregate 14
Capitol forms 302
Caps, conduit 505
Carborundum rub 52, 312
Card readers, parking 403
Carpenter, wage
  commercial & industrial 221
  residential 8
Carpenters iron 339
Carpentry
  cabinets 24, 406
  finish 24
  moulding 144
  plates 341
  rough 24, 340-342, 345-346
Carpet tile .379
Carpets 36-37
  access flooring 386
  adhesive .13
  Berber 37
  entrance 378-379
  mats .141, 407
  removal 55
Carports 37
Carriers, conveyors 415-416
Cars, elevator 73, 414
Cartridge fuses 538-539
Cartridges, respirator 247
Casement windows 215-216
Cases
  cabinets 23-24, 405-406
  china cabinets 24
  laboratory 406
  painting .154
  vanity 406
Cash dispenser 395
Casing
  aluminum siding 195
  beads 112
  door 59
  drywall 373
  moulding 144-145
  nails 148
  plastering 371
  wallboard 373
Cast iron
  bollards 280-281
  column bases 339
  fittings 435-436
  flanged pipe 434-435
  ornamental iron 151
  pipe 429-432, 434-435
  sections .339
  valve box 435
  valves 435-436
Cast-in-place concrete
  . . . . 45-52, 306, 308-309
Catch basins 269
  demolition 242
  entrance .407
  gratings .338
Cathodic protection 412
Catwalks, carpentry 26
Caulking 37-39, 355-356
  guns 39
Caution lights
  barricades 17
  signs 17
Cavity wall
  insulation 105-106, 140
  reinforcing 329
Cedar
  boards 39
  closet lining 39
  fencing 81-82
  lumber 118-120, 126
  paneling 157
  saunas 178
  shakes 173, 174
  shingles .174
  siding 126-127, 189-192
  veneer 124
  western red 118-120, 194
Ceiling
  access hatches 355
  adhesive .13
  assemblies, wood framed 27
  beams, carpentry 26, 341, 346
  demolition 54, 56, 244
  domes 39, 196
  drop 39-41
  fans 79
  grid 39, 375
  joists 24, 26-27
  lighting fixtures 39-41, 114-116
  materials 39-41
  metal pans 40
  paint 153
  panels 39-40
  radiant heating system 101
  shutters 79
  texture spray 96
  tile 40-41, 376
Ceilings
  acoustical 39-40, 375
  fire retardant 41
  integrated 408
  painting 154, 157
  stained glass look 40
  suspended 39, 375
  tin 41
Cellulose insulation 105
Celotex shingles 173
Cement 42
  adhesives 13-14
  base primer 151
  coloring 46, 139, 311
  concrete 46, 306, 324
  dampproofing 348
  duct, envelopes 502
  painting .155
  pargeting 332
  plaster 160, 370
  roofing 172
  stabilization, soil 269
  testing 227
  tile 176
  tile setting 374
  waterproof, hot mop 348
  white 46, 306, 374
Cement board siding 194
Cement mason, wage
  commercial & industrial 221
  residential 8
Cement, asbestos pipe 264
Cementitious
  dampproofing 348
  decks 324
  fireproofing 357-359
Central vacuum systems 210
Centrifugal chillers 487-488
Centrifugal pumps 454
Ceramic, facing tile 330
Ceramic tile 204-206, 373-374
  backer board 206
  demolition 55, 245
  mosaic 204-206
  veneer 330
Certainteed shingles 173
Cesspools, septic tanks 180-181
Chain link fence 79-80, 278
  demolition 242
Chain railing 334
Chain saw 78, 237
Chair rail 145
Chairs
  auditorium 396, 408
  school 398
Chalkboard 383
Chamfer strip 145
Chamfers
  cast in place 310
  tilt-up 318
Changes in costs 4
Channel hanger 516
Channel steel 136
Channel strap 516
Channelizing, earthwork 257
Charging, electric fence 80-81
Check desks 395
Check valves . 287, 448-450, 465, 480
Checkroom equipment 399
Cherry
  flooring 87-89
  lumber 123
  paneling 158
  veneer 124
Chestnut
  flooring 90
  paneling 158
  veneer 124
Chiller
  compressors 488
  controls .499
  monitor .499
Chillers
  absorption 488
  centrifugal 487-488
  reciprocating 487
Chimes, door
  transformer 54-57
  wiring 71
Chimney
  anchors 83
  caps 184
  firebrick .133
  flashing 184, 353
  flues 85, 137, 332
Chimneys
  fireplace 85, 332
  masonry 140, 332
  prefabricated 332, 488-489
China cases 24
Chippers 232
Chippers, rental 237
Chipping concrete 243
Chisels, rental 232
Chlordane .199
Church
  acoustical design fees 74
  equipment 396-397
  glass 395
  pews 396
Chutes
  demolition 245
  garbage .416
  linen 416
  mail 389
  trash 245
Cinder rock 258
Circline lighting 115
Circuit breaker panel 503
Circuit breakers 72
Circuits
  electrical 502-503, 512-513
  motor 528-529
Circulating pumps 453-454
City cost adjustments 10

Index

Civil engineering fees 74, 222
Cladding, preformed wall 350-351
Clamps
   column forms 310
   concrete, rental 234
   ground rod 527
Clamshell excavators 234
Clamshell work 231
Classrooms, manufactured .. 408-409
Clay
   backing tile 329
   brick 324-325
   flooring 133-134
   pipe 268
Clay tile
   load bearing 329
   roofing 176-177
Cleaning
   ball 268-269
   brick 331-332
   concrete 312
   glass 240
   glass block 330
   masonry 331-332
   new construction 240
   paint preparation 155
   pavement joints 273-274
   sandblasting 177
   stone 332
   terra cotta 332
   vacuum systems 210
Cleanout cover 459
Cleanout doors, fireplace 83
Cleanouts 459
Cleanup, overhead 222
Clearing
   brush 241
   excavation 75-76
Clerical help cost 222
Clevis hangers, pipe 445
Climbing crane 231
Clips
   conduit 516
   framing 206
   plywood 208
   timber connectors 206
Clock, receptacles 526
Clocks, wiring 70
Closers, door 57, 367
Closets
   cabinets 24, 405-406
   door systems 42-45
   lighting, wiring 70
   lining, cedar 39
   lockers 389
   racks 394
   shelves 186
   water 161-162, 464
Clothes chutes 416
Clothes dryers
   vent sets 184
   wiring 71
Clothesline units 45
Coal tar patch 172
Coat hooks and racks 394-395
Coatings
   concrete finishes 311-312
   dampproofing 348
   paint 151-155, 380-382
   roofing 171-172
   steel 339
   waterproofing 348
Coaxial cable 551
Cocks, gas 451
Coil units 483-484
Coils, heating & cooling ... 489-490
Cold box, insulation 350
Cold milling, asphalt 272
Collar beams, carpentry 28
Collar ties 28
Collection boxes, mail .. 129-130, 390
Colonial
   brick 132
   columns 45, 75
   doors 65
   entrances 74-75
Color coat, court 285

Coloring agents 311
   concrete 46, 52, 277, 306
   mortar 139, 275, 325
Column
   bases 45, 206-207
   clamps 234
   footings 49
   forms, cast-in-place .. 46-47, 49, 299
   forms, tilt-up 320
Columns
   carpentry, posts 28
   colonial 75
   concrete 299, 301-303
   decorative split 75
   fireproofing 359
   lally 31
   marble 331
   ornamental iron 151
   painting 156
   porch 45
   precast 315
   reinforcing 305
   structural steel 337
   tilt-up 320
   tube 309
   wood 28, 45
Combustion door 60-61
Combustion incinerators 408
Combustion train, boiler ... 470-471
Commercial
   brick 132
   hardware 366-368
   wage rate 221
   windows 366
Common brick 131, 140, 324-325
Communications cable 551
Communion rail 397
Compacting soil
   embankment 254
   equipment rental 77-78, 233
   excavation 75, 77, 252-253
Compaction testing 199
Compaction, roadway 257
Compactors, rental 78, 233
Compartments 384
   telephone enclosures 392
Composition
   flooring 378
   roof coating 152
   roofing 173, 351-352
Compression test 226
Compressor
   foundations 321
   paint 155
Compressors
   chiller 488
   rental 232
   service station 402
Computer flooring 385-386
Computer room HVAC 485-486
Concealed beds 19
Concrete
   additives 46, 306, 311-312
   adhesives 14
   aggregate 14, 306
   apron 277
   architectural 46
   beams 298
   buggies 233
   building walls 300, 307
   bush hammered 312
   caisson 261
   cast-in-place 46-47, 49-51,
                           306, 308-309
   catch basins 269, 407
   coatings 311-312
   coloring 46, 274, 306, 311-312
   column forms 301-303
   columns 299, 306, 309
   conduit envelopes 502
   conveyors 234
   core cutting (drilling) 250
   culvert, precast 266
   curb 47, 52, 276
   curing compounds 306-308
   curing paper 21
   cutting 250
   decking 324
   delivery charges 45-47

Concrete (continued)
   demolition 54-55, 57, 241
   driveways 45, 47
   dry pack 263
   embossed finish 52
   equipment foundations ... 321-323
   equipment rental 233-234
   etching 312
   expansion joints 46, 310
   exposed aggregate 52
   fibrous 273
   fill 320
   fill, columns 338
   finishing 52, 273, 311-312
   fireproofing 357-359
   flatwork 47-48, 52, 273, 306-307
   floor finishes 311-312, 324
   footings 49, 293
   form stakes 47
   forms 47-48, 122, 298, 300
   forms, paving 273
   foundations 47-51, 299
   grade beams 49-50
   grout 46, 311, 316, 327-328
   grouting 52
   gunite 312
   hardeners 311
   headwalls 271
   high early strength 46, 306
   inserts 311
   insulating deck 324
   joints 46, 273-274
   keyway 47
   labor placing 307-308
   lightweight 46, 306, 324
   lintels 331
   masonry units 134-137, 140, 326-328
   meter boxes 143
   mix 45-46
   mixers 233
   nails 149
   paint 152, 153
   paint preparation 156
   painting 155-157, 380
   patching 52
   pavers 136
   paving 160, 273
   piles 260
   pipe 265
   placing 49-52, 306-307
   plank 314
   plasticized 46
   pouring (placing) 49-52, 306-307
   precast 142, 312
   preparation 155
   primer 152
   pull & junction box 502
   pump mix 46
   pumping 237-238
   ready-mix 45-46, 306
   reinforcement 48-49, 305-306
   retaining walls 49-50, 300-301
   roof decks 324
   roof tiles 176
   sandblasting 177
   sawing 250
   saws 233
   screeds 234
   sealer 153, 312
   sidewalk 51, 277
   slab forms 299-300
   slab sawing 250
   slabs 51, 306-307, 314
   slip formed 309
   stairways 306, 309
   stamped finish 52
   steps 52, 309
   steps on grade 52
   structural 308, 316
   support poles 316
   suspended slabs 306-307
   testing 226
   ties 46, 311
   tile roofing 176
   tilt-up 316-321
   topping 52, 311-312
   underpinning 262
   unit masonry 134

Concrete (continued)
   valve boxes 143
   vibrators 233
   walks 51, 277
   wall saw, rental 233
   wall sawing 250
   walls 306, 318
   water stop 311
   waterproofing 211, 348
   work 45-52, 297, 316-322
Concrete angles 207
Concrete block
   demolition 54
   detailed 135-136
   glazed 326-327
   load bearing 326
   natural 134
   painting 155
   pavers 136
   primer 151
   screen 136, 327
   slump 136-137
   split face 136, 327
   testing 228
Concrete finishes 52
Concrete specialty finishes 52
Condensate pumps 475, 480
Condensers, boiler 475-476
Conditioner, limestone 291
Conductors, electric 519-521
Conduit 503-505
   aluminum 513-514
   EMT circuits 507-508
   flexible 514-515
   IMC 508-509
   lightweight 514-515
   PVC 509-512
   PVC circuits 512-513
   RSC 503, 512-513
   seal fittings 505
   steel 503-505
Conduit bodies
   aluminum 513-514
   PVC 509-510
   steel 504-505
Conduit clips 516
Cone flashing, plumbing vent ... 182
Cones, traffic control 17
Confessionals 396
Connections, sewer 181
Connectors
   Sealtite 514-515
   wood .. 108-109, 148-151, 206, 209
Construction
   barricades 17
   camps, manufactured 408-409
   elevators & hoists 231
   equipment rental 232-239
   keying 98
   photographs 230
   signs 230
   tool, rental 232
   tools, charge 232-237
Construction economics 219
Construction loans 223
Consultants fees 74, 225
Consumables, cost 222
Contact cement 14
Contactors, lighting 527
Containers, storage 240
Containment, asbestos 246
Contents of this book 2
Contingency 130-131, 222
Contractor insurance 106
Contractor's labor burden
   commercial & industrial 221
   residential 8
Control air handling 497-500
Control erosion 293
Control fund fee 223
Control panels
   alarm 548-550
   motor 529
Control valves 288
Controls
   boiler combustion 474
   building 497-500
   chiller 499

Index

Contols *(continued)*
  irrigation 282, 286, 288
  mechanical system 497-500
  motor 529
  pest (soil treatment) 199
  sprinkling 288
  thermostat 497-500
Conveying systems 413-416
Conveyors
  belt 415-416
  checkroom 399
  concrete, rental 234
  dish 401
Cooking equipment 170-171, 399-401
Cooktops 170-171
Coolers
  drinking water 163, 462
  HVAC 103
Cooling
  air conditioning 103, 484-488
  controls 499-500
  direct expansion 484-485
  equipment 484-488
  tower monitor 499
  tower screens 354
  towers 488
Coping
  masonry 137, 325, 331
  metal 352-353
Copper
  flashing 352
  grounding devices 527-528
  gutter 354
  pipe 436-439
  roof, cupola 53-54
  roofing 353
  wall ties 329
  wire 519
Cord sets 526
Core
  brick testing 228
  cutting, concrete 250
  sampling (soil testing) 199
Corian
  countertops 53
  vanity tops 53
Cork
  insulation 350
  sheathing 382
  tackboards 383
  tile flooring 90
  track, running 282
Cornalath 112
Corner
  bead 112, 145-146
  braces, metal 207
  cabinets 24
  guards, metal ... 320, 339, 353
  guards, plastic 384
  mould 144
  record 225
Corneraid 112
Cornerite lath 112
Cornices, painting 156
Corrugated
  fiberglass 82
  glass 369
  mats 141
  metal roofing 175, 350-351
  metal siding 194, 350-351
  pipe 266-267
  polyethylene pipe 264
Cost adjustments, city 10
Cost changes 4
Cost engineering 225
Costs listed in this book 3
Counter flashing 182, 352-353
Counters
  bank 395
  kitchen 399
  steel 399
Countertops
  Corian 53
  plastic 53, 406
  tile 206, 374
Couplings
  threaded 504
  victaulic 453

Court, surface primer 285
Courts
  basketball 282
  tennis 284-285
Cove
  flooring 86
  moulding 144
  terrazzo 375
Cover, ground 293-294
Coverage loss defined 3
Covers
  canvas 203
  joint 357
  manhole 271-272
  plastic 169, 203
Cowl caps 184
CPM scheduling 74, 223
CPVC pipe 440
Craft codes defined 5-7
Craft hours defined 3
Cranes
  concrete placing 306-307
  hoists 415
  jib 415
  pile driving 234
  rental 231, 234
  tower 231
  truck mounted 231
Crawl hole vents 184
Crawler tractors ... 77, 234, 252, 256
Credits 12
Creosote 128, 154
Cribbing & shoring 262
Critical path scheduling 223
Cross arms pole 539
Cross bridging 25-26, 108
Cross ties 295
Crossing, railroad 296-297
Crossings, grade 297
Crown moulding 144, 146, 148
Crushed
  rock 14, 174
  slag 14, 69
  stone 69
Cubicles 384
Cultivating 292-293
Cultured marble 212
Culvert pipe 264-266
Cup dispenser 399
Cup holder 18
Cupboards, painting 157
Cupolas
  aluminum roof 53
  copper roof 53-54
  redwood 53-54
  weather vanes 54
Cuprinol stain 154
Cuproligum 154
Curb
  asphalt 277
  demolition 55, 241
  granite 277
  inlets 271
  valve box 272
Curbs
  concrete 47, 52, 276-277
  elevated slab 304
  roof 500
Cure coat, concrete 312
Curing concrete 312, 317, 319
Curing papers 21
Curtain rod 19
Curtain wall 369
Curtains 69-70, 397-398, 407
  stage 397-398
Cushiontone 375
Cutting
  asphalt 251
  concrete 250
  masonry 139, 250, 328
  slabs 250
  steel 139
  walls 250
Cutting boards 399
Cylinder locks 98
Cypress flooring 90

D

Dampers, air regulating 498-499
  backdraft 498
  control 498-499
  fire 499
  fireplace 83
Dampproofing 348
Danger signs 230
Deadbolt locks 98
  sliding door 68
Deadening felt 21
Deadlocks 98
Debris removal 222, 241, 245
Deck
  insulation 350
  waterproofing 211
  yard and patio 54
Decking
  carpentry 33, 342
  cementitious 324
  concrete 324
  demolition 245
  fence 242
  fiber 324
  lumber 121
  manhole 242-243
  metal 335
  paving 241-242
  redwood 54
  roof 33, 324, 335
  wood 54
Decontamination, asbestos 247
Decorative glass 395
Delivery expense defined 3
Demobilization 226
Demolition 54-57, 241-245
  asbestos 246-249
  building 57
  concrete block 54
  curbs 55
  glazing, hack out 93
  joists 56
  roof 244-245
  stairs 57
  walls 56
  windows 56
  wood frame structures 57
Demountable partitions ... 390-391
Deposit boxes 395
Depository doors 395
Derricks, construction 231
Desks
  bank 395
  study 398
Destruction 54-57, 241-245
Detail steel 338-339
Detailed block 135
Detection systems 548-550
Detectors
  audio 179
  fire 179
  infrared 180
  security 548-550
  smoke 179
Detour signs, barricade 17
Dewatering 263
Diamond lath 112, 370
Diamonds, baseball 281-282
Dieldrin 199
Diffusers
  air 495-497
  floor 497
  light 540
Digging
  concrete work 297-298
  excavation 75-76, 251-252
Dimmer switches 72, 526
Direct expansion, cooling 484-485
Direct overhead 222
Directory boards 388
Disappearing stairs 202
Disconnect switches 539
Dish table 399

Dishwashers
  electrical 71, 164
  food service 399
  plumbing 164
  residential 164
Dispensers
  paper towel 393-394
  ticket 403
  toilet paper 392
Display case, food 399
Display shelf, kitchen 399
Disposal
  asbestos 246
  incinerators 408
Disposal bags, asbestos 249
Disposal fields 181
Disposers, garbage
  commercial 400
  plumbing 459
  residential 71, 163-164
Distribution
  electric 531-533
  panels 531-533, 535-536
  switchboards 531-533
Ditch Witch 254
Ditches, excavating 258
Ditching 75-76, 252, 298
Diverters, rain 183
Divider strip, brass 375
Dividers
  partitions 390-391
  redwood 277, 294
Dock
  bumpers 320, 404
  levelers 320, 403
  lifts 403
  shelters 404
Docks
  loading 309, 403
  marine 294
Documentation, photographs ... 230
Dolphins 294
Dome pan forms 304
Domes
  astronomy 410
  concrete form 304
  geodesic 410
  lighted ceilings 39
  playground 282
  skylight 196-197, 355
  slab form 304
  special construction 410
Domestic red oak 123
Door
  assembly, hollow metal 360
  bells, wiring 71
  bottom 368
  chimes 57
  closers 57, 367
  deadlocks 98
  demolition 55, 245-246
  entry locks 98-99
  frame protector 384
  frames 58, 74-75, 361-364
  frames, painting 157
  frames, steel 339
  framing, wood 28
  hardware 98-99, 366
  hinges 99
  jambs 58-59, 362
  kick plates 99
  knockers 100
  louver vents 385
  mail slot cut out 67
  mirror panels 43-45
  openers 91, 366
  openings, framing 28
  painting, exterior 156
  painting, interior 156-157, 380
  peep sight 100
  protectors 385
  pulls 100
  push plates 100
  release, electronic 108
  removal 55, 245-246
  security switch 549
  shoe 203-204
  sills 58

573

Index

Door *(continued)*
  stop 100, 368
  thresholds 144, 203-204, 368
  trim 59, 67, 144, 360
  weatherstripping 211-212, 368
  wood, paint grade 363
Door closers
  hydraulic 57
  pneumatic 57
  screen 57
  sliding door 57
Door seals
  loading dock 403
  weatherstrip 203-204
Doors 59-65, 360-369, 390
  access 181, 371, 458
  aluminum 60, 67-68, 364
  aluminum sliding 67-68
  ash 42, 44, 61-64
  bar 59-60
  basement 18
  bathtub 186-187
  birch 42, 44, 61-64
  by-passing 44
  cafe 59-60
  ceiling access 355
  cleanout, fireplace 83
  closet 42-45
  colonial 65
  combination 60
  demolition 55, 245-246
  depository 395
  double 66
  dutch 60, 391
  embossed 64
  exterior 60, 66-69, 361
  fire rated 60-61, 364
  fireplace 84-85
  flush 61-63, 66, 362
  folding 42-43, 64, 390
  freight 364
  French 64
  garage 92
  glass sliding 67-68
  hardboard 42, 44, 61-63
  hemlock 60, 64, 66
  hollow core 42, 44, 61-62
  interior 59-62, 65-66
  kalamein 361
  labor to hang 63
  lauan 44, 61-62
  louver 42, 64-65
  metal 360-362
  mirrored 43-45
  oak 44, 61-64, 66
  overhead 92, 333, 364
  panel 64
  partition 390
  patio 67-68
  pet 65
  pine 42-43, 59, 65-66
  pocket 58
  prefitting 63
  prehung 66-67, 362-364
  radiation shielding 410
  raised panel 65
  refrigerator, galvanized ... 364
  refrigerator, stainless steel 364
  revolving 364
  roll-up 364
  sash 64
  screen 67
  service 364
  shower 186-187
  single 67
  slatted 42, 64-65
  sliding, glass 67-68
  sliding, metal 67-68, 364, 390
  sliding, mirror 44
  sliding, wood 44, 68
  solid core 62, 66
  special 364
  spindle 60
  steel 66-67, 362
  steel, building 333
  storefront 365-366
  storm 60

Doors *(continued)*
  tub 186-187
  units, packaged 66, 360-361
  vault 365, 395
  wood panel 65
  wood sliding 44, 68
  X-ray protective 410
Dormer
  louvers 185
  studs 28
Dosch clay 133
Double tees, precast 315
Doughnut maker 400
Douglas fir
  flooring 89
  lumber 117-121
  siding 125, 192-193
Dove tail anchors 325
Dowel supports 299
Dowels 125
Downlight, baffle 544
Downspouts
  commercial 354
  painting 156
  residential 94-95
  steel buildings 334
Dozer
  clearing 241
  excavation 76, 256-257
  rental 77, 234
Drafting
  architectural 68-69
  plans 19
Drafting survey 224-225
Draftshield 212
Dragline
  excavation 254-255
  rental 234
Dragon teeth 403
Drain
  fields 181
  pipe 166, 264, 266-268
  pipe, plastic 69
  rock 259
  yard 272
Drain lines, demolition 242
Drainage
  fabric 259
  pipe 69, 264, 266-268
  rock fill 258
  slurry trench 259
  trenching 253
  tube 439
Drainage systems
  rock fill 69
  sand fill 69
Drainboards, countertops
  plastic 53, 406
  tile 205, 374
Drains
  area 459
  catch basins 269
  floor 459, 462
  foundation footing 69
  gratings 338
  piping 69
  roof 459
  shower 161, 459
  yard 272
Draperies 69-70, 407
Drawer bases 210
Drawers 22-23, 406
Drawers, stainless 399
Drawings
  architectural 15-16
  as-built 68-69
  record 68-69
  reproduction of 19-20
Dredging 294
Drench showers 462
Dressing cubicles 384
Drilled caissons 261
Drilling
  concrete 250
  post holes 277
  rock 251
  soil testing 199
  steel 139
  well 212

Drills, rental 232-237
Drinking fountains 163, 462
Drip cap
  door 203
  moulding, wood 144-146
  vinyl siding 195
Drip, eaves 183
Drive gates 79-80
Drive shoe 212
Driveway
  apron 277
  coating 152
  demolition 55, 241-242
  forms 47, 276
Driveways
  asphalt 159-160, 272
  concrete 51, 160, 273
Drop ceiling 39-41
Drop cloths 203
Drop poles, electrical 230
Drop siding 125-127
Dry pack grout 316
Dry pack, concrete 263
Dry packing, concrete 316
Dry pipe sprinklers 465
Dryer
  circuits 503
  receptacles 526
  vent sets 184
Dryers
  clothes, wiring 71
  hand 394
Dryvit, see EIFS 348
Drywall
  adhesive 13-14
  demolition 56, 244
  EIFS wall system 348
  expansion joints 356-357
  gypsum 95-96, 210, 372-373
  insulation board 346
  painting 154-155, 157
  wallboard 210
  wallboard assemblies 36
Drywall installer, wage
  commercial & industrial 221
  residential 8
Drywall taper, wage 8
Duct
  aluminum 494-495
  cable tray 518-519
  electrical 501-502
  fiberglass 103, 495
  flex 103
  flexible 495
  galvanized steel 494-495
  heaters 467-468
  insulation 495
  insulation removal 248-249
  insulation, asbestos .. 248-249
  sheet metal 103
  stainless steel 494
  underfloor 519
  vents 494
Ductile iron pipe 432-433
Ducting 170
Ducts, electric 501-502
Dumbwaiters 73, 413-414
Dummy knobs 99
Dump fees 241
  asbestos 249
Dump trucks
  debris hauling 245
  general hauling ... 76, 255-256
  rental 77-78, 235
Dumpsters, trash 240
Duplex nails 149
Duplex receptacles ... 72, 525-526
Dur-O-Wal 139
Durock backer board 206
Dust
  control 245, 269
  partition 245
Dutch doors 60

E

Earthwork 75-76, 252,
  .... 258-259, 272, 297-298
  air hammer 77
  backfill 75-76, 253, 297
  backhoes 76, 253-254
  basements 255
  blasting 251-252
  brush clearing 76-77, 241
  bulldozer 76, 256-257
  caissons 261
  clear and grub 77
  compaction ... 75-76, 252-253, 297
  dragline 254-255
  drilling 251, 261
  embankment 252
  equipment rental 77-78, 234
  erosion control 258, 269
  finish grade 75, 111, 159-160
  footings 75, 254-255, 297
  grading 75-76, 252, 297
  hand work 75-76, 258-259, 297
  hauling 255-256
  loaders 77, 255
  loading trucks 75, 77, 255
  pneumatic excavation 77
  ripping 251-252
  riprap 258, 294
  roadway 257-258
  rock 75, 251-252, 297
  scraper 257
  shovels 75
  slope protection 258
  spreading 75-76, 252
  sprinkling 77
  stripping soil 76, 252
  structural 297
  tamping 76, 77, 254, 297
  tractors 78
  tree removal 77, 241
  trenching 77, 253-254, 258-259
  trucks 76, 77, 255-256
  wheel loaders 255
Eave closure 177
Eave drip 183
Eaves, painting 156
Ebony veneer 124
Ecclesiastical equipment 396-397
Economics division 219
Edge forms 299-300
Edge polishing, glass 94
Edging
  redwood 112, 277, 294
  roof 183
  strip 86
Educational equipment 398
Efflorescence, testing 228
Eggcrate louvers 39-40
EIFS wall finish 348
Ejectors, sewage 460
Elasticity test 227
Elastomeric
  flashing 329
  membrane 211
  roofing 351-352
  waterproofing 211
Electric
  generator 199
  grounding 527-528
  heating 100-102
  incinerator 408
  motor connection 528-529
  motors 528
  outlets 72
  power 231
  receptacles 525-526
  sensor cable 549
Electric fence, charging unit ... 80
Electric service,
  boiler 472-473
  temporary 222, 230
Electrical conduit 503-505

Index

Electrical duct
concrete 502
PVC 500-501
steel 501-502, 517-518
surface 517-518
Electrical metallic tubing 505-508
Electrical outlet boxes 524-525
Electrical systems, solar 199
Electrical work 70-72, 500-552
alarm systems . . . 70, 179, 548-550
boxes 502, 524-525
circuit breakers . . . 72, 530, 536-537
conduit 503-505
controls 497
cutouts 539
ducts 500-502, 517-518
fuses 538-539
generators 547-548
generators, rental 237
heating 100-102, 466-467
junction boxes 524-525
lighting 70-71
loadcenters 536
locks 98-99
motors 528
outlets 72, 525-526
panels 534
poles 539
power, temporary 230
raceway 518
ranges 171
receptacles 72, 525-526
service 72, 230, 531-533
sprinkler controls 114, 288
subcontract 70-72
switches 526-527
switchgear 526-527, 531-533
temporary power 230
testing 552
transformers 534-535
tubing 505-507
valves 114
water heaters 72
wire 502-503, 511-513
wireway 517-518
Electrician, wage
commercial & industrial 221
residential 8
Electricity, solar 199-200
Elevated floors 385-386
Elevated water tanks 412
Elevators
commercial 414
construction type 231
hydraulic 73
residential 73
subcontract 73
Elm
lumber 123
veneer 124
Embankment grading 252
Embeco grout 311
Embedded steel 338
Embossed concrete 52
Emergency lighting 545
Emergency shower 462
Employer labor taxes
commercial & industrial 221
residential 8
EMT conduit 505-508
Emulsion wall coating . . . 171-172, 348
Enamel 152, 380-382
Encapsulation, asbestos 249
Enclosures
cubicles 384, 390
panel 524-525
sauna 178-179
shower and tub 161, 187, 384
sound deadening 412
telephone 392
temporary 238
Energy monitoring 499-500
Energy, solar 199-200
Engelmann spruce 119-120
Engineer, overhead 222
Engineering
fees 74, 225
testing 225, 229
tilt-up 318
Enthalpy sensor 499
Entrance
caps 505
carpeting 378-379
doors 55, 67
hoods 186
mats 141, 407
thresholds 203-204
Entrances
colonial 74-75
storefront 365-366
wood 74-75
Entry locks 98
Epoxy
bed 326
coated rebars 305
coatings 151
enamel 152
primer 151
terrazzo 378
Equipment
foundations 321-323
mobilization 226
overhead 222
pads 307
Equipment cost defined 3
Equipment rental 77-78, 232-237
paint sprayer 155
Erickson couplings 504
Erosion control 259, 269, 293
fabric 259
fence 258
landscaping 258, 293
soil covers 269
soil stabilization 269
Escalators 416
Escapes, steel 339
Estimate window 553
Estimate Writer
disk instructions 553
Estimating accuracy 4
Estimating costs 225-226
Etch, acid 285
Etching concrete 312
Evaporative coolers 103
Excavation . 75-77, 251-252, 258, 272 297-298,
air hammer 76
backfilling 75-76, 253, 297
backhoes 76, 255
basement 255
blasting 251-252
brushing 77, 241
bulldozer 76, 256-257
caissons 261
clearing 76
compaction . . . 75, 77, 252-253, 297
dragline 254-255
drilling 251, 261
embankment 252
equipment rental 77-78, 234
erosion control 258, 269
fine grading 75, 159-160
footings 75, 254-255, 297
foundation 297, 322-323
grading 75, 252, 298
hand work 75-76, 258, 297-298
hauling 255-256
loaders 77, 255
loading trucks 75, 77, 255
pneumatic excavation 77
ripping 251
riprap 258, 294
roadway 257-258
rock 75, 251-252, 297-298
scraper 257
shovels 75, 77
slope protection 258
spreading 75-76, 252
sprinkling 77
stripping soil 76, 252
tamping 76, 77, 254, 297
topsoil 76
tractors 78
tree removal 77, 241
trenching 77, 253, 258-259
trucks 76, 255-256
wheel loaders 255, 111
Excavation equipment rental 234
Exhaust
fans 78-79, 101, 491-492
fans, wiring 71
grilles 497
range hoods 170-171
roof 355
sets 491-492
tubeaxial 478
Exhaust fume system 402
Exhaust vents, dryer 184
Exit hardware 368
Exit signs 544-545
Expansion
shields 339, 517
strips 138
tanks 460
Expansion joints
composition 139
concrete 46, 310
covers 357
glass block 330
lath 112
masonry 139
pipe 452-453
plaster 371
preformed 356-357
sealants 359
Explosion-proof lights 545
Explosives, loading 251
Exposed aggregate finish 52, 311
Exterior door assembly, hollow metal 360
Exterior wall (EIFS) system 348
Exterior wall assemblies 36
Exterior wall system 350-351
Extinguishers, fire 386, 466
Eye wash shower 462

F

Fabric
draperies 69-70, 407
erosion control 258
stage curtain 398
wallcover 382
welded wire 305
Fabricated metals 338-339
Facade screen 354-355
Face brick 131, 140, 324
Faceted glass 395
Facia (see Fascia)
Facing tile 330
Facing, marble fireplace 130
Factory rep, cost 226
Fan coil
monitor 499
unit controls 499
units 483-484
Fan units, radiators 482-484
Fan wiring 71
Fans
attic 78
bathroom 78
decorative ceiling 78
electrical work 71
exhaust 78, 355, 491-492
kitchen 78, 170-171, 399
roof exhaust 79, 355, 491-492
supply 492
tubeaxial 478
Fascia (facia)
carpentry 28-29
forms 309
gutter 353-354
metal 353
soffit system 198
vinyl 351
wood 28-29
Fasteners
nails 148
shot 341
timber connectors 206, 209
Faucets 162-163, 451
Feather wedges 128
Featherock 258
Feeder section panels 534
Fees
acoustical design 74
appraisal 14-15
building permit 22
engineer 74, 225
estimating 225
overhead 222
scheduling 74, 223
sewer connection 223
specification 226
testing 226-229
valuation 14-15
water meter 224
Felt
asphalt 21, 173, 174, 351
netting 113
roofing 173, 174, 351
waterproofing 348
Felt underlay 86
Felton sand 312
Feminine napkin dispenser 392
Fence
athletic fields 281-282
barbed wire 278
chain link 79-80, 278
demolition 242
electric 80-81
erosion control 258
painting 156, 280
pet enclosure 80
post holes 277
sediment 258
swimming pool 151
temporary 238
tennis court 284-285
wire mesh 80-81
wood 81-82, 278-279
Fenders, marine 294
Fertilizer 291
Fescue seeding 293
Fiber
cant strips 352
forms 303
roof coating 172
Fiber drums, asbestos 249
Fiberboard backing 383
Fiberboard, wallboard 210
Fiberglass
bathtubs 161
blown 105
ceilings 40
duct 103, 495
formwork 304
insulation 104, 348
panels 82
pipe 442
pipe insulation 455-456
roofing 172, 173, 348
screen 179
septic tanks 180-181
shower receptors 187
showers 161
siding 194
whirlpool bath 161
FICA 203
Field office 239
Fields, playing 281-282
Fig veneer 124-125
Fill
caulking 37-39, 355-356
excavation 75, 253, 298
rock 258, 298
sand 298
slab base 298
Fill investigation
compacted 229
soil testing 199
Filler strips, fence 79
Film, polyethylene 169
Filter screen 212
Filters, air 493
Financing, project 223
Fine grading
embankment 252
excavation 75
hand 75
landscaping 111, 291-292
pavement 159-160, 272
slabs 252, 298

Index

Finish carpentry
  ceilings 39-41
  closet doors 42-45
  countertops 53
  cupolas 53-54
  door accessories 57
  doors 59
  entrances 74-75
  flooring 87-90, 121
  hardboard 97
  hardware 98
  lumber 144-148
  mantels 130
  medicine cabinets 141-142
  mouldings 144-148
  paneling 157-159
  sauna rooms 177-178
  shutters 188-189
  siding 125-127, 189-194
  skylights 196-198
  soffits 198
  stairs 200-202
  thresholds 203-204
  wallboard 210
  weatherstripping 211-212
  window sills 212
  windows ... 197-198, 212, 214-218
Finish grading ... 75, 111, 159-160, 53
Finish hardware 98, 101, 366-368
Finish, exterior 350-351
Finishers, rental 233
Finishes
  decks 54
  flooring 87, 377
  gypsum wallboard 96, 372
  paint 152
  plaster 160, 370
  wallpaper 210-211, 382
Finishing
  concrete ... 52, 307-308, 311-312
  nails 149
Fink domes 410
Fink truss 33-34, 346
Finned tube radiators 482
Fir
  flooring 89, 121
  lumber 117-121
  siding 125, 189-190, 192-193
Fire
  alarm breaker 531
  alarm systems 386, 550-551
  boxes 83
  brick 133
  dampers 83
  detectors 551
  doors 60-61, 364
  escapes 339
  extinguishers 386, 466
  hose cabinets 465
  hydrants 465
  hydrants, demolition 242
  protection equipment 464-466
  pumper connection 465
  retardant treatments 129
  sprinkler systems 83, 464-465
Fire pole tower 282
Fire rated
  doors 60-61, 364
  roofing shingles 174
  wallboard 95
  walls 95
Fire retardant wood 129
Fireblocks, carpentry 29
Fireclay 133
Fireplace
  accessories 83
  brick 133
  doors 84-85
  forms 83-84
  mantles 130
  marble 130
  masonry 140, 332
Fireplaces
  forced air 84
  heat circulating 84
  prefabricated 84-85, 332

Fireproofing
  columns 359
  concrete 359
  plaster 358-359
  spray-on 359-360
  steel 359
  testing 228
  vermiculite cement 359-360
  wood 129
First aid, overhead 222
Fitness trail 284
Fittings
  electrical 524-525
  pipe 416-417
Fixture lamps 543
Fixture valves 451
Fixtures
  electrical 70-71, 114-117
  lighting 114, 539-545
  plumbing 160-165
Flagpoles 386-387
  demolition 243
Flagstone
  floors 137, 141
  pavers 276
Flameproof draperies 407
Flanges
  pump 454
  welding 418-428
Flangeway 297
Flash cone 353
  plumbing vent 182
Flashing
  building paper 21
  chimney 184
  compound 172
  counter 182
  demolition 245
  masonry 328-329
  neoprene 352
  paper 21
  roof jack 459
  roofing 183, 352-353
  sheet metal ... 182-184, 352-353
  skylight 196-198
  through roof 182
  valley 183
  window 184
  Z-bar 184
Flat paint 152-153
Flexible conduit 514-515
Flexible duct 495
Flexure test 227
Flexwood 382
Flight, photography 225
Float finish
  concrete 52, 311
  glass 369
  plastering 371
Float finisher 233
Float switch 527
Float traps 479
Floodlights 545-546
Floor
  adhesive 13-14, 85, 86, 89
  assemblies, wood framed 30
  cleaning 240
  covering, carpet ... 36-37, 378-379
  decking, steel 335
  decking, subflooring ... 35, 342
  demolition 55, 244-245
  diffusers 497
  drains, plumbing 462
  furring 30-31
  grating 338
  joints 24, 29, 342, 357
  mats 141, 407
  paint, patio 153
  painting 154-155
  plank, precast 314
  plate, steel 338
  polishing 154-155
  sanding 154-155
  sheathing 342
  shellacking 154-155
  sinks, plumbing 462
  slab, concrete ... 51, 304, 306-308
  slab, precast 314

  staining 154-155
  varnishing 154-155
  waxing 154-155
Floor layer, wage
  commercial & industrial 221
  residential 8
Flooring
  access 385-386
  acoustic 377
  acrylic 378
  anti-fatigue 378
  asphalt tile 377
  brick 141
  carpet tile 379
  ceramic 206, 373-374
  composition 378
  conductive 378
  crossband parquet 89
  demolition 55, 244-245
  Douglas fir 89, 121
  fir 89, 121
  flagstone 137, 141
  gymnasium 376-378, 401
  hardwood 87-89
  marble 130, 331
  masonry 141, 326
  non-slip tiles 378
  parquet 88-89, 376
  pine 87, 90
  plank 87-88, 90
  plastic tiles 374, 378
  quarry tile 325-326, 374
  raised, installation 385-386
  removal 244-245
  resilient 90-91, 377-378
  rubber 90-91, 378
  safety tiles 378
  softwood 89-90, 376
  strip 87-90
  terrazzo 326, 374-375, 378
  tile ... 85-86, 206, 373-375, 379
  velvet tile 379
  vinyl 377
  wood 87-90, 376-377
Floors, removal 244-245
Flow meter 499
Flue
  fireplace 85, 332
  lining 137, 332
Fluorescent lighting . 114-116, 539-541
Flush
  bolts 368
  doors 61-63, 66, 362
Foam insulation 105
Foam panel adhesive 13-14
Foamglass insulation 349
Foil
  aluminum 21
  barrier 21
  wallcovering 210
Foil-backed gypsum 95, 372
Foil-faced insulation 104, 349
Folding beds 19
Folding doors 64, 390
  closet 42-43
Folding fire partitions 391
Folding stairs 202
Food
  mixers 400
  service equipment 399-401
  tables 401
  wells 399
Foot grilles, recessed 407
Football
  fields 281
  goal posts 284
Footing
  bollard 281
  concrete 46, 49, 307
  demolition 54-55, 244-245
  drains 69
  excavation ... 75, 254-255, 298-299
  fence 281
  forms 46, 298, 299
  keyway 47
  ties 46, 310
  wall 298

Footings
  bell 261
  caissons 261
  column 49
  concrete 46, 49, 307
  forming 299
  tilt-up 316
Forced air furnaces 102
Forklift, rental 235
Form
  board 47
  breaker 310
  clamps 311
  liner, architectural 320
  liners 300, 303, 320
  oil, concrete 311
  plywood 48, 168
  release 310
  spreader 310
  stakes 47
  stripping 48
  ties 46, 207
Formboard, fiber deck 324
Formica countertops 53
Forming
  architectural 300
  landing 305
  soffits 305
Forms
  beam 304
  column 301-303
  concrete 46-48, 298, 305
  curb 47, 276
  fireplace 83-84
  fireproofing 359
  girder 305
  paving 273
  plywood 301
  shoring 262
  sidewalk 47, 277
  slab 299-300, 304
  special 309
  stripping 48
  tilt-up 318, 321
  wall 300-301
Formwork
  foundation 322-323
  steps 309
  stripping 48
Fortifiber products 21
Foundation
  bolts 91, 339
  coating, asphalt 152
  demolition 54-55, 244-245
  drains 69
  excavation 298
  forms 47, 48, 299-300, 305
  investigation, soil testing ...199
  layout 114
  placing 307
  poisoning 199
  soil calculations 199
  stakes 127
  trenching 253, 258-259
  vents 184
  walls 49-50
  waterproofing 211
Foundation wall access doors 182
Foundations
  bell 261
  caissons 261
  concrete .. 47, 49-51, 299-300, 307
  equipment 321-323
  footings 47, 299-301
  injected 261
  pile 259
  tilt-up 316
Foundry iron 339
Fountains
  drinking 163, 462
  rough-in 463-464
  wash 463-464
Frame seal sets 212
Frames, door
  metal 361-362
  painting 156-157
  pocket 58
  steel 362
  wood 58, 74-75, 363-364

Index

Frames, manhole 270-271
Framing
adhesives 13
anchors 207, 341
canopy 338
carpentry 24, 34-35, 340-342
clips 169
demolition 56, 243, 244
door openings 31
hardware 108-109
lumber 117-121
metal stud 371
partitions, wood 34-35
steel 337-338
wall stud, wood 34-35
window openings 35-36
Free access flooring 385-386
Freezer, wiring 71, 72
Freezers, hospital 404-405
Freight elevators 414-415
French doors 64
Frequency meter 533
Fresnel globes 544
Frieze vents 355
Fringe benefits
commercial & industrial 221
residential 8
Fry pans .400
Fuel grates, fireplace 83
Fuel meters 458
Fuel storage tanks 466
Fuel train piping, boiler 470-472
Fume hoods 406
Fund control fee 223
Fungus treatment 128
Furan resin 374
Furnaces
electric work 71
floor 102-103
forced air 102
gas 102, 467-468
oil 102-103
wall 102, 468
Furnishings 405-408
Furring 30-31, 345, 370
nails 149
Fuses, cartridge 538-539
FUTA 203

G

Gable
truss 34, 346
vents 184, 385
Galvanized
fabricated metals 339
flashings 182-183
pipe 428
roofing .175
sheet metal 182-183
siding 350-351
steel 333-334
Galvanizing 339
Game equipment 282-284, 401
Gang shower 462
Garage door
hardware 100
openers, wiring 71
operators 91
seals 212
Garage doors 92
subcontract 92
Garage, carport 37
Garbage disposers . 163-164, 400, 459
electric work 71
Gas
chlorination 269
cocks 451
defroster 405
fireplace .85
fuel train, boiler 471-472
heating 467-468
incinerator 408
log lighter 83
meter boxes 272
meters 433
Gas *(continued)*
piping 171, 402, 433
regulators 433
stops 451
storage tanks 466
valves 83, 451
water heaters 164, 459-460
Gas station equipment 402
Gaskets, neoprene 357
Gasoline dispenser 402
Gate valves 447-448
Gates
fence 79-80, 81-82, 278
hardware 280
parking control 403
temporary 238
trash enclosure 321
wrought iron 340
Gauge, liquid level 481
Gauges
pressure 458
water meters 458
General contractor crews 4
General contractor markup 4
General requirements 222-240
Generator, electric 199
Generators
emergency electric 547-548
heat 467-468
hot water 468
rental 237
solar 199
steam 202
Geodesic domes 410
Geographic cost adjustments 10
Geotextiles 259
Ginnies 127
Girder forms 305
Girders
carpentry 31, 341
concrete 307
forms 305
precast .314
steel 337
Girts
fireproofing 360
steel 337
Glass
artwork .395
cleaning 222, 240
doors 64, 67-68, 366
mirrors 18, 43-45, 143-144
mosaic .395
rack 400
storefront 365-366
window 92-93, 368-369
Glass bead moulding 146
Glass block 330, 137-138
Glassweld panels 346
Glazed
block 326-327
brick 133, 140, 324
tile 204-205, 329-330
Glazier, wage
commercial & industrial 221
residential 8
Glazing 93-94, 368
sash 93-94
Globe valves 448, 449
Gloss paints 152
Glove bags 246
Glue, adhesives 13-14
Glulam beams 346-347, 410
Goal posts, football 284
Goals, soccer 284
Golf course
buildings 282
development costs 282
equipment 282
irrigation 282, 287
Grab bars, bath 18, 393
Gradall 235
trenching 253
Grade beam
concrete 47, 49
demolition 55
excavation 46
formwork 47-48, 298
reinforcing 48-49
Grade crossings 297
Graders, rental 235
Grading 252, 272, 298
embankment 252
excavation 75
fine 75, 111, 159-160
finish 111, 159-160, 253
hand 75, 259, 291
landscaping 111, 291
lumber 128
site 252
slab 299-300
tilt-up 316
Grandstands 402, 408
Granite
aggregate 306
bollards .281
curb 277
edging 294
masonry 331
paving 276, 326
terrazzo .326
tile 205
Grass
landscaping 111, 292-293
sod 111
stolons .293
Grasscloth 210, 382
Grates
catch basin 269
fireplace .83
Gratings, steel 338
Gravel
aggregate 14
bed, landscaping 294
riprap 258, 294
roofing 173, 352
stop 183, 352
Gravel stop, demolition 245
Grease interceptors 460
Greenboard 95
Greenhouses 410
Grey iron .339
Grid ceiling tiles 40-41
Griddles 400
Grill, kitchen 400
Grilles
air 495-497
aluminum 339, 364
door 364
ornamental 339-340
roll-up 364
ventilating 399
window .216
Grinders 232
Grinders, rental 237
Grinding concrete 312
Grip handle locks 99
Grits, concrete 311
Ground cover, landscape 109, 293
Ground fault interrupter . .71, 503, 534
Ground rod clamps 527-528
Grounding
devices 71, 527-528
electric 527-528
Grounds, wood 31
Groundwire 519-520
Grout
brick pavers 275
concrete 327-328
door frames 328
foundation 311
machine base 311
manhole 271
masonry 46, 135, 327-328
pressure injected 271, 312
testing .228
tile 205
Grout forms, tilt-up 319
Grout mix concrete 46
Grouting
concrete .52
pipe joints 271
pressure 271, 312
tilt-up 316
Grubbing 77, 241
Guard, dog 180
Guard, security 180
Guardrail
construction 238
demolition 242
handrails 334-335
roadway 280
temporary 222
Guards, corner 96, 339, 353, 384
Gunite 312
testing 228
Gutter
demolition 55, 241
ice melting cable 523-524
inlets 271
paint 152
wire, duct 518-519
Gutters
concrete 52, 276
rain 94-95, 354
vinyl 95
Gutting, demolition 244
Gym equipment 401
Gym floors 376, 401
Gymnasium joints 357
Gypboard demolition 244
Gypsum
adhesive 13-14
ceiling panels 40
decking, poured 324
demolition 56, 244
drywall 95-96, 210, 372-373
lath 112, 370
painting 154-155
plaster 160, 370
roofing 324
roofing, demolition 244
testing 228

H

Hack out, glazing 93
Hair interceptors 460
Half round 146
Hammering, concrete 312
Hand dryer 394
Hand excavation 75, 258-259
Hand mining 262
Handholes, electric 502
Handrails
iron 151
painting 154, 156
steel 334
wood 146
Handy box, electric 524
Hangers
channel 515-517
conduit 515-517
joist 108-109
pipe 445-447
rod 447
roof 343
timber 206-207
Hanging doors, labor 63
Hardboard
adhesive .97
carpentry 345
paneling 97, 158-159, 383
siding 189-182
wallcoverings 383
Hardener, concrete 312
Hardware
bath accessories . . .18-19, 392-394
builders 206-209
door 98, 366
exit 368
finish 98, 366-368
garage door 100
gate hardware 280
partition 390
rough 206-209
Hardwood
flooring 87-89, 376-377
flooring demolition 55
lumber 123-124, 144
mats 141
moulding 144
paneling 157-159
plywood 158
veneer 124-125

Index

Hasps ... 99
Hat channel ... 370
Hat racks ... 394-395
Hatches, roof ... 355
Hauling
   debris ... 245
   excavation ... 255-256
   truck ... 76, 255-256
Haunch forms ... 301
Hazard surveys ... 246
Hazard testing ... 21
Header hangers ... 207
Header pipe ... 263
Headers
   carpentry ... 31, 341
   yard ... 277
Heads
   masonry ... 325
   sprinkler ... 289-290
Headwalls, concrete ... 271
Hearths
   fireplace ... 332
   marble ... 130
Heat
   detectors ... 551
   exchangers, shell and tube ... 478
   pumps ... 103, 484
   recovery ... 487
   temporary ... 222
Heat exchanger foundations ... 321
Heaters ... 100-103
   baseboard ... 100, 468
   bathroom ... 101
   convection ... 100
   duct ... 467-468
   electrical work ... 71
   floor units ... 101
   food ... 400
   radiant ... 101
   solar ... 164-165, 413
   suspended units ... 102
   wall units ... 101
   water ... 164-165, 459-460
Heating
   boilers ... 468-477
   controls ... 497
   electric ... 101-102
   hydronic ... 470
   radiators ... 482
   solar ... 413
   steam ... 468-469
Heating and cooling ... 466-467
   computer room ... 485-486
Heating and venting ... 102-103, 466-467
Heating/cooling, roof-mounted ... 484-485
Hem-fir ... 28, 118
Hemlock ... 118, 158
Heptachlor ... 199
Hi-rise channel, masonry ... 139
Hide-a-beds ... 19
High bay lighting ... 543
High early cement ... 42
High early concrete ... 46
High intensity lighting ... 541-543
High pressure sodium ... 541-543
Highway
   concrete ... 310
   equipment rental ... 236-237
   stakes ... 127, 274
Hinges
   door ... 99, 367
   gate ... 280
Hip roof ... 33
Hip shingles ... 175
Hoists
   auto ... 402
   construction ... 231
   freight ... 415
Hoistway ... 73
Hollow metal
   doors ... 360-362
   frames ... 361-362
   studs ... 371-372
Hollow metal door assembly ... 360
Home inspection service ... 21
Home office cost ... 222

Hoods
   entrance ... 186
   laboratory ... 406
   range ... 170
Hook
   bar ... 394-395
   robe ... 18
Hooks, toilet ... 394
Horizontal ladder ... 283
Horns, paging ... 552
Hose
   air, rental ... 232
   bibs ... 167, 451
   clamps ... 184
   fire, cabinet ... 465
   reel ... 402
   rental ... 236
Hospital
   cubicles ... 384
   freezers ... 405
   refrigerators ... 405
Hospitals, concrete ... 310
Hot air dryers ... 394
Hot water
   boilers ... 468-477
   heaters ... 164-165, 459-460
Hotmopping
   dampproofing ... 348
   roofing ... 172
   waterproofing ... 172, 348
Hourly wages
   commercial & industrial ... 221
   residential ... 8
House numbers ... 100
House paint ... 152-153
Housing, manufactured ... 408-409
How to use this book ... 3
Hub and spigot pipe ... 428-429
Hubs, lumber ... 128
Humidifiers ... 488
   wiring ... 71
Humidistats ... 498
Humidity sensor ... 79
Humus, soil conditioners ... 291
Hydrants ... 465
   demolition ... 242
Hydrated lime
   builders ... 138
   fertilizer ... 291
Hydraulic door closures ... 57
Hydraulic elevators ... 73
Hydroblasting ... 177-178
Hydronic
   hot water generator ... 470
   monitor ... 499
   radiators ... 482
Hydroseeding ... 293
Hypalon ... 352
Hyperbolic domes ... 410

I

Ice maker ... 400
Ice melting cable ... 523-524
Ice plant ... 293
Idaho pine ... 119
Identifying devices, signs ... 388-389
Illustrations, architectural ... 15-16
IMC conduit ... 508-509
Incandescent lighting ... 117, 543-544
Incinerators ... 408
Incline lifts ... 73
Indicator valves ... 435, 450
Indicators, gauges ... 458
Indirect overhead ... 222
Industrial hygienist ... 247
Industrial wage rates ... 221
Inertia sensors ... 549
Infiltration barrier ... 20-21
Infrared heaters ... 101
Infrared thermography ... 106
Inlets, curb ... 271
Insect screens ... 179, 354-355
Inserts
   concrete ... 311
   tilt-up ... 319

Inspection
   building ... 21
   lumber ... 128
   pipe ... 269
   steel ... 229
Instructions
   use of Estimate Writer ... 553
   use of this book ... 3
Instrumentation, switchboard ... 533-534
Insulated siding ... 196, 351
Insulating
   board ... 104, 105, 348-350
   drywall ... 95
   glass ... 93, 369
   lath ... 112
   roof deck ... 324
   skylights ... 197-198
   tile ... 40
Insulation ... 104-106, 348-350
   analysis ... 106
   asbestos, removal ... 246
   blown ... 105
   board ... 346, 348-350
   board, demolition ... 56
   deck ... 324, 350
   demolition ... 245
   duct ... 495
   fiberglass ... 104
   foil ... 104
   foil faced ... 104
   jacket ... 164
   kraft faced ... 104
   masonry ... 105, 140
   panel adhesive ... 13-14
   pipe ... 455-458
   pouring ... 105
   sound control board ... 346, 372
   sprayed foam ... 105
   subcontract ... 105-106
   testing ... 21
   wall systems ... 349-350
   weatherstrip ... 368
Insurance ... 106-107
   overhead ... 222
Insurance included ... 3
Insurance, labor cost
   commercial & industrial ... 221
   residential ... 8
Integrated ceilings ... 408
Interceptors, plumbing ... 460
Intercom systems ... 107-108
Interest rate ... 223
Interior wall assemblies ... 36
Interlocking paving stones ... 326
Intermediate metal conduit ... 508-509
Inventory, lumber ... 128
Inverter system, solar ... 200
Investigation, soil ... 199
Iron
   cast ... 339
   compound, dampproofing ... 348
   embedded ... 310
   pipe ... 429-433
   wrought ... 339-340
Iron worker, wage ... 221
Ironing boards ... 108
Ironwork
   ornamental ... 151, 339-340
   railings ... 169, 335
   stairs (steel) ... 336, 339
   wrought iron ... 339-340
Irrigation control ... 114
Irrigation systems ... 286-289
   golf course ... 282, 287
   lawn sprinklers ... 113-114
   shrub sprinklers ... 289-290
Isolators, vibration ... 485
Ivy ... 293

J

Jack studs ... 340
Jacket insulation ... 164
Jackhammer ... 252
Jackhammers, rental ... 77, 233
Jacking, pipe ... 263

Jackposts ... 108
Jacks, roof ... 185
Jamb side stripping ... 212
Jambs, door ... 58-59, 390
   masonry ... 140, 325
   sets, door ... 362
Japanese shina ... 123
Jetting wellpoints ... 263
Jib cranes ... 415
Job site
   costs ... 226, 230
   engineer ... 225
   office ... 222, 239-240
   overhead ... 222
   phone ... 231
   security service ... 180
Job superintendent ... 226
Joint cleaning, pavement ... 273
Joint compound ... 96
Joint grouting, pipe ... 271
Joint tape ... 96
Joint treatment, paving ... 273
Joints
   brick ... 132-133
   expansion ... 46, 310, 356-357
   expansion, railroad track ... 296
   form ... 304
   masonry ... 326
   mortar ... 132-133
   pavement ... 273-274
   paving ... 273-274
   pipe ... 452-453
   precast concrete ... 314
   sandblasting ... 273-274
Joist
   bridges ... 25, 108
   hangers ... 108-109
Joists
   carpentry ... 24, 26-27, 29, 341
   ceiling ... 26-27
   concrete ... 309
   demolition ... 56
   floor ... 29
   roof ... 33-34
   steel ... 336-337
Jumbo brick ... 132
Jumbo Tex ... 21
Junction box, electric ... 502, 524-525
Junior mesh ... 112

K

Kalamein doors ... 361
Kalman course ... 312
Keenes cement ... 160, 370
Kettles, food ... 400
Key joints ... 298, 301, 304
Keyed wall form ... 301
Keyless locksets ... 98
Keys, construction ... 98
Keyway, concrete ... 47
Kick plates, hardware ... 100, 334, 368
Kitchen
   cabinets ... 22-24, 406
   fans ... 78
   fans, wiring ... 71
   sinks ... 162-163, 462-463
Kneelers ... 396
Knob, dummy ... 368
Knockers, door ... 100
Koa ... 123
Komar countertops ... 406
Kraft faced insulation ... 104
Kraft flashing ... 346
Kraft paper ... 21

L

L-straps ... 207
Lab refrigerators ... 405
Labor column defined ... 5
Labor cost defined ... 3
Labor costs, crews ... 5
Labor productivity ... 4

Index

Labor rates
   adjusting 9
   commercial & industrial 221
   residential 8
Labor, painting 154
Lacewood 124
Lacquer thinner 153
Ladders
   attic 202
   horizontal 283
   painting 380
   reinforcing 329
   rolling 401
   steel 336
Lag shields 517
Lally columns 31
Lamb's tongue 169
Laminated plastic
   cabinets 406
   countertops 53, 406
   wallcoverings 382
Laminated sandwich panels 351
Laminated wood
   cutting boards 399
   glued beams 346-347, 410
   timbers 346-347
Lamp post wiring 71
Lamps
   fixture 543
   fluorescent 114-116
Landing
   forming 305
   stair 202, 347
   steel 336
Landscape
   irrigation wiring 114
   lighting 117
   site curbs 52
Landscaping .. 109-111, 291, 293-294
   erosion control 258
   piles 294
Lane delineator 17
Lane spikes 403
Lap cement 172
Larch 117-120
Laser level, rental 237
Latch set 99, 367
Latches, gate 280
Latex
   caulk 37-38, 356
   paint 152-153
   primer 151-152
   stain 153-154
Lath 112, 122, 370
   demolition 244
   lead lined 410
   nails 149
   stakes 127
   wood 127
Lather, wage
   commercial & industrial 221
   residential 8
Lathing 112, 370
Lattice moulding 146
Lattice, painting 156
Lauan paneling 159
Laundry
   clothes chutes 416
   clothesline units 45
   sinks 163
Laurel 124
Lava rock 331
Lavatories
   china 162
   marble 130
Lavatory
   base 210
   cabinets 210
   rough-in 463
   vanities 210
Lawn
   lighting 117
   sprinklers .. 113-114, 287, 290-291
Lawns 111, 293-294
Layout
   engineer 222, 226
   fees 226
   foundation 114
   stakes 127-128
   survey 225
Leach fields 181
Lead lining 410
Lecterns 396
Lecture hall, acoustical design fees . 74
Ledgers 31, 345
Let-in
   bracing 25
   ribbons 33
Letter boxes 390
Letter drop plate 368
Letters, signs 388-389
Levee excavation 257
Levelers, dock 403
Liability insurance 106
Library ladders 401
Lift slabs 309
Lifting
   cranes 231, 234-236
   equipment 231, 234-236
   panels 319
Lifts
   dock 403
   elevators 73, 414-415
   incline 73
   rental 232, 234-236
   rolling platform 236
Light standards, demolition 243
Light tower, rental 237
Light-diffusing panels 39
Lighting 114-117, 539-545
   area 547
   ceiling 39
   court 284
   demolition 243
   emergency 545
   exit sign 544-545
   explosion proof 545
   fluorescent 114-116, 539-541
   high intensity 541-543
   incandescent 543-544
   integrated ceilings 408
   landscape 114-116
   metal halide 541-543
   outdoor 117, 541-542, 545-546
   panelboards 535-536
   poles 546-547
   street 545-547
   suspended 39
   switches 526-527
   temporary 222
   tennis court 284
   vandal resistant 544
   yard 545-547
Lighting circuits 502-503
Lighting contactors 527
Lighting fixtures 71, 114-117, 539-545
   emergency 545
   explosion proof 545
   exterior 541-542, 545-546
   fluorescent 114-116, 539-541
   high intensity 541-543
   incandescent 543-544
   indoor 539-545
   obstruction 544
   outdoor 117, 541-542, 545-546
Lighting systems, solar 200
Lightning
   arrestor 81
   protection 71
Lights, barricade 17
Lightweight concrete 46, 306, 324
Limba veneer 124
Lime
   builders 138
   fertilizer 291
   landscaping 111
   mason's 138
   soil conditioner 111
   spreading 291
   stabilization 291
Limestone
   conditioner 291
   masonry 331
   paving 276
Linen chutes 416
Liner, form 300, 303
Link mats 141, 407
Linoleum
   adhesive 14
   demolition 55
Linseed oil
   paints 152
   primer 152
Lintels
   masonry 331
   steel 339
Liquid level gauge 481
Liquid-tight conduit 514-515
Loaders
   excavation 255
   rental 235
Loaders, grading .. 77, 235, 256-257
Loading dock equipment 403-404
Loading docks, concrete 309
Loading trucks 75, 255
Loam topsoil 291, 292
Loan fees 223
Lockers 389
   demolition 246
Locks 98-99
Locksets 98-99, 364
   partition 390
Loudspeaker systems 551-552
Louver vents 385
Louvered
   closet doors 42
   doors 42-43
   shutters 188-189
   vent grating 340
   vents 185, 354
Louvers 354
   eggcrate 39-40
   parabolic 40
Low bay lighting 543
Lube rack 402
Lumber
   ash 123
   birch 123, 144
   building poles 123
   cedar 118-120, 126
   common lumber 117-122
   Douglas fir 117, 119, 121
   dowels 125
   Engelmann spruce 119-120
   finish lumber 121, 127, 144
   finish trim 144
   fir 117-118, 120
   flooring 117-121
   framing 117-121
   grading 128
   hardwood 123-124
   hardwood moulding 144
   hem-fir 118
   hemlock 118
   Idaho pine 119
   larch 117-120
   maple 123
   moulding 144
   oak 144
   ponderosa pine 118-119
   posts 123
   redwood 120, 123, 127
   selects 120
   siding 125-127
   southern pine 118, 120-121, 127
   spruce 118-120, 123, 127
   stakes 127-128
   studs 118
   teak 123
   trim 144-148
   veneer 124-125
   walnut 123-124
   west coast 117-121
   white fir 119
   willow 124
Lumber treatment 128
Luminaire 541-543
Luminous ceilings 39-40, 408

M

Machinery appraisal 15
Mahogany 123-124
   paneling 159
Mail
   chutes 389
   conveyors 415
   drop plate 368
Mail slot, cut out 67
Mailboxes 129-130, 389-390
Maintenance building, golf 282
Makore veneer 124
Man lifts 236
Manager cost 222
Manhole
   cut-in 224
   frames 270-271
   rings 338
Manholes
   demolition 242
   electric pull boxes 502
   precast 270
   repair 270-271
Manhour costs defined 5-7
Manhours defined 3
Mansards, sheet metal 352
Mantels 130
Manufactured housing 408-409
Manure 291
Map rail 383
Maple
   flooring 87-89, 376
   lumber 123
   paneling 158-159
   parquet 90
Mapping, aerial 225
Marble
   flooring 130
   masonry 331
   setting 130
   sills 212
   vanity tops 53
Marble setter, wage
   commercial & industrial 221
   residential 8
Margin, profit 130-131
Marine work 294
Markers, road 274
Market entry doors 366
Marking, pavement 274
Markup 222, 130-131
   contingency 130-131
   general contractor 4
   subcontractor 4
Marlite 97
Masonite
   hardboard 97
   siding 190-191
Masonry 131, 140, 322, 324-332
   acid-proof 325, 329
   anchors 517
   block 134-137, 140, 326-328
   brick veneer 324
   cleaning 331-332
   demolition 54, 57, 243
   fill 140
   furring 30
   insulation 105, 140
   joints 331-332
   materials 131-132, 140
   paint 153
   painting 155, 380
   pointing 331-332
   preparation 155-156
   primer 153
   reinforcing 139, 328-329
   sandblasting 177
   sawing 250
   saws 237
   testing 228
   toothing 331-332
   waterproofing 153
Mass excavation 257
Mastercure 52, 312
Mastic
   adhesives 14
   flooring 89
   roofing 172, 177
   tile 205-206
Mat, demolition 246
Material costs defined 3
Material grade defined 4

579

Index

Material handling systems 416
Material lifts 231, 234-236
Mats
   erosion control 259
   velvet tile 379
Mats and treads 141, 407
Mechanical 416-500
   controls 497-500
   drawing 68-69
Medical refrigerators 404-405
Medicare 203
Medicine cabinets .. 141-142, 393-394
Membrane
   poly film 169
   roofing 173, 351
   slab 51, 299, 308
   waterproofing 211
Mercury vapor lights 542-543
Mesh reinforcing 49, 305-306
Mesh, steel 305
Metal
   doors 360-363
   facing panels 350
   halide lights 541-543
   lath 112
   painting 155
   painting, preparation 156
   pans, ceiling 40
   studs 371-372
   studs, demolition 244
   windows 366
Metal work, aluminum 339-340
Metals 333-340
   fabricated 338-339
   ornamental 339-340
Meter boxes 142-143, 272
Meter fees 224
Meters
   electric 533-534
   fuel 458
   gas 433
   steam 458
   water 224, 458
Microphone 551
Microwave
   cabinets 24
   ovens 171
   sensors 549, 550
Midget louvers 185
Milling concrete 242
Millwork, treatment 128
Millwright, wage
   commercial & industrial 221
   residential 8
Mineral insulation 350
Mineral shingles 189
Mini-blinds 213, 407
Mini-brick 133
Mining camp, manufactured 409
Miradrain 259
Mirafi fabrics 259
Mirror panels, closet doors 44-45
Mirrors 93, 369, 143-144
   bath accessories 18, 393-394
Mission
   brick 132
   paver 133
   tile roofing 176
Mix design 227
Mixers, rental 233
Mixing boxes, air 493
Mixing valves 481
Mobilization 226
   cost defined 3
   piles 259
Modification factors, area 10
Modular
   brick 132
   cabinets 23-24
   carpet tile 379
Modulating valves 451
Moisture protection 169, 348, 355-356
Moisture resistant drywall 95
Moisture test, soil 199
Monitor panels 548-550
Monitoring, controls 497-500
Monitors, alarm 179
Mono rock, concrete 312

Monolithic
   curb 276
   topping, concrete 312
Monorail, hoist 416
Monuments, survey 225
Mop resilient floor 240
Mop sink 464
Moped racks 286
Morgue equipment 405
Mortar 132-133, 325
   bed 206
   block 135
   cement 42
   color 139, 275, 325
   glass block 138
   masonry 132-133, 138
   mix 42
   paving 276
   testing 228
   trowel-ready 139
Mortar joints, glass 138
Mortise
   hinges 99
   locks 98-99
Mosaic & terrazzo worker, wage
   commercial & industrial 221
   residential 8
Mosaic tile 204-206, 374
Motor
   connection 528-529
   controls 497-499
   graders, rental 235
   starters 529-530
Motorcycle racks 286
Motors, electric 528
Moulding 144-148
   astragal 145
   band 144
   base 144
   base shoe 144
   batten 144
   brick 144
   cap 144
   casing 144
   chair rail 144-145
   chalkboard 383
   chamfer 145
   corner 144
   cove 144
   crown 144
   door casing 144
   door stop 144
   door trim 144
   drip 144
   drip cap 144
   full round 147
   glass bead 146
   half round 146
   hand rail 146
   hardwood 144
   lattice 146
   metal 96
   mullion casing 146
   panel 146
   parting bead 146
   picture 146
   plaster 371
   plaster ground 146
   polystyrene 148
   quarter round 144, 146
   rail 144, 145
   rectangular 147
   redwood 144
   round edge stop 144
   S4S 145, 147
   sash bar 147
   screen bead 147
   shoe 145
   sill 145
   softwood 145
   sprung 146
   square 147
   stucco 145
   thresholds 144
   window 147, 189
   wood 144
Movable partitions 390-391
Moving stairs 416

Moving structures 240
Mud sill anchors 208
Mud sills 34, 340
Mulch 111
Mullion casing 146
Mullions, carpentry 341
Mushroom ventilators 355
Music systems 107-108

N

N.C.E. window 553
Nail base, sheathing 342
Nailer strip, carpentry 31
Nailers 32, 341
Nails
   aluminum 150
   blued box 148
   box 148
   bright 148
   casing 148
   cement coated 148
   common 148
   concrete 149
   dritite 150
   duplex head 149
   fiberglass panel 82
   finishing 149
   form 149
   furring 149
   galvanized 148
   hardboard 150
   lath 149
   plasterboard 149
   ring shank 149
   roofing 150
   scaffold 149
   screw shank 150
   sheetrock 149
   shingle 151
   siding 150-151
   sinkers 148, 151
   spikes 151
   timber connectors 206
   underlayment 151
NCX treatment 129
Needle gun finish 312
Neoprene
   flooring 378
   gaskets 352
   joints 357
   roofing 351-352
   tack coat 275
   weatherstrip 368
Nets
   safety 239
   tennis 284
Netting, stucco 112
Newel posts 169, 201
Nimmer detector 550
Nitrogen analyzer 499
No-hub pipe 429-432
Noise analysis 74
Non-metallic cable 522
Non-slip finishes, concrete 311
Non-slip strips 141, 378
Norman
   brick 131, 324
   pavers 134
Nosing
   roof edging 183
   stairs 339
   steel 339
Nozzle, sandblast 236
Numbers, house 100
Nylon carpet 37, 379

O

Oak
   brown 124
   flooring 87-89
   lumber 123
   moulding 144

Oak (continued)
   paneling 158-159
   parquet 88-89
   red 123, 124
   threshold 144
   white 123, 124
Obstruction lighting 544
Office
   clerk 222
   expenses 222
   job site 239-240
   manager costs 222, 226
   overhead 130-131
Office partitions 390
Office privacy locks 99
Offices
   concrete 310
   manufactured 408-409
   temporary 222, 239-240
Oil
   concrete form 311
   fuel train piping, boiler 472
   furnaces 102-103
   pump 455
   storage tanks 466
   tanks 466
Oil and water separators 272
Oil base
   caulk 38, 355
   paint 152
   primers 152
Openers
   entrance doors 365
   garage door 91
   storefront 366
Openings, wood framed
   doors 28
   windows 35-36
Operable walls 390
Operating engineer, wage 8
Option menu 553
Orchard clearing 241
Organic impurities 226
Oriental wood 124
Ornamental
   metals 169, 339-340
   railings 151, 335
   rock 258
Outdoor lighting 117, 541, 542, 545-546
Outlet circuit receptacle 503
Outlets, electric 72, 525-526
Oven cabinets 24
Ovens
   electrical work 72, 171
   food 400
   wall 171
Overhang, painting 156
Overhead
   costs 222
   mark-up 130-131
   office expense 222
   supervision 222
Overhead doors 364
Overhead electrical distribution .. 539
Oxygen analyzer 499

P

Padouk veneer 124
Padre
   brick 132
   pavers 134
Pads
   carpet 36-37, 379
   equipment 307
Paging systems 552, 107-108
Paint 151-154
   removal 240
   remover 153
   sandblasting 177
   spraying 154-156
   testing 21
   thinner 153
   wrought iron 154
Painter's caulk 37-38

Index

Painter, wage
commercial & industrial 221
residential 8
Painting 151, 154-157, 380-382
athletic courts 285
compressor 155
door frames 156-157
exterior 154
fence 280
interior 154-155
pavement 274
speed bumps 277
spray 155-156
steel 338, 381
Painting labor 154
Paints 151-154
Paldao veneer 124
Palm frond removal 241
Pan forms 304
Panel
adhesive 13-14
box, wiring 72
circuit breaker 503
doors 64
enclosures 524-525
moulding 146
Panelboards, electrical 535-536
Paneling
adhesive 96, 97
cedar 39, 157
chestnut 158
hardboard 97, 158-159
maple 158-159
pine 157-158
plastic 96
poplar 158
redwood 157
wallpapered 158
Z-brick 159
Panelized roof system 343-344
Panels
breaker 534
fiberglass 82
light-diffusing 39
wood walls 344
Panic buttons 179
Pans, sill 183
Pantry cabinets 24
Paper
building 21-22
roofing 174
Paper dispensers 392-394
Paperhanging 382, 210-211
Parabolic louvers 40
Parapet forms 309
Pargeting, masonry 332
Parking
bumper 274, 280
control equipment 403
garages, concrete 310
lot, paving 273
stripes 274
Parquet flooring 88-90, 376
Particleboard 91, 96-97, 168, 378
Parting bead 146
Partition framing 35-36
Partitions
cubicles 384, 390
demolition 244
dust 245
fire rated 391
folding doors 64, 390
load bearing 372
marble toilet 331
metal 384, 390-391
metal stud 371-372
non bearing 371-372, 390-391
office 390
toilet 384
wire 391
Passage hardware 99
Passive solar heat 164
Patching
concrete 52, 320
plastic 153
Path lighting 117
Patio
blocks 275
deck 54
doors 67-68
paint 153
pet door 65
stain 154
tile 141
Patiolife 154
Pavement
asphaltic 159-160, 272
brick 160, 274-275
concrete 45, 51, 160, 273-274
demolition 56, 241
joints 273-274
marking, removal 242
sawing 251
striping 274
sweeping 272
Paver tile 205-206
Pavers
brick 133-134, 274-275, 326
concrete 136, 294, 326
interlocking 326
landscaping 294
masonry 275-276
masonry grid 326
rubber 285
slate 326
terrazzo 326
Paving
asphalt 159-160, 272
breakers, rental 231
brick 160, 274-275
cold milling 272
concrete 45, 51, 160, 273-274
demolition 56, 241
joint treatment 273-274
machine 236
parking lot 273
planing 272
scarify 258
Paving stones 326
Payment bonds 107
Payroll taxes 203
Pea gravel
concrete 46
landscaping 293
riprap 258
Pearwood veneer 124
Peat humus 291
Pecan
paneling 157
veneer 124
Pedestal flooring 385-386
Pediment 74-75
Peel-off caulking 38
Peep sight 100
Pegboard, hardboard 97, 383
Pegging floors 90
Pension costs included 3
Penta treated post 123
Pentachlorophenol 154
Penthouse louvers 385
Percolation testing 180
Perfatape 96
Perforated pipe 264-265
Performance bonds 107
Perimeter door seal 404
Perling 32
Perlite
fertilizer 291
urethane board 348-350
Permits
building 22
well 212
Pest control
soil treatment 199
wood treatment 128
Pet
doors 65
enclosures 80
Petroleum storage tanks 466
Pews 396
Pharmacy pneumatic tube 416
Pharmacy refrigerators 405
Philippine mahogany 123
Phone support 3
Phones
temporary 222
wiring 551
Photo mapping 225
Photo micrographs 229
Photoelectric
sensors 549
switches 527
Photographs
aerial 225
construction 230
Photography, construction 230
Photovoltaic generator 199
Photovoltaic solar equipment 199
Picket fence 82, 151, 279-280
Pickup, superintendent overhead ..222
Picture moulding 146
Pictures, job site 230
Pier
blocks 32
caps 340
forms 304
pads 32
Piers, caissons 261
Pilaster forms, column ..301-302, 320
Pilasters
brick 325
painting 156
tilt-up 320
Pile caps, concrete 299
Pile driver 259
Pile driver, wage 221
Pile driving 234, 259
Pile foundations 259-261
Piles
landscaping 294
treatment 128
Piling, sheet steel 262
Pillars, colonial 75
Pine
building poles 123
door frames 58, 66
door jambs 58
doors 59, 65-66
flooring 87, 90
lumber 118-121
paneling 157-158
ponderosa 118-119
select 121
shutters 188-189
siding 127
southern 118, 120-121, 124
studs 118
veneer 124
white 119
yellow, southern 32, 118, 121
Pipe
ABS 441
asbestos cement 264, 433
asbestos removal 247-249
bellows 452
black steel 416-428
bollards 281
bumper 320
caps, conduit 505
cast iron ...264, 429-432, 434-435
clamps 445
clay 268
cleaning 269
columns 337-338
concrete, non-reinforced 265
concrete, reinforced 265
copper 166-167, 436-439
corrugated metal 266-267
CPVC 440
culvert 267
demolition 242
drainage ...264, 266-268, 439, 441
ductile iron 432-433
DWV 166, 439, 441
elliptical 265
fiberglass reinforced 442
galvanized 428
grouting 270
guides 446
hangers 445-447
hooks 446
hub and spigot 428-429
Pipe *(continued)*
inspection 270
insulation 455-458
insulation, asbestos 246
jacking 263
level 237
locating 181
markers 230
painting 381
perforated 264
piles 260-261
plastic 267-268, 439-441
polyethylene 264, 289, 441-442
PVC ..113, 167, 181, 267, 289, 439
railing 334
rolls 446
sewer 167, 181, 267-268
soil, cast iron 264
soil 167, 429-432
split 269
sprinkler 288-289
stainless steel 443-445
steel 416-428
sterilization 269
storm 267
storm drain 264
straps 446
supports 445-447
transite 264
troubleshooting service 181
valves 447-452
vitrified clay 181, 268
water 433-434, 439-441
Pipefitter, wage 221
Piping
combustion train, boiler ...470-472
demolition 242-243
drainage 69
fire protection 465
foundation drains 69
fuel train, boiler 470-472
gas 433, 471-472
Pitch pockets 353
Pits, excavation 75
Placing concrete 49-51, 306-307
Plan check fee 22
Planing, asphalt 272
Plank
flooring, unfinished 87-88, 90
precast concrete 314
scaffold 239
Plans
drafting 19
prints 19
Planting 109-111, 291-294
Plants 109-111, 291, 293-294
Plaques 389
Plaster
demolition 56, 244
ground 146
joints 357, 371
mixers 233
painting 154-157, 381
ring 524
soffits 160
Plasterboard nails 149
Plasterboard, drywall 372
Plasterer's helper, wage
commercial & industrial 221
residential 8
Plasterer, wage
commercial & industrial 221
residential 8
Plastering 160, 370-371
Plastic
countertops 53, 406
gutters 95
laminates 382, 399
panels 96-97
pipe 113, 181, 439-441
resin glue 14
shutters 188
siding 196
skylights 197, 355
vapor barrier 169, 308
wallcoverings 382, 384-385
Plastic drain pipe 69
Plasticized concrete 46

Index

Plate glass 92-93, 366
Plate straps 208
Plates
   carpentry 32, 34, 341
   door hardware 100
   switch 18
Platform lifts
   dock 403
   rental 236
Playground equipment 282-284
Playing fields, synthetic 282
Plexiglass 369
Plinth block 148
Plugs, electrical 525-526
Plumber's tape 446
Plumber, wage
   commercial & industrial 221
   residential 8
Plumbers flash cone 353
Plumbing
   equipment 160-165, 459-460
   fixtures 160-165, 461
   piping 461-462
   rough-in 166, 461
   single line system 164
   specialties 458-459
Plyclips 169, 208
Plyform 122, 168, 298, 301-302
Plywood 168-169, 342
   demolition 56
   fence 238
   forms 48, 299, 301
   framing clips 169, 208
   paneling 158-159, 343-344
   roof panels 343-344
   sheathing 33, 122, 342
   siding 122, 192-194
   softwood 168
   subfloor 35
   treatment 128
Pneumatic
   demolition 54-56, 243
   door closers 57
   excavation 77
   hammer 252
   tube conveyors 416
Pocket door frames 58
Pointing, masonry 331-332
Poisoning soil 199
Pole cross arms 539
Poles
   building 123
   concrete 316
   demolition 243
   flag 386-387
   job sign 230
   lighting 546-547
   lumber 123
   power distribution 539
   removal 243
   temporary utility power ... 230
   treatments 128
Polishing
   floors 154
   woodwork 157
Polybutene tape 357
Polyester
   fabric 172
   flooring 378
Polyethylene
   corrugated pipe 267
   culvert pipe 267
   film 169
   membrane, slab 51, 298, 308
   pipe 264, 267, 289, 441-442
   tilt-up 317
Polyisobutylene 357
Polystyrene
   insulation 104, 349-350
   moulding 148
   roof deck 349
Polysulfide
   caulk 356
   sealant 356
Polyurethane sealant 356
Ponderosa pine
   doors 59, 65
   lumber 118-119

mantels 130
Pool
   coating 153
   fences 151
   gunite 312
   heaters 165
   paint 153
Pools 412
Pop-up sprinklers 113
Poplar
   flooring 90
   lumber 123
   paneling 158
   veneer 124
Porcelain tile 204-205
Porch
   columns 45
   lifts 73
   painting 156
   posts 45
   rail 169
Portable structures 408-409
Portland cement 42, 138, 374
Post
   anchors 208
   bases 208
   caps 208
   column forms 301
   connectors 208
   indicator valve 435
Post hole diggers 237
Post holes
   fence 277
   footings 281
Postal specialties .. 129-130, 389-390
Posters 230
Posts
   athletic equipment 284
   barrier 404
   bollards 280-281
   building 123
   carpentry 32, 341
   fence 79-80, 81-82
   highway 280
   jackposts 108
   lumber 123
   parking barrier 403
   redwood 123
   sawn 123
   treated 123, 128
   treatment 128
   wood 341
   yellow pine 123
Pot rack 400
Pot wash sink 400
Pour strip, tilt-up 317
Pour-stop forms 299
Pouring
   concrete 49-50, 306
   insulation 105, 350
Power
   cable 521-524
   circuits 502-503, 512-513
   conduit 503-505
   duct 500-501
   EMT 505-507
   panels 533, 535-536
   temporary 222, 230
   transmission 539
Power factor meter 533
Power generation 547-548
   solar 199-200
Power shovel, rental 77
Power washer, rental 237
Pre-engineered buildings .. 333-334
Precast
   beams 314
   columns 315
   culvert 266
   double tees 315
   floor plank 314
   floor slab 314
   girders 314
   joists 314
   panels 312-313
   stairs 315
Precast concrete .. 266, 312-313, 315
Preformed siding 350-351

Prehung doors 66-67, 362
Preliminary soil investigation 199
Preparation, painting 155
Preservation, lumber 128-129
Preservative, wood 154
Pressure
   controls 458
   gauge 458
   grouting 312
   mats 180
   reducing valves ... 452, 479-481
   regulator valves 452
   treatment, lumber 128
Pressure regulators
   gas 433
   valves 452
Pressure vacuum breaker 224
Prestressed concrete 316
Prestressed, testing 229
Prima Vera veneer 124
Prime coat, pavement 272
Prime rate 223
Primer
   asphalt 285
   linseed 152
   painting 152, 381-382
Prints, plans 19
Privacy locks 99
Processed lime 138
Profit, markup 130-131, 222
Project
   cost adjustments 10
   financing 223
   manager cost 222
   scheduling 223
Propane
   furnaces 102
   tanks 466
Property boundaries 225
Public address systems ... 551-552
Pull box, electrical 502, 525
Pull plates 100, 368
Pull-up bar 283
Pullmans 211
Pulpits 396
Pump
   electrical connections 72
   foundations 321
   plumbing 460
   rental 236
Pump mix concrete 46
Pumper connection 465
Pumping
   concrete 46, 237-238, 306
   dewatering 263
   well 212
Pumps
   boiler feedwater 473
   centrifugal 454
   chemical feed 476
   chiller 488
   circulating 453-454
   condensate 480
   medical vacuum 455
   salt water 455
   sewage 455
   sump 72, 455
   turbine 454
   utility 455
   vacuum 455
Purling 32
Purlins
   carpentry 32, 341
   fireproofing 360
   hangers 208-209
   steel 337
Purpleheart veneer 124
Purse shelf 384
Push plates 100, 368
Putty, glass 94, 155
Puttying 157
PV solar equipment 199
PVC
   conduit 509-512
   duct 500-502
   pipe .. 113, 167, 181, 267, 289, 434
Pyro-Kure paper 21

Q

Quality control engineer 225
Quality inspection, lumber ... 128
Quantities defined 3
Quarry tile
   demolition 55
   flooring 325-326, 374
   masonry 141, 325-326
   wall tile 325
Quarry work 251-252
Quarter round mould 144, 146
Quartz heaters 101
Quick coupling valves 291

R

Raceway, electric 518
   concrete envelopes 502
Racks
   bicycle 286
   coat and hat 394-395
   kitchen 399-400
   motorcycle 286
   towel 19
Radial wall forms 301
Radiant heaters 101
Radiation protection 410
Radiators, heating 482
Radon gas, testing 21
Rafter vents 185
Rafters 24, 32-33, 342
   demolition 56
Rail
   communion 397
   demolition 243
   guard 238, 280, 384
   hand 151
   splice bars 295
   track 295
Rail fence 81-82, 279
Railing
   access flooring 386
   aluminum 334
   brass 334
   highway 280
   painting 154, 156
   pipe and chain 334
   porch, steel 169
   stair, steel 169, 334
   stair, wood 200-202, 347
   steel 169, 334
   temporary 238
   wrought iron 335
Railroad
   ballast 295
   crossing 297
   demolition 243
   track 295
   turnouts and switches ... 296
Railroad track demolition ... 243
Rain diverter 183
Rain drip cap 204
Rain gutters 94-95
Raised floors 385-386
Raised panel doors 42, 65
Rakers, steel 262
Rammer 254
Rammers, rental 233
Ramp forms, riser 305
Ramps, computer floor 386
Range circuit 503
Range hoods 400, 170-171
Ranges, built-in 170-171
   electrical work 72
Rattan black paper 21
Re-heat coils 493-494
Reactivity testing 227
Ready mix concrete 45-46, 306
Rebar
   concrete 48
   masonry 328-329
Rebound pad 378
Receptacle outlet circuit ... 503

Index

Receptacles, electric 72, 525-526
   wiring 72
Receptors, shower 187
Reciprocating chillers 487
Record drawings 15-16, 68-69
Recording meters 533
Recovery, heat 487
Red alkyd primer 152
Red cedar
   fence 81-82
   lumber 118-120
   posts 32
   roofing 174
   shakes 174
   shingles 174
   siding 126-127, 194
Red rosin sheathing 22
Redwood
   benderboard 111, 294
   clear 120, 125-126
   fence 81-82
   headers 277, 294
   lumber 29, 120
   oil stain 153
   paneling 157
   posts 32
   siding 125-126, 190, 193
   stakes 127
   strips, filler 79
   trim 144
Reflective stakes 274
Refrigeration
   air conditioning 103, 466-467
   doors 364
   electrical 72
   insulation 350
   medical equipment 404-405
   wiring 72
Refrigeration system, solar ... 200
Refrigerator doors 364
Refrigerators
   food service 400
   medical 404-405
   morgue 405
Registers 495-497
   furnace 102
Reglets
   parapet 320
   sheet metal 352, 353
Regulator valves, gas 451
Reinforcement testing 228-229
Reinforcing ... 139, 305-306, 328-329
   bars 48, 305, 328-329
   caissons 305
   columns 309
   demolition 243
   foundation 322-323
   gunite 312
   masonry 135, 139
   mesh 49, 160, 305-306
   pavement 160
   slab 305
   steel 48, 139
   tilt-up 316-317, 320
   walkways 277
   welding 305
Reinforcing ironworker, wage
   commercial & industrial ... 221
   residential 8
Reinforcing testing 228
Relays 498
Relief intake vents 492-493
Relief valves 452, 479-481
Relocatable structures ... 408-409
Removal
   asbestos 246-249
   clearing 241
   debris 245
   demolition 57, 246
   tree 77, 241
Renderings, architectural 15-16, 68-69
Rent, overhead 222
Rental
   air compressors 78
   barricade 17
   crane 231, 234
   equipment 77-78, 232-239
   lifting equipment .. 231, 234-236

power pole 230
scaffold 239
shoring 239
Report
   home inspection 20
   soil test 199
Repp cloth 398
Reproduction, plans 19
Residential rules of thumb 219
Residential wage rates 8
Resilient flooring 85-86, 377-378
   demolition 55, 246
Resorcinol glue 14
Respirator cartridges 247
Restaurant equipment ... 399-401
Retaining wall forms ... 48, 300-301
Reveal strips
   formwork 300
   tilt-up 318
Revolving doors 364
Ribbands 33-34
Ribbons, carpentry ... 33, 342, 345
Riblath 112
Richlite board 399
Ridge cap 175
Ridge framing 24
Ridge shingles 175
Ridge vents 185
Rigid insulation 104
Rigid steel conduit 503-505
Ring shank nails 149
Ripping, excavation ... 251-252, 258
Riprap 258, 294
   marine facilities 294
Riser
   clamps 445
   forms 48, 305
Risers
   concrete stair 52, 305
   flagstone 137
   forming 309
   masonry 137
   resilient 91, 378
   sprinkler head 114
   steel stairs 201
   wood stairs 200-202, 347
Road
   buttons 274
   paving 272
   rollers 233
   striping 274
Road closed sign 17
Roadway compaction 257
Roadway grading 252
Robe hooks 18, 394
Rock
   aggregate 14
   base 272
   cinder 258
   crushed, paving ... 174, 272
   drain rock fill 258
   drilling 251, 261
   drills, rental 232
   excavation 75, 251-252
   fill 258
   ornamental 258
   paving 272
   ripping 251-252, 258
   roofing 174, 352
Rock fill drainage systems 69
Rockwool insulation 105
Rod clamps 516
Rod hanger 447, 516
Rods
   lightning 71
   shower curtain 19
Roll insulation 104
Roll roofing 174, 351
Roll-up doors 364
Roller compaction 252-253
Rollers, rental 233
Rolling
   doors and grilles 364
   ladders 401
   platform lifts 236
   towers 236, 247
Rolling, compaction 252-253
Roman brick 324

Romex
   cable 70, 503, 522
   wire 522
Roof
   accessories 355
   cant strips 26, 105, 352
   coating 152, 171-172, 351
   coating, asphalt-asbestos ... 152
   curbs 500
   demolition 56, 244-245
   drains 459
   edging 183
   exhaust fans ... 79, 491-492
   framing 24, 33-34, 341-342
   hatches 355
   hip 33
   ice melting cable ... 523-524
   insulation 105
   insulation board ... 105, 348-349
   jacks 185, 459
   joints 342, 357
   painting 156, 381
   panels 343-344
   paper 21
   plank 315
   power ventilators 355
   safes 353
   scuttle 355
   sheathing 34, 342
   sump 354
   testing 228
   tile, concrete 176
   trusses 33-34, 346
   vapor barrier 348
   ventilators 79, 185, 491-492
   vents 185
   windows 196-198
Roof deck
   coatings 171-172
   concrete 324
   insulating 348-349
   steel 335
   wood 33, 342
Roof jack, flashing 459
Roof loading 177
Roof mounted heating\cooling 484-485
Roofer, wage
   commercial & industrial ... 221
   residential 8
Roofing 171-177
   bonds 352
   built-up 173, 351
   cant strips 26, 105, 352
   clay tile 173
   cold process built-up .. 173
   concrete tile 173
   corrugated 350-351
   demolition 56, 244-245
   elastomeric 351-352
   felt 173, 351
   fiberglass 351
   galvanized ... 175, 350-351
   gravel 173
   membrane 351
   nails 150
   preformed 350-351
   red cedar 174
   rock 174, 351
   rule of thumb costs ... 173
   shake 174-175
   sheet 175
   sheet metal ... 175, 352-353
   shingle 174-175
   slate 176
   specialties 351-352
   tear off 351
   tile 173
Roofing cement 172
Room air thermostat 499
Room dividers
   doors, folding 64
   partitions 64, 390-391
Rope net 282-283
Rosewood veneer 125
Rotary roof ventilators ... 185, 355
Rototilling 111, 292

Rough carpentry ... 24, 35, 340-342
   adhesive 13-14
   demolition 56-57
Rough hardware 206-209
   plywood clips 169
Rough sawn siding 192-193
Rough-in plumbing 166
Round edge stop 144, 147
Round moulding 147
Rounds, dowels 125
Router-groover, rental 236
RSC conduit 503, 512-513
Rub, Carborundum 52, 312
Rubber
   mats 141
   stair treads 91
   tile 90-91
Rubble, masonry 331
Rufon polyester fabric 172
Rugs, carpets ... 36-37, 378-379
Runners
   mats 141, 407
   vinyl 141
Running track 282
Rust resistant primer 152
Ryegrass seeding 293

S

Sack and patch 52, 320
Sack concrete 52, 322-323
Sack finish 52
Saddles, pipe 446
Safe deposit boxes 395
Safes
   recessed wall and floor ... 392
   residential 177, 392
   roof 353
Safety
   floor tiles 378
   glass 93
   nets 239
   overhead 222
   signs 389
   switches 530-531
Sag rods, steel 337
Salamanders 237
Salaries, overhead 222
Sales tax excluded 3
Salvage, demolition 245
Sampling concrete 229
Sand 69
   aggregate 14
   base 276
   bed, paving 160
   fill 51, 75, 299
   fill, drainage systems ... 69
   paving 276
   testing 229
Sandblast, masonry 140
Sandblasters, rental 237
Sandblasting, concrete ... 177, 312
   equipment rental 236
   joints 273-274
   masonry 136, 140, 177, 332
   steel 177
   tilt-up 320
Sandbox, playground 283
Sanders, rental 237
Sanding, floor ... 88-90, 155, 377
Sandstone, masonry 331
Sandwich panels 351
Sanitary piping, demolition .. 242
Sapele veneer 125
Sash
   bar moulding 147
   doors 64
   glazing 93-94
   painting 156, 381
Satinwood veneer 125
Sauna 177-178
   heaters, wiring ... 70, 177-178
   modular rooms 177-178
Saw, rental, concrete 233

583

Index

Sawing
   asphalt 251
   concrete 250, 317
   masonry wall 250
Sawn post, redwood 123
Saws, rental, wood 237
Scaffold 247
   jacks 311
   nails 149
Scaffolding 239
Scales, truck 391
Scarify pavement 258
Scheduling
   costs 223
   CPM 74, 223
   fees 74
   overhead 222
School equipment 398
School rooms, manufactured 409
Scissor
   lifts 403
   rental 236
Scoreboards, gym 401
Scoring concrete 311
Scotch marine boilers 468-469
Scrapers
   excavation 257
   rental 235
Scraping, paint 155
Scratch coat, plaster 160, 371
Screed
   lath 112
   rental 234
Screeds
   concrete 234, 317
   tilt-up 317
Screen
   architectural 354-355
   bead moulding 147
   bird 354-355, 491
   block, masonry 136, 327
   cooling tower 354
   insect 366
   ornamental 339
   projection 398
   sight 280, 384
   sun 340
   vent 184-186, 354-355, 385
   window 216-217, 366
   wire 179
Screen doors 60, 67
   closures 57
   hinges 99
Screens, urinal 384
Screws, self-tapping 82
Sculpture screens 136
Scuppers, sheet metal 354
Scuttle, roof 355
Seal coat
   asphalt 159-160, 272
   pavement 273
Seal fittings, conduit 505
Sealant pot, rental 237
Sealants 37-38, 355-356
   window 94
Sealer
   driveway 152
   paints 153
   sill 105
Seals, weatherstrip 211-212
Sealtite
   conduit 514-515
   connectors 515
Seat cover dispensers 392
Seating
   athletic 285
   gym 401
   school 398
   stadium 408
   theater 408
Security
   alarms 179, 548-550
   awning 16
   controls 549-550
   gratings 340
   guards 180
   locks 98
   racks 286
   systems 548-550

Sediment fence 258
Seeded aggregate finish 160, 277
Seeding
   grass 287, 292-293
   landscaping 111, 292-293
Seekure paper 22
Seesaws 283
Self-furring lath 112
Self-tapping screws 82
Selling price, residences 219
Semi-gloss paint 152-153
Sen hardwood 125
Sensor cable, electric 550
Sensors
   control 499
   humidity 79
   photoelectric 549
   security 549-550
   vibration 549
Separator, oil and water 272
Septic tanks 180-181, 461
Service doors 364
Service entrance
   connection 72, 503, 531-533
   wire 519-520
Service entrance connection,
   temporary 230
Service sections 531-533
Service sinks 163
Service station equipment 402
Serving fixtures 399
Settlement testing 199
Sewage ejectors 461
   pumps 181
Sewage leaks 181
Sewage pumps 455
Sewer
   demolition 242
   pipe, PVC 267-268
   pumps 181
   systems 180-181
   tanks 180-181
Sewer connection 181
   fees 223-224
Sewer lines 181, 429-432
   cleaning 269
   demolition 242
Shacks, storage 240
Shades, window 213, 407
Shaft drilling 261
Shaft walls 373
Shake felt 174
Shakertown siding 195
Shakes, roofing 174-175
Shale, excavation 252, 257
Shaping slopes 75
She bolts 311
Shear plates 209
Sheathing
   adhesive 13-14
   carpentry 34, 169, 342
   clips 169, 208
   cork 382
   demolition 56
   gypsum board 96
   gypsum board, exterior 348
   insulation board 105, 348-350
   lumber 122, 342
   paper 21
   plywood 169, 342
   roof 342
   skip sheathing 345
   wallboard 96, 210
   wood 340, 342
Sheds, storage 240
Sheepsfoot
   compaction 77, 253
   roller 77-78, 253
Sheer gate 435
Sheet flooring 85, 91
Sheet glass 92-93
Sheet metal 181-186, 352-353
   ducts 103, 493-494
   flashing 175, 182-184, 352-353
   gutters, downspouts 94-95, 354
   hoods 186
   roofing 175, 350-351
   siding 194-195, 350-351
   vents 184-186, 355

Sheet metal worker, wage
   commercial & industrial 221
   residential 8
Sheet piling 262
Sheet vinyl flooring 85, 91, 377
Sheeting (see also Sheathing)
   trench 262
Sheetrock, wallboard . . . 210, 372-373
   painting 154-155, 157
Sheets, polyethylene 169
Shelf lighting 116
Shell and tube heat exchanger . . . 478
Shellac 155
Shellacking 155
Shelters, bus stop 409
Shelves
   bathroom 393-394
   closet 186
   industrial 392
   library 392, 405
   linen cabinet 186
   storage 186, 392
   telephone directory 392
   toilet 384
   utility 186
Shelving, wood 186
Shina lumber 123
Shingle siding 195
Shingles
   asphalt 173
   cedar 173
   concrete 176
   demolition 56, 244-245
   fiberglass 173
   fire retardant 173
   mineral 189
   nails 151
   removal 244-245
   roofing 173-174
   siding 195
   slate 175
   stain 153
   tin 177
   wood 173
   wood fiber 174-175
Shoe moulding 144-145
Shores, rental 240
Shoring
   beam 304
   concrete 262
   forms 304
   heavy duty 239
   tilt-up 320
Shot fasteners 341
Shovels, earthwork 75, 77
Shower
   compartments 161, 384, 461
   doors 384
   drains 161, 459
   emergency 462
   plumbing 161, 461-462
   receptors 187, 384, 461
   rod 19, 161, 393-394
   stalls 161, 384, 461
   tile 206
   wallboard 96
Shrinkage defined 3
Shrubs
   landscaping 109-111
   planting 293-294
Shunt trip 534
Shutters 188-189
   ceiling 78
   louvered 188-189
Sidewalk
   demolition 54-55, 241
   forms 47, 277
Sidewalk cover 238
Sidewalk elevators 414
Sidewalks
   asphalt 277
   concrete 45, 51, 160, 273
   moving 416

Siding
   aluminum 194-195, 351
   anzac 126
   board 125-127
   cedar 126-127, 189-192
   cement board 194
   composition 189
   corrugated 194-195, 350-351
   demolition 56
   Douglas fir 125, 192-193
   fiberglass 194
   galvanized steel 195
   hardboard 189-192
   installation 127
   insulated 196
   lumber 125-127
   mineral fiber 189
   paint preparation 155
   painting 154, 156, 381-382
   plywood 122, 192-194
   preformed metal 350-351
   redwood 125-126, 190, 193
   rustic 125-126
   sandwich panels 351
   shingle 195
   shiplap 127
   spruce 127
   steel 350-351
   stone panels 194
   vinyl 196, 351
   yellow pine 127
Sieve analysis
   aggregate 229
   soil 226
Sight screen 280, 384
Sign lighting 541
Signal cable 551
Signs
   architectural signage 388
   display 388-389
   job type 230
   road closed 17
   safety 230, 389
   temporary 222
   traffic 170, 274
Silica sand 138
Silicone
   caulking 38, 356
   roofing 352
   sealant 356
Silicone carbide, sidewalk grits . . . 311
Sill
   anchors 91, 208, 340
   blocks 328
   pans 183
   plates 34
   sealer 105
   steel 339
Sills
   bolted 345
   carpentry 340
   cut 140
   door 55
   masonry 140
   mud 208, 341
   threshold 204
   window 145, 212
Sinkers, nails 148, 151
Sinks
   bar 162
   bases 210
   bathroom 162, 463
   cabinets 23, 210, 406
   demolition 246
   floor 462
   kitchen . 162-163, 400-401, 462-463
   laundry 163
   marble 130
   mop 464
   service 163, 464
   tops 53
Siphon breaker, irrigation 114
Sisalkraft 21
Site
   clearing 241
   grading 252
   utilities, accessories 271-272
   work 240-297

Index

Sizing ... 211
Skip sheathing ... 345
Skylights ... 196-198, 355
  roof windows ... 196-198
  trimmers ... 35
Slab
  aggregate base ... 51
  base ... 51
  coating ... 311-312
  concrete .. 45, 49-51, 304, 307, 311
  curing ... 312
  demolition ... 55, 241
  depression forms ... 300
  edge forms ... 299-300
  excavation ... 298
  finishing ... 52, 308
  forms ... 47, 304, 299-300
  grading ... 252
  membrane ... 211
  plank ... 314
  reinforcing ... 48-49, 305
  sand fill ... 51
  sawing ... 250
  tilt-up ... 317
Slag ... 14
  crushed ... 69
  stone ... 69
Slate
  flooring ... 91
  pavers ... 326
  paving ... 275-276
  roofing ... 176
  testing ... 228
Sleepers, carpentry ... 34
Sleeves, concrete ... 310
  tapping ... 435
  wall ... 447
Slide bolt lock ... 98
Slide, playground ... 282
Sliding doors
  closers ... 57
  deadbolt locks ... 68
  fire ... 364
  frames ... 58
  glass ... 67-68
  patio ... 65, 67-68
Sliding screen door, pet door ... 65
Slip formed walls ... 308
Slope protection ... 258
Slopes, shaping ... 75
Slump block ... 136
Slurry seal ... 272
Slurry trench, caissons ... 261
Small tool costs ... 222
Smoke detectors ... 70, 179, 499, 551
Smoke vents ... 355
Snap ties ... 311
Snow melting cable ... 523
Soap dishes ... 18, 392, 394
Soap dispensers ... 392, 394
  demolition ... 246
Soccer goals ... 284
Social security ... 203
Sod ... 111
  replacing ... 287
Sodding ... 111, 293
Sodium vapor lighting ... 541-543
Sodium, high pressure ... 541-543
Soffit
  forms ... 305
  systems ... 198
  vents ... 385
Soffits, plaster ... 160
Softball
  backstops ... 281
  fields ... 282
Softwood flooring ... 89-90
Softwood plywood ... 168
Soil
  cement ... 259
  compacting ... 77
  conditioner ... 291
  covers ... 269
  mixing ... 292
  pipe, cast iron ... 264, 429-432
  poisoning ... 199
  sealant ... 269, 293
  stabilization ... 259
  stacks ... 353
  testing ... 20, 226
  treatment ... 199, 259
Solar
  electrical systems ... 200
  heating ... 164-165
  photovoltaic systems ... 199
  pool heating ... 165
  power generating ... 199-200
  water heating ... 164-165, 413
Soldier beams, piles ... 262
Solid core doors ... 62, 66
Sonotube forms ... 320
Sound
  absorption walls ... 375
  board ... 350, 372
  control batts ... 105
  isolation rooms ... 412
  systems ... 108, 551-552
Sound trap, duct ... 495
Soundness testing ... 226
Soundproofing ... 412
Soundsoak ... 375
Southern pine ... 124
  siding ... 127
Space frame steel ... 335
Spackle ... 153
Spades, rental ... 232
Spandrel glass ... 369
Spanish tile roofing ... 176
Spark arrestor ... 85
Speaker outlets ... 108
Special construction ... 408-413
Special doors ... 364
Specific gravity ... 226
Specifications fee ... 226
Speed bumps, asphalt ... 277
Speed limit signs ... 274
Spikes ... 151
  lane ... 403
Spindle doors ... 60
Spirals, reinforcing ... 305
Splice bars, rail ... 295
Splices, piles ... 260
Splicing bars ... 305
Split face block ... 136, 327
Split post, redwood ... 123
Split rail fence ... 81
Split rib block ... 327
Split ring connectors ... 209
Split window ... 553
Split-wired outlets ... 72
Splitting rock ... 251-252
Spotlights ... 544
Spray heads, sprinkler ... 290-291
Spray painting ... 155
Spray texture, drywall ... 96
Spray-applied fireproofing ... 359-360
Sprayers, paint ... 155
Spraying, paint ... 155, 380-382
Spreading soil ... 75-76, 252, 291-292
Sprigging ... 293
Spring bronze ... 212
Sprinkler fitter, wage ... 221
Sprinkler systems
  trenching ... 287
  watering ... 286-289
Sprinklers
  control wiring ... 114
  controllers ... 114, 288
  fire ... 83, 464-465
  heads ... 113, 465
  irrigation ... 114, 289-290
  lawn ... 113-114, 290-291
  pipe ... 288-289
  shrub ... 289
  valves ... 287-288, 290-291
Sprinkling, excavation ... 77
Spruce ... 123
  lumber ... 118-120
  siding ... 127
Square foot costs, homes ... 219
Square moulding ... 147
Stabilization fabric ... 259
Stack
  boiler ... 477
  chimney ... 488-489
  economizer units ... 475
  masonry ... 140, 328
  prefabricated ... 85
Stadiums, concrete ... 310
Stage
  equipment ... 397-398
  scaffolding ... 239
  swinging ... 239
Stain, paint ... 153-157
Stained glass ... 395
Staining ... 154-157
Stainless steel
  fabrication ... 340
  pipe ... 443-445
Stair treads
  hardwood ... 347
  marble ... 331
  resilient ... 91, 378
  rubber ... 91, 141
  steel ... 336
Stairlifts ... 73
Stairs
  aluminum ... 201
  attic ... 202
  basement ... 200
  box ... 200
  carpeting ... 37
  concrete ... 52, 309
  curved ... 200
  demolition ... 57
  disappearing ... 202
  finishing ... 309
  folding ... 202
  forms ... 47, 305, 309
  hardwood ... 202, 347
  job built ... 202, 347
  moving ... 416
  nosing ... 339
  painting ... 157
  precast ... 315
  railing ... 169, 200-202, 334-335
  risers, resilient ... 91, 378
  spiral ... 201
  steel ... 201, 336
  stringer cover ... 378
  stringers ... 336
  wood ... 200-202, 347
Stakes
  concrete form ... 47
  form ... 128
  foundation ... 127
  ginnies ... 127
  lumber ... 127-128
  redwood ... 127
  reflector ... 274
  survey ... 127
Stalls
  shower ... 161, 384, 461
  toilet ... 384
Stamped concrete ... 52
Standard brick ... 131
Starter strip, aluminum siding ... 195
Starter switches ... 527
Starters, motor ... 529-530
Startup, boiler ... 475
State cost adjustments ... 10
Steam
  boilers ... 468-470
  gauge ... 458
  heater monitor ... 499-500
  heating ... 468-469, 482-484
  meters ... 458
  radiators ... 482
  traps ... 478-480
Steam bath, generators ... 202
  wiring ... 70, 202
Steam cleaning, masonry ... 332
Steam generators ... 202
Steel
  bridging ... 26
  buildings ... 333-334
  channel ... 136
  conduit ... 503-505
  curtain walls ... 369
  cutting ... 139
  decking ... 335
  doors ... 362
  drilling ... 139
  duct, electrical ... 501-502, 518-519
Steel *(continued)*
  fabricated ... 338-339
  fire hose cabinets ... 465
  fireproofing ... 359
  flagpoles ... 387
  flooring ... 385
  landings ... 336
  lath ... 112, 370
  lintels ... 339
  masonry ... 135, 139
  mesh ... 49, 305
  ornamental ... 340
  painting ... 338, 381
  piles ... 260-261
  pipe ... 416-428, 443-445
  plate cutting ... 243
  railing ... 169, 201
  rakers ... 262
  reinforcing ... 48-49, 52,
  ... 139, 305, 328-329
  reinforcing, masonry ... 328-329
  roof deck ... 335
  sag rods ... 337
  sandblasting ... 177
  securing gratings ... 340
  shoring ... 262
  stairs ... 201, 336
  storefronts ... 366
  structural ... 336-337
  testing ... 229
  tilt-up ... 319-320
  welding ... 139
Steel reinforcing ... 48
Steel shores, rental ... 235
Stem wall forms ... 298
Stenographer cost ... 222
Step lights ... 543
Step on grade, concrete ... 52
Stepping stones ... 111, 294
Steps
  brick ... 326
  carpeting ... 37
  computer floor ... 386
  concrete ... 52, 307, 309
  flagstone ... 137
  forming ... 305, 309
  manhole ... 270
  masonry ... 136
  painting ... 154-155, 157
Sterilization, pipe ... 269
Stock room door ... 365
Stockade fence ... 280
Stolons, grass ... 293
Stone
  aggregate ... 14
  base ... 51
  cleaning ... 331-332
  crushed ... 69
  masonry ... 136, 140, 331
  riprap ... 258, 294
  roofing ... 174, 352
  slate ... 91, 175
  stepping ... 111, 294
  veneer siding ... 194, 331
  veneer, fireplace ... 332
Stone slag ... 69
Stool
  carpentry ... 147
  marble ... 331
Stop
  door ... 100, 144-147, 368
  gravel ... 183, 352-353
  moulding ... 147
  plaster ... 371
  signs ... 274
  wallboard ... 373
Stops
  fixture ... 451
  plumbing ... 451-452
Storage
  bins ... 222
  overhead ... 222
  safes ... 177
  sheds ... 222, 240
Store front glass ... 94
Storefronts ... 365-366
Storm doors ... 60

Index

Storm drain 267
   demolition 242
   pipe 264
Storm sash 213
Stoves, ranges 170-171
   electrical work 72
Strainers, hydronic 481
Strand board 122
Strap, hanger 516
Straps
   hardware 206-209
   post anchors 139
   steel 99
Straw 111, 258
Strawberry 293
Street equipment rental .. 236-237
Street lights 545-546
Street marking 274
Street paving 272
Street sweeper 237, 272
Strip
   flooring 87-90
   lighting 539-541
Striper, rental 237
Striping
   athletic 285
   courts 285
   pavement 274
Striplights 540
Stripping
   forms 48
   soil 76, 252
Strongbacks 345
Structural engineering 74
Structural glazed tile ... 329-330
Structural steel 336-337
Structure moving 240
Structures, temporary 239-240
Stucco
   demolition 244
   mesh 370
   mix 42
   moulding 145
   netting 112
   paint 151, 154
   painting 154, 156, 381
   plastering 160, 370
   stone 331
   wall assembly, wood framed ... 36
Stud
   partitions, metal 371-372
   straps 208
   wall assemblies 36
   walls 24, 34-35, 340
Stud grade lumber 118
Studio equipment 397
Studs
   carpentry 34-35, 340
   demolition 56, 244
   hollow metal 371-372
Study carrels 398
Stump removal 77, 241
Sub panel connections 72
Subbase, pavement 272
Subcontract
   costs defined 4
   electrical work 70-72
   elevators 73
   garage doors 92
   insulation 105-106
Subcontractor markup 4
Subfloor
   adhesive 13-14
   plywood 35
Subflooring 13-14, 35, 168, 342
Subsurface exploration 199
Suits, protective 247
Sump pumps 72, 455, 460
Sumps, roof 354
Sun screens 340
Sunlamps, wiring 70
Superintendent 222, 226
Superphosphate 291
Supervision costs 226
Supervision costs excluded 3
Supply fans 492
Supply piping 166-167

Supports
   conduit 515-517
   fabricated metal 338
   pipe 445-447
Surface duct, electric ... 517-518
Surface prep, painting 379
Surface preparation 155
Survey stakes 127
Survey, aerial 225
Surveying 224-225
Suspended ceilings .. 39-41, 375
   demolition 244
Swales, excavating 258
Sweep pavement 272
Sweeper rental 237
Swimming pool 412
   fences 151
   gunite 312
   heaters, solar 165
   paint 153
Swing, tire 282
Swinging stage scaffold 239
Swings, playground 283-284
Switch box, electric 524
Switch circuits 503
Switch plates 18
Switchboards 531-533
Switches
   disconnect 539
   electric 72, 526-527, 530
   motor 530
   power distribution 531-533
   railroad 296
   safety 530-531
   security 549
   starter 527
   transfer 548
   wiring 72
Switchgear, electric 526-527
Sycamore veneer 125
Symbols 220
Synthetic surface, fields 282

T

T-bar ceiling 39-41, 408
T-hinges 99
T.V. pipe inspection 271
T.V. systems 551
Table of contents 2
Tables
   kitchen 399, 401
   school 398
Tack coat, pavement 272
Tackboard 383
Tamo veneer 125
Tampers, rental 232
Tamping, excavation 76-77, 254
Tanks
   demolition 242
   expansion 460
   foundations 321
   fuel 466
   oil 466
   painting 381
   propane 466
   septic 180-181, 461
   water 412, 460-461
Tape, thermoplastic 285
Taping compound 96
Taping wallboard 96, 372
Tapping
   pipe 436
   saddles 436
   sleeves 436
Tarkett flooring 85
Tarpaulins
   canvas 203
   plastic 169, 203
Taxable fringe benefits
   commercial & Industrial 221
   residential 8
Taxes & insurance 203, 222
Taxes included 3
Taxes, labor cost
   commercial & Industrial 221
   residential 8

Teak 89-90, 124
   lumber 123
   parquet 90
   veneer 125
Tear-off, demolition 244
   roof 351
Technical services 226-229
Tee hinges 99
Tee top flashing 182
Tees, precast concrete 315
Telephone
   answering equipment 108
   assistance 3
   enclosures 392
   outlets 551
   wiring 551
Television
   closed circuit 551
   equipment 398
   outlets 551
   stage equipment 398
   wiring 72
Teller
   automatic 395
   counters 395
   windows 395
Temperature
   controls 497-500
   gauge 458
Temporary
   caulking 38
   enclosures 238
   jackposts 108
   services 222
   structures 239-240, 408-409
   utilities 230-231
   wall bracing 25
Tennis courts
   installation 285
   striping 285
Tennis windbreak 285
Tensile test 228-229
Termite treatment 128, 199
Terra cotta, demolition 244
Terrazzo
   counter tops 374
   demolition 55, 245
   epoxy 374
   floors 374-375
   pavers 326
   tile 326, 374-375
   wainscot 374
Testimony
   appraisal 15
   lumber grading 128
Testing
   acoustical 74
   aggregate 226
   asbestos 21
   asphalt 227
   brick 228
   concrete 227
   electrical systems 552
   gunite 228
   gypsum 228
   hazards 21
   home 21
   lumber 128
   masonry 228
   noise 74
   percolation 180
   piles 261
   reinforcing 228-229
   soil 20, 180, 199, 226
   sound transmission 74
   steel 229
   thermal 106
   water 20
Tetherball post 284
Textolite counters 53
Texture paint 153
Textured wallboard ... 97, 210, 372
Theater equipment 397-398
Theater seating 408
Thermal analysis 106
Thermal and moisture protection 348-360
Thermoflow indicator 458
Thermography 106

Thermometer, steam 458
Thermoplastic marking 274
Thermoplastic tape 285
Thermostat controls 498-499
Thermostatic
   mixing valves 481
   traps 479-480
Thermostats 102-103, 498-499
Thinner
   lacquer 153
   paint 153
   swimming pool enamel 153
Threadolets 416-417
Thresholds
   aluminum 203-204, 368
   bronze 368
   marble 331
   wood 144
Thrust blocks
   concrete 435
   demolition 241
Ticket boxes 389
Tie beam forms 298
Tie straps 209
Tie wire 139
Tieback walls 262
Ties
   brick wall 206
   concrete 46, 207, 311
   cross 295
   form 46, 207
   railroad 295
Tile 204-206
   acoustical facing 330
   adhesive 14, 205-206
   aluminum 374
   backer board 95, 206
   ceiling 40-41, 376
   ceramic 204-206, 373-374
   clay 329
   concrete roof 176
   decorator panels 374
   demolition 55, 245
   epoxy bed 326
   flooring 85-86, 90-91
   glazed 204-205, 373-374
   marble 130, 205-206
   masonry 141
   paver 205-206
   plastic 374, 378
   quarry 205-206, 325-326, 374
   roofing 176-177
   security, glazed 330
   stainless steel 374
   structural facing 330
   wall 204-206
Tile layer, wage
   commercial & industrial 221
   residential 8
Tileboard 96
Tilt-up
   construction 316-321
   panels, sandblasting 177
   pilasters 320
Timber
   clips 209
   connectors 206-209
   construction 345-346
Timbers
   laminated 346-347
   lumber 119
Timekeeper costs 226
Timer switches 527
Timers 498
Tin
   ceilings 41
   flashing 183
   shingles 177
Tire changer 402
Tire swing 282
Tissue holder 18, 392
Toaster 401
Toggle switch 527
Toilet
   accessories 18, 392-394
   compartments 384
   overhead 222
Toilet partition, demolition .. 246

Index

Toilet room dispensers 392-394
Toilets 161-162, 464
  temporary 231
  trailer-mounted 231
Tool expense, overhead 222
Tool shed 240
Tools available 4
Tools, rental 232
Toothbrush holders 18
Toothed rings 209
Toothing masonry 331-332
Topping compound 96
Topping, concrete 52, 311-312
Tops
  bar 53
  sink 53
Topsoil 111
Topsoil excavation 76
Torch cutting 243
Total column defined 4
Towel
  bars 393-394
  dispensers 393-394
  racks 19
Tower, playground 282
Towers
  fire pole 282
  lifting 231
  scaffolding, rental 239
Track
  rail 295
  railroad 295
  rolling ladder 401
  running 282
  stage curtain 397
Track, railroad, demolition 243
Tractor 251-252
Tractor loader 255
Tractor operator, wage 221
Tractors, excavation 235
Traffic
  cones 17
  paint 153
  signs 274
  stripes 274
Trail, fitness 284
Trailer rental 240
Training, lumber grading 128
Transfer switches 548
Transformers 534-535
  pole mount 539
  switchboard 534
Transit mix concrete 45, 306
Transite pipe 264
Transmission, electric 539
Transom light 362
Traps
  bucket 478-479
  steam 478-48
Trash
  bins 240
  chutes 245, 416
  enclosure, tilt-up 321
  pump 236
  receptacles 393
  removal 222
Traverse rod 213
Treads
  concrete 312
  finishing 312
  flagstone 137
  marble 331
  resilient 91, 378
  rubber 141
  steel 336
  wood 202, 347
Treated poles 123
Treated wood 128
Treating
  fire retardant 129
  lumber 128
  soil 199
  wood preservative 154
Treatment, soil poisoning 199
Tree clearing 77, 241
Trees, planting 110-111, 293-294
Trench
  bracing 252-263
  covers 270, 338
  frame, manhole 270
  inlet 271
  shaping 253, 259
  sheeting 262
  shoring 262
  sprinkler pipe 287
Trenchers, rental 78, 235
Trenches, rock 252-253
Trenching
  backhoe 253-254, 297
  concrete 297
  excavation 76-77, 252-253, 259, 297
  hand 76, 113, 259
  sprinkler systems 113, 287
Trim
  bead 371
  carpentry 144
  door 59, 144
  lumber 144
  moulding 144
  painting 154, 156, 382
  pine 145
  vinyl 351
  wallboard 96
  window 144, 147
Trimmers 35
Trimming, trench bottom 76
Troffer fixtures 540-541
Trowel finishing, concrete 52, 311
Trowel-ready mortar 139
Trowels, rental 233
Truck
  bumpers 320
  cranes 231
  hauling 255-256
  scales 391
Truck driver, wage
  commercial & Industrial 221
  residential 8
Trucks
  excavation 77
  rental 235
Truss assembly 100
Truss joists 345
Truss reinforcing, masonry 329
Truss tee deck 324
Trusses
  fink 33-34
  gable 34
  roof 33-34
  steel 337
  timber 345
  wood 345
Tub access doors 182
Tub kits 96
Tube
  columns, steel 337-338
  conveyors 416
  copper 436-439
  electric, fluorescent 114-116
Tubeaxial fans 478
Tubs 161
  access doors 187
  caulking 38
  enclosures 187
  plumbing 161, 461
Turbine pumps 454
Turbine vents 185
Turf
  fields, grass 282
  fields, synthetic 282
  replacing 287
Turnouts, railroad 296
Turnstiles 389
Turntables, baggage 416
Turpentine 154
Twist straps 209
Tyvek housewrap 22
Tyvek suits 247

U

U bolts 447
UF cable 521-522
UF cable wire 521-522
Ultraflo water system 164
Ultrasonic sensors 550
Ultrasonic transceivers 549
Underfloor, duct 519
Underlayment 97, 121-122, 177
  carpet pad 379
  flooring 91
  hardboard 97
  nails 151
  plywood 168, 342
  tile backer board 95, 206
Underpinning, shoring 262-263
Unistrut
  embedded iron 311
  inserts 311
Unit column defined 4
Unit controls, fan coil 499
Unit heater monitor 499
Unit heaters 484
Uniturf 282
Urethane
  board 348-350
  caulk 356
  foam 105, 349
  sealer 88
Urinal bar 394
Urinal screens 246, 384
Urinals 462
Urns, ash 394
Using this book 3
Utilities
  overhead 222
  sewer connections 180-181
  temporary 230-231
Utility
  cabinets 23
  sets 492
  shelving, wood 186
  sinks 401

V

Vacation costs included 3
Vacuum breaker 287, 480
  hose bib 167
Vacuum systems 210
  wiring 72
Valance lighting 116, 142
Valley flashing 183, 352-353
Valley gutter, concrete 277
Valuation fees 14-15
Valuation, buildings 14-15
Valve box 143, 272, 435
Valves 114, 287-288, 447-452
  angle 450
  backflow 224
  balancing 480
  ball 451
  butterfly 448-449
  cast iron 435-436, 447
  check 448, 449-450, 480
  control 288
  gas regulator 451
  gate 435, 447-448
  globe 448, 449
  irrigation 288, 290-291
  mixing 481
  OS&Y 448
  pressure reducing 452, 479-481
  regulating 452
  relief 452, 479-481
  sprinkler 114, 288
  tapping 436
  wall fixture 451
  water control 452
Vanity bases 210
Vanity tops
  marble 53, 130
  plastic 53, 406
Vapor barriers
  building paper 21-22
  dampproofing 348
  polyethylene film 169
  slab 51, 308, 317
Varmeter 533
Varnish 154
Varnishing 154-155, 157

Vault doors 365, 395
Vegetable, peeling machines 401
Velour draperies 407
Velux
  roof windows 196-197
  skylights 196-198
Veneer
  brick 133, 140
  ceramic 330
  laurel 124
  lumber 124-125
  mahogany 124
  maple 124
  masonry 137
  southern pine 124
Venetian blinds 213, 407
Vent
  fans 78-79
  furnace 102-103
  hoods 170-171
  pipe 439, 441
Vent pipe flashing 182
Vent piping 164-165, 167-168
Ventilating
  doors 67
  skylights 196-198
Ventilation 489-494
Ventilators, intake 492-493
  relief 492-493
  roof 79, 185, 492
Vents
  attic 184, 355
  block 355, 385
  chimney 488-489
  clothes dryer 184
  door louver 354, 385
  foundation 184, 355
  frieze 355
  louver 184-185
  plaster 371
  plumbing 353
  rafter 185
  ridge 185
  roof 185
  screen 385
  sheet metal 184-186
  smoke 355
  soffit 385
  turbine 185
Vermiculite
  fertilizer 291
  insulation 105, 350
  roof deck 324
Vertical blinds 213
Vertical panels, precast tilt-up 318
Vibration isolators 453
Vibration sensors 549
Vibrator compaction 252-253
Vibrators, rental 233
Vibro plate, rental 78
Vibroflotation 259
Victaulic couplings 453
Viewer, door 368
Vinyl acrylic
  caulk 37
  primer 152
Vinyl covered wallboard 210
Vinyl drip cap 203
Vinyl flooring 85-86, 377
  demolition 55, 245
Vinyl siding 196, 351
Viscosity recorder 499
Visqueen 169
Vitrified clay
  drainage tile 267
  pipe 181, 268
Volcanic cinder track 282
Volcanic rock 258
Volleyball posts 284
Voltmeter 533

W

Waffle pan forms 304
Wage costs defined 3

587

Index

Wage rates
  adjusting 9
  commercial & industrial 221
  residential 8
Wagon drill 251
Wagon drill, rental 235
Wainscot
  cap 147
  ceramic tile 206, 373
  hardboard 96
  metal 351
  plywood paneling 157-159
  terrazzo 374
  tile 206
Waler brackets 311
Waler jacks 311
Walkway
  demolition 55, 241
  forms 47
  lighting 117
Walkways
  asphalt 159, 277
  concrete 51, 160, 277
  moving 416
  roof 351
  snow melting cable for 523
Wall
  anchors, glass block 330
  anchors, masonry 139, 329
  assemblies, wood framed 36
  beds 19
  braces 25, 209
  cladding 350-351
  coverings 210-211, 382-384
  demolition 54-56, 244
  footing 298
  forms, concrete 47-48, 300-301
  framing 24
  insulation and finish ... 348, 350
  jointing 310, 356-357
  louvers 385
  painting 154-155, 379-381
  panel systems 344, 350-351
  plates 32
  protection systems 384-385
  reinforcing 139, 329
  safes 177, 392
  sawing 250
  sheathing 34, 344
  studding 34-35, 340
  ties, masonry 139, 206
  valves 451
Wall ovens 171
Wall panels, precast 312
Wallboard
  adhesive 13-14
  bracing 25
  demolition 56, 244
  fire rated 95
  gypsum 95-96, 372-373
  hardboard 97, 383
  paint preparation 155, 379
  painting 154-155, 157, 381
  prefinished 210
  textured 210, 372
Wallcoverings . 210-211, 382-383, 385
Wallpaper 210-211, 382, 385
  adhesive 211
Walls
  brick 131
  concrete 49-50
  concrete block 134
  demolition 56
  glass block 137-138, 330
  masonry 131
  panelized wood 344
  precast panel 312
  retaining 48, 300-301
  tile 204-206, 373
  tilt-up 318-320
Walnut
  flooring 87-90
  lumber 124
  paneling 158-159

parquet 89
veneer 125
Wardrobe cabinets 398, 405
Warning signs 17, 274, 389
Wash
  concrete finishing 312
  fountains 462
Washed aggregate 311
Washers, clothes, wiring 71
Washers, foundation bolt 91
Washing
  concrete 312
  masonry 332
Waste
  asbestos disposal 249
  bins, toilet 393
  defined 3
  handling equipment 416
  pipe 166, 439, 441
Water
  blaster, rental 237
  blasting 177, 242, 332
  closets 161-162, 464
  control valves 452
  coolers 163, 401, 462
  fountains 163, 462
  hose reel 402
  meter boxes 142-143
  meter fee 224
  meters 224, 458
  overhead 222
  pipe 434, 441
  pumps 453-454, 460
  pumps, wiring 72
  reel 402
  repellents 154
  sealer 153
  softeners 165, 461
  stop, concrete 311
  storage tanks, solar 164
  tanks 412, 460-461
  temporary 231
  testing 20
  treatment 461
  truck, rental 237
  wells 212
Water capacity fees 224
Water heaters
  commercial 401
  residential 164-165
  rough-in 464
  safety pans 460
  solar 164-165, 413
  stands 460
Water pumping, solar 200
Water softening systems, boilers . 474
Water stop, concrete 311
Water table moulding 144-146
Waterblasting 177, 332
Watering
  embankment grading 252-253
  landscaping 289-291
  sprinkler systems 286-289
Waterproof adhesive 13-14
Waterproofing 211, 348
  acrylic 105
  baths 348
  concrete 299
  insulation 105
  masonry 153, 332
  roof coatings 171-172
  slab 299
Wattmeter 534
Waxing
  concrete 312
  floors 155, 377
  woodwork 157
Weather vanes, cupola 54
Weatherstripping 211-212, 368
Wedges 128
Weed control 296
Weld testing 229
Welded wire mesh 308

Welding
  flanges 418-428
  grounding 528
  machine, rental 237
  rebars 305
  reinforcing 139
  steel 139, 336
  tilt-up 319
Welfare costs included 3
Wellpoint pump rental 263
Wellpoints 263
Wells 212
Wenge veneer 125
West coast fir lumber 117-121
Western red cedar siding ... 126, 194
Wet patch 172
Wet sandblasting 177-178
Wharfs 294
Wheel corner guards 339
Wheel loaders 255
Wheelbarrow rental 237
Wheelchair lifts 73
Whirlpool baths, wiring 70
Whirlpool tub 161
Whirls, playground 284
White cement 42, 138, 312, 374
White pine 119
Willow 124
Wind clips 177
Windbreak, tennis 285
Window
  air conditioners 103
  awnings 16
  blinds 213, 407
  cleaning 240
  demolition 56, 246
  flashing 184
  glass 92-93
  glazing 93-94
  grating 340
  moulding 145, 147, 189
  openings, framed 35-36
  paint preparation 155
  painting 154, 156, 381
  reglazing 93-94, 155
  removal 56
  seals 212
  shades 213
  sills 145, 212
  stool, marble 331
  stool, wood 147
  treatment 69-70, 213, 407
  trim 147
  wall 369
  weatherstripping 212
Windows
  aluminum 217-218
  awning 213-214
  bank 395
  bay 214-215
  casement 215-216
  double hung 216
  louvered 218
  metal 213, 217-218, 366
  partition 390
  picture 216-217
  pine 213-214
  roof 197-198
  sliding 217
  steel 366
  teller 395
Wire
  aluminum 519-520
  brushing 380
  building 70-72, 519-521
  copper 519
  electric 70-72, 519-521
  lath 112
  mesh, gunite 312
  mesh, lath 370
  partitions 391
  Romex 522
  screen 179
  service entrance ... 72, 519-520

THW 520-521
UF cable 521-522
Wire brush 52
Wire mesh fence 80-81
Wiremold raceway 518
Wireway, electrical 517-518
Wiring
  alarms 179, 548-550
  electric 70-72
  fence, electric 80-81
  irrigation controls 114
  landscape sprinklers 114, 288
  telephone 551
Wonderboard, tile backer board .. 206
Wood
  and plastic 340-347
  ceiling 40
  doors 59-66
  fences 81-82
  fiber shingles 174
  floor finishing 377
  flooring 87-90
  hardwoods 123-124
  lumber 117-122, 129
  paint preparation 155
  panel doors 65
  piles 261
  preservatives 128-129, 154
  roofing 173-175
  sealer 154
  siding 125-127
  treatment 128-129
  veneer 124-125
Wood chip mulch 111
Wood frame structures
  demolition 56-57
Woodwork
  paint preparation 155
  painting 155-157
  waxing 157
Woodworking glue 13
Workers' comp insurance 106-107
Working conditions 4
Woven wire partitions 391
Wrecking, demolition 54-57, 241
Wrought iron 151
Wrought iron paint 154
Wrought washers 91

X

X-ray
  curtain 411
  protection 410-412
  room locks 99

Y

Yard
  decking 54
  drains 272
  lighting 545-546
Yardarm, nautical 386-387
Yellow pine
  lumber 118, 121
  posts 123
  siding 127
Yield signs 274

Z

Z-bar flashing 184
Z-brick 159
Zebra stripe veneer 125
Zinc chromate primer 152
Zinc dust primer 152
Zinc oxide primer 152
Zonolite, fill 324-325

Other Practical References

■ Construction Estimating Guides ■

■ Construction Estimating Reference Data
Provides the 300 most useful estimating reference tables. Labor requirements for nearly every type of construction, including: sitework, concrete work, masonry, steel, carpentry, thermal and moisture protection, doors and windows, finishes, mechanical and electrical. Each section details the work being estimated and gives the appropriate crew size and equipment needed. **368 pages, 8½ x 11, $26.00**

■ Building Cost Manual
Square foot costs for residential, commercial, industrial, and farm buildings. Quickly work up a reliable budget estimate based on actual materials and design features, area, shape, wall height, number of floors, and support requirements. Includes all the important variables that can make any building unique from a cost standpoint. **240 pages, 8½ x 11, $16.50. Revised annually**

■ Painting Cost Guide
A complete guide to estimating painting costs for just about any type of residential, commercial, or industrial painting, whether by brush, spray, or roller. Shows typical costs and bid prices for fast, medium, and slow work, including material costs per gallon; square feet covered per gallon; square feet covered per manhour; labor, material, overhead, and taxes per 100 square feet; and how much to add for profit. **448 pages, 8½ x 11, $27.50. Revised annually**

■ Berger Building Cost File
Labor and material costs needed to estimate major projects: shopping centers, hospitals, educational facilities, office complexes, industrial and institutional buildings, and housing projects. All cost estimates show both manhours required and typical crew needed. Figure the price and schedule the work quickly and easily. **304 pages, 8½ x 11, $30.00. Revised annually**

■ Estimating Tables for Home Building
Produce accurate estimates for nearly any residence in just minutes. This handy manual has tables you need to find the quantity of materials and labor for most residential construction. Includes overhead and profit, how to develop unit costs for labor and materials, and how to be sure you've considered every cost in the job. **336 pages, 8½ x 11, $21.50**

■ National Repair & Remodeling Estimator

The complete pricing guide for dwelling reconstruction costs. Reliable, specific data you can apply on every repair and remodeling job. Up-to-date material costs and labor figures based on thousands of jobs across the country. Provides recommended crew sizes; average production rates; exact material, equipment, and labor costs; a total unit cost and a total price including overhead and profit. Separate listings for high- and low-volume builders, so prices shown are accurate for any size business. Estimating tips specific to repair and remodeling work to make your bids complete, realistic, and profitable. *New this year! The complete book on a disk with a built-in estimating program for an IBM-compatible hard-drive computer. FREE on a 5¼" high-density (1.2 Mb) disk when you buy the book. (Add $10 for Repair & Remodeling Estimate Writer on extra 5¼" double density 360K disks or 3½" 720K disks.)* **288 pages, 11 x 8½, $29.50. Revised annually.**

■ Estimating Home Building Costs
Estimate every phase of residential construction from site costs to the profit margin you include in your bid. Shows how to keep track of manhours and make accurate labor cost estimates for footings, foundations, framing and sheathing finishes, electrical, plumbing, and more. Provides and explains sample cost estimate worksheets with complete instructions for each job phase. **320 pages, 5½ x 8½, $17.00**

■ Remodeling Guides ■

■ Profits in Buying & Renovating Homes

Step-by-step instructions for selecting, repairing, improving, and selling highly profitable "fixer-uppers." Shows which price ranges offer the highest profit-to-investment ratios, which neighborhoods offer the best return, practical directions for repairs, and tips on dealing with buyers, sellers, and real estate agents. Shows you how to determine your profit before you buy, what "bargains" to avoid, and how to make simple, profitable, inexpensive upgrades. **304 pages, 8½ x 11, $19.75**

■ Remodeling Contractor's Handbook
Everything you need to know to make a remodeling business grow: identifying a market, inexpensive sales and advertising techniques that work, making accurate estimates, building a positive company image, training effective sales people, placing loans for customers, and bringing in profitable work to keep your company growing. **304 pages, 8½ x 11, $18.25**

■ Audiotapes: Estimating Remodeling
Listen to the "hands-on" estimating instructions in this popular remodeling seminar. Make your own unit price estimate based on the prints enclosed. Then check your completed estimate with those prepared in the actual seminar. After listening to these tapes you will know how to establish an operating budget for your business, determine indirect costs and profit, and estimate remodeling with the unit cost method. *Includes seminar workbook, project survey and unit price estimating form, and six 20-minute cassettes,* **$65.00**

■ Manual of Professional Remodeling
The practical manual of professional remodeling that shows how to evaluate a job so you avoid 30-minute jobs that take all day, what to fix and what to leave alone, and what to watch for in dealing with subcontractors. Includes how to calculate space requirements; repair structural defects; remodel kitchens, baths, walls, ceilings, doors, windows, floors and roofs; install fireplaces and chimneys (including built-ins), skylights, and exterior siding. Includes blank forms, checklists, sample contracts, and proposals you can copy and use. **400 pages, 8½ x 11, $19.75**

Construction References

Fences & Retaining Walls

Everything you need to know to run a profitable business in fence and retaining wall contracting. Takes you through layout and design, construction techniques for wood, masonry, and chain link fences, gates and entries, including finishing and electrical details. How to build retaining and rock walls. How to get your business off to the right start, keep the books, and estimate accurately. The book even includes a chapter on contractor's math. **400 pages, 8½ x 11, $23.25**

Blueprint Reading for the Building Trades

How to read and understand construction documents, blueprints, and schedules. Includes layouts of structural, mechanical, HVAC and electrical drawings. Shows how to interpret sectional views, follow diagrams and schematics, and covers common problems with construction specifications. **192 pages, 5½ x 8½, $11.25**

Contractor's Guide to the Building Code Revised

This completely revised edition explains in plain English exactly what the Uniform Building Code requires. Based on the 1988 code, the most recent, it covers many changes made since then. Also covers the Uniform Mechanical Code and the Uniform Plumbing Code. Know how to design and construct residential and light commercial buildings that'll pass inspection the first time. Suggests how to work with an inspector to minimize costs, what common building shortcuts are likely to be cited, and where exceptions are granted. **544 pages, 5½ x 8½, $24.25**

Estimating Framing Quantities

Gives you hundreds of time-saving estimating tips. Shows how to make thorough step-by-step estimates of all rough carpentry in residential and light commercial construction: ceilings, walls, floors, and roofs. Lots of illustrations showing lumber requirements, nail quantities, and practical estimating procedures. **285 pages, 5½ x 8½, $34.95**

Carpentry Estimating

Simple, clear instructions on how to take off quantities and figure costs for all rough and finish carpentry. Shows how to convert piece prices to MBF prices or linear foot prices, use the extensive manhour tables included to quickly estimate labor costs, and how much overhead and profit to add. All carpentry is covered; floor joists, exterior and interior walls and finishes, ceiling joists and rafters, stairs, trim, windows, doors, and much more. Includes sample forms, checklists, and the author's factor worksheets. **320 pages, 8½ x 11, $25.50**

Rough Carpentry

All rough carpentry is covered in detail: sills, girders, columns, joists, sheathing, ceiling, roof and wall framing, roof trusses, dormers, bay windows, furring and grounds, stairs, and insulation. Explains practical code-approved methods for saving lumber and time. Chapters on columns, headers, rafters, joists, and girders show how to use simple engineering principles to select the right lumber dimension for any species and grade. **288 pages, 8½ x 11, $17.00**

Cabinetmaking From Design to Finish

Every aspect of cabinetmaking is covered from layout, through joinery, to finishing techniques. Gives illustrated instructions for designing cabinets to fit the kitchen workcenter, creating dado, mortise, tenon, lap & dowel joints, making frames and panels, to the actual construction of cabinets and installing cabinet hardware. **416 pages 5½ x 8½, $22.00**

Stair Builders Handbook

If you know the floor-to-floor rise, this handbook gives you everything else: number and dimension of treads and risers, total run, correct well hole opening, angle of incline, and quantity of materials and settings for your framing square for over 3,500 code-approved rise and run combinations – several for every ⅛ inch interval from a 3 foot to a 12 foot floor-to-floor rise. **416 pages, 5½ x 8½, $15.50**

How to Succeed With Your Own Construction Business

Everything you need to start your own construction business: setting up the paperwork, finding the work, advertising, using contracts, dealing with lenders, estimating, scheduling, finding and keeping good employees, keeping the books, and coping with success. If you're considering starting your own construction business, all the knowledge, tips, and blank forms you need are here. **336 pages, 8½ x 11, $19.50**

Video: Stair Framing

Shows how to use a calculator to figure the rise and run of each step, the height of each riser, the number of treads, and the tread depths. Then watch how to take these measurements to construct an actual set of stairs. You'll see how to mark and cut your carriages, treads, and risers, and install a stairway that fits your calculations for the perfect set of stairs. **60 minutes, VHS, $24.75**

Estimating Plumbing Costs

Offers a basic procedure for estimating materials, labor, and direct and indirect costs for residential and commercial plumbing jobs. Explains how to read and understand plot plans, design drainage, waste, and vent systems, meet code requirements, and make an accurate take-off for materials and labor. Includes sample cost sheets, manhour production tables, complete illustrations, and all the practical information you need. **224 pages, 8½ x 11, $22.50**

Handbook of Construction Contracting

Volume 1: Everything you need to know to start and run your construction business; the pros and cons of each type of contracting, the records you'll need to keep, and how to read and understand house plans and specs so you find any problems before the actual work begins. All aspects of construction are covered in detail, including all-weather wood foundations, practical math for the job site, and elementary surveying. **416 pages, 8½ x 11, $24.75**

Volume 2: Everything you need to know to keep your construction business profitable; different methods of estimating, keeping and controlling costs, estimating excavation, concrete, masonry, rough carpentry, roof covering, insulation, doors and windows, exterior finishes, specialty finishes, scheduling work flow, managing workers, advertising and sales, spec building and land development, and selecting the best legal structure for your business. **320 pages, 8½ x 11, $24.75**

Carpentry Layout

Explains the easy way to figure: cuts for stair carriages, treads and risers; lengths for common, hip, and jack rafters; spacing for joists, studs, rafters, and pickets; layout for rake and bearing walls. Shows how to set foundation corner stakes, even for a complex home on a hillside. Practical examples on how to use a hand-held calculator as a powerful layout tool. **240 pages, 5½ x 8½, $16.25**

■ **Roof Framing**
Shows how to frame any type of roof in common use today, even if you've never framed a roof before. Includes using a pocket calculator to figure any common, hip, valley, or jack rafter length in seconds. Over 400 illustrations cover every measurement and every cut on each type of roof: gable, hip, Dutch, Tudor, gambrel, shed, gazebo, and more. **480 pages, 5½ x 8½, $22.00**

■ **Masonry Estimating**
Step-by-step instructions for estimating nearly any type of masonry work. Shows how to prepare material take-offs, figure labor and material costs, add a realistic allowance for contingency, calculate overhead correctly, and build competitive profit into your bids. **352 pages, 8½ x 11, $26.50**

■ **Rafter Length Manual**
Complete rafter length tables and the "how to" of roof framing. Shows how to use the tables to find the actual length of common, hip, valley, and jack rafters. Explains how to measure, mark, cut and erect the rafters; find the drop of the hip; shorten jack rafters; mark the ridge and much more. Loaded with explanations and illustrations. **369 pages, 5½ x 8½, $14.25**

■ **Construction Surveying & Layout**
A practical guide to simplified construction surveying. How to divide land, use a transit and tape to find a known point, draw an accurate survey map from your field notes, use topographic surveys, and the right way to level and set grade. You'll learn how to make a survey for any residential or commercial lot, driveway, road, or bridge — including how to figure cuts and fills and calculate excavation quantities. Use this guide to make your own surveys, or just read and verify the accuracy of surveys made by others. **256 pages, 5½ x 8½, $19.25**

■ **Plumber's Exam Preparation Guide**
Hundreds of questions and answers to help you pass the apprentice, journeyman, or master plumber's exam. Questions are in the style of the actual exam. Gives answers for both the Standard and Uniform plumbing codes. Includes tips on studying for the exam and the best way to prepare yourself for examination day. **320 pages, 8½ x 11, $21.00**

■ **HVAC Contracting**
Your guide to setting up and running a successful HVAC contracting company. Shows how to plan and design all types of systems for maximum efficiency and lowest cost — and explains how to sell your customers on your designs. Describes the right way to use all the essential instruments, equipment, and reference materials. Includes a full chapter on estimating, bidding, and contract procedure. **256 pages, 8½ x 11, $24.50**

■ Electrical References ■

■ **Electrical Construction Estimator**
This year's prices for installation of all common electrical work: conduit, wire, boxes, fixtures, switches, outlets, loadcenters, panelboards, raceway, duct, signal systems, and more. Provides material costs, manhours per unit, and total installed cost. Explains what you should know to estimate each part of an electrical system. *New this year! The complete book on a disk with a built-in estimating program for an IBM-compatible hard-drive computer. Included FREE with the book on a 5¼" high-density (1.2 Mb) disk. (Add $10 for extra 5¼" double-density 360K disks or 3½" 720K disks.)* **416 pages, 8½ x 11, $28.50.** Revised annually.

■ **Estimating Electrical Construction**
Like taking a class in how to estimate materials and labor for residential and commercial electrical construction. Written by an A.S.P.E. National Estimator of the Year, it teaches you how to use labor units, the plan take-off, and the bid summary to make an accurate estimate, how to deal with suppliers, use pricing sheets, and modify labor units. Provides extensive labor unit tables and blank forms for your next electrical job. **272 pages, 8½ x 11, $19.00**

■ **Electrical Blueprint Reading Revised**
Shows how to read and interpret electrical drawings, wiring diagrams, and specifications for constructing electrical systems. Shows how a typical lighting and power layout would appear on a plan, and explains what to do to execute the plan. Describes how to use a panelboard or heating schedule, and includes typical electrical specifications. **208 pages, 8½ x 11, $18.00**

■ **Audiotapes: Estimating Electrical Work**
Listen to Trade Service's two-day seminar and study electrical estimating at your own speed for a fraction of the cost of attending the actual seminar. You'll learn what to expect from specifications, how to adjust labor units from a price book to your job, how to make an accurate take-off from the plans, and how to spot hidden costs that other estimators may miss. *Includes six 30-minute tapes, a workbook that includes price sheets, specification sheet, bid summary, and estimate recap sheet, blueprints used in the actual seminar, and blank forms for your own use.* **$65**

■ **Illustrated Guide to the National Electrical Code**
This fully-illustrated guide offers a quick and easy visual reference for installing electrical systems. Whether you're installing a new system or repairing an old one, you'll appreciate the simple explanations written by a code expert, and the detailed, intricately-drawn and labeled diagrams. A real time-saver when it comes to deciphering the current NEC. **256 pages, 8½ x 11, $24.00**

■ **Residential Electrician's Handbook**
Simple, clear instructions for wiring residences: understanding plans and specs, following the NEC, making simple load calculations, sizing wire and service equipment, installing branch and feeder circuits, and running wire. Explains how to estimate the cost of residential electrical systems, and speed up and simplify your estimates using composite unit prices. Includes forms and labor and material tables. **240 pages, 5½ x 8½, $16.75**

■ **Electrician's Exam Preparation Guide**
Need help in passing the apprentice, journeyman, or master electrician's exam? This is a book of questions and answers based on actual electrician's exams over the last few years. Almost a thousand multiple-choice questions — exactly the type you'll find on the exam — cover every area of electrical installation: electrical drawings, services and systems, transformers, capacitors, distribution equipment, branch circuits, feeders, calculations, measuring and testing, and more. It gives you the correct answer, an explanation, and where to find it in the Code. Also tells how to apply for the test, how best to study, and what to expect on examination day. **320 pages, 8½ x 11, $23.00**

■ Builders Office Manuals ■

■ Builder's Office Manual Revised

Explains how to create routine ways of doing all the things that must be done in every construction office – in the minimum time, at the lowest cost, and with the least supervision possible: organizing the office space, establishing effective procedures and forms, setting priorities and goals, finding and keeping an effective staff, getting the most from your record-keeping system (whether manual or computerized), and more. Loaded with practical tips, charts, and sample forms for your use. **192 pages, 8½ x 11, $15.50**

■ Contractor's Growth and Profit Guide

Step-by-step instructions for planning growth and prosperity in a construction contracting or subcontracting company. Explains how to prepare a business plan: select reasonable goals, draft a market expansion plan, make income forecasts and expense budgets, and project cash flow. You'll learn everything that most lenders and investors require, as well as the best way to organize your business. **336 pages, 5½ x 8½, $19.00**

■ Bookkeeping for Builders

Shows simple, practical instructions for setting up and keeping accurate records – with a minimum of effort and frustration. Explains the essentials of a record-keeping system: the payment, income, and general journals, and records for fixed assets, accounts receivable, payables and purchases, petty cash, and job costs. Shows how to keep I.R.S. records, and accurate, organized business records for your own use. **208 pages, 8½ x 11, $19.75**

■ Builder's Comprehensive Dictionary

Never let a construction term stump you again. Here you'll find almost 10,000 construction term definitions, over 1,000 detailed illustrations of tools, techniques, and systems, and a separate section of common legal, real estate, and management terms. **532 pages, 8½ x 11, $24.95**

■ Builder's Guide to Accounting Revised

Step-by-step, easy-to-follow guidelines for setting up and maintaining records for your building business. A practical, newly-revised guide to all accounting methods showing how to meet state and federal accounting requirements and the new depreciation rules. Explains what the 1986 Tax Reform Act can mean to your business. Full of charts, diagrams, blank forms, simple directions and examples. **304 pages, 8½ x 11, $20.00**

■ Contractor's Survival Manual

How to survive hard times and succeed during the up cycles. Shows what to do when the bills can't be paid, finding money and buying time, transferring debt, and all the alternatives to bankruptcy. Explains how to build profits, avoid problems in zoning and permits, taxes, time-keeping, and payroll. Unconventional advice on how to invest in inflation, get high appraisals, trade and postpone income, and stay hip-deep in profitable work. **160 pages, 8½ x 11, $16.75**

Craftsman Book Company
6058 Corte del Cedro
P.O. Box 6500
Carlsbad, CA 92018

In a hurry?
We accept phone orders
charged to your MasterCard, Visa
or American Express
Call 1-800-829-8123
FAX (619) 438-0398

Include a check with
your order and we pay shipping

Call 1-800-829-8123
for a FREE Full Color Catalog
with over 100 Titles

10 Day Money Back GUARANTEE

- ☐ 65.00 Audiotape: Estimating Electrical Work
- ☐ 65.00 Audiotape: Estimating Remodeling Work
- ☐ 30.00 Berger Building Cost File
- ☐ 11.25 Blueprint Reading for Building Trades
- ☐ 19.75 Bookkeeping for Builders
- ☐ 24.95 Builder's Comprehensive Dictionary
- ☐ 20.00 Builder's Guide to Accounting, Revised
- ☐ 15.50 Builder's Office Manual Revised
- ☐ 16.50 Building Cost Manual
- ☐ 22.00 Cabinetmaking: Design to Finish
- ☐ 25.50 Carpentry Estimating
- ☐ 16.25 Carpentry Layout
- ☐ 26.00 Construction Estimating Reference Data
- ☐ 19.25 Construction Surveying & Layout
- ☐ 19.00 Contractor's Growth & Profit Guide
- ☐ 24.25 Contractor's Guide to Building Code
- ☐ 16.75 Contractor's Survival Manual
- ☐ 18.00 Electrical Blueprint Reading Revised
- ☐ 28.50 Electrical Construction Estimator with free *Electrical Estimate Writer* on 5¼" (1.2Mb) disk. *Add $10 for extra* ☐ 5¼" (360K) or ☐ 3½" (720K) disks.
- ☐ 23.00 Electrician's Exam Preparation Guide
- ☐ 19.00 Estimating Electrical Construction
- ☐ 34.95 Estimating Framing Quantities
- ☐ 17.00 Estimating Home Building Costs
- ☐ 22.50 Estimating Plumbing Costs
- ☐ 21.50 Estimating Tables for Home Building
- ☐ 23.25 Fences & Retaining Walls
- ☐ 24.75 Handbook of Construction Contracting Vol. 1
- ☐ 24.75 Handbook of Construction Contracting Vol. 2
- ☐ 19.50 How to Succeed w/ Construction Business
- ☐ 24.50 HVAC Contracting
- ☐ 24.00 Illustrated Guide to National Electrical Code
- ☐ 19.75 Manual of Professional Remodeling
- ☐ 26.50 Masonry Estimating
- ☐ 29.50 National Repair & Remodeling Estimator w/ free Repair & Remodeling Estimate Writer on 5¼" (1.2Mb) disk. *Add $10 for extra* ☐ 5¼" (360K) or ☐ 3½" (720K) disks.
- ☐ 27.50 Painting Cost Guide
- ☐ 21.00 Plumber's Exam Preparation Guide
- ☐ 19.75 Profits in Buying & Renovating Homes
- ☐ 14.25 Rafter Length Manual
- ☐ 18.25 Remodeling Contractor's Handbook
- ☐ 16.75 Residential Electrician's Handbook
- ☐ 22.00 Roof Framing
- ☐ 17.00 Rough Carpentry
- ☐ 15.50 Stair Builder's Handbook
- ☐ 24.75 Video: Stair Framing
- ☐ 26.50 National Construction Estimator with free *Estimate Writer* on 5¼" (1.2Mb) disk. Add $10 for extra *Estimate Writer on either* ☐ 5¼" (360K) or ☐ 3½" (720K) disks.
- ☐ Free Full Color Catalog

FAX to (619) 438-0398

Total enclosed_____ (In Calif. add 7.25% tax)

Use your ☐ Visa ☐ MasterCard or ☐ American Express

Card # _____

Exp. date _____ Initials _____

Name _____

Company _____

Address _____

City/State/Zip _____